S0-BYH-691

Neuromethods

Series Editor
Wolfgang Walz
University of Saskatchewan
Saskatoon
SK, Canada

For other titles published in this series, go to
www.springer.com/series/7657

fMRI Techniques and Protocols

Edited by

Massimo Filippi

Neuroimaging Research Unit, Institute of Experimental Neurology, Division of Neuroscience, Scientific Institute and University Hospital, San Raffaele, Milan, Italy

Editor
Massimo Filippi
Scientific Institute and University Ospedale
San Raffaele, Milan, Italy
filippi.massimo@hsr.it

ISSN 0893-2336 e-ISSN 1940-6045
ISBN 978-1-60327-918-5 e-ISBN 978-1-60327-919-2
DOI 10.1007/978-1-60327-919-2
Springer Dordrecht Heidelberg London New York

Library of Congress Control Number: 2009931047

© Humana Press, a part of Springer Science+Business Media, LLC 2009
All rights reserved. This work may not be translated or copied in whole or in part without the written permission of the publisher (Humana Press, c/o Springer Science+Business Media, LLC, 233 Spring Street, New York, NY 10013, USA), except for brief excerpts in connection with reviews or scholarly analysis. Use in connection with any form of information storage and retrieval, electronic adaptation, computer software, or by similar or dissimilar methodology now known or hereafter developed is forbidden.
The use in this publication of trade names, trademarks, service marks, and similar terms, even if they are not identified as such, is not to be taken as an expression of opinion as to whether or not they are subject to proprietary rights.
While the advice and information in this book are believed to be true and accurate at the date of going to press, neither the authors nor the editors nor the publisher can accept any legal responsibility for any errors or omissions that may be made. The publisher makes no warranty, express or implied, with respect to the material contained herein.

Printed on acid-free paper

Springer is part of Springer Science+Business Media (www.springer.com)

Preface to the Series

Under the guidance of its founders Alan Boulton and Glen Baker, the Neuromethods series by Humana Press has been very successful since the first volume appeared in 1985. In about 17 years, 37 volumes have been published. In 2006, Springer Science + Business Media made a renewed commitment to this series. The new program will focus on methods that are either unique to the nervous system and excitable cells or which need special consideration to be applied to the neurosciences. The program will strike a balance between recent and exciting developments like those concerning new animal models of disease, imaging, in vivo methods, and more established techniques. These include immunocytochemistry and electrophysiological technologies. New trainees in neurosciences still need a sound footing in these older methods in order to apply a critical approach to their results. The careful application of methods is probably the most important step in the process of scientific inquiry. In the past, new methodologies led the way in developing new disciplines in the biological and medical sciences. For example, physiology emerged out of anatomy in the nineteenth century by harnessing new methods based on the newly discovered phenomenon of electricity. Nowadays, the relationships between disciplines and methods are more complex. Methods are now widely shared between disciplines and research areas. New developments in electronic publishing also make it possible for scientists to download chapters or protocols selectively within a very short time of encountering them. This new approach has been taken into account in the design of individual volumes and chapters in this series.

Saskatoon, SK, Canada *Wolfgang Walz*

0318870646

Preface

The past decade has witnessed a remarkable growth of fMRI as a tool for studying brain function. This is due to the appreciation of the potential for fMRI to provide an invaluable insight into the mechanisms through which the human brain works in healthy individuals and in patients with different neurological and psychiatric conditions. More recently, the potential of this technique to monitor the effect of treatment has also been shown. The scope of this book is to provide a complete and up-to-date review of the main methodological aspects of fMRI, as well as a state-of-the-art summary of the achievements obtained by its application to the study of central nervous system functioning in the clinical arena. The possible future evolution of fMRI is also discussed. The contributors of this volume are all worldwide renowned scientists and physicians with a broad experience in the technical development and clinical use of fMRI. Although the field is ample, based on a series of very different disciplines and expanding at a dramatic pace every day, I believe that this book provides an adequate background against which to plan and design new studies to advance our knowledge on the physiology of the normal human brain and its change following tissue injury.

Part I of the volume is aimed at providing the basic knowledge for the understanding of the technical aspects of fMRI. It covers the basic principles of MRI and fMRI, the different options that can be used to set up an fMRI experiment, and the steps of fMRI analysis, from the preparation of data to the achievement of interpretable results. This part is therefore essential to introduce the readers to the "fMRI world" and make them able to interpret with enough criticism the results of their own experiments. A chapter is devoted to the advantages, caveats, and pitfalls of fMRI data acquired using high-field MR scanners, since such scanners are increasingly available and likely to have an enormous impact on data acquisition and analysis in the near future. In addition, although still in its infancy, the assessment of brain connectivity with fMRI is considered at length, given its potential for improving the understanding of normal and pathological brain function.

Part II provides an overview of the main results derived from the application of fMRI to the study of healthy individuals. Given its noninvasiveness, safety, and repeatability, fMRI is rapidly replacing, whenever possible, other functional techniques, such as positron emission tomography, to image the function of the normal brain. In addition, due to its spatial resolution, fMRI is commonly preferred to neurophysiological techniques to locate with precision which areas are activated during the performance of experimental tasks. What has been achieved in the analysis of the main human functional systems with fMRI is illustrated, including, among many other aspects, behavior, language, memory, and emotion.

Part III is more clinically oriented and illustrates the main findings obtained by the application of fMRI to assess the role of brain plasticity in the major neurological and psychiatric conditions. The first chapter is devoted to fMRI studies of multiple sclerosis, since there is a growing body of evidence that brain functional reorganization has an important role, at least in same phases of the disease, in limiting the clinical consequences of MS-related irreversible tissue damage. Therefore, MS can be viewed as a "model" to understand how pathology can affect the patterns of brain recruitment. The results obtained in other white matter conditions, including isolated demyelinating myelitis and vasculitides, are then presented. The second chapter deals with stroke studies, which have shown consistently that reorganization of surviving neuronal networks is one of the key factors underlying recovery of function. The experimental caveats to be faced when studying patients with severe clinical impairment are also reviewed. The following two chapters cover psychiatric and neurodegenerative diseases, a field where fMRI is providing important pieces of information not only for the understanding of the mechanisms underlying disease pathophysiology and genesis of symptomatology, but also for planning and monitoring novel treatment strategies. Then, two conditions, i.e., epilepsy and tumors, where fMRI is gaining an important role in the presurgical evaluation of patients, are discussed. The last contribution of this part describes the potential and some preliminary, but nevertheless promising, results on the use of fMRI in the monitoring of pharmacological treatments and motor rehabilitation.

Part IV is a glimpse into the future and presents novel approaches for the integration of fMRI data with measures of damage assessed using structural MR techniques and the application of fMRI to image spinal cord function.

The hope that has inspired this book is that it will be of help to clinicians and researchers in their daily life activity by providing a "user-friendly" summary of the field and the necessary background against which to plan and carry out future and successful studies. This is, indeed, an ever-growing and exciting field of research, where we have reached a lot in the past few years, but where there is still a long journey ahead of us.

Milan, Italy *Massimo Filippi*

Contents

Contributors

FEDERICA AGOSTA • *Neuroimaging Research Unit, Institute of Experimental Neurology, Division of Neuroscience, Scientific Institute and University Hospital San Raffaele, Milan, Italy; Memory and Aging Center, UCSF Department of Neurology, San Francisco, CA, USA*

JOHN ASHBURNER • *Wellcome Trust Centre for Neuroimaging, UCL Institute of Neurology, London, UK*

ERIK B. BEALL • *Imaging Institute, Cleveland Clinic, Cleveland, OH, USA*

CHRISTIAN F. BECKMANN • *Division of Neuroscience and Mental Health, Imperial College London, Hammersmith Hospital, London, UK; Oxford University Centre for Functional Magnetic Resonance Imaging of the Brain (FMRIB), John Radcliffe Hospital, Oxford, UK*

TIMOTHY E.J. BEHRENS • *Centre for Functional MRI of the Brain and Department of Experimental Psychology, University of Oxford, Oxford, UK*

JEFFREY R. BINDER • *Language Imaging Laboratory, Departments of Neurology and Biophysics, The Medical College of Wisconsin, Milwaukee, WI, USA*

MICHAEL BUCHFELDER • *Neurochirurgische Klinik, Universität Erlangen-Nürnberg, Erlangen, Germany*

ANTHONY CHEN • *Helen Wills Neuroscience Institute, University of California, Berkeley, CA, USA*

STEVEN C. CRAMER • *Departments of Neurology and Anatomy & Neurobiology, University of California Irvine, Irvine, CA, USA*

MELISSA A. DANIELS • *McLean Hospital Brain Imaging Center and Department of Psychiatry, Harvard Medical School, Belmont, MA, USA*

RALF DEICHMANN • *Brain Imaging Center, Goethe University, Frankfurt/Main, Germany*

MARK D'ESPOSITO • *Helen Wills Neuroscience Institute, University of California, Berkeley, CA, USA*

BRADFORD C. DICKERSON • *Gerontology Research Unit, Massachusetts General Hospital, Charlestown, MA, USA*

EMMA G. DUERDEN • *Groupe de Recherche sur le Système Nerveux Central and Centre de Recherche de l'Institut Universitaire de Gériatrie de Montréal, Montreal, QC, Canada*

GARY H. DUNCAN • *Groupe de Recherche sur le Système Nerveux Central de l'Institut Universitaire de Gériatrie de Montréal, Montreal, QC, Canada; Département de Stomatologie, Université de Montréal, Montréal, QC, Canada; Neurology and Neurosurgery, McGill University, Montréal, QC, Canada*

SEAN P. FANNON • *Center for Mind and Brain, University of California, Davis, CA, USA; Folsom Lake College, Folsom, CA, USA*

MASSIMO FILIPPI • *Neuroimaging Research Unit, Institute of Experimental Neurology, Division of Neuroscience, Scientific Institute and University Hospital San Raffaele, Milan, Italy*

KARL FRISTON • *The Wellcome Trust Centre for Neuroimaging, Institute of Neurology, University College London, London, UK*
OLIVER GANSLANDT • *Neurochirurgische Klinik, Universität Erlangen-Nürnberg, Erlangen, Germany*
HUGH GARAVAN • *School of Psychology and Institute of Neuroscience, Trinity College Dublin, Dublin 2, Ireland*
JOY J. GENG • *Center for Mind and Brain, University of California, Davis, CA, USA*
MARIA LUISA GORNO-TEMPINI • *Memory and Aging Center, University of California San Francisco, San Francisco, CA, USA; Center for Mind/Brain Sciences (CIMeC), University of Trento, Trento, Italy*
WILLIAM GRISSOM • *Biomedical Engineering Department, University of Michigan, Ann Arbor, MI, USA*
PETER GRUMMICH • *Neurochirurgische Klinik, Universität Erlangen-Nürnberg, Erlangen, Germany*
ERIN L. HABECKER • *McLean Hospital Brain Imaging Center and Department of Psychiatry, Harvard Medical School, Belmont, MA, USA*
DEBORAH ANN HALL • *MRC Institute of Hearing Research, Nottingham, UK*
LUIS HERNANDEZ-GARCIA • *University of Michigan Functional MRI Laboratory and Biomedical Engineering Department, University of Michigan, Ann Arbor, MI, USA*
SAAD JBABDI • *Centre for Functional MRI of the Brain, University of Oxford, Oxford, UK*
HEIDI JOHANSEN-BERG • *Centre for Functional MRI of the Brain, University of Oxford, Oxford, UK*
ALAYAR KANGARLU • *Columbia University and New York State Psychiatric Institute, New York, NY, USA*
ANDREW KAYSER • *Helen Wills Neuroscience Institute, University of California, Berkeley, CA, USA*
ZOE KOURTZI • *School of Psychology, University of Birmingham, Birmingham, UK*
LOUIS LEMIEUX • *Department of Clinical and Experimental Epilepsy, UCL Institute of Neurology, Queen Square, London, UK; MRI Unit, National Society for Epilepsy, Chalfont St. Peter, Buckinghamshire, UK*
MARK J. LOWE • *Imaging Institute, Cleveland Clinic, Cleveland, OH, USA*
GEORGE R. MANGUN • *Center for Mind and Brain, University of California, Davis, CA, USA*
PAUL M. MATTHEWS • *Glaxo Smith Kline Clinical Imaging Centre, Hammersmith Hospital, London, UK; Department of Clinical Neurosciences, Imperial College, London, UK; Centre for Functional Magnetic Resonance Imaging of the Brain, University of Oxford, Oxford, UK*
EWALD MOSER • *MR Center of Excellence, Medical University of Vienna, Vienna, Austria; Center for Biomedical Engineering and Physics, Medical University of Vienna, Vienna, Austria; Department of Psychiatry, University of Pennsylvania Medical Center, Philadelphia, PA, USA*
KEVIN MURPHY • *Section on Functional Imaging Methods, Laboratory of Brain and Cognition, National Institute of Mental Health, NIH, Bethesda, MD, USA*

THOMAS E. NICHOLS • *GlaxoSmithKline Clinical Imaging Centre, Imperial College London, Hammersmith Hospital, London, UK*

CHRISTOPHER NIMSKY • *Neurochirurgische Klinik, Universität Erlangen-Nürnberg, Erlangen, Germany*

GUY A. ORBAN • *Laboratorium voor Neuro-en Psychofysiologie, KU Leuven Medical School, Leuven, Belgium*

ASPASIA ELENI PALTOGLOU • *MRC Institute of Hearing Research, Nottingham, UK*

SCOTT PELTIER • *University of Michigan Functional MRI Laboratory and Biomedical Engineering Department, University of Michigan, Ann Arbor, MI, USA*

MARTIN PEPER • *Philipps-Universität Marburg, Faculty of Psychology, General and Biological Psychology Section, Marburg, Germany*

ROBERT POWELL • *Department of Clinical and Experimental Epilepsy, UCL Institute of Neurology, Queen Square, London, UK; MRI Unit, National Society for Epilepsy, Chalfont St. Peter, Buckinghamshire, UK*

PERRY F. RENSHAW • *McLean Hospital Brain Imaging Center and Department of Psychiatry, Harvard Medical School, Belmont, MA, USA*

SIMON ROBINSON • *Center for Mind/Brain Sciences (CIMeC), Università di Trento, Mattarello (TN), Italy*

MARIA A. ROCCA • *Neuroimaging Research Unit, Institute of Experimental Neurology, Division of Neuroscience, Scientific Institute and University Hospital San Raffaele, Milan, Italy*

CLIFFORD D. SARON • *Center for Mind and Brain, University of California, Davis, CA, USA; M.I.N.D. Institute, University of California, Davis, CA, USA*

STEPHEN M. SMITH • *Oxford University Centre for Functional Magnetic Resonance Imaging of the Brain (FMRIB), John Radcliffe Hospital, Oxford, UK*

PATRICK W. STROMAN • *Centre for Neuroscience Studies, Department of Diagnostic Radiology, Queen's University, Kingston, ON, Canada*

RACHEL THORNTON • *Department of Clinical and Experimental Epilepsy, UCL Institute of Neurology, Queen Square, London, UK; MRI Unit, National Society for Epilepsy, Chalfont St. Peter, Buckinghamshire, UK*

ARTHUR TOGA • *Laboratory of Neuro Imaging, Department of Neurology, UCLA School of Medicine, University of California Los Angeles, Los Angeles, CA, USA*

JOHN DARRELL VAN HORN • *Laboratory of Neuro Imaging, Department of Neurology, UCLA School of Medicine, University of California Los Angeles, Los Angeles, CA, USA*

INDRE V. VISKONTAS • *Memory and Aging Center, University of California San Francisco, San Francisco, CA, USA*

NICK S. WARD • *Clinical Senior Lecturer and Honorary Consultant Neurologist Sobell Department of Motor Neuroscience, UCL Institute of Neurology, Queen Square, London, UK*

KATE E. WATKINS • *Centre for Functional MRI of the Brain and Department of Experimental Psychology, University of Oxford, Oxford, UK*

MARK W. WOOLRICH • *Oxford University Centre for Functional Magnetic Resonance Imaging of the Brain (FMRIB), John Radcliffe Hospital, Oxford, UK*

Part I

BOLD fMRI: Basic Principles

Chapter 1

Principles of MRI and Functional MRI

Ralf Deichmann

Summary

This chapter describes the basics of magnetic resonance imaging (MRI) and functional MRI (fMRI). It is aimed at beginners in the field and does not require any previous knowledge. Complex technical issues are made plausible by presenting plots and figures, rather than mathematical equations.

The part dealing with the basics of MRI covers spins, spin alignment in external magnetic fields, the magnetic resonance effect, field gradients, frequency encoding, phase encoding, slice selection, *k*-space, gradient echoes, and echo-planar imaging.

The part dealing with fMRI covers transverse relaxation times, the basics of the blood oxygen level-dependent contrast, and the haemodynamic response.

Key words: Spin, Field gradients, Frequency encoding, Phase encoding, Slice selection, *k*-Space, Gradient echo, Echo-planar imaging, Transverse relaxation time, Blood oxygen level dependent.

1. Basic Physical Principles

1.1. Spins in an External Magnetic Field

The first question arising is "What do we actually see in MRI?"

In general, we see protons. A proton is the nucleus of the hydrogen atom. Hydrogen is the most common element in tissue, so if we are able to detect the presence of protons and display them with a certain spatial resolution, it is fair to say that we can "see" tissue.

The detection of protons is based on a physical property called the "spin". A correct description of spins is only possible with quantum mechanics and would be beyond the scope of this book, so may it suffice to say that a spin is quite similar to a compass needle. In particular, a compass needle carries a "magnetisation"

M. Filippi (ed.), *fMRI Techniques and Protocols*, Neuromethods, vol. 41
DOI 10.1007/978-1-60327-919-2_1, © Humana Press, a part of Springer Science+Business Media, LLC 2009

which enables it to align in an external magnetic field and produce a magnetic field itself (for example, a compass needle can be used to attract small iron particles). A spin possesses an elementary magnetisation (albeit a tiny one), so it behaves in a similar way.

Let us consider a simple object containing protons, for example a container with water (**Fig. 1**). If there is no external magnetic field (usually labelled B), the spins will point in different directions, the contributions of their respective magnetisation vectors will cancel out, so there is no net magnetisation (usually labelled M). If, however, this object is placed into an external magnetic field B (e.g. into the bore of an MR scanner), the spins will align. According to the laws of quantum mechanics, this alignment is either parallel or anti-parallel to the external magnetic field, so once again one might assume that the single magnetisation vectors will cancel each other. However, a slight majority of spins prefers the parallel direction. This results in a macroscopic net magnetisation M which is parallel to B (**Fig. 1**, right). The idea is now: any measurement of M would correspond to the detection of the presence of protons in the object. If M is measured with a spatial resolution, we can display the result as an image. This is exactly what is done in MR imaging. The measurement of M is based on a physical effect which will be described in **Subheading 1.2**.

1.2. The Larmor Precession

Let us assume that we disturb the realigned spins in a way that (at least for a short time) the magnetisation vector M is not parallel to B but tilted by a certain angle (how this can be achieved will be discussed in **Subheading 1.3**). In this case, an interesting process begins: the tilted magnetisation vector rotates around the direction of the external magnetic field. This movement, which resembles closely the behaviour of a spinning top, is called precession (**Fig. 2**). The frequency is called *Larmor frequency*. It is

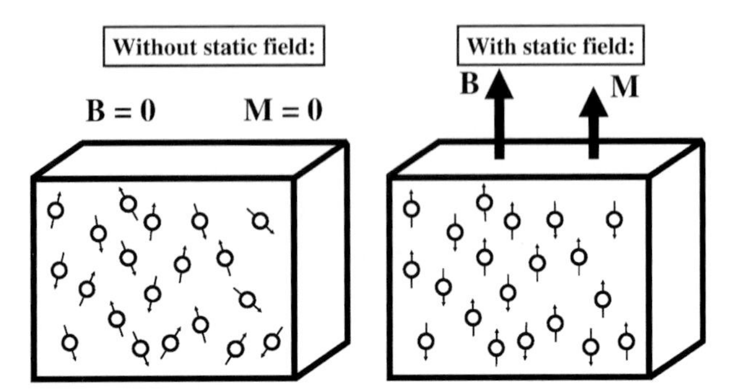

Fig. 1. (*Left*) Without an external magnetic field (labelled *B*) the spins point in different directions, the contributions of their respective magnetisation vectors cancel out, so there is no net magnetisation (labelled *M*). (*Right*) Inside an external magnetic field *B* the spins align in parallel or anti-parallel direction, with a slight majority of spins in parallel direction, giving rise to a net magnetisation *M*.

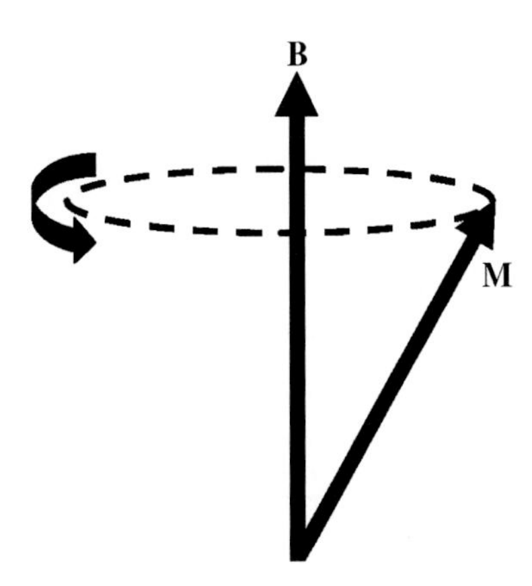

Fig. 2. Behaviour of the tilted magnetisation: Precession with the Larmor Frequency.

important to note that the Larmor frequency f is proportional to the external field B

$$f = \gamma B. \tag{1}$$

In this formula, γ is the gyromagnetic ratio with a value of 42.58 MHz/T for hydrogen.

This effect is of major interest because during the precession, the spins send out an electromagnetic wave with the Larmor frequency. For field strengths that are common on clinical MR scanners (1.5 and 3 T), the Larmor frequencies are about 63 MHz and 127 MHz, respectively. As the reader may know, the tuning on standard frequency modulation (FM) radios ranges from 88 to 108 MHz, so it is fair to say that the signal sent out by the spins is similar to an ordinary FM radio broadcasting signal. For this reason, it is also called *radiofrequency (RF) signal*. It can be detected with an appliance very similar to an FM tuner. This is exactly the way how the magnetisation M and thus the presence of protons are detected in MR scanners. In **Subheading 1.3**, it will be described how the magnetisation can be tilted.

1.3. The Magnetic Resonance Effect

The next physical effect is more or less the opposite of the one that has just been discussed. This time, the object is exposed to an external RF signal which has *exactly* the Larmor frequency and is produced by a kind of "inbuilt FM broadcasting station" (**Fig. 3**, part 1). This signal will tilt the magnetisation which subsequently starts to precede (**Fig. 3**, part 2). During precession, an RF signal is being sent out which can be detected with a kind of "FM tuner" (**Fig. 3**, part 3).

It should be noted that this concept only works if the incoming RF signal has *exactly* the spins' Larmor frequency. Otherwise, the magnetisation will not be tilted and it is impossible to detect

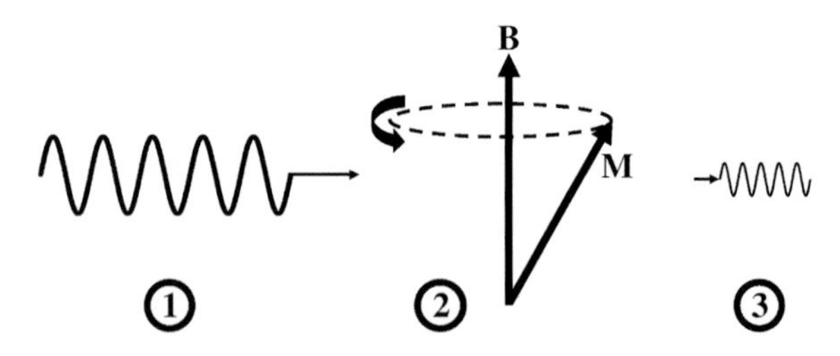

Fig. 3. An external radiofrequency (RF) pulse that has exactly the Larmor frequency tilts the magnetisation ***(1)***. The tilted magnetisation starts to precede ***(2)*** and sends out an RF pulse with the Larmor frequency itself ***(3)***.

a signal. Thus, we are dealing with a resonance effect, and this explains why the imaging technique based on this effect is called *magnetic resonance imaging* (MRI). The nuclear magnetic resonance effect was first described independently by Bloch and Purcell in 1946 *(1, 2)*.

We have now covered all the basic physical effects that are required to understand how MRI works. To summarise it, all MR experiments are based on the following principles:

- Put the object to be imaged into a strong external magnetic field B. The spins will align and create a net magnetisation M which is parallel to B.
- Knowing B, calculate the Larmor frequency $f = \gamma B$ and send an external RF pulse which has exactly this frequency. This will tilt the magnetisation. After a short time, the external RF has served its duty and can be switched off.
- The magnetisation vector now precedes. Switch on your FM tuner. If you detect a signal, you have detected the presence of protons.

Subheading 2 will deal with the question how spatial resolution can be achieved.

2. A One-Dimensional MR Experiment

Let us assume we have a closed box containing two glasses of water, one being half full and the other one full **(Fig. 4a)**.

We now have to answer the following questions:

1. Are there any glasses inside the box or is it empty?
2. How many glasses are inside the box?
3. What is the exact location of the glasses?
4. What are the relative filling levels?

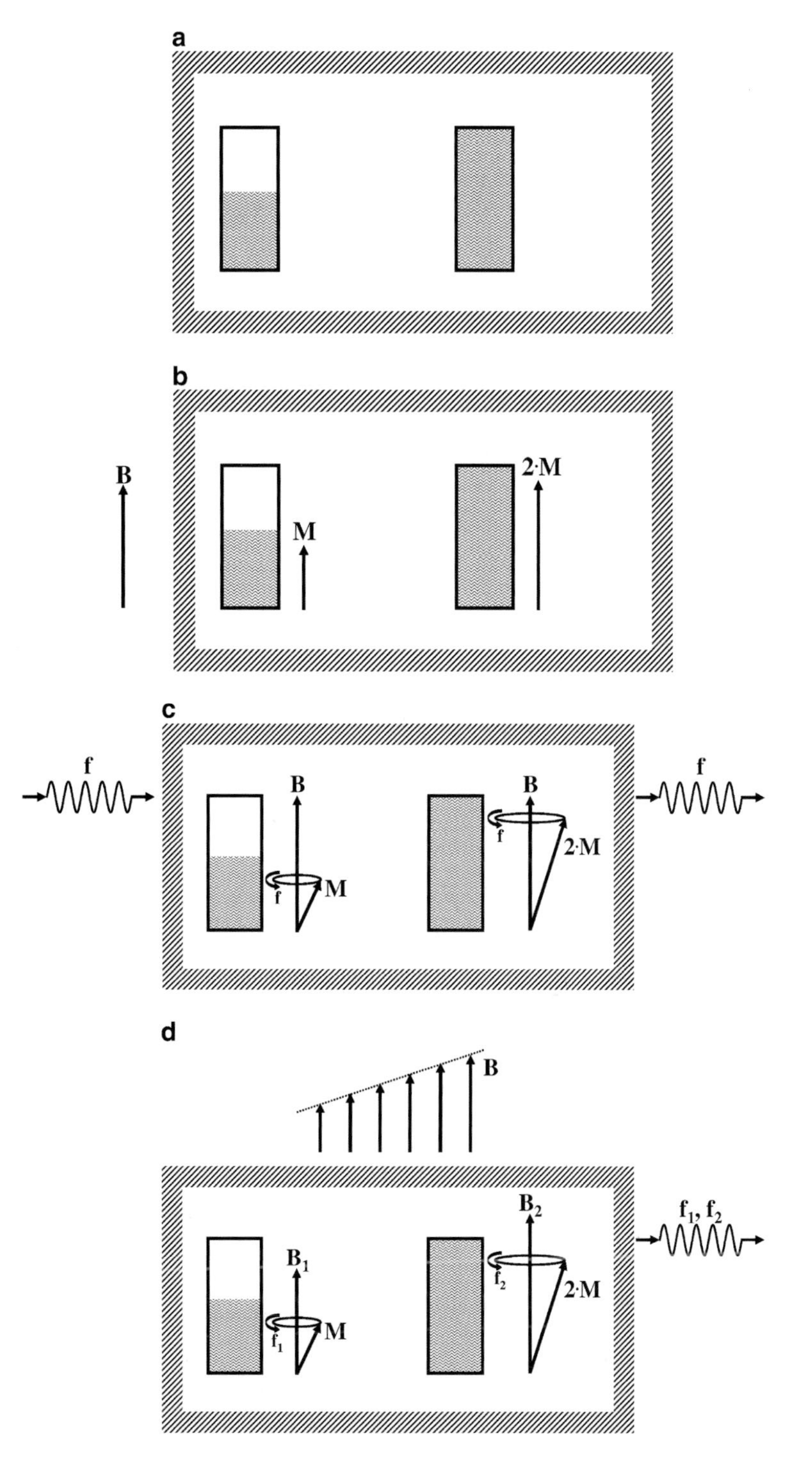

Fig. 4. **(a)** Two glasses of water inside a closed box. **(b)** Magnetisation inside both glasses after placing the box inside an external magnetic field. **(c)** Reaction to an external radiofrequency (RF) pulse: the magnetisation vectors in both glasses start to precede and send out RF signals with the Larmor frequency. **(d)** Effect of an external field gradient: the magnetic field strength at the position of both glasses is different, so radiofrequency signals with different Larmor frequencies are being sent.

Of course: *we are not allowed to open the box!* Our investigation must rely purely on the physical principles we have discussed so far.

The whole procedure will be discussed step by step as follows:

Step 1. We put the box into the MR scanner, i.e. into a strong magnetic field B. In both glasses, spins will align, resulting in a net magnetisation M in the first glass and, due to the larger number of protons, $2M$ in the second glass **(Fig. 4b)**.

Step 2. We calculate the Larmor frequency $f = \gamma B$ and send an external RF pulse which has exactly this frequency. This will cause a tilt of the magnetisation vectors in both glasses which consequently start to precede, sending out an RF signal with the same frequency f **(Fig. 4c)**.

This signal can be detected with a kind of FM tuner, so we can answer at least the first question: the box is sending out a signal, so it cannot be empty. However, so far we are not able to comment on the number and locations of the glasses, because they send out signals with the same frequency and we can only detect the *sum signal* outside the box.

Step 3. This is the crucial step: we switch on a so-called *gradient field.* This simply means that we modify for a certain time the external magnetic field B in a way that it is no longer constant across the box but increases (slightly) from the left-hand to the right-hand side **(Fig. 4d)**. In particular, the field strengths at the positions of the first and the second glass are now different, assuming the values B_1 and B_2, respectively. Since the Larmor frequency depends on the magnetic field strength, the magnetisation vectors continue their precession with different frequencies f_1 and f_2. Once again, we measure the sum signal outside the box.

As the next step, a *frequency analysis* of this signal is performed. This means that the signal is decomposed into its spectrum of frequency components. In the present case, this analysis yields the following results:

- The signal contains two frequency components, so there must be two glasses inside the box.
- We can measure the absolute values of the frequencies f_1 and f_2. According to **Eq. 1**, we can calculate from these frequencies the field strengths B_1 and B_2 at the positions of the first and the second glass, respectively. Since we know how we modified the magnetic field B, we can deduce from this the exact positions of the glasses.
- The amplitude of the frequency component f_2 is twice the amplitude of the frequency component f_1. From this we can deduce that there must be twice the number of spins in the

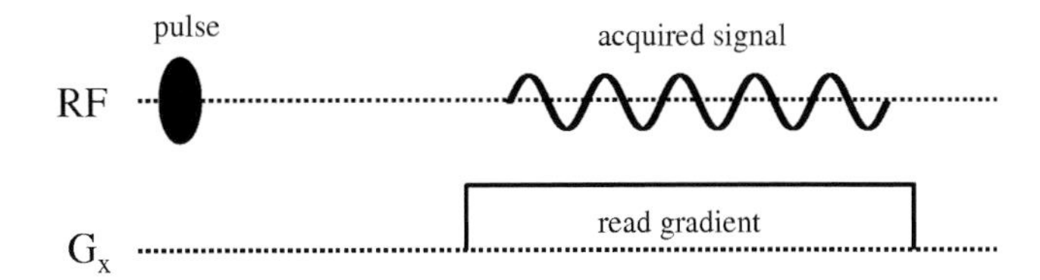

Fig. 5. Schematic plot of a one-dimensional MR experiment, showing a radiofrequency (RF) axis and a gradient (G_x) axis.

second glass, so we also obtain information about the relative filling levels.

In summary, we have answered all the questions above without opening the box, purely by using the concepts of magnetic resonance.

If you could follow this section, you have understood the basics of MR imaging.

The use of field gradients for spatial encoding was proposed by Lauterbur in 1973 *(3)*.

In textbooks and publications, MR experiments are usually described by special plots, the so-called *pulse diagrams*. For the experiment described above, the respective pulse diagram is shown in **Fig. 5**, comprising an *RF axis* and a *gradient axis*. The entries on the RF axis correspond to the initial RF pulse which tilts the magnetisation and the acquired signal. The entries on the gradient axis show that during signal acquisition the gradient G_x is switched on, i.e. during this time the external magnetic field is modified in a way that it increases linearly in a certain spatial direction (the *x*-direction of an arbitrarily chosen coordinate system). Because this gradient is switched on during the readout process, it is also referred to as *read gradient*.

3. The Fourier Transform

The frequency analysis of a signal is a mathematical process called *Fourier Transform*. A Fourier Transform yields the frequency spectrum of the signal, i.e. the amplitudes of the various frequency components. The frequency spectrum for the setup described above (two glasses in a box) is shown in **Fig. 6**.

As explained above, a spin's Larmor frequency depends on its local magnetic field strength and therefore on its position within the box, as long as the linear field gradient is switched on. It is therefore fair to say that the frequency spectrum shows a one-dimensional image of the scanned object. In the special case of **Fig. 6**, the two glasses, their respective positions, their relative filling levels, and even their diameters are clearly displayed.

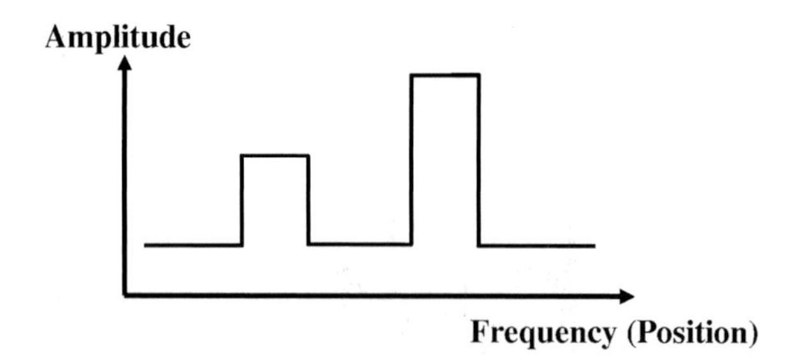

Fig. 6. Frequency spectrum resulting from the Fourier Transform. The frequency corresponds to the position of the originating spin.

4. The Gradient Echo

The *gradient echo* technique is of major importance in functional MRI (fMRI).

Let us consider the three experiments described in **Fig. 7**. Experiments will be discussed on a purely phenomenological basis first, in particular the experimental setup and the respective signal behaviours. The explanation will be given afterwards **(Fig. 8)**.

In the first experiment (**Fig.** 7, part 1), an initial RF pulse tilts the magnetisation. As expected, precession starts and a signal can be acquired immediately after sending the RF pulse. This signal has a relatively long duration.

In the second experiment (**Fig.** 7, part 2), the signal is acquired while a gradient is switched on. In this case, the signal decays much faster. The reason for this will be given below. For the time being it is sufficient to note that obviously inhomogeneities of the static magnetic field, as created by a gradient, cause a more rapid signal decay.

The third experiment (**Fig.** 7, part 3) starts off like the second one, resulting in the same fast signal decay. However, after a while the gradient is *inverted* (a negative gradient G_x means that the magnetic field strength decreases in x-direction, rather than increasing). This leads to a striking phenomenon: the signal which seemed to have disappeared completely, suddenly comes back. This effect is called *gradient echo*.

It is relatively simple to explain these effects. For illustration, **Fig. 8a** shows the precession of four individual spins at different positions in the first experiment after the initial RF pulse (the respective magnetisation vectors are seen "from top"). Although the spins are located at different positions, they are exposed to the same magnetic field because there are no gradients. Thus, they precede with the same Larmor frequency; their magnetisation vectors are always parallel and add up to a relatively strong net magnetisation.

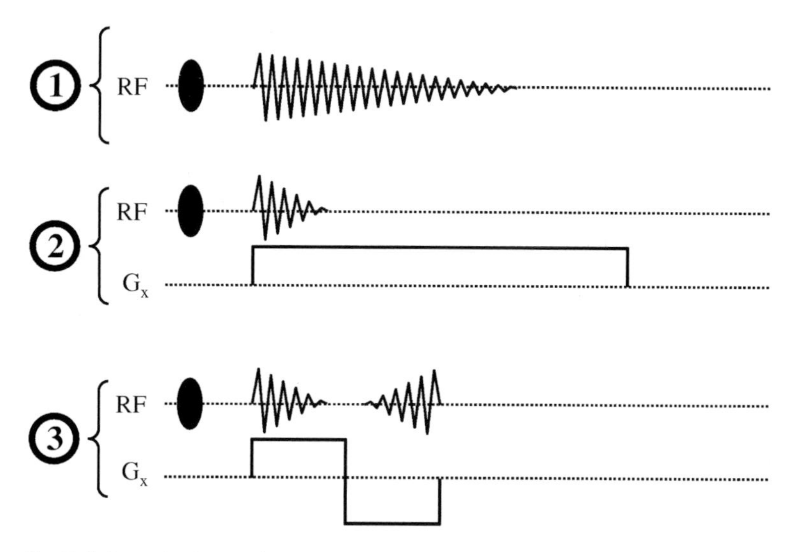

Fig. 7. Schematic sketch of three MR experiments: in the absence of any gradients, a long signal can be observed ***(1)***. In the presence of a gradient, the signal decays more rapidly ***(2)***. If the gradient is inverted, a gradient echo occurs ***(3)***.

In theory, this should go on forever. In practice, the signal decays due to transverse relaxation effects which will be discussed later.

Figure 8b shows the respective sketch for the second experiment. Due to the field gradient, the spins are exposed to different field strengths and precede with different Larmor frequencies. After a relatively short time, they are completely *dephased*, i.e. the magnetisation vectors cancel each other and there is no net magnetisation. The result is a fast signal decay, as depicted in **Fig.** 7 (part 2).

Figure 8c shows the respective sketch for the third experiment. Due to the gradient, the spins dephase, as described above. However, when the gradient is inverted the spins turn backwards, maintaining their frequencies. As a result, they rephase (i.e. they return to their original positions), resulting in a realignment which corresponds to a reappearance of the signal. This is the origin of the gradient echo.

The gradient echo helps to overcome a typical problem in MR imaging:

According to **Fig. 5**, a signal has to be acquired while the read gradient is switched on. However, due to technical limitations, gradients have a certain *rise time*, i.e. a certain time delay (typically several hundreds of μs or even several ms) is required to ramp up the gradient (**Fig. 9**, shaded area). This leads to a dilemma: on one hand, one has to wait for the gradient to reach its plateau level before signal acquisition can start, because only then there is a well-defined relationship between the position of a spin and its Larmor frequency, as required to deduce spatial information from the frequency spectrum. On the other hand, during the process of ramping up the gradient spins start to dephase, so

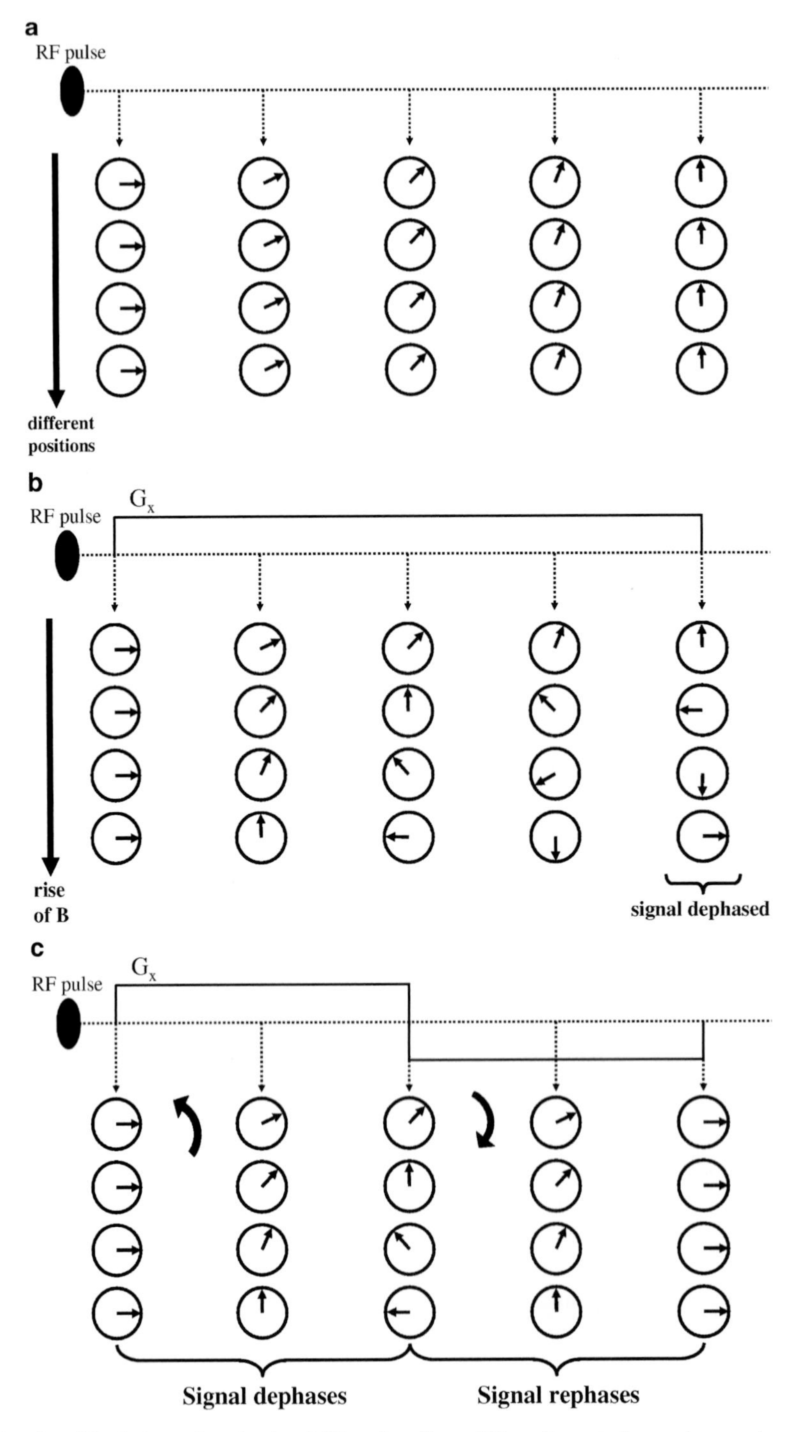

Fig. 8. ***(a)*** Explanation of the first experiment: spins at different positions still have the same Larmor frequencies as the magnetic field is homogeneous. As a consequence, their magnetisation vectors remain parallel and sum up to a strong net magnetisation over a relatively long time. ***(b)*** Explanation of the second experiment: spins at different positions have different Larmor frequencies due to the field gradient. Their magnetisation vectors dephase, reducing the duration of the signal. ***(c)*** Explanation of the third experiment: the gradient inversion leads to a change from anti-clockwise to clockwise rotation, so spins rephase. A strong signal, the so-called gradient echo can be observed when the magnetisation vectors are parallel again.

we will have lost a considerable part of the signal by the time the acquisition starts, resulting in a poor image quality.

This problem can be overcome by using the gradient echo concept as shown in **Fig. 10**: by means of an initial negative gradient, spins are dephased deliberately. Rephasing and the occurrence of a gradient echo take place during the plateau of the read gradient, so we can measure a strong signal at a time when the gradient is constant. Gradient echo sequences are widely used in MR imaging.

The experiment described in **Fig. 10** is one-dimensional, i.e. it allows for spatial resolution in one direction (the *x*-direction) only. If the gradient axes are chosen as shown in **Fig. 11** (left),

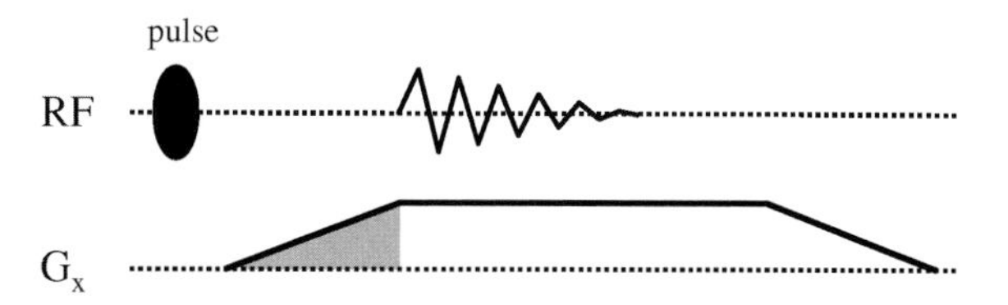

Fig. 9. Signal losses due to the finite gradient rise time: by the time the gradient has reached its full amplitude, the signal has decayed considerably.

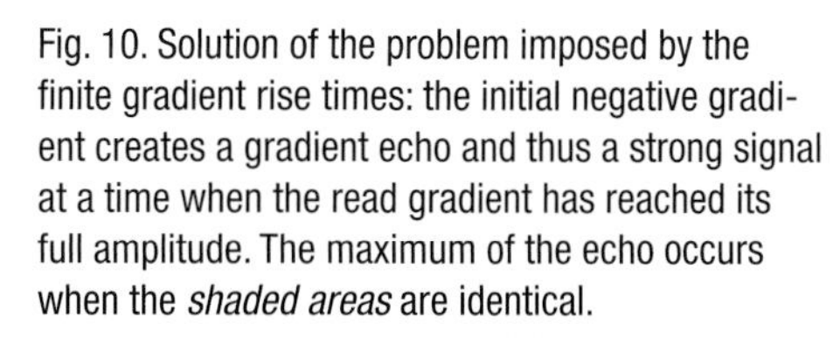

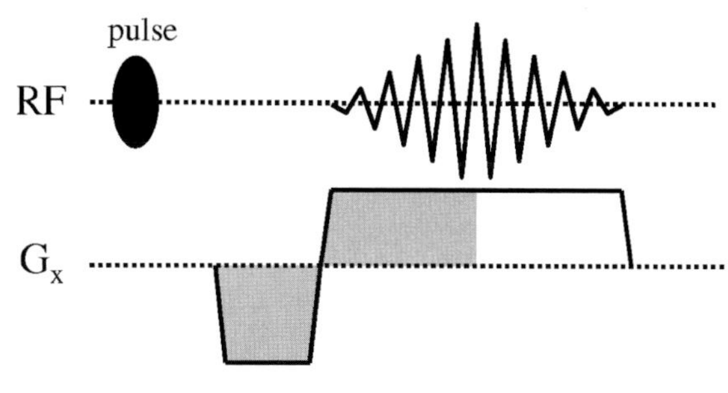

Fig. 10. Solution of the problem imposed by the finite gradient rise times: the initial negative gradient creates a gradient echo and thus a strong signal at a time when the read gradient has reached its full amplitude. The maximum of the echo occurs when the *shaded areas* are identical.

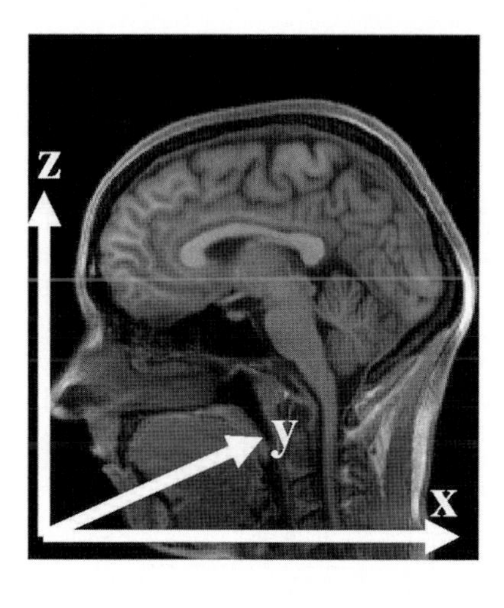

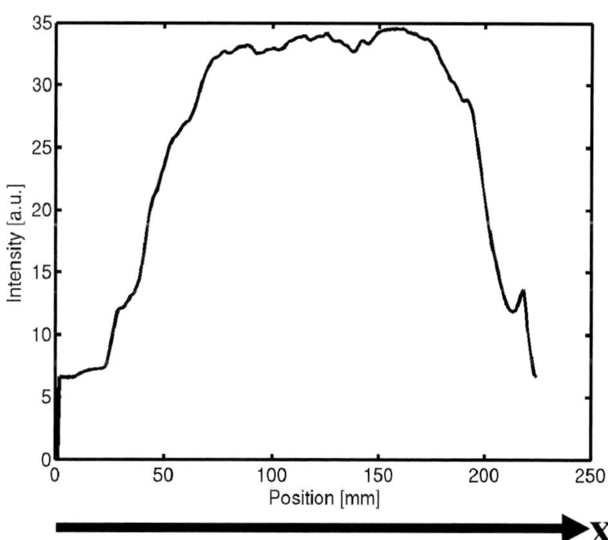

Fig. 11. If the gradient axes are chosen as shown on the *left-hand side*, the result of the one-dimensional experiment with a read gradient in *x*-direction is a profile in anterior/posterior direction.

the result is a profile in anterior/posterior direction (**Fig. 11**, right). The full extent of the imaged object in *y*- and *z*-direction is projected onto the *x*-axis.

5. The *k*-Space

Before we move on to two-dimensional experiments (which allow us to obtain real images), the so-called *k-space* will be introduced. The *k*-space is a very useful concept when it comes to describing and understanding MR sequences.

Imagine a gradient is turned on for a certain time. During this time, a single data point is acquired (**Fig. 12**). The *k-value* of this data point is the area under the *preceding* gradient (shaded), i.e. the area under the gradient up to the time point of acquisition. This is simply a definition.

Let us now analyse our one-dimensional MR experiment as depicted in **Fig. 10**. The signal acquisition consists basically of the acquisition of a series of discrete data points (**Fig. 13**) with different *k*-values.

The first data point is preceded by the negative dephasing gradient, so it has a negative *k*-value. The next data point "sees" a combined preparation: the negative dephasing gradient, followed by the first bit of the positive read gradient. Thus, its *k*-value (being the sum of individual areas under the preceding gradient)

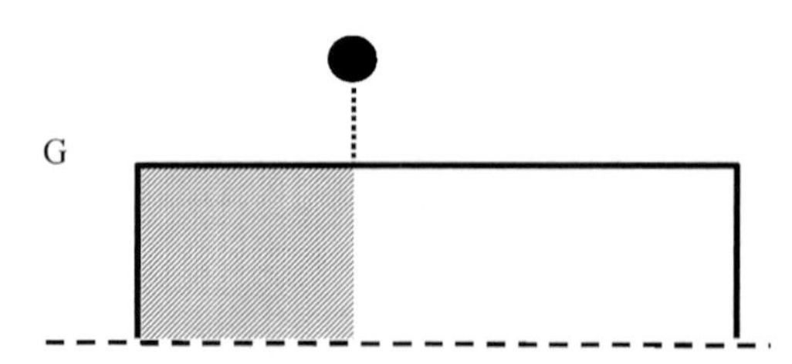

Fig. 12. Definition of a data point's *k*-value as the area under the gradient before the data point is sampled.

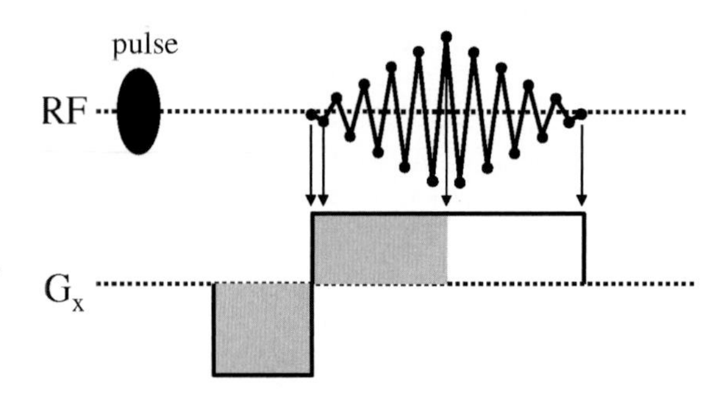

Fig. 13. Gradient echo experiment: the data points constituting the gradient echo have increasing *k*-values, ranging from a negative to a positive value. The *k*-value is zero for the central data point where the *shaded areas* are identical and the echo has maximum amplitude.

is still negative, but slightly higher than for the first data point. The subsequent data points have increasing *k*-values. The central data point is acquired when the shaded areas in **Fig. 13** are identical. Thus, its *k*-value is zero. As described above, this is where the centre of the gradient echo occurs, i.e. this data point will have the highest signal amplitude. The subsequent data points have increasing, positive *k*-values and the maximal *k*-value is attained for the last data point.

We learnt above that spatial resolution in one direction is achieved by acquiring a signal while a gradient in this direction is switched on. We further know that this gradient should be preceded by a negative dephasing gradient to obtain a gradient echo. The *k*-value concept allows us now to move on to an alternative formulation. However, please note that this formulation is basically identical to the previous one:

Spatial resolution in one direction is achieved by acquiring a series of data points with different *k*-values, ranging from a negative to a positive value. Maximum signal is attained for the data point for which the respective *k*-value is zero.

This concept was introduced because it makes it much easier to understand how two-dimensional imaging works. Basically, we need spatial resolution and thus gradients in two directions (the *x*- and the *y*-direction), so in general we have to attribute a k_x and a k_y value to each data point, corresponding to the areas under the respective preceding gradients **(Fig. 14)**.

To visualise these *k*-values, we can create a coordinate system with the axes k_x and k_y and insert the data point at the respective position **(Fig. 15)**. This is the so-called *k-space*. In the example of **Fig. 14**, there is a relatively large positive k_x and a relatively small k_y, so the position of this data point in *k*-space would be more or less as shown in **Fig. 15**.

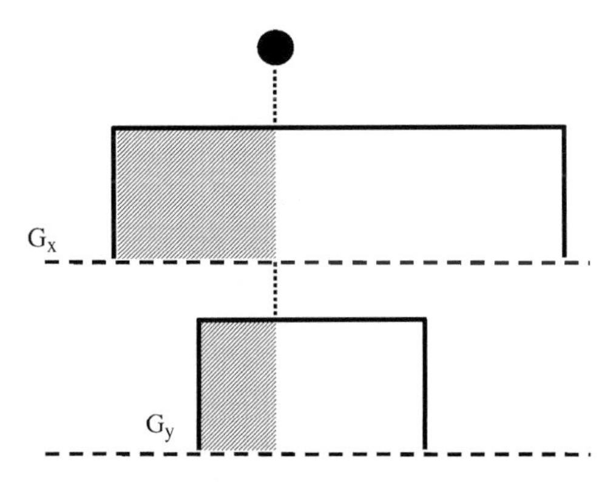

Fig. 14. Definition of a data point's k_x- and k_y-values as the areas under the respective gradients before the data point is sampled.

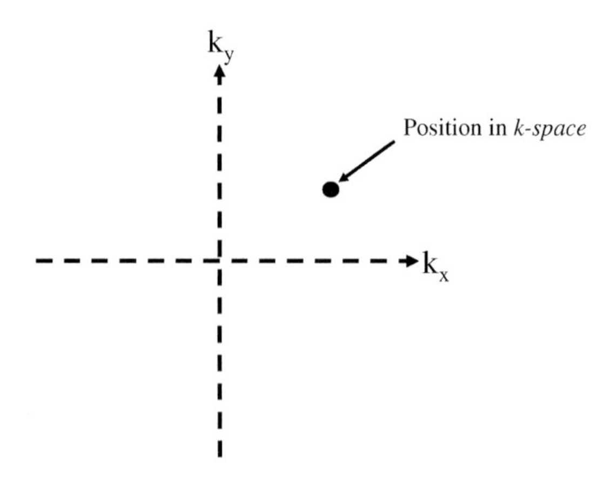

Fig. 15. Description of a data point's k_x- and k_y-values in *k*-space.

6. Two-Dimensional Acquisition

It is now easy to extend the one-dimensional concept described above to two dimensions:

Spatial resolution in two directions (x, y) is achieved by acquiring a series of data points with different combinations of k_x- and k_y-values, filling the two-dimensional *k*-space as shown in **Fig. 16**. Maximum signal is attained for the central data point for which both *k*-values are zero.

Let us now consider the experiment depicted in **Fig. 17**. It closely resembles the experiment discussed above: the RF axis and the G_x axis are identical. This means that the data points have increasing k_x-values, ranging from a negative to a positive value. However, there is now also a G_y axis, showing a negative gradient which is switched on and off before the acquisition starts. Since this gradient is off during acquisition, all data points have the same (negative) k_y-value, corresponding to the area under G_y. Thus, the k_y-value is constant, whereas the k_x-value increases. This means that the experiment acquires a single horizontal line in *k*-space.

To obtain spatial resolution in two dimensions, we have to acquire several horizontal lines in *k*-space. This means that we have to repeat the experiment from **Fig. 17** several times with different amplitudes of the gradient G_y. In textbooks and publications, this is usually depicted as shown in **Fig. 18**: the gradient G_y appears as a "ladder" with an arrow, which means that the acquisition is repeated several times with different discrete G_y values, stepping from a minimum to a maximum value.

It should be noted that the gradient G_y is also referred to as *phase gradient* or *phase encoding gradient*.

The experiment described in **Fig. 18** is two-dimensional, i.e. it allows for spatial resolution in two directions (the x- and y-direction). If the gradient axes are chosen as shown in **Fig. 19**

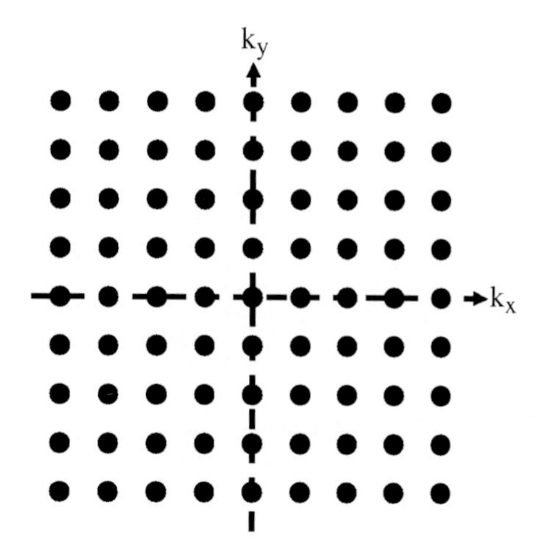

Fig. 16. The basis of two-dimensional imaging: several data points with different combinations of k_x- and k_y-values have to be sampled, filling the two-dimensional *k*-space.

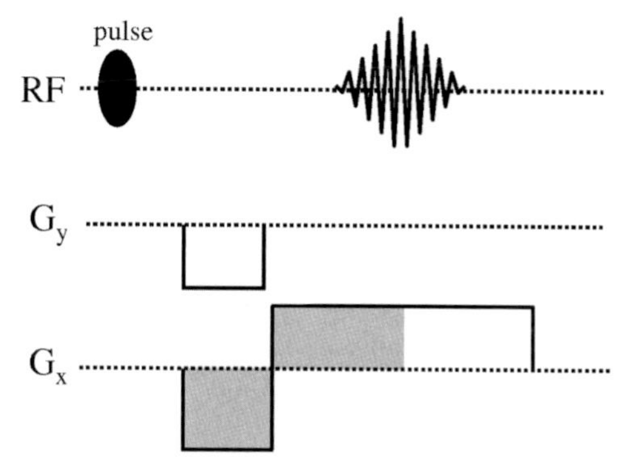

Fig. 17. A subset of a two-dimensional MR experiment. The echo covers a single horizontal line in *k*-space.

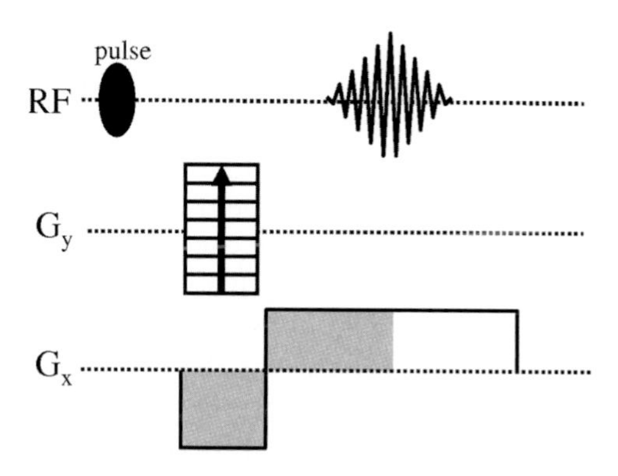

Fig. 18. Schematic description of a full two-dimensional MR experiment: the experiment shown in the previous figure is repeated several times with different values of the gradient G_y, so several *horizontal lines* in *k*-space are covered.

(left), the result is an image in the axial plane (**Fig. 19**, right). The full extent of the imaged object in *z*-direction is projected onto this plane, i.e. all axial slices are still added up.

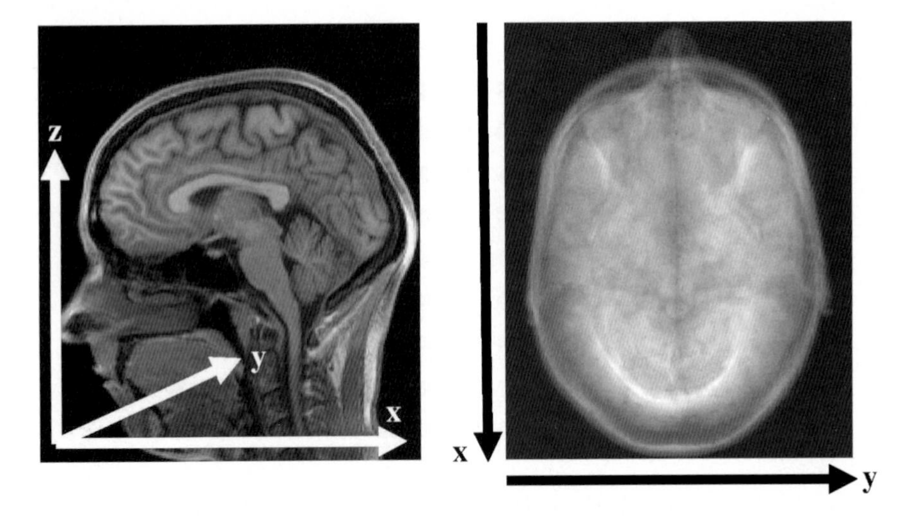

Fig. 19. If the gradient axes are chosen as shown on the left-hand side, the result of the two-dimensional experiment with a read gradient in *x*-direction and a phase gradient in *y*-direction is an image in the axial plane, showing an overlay of all axial slices.

7. Slice-Selective Excitation

So far, we have covered spatial resolution in two dimensions. To obtain spatial resolution in the third dimension, it would be useful to have a kind of "intelligent" excitation pulse which tilts the magnetisation only inside a slice of interest. This would mean that only spins within this slice could contribute to the signal, and we could subsequently employ the experiment described above to obtain spatial resolution in the remaining two directions. These *slice-selective excitation pulses* exist. They are based on the following principles. Let us assume we want to excite an axial slice through the brain (**Fig. 20**). As a first step, we switch on a gradient in the spatial direction perpendicular to this slice (*z*-direction). Consequently, the Larmor frequency of the spins will depend on their position: spins inside the slice of interest have a certain Larmor frequency f_0, whereas spins in upper/lower parts of the brain have higher/lower Larmor frequencies. If we send an RF pulse with the frequency f_0 while the gradient is switched on, it will tilt the magnetisation only inside the slice of interest (please remember that spin excitation is a resonance effect and affects only those spins whose Larmor frequency matches the frequency of the incoming RF pulse). In summary, we can say that an RF pulse which is transmitted while a field gradient is turned on causes a slice-selective excitation. After this special kind of excitation, we can proceed as shown in **Fig. 18** to achieve spatial resolution in two dimensions within the slice of interest.

The complete imaging experiment with spatial resolution in three dimensions is shown in **Fig. 21**. Basically, it is similar to **Fig. 18**, but comprises a slice selective excitation, including the *slice gradient* G_z.

The latter requires some further explanations. As shown in **Fig. 7** (part 2) and **Fig. 8b**, gradients cause dephasing of the spins and thus signal losses. The second half of the slice gradient (i.e. the part of the slice gradient that comes after sending the RF pulse) would have a similar effect. To avoid signal losses, a negative rephasing gradient has been added after the slice gradient which compensates this effect, as described for the gradient echo. Full compensation is approximately achieved if the shaded areas on the G_z axis in **Fig. 21** are identical.

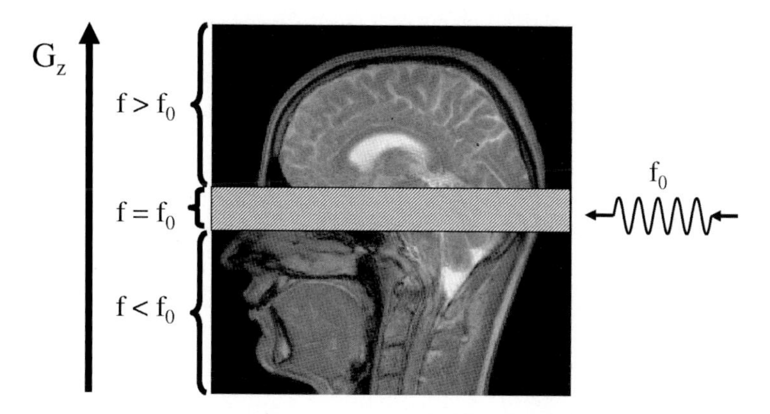

Fig. 20. The basis of slice selective excitation: due to the slice gradient G_z spins have spatially dependent Larmor frequencies. A radiofrequency pulse with frequency f_0 can only excite spins whose Larmor frequency corresponds to f_0. These spins are located in a plane perpendicular to the gradient direction.

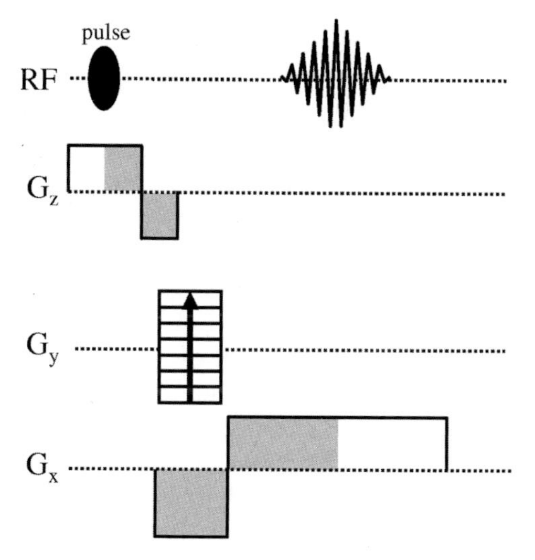

Fig. 21. Schematic description of a full three-dimensional MR experiment with slice selective excitation.

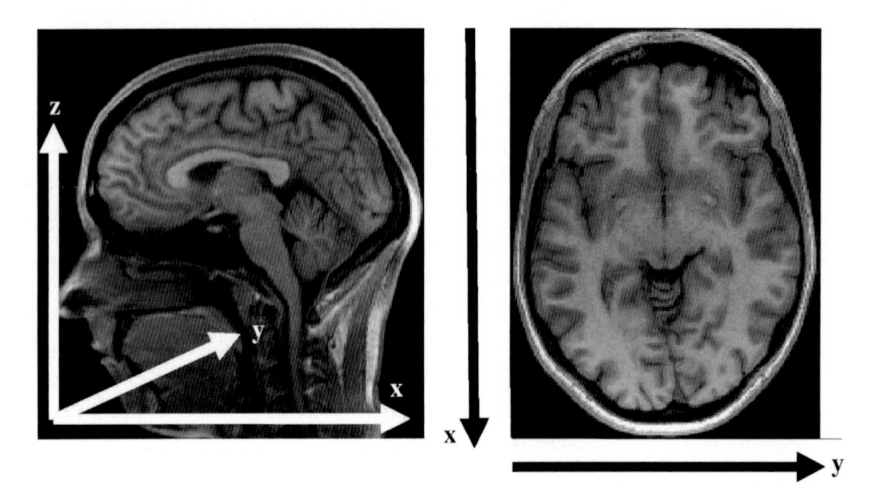

Fig. 22. If the gradient axes are chosen as shown on the *left-hand side*, the result of the three-dimensional experiment with a read gradient in *x*-direction, a phase gradient in *y*-direction, and a slice gradient in *z*-direction is an image in the axial plane with a finite slice thickness.

The experiment described in **Fig. 21** is three-dimensional, i.e. it allows for spatial resolution in all directions. If the gradient axes are chosen as shown in **Fig. 22** (left), the result is an image in the axial plane (**Fig. 22**, right) with a finite slice thickness.

8. Echo-Planar Imaging

The experiment described in **Fig. 21** can be relatively time consuming, because it requires an RF pulse for each single echo, i.e. for each line in *k*-space. Given the duration of RF pulses (typically several ms) it would be far more efficient to acquire all echoes after a single excitation pulse. One of these *single shot sequences*, dubbed echo-planar imaging (EPI), was developed by Mansfield *(4)* and is nowadays widely used in functional imaging experiments. It is based on the acquisition of multiple gradient echoes as described in **Fig. 23**.

The initial part of this experiment is identical to the one shown in **Fig. 13**: after the RF pulse, a negative gradient causes dephasing of the spins. During the subsequent positive read gradient, a gradient echo is acquired. The *k*-values of the data points constituting this echo increase, ranging from a negative to a positive value. Afterwards, the read gradient is *inverted*. It is obvious that this will result in another gradient echo with *decreasing k*-values, which means that the second echo covers the same *k*-values as the first one, only *in reverse order*. After a further gradient inversion, a third echo can be acquired which covers the *k*-values in exactly the same way as the first one. Of course, this is only a one-dimensional experiment, because no gradients in *y*-direction are used.

A two-dimensional expansion is shown in **Fig. 24**. As before, a series of gradient echoes is acquired by means of an oscillating read gradient. However, a phase gradient with negative amplitude

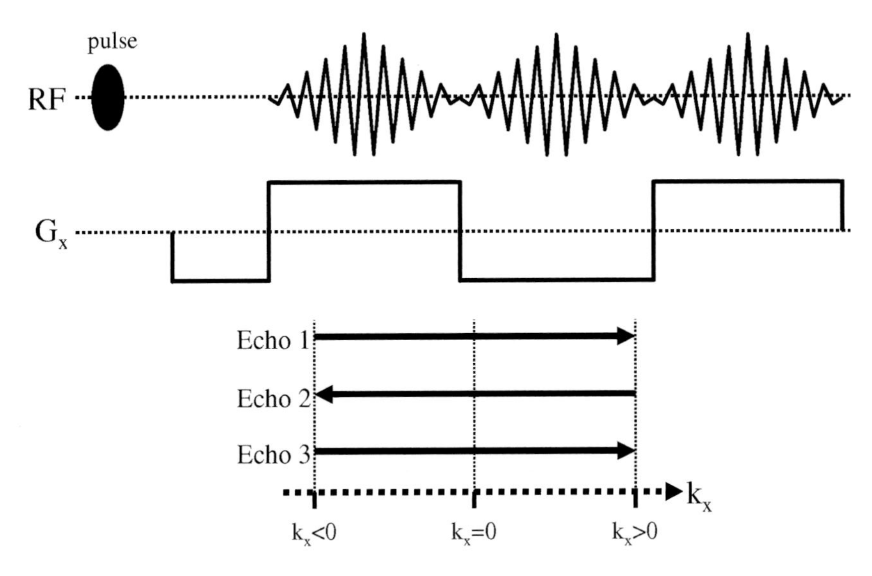

Fig. 23. Acquisition of multiple gradient echoes by successive read gradient inversion. All echoes cover the same *k*-values, but the order is reversed when comparing even and odd echoes.

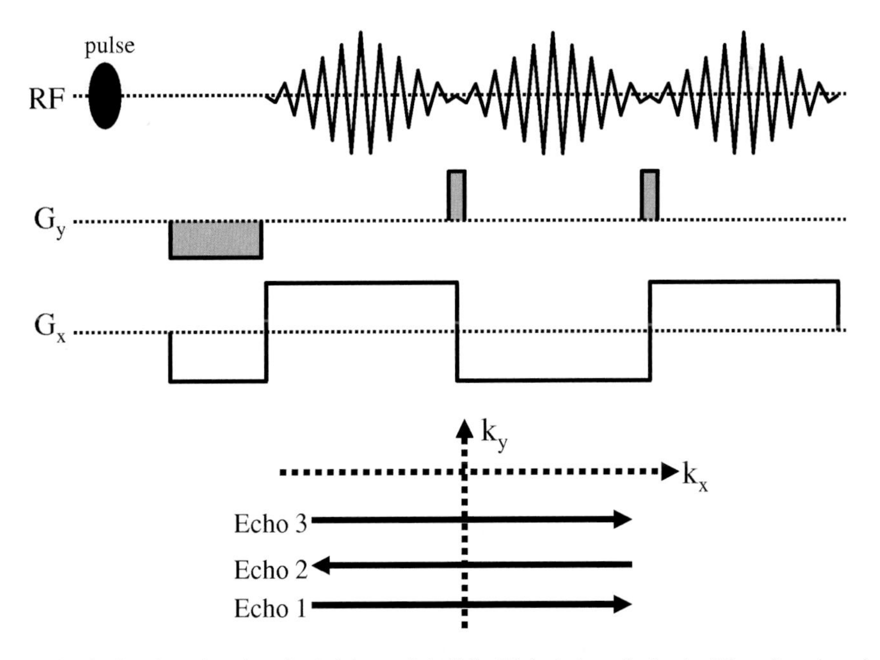

Fig. 24. The basis of echo planar imaging: due to intermediate "blips" (*shaded gradient pulses*) the echoes have increasing k_y-values, thus covering different lines in *k*-space.

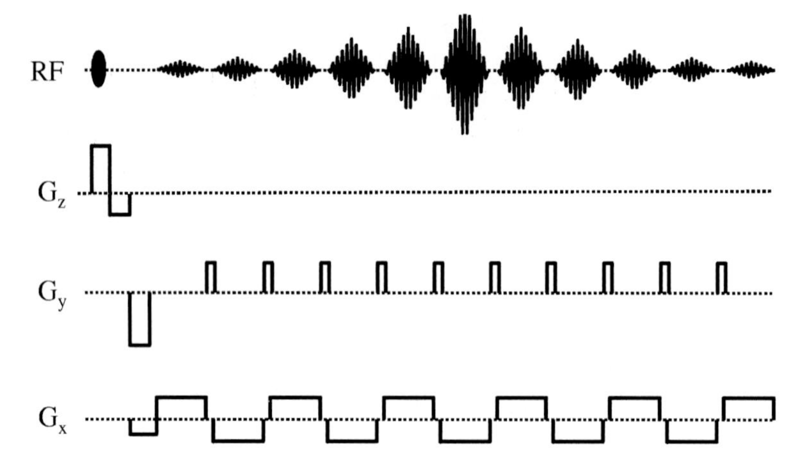

Fig. 25. A complete echo planar imaging experiment. The central echo which covers the centre of *k*-space has the highest amplitude.

is switched before the first echo (shaded), so all data points constituting the first echo have the same (negative) k_y-value, covering a horizontal line in *k*-space (**Fig. 24**, bottom). Between the acquisition of the first and the second echo, a very short phase gradient pulse is switched (a so-called *blip*, shaded in **Fig. 24**). The k_y-value of the second echo is determined by the sum of the areas under the initial phase gradient and this blip, so k_y is increased and the second echo covers another horizontal, slightly "higher" line in *k*-space in reverse direction. This concept of intermediate blips is maintained throughout the remaining acquisition, resulting in a meander-like journey through *k*-space.

Figure 25 shows the complete EPI experiment, including a slice gradient G_z (with a subsequent negative rephasing gradient) for slice-selective excitation. The echoes have different amplitudes due to their different positions in *k*-space, with the highest amplitude for the echo covering the centre of *k*-space, as described before.

The main advantage of the EPI sequence is its speed due to the use of a single RF pulse only, with typical acquisition times of 50–100 ms per slice. Another advantage is its susceptibility to the blood oxygen level-dependent (BOLD) effect *(5, 6)* which is exploited in the majority of functional imaging studies and will be discussed in a later section.

9. The Transverse Relaxation Times T_2 and T_2^*

The first experiment in **Fig. 7** (part 1) describes the acquisition of an MR signal after sending an excitation pulse, assuming the absence of any field gradients. According to **Fig. 8a**, spins at different locations will have the same Larmor frequency, so there are no dephasing effects, and in theory there should be no signal loss at all. However, the signal will still decay due to an effect called *transverse relaxation*. This is caused by the *spin-spin interaction*: in the classical view, the spins randomly exchange small amounts of energy, resulting in minor fluctuations of their Larmor frequencies and thus in gradual signal dephasing, even in otherwise perfectly homogeneous fields. The signal decays exponentially with a time constant called the *transverse relaxation time* T_2. In white matter and grey matter, T_2 has an approximate value of about 100 ms.

In reality, the signal would decay with a time constant even shorter than T_2. This is due to the following effect: tissue is not a homogeneous piece of matter, but consists of several, often microscopic components, for example arterioles and venules. In general, these components have slightly different magnetic properties, so they distort the magnetic field and create microscopic field gradients of varying amplitude and direction. As shown in **Fig. 7** (part 2) and **Fig. 8b**, the presence of field gradients speeds up the signal decay due to spin dephasing. Thus, the signal decays with the *effective transverse relaxation time* T_2^* which is shorter than T_2. This time constant depends on the scanner field strength. In white and grey matter, T_2^* amounts to about 70 ms at 1.5 T and 45 ms at 3 T.

For several applications (including the most common fMRI techniques), it is useful to acquire so-called T_2^* *weighted images*, i.e. images where the local signal intensity depends on the local T_2^* value. The degree of T_2^* weighting can be influenced by modifying a certain acquisition parameter, the *echo time* (TE).

10. The Echo Time

All MR experiments that were discussed so far are based on the same concept: an initial excitation pulse tilts the magnetisation. After a certain time (during which one or more gradients are switched on and off) a signal is acquired. The time delay between excitation and acquisition is called the echo time (TE) (**Fig. 26**).

To achieve a certain degree of T_2^* weighting, TE must be chosen carefully. This is depicted in **Fig. 27**, showing a fast (dashed line) and a slow (bold line) T_2^* decay, and three different choices for TE.

The first choice (TE much shorter than T_2^*) would be problematic, because signal amplitudes are almost identical, so T_2^* contrasts would be poor.

The second choice (TE similar to T_2^*) would yield a much better T_2^* contrast.

The third choice (TE much longer than T_2^*) is again problematic: signal has decayed in both compartments, so the image would show noise only, but hardly any structures.

As an example, **Fig. 28** shows T_2^* weighted brain images acquired with TE values of 10 ms, 50 ms, and 200 ms. At 10 ms, contrasts are relatively low. At 50 ms, a nice T_2^* contrast can be observed (arrow). This is due to an increased iron content which gives rise to magnetic field distortions and therefore reduces the T_2^* value. At a TE of 200 ms, signal has mostly decayed and only cerebrospinal fluid is visible due to its longer T_2^* value.

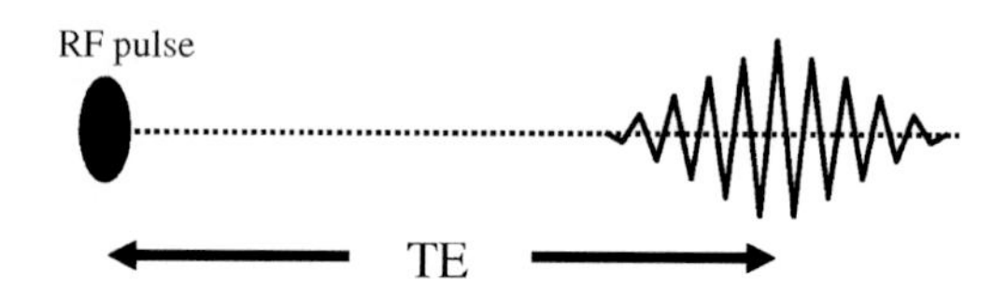

Fig. 26. Definition of the echo time (TE).

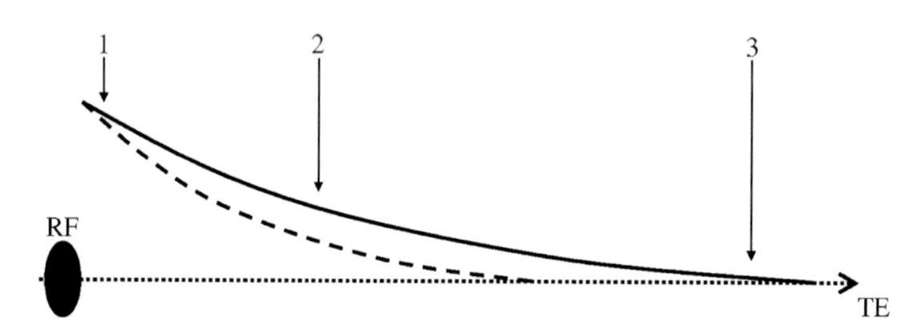

Fig. 27. Signal decay in compartments with a long T_2^* value (*solid line*) and a short T_2^* value (*dashed line*). If a short echo time (TE) is chosen ***(1)***, there is a high signal amplitude, but hardly any contrast between both compartments. For an intermediate TE ***(2)***, there is a good contrast and still a sufficient signal amplitude. For a long TE ***(3)***, the signal in both compartments has decayed.

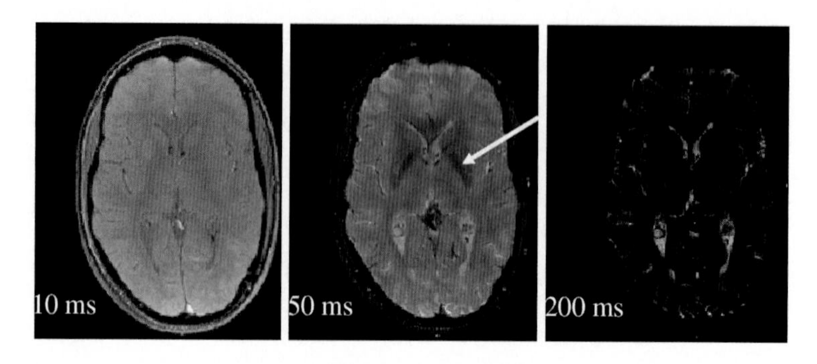

Fig. 28. T_2^* weighted brain images acquired with echo time values of 10 ms, 50 ms, and 200 ms. At 10 ms, contrasts are relatively low. At 50 ms, a nice T_2^* contrast can be observed (*arrow*). At 200 ms, signal has mostly decayed and only cerebrospinal fluid is visible.

11. The Basis of the BOLD Contrast

The majority of fMRI techniques are based on the BOLD contrast which will be explained in this section.

The BOLD contrast is closely linked to two physical phenomena, called "diamagnetism" and "paramagnetism". A full discussion would be beyond the scope of this book. To understand the BOLD effect, it is sufficient to know the following facts: if a diamagnetic substance is brought into an external magnetic field, it tends to decrease slightly this field, whereas a paramagnetic substance tends to increase it. This means that the close vicinity of paramagnetic and diamagnetic substances causes a local distortion of the magnetic field near the interface.

Tissue is mainly *diamagnetic*. In contrast, blood contains a certain level of deoxyhaemoglobin (i.e. haemoglobin that does not have oxygen attached) which is *paramagnetic*. Thus, the presence of blood in tissue means a close vicinity of substances with different magnetic properties, giving rise to microscopic field distortions. As explained above, the resulting field gradients cause spin dephasing and lower the T_2^* value. In summary one may say that due to the presence of deoxyhaemoglobin, the signal intensity of tissue is slightly reduced in T_2^* weighted images.

After neuronal activation, blood is locally hyperoxygenated, corresponding to a wash-out of deoxyhaemoglobin and an increased concentration of oxyhaemoglobin. In contrast to deoxyhaemoglobin, oxyhaemoglobin is diamagnetic, so it has similar magnetic properties as tissue, leading to a more homogeneous magnetic field and an increased signal intensity in T_2^* weighted images.

Thus, the basic concept of the BOLD effect can be summarised as follows: the *haemodynamic response* to brain activation consists in a *decrease in deoxyhaemoglobin* and an *increase in oxyhaemoglobin*, resulting in an increased field homogeneity and thus a higher signal intensity in a series of T_2^* weighted images. Quantification of this signal enhancement therefore allows for the detection of neuronal activation.

In reality, the physiology of the BOLD effect is more complex and depends on the following parameters: the cerebral blood flow (CBF), the cerebral blood volume (CBV), and the metabolic rate of oxygen consumption ($CMRO_2$). After a stimulus, the CBF goes up to deliver more oxygen to the site of neuronal activation, causing the BOLD effect as explained above. On the other hand, the $CMRO_2$ is increased, so more oxygen is consumed, which reduces the BOLD effect. The question arises if the first effect outpaces the second one, which is a prerequisite for blood hyperoxygenation and thus the detectability of the BOLD signal.

The exact physiology of the BOLD response is still controversial and several models have been proposed. For a more detailed overview, the reader is referred to the literature *(7)*. In the following section, one of the most common models will be explained.

Figure 29 is a simplified sketch, showing how the physiological parameters are affected by neuronal activation and how their interaction influences the signal intensity in T_2^* weighted images. Directly after the stimulus, $CMRO_2$ goes up, causing increased oxygen consumption and thus increased concentration of deoxyhaemoglobin. As explained above, deoxyhaemoglobin lowers the signal intensity in T_2^* weighted images, so there is an initial signal *reduction* with a duration of about 1 s, the so-called *initial dip*. It should be noted that this effect is small and not always present. About 1 s after the stimulus, the brain reacts by increasing the CBF, transporting oxygen to the site of activation. Fortunately, this effect outpaces the increase in $CMRO_2$, so blood becomes in fact hyperoxygenated. At the same time, the CBV is increased. As this parameter determines the total amount of blood, an increased CBV causes an increased quantity of deoxyhaemoglobin. However, this effect is again outpaced by the increase in CBF, and for a period of 4–6 s blood is hyperoxygenated, giving rise to a *positive BOLD response*. After this, $CMRO_2$ and CBF return to their baseline values. The relaxation of CBV is somewhat slower, so for a certain time there is an increased concentration of deoxyhaemoglobin, resulting in a *post-stimulus signal undershoot* with a duration of about 30 s.

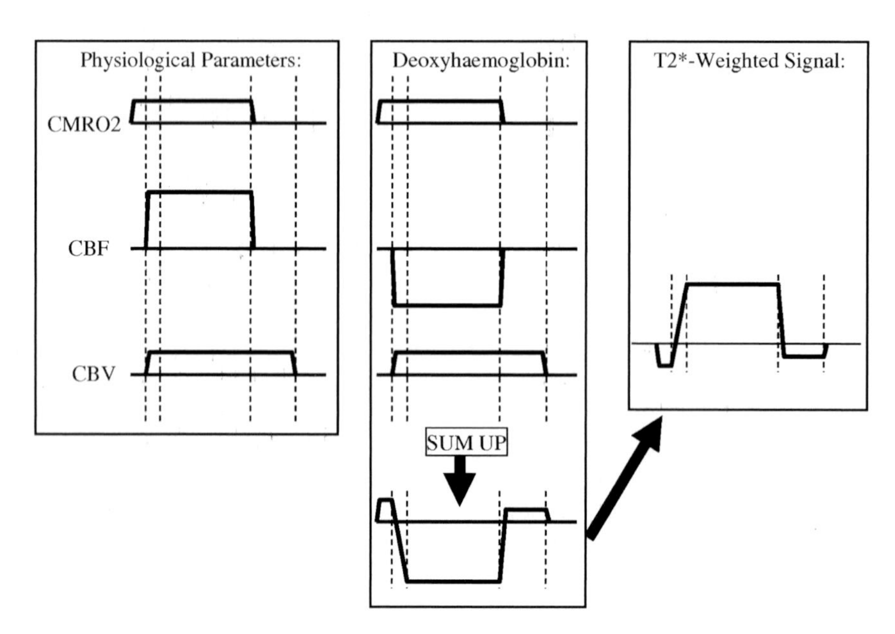

Fig. 29. Change of physiological parameters, the concentration of deoxyhaemoglobin, and the T_2^* weighted signal amplitude in response to neuronal activation.

In summary, the right-hand side of **Fig. 29** shows the complete signal behaviour following neuronal activation, the so-called *haemodynamic response function* (HRF): after the initial dip, there is a strong positive BOLD response, followed by a small negative signal undershoot.

12. Choice of TE in fMRI Experiments

The question arises, which value of TE should be chosen to maximise the BOLD signal, and how to set up the parameters of the EPI sequence described above to achieve this value.

Figure 30 shows the theoretical dependence of the BOLD sensitivity on TE (which is given in units of T_2^*). The results are not surprising and correspond to the previous discussion (*see* **Fig. 27**): at short TE, the BOLD sensitivity is low because there is hardly any T_2^* weighting. At long TE, the BOLD sensitivity goes down because the signal decays due to transverse relaxation effects. Maximum BOLD sensitivity is achieved when TE equals T_2^* (about 70 ms at 1.5 T and 45 ms at 3 T).

However, there is a fundamental problem with EPI sequences used in fMRI experiments: the magnetic field is usually distorted in brain areas that are close to air/tissue interfaces, in particular in the orbitofrontal cortex (due to the vicinity of the nasal sinus)

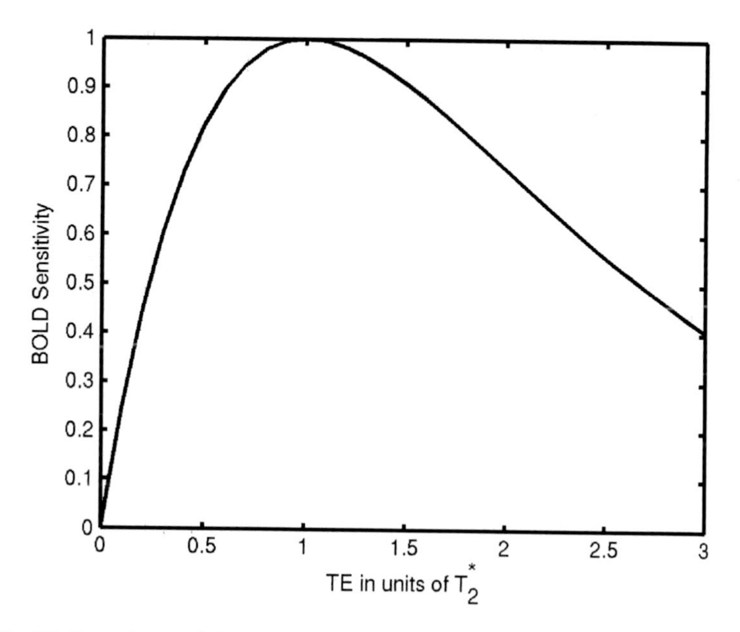

Fig. 30. Dependence of the blood oxygen level-dependent (BOLD) sensitivity on the chosen echo time (TE).

and the temporal lobes (due to the vicinity of the ear canals). In these areas, local macroscopic field gradients lower the T_2^* value, resulting in signal losses in EPI sequences with relatively long TE values. This shows the basic dilemma with fMRI experiments based on the BOLD effect: on one hand, T_2^* weighting is required to be able to detect the haemodynamic response to neuronal activation. On the other hand, T_2^* weighting causes severe signal losses in certain brain areas. Therefore, it is advisable to keep TE as short as possible to avoid signal losses in these areas, but still as long as necessary to detect a BOLD signal. According to **Fig. 30**, a decent BOLD sensitivity can still be achieved if TE corresponds to about 2/3 of T_2^*. The general advice is therefore to use a TE of about 50 ms at 1.5 T, and 30 ms at 3 T.

The next question is how TE can be defined for the EPI sequence. As shown in **Fig. 25**, EPI implies the acquisition of a series of gradient echoes after a single excitation pulse, i.e. each echo has a different echo time. However, as explained above these echoes have different amplitudes, with maximum signal strength for the echo that covers the centre of *k*-space (usually the central echo in the series if symmetric sampling is used). Therefore, the TE value of an EPI sequence is defined as the echo time of this central echo (**Fig. 31**). This shows that in EPI sequences TE is not simply a wasted waiting time (as one might have expected from **Fig. 26**), but can be used for the acquisition of the first half of the echo train, being another reason why EPI allows for a high temporal resolution in fMRI experiments.

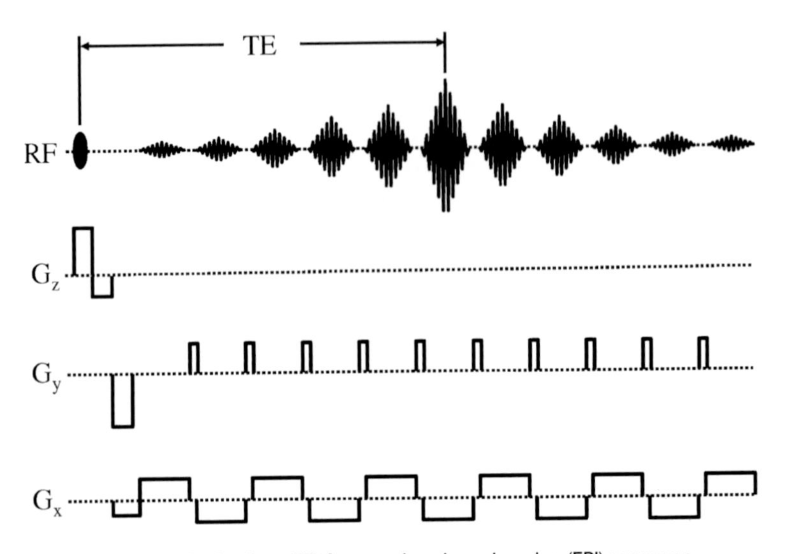

Fig. 31. Definition of echo time (TE) for an echo-planar imaging (EPI) sequence.

References

1. Bloch F, Hansen WW, Packard M. Nuclear induction. Phys Rev 1946;69:127.
2. Purcell EM, Torrey HC, Pound RV. Resonance absorption by nuclear magnetic moments in a solid. Phys Rev 1946;69:37–38.
3. Lauterbur PC. Image formation by induced local interactions: examples employing nuclear magnetic resonance. Nature 1973;242:190–191.
4. Mansfield P. Multiplanar image formation using NMR spin echoes. J Phys C 1977;10:L55–L58.
5. Ogawa S, Lee TM, Nayak AS, Glynn P. Oxygenation-sensitive contrast in magnetic resonance image of rodent brain at high magnetic fields. Magn Reson Med 1990;14:68–78.
6. Kwong KK, Belliveau JW, Chesler DA, et al. Dynamic magnetic resonance imaging of human brain activity during primary sensory stimulation. Proc Natl Acad Sci USA 1992;89:5675–5679.
7. Buxton RB, Uludag K, Dubowitz DJ, Liu TT. Modeling the hemodynamic response to brain activation. NeuroImage 2004;23:S220–S233.

Chapter 2

Introduction to Functional MRI Hardware

Luis Hernandez-Garcia, Scott Peltier, and William Grissom

Summary

This chapter discusses MRI hardware components, stimulus presentation devices, response, and physiological data collection systems. The general guidelines for MR-compatible hardware are also discussed. The chapter gives an overview of commonly used peripheral devices used in fMRI experiments, and it addresses the principles, performance aspects, and specifications of fMRI hardware. The target audience is quite broad and mathematical descriptions are kept to a minimum and qualitative descriptions are favored whenever possible.

Key words: Functional MRI, Hardware, Peripheral devices, MRI, Multimodal acquisition, Neuroimaging

1. Introduction

This chapter is concerned with both the MRI hardware components and the multitude of peripherals that are necessary for fMRI. Our primary aim is to describe the different components of each subsystem and to identify the important features and parameters. We hope this chapter will be of some use to those who want to get a deeper understanding of the electronics involved, but we mostly want to convey how each of the parts is responsible for the quality (or lack thereof) of the images and the experiment in general. Thus we will try to give minimum requirements for the performance of each component and describe what happens when those requisites are not met.

There is a myriad of subsystems to explore and we cannot possibly do justice to all of them in this chapter, so we will limit

M. Filippi (ed.), *fMRI Techniques and Protocols*, Neuromethods, vol. 41
DOI 10.1007/978-1-60327-919-2_2, © Humana Press, a part of Springer Science+Business Media, LLC 2009

ourselves to the main ones and to those that most commonly affect the performance of the system.

While this chapter primarily addresses the principles, performance aspects, and specifications of the fMRI hardware, it is important that we keep in mind that the objective is to carry out experiments on human subjects performing cognitive tasks. Thus we will also keep in mind the ergonomics and safety characteristics of the equipment.

Our target audience is fairly broad so we will try to keep mathematical descriptions to a minimum and give qualitative descriptions whenever possible. However, some of the descriptions and the requirements make a lot more sense in the context of the mathematical description of image acquisition and reconstruction. We also hope that this chapter can serve as a good introduction to each topic for those interested in the specific subjects. In a way, we approach this chapter as if we were trying to give advice to someone who is considering setting up a fMRI facility and/or establishing a set of quality control protocols for such a facility. We will be careful to leave our descriptions as general as possible and to avoid endorsing specific vendors or commercial products.

2. The MRI Scanner Environment

Let us begin by considering the surroundings of the MRI scanner. When deciding on the layout and location of an MRI scanner, there are several important questions one must ask. The first one is the scanner's purpose. Will it be used for clinical purposes or will it be dedicated to research on healthy subjects? In the case of research-dedicated scanners, one must think about a number of specific factors when designing the layout of the fMRI laboratory. This includes dressing rooms and testing rooms where the subjects can be trained on the experimental paradigm prior to scanning. Whenever possible, it is important to make sure that the control room is large enough to accommodate the needs of the researchers using it. It is not uncommon for fMRI experiments to require multiple pieces of custom stimulation/recording equipment in the control room and for several investigators to be present during the experiment, so that extra bench space and elbow room is very advantageous. When scanning clinical populations, there are additional precautions and considerations, like the presence of MR compatible first aid equipment. We will not address that aspect in detail, as it can be an extensive discussion that will vary from case to case.

The first thing one notices about an MRI scanner is how big it is (the magnet alone can take up a space of roughly $4 \times 4 \times 6$ m) and one must find a site with sufficient space for the scanner. Access is also important, as the main magnet is typically delivered and installed in one piece. Furthermore, MRI magnets are filled with cryogens, which are delivered in large dewars. Thus there must be a wide path to the loading dock that avoids stairs and is clear of obstacles. Typically, MRI scanners are housed in basements or ground floors near the loading docks of hospitals, although exceptions exist, as the one shown in **Fig. 1**.

Next one must consider how the magnetic field will affect its surroundings. Before the magnet is installed, one must consider how far the magnetic field will extend. Most people who work with MRI are familiar with the enormous forces that the main magnetic field can exert on objects in its proximity (i.e., the magnet room). Modern MRI scanners are actively shielded and contain the magnetic field fairly well within the magnet room. Even with shielding, sometimes a subtle but significant magnetic field can extend beyond the walls of the magnet room. Thus it is important to keep in mind two questions: how well the magnetic field is contained, and what sort of equipment is in the rooms adjacent to the magnet room. The first question is usually answered in terms of the location of the "5-Gauss line."

The United States FDA regulates that the general public (anyone not working with an MRI scanner or being scanned) not be exposed to static magnetic fields over 5 Gauss (5×10^{-4} T), and thus the 5-Gauss boundary must be contained inside the magnet room. MRI scanner vendors will typically provide contour plots of the magnetic field superimposed on the blueprints

Fig. 1. These photos were taken during the installation of an MRI scanner at Resurgens Orthopaedics in Atlanta, GA (courtesy of Resurgens Orthopaedics, Atlanta, GA).

of the room and provide consultation on the location of the scanner. One must realize, however, that smaller magnetic fields will extend beyond the walls of the magnet room, so it is important to note how quickly the field decays and what is located on those adjacent rooms. Subtle magnetic fields can affect electronic equipment in many ways. For example, a moving charge in the presence of a magnetic field will experience a force perpendicular to the magnetic field. This effect is most obvious in cathode ray tube monitors, whose images are skewed by the magnetic field, and in the performance of computer hard drives and magnetic media. In fact, magnetic media, including the magnetic strips on credit cards, floppy disks, etc., are typically erased when taken into the magnet room. Another example is that the lifespan of light bulbs near MRI scanners tends to be quite short because of the vibration of the filaments caused by switching the direction of the current in the presence of a large magnetic field. Hence, direct current (DC) lights are often used in MRI scanner rooms to avoid this problem. Pacemakers, neurostimulators, and implants must be kept outside the 5-Gauss line, unless they have been tested and classified as MR compatible. A number of publications exist ***(1–3)*** and are updated regularly with classification of MR compatible devices.

Containment of the magnetic field can be achieved by two different kinds of shielding. Passive shields consist of building a symmetric box around the magnet out of thick iron walls **(Fig. 2)** that contain the magnetic field. Alternatively, active shields can be built as secondary electromagnets concentrically placed around the main magnet. The shield magnets are built such that the field they produce points in the opposite direction of the main magnetic field. The strength of the shielding field is typically about half of the main field. For example, a 3.0-T actively shielded magnet can often be a 4.5-T magnet surrounded by a second, concentric, and opposing 1.5-T magnet. Such a system would have a 3.0-T field inside of the bore, and because of the inverse square dependence of the magnetic field on the distance to the coil, the field outside of the magnets bore is dramatically reduced. **Figure 3** shows a plot of the magnetic fields produced by the main magnet as a function of the distance to the center of the bore. The field produced by the shield and the net sum of both fields are superimposed on the same plot.

One must also consider how the environment will affect the MRI scanner. In this regard, the most important issue is the presence of electromagnetic noise sources. MRI scanners construct images from radio frequency (RF) electromagnetic signals. Thus, radio stations, cell phones, and other wireless communications will interfere with the MRI experiment and severely degrade image quality unless they are properly isolated. MRI rooms are usually encased in a copper shield box that blocks external RF

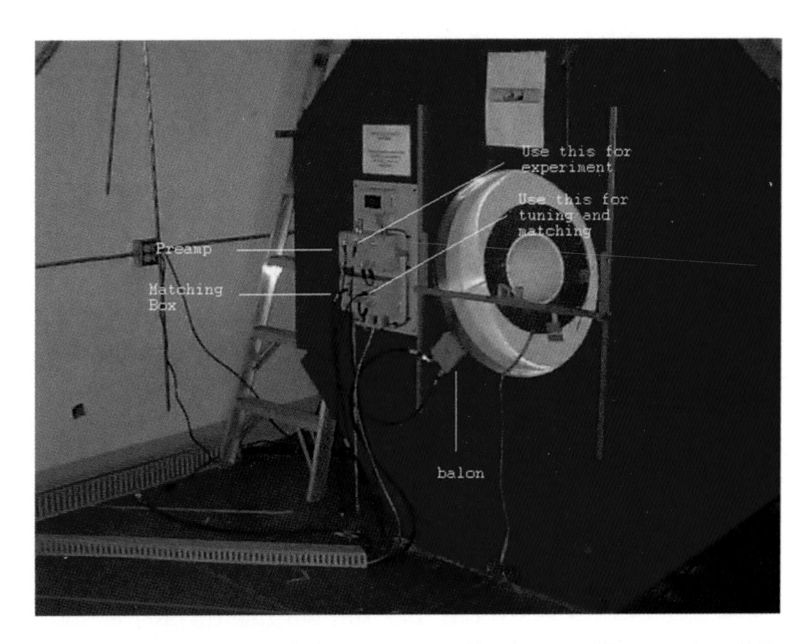

Fig. 2. A passively shielded 4.0-T magnet encased in a hexagonal iron cage to partially contain the main magnetic field.

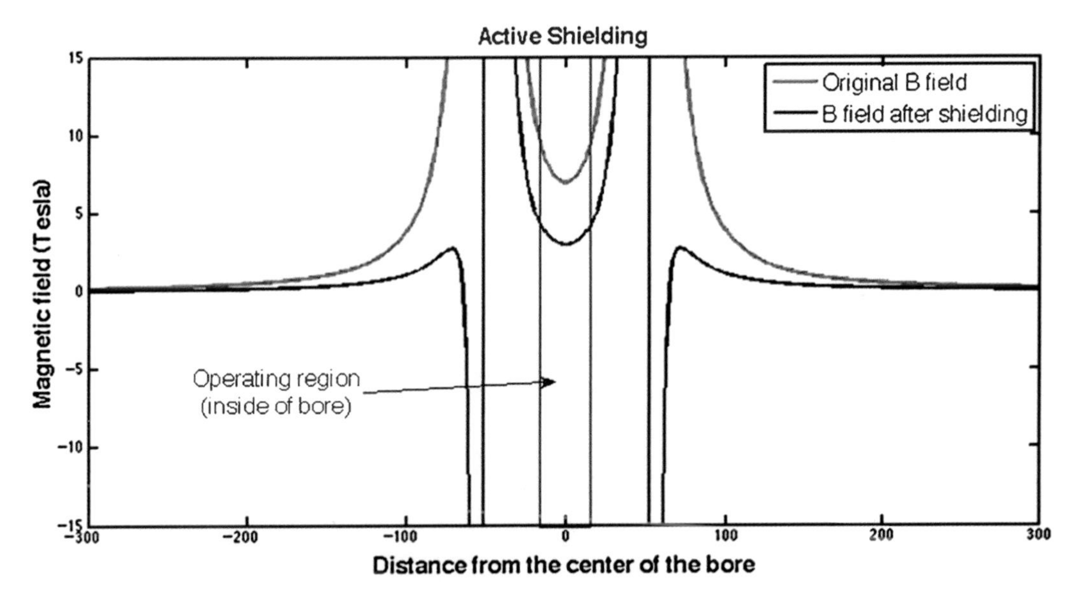

Fig. 3. A plot of the magnetic field strength along the radial direction. The *thick vertical lines* denote the location of the coil's windings. The *thinner line* denotes the operating region where the sample is placed. Ideally, this region should have a flat magnetic field, and additional fields (shims) are necessary.

radiation and contains the MRI's RF radiation as well. The quality of the shield is critically important to the performance of the scanner and it must be tested carefully before the magnet is ramped up to field. Typically attenuations for RF shielding are 100 dB at the operating frequency range. A slightly defective soldered seam between copper sheets, or a nail going through the copper sheeting, is sufficient to let RF noise into the room that ruins images, so testing of the room shielding must be stressed. **Figure 4** shows an example of the effects of an RF noise source on an MR image.

Good shielding of the room is not enough. MR-related electronic equipment produces RF noise and can act as an antenna that passively carries RF noise from the outside into the room. There are also a number of peripherals that are needed for fMRI in order to provide stimulation and record data from the subject (especially for fMRI! – *see* **Subheading 4**). It is preferable to keep all electronics out of the magnet room, but if the electronic equipment must be inside, it must be tested thoroughly for RF noise. If it is indeed noisy, care must be taken to shield the equipment to prevent image artifacts (Copper mesh is very useful for building RF shields).

Consider the case of a button response box. Typically, one keeps the bulk of the electronic components out of the magnet room, but the buttons themselves must be in the scanner and they need to communicate with the response recording electronics. Even if fiber-optic technology is used to carry signals into the room, RF noise can enter through the same opening as the fiber-optic cabling. The solution is to build a "penetration panel" into the shield. This is a panel on the wall that is outfitted with waveguides and filtered connectors. The role of a waveguide is to block any electromagnetic radiation that is not parallel with the direction of the waveguide, and at the same time, to guide the electromagnetic waves produced inside it. The filtered connectors typically remove high frequency radiation that may be carried by the cabling of the peripherals. Examples of penetration panels

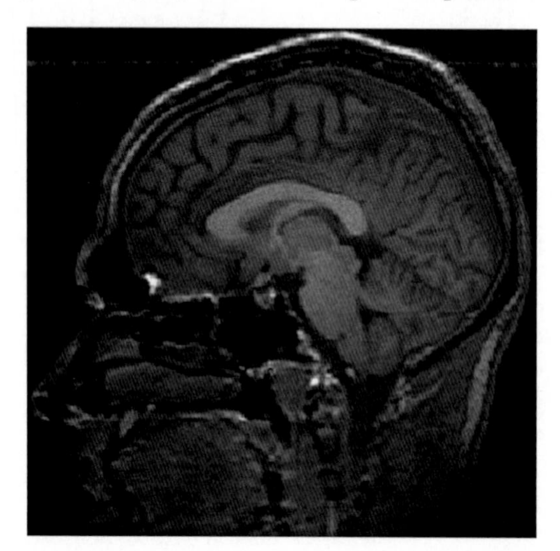

Fig. 4. The streaks in the image are caused by the presence of electromagnetic frequency noise at a single frequency. This is typically introduced by the presence of badly shielded electronic equipment in the room. The alternating current power supply running at 60 Hz to the device is in this case the culprit of the artifact.

used for scanner cabling and for general, user-specific peripherals are shown in **Fig. 5**. Additional precautions against RF contamination are the use of twisted-pair and coaxial cabling to contain the fields produced inside the transmission lines.

MR scanner equipment must be kept in stable temperature and humidity conditions, as these can change the performance of the electronics and the field strength. All the amplifiers and computer equipment required for MR imaging can generate a significant amount of heat, so it is important that the environmental temperature-regulating equipment be powerful enough to handle it. Changes in the scanner performance can mask changes in brain activity so the scanner's performance be maintained as constant as possible.

To put things in perspective, the MRI electronics equipment produces approximately 100,000 British Thermal Unit (BTU) per hour. Typical requisites for the temperature and humidity in the room are variability of less than 3°C per hour and 5% per hour, respectively. Normal operating ranges are in the 15–25°C temperature range and 30–75% humidity range. Of particular interest are the gradient coils, since they can heat up significantly as a result of the large currents that run through them. As gradient coils heat up, their performance is degraded and thus they require a cooling system to keep them stable. This system is usually a cooling loop involving a water chiller that can remove about 14,000 BTU per hour.[1]

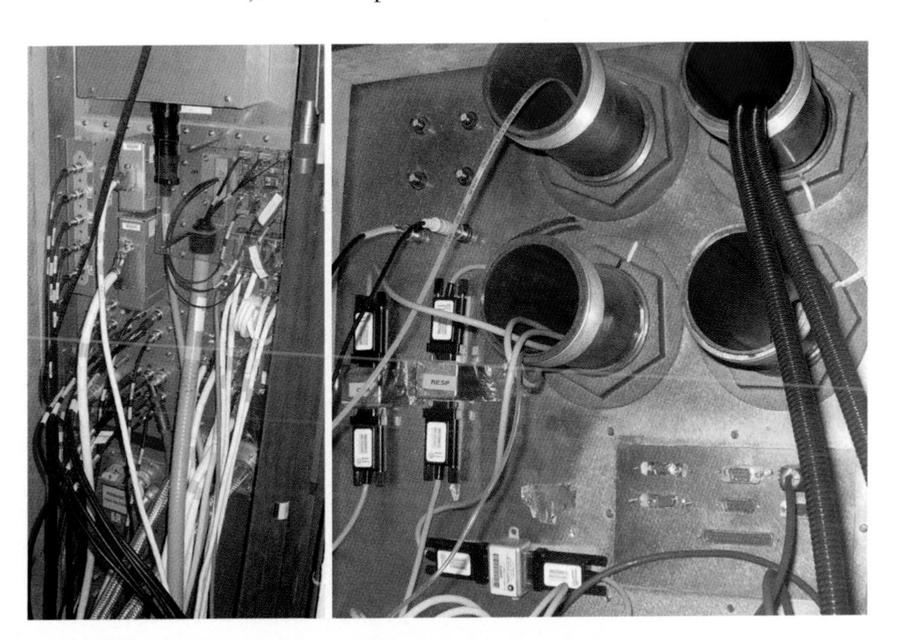

Fig. 5. On the *left* is the penetration panel that connects the MRI scanner electronics to the magnet hardware. The image on the *right* is a second penetration panel used for all the additional stimulus/response equipment needed for fMRI.

[1]These numbers are based on the specification of a 3.0-T scanner by General Electric.

It is also noteworthy that the scanner's performance can be affected by other unexpected environmental factors. Outside magnetic fields and vibrations can be an issue. For example, nearby construction can produce vibrations that will affect the MR signal's stability if the floor is not adequately mechanically damped. It is thus important to carry out vibration tests of the site before installation proceeds. Large moving objects, such as nearby trains can also generate magnetic fields that affect the scanner's stability *(4)*.

One other issue that can cause a great deal of grief to investigators is the production of small electromagnetic spikes inside the magnet room. These occur when metal objects in the room vibrate (typically because of the gradients) and bang against each other or when arcing occurs across badly soldered connections in homemade equipment. The RF receiver hardware is sensitive enough to pick up these spikes. Spikes in the *k*-space data translate into stripe patterns in the images **(Fig. 6)**. It is thus very important to make sure that all metal equipment is well secured.

A useful option to consider for a fMRI laboratory is a mock MRI scanner. This is advantageous since it allows subjects to get used to the fMRI experience prior to their actual scans, in a safe environment. This can help alleviate subjects' apprehension and claustrophobia, and lead to reduced head motion and improved task performance.

For completeness, replication of the entire MR suite would be preferable, but usually this is limited by available space and resources. The most important factors to reproduce are the spatial

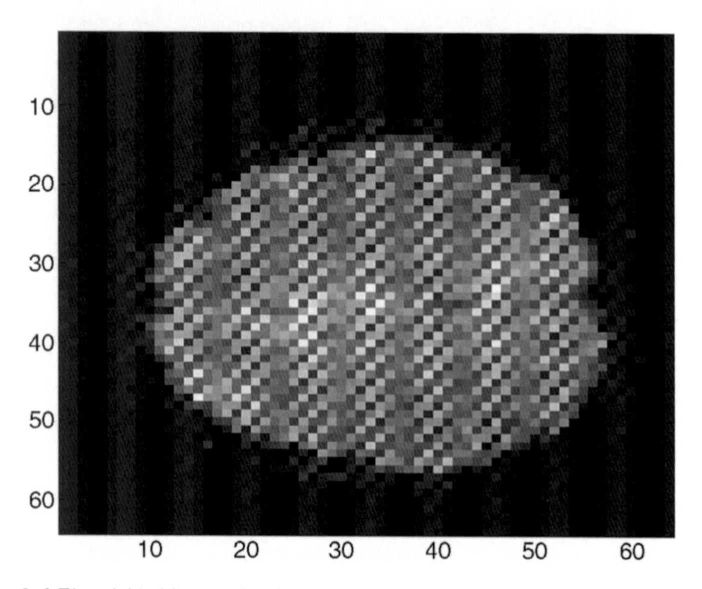

Fig. 6. A T_2^*-weighted image showing white pixel artifact. Spikes in *k*-space results in sinusoidal patterns in image spike.

dimensions, audio environment, and visual stimulus presentation of the MRI scanner.

It is important to have the subject familiarize themselves with the restrictive space in the MR scanner. This includes the inner diameters of both the bore of the main magnet and the head coil being used. These restrictions, combined with the distance the subject travels into the magnet from the home position of the patient bed, combine to give the overall physical experience. MR or CT patient beds and scanner housing may be recycled for this use, or the mock MR scanner can be built from scratch as the one shown in **Figs. 7** and **8**.

The audio environment of the scanner is also an important consideration. Having an audio recording of the actual scanner to play in the mock MR scanner will let the patient adjust to

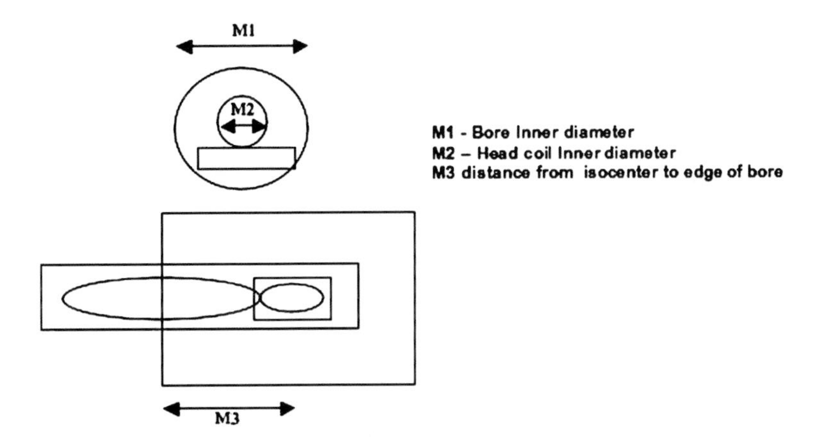

Fig. 7. Diagram of mock scanner showing critical dimensions.

Fig. 8. Example of a mock MRI scanner (University of Michigan).

the jarring transition when the scanner starts playing sequences and to the volume of this noise throughout the scan. Inclusion of audio feedback can also demonstrate to the subjects that the scanner operator will be able to communicate with them between scans.

A duplication of the fMRI visual stimulus presentation can also serve to acclimatize the subjects. The standard forward- or rear-projection systems used to present visual stimuli are relatively easy to replicate in a non-MR environment, and allow the subject to get used to task presentation during the scan.

An MR mock scanner can help to increase patient comfort and task performance, especially in some target populations (e.g., individuals with autism, children), and can be used for screening (e.g., individuals with claustrophobia, large physical dimensions). It can also be useful for the training of fMRI lab personnel or the designing, troubleshooting, or rehearsing of complicated fMRI experiments.

3. The MRI Scanner

We can divide the components of an MRI scanner into four categories: the magnet, the magnetic field gradients, the RF Excite/Receive hardware, and the data acquisition electronics.

3.1. The Magnet

We will begin by considering the magnet. Typical MRI magnets, like the one shown in **Fig. 9**, are large solenoid coils made of superconducting metal (niobium alloys, typically). They are kept cooled at approximately 4K by liquid Helium in order to achieve and maintain superconductivity.[2] The magnet is "ramped up" to field by introducing a current through a pair of leads that produces the desired magnetic field. Once the specified current has been built up, the circuit is closed such that the current "recirculates" through the coil constantly and there is no need to supply more power to it. It is crucial to maintain the low temperature to prevent the coil from resisting the current flow.

If the windings ever warm up and become resistive (i.e., they lose their superconducting state), they dissipate the electric power as heat. This very undesirable event is termed "quenching." As the magnet's windings become more resistive, the very high current circulating through the coil produces heat, such that the liquid helium that is responsible for maintaining the superconducting temperature boils off. This rapid boiling of the helium quickly

[2]For more information on superconductivity, see **ref.** *5*.

Fig. 9. A 3.0-T magnet during installation at the University of Michigan's FMRI Laboratory.

exacerbates the problem and the superconducting state is quickly lost. The very large currents can potentially produce enough heat to melt or severely damage the windings. The greatest danger, however, is that the rapid rate of helium boiling can build up a great deal of pressure in the magnet and also flood the room with helium gas and suffocate whoever is there. While helium is not toxic, it displaces the oxygen in the room. Thus, it is crucial that the magnet be outfitted with emergency vents (manufacturers of MRI scanners typically include emergency quench ventilation systems). Additionally, magnet rooms are outfitted with oxygen sensors that sound an alarm when the oxygen level falls below safe levels. It must be stressed that all personnel be trained in emergency quench procedures in case the emergency systems fail.

Having considered what can happen when the magnet fails, let us now return to the more cheerful subject of what the magnet can do.

The key parameter in the magnet is its field strength (B_0), as it determines many properties of the images. Primarily, field strength determines the amount of spins that align with and against the field. The higher the field strength, the larger the population of aligned spins that can contribute to the MR signal. More specifically, the population of spins aligned with the magnetic field (n_+) and against it (n_-) is described by the Boltzmann equation

$$\left(\frac{n_+}{n_-}\right) = \exp\left(\frac{\gamma\hbar B_0}{kT}\right), \tag{1}$$

where γ is the gyromagnetic constant for the material, h is Plank's constant, k is Boltzmann's constant, T is the temperature of the sample, and B_0 is the strength of the magnetic field. Hence, the higher the field, the more spins will contribute to the signal and thus yield a higher signal-to-noise ratio (SNR).

The field strength also determines the resonance frequency of the spins, ω_0, in a linear fashion according to the Larmor equation

$$\omega_0 = \gamma B_0, \tag{2}$$

where γ is again the gyromagnetic constant, which is specific for the nucleus in question. The resonant frequency will in turn determine the characteristics of the RF transmit and receive subsystem (*see* **Subheading 3.3**). Field strength also affects both the longitudinal and transverse relaxation rates of the materials via the resonant frequency as predicted by the equations

$$\frac{1}{T_1} \alpha \left\{ \frac{\tau_c}{1+\omega_0^2\tau_c^2} + \frac{4\tau_c}{1+4\omega_0^2\tau_c^2} \right\} \tag{3}$$

and

$$\frac{1}{T_2} \alpha \left\{ 3\tau_c + \frac{5\tau_c}{1+\omega_0^2\tau_c^2} + \frac{2\tau_c}{1+4\omega_0^2\tau_c^2} \right\}, \tag{4}$$

where τ_c is the correlation time (a measure of the tumbling rate and the frequency of collisions between molecules) of the species. Note that T_1 is more heavily dependent on B_0 than T_2.

While T_1 and T_2 typically get longer, T_2^* gets shorter at higher fields. Recall that T_2^* is the rate of transverse signal loss accounting for both T_2 and macroscopic field inhomogeneity, that is

$$\frac{1}{T_2^*} = \frac{1}{T_2} + \frac{1}{T'_2} = \frac{1}{T_2} + \gamma \Delta B_0, \tag{5}$$

where T'_2 is the relaxation purely due to the field inhomogeneity. This inhomogeneity in the magnetic field is usually produced by inhomogeneity in the magnetic susceptibility across the sample, e.g., the air in the ear canals has very different susceptibility than the water in brain tissue. The distortion of the magnetic field caused by magnetic susceptibility is described by

$$B'_0 = B_0(1-\chi), \tag{6}$$

where χ is the magnetic susceptibility of the sample, B_0 is the original magnetic field, and B'_0 is the resulting magnetic field after considering the magnetic susceptibility. Thus the change in the field gets worse as the magnetic field increases.

The implications for functional imaging are that most imaging and spectroscopy applications benefit in terms of SNR, and that the blood oxygen level-dependent (BOLD) effect is more pronounced at higher fields. But, as is common in MR, there is a tradeoff. As the field increases, and T_2^* effects get shorter, susceptibility artifacts get much more pronounced. This is particularly significant, as the BOLD effect is observed by T_2^* weighted imaging, which is very sensitive to susceptibility artifacts.

Arterial spin labeling techniques *(6)* also benefit from higher field strength, as the longer T_1 means a higher labeling efficiency. Another major practical implication of working at a higher field is that the resonant frequency of protons is proportionally higher, and thus RF pulses deposit more power into the subject. The US FDA regulates the amount of RF power that can be used on a human subject cannot exceed 1.5 W/kg.

The higher frequency of the pulses also means a shorter wavelength and the formation of standing waves in the sample during transmission. Hence, it is more challenging to achieve uniform excitation patterns across the imaging slice and parts of the imaging slice appear artificially brighter than others. **Figure 10** (left) shows an example of this phenomenon (usually referred to as "dielectric effects") in brain images at 3.0 T *(7–9)*. The right panel of the figure shows the corrected image.

To put things in context, at the time of this writing, T_2^* weighted imaging techniques required for BOLD fMRI are fairly challenging at 7 T and not many groups are doing human work at these high fields. Presently there are only two research groups that have 9-T human imaging systems. At this time, 7.0-T magnets

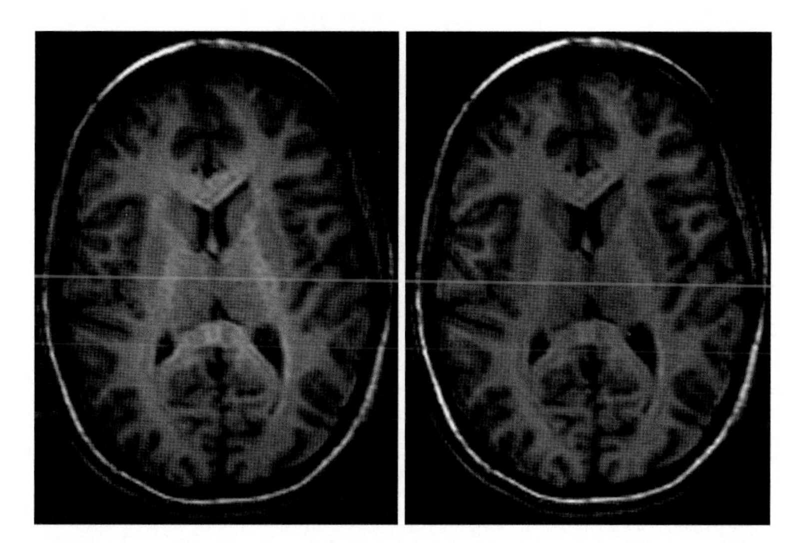

Fig. 10. Illustration of the dielectric effect on a high resolution, T_1 weighted, spoiled gradient recalled (SPGR) image. The center of the image appears brighter, because of the formation of standing waves in the radiofrequency pattern. As a result, the center of the field of view receives a higher flip angle than the periphery of the image. The image on the right has been corrected by removing the low frequency spatial oscillation with a 2D finite impulse response (FIR) filter.

are predominantly used for small animal research systems. The "de facto" standard field strength for human fMRI systems in the last few years has become 3.0 T, although many sites still use 1.5-T scanners for fMRI.

Besides the strength, one must consider the spatial homogeneity and temporal stability of the magnetic field. The magnetic field's homogeneity is crucial since the lack of it translates into severe image distortions. One challenge is that the shape of the magnetic field changes when an object (i.e., the subject) is introduced into the field. Consequently, MRI systems are usually outfitted with a set of small pieces of iron installed around the bore of the magnet, referred to as "passive shims." The location of the passive shims is carefully chosen to make the field more homogeneous. In order to adjust the field homogeneity for individual subjects, additional electromagnets can be used to superimpose additional fields that compensate inhomogeneities. These are referred to as active shims and the process of shaping the field is referred to as "shimming." There are many types of shim coils that are used to superimpose magnetic fields for shimming purposes. The shim coils are designed to produce spatial magnetic field gradients. These fields are typically shaped as linear, quadratic, and higher order functions of spatial position. While typical clinical scanning procedures require adjustments to the linear shims from patient to patient, it is our experience that T_2^* weighted (BOLD) fMRI benefits greatly from higher order shimming. Modern scanners are equipped with automatic shimming procedures *(10)* that can typically achieve homogeneities over a 1,500 cm^3 region of less than 20 Hz root-mean-square, approximately.

In addition to being homogeneous, it is quite important that the magnetic field be as constant as possible over time. The field tends to drift over time due to a number of factors, among them temperature of the room and the equipment, as mentioned previously. These drifts are typically subtle and slow enough that they do not affect clinical/structural imaging. FMRI, however, is based on subtle signal changes over time and therefore, drifts act as significant confounds, especially in slow paradigms. Statistical and signal processing tools do exist to reduce these drifts effects, but it is much more desirable that they be reduced during acquisition. Unfortunately, there are many sources of drift in the MRI hardware, so it is important that the magnet undergo extensive stability testing before it becomes operational and that quality control tests including stability measurements be performed regularly. The scanner's stability can be measured on a phantom over a small region of interest. **Figure 11** illustrates a typical stability test.

The physical configuration and shape of the magnet also plays an important role in many of these parameters. While "open" MRI systems exist and are used for oversized and claustrophobic subjects, their field strength is typically not sufficient for fMRI

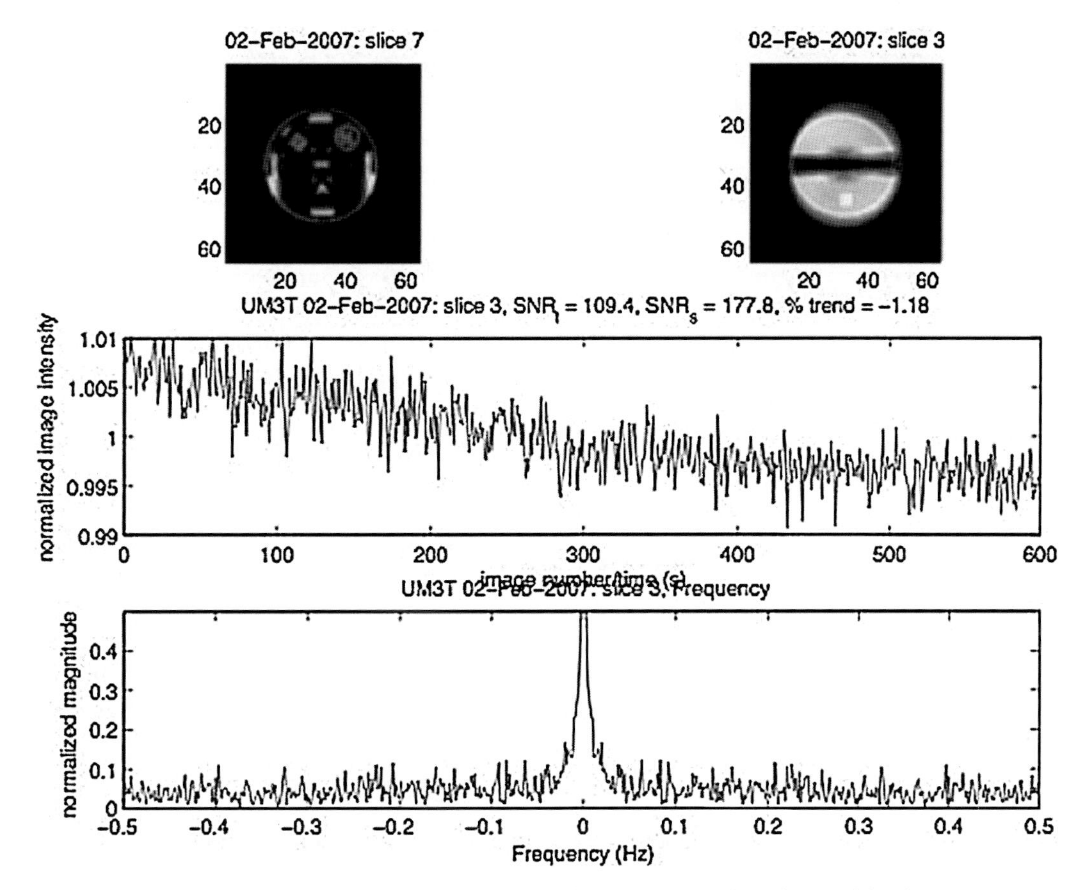

Fig. 11. A typical stability test showing the time course and its frequency content in a phantom (above).

applications and their use is limited to clinical applications that do not demand high quality imaging. Among the closed bore systems, one can choose between short- and long-bore systems. Short-bore systems are intended for head-only applications and can sometimes offer improved performance over smaller regions. Long-bore systems, although more cumbersome, achieve greater field homogeneity over a larger area, which is beneficial for some applications, such as arterial spin labeling.

When considering magnets, it is crucial that we also consider safety issues. The most obvious issue is the very powerful force that magnetic fields of the magnitude required for MRI exert on ferromagnetic objects. These forces are inversely proportional to the square of the distance between the object and the dipole, and are directly proportional to the mass of the metal in question. Recall that the magnetic field produced by a current is described by the Biot-Savart law:

$$\bar{B} = \frac{\mu I \oint \bar{l} \times \mathrm{d}\bar{L}}{4\pi r^2}, \tag{7}$$

where I is the current, L is a unit vector in the direction of the current, l is another unit vector pointing from the location of the wire to the location of interest, and r is the distance from the location of interest to the current source.

One must obviously be very careful to keep ferromagnetic objects out of the magnet room. Typical accidents occur when someone forgets about small metallic objects in their pocket and they fly out of their pocket and strike someone. Accidents sometimes happen because there may be a very subtle force on the object at a specific location in the room leading an unsuspecting investigator to believe that the object is not ferromagnetic. However, moving the object a very small amount toward the magnet can translate into a very rapid increase of the magnetic field, since the magnetic field increases with the inverse of the square of the distance, as mentioned. A small step in the wrong direction while carrying a ferromagnetic object can be the difference between a gentle tug on the object and the object being launched into the bore of the magnet, to the horror of the investigator and the subject. It is thus paramount that strict screening procedures be followed before allowing people into the magnet room. Sometimes small bar magnets and airport security style metal detectors are used to verify the absence of ferromagnetic objects on the subject's body or to test allegedly MR compatible equipment.

These forces can also affect metal implants in the subject's bodies. Pacemakers, neurostimulators, and other implanted electronic devices are likely to malfunction putting the subject at great risk. It is thus crucial that subjects be thoroughly screened for the presence of implants, shrapnel, or other metal sources in their bodies. Having said that, many modern implants are built of titanium and nonreactive materials that are not ferromagnetic and are therefore "MR compatible". A number of publications exist cataloging medical devices and their MR compatibility according to model and manufacturer *(1, 2)* (and on the Web: http://www.mrisafety.com/).

3.2. Magnetic Field Gradients

In order to produce images, an MRI scanner needs a spatially varying magnetic field under tight control by the user. This is accomplished by using an additional set of coils that add extra magnetic fields to the main field. A set of such coils is shown in **Fig. 12**. By supplying customized current waveforms to these coils, the user can change the distribution of the magnetic field's shape at will. In broad terms, by varying gradient's strength over time during the pulse sequence, one can obtain MR signals whose phase distribution is a function of the spatial distribution of the sample.

The ideal gradient set is capable of quickly changing the magnetic field as a linear function of spatial location along each of the Cartesian axes. Typical gradients in clinical and fMRI are

Fig. 12. Gradient coils from Doty Scientific (reproduced with permission of Doty Scientific).

between 10 and 40 mT/m, but specialized gradient inserts exist that can produce larger gradients (in the range of 100 mT/m). Small-bore animal systems can be outfitted with more powerful gradients (up to approximately 400 mT/m). The main challenges in MRI gradient design and construction usually consist of producing linear gradients in space and time, and the production of eddy currents.

The spatial linearity of the gradients must be maintained over the volume of the sample to be imaged, or the images will appear warped (although these distortions can be corrected during reconstruction if the true shape of the gradient is known). The spatial linearity of the gradient fields is primarily a function of the shape of the gradient coils, and a great deal of effort goes into their design and construction (we will not go into those details here). Typical MRI scanner head gradients can maintain 95% linearity over 30 cm.

The gradient coils must also be able to produce magnetic field gradients very quickly and accurately. The inductive nature of the coils causes their response to the input currents to be severely dampened. In order to correct this problem, typical gradient currents are "precompensated" in order to produce the desired waveform *(11, 12)*. An example of precompensation is illustrated in **Fig. 13**. Most maintenance or quality assurance protocols include gradient linearity and precompensation.

The rate at which a gradient is achieved is referred to as the slew rate. Slew rates of about 200 T/m/s can be generally achieved. However, there are FDA limitations (These are determined by the imaging sequence type and the duration of the stimulation. Typically the maximum allowed rate of change in magnetic field is approximately 20 T/s.), since the sudden

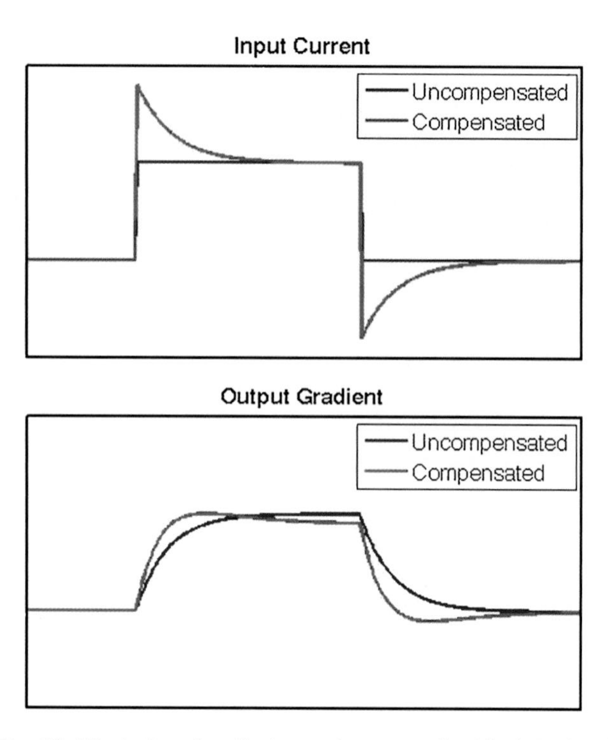

Fig. 13. Simplified illustration of gradient current compensation. The inductive effects of the gradient coils are partially corrected by modifying the input currents to the coil.

changes in the magnetic field can induce currents in the peripheral nervous system, causing muscle contractions and unpleasant or even painful sensations in the patient. This phenomenon is referred to as peripheral nerve stimulation (PNS).

Active shielding of gradients is necessary to contain the gradient fields and reduce the interactions between gradient coils and other conductors in the scanner. The principles are the same as the active shielding of the main magnetic field (*see* **Subheading 3.1**). In other words, a coil producing an opposite gradient field is built around the main gradients in order to cancel the fields outside the area of interest *(13, 14)*.

Gradient vibration and noise are another issue to consider in gradient design. As the gradients are rapidly turned on and off, especially in echo-planar sequences, they experience torques due to the presence of the main magnetic field. Thus the coils vibrate violently thus causing the familiar banging MRI sounds. The sound levels are quite loud and require that the subject's wear ear plugs. Active shielding can help reduce the vibrations and acoustic noise produced by the gradients *(14)*.

3.3. RF Hardware

MRI scanners employ RF hardware to generate oscillating magnetic fields that cause the magnetization vector to tip into the transverse plane, and for signal reception. To create signal, an MRI scanner uses a powerful amplifier (generally 1–25 kW) to drive an excitation coil with a large pulse of electric current. In the reception stage, the receive coil is used to pick up the MR signal, which is then processed to create an image. The components of the RF chain that an fMRI user should pay attention to are the transmit coils and the receive coils. Commonly, the transmit and receive coils are the same, but since the choice of RF coils will influence image characteristics such as noise level and image homogeneity, sometimes different coils are used for transmit and receive. Furthermore, a certain class of RF coils allows the use of recently-developed "parallel imaging" techniques *(15)* that aim to improve image quality via reduced acquisition time.

Motivated by the inherently low SNR of the BOLD fMRI signal, and by a desire to minimize RF energy deposited in the body, fMRI experiments forego the use of the body RF coil built into MR scanners, in favor of coils that sit close to the head. Head coils can take many shapes and functional forms; however, for the purposes of SNR and image homogeneity comparisons, there are two main classes of coils: single-channel and phased-array coils.

An RF coil is a resonant circuit, and can be modeled as a simple loop containing an inductor, a capacitor, and a resistor (**Fig. 14**). The inductor and the capacitor represent actual circuit components lumped together with the inductance and the capacitance of the sample. When a source of electrical current that oscillates at the circuit's resonant frequency is placed across its terminals, the impedances of the inductor and the capacitor cancel, and the coil delivers the maximal amount of energy to the sample. In a reciprocal manner, current measured at the coil terminals due to energy radiated by the sample will be of maximum amplitude when that energy oscillates at the coil's resonant frequency. Because the frequency at which biological spins oscillate is determined by the main magnetic field strength via the Larmor relationship, an RF coil must be "tuned" to resonate at this frequency.

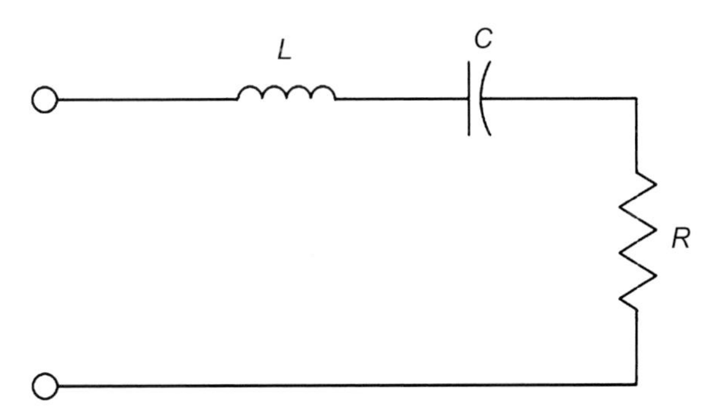

Fig. 14. Series Resistor–Inductor–Capacitor (RLC) circuit representing an RF coil and a biological sample.

Single channel surface coils are the simplest RF coil design. These are typically used to image small areas close to the surface of the skin, as their penetration depth drops sharply with distance from the coil. They consist of a simple loop that induces an oscillating magnetic field when a current is passed through it.

Although the phase information is routinely discarded in image reconstruction, MR signals are inherently vector quantities. RF coils are used to generate and oscillate magnetic fields, and the signal emitted back by the sample is a rotating electromagnetic field. However, a simple surface coil can only produce and detect only one component of that vector: the component that is perpendicular to the plane of the coil. Quadrature coils typically consist of two perpendicular coils that transmit at 90° phase from each other. The resulting magnetic field is the vector sum of the two perpendicular fields. Quadrature reception can also be achieved with the same coils by adding phase to one of the channels in the receive chain.

The most popular quadrature-channel head coils generally take on a "birdcage" design **(Fig. 15a)**. This classic design provides good SNR and image homogeneity characteristics, owing to the unique nature of the magnetic field that it creates *(16)*. Another type of single-channel coil commonly used in studies of the occipital cortex is a quadrature occipital coil **(Fig. 15b)**, which provides high SNR in this localized region of the brain, and has a compact design compared to full head coils, allowing the experimenter greater freedom in stimulus hardware setup.

In contrast, phased-array coils **(Fig. 15c)** are composed of a set of discrete and independent "surface" coils, arranged around the head. Generally, phased-array coils can be brought into a tighter conformation around the head, which improves SNR. Taken alone, images obtained with individual surface coils will possess lower SNR and poor image homogeneity compared to a birdcage coil; however, when images from the coils are combined in a sum-of-squares reconstruction, excellent SNR can be achieved *(17)*, though image homogeneity will still be worse than for a birdcage coil. Furthermore, most phased-array coils can be used only for reception, so a separate coil (usually the scanner's body coil) must be employed for excitation, which can result in degraded image homogeneity, as well as increased specific absorption rate (SAR) in areas outside the head. The central advantage to phased-array coils is that they allow the use of parallel imaging techniques, such as sensitivity encoding (SENSE) *(18)* and generalized autocalibrating partially parallel acquisitions (GRAPPA) *(19)*, when paired with RF signal chains capable of receiving multiple channels simultaneously, which are now commonly available from most MR vendors. Parallel imaging exploits the inhomogeneity of images obtained with surface coils to accelerate image acquisition, which results in a reduction of artifacts

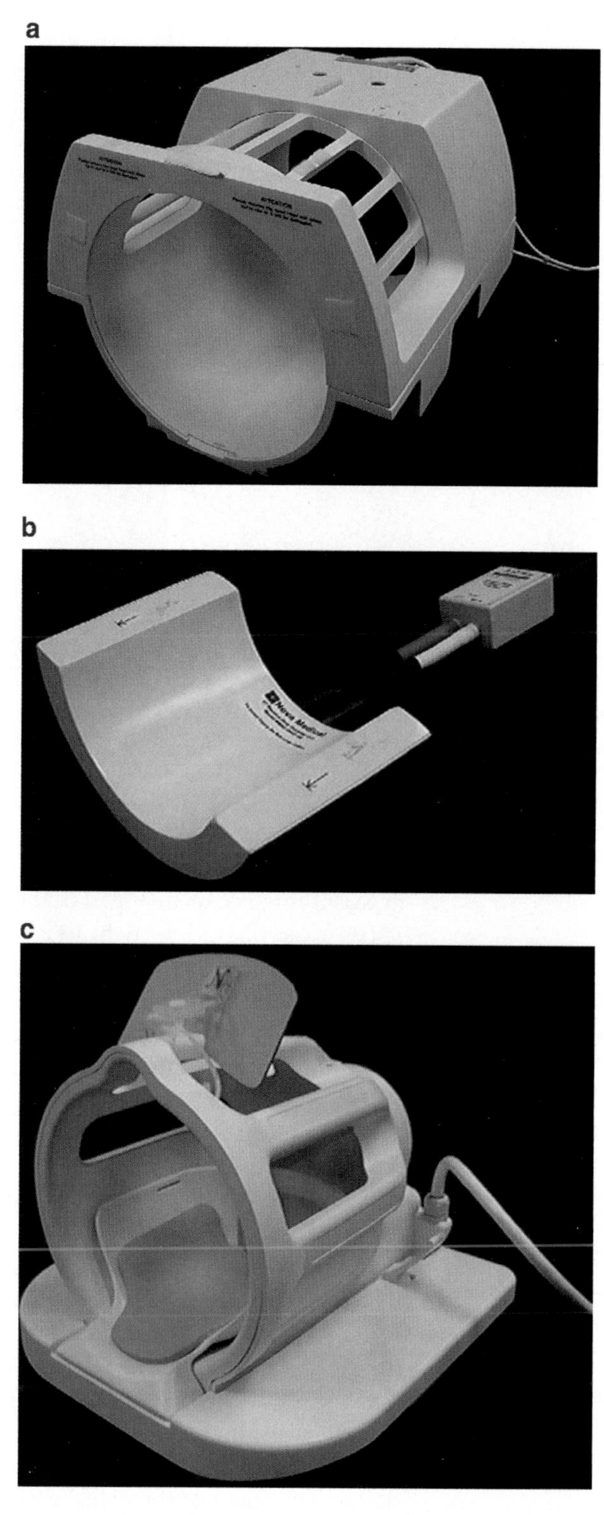

Fig. 15. (**a**) Transmit/receive birdcage head coil. (**b**) Receive-only quadrature occipital coil. (**c**) Eight-channel phased array coil.

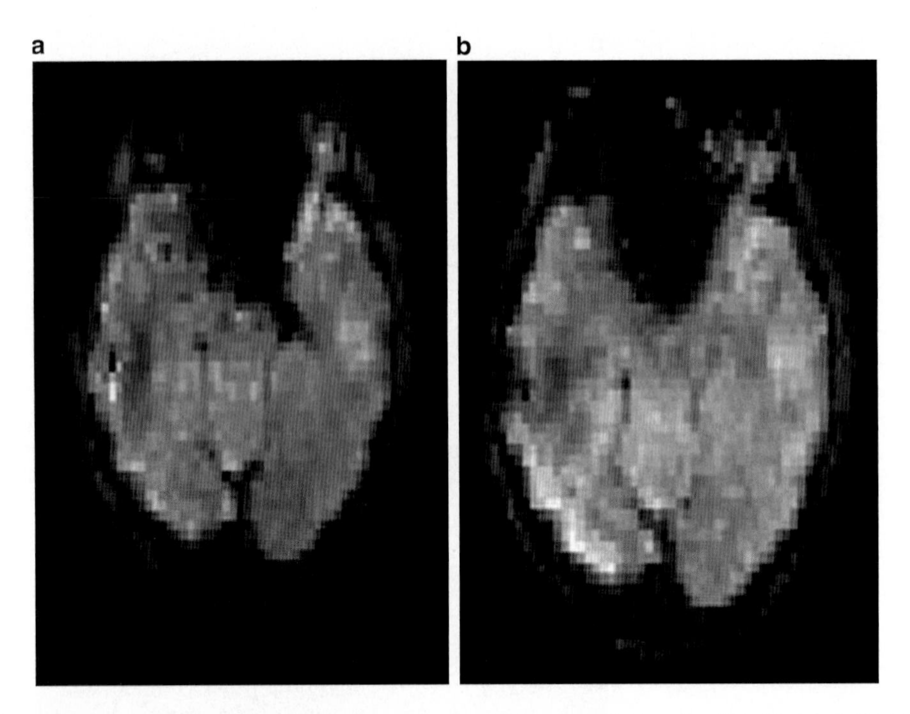

Fig. 16. Comparison of fMRI images obtained using conventional (**a**) and parallel imaging (**b**) in the inferior brain. In this example, parallel imaging with an eight-channel phased array coil was used to reduce data readout time by a factor of 2. This reduced signal loss and image distortions due to susceptibility, particularly in the region indicated by the *arrows* (courtesy of Yoon Chung Kim, University of Michigan Functional MRI Laboratory).

in fMRI images, as in **Fig. 16**. This comes at the cost of reduced SNR. Most phased-array head coils have eight individual coil elements, and can be effectively used to reduce image acquisition time by a factor of 2–4.

However, a birdcage coil generally provides a lot of flexibility in stimulus presentation setup, due to the large amount of room within the coil. One can use goggles, projector/mirror systems, and a range of other solutions in conjunction with a birdcage coil (as we discuss later). In comparison, current phased-array coil designs limit presentation options, since the coils forming it are placed closer to the head, which may prevent the use of visors. However, phased array coils are generally compatible with the most commonly used projector-based visual stimulus setups. When using visor-based stimulus systems, one must ensure that the electronics of the visor are properly shielded to prevent image artifacts of the type shown in **Fig. 4**. The "streaking" artifact shown in this image was caused by a visor with an electromagnetic "leak," that coupled to the receive coil and manifested as a false MR signal. The form of the artifact will depend on the pulse sequence used; in this example the streak is localized, while in another pulse sequence the artifact may be spread over the entire image.

3.4. Computing Resources

FMRI experiments generate large datasets. The raw data alone for a single subject, scanned for 1 h, weighs in between 100 and 200 MB. Combine this with the space needed for image reconstructions and analysis and one should budget storage and scratch space of around 2 GB for every hour of scanning. In this section, we will provide some guidelines and suggestions on how to set up a computing environment to handle all this data. The two major factors influencing the design of your environment will be (1) money, and (2) the expertise available, in terms of computer systems and MRI data processing.

The data stream in a typical laboratory consists of four main stages. The first stage is the MRI scanner, which produces either raw, unreconstructed data, or reconstructed images that are ready for postprocessing and analysis. If a laboratory has an MR physicist at its disposal, then the former case is often true, since an MR physicist may develop image reconstruction codes that provide improved image quality over vendor-provided software, and that can add in improved reconstruction techniques as they are developed. Assuming images have been reconstructed, the second stage consists of slice timing correction, realignment, coregistration, warping, and smoothing, which are all operations that prepare the dataset for statistical modeling and analysis. The third stage is statistical modeling and analysis. The fourth stage is data backup, though it is advisable to make backups of data at more than one point in the stream.

The majority of fMRI laboratories maintain one powerful workstation or computer cluster to do the bulk of their processing. Users can log into this computer from their own machines to initiate processing or view and download the preprocessed data for local analysis. The central advantage to this model is that the large and complicated software packages used to process fMRI need only be maintained on one machine, which greatly simplifies maintenance. A secondary advantage is that this model allows users greater flexibility in choosing the operating system of the computers they use; the workstation can be running a Unix derivative, which has benefits in terms of networking, stability, and the availability of fMRI software packages, while the users can be using Windows PCs or Macintoshes, which are generally more user-friendly systems. All computers that will be involved in processing or storing data should be connected with the fastest network possible, such as a gigabit network.

The central processing workstation should have as much RAM as possible. By today's computing standards, each CPU should have at least 2 GB RAM available to it, so that it may store the entire dataset and related files. The reason for this is to minimize the frequency with which the computer accesses its hard disk processing, which costs large amounts of time. The second main consideration is the number of CPUs the computer should have. This will largely depend on the number of simultaneous processing streams one expects to have running on the computer. The third consideration is storage. The main computer should be connected to a large data storage device, so that all current experiments can be instantly accessed, without requiring reloading of backed-up data. This device should be a set of hard disks configured in a redundant manner, such as a redundant array of independent devices array.

One of the central computing dilemmas that an fMRI laboratory will continually deal with is making data backups. There are two main questions to answer here (1) at what points in the processing stream should backups be made, and (2) what form should backups take? Backups of the initial MR data are absolutely necessary, since an experimenter may be asked to reproduce their results at a later date, or bugs may be found in the postprocessing stream (stage two), which will require reprocessing the original data. After this stage in the stream, the choice of where to do backups will depend on the amount of backup space available and the speed with which the second and third stages may be executed, should the analyzed data be lost. The more points at which backups are made, the more quickly an experimenter could recover after a data loss or processing interruption. A laboratory also has many options in choosing backup forms, and it may be best to use a combination of them. Perhaps the simplest backup form is mirroring the data, i.e., storing the data in another set of hard disks whose sole purpose is to store backed-up and archived data. This solution is particularly simple in that backups can be fully automated and instantly accessed. The other two main options are tape storage and optical media (DVD). While automated machines may be purchased to manage tape backups, hard disk capacity is rapidly outstripping tape capacity, and the mechanical nature of these machines makes them prone to frequent failure. On the other hand, DVD backups are more reliable and cheap, but they require human interaction to load DVDs and execute burning software. A backup schedule will also have to be worked out by the laboratory.

Because computer components (i.e., disks, CPUs, and video cards) frequently fail or need replacement, it is advisable to set aside part of the laboratory's initial capital for yearly computer maintenance. It is also advisable to purchase extended warranties for the computers, as they will be heavily used and if they fail, this can save the laboratory a significant amount of money in the long run. Furthermore, as technology advances, the laboratory should build into their operational budget the cost for new machines every few years.

4. Stimulus Presentation and Behavioral Data Collection Devices

There is a small but increasing number of manufacturers of stimulus presentation hardware and software. We will not review specific products or vendors but limit ourselves to describe the important characteristics of these devices.

Some components of the stimulation and response recording equipment must go inside and/or near the magnet. These must not be susceptible to magnetic forces for obvious safety reasons. Additionally, the presence of metals in headphones and head mounted displays can generate field distortions that cannot be compensated by shimming, even if they are not ferromagnetic. This results in severe image degradation and it is thus important that the devices be thoroughly tested on phantoms for image degradation.

As we alluded to before, electronic equipment in the scanner room must be adequately shielded to avoid introducing RF noise into the system. For example, LCD displays for visual stimulation inside the magnet are typically encased in a fine wire mesh that acts as a Faraday cage to contain RF leakage. Other audiovisual electronic equipment used in the MR environment, such as projectors and button response units (BRU) are typically encased in brass or aluminum for the same reasons. Regardless of the manufacturers' best intentions, sometimes the shielding is not adequate or becomes damaged over time in subtle ways. Just as in the case of the room's RF shielding, an exposed wire or bad shielding connections can produce severe RF contamination of the images. Thus, it is paramount that the stimulation devices be checked upon purchase and periodically for RF leaks that may develop during delivery, installation, or daily use.

There are different technologies commercially available for MR compatible visual stimulation. The simplest approach is perhaps an LCD projector outfitted with narrow focus lenses that project the images into a back projection screen placed inside the bore of the magnet. The subject then can see the display through a set of mirrors that are mounted on the head coil assembly. The main advantages of this approach are simplicity and lower cost. The disadvantages are related to positioning issues and a reduced visual field for the subject.

Another approach to MR compatible visual stimulation is fiber-optic display visors. This sort of display system is based on an optical signal converter that carries a super video graphics adapter quality image through an array of micro-optic fibers to a head-mounted display inside the scanner's head coil. This approach is very attractive in that there are no electronic components that need to be installed inside the scanner room and the display can be placed very accurately in front of the subject's eyes, maximizing the available visual field. The drawbacks are that in additional to being expensive, the fiber optics used in the array are very fine and brittle, so that regardless of the high quality of fabrication, there will always be a small number of broken fibers that result in dead pixels or small streaks in the image.

One of the more popular approaches is to display the images on a shielded LCD screen mounted in front of the subject's head. This screen can be either a large one that is mounted outside the scanner's RF coil, or a small one in a visor that the subject wears inside the coil. The advantages of this are the large visual field and ease of use of the system. The drawbacks are the high cost and the interactions between the display electronics and the magnet. Some of these devices become dimmer when placed inside the magnetic field. Additionally, if any RF leaks develop, they severely degrade the images, especially in the visor type systems since they sit inside of the RF coil.

Auditory stimulation is typically performed in the MR environment through two different kinds of headphones: pressure waveguide types and shielded piezoelectrics. Both of these are highly effective devices. The pressure waveguide headphones keep the speakers outside of the magnet's bore and the sound is carried through rigid tubing into the headphones. The piezoelectric headphones are akin to standard speaker technology but use piezoelectrics to produce the vibrations. They require RF shielding of the cables and the electronics to prevent artifacts. Perhaps the biggest challenge for auditory stimulation is reduction of the MRI scanner's noise. There is very limited space inside the scanner's head coil for building an effective muffler into the headphones but fairly effective noise reduction (typically around 30 dB) can be achieved. The headphones' acoustic insulation is sometimes achieved by gel padding that attenuates the sound very effectively by forming a tight seal around the ear. Caution must be used as the gel in the padding produces an MR signal and is visible in the images so it must be taken into consideration during registration and normalization of structural images. The gel's resonant frequency is typically not the same as water and produces some off-resonance artifacts, but these tend to be mild.

MR compatible microphones for patient communication and verbal response recording are typically based on piezoelectric technology and require both electronic and acoustic shielding to reduce the scanner sound. To our knowledge of the present state of the art, the acoustic shielding of the microphone from the scanner sound is somewhat effective, but communication with the subject during the scan is still challenging. Some systems are equipped with active noise cancellation with limited success.

Consequently, investigators often use pulse sequences with "quiet" (i.e., no gradient pulses) periods during the subject response time instead *(20, 21)*.

Other response recording devices are primarily "button response units (BRU)" that are built into hand rests and strapped to the subject's hands. They typically carry only DCs through twisted pair cables and RF noise is not an issue as the electronics to drive the system are kept outside the scanner room. There are a number of other response units, such as MR compatible joysticks and keyboards that are manufactured by small companies. While these are typically safe and effective, one should test all such equipment immediately upon purchase not only for functionality but also for RF leakage and ferromagnetic forces. Periodic RF testing of peripherals should be an integral of the fMRI facility's quality assurance procedures.

5. Subject Monitoring

When running an fMRI experiment, it can be desirable to monitor and record subjects' status and peripheral signals during an fMRI experiment to use as correlates of subject behavior or as nuisance signals in data correction. Some possibilities include monitoring cardiorespiratory rhythms, galvanic skin response (GSR), head motion, or eye-tracking. As stressed in the previous sections, all these considerations should fit with the comfort and safety of the subject.

In general, when considering recording peripheral signals on fMRI subjects, one should pay attention to synchronization with the MR scanner, adequate sampling of the signal in question, and avoiding introducing signal noise in both the MR data and the recorded peripheral signals.

5.1. Scanner Synchronization

In order to match the recorded external signals with the fMRI data being recorded, synchronization with the start of the scan must be achieved. This can be done using a TTL pulse to/from the scanner from/to the external device or recording media. For instance, a logic pulse from the MR scanner to the computer recording physiological noise can be set to trigger the recording sequence. Commercial MR scanners from the main vendors (GE, Siemens, and Philips) all have the capability to send or receive TTL sync pulses.

5.2. Physiological Monitoring

A limit to the effectiveness of fMRI in detecting activation is the presence of physiological noise, which can equal or exceed the desired signal changes in an fMRI experiment *(22)*. These physiological fluctuations that are present during an fMRI scan can obscure the BOLD activity that the researcher is trying to detect. In addition, monitoring physiological rates can help as secondary reaction measures (such as monitoring the cardiac rate variability during a stress experiment).

5.3. Cardiac Monitoring

Monitoring the cardiac waveform can be achieved in several ways. The most common solutions are pulse oximeters or ECG patches. The primary cardiac harmonic frequency lies in the 0.5–2.0 Hz range, with both the first and secondary harmonics shown to affect the fMRI signal *(23)*.

Pulse oximetry refers to indirectly monitoring the oxygen levels in the extremities to monitor the cardiac waveform. This is most often accomplished in fMRI laboratories by using a light emitting diode and photodiode that clips to the subject's finger, connected to a data acquisition board **(Fig. 17)**. Several MR scanner vendors offer this as part of the MR system (GE and Siemens), and stand-alone monitoring units from commercial vendors are also available (Invivo and Biopak). Normal setup with compliant subjects allows adequate sampling of cardiac rhythm as seen in **Fig. 18**. Drawbacks include the fact that subject motion

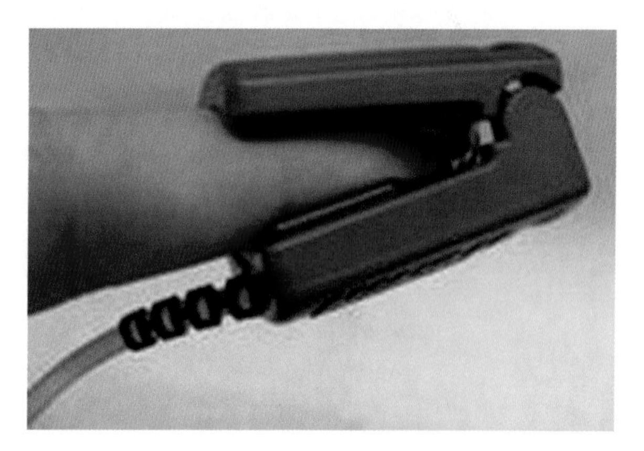

Fig. 17. Pulse oximeter for indirect measure of cardiac waveform.

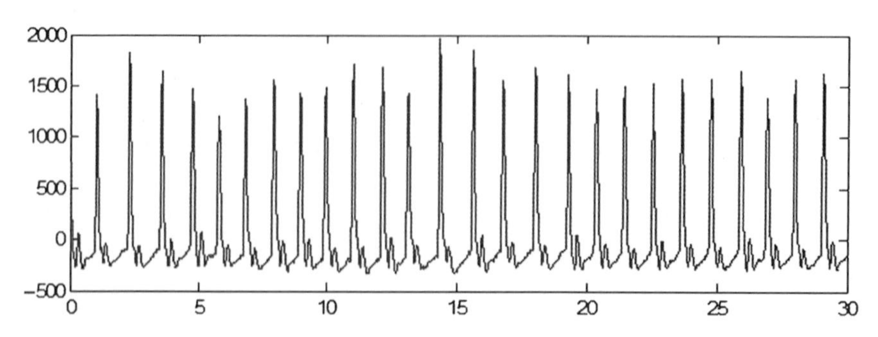

Fig. 18. Cardiac waveform acquired during an fMRI scan (data acquired on a 3.0-T GE scanner, using a pulse oximeter with a sampling rate of 40 Hz).

may corrupt the signal, and motor tasks may be impeded with the oximeter placed on the finger (alternative placement on the ear or toe is possible).

ECG patches located over the heart allow high-fidelity monitoring of cardiac electrical activity. This allows identification of features beyond the simple cardiac peaks, such the QRS complex during the depolarization of the ventricles (**Fig. 19**). Disadvantages of ECG recording include increased setup complexity and patient comfort.

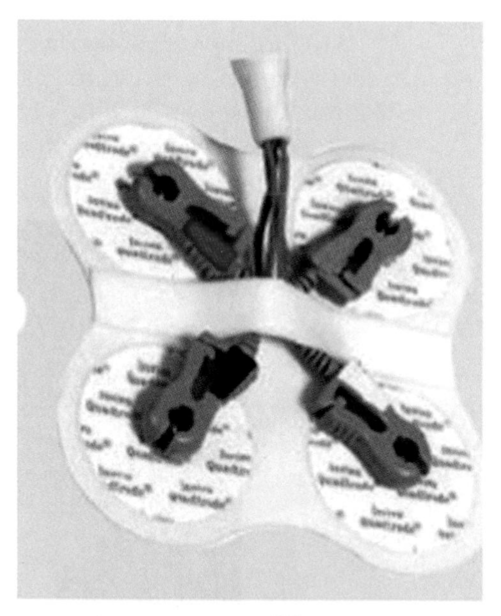

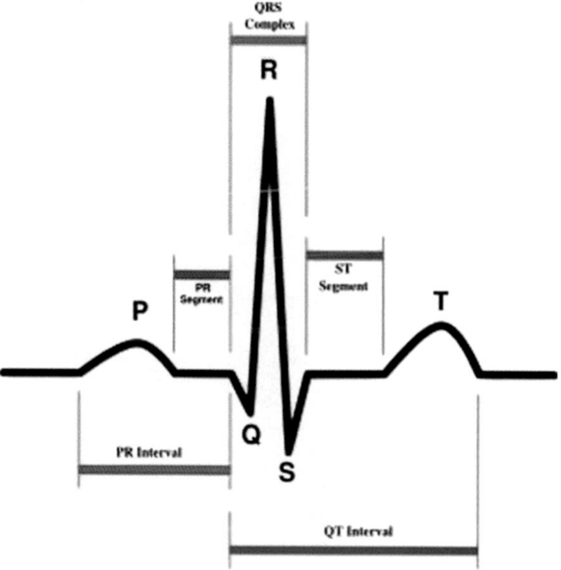

Fig. 19. Example of ECG patch (Courtesy of Invivo, http://www.invivocorp.com). With a schematic of a typical QRS waveform.

5.4. Respiratory Monitoring

The respiratory rhythm has a normal frequency range of 0.1–0.5 Hz. Motion of the chest during respiration combined with the changes in oxygen saturation in the lungs lead to a modulation of the local magnetic field that can affect the phase of the MR signal at the position of the head during scanning (**Fig. 20**), which can lead to modulation of the recorded MR signal intensity. Mitigation of respiratory effects on the MR signal can include scanning during breath-holds, modified pulse sequences to sample and account for the modulation in magnetic field *(24)*, and recording of the respiratory signal to use as nuisance covariates in post-processing analysis *(23)*.

Monitoring respiratory rhythm can be accomplished by using a plethysmograph (pressure belt) around the waist of a subject, like the one shown in **Fig. 21**, or by using a nasal cannula to monitor expired CO_2 concentration. A sample respiratory waveform is shown in **Fig. 22**.

Fig. 20. Phase difference between inspiration and expiration for a coronal slice.

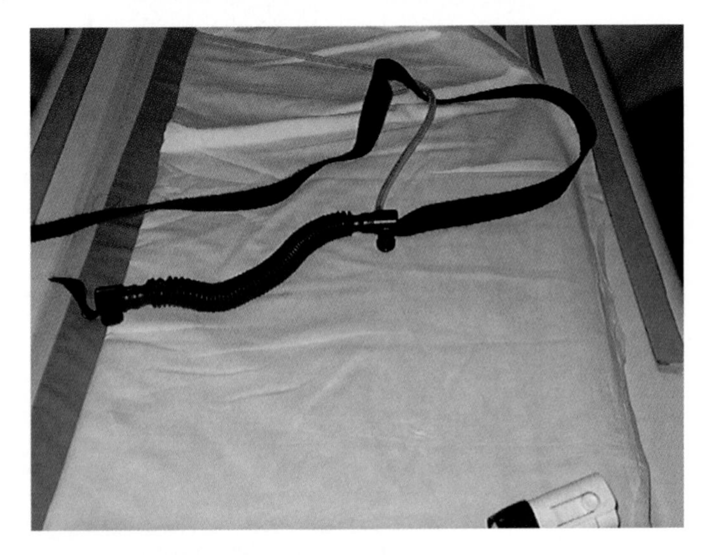

Fig. 21. Plethysmograph belt for measuring respiration.

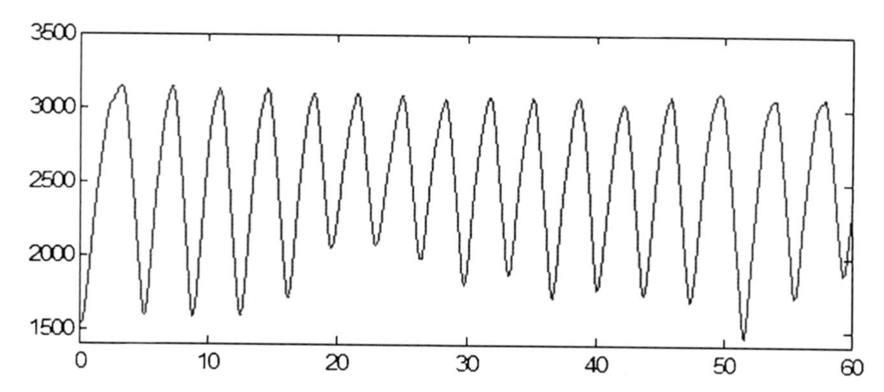

Fig. 22. Respiratory waveform acquired during an fMRI scan (data acquired on a 3.0-T GE scanner, using a pulse oximeter with a sampling rate of 40 Hz).

5.5. Galvanic Skin Response

GSR is a measure of the electrical resistance of the skin, a physical property that has been shown to increase in response to subject arousal, mental effort, or stress. It is monitored by measuring a voltage drop across the skin using paired electrodes. This can have the same motion sensitivity and motor task complication as the pulse oximeter if the electrodes are placed on the fingers. These problems are usually reduced by placing the electrodes between the second and third knuckles, instead of on the fingertip.

5.6. Head Motion Tracking

Severe head motion during an fMRI scan can severely corrupt the data. Besides minimizing patient motion using cushioning and restraints, a measure of head motion may also be collected to help correct data in postprocessing. This may be done using modified pulse sequences (e.g., Prospective Acquisition CorrEction, PACE), or by using external motion tracking *(25–27)*. Again, patient comfort and visual path should be taken into consideration.

5.7. Eye Tracking

Patient gaze and fixation time is important for several types of fMRI paradigms and pathological types. In addition, eye motion can be a source of variance in fMRI scans *(28)*. Thus, tracking eye position can be desirable. The common method is to monitor the position of infrared (IR) light that is reflected off the eye of the subject. This involves transmitting IR light to the subject's eye, and then recording it, using MR-compatible equipment. Several vendors provide hardware solutions including both long-range and short-range cameras. These systems typically have on the order of 0.1–1° spatial resolution and accuracy, with working ranges of 10–25° horizontally and vertically, and 60–120 Hz sampling rate.

In acquiring a physical eye-tracking system for an fMRI laboratory, consideration should be given to the optical path for the eye-tracking system, taking into account the MR bore, head coil, and visual stimulus presentation system; signal integrity of the MR data; and ease of setup for the MR technicians and scanners. This will also involve peripheral equipment located in the scanner room or control room: usually a camera, power supply, video monitor for real-time display of the subject's eye, and a PC.

6. Multimodal fMRI

6.1. Fmri-eeg

Acquiring other neurophysiological measures with fMRI data can complement the excellent spatial resolution and depth penetration of fMRI with modalities that have superior temporal resolution and different sensitivities to the underlying neuronal activity. In particular, the simultaneous acquisition of electroencephalography (EEG) data with fMRI data allows the higher temporal sampling rate of EEG (~5,000 Hz) to be combined with the superior resolution (~mm) and depth penetration of fMRI.

Several factors must be accounted for when setting up a simultaneous EEG/fMRI acquisition, chief among them safety and signal quality. In the following, we will try to touch on most of the considerations one will make when selecting and setting up EEG-fMRI hardware.

Several companies have MR-compatible EEG hardware commercially available (Brain Products and Neuroscan). A common EEG setup includes an electrode cap, connected to a signal amplifier and recording device. Any part of this setup that is inside the MR scanner room within the 5-Gauss line must be nonferrous, and the amount of metal must be kept to a minimum, with care exercised around all metallic components (this includes electrode leads on the cap and skin, batteries in amplifiers, etc.). Fiber optics can be used for signal transmission after amplification, with recording devices located in the scanner control room.

Much planning is required to integrate the EEG with the MR. The head coil used for MR acquisition may affect the physical setup. For instance, an open birdcage coil may allow placement of the EEG cap cord and amplifier above the head of the subject, with no obstruction of the subject's field of view. With an alternative phased array coil that is closed at one end, this setup may not be possible.

Also, for time synchronization, it may be necessary to use the TTL pulse from the scanner (mentioned in **Subheading 5.1**) to trigger the EEG recording device at each TR.

Finally, for fusion of the fMRI and EEG data, accurate electrode locations on the head should be recorded and transferred to the structural MR images. Commercial head position recording systems are available (Brainsight). These systems can record points on the subject's head using infrared positional markers, and coregister these coordinates with the structural MR scans of the subject. Subsequent coregistration of the fMRI and structural MR data allows direct overlay and source localization using both between the MR, fMRI, and EEG data.

The EEG equipment should not adversely affect the MR data, if proper materials and shielding are used. Images with and without the EEG equipment should be inspected for any introduction of alternating current line noise (~60 Hz) or localized variance in the structural and fMRI images.

The MR equipment will affect the EEG recording, due to the high field environment, and the application of gradients during the MR acquisition (**Fig. 23**). However, the MR gradient artifact can be corrected for in postprocessing, using either available software (EEGLAB, http://www.sccn.ucsd.edu/eeglab/) or adapting techniques such as principal component analysis or independent component analysis.

6.2. Transcranial Magnetic Stimulation and fMRI

Transcranial magnetic stimulation (TMS) has great potential not only as a research tool but also as a therapeutic device *(29–31)*. The principle behind TMS is that a large current waveform is driven through a coil placed adjacent to the tissue of interest. The current in turn induces an electromagnetic field depending on the rate of change of the current, as predicted by classical electrodynamics. The induced electric field penetrates the tissue and induces eddy currents on conductors, such as nerve fibers. When a nerve fiber is aligned with the direction of the electric field, a large current is induced in the axon which causes its membrane to depolarize, effectively causing the transmission of an action potential *(32–34)*. After depolarization of the axonal membrane, the sodium–potassium pumps rebuild the membrane potential and the nerve fibers return to their original state within seconds.

The effects of these induced discharges are complex, depending upon the magnitude and timing of the TMS pulse affecting inhibitory and/or excitatory neuronal populations (for recent reviews, *see* **refs.** *35–37)*. If a single TMS pulse is applied in the hand region of the motor cortex, for example, motor neurons depolarize and the hand will twitch (motor evoked potential, MEP). Subthreshold stimulation, followed by suprathreshold stimulation, can inhibit or facilitate the MEP, which varies with the distance and location of the subthreshold stimulus *(38)*.

By selectively applying a TMS pulse during the performance of a task, neural circuits are effectively jammed, and performance interruptions can be observed. In effect, one creates a controlled, completely reversible "lesion," enabling the study of brain function through perturbation of neuronal activity *(35, 39)*. Although these studies have rapidly become a popular investigative tool for cognitive neuroscientists, one important limitation stems from

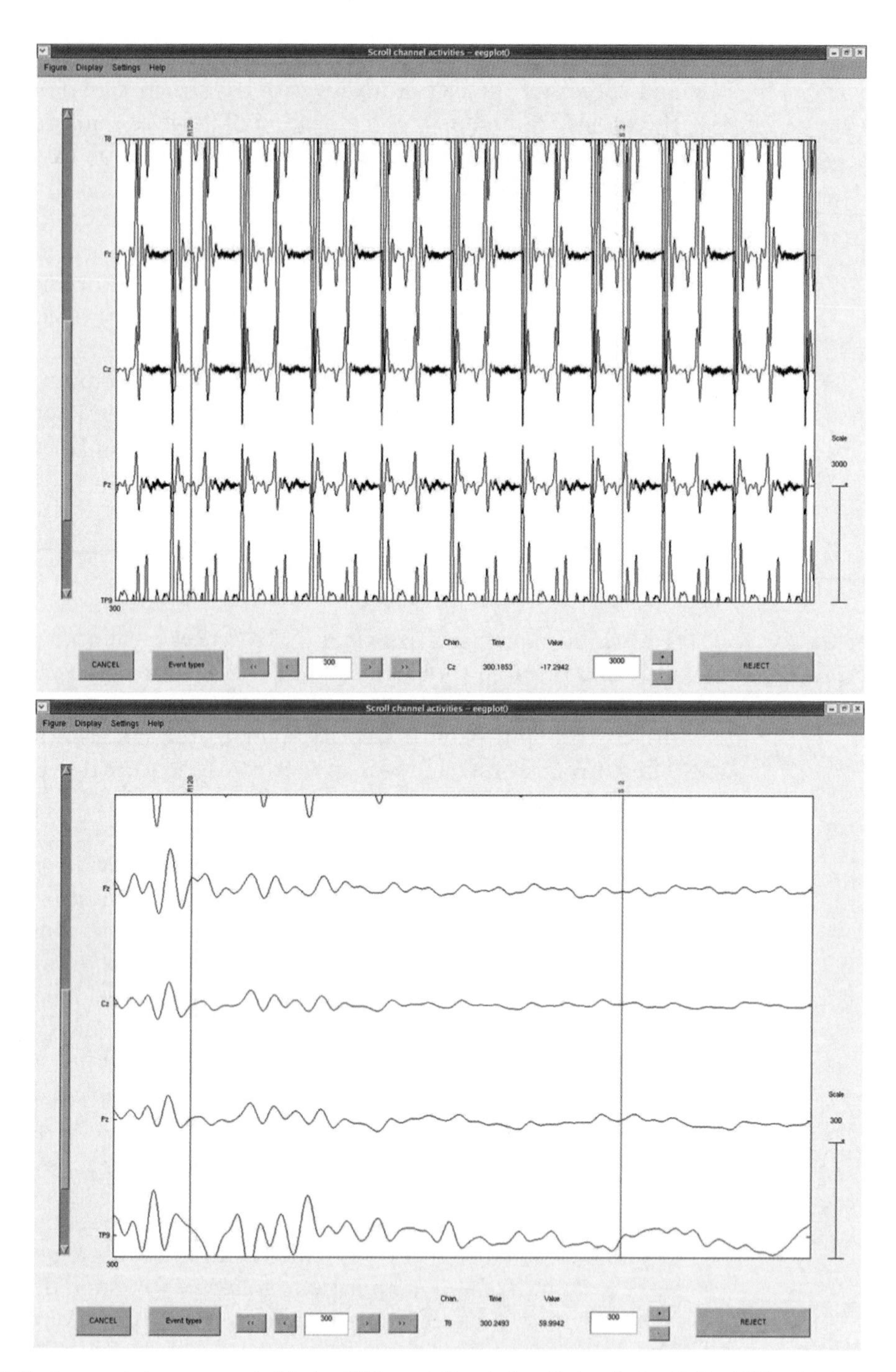

Fig. 23. Electroencephalogram recorded during fMRI acquisition, before (*top*) and after (*bottom*) MR artifact correction.

inadequate knowledge about the shape and magnitude of the induced current fields that introduce the perturbation. Hence, TMS and fMRI can be complementary for the study of brain function. TMS can interfere, or modulate the cognitive process under scrutiny by locally altering the responsiveness of the tissue, while fMRI can allow the investigators to precisely map out these effects. While TMS and fMRI experimental data can be coregistered and integrated after each experiment has been carried out separately *(40)*, it is desirable to be able to observe the BOLD responses to TMS.

However, there are some clear challenges to carrying out joint TMS and fMRI experiments. Such experiments require TMS coils that contain no ferromagnetic parts and extra long cabling so that the amplifier/capacitor bank can be kept outside of the 5-Gauss line. Specialized holders must be constructed to hold the TMS coil in the appropriate position during the duration of the scanning session (**Fig. 24**). Like all electronic equipment, the TMS hardware must be shielded in a Faraday cage to prevent RF contamination of the MRI signal *(41)*. Commercial TMS coils currently have these characteristics as optional features.

Another important aspect is the synchronization of TMS pulses and image acquisition. It is very important that the scanner's RF receiver chain be switched off while the stimulator is pulsing in order to protect it from the large signals that may severely damage it. At the same time the scanner's pulses may induce currents in the TMS hardware (although these are reportedly not harmful to the hardware). Isolation between the two is achieved by synchronizing the TMS pulses with the MRI scanner via transistor-transistor logic (TTL) pulses. The imaging pulse sequence is typically designed with long gaps for the TMS pulses.

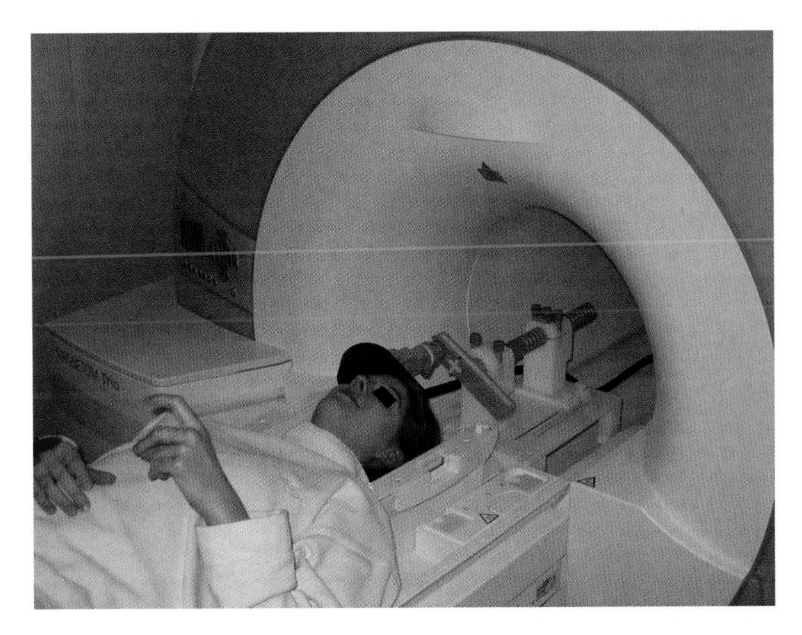

Fig. 24. MRI-compatible transcranial magnetic stimulation (TMS) coil and holder apparatus (courtesy of Dr. A. Thielscher of the Max Plank Institute for Biological Cybernetics http://www.kyb.mpg.de/).

These gaps include about 0.1 s to allow TMS-induced eddy currents inside the bore to decay. There is also the challenge of the large torques that a large, sudden dipole exerts when in the presence of a large magnetic field. However, in the case of figure-eight coils, these torques cancel since the two "wings" of the coil are torqued in opposite directions. Furthermore, rapid biphasic pulses also cancel these torques and the subject does not perceive any such effects. The coil, however, experiences internal stresses from the magnetic forces *(41–43)*.

7. Conclusions

The hardware used in MRI is quite extensive and we hope to have provided an adequate overview of the subsystems involved in the generation of MRI images. fMRI requires additional hardware for collection of behavioral data and stimulation of the subject while collecting the functional images. The greatest challenge is perhaps to coordinate all these devices while being mindful of the interactions between the devices and the MRI scanner. Failure to do so often result in severe artifacts in the desired measurements, or worse, the subject could be severely injured. In this chapter, we have also explored the hardware requirements for multimodal imaging, such as EEG-fMRI or TMS-fMRI.

References

1. Shellock FG. Reference manual for magnetic resonance safety, implants, and devices. Oxford, UK: W.B. Saunders, 2002.
2. Shellock FG, Crues JV III. MR safety and the American College of Radiology White Paper. AJR Am J Roentgenol 2002;178:1349–1352.
3. Train JJ. Magnetic resonance compatible equipment. Anaesthesia 2003;58:387.
4. Durand E, van de Moortele PF, Pachot-Clouard M, Le Bihan D. Artifact due to B(0) fluctuations in fMRI: correction using the k-space central line. Magn Reson Med 2001;46:198–201.
5. Tinkham M. Introduction to superconductivity (2nd ed.). Dover Books on Physics. New York, NY: Dover, 2004.
6. Williams DS, Detre JA, Leigh JS, Koretsky AP. Magnetic resonance imaging of perfusion using spin inversion of arterial water. Proc Natl Acad Sci USA 1992;89:212–216.
7. Yang QX, Wang J, Zhang X, et al. Analysis of wave behavior in lossy dielectric samples at high field. Magn Reson Med 2002;47:982–989.
8. Collins CM, Liu W, Schreiber W, Yang QX, Smith MB. Central brightening due to constructive interference with, without, and despite dielectric resonance. J Magn Reson Imaging 2005;21:192–196.
9. Tropp J. Image brightening in samples of high dielectric constant. J Magn Reson 2004;167:12–24.
10. Schneider E, Glover G. Rapid in vivo proton shimming. Magn Reson Med 1991;18:335–347.
11. Gach HM, Lowe IJ, Madio DP, et al. A programmable pre-emphasis system. Magn Reson Med 1998;40:427–431.
12. Wysong RE, Madio DP, Lowe IJ. A novel eddy current compensation scheme for pulsed gradient systems. Magn Reson Med 1994;31:572–575.
13. Mansfield P, Chapman B. Active magnetic screening of coils for static and time-dependent magnetic field generation in NMR imaging. J Phys E Sci Instrum 1986;19:540–545.

14. Edelstein WA, Kidane TK, Taracila V, et al. Active-passive gradient shielding for MRI acoustic noise reduction. Magn Reson Med 2005;53:1013–1017.
15. Blaimer M, Breuer F, Mueller M, et al. SMASH, SENSE, PILS, GRAPPA: how to choose the optimal method. Top Magn Reson Imaging 2004;15:223–236.
16. Hoult DI, Chen CN, Sank VJ Quadrature detection in the laboratory frame. Magn Reson Med 1984;1:339–353.
17. Roemer PB, Edelstein WA, Hayes CE, Souza SP, Mueller OM. The NMR phased array. Magn Reson Med 1990;16:192–225.
18. Pruessmann KP, Weiger M, Scheidegger MB, Boesiger P. SENSE: sensitivity encoding for fast MRI. Magn Reson Med 1999;42:952–962.
19. Griswold MA, Jakob PA, Heidemann RM, et al. Generalized autocalibrating partially parallel acquisitions (GRAPPA). Magn Reson Med 2002;47:1202–1210.
20. Jakob PM, Schlaug G, Griswold M, et al. Functional burst imaging. Magn Reson Med 1998;40:614–621.
21. Edmister WB, Talavage TM, Ledden PJ, Weisskoff RM. Improved auditory cortex imaging using clustered volume acquisitions. Hum Brain Mapp 1999;7:89–97.
22. Noll DC, Schneider W. Theory, simulation, and compensation of physiological motion artifacts in functional MRI. Image Processing, 1994. Proceedings of ICIP-94, IEEE International Conference 1994;3:40–44.
23. Hu X, Le TH, Parrish T, Erhard P. Retrospective estimation and correction of physiological fluctuation in functional MRI. Magn Reson Med 1995;34:201–212.
24. Pfeuffer J, Van de Moortele PF, Ugurbil K, Hu X, Glover GH. Correction of physiologically induced global off-resonance effects in dynamic echo-planar and spiral functional imaging. Magn Reson Med 2002;47:344–353.
25. Tremblay M, Tam F, Graham SJ. Retrospective coregistration of functional magnetic resonance imaging data using external monitoring. Magn Reson Med 2005;53:141–149.
26. Zaitsev M, Dold C, Sakas G, Hennig J, Speck O. Magnetic resonance imaging of freely moving objects: prospective real-time motion correction using an external optical motion tracking system. Neuroimage 2006;31:1038–1050.
27. Thesen S, Heid O, Mueller E, Schad LR. Prospective acquisition correction for head motion with image-based tracking for real-time fMRI. Magn Reson Med 2000;44:457–465.
28. Chen W, Zhu XH. Suppression of physiological eye movement artifacts in functional MRI using slab presaturation. Magn Reson Med 1997;38:546–550.
29. Barker AT. An introduction to the basic principles of magnetic nerve stimulation. J Clin Neurophysiol 1991;8:26–37.
30. Barker AT. The history and basic principles of magnetic nerve stimulation. Electroencephalogr Clin Neurophysiol 1999;Suppl 51:3–21.
31. Jalinous R. Technical and practical aspects of magnetic nerve stimulation. J Clin Neurophysiol 1991;8:10–25.
32. Ruohonen J, Ravazzani P, Tognola G, Grandori F. Modeling peripheral nerve stimulation using magnetic fields. J Peripher Nerv Syst 1997;2:17–29.
33. Ilmoniemi RJ, Virtanen J, Ruohonen J, et al. Neuronal responses to magnetic stimulation reveal cortical reactivity and connectivity. Neuroreport 1997;8:3537–3540.
34. Berne RM, Levy MN. Physiology. St. Louis, MO: Mosby, 1993.
35. George MS, Nahas Z, Lisanby SH, Schlaepfer T, Kozel FA, Greenberg BD. Transcranial magnetic stimulation. Neurosurg Clin N Am 2003;14:283–301.
36. Paus T. Inferring causality in brain images: a perturbation approach. Philos Trans R Soc Lond B Biol Sci 2005;360:1109–1114.
37. Pascual-Leone A, Walsh V, Rothwell J. Transcranial magnetic stimulation in cognitive neuroscience – virtual lesion, chronometry, and functional connectivity. Curr Opin Neurobiol 2000;10:232–237.
38. Rothwell JC. Paired-pulse investigations of short-latency intracortical facilitation using TMS in humans. Electroencephalogr Clin Neurophysiol 1999;Suppl 51:113–119.
39. Ilmoniemi RJ, Ruohonen J, Karhu J. Transcranial magnetic stimulation - a new tool for functional imaging of the brain. Crit Rev Biomed Eng 1999;27:241–284.
40. Bastings EP, Gage HD, Greenberg JP, et al. Co-registration of cortical magnetic stimulation and functional magnetic resonance imaging. Neuroreport 1998;9:1941–1946.
41. Bohning DE, Shastri A, Nahas Z, et al. Echoplanar BOLD fMRI of brain activation induced by concurrent transcranial magnetic stimulation. Invest Radiol 1998;33:336–340.
42. Bohning DE, Shastri A, McConnell KA, et al. A combined TMS/fMRI study of intensity-dependent TMS over motor cortex. Biol Psychiatry 1999;45:385–394.
43. Bohning DE, Shastri A, Wassermann EM, et al. BOLD-f MRI response to single-pulse transcranial magnetic stimulation (TMS). J Magn Reson Imaging 2000;11:569–574.

Chapter 3

Selection of Optimal Pulse Sequences for fMRI

Mark J. Lowe and Erik B. Beall

Summary

In this chapter, we discuss technical considerations regarding pulse sequence selection and sequence parameter selection that can affect fMRI studies. The major focus is on optimizing MRI data acquisitions for blood oxygen level-dependent signal detection. Specific recommendations are made for generic 1.5-T and 3.0-T MRI scanners.

Key words: MRI, fMRI, Pulse sequences, Blood oxygen level dependent, Echo-planar imaging, Spiral imaging

MRI signals are generated by exposing nuclei placed in a static magnetic field to radiofrequency (RF) pulses in the presence of rapidly switching magnetic field gradients. These patterns of RF pulses and magnetic field gradients are referred to as *pulse sequences.* Pulse sequences dictate the contrast that will be present in MR images.

The issue of optimal pulse sequences, or more generally, optimal data acquisition strategies for functional neuroimaging, is a complex one. One cannot categorically say that a particular approach is superior to any other in all cases. In this chapter, we will examine the issues that affect the detection sensitivity of neuronal activation in MRI and discuss relevant data acquisition strategies that can be optimal in each situation. For those readers who are not interested in the technical details involved in fMRI pulse sequence optimization and who wish to simply read a summary of recommended pulse sequence strategies for blood oxygen level-dependent (BOLD) fMRI, it is recommended that they skip to **Section 5**, which summarizes the

M. Filippi (ed.), *fMRI Techniques and Protocols*, Neuromethods, vol. 41
DOI 10.1007/978-1-60327-919-2_3, © Humana Press, a part of Springer Science+Business Media, LLC 2009

issues and presents recommendations and caveats for each relevant sequence parameter.

1. Physics of Functional Contrast in MRI

With a few notable exceptions, such as diffusion-weighted MRI, MRI contrast stems from taking advantage of the different MR relaxation rates in different tissues and in the presence of pathology. Felix Bloch phenomenologically characterized the dynamic evolution of spin magnetization with two time constants, referred to as T_1 and T_2. An in-depth discussion of the Bloch equations is beyond the scope of this chapter, but for the purposes of understanding the interaction of pulse sequence parameters and functional contrast in MRI, it is useful to briefly describe the processes associated with these relaxation time constants.

T_1 relaxation is taken to be the time constant of the return of an excited ensemble of nuclei to the equilibrium state of the "lattice" or surroundings. So, before excitation, the ensemble will generally be in equilibrium with its surroundings. After excitation, T_1 governs the time for it to return to the state of equilibrium with the lattice. This is sometimes referred to as spin–lattice relaxation.

T_2 relaxation, which is technically an enhancement of T_1 relaxation (i.e., the upper limit of T_2 is T_1), is the time constant for an excited ensemble of nuclei to lose phase coherence through interactions with each other. This is sometimes referred to as spin-spin relaxation. T_2 relaxation in solids and tissue is typically much faster than T_1.

For functional imaging, another important parameter governing relaxation is T_2^*, which is an enhancement of T_2 caused by magnetic field gradients inhomogeneities. T_2^* is defined as

$$\frac{1}{T_2^*} = \frac{1}{T_2} + \frac{1}{T_2'}, \qquad (1)$$

where T_2' is the additional relaxation contribution from field inhomogeneities.

These relaxation processes are sensitive to the chemical environment of the nuclei. MRI utilizes this fact to produce images whose contrast is based on the different relaxation rates in different tissues.

1.1. Magnetic Resonance Relaxometry

Exposing nuclei in a static magnetic field to RF radiation at the Larmor frequency, given by

$$\nu_L = \gamma B \tag{2}$$

will result in the absorption of energy by the nuclei. In **Eq. 2**, γ is the gyromagnetic ratio and is a property of the nucleus. Since B is the static field strength, we see from **Eq. 2** that the Larmor frequency will rise linearly with field strength. For protons, $\gamma = 42.58$ MHz/T so the Larmor frequency at 1.5 T is approximately 64 MHz.

When the RF radiation is stopped, the nuclei will gradually release the energy into the surrounding material until they return to the pre-excited state of equilibrium with their surroundings.

An MR pulse sequence is characterized mainly by two parameters that control the contrast of the acquired data. The first is called the repetition time, or TR, which dictates how frequently the nuclei in a particular location are excited. If they are excited much more rapidly than the T_1 relaxation rate of the tissue, the protons will not recover to equilibrium between excitations. After a few excitations, the nuclei in a given location will approach a steady state. **Figure 1** shows an example of the signal evolution in tissue with different T_1 as a function of TR.

The other important parameter that is used to control contrast is the echo time, or TE. This is the time after excitation that the observed signal is spatially encoded. The amount of signal that can be spatially encoded is dictated by both TR and TE. Two of the most common methods for refocusing MR signal to allow spatial encoding are the spin echo (SE) and the gradient recalled echo (GRE) methods.

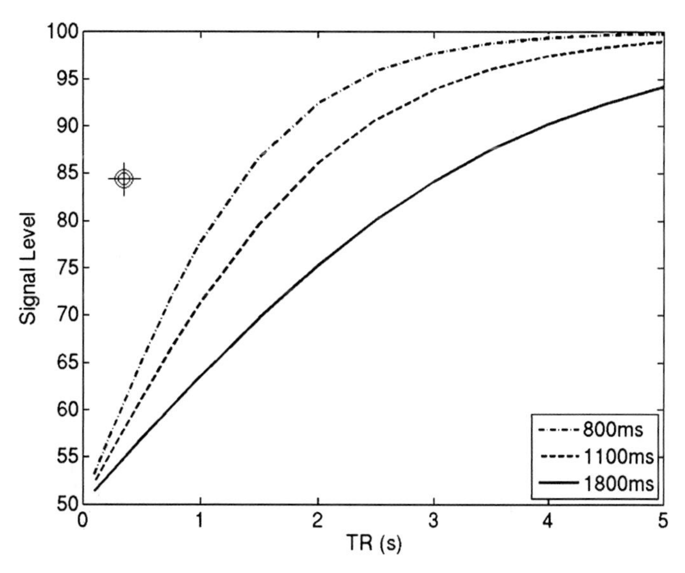

Fig. 1. Steady-state MR signal as a function of repetition time (TR) for tissue with three different T_1 relaxation times.

1.1.1. The Spin Echo

The time evolution of MR signal immediately after excitation is referred to as free induction decay (FID) in MR. During the FID, the processes governing the loss of signal coherence are a combination of T_1, T_2, and T_2'-related processes. In tissue, T_2' will have a large effect on the loss of signal. T_2' effects are what are referred to as reversible processes. The loss of phase coherence from these effects can be reversed by applying a refocusing RF pulse. **Figure 2** illustrates the sequence timing, using a pulse sequence timing diagram. Application of a refocusing RF pulse at a time t after the initial excitation pulse will result in a complete refocusing of the reversible dephasing effects at a time $2t$. This is referred to as a spin echo. The time $2t$ is usually called the echo time or TE. The MR signal from a given pulse sequence can be derived from the Bloch equations. For a SE acquisition, the MR signal will be given by

$$S_{\mathrm{SE}} \propto \exp\left(\frac{-\mathrm{TE}}{\mathrm{T}_2}\right)\left\{1-2\exp\left(-\frac{\mathrm{TR}-\frac{\mathrm{TE}}{2}}{\mathrm{T}_1}\right)+\exp\left(\frac{-\mathrm{TR}}{\mathrm{T}_1}\right)\right\}. \quad (3)$$

As is clear from **Eq. 3**, the MR signal from a SE acquisition is moderated by T_1 and T_2. From this we can see that the T_2 will affect the encoded signal if the TE is comparable to, or longer than T_2. **Figure 3** shows an example of the signal evolution for different TEs and T_2s.

1.1.2. The Gradient Recalled Echo

It is possible to perform the spatial encoding for MRI during the FID. An echo can be created by increasing the dephasing

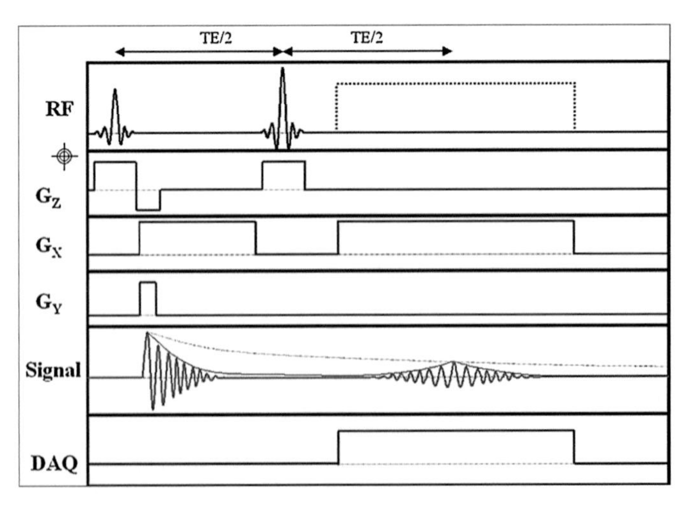

Fig. 2. Pulse sequence diagram of a spin-echo acquisition. The envelope of the signal indicates the free induction decay (FID), while the peak of the echo is modulated by T_2 according to **Eq. 3**.

through application of a brief field gradient along a particular direction and then reversing it while acquiring the signal data. This is called a GRE or a field echo.[1] The sequence diagram for this technique is shown in **Fig. 4**. The signal obtained from a GRE acquisition is given by

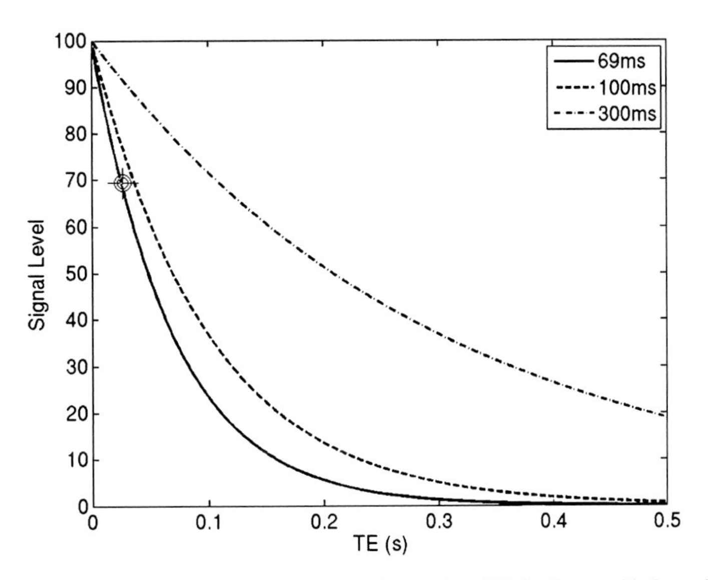

Fig. 3. Steady-state MR signal as a function of echo time (TE) for tissue with three different T_2 relaxation times.

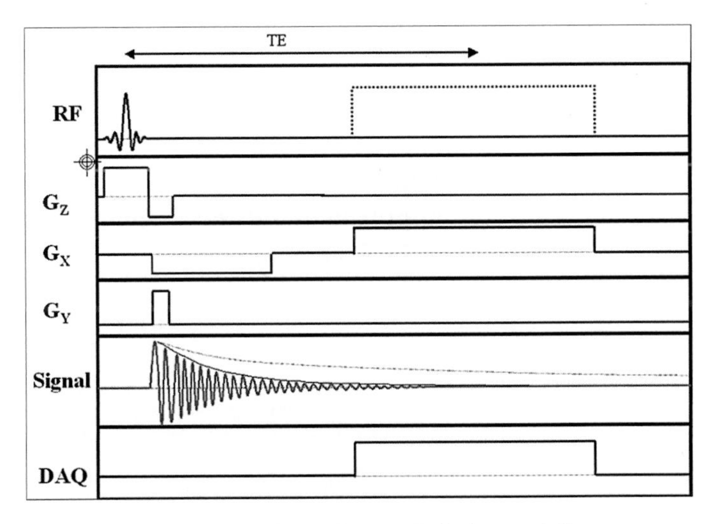

Fig. 4. Pulse sequence diagram of a gradient recalled echo acquisition.

[1] This is sometimes shortened to gradient-echo (GE).

$$S_{\text{GRE}} \propto \exp\left(\frac{-\text{TE}}{\text{T}_2^*}\right)\sin\alpha\frac{1-\exp\left(\frac{-\text{TR}}{\text{T}_1}\right)}{1-\cos\alpha\exp\left(\frac{-\text{TR}}{\text{T}_1}\right)}, \tag{4}$$

where α is the flip angle, which is a sequence parameter that is a function of the amount of transmitted power.[2]

It can be shown from **Eqs. 3** and **4** that the TE that will maximize the difference in signal between two tissues with different T_2 or T_2^* is approximately the average of the T_2s or T_2^* of the two tissue types. **Table 1** lists the T_1 and T_2 of gray matter and white matter in the human brain at 1.5 T and 3.0 T.

Historically, evidence of regionally specific functional contrast using MRI was first observed using an exogenous contrast agent *(1)*. However, this observation was very quickly followed by several groups, employing the phenomenon of BOLD contrast observed by Ogawa and colleagues *(2)*, utilizing endogenous contrast to observe brain activation in several different brain regions *(3–6)*.

1.2. Exogenous Functional Contrast

It is possible to generate dynamic MR images with functional contrast by utilizing the fact that regional blood flow increases proximal to activated neurons. This is typically done by using methods similar to those used to measure regional blood perfusion with MRI. Gadolinium chelates will not cross the blood-brain barrier. Thus, a bolus injection of such a paramagnetic material will cause a transient change in the T_2 relaxation near arterial blood vessels that are perfusing brain tissue. If this is done while rapidly acquiring T_2^*-weighted MR images of the brain region that is active, one will observe a decrease in the measured signal intensity that is monotonically related to the volume of gadolinium passing

Table 1
Approximate relaxation times for gray and white matter at 1.5 T and 3.0 T

	1.5 (T)			3.0 (T)		
	T_1(ms)	T_2(ms)	T_2^*(ms)	T_1(ms)	T_2(ms)	T_2^*(ms)
White matter	600	80	70	800	70	60
Gray matter	900	100	60	1,100	90	50

[2] The flip angle can also affect the contrast of the generated images, but for simplicity, we focus here on the more intuitive parameters TE and TR.

through. One can infer directly the volume of blood perfusing this region from the signal decrease.

Functional contrast can thus be obtained in MRI by comparing the regional perfusion, measured with bolus contrast injection, while performing a task to that measured while at rest. **Figure 5** is an example of the difference in the MRI signal evolution from the same brain region in visual cortex while undergoing photic stimulation and in darkness. One can see that the area under the curve for photic stimulation is larger than that for rest, indicating that the volume of blood perfusing the tissue was increased during stimulation.

In this manner, one can produce voxel level comparisons of the area under the bolus passage curve in the MR time courses and determine those whose measured volume change was statistically significant.

Due to the invasive nature of the necessary bolus injection, the (albeit low level) risk of adverse reaction to contrast agents, the lower signal-to-noise ratio (SNR) of perfusion measurement techniques, and to a lesser extent, the limited volume coverage of perfusion measuring techniques, exogenous contrast-enhanced fMRI is only rarely performed and usually for reasons specific to a particular experimental design.

1.3. Endogenous Functional Contrast

There are two principal mechanisms for generating contrast in MR images using endogenous features related to neuronal activation. Both of these mechanisms are related to the hemodynamic response to an increase in neuronal activation. One of these is a regional increase in blood flow and the other is a concomitant

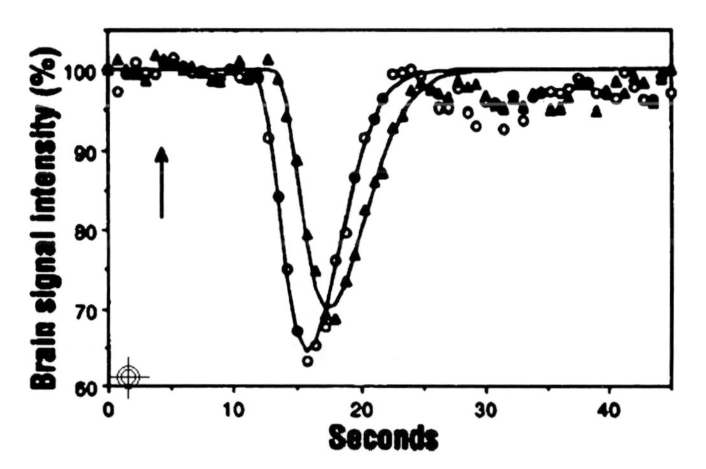

Fig. 5. Changes in MR brain signal intensity during the first-pass transit of intravenously administered paramagnetic contrast agent. *Open triangle* symbols represent the time course of signal during photic stimulation and *open circle* symbols represent the time course during rest (darkness). Reproduced with permission from *(1)*.

increase in the oxygenation content of the blood perfusing tissue near activated neurons.

Far and away the most commonly employed fMRI acquisitions utilize the fact that regional brain activation results in a local increase in blood oxygenation. This is called BOLD contrast. The contrast in BOLD stems from the fact that oxygenated hemoglobin is a weakly diamagnetic molecule, while deoxygenated hemoglobin is a strongly paramagnetic molecule. The relative increase in the concentration of oxygenated hemoglobin in the vessels perfusing activated tissue results in an increase in the T_2 and T_2^* relaxation times in the affected brain regions. Thus, methods utilizing BOLD contrast for fMRI employ acquisition techniques that are sensitive to changes in T_2 and T_2^*. Because of the flexibility of T_2 and T_2^* acquisition methods, this has become the contrast of choice for the vast majority of fMRI experiments. For this reason, the remainder of this chapter will focus on acquisition strategies to acquire BOLD-weighted MRI data and we will discuss methods to optimize these depending on experimental needs.

2. Ultrafast Spatial Encoding

It is possible to generate MR images that will demonstrate a change in signal in brain regions that transition from the inactive to active state. These transitions are typically very rapid and the advantage of MRI over other imaging techniques is the ability to acquire even whole brain images very rapidly. In this section, we introduce the concept of spatial encoding in MRI and discuss the most common ultrafast imaging pulse sequences used in fMRI. For simplicity, throughout this section, we refer to the net magnetization within a sample as "spin."

2.1. The Pulse Sequence

As stated above, the pulse sequence refers to the specific acquisition strategy in which spatial encoding and magnetization readout is performed, providing the basic structure of the RF pulses and field gradients used. Because conventional MRI is based on Fourier spatial encoding, it has become convention within MRI to discuss pulse sequences in the context of *k*-space, another name for the Fourier conjugate of coordinate space. *k*-Space is essentially the image in the spatial frequency domain, and most pulse sequences acquire image data in this domain. There are a variety of advantages to this; most importantly that a coordinate space image can be produced simply by performing a two-dimensional Fourier transform (typically computed using the Fast Fourier Transform, or FFT) on sequentially acquired MRI data.

A pulse sequence for reading one arbitrary line of *k*-space with a gradient-recalled echo is shown in **Fig. 6**.

Starting with stage (1), waveforms are played out on the *z*-direction gradient, G_z, and the RF transmit channel to excite a slice of proton spins. During stage (2), the readout gradient (G_x) prewind and phase-encode gradient selection (G_y) is performed while rephasing spins across the slice/slab with G_z. During stage (3), the readout gradient is switched on while the emitted RF signal from the sample is recorded, denoted by the block of dotted lines. The diagrams shown are simplifications, where the timing and form of the gradients are changed according to various design considerations. The corresponding traversal of *k*-space for the pulse sequence in **Fig. 6** is shown in **Fig. 7**.

Typical conventional (i.e., not single-shot) sequences repeat this process, for different lines of k_y, or phase-encoding positions. This is shown in **Fig. 6**, step 2 with the G_y gradient at multiple possible values containing the variation in the repeated lines of *k*-space sampling. It should be noted that the distance traveled in *k*-space is proportional to the time integral of the gradient strength in space, so it is possible to use a larger amplitude to shorten the time taken to traverse *k*-space.[3]

The sequence described above pertains to a GRE, but without loss of generality, the same sequence applies to a SE sequence. A slice selective RF excitation pulse for both GRE and SE is typically a sinc function-shaped pulse, with amplitude set to rotate

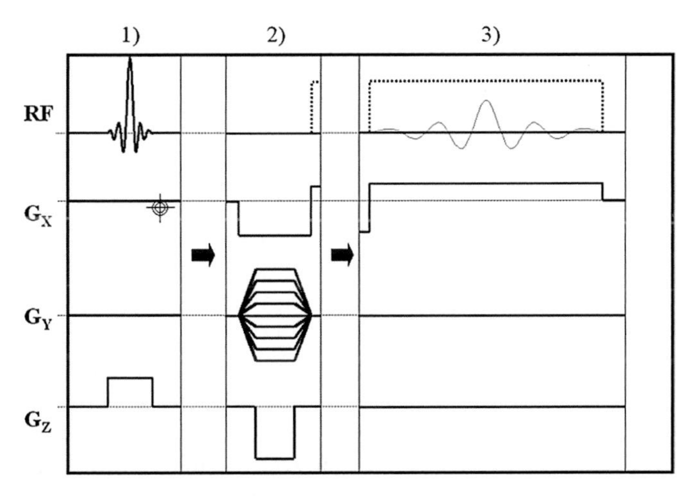

Fig. 6. Pulse sequence diagram for reading one line of *k*-space, time increases from *left* to *right*. The proton echo from the sample is shown here in *light gray* during and under the readout window, which will be sampled by a receive coil.

[3] Up to the limits of the gradient hardware and not without various drawbacks.

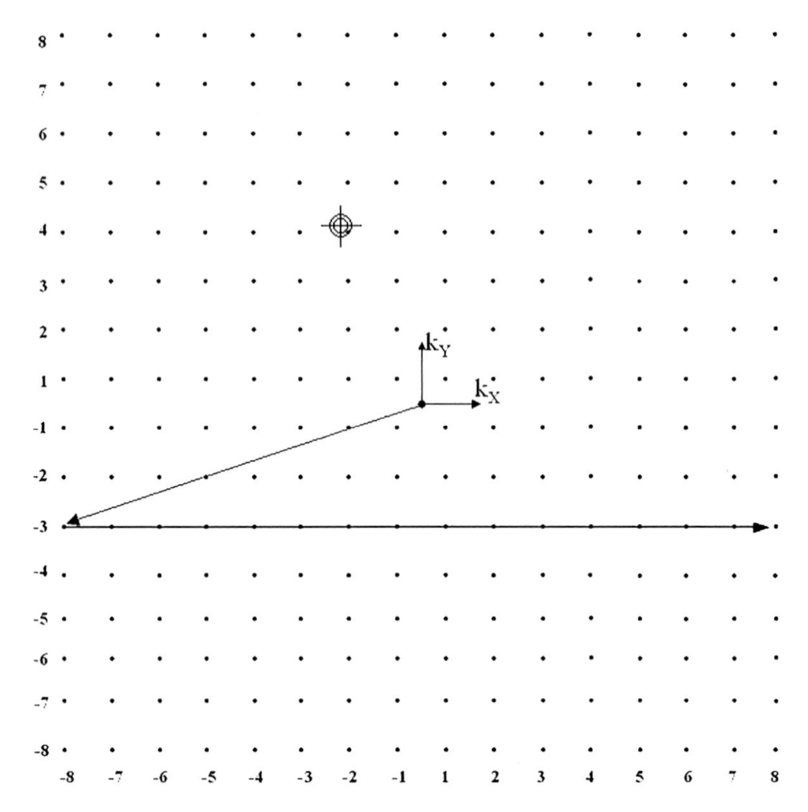

Fig. 7. *k*-Space diagram showing 16 × 16 matrix of data sampling points and trajectory of pulse sequence diagrammed in **Fig. 6**. *k*-Space is first prewound in step 2 (G_x moves position in k_x from 0 to −8, G_y moves position in k_y from 0 down to line −3), then k_x is traversed in step 3 while sampling from −8 to +8.

the slice magnetization 90° from the longitudinal magnetization direction into the plane transverse to the static field. A SE sequence is very similar, but with the addition of a refocusing pulse set at 180°, timed to play out midway between the centers of the RF excitation pulse in step 1 and the readout window in step 3 in **Fig. 6**. The differences between SE and GRE can be seen in **Fig. 8**. The timing of the inversion pulse after the excitation pulse dictates the TE, so a short TE can preclude the SE method. SE can be advantageous because it can cancel dephasing due to local field inhomogeneities, leading to an improved SNR, but single-shot sequences get less of this benefit due to an effective spread of TEs which will be discussed later.

Strategies for what is referred to as "single-shot" imaging are critical to the high sampling rates necessary for dynamic imaging techniques such as fMRI. Echo-planar imaging (EPI) and spiral imaging are the most widely used of these. In addition, parallel imaging techniques, combined with the increasing use of multichannel coil technology, are playing an increasing role in fMRI.

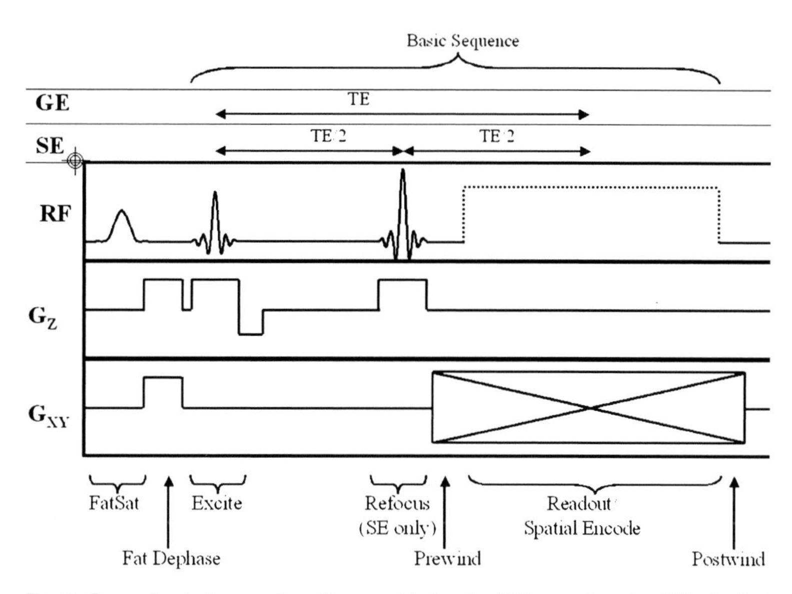

Fig. 8. Generalized diagram for either gradient-echo (GE) or spin-echo (SE) ultrafast pulse sequences specific to ultrafast imaging. A fat saturation pulse (described later), followed by excitation, then a refocus pulse (if spin-echo only), prewind gradients to set position in *k*-space, then the readout and spatial encoding gradients. Finally there may be postwind spoiler/crusher gradients (RF may also be used at the end) to dephase residual signal.

Single-shot sequences differ from conventional pulse sequences in that the data for an entire slice are acquired in one readout window after one excitation. This has been made possible by fast gradient switching technologies, and single-shot sequences are available on all modern MRI scanners. Common to all fast imaging sequences are higher demands on the hardware, which increase vibration and heating of the scanner, leading to increasing inhomogeneity and field drift over time during long scans *(7, 8)*. Parallel imaging is a recent development, which reduces the readout time by acquiring data from multiple coils. These imaging strategies have various artifacts and tradeoffs, which will be discussed below following an introduction to the most common strategies.

2.2. Echo-Planar Imaging

EPI follows the basic strategy of excitation of a slice or slab followed by readout of one line (in the readout direction, or k_x) in *k*-space. The GRE sequence was shown in the pulse sequence timing diagram in **Fig. 8** without the SE refocusing pulse, which was also shown in parts in **Fig. 6**. With the fast gradient switching speeds available in recent years, it has become possible to spatially encode an entire slice in one echo by performing multiple readouts and phase-encoding steps after a single excitation. The most common implementation, known as “Blipped EPI,” involves excitation of a slice followed by readout of k_x line like the

GRE sequence. The sequence continues, however, after an increment, or "blip," of the position in *k*-space in the other dimension using a short-duration gradient pulse (in the phase-encode direction, or k_y). Readout continues when the readout gradient is reversed to read another k_x line in *k*-space immediately adjacent to the first line sampled but in the opposite direction. This is shown in **Fig. 9**.

These reversals and blips are repeated to adequately sample *k*-space and the resulting data can be treated in the same manner as multishot imaging, with the full readout of the slice or slab centered on the TE. The trajectory in *k*-space is shown in **Fig. 10**.

Three-dimensional acquisitions can be performed using an additional increment in the perpendicular dimension, or k_z, although most blipped-EPI sequences are two dimensional only due to the constraint of a shorter TE required. There exist many modifications to this basic structure, but all EPI strategies contain a fast back-and-forth cycling of the gradients to produce a GRE train. The blipped EPI strategy is commonly also referred to as Cartesian imaging, due to the rectangular trajectory of readout in *k*-space. We will not discuss other non-Cartesian strategies that are no longer common such as constant-phase encode (PE) EPI or square-spiral EPI. The signal generation stage before the spatial encoding can include a refocusing pulse or not, depending on whether T_2 (SE-EPI) or T_2^* (GRE-EPI) weighting is desired.

2.3. Spiral Imaging

Another common strategy for single-shot imaging is spiral imaging *(9)*. In this scheme, *k*-space is sampled in a spiral or circular

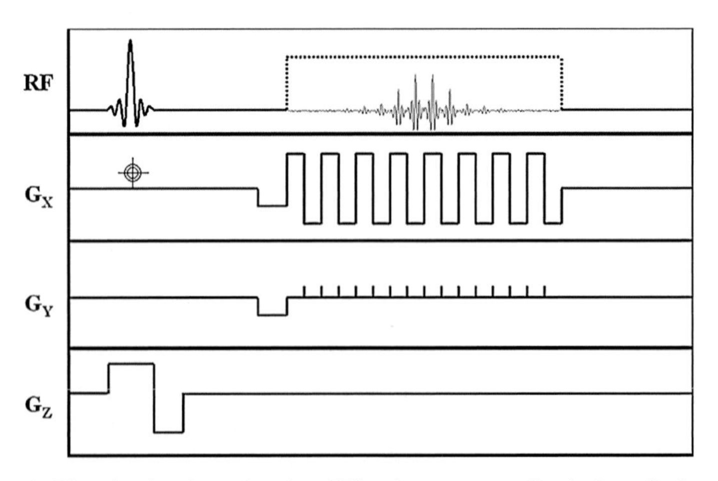

Fig. 9. Blipped echo-planar imaging (EPI) pulse sequence. Readout gradients are reversed following readout of each k_x line, along with a small increment of *k*-space in k_y direction, or a "blip" in G_y. Sixteen k_y lines are read out, corresponding to the *k*-space diagram in **Fig. 7**. Gradient-stimulated echo train is shown in *light gray*, which becomes stronger closer to center of *k*-space, and at center of each k_x-readout.

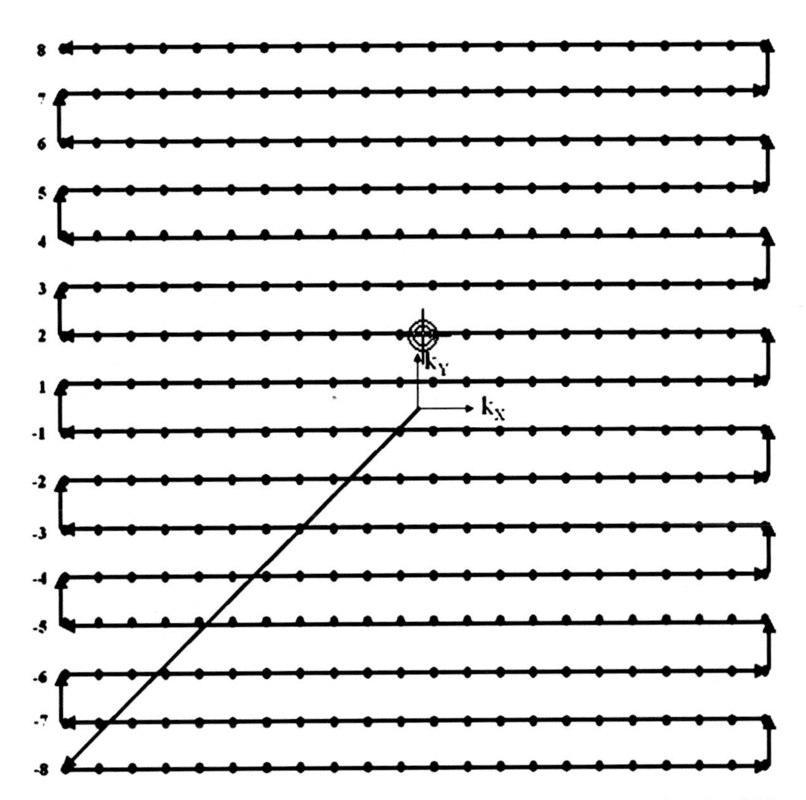

Fig. 10. Cartesian trajectory in *k*-space for one-shot blipped echo-planar imaging (EPI) sequence shown in **Fig. 5**. One-shot means full coverage of *k*-space is accomplished during echo of one excitation.

manner, such as in **Fig. 11**, with less asymmetry between the rate of sampling in k_x- and k_y-space.

By applying sinusoidal gradients 90° out of phase to the read- and phase-encode gradients, *k*-space can be traversed in a circular manner by increasing the amplitude of the sinusoidal gradients. A typical sequence for spiral acquisitions is shown in **Fig. 12**.

This process continues until *k*-space is adequately sampled. There are many trajectory modifications to this scheme, but all have the basic property that the sampling of *k*-space is not Cartesian, but instead approximately radial symmetric. Reconstruction of image space from *k*-space may be done using a Fourier transform after resampling the *k*-space data to a Cartesian grid, but there are implementations that reconstruct the image data using the discrete Fourier transform *(10, 11)*.

Spiral imaging has advantages compared with blipped EPI, mostly related to the lower overall demand on the gradients. These include reduced gradient noise, improved SNR, lower induced eddy currents,[4] and different geometric distortion artifacts *(12)*.

[4] Eddy currents are currents induced in gradient coils and other scanner components from the rapidly changing fields generated by the gradient coils.

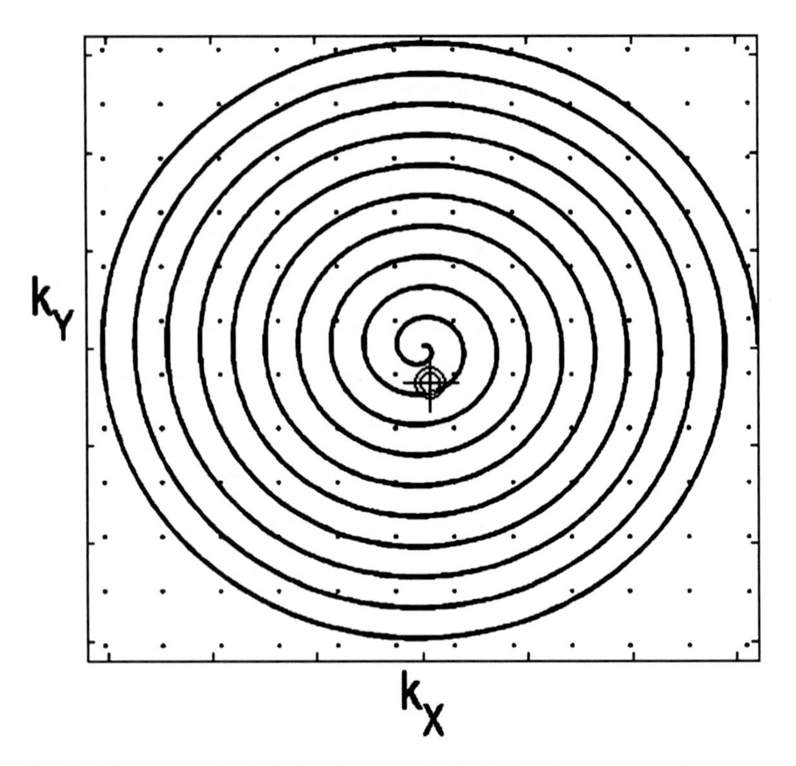

Fig. 11. *k*-Space sampling trajectory for a one-shot spiral imaging sequence shown over a rectangular grid. Central *k*-space is sampled first. Prior to fast Fourier transform (FFT) reconstruction, the data must be resampled from spiral grid to Cartesian grid.

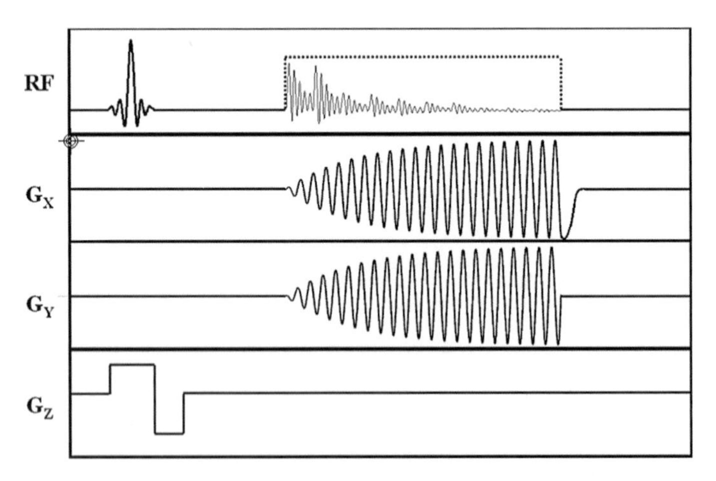

Fig. 12. Pulse sequence timing diagram for spiral acquisition. Gradients during readout window are 90° phase-offset ramped sinusoids.

Increased SNR is due to the earlier sampling of the center of *k*-space, but the correspondingly later acquisition of the outer regions of *k*-space mean that the higher spatial frequencies have lower specificity than the readout direction of a blipped-EPI image. The typically sinusoidal gradient play-out means the gradients are switched at a lower rate of change, which reduces the induced eddy currents and gradient noise. Since most of the magnetization signal naturally lies near the center of *k*-space, which is sampled early, it is preferable to start the readout window at the TE. This modifies the timing from the Cartesian EPI sequence where the readout is centered on the TE, although newer spiral sequences (such as spiral-in/out) are available which also center the TE in the readout window *(13)*. A spiral-in/out sequence is shown in **Fig. 13**.

A spiral-in/out sequence reduces the effects of the echo shifting by centering the readout at the TE as in EPI. Readout begins prior to the TE, starting near the edges of *k*-space and spiraling into the center, which is reached at the TE, before spiraling back out over new data points. After resampling the grid, every point in *k*-space now has two samplings symmetrically spaced about the TE, which are passed through FFT to give two images. These two images can be combined and the result is a reduced sensitivity to susceptibility signal loss and image data with an effective TE closer to that specified *(14)*. This is more demanding on the gradient hardware than spiral imaging and there can be drawbacks in image quality.

2.4. Parallel Imaging

Multiple receive coils have become a popular and widely available means to increase image SNR by providing multiple samples of a *k*-space trajectory. Because the coils cannot be located in the same place, they have varying spatial sensitivities to the tissue, which is maximal at the tissue nearest each coil element. This provides an alternative spatial encoding mechanism, where

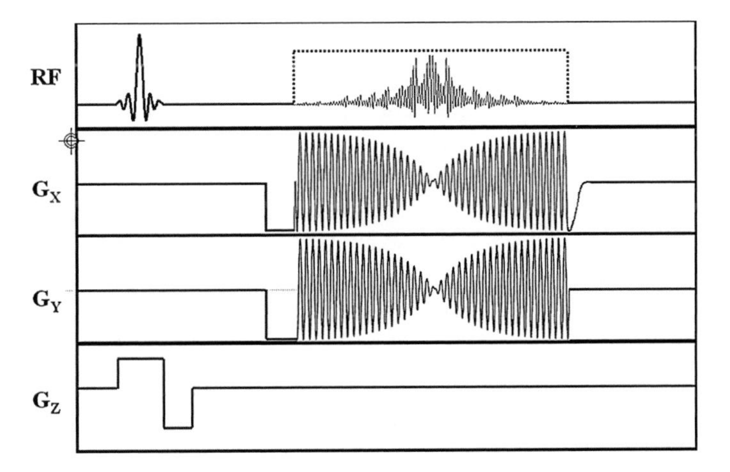

Fig. 13. Spiral-in/out sequence acquires full *k*-space data prior to echo time and a second acquisition of *k*-space after echo time. The data from these echoes are combined in reconstruction.

one sample of several parallel coil elements provides information about the magnetization density over several regions of tissue. Parallel imaging combines this spatial encoding with the gradient-mediated spatial encoding to skip some gradient-encoded lines in *k*-space and replace those gaps with information derived from the parallel coil elements *(15)*. The skipped lines in *k*-space reduce the field of view (FOV) seen by the coils by a reduction factor. The individual coils, if reconstructed with only the data acquired, would see the nearest portion of tissue inside that coil's FOV, but with aliased image overlap with other portions of tissue further away from the coil. The methods used to un-alias the data can be separated into two strategies: image-space unfolding and *k*-space interpolation. While there are many methods, and more than a few hybrids, the most common implementations of each [sensitivity encoding, SENSE *(16)*, and generalized autocalibrating partially parallel acquisitions, GRAPPA *(17)*] will be discussed, along with benefits/drawbacks.

2.4.1. SENSE

SENSE performs the reconstruction of parallel images in image space in an iterative manner using a seeded coil sensitivity matrix. Prior to the parallelized scan, the sensitivity of each coil in the full FOV is measured. These sensitivity maps are used as an initial guess for the "unfolding" matrix.

The undersampling of *k*-space shown in **Fig. 14** leads to image aliasing when reconstructed. However, if the multiple coils are sensitive to spins from different aliased regions, then the portion of signal aliased or unaliased in each image can be differentiated using the sensitivity of the multiple coils to the different regions. The matrix inversion is performed iteratively after preprocessing the data to handle the problem of nonideal coil geometry.

2.4.2. GRAPPA

GRAPPA is a regenerative *k*-space method, using measured data to calculate missing phase-encoding lines. Outer regions of *k*-space have reduced sampling. The acceleration factor defines the number of lines skipped per line acquired. The central *k*-space lines, or autocalibration signal (ACS) lines, are fully sampled, which is shown in **Fig. 15**.

The ACS lines are used to interpolate the nonacquired lines of *k*-space by fitting the acquired lines to the ACS data. This is performed separately for each coil used, leading to weights specific to each ACS line, for each coil. So for N coils, there will be N^2 weights resulting from the fitting procedure to use in interpolating the nonacquired lines. A particular coil's matrix is based on all coil signals, but masks out, or de-weights, signal from other regions outside the FOV of that coil, in *k*-space. The matrix weighting removes the aliasing seen in the original, undersampled images. The final, unaliased image data for each coil is combined by sum-of-square. The greatest advantage of GRAPPA

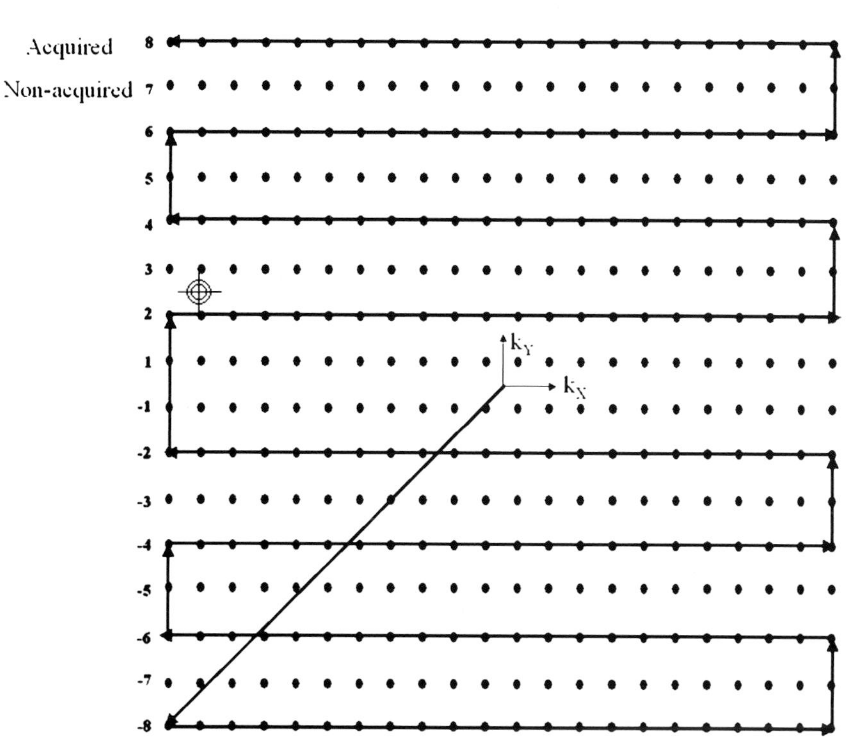

Fig. 14. Sensitivity encoding (SENSE) *k*-space traversal for acceleration factor 2. Odd lines of *k*-space are missed, even lines acquired. Acceleration factor equals acquired plus nonacquired number of lines, divided by acquired lines, in this case the full *k*-space matrix would have twice the number of lines as were actually acquired.

over image-space methods is the determination of sensitivity from the *k*-space data itself, which is useful in images containing regions with poor homogeneity or low signal, both of which are the case with ultrafast imaging *(18)*.

2.4.3. Tradeoffs

The primary benefit of parallel imaging is a reduction of the time spent spatially encoding (the readout window), but at a cost of SNR compared to the same sequence with a fully gradient-based spatially encoded image using the average signal from the parallel coils *(19, 20)*. The reduction in SNR is due to reduced coverage of *k*-space, or the square root of the acceleration factor. An additional cost for any parallel imaging is the coil geometry coverage or *g*-factor. In SENSE imaging, the *g*-factor directly relates to the invertibility of the sensitivity matrix *(21)*. Since most ultrafast parallel blipped-EPI sequences are acquired in two dimension only, the coil coverage should be optimized in the phase encoding direction *(22)*. Spiral-EPI with parallel imaging is more complicated than blipped-EPI and much more time consuming but recent advances have reduced the reconstruction time for parallel spiral-EPI *(23–25)*.

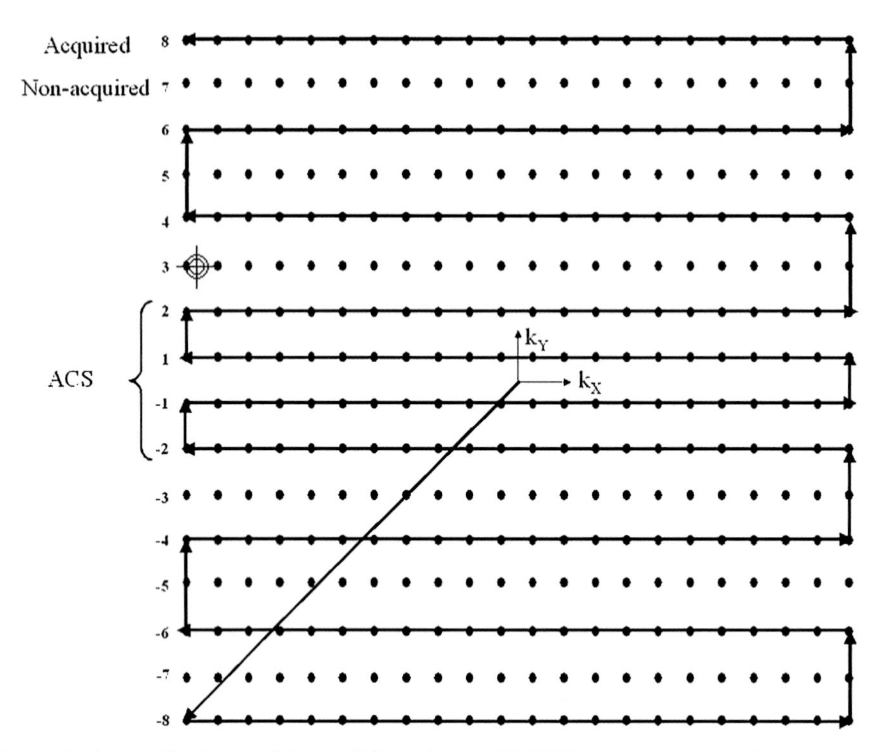

Fig. 15. Generalized autocalibrating partially parallel acquisitions (GRAPPA) *k*-space traversal. Central *k*-space is fully sampled to provide ACS lines. Outer *k*-space is undersampled in phase-encoding direction by acceleration factor. Acceleration factor here is 2, so every other line is acquired. Same sampling is acquired for other coils.

2.4.4. Artifacts

Residual alias in the image is a common artifact seen with parallel imaging. The SENSE method requires the full image FOV to be greater than the object of interest in any accelerated directions; otherwise reconstruction will fail resulting in considerably aliased images. The only current solution with SENSE is to expand the FOV so that there is no image wrapping *(26)*. This is not an issue with GRAPPA, because the spacing of *k*-space lines determines the FOV, but not the signal of a particular line. Therefore, *k*-space fitting under GRAPPA is not compromised by a smaller FOV, while image space fitting would be compromised by the aliasing image. With an ideal sensitivity map, SENSE gives better results than GRAPPA, however, the accuracy of the maps are highly dependent on local field homogeneity and subject motion can invalidate them. There are now several sensitivity map methods, including ones based in part on GRAPPA autocalibrating methods that derive the maps from the data to get around these problems *(27)*. Finally, there is the issue of fitting the systems of equations in the presence of incomplete coil coverage. The SENSE method requires the solution of an inverse problem, but if there are regions of tissue that no coil has adequate sensitivity to, this is an ill-conditioned problem that cannot be exactly solved. All implementations of SENSE regularize or condition the data to work around this, but it means that the final reconstructed image may have local noise enhancements *(28–30)*. GRAPPA is also sensitive to this problem, but because fitting is done with *k*-space data, the noise enhancement is global rather than local *(21)*.

2.4.5. Limitations

There is a practical limit to the number of coils and the acceleration factor, because adjacent coils will overlap spatially in their sensitivity and coverage of magnetized spins, reducing the ability to separate aliased signals. In typical applications of ultrafast imaging, the use of two-dimensional EPI sequences leads to a limit on the acceleration factor of between 4 and 5 *(19)*. While not every scanner has multiple channels, it is becoming the standard for vendors to offer such capability. However, most sites are unable to use parallel imaging due to higher than expected SNR loss at even the lowest acceleration factor *(31)*. Future implementations of parallel imaging promise higher acceleration factors with less loss of SNR using hybrids of *k*-space and image-space methods with dynamically changing undersampling strategies, such as *k–t* SENSE *(32)* or *k–t* GRAPPA *(33)*.

2.5. Partial Fourier Imaging

Ideal *k*-space data has complex conjugate symmetry, which can be exploited to reduce the acquisition time. Up to half of *k*-space can be interpolated from symmetry with the other half. This is referred to as partial Fourier imaging. The symmetry is only approximately true in real data due to scanner and tissue nonidealities, so algorithms to take advantage of this fact must use low-resolution approximations to account for nonzero phases in regions breaking this symmetry *(34, 35)*. With the use of partial Fourier acquisition, a higher spatial resolution can be acquired with less signal loss and blurring, with the result that the SNR does not drop along with the reduced acquisition time *(36, 37)*.

3. Artifacts

3.1. Nonphysiologic

There are several potential artifacts from single-shot imaging techniques due to hardware realities, such as chemical shift (fat) artifact, eddy current artifacts induced by the fast gradient switching, imperfections in gradient ramping waveforms, and both blurring and signal loss due to nonuniform TEs combined with static-field inhomogeneity.

3.1.1. Water-Fat Shift

Water-fat shift image artifact is a consequence of the off-resonance frequency of body fat that shifts the fat signal mostly in the phase-encode direction, misplacing it across the image. Fat suppression with an RF pulse at the resonance frequency of fat, often called chemical saturation, followed by a strong dephasing gradient is the standard countermeasure on EPI sequences. This RF pulse is done immediately before the initial excitation pulse and, due to the fact that the longitudinal signal of the fat is saturated, water protons in fat will experience no excitation. One immediate consequence of this approach is an increase in the time taken by the sequence, as this off-resonance pulse must be performed once before every excitation pulse. Because the resonance frequency of fat protons is only 3.35 parts per million (ppm) in frequency away from water protons, the homogeneity of the static field must be very good to help ensure that the suppression pulse acts only on fat protons and the excitation pulse acts only on water protons.

An alternative strategy to chemical saturation is the use of spatial-spectral RF excitation pulses *(38)*. These are patterned RF and gradient pulses played out over many milliseconds. Properly designed, the aggregate affect of the ensemble of pulses is to create discrete regions of excitation in space and frequency. It is possible to design these pulses such that the excited regions are separated by more than 3.35 ppm in frequency, such that water protons in tissue in a given slice will be excited and water protons in fat will not. This strategy requires less field homogeneity than chemical saturation, but they tend to have a poor slice profile and can take up to twice as long as a good chemical saturation pulse to achieve the same result.

3.1.2. Gradient Nonidealities

Time-varying magnetic gradients induce eddy currents in nearby electrical conductors, such as the magnet cryostat. These eddy currents create magnetic fields that partially cancel the effect of the applied gradients. Fast gradient ramping is limited in hardware by the reactance of the gradient coils, creating effective upper limits on gradient switching that perturb the intended ramping waveform that the gradient coil is driven with. These waveform perturbations increase as gradient switching time decreases. The induced eddy currents and gradient ramping imperfections create phase errors in *k*-space magnetization readout, which produces different artifacts depending on the *k*-space trajectory.

In blipped-EPI, artifact is magnified in the phase-encode direction. This creates what is known as a "phase ghost" or "N/2 ghost," an identical image at 2–5% of the original image signal level but offset by 90° in the phase-encode direction. The ghost can have some overlap with the image of interest. In spiral EPI, the artifact is not as simple, but will result in an increase in noise level.

The corrective methods used vary by scanner manufacturer, but there are some commonalities. The first line of defense is in the screening of the gradient coils to reduce the field change and concomitant eddy currents. A second method employed is calibration of the gradient waveforms. To an extent, eddy currents can be predicted and compensated for by pre-emphasis of the gradient waveforms. This is shown in **Fig. 16**.

Generally, the effect of coil reactance is to dampen the intended gradient waveform by providing a resistance to it, so the waveform to be played out on each gradient coil is modified by a predetermined calibration. Changing the configuration of conductive objects in the scanner room can make this calibration obsolete, if they are near and large enough to be affected by the gradient fields. This could show up as a sudden increase in *N/2* ghosting in blipped-EPI images, requiring a recalibration of the gradient waveform perturbation. Further antighost calibrations to account for system timing offsets and residual eddy current effects may be performed, such as phase line correction. A calibration is typically taken just prior to the readout window in the blipped-EPI sequence by sampling forward and backward across the middle of *k*-space. Eddy current and timing offsets result in a nonideal *k*-space trajectory that can be approximated as a simple shifting of each line of *k*-space forward and backward depending on the direction of traversal. The calibration lines are used to resample every readout k_y line to center the received echo *(39, 40)*. A failure of this online phase ghost correction algorithm would show up as a dramatic increase in phase ghost signal level, to a level comparable to the image of interest. An example of a failure of online phase ghost correction is shown in **Fig. 17**. In this case, signal changes seen as a result of phase ghost correction overwhelm the BOLD effect.

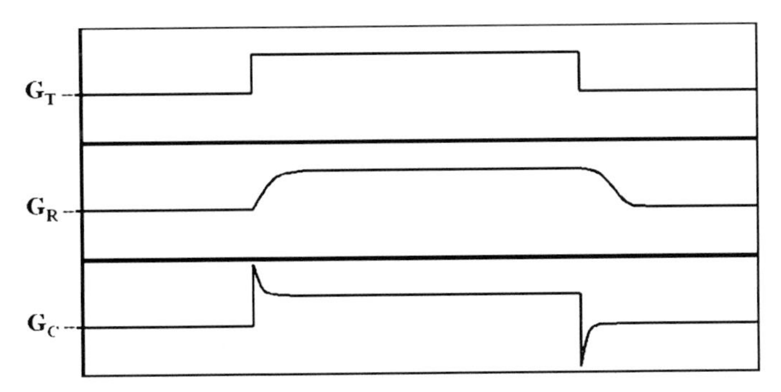

Fig. 16. Gradient waveform calibration. G_T shows the theoretical, intended gradient, while G_R shows the real waveform due to eddy current damping the intended waveform. G_C shows a calibrated waveform to be played out on the gradient coils to produce the intended gradient despite the presence of eddy current.

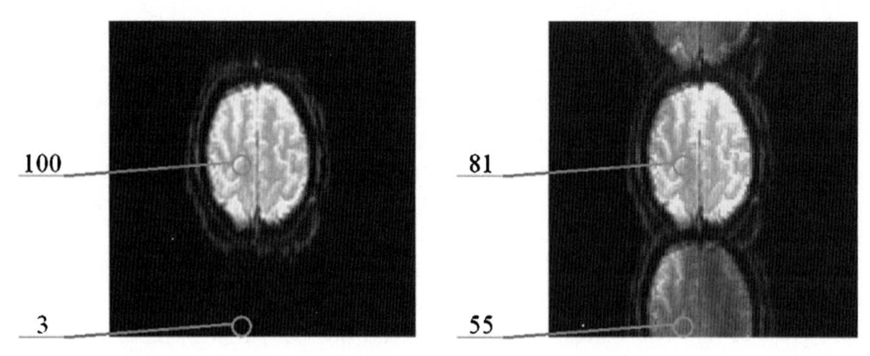

Fig. 17. Phase ghost typical level on *left*. Phase ghost correction algorithm failure on *right*. All measured signal values normalized to first brain tissue measurement on *top left*.

The effect of eddy current on spiral imaging is smaller since dB/dt is lower, but if uncompensated will result in warping, because it warps the k-space trajectory in both dimensions. Measuring the actual trajectory taken in k-space can be used in the resampling portion of reconstruction to correct this image warping similar to the blipped-EPI phase line correction *(41)*.

3.1.3. Echo Shifting

The long readout time employed leaves single-shot images highly sensitive to static field homogeneity, leading to image distortion artifact in regions of inhomogeneity. The off-resonance frequency in these regions causes an accumulation of phase errors in those regions over the readout time. Phase errors specific to a region result in spatial encoding errors, which manifest as signal misplacement from that region. For blipped-EPI images, this is insignificant in the readout, or k_x direction because it is sampled so quickly, but the phase-encode direction is sampled more slowly, resulting in spatial distortion, or blurring in the phase-encode direction in those regions. Spiral imaging samples the k_x and k_y dimensions at approximately the same rate, but the radial dimension is sampled more slowly, akin to the phase-encode direction in EPI. Spiral images are therefore blurred across both dimensions *(42)*. The geometric distortion can be "unwarped" from the images using the calculated pixel shifts from an acquired fieldmap for both blipped-EPI *(7, 43)* and spiral imaging *(35)* (**Fig. 18**).

3.1.4. Signal Loss

Signal loss, or slice dropout, is caused by through-slice dephasing after the RF excitation. This signal loss cannot be recovered without modifying the pulse sequence. Strategies for overcoming this include: use of SE to refocus the dephasing effects, reducing the TE, reducing slice thickness and/or in-plane voxel size, and changing the scan plane. If hardware permits, the use

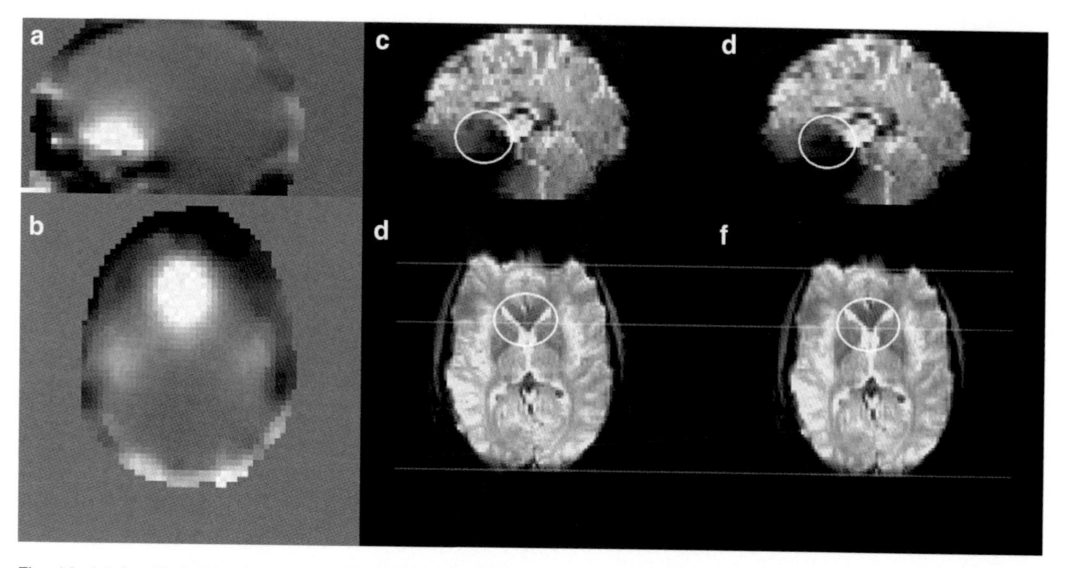

Fig. 18. (**a**) Sagittal, (**b**) axial views of fieldmap. Other pictures show (**c**) and (**d**) blurred blipped echo-planar imaging (EPI) signal and (**e**) and (**f**) unwarped images of the same using pixelshifts calculated from the fieldmap. Anterior regions shifted roughly 1–2 pixels, but signal loss near the frontal sinuses cannot be recovered, leaving a signal void.

of high order gradient shims and image-based shimming can help *(44–46)*.

Another common method is *z*-shimming to unwrap some of the dephasing, but this has the disadvantage of reducing the SNR in unaffected regions. *z*-Shim methods rely on acquiring a fieldmap to estimate the gradient across a slice, and then applying an opposing (negative) slice gradient that is equal to the fieldmap measured gradient in the high susceptibility region *(47, 48)*. This reduces the dephasing in the high susceptibility region, but at the cost of increasing dephasing in other less-affected regions. To deal with this problem, typically two images are taken; one with the *z*-shim and one without and then these are added together to restore all signal loss, but at the cost of nearly doubled imaging time. This method is approximately equivalent to an older technique of tailoring RF pulses to apply a set dephase across the slice at the time of excitation *(49)*. Alternatively, several images with a range of *z*-shim gradients linearly spaced between zero and the maximum measured gradient are taken, and these are averaged; however, this is more time consuming with little benefit over the more common method with two images.

3.1.5. Spiral Regridding

Apart from lower spatial specificity and increased acquisition time, problems typically associated with spiral imaging lie in the regridding of the spiral trajectory to a Cartesian coordinate system prior to Fourier transform. Regridding introduces subtle artifacts and reduces SNR, and is computationally intensive compared with the reconstruction methods used in blipped EPI, although computational improvements have been made ***(50, 51)***. Until recently, the reconstruction had to be performed offline on a separate image-processing computer after the scan, which made it difficult to validate data online or use prospective motion correction. There are now online versions of spiral reconstruction, such that prospective motion implementations for spiral imaging have been used to monitor subject motion ***(52)***.

3.2. Physiologic Artifacts

There is another class of image artifacts present in functional neuroimaging that have physiologic origins. Head motion ***(53–55)*** and physiologic noise from the heart and breathing cycles ***(56–58)*** are unavoidable nonneuronal sources of variance, changing underlying statistical distributions and introducing possible systematic effects in population studies. Accounting for these artifacts requires care in the acquisition of data and several stages of postprocessing of data (retrospective techniques will not be discussed here) after collection is complete. Preventive measures include head restraints or navigator echoes to reduce the effects of head motion, and routine scanner quality assurance measures to track the stability of the scanning hardware ***(59)***. In addition, the collection of parallel measures of state during the image acquisition may be useful for artifact removal during postprocessing. These parallel measurements can include online motion detection parameters from navigator echo or prospective motion correction along with signals representing physiologic cardiac and respiratory cycles.

3.2.1. Head Motion

Due to the fact that it is desired to maintain a high temporal resolution, the sampling rate in most fMRI acquisitions is fast compared to the T_1 of brain tissue. The consequence of this is that, after equilibrium is achieved after acquisition of a few volumes, the tissue is in a saturated state. This means that the magnetization is not completely recovered between excitations of a given slice. If a subject moves such that tissue from one slice moves into an adjacent slice, the tissue will, for the first excitation after the motion, be in a different state of saturation than the rest of the tissue in that slice. This leads to a signal change that is correlated with the motion, but will not be corrected by the traditional technique of retrospective realignment of the images. Prospective motion correction techniques are intended to deal with this problem in real-time.

Navigator echoes can be used to obtain motion information during the acquisition of data ***(60, 61)***. This technique uses the fact that the phase of MR data is sensitive to motion. Typically, low-power RF pulses are interspersed with the fMRI data acquisition and the phase information from the readout of the signal from these pulses is used to infer motion along a given direction. The drawbacks of the navigator echo approach is that, unless the power is very low, in which there is a limited ability to determine phase changes, the excitation pulses will affect the spin history of the fMRI data. Nevertheless, these approaches have been used with some success in fMRI.

An alternative approach that has been applied is to use real-time coregistration of a reference volume to the current volume to determine motion. The motion parameters are determined from the coregistration and are applied to the imaging system prior to acquiring the next volume *(52, 62)*. The drawback of this approach is that it is more computationally intensive than the navigator echo approach and it does not update the slice locations until after the motion occurs. Thus, registration-based prospective motion correction must be coupled with postprocessing motion correction.

3.2.2. Physiologic Noise

Ongoing physiologic processes in living subjects present an additional potential artifact. Effects due to the cardiac and respiratory cycles have been identified as being significantly coupled to BOLD-weighted MR signal in voxels in the brain and the spinal cord. The primary effect of the respiratory cycle on blipped-EPI fMRI data is an apparent shift in image position in the phase-encode direction. This is due to shifting of the resonant water frequency as the main field drifts due to chest expansion and contraction *(63, 64)*. The primary effect of the cardiac cycle is pulsatility artifact with each heartbeat, although the structure and timing of the artifact may vary across the brain due to the range of vessel sizes, stage in the vessel network, and location in the brain *(65, 66)*. The cardiac effects are more pronounced in certain regions such as the insula and brainstem, while respiratory effects are more global (**Fig. 19**).

Correcting for the effect of respiration can be accomplished using navigator echoes or off-resonance detection to follow the field shift during the scan, in the same way as described above for gross head motion. Gating acquisitions on either cardiac or respiratory rates is another method that has been used to reduce the variability in fMRI data *(67)*. This approach necessitates a correction for T_1 effects introduced by the variable TR.

Use of parallel measures can be used to effectively remove physiologic noise sufficient for most purposes in fMRI *(68, 69)*. Therefore, acquiring a pair of signals representing the cardiac cycle and respiratory cycle during the scan is desirable, although recent methods have been developed to estimate equivalent signals from the echo-planar data itself *(70)*.

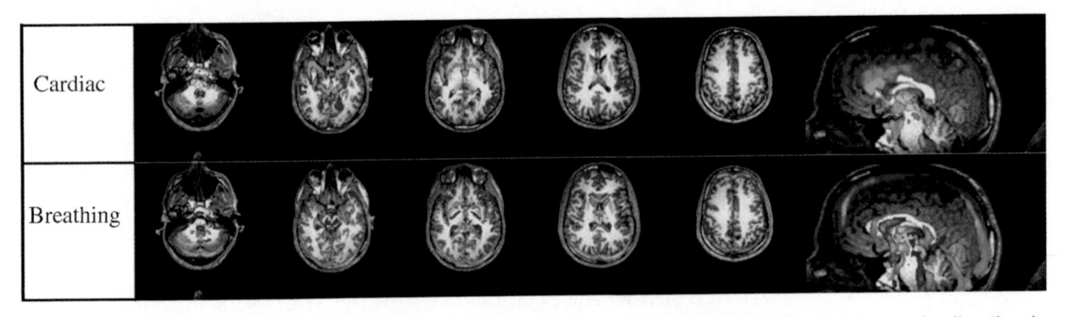

Fig. 19. Averaged physiologic coupling maps in blipped echo-planar imaging (EPI) with phase-encode direction in anterior–posterior axis, determined by temporal independent component analysis (ICA). Cardiac coupling overlain on anatomy is shown in *top row* and respiratory coupling is shown in *bottom row*.

4. Optimization of Sequence Parameters

In order to generate MRI data with functional contrast, there are a number of experimental issues that need to be considered. Generally speaking, if the issues were simply rapidly generating images with BOLD contrast then the procedure would be to simply select sequence parameters such that the TR is as short as possible and the TE such that the expected changes in capillary or venous oxygenation result in a maximal signal change between rest and active neuronal state. However, the choice of optimal acquisition encompasses many other experimental issues and these should all be considered.

4.1. Relaxation Parameters and Functional Contrast

As described above, BOLD contrast in MRI is produced either through changes in T_2 or T_2^*. We are interested in optimizing MR signal differences between two states, rest and active. In this section, we discuss scanner and sequence issues that can affect the detection of these two states.

4.1.1. Field Strength and Relaxation Parameters

The effect of static field strength on T_2 relaxation in brain tissue, as evidenced from examination of **Table 1**, is generally that it is reduced. The effect of field strength on the BOLD signal is complex, and depends on the nature of the proton transport mechanism in the presence of the field defects introduced by the deoxygenated hemoglobin. Recent studies suggest that this mechanism is largely diffusive in nature at clinical field strengths, which would suggest a linear dependence of the BOLD signal on field strength. Experimental data bear out the linearity of the dependence of BOLD contrast on field strength *(71)*. Thus, BOLD contrast from extravascular protons can be taken to increase approximately linearly in the regime used by most commonly available MRI scanners (i.e., 0.3–3.0 T).

The intravascular contribution to BOLD signal stems from the impact of the change in oxygenated hemoglobin concentration within the vessels and the consequent change in T_2 of the blood. The effect of this at the voxel level has been more difficult to describe than the extravascular effect due to the dependence on many factors such as blood volume, vessel size, and volume fraction.

4.2. Experimental Design

The goal in experimental design of fMRI studies is to take advantage of the fact that regional changes in blood flow and oxygenation result proximal to regions of increased neuronal activation. Historically, the basic methodology has been to acquire properly weighted MRI data of the brain regions of interest while a subject repeatedly performs tasks related to the brain function of interest. Initial methodology took advantage of the observation that continuously repeating a task during short intervals leads to an accumulation of signal from the overlapping of events in a time short compared to the hemodynamic response. This is a feature of the general linear model (GLM) of functional imaging *(72)*.

Blocking activation in bursts of extended activity over many seconds, interleaved with long period of rest, leads to up to a much higher increase in hemodynamic response than short, isolated events. This fact makes it desirable, when possible, to use what is typically referred to as a *block design*.

In 1997, Josephs and colleagues proposed an alternative experimental design, intended to more specifically detect the MRI signal associated with neuronal events *(73)*. This experimental design takes advantage of the fact that, through synchronization of the time of stimuli and measurement of behavioral responses, functional imaging data can be analyzed for signal fluctuations correlated with brief, temporally separated neuronal events. This type of experimental design is referred to as *event-related* fMRI. Due to its suitability to address more complex neuroscience questions regarding brain activation and interactions, this has become a preferred experimental design among neuroscience researchers.

Since these two experimental approaches have different analysis strategies, the issues with regard to optimizing pulse sequences are different between them. In the sections below, we separately discuss these issues.

4.2.1. Block Design fMRI

As stated above, block design fMRI experiments are designed to create a large aggregate signal from activated neurons extended in time, interspersed with long periods of rest, or alternate task performance. Analysis of this type of data is typically performed with what is referred to as a reference function. The simplest method for analyzing these data is simply to calculate the cross correlation of the experimental reference function with the time series at each voxel *(74)*. Although more sophisticated methods have been developed that allow more complex analyses, accounting for nuisance effects and systematic effects of no interest, for purposes of pulse sequence optimization, a correlation approach is sufficient to illustrate the issues.

Figure 20 shows an example of the signal time course from a voxel in response to a block design paradigm. The issues with regard to pulse sequence optimization are contrast-to-noise ratio (CNR), sampling rate, and number of samples. In principle, the TE and TR will control the CNR for a given pulse sequence (i.e., EPI, spiral, etc.). The sampling rate is the inverse of the TR. The detection efficiency of a pulse sequence will be determined by these and the number of samples. As an illustration of this, **Fig. 21** shows the probability of getting a type II error at a false positive rate of 0.01 as a function of the number of samples.

4.2.2. Event-Related fMRI

Event-related fMRI relies on a different analysis strategy than block design fMRI. The principal difference as it relates to choice of pulse sequence is temporal resolution. A typical method for analyzing event-related fMRI is *deconvolution*. Deconvolution is an analysis method where rapidly repeated, although temporally separated, events can be extracted if the timing of the onset of the signal and either the duration or the hemodynamic response function is known. A detailed discussion of deconvolution techniques is beyond the scope of this chapter. The reader is referred to chapter 7 for a more complete treatment of the analysis of event-related fMRI.

Figure 22 shows a typical timing and signal response for a rapidly presented event-related fMRI experiment. Studies on the

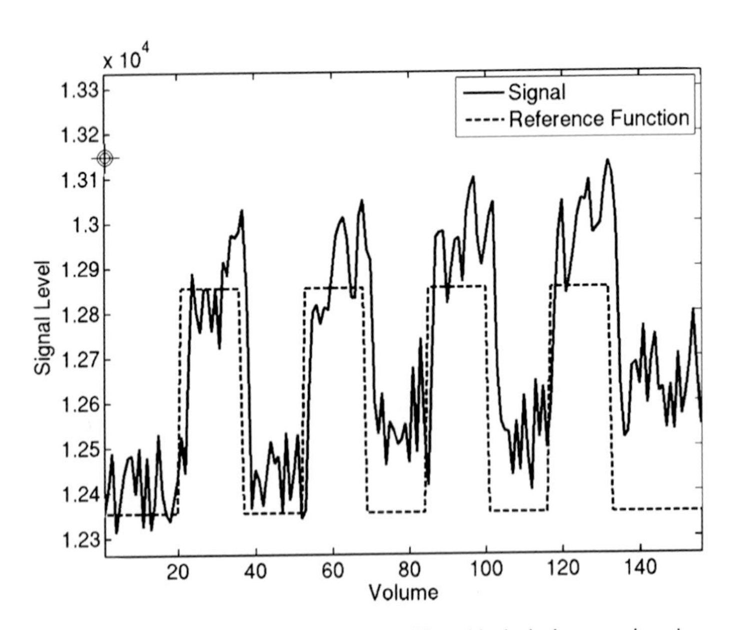

Fig. 20. Example of fMRI timecourse from a voxel for a block-design experiment.

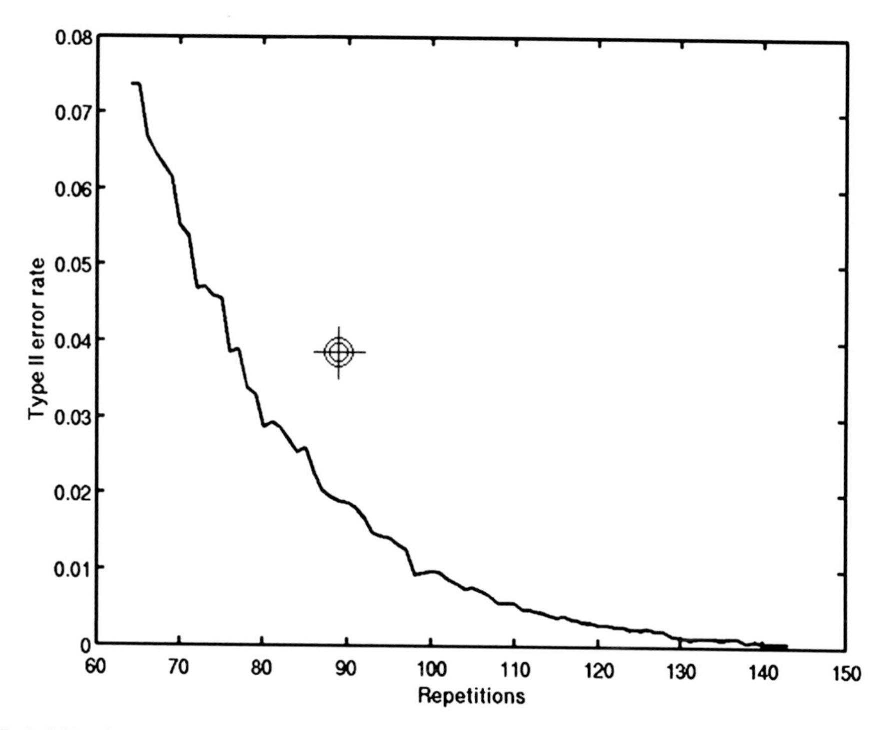

Fig. 21. Probability of rejecting a true event (type II error) as a function of the number of samples at false positive rate of 0.01 for a two cycle block design fMRI experiment. Result is from simulating image (signal-to-noise ratio = 50 with a 2% blood oxygen level-dependent signal change).

ability of deconvolution techniques to resolve neuronal timing shifts indicate that volume sampling rates (i.e., TR) of up to 3 s permit identification of neuronal timing shifts of order 100 ms *(75)*. Therefore, if a goal of an experiment is to study relative timing of events, TRs of up to 3 s should be sufficient.

Another pulse sequence issue that should be of concern to researchers employing event-related experimental designs is SNR. As stated above, the CNR of event-related design is much lower than block design experiments. Therefore, it is recommended that researchers choose their acquisition strategy with this in mind. For instance, if a local cortical region is of interest, a surface coil array could be adopted to significantly increase SNR. Pulse sequence choices should be made carefully to avoid loss of SNR (shortest TE permissible for BOLD contrast, for example). In **Section 5**, issues with regard to SNR for the common pulse sequence parameters will be discussed in detail.

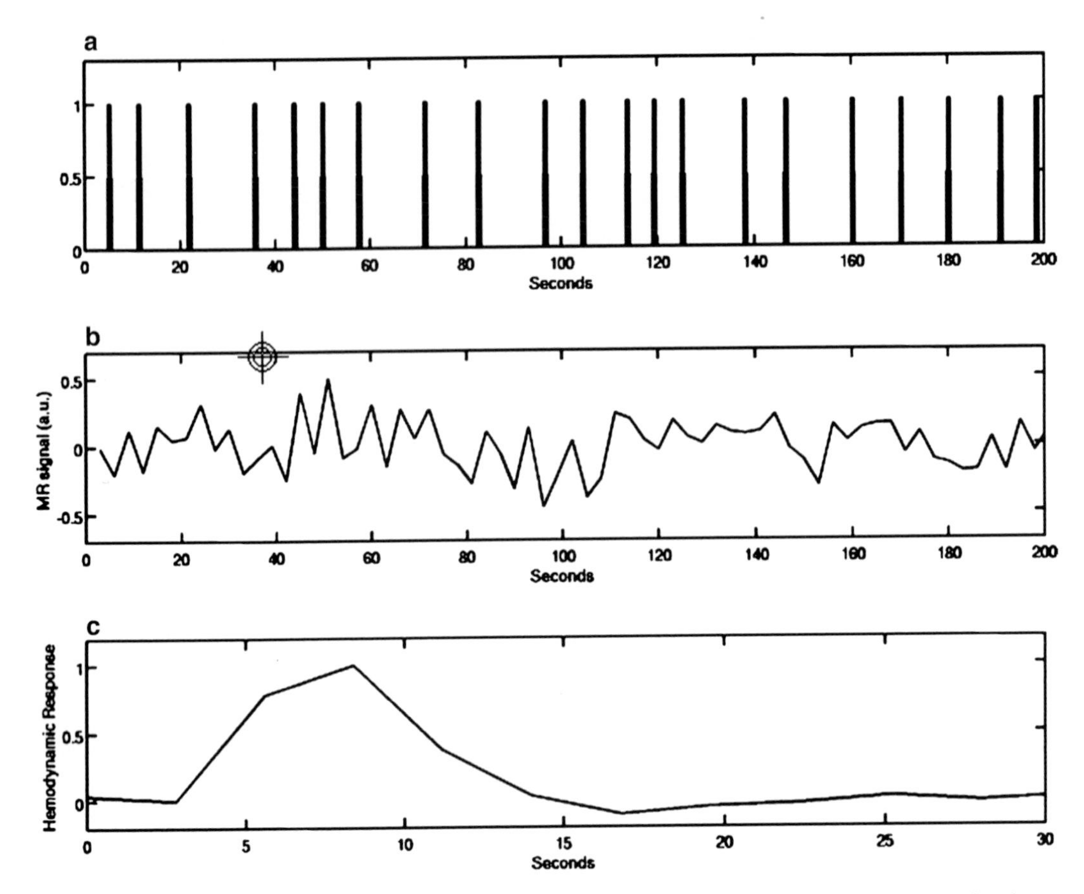

Fig. 22. Example of event-related fMRI experiment. Rapid presented stimuli (**a**) results in single-voxel time series shown in (**b**). Hemodynamic response (**c**) is deconvolved from signal average over 100 voxels with temporal resolution 3 s.

5. Summary Recommendations for Optimal BOLD fMRI

In this section, we will summarize the issues with regard to pulse sequence optimization in the performance of fMRI experiments. As stated above, currently, BOLD-weighted fMRI is the overwhelming method chosen for fMRI. For that reason, the recommendations presented in this summary will focus on BOLD-weighted acquisitions. In most cases, the technical discussions will be relevant for other types of image weighting.

In the context of the information presented previously in this chapter, we will present recommended acquisition strategies for a given field strength and we will then expand on the optimization issues with regard to each of the sequence parameters.

Table 2 lists basic pulse sequence recommendations for two field strengths and two experimental design strategies.

Table 2
Basic sequence parameters for BOLD fMRI acquisition on most clinical MRI scanners

Sequence parameter	Field strength	
	1.5 (T)	3.0 (T)
Sequence	GRE-EPI	GRE-EPI
Scan plane	Axial	Axial
FOV (mm)	256 × 256	256 × 256
Matrix	64 × 64	128 × 128
TE (ms)	50	30
TR (ms)	2,000	2,800
Flip angle	77°	80°
Receiver bandwidth (total) (kHz)	125	250
Slice thickness (mm)	7	4
Slices (for whole-brain coverage)	18	32

FOV Field of view, *GRE-EPI*, gradient recalled echo–echo planar imaging, *TE* echo time, *TR* repetition time

The recommendations in **Table 2** should be possible in almost any recently installed clinical MR scanner of the indicated field strength. Further recommendations made below may require special modifications to supplied clinical pulse sequences and it is strongly recommended that researchers seek the input of an MRI physicist experienced in fMRI.

5.1. Pulse Sequence

Various pulse sequences that have been proposed for fMRI acquisition were outlined in **Section 2**. Issues with regard to optimization of fMRI experiments are discussed here.

5.1.1. Echo-planar Imaging

This is now the most commonly available single-shot imaging sequence available on MRI scanners. Data can be acquired in GRE mode and SE mode.[5] The issues with regard to optimization

[5] In addition, a mixed mode EPI has been used in the literature known as asymmetric spin echo (ASE). This is a SE EPI with the acquisition window shifted to be centered on a time point early on in the SE evolution. The result is an acquisition that has, in a sense, *adjustable* sensitivity to capillary and venous signal.

are that SE EPI employs, by necessity, a longer TE that will result in a smaller intravascular contribution, particularly at 3.0 T and higher, and refocuses dephasing effects from the static dephasing component of the extravascular signal that is specific to larger, distal vessels. The result is that SE EPI can be more spatially specific to the localization of neuronal activation. However, the SNR is much lower than GRE EPI. SE EPI is not recommended for field strengths below 3.0 T because the T_2 of blood is not short enough to be of benefit with regard to the intravascular BOLD signal and the SNR is too low to employ high enough spatial resolution for the spatial specificity to have a significant impact.

5.1.2. Spiral Imaging

Spiral imaging is becoming more common and is available as a product sequence on some clinical MRI scanners. As outlined above, the major advantage of spiral imaging with respect to EPI is that it is less demanding on the imaging gradients. Thus, there will be reduced image warping from eddy current effects. In addition, the nonuniform sampling of the Fourier domain image will lead to a different, possibly lower, sensitivity to motion effects and even some types of physiologic noise. Variants of the spiral technique have been proposed that are specifically designed to be more sensitive to the characteristics of the BOLD signal. This sequence is recommended for researchers employing systems with underpowered gradient systems and for situations where motion and/or physiologic noise or other types of image artifact, as discussed above, are a concern and alternate methods of addressing these are not available.

5.1.3. Parallel Imaging

MRI scanner manufacturers are increasingly moving to the use of head array RF coils in lieu of circularly polarized quadrature head RF coils for MRI. The cost, particularly at 1.5 T, is uniformity of SNR, and thus fMRI signal detection efficiency across the brain. This is less of an issue at high field strength, since dielectric effects in this frequency regime reduce the uniformity of the quadrature coil anyway. The advantage of head arrays is that parallel imaging methods can be used to accelerate spatial encoding of images. The result dramatically increased image quality in regions where single-shot imaging methods have historically been very poor in quality (e.g., orbitofrontal regions, mesial temporal lobe, and brainstem). Some of these brain regions have important and interesting functions. Parallel imaging techniques with the head arrays available to most researchers will typically result in lower SNR throughout most of the brain, but these can still be effectively employed in situations where image artifact severely limits experimental options.

5.2. Scan Plane

Issues with regard to scan plane are largely esthetic. However, there are some technical issues that are worth discussing here.

Perhaps the most important issue is brain tissue coverage. The scan plane of choice can affect the coverage of brain tissue. Given a TE, receiver bandwidth (RBW), and TR, the number of slices available to be acquired in one TR is fixed on a given scanner.[6] The most efficient scan plane for acquiring most human brains is the sagittal plane. The brain in most adults is shortest in the right/left dimension, and so fewer slices will be required to cover the entire brain.

An axial acquisition plane is recommended in **Table 2** due to the fact that it is a more intuitive scan plane to work in, both anatomically and from a physics perspective. Eddy current and higher order artifacts stemming from gradient and shim coil interactions, that are essentially related to coil geometry, are more easily understood in the axial plane. With that said, it is a simple extension to understand these effects in other scan planes. Axial imaging has the added advantage over sagittal and coronal imaging planes in that lateral, frontal, prefrontal, and posterior regions of the brain can be imaged entirely within a relatively few slices (i.e., the very top and very bottom of the brain are considered by many to be more "expendable" than these other regions). The axial plane is a very common imaging choice in fMRI and thus is listed in **Table 2**.

5.3. Field-of-View

FOV has an impact on fMRI signal optimization in three ways (1) together with image matrix, it determines the in-plane voxel size and there are a number of issues with regard to this that will be outlined below, (2) image artifact reduction, particularly in the phase direction, with volume RF coils, and (3) brain coverage.

The last of these points is rather trivial; however, the other two are important with regard to general acquisition strategy.

5.3.1. In-Plane Voxel Size

In-plane voxel size affects fMRI acquisition SNR (and thus fMRI signal detection efficiency) and image quality. At lower field strengths, the SNR issue will dominate and thus it is recommended to use a larger voxel size at the expense of spatial resolution in order to enhance signal detection efficiency. Further signal enhancement can be attained with minimal loss of spatial resolution at 1.5 T through special spatial filtering techniques *(76, 77)*.

At field strengths of 3.0 T and higher, voxel size has an interaction with physiologic noise from cardiac and respiratory sources that can be detrimental to fMRI signal detection *(58)*. It is recommended that smaller voxels be employed at 3 T and higher

[6] There are, of course, other parameters that can affect this, such as gradient slew rate, partial Fourier, and/or field-of-view acquisition, etc.

to limit the impact on BOLD CNR from physiologic noise. If spatial resolution is not a concern, it is still recommended that data be acquired at a higher spatial resolution (i.e., smaller voxel size) and retrospective spatial filtering be employed to further increase the CNR *(77)*.

5.3.2. Image Artifact Reduction

Due to the fact that the spatial encoding in the phase direction (i.e., the direction encoded using, for instance, the phase blipping described in **Section 2.2**) is not bandwidth limited, tissue outside of the FOV in the phase direction that experiences RF excitation will be appear wrapped into the FOV with a signal intensity related to the leakage RF experienced by that tissue. For that reason, it is important for most acquisitions that the FOV in the phase direction is adequate to contain the entire brain volume and is oriented such that other tissue is not proximal to the FOV. An example would be a coronal plane acquisition with the phase-encode direction in the inferior/superior direction. In this acquisition, even if the entire brain volume is within the FOV, tissue from the neck and the trunk of the body that is within the sensitive volume of transmit and receive RF coils will appear aliased into the top of the FOV. A more common problem is that the FOV is chosen too small and one side of the brain is wrapped into the other side of the brain **(Fig. 23)**. This is avoided most simply by adopting a large enough FOV in the phase direction. The recommendation in **Table 2** is sufficient for most adults.

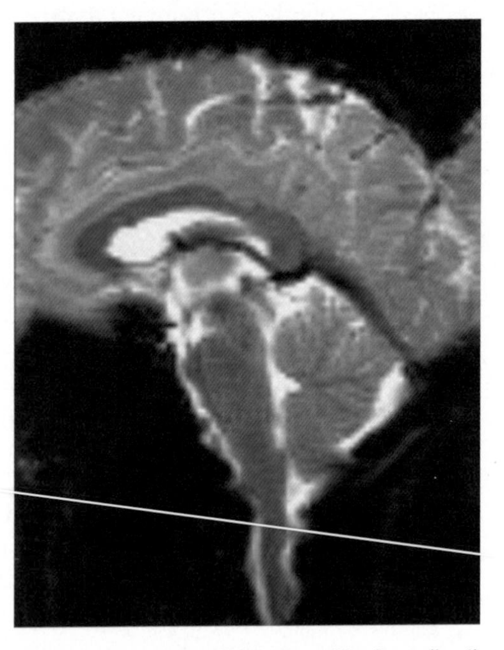

Fig. 23. Sagittal echo-planar imaging (EPI) image with phase direction too small for the brain dimension in the anterior–posterior direction. The anterior portion is phase-wrapped into the tissue at the posterior part of the brain.

5.4. Image Matrix

As stated above, the principal impact of image matrix is on voxel size and these issues are outlined above. However, it will also affect the duration of the data acquisition for a single slice in single-shot imaging. This duration, as discussed in **Section 3.1.3**, can have a detrimental effect on image quality in ultrafast imaging. Generally, the total readout time should be much less than T_2 or T_2^*. Typical methods of decreasing the scan duration while maintaining good spatial resolution are partial Fourier and partial FOV techniques, discussed briefly in **Section 2.2**. These are very commonly employed and can result in acceptable SNR tradeoffs that allow good image quality. For the 3.0-T acquisition recommended in **Table 2**, it is typical to use a partial Fourier, or partial echo, acquisition strategy. These are recommended such that reasonable SNR is maintained.

5.5. Echo Time

The TE is probably the most important consideration in optimizing a pulse sequence for BOLD contrast (or any T_2 or T_2^* contrast for that matter). As discussed in **Section 3.1**, optimal BOLD contrast is obtained by selecting a TE that is the average of the T_2 or T_2^* of the tissue in the active and inactive states. This will necessarily depend on the tissue characteristics and the field strength. The TEs recommended in **Table 2** are typical TEs for a GRE-EPI acquisition that have a reasonable balance between tissue sensitivity and specificity. Adopting a longer TE can result in less sensitivity to intravascular signal, especially at 3.0 T and higher, while a shorter TE can result in higher SNR. Ranges of TE for T_2^* BOLD imaging at 1.5 T include 40–65 ms, while ranges used experimentally at 3.0 T can range from 25 to 40 ms. One should be careful when adopting TEs outside of these ranges as BOLD contrast can be severely attenuated.

5.6. Repetition Time

With regard to BOLD contrast, TR has the fairly simple effect of increasing or lowering SNR based on the T_1 of the tissue. For a TR short with respect to the T_1 of the tissue of interest, the MR signal will be saturated. This will be discussed in some detail in the section regarding the flip angle. Here, we will simply point out that very short TRs can result a significant reduction in SNR, and can subsequently also result in a significant contribution from flow contrast, depending on the saturated state and whether a slice gap is included in the prescription.

Increased blood flow results in an apparent shortening of the T_1 relaxation time due to the effect of infusing blood on the net saturated state of the protons in a given voxel. Generally, the effect of increased flow on the T_1 of a given voxel can be expressed as

$$\frac{1}{T_{1\text{eff}}} = \frac{1}{T_1} + \frac{f}{\lambda}, \tag{5}$$

where T_{1eff} is the observed T_1 in the presence of flow. λ is the tissue blood volume fraction and f is the rate of flow. Thus, an *increase* in blood flow will result in an apparent shortening of the observed T_1 in the affected brain region.

With regard to fMRI, the TR normally determines the sampling rate. This, combined with experimental design (stimulus presentation, duration, number of samples, etc.) determines the detection efficiency for BOLD signals (see also **Section 4.2.1**).

A note with regard to TR and fMRI is that sensitivity to out-of-plane motion, discussed above in **Section 3.2.1**, is a consequence of spin saturation in two-dimensional single-shot MRI. Longer TRs will lead to lower saturation, and thus lower sensitivity to out-of-plane motion.[7]

5.7. Flip Angle

Technically, the flip angle relates to the amount of RF power applied at the excitation stage of a pulse sequence. For a given tissue type (i.e., T_1) and TE, MR signal is optimized at a flip angle referred to as the Ernst angle. The formula for the Ernst angle is given by

$$\alpha_E = \cos^{-1}\left\{\exp\left(\frac{-TR}{T_1}\right)\right\}. \tag{6}$$

Since the flip angle controls the amount of RF power transmitted to the tissue, reducing the flip angle in acquisitions with a short TR can reduce the amount of flow contribution observed in a BOLD-weighted acquisition, possibly increasing spatial specificity of detected neuronal activation.

5.8. Receiver Bandwidth

RBW, in a conventional MRI sequence, has an easily interpreted impact on images: since the RBW is essentially the speed at which the MR signal is digitized, lower bandwidth results in higher SNR, but the resulting longer data acquisition can impact image quality. With single-shot imaging techniques, the effect of RBW is not as straightforward. At low RBW, the readout window length can be long enough such that SNR is reduced. At higher RBW, the reduced artifact from shorter readout time lessens or even completely eliminates the expected reduction in SNR. It is difficult to recommend an exact RBW for these types of acquisitions, since the optimal operating point will depend on scanner hardware characteristics such as slew rate that is highly variable between scanners. The parameters indicated in **Table 2** should give reasonable results in most clinical MRI scanners. Increasing RBW can help to reduce susceptibility artifacts, such as image

[7] More accurately, retrospective motion correction techniques will be more effective.

warping, in orbitofrontal or other regions in much the same way as discussed above under parallel imaging.

5.9. Slice Thickness

Choice of slice thickness has generally the same effect as spatial resolution mentioned earlier. Principal effects are brain coverage, SNR, and image quality. The recommended slice thickness in **Table 2** should give nearly whole-brain coverage in most adults, with acceptable SNR and image artifact given field strength limitations.

As a further note, there is no mention of a slice gap in **Table 2**. A slice gap is not recommended in fMRI studies where the entire brain is desired. The RF excitation for a given slice will not be perfect, so if a small gap (<10% of slice thickness) between slices were permitted, the tissue in this gap would still be sampled, although less than if the slices were simply made thicker. Historically, slice gaps were included to improve SNR in two-dimensional acquisition by reducing RF crosstalk between adjacent slices. The longer TR recommended in **Table 2**, along with an interleaved style pattern of slice excitation, should be sufficient to make this a negligible effect in most clinical scanners.

5.10. Number of Slices

The desired number of slices in an fMRI acquisition will affect temporal resolution and brain coverage. The recommended number of slices in **Table 2** should permit whole brain coverage in most situations. Issues with regard to TR are discussed earlier.

6. Concluding Remarks

As stated at the beginning, and illustrated throughout this chapter, there are many experimental design issues in fMRI that will affect the exact pulse sequence prescription adopted for a particular study. The intent of this chapter is to give an overview of these issues and recommend a starting point for a sequence prescription that will be generally feasible on most modern MRI scanners, along with a sense of the impact of each of the sequence features. There are two caveats with regard to the content of this chapter (1) it is strongly recommended that fMRI researchers work closely with an MR physicist experienced in fMRI when there are specific issues that may effect acquisition choices and (2) ideally, pulse sequence prescription and paradigm design should both be considered together when designing an fMRI experiment. Informing the data acquisition design based on the needs with regard to the experimental hypothesis or analysis methods is critical to a successful fMRI experiment.

References

1. Belliveau JW, Kennedy DN, Jr., McKinstry RC, et al. Functional mapping of the human visual cortex by magnetic resonance imaging. Science 1991;254:716–719.
2. Ogawa S, Lee TM, Nayak AS, Glynn P. Oxygenation-sensitive contrast in magnetic resonance image of rodent brain at high magnetic fields. Magn Reson Med 1990;14:68–78.
3. Bandettini PA, Wong EC, Hinks RS, Tikofsky RS, Hyde JS. Time course EPI of human brain function during task activation. Magn Reson Med 1992;25:390–397.
4. Kwong KK, Belliveau JW, Chesler DA, et al. Dynamic magnetic resonance imaging of human brain activity during primary sensory stimulation. Proc Natl Acad Sci USA 1992;89:5675–5679.
5. Ogawa S, Tank DW, Menon R, et al. Intrinsic signal changes accompanying sensory stimulation: functional brain mapping with magnetic resonance imaging. Proc Natl Acad Sci USA 1992;89:5951–5955.
6. Frahm J, Bruhn H, Merboldt KD, Hanicke W. Dynamic MR imaging of human brain oxygenation during rest and photic stimulation. J Magn Reson Imaging 1992;2:501–505.
7. Weisskoff RM, Davis TL. Correcting gross distortion on echo planar images, Society of Magnetic Resonance in Medicine 11th Annual Meeting, Berlin, 1992.
8. Foerster BU, Tomasi D, Caparelli EC. Magnetic field shift due to mechanical vibration in functional magnetic resonance imaging. Magn Reson Med 2005;54:1261–1267.
9. Ahn CB, Kim JH, Cho ZH. High-speed spiral-scan echo planar NMR imaging-I. IEEE Trans Med Imaging 1986;5:2–7.
10. Bruder H, Fischer H, Reinfelder HE, Schmitt F. Image reconstruction for echo planar imaging with nonequidistant k-space sampling. Magn Reson Med 1992;23:311–323.
11. Pipe JG, Duerk JL. Analytical resolution and noise characteristics of linearly reconstructed magnetic resonance data with arbitrary k-space sampling. Magn Reson Med 1995;34: 170–178.
12. Bornert P, Schomberg H, Aldefeld B, Groen J. Improvements in spiral MR imaging. Magma 1999;9:29–41.
13. Glover GH, Law CS. Spiral-in/out BOLD fMRI for increased SNR and reduced susceptibility artifacts. Magn Reson Med 2001;46: 515–522.
14. Preston AR, Thomason ME, Ochsner KN, Cooper JC, Glover GH. Comparison of spiral-in/out and spiral-out BOLD fMRI at 1.5 and 3 T. Neuroimage 2004;21:291–301.
15. Sodickson DK, Manning WJ. Simultaneous acquisition of spatial harmonics (SMASH): fast imaging with radiofrequency coil arrays. Magn Reson Med 1997;38:591–603.
16. Pruessmann KP, Weiger M, Scheidegger MB, Boesiger P. SENSE: sensitivity encoding for fast MRI. Magn Reson Med 1999;42: 952–962.
17. Griswold MA, Jakob PM, Heidemann RM, et al. Generalized autocalibrating partially parallel acquisitions (GRAPPA). Magn Reson Med 2002;47:1202–1210.
18. Heidemann RM, Griswold MA, Kiefer B, et al. Resolution enhancement in lung 1H imaging using parallel imaging methods. Magn Reson Med 2003;49:391–394.
19. Ohliger MA, Grant AK, Sodickson DK. Ultimate intrinsic signal-to-noise ratio for parallel MRI: electromagnetic field considerations. Magn Reson Med 2003;50:1018–1030.
20. Wiesinger F, Boesiger P, Pruessmann KP. Electrodynamics and ultimate SNR in parallel MR imaging. Magn Reson Med 2004;52: 376–390.
21. Blaimer M, Breuer F, Mueller M, Heidemann RM, Griswold MA, Jakob PM. SMASH, SENSE, PILS, GRAPPA: how to choose the optimal method. Top Magn Reson Imaging 2004;15:223–236.
22. Ohliger MA, Sodickson DK. An introduction to coil array design for parallel MRI. NMR Biomed 2006;19:300–315.
23. Pruessmann KP, Weiger M, Bornert P, Boesiger P. Advances in sensitivity encoding with arbitrary k-space trajectories. Magn Reson Med 2001;46:638–651.
24. Weiger M, Pruessmann KP, Osterbauer R, Bornert P, Boesiger P, Jezzard P. Sensitivity-encoded single-shot spiral imaging for reduced susceptibility artifacts in BOLD fMRI. Magn Reson Med 2002;48:860–866.
25. Heidemann RM, Griswold MA, Seiberlich N, et al. Direct parallel image reconstructions for spiral trajectories using GRAPPA. Magn Reson Med 2006;56:317–326.
26. Griswold MA, Kannengiesser S, Heidemann RM, Wang J, Jakob PM. Field-of-view limitations in parallel imaging. Magn Reson Med 2004;52:1118–1126.
27. Griswold MA, Breuer F, Blaimer M, et al. Autocalibrated coil sensitivity estimation for parallel imaging. NMR Biomed 2006;19: 316–324.

28. Sodickson DK. Tailored SMASH image reconstructions for robust in vivo parallel MR imaging. Magn Reson Med 2000;44:243–251.
29. Sanchez-Gonzalez J, Tsao J, Dydak U, Desco M, Boesiger P, Paul Pruessmann K. Minimum-norm reconstruction for sensitivity-encoded magnetic resonance spectroscopic imaging. Magn Reson Med 2006;55:287–295.
30. Lin FH, Kwong KK, Belliveau JW, Wald LL. Parallel imaging reconstruction using automatic regularization. Magn Reson Med 2004;51:559–567.
31. Block KT, Frahm J. Spiral imaging: a critical appraisal. J Magn Reson Imaging 2005;21:657–668.
32. Tsao J, Boesiger P, Pruessmann KP. k–t BLAST and *k–t* SENSE: dynamic MRI with high frame rate exploiting spatiotemporal correlations. Magn Reson Med 2003;50: 1031–1042.
33. Huang F, Akao J, Vijayakumar S, Duensing GR, Limkeman M. *k–t* GRAPPA: a k-space implementation for dynamic MRI with high reduction factor. Magn Reson Med 2005;54:1172–1184.
34. Cuppen JJ, Groen JP, Konijn J. Magnetic resonance fast Fourier imaging. Med Phys 1986;13:248–253.
35. Noll DC, Nishimura DG, Macovski A. Homodyne detection in magnetic resonance imaging. IEEE Trans Med Imaging 1991;10:154–163.
36. Jesmanowicz A, Bandettini PA, Hyde JS. Single-shot half k-space high-resolution gradient-recalled EPI for fMRI at 3 Tesla. Magn Reson Med 1998;40:754–762.
37. Hyde JS, Biswal BB, Jesmanowicz A. High-resolution fMRI using multislice partial k-space GR-EPI with cubic voxels. Magn Reson Med 2001;46:114–125.
38. Meyer CH, Pauly JM, Macovski A, Nishimura DG. Simultaneous spatial and spectral selective excitation. Magn Reson Med 1990;15: 287–304.
39. Zhou XJ, Du YP, Bernstein MA, Reynolds HG, Maier JK, Polzin JA. Concomitant magnetic-field-induced artifacts in axial echo planar imaging. Magn Reson Med 1998;39: 596–605.
40. Reeder SB, Atalar E, Faranesh AZ, McVeigh ER. Referenceless interleaved echo-planar imaging. Magn Reson Med 1999;41:87–94.
41. Duyn JH, Yang Y, Frank JA, van der Veen JW. Simple correction method for k-space trajectory deviations in MRI. J Magn Reson 1998;132:150–153.
42. Yudilevich E, Stark H. Spiral sampling in magnetic resonance imaging-the effect of inhomogeneities. IEEE Trans Med Imaging 1987;6:337–345.
43. Jezzard P, Balaban RS. Correction for geometric distortion in echo planar images from B0 field variations. Magn Reson Med 1995;34:65–73.
44. Blamire AM, Rothman DL, Nixon T. Dynamic shim updating: a new approach towards optimized whole brain shimming. Magn Reson Med 1996;36:159–165.
45. Wilson JL, Jenkinson M, de Araujo I, Kringelbach ML, Rolls ET, Jezzard P. Fast, fully automated global and local magnetic field optimization for fMRI of the human brain. Neuroimage 2002;17:967–976.
46. Ward HA, Riederer SJ, Jack CR, Jr. Real-time autoshimming for echo planar time-course imaging. Magn Reson Med 2002;48: 771–780.
47. Yang QX, Williams GD, Demeure RJ, Mosher TJ, Smith MB. Removal of local field gradient artifacts in $T2^*$-weighted images at high fields by gradient-echo slice excitation profile imaging. Magn Reson Med 1998;39:402–409.
48. Constable RT, Spencer DD. Composite image formation in z-shimmed functional MR imaging. Magn Reson Med 1999;42:110–117.
49. Chen N, Wyrwicz AM. Removal of intravoxel dephasing artifact in gradient-echo images using a field-map based RF refocusing technique. Magn Reson Med 1999;42:807–812.
50. Oesterle C, Markl M, Strecker R, Kraemer FM, Hennig J. Spiral reconstruction by regridding to a large rectilinear matrix: a practical solution for routine systems. J Magn Reson Imaging 1999;10:84–92.
51. Moriguchi H, Duerk JL. Modified block uniform resampling (BURS) algorithm using truncated singular value decomposition: fast accurate gridding with noise and artifact reduction. Magn Reson Med 2001;46: 1189–1201.
52. Nehrke K, Bornert P. Prospective correction of affine motion for arbitrary MR sequences on a clinical scanner. Magn Reson Med 2005;54:1130–1138.
53. Hajnal JV, Myers R, Oatridge A, Schwieso JE, Young IR, Bydder GM. Artifacts due to stimulus correlated motion in functional imaging of the brain. Magn Reson Med 1994;31: 283–291.
54. Friston KJ, Williams S, Howard R, Frackowiak RS, Turner R. Movement-related effects in fMRI time-series. Magn Reson Med 1996;35:346–355.
55. Bullmore ET, Brammer MJ, Rabe-Hesketh S, et al. Methods for diagnosis and treatment of

stimulus-correlated motion in generic brain activation studies using fMRI. Hum Brain Mapp 1999;7:38–48.

56. Jezzard P, LeBihan D, Cuenod D, Pannier L, Prinster A, Turner R. An investigation of the contribution of physiological noise in human functional MRI studies at 1.5 Tesla and 4 Tesla, Society of Magnetic Resonance in Medicine 12th Annual Meeting, New York, NY, 1992.
57. Lowe MJ, Mock BJ, Sorenson JA. Functional connectivity in single and multislice echoplanar imaging using resting-state fluctuations. Neuroimage 1998;7:119–132.
58. Triantafyllou C, Hoge RD, Krueger G, et al. Comparison of physiological noise at 1.5 T, 3 T and 7 T and optimization of fMRI acquisition parameters. Neuroimage 2005;26:243–250.
59. Friedman L, Glover GH. Report on a multicenter fMRI quality assurance protocol. J Magn Reson Imaging 2006;23:827–839.
60. Fu ZW, Wang Y, Grimm RC, et al. Orbital navigator echoes for motion measurements in magnetic resonance imaging. Magn Reson Med 1995;34:746–753.
61. Lee CC, Jack CR, Jr., Grimm RC, et al. Real-time adaptive motion correction in functional MRI. Magn Reson Med 1996;36:436–444.
62. Thesen S, Heid O, Mueller E, Schad LR. Prospective acquisition correction for head motion with image-based tracking for real-time fMRI. Magn Reson Med 2000;44:457–465.
63. Zhao X, Bodurka J, Jesmanowicz A, Li SJ. B(0)-fluctuation-induced temporal variation in EPI image series due to the disturbance of steady-state free precession. Magn Reson Med 2000;44:758–765.
64. Raj D, Anderson AW, Gore JC. Respiratory effects in human functional magnetic resonance imaging due to bulk susceptibility changes. Phys Med Biol 2001;46:3331–3340.
65. Dagli MS, Ingeholm JE, Haxby JV. Localization of cardiac-induced signal change in fMRI. Neuroimage 1999;9:407–415.
66. Bhattacharyya PK, Lowe MJ. Cardiac-induced physiologic noise in tissue is a direct observation of cardiac-induced fluctuations. Magn Reson Imaging 2004;22:9–13.
67. Guimaraes AR, Melcher JR, Talavage TM, et al. Imaging subcortical auditory activity in humans. Hum Brain Mapp 1998;6:33–41.
68. Hu X, Le TH, Parrish T, Erhard P. Retrospective estimation and correction of physiological fluctuation in functional MRI. Magn Reson Med 1995;34:201–212.
69. Glover GH, Li TQ, Ress D. Image-based method for retrospective correction of physiological motion effects in fMRI: RETROICOR. Magn Reson Med 2000;44:162–167.
70. Beall EB, Lowe MJ. Isolating physiologic noise sources with independently determined spatial measures. Neuroimage 2007;37:1286–1300.
71. Stefanovic B, Pike GB. Human whole-blood relaxometry at 1.5 T: Assessment of diffusion and exchange models. Magn Reson Med 2004;52:716–723.
72. Friston KJ, Holmes AP, Worsley KJ, Poline J-B, Frith CD, Frackowiak R. Statistical parametric mapping in functional imaging: A general linear approach. Hum Brain Mapp 1995;2:189–210.
73. Josephs O, Turner R, Friston KJ. Event-related fMRI. Hum Brain Mapp 1997;5:243–248.
74. Bandettini PA, Jesmanowicz A, Wong EC, Hyde JS. Processing strategies for time-course data sets in functional MRI of the human brain. Magn Reson Med 1993;30:161–173.
75. Miezin FM, Maccotta L, Ollinger JM, Petersen SE, Buckner RL. Characterizing the hemodynamic response: effects of presentation rate, sampling procedure, and the possibility of ordering brain activity based on relative timing. Neuroimage 2000;11:735–759.
76. Lowe MJ, Sorenson JA. Spatially filtering functional magnetic resonance imaging data. Magn Reson Med 1997;37:723–729.
77. Triantafyllou C, Hoge RD, Wald LL. Effect of spatial smoothing on physiological noise in high-resolution fMRI. Neuroimage 2006;32:551–557.

Chapter 4

High-Field fMRI

Alayar Kangarlu

Summary

Imaging of human brain function is possible only through a few techniques of which magnetic resonance is the safest and most widely used. The soft tissue contrast and high resolution functional maps of the human brain are making a profound contribution to our understanding of the brain function. Resolution and strength of activation signal in fMRI images depends on the static magnetic field. This fact must be fully exploited by availing the highest field fMRI scanners for neurofunctional studies. During the last decade of the last century, field strength of human imaging was raised to 8 T. As a result, today commercial 7-T MRI scanners are available for harnessing the strongest possible signal from the brain neuronal response to external stimulations. Such achievement has involved development of many ancillary technologies and overcoming of many inherent technological and scientific challenges. In this chapter, advantages and challenges of the use of high-field scanners for fMRI studies involving human subjects are discussed. Among challenges, susceptibility artifacts, relaxation rates, and RF coil designs are discussed in details.

Key words: High Field, fMRI, Neuroimaging, Magnetic field, High resolution

1. Introduction

MRI is a diagnostic tool with capability of acquiring signal from biological tissues with sufficient contrast for visualization of soft tissues at high resolution *(1, 2)*. In addition to structures, MRI can detect changes in signal as a result of altered physiological conditions *(3)*. fMRI has become a powerful tool for the study of neuroscience, since it provides access to inner working of the brain during the neuronal response to a specific stimulation designed to activate specific areas of the brain *(4–10)*. The accuracy of brain function detected by fMRI depends on sensitivity to changes induced in the brain in response to external stimulations

Massimo Filippi (ed.), *fMRI Techniques and Protocols*, Neuromethods, vol. 41
DOI 10.1007/978-1-60327-919-2_4, © Humana Press, a part of Springer Science+Business Media, LLC 2009

of neurons. Since such responses are spatially localized for neurons performing similar tasks, fMRI must optimize its sensitivity to better detect these neuronal clusters. Therefore, enhancing sensitivity is a permanent aim of neuroscientists in order to better visualize intricacies of the structure and function of the brain. To acquire functional images, signal must be acquired from the neuronal activity or their consequences. While direct neuronal currents have not yet been detected, the physiological correlates of neuronal activity, i.e., the hemodynamics causes disturbance in magnetic field that makes it visible in MRI *(3)*. However, the relationship between the hemodynamic response and the neuronal activity is still sketchy *(11)*. The hemodynamic signals detected by fMRI are reliable, reproducible, and consistent with the pattern of sensory, motor, or cognitive activations. As such, fMRI has turned into a trusted tool for neuroscience and psychiatry and is expanding its circle of influence into neurology, neurosurgery, and other areas of mainstream medicine as well.

To detect signal from brain function requires sensitivity for the changes caused by the brain activity and having a high structural and temporal resolution. This means an ability to produce images of the brain structure combined with sensitivity for functional activation over the entire brain. This is made more challenging due to the fact that functional organization is not exactly commensurate with anatomical organization. But, MRI is uniquely capable of visualizing both of these organizations on the same subject in one study. In fact, the power of fMRI to detect various regions involved in a specific function while at the same time visualizing the cortical anatomy enabling precise identification of the functional regions makes it an exceptional tool for functional brain studies. On the other hand, fMRI provides an indirect measure of neuronal activity. The unique sensitivity of MRI to paramagnetic entities, however, brings hemodynamic and its coupling to neuronal activity to rescue *(3, 11)*. Diamagnetic nature of biological tissues makes blood with its rich iron content an ideal medium for the detection of physiological changes. The blood oxygen level dependent (BOLD) is an effect that measures changes in MR signal from deoxygenated hemoglobin (dHb) to oxygenated hemoglobin (O_2Hb) required by neuronal activation which modifies the magnetic field around the regions of oxygenation to the extent that changes in MR signal-to-noise ratio (SNR) can be measured in a comparative measurement. In fMRI studies, this change in signal ($\Delta R/R$) is taken as accurately representing neuronal activity. Higher magnetic susceptibility (χ) of dHb compared to that of O_2Hb is enough at fields above 1 Tesla (T) to raise $\Delta R/R$ to about 1%/T which with modern instrumentation is detectable. Dependence of BOLD strength on the static magnetic field ($\mathbf{B}_0$) is a valuable attribute of fMRI, which presently is benefiting from availability of 7.0 T whole body magnets for the study on human subjects.

The BOLD effect, however, depends on a number of physiological factors. The primary ones being cerebral blood flow (CBF), cerebral blood volume (CBV), and cerebral metabolic rate of oxygen (CMRO2). BOLD-based fMRI studies operate on the bases of the assumption that all three mechanisms CBF, CBV, and CMRO2 are caused by changes in neuronal activities. To use fMRI in any capacity, i.e., for better understanding of it as a technique or to use it as a tool in neuroscience research, the underlying mechanisms producing BOLD must be understood. In this regard, the role of high magnetic field in shaping BOLD signal is important and will be discussed in the context of the nature of MR signal.

2. MR Signal

Magnetic field used in MRI induces precession in protons that requires electromagnetic waves in the radio frequency (RF) range to excite them *(12, 13)*. RF waves of frequencies in 100 MHz range (10^8 Hz) will transfer energy into the protons that will be deflected from parallel alignment with $\mathbf{B}_0$. Each proton possessing a magnetic dipole moment (μ) which in returning to their equilibrium position they will induce an electric current in the RF coil *(15, 16)*. The oscillating magnetic field ($\mathbf{B}_1$) of the RF wave induced into the coil circuit is modulated by the environment of the proton that produces it. These effects at a frequency of ω with a group of protons of μ magnetic moment are to be detected by an RF coil, either the same or different from the exciting coil. Detection of signal by the RF coil is affected by the narrow resonance condition that is set up by MR. The design and construction of RF coils are discussed in the literature and a brief description is included later in this section. Suffice to state that RF coil should be capable of delivering a large signal (kW range) to the sample and within less than a few milliseconds be ready to receive a signal that could be 100,000 times weaker.

The strength of the signal is a function of some tissue parameters that determine realignment of μ with $\mathbf{B}_0$ through spin–lattice (T_1) relaxation and spin–spin (T_2) relaxation *(17)*. In order to produce images where the tissue intensity represents relative T_1 values, T_1-weighted (T1W) images, the T_2 relaxation must be suppressed. Since T_2 relaxation is a process of rapid signal loss compared to T_1, T1W images are produced at very short echo times (TE). A typical clinical imaging of the whole head with 256

× 256 matrix requires about 4 min to be acquired. The acquisition time is primarily due to the slow T_1-based realignment process. One way to accelerate data acquisition is to acquire all lines of the *k*-space during one excitation. But, the other relaxation process in MRI, spin-spin relaxation or T_2 decay, is also a process that must be taken into account for image production from MR signals. The detectable entity in MRI is magnetic moment of all the protons within a voxel which is referred to as magnetization vector or **M**. The magnitude of **M** is a function of μ population within the voxel. At high field, the net number of μs participating in MR signal formation increases for a group of water molecules. This is one reason why strong signal can be received from smaller voxels, hence resulting in higher resolution. In fMRI, the dominant contrast is T_2^*, which is an inverse of sum of signal decay rates caused by spin-spin as well as due to local inhomogeneities in $\mathbf{B}_0$. These inhomogeneities cause a faster loss of coherent precession of **M** about $\mathbf{B}_0$ lost over time *(17, 18)*. All gradient echo (GE)-based BOLD studies have their fMRI contrast-to-noise ratio (CNR) as a direct function of the loss of coherence that occurs at a rate of T_2^*, which is much shorter than T_2 at high fields. T_2^*-weighted (T_2^*W) images in fMRI are rich with information about local sources of magnetic field inhomogeneity but in high fields fMRI many hardware challenges are involved in converting that information into scientifically interpretable data *(8, 9)*.

3. Static Magnetic Field Effects

MR sensitivity depends on $\mathbf{B}_0$, RF coil design, and acquisition parameters. $\mathbf{B}_0$ is the most important factor as it determines the number of protons participating in the MR signal formation and is tied to the largest and costliest component of a scanner (**Fig. 1**). While all other parameters impacting image quality and SNR could be optimized, $\mathbf{B}_0$ is the single parameter that a researcher must think hard and long to choose its optimum value for his/her applications after which he/she will have to live by. Given the nature of MR and the fact that every aspect of the technique such as relaxations, susceptibility, CNR, and hardware have advantages and disadvantages at high field, the choice of premium field strength for fMRI studies is a complex venture. This point has been highlighted in all sections of this chapter. One prominent example is magnetic susceptibility at high-field strength which has been considered to be a blight rather than blessing but its implications on fMRI are immense. These aspects of fMRI are analyzed later.

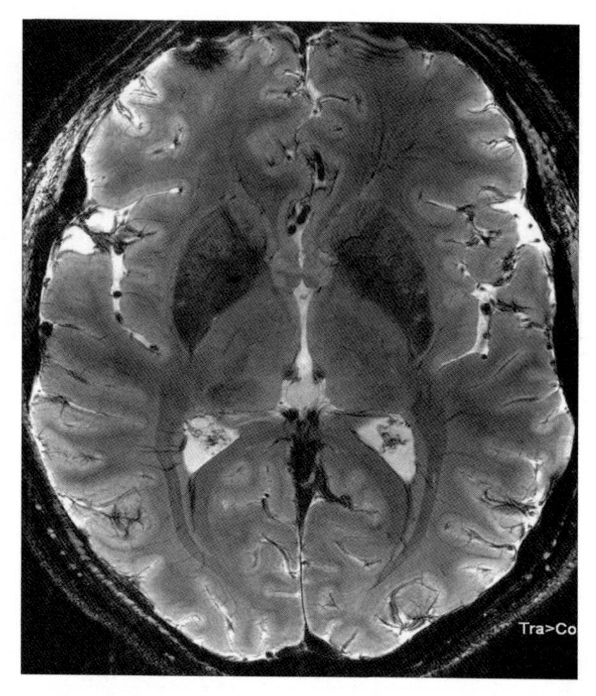

Fig. 1. Susceptibility weighted (two-dimensional T_2^*) images acquired using a Siemens 7.0-T scanner. Acquisition parameters are: FLASH 500 ms/25 ms/35°, acquisition time = 8:30 min, 200 μm × 200 μm × 1 mm, matrix = 1024 × 1024. Courtesy of Prof. L.L. Wald of MGH Martinos Center, Boston, USA.

4. Relaxation Effects

At 7.0 T, the SNR is more than twice that of 3.0 T and about five times higher than 1.5 T (**Fig. 2**). Typical voxel size at 1.5 T is about 5 mm^3, while the SNR at 7.0 T and higher fields can produce signals from typical biological cells to generate images with 0.5-mm^3 resolution. Of course, this gain is also a function of the relaxation values which will modulate the magnetization vectors after excitation. The fMRI resolutions are still an open question. The need to phase encode the signal to cover the entire *k*-space in one shot for fast imaging, such as echo-planar imaging (EPI), makes relaxations critical parameters in fMRI studies at high field. T_1 values reported for gray matter (GM) at 7.0 T are about 2 s and that of white matter is about 1.3 s *(18–21)*. At 3.0 T, T_1 values for GM are approximately 1.5 s and for white matter is about 0.8 s *(19–21)*. At 1.5 T, GM has a T_1 of about 1.2 s while that of white matter is about 0.6 s *(22)*. While T_1 values at various fields have a minimal effect on fMRI images, T_2^* relaxation has a much more direct effect. This is especially true due to drastic change in T_2^* as a function of field strength. A recent work *(18)* has reported T_2^* for GM, white matter, and putamen, respectively, to be 84.0, 66.2,

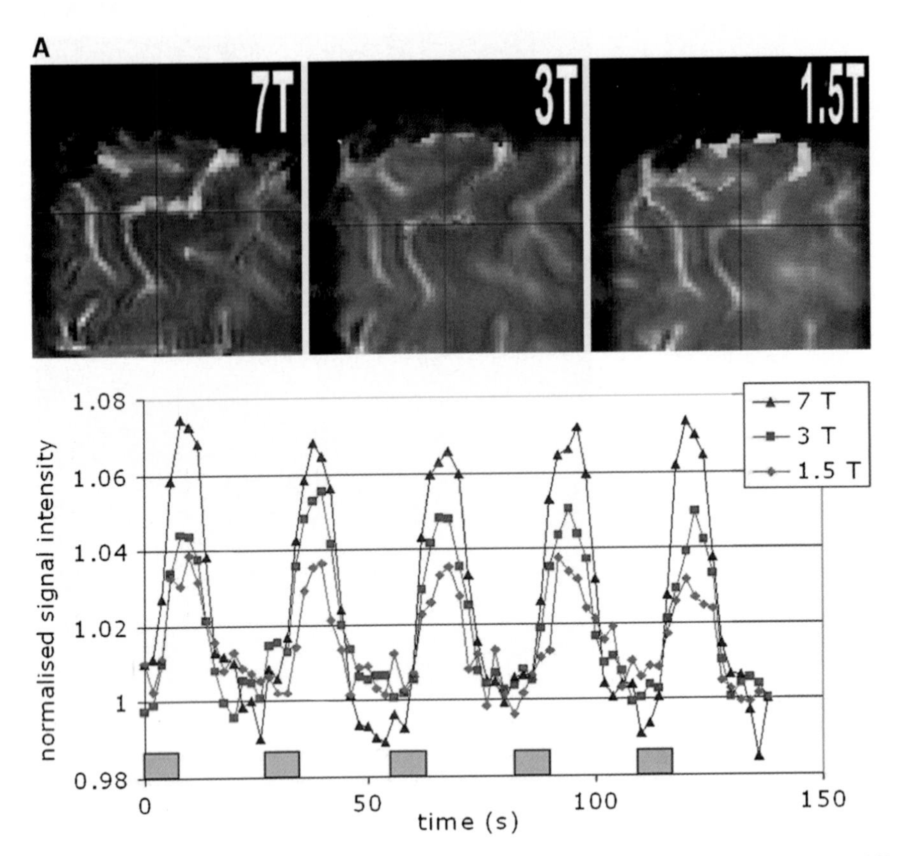

Fig. 2. fMRI study at 1.5, 3.0, and 7.0 T performed on Philips Achieva. Courtesy of University of Nottingham, UK.

and 55.5 ms at 1.5 T; 66.0, 53.2, and 31.5 ms at 3.0 T; and 33.2, 26.8., and 16.1 ms at 7.0 T. This shows that at an 18 ms difference between T_2^* of GM/white matter at 1.5 T has reduced to 7 ms at 7.0 T. While this shows a drop in T_2^* of these two tissues by a factor of almost 2.5 as field strength increases from 1.5 to 7.0 T, there are profound implications on the implementation of fast sequences at high field. This is due to the fact that as T_2^* of some tissues, such as putamen, drops to numbers near 10 ms, the possibility of phase encoding the structure in a single-shot image for high resolution images is reduced. This warrants a better understanding of the role of high field in producing high resolution functional images.

There are needs for higher gradient amplitudes at high fields. At 7.0 T, gradient fields of about 5 Gauss/cm are available on commercial scanners today which reduce the encoding time for a resolution of 0.5 mm to be 1 ms/line. This makes the total encoding time for 256 phase encoding steps to be 256 ms. At 7.0 T, T_2^* of the human brain is about 30 ms while T_2 of GM and white matter are 93 ms and 76 ms, respectively. Thus, readouts of about 100 ms might be needed for SE EPI or partial *k*-space filling in GE EPI. Even such strong magnitudes are not going to be able to read a signal that lasts 30 ms. Such fast decay during readout causes blurring of images that makes high resolution EPI challenging at high field. As sensitivity and specificity of BOLD fMRI images improve at high field, the additional information provided by BOLD images would require demanding hardware for accurate localization and coregistration with high resolution anatomical images. The increase in BOLD sensitivity with higher fields is not fully utilized unless high resolution structural images are produced at a reasonable acquisition time from the same subject during the same study session.

5. Imaging the Brain Function

An imaging technique capable of directly detecting neuronal activity has yet not been devised. In the absence of such a direct observation of brain function, techniques such as BOLD have enabled us to measure changes in hemodynamics as a result of controlled neuronal activation. It should be kept in mind, however, that for BOLD to provide a measure of neural activity, one should rely on intact communication between neurons, glia, and blood vessels. In addition, a tight level of vascular reactivity must exist between the vascular system and the neuronal network.

In order to use fMRI data for understanding of the brain function, the mechanism of neurovascular coupling to neuronal activity and how it affects hemodynamic response should be explored. Once such a relationship is established, pharmacological or disease-induced modulations of neurovascular coupling could use BOLD signal changes, or perfusion, for the assessment of drug efficacy and physiological implications of pathology. These utilities of imaging of brain function are in addition to the role that fMRI plays in psychology, psychiatric studies, and basic cognitive neuroscience research *(11)*.

Given the organization of the brain, functional studies can help identify the neuronal basis of behavior. This is best achieved by being able to map the functional units involved in various brain systems *(23–25)*. The units such as cortical columns made up of neuronal pathways involved in the implementation of a specific function form an organized structure that interact with other units of the system *(26)*. It would be ideal to be able to detect the activation of these units as well as their collective response to specific external stimulations. This way the extent of spatial localization of functional units can be determined. In case of interaction between isolated clusters of columns involved in implementation of a function, such functional connectivity could be studied by fMRI. In addition to these, fMRI ability of whole brain imaging provides a precious opportunity of simultaneously detect functional synchrony between various system subunits. High field has shown to have the potential to further empower fMRI with high resolution maps of functional units of the brain *(5–8, 23)*.

Another great advantage of high-field fMRI is that it allows the study of such profound issues noninvasively with a high spatial and temporal resolution too. The fact that fMRI is capable of visualizing the entire brain networks based on experiments that measure brain activation due to the execution of a specific task, avails the entire brain to investigation at once. It is important that as field strength increases, the ability of whole brain imaging is not compromised. Expectations for a robust functional mapping tool require spatiotemporal resolution improvement uniformly over the whole brain as field strength increases. This poses some challenges as brain regions close to tissue/air interfaces are exceedingly vulnerable to susceptibility-induced signal dropouts, an artifact that worsens as the field strength increases. And, in spite of all its capabilities a functional measurement that is sensitive only to hemodynamics must be combined with reliable models of neuronal underpinning of the detectable mechanism, i.e., BOLD or perfusion, with as high of spatial specificity and temporal resolution as possible. This is where high field can play another important role than just improving resolution. In fact, since high-field MRI can boost both spatial and temporal resolution of functional brain studies and considering availability of viable solutions to problems such as susceptibility, one should strive for the highest feasible field strength for the study of brain functions.

Functional imaging at 7.0 T **(Fig. 2)** has shown tremendous potential to convert the magnetic susceptibility and consequently the BOLD effect to more information content *(27)*. Such advantage could increase linearly or even more than linearly with $\mathbf{B}_0$ as more technology in hardware and software is developed for high field. For example, the high SNR of SE BOLD could eliminate contributions to the signal from large draining veins at fields 3.0 T and higher. This way, BOLD from microvascular networks directly on or near the site of neuronal activity could be detected *(28)*. Such tool is capable of dealing with more fundamental question in quantification of the signal. An fMRI signal with resolution high enough to consistently quantify blood flow and energy consumption provides a valuable insight into the relationship between neuroenergetics and neuronal activity. Such relationship has not received attention in fMRI studies. Most of fMRI studies, instead, have concentrated on experimentally proving cognitive neuroscience theories. High field can provide more powerful tools and quantifiable measures for such endeavors.

Arterial spin labeling (ASL) based sequences have provided reproducible measurements of CBF. This feature makes perfusion-based fMRI as a complement to BOLD. A version of ASL called continuous or CASL has shown particular potential to take advantages of high-field strengths to obtain high SNR and CNR. ASL is implemented by tagging (normally with inversion RF pulse) the blood flowing to the brain in the neck. After a delay time, the slice select RF pulse is followed by an acquisition sequence. The blood with its water magnetically labeled flows into the brain has its transverse magnetization decaying at the rate of T_1. So, T_1 duration is important in detection of tissue perfusion. The signal in perfusion imaging is a function of regional blood flow and the longitudinal relaxation time T_1. The T_1-dependent part of perfusion signal makes perfusion SNR a function of magnetic field. At high field, perfusion will benefit from increase in T_1 as it provides more transverse magnetization in the image slice. At high field, perfusion can provide quantitative measures of absolute CBF, a more direct representation of neuronal activity than the BOLD signal.

5.1. Fast Imaging

Image acquisition in MRI is slower than in other techniques such as computed tomography and positron emission tonography. This is mostly due to the relaxation phenomenon. Fast imaging techniques are not widely popular for structural imaging due to the poor image qualities and technical limitations. Relaxations and dephasing requires refocusing of signals in the intervals of the order of TE and realignment of spins with the main magnetic field every TR seconds, where TR is called the repetition time. Refocusing can be achieved by gradient reversal or RF pulses. Depending on the acceleration rate and safety concerns, one or the other method can be used. For detection of physiological signals, however, the image acquisition rate should match the rate of physiological event. For brain functional imaging, this rate is of the order of a second. So, there is a need for imaging the entire brain within that timescale. For resolutions of the order of 5 × 5 × 5 mm, the whole brain coverage requires 30-40 slices for whole brain with 5mm slice. For an image with 64 phase-encoding steps there is only 300 ms for refocusing and readout. These facts leave very few sequences for imaging at such rate. EPI is one such sequence. Its sequence details and implications of its execution at high fields needs close scrutiny in order to fully exploit its potentials in high-field functional imaging studies.

5.1.1. Echo-Planar Imaging

Fast imaging techniques achieve their speed by multiple refocusing of the spin ensemble during one TR. EPI as a GE-based technique is the fastest sequence and has a very low RF power content *(29)*. This aspect of EPI makes it suitable for high-field applications as RF absorption increases at high fields increasing the RF requirements. On the other hand, other aspects of EPI such as geometric distortion, blurring artifacts, and T_2^* signal loss are aggravated at higher fields *(30, 31)*. For instance, the geometric distortion that is caused by off-resonance effects will be further aggravated by long readout train of EPI. A phase offset that increases with TE will be created that will establish a linear phase gradient over *k*-space in the phase-encoding direction *(32)*. The image signal from these spins will get shifted as image is reconstructed. At high fields, this effect is proportionally stronger resulting in larger frequency shifts. However, the effect of long readouts can be drastically reduced by using parallel imaging. This will reduce geometrical distortions, but the T_2^* signal decay and blurring on the images will still remain. Other techniques have been introduced to deal with T_2^* relaxation causing distortion in images due to the decay in the signal along the *k*-space trajectory. Minimization of magnetic field inhomogeneities and susceptibility-induced effects requires the choice of TE close to T_2^*. As $\mathbf{B}_0$ increases, T_2^* decreases and hardware and safety considerations often makes the minimum TE of single-shot EPI longer compared to T_2^*, which causes signal loss due to the phase dispersion caused by such choices of TE. Higher bandwidth could alleviate this problem but possibility of peripheral nerve stimulation will limit the use of much stronger gradients to achieve this. Other techniques have been proposed that will effectively restore T_2^* relaxation-induced signal loss and blurring. GE slice excitation profile imaging (GESEPI) is one such method that combined with multichannel parallel receiver technology, such as sensitivity encoding (SENSE), will significantly enhance high-field EPI image qualities *(33, 34)*.

Other EPI artifacts such as Nyquist ghost are independent of field strength and are inherent to the sequence *k*-space trajectory with various solutions applicable to their minimization at all field strengths *(35)*. Nyquist artifact is due to the time-reversal asymmetry of even and odd echoes and its ghosts overlap with the image causing a reduction in EPI SNR.

Next to its speed, the most important characteristic of EPI is the high magnetic susceptibility weighting it casts on images (**Figs. 2** and **3**). In fact, fMRI as the most important application of EPI takes advantage of EPI sensitivity to susceptibility change due to blood oxygenation. Unlike fMRI applications in which susceptibility enhances BOLD contrast, susceptibility weighting of EPI is not considered an advantage in applications such as diffusion-weighted imaging. As such, understanding of susceptibility is essential in enhancing its role where it helps fMRI and suppressing its undesirable aspects where it hurts data quality. A brief account of magnetic susceptibility of biological tissues is presented here to help appreciate the role of susceptibility in EPI-based BOLD signal changes.

5.2. Magnetic Susceptibility

Magnetic susceptibility, χ, is at the core of BOLD-based fMRI studies. When matter is exposed to strong magnetic field it will be magnetized *(36)*. In formation of χ, magnetic field (**H**),

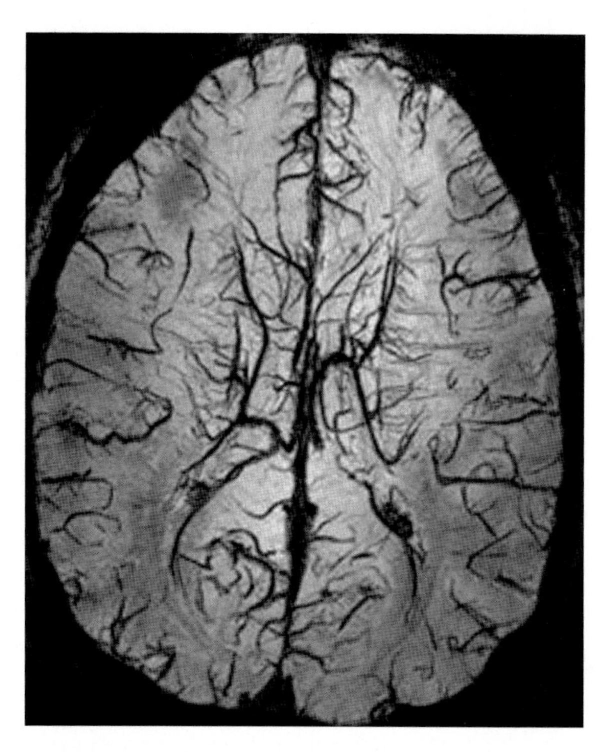

Fig. 3. An axial image acquired at 7.0 T using susceptibility enhanced sequence to produce an MRI venographic image of a human subject. The sequence parameters are three-dimensional T_2^* gradient echo, TR/TE/FA = 30/14.8/18 ms, resolution = 0.5 × 0.5 × 1.0 mm, 30 slices, acquisition time = 7:18 min. Study was performed on a 7T Phillips Achieva in Vanderbilt University. Courtesy of Philips Medical Systems.

magnetic induction (**B**), and magnetization (**M**) play roles. **H** is the entity that exists in vacuum and its penetration through space, i.e., free space of permeability $\mu_0 = 4\pi \times 10^{-7}$ Henry/m, is given by $\mathbf{B} = \mu_0\mathbf{H}$. The magnetization, **M**, represents the total magnetic moments per unit volume $\mathbf{M} = \sum\mu/v$. **M** is caused by **H** according to $\mathbf{M} = \chi\mathbf{H}$. **B** and **H** in SI unit system have units of Tesla and Ampere/meter, respectively. Inside a body placed in a magnetic field a magnetization **M** is generated that will produce a magnetic field of $\mathbf{B} = \mu_0(\mathbf{M} + \mathbf{H})$. Replacing **M** in this expression will yield $\mathbf{B} = \mu\mathbf{H}$ where $\mu = \mu_0(1 + \chi)$ will be the magnetic permeability of matter. As such, susceptibility of an object is a measure of enhancement of the magnetic field within its volume. This is important as it will determine how uniform a magnetic field (**B0**) can be established inside the body in MRI. In μ are hidden the characteristics of the free space and how its magnetic properties are modulated by matter through χ. $\mathbf{B}_0$ in turn, changes locally by χ causing the so-called susceptibility artifacts in MRI particularly in the areas of air/tissue interface *(36, 37)*. This effect causes a change in magnetic field as it is sensed inside a tissue and for heterogeneous tissues a contrast is generated between tissues, which are $\mathbf{B}_0$ dependent. Difference in susceptibility, $\Delta\chi$, between adjacent tissues are small at low fields. If susceptibility-based inhomogeneity is smaller than inherent $\mathbf{B}_0$ inhomogeneity it could be used for generating contrast for better visualization of tissues such as GM. High $\Delta\chi$ as exists at the air/tissue interfaces causes large variation in $\mathbf{B}_0$ that is responsible for signal drop-outs interfering with studies focused on these regions *(31)*. fMRI studies of regions near the ear canal, nasal cavity, and dorsolateral prefrontal cortex (DPFC) suffer from this phenomenon.

The most distinct role of susceptibility effect in MRI is in fMRI. It is based on the fact that **M** within a voxel is linearly proportional to $\mathbf{B}_0$ determining the role of high field in susceptibility-based enhancements of CNR. Specifically, T_2^* values decrease allowing paramagnetic molecules such as dHb to generate more dephasing in collective proton precession at high magnetic fields. **Figures 2** and **3** show how high activation induces high functional CNR on the images taken at 7.0 T compared to lower fields. The short T_2^* values due to paramagnetic properties of dHb causes the veins and structures with high density vasculatures to have their dimensions exaggerated as shown in **Fig. 1**. This mechanism affects $\mathbf{B}_0$ through χ making a larger variation in susceptibility, $\Delta\chi$, in brain tissues around activated neurons at high field subsequent to a perturbation. In brain activation studies, the stimulus causes change in volume and flow of oxygenated blood in the near proximity of activated brain regions. For the same activation, high field will use higher $\Delta\chi$ for better visualization of vasculature network which is coupled into the neuronal system in the brain.

Furthermore, high-field SNR allows the use of in vivo vascular imaging in establishing a relationship between brain tissue vascular density and functional imaging results. Independent information from vascular density could be attained from MR angiography to help better analysis of the fMRI data. In addition, such vascular density information could be used for the study of various topics from brain development and brain tumor staging to multiple sclerosis (MS). High-field fMRI in MS could better assess the effect of any changes in cortical activation during a particular task such as attention, memory, motor, etc. As high field enables better spatialized maps of the response to stimuli, fMRI could help assess the extent of neuropsychological problems. T_2-weighted images have so far been considered to best lend their contrast mechanism to changes due to demyelination *(38)*. However, as fMRI becomes faster, more detailed brain activation in MS patients could be used to assess their normal motor function as is done clinically today. As the strength of response signal varies depending on the activated region of the brain and the accessibility to the detector, i.e., RF coil, high field could allow a wide range of regions and paradigms to be designed to compare performance of MS patients with healthy controls.

Such quantitative assessment of functional performance of the brain will provide a valuable tool in enhancing the disease management. Structural MRI, however, has had great success in visualizing lesions of demyelination *(38–40)*. But, lack of specificity has prevented MRI from being established as a reliable one-shop stop for diagnosis of MS. A reliable fMRI technique with resolution to reveal accurate cerebral functional response to controlled stimulations will complement the existing structural MRI tools in better understanding of MS. Such potential is entirely due to inherent sensitivity of fMRI to hemodynamics induced by cognitive perturbation and will provide information independent of structural changes of the disease. In this regard, a unique aspect of high field, i.e., high susceptibility and SNR, offers a tool that, although is MR in nature, its attributes are not equally available at lower fields. The $\mathbf{B}_0$-dependent susceptibility contrast, furthermore, provides potential for depiction of microvascular structures that will further enrich the tool box of high-field magnets *(41)*.

Magnetic susceptibility of blood is governed by the same effects as discussed above. It is high enough to generate the BOLD effect just due to the change in its oxygenation state. A dHb molecule contains four paramagnetic iron ions. During oxygenation, dHb combines with four oxygen molecules which results in an O_2Hb molecule, which normally has no net paramagnetic moment. O_2Hb is in fact slightly diamagnetic. This will cause the magnetic susceptibility of blood to change by about 10^{-6} if the blood is fully oxygenated. Taking the susceptibility of O_2Hb as zero, then change in blood magnetic susceptibility with oxygenation constitutes the basis of BOLD contrast in fMRI. A detailed account of the effect of $\mathbf{B}_0$ on BOLD through the susceptibility mechanism will further elucidate the impact of high field in fMRI.

5.3. Blood Oxygen Level Dependent

A change in magnetic susceptibility of $\Delta\chi = 10^{-6}$ (SI system) in blood as a result of oxygenation is possible which forms the basis of fMRI. Through $\mu = \mu_0(1 + \chi)$, magnetic dipole strength of a voxel changes by $\Delta\chi$ and results in change in magnetization which constitute the basis of NMR signal. The maximum possible change in susceptibility due to blood oxygenation change is about 1 unit in SI. Assuming that $\Delta\chi = 10^{-6}$ is achieved during the activation, a corresponding 1.0×10^{-6} or 1 ppm change will result in magnetic field inhomogeneity. While at 1.5 T, 1 ppm inhomogeneity corresponds to about 63 Hz, at 7.0 T it could produce frequency shift of about 300 Hz. Such $\mathbf{B}_0$ inhomogeneity will induce dephasing of spin coherence which will reduce the signal causing dark regions on T_2^*-weighted EPI images. Even spin-echo sequence will bear reminiscent of such susceptibility-induced signal loss near the veins. While for stationary tissues RF does refocus the resulted dephasing of spins, for moving water molecules in veins protons rephrasing is not complete making BOLD effective as a T_2 as well as T_2^* effect.

fMRI signal is believed to largely originate from BOLD effect around small vessels, i.e., arterioles, capillaries, and venules *(27)*. The extravascular areas surrounding the small vessels represent loci of neuronal activity. But, there are contributions from large vessels to the BOLD signal as well. Such contribution must be quantified to ensure an accurate account of the role of small vessels *vs.* large vessels in fMRI. High magnetic fields provide a powerful tool in this regard. A known magnetic field at any position puts spins in a well-defined precession whose frequency provides knowledge of its location to produce a map of proton density. Spatial homogeneity and temporal stability of the field are important requirements for creating images faithful to the structures being studied. $\mathbf{B}_0$ field homogeneity of high-field magnets is around 0.5 ppm that using high order shimming could improve it to about 0.1 ppm over the head. Beyond this, as it was discussed, dHb produces high magnetic susceptibilities leading into comparable local inhomogeneities in the static field within the brain. At high field, regions in the brain, such as temporal lobes and basal ganglia, demonstrate high magnetic susceptibility providing a high contrast from the surrounding tissues *(42)*. Different scenarios for change in T_2^* are possible depending on the occupation of the voxel by capillaries, large vessels, and extravascular and intravascular BOLD *(8)*. In general, it can be stated that T_2^* signal differential between activation and rest period from these regions increases as a function of magnetic field. For example, if typical acquisition parameters for fMRI studies are receiver bandwidth of 2 kHz/pixel; TR 4,000 ms; TE 40 ms; FOV 190 × 190 mm^2; 30-40 slices, slice thickness, 5 mm; then implications of these parameters at 7.0 T can be contrasted to 1.5 T through a simple frequency shift. A typical BOLD effect of 0.5 ppm or 150 Hz frequency shift at 7.0 T could result in as high as 7% change in signal. Considering that BOLD has typically produced SNR, $\Delta R/R$, of the order of 1% at 1.5 T, this fact indicates that a linear increase in $\Delta R/R$ with $\mathbf{B}_0$ is possible.

BOLD contrast acts as a change in T_2^* rate, ΔR_2^*. What are the factors affecting ΔR_2^*? First, ΔR_2^* is directly influenced by the change in concentration of dHb. In fact, the volume susceptibility is directly proportional to volume of dHb and as such on ΔR_2^* *(36)*. Assuming that dHb is proportional to blood volume, the fraction of the blood volume f_{dHb} occupied by dHb will have direct effect on the signal. Models have been proposed that assign dependence of magnetic susceptibility difference between blood O_2Hb and dHb, $\Delta\chi$, raised to a power between 1 and 2. For a venous oxygen saturation increasing from 60 to 95%, Davis et al. found a power of 1.5 fitting the simulated ΔR_2^* vs. $\Delta\chi$ curve best *(43)*. Such studies measure the oxygen consumption increase vs. blood flow increase as a result of functional stimulation of the brain, visual cortex in this case *(43)*. So, while there could be considerable differences between the increase in blood flow and oxygen consumption, there is a unanimous consent on the role of oxidative metabolism as a significant component of the metabolic response of the brain to externally induced neuronal activation.

A key role for oxidative metabolism during neuronal activation makes the role of high field more momentous in both settling such issues and enhancing the fMRI SNR. One needs to determine with certainty where the changes of the blood activation come from. They could come from the brain tissue or from the draining veins near the activated region. Many fMRI studies do not make any distinctions between these two contributions. This is partially due to the challenges involved in addressing the issue. As it was mentioned earlier, spin echo and diffusion weighting are used to differentiate contributions from different-sized vessels. Considering the small BOLD effect at low field, about 1% change in signal, an increase in fMRI signal is essential to enable suppression of BOLD signal through SE or diffusion in order to accurately investigate the source of activation. This is owing to the fact that SE EPI has more T_2 than T_2^* weighting reducing sensitivity to local susceptibility-based changes. The GE readout is responsible for the T_2^* contrast. The extent of T_2^* overlay on T_2 contrast of SE EPI is field dependent and drastic difference between T_2 and T_2^* at high field makes EPI readout in BOLD fMRI a good tool to investigate the exact location of the activated region.

Furthermore, changes in oxygenation induced by neuronal activation are complex. In the early stage of response, within the first 2–3 s, an increase in dHb is observed, which is called the "initial dip" *(44)*. At the end of this stage a decrease in dHb and an increase in O_2Hb are observed. High field can refine this hemodynamic behavior. The initial dip has not been so conspicuous at 1.5 T and as such not well documented. The strength of the initial dip has been reported to be more than five times stronger at 7.0 T compared to 1.5 T. Furthermore, the nature of the initial dip provides insight into the mechanism of oxygen utilization vs. cerebral blood flow. In this regard, the initial dip could be used as another tool at high field to study the correlation between hemodynamics and neuronal activities.

5.4. Physiological Noise

In the absence of physiological noise, fMRI at high field could produce functional maps with even higher resolution from the brain regions *(45)*. Resolutions that already acquire submillimeter images at 7.0 T will approach microscopic resolution in the absence of the limiting noise. Unlike thermal noise which is temperature dependent, flat in frequency, not encoded by gradients, and hence constant at room temperature, physiological noise is a function of biological activities with relatively strong MR implications. As acquisition time of individual slices is around 10 ms, physiological noise during that time is not as debilitating as it is in the time series. Intensity of this time series noise in fMRI has shown variations in the signal which correlate with the respiratory and cardiac cycles indicating physiological modulation of BOLD by lung and cardiac function. In fact, these signals have variations independent of stimulation paradigm and thermal noise. Physiological noises and their correlation with various physiological functions are independent of field strength. Nevertheless, there are some indications that physiological noise might have some components in brain activation as well *(46, 47)*. Nevertheless, physiological noises have BOLD-like signal with low-frequency and TE-dependent variations *(48)*. It has also been shown that physiological noise could be dependent on the signal strength and its brain regional dependence. In this regard, it has been shown to have greater magnitude in cortical GM than in white matter *(47)*. The possibility of physiological noise dependence on the signal strength could not be related to magnetic field strength. However, conversion of brain metabolism into MR signal might produce a "resting-state" signal that will not correlate with external stimulations and consequently degrades the fMRI SNR. It has been proposed that relative strength of physiological noise could also be due to the choice of imaging resolution. This could be caused by a large voxel size which results in an increase in physiological noise which in turn degrades the activation signal *(49)*. These optimum voxels become smaller as field strength increases. However, if there is any vascular cause of physiological noise the inverse relation between the field strength and optimum voxel size will be limited.

6. High-Resolution fMRI

High spatial resolution (submillimeter voxels) is an expected outcome of imaging at high magnetic fields. The Information content of fMRI data can best be extracted by using an accurate account of the effect of neural activity on fMRI signals. In order to make fMRI images directly depicting cortical information, it is crucial to image at the scale of functional units of cortical structure, i.e., cortical columns *(50)*. Details of structures of cortical columns are the most prominent features of the architecture of the cortex. The cortex is organized in layers parallel to its outer surface (horizontal layers). Layers are specialized in the cell types they contain. Both the cell types and their connections with other neurons are unique in each layer. Nevertheless, there are distinct units connecting neurons in the vertical direction (perpendicular to the outer surface). From the outer surface of the cortex inward, these neuronal units are piled up deep into the cortex and participate in producing response to the same external stimulations. The fact that these vertical structures penetrate though the entire cortical thickness gives them the attribute of cortical columns. The cortex is made up of about 20 billion neurons and contains progenitor cells and glial cells and their structure is organized in units of minicolumn each constituted of about 100 neurons. These minicolumns are tied to each other to form cortical columns *(50)*. fMRI at high field is capable of visualizing these columnar units.

To date, high-field fMRI of columnar organization has been concentrated mostly on the visual systems. Since neurons involved in the specific functions are incorporated in the same columns with average dimension of 0.5 mm along cortex surface, fMRI resolutions comparable to this dimension is essential for their observation. At lower fields, fMRI has shown to be able to detect site of the BOLD signal to within 5 mm at 1.5 T down to <1 mm at 7.0 T. But the point spread functions make the relationship between susceptibility-based BOLD and loci of neuronal activity a function of correlation between hemodynamics and neuronal response which is not known with certainty *(51)*. It has been reported that submillimeter in-plane resolution and the negative bold response (NBR) or "initial dip" can be used to locate the site of neural activation in the visual cortex (V2) of anesthetized cats at 7.0 T *(52, 53)*. Such findings at the columnar level will bestow fMRI a new capability in functional mapping of the brain. Also, it is clear that low-field fMRI cannot achieve similar results due to lower susceptibility and poor SNR and CNR in reduced voxel volumes. The spatial resolutions required for positive identification of sites of neuronal activities require resolutions in hundreds of microns range which are only possible at high magnetic field, i.e., >4.0 T. The neurophysiology of neuronal columns has to be reflected in BOLD response in a way to increase specificity and spatial resolution of fMRI. This places the spotlight on high magnetic fields. One major requirement of an imaging technique that is to elucidate the neurophysiology of the central nervous system using the BOLD dynamics is to reach high spatial and temporal resolution at the same time. High-field MRI has shown to have such potential.

6.1. RF and Gradient Coil Technology

Gradients and RF coils are the two components of MRI scanners at the forefronts of signal generation and detection. As such, their less than ideal performance is the source of great many nuisances collectively referred to as artifacts *(54)*. To eliminate the high-field distortions of fMRI images, a variety of solutions are available *(55)*. Postprocessing techniques are proposed to correct for some of the distortions with known origins. Strong gradients also help reduce distortions as they increase the receiver bandwidth which in comparison, susceptibility induced changes will reduce. High-field works have also shown that multishot EPI has been able to reduce distortions causing an increase in acquisition time. For those measurements in which temporal resolution and spatial resolution do not have to be at maximum multiecho EPI is a viable approach. However, due to the low frequency nature of physiological noise, longer acquisition will increase the signal variations of physiological functions.

RF coil technology appropriate for high field has many design aspects in common with coils used in lower fields *(16)*. However, due to the nature of RF distribution at higher field, RF engineering needs many advances for adaptation to high field *(56–58)*. The popular bird-cage designs will be unable to take full advantage of high field. In particular, lumped element technology in which capacitors and inductors are used is a design based on circuit analysis using quasistatic field approximations *(16)*. But this analysis is only valid at low fields since the RF wavelength required for spin excitation decreases as the field strength increases. Specifically, RF wavelength in air at 1.5 T, 3.0 T, and 7.0 T are about 5 m, 2.5 m, and 1 m long. Taking into account the dielectric constant of biological tissues which are around 80, the wavelength inside the body reduces by a factor of inverse of square root of dielectric constant to around 50 cm, 25 cm, and 10 cm, respectively. Comparing the typical dimension of a RF coil, say 20 cm diameter, with these numbers makes it clear that quasistatic approximations are only valid for fields below 1.5 T where the RF coil dimension is much smaller than the wavelength of the RF field. At 7.0 T, the resonance frequency of 300 MHz makes the in-tissue wavelength to be about 10 cm. Since typical dimensions of the human head are comparable to this wavelength, the wave nature of the RF pulse becomes dominant within the head. Consequently, full wave Maxwell equation solutions are required to estimate the magnetic field ($\mathbf{B}_1$) and electric field (E_1) of the RF as it penetrates into the body during the spin excitations *(57)*. Such solutions are only possible through the use of sophisticated numerical computations, such as finite difference time domain (FDTD). This approach treats the RF coil interaction with the human body as a full wave electromagnetic modeling that not only provides an accurate map of distribution of $\mathbf{B}_1$ field over the subject but also offers a precise measure of specific absorption rate (SAR) which is an important indication of RF heating.

Inhomogeneous images acquired at high field (**Fig. 1**) demonstrate the effectiveness of the techniques developed for alleviation of inherent inhomogeneities of high-field images. These images point to a need for change in paradigm in the use of RF in high-field MRI. Use of computational tools for coil design is one pillar of the new paradigm. In addition, potential for excessive heat deposition predicted early in the history of MRI due to RF power constitutes a major safety issue that high field will have for a long time. Another issue that is highlighted at high field is dielectric effects that have revealed their presence in high-field images due to focusing of RF power at the central regions of the imaged body *(59)*. In studies at high field which are mostly done on the human heads, this effect shows strong inhomogeneous spread of RF reducing the power required in the peripheral regions for spin excitation. Dielectric effects or dielectric resonance problems at high field is an issue of concern for coil designers and must be taken into account in the use of high-field scanners and analysis of the data acquired by these systems.

Another pillar of the new paradigm is parallel imaging. Recent techniques for acceleration of image acquisition based on parallel imaging, SMASH-like methods and SENSE-like methods *(60, 61)*, have shown promise in alleviating RF inhomogeneities at high field. Both methods use surface like coil element which have an RF profile stronger in the proximal regions than in the deep regions of the body. While the immediate use of multichannel coil technology (parallel imaging) is in the receiver mode to accelerate signal reception, parallel transmit will also play an important role by restoring RF distribution over the whole head *(62)*. Possibility of the use of multichannel receive and transmit technology will allow high-field fMRI to further accelerate and enhance image qualities with potential to achieve microscopic resolution BOLD and perfusion-based images with high temporal and spatial resolution.

Need for more powerful gradients is another necessity of high field which are more highlighted recently as the receiver bandwidth increases in high-field scanners. Although modern scanners are equipped with more robust gradients, the increase in gradient strength and slew rate continues. During the 1980s, clinical scanners were equipped with gradients of 20 mT/m strength and 50 T/m/s slew rate. Today, 40-mT/m gradients with 150 T/m/s slew rate are available in most clinical scanners. Such hardware has helped many fMRI studies at 3.0 T and has helped research in the development and use of more powerful gradients.

Gradients also are the source of many image artifacts. At high field, artifacts due to EPI are aggravated and research has achieved many successes in minimizing image artifacts. Advances have been also achieved in gradient coil design and gradient amplifiers. Technologies such as active shielding (AS) of gradient have been realized. Considerable reduction in eddy current and its artifacts are reported by the use of AS gradients. Improved technology in pre-emphasis also has contributed in making modern gradients capable of higher performance even compared to the recent generations.

Manufacturers of specialty high-field gradients offer products with strengths of 50–100 mT/m with capability of 150–300 μs switch time. Such gradients can clock slew rate up to 200–500 T/m/s. High-field fMRI is the primary beneficiary of this technology as strong gradients capable of faster switching rates can be used to recover signal losses due to inhomogeneities through suppression of T_2^* artifacts. There are, however, disadvantages such as dB/dt which emerge as switching time reduces in gradients. Faster switching increases dB/dt which induces stronger electric fields in conducting tissues of the body causing nerve stimulation in the subject. Both designers and users of MRI scanners are made aware of the potential hazards of high gradient-induced electric fields and their use is governed by software and hardware safety supervisors to prevent incidents, such as ventricular fibrillation. Fortunately, fMRI uses sequences such as EPI which is very similar to the conventional gradient recalled echo sequence and it acquires the entire image in a single shot. The artifacts due to susceptibility and fast switching have been addressed in solutions, such as interleaved EPI and discontinuity in *k*-space, which have been dealt with by flip angle adjustments. In all, solutions in cleaning up EPI and other fast imaging techniques are making strong gradients more useful for application in high-field fMRI studies.

Another approach for using strong gradients without their undesirable side effects is through asymmetric designs, where the gradient field is produced only over the intended body part. For fMRI of the brain, this is particularly useful as it allows the establishment of stronger and faster gradients while at the same time keeping the heart isolated from induced electric fields. As the field strength increases, head-only scanners are gaining more attention. While high-field advantage of SNR is independent of higher gradient strength or slew rate, the additional in-plane resolution and slice thickness that can be achieved by using powerful gradients will help achieve isotropic voxels and ultimately microscopic map of brain function.

7. Conclusion

Functional imaging has achieved much success due to the MRI inherent soft-tissue contrast and its capability of detecting paramagnetic-based brain activation signals. The proportionality of SNR with field strength is an opportunity that has the potential of achieving microscopic brain mapping. High-field fMRI uses the SNR currency to enhance sensitivity and specificity in probing neurophysiology. Many high-field advantages can be utilized through realizable improved ancillary technologies such as RF coils, new excitation/detection schemes, artifact reduction, gradient technology, and parallel imaging. Low-field fMRI has already produced data from brain function that allows much insight into cognitive neuroscience. High field, in turn, has shown potential of further unraveling brain mysteries by detecting activation caused by controlled external stimulations with resolution that is approaching dimensions of functioning units of the brain. Such is the true potential of high-field fMRI.

References

1. Lauterbur PC. Image formation by induced local interactions: Example employing nuclear magnetic resonance. Nature 1973;242:190–191.
2. Hoult DI, Lauterbur PC. The sensitivity of the zeumatographic experiment involving human samples. J Magn Reson 1979;34:425–433.
3. Ogawa S, Tank DW, Menon R, Ellermann JM, Kim SG, Merkle H, Ugurbil K. Intrinsic signal changes accompanying sensory stimulation: Functional brain mapping with magnetic resonance imaging. Proc Natl Acad Sci USA 1992;89:5951–5955.
4. Sadek JR, Hammeke TA. Functional neuroimaging in neurology and psychiatry. CNS Spectr 2002;7:286–290, 295–299.
5. Yacoub E, Van De Moortele PF, Shmuel A, Uğurbil K. Signal and noise characteristics of Hahn SE and GE BOLD fMRI at 7 T in humans. Neuroimage 2005;24:738–750.
6. Duong TQ, Yacoub E, Adriany G, Hu X, Ugurbil K, Vaughan JT, Merkle H, Kim SG. High-resolution, spin-echo BOLD, and CBF fMRI at 4 and 7 T. Magn Reson Med 2002;48:589–593.
7. Pfeuffer J, Adriany G, Shmuel A, Yacoub E, Van De Moortele PF, Hu X, Ugurbil K. Perfusion-based high-resolution functional imaging in the human brain at 7 Tesla. Magn Reson Med 2002;47:903–911.
8. Uğurbil K, Hu X, Chen W, Zhu XH, Kim SG, Georgopoulos A. Functional mapping in the human brain using high magnetic fields. Philos Trans R Soc Lond B Biol Sci 1999 29;354:1195–1213.
9. Logothetis NK. What we can do and what we cannot do with fMRI. Nature 2008 12;453:869–878.
10. Goense JB, Zappe AC, Logothetis NK. High-resolution fMRI of macaque V1. Magn Reson Imaging 2007;25:740–747.
11. Shulman RD. Functional imaging studies: Linking mind and basic neuroscience. Am J Psychiatry 2001;158:11–20.
12. Bloch F. Nuclear induction. Phys Rev 1946;7: 460–473.
13. Pourcell EM, Torrey HC, Pound RV, Resonance absorption by nuclear magnetic moments in a solid. Phys Rev 1946;69:37–38.
14. Boskamp EB. Improved surface coil imaging in MR: Decoupling of the excitation and receiver coils. Radiology 198;157:449–452.
15. Hoult DI, Richards RE. The signal-to-noise ratio of nuclear magnetic resonance experiment. J Magn Reson 1976;24:71–85.
16. Tropp J. The theory of the bird-cage resonator. J Magn Reson 1989;82:51–62.
17. Bloembergen, Purcell EM, Pound RV. Relaxation effects in nuclear magnetic resonance absorption. Phys Rev 1948;73:679–746.
18. Peters AM, Brookes MJ, Hoogenraad FG, Gowland PA, Francis ST, Morris PG, Bowtell R. T2* measurements in human brain at 1.5, 3 and 7 T. Magn Reson Imaging 2007;25:748–753.
19. Wansapura JP, Holland SK, Dunn RS, Ball WS Jr. NMR relaxation times in the human brain at 3.0 Tesla. J Magn Reson Imaging 1999;9:531–538.
20. Vymazal J, Righini A, Brooks RA, Canesi M, Mariani C, Leonardi M, Pezzoli G. T1 and T2 in the brain of healthy subjects, patients with Parkinson disease, and patients with multiple system atrophy: Relation to iron content. Radiology 1999;211:489–495.
21. Liu F, Garland M, Duan Y, Stark RI, Xu D, Dong Z, Bansal R, Peterson BS, Kangarlu A. Study of the development of fetal baboon brain using magnetic resonance imaging at 3 Tesla. Neuroimage 2008;40:148–159.
22. Wright PJ, Mougin OE, Totman JJ, Peters AM, Brookes MJ, Coxon R, Morris PE, Clemence M, Francis ST, Bowtell RW,

Gowland PA. Water proton T (1) measurements in brain tissue at 7, 3, and 1.5T using IR-EPI, IR-TSE, and MPRAGE: Results and optimization. MAGMA 2008;21:121–130.

23. Kim SG, Ugurbil K. High-resolution functional magnetic resonance imaging of the animal brain. Methods 2003;30:28–41.
24. Meltzer HY, McGurk SR. The effects of clozapine, risperidone, and olanzapine on cognitive function in schizophrenia. Schizophr Bull 1999;25:233–255.
25. Sadek JR, Hammeke TA. Functional neuroimaging in neurology and psychiatry. CNS Spectr 2002;7:286–290, 295–299.
26. Kim SG, Fukuda M, Lessons from fMRI about mapping cortical columns. Neuroscientist 2008;14:287–299.
27. Yacoub E, Shmuel A, Logothetis N, U urbil K. Robust detection of ocular dominance columns in humans using Hahn Spin Echo BOLD functional MRI at 7 Tesla. Neuroimage 2007;37:1161–1177.
28. Yacoub E, Shmuel A, Pfeuffer J, Van De Moortele PF, Adriany G, Andersen P, Vaughan JT, Merkle H, Ugurbil K, Hu X. Imaging brain function in humans at 7 Tesla. Magn Reson Med 2001;45:588–594.
29. Mansfield P, Pykett IL, Morris PG. Human whole body line-scan imaging by NMR. Br J Radiol 1978;51:921–922.
30. Goense JB, Logothetis NK. Neurophysiology of the BOLD fMRI signal in awake monkeys. Curr Biol 2008;18:631–640.
31. Goense JB, Ku SP, Merkle H, Tolias AS, Logothetis NK. fMRI of the temporal lobe of the awake monkey at 7 T. Neuroimage 2008;39:1081–1093.
32. Farzaneh F, Riederer SJ, Pelc NJ. Analysis of T2 limitations and off-resonance effects on spatial resolution and artifacts in echo-planar imaging. Magn Reson Med 1990;14: 123–139.
33. Yang QX, Smith MB, Briggs RW, Rycyna RE. Microimaging at 14 Tesla using GESEPI for removal of magnetic susceptibility artifacts in T(2)(*)-weighted image contrast. J Magn Reson 1999;141:1–6.
34. Yang QX, Wang J, Smith MB, Meadowcroft M, Sun X, Eslinger PJ, Golay X. Reduction of magnetic field inhomogeneity artifacts in echo planar imaging with SENSE and GESEPI at high field. Magn Reson Med 2004;52: 1418–1423.
35. Chen NK, Wyrwicz AM. Removal of EPI Nyquist ghost artifacts with two-dimensional phase correction. Magn Reson Med 2004;51:1247–1253.
36. Schenck JF. The role of magnetic susceptibility in magnetic resonance imaging: MRI magnetic compatibility of the first and second kinds. Med Phys 1996;23:815–850.
37. Callaghan PT. Susceptibility-limited resolution in nuclear magnetic resonance microscopy. J Magn Reson 1990;87:304–318.
38. Kangarlu A, Bourekas EC, Ray-Chaudhury A, Rammohan KW. Cerebral cortical lesions in multiple sclerosis detected by MR imaging at 8 Tesla. AJNR Am J Neuroradiol 2007;28:262–266.
39. Filippi M, Rocca MA. Conventional MRI in multiple sclerosis. J Neuroimaging 2007;17 (Suppl 1):3S–9S.
40. Fazekas F, Soelberg-Sorensen P, Comi G, Filippi M. MRI to monitor treatment efficacy in multiple sclerosis. J Neuroimaging 2007;17(Suppl 1):50S–55S.
41. Christoforidis GA, Bourekas EC, Baujan M, Abduljalil AM, Kangarlu A, Spigos DG, Chakeres DW, Robitaille PM. High resolution MRI of the deep brain vascular anatomy at 8 Tesla: Susceptibility-based enhancement of the venous structures. J Comput Assist Tomogr 1999;23:857–866.
42. Bourekas EC, Christoforidis GA, Abduljalil AM, Kangarlu A, Chakeres DW, Spigos DG, Robitaille PM. High resolution MRI of the deep gray nuclei at 8 Tesla. J Comput Assist Tomogr 1999;23:867–874.
43. Davis TL, Kwong KK, Weisskopff RM, Rosen BR, Calibrated functional MRI: Mapping the dynamics of oxidative metabolism (hypercapniaycerebrovascular reactivity). Proc Natl Acad Sci USA 1998;95:1834–1839.
44. Yacoub E, Shmuel A, Pfeuffer J, Van De Moortele PF, Adriany G, Ugurbil K, Hu X. Investigation of the initial dip in fMRI at 7 Tesla. NMR Biomed 2001;14:408–412.
45. Krüger G, Glover GH. Physiological noise in oxygenation-sensitive magnetic resonance imaging. Magn Reson Med 2001;46: 631–637.
46. Wang SJ, Luo LM, Liang XY, Gui ZG, Chen CX. Estimation and removal of physiological noise from undersampled multi-slice fMRI data in image space. IEEE-EMBS 2005;27:1371–1373.
47. Krüger G. Glover GH, Physiological noise in oxygenation-sensitive magnetic resonance imaging. Magn Reson Med 2001;46:631–637.
48. Hyde JS, Biswal BB, Jesmanowicz A. High-resolution fMRI using multislice partial k-space GR-EPI with cubic voxels. Magn Reson Med 2001;46:114–125.

49. Glover GH, Krüger G. Optimum voxel size in BOLD fMRI. Proc Intl Soc Magn Reson Med 2002;10:1395.
50. Mountscale VB. The columnar organization of the neocortex. Brain 1997;120:701–722.
51. Triantafylloua C, Hogea RD, Wald LL. Effect of spatial smoothing on physiological noise in high-resolution fMRI. Neuroimage 2006;32:551–557.
52. Duong TQ, Kim DS, Ugurbil K, Kim SG. Localized cerebral blood flow response at submillimeter columnar resolution. Proc Natl Acad Sci USA 2001;98:10904–10909.
53. Kim DS, Duong TQ, Kim SG. High-resolution mapping of isoorientation columns by fMRI. Nat Neurosci 2000;3:164–169.
54. Jezzard P, Clare S. Sources of distortion in functional MRI data. Hum Brain Mapp 1999;8:80–85
55. Speck O, Stadler J, Zaitsev M. High resolution single-shot EPI at 7T. MAGMA Magn Reson Mater in Phys Biol Med 2008;21:73–86.
56. Baertlein BA, Ozbay O, Ibrahim T, Lee R, Yu Y, Kangarlu A, Robitaille PM. Theoretical model for an MRI radio frequency resonator. IEEE Trans Biomed Eng 2000;47: 535–546.
57. Ibrahim TS, Lee R, Baertlein BA, Kangarlu A, Robitaille PL. Application of finite difference time domain method for the design of birdcage RF head coils using multi-port excitations. Magn Reson Imaging 2000;18: 733–742.
58. Ibrahim TS, Kangarlu A, Chakeress DW. Design and performance issues of RF coils utilized in ultra high field MRI: Experimental and numerical evaluations. IEEE Trans Biomed Eng 2005;52: 1278–1284.
59. Kangarlu A, Baertlein BA, Lee R, Ibrahim T, Yang L, Abduljalil AM, Robitaille PM. Dielectric resonance phenomena in ultra high field MRI. J Comput Assist Tomogr 1999;23: 821–831.
60. Pruessmann KP, Weiger M, Scheidegger MB, Boesiger P. SENSE: Sensitivity encoding for fast MRI. Magn Reson Med 1999;42: 952–962.
61. Sodickson DK, Manning WJ. Simultaneous acquisition of spatial harmonics (SMASH): Fast imaging with radiofrequency coil arrays. Magn Reson Med 1997;38:591–603.
62. Katscher U, Börnert P, Leussler C, van den Brink JS. Transmit SENSE. Magn Reson Med 2003;49:144–150.

Chapter 5

Experimental Design

Hugh Garavan and Kevin Murphy

Summary

This chapter addresses issues particular to the optimal design of fMRI experiments. It describes procedures for isolating the psychological process of interest and gives an overview of block, event-related, and participant-response dependent designs. An additional focus is placed on data analysis with emphasis on optimizing and isolating the neuroimaging signal in activated brain regions. Finally, the chapter addresses a number of practical matters including optimal sample sizes and trial durations that confront all researchers when designing their experiments.

Key words: Task design, Sample size, Scan durations, Analysis, Regression, Efficiency, Frequency

1. Overview

Non-invasive functional neuroimaging techniques enable researchers to study the neurobiological substrates of psychological processes. The increasingly large body of neuroimaging research has two fundamental purposes. The first is to identify the brain regions that underlie a particular psychological process while the second seeks to identify differential responses of these regions to various stimuli or task challenges. The latter focus yields insights into both how the brain accommodates varying task demands and how differences between individuals or between clinical and healthy comparison groups might be explained by differences in neurobiological functioning. To achieve these goals, it is essential that one be able to isolate the psychological process of interest and how best to do so, with particular regard to experimental design, is the focus of this chapter. **Part II** describes issues particular to psychological experimental design, that is, experimental

M. Filippi (ed.), *fMRI Techniques and Protocols*, Neuromethods, vol. 41
DOI 10.1007/978-1-60327-919-2_5, © Humana Press, a part of Springer Science+Business Media, LLC 2009

control over the cognitive or emotional process of interest. **Part III** focuses on data analysis with emphasis on optimizing and isolating the neuroimaging signal in the activated brain regions. It also addresses a number of practical matters that confront all researchers when designing their experiments.

The distinction between isolating the psychological process of interest and isolating the signal associated with that process is made for pedagogical purposes. In practice, the two considerations are closely intertwined in that the experiment must be designed with a view to how the data are to be analysed. In brief, a typical analysis decomposes the time-series data into their contributing sources of variance. These sources of variance, which are generally assumed to be linearly additive, can include signals of interest such as task-induced brain activity as well as nuisance signals such as those created by head-movement, scanner signal drift, or the intrusion of extraneous psychological processes. The most common method for analysing these data is a linear decomposition of the various signal sources using, for example, a multiple regression in which separate regressors and planned contrasts between regressors capture both the unwanted variance and the variance of interest. Clearly, the design of the experiment needs to take into consideration what regressors and what contrasts of interest will be included in the analyses in order to ensure that the final brain activation map can be attributed to the psychological process of interest.

2. Task Design

As fMRI data are inherently noisy, it is important to induce as strong a signal as possible. This serves to maximize the contrast between the active task state and a comparison state (e.g. between a cognitive task and a visuomotor control task). In addition to maximizing contrast within an individual, it is also important to maximize contrast between individuals (e.g. between a clinical group and healthy controls) or between two times of testing (e.g. a pre-post comparison of treatment effects). To maximize the contrast between groups, it is advisable to isolate the psychological process that best discriminates the two groups. In this regard, neuroimaging researchers would be well served by grounding their experimental methods in the relevant psychological literature that identifies the key functions that distinguish the clinical and control groups and that provides a wealth of research methods detailing how to isolate those functions experimentally.

Experimental designs in fMRI can be categorized into block, event-related, and a third, broader category, labelled participant-response dependent, in which a continuous measure obtained from the participant provides a regressor for probing brain activity. The block design averages brain activation over a sustained period of time (20–30 s would be typical durations) and contrasts this with similar periods of either a resting state or a comparison task which is typically chosen to contain all task demands bar the psychological process of interest. Brain regions that differ between these conditions may then be attributed to the psychological process. In an effort to exclude signals associated with confounding physiological processes (described in detail below), aperiodic block durations, in which the alternating ON and OFF periods vary in durations, may be advisable.

This standard block design can be supplemented by including gradations of task challenge. This type of parametric manipulation can be quite advantageous: whereas the standard two-condition comparison (e.g. task A vs. task B or task A vs. rest) is open to the criticism of pure insertion (i.e. whether it is possible to selectively include and exclude a psychological function without affecting other task-related processes), the parametric manipulation assumes that the process is always present but to varying degrees in accordance with the demands placed on that process. Examples would include presenting various intensity levels of a sensory stimulus *(1)* or manipulating the number of memoranda in a working memory task *(2)*. Block designs can also be enhanced by a sort of psychological triangulation in which the conjunction of distinct block design contrasts allows one to isolate a psychological process that can be separated from irrelevant surface features of the tasks *(3)*. For example, if one wishes to isolate the neuroanatomy of the mental rehearsal component of verbal working memory, one might design an experiment using quite distinct classes of stimuli with each class accompanied by its own control comparison. One task might require participants to store a list of common nouns over a rehearsal period and recall the words after that rehearsal period. A reasonable control condition for this task might be one in which the word list remains on-screen for the duration of the rehearsal period and participants read, rather than recall, the words after the rehearsal period. The second task might present a list of nonsense syllables through earphones. At the end of the rehearsal period, a single nonsense syllable is presented and participants report, using a button box, if the single item was one of the rehearsed items. A control condition for this second task might simply prompt the participant to make a predetermined button press response at the end of a delay period that was of similar duration to the rehearsal period. The conjunction between the two activation maps, in which activation for each task is first subtracted from its control condition, may be argued to represent core regions responsible for verbal working memory for which the influence of extraneous task features (e.g. linguistic stimulus properties, response modalities, recall vs. recognition) is minimized. This strength of the conjunction approach, however, may often need to be balanced against the time costs involved in testing all the required conditions.

In circumstances in which a psychological process can be isolated temporally, then event-related designs are particularly useful. Here, brain activation time-locked to the events of interest can be selectively averaged enabling the researcher to embed trials of interest amidst other control trials and to categorize the trials after the participant has completed the experiment. Error trials can be excluded (or averaged separately) and events can be coded by whether a participant detected a target or not, responded relatively fast or not, produced a subsequent behaviour or not, and so on *(4)*. This affords the researcher increased flexibility in probing the dataset and has obvious advantages over the block design in circumstances in which the psychological process cannot be presented in blocks as in, for example, an oddball paradigm in which the nature of the phenomenon mandates that events are infrequent and unpredictable. The block and event-related designs can also be combined such that events of interest during an active task period can be isolated while the task period itself can be simultaneously contrasted against a control period *(5)*. This type of mixed design provides additional information in which one can determine the inter-relationships between tonic activity levels (e.g. sustained attention or an induced emotional state) and the processing of a discrete trial (e.g. detection of a fearful face).

A final category of experimental design is what we have labelled participant-response dependent. Here, the participant provides a continuous measure that can, for example, be used to generate a regressor to correlate against brain activity measures. Despite a loss of experimental control over the participant's behaviour, this category of design affords much flexibility when the phenomenon of interest is either not strictly task-dependent or is difficult to experimentally induce. Examples include resting-state acquisitions (in which correlated patterns of brain activity can be detected while the participant simply rests) *(6)*, biofeedback (in which, for example, a participant learns to control their level of brain activity) *(7)*, passive viewing of a movie clip (in which there may be multiple sources of stimulation with each varying with a different time course) *(8)*, or in which performance varies in an unpredictable manner *(9)*. Performance modulations for which one could assess brain activation changes can be quite wide-ranging including response times or response time variability on a continuous performance task *(10)*, frequently sampled self-report measures of mood *(11)*, and physiological measures such as heart rate or pupil diameter *(12)*. In these examples, the discrete measurements can be interpolated to provide a continuous time series that can be correlated with the brain activation time-series data.

An important consideration permeating all experimental designs is the choice of baseline against which activation is contrasted. These baselines can be explicit as in the block design in which specific blocks are chosen for comparison or implicit as in the event-related and participant-response dependent designs in which the baseline is all task-related activity that is not accommodated by regressors in the data analysis. The choice of baseline determines the interpretation of what processes are captured in an activation map and requires very careful consideration by the experimenter.

2.1. Choosing an Experimental Design

The choice of which design to employ will be dictated by the particulars of the psychological process to be investigated and how easy it is to isolate. Block designs can be employed if the psychological function is easy to isolate or if it is of particular interest to compare two tasks. A simple example would be a contrast between unilateral and bilateral finger movements. Here, blocks of finger movement in just one hand could be alternated with blocks of finger movements in two hands. Rest periods might also be included in order to provide a low-level baseline against which any task-related activity could be assessed. The inclusion of a resting-state baseline is generally advantageous as contrasts between two task-active periods can often be ambiguous in that greater activation in condition A relative to condition B could result from either more positive activation in A or a greater deactivation in B. A resting-state baseline allows one to resolve this ambiguity by showing if activation increases or decreases in any one condition relative to the resting baseline.

If the psychological function is not easily isolated, then a conjunction analysis may be useful. As can be seen in the verbal working memory example given earlier, the conjunction design enables the researcher to identify the core functional neuroanatomy that is common across different operationalizations of a psychological process. In addition, it can also reveal task-specific activations enabling, for example, one to determine how verbal working memory rehearsal for linguistic information differs to that of non-linguistic information. An alternative approach may parametrically manipulate verbal working memory demands by asking participants to rehearse items of varying set sizes. The presumption here is that more items will engage verbal working memory rehearsal to a greater extent resulting in changes in activation corresponding to the increased memory loads.

Although block designs suffer from an inability to isolate cognitive events that are temporally proximal by virtue of averaging over a prolonged duration, and may provide activation measures contaminated by extraneous tonic processes or isolated events (e.g. errors), they nonetheless have some advantages. For example, if the psychological process of interest by its very nature exists over a prolonged duration (e.g. sustained attention) or does not exist as a temporally discrete event (e.g. an emotional reaction), then it may be assayed best by a block design.

Conversely, the event-related design is particularly useful if one's goal is to isolate distinct cognitive events. In between-group comparisons (or time 1 vs. time 2 comparisons), the event-related design has the added advantage of being able to equate performance levels by comparing the groups on correct trials only. That is, one can compare correct performance trials of one group against the correct performance trials of the second group even if the absolute numbers of correct trials differ between the groups. In this regard, contamination from activity specific to error-related processes will not confound the between-group comparison *(13)*. In a similar manner, selective averaging of trials may make it possible to eliminate other group differences (e.g. response speed) assuming that there are sufficient numbers of trials for this type of a matched-trial analysis. This is a particularly welcome feature as activation differences between groups that one may wish to attribute to a psychological difference can be confounded by secondary behavioural or performance differences *(14)*. Indeed, the relationship between performance and activation levels is not straightforward. Often, researchers wish to ensure that the task produces performance differences between groups (or within a group following some experimental manipulation) in order to justify that choice of task or the focus on the psychological process engaged by the task. Why study the neurobiology of attention between healthy controls and children with attention deficit hyperactivity disorder (ADHD), if the latter are not shown to be worse on the attention task? However, this can be a double-edged sword in which performance differences and knock-on effects such as differences in frustration or anxiety levels can confound interpretation of activation levels. One proposed solution is to administer a task that is within the level of competence of all participants and which, therefore, may not produce group differences in performance. Such a task can be considered a probe of the neurocognitive functioning of the groups and substantial empirical evidence shows that brain activation differences are often observed in the absence of performance differences. The typical interpretation of activation differences when there are no performance differences is that reduced activation reflects better neural efficiency and less "effort". This interpretation is supported by studies showing greater levels of activity as task difficulty increases or those that show reduced activation following practice of a psychological process *(15)*.

Finally, the participant-dependent response design may be a sensible choice when one can obtain a continuous measurement from the participant but cannot exercise full experimental control over behaviour. For example, although emotional states are difficult to induce (and extinguish) experimentally, a physiological, self-report, or task-induced measure can provide a time-course of that emotional state that can be used to detect correlated brain regions.

3. Optimizing Experimental Task Designs

The key issue when optimizing experimental task design in fMRI is statistical power. There are many important basic variables to be chosen which, if selected wisely, will lead to high power and thus robust and reliable results. Too often these variables are chosen arbitrarily leading to poor experimental designs that fail to yield the expected outcomes.

When designing a task one needs to consider practical issues such as the number of participants or events required to give reliable results along with more analytic issues such as the estimation efficiency of the task design. These pragmatics are often dictated by feasibility constraints such as the availability of participants (e.g. how much access to the clinical population under study does one have?), the cost of scan time, or the amount of available time in which the participant will remain comfortable and compliant. When it comes to the practical issues, the real question researchers have is not "How many participants/events does my study require?", but "How few can I get away with?" Despite the ubiquity of these concerns, surprisingly few studies have addressed them and, instead, more emphasis has been placed on the analytic issues of presentation rate, duty cycles, sampling procedures, detectability of activation, and efficiency of response estimation. These analytic issues relate the task that will be performed to the analysis methods that will be employed and provide guidance on the design details of an experiment. It is important, however, that analytic considerations are not allowed to dictate the design of a task such that it is no longer appropriate for measuring/engaging the psychological process under study.

3.1. Practical Issues

Only a handful of studies have addressed how many participants are required to yield stable activation maps. The first paper addressing this issue showed that conjunction analysis with a fixed-effect model is sufficient to make inferences about population characteristics thus reducing the number of participants required to infer differences between populations *(16)*. Although quite useful, this conclusion does not give a clear indication of the number of participants required. By estimating the mean differences and variability between two block conditions, Desmond and Glover were able to perform simulation experiments generating power curves from which they could calculate the required number of participants *(17)*. They found that for a liberal threshold of $p = 0.05$, 12 participants were required to yield 80% power in a single voxel for typical block design activation levels. However, in fMRI the multiple comparisons problem and the associated potential for high levels of false positives require us to go to stricter thresholds where they demonstrated that twice the number of participants would be needed to maintain the same level of statistical power. This recommended number of participants is higher than the majority of fMRI studies, but is similar to independent assessments based on empirical data from a visual/audio/motor task *(18)* and from an event-related cognitive task *(19)*. The Murphy and Garavan study *(19)* found that statistical power is surprisingly low at typical sample sizes ($n < 20$), but that voxels that were significantly active from these smaller sample sizes tended to be true positives. Although voxelwise overlap may be poor in tests of reproducibility, the locations of activated areas provide some optimism for studies with typical sample sizes. It was found that the similarity between centres-of-mass for activated regions does not increase after more than 20 participants are included in the statistics. The conclusion can be drawn from this paper that a study with fewer numbers of participants than Desmond and Glover propose is not necessarily inaccurate but it is incomplete; activated areas are likely to be true positives but there will be a sizable number of false negatives. Needless to say, the required number of participants is influenced by the effect size which, in turn, is affected by the sensitivity of the experiment (e.g. the strength of the experimental manipulation, the quality of the data acquisition, and the accuracy of the data analyses). These considerations may be even more important if one's intention is to detect what is likely to be an even smaller effect size of a between-group comparison.

Little research has addressed the optimal number of scans/events needed for a successful fMRI study. A simple reason for this is that there is no standard metric for determining the required number of scans/events and no gold standard for determining when the optimal number of events has been reached. One metric that has been utilized is the spatial extent of activation under the assumption that as more scans/events are included in the analysis, the spatial extent of activation will increase until all activated cortex is deemed above significance. When this occurs the spatial extent should asymptote providing an estimate of the required number of scans/events. Using this approach in a block design experiment, Saad et al. demonstrated that the spatial extent of activation increased monotonically with the number of scans included in the analysis and failed to asymptote after twenty-two 200 s long scans *(20)*. Similarly, Huettel and McCarthy found that the spatial extent of activation failed to asymptote even after 150 events in an event-related design *(21)*. However, this failure to asymptote may be a consequence of the analysis method employed *(22)*. The correlation method does not asymptote because the goodness-of-fit to the regressor continues to rise with increasing degrees-of-freedom (df), which implies that the correlation measure will never plateau by adding more time points. The Huettel and McCarthy result *(21)* was replicated by Murphy and Garavan *(22)*, but they also demonstrated that when using a standard general linear modelling (GLM) analysis rather than a correlation, the spatial extent of activation asymptotes after roughly 25 events in a properly jittered event-related design. This is certainly a more attainable number of events in the available scan time of standard fMRI studies. It can be assumed that at least 25 of each type of event are needed if there is more than one psychological process under study. Also, these results have been derived from primary sensorimotor processes in the brain so that it is unclear whether they will still hold for more subtle cognitive activations. Again, differences in activation have not been addressed either; it is quite possible that many more events would be required to distinguish two processes with slightly varying activation levels since these differences could be dominated by noise.

A related concern is the optimal duration of a scan. How long a scan should last is obviously dependent on how densely the required number of events can be distributed. For example, a GO/NOGO task must sparsely distribute NOGO events due to the need to build up a prepotency to respond while a simple motor response task can present the events more frequently. Other issues that limit how long a scan can last include participant comfort and ability to stay engaged in the task along with technical concerns such as throughput of data and image reconstruction times. For these reasons and more, it is common to split a scanning session into separate scans lasting 5–10 min each after which they can be concatenated into one single time series and treated as a single scan in the analyses. However, breaks in scanning reduce the efficiency of any temporal filtering that is used and can also introduce unwanted session effects. If the goal is to detect activation then a block design is the most efficient approach (see later). In this case, the length of the scan is dependent on the amount of noise in the time series (which can be measured by calculating the temporal signal-to-noise ratio (TSNR) defined as the mean of the time series divided by its standard deviation), the size of the effect to be measured (eff), and the significance level (P) at which the activation is to be detected *(23)*. These authors derive an equation that determines how long one needs to scan to detect activation with a block design for volumes with high spatial resolution and suggest how this can be extended to an event-related design:

$$N_G = 8\left[1.5\left(1+e^{\log_{10}\frac{P}{2}}\right)\left(\frac{\text{erfc}^{-1}(P)}{(\text{TSNR})(\text{eff})}\right)\right]^2 \quad (1)$$

where N_G is the number of time points required for activation detection. Estimates of the size of the effect can be obtained from previous studies and TSNR measurements can be made using a short resting scan. Since these variables differ widely across types of task, brain regions, and scanners, it is impractical to suggest an optimal scan duration here.

3.2. Analytic Issues

The purpose of an experimental design is to alter neural activity, and hence the blood oxygen level dependent (BOLD) signal, as effectively as possible and in a predicted way, thereby enabling the researcher to detect the resulting brain changes. Using this prediction, one looks for corresponding patterns in the fMRI time series to determine which voxels were engaged in the task. A simple reference time series can be produced by convolving the stimulus timing function (which is equal to 0 when no stimulus is applied and 1 when a stimulus is presented) with a haemodynamic response function (HRF) that accurately represents the shape of the BOLD response after a single event. The gamma-variate function, $y(t) = t^r e^{-t/b}$, has been shown to effectively model the haemodynamic response to brief stimuli *(24)*, with parameters $r = 8.6$ and $c = 0.51$, and is a popular choice for modelling the haemodynamic shape. The difference between two gamma variates is also used in order to model the post-stimulus undershoot. It is important that the chosen HRF model accurately reflects the true shape of the response. If, for some reason, the haemodynamic shape of a participant is atypical (e.g. following treatment with a substance that directly affects the vasculature or a patient group with vascular damage), then the results of the analysis could be confounded by this difference in shape. (It should be noted that there are more advanced approaches to reference time-series formation, such as ones that use basis functions rather than a predetermined HRF shape and these are addressed in a later chapter.)

The simplest type of analysis is a linear least squares regression of the equation:

$$y(t) = \beta \cdot x(t) + \alpha + \varepsilon(t) \quad (2)$$

where $y(t)$ is the voxel time-series data, $x(t)$ is the reference function (i.e. the expected BOLD response to the stimulus) with β its scaling factor, α is a constant, and $\varepsilon(t)$ is a random Gaussian white-noise term. Both β and α are unknown parameters that are fit by the linear regression method. This equation can be extended to include extra regressors to remove unwanted trends in the data, such as baseline drift, whilst simultaneously computing the scaling factor. This scaling factor, β, can then be used as an activation measure for each voxel.

Multiple reference waveforms can easily be included in this type of analysis, denoted by the term *multiple linear regression*. In an experiment with two active conditions (1 and 2), the equation:

$$y(t) = \beta_1 \cdot x_1(t) + \beta_2 \cdot x_2(t) + \alpha + \delta \cdot t + \varepsilon(t) \quad (3)$$

is fitted to the data. For this model, four parameters are estimated, the two scaling factors β_1 and β_2 and the baseline α and also a baseline drift rate δ which accommodates for linear changes in the baseline over time. It is possible to investigate whether β_1 or β_2 are non-zero and whether β_1 is different from β_2 with statistical significance calculated using *F*-tests. This method allows one to identify active areas in the brain and calculate if an area is more active in one condition than another, thereby satisfying the two primary purposes of fMRI. This equation can be further generalized to $\mathbf{Y} = \mathbf{X}\boldsymbol{\beta} + \boldsymbol{\varepsilon}$, where $\mathbf{Y}$ is a column vector of the voxel's time-series data, $\mathbf{X}$ is the design matrix, $\boldsymbol{\beta}$ is a column vector of scaling factors, and $\boldsymbol{\varepsilon}$ is a column vector of Gaussian white-noise terms *(25, 26)*. This equation is called the *general linear model* (GLM) and is the basis for most fMRI analytic techniques. The columns of the design matrix $\mathbf{X}$ model the effects of interest and also confounding variables and are, in essence, the reference waveforms mentioned earlier.

When designing an experimental task, one is essentially specifying these reference waveforms/regressors. To maximize statistical power, these regressors must be chosen wisely. For example, *F*-tests are used to determine if there are significant differences between the regressors. To increase statistical power, one can increase the df by lengthening the task (for long TRs, each additional time point adds a new df). It might seem like a good idea to use extremely short TRs to increase the number of time points and hence the statistical power. However, to gain an extra df for each additional time point, each time point must be statistically independent from every other. Unfortunately, due to autocorrelations introduced into the fMRI data by physiological noise and scanner drifts, this is not the case. This example demonstrates that knowledge of the underlying mechanisms of fMRI along with the analysis methods is required when choosing even the simplest variables (such as the number of time points and the TR) for experimental design.

Efficiency of a task design is a measure of how accurately the GLM can estimate the β weights for each of the regressors, that is, how small the predicted variance of the β estimates will be. For example, assume a block design task that induces exactly a 2% signal change in hundreds of voxels, all with different noise properties but with the same noise variance. If a GLM analysis is performed, the β estimate for every voxel will be approximately 2% for all voxels with very little deviation. Since the variance of the estimates does not differ widely with noise distributions, this would be considered an efficient design. However, matters become more complicated when there is more than one regressor. Assume that there are two block conditions, A and B, where A and B are identical with the exception that B is delayed with respect to A by one TR. These two regressors are highly correlated so if a voxel responds only to condition A, it will be extremely difficult for the GLM to distinguish this from a voxel that responds only to B. For this reason, the β estimates for each of the conditions will vary substantially and this would be considered an inefficient design. If the conditions are designed so that they have zero correlation (this is achieved by delaying B by half a block length relative to A), it would be very easy to distinguish voxels that respond to each of the conditions individually or both of the conditions together. Therefore, the variance of the β estimates would be quite small and so the design is efficient. These simple examples show that efficient task designs come from regressors that are not correlated with each other. This can be slightly complicated by the contrasts of interest. For example, say we have a jittered event-related design where conditions A and B are randomly presented. If we want to find voxels that respond only to A (i.e. a contrast matrix of $C = [1\ 0]$), only to B (i.e. $C = [0\ 1]$), or differ in their response from A to B (i.e. $C = [1\ -1]$), this design is very efficient. However, if we want to determine voxels that respond equally to both A and B (i.e. $C = [1\ 1]$), then the design is very inefficient because such a voxel will always have an elevated activation level and therefore will be indistinguishable from a voxel that does not respond to either task. The simple idea that regressors must be minimally correlated becomes more complicated when multiple conditions, nuisance regressors, and contrasts are placed into a GLM analysis. The efficiency of a task is related to the covariance of the design matrix X (i.e. all regressors expressed as columns of a matrix) and is given the formula:

$$e = \text{trace}(C' \times (X'X)^{-1} \times C)^{-1} \qquad (4)$$

where C is the matrix of contrast weights and $'$ denotes the transpose of a matrix. Efficiency calculations should be carried out on all experimental designs before scanning to check that the regressors are sufficiently independent. A paper by Smith et al. argues against this efficiency calculation since it relates to computational precision rather than image noise *(27)*. This paper formulates the standard efficiency equations in terms of the required BOLD effect which takes into account the strength and smoothness of the time-series noise.

The question "how do we design a good fMRI task?" is really asking "what experimental timing will produce the most efficient design?". There are two variables under our control, the stimulus duration (SD: defined as the length of time the stimulus is displayed) and the interstimulus interval (ISI: defined as the length of time between the offset of one event and the onset of another). Another common term is stimulus onset asynchrony (SOA) defined as SOA = SD + ISI. Sometimes, ISI is used to mean SOA so that care must be taken to understand the true meaning when reading the literature. To maximize efficiency (i.e. minimize correlations between regressors by ensuring a clear temporal separation between the event types) one can use either a fixed ISI, but vary the order of events from different conditions or one can fix the order of the conditions and vary the ISI. For example, if an event from either condition A or condition B is to be presented every repetition time (TR), it is very inefficient to present the events in an alternating fashion A, B, A, ... However, efficiency is increased if the order is randomized. On the contrary, if B must follow A (e.g. A is a picture of an object and the participant must respond to B, a word, deciding whether it matches the object or not), then randomizing the order is not possible. Therefore, we must vary the ISI between successive As and Bs to increase the efficiency of the design.

The issue of experimental timing is very important in fMRI tasks due to the relatively poor temporal resolution of the technique. Bandettini and Cox have shown that with a 2 s SD the optimal ISI is 12–14 s when the ISI is kept constant *(28)*. At this optimal ISI, the experimentally determined functional contrast (i.e. the ability to detect activation) of an event-related task is only 35% lower than that of a block design (which, as explained below, is the most efficient design). Simulations assuming a linear system showed that this should be 65% lower suggesting the HRF is a non-linear system. Most techniques in event-related fMRI analysis assume that the haemodynamic shape of the BOLD signal is linearly additive. It has also been shown that when the ISI is allowed to vary, the haemodynamic response shows a 17–25% reduction in amplitude when trial onsets are spaced (on average) 5 s apart compared to those spaced 20 s apart *(29)*. However, power analysis indicated that the increased number of trials at fast rates outweighs this decrease in amplitude if statistically reliable response detection is the goal. So, despite the HRF being non-linear at fast presentation rates, the mismatch with the regressor is compensated by the increase in trial numbers. Dale also demonstrated that if the ISI varies, the statistical efficiency improves monotonically with decreasing mean ISI and that the efficiency can be up to ten times greater than that of a fixed ISI design *(30)*. These lessons on stimulus timing suggest that even though the HRF is non-linear at short ISIs, closely packed, randomly presented events produce highly efficient designs.

There are two fundamentally different goals when analysing event-related fMRI tasks: detection of signal change (which has been the focus thus far) and estimation of the HRF. Detection of the signal change involves determining one variable: the amplitude of the haemodynamic response. More information can be gleaned by estimating the HRF (e.g. time to onset, rise time, fall time, area under the curve), which can be used to determine subtle differences between groups or conditions that may not show up in an amplitude measure. However, this information comes at a cost: the experimental task can be optimized for either detection or estimation, but not both. Birn et al. showed that the estimation of the HRF is optimized when stimuli are frequently alternated between task and control states, whereas detection of activated areas is optimized by block designs *(31)*. Liu et al. have developed a method that can simultaneously achieve the estimation efficiency of randomized designs and the detection power of block designs at a cost of increasing the length of the experiment by less than a factor of two *(32)*. There are many programs that allow one to randomly (or not so randomly) generate thousands of task designs in order to choose the most efficient for the task at hand, be it detection or estimation. Genetic algorithms (optimization algorithms that code different designs like chromosomes and allow them to "crossover" and "point mutate" as they "replicate") that can produce designs that outperform random designs on estimation efficiency, detection efficiency, and design counterbalancing have also been developed *(33)*. Further work has also shown that using advanced mathematical techniques, block designs, rapid event-related designs, *m*-sequence designs (reference time series with an autocorrelation of zero), and mixed designs can nearly achieve their theoretically predicted efficiency and can be used in practice to obtain advantageous trade-offs between efficiency and detection power *(34)*. It is important when using programs to design experiments to realize that they may converge on a structure that may be problematic for the psychological process under investigation (e.g. the most efficient task for detecting activation is a block design; however, if we want to design an oddball study the oddball events of interest should not occur in a block).

When designing a task, one must also consider the frequencies at which the events of interest are presented. Analysis packages often perform high-pass filtering to remove low-frequency drifts from the data. If all frequencies below the limit of, say, 0.01 Hz are removed, then the activation to a task with a block lasting greater than 100 s will also be removed. Similarly, this would be true for event-related tasks if the events were presented at the same low frequency. Other frequencies exist in the data that one must be aware of. It is possible to remove the influence of physiological noise from fMRI data using techniques such as RETROICOR *(35)*. These physiological noise sources are known to produce fluctuations in the data at the cardiac frequency ~1.1 Hz, at the respiration frequency ~0.3 Hz, and also at the respiration volume variation frequency ~0.03 Hz *(36)*. If these techniques are to be used and the task predominantly displays power at one of these frequencies (e.g. blocks lasting 33 s have a frequency of 0.03 Hz), then the correction techniques may remove the activations of interest and not just the fluctuations due to unwanted physiological processes. Conversely, if these corrections are not used, then the GLM may denote these physiological fluctuations as activations (if the phase of the fluctuations matches the phase of the task). One must also bear in mind that when using a long TR, all frequencies will be aliased into a narrow frequency band (e.g. with a TR of 2 s all frequencies above 0.25 Hz will be aliased into the range of 0–0.25 Hz). This means that although the frequencies may seem far apart, the task and the physiological noise may alias to the same frequency (e.g. for a TR of 2 s, the respiratory frequency 0.3 Hz will be aliased to 0.2 Hz as will a task frequency of 0.7 Hz, that is, one event every 1.4 s). To avoid this problem it is best not to have the events regularly spaced so that they reside at one frequency but to have random ISIs, thus spreading the power to different frequencies. The most efficient tasks are ones whose power is spread widely across the whole available frequency spectrum.

4. Conclusions

Designing fMRI tasks can be difficult with logistical constraints (e.g. how many participants and how much time per participant can one afford) obliging the experimenter to optimize the study design. The emphasis here has been on the experimental and analytic means of isolating a psychological process and its associated fMRI signal. Both considerations are central: optimal efficiency is of little comfort if one measures the wrong thing, but there is little to be gained from an inaccurate measurement of a robust psychological phenomenon. General recommendations include the importance of grounding one's experiment in the appropriate theoretical framework and using appropriate experimental methods, generating designs that are tested for their efficiency prior to data collection, ensuring that a sufficiently large sample is tested and being clear on whether one's goal is the detection of a response or the estimation of that response.

References

1. Helmchen C, Mohr C, Erdmann C, Binkofski F, Buchel C. Neural activity related to self- versus externally generated painful stimuli reveals distinct differences in the lateral pain system in a parametric fMRI study. Hum Brain Mapp 2006;27:755–765.
2. Braver TS, Cohen JD, Nystrom LE, Jonides J, Smith EE, Noll DC. A parametric study of prefrontal cortex involvement in human working memory. Neuroimage 1997;5:49–62.
3. Price CJ, Friston KJ. Cognitive conjunction: a new approach to brain activation experiments. Neuroimage 1997;5:261–270.
4. Garavan H, Ross TJ, Murphy K, Roche RA, Stein EA. Dissociable executive functions in the dynamic control of behavior: inhibition, error detection, and correction. Neuroimage 2002;17:1820–1829.
5. Donaldson DI, Petersen SE, Ollinger JM, Buckner RL. Dissociating state and item components of recognition memory using fMRI. Neuroimage 2001;13:129–142.
6. Margulies DS, Kelly AM, Uddin LQ, Biswal BB, Castellanos FX, Milham MP. Mapping the functional connectivity of anterior cingulate cortex. Neuroimage 2007;37:579–588.
7. Weiskopf N, Veit R, Erb M, et al. Physiological self-regulation of regional brain activity using real-time functional magnetic resonance imaging (fMRI): methodology and exemplary data. Neuroimage 2003;19:577–586.
8. Hasson U, Nir Y, Levy I, Fuhrmann G, Malach R. Intersubject synchronization of cortical activity during natural vision. Science 2004;303:1634–1640.
9. Slotnick SD, Yantis S. Common neural substrates for the control and effects of visual attention and perceptual bistability. Brain Res Cogn Brain Res 2005;24:97–108.
10. Hahn B, Ross TJ, Stein EA. Cingulate activation increases dynamically with response speed under stimulus unpredictability. Cereb Cortex 2007;17:1664–1671.
11. Risinger RC, Salmeron BJ, Ross TJ, et al. Neural correlates of high and craving during cocaine self-administration using BOLD fMRI. Neuroimage 2005;26:1097–1108.
12. Kampe KK, Frith CD, Frith U. "Hey John": signals conveying communicative intention toward the self activate brain regions associated with "mentalizing," regardless of modality. J Neurosci 2003;23:5258–5263.
13. Murphy K, Garavan H. Artifactual fMRI group and condition differences driven by performance confounds. Neuroimage 2004;21:219–228.
14. Poldrack RA. Imaging brain plasticity: conceptual and methodological issues – a theoretical review. Neuroimage 2000;12:1–13.
15. Kelly AM, Garavan H. Human functional neuroimaging of brain changes associated with practice. Cereb Cortex 2005;15:1089–1102.
16. Friston KJ, Holmes AP, Worsley KJ. How many subjects constitute a study? Neuroimage 1999;10:1–5.
17. Desmond JE, Glover GH. Estimating sample size in functional MRI (fMRI) neuroimaging studies: statistical power analyses. J Neurosci Methods 2002;118:115–128.
18. Thirion B, Pinel P, Meriaux S, Roche A, Dehaene S, Poline JB. Analysis of a large fMRI cohort: statistical and methodological issues for group analyses. Neuroimage 2007;35:105–120.
19. Murphy K, Garavan H. An empirical investigation into the number of subjects required for an event-related fMRI study. Neuroimage 2004;22:879–885.
20. Saad ZS, Ropella KM, DeYoe EA, Bandettini PA. The spatial extent of the BOLD response. Neuroimage 2003;19:132–144.
21. Huettel SA, McCarthy G. The effects of single-trial averaging upon the spatial extent of fMRI activation. Neuroreport 2001;12:2411–2416.
22. Murphy K, Garavan H. Deriving the optimal number of events for an event-related fMRI study based on the spatial extent of activation. Neuroimage 2005;27:771–777.
23. Murphy K, Bodurka J, Bandettini PA. How long to scan? The relationship between fMRI temporal signal to noise ratio and necessary scan duration. Neuroimage 2007;34: 565–574.
24. Cohen MS. Parametric analysis of fMRI data using linear systems methods. Neuroimage 1997;6:93–103.
25. Friston KJ, Holmes AP, Poline JB, et al. Analysis of fMRI time-series revisited. Neuroimage 1995;2:45–53.
26. Worsley KJ, Friston KJ. Analysis of fMRI time-series revisited – again. Neuroimage 1995;2:173–181.
27. Smith S, Jenkinson M, Beckmann C, Miller K, Woolrich M. Meaningful design and contrast estimability in FMRI. Neuroimage 2007;34:127–136.
28. Bandettini PA, Cox RW. Event-related fMRI contrast when using constant interstimulus interval: theory and experiment. Magn Reson Med 2000;43:540–548.

29. Miezin FM, Maccotta L, Ollinger JM, Petersen SE, Buckner RL. Characterizing the hemodynamic response: effects of presentation rate, sampling procedure, and the possibility of ordering brain activity based on relative timing. Neuroimage 2000;11:735–759.
30. Dale AM. Optimal experimental design for event-related fMRI. Hum Brain Mapp 1999;8:109–114.
31. Birn RM, Cox RW, Bandettini PA. Detection versus estimation in event-related fMRI: choosing the optimal stimulus timing. Neuroimage 2002;15:252–264.
32. Liu TT, Frank LR, Wong EC, Buxton RB. Detection power, estimation efficiency, and predictability in event-related fMRI. Neuroimage 2001;13:759–773.
33. Wager TD, Nichols TE. Optimization of experimental design in fMRI: a general framework using a genetic algorithm. Neuroimage 2003;18:293–309.
34. Liu TT. Efficiency, power, and entropy in event-related fMRI with multiple trial types. Part II: Design of experiments. Neuroimage 2004;21:401–413.
35. Glover GH, Li TQ, Ress D. Image-based method for retrospective correction of physiological motion effects in fMRI: RETROICOR. Magn Reson Med 2000;44:162–167.
36. Birn RM, Diamond JB, Smith MA, Bandettini PA. Separating respiratory-variation-related fluctuations from neuronal-activity-related fluctuations in fMRI. Neuroimage 2006;31: 1536–1548.

Chapter 6

Preparing fMRI Data for Statistical Analysis

John Ashburner

Summary

This chapter describes the procedures applied to fMRI data prior to their statistical analysis. This usually begins with converting the data from original MR format to a form that can be used by the analysis software. The data are then motion corrected. If an anatomical scan is collected for the subject, then it would be co-registered with the fMRI, and may serve to estimate the warps needed to spatially normalise the fMRI to some standard space. The final pre-processing step is usually to smooth the data.

Key words: Generative Model, fMRI, Registration, Artefact Correction, Spatial Normalisation, Smoothing

1. Introduction

This chapter provides a brief overview of the pre-processing steps currently used for transforming fMRI data into a form suitable for statistical analysis. Processing strategies for fMRI data are not fixed. The particular procedures used depend on the data and the aims of the analysis. Pragmatic motivations, such as software availability and ease of use, also play a major role in determining how fMRI data are pre-processed. This chapter focusses on the main processing steps that are usually applied to the data, prior to performing statistical analysis. Most of the procedures dealt with are various forms of image registration. The first step is usually to convert from DICOM format to a file format that is more manageable. This is followed by motion correcting the data, which may include a distortion correction procedure. Often, there is also an anatomical scan collected for each subject, and this would be brought into alignment with the fMRI by a co-registration step.

M. Filippi (ed.), *fMRI Techniques and Protocols*, Neuromethods, vol. 41
DOI 10.1007/978-1-60327-919-2_6, © Humana Press, a part of Springer Science+Business Media, LLC 2009

The anatomical image is useful in order to spatially normalise the fMRI data. The warps, needed to deform the fMRI to some standard space, can be estimated using the anatomical image. Once these warps have been estimated, they can be applied to the motion-corrected fMRI data, to spatially normalise them. The final step, before statistical analysis, is usually to smooth the data.

There are many variations on this sequence of operations. For example, a two-level approach for multi-subject analysis (random effects model) may be performed by generating parameter-estimate images from fMRI data that have not been spatially normalised. These parameter images could then be warped to the standard space and the statistical analysis performed on them. The smoothing step would be omitted if the statistical analysis included a model of spatial smoothness. This chapter says nothing about "slice-time correction," and assumes that the model used in the subsequent statistical analysis accounts for the fact that slices of fMRI data are not acquired simultaneously. Surface-based approaches, in which the fMRI data is projected on to a representation of the cortical surface, are not covered.

2. File Format Conversion

Most MRI scanners produce image data in a format that conforms to the *DICOM Standard*. This stands for "*Digital Imaging and Communications in Medicine*," and it is the standard used in virtually all hospitals worldwide. To keep up with technological advances, the DICOM Standard is re-published approximately every year or two. The standard is also extensible, and scanner manufacturers customise file formats to suit their own particular needs. Full details, along with several hundred pages of documentation describing the basic file format, are available from http://dicom.nema.org/

Most fMRI analysis is currently performed within an academic setting, where the complexity of DICOM is unnecessary. Neuroimaging analysis tools are written by scientists and engineers who wish to avoid working with complex and difficult formats. As a result, several different file formats for fMRI data arose, many of which were variants of the ANALYZE™ 7.5 format, which consists of an ".img" file containing the image data itself, plus a ".hdr" file containing various pieces of descriptive information. For a number of years, various fMRI analysis software developers used the ANALYZE format in slightly different ways, or had their own file formats. This made inter-operability among packages very difficult, which precluded the use of tools developed at one site with tools developed at another. One classic example of such problems was the different ways in which the *voxels*[1] of an image are ordered, which often caused uncertainty about the laterality of the brain.

The *NIfTI-1* data format was recently developed in order to facilitate inter-operability among fMRI data analysis packages.[2] Standards have been agreed on how the data format should be used, with the aim of making it easier to mix different software packages. Providing that only NIfTI-1 compliant software is used, there should no longer be any confusion about the orientation of the brains within the images.

Images are usually treated as an array of voxels. For example, an anatomical image is generally treated as a 3D array, and most packages require this volume to be stored in a single file. DICOM usually stores each slice separately (as a series of 2D arrays), but most file format conversion routines will stack these slices together into a 3D volume. A run of fMRI data is usually considered as a 4D array, although for many procedures, it is often convenient to treat it as a time series of 3D arrays. Some packages assume that the entire run is saved in a single file, whereas other packages treat the data as a series of files containing 3D volumes.

The NIfTI format allows storage on disk to be in either a left- or a right-handed co-ordinate system. However, the format includes an implicit spatial transformation into a right-handed co-ordinate system. This transform maps from data co-ordinates (e.g., column i, row j, slice k) into some real-world (x,y,z) positions in space. These positions could relate to Talairach & Tournoux (T&T) space *(1)*, Montreal Neurological Institute (MNI) space *(2, 3)*, or patient-based scanner co-ordinates. For T&T and MNI co-ordinates, x increases from left to right, y increases from posterior to anterior, and z increases from inferior to superior direction. Directions in the scanner co-ordinate system are similar. MRI data are usually exported as DICOM format, which encodes the positions and orientations of the slices. When data are converted from DICOM to NIfTI-1 format, the relevant position and orientation information can be determined from the "*Pixel Spacing*," "*Image Orientation*," and "*Image Position*" fields of the DICOM files.

Terms such as "neurological" and "radiological convention" relate only to visualisation of axial images. They are unrelated to how the data are stored on disk, or even how the real-world co-ordinates are represented. It is more appropriate to consider

---

[1] A voxel is a three-dimensional pixel, and can be thought of as a "volume element," as opposed to a "picture element."

[2] See http://nifti.nimh.nih.gov/

whether the real-world co-ordinates system is left- or right-handed. T&T use a right-handed system, whereas the storage convention of ANALYZE files is usually considered as left-handed (x increases from right to left). These co-ordinate systems are mirror images of each other, so transforming between left- and right-handed systems involves flipping, and cannot be done by rotations alone.

3. Corrections to fMRI Data

Most pre-processing of fMRI data involves some form of spatial registration. The head of a single individual is generally considered to be fairly rigid, so pre-processing generally tries to bring the image volumes of each individual subject into alignment, prior to any inter-subject registration. Various artefact corrections may be incorporated within the intra-subject registration procedures.

3.1. Artefact Correction

There are a number of image artefacts that result from the very fast acquisition times required for fMRI. Many of these have detrimental effects if not properly modelled. Some groups have developed in-house software to improve on the algorithms supplied by scanner manufacturers for reconstructing images from the original k-space data. The objectives of these custom reconstruction algorithms include reducing Nyquist ghosting artefacts and ensuring that the model uses a better trajectory through k-space. Sites that perform their own image reconstruction require the original complex k-space data, for which there is no clearly defined DICOM standard.

The introduction of a subject into the scanner causes distortions of the magnetic field *(4)*. The field may be uniform when there is no subject present; but with a subject in the scanner, the field is influenced (through Maxwell's equations) by the varying magnetic susceptibilities of the different tissues within the field of view. These effects are especially prominent at the interface between tissue and air, resulting in, for example, dropout and distortion in the frontal lobe in regions close to the nasal sinuses. For echo-planar images, the main effects of magnetic field inhomogeneity are spatial distortions in the phase-encoding direction of the images and dropouts (signal loss) that arise through through-plane de-phasing. Some of the distortions can be reduced by active shimming (changing the field of the scanner via the shim coils) or passive shimming (introducing diamagnetic material into the orifices of the subject), but these measures only reduce the effects of distortions and dropouts, and cannot completely counteract them.

The models used for intra-subject registration of the head often assume rigid-body movement. Obtaining accurate alignment of a relatively distortion-free anatomical image with highly distorted fMRI data is not possible, unless the geometric distortions are corrected. Therefore, one of the first steps is often a correction for these distortions. Retrieving signal that is lost in dropouts is not possible, but there are a number of post hoc approaches for correcting geometric distortions in the images.

- It is possible to generate field maps from additional scans that are normally collected just prior to the fMRI runs *(4)*. This is the most common approach and generally works well (**Fig. 1**).
- If the air, bone, and other tissue can be segmented from the anatomical images, then it becomes possible to simulate field maps *(5)* by solving Maxwell's equations. Separating air from bone is quite difficult from MRI scans, as both generally appear dark. Air appears dark because of its low proton density, whereas hard tissue such as bone has a very short T2 relaxation time, so most of the signal has decayed before it is detected. Additional prior information generated from computed tomography scans will probably be needed in order to attempt such segmentation. For this reason, the approach has not been widely adopted.

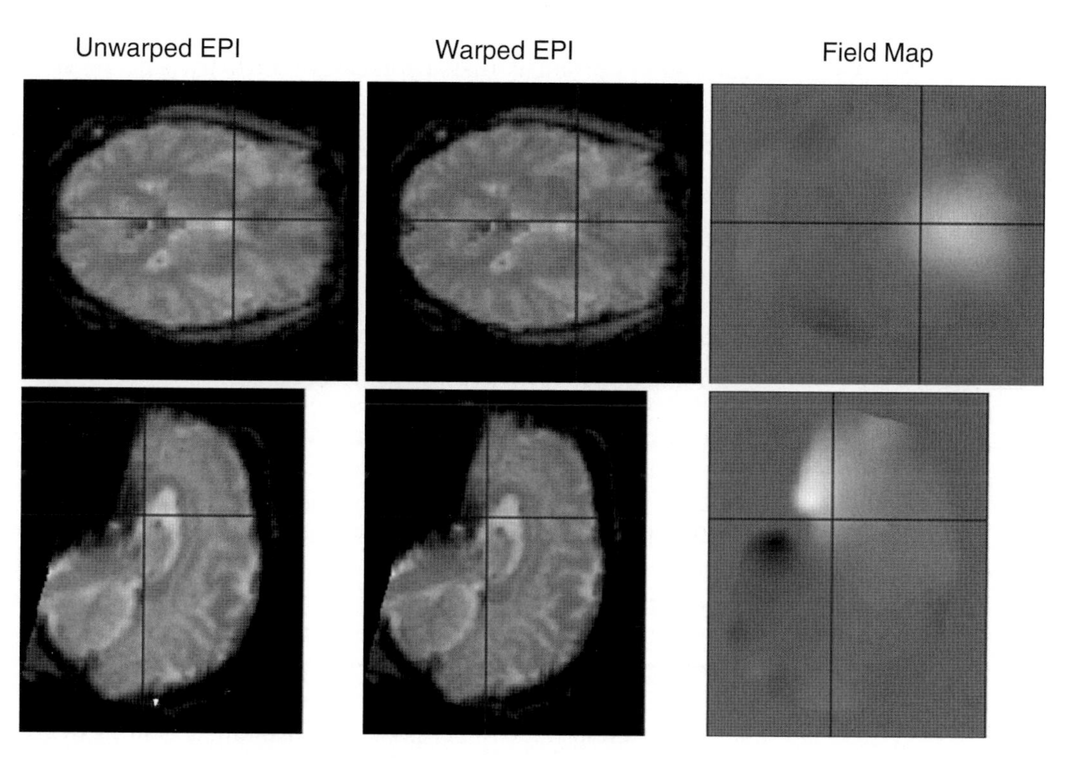

Fig. 1. Distortions in echo-planar imaging (*EPI*) can be corrected by field maps.

- Image registration procedures can be used to estimate the warps that align a distorted fMRI scan with a distortion-free anatomical image *(6, 7)*. Contrast differences between the images mean that some form of information-theoretic objective function is required, and the effects of signal dropout in the fMRI should also be taken into consideration.

A correction approach that combines all of the above strategies into a single model is likely to be the most accurate.

3.2. Motion Correction

The most common application of within-modality registration in functional imaging is to reduce motion artefacts by re-aligning the volumes in the image time series. The objective of re-alignment is to determine the rigid-body transformations that best map the series of functional image volumes to the same space. Blood oxygen level dependent (BOLD) signal changes elicited by the haemodynamic response tend to be small compared with apparent signal differences that can result from subject movement *(8)*. Subject head movement in the scanner cannot be completely eliminated, so retrospective motion correction is usually performed as a pre-processing step. This is especially important for experiments where subjects may move in a way that is correlated with the different experimental conditions. Even tiny systematic differences can result in a significant signal accumulating over numerous scans. Without suitable corrections, artefacts arising from subject movement, which correlate with the experiment, may appear as activations. A second reason why motion correction is important is that it increases sensitivity. The *t*-test is based on the signal change relative to the residual variance. The residual variance is computed from the sum of squared differences between the data and the linear model to which it is fitted; movement artefacts add to this residual variance, and so reduce the sensitivity of the test to true activations.

At its simplest, image registration involves estimating a mapping between a pair of images. One image is assumed to remain stationary (the reference image), whereas the other (the source image) is spatially transformed to match it. In order to transform the source to match the reference, it is necessary to determine a mapping from the location of each voxel in the reference to a corresponding location in the source. The source is then re-sampled at the new locations. The mapping can be thought of as a function of a set of estimated transformation parameters. The shape of a human brain changes very little with head movement, so rigid-body transformations can be used to model different head positions and orientations of the same subject. Matching of two images is performed by finding the spatial transformation (mapping) that optimises some mutual function of the images. For the case of a rigid-body transformation in three dimensions, the mapping is defined by six parameters: three translations and three rotations.

There are two steps involved in registering a pair of images together. There is the *registration* itself, whereby the set of parameters describing the mapping is estimated. Then there is the *transformation*, where one of the images is transformed according to the estimated parameters. For rigid registration, this step is often referred to as "re-slicing."

Registration involves estimating the parameters of a spatial transformation that "best" match the images. The quality of the match is based on an *objective function*, which is maximised or minimised using some *optimisation algorithm*. Performing the registration normally involves iteratively transforming the source image many times, using different parameters, until the objective function can no longer be improved upon. Alignment of fMRI data is usually achieved by minimising the mean-squared difference between each of the images and a reference image, where the reference image could be one of the images in the series. For slightly better results, this procedure could be repeated, but instead of matching to one of the images from the series, the images would be registered to their mean (after first-pass re-alignment).

Even after rigid re-alignment, there may still be some motion-related artefacts remaining in the data. There are many sources of such residual artefacts, and the most obvious ones are:

- Interpolation error from the re-sampling algorithm *(9)* used to transform the images can be a source of motion-related artefacts. For this reason, and also for speed and efficiency, some registration algorithms use a Fourier interpolation method *(10, 11)*.
- When MR images are reconstructed, the final images are usually the modulus of the initially complex data. This results in voxels that should be negative being rendered positive. This has implications when the images are re-sampled because it leads to errors at the edge of the brain that cannot be corrected, however good the interpolation method is. Possible ways to circumvent this problem are to work with complex data or apply a low-pass filter to the complex data before taking the modulus.
- The sensitivity (slice-selection) profile of each slice also plays a role in introducing artefacts *(12)*. Gaps between slices are not accounted for within the current framework.
- fMRI images are spatially distorted and the amount of distortion depends partly upon the position of the subject's head within the magnetic field. Interactions between image distortion and the orientation of a subject's head in the scanner can also cause other problems because purely rigid alignment does not take this into account. Relatively large subject movements result in the brain images changing shape, and these shape changges

cannot be corrected by a rigid-body transformation alone. The interaction between image distortion and head orientation illustrates a limitation of conceptualising pre-processing as the application of a series of tools to the data. These issues are better resolved by a generative model that combines both a model for image distortions and a model of subject motion *(13)*.

- Each volume of a series of fMRI data is currently acquired a plane at a time over a period of about a second. Subject movement between acquiring the first and last plane of any volume is another reason why the image volumes may not strictly obey the rules of rigid-body motion *(14)*. A better model would allow each slice to move separately – but it may lead to technical problems if there was too much movement. For example, the model would allow some points in the brain to be scanned more than once during the acquisition of a volume, and some points not to be scanned at all. Filling in the appropriate values in the corrected images is difficult if there is no actual data to sample.
- After a slice is magnetised, the excited tissue takes time to recover to its original state, and the amount of recovery that has taken place will influence the intensity of the tissue in the image. This effect can be seen in the first few scans of an fMRI time series, and is the reason why a few "dummy scans" are collected at the start of an fMRI run in order for the intensities to stabilise. Out of plane movement will result in a slightly different part of the brain being excited during each repeat. This means that the spin-excitation will vary in a way that is related to head motion, and so leads to more movement-related artefacts *(15)*.
- Nyquist ghost artefacts in MR images do not obey the same rigid-body rules as the head, so a rigid rotation to align the head will not mean that the ghosts are aligned. The same also applies to other image artefacts such as those arising due to chemical shifts.
- The accuracy of the estimated registration parameters is normally in the region of tens of micrometers. This is dependent upon many factors, including the effects just mentioned. Even the signal changes elicited by the experiment can have a slight effect (a few micrometer) on the estimated parameters *(16)*, so this in turn may have consequences in terms of how significant differences are interpreted.

These problems cannot be corrected by simple rigid-body re-alignment, and so may be sources of stimulus-correlated motion artefacts. Systematic movement artefacts resulting in a signal change of only one or two percent can lead to highly significant false positives over an experiment with hundreds of scans.

This is especially important for experiments where some conditions may cause slight head movements (such as motor tasks or speech) because these movements are likely to be highly correlated with the experimental design. In cases like this, it is difficult to separate true activations from stimulus-correlated motion artefacts. All that can be concluded is whether there is a difference among the data. The specific causes of any difference remain unknown, but it is generally hoped that they relate to BOLD signal changes. Providing there are enough images in the series and the movements are small, some of these artefacts can be removed during the subsequent statistical analysis by regressing out any signal that is correlated with functions of the estimated movement parameters *(15)*. However, when the estimates of the movement are related to the experimental design, it is likely that much of the interesting BOLD signal will also be regressed out of the data. These issues are and will remain unresolved until interactions among processing steps (including the statistical analysis) are properly modelled.

4. Inter-modality Registration

For studies of a single subject, sites of activation can be localised more clearly by superimposing them on a high-resolution anatomical (structural) image of the subject (typically a T1-weighted MRI). This requires registration of the functional images with the anatomical image. A further use for this registration is that a more precise spatial normalisation can be achieved by estimating the requisite warps from a more detailed anatomical image. If the functional and anatomical images are in register, then a warp estimated from the anatomical image can also be applied to the functional images.

As in the case of movement correction, this registration is normally performed by optimising a set of parameters describing a rigid-body transformation, but the matching criterion needs to be more complex because the anatomical and functional images normally have very different patterns of intensity. A simple mean-squared difference model will no longer be effective, so alternative objective functions (similarity measures) are needed. Inter-modal registration approaches initially involved the use of landmarks, which were manually defined on the images. The images were registered by bringing the landmarks into alignment. An early automated approach based on a similarity measure between images was the automated image registration algorithm *(17)*. The objective function was obtained by dividing the intensities of one image into a number of bins. For the voxels associated with each bin, the idea was to minimise the variance of the corresponding voxel intensities of the other image. The algorithm was originally intended for registering positron emission tomography and anatomical MRI and worked well – providing the MRI had non-brain tissue removed.

The more recent similarity measures used for inter-modal [as well as intra-modal *(16, 18)*] registration are based on *information theory*. These measures are based on joint probability distributions of intensities in the images, usually discretely represented in the form of 2D joint histograms, which are normalised to sum to one. The most commonly used measure of image alignment is *mutual information* (MI) *(19, 20)* (also known as *Shannon information*). MI is a measure of dependence of one image on the other, and can be considered as a distance (*Kullback-Leibler divergence*) between the joint distribution and the equivalent distribution assuming complete independence. Another perspective is that MI is a measure of the reduction of uncertainty about one image given the other. Registration algorithms work under the assumption that the MI between the images is maximised when they are in register (**Fig. 2**). A number of other information theoretic measures have since been devised *(21, 22)*, and a more complete review of information theoretic image registration approaches is given by Pluim et al. *(23)*. A number of inter-modality registration algorithms have been thoroughly evaluated on the same data *(24)*, and alignment accuracy is generally found not to be as high as that obtained by within modality registration.

5. Spatial Normalisation

Currently, the main application for non-linear image registration within imaging neuroscience is the procedure known as *spatial normalisation*. This involves warping the brain images from different subjects in a study into roughly the same standard space to allow signal averaging across subjects. In functional imaging studies, spatial normalisation is useful for determining what happens generically over individuals. A further advantage of using spatially normalised images is that activation sites can be reported according to their Euclidean co-ordinates within a standard co-ordinate system *(25)*. The most commonly adopted co-ordinate system within the brain imaging community is that described by T&T *(1)*, although new standards are also emerging, which are based on digital atlases *(2, 3, 26)*.

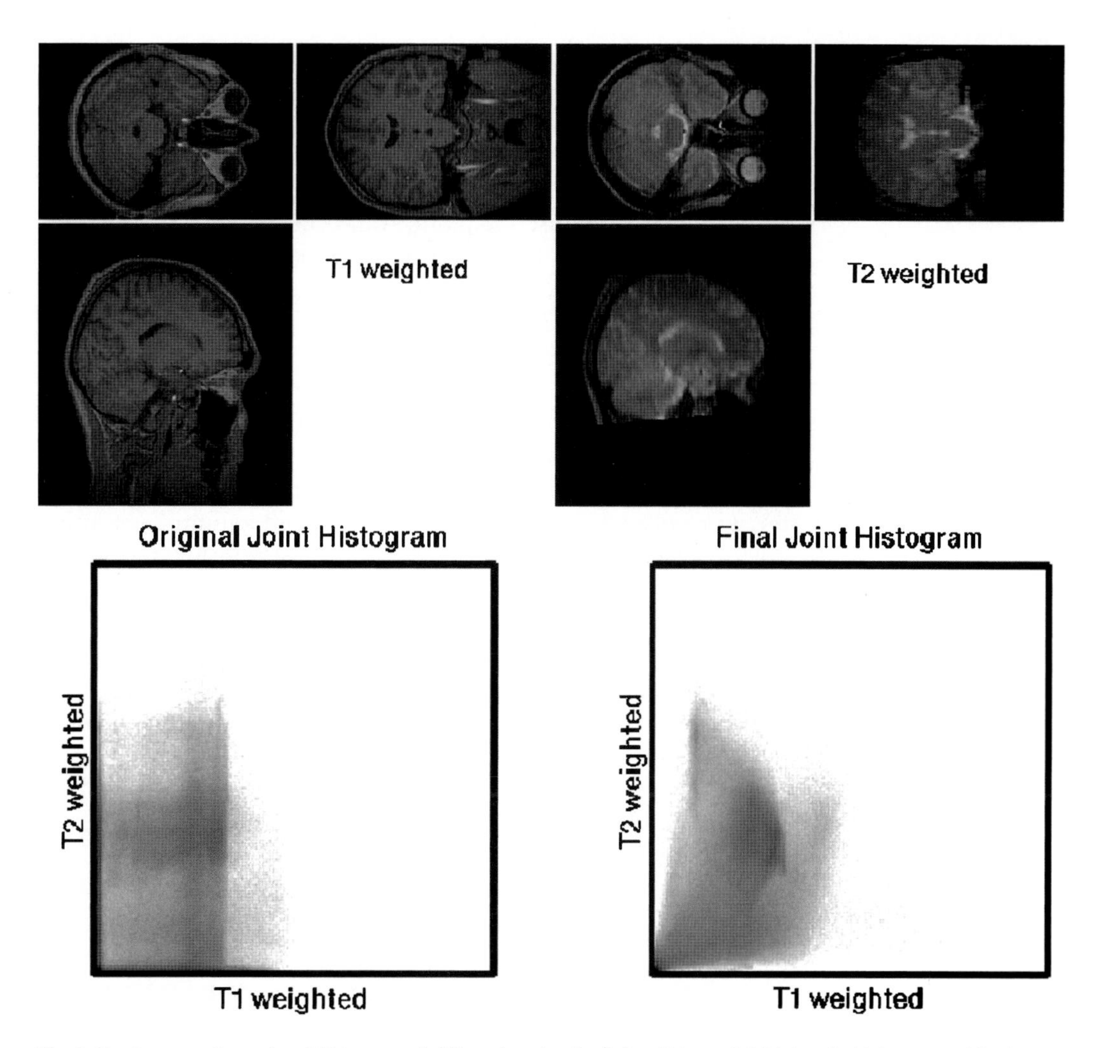

Fig. 2. The top row shows two MR images of different contrasts. Below this are joint intensity histograms of the image pair, both before and after image registration (note that the pictures show log(1 ± N), where N is the count in each histogram bin).

This section provides an overview of the ideas underlying non-linear image registration. The previous sections described rigid-body approaches for registering brain images of the same subject, where it was assumed that there are no differences among the shapes of the brains. This is often a reasonable assumption to make for intra-subject registration, but this model is not appropriate for aligning brain images of different subjects. In addition to estimating an unknown pose and position, inter-subject registration approaches also need to model the different sizes and shapes of the subjects' heads.

Methods of spatially normalising images can be divided broadly into *label based* and *intensity based*. Label-based techniques involve identifying features (labels) in a subject's image, and then bringing these into alignment with the appropriate location in some atlas. The original strategy proposed by T&T involved matching discrete points, but other forms of label could also be used such as lines or surfaces. Homologous features are often identified manually, but this process is time consuming and subjective. Another disadvantage of using points as landmarks is that there are very few readily identifiable discrete points in the brain, so the registration accuracy in regions away from those points is likely to be limited. The required transformation at the defined features is known, but the deforming behaviour in regions distant from the features can only be estimated, so it is usually forced to be as smooth as possible. There are a number of interpolation methods that ensure smooth spatial transforms, but the most commonly used approaches generally involve some form of radial basis functions, which are centred at the landmarks.

Intensity-based approaches operate by identifying a spatial transformation that optimises some voxel-similarity measure between a subject's image and a template image. The template image defines the standard space to which all the subjects' data is warped during spatial normalisation. Typically, the spatial transformation that best matches a subject's anatomical image with the template is estimated using an iterative optimisation procedure.

Image registration uses a mathematical model to explain the data. Such a model will contain a number of unknown parameters that describe how an image is deformed, or warped. The objective is usually to determine the best possible values for these parameters by optimising some objective function; in other words, to find the single most probable deformation, given the data. In such cases, the objective function can be considered as a measure of this probability. A key element of probability theory is Bayes' theorem:

$$p(\theta \mid D) = p(D \mid \theta)\frac{p(\theta)}{p(D)}$$

This *posterior probability* of the parameters, given the image data ($p(\theta|D)$) is proportional to the probability of the image data given the parameters ($p(D|\theta)$ – the *likelihood*), times the *prior probability* of the parameters ($p(\theta)$). The probability of the data ($p(D)$) is treated as a constant because the data are fixed and known. The objective is to find the most probable parameter values, and not the actual probability density, so this factor can be ignored. The most probable set of values for the parameters is known as the *maximum a posteriori* estimate. In practice, the objective function is normally the logarithm of the posterior probability (which is maximised) or the negative logarithm (which is minimised).

The objective function can therefore be considered as the sum of two terms: a likelihood term and a prior term:

$$-\log p(\theta, D) = -\log p(\mathrm{D} \mid \theta) - \log p(\theta)$$

The likelihood term is a measure of the probability of observing an image, given some set of model parameters. A simple example would be where an image is modelled as a warped version of a template image, but with Gaussian random noise added. Such a model reduces to minimising the sum of squared differences between the image and warped template. It is possible that parts of the image correspond to a region that falls outside the field of view of the template, so it is usual to simply use the mean-squared difference in the overlapping region.

The prior term reflects the prior probability of a deformation occurring – effectively biasing the deformations to be realistic. If one considers a model whereby each voxel can move independently in three dimensions, then there would be three times as many parameters to estimate as there are observations. This would simply not be achievable without *regularising* the parameter estimation by modelling a prior probability.

Registration is usually considered as an optimisation procedure which involves finding the model parameters that maximise or minimise the objective function. If the registration is based on matching landmarks together, then it is often possible to register the images in a single step because the problem can be solved by a single matrix inversion. In contrast, if image registration is based on matching intensities, then some form of iterative scheme is required. These procedures are usually very susceptible to poor starting estimates, so a number of hybrid approaches have emerged that combine intensity-based methods with feature matching (typically sulci). Registration methods usually attempt to find the single most probable realisation of all possible transformations. Robust methods that almost always find the global optimum would take an extremely long time to run with a model that uses millions of parameters, so these methods are simply not feasible for problems of this scale. However, if sulci and gyri can be labelled easily from the brain images, then these features can be used to bias the registration *(27)*, therefore increasing the likelihood of obtaining a more globally optimal solution.

In practice, the parameters describing the spatial transformations, which warp a subject's images to a standard co-ordinate system, are usually estimated by matching an anatomical image with a template. Providing the fMRI data are in accurate alignment with the anatomical image, then the spatial transformation that would warp them to the standard co-ordinate system is also known. Spatially normalised fMRI can simply be created using the same set of estimated parameters.

5.1. Matching Criteria

The matching criterion is often based on minimising the mean-squared differences or maximising the correlation between the image and the template. For this criterion to be successful, it requires the individual's image to have the visual appearance of a warped version of the template. In other words, there must be correspondence in the grey levels of the different tissue types between the images. The mean-squared difference objective function makes a number of assumptions about the data. If the data do not meet these assumptions, then the objective function may not accurately reflect the goodness of fit, and the estimated deformations will be sub-optimal. Under some circumstances, it may be better to weight different regions to a greater or lesser extent. For example, when matching a template to a brain image containing a lesion, the mean-squared difference around the lesion should contribute little or nothing to the objective function *(28)*. This is achieved by assigned lower weights for the matching criterion in these regions, so that they have much less influence on the final solution.

In addition to modelling non-linear deformations of the template, there may also be extra parameters within the model that describe intensity variability. A very simple example would be the inclusion of an additional intensity scaling parameter, but the models can be much more complicated. There are many possible objective functions, each making a different assumption about the data and requiring different parameterisations of the template intensity distribution. There is no single universally best criterion to use for all data. For example, matching can be based on feature vectors derived from the images *(29)*, or can rely on some information theoretic model *(30)*.

Often, it is necessary to process the anatomical images prior to any attempt to align them with a template image. This may involve stripping out non-brain tissue from the image, which can improve the accuracy with which the brains themselves are registered. Non-linear registration methods that use only a small number of parameters to model the deformations are likely to be heavily influenced by any variability of tissue outside the skull. The alignment of the subject's brains is compromised by also having to model the enormous variability among the faces and necks of different subjects. Because the interesting signal predominantly arises in grey matter, another strategy for increasing spatial normalisation accuracy is to simply spatially normalise the images by aligning grey matter with a grey matter template image. There are a number of readily available tissue segmentation algorithms that can be used for identifying grey matter in brain MRI.

Another strategy that can be of great benefit for increasing registration accuracy is the correction of intensity non-uniformity artefact (also known as "bias" or "inhomogeneity"), which would otherwise prevent accurate alignment. Such correction algorithms may be standalone *(31)*, or they may be incorporated within tissue segmentation procedures *(32, 33)*. Some non-linear registration procedures explicitly incorporate bias correction *(34)*, but recent developments combine bias correction, non-linear warping and tissue segmentation into the same model *(35–38)*. Within such unified generative models, the registration and bias correction inform the tissue segmentation, and the tissue segmentation informs the registration and bias correction.

Currently, most spatial normalisation algorithms use only a single image from each subject, which is typically a T1-weighted MR image. Such images only really delineate different tissue types. Further information that may help the registration could be obtained from other data such as diffusion-weighted images *(39)*. These provide anatomical information more directly related to connectivity and implicitly function, possibly leading to improved registration of functionally specialised areas *(40)*. Matching diffusion images of a pair of subjects together is likely to give different deformation estimates than that would be obtained through matching T1-weighted images of the same subjects. The only way to achieve an internally consistent match is through performing the registrations simultaneously, within the same model. Similarly, the patterns of activation across subjects could, in principle, be used to drive the registration – although a naïve implementation of such an approach may cause problems for interpreting group results.

The choice of template used for spatial normalisation is important. It is sometimes tempting to base the template on the brain of a single individual, but such a procedure would produce different results depending upon the choice of individual. One could consider an optimal template being some form of average *(41–44)*. Registering such a template with a brain image generally requires smaller (and therefore less error prone) deformations than would be necessary for registering with an unusually shaped template. Such averages generally lack some of the detail present in the individual subjects. Structures that are more difficult to match are generally slightly blurred in the average, whereas the structures that can be more reliably matched are sharper. Such an average generated from a large population of subjects would be ideal for use as a general purpose template. Another reason for using a template that better represents the study population is that it does not bias the results more towards some brain regions than others. During spatial normalisation of a brain image, some regions need to be expanded and other to be contracted in order to match the template. If some structure is especially small in the template brain, then this region will be contracted in the brains in the study, leading to a systematic reduction in the amount of BOLD signal being detected from this brain region.

5.2. Deformation Models

At its simplest, non-linear image registration involves estimating a smooth, continuous mapping between the points in one image and those in another. This mapping allows one image to be re-sampled so that it is warped (deformed) to match another (**Fig. 3**). There are many ways of modelling such mappings, but these fit into two broad categories *(45)*.

- The *small-deformation* framework does not necessarily preserve topology[3] – although if the deformations are relatively small, then there may still be a one-to-one mapping between the images. This framework usually models deformations by a smooth displacement field.

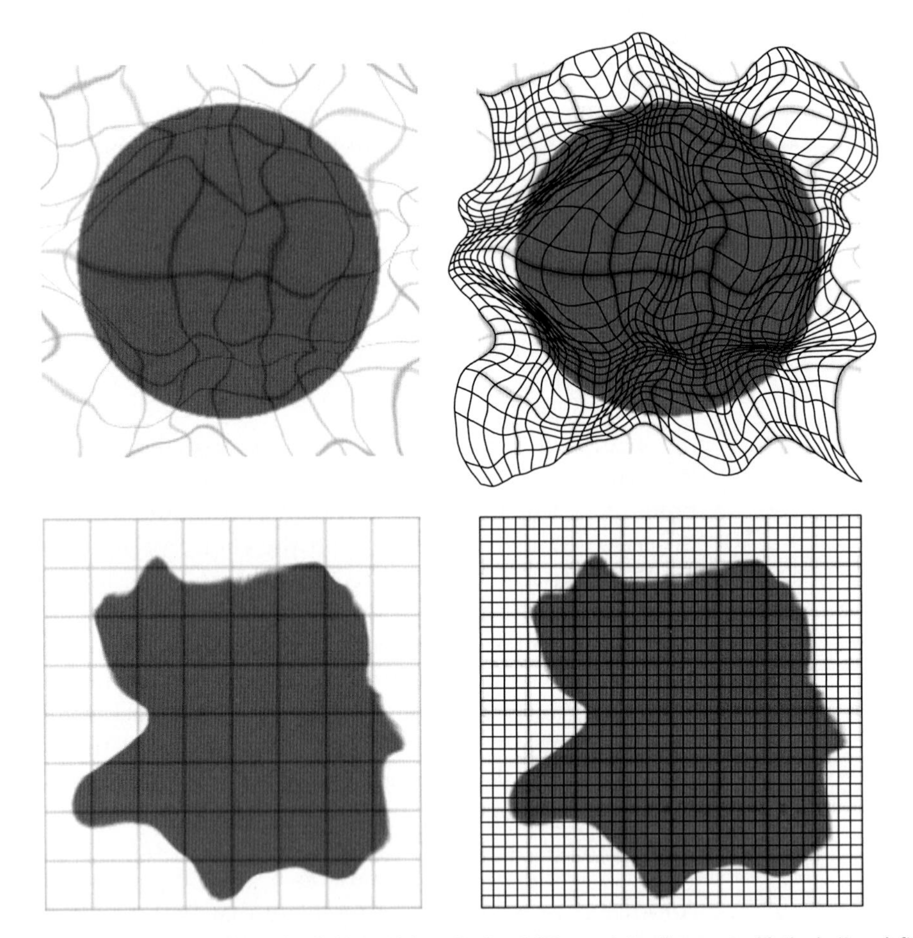

Fig. 3. This figure illustrates a deformation field that brings the top-left image into alignment with the bottom-left image. At the top-right is the image with the deformation field overlaid, and at the bottom-right is this image after it has been warped. Note that in this example, the deformation wraps around at the boundaries.

[3] The word "topology" is used in the same sense as in "Topological Properties of Smooth Anatomical Maps" *(46)*. If spatial transformations are not one-to-one and continuous, then the topological properties of different structures can change.

- The *large-deformation* framework generates deformations (*diffeomorphisms*) that have a number of elegant mathematical properties, such as enforcing the preservation of topology *(47)*. Within this framework, the models are parameterised in terms of smooth velocity fields.

Images can be treated as continuous functions of space. Reading off the value at some arbitrary point involves interpolating between the original voxels. For many interpolation methods, the functions are parameterised by linear combinations of basis functions, such as B-spline bases, centred at each original voxel. Similarly, the deformations themselves can also be parameterised by linear combinations of smooth, continuous basis functions.

A potentially enormous number of parameters are required to describe the non-linear transformations that warp two images together (i.e., the problem can be very high-dimensional). However, much of the spatial variability can be captured using just a few parameters. Sometimes only an affine transformation is used to approximately register images of different subjects. This accounts for differences in position, orientation, and overall brain dimensions, and often provides a good starting point for higher-dimensional registration models.

Low spatial frequency global variability of head shape can be accommodated by describing deformations by a linear combination of a few low-frequency basis functions (**Fig. 4**). One widely used basis function registration method is part of the automated image registration package *(48, 49)*, which uses polynomials (**Fig. 5**) to model shape variability. These functions are a simple extension to those used for parameterising affine transformations. Other models parameterise a displacement field, which is added to an identity transform. In such parameterisations, the inverse transformation is sometimes approximated by subtracting the displacement. It is worth noting that this is only a very approximate inverse, which fails badly for larger deformations (**Fig. 6**). Families of basis functions for such models include Fourier bases *(50)*, sine and cosine transform basis functions used by statistical parametric mapping *(51, 52)* (**Fig. 7**). These models usually use in the order of about 1,000 parameters. The small number of parameters will not allow every feature to be matched exactly, but it will permit the global head shape to be modelled rapidly.

Radial basis functions are another family of parameterisations, which are often used in conjunction with an affine transformation. Each radial basis function is centred at some point and the amplitude is then a function of the distance from that point. Thin-plate splines are one of the most widely used radial basis functions for image warping and are especially suited to manual landmark matching *(53, 54)*. The landmarks may be known, but interpolation is needed in order to define the mapping between these known points. By modelling it with thin-plate splines, the

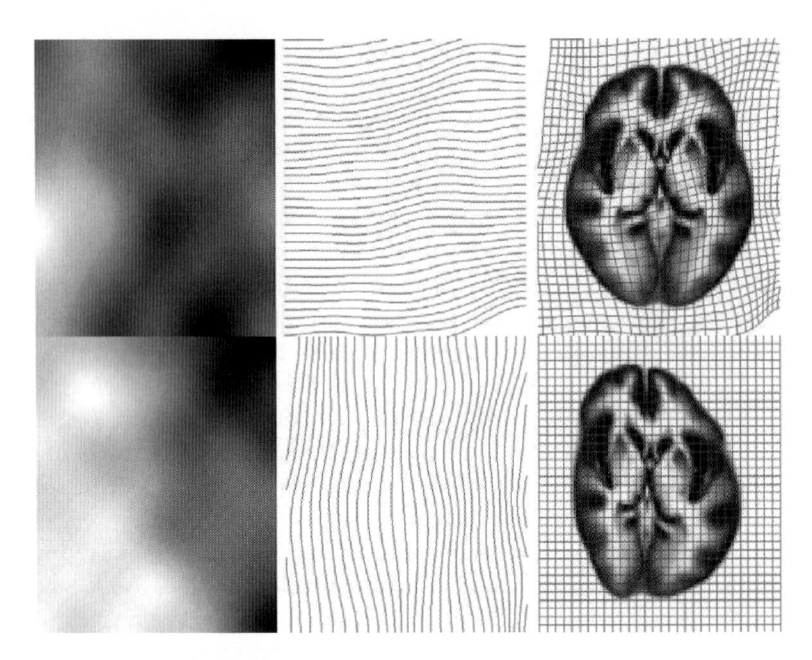

Fig. 4. This figure illustrates the effect of different types of regularisation. The top row on the left shows simulated 2D images of a circle and a square. Below these is the circle after it has been warped to match the square, using both membrane and bending energy priors. These warped images are almost visually indistinguishable, but the resulting deformation fields using these different priors are quite different. These are shown on the right, with the deformation generated with the membrane energy prior shown above the deformation that used the bending energy prior.

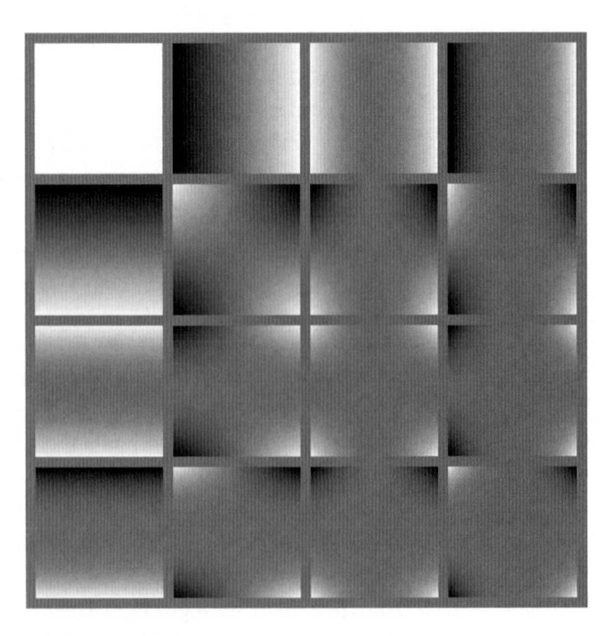

Fig. 5. Polynomial basis functions.

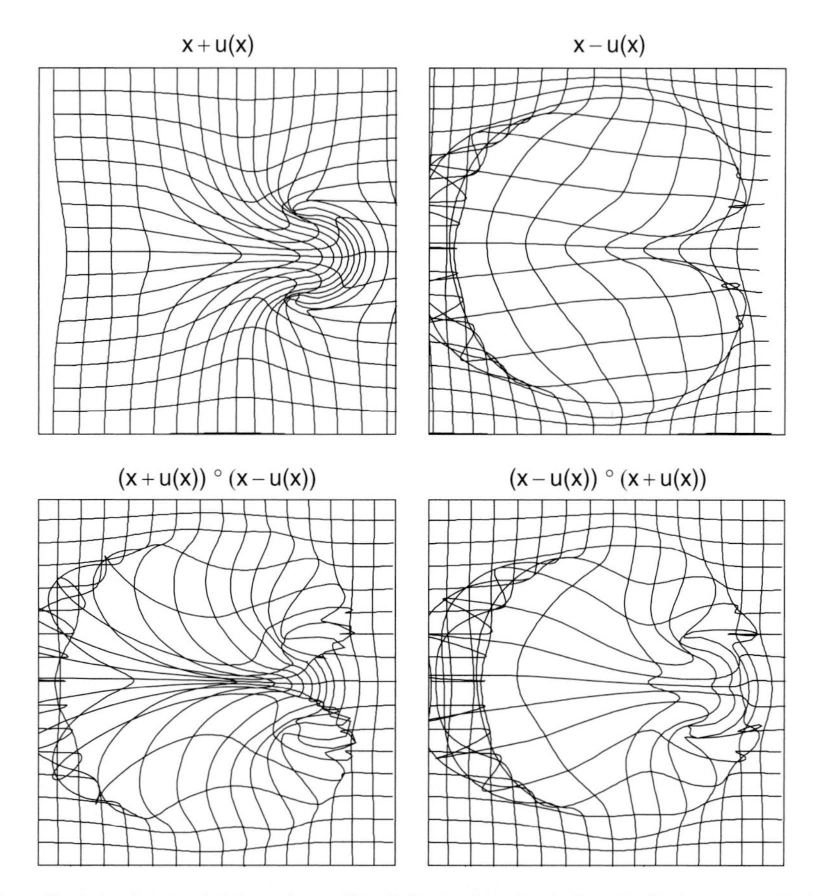

Fig. 6. This figure illustrates the small-deformation setting. At the top is an illustration of a displacement added to an identity transform. Below this is a forward and inverse deformation generated within the small-deformation setting. Note that the one-to-one mapping is lost because the displacements are too large and the mappings are not accurate inverses of each other.

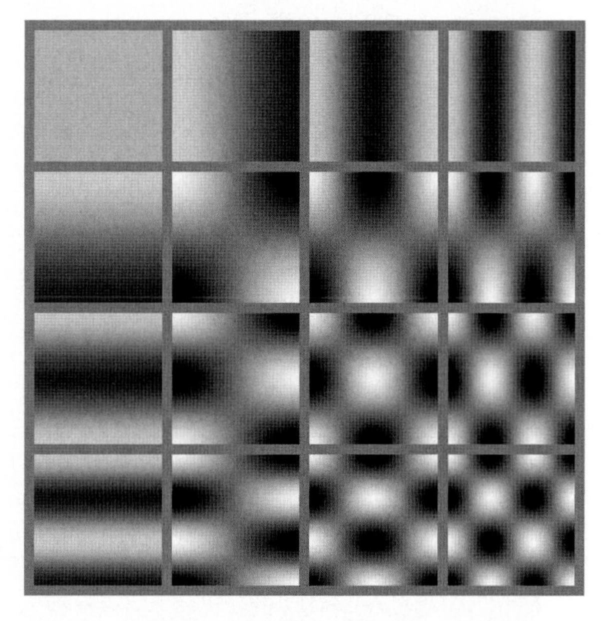

Fig. 7. This figure illustrates 2D displacements generated from two scalar fields. On the left are the displacements represented as images, which are modelled by linear combinations of basis functions. In the centre column is a different representation. The resulting vector field is overlaid on the top-right image, in order to deform the image as shown at the bottom-right.

mapping function has the smallest bending energy. Other choices of basis function reduce other energy measures, and these functions relate to the convolution filters that are sometimes used for fast image matching *(55–57)*.

B-spline bases are also used for parameterising displacements *(6, 58)* (**Fig. 8**). They are related to the radial basis functions in that they are centred at discrete points, but the amplitude is the product of functions of distance in the three orthogonal directions (i.e., they are *separable*). The separability and local support of these basis functions confers certain advantages in terms of being able to rapidly generate displacement fields through a convolution-like procedure. Very detailed displacement fields can be generated by modelling an individual displacement at each voxel *(59)*. This may not appear to be a basis function approach, but the assumptions within such models are often that the fields are tri-linearly interpolated. This is the same as a first degree B-spline basis function model.

Regularisation is generally based on some measure of deformation smoothness. Smoother deformations are deemed to be more probable, a priori, than deformations containing a great deal of detailed information. The regularisation term (prior term) of the objective function is often thought of as an "*energy density*."

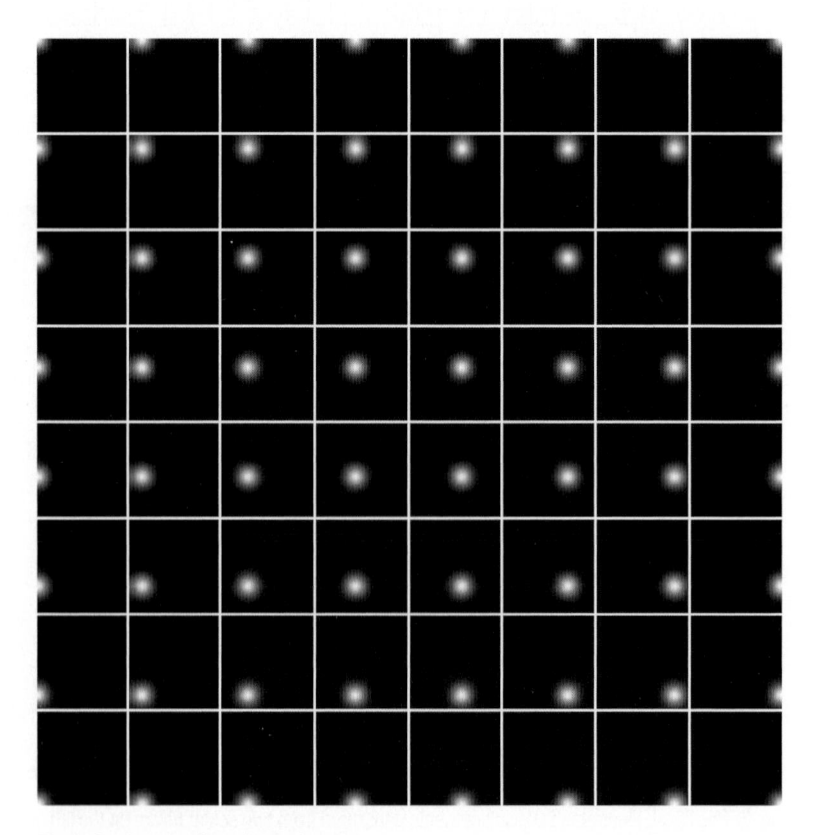

Fig. 8. Cosine transform basis functions.

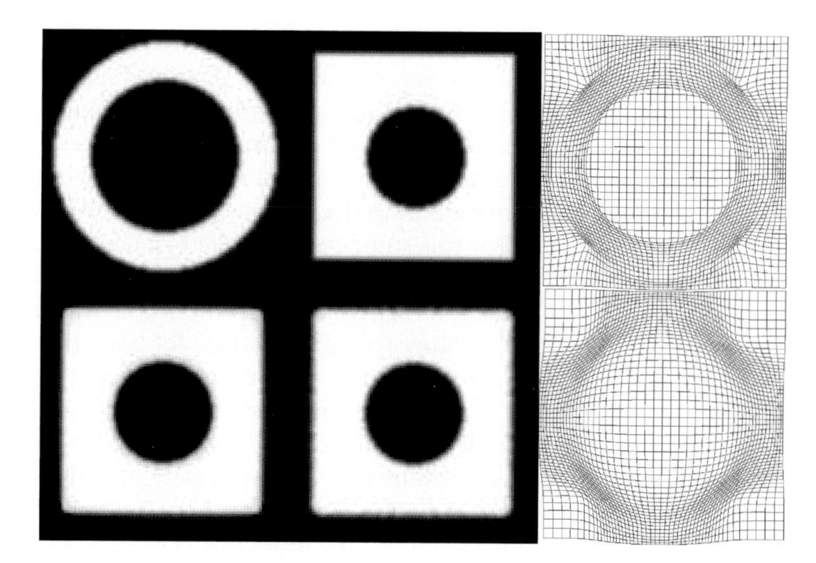

Fig. 9. B-spline basis functions.

Commonly used forms for this are the *membrane energy*, *bending energy*, or *linear-elastic energy*. The form of the prior used by the registration will influence the estimated deformations. This is illustrated in **Fig. 9**.

The key concept for the large-deformation or *diffeomorphic*[4] setting is that the deformations are generated by the composition of a number of small deformations (i.e., warped warps). For deformations, the composition operation is achieved by re-sampling one deformation field by another. Providing the original deformations are small enough, they are likely to be one-to-one. A composition of a pair of one-to-one mappings will produce a new mapping that is also one-to-one. Multiple nesting can be achieved, so that large one-to-one deformations can be obtained from the composition of many very small deformations. From a mathematical perspective, diffeomorphic deformations are the result of integrating differential equations over a unit of time, in which the deformations are a function of smooth continuous velocity fields. The composition of a series of small deformations can be viewed as an *Euler integration* of these differential equations.

Early diffeomorphic registration approaches were based on the *greedy* "*viscous fluid*" registration method of Christensen and Miller *(51, 60)*. In these models, finite difference methods are used to solve the partial differential equations that model one image as it "flows" to match the shape of the other. The advantage of these methods is that they can account for large

[4]A diffeomorphism is a globally one-to-one (bijective) smooth and continuous mapping with derivatives that are invertible (i.e., non-zero Jacobian determinant).

displacements while ensuring that the topology of the warped image is preserved. The disadvantage is that they are not formulated to find the smoothest deformation. At each time step, one image is warped so that it slightly better matches the other. This is repeated over a number of time steps, building up the final deformation from the composition of a number of small deformations. For this strategy, the trajectory taken by the deforming image is not the shortest[5] because it does not consider the inverse deformation that would warp the second image to match the first.

More recent algorithms for large-deformation registration do aim to find the smoothest solution. For example, the *LDDMM* (large-deformation diffeomorphic metric mapping) algorithm *(57)* does not fix the deformation parameters once they have been estimated. It continues to update them such that the objective function is properly optimised. Such approaches essentially parameterise the model by velocities, and compute the deformation as the medium warps over unit time *(41, 45, 47, 61)*.

Diffeomorphic warps can be inverted (**Fig. 10**). If each constituent small deformation is generated by adding a small displacement (velocity) to an identity transform, then their approximate inverses can be derived by subtracting the same displacement. A composition of these small deformation inverses will then produce the inverse of the large deformation.

Optimisation problems for complex non-linear models, such as those used for image registration, can easily get caught in local optima; so there is no guarantee that the estimate determined by the algorithm is globally optimal. If the starting estimates are sufficiently close to the global optimum, then a local optimisation algorithm is more likely to find the globally optimal solution. Therefore, the choice of starting parameters can influence the accuracy of the final registration result. One method of increasing the likelihood of achieving a good solution is to gradually reduce the amount of regularisation (cf, simulated annealing). Registration is first performed using heavy regularisation. Once this solution is found, then the procedure is repeated using less regularisation and so on. This has the effect of making the registration estimate the more global deformations before estimating more detailed ones. Most shape variability is of low frequency, so an algorithm can get reasonably close to a good solution using a relatively large amount of regularisation. This also reduces the number of local minima for the early iterations. The images could also be smoother for the earlier iterations in order to reduce the amount of confounding information and the number of local optima. A review of such approaches can be found in *(62)*.

---

[5]Here, the length is derived by integrating a smoothness measure for each of the component small deformations.

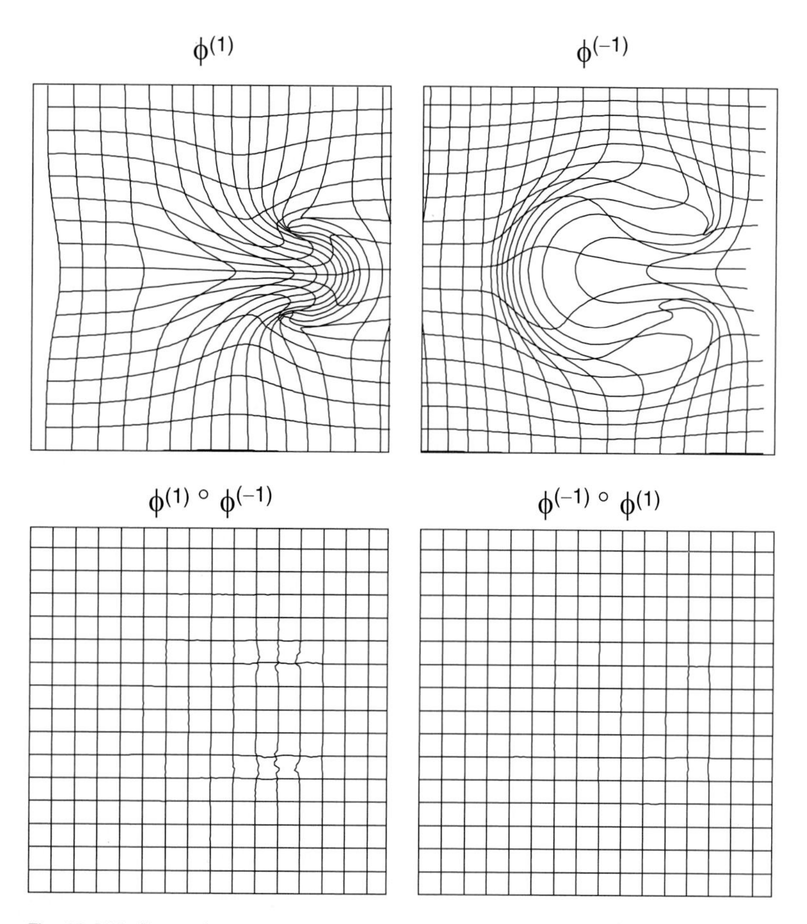

Fig. 10. This figure shows an example of a forward and inverse deformation in a large-deformation setting. Unlike those shown in **Fig. 6**, these are one-to-one mappings, and the transforms are actually inverses of each other.

6. Smoothing

Usually, the final step of the pre-processing pipeline is to smooth the images, which involves convolving the data with a 3D Gaussian kernel (**Figs. 11** and **12**). The amount of smoothing is defined by the *full width at half maximum* (*FWHM*) of the smoothing kernel. A broader FWHM produces smoother results, and the choice of FWHM is determined by many factors.

More smoothing is usually used prior to statistical analyses of group studies than would be used for studies of single individuals. Inter-subject registration is generally less accurate than the rigid-body registration that is used within subject. If homologous functional regions in different subjects are not well aligned, then the activations will appear in different places in different subjects. Smoothing is used in order to increase the amount of overlap

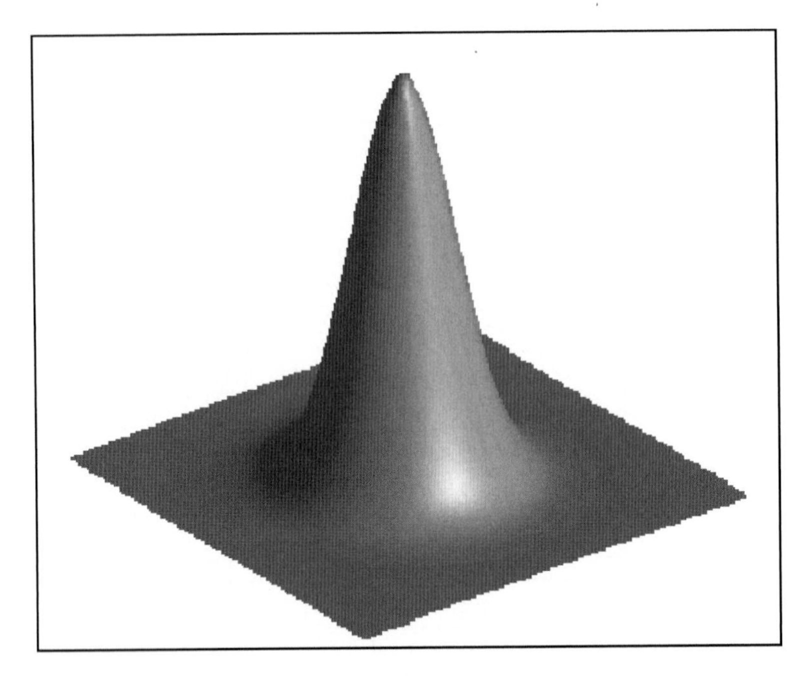

Fig. 11. A two-dimensional Gaussian function.

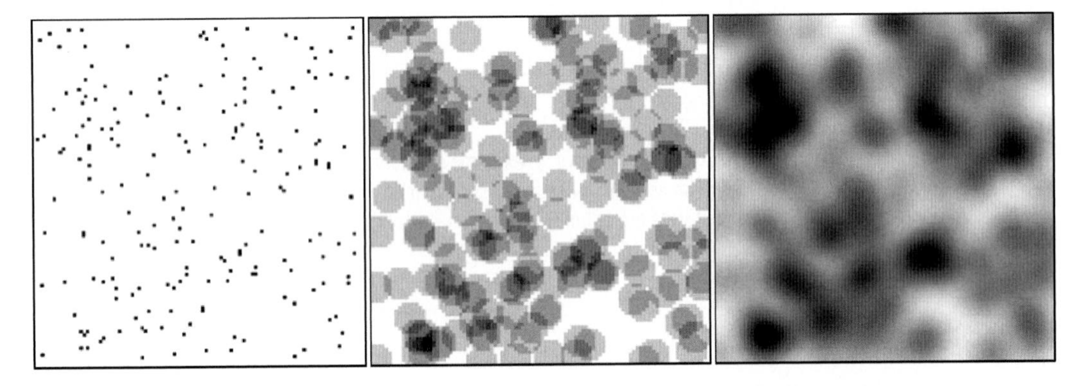

Fig. 12. This figure shows the effect of convolving the image on the left with different kernels. In the centre, the image has been convolved with a circular kernel. The result is an image in which each pixel is the average of the values from the original image within the radius of the kernel. At the right is a result from convolving with a Gaussian kernel. Pixels in this image are weighted averages, where the weights depend on the distance from the centre of the kernel.

across subjects, and so increases the significance of the results. The optimal amount of smoothing therefore depends upon the alignment accuracy.

The current norm within the field is to smooth the spatially normalised images, as opposed to smoothing the native images and then warping them. The two procedures lead to subtly different results. When an image is spatially normalised, some regions will contract, whereas others expand. It is worth considering what happens to the signal from two adjacent regions, in which one expands and the other contracts. A voxel in the smoothed image will contain proportionally more signal from the expanded region than the contracted one, which is not a desirable result.

Less smoothing would typically be used for single-subject analyses. There are various reasons for smoothing such data, but the main one relates to the spatial frequencies of the interesting signal, compared with the noise. If the noise contains proportionally more high frequency than the signal, then it makes sense to remove some of the high frequencies by smoothing. Another reason for smoothing is that it reduces the effective number of independent statistical tests that are performed. This can lead to greater sensitivity in results that are corrected for multiple comparisons. The disadvantage would be that localisation of activations is less precise.

Some of the more recent approaches for the analysis of fMRI data involve spatiotemporal models of activation. Such analyses do not require the data to be smoothed, as the generative models themselves deal with this issue.

7. Summary and Conclusions

The current fMRI data analysis paradigm involves applying a series of tools to the data. One procedure is applied to the images to produce some output. Then another procedure is applied to the output of that, and so on. The end result is a version of the data that has been massaged into a form suitable for applying simple statistical tests to. A number of centres have developed pipeline environments *(63–65)* to facilitate such processing streams.

An alternative and more principled approach is to consider a full model of how the data could have arisen. Such a *generative model* would involve components for modelling the physics of the scanner, the motion of the subjects, the brain-shape variability among the population, and models for how the experiment elicits changes in the data. It would then be used to model the raw data, in order to make the kinds of inferences in which neuroscientists are interested. A full model for everything is a long way off, but scientists are making progress in terms of simulating fMRI data in individual subjects *(66)*. Once a simulation model can be made, then it is simply a matter of inverting it, to make the necessary inferences. Such models usually include a number of parameters that influence the data, but are of no interest. Accurate model inversion would require the effects of these uninteresting variables to be "integrated out," which is not a straightforward procedure and is an area that occupies much of current methodological research. Choosing an optimal model for a data set would essentially be a form of *model selection*, and could be done empirically by determining which model has the greatest evidence.

Many of the issues that concern users are related to the unmodelled interactions that arise through the sequential application of tools to the data. For example, when motion correction is applied to an fMRI time series, the interesting signal elicited by the experiment will influence how the algorithm estimates subject motion. Similarly, interactions between image artefacts and subject movement can also lead to problems, which can only be reduced by including artefact correction within the motion correction. There are many other examples of where combining procedures could produce similar benefits.

References

1. Talairach J, Tournoux P. Coplanar stereotaxic atlas of the human brain. New York: Thieme Medical, 1988.
2. Evans C, Collins DL, Milner B. An MRI-based stereotactic atlas from 250 young normal subjects. Society of Neuroscience Abstracts 1992;18:408.
3. Evans C, Collins DL, Mills SR, Brown ED, Kelly RL, Peters TM. 3D statistical neuroanatomical models from 305 MRI volumes. Proc IEEE Nuclear Science Symposium and Medical Imaging Conference, 1993:1813–1817.
4. Jezzard P, Clare S. Sources of distortion in functional MRI data. Hum Brain Mapp 1999;8:80–85.
5. Jenkinson M, Wilson J, Jezzard P. A perturbation method for magnetic field calculations of non-conductive objects. Magn Reson Med 2004;52:471–477.
6. Studholme C, Constable RT, Duncan JS. Accurate alignment of functional EPI data to anatomical MRI using a physics-based distortion model. IEEE Trans Med Imaging 2000;19:1115–1127.
7. Kybic J, Thévenaz P, Nirkko A, Unser M. Unwarping of unidirectionally distorted EPI images. IEEE Trans Med Imaging 2000; 19:80–93.
8. Hajnal JV, Mayers R, Oatridge A, Schwieso JE, Young JR, Bydder GM. Artifacts due to stimulus correlated motion in functional imaging of the brain. Magn Reson Med 1994;31:289–291.
9. Thévenaz P, Blu T, Unser M. Interpolation revisited. IEEE Trans Med Imaging 2000;19:739–758.
10. Eddy WF, Fitzgerald M, Noll DC. Improved image registration by using Fourier interpolation. Magn Reson Med 1996;36:923–931.
11. Cox RW, Jesmanowicz A. Real-time 3D image registration for functional MRI. Magn Reson Med 1999;42:1014–1018.
12. Noll DC, Boada FE, Eddy WF. A spectral approach to analyzing slice selection in planar imaging: optimization for through-plane interpolation. Magn Reson Med 1997;38: 151–160.
13. Andersson JLR, Hutton C, Ashburner J, Turner R, Friston KJ. Modeling geometric deformations in EPI time series. NeuroImage 2001;13:903–919.
14. Bannister PR, Brady JM, Jenkinson M. Integrating temporal information with a non-rigid method of motion correction for functional magnetic resonance images. Image and Vision Computing 2007;25:311–320.
15. Friston KJ, Williams S, Howard R, Frackowiak RSJ, Turner R. Movement-related effects in fMRI time-series. Magn Reson Med 1996;35:346–355.
16. Freire L, Mangin JF. Motion correction algorithms of the brain mapping community create spurious functional activations. In: Insana MF, Leahy RM, eds. Proc Information Processing in Medical Imaging (IPMI). Lecture Notes in Computer Science, vol 2082. Berlin-Heidelberg: Springer-Verlag, 2001:246–258.
17. Woods RP, Mazziotta JC, Cherry SR. MRI-PET registration with automated algorithm. J Comput Assist Tomogr 1993;17:536–546.
18. Holden M, Hill DLG, Denton ERE, Jarosz JM, Cox TCS, Rohlfing T, Goodey J, Hawkes DJ. Voxel similarity measures for 3-D serial MR brain image registration. IEEE Trans Med Imaging 2000;19:94–102.
19. Collignon A, Maes F, Delaere D, Vandermeulen D, Suetens P, Marchal G. Automated

multi-modality image registration based on information theory. In: Bizais Y, Barillot C, Di Paola R, eds. Proc Information Processing in Medical Imaging (IPMI). Dordrecht, The Netherlands: Kluwer Academic Publishers, 1995:263–274.

20. Wells WM III, Viola P, Atsumi H, Nakajima S, Kikinis S. Multi-modal volume registration by maximisation of mutual information. Med Image Anals 1996;1:35–51.
21. Maes F, Collignon A, Vandermeulen D, Marchal G, Seutens P. Multimodality image registration by maximisation of mutual information. IEEE Trans Image Process 1997;16:187–197.
22. Studholme C, Hill DLG, Hawkes DJ. An overlap invariant entropy measure of 3D medical image alignment. Pattern Recognition 1999;32:71–86.
23. Pluim JPW, Maintz JBA, Viergever MA. Mutual-information-based registration of medical images: a survey. IEEE Trans Med Imaging 2003;22:986–1004.
24. West J, Fitzpatrick JM, Wang MY, Dawant BM, Maurer CR, Kessler RM, Maciunas RJ et al. Comparison and evaluation of retrospective intermodality brain image registration techniques. J Comput Assist Tomogr 1997;21:554–566.
25. Fox PT. Spatial normalization origins: objectives, applications, and alternatives. Hum Brain Mapp 1995;3:161–164.
26. Mazziotta JC, Toga AW, Evans A, Fox P, Lancaster J. A probabilistic atlas of the human brain: theory and rationale for its development. NeuroImage 1995;2:89–101.
27. Joshi SC, Miller MI. Landmark matching via large deformation diffeomorphisms. IEEE Trans Med Imaging 2000;9:1357–1370.
28. Brett M, Leff AP, Rorden C, Ashburner J. Spatial normalization of brain images with focal lesions using cost function masking. NeuroImage 2001;14:486–500.
29. Shen D, Davatzikos C. HAMMER: hierarchical attribute matching mechanism for elastic registration. IEEE Trans Image Process 2002;21:1421–1439.
30. D'Agostino E, Maes F, Vandermeulen D, Suetens P. Non-rigid atlas-to-image registration by minimization of class-conditional image entropy. In: Barillot C, Haynor DR, Hellier P, eds. Proc Medical Image Computing and Computer-Assisted Intervention (MICCAI). Lecture Notes in Computer Science, vol 3216. Berlin-Heidelberg, Springer-Verlag, 2004:745–753.
31. Sled JG, Zijdenbos AP, Evans AC. A non-parametric method for automatic correction of intensity non-uniformity in MRI data. IEEE Trans Med Imaging 1998;17:87–97.
32. van Leemput K, Maes F, Vandermeulen D, Suetens P. Automated model-based bias field correction of MR images of the brain. IEEE Trans Med Imaging 1999;18:885–896.
33. Wells WM III, Grimson WEL, Kikinis R, Jolesz FA. Adaptive segmentation of MRI data. IEEE Trans Med Imaging 1996;15: 429–442.
34. Studholme C, Cardenas V, Song E, Ezekiel F, Maudsley A, Weiner M. Accurate template-based correction of brain MRI intensity distortion with application to dementia and aging. IEEE Trans Med Imaging 2004:23:99–110.
35. Fischl B, Salat DH, Busa E, Albert M, Dieterich M, Haselgrove C, van der Kouwe A, Killiany R, Kennedy D, Klaveness S, Montillo A, Makris N, Rosen B, Dale AM. Whole brain segmentation: automated labeling of neuroanatomical structures in the human brain. Neuron 2002;33:341–355.
36. Fischl B, Salat DH, van der Kouwe AJW, Makris N, Ségonne F, Quinn BT, Dale AM. Sequence-independent segmentation of magnetic resonance images. NeuroImage 2004; 23:S69–S84.
37. Ashburner J, Friston KJ. Unified segmentation. NeuroImage 2005;26:839–851.
38. D'Agostino E, Maes F, Vandermeulen D, Suetens P. A unified framework for atlas based brain image segmentation and registration. In:Pluim JPW, Likar B, Gerritsen FA, eds. Proc Third International Workshop on Biomedical Image Registration (WBIR). Lecture Notes in Computer Science, vol 4057. Berlin-Heidelberg, Springer-Verlag, 2006:136–143.
39. Zhang H, Yushkevich PA, Gee JC. Registration of diffusion tensor images. Proc IEEE Computer Society Conference on Computer Vision and Pattern Recognition (CVPR'04) 2004;1:842–847.
40. Behrens TEJ, Jenkinson M, Robson MD, Smith SM, Johansen-Berg H. A consistent relationship between local white matter architecture and functional specialisation in medial frontal cortex. NeuroImage 2006;30: 220–227.
41. Joshi S, Davis B, Jomier M, Gerig G. Unbiased diffeomorphic atlas construction for computational anatomy. NeuroImage 2004;23: S151–S160.
42. Avants B, Gee JC. Geodesic estimation for large deformation anatomical shape averaging and interpolation. NeuroImage 2004;23:S139–S150.
43. Davis B, Lorenzen P, Joshi S. Large deformation minimum mean squared error template estimation for computational anatomy. Proc IEEE International Symposium on Biomedical Imaging (ISBI) 2004:173–176.

44. Lorenzen P, Davis B, Gerig G, Bullitt E, Joshi S. Multi-class posterior atlas formation via unbiased kullback-leibler template estimation. In: Barillot C, Haynor DR, Hellier P, eds. Proc Medical Image Computing and Computer-Assisted Intervention (MICCAI). Lecture Notes in Computer Science, vol 3216. Berlin-Heidelberg, Springer-Verlag, 2004: 95–102.
45. Miller MI. Computational anatomy: shape, growth, and atrophy comparison via diffeomorphisms. NeuroImage 2004;23:S19–S33.
46. Christensen GE, Rabbitt RD, Miller MI, Joshi SC, Grenander U, Coogan TA, Van Essen DC. Topological properties of smooth anatomic maps. In: Bizais Y, Barillot C, Di Paola R, eds. Proc Information Processing in Medical Imaging (IPMI). Edited by. Dordrecht, The Netherlands: Kluwer Academic Publishers, 1995:101–112.
47. Miller MI, Younes L. Group actions, homeomorphisms, and matching: a general framework. International Journal of Computer Vision 2001;41:61–84.
48. Woods RP, Grafton ST, Holmes CJ, Cherry SR, Mazziotta JC. Automated image registration: I. General methods and intrasubject, intramodality validation. J Comput Assist Tomogr 1998;22:139–152.
49. Woods RP, Grafton ST, Watson JDG, Sicotte NL, Mazziotta JC. Automated image registration: II. Intersubject validation of linear and nonlinear models. J Comput Assist Tomogr 1998;22:153–165.
50. Christensen GE. Consistent linear elastic transformations for image matching. In: Kuba A, Sámal M, Todd-Pokropek A, eds. Proc Information Processing in Medical Imaging (IPMI). Lecture Notes in Computer Science, vol 1613. Berlin-Heidelberg: Springer-Verlag, 1999:224–237.
51. Christensen GE. Deformable shape models for anatomy. Doctoral Thesis. Washington University, Sever Institute of Technology, 1994.
52. Ashburner J, Friston KJ. Nonlinear spatial normalization using basis functions. Hum Brain Mapp 1999;7:254–266.
53. Bookstein FL. Principal warps: thin-plate splines and the decomposition of deformations. IEEE Transactions on Pattern Analysis and Machine Intelligence 1989;11: 567–585.
54. Bookstein FL. Quadratic Variation of Deformations. In: Duncan J, Gindi G, eds. Proc Information Processing in Medical Imaging (IPMI). Lecture Notes in Computer Science, vol 1230. Berlin-Heidelberg: Springer-Verlag, 1997:15–28.
55. Bro-Nielsen M, Gramkow C. Fast fluid registration of medical images. In: Höhne KH, Kikinis R, eds. Proc Visualization in Biomedical Computing (VBC). Lecture Notes in Computer Science, vol 1131. Berlin-Heidelberg: Springer-Verlag, 1996: 267–276.
56. Thirion JP. Fast non-rigid matching of 3D medical images. Technical report no 2547. Institut National de Recherche en Informatique et en Automatique, 1995.
57. Beg MF, Miller MI, Trouvé A, Younes L. Computing large deformation metric mappings via geodesic flows of diffeomorphisms. International Journal of Computer Vision 2005;61:139–157.
58. Thévenaz P, Unser M. Optimization of mutual information for multiresolution image registration. IEEE Trans Image Process 2000;9:2083–2099.
59. Rueckert D, Sonoda LI, Hayes C, Hill DLG, Leachand MO, Hawkes DJ. Nonrigid registration using free-form deformations: application to breast MR images. IEEE Trans Image Process 1999;18:712–721.
60. Christensen GE, Rabbitt RD, Miller MI. Deformable templates using large deformation kinematics. IEEE Trans Image Process 1996;5:1435–1447.
61. Miller MI, Trouvé A, Younes L. Geodesic shooting for computational anatomy. J Mathematical Imaging and Vision, Norwell, MA, USA: Kluwer Academic Publishers, 2006;24:209–228.
62. Lester H, Arridge SR. A survey of hierarchical non-linear medical image registration. Pattern Recognition 1999;32:129–149.
63. Fissell K, Tseytlin E, Cunningham D, Carter CS, Schneider W, Cohen JD. Fiswidgets: a graphical computing environment for neuroimaging analysis. Neuroinformatics 2003;1: 111–125.
64. Rex DE, Maa JQ, Toga AW. The LONI pipeline processing environment. NeuroImage 2003;19:1033–1048.
65. Zijdenbos P, Forghani R, Evans AC. Automatic 'pipeline' analysis of 3-D MRI data for clinical trials: application to multiple sclerosis. IEEE Trans Med Imaging 2002;21: 1280–1291
66. Drobnjak I, Gavaghan D, Suli E, Pitt-Francis J, Jenkinson M. Development of a fMRI simulator for modelling realistic rigid-body motion artifacts. Magn Reson Med 2006;56:364–380.

Chapter 7

Statistical Analysis of fMRI Data

Mark W. Woolrich, Christian F. Beckmann, Thomas E. Nichols, and Stephen M. Smith

Summary

fMRI is a powerful tool used in the study of brain function. It can non-invasively detect signal changes in areas of the brain where neuronal activity is varying. This chapter is a comprehensive description of the various steps in the statistical analysis of fMRI data. This will cover topics such as the general linear model (including orthogonality, haemodynamic variability, noise modelling, and the use of contrasts), multi-subject statistics, and statistical thresholding (including random field theory and permutation methods).

Key words: fMRI, Analysis, Statistics, General linear model (GLM), Multi-subject statistics, Statistical thresholding

1. Introduction

fMRI is a powerful tool used in the study of brain function. It can non-invasively detect signal changes in areas of the brain where neuronal activity is varying. fMRI can therefore give high-quality visualisation of the location of activity in the brain resulting from sensory stimulation or cognitive function. It allows, for example, the study of how the healthy brain functions, how different diseases affect the brain, or how drugs can modulate activity or post-damage recovery.

After an fMRI experiment has been designed and carried out, the resulting data must be passed through various analysis steps before the experimenter can get answers to questions about experimentally related activations at the individual or multi-subject level. This chapter focuses on the statistical aspects of such analysis.

M. Filippi (ed.), *fMRI Techniques and Protocols*, Neuromethods, vol. 41
DOI 10.1007/978-1-60327-919-2_7, © Humana Press, a part of Springer Science+Business Media, LLC 2009

We need a statistical approach for two reasons. First, fMRI data is very noisy. The noise is often of the same order of magnitude as the fMRI signal changes we are trying to detect, and as such we can only approximately estimate the signal changes. Statistics are therefore needed to ask if the estimated signal changes are significant, given the quality of the approximation. Second, many fMRI studies are carried out with the intention of answering some question about a population of individuals. For example, we might want to know what the difference in neural activation is between a patient and control group. Such population differences in neural activation can only ever be approximated, because not only is the fMRI data noisy but also we only ever have a sample of subjects from the populations. This issue is somewhat exacerbated in fMRI studies as the number of subjects sampled is typically quite small! Statistics are needed to see if the approximated population differences are significant given the quality of those approximations.

Figure 1 shows an illustration of the main analysis steps carried out in a typical fMRI study. There are three main components in the process. First, the individual subjects' fMRI data must be processed. Second, the information gleaned from this about the effect sizes (the size of the fMRI signal change in response to the experimental task) for each subject is then combined in a group analysis. Finally, the group effect sizes are statistically thresholded to ask questions such as "Where is there significant activity in response to the experimental task for the population?" or "Where are there significant differences between populations (e.g., controls versus patients)?." This final thresholding is carried out on statistic images as it takes into account the spatial characteristics of the data. Note that one could also perform thresholding on the effect size statistics from a single-subject's analysis, allowing one to ask questions such as "Where is there significant activity in response to the experimental task for this subject?" The various steps in the analysis will be described in detail throughout the chapter.

2. Statistical Analysis of a Single fMRI Dataset

Later in the chapter, we will see how we go about asking statistical questions about populations of subjects. However, before this can be done, the individual subject fMRI data must be analysed. The most common way of statistically analysing fMRI data is by using a general linear model (GLM). As we shall see, this is a powerful framework that allows a wide range of different statistical questions to be asked about the data.

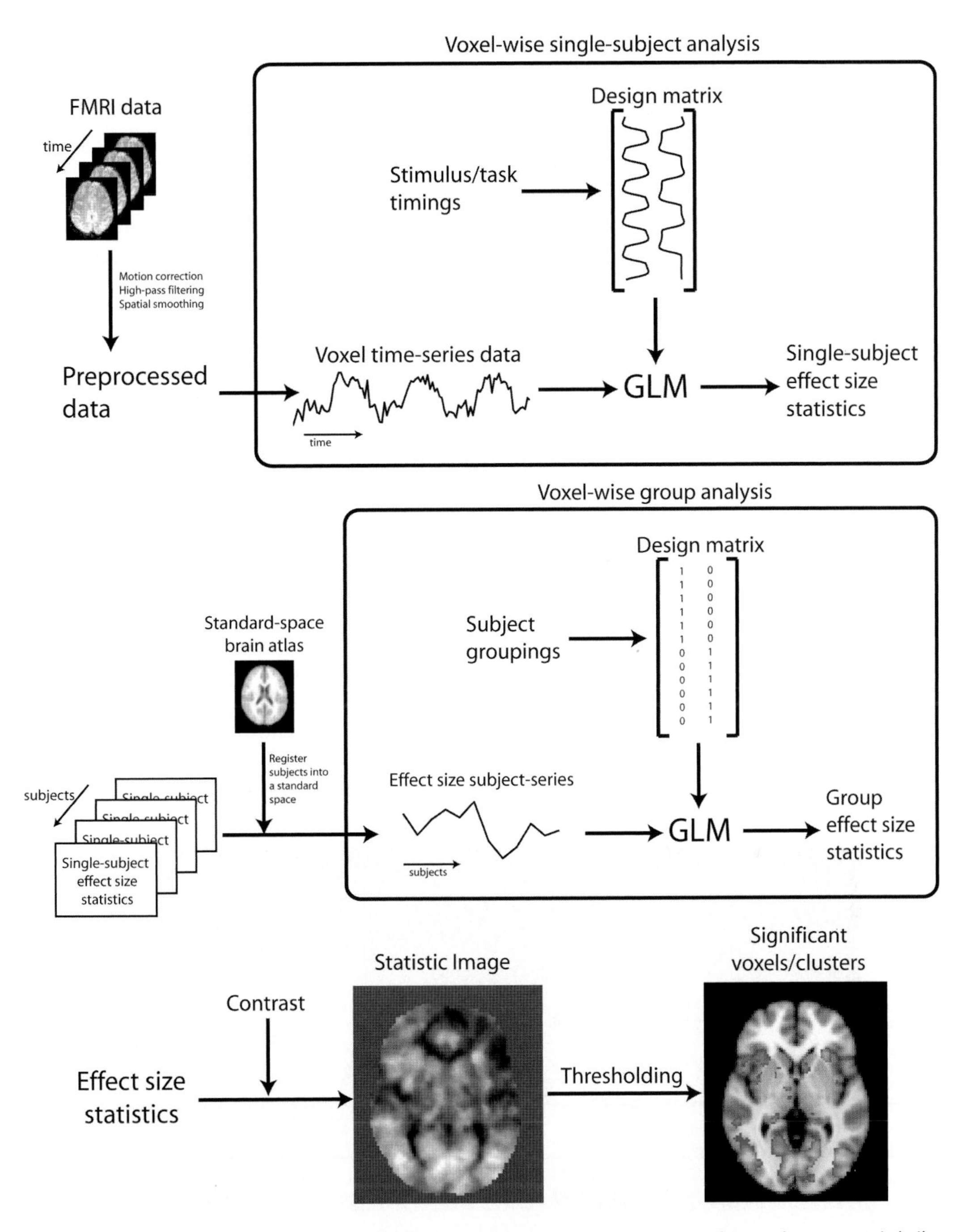

Fig. 1. Illustration of the analysis steps carried out in a typical fMRI group study. There are three main components in the process. First, the individual subjects' fMRI data must be processed (*top*). The information gleaned from this about the effect sizes (the size of the fMRI signal change in response to the experimental task) for each subject are then combined in a group analysis (*middle*). The group effect sizes statistic images are then statistically thresholded to find significant brain areas (*bottom*).

2.1. fMRI Data

In a typical fMRI imaging session, a low-resolution functional volume is acquired every few seconds (MR volumes are often also referred to as "images" or "scans"). Over the course of the experiment, 100 volumes or more are typically recorded. In the simplest possible experiment, some images will be taken while stimulation[1] is applied, and some will be taken with the subject at rest. Because the images are taken using an MR sequence which is sensitive to changes in local blood oxygenation level, parts of the images taken during stimulation should show increased intensity, compared with those taken while at rest. The parts of these images that show increased intensity should correspond to the brain areas that are activated by the stimulation. The goal of fMRI analysis is to detect, in a robust, sensitive, and valid way, those parts of the brain that show changes in intensity at the points in time that the stimulation was applied.

A single volume is made up of individual cuboid elements called voxels **(Fig. 2)**. An fMRI dataset from a single session can be thought of either as t volumes, one taken every few seconds, or as v voxels, each with an associated time series of t time points. It is important to be able to conceptualise both of these representations, as some analysis steps make more sense when thinking of the data in one way and others make more sense the other way.

An example time series from a single voxel is shown in **Fig. 3**. Image intensity is shown on the y-axis and time (in scans) on the x-axis. As described above, for some of the time points, stimulation was applied (the higher intensity periods), and at some time

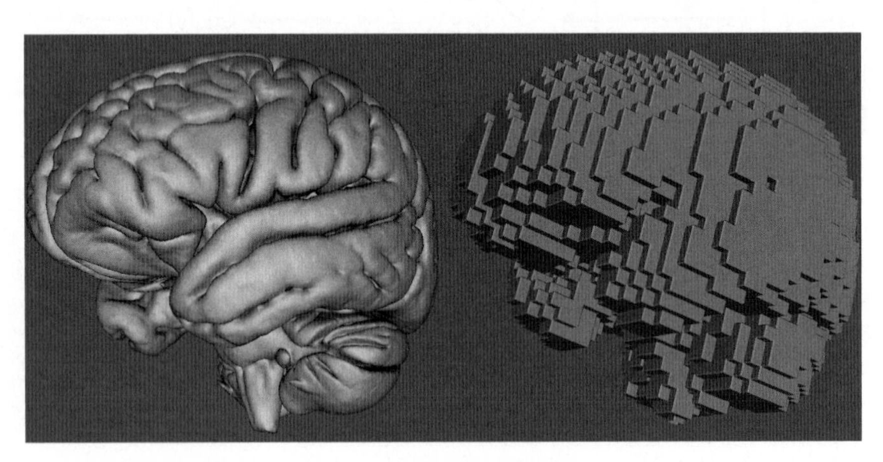

Fig. 2. What are voxels? Shown here are surface renderings of 3D brain images. On the left is a high-resolution image, with small (0.5 × 0.5 × 0.5 mm) voxels; the voxels are too small to see. On the right is a low-resolution image of the same brain, with large (5 × 5 × 5 mm) voxels, clearly showing the voxels making up the image.

[1] For the remainder of this chapter, reference to "stimulation" should be taken to include also the carrying out of physical or cognitive activity.

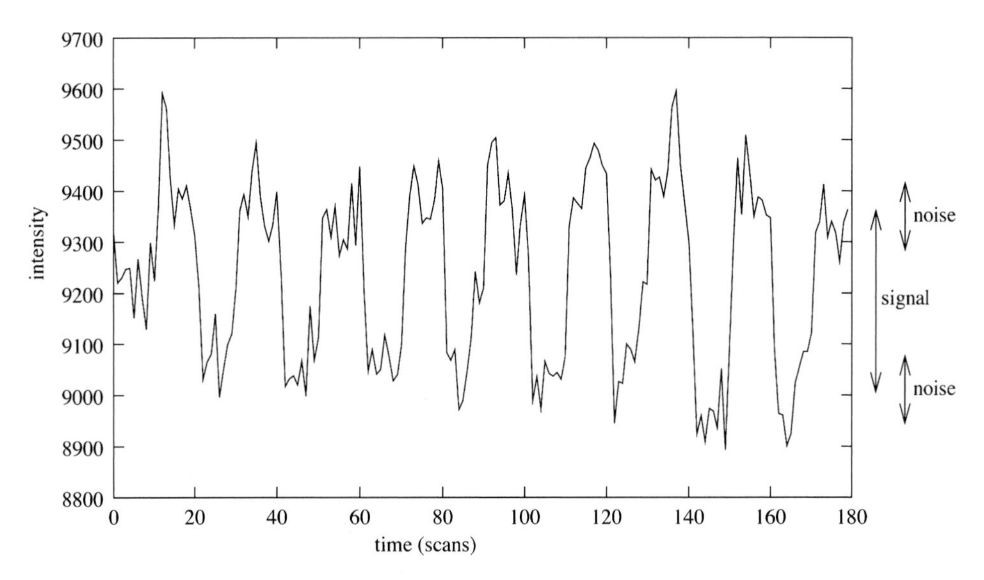

Fig. 3. An example time series at a strongly activated voxel from a visual stimulation experiment. Here the signal is significantly larger than the noise level. Periods of stimulation are alternated with periods of rest – a complete stimulation – rest cycle lasts 20 scans.

points, the subject was at rest. As well as the effect of the stimulation being clear, the high-frequency noise is also apparent. The aim of fMRI analysis is to identify in which voxels' time series the signal of interest is significantly greater than the noise level.

2.2. Preparing fMRI Data for Statistical Analysis

Initially, a 4D dataset is pre-processed. This pre-processing is aimed at not only removing artefacts and reducing noise but also conditioning the data so that it is more amenable to the statistical analysis that is to follow.

The most basic required steps that will be typically carried out are as follows. Once data has been acquired by the MR scanner, the pre-processing starts by *reconstructing* the raw "k-space" data into images that actually look like brains. The data is then *motion corrected*, where each volume is transformed (using rotation and translation) so that the image of the brain within each volume is aligned with that in every other volume. *Spatial smoothing* is then carried out, principally to reduce noise, hopefully without significantly affecting the activation signal. Finally, each voxel's time series is *temporally high-pass filtered* with a filter designed to remove the large amount of low-frequency temporal noise found in FMRI data, without removing the signal of interest.

Chapter 6 has already covered fMRI pre-processing in much more detail, including other optional steps that have not been mentioned here.

2.3. Predicting the Response

The first step in the statistical analysis is to come up with a good prediction, or model, of what we think the measured fMRI signal response will look like in voxels that are active. In the simplest type of fMRI experiment, we alternate periods of stimulation with periods of rest, in what we will refer to as a square-wave block design, as shown in **Fig. 4** (left). We expect that a voxel which is active in response to the stimulus will contain an fMRI signal that generally fluctuates up and down with a time course that is similar to the stimulus time course (**Fig. 4** left), whereas an inactive voxel will not.

2.4. Haemodynamic Response Function

However, can we come up with a better prediction of the fMRI signal than **Fig. 4** (left)? In particular, we know, from experiment, that the response to a very short stimulus looks like the curve shown in **Fig. 5**. We refer to this response to an impulse of stimulus as the haemodynamic response function (HRF), and it is basically a delayed and blurred version of the short stimulus burst. This is because the variations that we can detect in blood oxygen level dependent (BOLD) fMRI signal are due to processes taking place in the vasculature: things such as the amount of blood oxygenation, blood flow, and blood volume change when neural activation increases or decreases. Unsurprisingly, these vascular changes occur on a slower timescale than the neural activity. Note that there are also more subtle characteristics of the HRF. For example, as can be seen in **Fig. 5**, there can be a post-stimulus undershoot as the HRF temporarily drops below baseline before rising back to zero. The HRF in **Fig. 5** is a commonly used HRF, and is a double-gamma function of the form:

$$h(t) = \frac{G(\mu_1, \sigma_1^2) - G(\mu_2, \sigma_2^2)}{\rho} \tag{1}$$

where $h(t)$ is the HRF as a function of time, $G(\mu, \sigma^2)$ is a Gamma distribution parameterised by its mean, μ, and variance, σ^2 (note

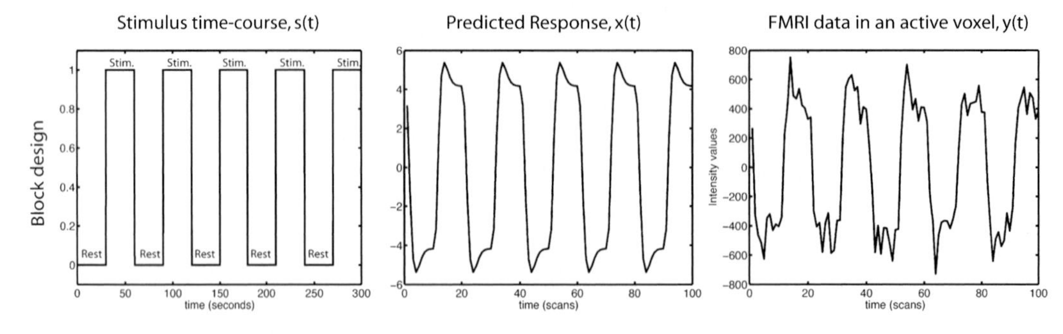

Fig. 4. Predicting the response using the known stimulus timings. This example is a square-wave block design where blocks of stimulation are alternated with blocks of rest. The square-waveform (*left*) describes the input stimulus timing; the predicted response (*middle*) results from convolving the stimulus time course with the haemodynamic response function and then sampling it at the temporal resolution of the experiment. This experiment has a repetition time (*TR*) of 2 s. This process produces a model, or predicted response, that looks much more like the data measured in voxels that are responding to the stimulus (*right*).

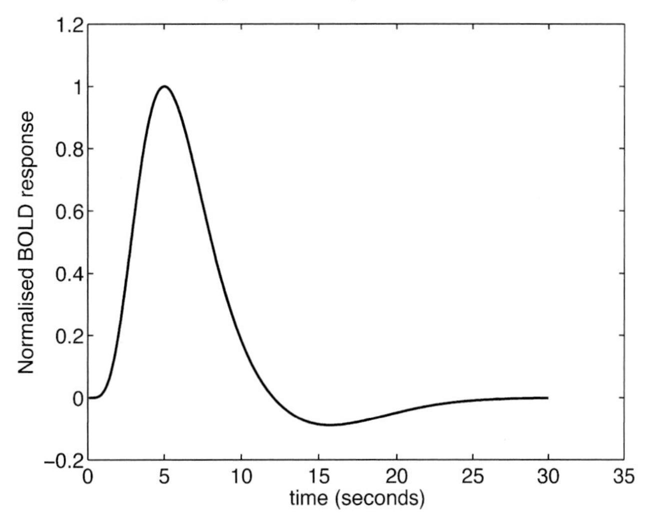

Fig. 5. The haemodynamic response function. A brief impulse of stimulation at t = 0 s causes a blood oxygen level dependent (*BOLD*) signal that is delayed and blurred. Here it is modelled as a double-gamma function.

that we can convert these to the traditional Gamma distribution parameters using $\alpha = \mu^2/\sigma^2$ and $\beta = \sigma^2/\mu$), and ρ is the ratio of the height of the positive Gamma to the negative Gamma.[2]

The most straightforward way of incorporating the HRF into our predicted response is to apply its delaying and blurring effect to the raw stimulus time course that we have in our fMRI experiment. This is achieved by the mathematical operation of convolution:

$$x(t) = \int_0^\infty h(\tau)s(t-\tau)\mathrm{d}\tau \qquad (2)$$

This essentially assumes that the effects of the different impulses that make up the stimulus time course add together in an additive, linear fashion *(1–4)*. **Figure 4** (middle) shows the result of convolving the HRF in **Fig. 5** with the square-wave stimulus in **Fig. 4** (left) to form our new improved predicted response. Strictly speaking, making the assumption of linearity is incorrect. We will see later how we can address this issue, and also discuss how to tell if that is necessary.

For now, using the HRF and convolving it with the stimulus time course provides us with a way in which we can come up with a reasonable prediction of the response for any general stimulus type. For example, **Fig. 6** shows the predicted response for a sparse single-event design and a dense randomised single-event

[2] The particular HRF in **Fig. 5** has parameter values $\mu_1 = 6$s, $\sigma_1 = 2.45$s, $\mu_2 = 16$s, $\sigma_2 = 4$s, and $\rho = 6$.

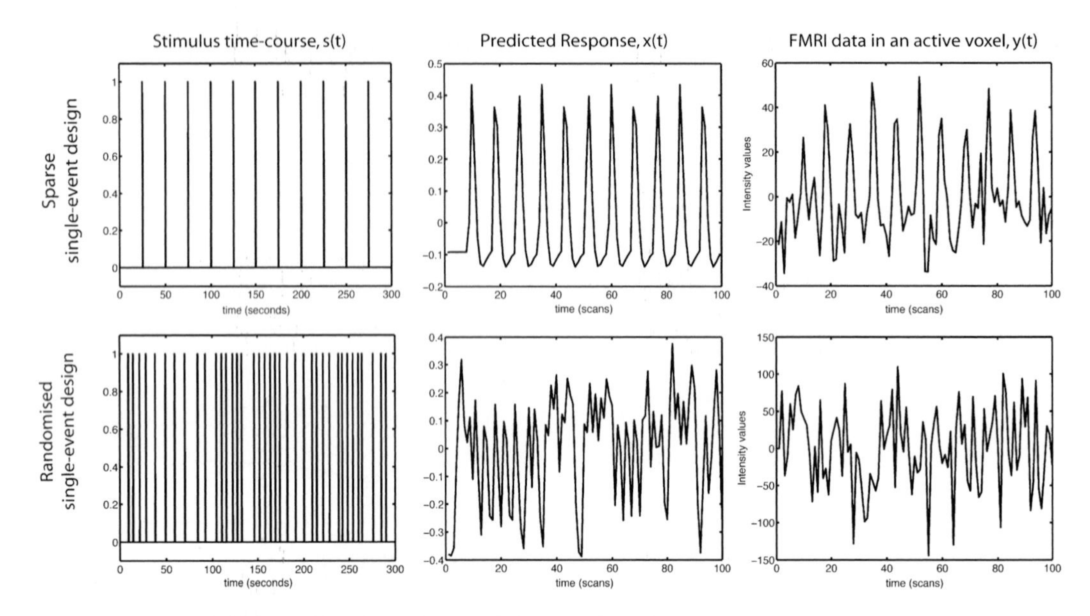

Fig. 6. Using the haemodynamic response function (*HRF*) and convolving it with the stimulus time course provides a way in which we can predict the response for any general stimulus type. Here we can see the predicted response for a sparse single-event design, where short bursts of 0.1 s stimulation are 25 s apart (*top*), and for a randomised single-event design, where short bursts of 0.1 s stimulation are separated with inter-stimulus intervals (*ISIs*) sampled from a Poisson distribution with mean of 7 s (*bottom*). Randomised single-event designs are an excellent way of working with stimuli that by their nature need to be single events, as they generally have better sensitivity than sparse single-event designs ***(5)***.

design. Armed with our predicted response, we can then look to find those voxels that have fMRI time courses that match the predicted response well and, if they pass a statistical test, label them as being active voxels.

3. General Linear Modelling

We have so far discussed how to come up with a prediction of the fMRI signal in response to an experimental stimulus. However, what do we do when we have more than one stimulus switching on or off throughout the experiment? The answer is to use a GLM. This assumes that each of the stimuli have their own predicted response, and that these predicted responses then add together linearly in some combination unique to each voxel, to explain the data measured in that voxel. For example, consider that we have two experimental stimuli: one auditory and one visual. Both are square-wave block designs, but they switch on and off at different times. The overall predicted response is given by a linear combination of the predicted responses:

$$y(t) = \beta_1 x_1(t) + \beta_2 x_2(t) + c + e(t) \quad (3)$$

where $y(t)$ is the data in one voxel, and is a 1D vector (time course) of intensity values with one value for each time point. $x_1(t)$ and $x_2(t)$ are the predicted responses for our auditory and visual experimental stimuli, respectively, and both are also 1D time courses with one value for each time point. c is a constant and would correspond to the mean intensity value in the data. The linear combination of the predicted responses needed to explain the data in a particular voxel is described by the parameters β_1 and β_2. $e(t)$ models the noise that is present in fMRI data.

Model fitting involves adjusting the mean level, c, and the parameters β_1 and β_2, to best fit the data. For example, if a particular voxel responds strongly to model x_1, the model-fitting will find a large value for β_1; if the data instead looks more like the second model time course, x_2, then the model-fitting will give β_2 a large value. The GLM is used to analyse each voxel's time series independently. This is often referred to as a mass univariate analysis, and outputs statistics independently at each voxel.

There are a number of different of names that get used to describe the different components of the GLM. The predicted responses within a GLM are often referred to as *explanatory variables (EVs)*, as they explain different processes in the data. They can also be referred to as *regressors*, and the βs as *regression parameters*, as we are performing what is also known as a multiple regression. The regression parameters, β, are also sometimes referred to as *effect sizes*, as they describe the size of the response to the corresponding underlying experimental stimuli.[3]

3.1. Design Matrix

The GLM is often formulated in matrix notation. All of the parameters are grouped together into a $P \times 1$ vector β (where P is the number of EVs) and all of the EVs are grouped together into an $N \times P$ matrix X, often referred to as the design matrix, where N is the number of time points in the experiment. This gives us the GLM in the following form:

$$Y = X\beta + e \quad (4)$$

where Y is the $N \times 1$ vector of intensity values in the data, and e is the $N \times 1$ noise vector. You may wonder what has happened to the mean parameter, c, in this new equation. There are two

[3] Note that this common usage is slightly different from the definition sometimes used in the statistics literature, where effect size means β divided by the noise level.

common ways in which this is handled. The first is to remove the mean, or de-mean, the data Y, and to also separately de-mean all of the EVs in the design matrix. This is appropriate as there is no information in the mean signal intensity of the fMRI data that can aid us in our statistical analysis. The second is to leave the mean as part of the GLM by creating an EV that has the value of 1 at every time point. The β or regression parameter for this EV will be determined when we fit the GLM to the data and will relate to the estimate of the mean parameter.

Figure 7 shows a design matrix for our example experiment with two stimuli (auditory and visual). Each column in the design matrix is a different part of the model. The left column (x_1 or EV1) models the auditory stimulation, and the right column (x_2 or EV2) models the visual stimulation.

3.2. Fitting the GLM to the Data

The GLM is fit to the data at each voxel separately. This is achieved by adjusting the estimates of the regression parameters to find the best fit of the model to the data. Typically, it is assumed that the fMRI noise, e, is well modelled by a Gaussian distribution with a standard deviation, σ, unique to each voxel. When this assumption is made, the best fit of the model to the data is equivalent to minimising the sum (over all time points) of the squared difference between the data, Y, and the signal model, $X\beta$. That is we choose the β values that minimise:

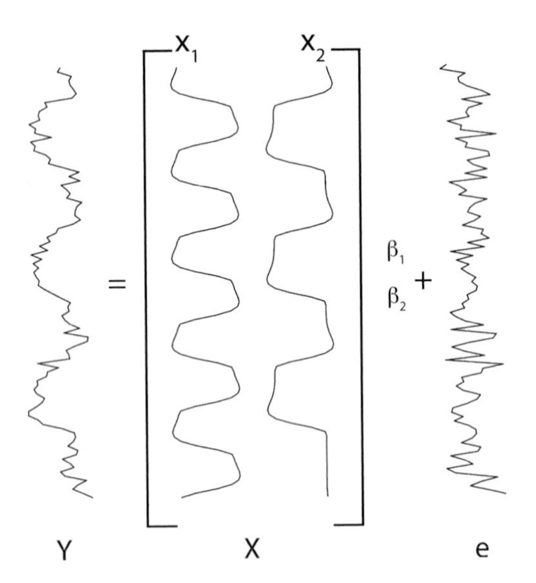

Fig. 7. Example of the general linear model (*GLM*) for an experiment containing auditory and visual stimuli which have different stimulus timings. The design matrix, *X*, contains two predicted responses (also known as regressors or explanatory variables): x_1 for the auditory stimulus and x_2 for the visual stimulus – *see* **Eq. 4**. In this, visualisation time is running downwards. It can be seen that the particular voxel data shown, *Y*, is from a voxel that is strongly activating in response to the visual stimulus modelled by x_2, but not to the auditory stimulus modelled by x_1.

$$\sum_t (Y_t - X_t\beta)^2 \tag{5}$$

Mathematically, it can be shown that this is minimised when we estimate the βs as:

$$\hat{\beta} = (X^T X)^{-1} X^T Y \tag{6}$$

where $\hat{\beta}$ is our regression parameter estimate. In the example visual/auditory experiment in **Fig. 7**, it can be seen that the particular voxel data shown, Y, is from a voxel that is strongly activating in response to the visual stimulus modelled by x_2, but not to the auditory stimulus modelled by x_1. This would result in a large value for β_2 and a low value for β_1 when the GLM is fit to this data, suggesting that there is visual activation but no auditory activation in this voxel.

3.3. Temporal Autocorrelation

Until now we have considered a rather simple approach to dealing with the noise that is present in fMRI, by assuming that it is well modelled as coming from a Gaussian distribution. Unfortunately, in practice, this is not the whole story. This is because fMRI noise, that is, the signal we record in the absence of any stimulation, is temporally autocorrelated. In particular, in the grey matter, this corresponds to the fMRI noise being temporally smooth. This is because the nature of the many artefacts that make up fMRI noise, for example, thermal noise, cardiac and respiratory rhythms, auto-regulatory oscillations, and networks of spontaneous neural activity, tend to occur more at low frequency than at high frequency. We can see this imbalance between low and high frequency if we look at a plot of the power spectrum (the absolute value of a Fourier transform) of fMRI noise, an example of which is shown in **Fig. 8a**.

The presence of temporal autocorrelation is a concern because it affects the choice of the optimal estimation (model fitting) method, and perhaps more important, the accuracy of the subsequent statistical tests. For example, if we ignore the fact that there is an increased amount of noise at low frequency, then we can underestimate the variability in the data and produce under-conservative statistical tests.

A number of strategies have been proposed for dealing with the problem of temporally autocorrelated noise. Traditionally, the most commonly used approach is pre-whitening *(6–8)*. The process of pre-whitening is summarised in **Fig. 8**. The first step is to estimate the temporal autocorrelation on the residuals, $r = Y - X\hat{\beta}$, from an initial GLM fit. The temporal autocorrelation estimate is then used to construct a pre-whitening temporal filter designed to "undo" the autocorrelation. In other words,

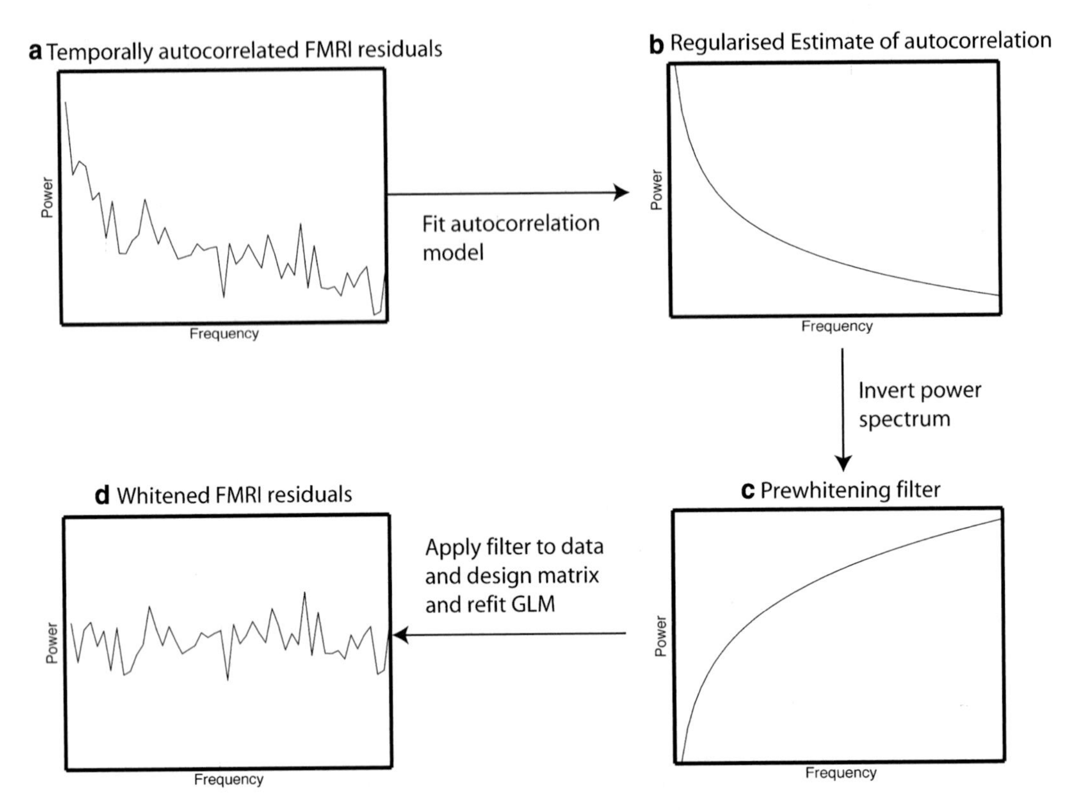

Fig. 8. A summary of the process of pre-whitening. (**a**) Plot of the power spectrum of the residuals from an initial general linear model (*GLM*) fit to the data from one voxel. (**b**) Estimated power spectrum (representing the temporal autocorrelation estimate) obtained from fitting an autocorrelation model. (**c**) This spectrum is inverted to create the frequency characteristics of a temporal filter designed to "undo" the autocorrelation. (**d**) The pre-whitening temporal filter is applied to both the data and the explanatory variables (*EVs*) in the design matrix, and then this pre-whitened GLM is refit to the pre-whitened data. The residuals that result from this refit of the GLM should now be approximately white, that is have a flat power spectrum.

the filter is designed to re-dress the imbalance between high and low frequency in the fMRI noise power spectrum, so that the new power spectrum has equal power at all frequencies. By analogy, since white light is a result of there being equal amounts of light at all frequencies/colours, we refer to noise with equal power at all frequencies as white noise – and therefore to the whole approach as pre-whitening. The pre-whitening temporal filter is applied to both the data and the EVs in the design matrix, and then this pre-whitened GLM is refitted.

A crucial step in pre-whitening is the estimation of the temporal autocorrelation. A wide range of approaches have been proposed, including the use of auto-regressive (AR) models *(6, 9)*, AR plus white noise models *(10)*, spectral smoothing *(8, 11)*, and spatial regularisation of autocorrelation estimates *(8, 12, 13)*.

The pre-whitening approach described above is one that was designed to work in a classical statistical ("frequentist") framework.

It is possible that inferring the autocorrelation on the residuals from an initial GLM fit can introduce inaccuracies, since the residuals, r, only serve as an approximation to the true error, e. More recently, alternative Bayesian strategies have been developed to deal with this problem *(14, 15)*. These have the advantage of inferring the autocorrelation characteristics at the same time as the GLM regression parameters, and to take into account the uncertainty in the temporal autocorrelation estimation. However, these issues aside, they are essentially performing the same pre-whitening approach we have already described. Although computationally more demanding these techniques are being increasingly used.

3.4. Inferring Neural Activity

When we fit a particular GLM to a particular voxel's data, we get regression parameter estimates that indicate how much of each EV is needed to explain what we see in the data. If the parameter estimate of β for any particular EV is non-zero, then it might seem reasonable to assume that the voxel in question is neuronally responding to the stimulus that the EV represents. However, we only have estimates/approximations of the true β obtained from a limited amount of noisy fMRI data. So, given the amount of noise and the estimate of β obtained, how much can we trust that any particular β is non-zero? It is only those voxels where we can satisfy ourselves in a statistical manner that this it the case, that we label as being active. The first step towards this is to convert the parameter estimates of β into a useful statistic. Most commonly, we use a T-statistic, given by:

$$t = \frac{\hat{\beta}}{\mathrm{std}(\hat{\beta})} \tag{7}$$

where the denominator is the standard deviation (uncertainty) of our parameter estimate. If the parameter estimate is low relative to its estimated uncertainty, the T-statistic, t, will be low, implying that β is unlikely to be significantly non-zero (and vice versa). We will see later what the standard deviation of our parameter estimate depends upon.

The question remains as to how we determine that a T-statistic is significantly non-zero. This is achieved by comparing the calculated T-statistic to the distribution of T-statistics we would expect to get if the true β value was zero. This is a *null hypothesis test* (the null hypothesis is that β is zero).

As mentioned earlier, we typically assume that the noise in fMRI is Gaussian distributed. This means that our expected distribution of T-statistics under the null hypothesis is T-distributed. This is a standard statistical distribution for which the probability, or P-value, of getting a T-value greater than the

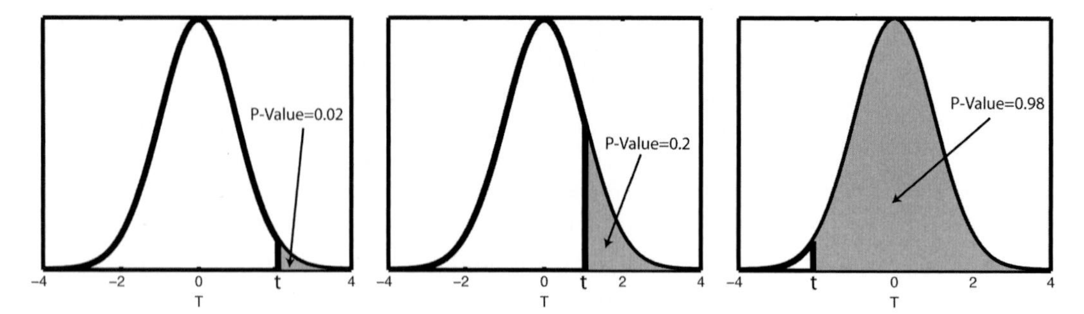

Fig. 9. Performing a *T*-test. The *T*-statistic we calculate for each regression parameter estimate is compared with the distribution of *T*-statistics we would expect if the true regression parameter were zero. This "null distribution" is a standard *T*-distribution. Here we are using a *T*-distribution with 100 DOF (this would correspond to about 100 time points in the data). We calculate a probability, or *P*-value, as the proportion of the area under the curve in the positive tail of the distribution defined by our *T*-statistic, *t*. A low probability, or *P*-value, (*left plot*) of the null hypothesis being true means that we can more confidently reject the null hypothesis and label the voxel as having a non-zero β, and therefore as being neuronally activated by the stimulus that the β in question represents. *T*-statistics that have larger *P*-values (*middle plot*) or are deep into the negative tail of the distribution (*right plot*) have high *P*-values. In the latter case, this might seem a bit counterintuitive as we have a *T*-statistic that is in the extremities of the null distribution. However, this is because the test is directional and so we calculate the *P*-value by looking in the positive tail of the distribution only.

one we have calculated if the null hypothesis were true, can easily be calculated. As illustrated in **Fig. 9**, a low probability (low *P*-value) of the null hypothesis being true means that we can more confidently reject the null hypothesis and label the voxel as having a non-zero β. We will see in **Subheading 7** how we choose a threshold for the *P*-values.

3.5. Contrasts

So far, we have addressed how we might go about producing *T*-statistics and *P*-values, which describe how strongly each voxel is related to each EV in our design matrix. Contrasts are a framework whereby we can ask not just questions about each EVs parameter estimate (PE, or β) in isolation, but also a wider range of questions that compare the different parameter estimates with each other.

In general, a contrast is defined by a $P \times 1$ vector, c, (recall that P is the number of EVs in the design matrix). This is multiplied by the $P \times 1$ vector of parameter estimates, $\hat{\beta}$, to give what is known as a contrast of parameter estimates (COPEs), $c^{\mathrm{T}}\hat{\beta}$. The COPE is therefore just a linear combination of the parameters estimates; it is equal to the sum of each PE multiplied by the relevant number in the contrast vector.

For example, it may be desirable to compare two different PEs to test directly whether one EV is more "relevant" to the data than another EV. In our combined auditory and visual experiment, this would be asking "Where does the brain respond more strongly to the auditory stimulus compared with the visual stimulus?". In this example, we have two EVs, one for the auditory stimulus and one for the visual (recall **Fig. 7**). So our contrast to answer this

ITEM ON HOLD

Titl fMRI basics and clinical applica
tions
Auth
Call 62709212 ocm
Enum
Chro
Copy:
Item *89108293796*

Item
Patr YELENA GULLER

Hold 3/6/2010
Pick ERLING CIRC DESK

ld be c^{T} = [1 −1] (c is a column vector, hence the
:), resulting in a COPE, $c^{T}\beta = 1\beta_1 - 1\beta_2$.
'E is then simply treated as if it were itself an indi-
ion parameter estimate. In other words, in the same
ı. 7, we calculate a T-statistic by dividing the COPE
'd deviation:

$$t = \frac{c^{T}\hat{\beta}}{\sqrt{\mathrm{var}(c^{T}\hat{\beta})}} \qquad (8)$$

T-distribution in question has degrees of freedom
$[- P$ (recall that N is the number of time points in the
), and:

$$\mathrm{var}(c^{T}\hat{\beta}) = \hat{\sigma}^{2} c^{T} (X^{T}X)^{-1} c \qquad (9)$$

the estimate of the variance of the fMRI noise:

$$\hat{\sigma}^{2} = \frac{r^{T}r}{(N-P)} \qquad (10)$$

the residual, that is, an estimate of the error, e, and is
t over after the model is fit to the data, and is given by
$\hat{\beta}$.

th the T-test on a single PE, we can calculate a probabil-
value, of getting the calculated T-statistic under the null
s in the same manner as in **Fig. 9**. The null hypothesis
COPE = 0, so if the calculated P-value is low then this
that the it is unlikely that the COPE = 0. If we apply a
1] contrast in the auditory/visual experiment, then this is equivalent to saying "the voxel is responding more strongly to the auditory stimulus compared with the visual stimulus." We can infer that it is the auditory that is stronger than the visual, and not vice versa, because these T-tests on these contrasts are directional. Technically, this is because we are doing the null-hypothesis test in **Fig. 9** on one tail (in particular, the right-hand tail) of the distribution only. As shown in **Table 1**, if we want to ask "Where does the brain respond more strongly to the visual stimulus compared with the auditory stimulus?," then we would use c^{T} = [−1 1]. All that remains to complete the null hypothesis test is a choice of threshold, such that if the calculated P-value drops below that threshold, we reject the null hypothesis and say that the contrast is significant. We shall see later in **Subheading 7** how we go about doing this while also taking into account the spatial nature of the data.

Even in the relatively simple auditory/visual experiment, there are a number of different questions that can be answered. *See* **Table 1** for some other examples.

Table 1
Examples of contrasts that might be used in the two stimulus auditory/visual experiment

Contrast, c^T	COPE, $c^T\beta$	Meaning
[1 0]	β_1	Where is there significant auditory activation?
[0 1]	β_2	Where is there significant visual activation?
[-1 0]	$-\beta_1$	Where is there significant negative auditory activation?
[0 -1]	$-\beta_2$	Where is there significant negative visual activation?
[1 -1]	$\beta_1 - \beta_2$	Where is there auditory activation significantly greater than visual activation?
[-1 1]	$\beta_2 - \beta_1$	Where is there visual activation significantly greater than auditory activation?
[1 1]	$\beta_1 + \beta_2$	Where is there significant activation averaged across both conditions?

Even in this relatively simple experiment, there are a number of different questions that can be answered. In particular, it is important to remember that the directionality of the *T*-test is important. Note that:

EV1 is the auditory predicted response and EV2 is the visual predicted response.

COPE stands for contrast of parameter estimate and *EV* stands for explanatory variable

3.5.1. F-Tests

The last example contrast in **Table 1** was a [1 1] contrast. It is a common misconception that such a contrast asks "Where is there significant activation due to either the visual or the auditory stimulation?." In fact, this contrast calculates COPE = $\hat{\beta}_1 + \hat{\beta}_2$, which is proportional to the average value of the two regression parameters, $(\hat{\beta}_1 + \hat{\beta}_2)/2$, and hence is actually asking "Where is there significant activation averaged across both conditions?."

So how do we go about asking the question "I want to find where there is significant activity due to either the visual or the auditory stimulation"? The answer is to use *F*-tests. An *F*-test is defined by specifying a set of contrasts that we want to test simultaneously. This then tests the null hypothesis that all of the COPEs that are in the *F*-test are equal to zero. Therefore, we can find significance (reject the null hypothesis) if any of the COPEs is non-zero. Another perspective is that the *F*-test will find significance if there is any linear combination of the COPEs that explains a significant amount of variance in the data.

So to ask "Where is there significant activity due to either the visual or the auditory stimulation?," we simply need to include in an F-test the contrast that asks where there is significant activity due to the auditory stimulation, [1 0], along with the contrast that asks where there is significant activity to the visual stimulation, [0 1]. Formally, this is done with an F-test contrast matrix, c, that contains both of these contrasts:

$$c^T = \begin{pmatrix} 1 & 0 \\ 0 & 1 \end{pmatrix}.$$

Note that in the T-tests, our contrasts were $P \times 1$ vectors. Now, F-tests are generally described as $P \times K$ contrast matrices, c, where K is the number of contrasts in the F-test, and recall that P is the number of regression parameters in the GLM. Using this contrast matrix, we can then calculate an F-statistic:

$$f = \frac{\hat{\beta}^{\mathrm{T}} c\, \mathrm{var}(c^{\mathrm{T}} \hat{\beta}) c^{\mathrm{T}} \hat{\beta}}{K}. \tag{11}$$

Analagous to how the T-statistics were T-distributed under the null hypothesis that the COPE is zero, this F-statistic is F-distributed (with DOF K and $N - P$) under the null hypothesis that all of the contrasts in the F-test are zero $(c^{\mathrm{T}} \hat{\beta} = 0)$. As such, we can proceed with a null-hypothesis test in exactly the same manner as we did with the T-test.

An important characteristic of F-tests is that they are blind to the directionality of the contrasts that make up the test. In other words, in our auditory/visual experiment example, the following F-tests are all equivalent:

$$\begin{pmatrix} 1 & 0 \\ 0 & 1 \end{pmatrix}, \begin{pmatrix} -1 & 0 \\ 0 & -1 \end{pmatrix}, \begin{pmatrix} -1 & 0 \\ 0 & 1 \end{pmatrix}, \begin{pmatrix} 1 & 0 \\ 0 & -1 \end{pmatrix}.$$

This is because F-tests are testing if there is a linear combination of the COPEs that explain a significant amount of variance in the data, and when we consider this variance, we are ignoring the sign of the COPEs.

It is instructive to consider how F-tests relate to T-tests by considering an F-test that consists of just one contrast. As illustrated in **Fig. 10**, such an F-test is equivalent to a two-tailed T-test on the contrast in question, with the relationship $f = t^2$. Hence, F-contrasts containing single contrasts can be used to mimic two-tailed T-tests. The F-test's blindness to the directionality of the contrasts is readily apparent when we consider the equivalent two-tailed T-test. It is because we can get significance with either a significantly positive or negative COPE in either tail. As a result

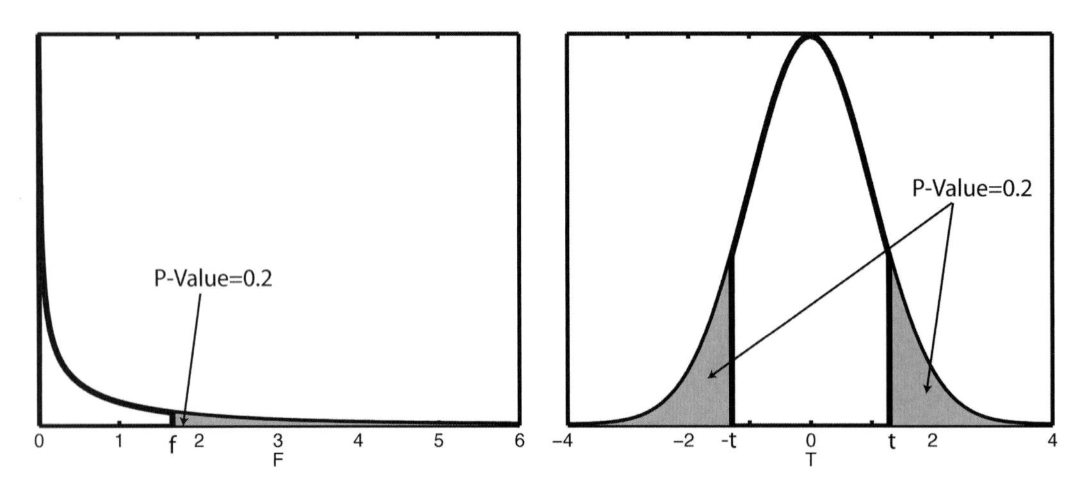

Fig. 10. Here we are considering an *F*-test that consists of just one contrast. In this case, the *F*-test (*left*) is equivalent to a two-tailed *T*-test (*right*) on the same contrast, with the relationship $f = t^2$.

of calculating the *P*-value under both tails, a two-tailed *T*-test (or equivalently the *F*-test) is more conservative (with respect to positive activation) than the one-tailed *T*-test on the same contrast.

3.6. Interaction Example

It is possible that the response to two different stimuli, when applied simultaneously, is greater than that predicted by adding up the responses to the stimuli when applied separately. If this is the case, then such "non-linear interactions" may need to be allowed for in the model. The simplest way of doing this is to set up the two originals EVs, and then add an interaction EV, which will only be "up" when both of the original EVs are "up" and "down" otherwise. In **Fig. 11**, EV1 could represent the application of a drug and EV2 could represent visual stimulation. EV3 will model the extent to which the response to drug + visual is greater than the sum of drug-only and visual-only. A contrast of [0 0 1] will show this measure, whereas a contrast of [0 0 −1] shows where negative interaction is occurring. An F-contrast of

$$\begin{pmatrix} 1 & 0 & 0 \\ 0 & 1 & 0 \end{pmatrix}$$

will ask where is there significant activity to either drug-only or visual-only.

3.7. Converting T- and F-Statistics into Z-Statistics

T- and *F*-statistics can be converted to *Z*-statistics, that is, statistics that are distributed as a standardised Normal (Gaussian) distribution. This is simply achieved by ensuring that the *P*-value is the same regardless of which statistic is used, so to convert from

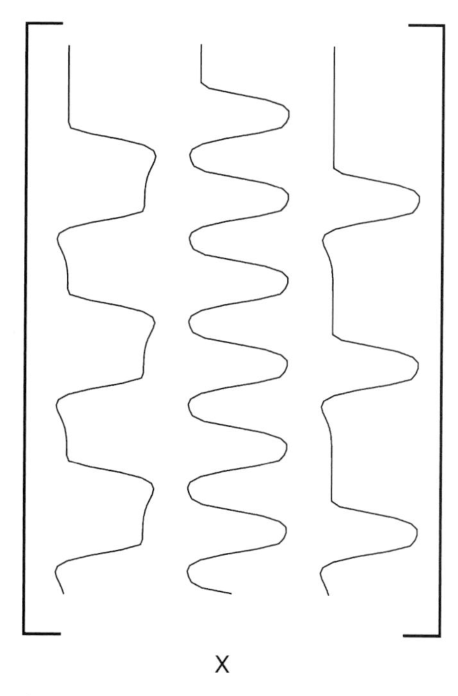

Fig. 11. Example of modelling a non-linear interaction between stimuli. The first two explanatory variables (*EVs*) model the separate stimuli and the third models the interaction, that is, accounts for the "extra" response when both stimuli are applied together.

a *T*- to *Z*-statistic, we calculate the *P*-value for the given *T*-statistic and then determine the *Z*-statistic as being the one that gives the same *P*-value. One reason for doing this is so that we can compare statistics using a common currency. Another reason is so that we can perform generic thresholding techniques (such as described in **Subheading 7**), using *Z*-statistic maps, regardless of whether we have done *T*- or *F*-tests.

3.8. Percent Signal Changes

It is useful to be able to convert regression parameter estimates into percent BOLD changes. This is because percent BOLD change can be a common currency across different experiments (though BOLD is not a quantitative measure and depends on many experimental factors, and so should not be treated as comparable without a good deal of care). It does not depend on things like the arbitrary scaling of intensity values output from the scanner or arbitrary scaling of the EVs. As illustrated in **Fig. 12**, this is simple to do and just requires that we have an estimate of the baseline signal intensity, *C*, from the voxel in question, and that we know the peak-to-peak height, *H*, of the relevant EV. The percent BOLD change is then calculated as:

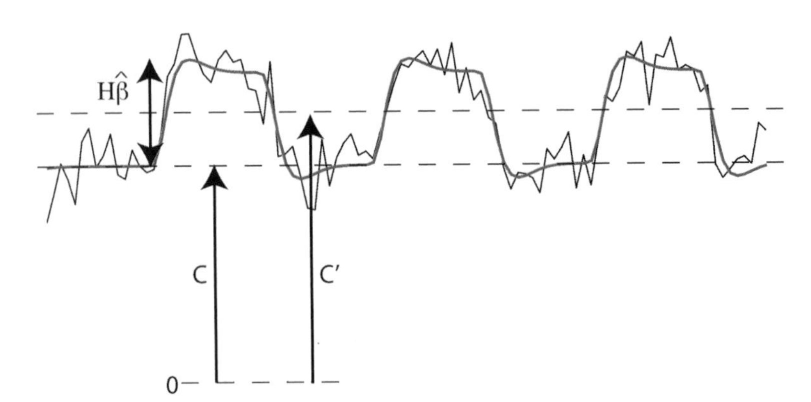

Fig. 12. Illustration of the calculation of percent blood oxygen level dependent (*BOLD*) signal change from a general linear model (*GLM*) fit. % change = $100H\hat{\beta}/C$, where $\hat{\beta}$ is the regression parameter estimate (effect size) and *H* is the peak-to-peak height for the relevant explanatory variable (*EV*). *C* is the baseline intensity of the fMRI data, but is common to approximate this with the mean of the time series, *C′*.

$$\%\,change = 100\frac{H\hat{\beta}}{C}. \quad (12)$$

Since the signal fluctuation $\hat{\beta} \times \mathrm{H}$ is always small with respect to the baseline *C*, it is common to approximate *C* as more simply the mean of the time series, *C′*.

It is possible to do this for contrasts as well. However, it is not immediately obvious what the peak-to-peak height is in the context of a general contrast. This can be dealt with by determining the effective regressor for a contrast. The effective regressor is the regressor in a new version of the design matrix whose regression parameter estimate (and its variance) is equal to the original COPE. This is given by *(16)*:

$$X_{\mathrm{eff}} = XQc(c^{\mathrm{T}}Qc)^{-1}, \text{where } Q = (X^{\mathrm{T}}X)^{-1} \quad (13)$$

The peak-to-peak height of this effective regressor can then be used in the percent signal change calculation above.

3.9. Issues with Orthogonality and Estimating Contrasts

We described in **Subheading 3.2** how we obtain regression parameter estimates by finding the best fit of the GLM to the data (by using **Eq. 6**). The regression parameter estimates describe how much we need of each EV to explain what we see in the data. However, consider what would happen in a poorly designed experiment, where two different stimuli are switched on and off at very similar times. The resulting predicted responses (EVs) are highly correlated. Note that we use the terms correlated and non-orthogonal (similarly uncorrelated and orthogonal) interchangeably. The top of **Fig. 13** shows a design matrix containing our two very similar EVs. The model fitting will determine the regression parameter estimates $\hat{\beta}_1$ and $\hat{\beta}_2$ and describe how much we need of each EV to explain what we see in the data. However, because EV1 and EV2 are so similar, we can equally well use either EV1 or EV2 to explain what we see in the data. The result is that we cannot estimate either $\hat{\beta}_1$ or $\hat{\beta}_2$, separately from each other, very well. Mathematically, we say that the design matrix is not of "full rank," or that it is "rank deficient."

To understand this, consider solving two simultaneous equations. If we have two unknowns to solve for, then we need two equations to solve for them. However, if the two equations are the same, then we really have only one equation and we cannot solve for the two unknowns.

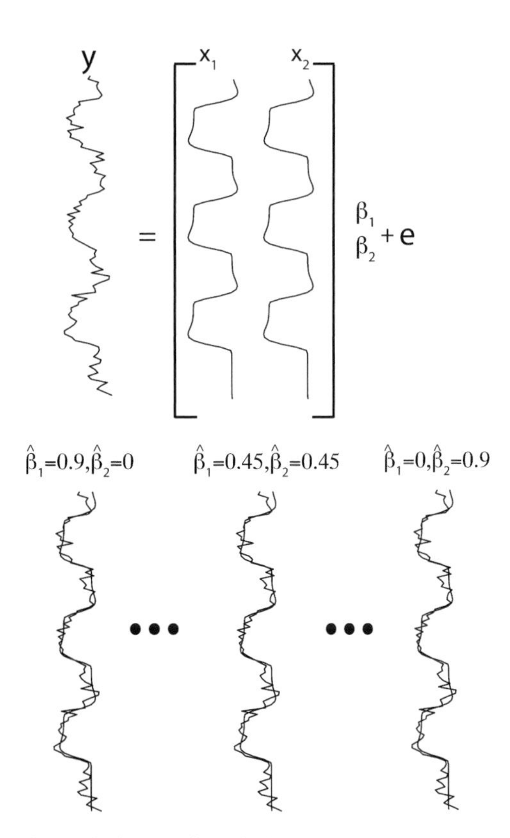

Fig. 13. Rank-deficient design matrices. At the top, we can see a general linear model (*GLM*) with a design matrix containing two identical explanatory variables (*EVs*). Since EV1 and EV2 are so similar, we can equally well use either EV1 or EV2 to explain what we see in the data. At the bottom of the figure, we can see examples of identically good model fits with any linear combination of the EVs as long as $\hat{\beta}_1 + \hat{\beta}_2 = 0.9$. The result is that we cannot estimate either $\hat{\beta}_1$ or $\hat{\beta}_2$ with any certainty (corresponding to [1 0] or [0 1] contrasts), but we can estimate $\hat{\beta}_1 + \hat{\beta}_2$ (corresponding to a [1 1] contrast).

The good news is that our statistical tests (T- and F-tests) take this all into account. When two EVs are highly correlated, the appropriate variances of the regression parameter estimates (*see* **Eq. 9**) are automatically increased – acknowledging the fact that we cannot determine which EV should explain what in the data. So, even though the statistics accounts for orthogonality, it is clear that when we design our experiments, we want to avoid this being an issue whenever possible. This can be achieved by using approaches that assess the efficiency of experimental designs such as *(16–18)*.

In **Fig. 14**, we can see three different design matrices. The first contains two EVs that are highly correlated, the second shows two EVs that are partly correlated, and the third shows two EVs that are completely uncorrelated. As discussed, the first design matrix is "rank deficient." The third design matrix is the ideal scenario in that the EVs are uncorrelated and there is no ambiguity in how to determine which EV explains what in the data. We refer to this as a "well-conditioned" design matrix. But what happens in the case of the second design matrix where the EVs are partially correlated? It is useful to think of the EVs as having uncorrelated (orthogonal) and correlated (non-orthogonal) components. The correlated (or non-orthogonal) components are of no use, as they are, by definition, identical and cannot be used to disambiguate which EV explains what in the data. Hence, the model fitting can only be driven by the uncorrelated, or orthogonal, components of the EVs. As long as there is a

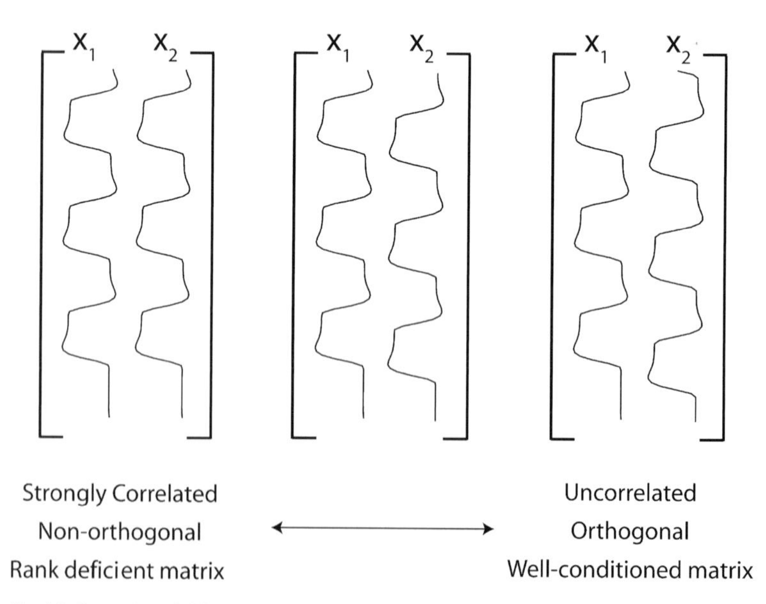

Fig. 14. Examples of different design matrices. Design matrix with two explanatory variables (*EVs*) that are highly correlated (*left*). Design matrix with two EVs that are partially correlated (*middle*). Design matrix with two EVs that are uncorrelated (*right*).

substantial orthogonal component, then there is sufficient information to get an efficient estimate of the regression parameters, and we can successfully infer on such GLMs.

The idea that the model fitting can only be driven by the uncorrelated, or orthogonal, components of the EVs is a crucial one, and is well illustrated by the following example. What happens to the regression parameter estimates when we have two partially correlated EVs, and orthogonalise one with respect to the other? **Figure 15** illustrates such a case. On the right, EV2 has been "orthogonalised with respect to" EV1, which just means that the part of EV2 which is correlated with EV1 has been subtracted from it. The counterintuitive result is that, even though it is EV2 that has been changed and EV1 has remained the same, it is β_1 that has changed and β_2 that remains the same. To understand this, remember that the model fitting can only be driven by the orthogonal components of the EVs. Although EV2 has changed, it has changed to be equal to the original orthogonal component and hence its orthogonal component is unchanged; subsequently, its regression parameter estimate is still the same. In contrast, although EV1 has not changed, because EV2 has

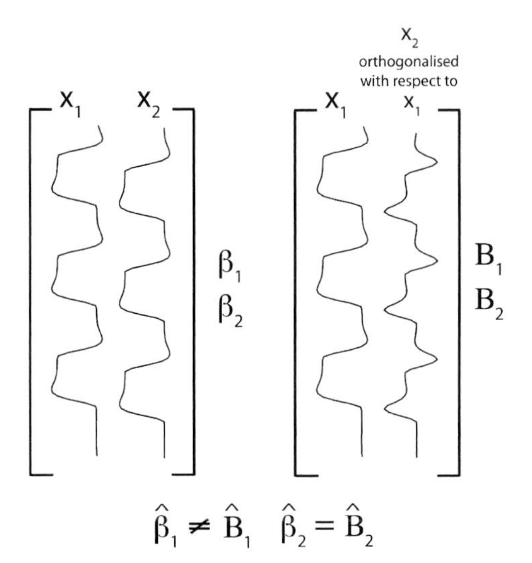

Fig. 15. Effects of orthogonalisation. We, first, fit the design matrix (containing two partially correlated explanatory variables, *EVs*) on the left to the data from a voxel, and obtain regression parameter estimates $\hat{\beta}_1$ and $\hat{\beta}_2$. We then construct a new design matrix, shown on the right, where EV1 is unchanged and EV2 is the old EV2 orthogonalised with respect to EV1. We then fit this new design matrix to the same data and obtain new regression parameter estimates $\hat{\beta}_1$ and $\hat{\beta}_2$. The counterintuitive result is that even though it is EV2 that has been changed and EV1 that has remained the same, it is $\hat{\beta}_1$ that has changed and $\hat{\beta}_2$ that remains the same. The underlying reason for this is that the model fitting can only ever be driven by the orthogonal components of the EVs.

been orthogonalised with respect to EV1, the orthogonal component of EV1 has changed; subsequently, its regression parameter estimate is different.

Everything we have considered up to now has been within the context of considering problems where we have two (or more) EVs that are correlated with one another. In fact, the problem is more general than this. We can see this by extending the analogy of solving simultaneous equations. In general, one encounters the same problems whenever it is possible to find a linear combination of the equations that is equal to another of the equations. For example, consider that we have three equations to solve for three unknowns, but it turns out that if we take two times equation 1 and subtract equation 2, then we get exactly equation 3. In that case, we actually only really have two equations to solve for our three unknowns, and so we are in trouble. In the GLM, the EVs are analogous to the equations, and the regression parameters are the unknowns. And so we have a problem if any EV is the same (or close to being the same) as a weighted sum of the other EVs in the design matrix. Again, we describe such a design matrix as being (or close to being) "rank deficient."

Even if we do have a design matrix that is close to being rank deficient, and therefore, there are some regression parameters that cannot be very well estimated, there may well be other regression parameters that can be estimated. There may even be contrasts that actually include the hard-to-estimate regression parameters, but that can still be estimated *(16)*. At first glance, this may seem a little counterintuitive. However, things should become clear if we consider a simple example of this. This occurs when we have the situation shown in **Fig. 13** where we had a design matrix with two very similar EVs. As already discussed, we cannot estimate at all well the individual parameters β_1 and β_2 with [1 0] and [0 1] contrasts. However, we can estimate a [1 1] contrast since this does not require us to separate out which EV explains what in the data. Again the simultaneous the first equation analogy is useful. Consider that we have two equations, $2x + 2y = 2$ and $4x + 4y = 4$. The second equation is simply two times of the first equation; and so we have a problem, and we cannot solve for x and y individually. However, we can solve for $x + y$; it is equal to one. Solving for $x + y$ is analogous to estimating the [1 1] contrast in our GLM.

4. Modelling Haemodynamic Variability

Up to this point, we have been working under the assumption that the HRF is known. However, it is well established that the HRF varies between brain regions and subjects *(19)* and so in practice, it is necessary to incorporate into our modelling of the fMRI data some flexibility in the HRF. One option is to have a parameterised model of the HRF and then estimate the HRF shape parameters at the same time as we estimate the GLM regression parameters that represent the size of the response. For example, we could use the double gamma HRF illustrated in **Fig. 5**, but instead of fixing the five parameters that describe the shape, we now look to estimate those parameters from the fMRI data. The problem with this approach is that it is not straightforward to estimate these HRF shape parameters within the GLM framework, as they generally require non-linear estimation approaches. A number of these approaches have been proposed, predominantly using Bayesian techniques *(20–24)*. However, these approaches are computationally demanding and are not yet in common use.

4.1. HRF Basis Sets

A popular alternative is to use the approach of basis functions. These allow HRF modelling flexibility but within the computationally undemanding GLM framework *(19)*. **Figure 16a** shows just one example of an HRF basis set that contains three basis functions. The choice of basis set is clearly important and we will come to that later. Whatever basis set is used, the principle is the same: different linear combinations of the basis sets can be used to give different HRF shapes. This is illustrated in **Fig. 16b**.

But how do we use these HRF basis functions in combination with our known stimulus timings to create predicted responses

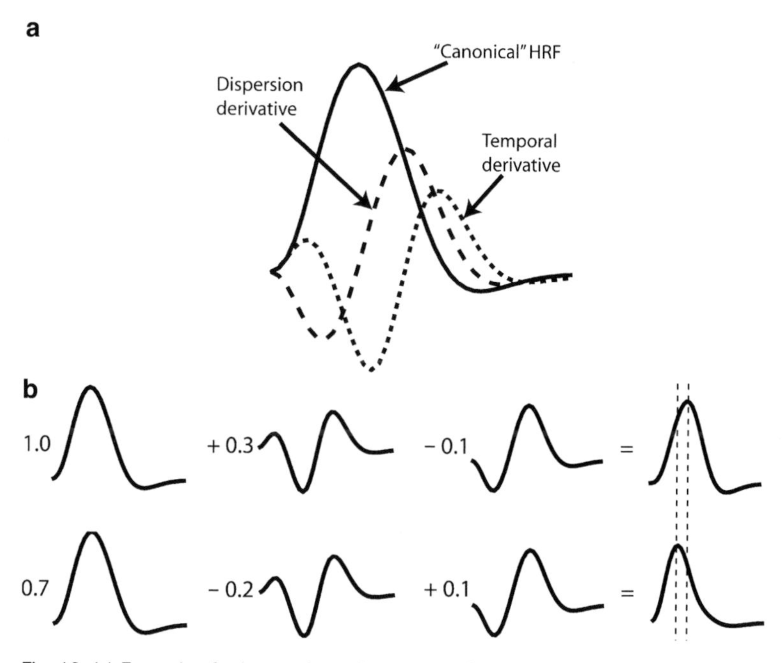

Fig. 16. (**a**) Example of a haemodynamic response function (*HRF*) basis set containing three basis functions. (**b**) Different linear combinations of the basis functions in the basis set can be used to obtain different of HRF shapes.

that can be used in the GLM? The answer is to separately convolve each of the HRF basis functions with the stimulus function to create an EV for each of the basis functions, as shown in **Fig. 17**. When the resulting design matrix is fit to the fMRI data, the required linear combination of these EVs is determined. If desired, the same linear combination can then be applied to the HRF basis functions to show the implied HRF shape.

The question remains as to how we set up a statistical test to ask, for example, "Where is there significant activation due to condition A?" when we model the response to condition A using HRF basis functions. One answer is to use an *F*-test. Recall that one way to think about an *F*-test is that it will find significance if there is any linear combination of the contrasts (in the *F*-test) that explain a significant amount of variance in the data. So if we simply create an *F*-test made from the contrasts that pick out each of the regression parameter estimates for each of our basis function EVs for condition A, then we will find where there are any linear combinations of the basis set EVs that can be used to give HRF shapes that explain significant amounts of variation in the data. In other words, if we have an experiment with just condition A, and we model it using three basis functions (**Fig. 17**), then we can ask "Where is there significant activation to condition A?" with the *F*-test contrast matrix:

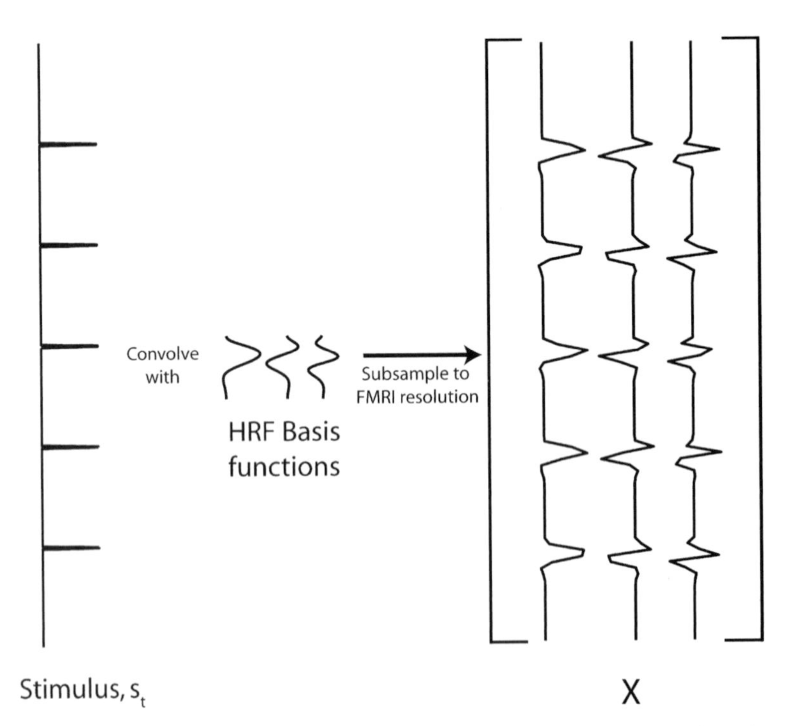

Fig. 17. Setting up design matrix explanatory variables (*EVs*) using a haemodynamic response function (*HRF*) basis set. Each HRF basis function is separately convolved with the stimulus function to create an EV for each of the basis functions.

$$c^T = \begin{pmatrix} 1 & 0 & 0 \\ 0 & 1 & 0 \\ 0 & 0 & 1 \end{pmatrix}.$$

An important point to remember is that, as we are using an *F*-test, we lose directionality of the test. So we cannot tell if we are finding significance with a positive or negative response. However, it is possible to recover this post hoc if we are using a basis set that contains a canonical HRF, by looking at the sign of the regression parameter estimate for the corresponding canonical HRF EV. We can also make comparisons between two different conditions by pairing up the corresponding HRF EVs for the two conditions and setting up an *F*-test that looks for linear combinations of differences between corresponding HRF EV regression parameters. In other words, with three basis functions in our basis set, EVs 1–3 model the HRF EVs for condition A, and EVs 4–6 model the HRF EVs for condition B, then we can ask "Where is there significantly different activation between condition A and condition B" with the *F*-test contrast matrix:

$$c^T = \begin{pmatrix} 1 & 0 & 0 & -1 & 0 & 0 \\ 0 & 1 & 0 & 0 & -1 & 0 \\ 0 & 0 & 1 & 0 & 0 & -1 \end{pmatrix}.$$

More precisely, this is looking for where there is a significant amount of variance being explained by any linear combination of differences between condition A and condition B for corresponding basis function EVs. This means that we can find a significant difference due to either a "shape" or "size" change. Note that for this approach to be sensible, we need to use the same basis set for both conditions.

4.1.1. Choosing a Basis Set

Figure 16a showed an example basis set that had been derived from a parameterised HRF that was made up of a series of half-cosine functions *(15)*. This was obtained by sampling thousands of example HRFs from within a range of plausible parameter values for the HRF model (**Fig. 18**), and then a principal component analysis was carried out on these samples to determine the principal modes/components of variation in the HRF shape. The three highest principle components were then used as the basis set. Note that this can equally well be done on any form of parameterised HRF, for example, a double gamma HRF or a biophysical model such as the balloon model *(25)*.

When this approach is taken with any plausible HRF model, it is typical for the first basis function to turn out to be the mean HRF shape, or a "canonical" HRF, for the second to approximate

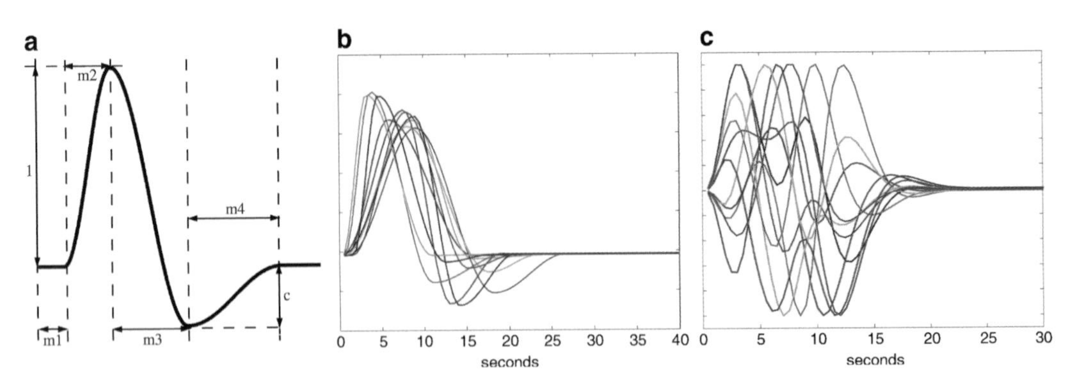

Fig. 18. (**a**) Parameterised haemodynamic response function (*HRF*) model. (**b**) Example HRFs sampled from this parameterised model of the HRF for plausible parameter values. (**c**) Samples of the HRF obtained from random linear combinations of the basis set shown in **Fig. 16a**.

the temporal derivative (i.e., linear combinations of the first and second basis functions result in versions of the canonical HRF shifted in time), and for the third to approximate a dispersion derivative (i.e., linear combinations of the first and third basis functions result in versions of the HRF with different widths of the main positive response). Although other basis sets that have been proposed (e.g., sets of Gamma functions and finite impulse response functions), this kind of basis set is highly recommended since it parsimoniously captures shape variations. It also has the advantage that the first (canonical) basis function will tend to dominate the fit to the data and can then be used to determine the positivity or negativity of the HRF. Note that it is quite common for people to neglect the dispersion derivative and use just a temporal derivative since temporal shifts represent the most important variation in the HRF shape, particularly when working with boxcar stimuli.

Thus far, we have considered basis sets made up of two or three basis functions. But why not use more? The reason for this, and in general the reason why basis sets with large numbers of basis functions (e.g., finite impulse response basis sets) are sub-optimal, is that the GLM becomes unrealistically flexible. This problem is evident even with just three basis functions. **Figure 18c** shows possible HRFs that result from random linear combinations of the basis set in **Fig. 16a**. Clearly, many of these HRFs are non-sensical. The problem is that random fluctuations in the fMRI noise can, by chance, look like these non-sensical HRFs and so we "over-fit" the model to the noise. The statistical inference (e.g., via *F*-tests across the basis functions) is still valid, but we lose sensitivity; it becomes harder to detect genuine activations. On the contrary, if we use too few basis functions then we can fail to estimate the true HRF and our model is then a poor match to the data and again we suffer a reduction in sensitivity. So there is a trade-off between providing enough basis functions to provide enough HRF variability, while not having so many that over-fitting becomes a problem. A general rule of thumb is that three basis functions are good for single-event designs and two basis functions are good for boxcar designs.

An increasingly used approach to overcome this problem is to infer on models that incorporate HRF variability using a Bayesian framework. One advantage of a Bayesian approach is that prior information can be included. Priors can be used that prohibit non-sensical HRFs. Subsequently, more flexibility can be allowed while protecting against over-fitting *(15, 22–24)*.

4.1.2. Basis Functions and Group Inference

In **Subheading 6**, we will discuss how we model multi-session/subject fMRI data. However, it is worth mentioning how basis functions are best used when we are ultimately doing a group analysis. In particular, we consider this in the context of the simple case of inferring a population group mean. One option might be to pass up the regression parameter estimates for all basis functions into the higher-level group analysis, obtain the group average for each basis function separately, and then perform an *F*-test across them at the group level (in the same manner as we would do in a single-session analysis). However, it is not clear what benefit there would be of doing this. When our basis set contains a "canonical" HRF, the other basis functions, such as the temporal and dispersion derivatives, tend to average out to zero at the group level due to the different subject HRF shape variations. Subsequently, an often-recommended approach is to only pass up to the group level the canonical HRF regression parameter estimates. This makes for a simple group analysis, and the benefits of including the basis function at the first level are still felt in terms of accounting for HRF variability that would otherwise cause increased noise in the first-level analysis.

Another option that can be taken is to calculate a size summary statistic from the single-session analyses (e.g., the root mean square of the basis function regression parameter estimates), and pass that up to the group level. However, it is important to note that this would then require different inference methods at the group level (e.g., *see* **Subheading 7.4.2**) than is generally used, as the population distribution of such a summary statistic is likely to be non-Gaussian.

4.2. Non-linearities

So far we have assumed linearity of the HRF. That is, we have assumed that the response to a stimulus is well modelled by (linear) convolution of the stimulus with the HRF. Typically, this is the approach that people take in the majority of fMRI analyses. However, it has been shown that this assumption is poor in certain situations. For example, it can be shown that the response to a prolonged stimulus is not as large as the one we would predict from extrapolating results from applying a short stimulus *(26, 27)*, and non-linearities are predominant when there are short separations (less than ~3 s) between stimuli *(28)*. Normally, these situations are intentionally avoided by designing experiments appropriately. For example, we avoid experiments where single events are occurring less than approximately 3 s apart, or experiments that require comparisons between a mix of short (e.g., single-event) and prolonged (e.g., boxcar) stimuli. However, if these situations are unavoidable, then it becomes necessary to model the non-linearities.

Such non-linearities are predicted by non-linear biophysical models, for example, the balloon model *(25)*. Hence, one solution is to model fMRI data using these non-linear biophysical models *(22)*. Another approach that can be used in the GLM setting is to extend the idea of convolution to include second-order non-linear terms using Volterra kernels *(28)*.

4.2.1. Volterra Kernels

Volterra kernels are a generalisation of convolution to include higher order non-linear terms. In fMRI, we need only to add the second-order terms to the first-order convolution terms to get the most important non-linear behaviour. A second-order Volterra kernel model is given by:

$$x(t) = \int_0^\infty h_1(\tau)s(t-\tau)\mathrm{d}\tau + \int_0^\infty \int_0^\infty h_2(\tau_1,\tau_2)s(t-\tau_1)s(t-\tau_2)\mathrm{d}\tau_1\mathrm{d}\tau_2 \tag{15}$$

where $s(t)$ is the stimulus, the first term contains the traditional linear HRF first-order kernel, $h_1(\tau)$, and the second term includes the second order kernel, $h_2(\tau_1, \tau_2)$.

Many of the issues surrounding the use of second-order Volterra kernel basis functions are the same as they are for linear basis functions. For example, Volterra kernels can be determined empirically *(25, 28)*, or derived from non-linear biophysical models *(22)*. Either way, as with linear basis functions, there is variability in the response between different subjects and brain regions, and this variability can be parsimoniously captured within the GLM by using basis functions. We can obtain parsimonious basis sets for first- and second-order kernels by using principle component analysis on samples of the response from empirical data or from parameterised models. Furthermore, we can infer on the Volterra kernels in a Bayesian framework with priors that prohibit nonsensical responses, so that more flexibility can be allowed while protecting against over-fitting *(22)*.

5. De-noising fMRI Data

fMRI data are inherently noisy and contain a variety of fluctuations induced by processes beyond the control of the experimenter. Examples of such effects include artefacts related to the MR physics (such as slice-dependent signal dropout due to imperfect switching of the slice-select gradients, EPI "ghosting," and thermal noise), subjects' head motion effects, fluctuations induced by the cardiac and respiratory cycles, and spontaneous low-frequency fluctuations of the baseline signal.

Under the assumptions of the GLM, any fluctuation in the measured BOLD signal that is not modelled by the EVs in the design matrix is deemed to be noise. In **Subheading 3.3**, we discussed how such artefacts are more likely to occur at low frequencies and how we can use pre-whitening to deal with this. However, this assumed that all these fluctuations are stochastic, and that within the framework of the GLM, this stochastic noise is well described by a Gaussian distribution.

In practice, however, some of the underlying random noise fluctuations will have very distinct spatial and/or temporal structure. As an example, **Fig. 19** shows a variety of such structured noise components identified from a single fMRI dataset using an independent component analysis (ICA) decomposition *(29)*.

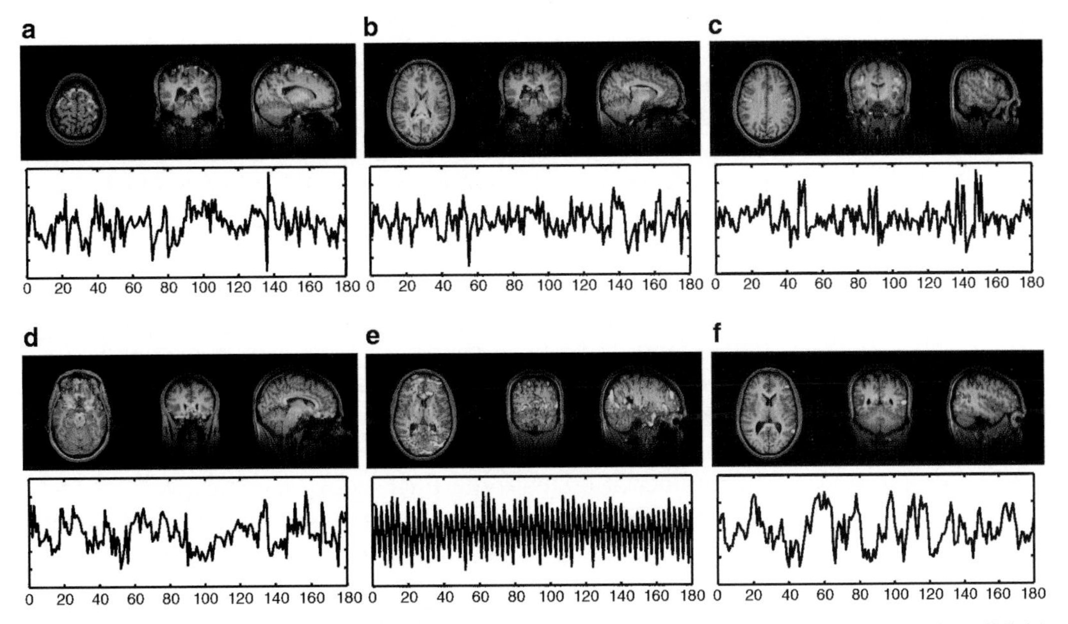

Fig. 19. Examples of "structured noise" identified in a single fMRI dataset using independent component analysis *(29)*: (**a**) residual head motion, (**b**) signal fluctuations in the ventricles, (**c**) spontaneous fluctuations in the bilateral sensory motor cortex, (**d**) fluctuations close to the sinuses (likely due to interactions between B_0 field inhomogeneities and head motion), (**e**) high-frequency image ghosting, and (**f**) more spontaneous low-frequency fluctuations.

This suggests that such effects are structured rather than random stochastic noise and the challenge is to account for their existence in order to obtain optimal estimates of the GLM model parameters.

5.1. Structured Noise and the GLM

The main problem with such structured noise artefacts is that they can severely impact our GLM-based analysis. The part of the artefact that is orthogonal (uncorrelated) with all of the EVs will simply not be modelled by the design matrix regressors, and therefore, the presence of the structured noise effect will be reflected by an increase in the residual GLM noise variance. This, in turn, will decrease any T- or F-statistics value, making it harder for us to detect true activations. The non-orthogonal (correlated) part of such an artefact, however, will result in wrong parameter estimates for those EVs that correlate with the artefact. If the correlation is positive, we will overestimate the parameter for the EV, making it more likely that we wrongly detect activations where there are none. If, on the other hand, the correlation is negative, then estimated effect sizes will be underestimated, making it harder to detect true activations.

In short, the impact on the GLM estimates and the statistics values can be profound. Therefore, if we can characterise the spatial and/or temporal structure of such artefacts, it is desirable to incorporate this knowledge into the data analysis and to explicitly account for the presence of these effects in the data.

5.2. Nuisance Regressors in the GLM

One possible way of correcting for the negative impact on GLM statistics is to introduce additional "nuisance" or "confound" regressors in the GLM design matrix. Remember from **Subheading 3.9** that in the case of multiple regressors, the parameter estimates for each of the EVs can only be driven by the uncorrelated (orthogonal) component of an EV. If we can find a suitable characterisation of the temporal structure of an artefact, we can add this as a new regressor to the design matrix in order to use this to "explain" some of the measured variation in the data. The parameter estimate for EVs of interest will then only reflect the amount of variation that the EV can explain over and above what can already be explained by nuisance variables. Note, however, that we do need to pay a price for the use of nuisance regressors as part of the design matrix: every new regressor does reduce the number of DOF for our final statistical comparison. As such, it is desirable to keep the number of nuisance regressors to a minimum while trying to maximise the amount of structured noise variance captured by these regressors.

Figure 20 illustrates the use of a nuisance regressor. In this example, an fMRI time series exhibits an intensity jump due to the presence of a scanner-induced image artefact such that during one of the TRs, the measured image intensity for this time

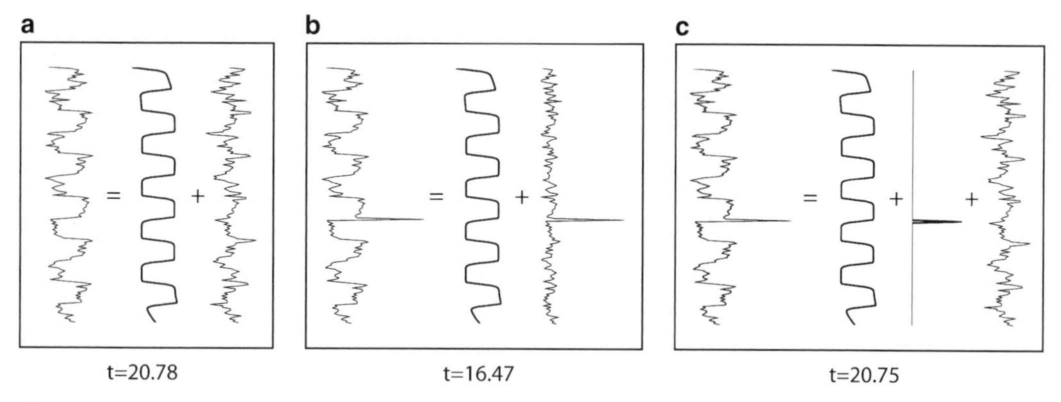

Fig. 20. Example of the utility of nuisance regressors in the general linear model (*GLM*): (**a**) data without artefact regressed against a single explanatory variables (*EVs*), (**b**) data with confound analysed in a GLM without nuisance regressor, and (**c**) confounded data analysed using both the EV of interest and a nuisance regressor.

point is about 10% above the mean intensity level. In **Fig. 20a**, the voxel's time series without the artefact is analysed using a simple GLM with one single EV. The analysis represents the data as a linear combination of the EV of interest and residual noise. The level of activation in this voxel is high, resulting in a very significant T-statistic. In **Fig. 20b**, the same GLM design matrix is now fitted to the voxel's time series with the artefact present. The timing of the artefact is almost entirely uncorrelated with the primary EV and the presence of the artefact will therefore result in an inflated residual variance, causing a significant drop in the T-statistics value of more than 20%. If information about the temporal characteristics of the artefact is available, then we can model the intensity variation at the specific time of the artefact by introducing a nuisance variable into the GLM design.

5.2.1. Deriving Nuisance Regressors

There are various ways of deriving useful nuisance variables. In general, these additional EVs should reflect the temporal characteristics of structured noise that is thought to exist in the data.

Motion of the subject in the scanner is a typical problem in fMRI, and there are often intensity fluctuations related to head motion still present in the data even after alignment-based motion correction. A common set of nuisance regressors (used to help model out such residual effects of motion) is the set of six time series obtained as the parameters of the head motion correction procedure. In this case, the intensity variations in the data are expected to correlate with the size of the three translations and three rotations. When included, these regressors can jointly "explain" any signal variation in the data correlated with head motion.

Other sources of structured noise effects are the subjects' cardiac and respiratory cycles. A popular approach is to use retrospective image correction [RETROICOR *(30)*] in order to correct for these effects. Using additional measurements of the heart and respiration cycles, one can derive nuisance regressors that permit one to remove all signal variation in the fMRI data that temporally correlates with the relative phase of these physiological cycles. The regressors are based on these additional measurements as low-order Fourier terms, and can significantly reduce the amount of structured noise induced by physiological fluctuations.

5.2.2. ICA-Based De-noising

The ability to correct for additive structured noise depends on the ability to characterise these noise components in terms of their temporal evolution. In the previous two examples, this was obtained by accurately estimating rigid-body motion or by using secondary measurements of physiological processes. For other types of noise, it is often not easy to predict such nuisance regressors based on the understanding of the biophysics and of the imaging process. One possibility is to use a model-free data analysis approach, such as ICA, in order to identify structured noise effects in the data prior to the GLM analysis. ICA and related techniques decompose the fMRI data into modes of variation that define the spatial and temporal extent of underlying fluctuations *(29)*. The estimated time courses of a component can then be used as nuisance regressors as part of a GLM analysis. An alternative is to explicitly regress out such effects prior to a GLM-based analysis, effectively running the model-based analysis on the residuals of a prior linear regression model designed to de-noise the data. Currently, as no well-established techniques exist for automatically identifying such noise components, such an approach relies on the experimenter to visually inspect all components. Further research is required to integrate such a model-free identification (e.g., using ICA) into the standard GLM in an unbiased objective way.

5.3. Example

Figure 21 demonstrates the impact of structured noise on the GLM and highlights the utility of such a de-noising approach. Subjects were requested to perform simple finger tapping using either the left or right hand. All subjects were right-handed and one of the contrasts of interest involved the left-right comparison "Where is the activity larger when using the left hand when compared with the right hand?." Prior to the noise removal, the histogram of the *Z*-statistic values for this contrast is highly non-Gaussian. The contrast map itself did not reveal any significant differences when using standard thresholding. After de-noising, the *Z*-statistic image of this contrast identifies significant differences in the BOLD, particularly in right motor cortical areas.

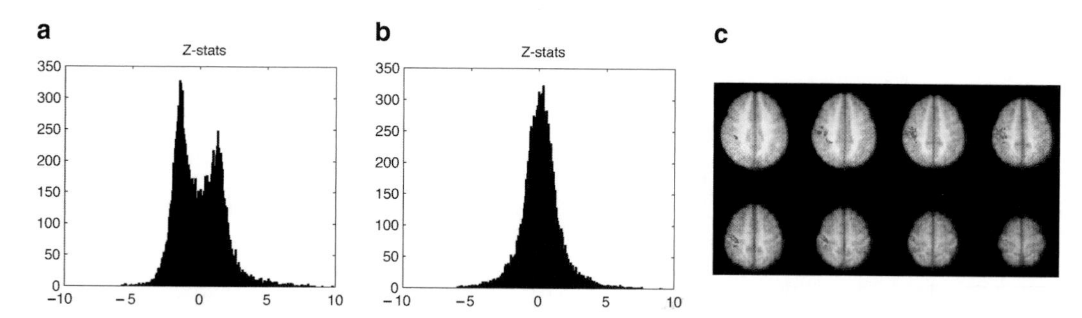

Fig. 21. Example of the effect of fMRI de-noising in a simple finger tapping experiment: (**a**) histogram of the *Z*-statistic image for the differential left- versus right-hand finger tapping contrast. Because of the presence of structured noise, the histogram is far from being Gaussian distributed; (**b**) after regressing out a variety of structured noise effects, the histogram of *Z*-statistic values becomes unimodal and much closer to a Gaussian distribution; and (**c**) map of significant voxels after regressing nuisance effects out of the data.

6. Multi-Subject Statistics

We have so far only focused on ways of modelling and fitting the (time series) signal and residual noise at the individual single-session level, in order to derive effect size estimates from a single fMRI dataset. The majority of fMRI studies, however, are used to address questions about activation effects in populations of subjects. This generally involves a multi-subject and/or multi-session approach where data are analysed in such a way as to allow for hypothesis tests at the group level *(12, 31)*, for example, in order to assess whether the observed effects are common and stable across or between groups of interest.

Figure 22 illustrates an example scenario where the question of interest involves estimating the difference in activation between two groups of subjects. This question is addressed by having different GLMs at the session, subject, and group level in a hierarchical fashion. At the lowest level of the analysis, the single-session time series data is modelled in the way described previously. At the subject and group level, there are GLMs that, for example, model the subject (cross-session) mean and group (cross-subject) mean effect sizes, respectively. At the top-level of the hierarchy, a set of statistic images is created that can be used for final statistical inference.

6.1. Brain Atlases

Registration (aligning different brain images) is typically used when combining fMRI data from different sessions or subjects in a multi-subject analysis. This allows us to assume that the data we are comparing across sessions or subjects come from approximately corresponding areas of the brain. In doing this, it is typical to transform the data into a common "standard brain space," for example, the co-ordinate system specified by Talairach and Tournoux *(32)*. These standard spaces can be either what are known as templates or atlases.

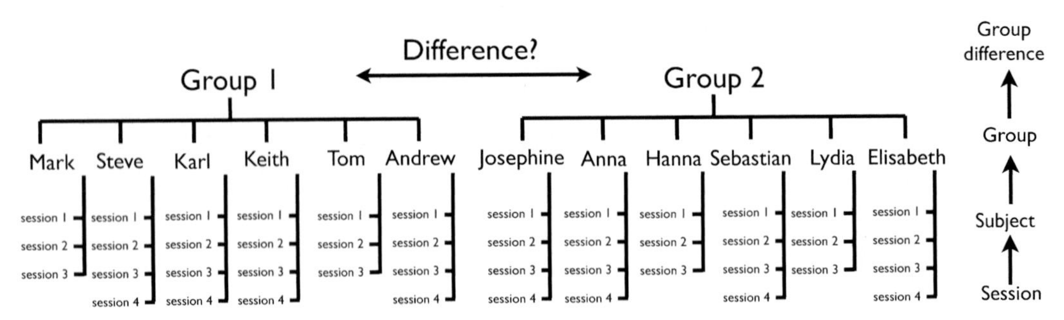

Fig. 22. Hierarchical general linear model (*GLM*) for the analysis of group fMRI data. Within a summary statistics approach, the GLMs are estimated one level at a time and summary statistics are passed up to the next level of the hierarchy.

A *template* is typically an average of many brains, all registered into a common co-ordinate system. An example is the MNI 305 average *(33)*. An *atlas* is also based in a common co-ordinate system, but contains richer information about the brain at each voxel, for example, information about tissue type, local brain structure, or functional area. Atlases can inform interpretation of fMRI experiments in a variety of ways, helping the experimenter gain the maximum value from the data.

6.2. Fixed- Versus Mixed-Effects Models

An important question is that of how to model and estimate effects at the intermediate and higher level of the hierarchy. If we were only concerned about the particular set of subjects in our study, then we would use a fixed-effects model. More typically, however, we would want to generate results that extend beyond the particular population of subjects scanned as part of the study, into the wider population. In this case, we also need to account for the fact that the individual subjects themselves are sampled from the wider population and thus are random quantities with associated variances. It is exactly this step that marks the transition from a simple fixed-effects model to a mixed-effects model and it is imperative to formulate a model at the group level that allows for the explicit modelling and estimation of these additional variance terms.

As an example, consider the simplest case of estimating the effect size of a group of M subjects, where for each subject k, the pre-processed fMRI data is Y_k, the design matrix is X_k and the parameter estimates are β_k (for $k = 1,\ldots, M$). The individual first-level GLMs relate first-level regression parameters to the M individual datasets: $Y_k = X_k\beta_k + \varepsilon_k$, where ε_k specifies the single-subject residuals. If we are only concerned about the exact population of subjects scanned under our fMRI paradigm, then the estimate of the group mean effect size is simply the average over all the lower-level parameter estimates: $\beta_g = (1/M)\sum_{k=1}^{M}\beta_k$, that is, our second-level GLM is simply

$$\beta_k = X_g \beta_g \tag{16}$$

where $X_g = [1/M, \ldots, 1/M]^T$, and we have concatenated all of the first-level regression parameters into one vector:

$$\beta_k = \begin{bmatrix} \beta_1 \\ \beta_2 \\ \vdots \\ \beta_2 \end{bmatrix}.$$

In this case, we simply need to average the first-level regression parameters and the only variance to consider is the average first-level variance.

If, however, we want to generalise our findings to the wider population then the second-level analysis needs to account for the sampling of the subjects, and we need to proceed by modelling the group effect size of interest as

$$\beta_k = X_g \beta_g + \varepsilon_g \tag{17}$$

where ε_g accounts for the variation of the different subjects' means from the overall group mean. Both the within- and the between-subject variations contribute to the total mixed-effects variance against which the mean effect size is tested during the statistical inference procedure.

The difference between the two approaches is illustrated in **Fig. 23**. In the fixed-effects model (**a**), only the first-level variances need to be considered, whereas in the case of the mixed-effects analysis (**b**), both the first-level fixed-effects variances and the higher-level random-effects variance contribute to the total mixed-effects variance used for inference. The between-subject

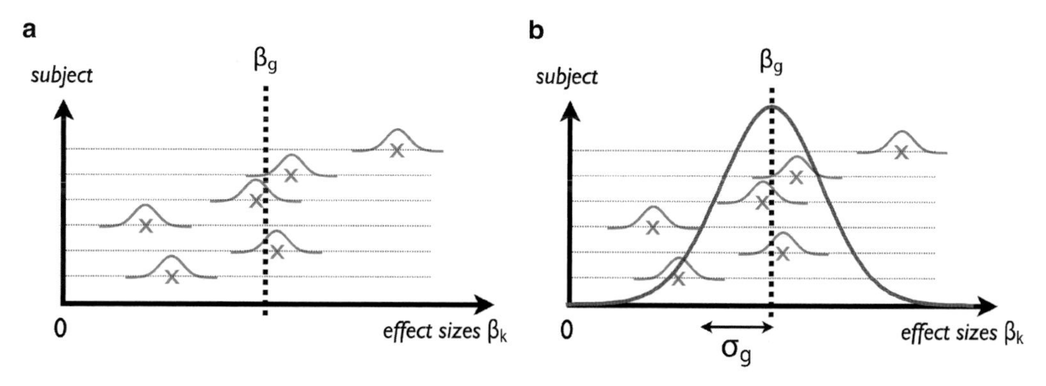

Fig. 23. Illustration of a simple group analysis using the fixed-effect and the mixed-effects models: (**a**) in the fixed-effects analysis, the only variance contribution to consider is the lower-level within-subject variance; (**b**) in the mixed-effects analysis, the between-subject random-effects variance, σ^2, accounts for the random sampling of the subjects themselves and contributes to the overall mixed-effects variance.

variance, σ_g^2, then accounts for the random sampling of the particular subjects from the wider population.

In the following sections, we assume that a mixed-effects analysis is being performed, as this is typically what is required. Fixed-effect analyses are performed in a similar manner but without the complication of needing to estimate the random-effect variances.

6.3. Summary Statistics Approach

We now come to the question of how we infer on the multilevel hierarchy of a group study (an example of which was shown in **Fig. 22**), in order to ask questions such as "Where is there significant activity in response to the experimental task for the population?" or "Where is there significant differences between populations (e.g., controls versus patients)?" Recall that each level in the hierarchy is represented by its own GLM. Hence, one approach is to formulate a single complete GLM that combines together the first-level and higher-level GLMs. An example of such an approach is presented in Friston et al. *(34)* where the group analysis is carried out "all-in-one" using the within-session fMRI time series data as input. However, in fMRI, where the human and computational costs involved in data analysis are relatively high, it is desirable to be able to make group-level inferences using the results of separate first-level analyses. This approach is commonly referred to as the *"summary statistics"* approach to fMRI analysis *(31)*. Within such an approach, group parameters of interest can easily be refined as more data become available.

In Holmes and Friston *(31)*, the regression parameter estimates were used as summary statistics. The regression parameter estimates from the lower level are used as the "data" at the next level up. For example, the estimates of β_k are used in place of β_k in **Eq. 17**. This approach was shown to be equivalent to inferring all-in-one under certain conditions *(31)*. For example, it requires balanced designs, that is, all lower-level design matrices need to be identical, preventing the use of behavioural scores or subject-specific confound regressors.

However, top-level inference using the summary statistics approach can be made equivalent to the all-in-one approach without such restrictions *(35, 36)*, if we pass up the correct summary statistics. In particular, it is important to pass up information about not only the effect sizes from the lower levels, but also their variances. We shall explore the benefits of doing this in **Subheading 6.4**.

6.4. Estimation of the Mixed-Effects Model

When we use a summary statistic approach, we infer on each level of the hierarchy one at a time. The first-level inference is as we described earlier in the chapter. At higher levels in the hierarchy, however, estimating the regression parameters and variances of group-level GLMs offers a different set of challenges. At the first level, there typically exists a large number of observations (typically more than 100), so that relevant parameters and variances can be estimated with high DOF. In contrast, group-level variance component estimation is typically troubled by having very few observations (i.e., low DOF).

A key issue in estimating mixed-effects models within the "summary statistics" approach is whether the variance information from the lower levels (e.g., first level) is used at the higher levels (e.g., group level). Approaches that use the variance information from the lower levels have a number of substantial advantages. First, such approaches do not require balanced designs and so permit the analysis of fMRI data where the first-level design matrices have different structure from each other (e.g., contain behavioural scores as regressors) or where the data contains different numbers of observations (e.g., different numbers of sessions for each subject). Second, they provide more accurate variance estimation (and therefore more accurate inference) by ensuring that at every level, only positive estimates of the random-effects variances contribute to the overall mixed-effects variance. Finally, such approaches increase the ability to detect real activation, by weighting the different contributions from the lower levels by using the lower-level variance information. For example, effect sizes from subjects with high first-level variance get down-weighted compared with those with low first-level variance, when inferring at the group level.

Estimating mixed-effects models when the lower-level variance information is ignored can be carried out easily using ordinary least squares *(31)*. Approaches that use the lower-level variance information and provide the advantages described above are a little more involved. For example, Worsley et al. *(12)* used an expectation maximisation approach. Woolrich et al. *(36)* used a fully Bayesian framework using appropriate non-informative priors. This approach had the added advantage that one can model different variance components for different groups. For example, one can contrast effect sizes in a population of patients relative to a population of controls under the assumption that these two groups have different within-group variance. This is important as, empirically, patient populations exhibit larger within-group variability than a carefully selected population of controls.

6.5. Handling Outlier Subjects

The approaches described so far assume that the population distributions of the effect sizes are well modelled using a Gaussian distribution. However, in practice group studies can include "outlier" subjects whose effect sizes are completely at odds with the general population for reasons that are not of experimental interest. For example, it could be due to excessive subject motion or misunderstanding by the subject of the experiment instructions. The estimate of the population variance can be inflated by outlier subjects, and the population mean estimates can be under- or over-estimated. This is analogous to the presence of structured noise in the context of single-session analysis, as discussed in **Subheading 5**.

A number of approaches have been proposed to deal with this problem. One option is to visually inspect both the data and the results of a group analysis to deduce outliers. These outlier subjects can then be removed and the group study re-analysed without them. Although useful exploratory approaches have been proposed that aid in this process *(37–39)*, human intervention is often still required. It is preferable to use approaches that are automatic, and soft-assign outlier behaviour in a spatially localised manner *(40, 41)*. The approach of Woolrich *(41)* has the added benefit that it uses the lower-level variance information.

Another possibility to dealing with outliers is to use non-parametric statistics (*see* **Subheading 7.4.2**). For example, Meriaux et al. *(42)* and Roche et al. *(43)* use permutation tests that take advantage of the lower-level variance information. These approaches protect the validity of the statistics but can be less sensitive compared with techniques that explicitly model the outliers *(41)*. However, they are also potentially able to handle other deviations from Gaussian population distributions (e.g., populations with two sub-populations), and there is evidence that non-parametric statistics in general handle the multiple comparison problem better than random field theory (RFT) *(44)* (this is discussed further in **Subheading 7.4.2**).

6.6. Creating Higher-Level GLMs

When creating higher-level GLM design matrices, it is important to ensure that the design matrix at least models the cross-subject mean effect. This is different from a first-level analysis where the overall time series mean is normally not of interest and might actually be removed prior to the first-level GLM. In the case of a higher-level analysis, however, the mean lower-level effect often is exactly what is of interest and therefore needs to be explicitly modelled as part of the design matrix, either as a single EV or as a linear combination of EVs.

Figure 24 gives a selection of typical fMRI higher-level designs. In the simplest case (**a**), the higher-level design only models a single group mean effect and a simple [1] contrast then tests if the mean effect is greater than 0.

In some cases, additional subject-specific behavioural scores need to be included as additional EVs (**b**). This can be either to remove some higher-level variation of no interest by including these EVs as nuisance regressors (e.g., by regressing out subjects' age or gender or reaction time), or because these regressors are part of a testable hypothesis and need to be included in a contrast of interest (e.g., a researcher might be interested in effects which correlate significantly with duration of treatment in

a $\begin{bmatrix} 1 \\ 1 \\ 1 \\ 1 \\ 1 \\ 1 \\ 1 \\ 1 \end{bmatrix}$ **b** $\begin{bmatrix} 1 & -2 \\ 1 & -7 \\ 1 & 4 \\ 1 & 0 \\ 1 & 3 \\ 1 & 4 \\ 1 & -5 \\ 1 & 3 \end{bmatrix}$ **c** $\begin{bmatrix} 1 & 0 \\ 1 & 0 \\ 1 & 0 \\ 1 & 0 \\ 0 & 1 \\ 0 & 1 \\ 0 & 1 \\ 0 & 1 \end{bmatrix}$ **d** $\begin{bmatrix} 1 & 1 & 0 & 0 & 0 \\ -1 & 1 & 0 & 0 & 0 \\ 1 & 0 & 1 & 0 & 0 \\ -1 & 0 & 1 & 0 & 0 \\ 1 & 0 & 0 & 1 & 0 \\ -1 & 0 & 0 & 1 & 0 \\ 1 & 0 & 0 & 0 & 1 \\ -1 & 0 & 0 & 0 & 1 \end{bmatrix}$

Fig. 24. Typical higher-level general linear model (*GLM*) design matrices: (**a**) group mean effect size over eight subjects, (**b**) group mean and confounds over eight subjects, (**c**) unpaired group difference over two groups of four subjects, (**d**) paired group difference test over two conditions for four subjects. Note that for the sake of space, the number of subjects assumed here is lower than what would be typically expected in a group study.

a clinical population). In both cases, the additional EVs would need to be orthogonalised relative to the EV that is modelling the group mean. This is so that the regression parameter for the group mean EV can indeed be interpreted as the overall group mean effect.

The simplest multiple-group design involves just two groups where the question of interest involves assessing the between-group difference (**c**). In this case, each group's mean effect is modelled using a separate EV, a [1 –1] contrast can then be used to assess A > B differences; the negative [–1 1] contrast tests for B > A differences.

Another typical design involves testing for differences between a set of data generated under different conditions A and B in the same population (**d**), for example, where subjects get scanned before and after a period of learning. This is often referred to as a *paired T-test.* Every subject has two observations and we need to account for the within-subject covariance by means of subject-specific confound regressors. In this case, the first EV models the A – B differences for the *M* subjects, while the *M* additional EVs account for the subject-specific mean effects. For example, the second EV in **Fig. 24d** models the mean effect for the first subject. A [1 0 0 0 0] contrast can then be used to assess the A – B paired difference.

7. Inference ("Thresholding")

As we saw in **Subheading 3.5**, a result of fitting a GLM is a *T*-statistic image for each contrast, where the intensity at each voxel assesses the evidence for a non-zero effect. Ideally, the statistic image would be zero where there is no effect and very large where there is an effect. Of course, because of the noise, this is not the case, and we must make inference - a statistically calibrated decision – on which voxels exhibit a signal and which voxels are just consistent with noise. Here we are talking about inferring on *T*-statistic images; however, the issues involved with *F*-statistic images generated from *F*-tests, or *Z*-statistic images (*see* **Subheading 3.7**), are similar.

The simplest inference method is voxel-wise thresholding. If the value of a *T*-statistic image is t at a given voxel, then we reject the null hypothesis of no experimental effect if $t > u$ (where u is a significance threshold). However, one should first ask: Why threshold?

7.1. Inference with the Mass Univariate Model

The natural questions that any user of FMRI has of their data are: "What is the location of my signal?," "What is the extent of my signal about that location?," "What is the magnitude of my signal?." For each of these, our statistical model should provide an estimate, a measure of uncertainty of the estimate (i.e., a standard error or a confidence interval), and a significance measure, like a *P*-value (i.e., could our result be explained by chance alone). For example, if a visual effect produced a cluster (a contiguous groups of supra-threshold voxels, more on this in **Subheading 7.2**) with a peak at a certain location, how certain can I be that the true centre of activation is near that location? Or, if a cluster had a volume of 500 voxels, what is my confidence that the true signal extent is 500 voxels?

Surprisingly, such basic questions cannot be answered with standard fMRI methods. In fMRI, we generally use a mass univariate model, where a GLM is fit independently at each voxel. No information is shared over space, and, specifically, no explicit spatial model is used to express the extended signals that we expect. While more advanced methods that address these issues exist *(45)*, the only inferential questions that a mass univariate model can answer are *(a)* "What is the signal magnitude at each voxel (with standard errors and *P*-values)?" and *(b)* "What is the signal extent for a given cluster-defining threshold (*P*-values only)?." However, standard errors and *P*-values on locations (e.g., confidence intervals on local maxima or centre of mass of a cluster) are not available.

The remainder of **Subheading 7.2** focuses on these two types of inferences: voxel-wise and cluster-wise.

7.2. Voxel-Wise Versus Cluster-Wise Inference

The result of applying a contrast to the GLM fit at each voxel is a statistic image. This is anywhere from I = 20,000 to 100,000 brain voxels in a statistic image. The value in the image at each voxel is a *T*-, *F*-, or *Z*-statistic that measures the evidence for an effect defined by the contrast. The process of applying threshold u, and retaining all voxels with statistic value greater than u is known as voxel-wise inference. Precisely, we are performing I statistical tests of significance, rejecting the null hypothesis at voxel i if $t_i \geq u$, where t_i is the statistic value at voxel i.

Alternatively, we can apply a cluster-forming threshold, u_{clus}, create a binary image of voxels $t_i \geq u_{clus}$, and identify clusters, that is, contiguous groups of supra-threshold voxels, with cluster having size S. The process of retaining all clusters with size greater than k is known as cluster-wise inference. Precisely, we are performing L statistical tests, one for each of the L clusters in the image, rejecting the cluster null hypothesis if $S \geq k$. The cluster null hypothesis is that all of the voxels in cluster have no signal, and the inference is cluster-by-cluster. Hence, an unusually large cluster extent only tells us that there exists one or more signal voxels within the cluster, but not which voxels within the cluster have signal.

Figure 25 illustrates the difference between voxel- and cluster-wise inference. While voxel-wise inference rejects the null hypothesis for individual voxels, cluster inference jointly rejects the null hypothesis for a set of voxels within a cluster as significant. As such, we say that cluster-wise inference has less spatial specificity than voxel-wise inference. On the contrary, since there are many fewer clusters than voxels, the multiple testing problem (*see* **Subheading 7.3**) is more severe with voxel-wise inference.

What cluster-forming threshold should be used? Using a relatively low threshold will allow clusters with small statistic values to be formed, but the clusters may not be significant, as a low threshold will also result in large clusters by chance alone. A high threshold ensures that clusters due to chance noise are small, but may then miss true signals that have relatively small magnitude. Below we will introduce two methods for finding P-values for cluster size, one of which requires relatively high thresholds: Random Field Theory requires relatively high u_{clus} values (uncorrected P-value of 0.001 or smaller) to give accurate inferences *(46)*, while permutation is valid with any chosen u_{clus} threshold. Regardless of which threshold is chosen, the threshold should be picked before examining the data. Trying many thresholds introduces a multiplicity that is not easily accounted for, and will reduce the confidence of any significant findings. Later, in **Subheading 7.5**, we will consider the approach of threshold-free cluster enhancement, which provides cluster-like inference without the dependence on u_{clus}.

Should we be using voxel-wise or cluster-wise inference, or both? If one examines both cluster-wise and voxel-wise results yet another multiple testing problem is introduced, and so one method does need to be chosen a priori. Friston et al. *(47)* shows that when the anticipated signals are broad or spatially extended, cluster size inference is best, and when focal, intense signals are anticipated, voxel-wise inference is best. Both methods are widely

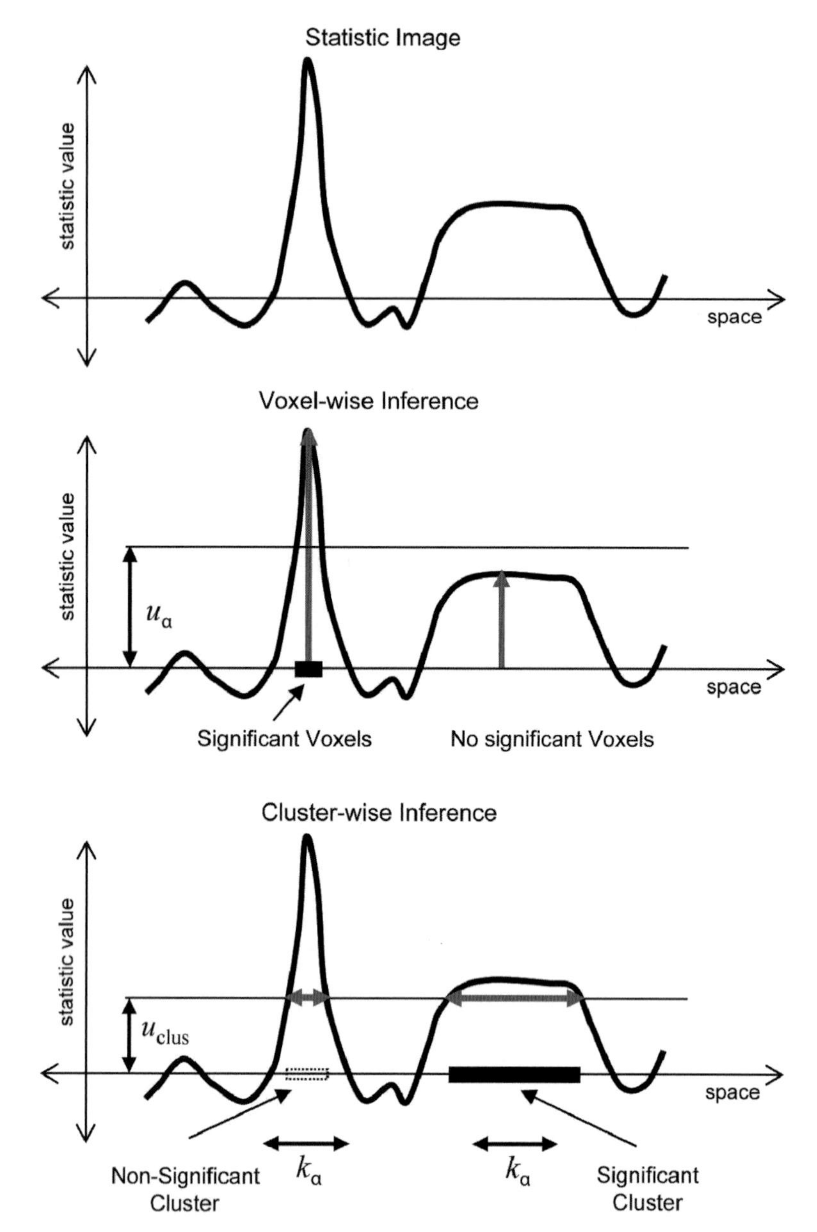

Fig. 25. Voxel-wise versus cluster-wise inference. Inferences on fMRI statistic images are made either through voxel-wise or through cluster-wise methods. The top panel illustrates the values in a statistic image through one line of space, where large values indicate evidence for an experimental effect. The middle panel illustrates voxel-wise inference, where a significance threshold u_α is applied to the image, and voxels above that threshold are labelled as significant. The advantage of voxel-wise inference is that individual voxels are marked as significant, but the disadvantage is no spatial information is considered, and unusually expansive effects may be missed. The bottom panel illustrates cluster-wise inference, where a cluster-forming threshold u_{clus} is applied to the image, and contiguous voxels are formed into clusters. Clusters that exceed a significance threshold k_α in size are marked as significant. The advantage of cluster-wise inference is that low, spatially extended signals can be detected. The disadvantage is that clusters as a whole are marked as significant, and individual voxels within a cluster cannot be marked individually as significant.

used, with cluster-wise inference being slightly more common, probably due to the large extent of effects typical after spatial smoothing.

7.3. Correction for Multiple Tests

Statistical hypothesis testing provides a means to test a default, or null, hypothesis with a pre-specified false positive rate. At a single voxel, a test statistic can be compared to an $\alpha = 0.05$ significance threshold, denoted u_α, where α is the allowed risk of false positives. For example, a Z-statistic at one voxel will, with many repetitions of a null experiment, exceed $u_\alpha = 2$ about 2% of the time.

For a particular observed statistic value, say $t = 3.3$, we measure the evidence against the null hypothesis with a P-value, here $p = 0.0005$, which is the chance of obtaining, over repeated null experiments, a result greater or equal to $t = 3.3$. (Take care not to confuse P-values with posterior probabilities: Bayesian methods give the posterior probability that the null is true, conditional on the data; classical P-values, in contrast, are the probability of the data, conditional on the null hypothesis).

But if $I = 10{,}000$ voxels (or $K = 100$ clusters) are tested at level $\alpha = 0.05$, the false positive risk is only controlled at each voxel. Then over all voxels, a total of $I \times \alpha = 500$ false positive voxels (or five false positive clusters) would be expected. The problem is that standard hypothesis testing only accounts for the risk of false positives for one test. If multiple tests are to be considered, some measure of the risk of false positives over multiple tests is required.

7.3.1. Measures of Multiple False Positives: FWE and FDR

To carefully define different types of false positive measures, **Table 2** shows a cross classification of each of the I voxels in an image. Voxels can be truly null, or truly have non-zero signal, and additionally each voxel can be detected by some (imperfect) thresholding method, or fail to be detected.

Table 2
Cross-tabulation of the number voxels in difference inference categories

	Voxels not detected	Voxels detected	
True null voxels	I_{N-}	I_{N+}	I_N
True signal voxels	I_{S-}	I_{S+}	I_S
	I_-	I_+	I

To fill in the values for true null and true signal voxels, we need to somehow know the underlying truth. Voxels that are marked as significant are "detected" and voxels that are not marked as significant are "not detected"

The standard measure of false positives in multiple testing is the *family-wise error (FWE) rate*, that is, the chance of one or more false positive voxels (or clusters) anywhere in the image [FWE = $P(I_{N+} > 0)$]. *Bonferroni* is the most widely known FWE method, which produces critical thresholds u^{FWE} that controls the FWE. This method simply divides the FWE rate (e.g., 0.05) by the number of tests (e.g., 10,000), to give a voxel-wise *P*-value threshold α (0.000005) that, when applied voxel-wise, results in the originally desired FWE control. Regardless of the method, if an $\alpha^{FWE} = 0.05$ threshold is used, one can be 95% confident that there are no false positives in the image at all.

A more recent measure of false positives is the false discovery rate (FDR), the expected false discovery proportion (FDP), where FDP is the proportion of false positives among all reported positives. Precisely, if I_+ voxels are detected, FDP is the proportion of these that are false "discoveries," FDP = I_{N+}/I_+, where FDP is defined to be 0 if no voxels are detected. FDP is a random quantity that cannot be known for any particular real dataset, so the FDR is defined as the expected value of FDP over many datasets, FDR = *E*(FDP). **Figure 26** shows the difference between FDR and FWE inference. Imagine the ten images shown in the figure as the next ten experiments you will analyse; of course, you only consider a single dataset at a time, but this illustrates how the methods are calibrated over many (idealised) repetitions of an experiment.

FDR is a more lenient measure of false positives allowing some false positives – as a fraction of the number of detections – while FWE regards any false positives as an error. One special case is notable when the definitions of FDR and FWE coincide. If there are no signal voxels at all ($I_S = 0$), then FDP is 1 whenever there is an FWE, and the two methods give the same control of false positives. (In technical terms, it is said that FDR has weak control of FWE). This hints at the adaptive nature of FDR: when there is no signal, it behaves like FWE; as there are more and more signal voxels, it admits more and more false positives, yielding increased power while still controlling false positives in proportion.

The voxel-wise *P*-values referred to in previous sections are more accurately referred to as "uncorrected *P*-values," as they do not account for the multiple testing problem. "Corrected *P*-values" refer to the FWE rate (or FDR if this is used instead).

7.4. Corrected Inference Methods

So far we have only defined measure of false positives, but we have not described how we obtain thresholds that control these false positive measures. In addition to Bonferroni, there are two further methods that are commonly used in fMRI for controlling FWE; these are RFT and permutation, whereas there is generally just a single method for FDR.

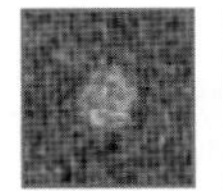
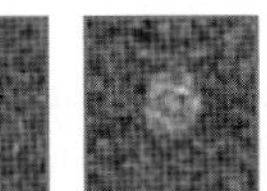
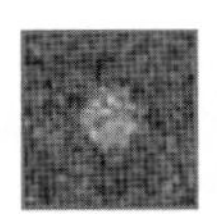
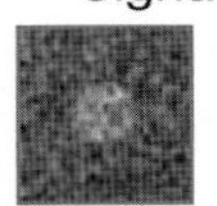

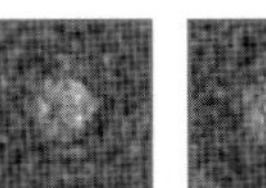

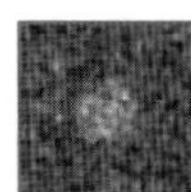
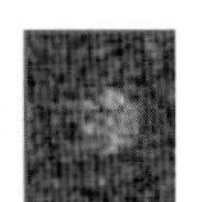

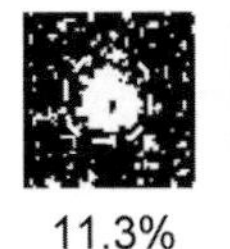
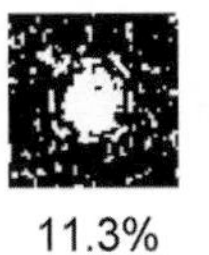

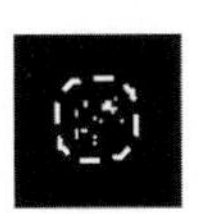
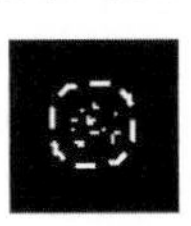

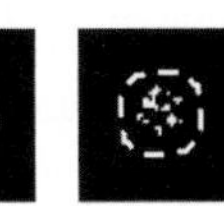
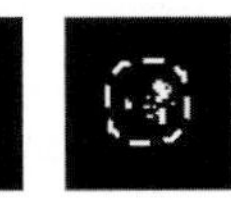
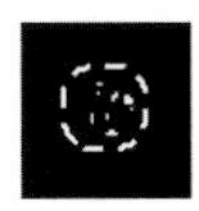

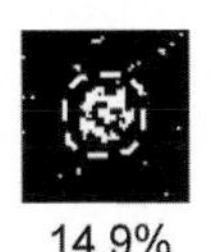

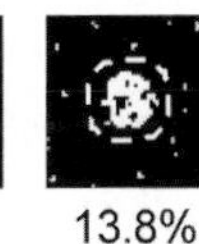
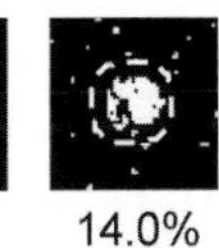
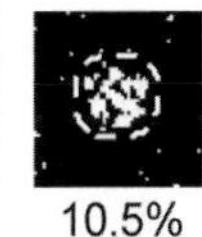
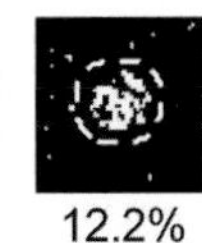

Fig. 26. Comparison of uncorrected versus family-wise error (*FWE*)-corrected versus false discovery rate (*FDR*)-corrected inferences. The top rows shows ten realisations of a central, circular signal added to smooth noise, and the next three rows show different possible voxel-wise thresholding methods applied to each realisation. The dashed circles indicate the extent of the signal. The second row shows the result of an $\alpha = 10\%$ uncorrected threshold; most of the signal is correctly detected, but much of the background is also incorrectly detected. On average, 10% of the background consists of false positives, but notice that the exact proportion of false positives varies from realisation to realisation. The third row shows the result of using an $\alpha^{\mathrm{FWE}} = 10\%$ threshold (e.g., a threshold from Bonferroni or random field theory). Much less of the signal is detected, but there are many fewer false positives, only 1-in-10 of the datasets considered had any false positives, this is a FWE. Of course in practice, we never know if the dataset in our hands is the 1-in-10 (or 20) that contains a family-wise error. The bottom row shows the result of using an $\alpha^{\mathrm{FDR}} = 10\%$ threshold. While we are guaranteed that the percentage of detected voxels that are false positive does not exceed 10% on average, the actual percentage can vary considerably.

7.4.1. Controlling FWE with RFT

The Bonferroni method for controlling FWE uses a significance threshold corresponding to $\alpha = \alpha^{\mathrm{FWE}}/I$, the nominal FWE test level divided by the number of tests. Bonferroni becomes quite conservative when the data is smooth, and has no way to adapt to the data in anyway. For example, imagine an extreme case where FMRI data is smoothed with a 1 metre wide Gaussian smoothing kernel; such data will produce a statistic image with essentially a single constant value, meaning there is no multiple testing problem any more; however, the Bonferroni threshold will still prescribe dividing α^{FWE} by, say, 10,000.

RFT uses the smoothness of the data to adjust the significance threshold while still controlling FWE. The mathematics involved are elegant yet quite involved, and in what follows, we only give the most cursory review. For a more detailed review, see *(48)*, or for a more technical overview, see *(49)*. The original Gaussian RFT paper for PET imaging remains a useful introduction *(50)*, though also see *(51)* for more up-to-date results including *T*- and *F*-statistic RFT results.

To use RFT results, we must know the smoothness of the data, precisely the smoothness of the standardised noise images (ε/σ in the notation from **Subheading 3**). Smoothness is parameterised by the full width at half maximum (FWHM) of the Gaussian kernel required to simulate images with the same apparent spatial smoothness as our data (**Fig. 27**). For example, if we say that our data has 6-mm FWHM smoothness, it means that if we were to simulate our data, we would generate noise data with no correlation and then convolve it with a Gaussian kernel with FHWM of 6 mm. The exact form of the spatial dependence of our data does not have to follow a Gaussian kernel, but for convenience, we describe the strength of the dependence in terms of the size a Gaussian kernel.

It may seem that if we take our fMRI data fresh from the scanner and convolve it with a 6-mm Gaussian kernel, our FWHM for RFT would be 6 mm. This is incorrect, however, as the noise smoothness includes both intrinsic sources of smoothness (imperfect MRI resolution, physiological artefacts, etc.) and smoothness induced by the applied smoothing. As a result, the smoothness for RFT is not a user-specified parameter, but rather estimated from the residuals of the GLM $(Y - X\hat{\beta})$.

RESELs

The definition of FWHM smoothness creates a notion of a smoothness-equivalent volume, a resolution element or RESEL. If the smoothness of the data is $FWHM_x$, $FWHM_y$, $FWHM_z$ in each of the principal directions, then a volume of space with

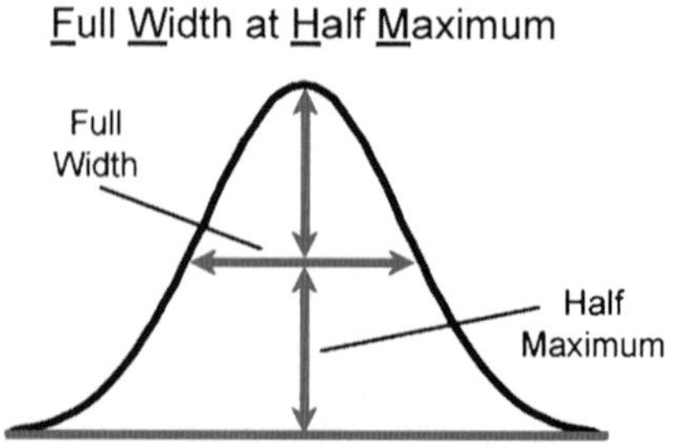

Full Width at Half Maximum

Full Width

Half Maximum

Fig. 27. Full width at half maximum (*FWHM*) is a generic way to describe the spread of a distribution, and is the way that smoothness is measured for random field theory.

dimensions $\text{FWHM}_x \times \text{FWHM}_y \times \text{FWHM}_z$ is one RESEL. In very approximate terms, the RESEL count captures the amount of independent information in the image; fewer RESELs = smoother data = less information = less severe multiple testing problem; more RESELS = rougher data = more severe multiple testing problem. The total RESEL count for a search volume is:

$$\text{RESELcount} = I / (\text{FWHM}_x \times \text{FWHM}_y \times \text{FWHM}_z) \quad (18)$$

where I is the total number of voxels in the brain and FHWM is expressed in units of voxels. The RESEL count is important because it is the summary measure that determines the RFT threshold, as illustrated next.

RFT-Corrected *P*-Values

RFT can be used with any type of statistic image a GLM can create, including *T*-, *F*-, and *Z*-statistic images *(51)*. The formulas provide FWE-corrected *P*-values for voxel-wise thresholds and cluster sizes and include corrections to account for edge effects (i.e., when blobs touch the edge of the search volume). The simplest result, for *Z*-statistic images with no edge corrections, can be used to gain some insight into the method.

For voxel-wise inference, a voxel with value z has a corrected *P*-value of:

$$P_{\text{vox}}^{\text{FWE}}(z) = \text{RESELcount}\,(2\pi)^{-2}(z^2 - 1)\exp\left(-\frac{z^2}{2}\right).$$

This shows that as z grows, the corrected *P*-value shrinks (the exponential term dominates), which of course makes sense, as larger statistic values should produce smaller *P*-values. As the RESEL count grows, the corrected *P*-value grows. The RESEL count can increase because the search volume increases, which again is sensible, as a greater search volume demands a greater correction for multiple testing, and hence a less significant *P*-value. The RESEL count can also increase if the smoothness decreases, as per *(18)*, which increases the amount of information in the image, again demanding a greater correction for multiple testing. For cluster-wise inference, the equation for the FWE-corrected *P*-value is more involved, but it also accounts for the image search space and smoothness.

Small-Volume Correction

The first RFT results published (and the equation shown above) assumed that the search region was large relative to the smoothness of the image. This assumption was needed to avoid dealing with the case when a cluster touches the boundary of the search region. To see why this could be a problem, imagine two statistic images with the same smoothness, one the size and shape of the brain, the other with the same total volume but the shape of a long, narrow sausage. In the latter case, it is more likely that clusters will touch the edge of the image, and, relative to other clusters, have smaller volume. The results in *(51)* can be used to produce *P*-values that are accurate even with small search regions. When these results are used, they are some times referred to as "small volume correction."

RFT Assumptions

The use of RFT results is based on several assumptions and approximations. The essential assumptions are:

1. *Gaussian data.* For any collection of voxels, the distribution of the data is multivariate Gaussian.
2. *Sufficient smoothness.* The data must be sufficiently smooth to approximate continuous random fields (upon which the theory is based).
3. *Known smoothness.* The results assume that the FWHM smoothness parameters are exact and contain at most negligible error.
4. *Constant smoothness for cluster-wise inference only.* The standard cluster-wise results assume that the smoothness is the same everywhere in the image. If the data are "non-stationary," regions of the brain that are smoother than others will generate large clusters just by chance. Updated methods are available *(52)* which account for varying non-stationarity, but they have reduced sensitivity unless the non-stationarity is severe; hence, this is principally suitable for voxel-based morphometry data. Generally, FMRI data does not exhibit severe non-stationarity.

In the light of these extensive assumptions, there can be good reason to seek alternative methods that do not require as many assumptions.

7.4.2. Controlling FWE with Permutation

Non-parametric methods are generally used when the standard parametric assumptions are known to be false or cannot be verified. In the case of RFT, the assumptions are nearly impossible to verify, but more importantly, it has been found that voxel-wise RFT results are quite conservative for small group studies [e.g., when the number of subjects is less than about 40; *(48, 53)*]. Hence, there has been interest in using alternative methods.

Instead of making assumptions about the distribution of the data, permutation testing uses the distribution of the data itself to find *P*-values and thresholds. **Figure 28** illustrates the reasoning of the permutation test in the two-group setting.

While non-parametric tests are sometimes referred to as assumption-free, in fact they also have assumptions, just much weaker ones than standard parametric methods. The essential assumption for the permutation test is exchangeability under the null hypothesis. Exchangeability means that the data can be permuted (relative to the model) without altering its joint

Illustration of Permutation Inference: Two-Sample T

a Data. Single voxel %BOLD change for 6 subjects, 3 from each group

Group A			Group B		
2.60	1.99	1.95	0.10	−0.43	1.20

b Possible permutations. Twenty possible ways of shuffling A and B group labels. Under the null hypothesis group labels are meaningless, and all of these labellings are equivalent.

AAABBB	ABABAB	BAAABB	BABBAA
AABABB	ABABBA	BAABAB	BBAAAB
AABBAB	ABBAAB	BAABBA	BBAABA
AABBBA	ABBABA	BABAAB	BBABAA
ABAABB	ABBBAA	BABABA	BBBAAA

c Permutation distribution. Twenty T statistic values from each possible permutation. Correctly labelled data gives T = 3.61. Only that permutation is as large or larger, so P-value is 1/20 = 0.05

AAABBB	3.61	ABABAB	0.26	BAAABB	0.21	BABBAA	−0.61
AABABB	0.64	ABABBA	1.63	BAABAB	−0.12	BBAAAB	−1.68
AABBAB	0.28	ABBAAB	−0.99	BAABBA	0.99	BBAABA	−0.28
AABBBA	1.68	ABBABA	0.12	BABAAB	−1.63	BBABAA	−0.64
ABAABB	0.61	ABBBAA	−0.21	BABABA	−0.26	BBBAAA	−3.61

d Permutation distribution in histogram form.

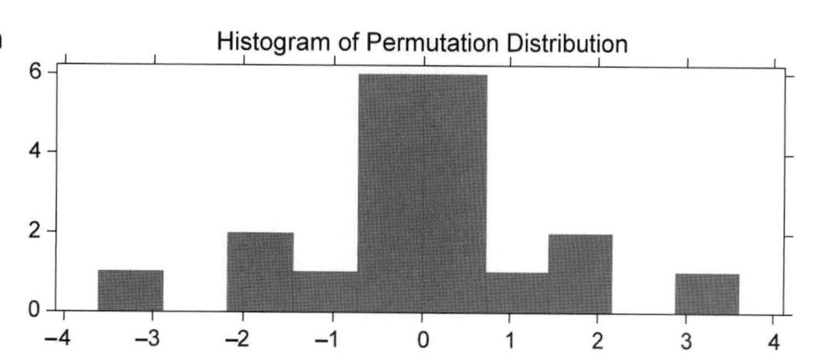

Fig. 28. Permutation test applied to data from a single voxel for a hypothetical two-group fMRI second-level analysis.

distribution. fMRI data presents a challenge for permutation testing. At the first level, temporal autocorrelation renders the data non-exchangeable and permutation methods cannot be directly applied (the data must be de-correlated, then permuted, and then re-correlated [*see* (*6, 54*) for more details]). However, in second level analyses, exchangeability is generally not a problem. In the example in the figure, we assume that, under the null hypothesis, all six subjects are exchangeable – this is very reasonable because if there is no group effect (this is the null hypothesis), then there is nothing special about the first three subjects versus the last three. For this example, there are 20 possible ways of permuting the groups (including the correct labelling). For an arbitrary dataset with group sizes n_1 and n_2, the number of possible permutations is $(n_1 + n_2)!/(n_1!n_2!)$.

By permuting the data many times, and for each permutation, assuming that the resulting test statistic (in this case, the group-difference *T*-statistic) is an sample of what we would see if there were no real effect present, we build up a histogram of test statistic values that will serve as the null distribution of that test statistic. We can then look to see how much "area under the tail" lies to the right of the actual test statistic value that we originally observed (under the correct labelling of the data), and hence estimate our *P*-value; this is the same principle for relating the null distribution of the test statistic to the *P*-value as we saw in **Fig. 9**, but in this case, the null distribution has been generated via a completely different methodology.

The permutation test for the two group case can be generalised to three or more groups. In that case, we are testing the null hypothesis that all groups are the same, and use an *F*-test to measure the evidence of any difference. Under the null hypothesis, all subjects can be freely permuted. Likewise for a simple correlation model, the null hypothesis of no association justifies the free permutation of all of the subjects. The permutation test for the one group case, however, is problematic without further assumptions: If all we have is a single group, what is there to permute? Shuffling the order of subjects will not change the value of a one-sample *T*-test.

In a second-level single group mean, the COPE images are always created as relative differences between baseline and active data. Even if an event-related design is used, and a contrast selects a single predictor, the effective predictor is a subtraction of event and baseline data. This is the case due to the relative, non-quantitative nature of the BOLD signal. As a result, we use here a slightly different assumption to generate "permutations." Under the null hypothesis, we assume that each individual's second-level COPE data are mean zero and have a symmetric distribution. Assuming mean zero data is reasonable, as, under the null, we expect no activation, positive or negative. Assuming a symmetric distribution is a weakened form of normality, and is exactly satisfied for any balanced effect (i.e., a COPE constructed as difference of two averages, where an equal number of scans contributed to each average).

The one-sample, group-level fMRI permutation thus works as follows. Assuming mean zero, symmetrically distributed COPE data, we randomly multiply each subject's data by 1 or -1, or, equivalently, randomly flip the signs of each subject's COPE image. Since the data are symmetrically distributed about zero, multiplication by -1 does not alter the distribution, and we generate a realisation that is equivalent to the original data. If there are n subjects in the analysis, there are 2^n possible ways to flip the signs of the group-level data.

The permutation methods described so far will create uncorrected P-values at each voxel. Control of the FWE rate is easily obtained with permutation testing via the following observation: In complete-null data, an FWE occurs whenever one or more voxels exceed the threshold, which occurs exactly when the voxel with the largest statistic exceeds the threshold. Hence, inference based on the permutation distribution of the largest (maximum) statistic provides valid FWE inferences. Specifically, at each permutation, the maximum statistic value (across all voxels in the brain) is noted, creating a null distribution of the maximum-across-space test statistic. The 95th percentile of that distribution gives an FWE-corrected threshold ("$p < 0.05$, corrected"), and any particular statistic value can be compared to the maximum permutation distribution to obtain an FWE-corrected P-value. Similarly, the maximal cluster size distribution can be created to provide FWE cluster-wise inferences.

It is important to note that, while permutation may appear to be a completely different approach than those we described earlier, in fact all pre-processing and modelling are generally the same, and it is only the P-value computation that differs. This is because the standard statistical models used generally have good sensitivity and robustness properties and should be used unaltered. For example, above we only discussed one- and two-sample T-tests, and did not mention other traditional non-parametric test statistics based on ranks, like the Wilcoxon Mann-Whitney test, as they often have much less power. There is an exception, however with small group data, with 20 or fewer subjects. With such small group data, there can be substantial sensitivity gains by using a non-standard statistic, namely the smoothed variance T-test. As the fMRI data is generally smoothed before statistical modelling, we expect the variance image to be smooth as well. However, when the number of subjects is very small, the DOF available to estimate the variance is very low, and this can result in a noisy sample variance image. Smoothing regularizes the estimated variance image, effectively increasing the DOF and increasing sensitivity. While the null distribution of the smoothed estimated variance T-statistic image is not known, and so parametric methods cannot be used, non-parametric permutation methods can easily be used to generate FWE inferences based on the smoothed variance results.

Finally, if the number of possible permutations is very large, it can be impossible to compute them all. For example, for a 20 subject one-group analysis, there are over one million possible sign-flips of the data. In fact, it is sufficient to run a random subset of all possible permutations. If only k of a large number of possible permutations is used, the margin of error on the P–values is approximately $\pm 2^{p} p(1-p)/k$, where p is the true P-value. For a nominal p of 0.05, this suggests that 1,000 permutations is nearly sufficient (ME = ± 0.014, or 28% of 0.05), while 10,000 is probably more than enough (ME = ± 0.0044, or 8.7% of 0.05).

7.4.3. Controlling FDR

The method for finding a threshold that controls FDR is surprisingly simple. It is based only on the uncorrected voxel-wise P-values in the statistic image. Let P_i be the P-value at voxel i, and $P_{(i)}$ be the ordered P-values, $P_{(1)} \le P_{(2)} \le \cdots \le P_{(I)}$. Then the largest index i that satisfies

$$P_{(i)} \le \frac{i}{I} \alpha^{\mathrm{FDR}} \tag{19}$$

defines the FDR threshold as $P_{(i)}$ *(55)*. This method works even when there is positive dependence between voxels *(56, 57)*.

7.4.4. Controlling False Positives and True Negatives: Mixture Modelling

So far we have considered techniques that control the rate of false positives. This depends on knowing the null distribution (or non-activation distribution) for relevant statistics under the null hypothesis. In contrast, mixture modelling provides us with a way of estimating the "activating" and "non-activating" distributions from the data itself. For example, non-activating voxel statistics may be modelled as coming from a (zero, or close-to-zero, mean) Gaussian distribution, activating voxels statistics as coming from a Gamma distribution, and de-activating voxels statistics as coming from a negative Gamma distribution *(58–60)*. The means and variances of these distributions are estimated from the whole statistic image. An example is shown in **Fig. 29**.

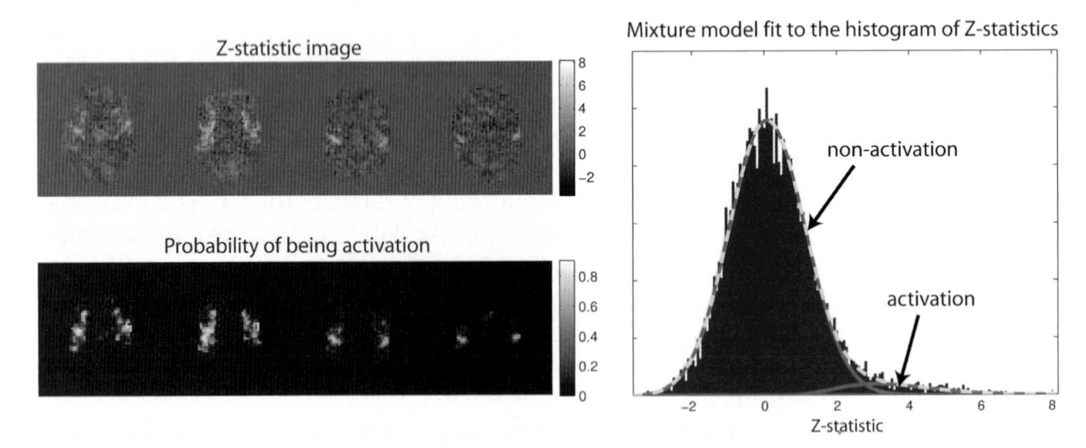

Fig. 29. Mixture modelling of a *Z*-statistic image. *Top-left*: Four example slices of a *Z*-statistic image obtained from fitting a general linear model (*GLM*) to the fMRI data at each voxel from an individual subject. The experiment was a pain stimulus applied using a sparse single-event design. *Right*: Mixture model fit to the histogram of *Z*-statistics. Non-activating voxel are modelled as coming from a close-to-zero mean Gaussian distribution, activating voxels as coming from a Gamma distribution, and de-activating voxels as coming from a negative Gamma distribution. However, note that there were found to be no de-activating voxels in this case. *Bottom-left*: Image showing the probability that a voxel is activating - this information that can be extracted from the mixture model fit and can be used in thresholding to approximately control the true positive rate (*TPR*) as an alternative to null hypothesis testing.

Note that the mixture modelling approach is similar to the permutation approach (discussed in **Subheading 7.4.2**) in that both approaches extract information about the null distribution (or non-activation distribution) from the data itself. However, permutation methods extract information about only the null distribution without making strong distributional assumptions, whereas mixture modelling extracts information about both the non-activating and activating distributions by making strong distributional assumptions.

Mixture modelling can provide a number of advantages over null hypothesis testing. First, there is a well-known problem in null hypothesis testing of FMRI in that if enough observations (e.g., time points) are made, then every voxel in the brain will reject the null hypothesis *(34)*. This is because in practice no voxels will show completely zero response to the stimulus, if only due to modelling inadequacies such as unmodelled stimulus-correlated motion or the point spread function of the scanner. By doing mixture modelling, we can overcome this by instead of asking the question "Is the activation zero or not?," we ask the question "Is the activation bigger than the overall background level of 'activation'?."

Mixture modelling also provides inference flexibility. Because we have both the "activating" and "non-activating" distributions, we can calculate the probability of a voxel being "activating" and the probability of a voxel being "non-activating." This provides us with far more inference flexibility compared with null hypothesis testing. We can still look to control the FPR by thresholding using the probability of a voxel being "non-activating." But now we could also look to approximately control the true positive rate (TPR) by thresholding using the probability of a voxel being "activating." Controlling the TPR may be of real importance when using fMRI for pre-surgery planning *(61)*.

The wider-spread use of mixture modelling is currently somewhat hampered by the violation of the strong distributional assumptions that need to be made. In the future, this may be alleviated by the improvement of techniques such as ICA de-noising (*see* **Subheading 5.2.2**) rendering the distributional assumptions valid.

7.5. Enhancing Statistic Images

Aside from thresholding the final statistic images (either voxel-wise or cluster-wise), it is not advisable to make image-processing adjustments to statistic images. For example, smoothing a *T*-statistic image would be disastrous: While a *T*-statistic has approximately unit variance and follows a particular null distribution, a smoothed *T*-statistic image will have dramatically reduced variance with no particular distribution. Two exceptions to this are wavelet de-noising methods and a recently proposed threshold-free cluster enhancement (TFCE) method.

Wavelet methods transform the data in a scale-dependent fashion, so that all of the large-scale information is segregated from the fine-scale information. Since we generally expect the signals of interest to be spatially extended, wavelet methods can be used to "shrink" variation associated with the finest scales, "de-noising" the image, while preserving the large-scale structure. For an overview of wavelet methods applied to fMRI, see *(62)*.

Cluster-wise inference also tries to capture spatially extended signals, but requires the specification of an arbitrary cluster-forming threshold u_{clus}. TFCE removes this dependence by, in essence, using all possible u_{clus} values and then merging all the results into a single image. Specifically, at each voxel, let $e_i(h)$ be the extent of the cluster that voxel i belongs to with cluster-forming threshold h (or 0 if $t_i < h$). Then, the TFCE image is defined by $\sum_{h>0} e_i(h)^E h^H$, where the sum is computed for a discrete set of h values, from 0 to the maximum statistic value, and E and H are tuning parameters. In *(63)*, $E = 0.5$ and $H = 2$ were found to give generally good performance for a range of classes of signals. TFCE seems to succeed in matching or exceeding the sensitivity of optimised cluster-based thresholding without the arbitrariness and instability of the smoothing and initial thresholding. There is no known distribution for the TFCE image, and so permutation testing is used to convert the TFCE image into P-values.

References

1. Friston K, Worsley K, Frackowiak R, Mazziotta J, Evans A. Assessing the significance of focal activations using their spatial extent. Hum Brain Mapp 1994;1:214–220.
2. Hykin J, Bowtell R, Glover P, Coxon R, Blumhardt L, Mansfield P. Investigation of the linearity of functional activation signal changes in the brain using echo planar imaging (EPI) at 3.0 T. In: Proc of the SMR and ESMRB Joint Meeting 1995, page 795.
3. Cohen M. Parametric analysis of fMRI data using linear systems methods. NeuroImage 1997;6:93–103.
4. Dale A, Buckner R. Selective averaging of rapidly presented individual trials using fMRI. Hum Brain Mapp 1997;5:329–340.
5. Burock MA, Buckner RL, Woldorff MG, Rosen BR, Dale AM. Randomized event-related experimental designs allow for extremely rapid presentation rates using functional MRI. NeuroReport 1998;9:3735–3739.
6. Bullmore E, Brammer M, Williams S, et al. Statistical methods of estimation and inference for functional MR image analysis. Magn Reson Med 1996;35:261–277.
7. Friston K, Josephs O, Zarahn E, Holmes A, Rouquette S, Poline J-B. To smooth or not to smooth? NeuroImage 2000;12:196–208.
8. Woolrich M, Ripley B, Brady J, Smith S. Temporal autocorrelation in univariate linear modelling of FMRI data. NeuroImage 2001; 14:1370–1386.
9. Locascio J, Jennings P, Moore C, Corkin S. Time series analysis in the time domain and resampling methods for studies of functional magnetic resonance brain imaging. Hum Brain Mapp 1997;5:168–193.
10. Purdon P, Weisskoff R. Effect of temporal autocorrelation due to physiological noise and stimulus paradigm on voxel-level false-positive rates in fMRI. Hum Brain Mapp 1998;6: 239–249.
11. Marchini J, Ripley B. A new statistical approach to detecting significant activation in functional MRI. NeuroImage 2000;12:366–380.
12. Worsley K, Liao C, Aston J, et al. A general statistical analysis for fMRI data. NeuroImage 2002;15:1–15.
13. Gautama T, Van Hulle MM. Optimal spatial regularisation of autocorrelation estimates in

fMRI analysis. Neuroimage 2004;23:1203–1216.

14. Penny W, Kiebel S, Friston K. Variational Bayesian inference for fMRI time series. NeuroImage 2003;19:1477–1491.
15. Woolrich M, Behrens T, Smith S. Constrained linear basis sets for HRF modelling using Variational Bayes. NeuroImage 2004; 21:1748–1761.
16. Smith S, Jenkinson M, Beckmann C, Miller K, Woolrich M. Meaningful design and contrast estimability in fMRI. NeuroImage 2007;34: 127–136.
17. Dale A, Greve D, Burock M. Optimal stimulus sequences for event-related fMRI. NeuroImage 1999;9:S33.
18. Wager T, Nichols T. Optimization of experimental design in fMRI: A general framework using a genetic algorithm. Neuroimage 2003; 18:293–309.
19. Josephs O, Turner R, Friston K. Event-related fMRI. Hum Brain Mapp 1997;5:1–7.
20. Lange N, Zeger S. Non-linear Fourier time series analysis for human brain mapping by functional magnetic resonance imaging. Applied Statistics 1997;46:1–29.
21. Genovese C. A Bayesian time-course model for functional magnetic resonance imaging data (with discussion). J Am Stat Assoc 2000; 95:691–703.
22. Friston KJ. Bayesian estimation of dynamical systems: An application to fMRI. NeuroImage 2002;16:513–530.
23. Marrelec G, Benali H, Ciuciu P, Pélégrini-Issac M, Poline, J-B. Robust Bayesian estimation of the hemodynamic response function in event-related BOLD MRI using basic physiological information. Hum Brain Mapp 2003; 19:1–17.
24. Woolrich M, Jenkinson M, Brady J, Smith S. Fully Bayesian spatio-temporal modelling of FMRI data. IEEE Trans Med Imaging 2004; 23:213–231.
25. Buxton R, Uludag K, Dubowitz D, Liu T. Modeling the hemodynamic response to brain activation. NeuroImage 2004;23(S1):220–233.
26. Boynton G, Engel S, Glover G, Heeger D. Linear systems analysis of functional magnetic resonance imaging in human V1. J Neurosci 1996;16:4207–4221.
27. Glover G. Deconvolution of impulse response in event-related BOLD fMRI. NeuroImage 1999;9:416–429.
28. Friston K, Josephs O, Rees G, Turner R. Nonlinear event-related responses in fMRI. Magn Reson Med 1998;39:41–52.
29. Beckmann C, Smith S. Probabilistic independent component analysis for functional magnetic resonance imaging. IEEE Trans Med Imaging 2004;23:137–152.
30. Glover G, Li T, Ress D. Image-based method for retrospective correction of physiological motion effects in fMRI: Retroicor. Magn Reson Med 2000;44:162–167.
31. Holmes A, Friston K. Generalisability, random effects \amp population inference. In: Fourth Int Conf on Functional Mapping of the Human Brain. NeuroImage, 1998;7:S754.
32. Talairach J, Tournoux P. Co-planar Stereotaxic Atlas of the Human Brain. Thieme Medical Publisher, Inc., New York, 1988.
33. Collins D, Neelin P, Peters T, Evans A. Automatic 3D intersubject registration of MR volumetric data in standardized Talairach space. J Comput Assist Tomo 1994;18:192–205.
34. Friston KJ, Penny W, Phillips C, Kiebel S, Hinton G, Ashburner J. Classical and Bayesian inference in neuroimaging: Theory. NeuroImage 2002;16:465–483.
35. Beckmann C, Jenkinson M, Smith S. General multi-level linear modelling for group analysis in FMRI. NeuroImage 2003;20:1052–1063.
36. Woolrich M, Behrens T, Beckmann C, Jenkinson M, Smith S. Multi-level linear modelling for FMRI group analysis using Bayesian inference. NeuroImage 2004;21:1732–1747.
37. Kherif F, Poline J-B, Meriaux S, Benali H, Flandin G, Brett M. Group analysis in functional neuroimaging: Selecting subjects using similarity measures. Neuroimage 2003;20: 2197–2208.
38. Luo W-L, Nichols TE. Diagnosis and exploration of massively univariate neuroimaging models. Neuroimage 2003;19:1014–1032.
39. Seghier M, Friston K, Price C. Detecting subject-specific activations using fuzzy clustering. Neuroimage 2007;36:594–605.
40. Wager T, Keller M, Lacey S, Jonides J. Increased sensitivity in neuroimaging analyses using robust regression. NeuroImage 2005; 26:99–113.
41. Woolrich M. Robust group analysis using outlier inference. NeuroImage 2008;41:286–301.
42. Meriaux S, Roche A, Dehaene-Lambertz G, Thirion B, Poline J. Combined permutation test and mixed-effect model for group average analysis in fMRI. Hum Brain Mapp 2006;27: 402–410.
43. Roche A, Meriaux S, Keller M, Thirion B. Mixed-effect statistics for group analysis in fMRI: A nonpara-metric maximum likelihood approach. Neuroimage 2007;38:501–510.

44. Thirion B, Pinel P, Mriaux S, Roche A, Dehaene S, Poline J. Analysis of a large fMRI cohort: Statistical and methodological issues for group analyses. Neuroimage 2007;35: 105–120.
45. Hartvig NV, Jensen JL. Spatial mixture modeling of fMRI data. Hum Brain Mapp 2000; 11:233–248.
46. Hayasaka S, Nichols TE. Validating cluster size inference: Random field and permutation methods. NeuroImage 2003;20:2343–2356.
47. Friston KJ, Holmes A, Poline J-B, Price CJ, Frith CD. Detecting activations in PET and fMRI: Levels of inference and power. NeuroImage 1996;4:223–235.
48. Nichols TE, Hayasaka S. Controlling the familywise error rate in functional neuroimaging: A comparative review. Stat Methods Med Res 2003;12:419–446.
49. Cao J, Worsley KJ. Applications of random fields in human brain mapping. In: Moore M, editor, Spatial Statistics: Methodological Aspects and Applications, volume 159 of Springer Lecture Notes in Statistics, pages 169–182. Springer, 2001.
50. Worsley KJ, Evans AC, Marrett S, Neelin P. Three-dimensional statistical analysis for cbf activation studies in human brain. J Cerebr Blood F Met 1992;12:900–918.
51. Worsley KJ, Marrett S, Neelin P, Vandal AC, Friston KJ, Evans AC. A unified statistical approach for determining significant signals in images of cerebral activation. Hum Brain Mapp 1996;4:58–73.
52. Hayasaka S, Luan Phan K, Liberzon I, Worsley KJ, Nichols TE. Nonstationary cluster-size inference with random field and permutation methods. NeuroImage 2004;22:676–687.
53. Nichols T, Holmes A. Nonparametric permutation tests for functional neuroimaging: A primer with examples. Hum Brain Mapp 2001;15:1–25.
54. Bullmore E, Long C, Suckling J, et al. Colored noise and computational inference in neurophysiological (fMRI) time series analysis: Resampling methods in time and wavelet domains. Hum Brain Mapp 2001; 12:61–78.
55. Benjamini Y, Hochberg Y. Controlling the false discovery rate: A practical and powerful approach to multiple testing. Journal of the Royal Statistical Society, Series B, Methodological 1995;57:289–300.
56. Genovese C, Lazar N, Nichols T. Thresholding of statistical maps in functional neuroimaging using the false discovery rate. NeuroImage 2002;15:870–878.
57. Benjamini Y, Yekutieli D. The control of the false discovery rate in multiple testing under dependency. Ann Stat 2001;29: 1165–1188.
58. Everitt B, Bullmore E. Mixture model mapping of brain activation in functional magnetic resonance images. Hum Brain Mapp 1999;7: 1–14.
59. Hartvig N. A stochastic geometry model for fMRI data. Technical Report 410, Department of Theoretical Statistics, University of Aarhus, 2000.
60. Woolrich M, Behrens T. Variational Bayes Inference of Spatial Mixture Models for Segmentation. IEEE Trans Med Imaging 2006; 25:1380–1391.
61. Bartsch A, Homola G, Biller A, Solymosi L, Bendszus M. Diagnostic functional MRI: Illustrated clinical applications and decision-making. J Magn Reson Imaging 2006;23: 921–932.
62. Van De Ville D, Blu T, Unser M. Surfing the brain – an overview of wavelet-based techniques for fMRI data analysis. IEEE Eng Med Biol 2006;25:65–78.
63. Smith SM, Nichols TE. Threshold-free cluster enhancement: Addressing problems of smoothing, threshold dependence and localisation in cluster inference. NeuroImage 2008 (in press) [Epub ahead of print April 11, 2008; doi: 10.1016/j.neuroimage.2008.03.061].

Chapter 8

Dynamic Causal Modelling of Brain Responses

Karl J. Friston

Summary

This chapter is about modelling-distributed brain responses and, in particular, the functional integration among neuronal systems. Inferences about the functional organisation of the brain rest on models of how measurements of evoked responses are caused. These models can be quite diverse, ranging from conceptual models of functional anatomy to mathematical models of neuronal and haemodynamics. The aim of this chapter is to introduce dynamic causal models. These models can be regarded as generalisations of the simple models employed in conventional analyses of regionally specific brain responses. In what follows, we will start with anatomical models of functional brain architectures, which motivate some of the basic principles of neuroimaging. We then review briefly statistical models (e.g., the general linear model) used for making classical and Bayesian inferences about *where* neuronal responses are expressed. By incorporating biophysical constraints, these basic models can be finessed and, in a dynamic setting, rendered causal. This allows us to infer *how* interactions among brain regions are mediated. This chapter focuses on causal models for distributed responses measured with fMRI and electroencephalography. The latter is based on neural-mass models and affords mechanistic inferences about how evoked responses are caused, at the level of neuronal sub-populations and the coupling among them.

Key words: Functional connectivity, Effective connectivity, Dynamic causal modelling, Causal, Dynamic, Non-linear

1. Introduction

Neuroscience depends on conceptual, anatomical, statistical, and causal models that link ideas about how the brain works to observed neuronal responses. Here, we highlight the relationships among the sorts of models that are employed in imaging, with a special focus on dynamic causal models of functional brain architectures. We will show how simple statistical models used

M. Filippi (ed.), *fMRI Techniques and Protocols*, Neuromethods, vol. 41
DOI 10.1007/978-1-60327-919-2_8, © Humana Press, a part of Springer Science+Business Media, LLC 2009

to identify where evoked brain responses are expressed (cf, neo-phrenology) can be elaborated to provide models of how neuronal responses are caused (e.g., dynamic causal modelling – DCM). We will review a series of models that range from conceptual models, motivating experimental design, to detailed biophysical models of coupled neuronal ensembles that enable questions to be asked, at a physiological and computational level.

Anatomical models of functional brain architectures motivate the fundaments of neuroimaging. In the first section, we review the distinction between functional *specialisation* and *integration* and how these principles serve as the basis for most models of neuroimaging data. The next section turns to simple statistical models (e.g., the general linear model – GLM) used for making classical and Bayesian inferences about functional specialisation, in terms of where neuronal responses are expressed. By incorporating biological constraints, simple observation models can be made more realistic and, in a dynamic framework, causal. This section concludes by considering the biophysical modelling of haemodynamic responses. All the models considered in this section pertain to regional responses. In the final section, we focus on models of distributed responses, where the interactions among cortical areas or neuronal sub-populations are modelled explicitly. This section covers the distinction between *functional* and *effective connectivity* and reviews DCM of functional integration, using fMRI and electroencephalogram (EEG). We conclude with an example from event-related potential (ERP) research and show how the mismatch negativity (MMN) can be explained by changes in coupling among neuronal sources that may underlie perceptual learning.

2. Anatomical Models

2.1. Functional Specialisation and Integration

From a historical perspective, the distinction between functional specialisation and functional integration relates to the dialectic between *localisationism* and *connectionism* that dominated thinking about brain function in the nineteenth century. Since the formulation of phrenology by Gall, who postulated fixed one-to-one relations between particular parts of the brain and specific mental attributes, the identification of a particular brain region with a specific function has become a central theme in neuroscience. Somewhat ironically, the notion that distinct brain functions could be localised in the brain was strengthened by early scientific attempts to refute the phrenologists' claims. In 1808, a scientific committee of the Athénée at Paris, chaired by Cuvier, declared that phrenology was an unscientific and invalid theory *(1)*. This conclusion, which was not based on experimental results, may have been enforced by Napoleon Bonaparte (who, allegedly, was not amused after Gall's phrenological examination of his own skull did not give the flattering results expected). During the following decades, lesion and electrical stimulation paradigms were developed to test whether functions could indeed be localised in animal models. Initial lesion experiments by Flourens on pigeons were incompatible with phrenologist predictions, but later experiments, including stimulation experiments in dogs and monkeys by Fritsch, Hitzig, and Ferrier, supported the idea that there was a relation between distinct brain regions and certain cognitive or motor functions. Additionally, clinicians like Broca and Wernicke showed that patients with focal brain lesions in particular locations showed specific impairments. However, it was realised early on that, in spite of these experimental findings, it was generally difficult to attribute a specific function to a cortical area, given the dependence of cerebral activity on the anatomical connections between distant brain regions; for example, a meeting that took place on 4 August 1881 addressed the difficulties of attributing function to a cortical area, given the dependence of cerebral activity on underlying connections *(2)*. This meeting was entitled "localisation of function in the cortex cerebri." Goltz *(3)*, although accepting the results of electrical stimulation in dog and monkey cortex, considered that the excitation method was inconclusive, in that movements elicited might have originated in related pathways, or current could have spread to distant centres. In short, the excitation method could not be used to infer functional localisation because localisationism discounted interactions or functional integration among different brain areas. It was proposed that lesion studies could supplement excitation experiments. Ironically, it was observations on patients with brain lesions some years later (*see* **ref.** *4)* that led to the concept of *disconnection syndromes* and the refutation of localisationism as a complete or sufficient explanation of cortical organisation. Functional localisation implies that a function can be localised in a cortical area, whereas specialisation suggests that a cortical area is specialised for some aspects of perceptual or motor processing, and that this specialisation is anatomically *segregated* within the cortex. The cortical infrastructure supporting a single function may then involve many specialised areas whose union is mediated by the functional integration among them. In this view, functional specialisation is only meaningful in the context of functional integration and vice versa.

2.2. Functional Specialisation and Segregation

The functional role of any component (e.g., cortical area, subarea, or neuronal population) of the brain is defined largely by its connections. Certain patterns of cortical projections are so common that they could amount to rules of cortical connectivity. "These rules revolve around one, apparently, over-riding strategy that the cerebral cortex uses – that of functional segregation" *(5)*. Functional segregation demands that cells with common functional properties be grouped together. This architectural constraint necessitates both convergence and divergence of cortical connections. Extrinsic connections among cortical regions are not continuous but occur in patches or clusters. This patchiness has, in some instances, a clear relationship to functional segregation. For example, when recordings are made in V2, directionally selective (but not wavelength or colour selective) cells are found exclusively in its thick stripes. Retrograde (i.e., backward) labelling of cells in V5 is limited to these thick stripes; all the available physiological evidence suggests that V5 is a functionally homogeneous area that is specialised for visual motion. Evidence of this nature supports the notion that patchy connectivity is the anatomical infrastructure that mediates functional segregation and specialisation. If it is the case that neurons in a given cortical area share a common responsiveness, by virtue of their extrinsic connectivity, to some sensorimotor or cognitive attribute, then this functional segregation is also an anatomical one.

In summary, functional specialisation suggests that challenging a subject with the appropriate sensorimotor attribute or cognitive process should lead to activity changes in, and only in, the specialised areas. This is the anatomical and physiological model upon which the search for regionally specific effects is based. We will deal briefly with models of regionally specific responses and return to models of functional integration.

3. Statistical Models

3.1. Statistical Parametric Mapping

Functional mapping studies are usually analysed with some form of statistical parametric mapping (SPM). SPM entails the construction of continuous statistical maps (e.g., *t*-maps) to test hypotheses about regionally specific effects *(6)*. SPM uses the GLM and random field theory (RFT) to analyse and make classical inferences about brain responses. Parameters of the GLM are estimated in exactly the same way as in conventional analysis of discrete data. RFT is used to resolve the multiple-comparisons problem induced by making inferences over a volume of the brain. RFT provides a method for adjusting *p*-values for the search volume of an SPM to control false positive rates. It plays the same role for continuous data (i.e., images or time series) as the Bonferroni correction for a family of discontinuous or discrete statistical tests.

There is a Bayesian alternative to classical inference with SPMs. This rests on conditional inferences about an effect, given the data, as opposed to classical inferences about the data, given the effect is zero. Bayesian inferences about effects that are continuous in space use posterior probability maps (PPMs). Although less established than SPMs, PPMs are potentially useful, not least because they do not have to contend with the multiple-comparisons problem induced by classical inference (see **ref.** *7)*. In contradistinction to SPM, this means that inferences about a given regional response do not depend on inferences about responses elsewhere. Bayesian inference is particularly relevant to dynamic casual modelling because the Bayesian formulation is an essential part of model specification and inversion. Before looking at the models underlying Bayesian inference, we briefly review estimation and classical inference in the context of the GLM and show how this can be generalised to give a Bayesian approach.

3.2. General Linear Model

The GLM is a simple equation

$$y = X\beta + \varepsilon \tag{1}$$

that expresses an observed response y in terms of a linear combination of explanatory variables in the design matrix X, plus a well-behaved error term. The GLM is variously known as analysis of variance or multiple-regression and subsumes simpler variants, like the t-test for a difference in means, to more elaborate linear convolution models (see below). Each column of the design matrix models a cause of the data. These are referred to as explanatory variables, covariates, or regressors. Sometimes the design matrix contains covariates or indicator variables that take values of zero or one, to indicate the presence of a particular level of an experimental factor (cf, analysis of variance). The relative contribution of each of these columns to the response is controlled by the parameters, β. Inferences about the parameter estimates are made using t or F-statistics, as in conventional statistics. Having computed the statistic, RFT is used to assign adjusted p-values to topological features of the SPM, such as the height of peaks or the spatial extent of blobs. This p-value is a function of the search volume and smoothness. The intuition behind RFT is that it controls the false positive rate of peaks corresponding to regional effects. A Bonferroni correction would control the false positive rate of voxels, which is inexact and unnecessarily severe. The p-value is the probability of getting a peak in the SPM, or higher, by chance over the search volume. If sufficiently small (usually less than 0.05), the regional effect is declared significant.

3.3. Classical and Bayesian Inference

Inference in neuroimaging is restricted largely to classical inferences based upon SPMs. The statistics that comprise these SPMs are essentially functions of the data. The probability distribution of the chosen statistic, under the null hypothesis (i.e., the null distribution), is used to compute a *p*-value. This *p*-value is the probability of obtaining the statistic, or the data, given that the null hypothesis is true. If sufficiently small, the null hypothesis is rejected and an inference is made. The alternative approach is to use Bayesian or conditional inference based upon the posterior distribution of the activation given the data *(8)*. This necessitates the specification of priors (i.e., the probability distribution of the activation or model parameter). Bayesian inference requires the conditional or posterior distribution and therefore rests upon a posterior density analysis. A useful way to summarise this posterior density is to compute the probability that the activation exceeds some threshold. This represents a Bayesian inference about the effect, in relation to the specified threshold. By computing posterior probability for each voxel, we can construct PPMs that are a useful complement to classical SPMs.

The motivation for using conditional or Bayesian inference is that it has high face validity. This is because the inference is about an effect, or activation, being greater than some specified size that has some meaning in relation to underlying neurophysiology. This contrasts with classical inference, in which the inference is about the effect being significantly different than zero. The problem for classical inference is that trivial departures from the null hypothesis can be declared significant, with sufficient data or sensitivity. From the point of view of neuroimaging, posterior inference is especially useful because it eschews the multiple-comparisons problem. In classical inference, one tries to ensure that the probability of rejecting the null hypothesis incorrectly is maintained at a small rate, despite making inferences over large volumes of the brain. This induces a multiple-comparisons problem that, for spatially continuous data, requires an adjustment or correction to the *p*-value using RFT as mentioned earlier. This correction means that classical inference becomes less sensitive or powerful with large search volumes. In contradistinction, posterior inference does not have to contend with the multiple-comparisons problem because there are no false-positives. The probability that activation has occurred, given the data, at any particular voxel is the same, irrespective of whether one has analysed that voxel or the entire brain. For this reason, posterior inference using PPMs represents a relatively more powerful approach than classical inference in neuroimaging.

3.3.1. Hierarchical Models and Empirical Bayes

PPMs require the posterior distribution or conditional distribution of the activation (a contrast of conditional parameter estimates), given the data. This posterior density can be computed, under Gaussian assumptions, using Bayes rule. Bayes rule requires the specification of a likelihood function and the prior density of the model parameters. The models used to form PPMs and the likelihood functions are exactly the same as in classical SPM analyses, namely, the GLM. The only extra information that is required is the prior probability distribution of the parameters. Although it would be possible to specify those using independent data or some plausible physiological constraints, there is an alternative to this fully Bayesian approach. The alternative is *empirical Bayes* in which the prior distributions are estimated from the data. Empirical Bayes requires a *hierarchical observation model* where the parameters and hyper-parameters at any particular level can be treated as priors on the level below. There are numerous examples of hierarchical observation models in neuroimaging. For example, the distinction between fixed- and mixed-effects analyses of multi-subject studies relies upon a two-level hierarchical model. However, in neuroimaging, there is a natural hierarchical observation model that is common to all brain mapping experiments. This is the hierarchy induced by looking for the same effects at every voxel within the brain (or grey matter). The first level of the hierarchy corresponds to the experimental effects at any particular voxel and the second level comprises the effects over voxels. Put simply, the variation in a contrast, over voxels, can be used as the prior variance of that contrast at any particular voxel. Hierarchical linear models have the following form:

$$\begin{aligned} y &= X^{(1)}\beta^{(1)}\varepsilon^{(1)} \\ \beta^{(1)} &= X^{(2)}\beta^{(2)}\varepsilon^{(2)} \\ \beta^{(2)} &= \ldots \end{aligned} \qquad (2)$$

This is exactly the same as **Eq. 1** but now the parameters of the first level are generated by a supra-ordinate linear model and so on to any hierarchical depth required. These hierarchical observation models are an important extension of the GLM and are usually estimated using expectation maximisation (EM) *(9)*. In the present context, the response variables comprise the responses at all voxels and $\beta^{(1)}$s are the treatment effects we want to make an inference about. Because we have invoked a second level, the first-level parameters embody random effects and are generated by a second-level linear model. At the second level, $\beta^{(2)}$ is the average effect over voxels and $\varepsilon^{(2)}$ is its voxel-to-voxel variation. By estimating the variance of $\varepsilon^{(2)}$, one is implicitly estimating an empirical prior on the first-level parameters at each voxel. This prior can then be used to estimate the posterior probability of $\beta^{(1)}$ being greater than some threshold at each voxel. An example of the ensuing PPM is provided in **Fig. 1** along with the classical SPM.

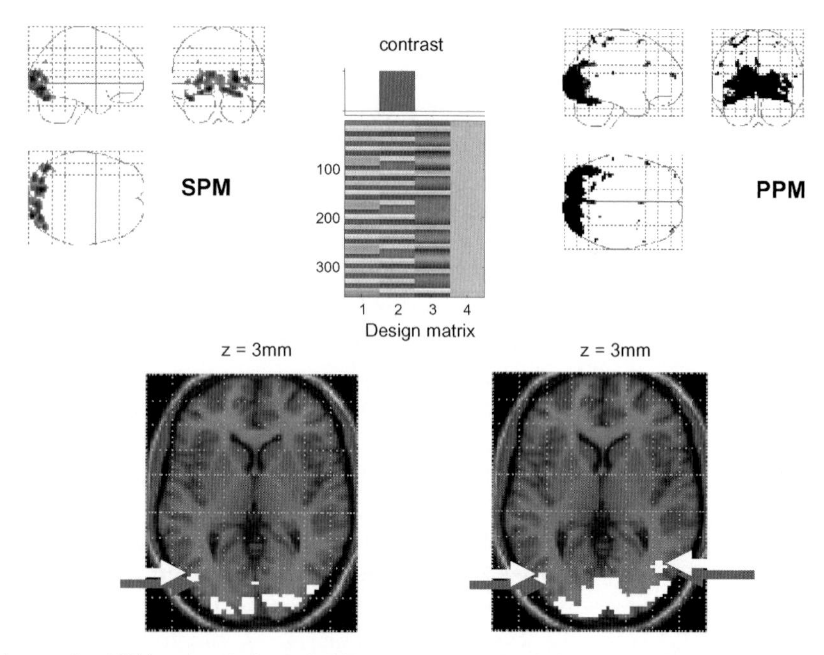

Fig. 1. Statistical parametric mapping (*SPM*) and posterior probability map (*PPM*) for an fMRI study of attention to visual motion. The display format (*lower panel*) uses an axial slice through extra-striate regions but the thresholds are the same as employed the in maximum-intensity projections (*upper panels*). *Upper right*: The activation threshold for the PPM was 0.7 au, meaning that all voxels shown had a 90% chance of an activation of 0.7% or more. *Upper left*: The corresponding SPM using an adjusted threshold at $p = 0.05$. Note the bilateral foci of motion-related responses in the PPM that are not seen in the SPM (*grey arrows*). As can be imputed from the design matrix (*upper-middle panel*), the statistical model of evoked responses comprised boxcar regressors convolved with a canonical haemodynamic response function. The middle column corresponds to the presentation of moving dots and was the stimulus attribute tested by the contrast.

In summary, we have seen how the GLM can be used to test hypotheses about brain responses and how, in a hierarchical form, it enables empirical Bayesian or conditional inference. Then, we deal with the dynamic systems and how they can be formulated as GLMs. These dynamic models take us closer to how brain responses are actually caused by experimental manipulations and represent the next step towards dynamic causal models of brain responses.

3.4. Dynamic Models

3.4.1. Convolution Models and Temporal Basis Functions

In Friston et al. *(10)*, the form of the impulsed haemodynamic response function (HRF) was estimated using a least squares deconvolution and a linear time invariant model, where evoked neuronal responses are *convolved* or smoothed with an HRF to give the measured haemodynamic response (*see also* **ref.** *11)*. This simple linear convolution model is the cornerstone for making statistical inferences about activations in fMRI with the GLM. An impulse response function is the response to a single impulse, measured at a series of times after the input. It characterises the input-output behaviour of the system (i.e., voxel) and places important constraints on the sorts of inputs that will excite a response.

Knowing the form of the HRF is important for several reasons, not least because it furnishes better statistical models of the data. The HRF may vary from voxel to voxel and this has to be accommodated in the GLM. To allow for different HRFs in different brain regions, temporal basis functions were introduced *(12)* to model evoked responses in fMRI and applied to event-related responses in Josephs et al. *(13)* (*see also* **ref.** *14)*. The basic idea behind temporal basis functions is that the haemodynamic response, induced by any given trial type, can be expressed as the linear combination of (basis) functions of peri-stimulus time. The convolution model for fMRI responses takes a stimulus function encoding the neuronal responses and convolves it with an HRF to give a regressor that enters the design matrix. When using basis functions, the stimulus function is convolved with each basis function to give a series of regressors. Mathematically, we can express this model as

$$\begin{matrix} y(t) = X\beta + \varepsilon \\ X_i = T_i(t) \otimes u(t) \end{matrix} \Leftrightarrow \begin{matrix} y(t) = (t) \otimes h(t) \\ h(t) = \beta_1 T_1(t) + \beta_2 T_2(t) + \ldots \end{matrix} \quad (3)$$

where $\otimes$ means convolution. This equivalence shows how any convolution model (right) can be converted into a GLM (left), using temporal basis functions. The parameter estimates β^i are the coefficients or weights that determine the mixture of basis functions of time $T_i(t)$ that models $h(t)$, the HRF for the trial type and voxel in question. We find the most useful basis set to be a canonical HRF and its derivatives with respect to the key parameters that determine its form (see below). Temporal basis functions are important because they provide a graceful transition between conventional multi-linear regression models with one stimulus function per condition and finite impulse response (FIR) models with a parameter for each time point, following the onset of a condition or trial type. **Figure 2** illustrates this graphically. In short, temporal basis functions offer useful constraints on the form of the estimated response that retain the flexibility of FIR models and the efficiency of single regressor models.

3.5. Biophysical Modelsa

3.5.1. Input-State-Output Systems

By adopting a convolution model for brain responses in fMRI, we are implicitly positing a dynamic system that converts neuronal responses into observed haemodynamic responses. Our understanding of the biophysical and physiological mechanisms that underpin the HRF has grown considerably in the last few years [e.g., *(15–17)*]. **Figure 3** shows some simulations based on the haemodynamic model described in Friston et al. *(18)*. Here, neuronal activity induces some auto-regulated vasoactive signal that causes transient increases in regional cerebral blood

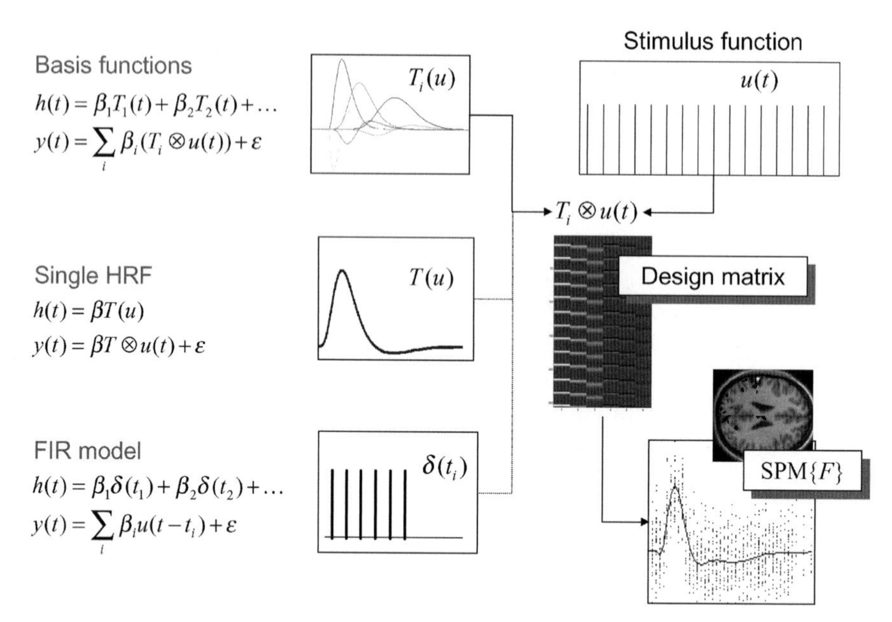

Fig. 2. Temporal basis functions offer useful constraints on the form of the estimated response that retain the flexibility of finite impulse response (*FIR*) models and the efficiency of single regressor models. The specification of these constrained FIR models involves setting up stimulus functions $u(t)$ that model expected neuronal changes, for example, boxcar-functions of epoch-related responses or spike-(δ)-functions at the onset of specific events or trials. These stimulus functions are then convolved with a set of basis functions $T_i(t)$ of peri-stimulus time that, in some linear combination, model the HRF. The ensuing regressors are assembled into the design matrix. The basis functions can be as simple as a single canonical HRF (*middle*), through to a series of top-hat-functions $\delta_i(t)$ (*bottom*). The latter case corresponds to an FIR model and the coefficients constitute estimates of the impulse response function at a finite number of discrete sampling times. Selective averaging in event-related fMRI ***(39)*** is mathematically equivalent to this limiting case.

flow (rCBF). The resulting flow increases dilate a venous balloon, increasing its volume, and diluting venous blood to decrease deoxyhaemoglobin content. The blood oxygen level dependent (BOLD) signal is roughly proportional to the concentration of deoxyhaemoglobin and follows the rCBF response with about a second delay. The model is framed in terms of differential equations, examples of which are provided in left panel.

Note that we have introduced variables like volume and deoxyhaemoglobin concentrations that are not actually observed. These are referred to as the *hidden states* of input-state-output models. The state and output equations of any analytic dynamical system are

$$\begin{aligned} \dot{x}(t) &= f(x, u, \theta) \\ y(t) &= g(x, u, \theta) + \varepsilon \end{aligned} \tag{4}$$

The first line is an ordinary differential equation and expresses the rate of change of the states as a parameterised function of the

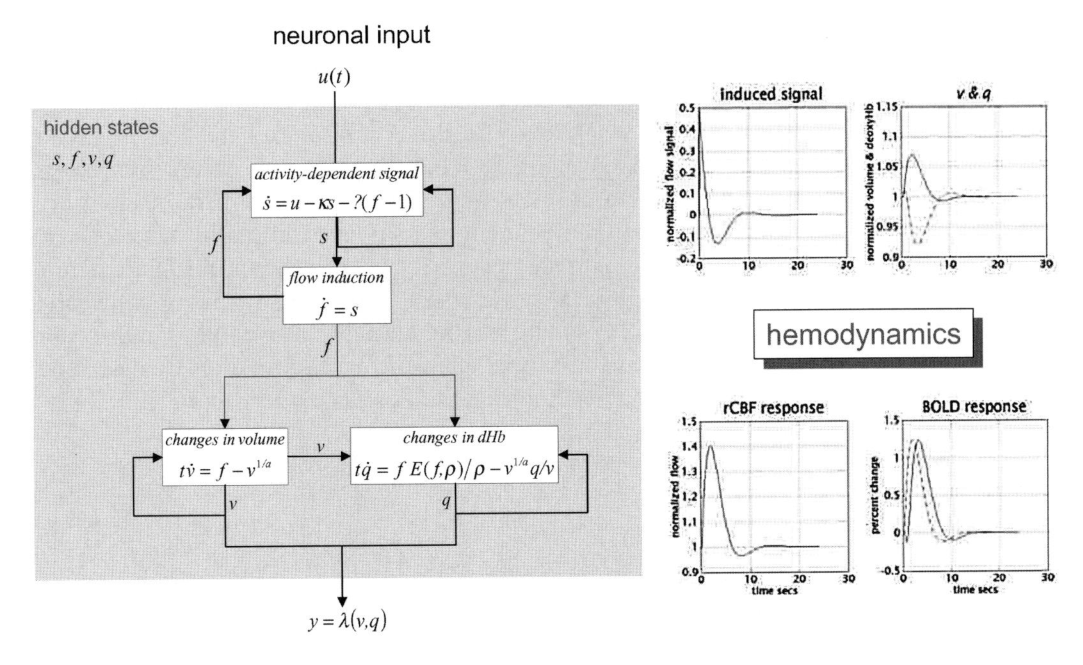

Fig. 3. *Right*: Haemodynamics elicited by an impulse of neuronal activity as predicted by a dynamical biophysical model (*left*). A burst of neuronal activity causes an increase in flow-inducing signal that decays with first-order kinetics and is down-regulated by local flow. This signal increases regional cerebral blood flow (*rCBF*), which dilates the venous capillaries, increasing volume *v*. Concurrently, venous blood is expelled from the venous pool decreasing deoxyhaemoglobin content *q*. The resulting fall in deoxyhaemoglobin concentration leads to a transient increases in blood oxygen level dependent (*BOLD*) signal and a subsequent undershoot. *Left*: Haemodynamic model; on which these simulations were based.

states and inputs. Typically, the inputs $u(t)$ correspond to designed experimental effects (e.g., the stimulus function in fMRI). There is a fundamental and causal relationship *(19)* between the outputs and the history of the inputs in **Eq. 4**. This relationship conforms to a Volterra series, which expresses the output as a generalised convolution of the input, critically without reference to the hidden states $x(t)$. This series is simply a functional Taylor expansion of the outputs with respect to the inputs *(20)*. The reason for it is a *functional* expansion that the inputs are a function of time.[1]

$$y(t) = \sum_i \int_0^t \ldots \int_0^t \kappa_i(\sigma_1, \ldots, \sigma_i) u(t-\sigma_1), \ldots, u(t-\sigma_i)$$
$$d\sigma_1, \ldots, d\sigma_i \;\; \kappa_i(\sigma_1, \ldots, \sigma_i) = \frac{\partial^i y(t)}{\partial u(t-\sigma_1), \ldots, \partial u(t-\sigma_i)} \qquad (5)$$

[1] For simplicity, here and in **Eq. 7**, we deal with only one experimental input.

where $\kappa_i(\sigma_1,\ldots,\sigma_i)$ is the ith-order kernel. In **Eq. 5**, the integrals are restricted to the past. This renders the system causal. The key thing here is that **Eq. 5** is simply a convolution and can be expressed as a GLM as in **Eq. 3**. This means that we can take a neurophysiologically realistic model of haemodynamic responses and use it as an observation model to estimate parameters using observed data. Here the model is parameterised in terms of kernels that have a direct analytic relation to the original parameters θ of the biophysical system. The first-order kernel is simply the conventional HRF. High-order kernels correspond to high-order HRFs and can be estimated using basis functions as described above. In fact, by choosing basis functions according to

$$A(\sigma)_i = \frac{\partial \kappa(\sigma)_1}{\partial \theta_i}, \qquad (6)$$

one can estimate the biophysical parameters because $\beta_i = \theta_i$ to a first-order approximation. The critical step we have taken here is to start with a dynamic causal model of how responses are generated and construct a general linear observation model that allows us to estimate and infer things about the parameters of that model. This is in contrast to the conventional use of the GLM with design matrices that are not informed by a forward model of how data are caused. This approach to modelling brain responses has a much more direct connection with underlying physiology and rests upon an understanding of the underlying system.

3.5.2. Non-linear System Identification

Once a suitable causal model has been established (e.g., **Fig. 3**), we can estimate second-order kernels. These kernels represent a non-linear characterisation of the HRF that can model interactions among stimuli in causing responses. One important manifestation of the non-linear effects, captured by the second-order kernels, is a modulation of stimulus-specific responses by preceding stimuli that are proximate in time. This means that responses at high-stimulus presentation rates saturate and, in some instances, show an inverted U behaviour. This behaviour appears to be specific to BOLD effects (as distinct from evoked changes in CBF) and may represent a *haemodynamic refractoriness*. This effect has important implications for event-related fMRI, where one may want to present trials in quick succession.

In summary, we started with models of regionally specific responses, framed in terms of the GLM, in which responses were modelled as linear mixtures of designed changes in explanatory variables. Hierarchical extensions to linear observation models enable random-effects analyses and, in particular, empirical Bayes. The mechanistic utility of these models is realised though the use of forward models that embody causal dynamics. Simple variants of these are the linear convolution models used to construct explanatory variables in conventional analyses of fMRI data. These are a special case of generalised convolution models that are mathematically equivalent to input-state-output systems comprising hidden states. Estimation and inference with these dynamic models tells us something about *how* the response was caused, but only at the level of a single voxel. **Subheading 4** retains the same perspective on models, but in the context of distributed responses and functional integration.

4. Models of Functional Integration

4.1. Functional and Effective Connectivity

Imaging neuroscience has established functional specialisation as a principle of brain organisation in man. The integration of specialised areas has proven more difficult to assess. Functional integration is usually inferred on the basis of correlations among measurements of neuronal activity. Functional connectivity is defined as statistical dependencies or correlations *among remote neurophysiological events*. However, correlations can arise in a variety of ways. For example, in multi-unit electrode recordings, they can result from stimulus-locked transients evoked by a common input or reflect stimulus-induced oscillations mediated by synaptic connections *(21)*. Integration within a distributed system is usually better understood in terms of effective connectivity. Effective connectivity refers explicitly to *the influence that one neural system exerts over another*, either at a synaptic (i.e., synaptic efficacy) or at a population level. It has been proposed that "the (electrophysiological) notion of effective connectivity should be understood as the experiment- and time-dependent, simplest possible circuit diagram that would replicate the observed timing relationships between the recorded neurons" *(22)*. This speaks about two important points: (a) effective connectivity is dynamic, that is, activity-dependent and (b) it depends upon a model of the interactions. The estimation procedures employed in functional neuroimaging can be divided into linear non-dynamic models [e.g., *(23)*] or non-linear dynamic models.

There is a necessary link between functional integration and multi-variate analyses because the latter are necessary to model interactions among brain regions. Multi-variate approaches can be divided into those that are inferential in nature and those that are data-led or exploratory. We will first consider multi-variate approaches that look at functional connectivity or covariance patterns (and are generally exploratory) and then turn to models of effective connectivity (that allow for inference about their parameters).

4.1.1. Eigenimage Analysis and Related Approaches

In Friston et al. *(24)*, we introduced voxel-based principal component analysis (PCA) of neuroimaging time series to characterise distributed brain systems implicated in sensorimotor, perceptual, or cognitive processes. These distributed systems are identified with principal components or *eigenimages* that correspond to spatial modes of coherent brain activity. This approach represents one of the simplest multi-variate characterisations of functional neuroimaging time series and falls into the class of exploratory analyses. Principal component or eigenimage analysis generally uses singular value decomposition to identify a set of orthogonal spatial modes that capture the greatest amount of variance expressed over time. As such, the ensuing modes embody the most prominent aspects of the variance–covariance structure of a given time series. Noting that covariance among brain regions is equivalent to functional connectivity renders eigenimage analysis particularly interesting because it was among the first ways of addressing functional integration (i.e., connectivity) with neuroimaging data. Subsequently, eigenimage analysis has been elaborated in a number of ways. Notable among these is canonical variate analysis (CVA) and multi-dimensional scaling *(25, 26)*. CVA was introduced in the context of multiple analysis of covariance and uses the generalised eigenvector solution to maximise the variance that can be explained by some explanatory variables relative to error. CVA can be thought of as an extension of eigenimage analysis that refers explicitly to some explanatory variables and allows for statistical inference.

In fMRI, eigenimage analysis [e.g., *(27)*] is generally used as an exploratory device to characterise coherent brain activity. These variance components may, or may not, be related to experimental design. For example, endogenous coherent dynamics have been observed in the motor system at very low frequencies *(28)*. Despite its exploratory power, eigenimage analysis is limited for two reasons. First, it offers only a linear decomposition of any set of neurophysiological measurements and second, the particular set of eigenimages or spatial modes obtained is determined by constraints that are biologically implausible. These aspects of PCA confer inherent limitations on the interpretability and usefulness of eigenimage analysis of biological time series and have motivated the exploration of non-linear PCA and neural network approaches.

Two other important approaches deserve to be mentioned here. The first is independent component analysis (ICA). ICA uses entropy maximisation to find, using iterative schemes, spatial modes or their dynamics that are approximately *independent*. This is a stronger requirement than *orthogonality* in PCA and involves removing high-order correlations among the modes (or dynamics). It was initially introduced as *spatial* ICA *(29)*, in which the independence constraint was applied to the modes (with no constraints on their temporal expression). More recent approaches use, by analogy with magneto- and electrophysiological time series analysis, *temporal* ICA where the dynamics are enforced to be independent. This requires an initial dimension reduction (usually using conventional eigenimage analysis). Finally, there has been an interest in cluster analysis *(30)*. Conceptually, this can be related to eigenimage analysis through multi-dimensional scaling and principal co-ordinate analysis.

All these approaches are interesting, but they are not used very much. This is largely because they tell you nothing about how the brain works or allow one to ask specific questions. Simply demonstrating statistical dependencies among regional brain responses or endogenous activity (i.e., demonstrating functional connectivity) does not address how these responses were caused. To address this, one needs explicit models of integration or more precisely, effective connectivity.

4.2. Dynamic Causal Modelling with Bilinear Models

This section is about modelling interactions among neuronal populations, at a cortical level, using neuroimaging time series and dynamic causal models that are informed by the biophysics of the system studied. The aim of DCM *(31)* is to estimate, and make inferences about, the coupling among brain areas and how that coupling is influenced by experimental changes (e.g., time or cognitive set). The basic idea is to construct a reasonably realistic neuronal model of interacting cortical regions or nodes. This model is then supplemented with a forward model of how neuronal or synaptic activity translates into a measured response (see previous section). This enables the parameters of the neuronal model (i.e., effective connectivity) to be estimated from observed data.

Intuitively, this approach regards an experiment as a designed perturbation of neuronal dynamics that are promulgated and distributed throughout a system of coupled anatomical nodes to change region-specific neuronal activity. These changes engender, through a measurement-specific forward model, responses that are used to identify the architecture and time constants of the system at a neuronal level. This represents a departure from conventional approaches [e.g., structural equation modelling and auto-regression models; *(32, 33)*], in which one assumes that the observed responses are driven by endogenous or intrinsic noise (i.e., innovations). In contrast, dynamic causal models assume that the responses are driven by designed changes in inputs. An important conceptual aspect of dynamic causal models pertains to how the experimental inputs enter the model and cause neuronal responses. Experimental variables can elicit responses in one of two ways. First, they can elicit responses through direct influences on specific anatomical nodes. This would be appropriate, for example, in modelling sensory-evoked responses in early visual cortices. The second class of input exerts its effect vicariously, through a modulation of the coupling among nodes. This sort of experimental variable would normally be more enduring, for example, attention to a particular attribute or the maintenance of some perceptual set. These distinctions are seen most clearly in relation to particular forms of causal models used for estimation, for example, the bilinear approximation

$$\begin{aligned} \dot{x} &= f(x,u) = Ax + uBx + Cu \\ y &= g(x) + \varepsilon \\ A &= \frac{\partial f(0,0)}{\partial x} \quad B = \frac{\partial^2 f(0,0)}{\partial x \partial u} \quad C = \frac{\partial f(0,0)}{\partial u} \end{aligned} \tag{7}$$

where $\dot{x} = \partial x / \partial t$. This is an approximation to any model of how changes in neuronal activity in one region x_i are caused by activity in the other regions. Here the output function $g(x)$ embodies a haemodynamic convolution, linking neuronal activity to BOLD, for each region (e.g., that in **Fig. 3**). The matrix A represents the coupling among the regions in the absence of input $u(t)$. This can be thought of as the endogenous coupling in the absence of experimental perturbations. The matrix B is effectively the change in coupling induced by the input. It encodes the input-sensitive changes in A or, equivalently, the modulation of coupling by experimental manipulations. Because B is a second-order derivative, it is referred to as *bilinear*. Finally, the matrix C embodies the exogenous influences of inputs on neuronal activity. The parameters $\theta = A, B$, and C are the connectivity or coupling matrices that we wish to identify and define the functional architecture and interactions among brain regions at a neuronal level. They play the same role as rate constant in kinetic models and therefore have units of Hertz or per second.

Because **Eq.** 7 has exactly the same form as **Eq. 4**, we can express it as a GLM and estimate the parameters using EM in the usual way (*see* **ref.** *31)*. Generally, estimation in the context of highly parameterised models like DCMs requires constraints in the form of priors. These priors enable conditional inference about the connectivity estimates. The sorts of questions that can be addressed with DCMs are now illustrated by looking at how attentional modulation is mediated in sensory processing hierarchies in the brain.

4.2.1. DCM and Attentional Modulation

It has been established that the superior parietal cortex (SPC) exerts a modulatory role on V5 responses using Volterra-based regression models *(34)* and that the inferior frontal gyrus (IFG) exerts a similar influence on SPC using structural equation modelling *(32)*. The example here shows that DCM leads to the same conclusions but starting from a completely different construct.

The experimental paradigm and data acquisition are described in the legend to **Fig. 4**. This figure also shows the location of the regions that entered the DCM. These regions were based on maxima from conventional SPMs testing for the effects of photic stimulation, motion, and attention. Regional time courses were taken as the first eigenvariate of 8-mm-spherical volumes of interest, centred on the maxima shown in the figure. The exogenous inputs, in this example, comprise one sensory perturbation and two contextual inputs. The sensory input was simply the presence of photic stimulation and the first contextual one was presence of motion in the visual field. The second contextual input, encoding attentional set, was one during attention to speed changes and

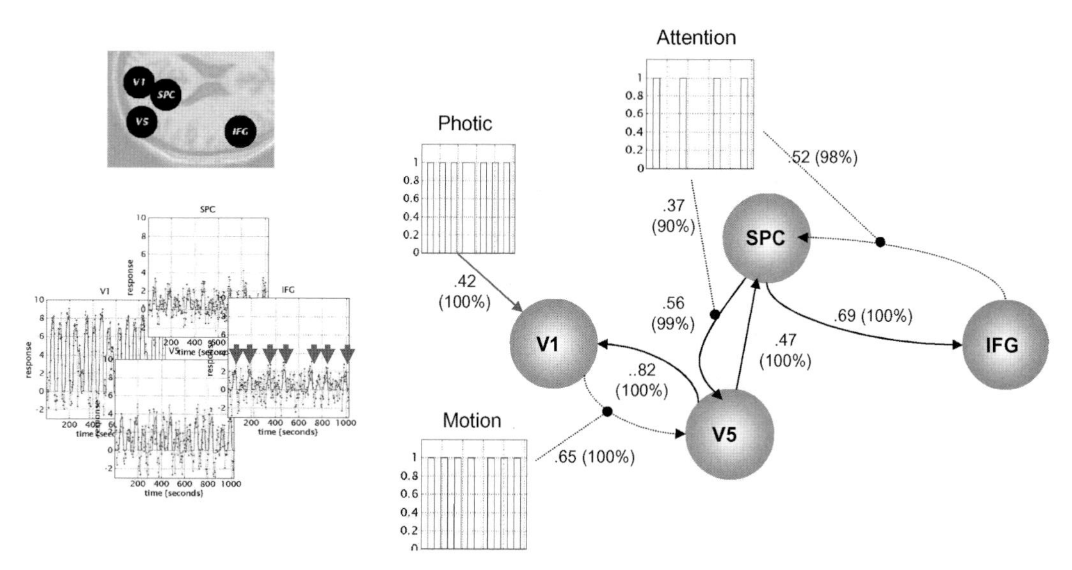

Fig. 4. Results of a dynamic causal modelling (*DCM*) analysis of attention to visual motion with fMRI. *Right panel*: Functional architecture based upon the conditional estimates shown alongside their connections, with the percent confidence that they exceeded threshold in brackets. The most interesting aspects of this architecture involve the role of motion and attention in exerting bilinear effects. Critically, the influence of motion is to enable connections from V1 to the motion-sensitive area V5. The influence of attention is to enable backward connections from the inferior frontal gyrus (*IFG*) to the superior parietal cortex (*SPC*). Furthermore, attention increases the influence of SPC on V5. Dotted arrows connecting regions represent significant bilinear effects in the absence of a significant intrinsic coupling. *Left panel*: Fitted responses based upon the conditional estimates and the adjusted data are shown for each region in the DCM. The insert (*upper left*) shows the location of the regions.

Subjects were studied with fMRI under identical stimulus conditions (visual motion subtended by radially moving dots) while manipulating the attentional component of the task (detection of velocity changes). The data were acquired from a normal subject at 2 Tesla. Each subject had four consecutive 100-scan sessions comprising a series of ten-scan blocks under five different conditions D F A F N F A F N S. The first condition (D) was a dummy condition to allow for magnetic saturation effects. F (fixation) corresponds to a low-level baseline where subjects viewed a fixation point at the centre of a screen. In condition A (attention), subjects viewed 250 dots moving radially from the centre at 4.7°/second and were asked to detect changes in radial velocity. In condition N (no attention), the subjects were asked simply to view the moving dots. In condition S (stationary), subjects viewed stationary dots. The order of A and N was swapped for the last two sessions. In all conditions, subjects fixated the centre of the screen. During scanning, there were no speed changes. No overt response was required in any condition.

zero otherwise. The outputs corresponded to the four regional eigenvariates in **Fig. 4** (left panel). The intrinsic connections were constrained to conform to a hierarchical pattern in which each area was reciprocally connected to its supra-ordinate area. Photic stimulation entered at, and only at, V1. The effect of motion in the visual field was modelled as a bilinear modulation of the V1 to V5 connectivity and attention was allowed to modulate the backward connections from IFG and SPC.

The results of the DCM are shown in **Fig. 4** (right panel). Of primary interest here is the modulatory effect of attention that is expressed in terms of the bilinear coupling parameters for this input. As expected, we can be highly confident that attention modulates the backward connections from IFG to SPC and from SPC to V5. Indeed, the influences of IFG on SPC are negligible in the absence of attention (dotted connection). It is important to note that the only way that attentional manipulation can affect brain responses was through this bilinear effect. Attention-related responses are seen throughout the system (attention epochs are marked with arrows in the plot of IFG responses in the left panel). This attentional modulation is accounted for, sufficiently, by changing just two connections. This change is, presumably, instantiated by instructional set at the beginning of each epoch.

The second thing, this analysis illustrates, is how functional segregation is modelled in DCM. Here one can regard V1 as "segregating" motion from other visual information and distributing it to the motion-sensitive area V5. This segregation is modelled as a bilinear "enabling" of V1 to V5 connections when, and only when, motion is present. Note that in the absence of motion, the intrinsic V1 to V5 connection was trivially small (in fact the estimate was –0.04 Hz). The key advantage of entering motion through a bilinear effect, as opposed to a direct effect on V5, is that we can finesse the inference that V5 shows motion-selective responses with the assertion that these responses are mediated by afferents from V1. The two bilinear effects above represent two important aspects of functional integration that DCM is able to characterise.

4.2.2. Structural Equation Modelling as a Special Case of DCM

The central idea, behind DCM, is to treat the brain as a deterministic non-linear dynamic system that is subject to inputs and produces outputs. Effective connectivity is parameterised in terms of coupling among unobserved brain states (e.g., neuronal activity in different regions). The objective is to estimate these parameters by perturbing the system and measuring the response. This is in contradistinction to established methods for estimating effective connectivity from neurophysiological time series, which include SEM and models based on multi-variate auto-regressive processes. In these models, there is no designed perturbation and the inputs are treated as unknown and stochastic. Furthermore, the inputs are often assumed to express themselves instantaneously such that, at the point of observation, the change in states is zero. From **Eq. 7**, in the absence of bilinear effects, we have

$$\dot{x} = 0 = Ax + Cu$$
$$x = -A^{-1}Cu \tag{8}$$

This is the regression equation used in SEM where $A = D-I$ and D contains the off-diagonal connections among regions. The key point here is that A is estimated by assuming that $u(t)$ is some random innovation with known covariance. This is not really tenable for designed experiments when $u(t)$ represent carefully structured experimental inputs. Although SEM and related auto-regressive techniques are useful for establishing dependence among regional responses, they are not surrogates for informed causal models based on the underlying dynamics of these responses.

In this section, we have covered multi-variate techniques ranging from eigenimage analysis that does not have an explicit forward or causal model to DCM that does. The bilinear approximation to any DCM has been illustrated though its use with fMRI to study attentional modulation. The parameters of the bilinear approximation include first-order effective connectivity A and its experimentally induced changes B. Although the bilinear approximation is useful, it is possible to model coupling among neuronal sub-populations explicitly. We conclude with a DCM that embraces a number of neurobiological facts and takes us much closer to a mechanistic understanding of how brain responses are generated. This example uses responses measured with EEG.

4.3. Dynamic Causal Modelling with Neural Mass Models

ERPs have been used for decades as electrophysiological correlates of perceptual and cognitive operations. However, the exact neurobiological mechanisms underlying their generation are largely unknown. In this section, we use neuronally plausible models to understand event-related responses. Our example shows that changes in connectivity are sufficient to explain certain ERP components. Specifically, we will look at the MMN, a component associated with rare or unexpected events. If the unexpected nature of rare stimuli depends on learning which stimuli are frequent, then the MMN must be due to plastic changes in connectivity that mediate perceptual learning. We conclude by showing that advances in the modelling of evoked responses now afford measures of connectivity among cortical sources that can be used to quantify the effects of perceptual learning.

4.3.1. Neural Mass Models

The minimal model we have developed *(35)* uses the connectivity rules described by Felleman and Van Essen *(36)* to assemble a network of coupled sources. These rules are based on a partitioning of the cortical sheet into supra-, infra-granular, and granular layer (layer 4). Bottom-up or forward connections originate in agranular layers and terminate in layer 4. Top-down or backward connections target agranular layers. Lateral connections originate in agranular layers and target all layers. These long-range or extrinsic cortico-cortical connections are excitatory and arise from pyramidal cells.

Each region or source is modelled using a neural mass model described by David and Friston *(35)*, based on the model of Jansen and Rit *(37)*. This model emulates the activity of a cortical area using three neuronal sub-populations, assigned to granular and agranular layers. A population of excitatory pyramidal (output) cells receives inputs from inhibitory and excitatory populations of inter-neurons, via intrinsic connections (intrinsic connections are confined to the cortical sheet). Within this model, excitatory inter-neurons can be regarded as spiny stellate cells found predominantly in layer 4 and in receipt of forward connections. Excitatory pyramidal cells and inhibitory inter-neurons are considered to occupy agranular layers and receive backward and lateral inputs (**Fig. 5**).

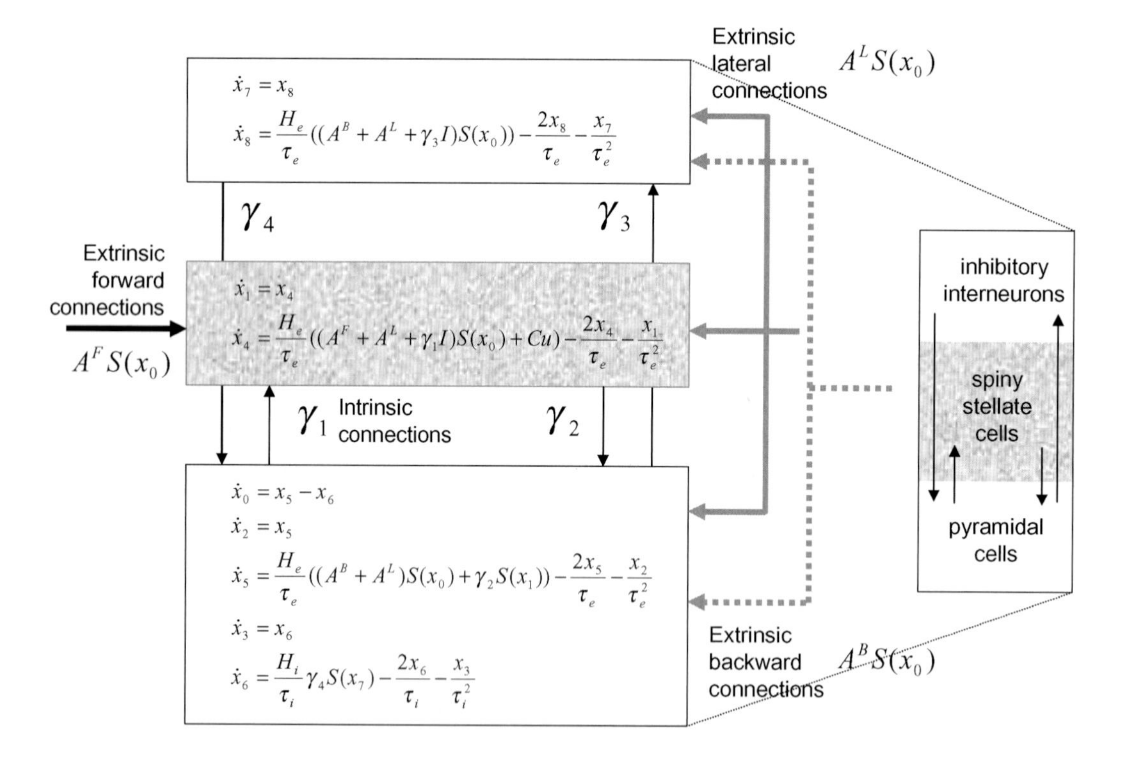

Fig. 5. Schematic of the dynamic causal modelling (*DCM*) used to model electrical responses. This schematic shows the state equations describing the dynamics of sources or regions. Each source is modelled with three sub-populations (pyramidal, spiny stellate, and inhibitory inter-neurons) as described in the main text. These have been assigned to granular and agranular cortical layers that receive forward and backward connections, respectively.

To model event-related responses, the network receives inputs via input connections. These connections are exactly the same as forward connections and deliver inputs to the spiny stellate cells in layer 4. In the present context, inputs $u(t)$ model subcortical auditory inputs. The vector C controls the influence of the input on each source. The lower, upper, and leading diagonal matrices A^F, A^B, A^L encode forward, backward, and lateral connections, respectively. The DCM here is specified in terms of the state equations shown in **Fig. 5** and a linear output equation

$$\begin{aligned} \dot{x} &= f(x, u) \\ y &= Lx_0 + \varepsilon \end{aligned} \tag{9}$$

where x_0 represents the trans-membrane potential of pyramidal cells and L is a lead field matrix coupling electrical sources to the EEG channels. This should be compared with the DCM above for haemodynamics; here, the equations governing the evolution of neuronal states are much more complicated and realistic, as opposed to the bilinear approximation in **Eq.** 7. Conversely, the output equation is a simple linearity, as opposed to the non-linear observer used for fMRI. As an example, the state equation for the inhibitory sub-population is[2]

$$\begin{aligned} \dot{x}_7 &= x_8 \\ \dot{x}_8 &= \frac{H_e}{\tau_e}[(A^B + A^L + \gamma_3 I)S(x_0)] - \frac{2x_8}{\tau_e} - \frac{x_7}{\tau_e^2} \end{aligned} \tag{10}$$

Within each sub-population, the evolution of neuronal states rests on two operators. The first transforms the average density of presynaptic inputs into the average postsynaptic membrane potential. This is modelled by a linear transformation with excitatory and inhibitory kernels parameterised by $H_{e,i}$ and $\tau_{e,j}$. $H_{e,j}$ controls the maximum post-synaptic potential and $\tau_{e,j}$ represents a lumped rate constant. The second operator S transforms the average potential of each sub-population into an average firing rate. This is assumed to be instantaneous and is a sigmoid function. Interactions, among the sub-populations, depend on constants, $\gamma_{1,2,3,4}$, which control the strength of intrinsic connections and reflect the total number of synapses expressed by each sub-population. In **Eq. 10**, the top line expresses the rate of change of voltage as a function of current. The second line specifies how current changes as a function of voltage, current, and pre-synaptic input from extrinsic and intrinsic sources. Having specified the DCM in terms of these equations, one can estimate the coupling parameters from empirical data using *EM* as described above. See reference *(38)* for more details.

[2] Propagation delays on the extrinsic connections have been omitted for clarity here and in **Fig. 5**.

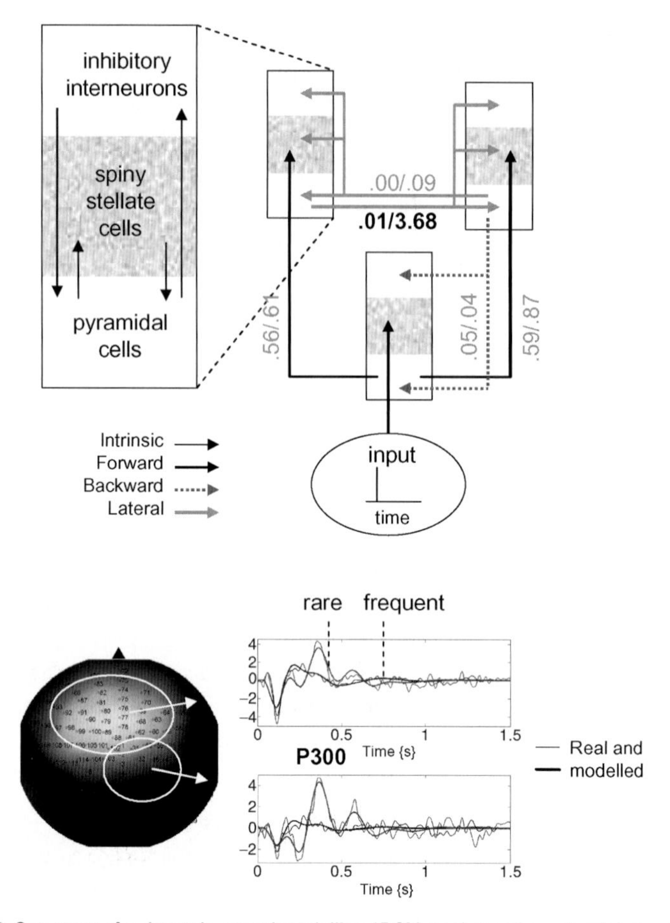

Fig. 6. Summary of a dynamic causal modelling (*DCM*) analysis of event-related potentials (*ERPs*) elicited during an auditory oddball paradigm, employing rare and frequent pure tones. *Upper panel*: Schematic showing the architecture of the neuronal model used to explain the empirical data. Sources were coupled with extrinsic cortico-cortical connections following the rules of Felleman and van Essen. The free parameters of this model included intrinsic and extrinsic connection strengths that were adjusted to best explain the data. In this example, the lead field was also estimated, with no spatial constraints. The parameters were estimated for ERPs recorded during the presentation of rare and frequent tones and are reported beside their corresponding connection (frequent/rare). The most notable finding was that the mismatch response could be explained by a selective increase in lateral connection strength from 0.1 to 3.68 Hz (highlighted in bold). *Lower panel*: The channel positions (*left*) and ERPs (*right*) averaged over two subsets of channels (*circled on the left*). Note the correspondence between the measured ERPs and those generated by the model. Auditory stimuli, 1,000 or 2,000 Hz tones with 5 ms rise and fall times and 80 ms duration, were presented binaurally. The tones were presented for 15 min, every 2 s in a pseudo-random sequence with 2,000-Hz tones occurring 20% of the time and 1,000-Hz tones occurring 80% of the time. The subject was instructed to keep a mental record of the number of 2,000-Hz tones (non-frequent target tones). Data were acquired using 128 EEG electrodes with 1,000 Hz sample frequency. Before averaging, data were referenced to mean earlobe activity and band-pass filtered between 1 and 30 Hz. Trials showing ocular artefacts and bad channels were removed from further analysis.

4.3.2. Perceptual Learning and the MMN

The example shown in **Fig. 6** is an attempt to model the MMN in terms of changes in backward and lateral connections among cortical sources. In this example, two (averaged) channels of EEG data were modelled with three cortical sources. Using this generative or forward model, we estimated differences in the strength of these connections for rare and frequent stimuli. As expected, we could account for detailed differences in the ERPs (the MMN) by changes in connectivity (see figure legend for details). Interestingly, these differences were expressed selectively in the lateral connections. If this model is a sufficient approximation to the real sources, these changes are a non-invasive measure of plasticity, mediating perceptual learning, in the human brain.

5. Conclusion

In this chapter, we have reviewed some key models that underpin image analysis and have touched briefly on ways of assessing specialisation and integration in the brain. These models can be regarded as a succession of modelling endeavours, that drawing more and more on our understanding of how brain-imaging signals are generated, both in terms of biophysics and the underlying neuronal interactions. We have seen how hierarchical linear observation models encode the treatment effects elicited by experimental design. GLMs based on convolution models imply an underlying dynamic input-state-output system. The form of these systems can be used to constrain convolution models and explore some of their simpler non-linear properties. By creating observation models based on explicit forward models of neuronal interactions, one can model and assess interactions among distributed cortical areas and make inferences about coupling at the neuronal level. The next years will probably see an increasing realism in the dynamic causal models introduced above (*see* **ref.** *39)*. These endeavours are likely to encompass fMRI signals enabling the conjoint modelling, or fusion, of different modalities and the marriage of computational neuroscience with the modelling of brain responses.

References

1. Staum M. Physiognomy and phrenology at the Paris Athénée. J Hist Ideas 1995;6: 443–462.
2. Phillips CG, Zeki S, and Barlow HB. Localisation of function in the cerebral cortex: Past present and future. Brain 1984;107: 327–361.
3. Goltz F. In “Transactions of the 7th international medical congress” (W. MacCormac, Ed.), Vol. I, JW Kolkmann: London, 1881:218–228.
4. Absher JR and Benson DF. Disconnection syndromes: An overview of Geschwind’s contributions. Neurology 1993;43:862–867.

5. Zeki S. The motion pathways of the visual cortex. In "Vision: Coding and efficiency" (C. Blakemore, Ed.), Cambridge University Press, UK, 1990:321–345.
6. Friston KJ, Frith CD, Liddle PF, and Frackowiak RSJ. Comparing functional (PET) images: The assessment of significant change. J Cereb Blood Flow Metab 1991;11:690–699.
7. Berry DA and Hochberg Y. Bayesian perspectives on multiple comparisons. J Statistical Planning and Inference 1999;82:215–227.
8. Holmes A and Ford I. A Bayesian approach to significance testing for statistic images from PET. In "Quantification of brain function, tracer kinetics and image analysis in brain PET" (K. Uemura, N.A. Lassen, T. Jones, and I. Kanno, Eds.), Excerpta Medica, Int. Cong. Series No. 1993;1030:521–534.
9. Dempster AP, Laird NM, and Rubin. Maximum likelihood from incomplete data via the EM algorithm. J Roy Stat Soc 1977;Series B 39:1–38.
10. Friston KJ, Jezzard P, and Turner R. Analysis of functional MRI time series. Human Brain Map 1994;1:153–171.
11. Boynton GM, Engel SA, Glover GH, and Heeger DJ. Linear systems analysis of functional magnetic resonance imaging in human V1. J Neurosci 1996;16:4207–4221.
12. Friston KJ, Frith CD, Turner R, and Frackowiak RSJ. Characterising evoked hemodynamics with fMRI. NeuroImage 1995;2:157–165.
13. Josephs O, Turner R, and Friston KJ. Event-related fMRI Hum. Brain Mapp 1997;5: 243–248.
14. Lange N and Zeger SL. Non-linear Fourier time series analysis for human brain mapping by functional magnetic resonance imaging (with discussion). J Roy Stat Soc Ser C 1997;46:1–29.
15. Buxton RB and Frank LR. A model for the coupling between cerebral blood flow and oxygen metabolism during neural stimulation. J Cereb Blood Flow Metab 1997;17:64–72.
16. Mandeville JB, Marota JJ, Ayata C, Zararchuk G, Moskowitz MA, Rosen B, and Weisskoff RM. Evidence of a cerebrovascular postarteriole Windkessel with delayed compliance. J Cereb Blood Flow Metab 1999;19:679–689.
17. Hoge RD, Atkinson J, Gill B, Crelier GR, Marrett S, and Pike GB. Linear coupling between cerebral blood flow and oxygen consumption in activated human cortex. Proc Natl Acad Sci 1999;96:9403–9408.
18. Friston KJ, Mechelli A, Turner R, and Price CJ. Nonlinear responses in fMRI: The Balloon model, Volterra kernels, and other hemodynamics. NeuroImage 2000;12:466–77..
19. Fliess M, Lamnabhi M, and Lamnabhi-Lagarrigue F. An algebraic approach to nonlinear functional expansions. IEEE Trans Circuits Syst 1983;30:554–570.
20. Bendat JS. Nonlinear system analysis and identification from random data. John Wiley and Sons, New York USA, 1990.
21. Gerstein GL and Perkel DH. Simultaneously recorded trains of action potentials: Analysis and functional interpretation. Science 1969;164:828–830.
22. Aertsen A and Preißl H. Dynamics of activity and connectivity in physiological neuronal Networks. In "Non linear dynamics and neuronal networks" (H.G. Schuster, Ed.), VCH publishers, Inc., New York NY USA, 1991:281–302.
23. McIntosh AR and Gonzalez-Lima F. Structural equation modelling and its application to network analysis in functional brain imaging. Hum Brain Mapp 1994;2:2–22.
24. Friston KJ, Frith CD, Liddle PF, and Frackowiak RSJ. Functional connectivity: The principal component analysis of large data sets. J Cereb Blood Flow Metab 1993;13:5–14.
25. Friston KJ, Poline J-B, Holmes AP, Frith CD, and Frackowiak RSJ. A multivariate analysis of PET activation studies. Hum Brain Mapp 1996;4:140–151.
26. Friston KJ, Frith CD, Fletcher P, Liddle PF, and Frackowiak RSJ. Functional topography: Multidimensional scaling and functional connectivity in the brain. Cerebral Cortex 1996;6:156–164.
27. Sychra JJ, Bandettini PA, Bhattacharya N, and Lin Q. Synthetic images by subspace transforms. I. Principal component images and related filters. Med Physics 1994;21: 193–201.
28. Biswal B, Yetkin FZ, Haughton VM, and Hyde JS. Functional connectivity in the motor cortex of resting human brain using echo-planar MRI. Mag Res Med 1995;34:537–541.
29. McKeown M, Jung T-P, Makeig S, Brown G, Kinderman S, Lee T-W, and Sejnowski T. Spatially independent activity patterns in functional MRI data during the Stroop colour naming task. Proc Natl Acad Sci 1998;95: 803–810.
30. Baumgartner R, Scarth G, Teichtmeister C, Somorjai R, and Moser E. Fuzzy clustering of gradient-echo functional MRI in the human visual cortex. Part 1: Reproducibility. J Mag Res Imaging 1997;7:1094–1101.

31. Friston KJ, Harrison L, and Penny W. Dynamic causal modelling. NeuroImage 2003;19:1273–1302.
32. Büchel C and Friston KJ. Modulation of connectivity in visual pathways by attention: Cortical interactions evaluated with structural equation modelling and fMRI. Cerebral Cortex 1997;7:768–778.
33. Harrison LM, Penny W, and Friston KJ. Multivariate autoregressive modelling of fMRI time series. NeuroImage 2003;19:1477–1491.
34. Friston KJ and Büchel C. Attentional modulation of effective connectivity from V2 to V5/MT in humans. Proc Natl Acad Sci U S A 2000;97:7591–7596.
35. David O and Friston KJ. A neural mass model for MEG/EEG: coupling and neuronal dynamics. NeuroImage 2003;20:1743–1755.
36. Felleman DJ and Van Essen DC. Distributed hierarchical processing in the primate cerebral cortex. Cereb Cortex 1992;1:1–47.
37. Jansen BH and Rit VG. Electroencephalogram and visual evoked potential generation in a mathematical model of coupled cortical columns. Biol Cybern 1995;73:357–366.
38. David O, Kiebel SJ, Harrison LM, Mattout J, Kilner JM, and Friston KJ. Dynamic causal modelling of evoked responses in EEG and MEG. NeuroImage 2006;30:1255–1272.
39. Horwitz B, Friston KJ, and Taylor JG. Neural modelling and functional brain imaging: an overview. Neural Networks 2001;13:829–846.

Chapter 9

Brain Atlases: Their Development and Role in Functional Inference

John Darrell Van Horn and Arthur W. Toga

Summary

Imparting functional meaning to neuroanatomical location has been among the greatest challenges to neuroscientists. The characterization of the brain architecture responsible in human cognition received a boost in momentum with the emergence of in vivo functional and structural neuroimaging technology over the past 20 years. Yet, individual variability in cortical gyrification as well as the pattern of blood flow-related activity measured using fMRI and positron emission tomography complicated direct comparisons across subjects without spatially accounting for overall brain size and shape. This realization resulted in considerable effort now involving the collective efforts of neuroscientists, computer scientists, and mathematicians to develop common brain atlas spaces against which the regions of activity may be accurately referenced. We examine recent developments in brain imaging and computational anatomy that have greatly expanded our ability to analyze brain structure and function. The enormous diversity of brain maps and imaging methods has spurred the development of population-based digital brain atlases. These atlases store information on how the brain varies across age and gender, across time, in health and disease, and in large human populations. We describe how brain atlases, and the computational tools that align new datasets with them, facilitate comparison of brain data across experiments, laboratories, and from different imaging devices. The major philosophies are presented that underlie the construction of probabilistic atlases, which store information on anatomic and functional variability in a population. Algorithms that create composite brain maps and atlases based on multiple subjects are examined. We show that group patterns of cortical organization, asymmetry, and disease-specific trends can be resolved that may not be apparent in individual brain maps. Finally, we describe the development of four-dimensional maps that store information on the dynamics of brain change in development and disease.

Key words: Brain atlases, Neuroanatomy, Diffeomorphism, Warping, Functional activity, Inference

M. Filippi (ed.), *fMRI Techniques and Protocols*, Neuromethods, vol. 41
DOI 10.1007/978-1-60327-919-2_9, © Humana Press, a part of Springer Science+Business Media, LLC 2009

1. Introduction

Over a century ago, in a horrific accident, damage to the frontal lobe of Phineas Gage produced profound changes in his personality and cognitive function, the beginning of what may be considered as the modern era of the localization of brain function *(1)*. Nearly a decade later, the examination by Paul Broca of aphasic patients having damage to the left inferior frontal areas solidified the notion that function could be linked to specific areas of cortical tissue *(2)*. At the turn of the twentieth century, Dr. Alois Alzheimer, a German psychiatrist, identified the first case of what became known as Alzheimer's disease (AD) *(3)*. Since those early reports, a principle goal in neuroscience has been to classify the specific brain regions possessing unique functional components of complex thought and how these functions might be altered in response to injury or as a result of disease.

Over the past two decades, with the emergence of neuroimaging as the primary tool for the examination of the brain in vivo during cognitive task performance, the mapping of mental function has given rise to an explosion of functional experimentation and an ever-widening interest in understanding brain processes from fields beyond traditional neuroscience (e.g., economics, criminology, social science, etc.). This intense effort and expanse of data has emphasized the realization that there is considerable individual variation in brain size and shape that must be accounted for in the processing of brain imaging data and the assignment of functional significance. Evaluation and comparison of brain imaging data with respect to and against well-defined anatomical references is now a critical element in the localization of essential cognitive functions in nearly all functional imaging investigations.

Brain atlases can now comprise imaging data describing multiple aspects of brain structure or function at different scales from different subjects, yielding a truly integrative and comprehensive description of this organ in health and disease *(4, 5)*. The complexity and variability of brain structure, especially in the gyral patterns of the human cortex, present challenges in creating standardized brain atlases that reflect the anatomy of a population *(6)*. This chapter discusses the concepts behind population-based, age- and disease-specific brain atlas construction that can be used to reflect the specific anatomy and physiology of a particular clinical subpopulation. Based on well-characterized subject groups, age-specific atlases can potentially contain thousands of structure models, composite maps, average templates, and visualizations of structural variability, asymmetry, and group-specific differences. They correlate the structural, metabolic, molecular, and histologic hallmarks of the disease *(7, 8)*. Rather than simply fusing information from multiple subjects and sources, new mathematical strategies are introduced to resolve group-specific features not apparent in individual scans *(9)*. High-dimensional elastic mappings, based on covariant partial differential equations, are developed to encode patterns of cortical variation *(10–12)*. In the resulting brain atlas, age-stratified features and regional asymmetries emerge that are not apparent in individual anatomies. The resulting probabilistic atlas spaces can be used to identify patterns of altered structure and function, and can guide algorithms for knowledge-based image analysis, automated image labeling, tissue classification, data mining, and functional image analysis. These integrative techniques have provided significant motivation for human brain mapping initiatives and have important applications in health and disease.

2. Methods for Brain Atlas Construction

Creating atlases relies on the accumulation and compilation of many image sets along with appropriate registration and warping strategies, indexing schemes, and nomenclature systems. The processing of multi-modal brain images in the context of an atlas enables a more meaningful interpretation (**Fig. 1**). The complexity and variability of human brain (as well as other species) across

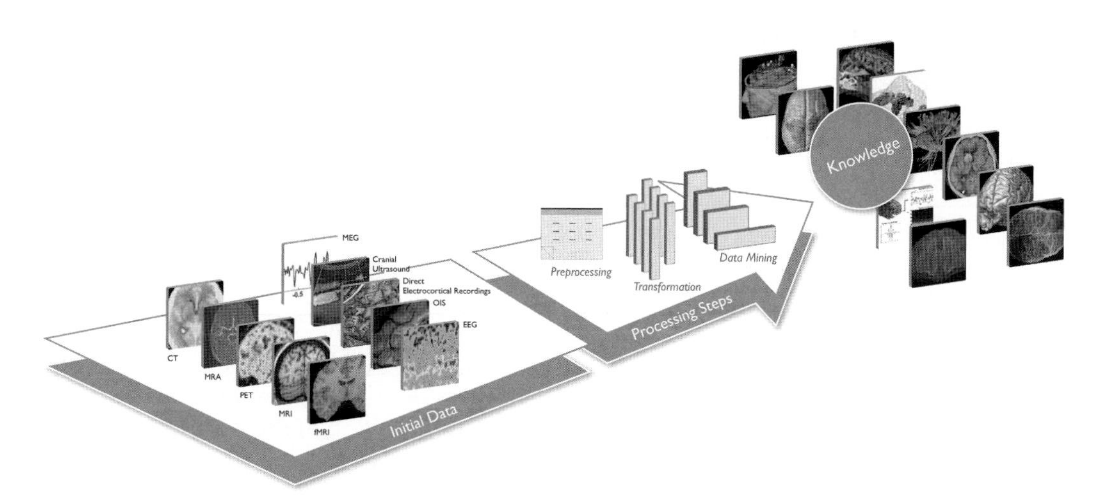

Fig. 1. A variety of neuroimaging methods permit the acquisition of brain data over time and space having a range of resolution granularity. Moreover, variation across individuals and how this changes over the lifespan must be accounted for in statistical examination of the data. Mapping these data to known spatial coordinate systems enables highly accurate inference concerning the brain's structural change over time, between populations, or in terms of localizing functional change. Atlases denoting this variation after spatial warping, the characterization of shape, three-dimensional (3D) distortion, etc. will be essential in describing structural and functional alteration associated with normal aging as well as in disease.

subjects is so great that reliance on atlases is essential to effectively manipulate, analyze, and interpret brain data. Central to these tasks is the construction of averages, templates, and models to describe how the brain and its component parts are organized. Design of appropriate reference systems for human brain data presents considerable challenges, since these systems must capture how brain structure and function vary in large populations, across age and gender, in different disease states, across imaging modalities, and even across species.

2.1. Basic Image Registration

Image registration is elemental to many of the challenges in brain imaging today *(13)*. Initially developed as an image processing subspecialty to spatially align one image to match another, image registration now has a vast range of applications, such as automated image labeling and for pathology detection in individuals or groups *(14)*. Registration algorithms can encode patterns of anatomic variability in large human populations, and can use this information to create disease-specific, population-based brain atlases *(15)*. They may also blend data from multiple imaging devices to correlate different measures of brain structure and function. Finally, registration algorithms can serve as a basic measure for patterns of structural change during brain development, tumor growth, or degenerative disease processes *(16)*.

2.2. Geodesic Averaging of Brain Shape

The objective in geodesic approaches has been to encourage variational methods for anatomical averaging that operate within the space of the underlying image registration problem *(17)*. This approach is effective when using the large deformation viscous framework, where linear averaging might not be appropriate. The theory behind it is similar to registration-based techniques but with single image force replaced by the average forces from multiple sources. These group forces drive an average transport ordinary differential equation allowing one to estimate the geodesic that moves an image toward the mean shape configuration. This model provides large deformation atlases that are optimal with respect to the shape manifold as defined by the data and the image registration assumptions. These procedures generate refined average representations of highly variable anatomy from distinct populations. For example, the population statistics have been used to show a significant doubling of the relative prefrontal lobe size in humans, as compared to chimpanzees *(18)*.

2.3. Density-Based Atlases

Initial approaches to population-based atlasing concentrated on generating mean representations of anatomy through the "intensity pooling" of multiple MRI scans. This involves large number of MRI scans that are each linearly transformed into stereotaxic space, intensity-normalized, and averaged on a voxel-by-voxel basis, producing an average intensity MRI dataset. The average brains that result have large areas, especially at the cortex, where individual structures are blurred because of spatial variability in the population. While this blurring limits their usefulness as a quantitative tool, the templates can be used as targets for the automated registration and mapping of MR and co-registered functional data into stereotaxic space *(19)*.

2.4. Label-Based Atlases

In label-based approaches, large ensembles of brain data are labeled or "segmented" by a human operator into sub-volumes after mapping individual datasets into stereotaxic space. A probability map is then constructed for each segmented structure, by determining the proportion of subjects assigned a given anatomic label at each voxel position in stereotactic space. The prior information which these probability maps provide on the location of various tissue classes in stereotactic space has been useful in designing automated tissue classifiers and approaches to correct radio-frequency and intensity inhomogeneities in MR scans. Statistical data on anatomic labels and tissue types normally found at given positions in stereotactic space provide a vital independent source of information to guide and inform mathematical algorithms, which analyze neuroanatomical data in stereotactic space.

2.5. Encoding Brain Variation

Measuring and accounting for the considerable variability in brain shape across human populations necessitates realistically complex mathematical strategies to encode comprehensive information on structural variability *(20)*. Particularly relevant is a three-dimensional (3D) statistical information on group-specific patterns of variation and how these patterns are altered in disease. This information can be represented such that it can be exploited by expert diagnostic systems, whose goal is to detect subtle or diffuse structural alterations in disease *(21)*. Strategies for detecting structural anomalies can leverage information in anatomical databases by invoking encoded knowledge on the variations in geometry and location of neuroanatomic regions and critical functional interfaces, especially at the cortex.

2.6. Shape and Pattern Theory

Of particular relevance in dealing with brain substructures are methods used to define a mean shape in such a way that departures from this mean shape can be treated as a linear process *(22)*. Linearization of the pathology detection problem, by constructing various shape manifolds and their associated tangent spaces, allows the use of conventional statistical procedures and linear decomposition of departures from the mean to characterize shape change. These approaches have been applied to detect structural anomalies in schizophrenia by identification of statistical differences in mean shape of brain structures *(23–25)* (**Fig. 2**).

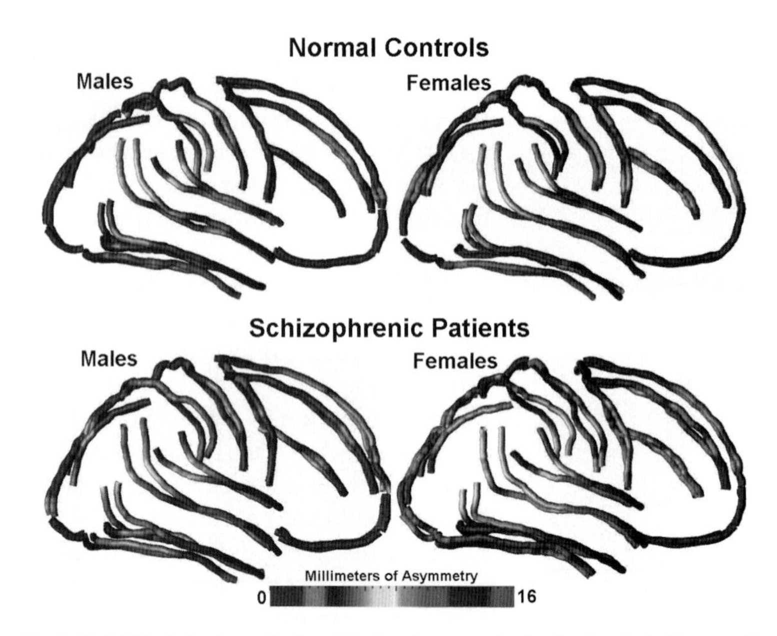

Fig. 2. Variability in brain cortical architectural asymmetry leads directly to the consideration of probabilistic atlases and the use of cortically derived landmarks, for example, sulcal lines. In this figure adapted from ***(23)***, sulcal lines were manually determined for male and female normal subjects and those diagnosed with schizophrenia. Asymmetry maps were created in each group as defined by sex and diagnosis (*NC* normal controls, *SZ* schizophrenic patients). Sulcal mesh averages for each hemisphere were subtracted from a reflected version of the same structure in the other hemisphere to create displacement vectors. Thus, the color mapping indicates the degree of lateralization in terms of millimeters of displacement for the line under diffeomorphic and atlas space constraints. These maps represent in color the magnitude of average asymmetry in sulcal anatomy between the two hemispheres between males and females and schizophrenic and normal subjects. [Figure adapted from Narr et al. ***(72)***].

2.7. Deformation Atlases

When applied to two different 3D brain scans, a non-linear registration or warping algorithm calculates a deformation map that matches up brain structures in one scan with their counterparts in the other. The deformation map indicates 3D patterns of anatomic differences between the two subjects or populations *(26)*. In probabilistic atlases based on deformation maps, statistical properties of these deformation maps are encoded locally to determine the magnitude and directional biases of anatomic variation *(27)*. Encoding of local variation can then be used to assess the severity of structural variants outside of the normal range, which may be a sign of disease. A major goal in designing this type of pathology detection system is to recognize that both the magnitude and local directional biases of structural variability in the brain may be different at every single anatomic point. Such atlases are not only cortically based but can also be done, for instance, on substructures and cerebellum *(28)*.

2.8. Disease-Specific Atlases

Disease-specific atlases are designed to reflect the unique anatomy and physiology of a particular clinical subpopulation. Based on well-characterized patient groups, these atlases contain thousands of structure models, as well as composite maps, average templates, and visualizations of structural variability, asymmetry, and group-specific differences. They act as a quantitative framework that correlates the structural, metabolic, molecular, and histologic hallmarks of the disease. Because they retain information on group anatomical variability, disease-specific atlases are a type of probabilistic atlas specialized to represent a particular clinical group. The resulting atlases can identify patterns of altered structure or function, and can guide algorithms for knowledge-based image analysis, automated image labeling, tissue classification, and functional image analysis.

2.9. Genetic Atlases

Inclusion of genetic data in an atlas makes it possible to go beyond simply describing the effects of a disease on the brain to investigating its fundamental causes. This not only allows the direct mapping of genetic influences on brain structure, but also allows us to quantify heritability for different features of the brain. Familial, twin, and genetic linkage studies have recently begun to expand the atlas concept to tie together genetic and imaging studies of disease *(7, 29)*. Atlases that contain genetic brain maps, and a means to analyze them, can help screen relatives for inherited disease. They also offer a framework to mine large imaging databases for risk genes and quantitative trait loci, as well as genetic and environmental triggers of disease.

2.10. Age and Developmental Stratification

The brain changes remarkably in its size and complexity over the lifespan. There is considerable need to account for the age of particular populations in the context of brain maturation and the development of age-stratified normal brain atlas spaces *(30)*. People who are mildly cognitively impaired, for instance, are at a fivefold increased risk of imminent conversion to dementia, and present specific structural brain changes that are predictive of imminent disease onset *(31, 32)*. Language impairment in AD patients is also correlated with cortical atrophy in the left temporal and parietal lobes, bilateral frontal lobes, and the right temporal pole *(33)*. However, characterizing such change presents particular computational challenges. The fitting of brain anatomy to a single template of undetermined age specification may lead to errors in inference about brain morphometry of function relative to an inappropriate underlying template. Alternative approaches can also be fruitful and metrics, such as shape *(34)*, cortical thickness mapping, tensor-based morphometry (TBM), may be better suited for shedding light on the neuroscience of aging and brain degeneration in AD and mild cognitive impairment (MCI) *(35)*.

3. Openly Available Atlases of the Brain

An increasing number and variety of brain atlases for humans, as well as other species, are being made openly available online for the neuroscience community to use as authoritative references, for inclusion in data processing pipelines, or for the display of results. These include probabilistic anatomical atlases *(15,36, 37)*, white matter fiber atlases *(38)*, and cortical surface atlases *(39)*. A brief listing of several from human, non-human primate, and the mouse are provided in **Table 1**.

4. Applications of Atlases for Regional Parcellation and Functional Inference

Without reference to known geometries or atlas spaces, functional localization is not formally possible at the population level. For instance, in positron emission tomography (PET)/fMRI studies of human cognition analyzed using the statistical parametric mapping (SPM) software package relying on the Montreal Neurological Institute (MNI) atlas as the basis for within group and between group statistical comparisons, typically, each subject's high resolution anatomical image is warped to the MNI multi-subject T1 whole brain template using non-linear and affine methods *(40)*. This transformation is then applied to the collection of linearly aligned fMRI time series or task-condition-specific PET images *(41)*. Employing regional labeling based upon the chosen atlas space, the process of localization analysis is enhanced by reference to known anatomical delineations. The process is typically decomposed into a series of steps: data warping, feature extraction, identification of loci, fitting of labels, and region activity value extraction (*(42)*, for discussion).

Obtaining reliable spatial registration with the chosen atlas space is essential to accurate localization of functional activity. The alignment accuracy and impact on functional maps of four spatial normalization procedures have been compared using a set of high-resolution brain MRIs and functional PET volumes *(43)*, suggesting that the functional variability is much larger than that comprised anatomically and that precise alignment of anatomical features has low influence on the resulting inter-subject functional maps. At larger spatial resolution, however, differences in localization of activated areas appear to be a consequence of the particular spatial normalization procedure employed. Despite these concerns, however, for typical sample sizes and numbers of observations per subject, reliable functional localization is achieved when performed for each individual using data in atlas space *(44)*.

Table 1
Human, non-human primate, and mouse probabilistic anatomical atlases

Name	URL(s)	Comment
Adult C57BL/6J Mouse Brain	http://www.bnl.gov/CTN/mouse/	Anatomical atlas of the C57BL/6J mouse brain
Atlases of the Brain	http://library.med.utah.edu/kw/brain_atlas/	Anatomical sections and MR image data of human brain
The Allen Brain Atlas	http://www.alleninstitute.org/ http://www.brain-map.org/	Atlas of gene expression in the mouse
Brainmaps.org	http://brainmaps.org/	Digitally scanned images of serial sections of both primate and non-primate brains
Comparative Mammalian Brain Collections	http://www.brainmuseum.org/	Digital photos of whole brain and serial sections from a range of primate brains
Digital Anatomist	http://da.biostr.washington.edu/da.html	Interactive brain atlas surface models and neuroimaging data
The Human Brain Atlas	http://www.msu.edu/%7ebrains/brains/human/index.html	Images of stained sections of human brain
ICBM/LONI Probabilistic Atlas Series	http://www.loni.ucla.edu/Atlases	Human brain atlases based upon probabilistic metrics of regional location
The Mouse Brain Library	http://www.mbl.org/	High-resolution images and database of brains from many genetically characterized strains of mice
Mouse Lemur Brain	http://atlasserv.caltech.edu/Lemur/Start_lemur.html	Downloadable MR image volumes of the mouse lemur (*Microcebus murinus*) brain
Surface Data Management System (SuMs) Atlases	http://sumsdb.wustl.edu:8081/sums/humanpalsmore.do	Standardized brain surface models with links to associated functional data
White Matter Atlas	http://www.dtiatlas.org/	Human brain white matter maps obtained from diffusion tensor imaging (DTI)
The Whole Brain Atlas	http://www.med.harvard.edu/AANLIB/home.html	Human brain atlas including images from post-mortem serial sections and MRI. Includes aging brain images

The use of atlases provides more than just a space in which to morph images for computing population averages and inferential statistics on function, but is also useful for the precise labeling of cortical regions. **Figure 3** shows an example of a LONI Pipeline

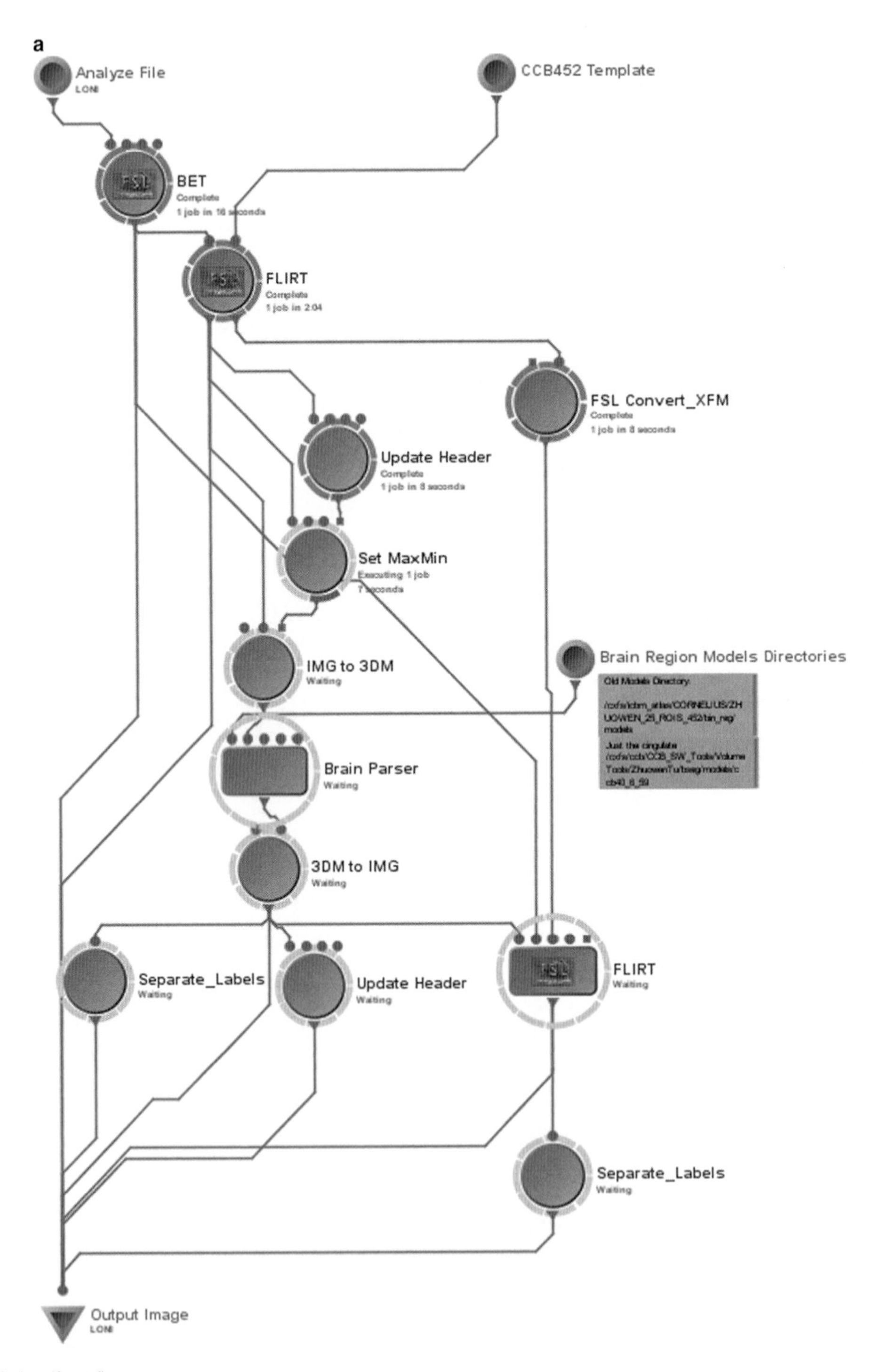

Fig. 3. (continued)

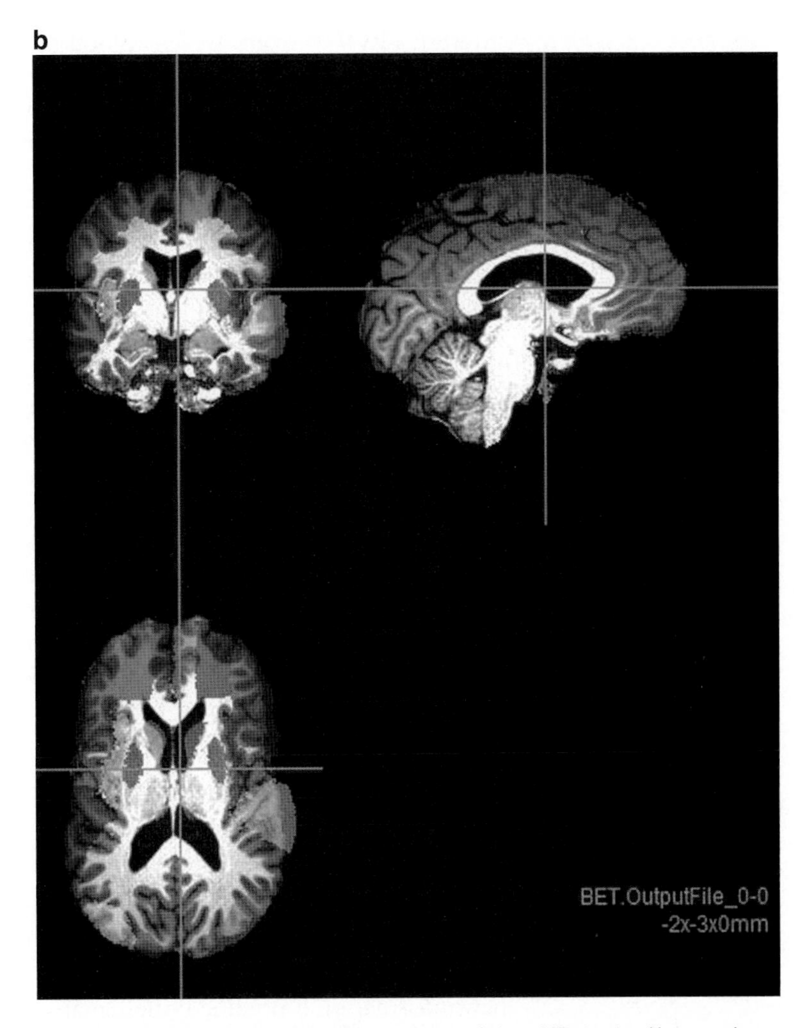

Fig. 3. (**a**) A LONI Pipeline workflow diagram that reads in an MR structural image volume, performs skull stripping, fits the data to the ICBM452 standardized atlas, performs Bayesian boost-tree region classification, and (**b**) returns the regional delineation results to the original MR image space. Once obtained, the region labeling can be used to mask functional and/or extract functionally related signal from blood oxygen level dependent data. Such automated labeling of brain regions would be made considerably more challenging without the use of standard atlas spaces.

diagram for the process of brain extraction using FSL's BET followed by warping to the ICBM452 atlas using FLIRT, and brain surface parcellation using the Brain Parser algorithm developed at UCLA *(45)*. Unlike alternative methods for detecting the major cortical sulci, which use a set of predefined rules based on properties of the cortical surface such as the mean curvature, this approach learns a discriminative model using the probabilistic boosting tree algorithm (PBT), a supervised learning approach which selects and combines hundreds of features at different scales, such as curvatures, gradients, and shape index *(46)*. Example output from this method can then be used as regions of interest (ROIs) from which blood oxygen level dependent (BOLD) values may then be obtained.

Historically, standardized atlases have demonstrated their greatest utility in the fitting of experimental data from functional imaging studies with PET *(40)* and fMRI *(47)*. Population-based inference *(48)*, the identification of individual differences *(49)*, meta-analytic comparisons *(50)*, and other applications performed across subjects depend upon accurate atlas-based normalization. There is no doubt that this is the most scientifically beneficial justification of altas construction and demonstrates their ability to form the spatial basis for comparing subjects with respect to cognitive operations or in comparisons between patient samples.

Continued improvement and enhancement of extant atlases, such as the Talairach atlas *(51)*, the several iterations of the MNI atlas, cytoarchitectonic atlases *(52)*, as well as those of the ICBM probabilistic atlas *(53)*, provide greater accuracy with respect to functional data and, hence, localization power. Improvement and enhancement of MNI or Talairach landmarks, for example, may enable more rapid calculation of spatial transformations, thereby providing flexibility for specific applications *(54)*.

5. Brain Atlas Revision and Evolution

Spatial mapping of any form is an ongoing process of determining accurate coordinate locations for content appropriate to that mapping's purpose. As previous content in a map changes or as new information is obtained, these maps need to be updated, corrected, and/or modified to reflect these changes in knowledge and the importance of what new knowledge is being conveyed. For instance, the information contained in US aeronautical charts is re-published approximately every 3 months partly to reflect changes in the Earth's magnetic field isogonic declination lines that point toward the magnetic North Pole. These field lines vary across North America, from approximately –19° in the Western US through to nearly +20° in the East, and must be accounted for to locate the direction of the true North Pole when charting a navigational course. However, the Earth's magnetic field has been drifting slowly westward since measurements began around 1850, shifting roughly 0.1° or 40 km/year. Failure to periodically update published maps that incorporate this shift in magnetic declination, as well as other information concerning changes in the Earth's topographical features, the construction of tall buildings in urban areas, alterations in air traffic routing, errors in earlier revisions, etc., could result in pilots or computers making substantial navigational errors due to improper compass and course directional settings in the absence of global positioning system (GPS) satellite information.

Errors often appear in maps resulting from the information that was used to create them. Landmarks of note may be mislocated and place names misspelled or mistranslated. This would also be true for brain mapping atlases where previous inaccuracies must be addressed, additional data included, or data from other modalities considered. Cytoarchitectonic maps from the classical period of describing brain anatomy have been noted as failing to incorporate sulcal pattern, variation in cell orientation, and being presented as idealized versions of brain structure *(15)*.

However, even modern approaches, using multi-modal methods, large databases, and sophisticated computer methods, are not immune from introducing errors. Data misaligned with respect to a standard atlas space may result in gross inaccuracies and considerable problems when trying to make inferences between diagnostic groups. For instance, the statistics of resulting voxel-based morphometric comparisons may be uninformative about group differences wherever the spatial normalization algorithm has failed to register on any robustly appearing image gradient *(55)*. This has severe consequences for random-effects-based analyses of morphometric changes due to disease or clinical outcome.

Electronic versions of the atlas of Talairach and Tournoux *(56)*, including the Talairach Daemon (http://ric.uthscsa.edu/projects/talairachdaemon.html) and the official versions published by Thieme, have been found to contain a discrepant region of the precentral gyrus on axial slice +35 mm that extends far forward into the frontal lobe. This region has been found to be anatomically incorrect and internally inconsistent within the digital atlas software applications that employ multi-planar cross-referencing tools *(57)*. This may be a case of simple mislabeling but other forms of atlas warping are known to result in distortions which must be predicated in context with the accurate interpretation of location. As new data are included and novel techniques are developed to inform atlases that are open to scrutiny by researchers, with ongoing updates and corrections, will they become most widely valuable.

Workers in our lab have recently described the construction of an improved digital brain atlas composed of data from manually delineated high-resolution MRI *(58)*. A total of 56 structures were labeled on the MR volumes of 40 normal healthy volunteers. The labeling was performed according to a set of protocols developed specifically for this project. In brief, pairs of raters were assigned to each structure and trained on the protocol for delineating that structure. Each rater pair was tested for concordance on 6 of the 40 brains; once they had achieved reliability standards, they divided the task of delineating the remaining

34 brains. The data were then spatially normalized to well-known atlas-based templates using each of three popular algorithms: AIR5.2.5's non-linear warp *(59)* paired with the ICBM452 Warp 5 atlas *(60)*, FSL's FLIRT *(61)* was paired with its own template, a skull stripped version of the ICBM152 T1 average; and SPM5's unified segmentation method *(62)* was paired with its canonical brain, the whole head ICBM152 T1-weighted average. In the end, these approaches produced three variants of a resultant atlas, where each was constructed from 40 representative samples of a data processing stream that one might use for analysis. For each normalization algorithm, the individual structure delineations were then re-sampled according to the derived transformations and computed averages were obtained at each voxel location to estimate the probability of that voxel belonging to each of the 56 structures. Each version of the atlas contains, for every voxel, probability densities for each region, thereby providing a resource for automated probabilistic labeling of external data types registered into standard spaces. Additionally, computed average intensity images and tissue density maps based on the three methods and target spaces were also obtained. These atlases are publicly available on the LONI Web site (http://loni.ucla.edu) and, we believe, will serve as critical resources for diverse applications including meta-analysis of functional and structural imaging data and other bioinformatics applications where display of arbitrary labels in probabilistically defined anatomic space will facilitate both knowledge-based development and visualization of findings from multiple disciplines. However, in time these, too, will be replaced by still more accurate atlases of larger sample size, improved spatial resolution, and anatomical detail.

6. Conclusions

The evolution of brain atlases has seen tremendous advances; they can now accommodate observations from multiple modalities and from populations of subjects collected at different laboratories. The probabilistic systems described here show promise for identifying patterns of structural, functional, and molecular variation in large image databases for pathology detection in individuals and groups and for determining the effects of age, gender, handedness, and other demographic or genetic factors on brain structures in space and time. Integrating these observations to enable statistical comparison has already provided a deeper understanding of the relationship between brain structure and function. Importantly, the utility of an atlas depends on appropriate coordinate systems, registration, and deformation methods to allow the statistical combination of multiple observations in an agreed, but expandable, digital reference framework. In this review, we highlighted two sources of data that will have an increasingly important role in integrative brain atlases: molecular architectonics and diffusion tensor imaging (DTI). Once stored in a population-based atlas, information from these techniques can help to interpret more conventional functional and structural brain maps by integrating them with data on molecular content, physiology, and fiber connections – a development that can help to formulate and test new types of neuroscientific models. A goal of systems neuroscience is to establish brain systems that underlie cognitive processes and the factors that influence them. DTI data on fiber connectivity, stored in an atlas coordinate system, can offer a rigorous computational basis to test how identifiable anatomical systems (e.g., visual, limbic, or corticothalamic pathways) interact. This atlas information can be invoked as ROI that are incorporated into the statistical design of functional brain mapping studies (e.g., with fMRI or electroencephalography), even when underlying fiber connections are not evident in the data being collected for a particular study. Molecular architectonic mapping also provides a complementary perspective in which known neurotransmitter and receptor pathways – the physiology and molecular features of which are now well understood – can be associated with functional subdivisions of the cortex, identified with tomographic imaging. For example, an fMRI study of inhibitory cognitive processes in drug abusers might be informed by other modalities of data on limbic–prefrontal connectivity (from DTI), or on cortical monoamine receptor distributions (from architectonic mapping). In each of these contexts, the coordinate system of the atlas, and the transformations that equate different modality data in the same reference frame, provide the means to build and test system-level models of cognition or disease, incorporating data from traditionally separate domains of neuroscience.

As brain atlases begin to incorporate data from thousands of subjects, new questions in basic and clinical neuroscience can be addressed that were previously out of reach. For example, quantitative genetic studies are underway to link functional, structural, and connectivity information with variations in candidate genetic polymorphisms that could influence them. As polygenic disorders involve the interaction of multiple genetic variations, each with a small effect on the overall phenotype, digital atlases provide the ideal setting to mine large numbers of images computationally with hybrid techniques from computational anatomy and quantitative genetics (such as linkage and association studies in which a statistic is computed at each voxel location in the brain).

Should atlases be constructed specific to different age groups or different age-related diseases? Numerous other papers *(30, 63)* have come to the same conclusion and many population-based atlases have emerged in response to this need *(64)*. But the same logic can be carried to the next level by creating many different population-based atlases, each specific to the group demographics, disease, age, or other characteristics of the subjects being studied. These provide, not only population statistics within the map, but arguably better represent the morphological signature of that particular cohort. What must be included in all analyses are confidence statistics on where the activity takes place. Whether this entails a statistic on probability, percentile, or other metric may depend on the experimental design and other factors. Adoption of a single normal atlas, even a probabilistic version, for all subject studies provides for the nominal capability for easier comparisons but in doing so fails to adequately measure the nuances within or between each group **(Fig. 3)**. It, therefore, seems that it might be prudent to avoid dependency on a single modality, single group representation for every study. Any given imaging experiment will be better served by mapping to a population-based atlas that closely resembles the cohort under study. We suggest that population atlases for groups, such as AD *(35,65)*, schizophrenia *(66,67)*, pediatric populations *(68,69)*, autism *(70)*, even decades of life *(71)*, should be utilized, as appropriate, for the subject group.

The next generation of population-based atlases will provide the necessary statistical power to identify demographic, genetic, and environmental factors that influence therapeutic response. These will be essential in the study of normal and abnormal human brain aging. Most important of all, brain atlases are now being enriched with data from newer technologies, such as DTI, fMRI, and modern high-throughput cytoarchitectural methods. These efforts are yielding whole new avenues of research into the functional organization of the brain and how this is altered as we age will be of interest not only just to specialists in neuroimaging, but also to all basic and clinical neuroscientists.

References

1. Haas LF. Phineas Gage and the science of brain localisation. J Neurol Neurosurg Psychiatry 2001;71:761.
2. Cowie SE. A place in history: Paul Broca and cerebral localization. J Invest Surg 2000; 13:297–8.
3. Goedert M, Ghetti B. Alois Alzheimer: his life and times. Brain Pathol 2007;17:57–62.
4. Roland PE, Zilles K. Brain atlases – a new research tool. Trends Neurosci 1994;17: 458–67.
5. Toga AW, Thompson PM. Maps of the brain. Anat Rec 2001;265:37–53.
6. Toga AW, Thompson PM. New approaches in brain morphometry. Am J Geriatr Psychiatry 2002;10:13–23.
7. Thompson P, Cannon TD, Toga AW. Mapping genetic influences on human brain structure. Ann Med 2002;34:523–36.
8. Narr KL, Thompson PM, Sharma T, Moussai J, Cannestra AF, Toga AW. Mapping morphology of the corpus callosum in schizophrenia. Cereb Cortex 2000;10:40–9.
9. Davatzikos C. Spatial normalization of 3D brain images using deformable models. J Comput Assist Tomogr 1996;20:656–65.

10. Davatzikos C. Spatial transformation and registration of brain images using elastically deformable models. Comput Vis Image Underst 1997;66:207–22.
11. Thompson PM, Woods RP, Mega MS, Toga AW. Mathematical/computational challenges in creating deformable and probabilistic atlases of the human brain. Hum Brain Mapp 2000;9:81–92.
12. Weaver JB, Healy DM, Jr., Periaswamy S, Kostelec PJ. Elastic image registration using correlations. J Digit Imaging 1998;11:59–65.
13. Barillot C, Lemoine D, Le Briquer L, Lachmann F, Gibaud B. Data fusion in medical imaging: merging multimodal and multipatient images, identification of structures and 3D display aspects. Eur J Radiol 1993;17:22–7.
14. Woods RP, Grafton ST, Holmes CJ, Cherry SR, Mazziotta JC. Automated image registration. I. General methods and intrasubject, intramodality validation. J Comput Assist Tomogr 1998;22:139–52.
15. Toga AW, Thompson PM, Mori S, Amunts K, Zilles K. Towards multimodal atlases of the human brain. Nat Rev Neurosci 2006;7:952–66.
16. Woods RP. Characterizing volume and surface deformations in an atlas framework: theory, applications, and implementation. Neuroimage 2003;18:769–88.
17. Avants B, Gee JC. Geodesic estimation for large deformation anatomical shape averaging and interpolation. Neuroimage 2004;23(Suppl 1): S139–50.
18. Avants BB, Schoenemann PT, Gee JC. Lagrangian frame diffeomorphic image registration: morphometric comparison of human and chimpanzee cortex. Med Image Anal 2006;10:397–412.
19. Evans AC, Collins DL, Milner B. An MRI-based stereotactic atlas from 250 young normal subjects. J Neurosci Abstracts 1992;18:408.
20. Durrleman S, Pennec X, Trouve A, Ayache N. Measuring brain variability via sulcal lines registration: a diffeomorphic approach. Med Image Comput Comput Assist Interv Int Conf Med Image Comput Comput Assist Interv 2007;10(Pt 1):675–82.
21. Alayon S, Robertson R, Warfield SK, Ruiz-Alzola J. A fuzzy system for helping medical diagnosis of malformations of cortical development. J Biomed Inform 2007;40:221–35.
22. Rohlfing T, Maurer CR, Jr. Shape-based averaging. IEEE Trans Image Process 2007;16:153–61.
23. Narr KL, Bilder RM, Luders E, et al. Asymmetries of cortical shape: effects of handedness, sex and schizophrenia. Neuroimage 2007;34:939–48.
24. Thompson PM, Giedd JN, Woods RP, MacDonald D, Evans AC, Toga AW. Growth patterns in the developing brain detected by using continuum mechanical tensor maps. Nature 2000;404:190–3.
25. Corouge I, Dojat M, Barillot C. Statistical shape modeling of low level visual area borders. Med Image Anal 2004;8:353–60.
26. Cardenas VA, Boxer AL, Chao LL, et al. Deformation-based morphometry reveals brain atrophy in frontotemporal dementia. Arch Neurol 2007;64:873–7.
27. Leow AD, Klunder AD, Jack CR, Jr., et al. Longitudinal stability of MRI for mapping brain change using tensor-based morphometry. Neuroimage 2006;31:627–40.
28. Diedrichsen J. A spatially unbiased atlas template of the human cerebellum. Neuroimage 2006;33:127–38.
29. Toga AW, Thompson PM. Genetics of brain structure and intelligence. Annu Rev Neurosci 2005;28:1–23.
30. Toga AW, Thompson PM, Sowell ER. Mapping brain maturation. Trends Neurosci 2006;29:148–59.
31. Apostolova LG, Thompson PM. Brain mapping as a tool to study neurodegeneration. Neurotherapeutics 2007;4(3):387–400.
32. Apostolova LG, Akopyan GG, Partiali N, et al. Structural correlates of apathy in Alzheimer's disease. Dement Geriatr Cogn Disord 2007;24:91–7.
33. Apostolova LG, Lu P, Rogers S, et al. 3D mapping of language networks in clinical and pre-clinical Alzheimer's disease. Brain Lang 2008;104:33–41.
34. Scher AI, Xu Y, Korf ES, et al. Hippocampal shape analysis in Alzheimer's disease: a population-based study. Neuroimage 2007;36:8–18.
35. Thompson PM, Hayashi KM, Dutton RA, et al. Tracking Alzheimer's disease. Ann N Y Acad Sci 2007;1097:183–214.
36. Mazziotta JC, Toga AW, Evans AC, Fox PT, Lancaster JL. Digital brain atlases. Trends Neurosci 1995;18:210–1.
37. Toga AW, Thompson PM, Mega MS, Narr KL, Blanton RE. Probabilistic approaches for atlasing normal and disease-specific brain variability. Anat Embryol (Berl) 2001;204:267–82.
38. Wakana S, Jiang H, Nagae-Poetscher LM, van Zijl PC, Mori S. Fiber tract-based atlas of human white matter anatomy. Radiology 2004;230:77–87.
39. Van Essen DC. A Population-Average, Landmark- and Surface-based (PALS) atlas of human cerebral cortex. Neuroimage 2005;15:635–62.

40. Fox PT, Perlmutter JS, Raichle ME. Stereotactic method for determining anatomical localization in physiological brain images. J Cereb Blood Flow Metab 1984;4:634.
41. Evans AC, Marrett S, Neelin P, et al. Anatomical mapping of functional activation in stereotactic coordinate space. Neuroimage 1992;1:43–53.
42. Nowinski WL, Thirunavuukarasuu A. Atlas-assisted localization analysis of functional images. Med Image Anal 2001;5:207–20.
43. Crivello F, Schormann T, Tzourio-Mazoyer N, Roland PE, Zilles K, Mazoyer BM. Comparison of spatial normalization procedures and their impact on functional maps. Hum Brain Mapp 2002;16:228–50.
44. Swallow KM, Braver TS, Snyder AZ, Speer NK, Zacks JM. Reliability of functional localization using fMRI. Neuroimage 2003;20:1561–77.
45. Tu Z, Zheng S, Yuille AL, et al. Automated extraction of the cortical sulci based on a supervised learning approach. IEEE Trans Med Imaging 2007;26:541–52.
46. Luders E, Thompson PM, Narr KL, Toga AW, Jancke L, Gaser C. A curvature-based approach to estimate local gyrification on the cortical surface. Neuroimage 2006;29: 1224–30.
47. Ashburner J, Friston KJ. Nonlinear spatial normalization using basis functions. Hum Brain Mapp 1999;7:254–66.
48. Friston KJ, Stephan KE, Lund TE, Morcom A, Kiebel S. Mixed-effects and fMRI studies. Neuroimage 2005;24:244–52.
49. Miller MB, Van Horn JD, Wolford GL, et al. Extensive individual differences in brain activations associated with episodic retrieval are reliable over time. J Cogn Neurosci 2002;14:1200–14.
50. Fox PT, Parsons LM, Lancaster JL. Beyond the single study: function/location metanalysis in cognitive neuroimaging. Curr Opin Neurobiol 1998;8:178–87.
51. Nowinski WL. The cerefy brain atlases: continuous enhancement of the electronic talairach-tournoux brain atlas. Neuroinformatics 2005;3:293–300.
52. Amunts K, Schleicher A, Zilles K. Cytoarchitecture of the cerebral cortex – more than localization. Neuroimage 2007;37:1061–5; discussion 6–8.
53. Mazziotta J, Toga AW, Evans A, et al. A probabilistic atlas and reference system for the human brain: International Consortium for Brain Mapping (ICBM). Philos Trans R Soc Lond B Biol Sci 2001;356:1293–322.
54. Nowinski WL. Modified Talairach landmarks. Acta Neurochir (Wien) 2001;143:1045–57.
55. Bookstein FL. "Voxel-based morphometry" should not be used with imperfectly registered images. Neuroimage 2001;14:1454–62.
56. Talairach J, Tournoux P. Co-Planar Stereotactic Atlas of the Human Brain. Tieme, New York 1988.
57. Maldjian JA, Laurienti PJ, Burdette JH. Precentral gyrus discrepancy in electronic versions of the Talairach atlas. Neuroimage 2004;21:450–5.
58. Shattuck DW, Mirza M, Adisetiyo V, et al. Construction of a 3D probabilistic atlas of human cortical structures. Neuroimage 2008;39: 1064–80.
59. Woods RP, Grafton ST, Watson JD, Sicotte NL, Mazziotta JC. Automated image registration. II. Intersubject validation of linear and nonlinear models. J Comput Assist Tomogr 1998;22:153–65.
60. Rex DE, Ma JQ, Toga AW. The LONI Pipeline Processing Environment. Neuroimage 2003;19:1033–48.
61. Smith SM, Jenkinson M, Woolrich MW, et al. Advances in functional and structural MR image analysis and implementation as FSL. Neuroimage 2004;23(Suppl 1):S208–19.
62. Ashburner J, Friston KJ. Unified segmentation. Neuroimage 2005;26:839–51.
63. Van Essen DC. Windows on the brain: the emerging role of atlases and databases in neuroscience. Curr Opin Neurobiol 2002;12:574–9.
64. Mazziotta J, Toga A, Evans A, et al. A four-dimensional probabilistic atlas of the human brain. J Am Med Inform Assoc 2001;8: 401–30.
65. Mega MS, Dinov ID, Mazziotta JC, et al. Automated brain tissue assessment in the elderly and demented population: construction and validation of a sub-volume probabilistic brain atlas. Neuroimage 2005;26:1009–18.
66. Yoon U, Lee JM, Koo BB, et al. Quantitative analysis of group-specific brain tissue probability map for schizophrenic patients. Neuroimage 2005;26:502–12.
67. Cannon TD, Thompson PM, van Erp TG, et al. Mapping heritability and molecular genetic associations with cortical features using probabilistic brain atlases: methods and applications to schizophrenia. Neuroinformatics 2006;4:5–19.
68. Wilke M, Schmithorst VJ, Holland SK. Assessment of spatial normalization of whole-brain magnetic resonance images in children. Hum Brain Mapp 2002;17:48–60.

69. Jelacic S, de Regt D, Weinberger E. Interactive digital MR atlas of the pediatric brain. Radiographics 2006;26:497–501.
70. Joshi S, Davis B, Jomier M, Gerig G. Unbiased diffeomorphic atlas construction for computational anatomy. Neuroimage 2004;23(Suppl 1):S151–60.
71. Mazziotta J, Toga A, Evans A, et al. A four-dimensional probabilistic atlas of the human brain. J Am Med Inform Assoc 2001;8:401–30.
72. Narr K, Thompson P, Sharma T, et al. Three-dimensional mapping of gyral shape and cortical surface asymmetries in schizophrenia: gender effects. Am J Psychiatry 2001;158:244–55.

Part II

fMRI Application to Measure Brain Function

Chapter 10

fMRI: Applications in Cognitive Neuroscience

Mark D'Esposito, Andrew Kayser, and Anthony Chen

Summary

Neuroimaging has, in many respects, revolutionized the study of cognitive neuroscience, the discipline that attempts to determine the neural mechanisms underlying cognitive processes. Early studies of brain–behavior relationships relied on a precise neurological exam as the basis for hypothesizing the site of brain damage that was responsible for a given behavioral syndrome. The advent of structural brain imaging, first with computerized tomography and later with magnetic resonance imaging, paved the way for more precise anatomical localization of the cognitive deficits that manifest after brain injury. In recent years, functional neuroimaging, broadly defined as techniques that provide measures of brain activity, has further increased our ability to study the neural basis of behavior. fMRI, in particular, has rapidly emerged as an extremely powerful technique that affords excellent spatial and temporal resolution. This chapter focuses on the principles underlying fMRI as a cognitive neuroscience tool for exploring brain-behavior relationships.

Key words: Functional MRI, Cognitive neuroscience, Hemodynamic response, Experimental design, Statistics

1. Introduction

Cognitive neuroscience is a discipline that attempts to determine the neural mechanisms underlying cognitive processes. Specifically, cognitive neuroscientists test hypotheses about brain–behavior relationships that can be organized along two conceptual domains: *functional specialization* – the idea that functional modules exist within the brain, that is, areas of the cerebral cortex that are specialized for a specific cognitive process, and *functional integration* – the idea that a cognitive process can be an emergent property of interactions among a network of brain regions which suggests that a brain region can play a different role across many functions.

M. Filippi (ed.), *fMRI Techniques and Protocols*, Neuromethods, vol. 41
DOI: 10.1007/978-1-60327-919-2_10, © Humana Press, a part of Springer Science+Business Media, LLC 2009

Early investigations of brain-behavior relationships consisted of careful observation of individuals with neurological injury resulting in focal brain damage. The idea of functional specialization evolved from hypotheses that damage to a particular brain region was responsible for a given behavioral syndrome that was characterized by a precise neurological examination. For instance, the association of nonfluent aphasia with right-sided limb weakness implicated the left hemisphere as the site of language abilities. Moreover, upon the death of a patient with a neurological disorder, clinicopathological correlations provided confirmatory information about the site of damage causing a specific neurobehavioral syndrome such as aphasia. For example, in 1861 Paul Broca's observations of nonfluent aphasia in the setting of a damaged left inferior frontal gyrus (IFG) cemented the belief that this brain region was critical for speech output *(1)*. The introduction of structural brain imaging more than 100 years after Broca's observations, first with computerized tomography (CT) and later with magnetic resonance imaging (MRI), paved the way for more precise anatomical localization in the living patient of the cognitive deficits that develop after brain injury. The superb spatial resolution of structural neuroimaging has reduced the reliance on the infrequently obtained autopsy for making brain–behavior correlations.

Functional neuroimaging, broadly defined as techniques that measure brain activity, has expanded our ability to study the neural basis of cognitive processes. One such method, fMRI, has emerged as an extremely powerful technique that affords excellent spatial and temporal resolution. Measuring regional brain activity in healthy subjects while they perform cognitive tasks links localized brain activity with specific behaviors. For example, functional neuroimaging studies have demonstrated that the left IFG is consistently activated during the performance of speech production tasks in healthy individuals *(2)*. Such findings from functional neuroimaging are complementary to findings derived from observations of patients with focal brain damage. This chapter focuses on the principles underlying fMRI as a cognitive neuroscience tool for exploring brain-behavior relationships.

2. Inference in Functional Neuroimaging Studies of Cognitive Processes

Insight regarding the link between brain and behavior can be gained through a variety of approaches. It is unlikely that any single neuroscience method is sufficient to fully investigate any particular question regarding the mechanisms underlying cognitive function. From a methodological point of view, each method will offer different temporal and spatial resolution. From a conceptual point of view, each method will provide data that will support different types of inferences that can be drawn from it. Thus, data obtained addressing a single question but derived from multiple methods can provide more comprehensive and inferentially sound conclusions.

Functional neuroimaging studies support inferences about the association of a particular brain system with a cognitive process. However, it is difficult to prove in such a study that the observed activity is necessary for an isolated cognitive process because perfect control over a subject's cognitive processes during a functional neuroimaging experiment is never possible. Even if the task performed by a subject is well designed, it is difficult to demonstrate conclusively that he or she is differentially engaging a single, identified cognitive process. The subject may engage in unwanted cognitive processes that either have no overt, measurable effects or are perfectly confounded with the process of interest. Consequently, the neural activity measured by the functional neuroimaging technique may result from some confounding neural computation that is itself not necessary for executing the cognitive process seemingly under study. In other words, functional neuroimaging is an observational, correlative method *(3)*. It is important to note that the inferences that can be drawn from functional neuroimaging studies such as fMRI apply to all methods of physiological measurement (e.g., electroencephalography, EEG, or magnetoencephalography, MEG).

The inference of necessity cannot be made without showing that inactivating a brain region disrupts the cognitive process in question. However, unlike precise surgical or neurotoxic lesions in animal models, lesions in patients are often extensive, damaging local neurons and "fibers of passage." For example, damage to prominent white matter tracts can cause cognitive deficits similar to those produced by cortical lesions, such as the amnesia resulting from lesions of the fornix, the main white matter pathway projecting from the hippocampus *(4)*. In addition, connections from region "A" may support the continued metabolic function of region "B," but region A may not be computationally involved in certain processes undertaken by region B. Thus, damage to region A could impair the function of region B via two possible mechanisms: (1) diaschisis *(5)* and (2) retrograde transsynaptic degeneration. Consequently, studies of patients with focal lesions cannot conclusively demonstrate that the neurons within a specific region are themselves critical to the computational support of an impaired cognitive process.

Empirical studies using lesion and electrophysiological methods demonstrate these issues regarding the types of inferences that can be logically drawn from them. For example, in monkeys, single-unit recording reveals neurons in the lateral prefrontal cortex (PFC) that increase their firing during the delay between the presentation of information to be remembered and a few seconds later when that information must be recalled *(6, 7)*. These studies are taken as evidence that persistent neural activity in the PFC is involved in temporary storage of information, a cognitive process known as working memory. The necessity of PFC for working memory was demonstrated in other monkey studies showing that prefrontal lesions impair performance on working memory tasks, but not on tasks that do not require temporarily holding information in memory *(8)*. Persistent neural activity during working memory tasks is also found in the hippocampus *(9, 10)*. Hippocampal lesions, however, do not impair performance on most working memory tasks *(11)*, which suggests that the hippocampus is *involved* in maintaining information over short periods of time, but is not *necessary* for this cognitive operation. Observations in humans support this notion. For example, the well-studied patient H.M., with complete bilateral hippocampal damage and the severe inability to learn new information, could nevertheless perform normally on working memory tasks such as digit span *(12)*. The hippocampus is implicated in long-term memory especially when relations between multiple items and multiple features of a complex, novel item must be retained. Thus, the hippocampus may only be engaged during working memory tasks that require someone to subsequently remember novel information *(13)*.

When the results from lesion and functional neuroimaging studies are combined, a stronger level of inference emerges. As in the examples of Broca's aphasia or working memory, a lesion of a specific brain region causes impairment of a given cognitive process and when engaged by an intact individual, that cognitive process evokes neural activity in the same brain region. Given these findings, the inference that this brain region is computationally necessary for the cognitive process is stronger than the data derived from each study performed in isolation. Thus, lesion and functional neuroimaging studies are complementary, each providing inferential support that the other lacks.

Other types of inferential failure can occur in the interpretation of functional neuroimaging studies when other common assumptions do not hold true. First, it is assumed that if a cognitive process activates a particular brain region (evoked by a particular task), the neural activity in that brain region must depend on engaging that particular cognitive process. For example, a brain region showing greater activation during the presentation of faces than to other types of stimuli, such as photographs of cars or buildings, is considered to engage face perception processes. However, this region may also support other higher level cognitive processes such as memory processes, in addition to lower level perceptual processes *(14)*. See **ref.** ***15*** for a further discussion of this issue.

The opposite type of inference is made when it is assumed that if a particular brain region is activated during the performance of a cognitive task, the subject must have engaged the cognitive process supported by that region during the task (referred to as a "reverse inference"). For example, observing activation of the frontal lobes during a mental rotation task, it was proposed that subjects engaged working memory processes to recall the identity of the rotated target *(16)*. (They derived this assumption from other imaging studies showing activation of the frontal lobes during working memory tasks.) However, in this example, because some other cognitive process supported by the frontal lobes could have activated this region *(17)*, one cannot be sure that working memory was engaged leading to the activation of the frontal lobes. Unfortunately, this potentially faulty logic is a fairly common practice in fMRI studies. See **ref.** ***18*** for a further discussion of this issue.

In summary, interpretation of the results of functional neuroimaging studies attempting to link brain and behavior rests on numerous assumptions. Familiarity with the types of inferences that can and cannot be drawn from these studies is helpful for assessing the validity of the findings reported by such studies.

3. Functional MRI as a Cognitive Neuroscience Tool

Functional MRI has become the predominant functional neuroimaging method for studying the neural basis of cognitive processes in humans. Compared to its predecessor, positron emission tomography (PET) scanning, fMRI offers many advantages. For example, MRI scanners are much more widely available, and imaging costs are less expensive since MRI does not require a cyclotron to produce radioisotopes. MRI is also a noninvasive procedure since there is no requirement for injection of a radioisotope into the bloodstream. Also, given the half-life of available radioisotopes, PET scanning is unable to provide comparable temporal resolution to that of fMRI which can provide images of behavioral events occurring on the order of seconds rather than the summation of many behavioral events over tens of seconds.

In selected circumstances, however, PET can provide an advantage over fMRI for studying certain questions concerning the neural basis of cognition. For example, at present, fMRI does not adequately image the regions within the orbitofrontal cortex and the anterior or inferior temporal lobe because of the susceptibility artifact near the interface of the brain and sinuses. These artifacts worsen at higher magnetic fields (i.e., 3.0 or 4.0 T), and such scanners are becoming commonly available and increasingly utilized by cognitive neuroscientists. Improvements in pulse sequences for acquiring fMRI data, development of algorithms for limiting susceptibility loss of signal and distortion correction of images should eventually eliminate or reduce these artifacts *(19–21)*. Currently, however, such sequences and methods are not widely available and implemented.

PET scanning may remain desirable or necessary when studying certain populations of individuals. For example, amnesic patients resulting from cerebral anoxia often have implanted cardiac pacemakers precluding them from having an MRI scans due to the magnetic field. However, PET scanning is unacceptable for studies of children due to the radiation exposure. A particular advantage of PET scanning in the study of cognition that can nicely complement fMRI studies is its ability to assess neurochemical (neurotransmitter and neuromodulator) systems. Radioactively labeled ligands may be used to directly measure density and distribution of particular receptors and even receptor subtypes, distribution of presynaptic terminals or enzymes involved in the production or breakdown of particular neurochemicals *(22)*.

The MRI scanner, compared to a behavioral testing room, is less than ideal for performing most cognitive neuroscience experiments. Experiments are performed in the awkward position of lying on one's back, often requiring subjects to visualize the presentation of stimuli through a mirror, in an acoustically noisy environment. Moreover, most individuals develop some degree of claustrophobia due to the small bore of the MRI scanner and find it difficult to remain completely motionless for a long duration of time that is required for most experiments (e.g., usually 60–90 min). These constraints of the MRI scanner make it especially difficult to scan children or certain patient populations (e.g., Parkinson's disease patients), which has resulted in many fewer fMRI studies involving children than adults and neurological patients in general. However, mock scanners have been built in many imaging centers, with motion devices, which acclimate children to the scanner environment before they participate in an fMRI study. This approach has led to an increasing number of fMRI studies of children being reported in the literature that are providing tremendous insight regarding the mechanisms underlying the developing brain *(23, 24)*.

All sensory systems have been investigated with fMRI including the visual, auditory, somatosensory, olfactory, and gustatory systems. Each system requires different technologies for successful presentation of relevant stimuli within an MRI environment. At the time of this writing, very few off-the-shelf commercial products exist that are MRI-compatible and most in use today have been engineered locally by individual laboratories. Most published fMRI studies have utilized visual stimuli although great strides have been made to allow the presentation of other types of stimuli. Details regarding the issues related to presenting visual and auditory stimuli in the MRI environment can be further reviewed in a comprehensive chapter on the topic by Savoy et al. *(25)*. In brief, the most common means of presenting visual stimuli is via an LCD projector system with the sophistication of the system depending on the quality of image resolution required for the experiment. Several options exist for auditory stimuli such as piezoelectric or electrostatic headphones. However, the biggest challenge is the acoustically noisy scanner environment. The pulsing of the fMRI gradient coils is the source of such noise making the study of auditory processes challenging *(26, 27)*. For example, during echoplanar imaging within a 4.0 T magnet using a high-performance head gradient set, sound levels can reach 130 dB. As a reference point, FDA safety regulations require no greater than an average of 105 dB for 1 h. With placement of absorbing materials within the scanner and on the walls of the room, as well as a fiberglass bore liner surrounding the gradient set, we have been able to reduce sound levels by about 25 dB. One of the biggest technical challenges within an MRI scanner has been the ability to present olfactory stimuli. However, sophisticated MR compatible olfactometers have been designed and utilized successfully. Such methods use a nasal mask in which the change from odorant to no-odorant conditions occurs within a few milliseconds *(28, 29)*.

Acquiring ancillary electrophysiological data such as electromyographic recordings to measure muscle contraction or electrodermal responses to measure autonomic activity enhances many cognitive neuroscience experiments. Devices have been developed that are MR compatible for these types of measurements as well as other physiological measures such as heart rate, electrocardiography, oxygen saturation, and respiratory rate. The recording of eye movements is becoming commonplace in MRI scanners predominantly with the use of infrared video cameras equipped with long range optics *(30, 31)*. Video images of the pupil–corneal reflection can be sampled at 60/120/240 Hz allowing for the accurate ($<1°$) localization of gaze within 50 horizontal and 40 vertical degrees of visual angle. Although most behavioral tasks used in cognitive neuroscience experiments rely on collecting manual responses, the ability to reliably collect verbal responses without significant artifact introduced into the data has been demonstrated by several laboratories *(32–34)*.

EEG recordings have also been successfully performed during MRI scanning (e.g., **refs.** *35, 36)*. However, the recording of event-related potentials (ERP), a signal that is much smaller in amplitude than the signal in EEG, can be more difficult in a magnetic field due to artifacts induced by gradient pulsing and head movement from cardiac pulsation. New monitoring devices and algorithms to remove artifact are being developed allowing for reliable measurements of ERPs during MRI scanning *(37, 38)*. In summary, most initial challenges facing cognitive experiments within the MRI environment have been overcome, creating an environment that is comparable to standard psychophysical testing labs outside of a scanner. Although individual laboratories have achieved most of these advancements, MRI scanners originally designed for clinical use by manufacturers are now being designed with consideration of many of these research-related issues.

3.1. Temporal Resolution

Two types of temporal resolution need to be considered for cognitive neuroscience experiments. First, what is the briefest neural event that can be detected as an fMRI signal? Second, how close together can two neural events occur and be resolved as separable fMRI signals?

The time scale on which neural changes occur is quite rapid. For example, neural activity in the lateral intraparietal area of monkeys increases within 100 ms of the visual presentation of a saccade target *(39)*. In contrast, the fMRI signal gradually increases to its peak magnitude within 4–6 s after an experimentally induced brief (<1 s) change in neural activity, and then decays back to baseline after several more seconds *(40–42)*. This slow time course of fMRI signal change in response to such a brief increase in neural activity is informally referred to as the blood oxygen level dependent (BOLD) fMRI hemodynamic response or simply, the hemodynamic response (**Fig. 1**). Thus, neural dynamics and neurally evoked hemodynamics, as measured with fMRI, are on quite different time scales.

The sluggishness of the hemodynamic response limits the temporal resolution of the fMRI signal to hundreds of milliseconds to seconds as opposed to the millisecond temporal resolution of electrophysiological recordings of neural activity, such

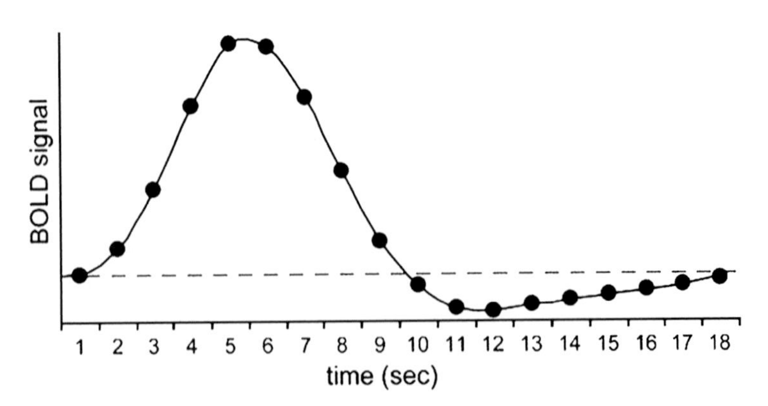

Fig. 1. A typical hemodynamic response (i.e., fMRI signal change in response to a brief increase of neural activity) from the primary sensorimotor cortex. The fMRI signal peaked approximately 5 s after the onset of the motor response (at time zero).

as from single-unit recording in monkeys and EEG or MEG in humans. However, it has been clearly demonstrated that brief changes in neural activity can be detected with reasonable statistical power using fMRI. For example, appreciable fMRI signal can be observed in sensorimotor cortex in association with single finger movements *(43)* and in visual cortex during very briefly presented (34 ms) visual stimuli *(44)*. In contrast, the temporal resolution of fMRI limits the detection of sequential changes in neural activity that occur rapidly with respect to the hemodynamic response. That is, the ability to resolve the changes in the fMRI signal associated with two neural events often requires the separation of those events by a relatively long period of time compared with the width of the hemodynamic response. This is because two neural events closely spaced in time will produce a hemodynamic response that reflects the accumulation from both neural events, making it difficult to estimate the contribution of each individual neural event. In general, evoked fMRI responses to discrete neural events separated by at least 4 s appear to be within the range of resolution *(45)*. However, provided that the stimuli are presented randomly, studies have shown significant differential functional responses between two events (e.g., flashing visual stimuli) spaced as closely as 500 ms apart *(46–48)*. The effect of fixed and randomized intertrial intervals on the BOLD signal is illustrated in **Fig. 2**.

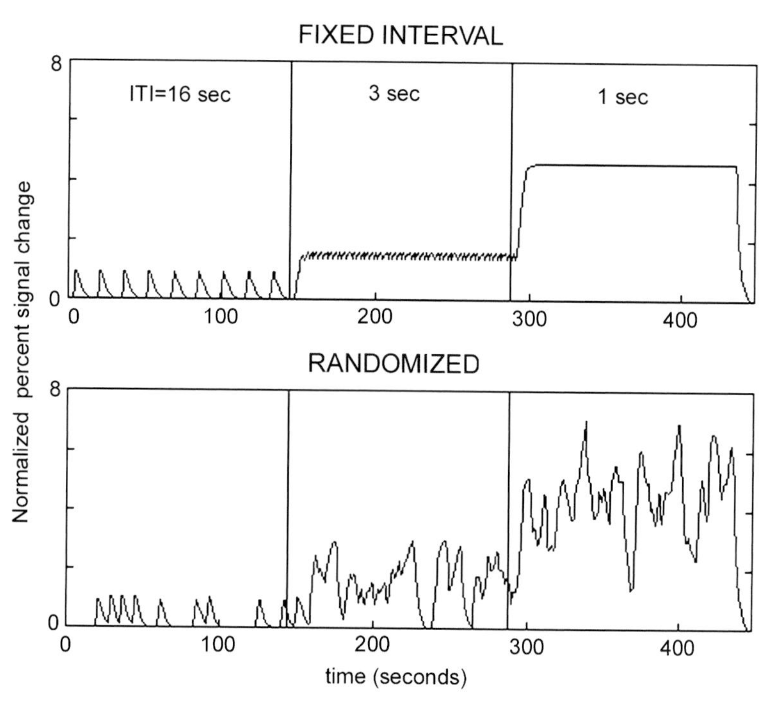

Fig. 2. Effect of fixed vs. randomized intertrial intervals on the blood oxygen level dependent (BOLD) fMRI signal ***(46)***.

In some tasks, the order of individual trial events cannot be randomized. For example, in certain types of working memory tasks, the presentation of the information to be remembered during the delay period, and the period when the subject must recall the information are individual trial events whose order cannot be randomized. In these types of tasks, short time scales (<4 s) cannot be temporally resolved. These temporal resolution issues in fMRI have been extensively considered regarding their impact on experimental design *(49, 50)*.

3.2. Spatial Resolution

It is yet to be determined how precisely the measured BOLD fMRI signal, which arises from the vasculature, reflects adjacent neural activity. Thus, the ultimate spatial resolution of BOLD fMRI is unknown. Functional MRI studies in both monkey and man at high field (4.0–4.7 T) have demonstrated that BOLD signal can be obtained with high spatial resolution – approximately $0.75 \times 0.75\ \text{mm}^2$ in-plane resolution *(51, 52)*. In monkeys, with novel approaches such as using a small, tissue-compatible, intraosteally implanted radiofrequency coil, ultra high spatial resolution of $125 \times 125\ \text{mm}^2$ has been obtained *(53)*. Using this method, Logothetis et al. demonstrated cortical lamina-specific activation in a task that compared responses to moving stimuli with those elicited by flickering stimuli. This contrast elicited BOLD signal mostly in the granular layers of the striate cortex of the monkey, which are known to have a high concentration of directionally selective cells. Advances in such methods would allow for imaging of hundreds of neurons per voxel as opposed to hundreds of thousands of neurons per voxel, which is more typical for a human cognitive neuroscience fMRI experiment.

Virtually all fMRI studies model the large BOLD signal increase, which is due to a local low-deoxyhemoglobin state, in order to detect changes correlating with a behavioral task. However, optical imaging studies have demonstrated that preceding this large positive response there is an initial negative response reflecting a localized increase in oxygen consumption causing a high-deoxyhemoglobin state *(54)*. This early hemodynamic response is called the "initial dip" and is thought to be more tightly coupled to the actual site of neural activity evoking the BOLD signal as compared to the later positive portion of the BOLD response. For example, Kim et al., scanning cats in a high field scanner, demonstrated that the early negative BOLD response (e.g., initial dip) produced activation maps that were consistent with orientation columns within visual cortex. This finding is quite remarkable given that the average spacing between two adjacent orientation columns in cortex is approximately 1 mm. In contrast, the activation maps produced by the delayed positive BOLD response appeared more diffuse and cortical columnar organization could not be identified *(55)*.

Thus, empirical evidence suggests that deriving activation maps by correlating behavioral responses with the initial dip may markedly improve spatial resolution. However, it is important to note that observation of the initial dip of the BOLD signal has been inconsistently observed in humans across laboratories for reasons that are still unclear. Several groups, however, were able to detect columnar architecture (in this case ocular dominance columns) by modeling the positive BOLD response in humans scanning at 4.0 T *(52, 56)*. These investigators attributed their success to optimized radiofrequency coils, limiting head motion, optimizing slice orientation, and the enhanced signal-to-noise ratio (SNR) provided by a high magnetic field.

Another unique method for improving spatial resolution has been called functional magnetic resonance-adaptation (fMR-A), which could provide a means for identifying and assessing the functional attributes of sharply defined neuronal populations within a given region of the brain *(57)*. Even if the spatial resolution of fMRI evolves to the point of being able to resolve a population of a few hundred neurons within a voxel, it is still likely that this small population will contain neurons with very different functional properties that will be averaged together. The adaptation method is based on several basic principles. First, repeated presentation of the same type of stimuli (i.e., a picture of the one object) causes neurons to adapt to those stimuli (i.e., neuronal firing is reduced). Second, if these neurons are then exposed to a different type of stimulus (i.e., a picture of another object) or a change in some property of the stimulus (i.e., the same object in a different orientation), then recovery from adaptation can be assessed (i.e., whether or not the BOLD signal returns to its original state). If the signal remains adapted it implies that the neurons are invariant to the attribute that was changed or if the signal recovers from the adapted state it would imply that the neurons are sensitive to that attribute. For example, Grill-Spector et al. demonstrated that an area of lateral occipital cortex thought to be important for object recognition was less sensitive to changes in object size and position as compared to changes in illumination and viewpoint *(58)*. Thus, with this method it is possible to investigate the functional properties of neuronal populations with a level of spatial resolution that is beyond that obtained from conventional fMRI data analysis methods.

Considering all the neuroscientific methods available today for studying human brain–behavior relationships, fMRI provides an excellent balance of temporal and spatial resolution. Improvements on both fronts will clearly add to the increasing popularity of this method.

4. Issues in Functional MRI Experimental Design

Numerous options exist for designing experiments using fMRI. The prototypical fMRI experimental design consists of two behavioral tasks presented in blocks of trials alternating over the course of a scanning session, and the fMRI signal between the two tasks is compared. This is known as a blocked design. For example, a given block might present a series of faces to be viewed passively, which evokes a particular cognitive process, such as face perception. The "experimental" block alternates with a "control" block that is designed to evoke all of the cognitive processes present in the experimental block except for the cognitive process of interest. In this experiment the control block may comprise a series of objects. In this way, the stimuli used in experimental and control tasks have similar visual attributes, but differ in the attribute of interest (i.e., faces). The inferential framework of "cognitive subtraction" *(59)* attributes differences in neural activity between the two tasks to the specific cognitive process (i.e., face perception). Cognitive subtraction was originally conceived by Donders in the late 1800s for studying the chronometric substrates of cognitive processes *(60)* and was a major innovation in imaging *(59, 61)*.

The assumptions required for cognitive subtraction may not always hold and could produce erroneous interpretation of functional neuroimaging data *(45)*. Cognitive subtraction relies on two assumptions: "pure insertion" and linearity. Pure insertion implies that a cognitive process can be added to a preexisting set of cognitive processes without affecting them. This assumption is difficult to prove because one needs an independent measure of the preexisting processes in the absence and presence of the new process *(60)*. If pure insertion fails as an assumption, a difference in the neuroimaging signal between the two tasks might be observed, not because a specific cognitive process was engaged in one task and not the other, but because the added cognitive process and the preexisting cognitive processes interact.

An example of this point is illustrated in working memory studies using delayed-response tasks *(62)*. These tasks *(63)* typically present information that the subject must remember (engaging an *encoding* process), followed by a delay period during which the subject must hold the information in memory over a short period of time (engaging a *memory* process), followed by a probe that requires the subject to make a decision based on the stored information (engaging a *retrieval* process). The brain regions engaged by evoking the *memory* process theoretically are revealed by subtracting the BOLD signal measured by fMRI during a block of trials that the subject performs that do not have a delay period (only engaging the *encoding* and *retrieval* processes) from a block of trials with a delay period (engaging the *encoding*, *memory*, and *retrieval* processes). In this example, if the addition or "insertion" of a delay period between the *encoding* and *retrieval* processes affects these other behavioral processes in the task, the result is failure to meet the assumptions of cognitive subtraction. That is, these "nonmemory" processes may differ in delay trials and no-delay trials, resulting in a failure to cancel each other out in the two types of trials that are being compared.

Empirical evidence of such failure exists *(64)*. For example, **Fig. 3** demonstrates BOLD signal derived from the PFC from a subject performing a delayed-response task similar to the tasks described above. The left side of the figure illustrates BOLD signal consistent with delay period activity whereas the right side of the figure illustrates BOLD signal from another region of PFC that did not display sustained activity during the delay yet showed greater activity in the delay trials as compared to the trials without a delay. In any blocked functional neuroimaging study that compares delay vs. no-delay trials with subtraction, such a region would be detected and likely assumed to be a "memory" region. Thus, this result provides empirical grounds for adopting a healthy doubt regarding the inferences drawn from imaging studies that rely exclusively on cognitive subtraction.

In functional neuroimaging, the transform between the neural signal and the hemodynamic response (measured by fMRI) must also be linear for the cognitive subtractive method to yield

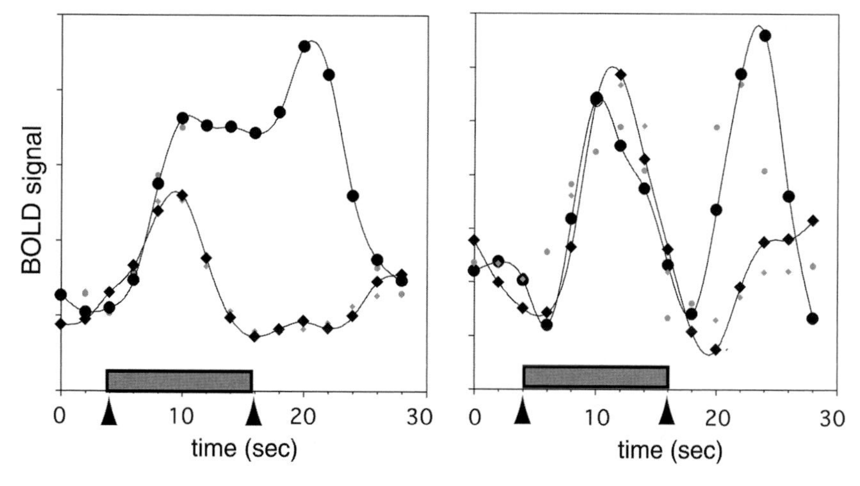

Fig. 3. Data derived from the performance of a normal subject on a spatial delayed-response task ***(64)***. This task comprised both delay trials (*circles*) as well as trials without a delay period (no-delay trials; *diamonds*). (**a**) Trial averaged fMRI signal from prefrontal cortex that displayed delay-correlated activity. The gray bar along the *x*-axis denotes the 12 s delay period during delay trials. The delay trials display a level of fMRI signal greater than baseline throughout the period of time corresponding to the retention delay (taking into account the delay and dispersion of the fMRI signal). The peaks seen in the signal correspond to the encoding and retrieval periods. (**b**) Trial averaged fMRI signal from a region in prefrontal cortex that did not display the characteristics of delay-correlated activity. This region displays a significant functional change associated with the no-delay trials, and a significant functional changes associated with the encoding and retrieval periods of the delay trials, but not one associated with the retention delay of delay trials.

valid results. In other words, it is assumed that the BOLD signal being measured is approximately proportional to the local neural activity that evokes it. Surprisingly, although thousands of empirical studies using fMRI to study brain-behavior relationships have been published, only a handful exist that have explored the neurophysiological basis of the BOLD signal (for reviews *see* **refs.** ***65, 66)***. In several studies, linearity did not strictly hold for the BOLD fMRI system but the linear transform model was reasonably consistent with the data. For example, Boynton et al. tested whether BOLD signal in response to long duration stimuli can be predicted by summing the responses to shorter duration stimuli ***(42)***. Using pulses of flickering checkerboard patterns and measuring within human primary visual cortex, these investigators found that the BOLD signal response to various durations of stimulus presentation (6, 12, or 24 s) could be predicted from the responses they obtained from shorter stimulus presentations. For example, the BOLD signal response to a 6 s pulse could be predicted from the summation of the BOLD signal response to the 3 s pulse with a copy of the same response delayed by 3 s. However, temporal summation did not always hold, and there are clearly nonlinear effects in the transform of neural activity to a hemodynamic response that must be considered ***(67–70)***. If these nonlinearities lead to saturation of the BOLD effect at a certain stimulus intensity, erroneous interpretation of particular results of fMRI experiments may occur.

Another class of experimental designs, called event-related fMRI, attempts to detect changes associated with individual trials, as opposed to the larger unit of time comprising a block of trials ***(71, 72)***. Each individual trial may be composed of one behavioral "event," such as the presentation of a single stimulus (e.g., a face or object to be perceived) or several behavioral events such as in the delayed-response task described above (e.g., an item to be remembered, a delay period, and a motor response in a delayed-response task). For example, with an event-related design, activity within the PFC has consistently been shown to correlate with the delay period, supporting the role of the PFC in temporarily maintaining information ***(64)***. This finding is consistent with single-neuron recording studies in the PFC of monkeys ***(7)***. An event-related design offers numerous advantages. For example, it allows for stimulus or trial randomization avoiding the behavioral confounds of blocked trials. It also permits the separate analysis of functional responses that are identified only in retrospect (i.e., trials on which the subject made a correct or incorrect response). Of course, an experiment does not have to be limited to either a block or event-related designs – a mixed-type (both event-related and blocked) design where particular trial types are randomized within a block is perfectly feasible. In this type of design, both item-related processes (e.g., transient responses to stimuli) as well as state-related processes (processes sustained throughout a block of trials or a task) are perfectly feasible *(73, 74)*.

Overall, much flexibility exists in the type of experimental design that can be utilized in fMRI experiments and continued innovation in this area will greatly expand the types of neuroscientific questions that can be addressed.

5. Issues in Interpretation of fMRI Data

5.1. Statistics

Many statistical techniques are used for analyzing fMRI data, but no single method has emerged as the ideal or "gold standard." The analysis of any fMRI experiment designed to contradict the null hypothesis (i.e., there is no difference between experimental conditions) requires inferential statistics. If the difference between two experimental conditions is too large to be reasonably due to chance, then the null hypothesis is rejected in favor of the alternative hypothesis, which typically is the experimenter's hypothesis (e.g., the fusiform gyrus is activated to a greater extent by viewing faces than objects). Unfortunately, since errors can occur in any statistical test, experimenters will never know when an error is committed and can only try to minimize them *(75)*. Knowledge of several basic statistical issues provides a solid foundation for the correct interpretation of the data derived from fMRI studies.

Two types of statistical errors can occur. A type I error is committed when the null hypothesis is falsely rejected, that is, a difference between experimental conditions is found but a difference does not truly exist. This type of error is also called a false-positive error. In an fMRI study, a false-positive error would be finding a brain region activated during a cognitive task, when actually it is not. A type II error is committed when the null hypothesis is accepted when it is false, that is, no difference between experimental conditions exists when a difference does exist. This type of error is also called a false-negative error. A false-negative error in an fMRI study would be failing to find a brain region activated during the performance of a cognitive task when actually it is. The concept of type II error is closely related to the idea of statistical power. If the false-negative rate for a given study design is 20%, for instance, then the "power" of that design to detect an activation is 100 – 20% or 80%.

In cognitive neuroscience studies, much emphasis has been placed on avoiding type I errors. The negative effects of incorrectly identifying a brain region as task-active include the expenditures of time, money, and effort spent in replicating and/or expanding upon a false positive result. Type II error, on the other hand, is seen as less damning; failure to detect brain activity in a research study has fewer implications for future research, provided that one is careful to interpret so-called "null" results correctly. For example, cognitive neuroscience studies (due to factors such as the expense and the difficulty of finding research participants, for example) tend to employ a small number of subjects – 15 would not be atypical - and therefore frequently lack power to detect significant brain activations. One must consequently be careful to avoid interpreting a lack of activation in one part of the brain as true inactivity during the task.

In a clinical research study, on the contrary, the emphasis may be different, especially when fMRI studies are being used diagnostically in single patients. A type II error - failing to detect active brain regions related to movement or language in the vicinity of a brain tumor, for example - may lead to a larger surgical resection that leaves the patient with avoidable residual deficits. On the contrary, a type I error - for example, identifying motor activity adjacent to a tumor when in fact none exists - may erroneously lead to a more cautious surgical resection, or to use of a different treatment modality. Which error is deemed more tolerable may depend on the clinical situation.

In fMRI experiments, like all experiments, a tolerable probability for type I error, typically less than 5%, is chosen for adequate control of specificity, that is, control of false-positive rates. Two features of fMRI data can cause unacceptable false-positive rates, even with traditional parametric statistical tests. First, there is the problem of multiple comparisons. For the typical resolution of images acquired during fMRI scans, the full extent of the human brain could comprise as many as 15,000 voxels. Thus, with any given statistical comparison of two experimental conditions, there are actually 15,000 statistical comparisons being performed. With such a large number of statistical tests, the probability of finding a false-positive activation, that is, committing a type I error, somewhere in the brain increases. Several methods exist to deal with this problem. One method, a Bonferroni correction, assumes that each statistical test is independent and calculates the probability of type I error by dividing the chosen probability ($p = 0.05$) by the number of statistical tests performed. Another method is based on Gaussian field theory *(76)*, and calculates the probability of type I error when imaging data are spatially smoothed. Many other methods for determining thresholds of statistical maps are proposed and utilized *(77, 78)* but unfortunately, no single method has been universally accepted. Nevertheless, all fMRI studies must apply some type of correction for multiple comparisons to control the false-positive rate.

The second feature that might increase the false-positive rate is the "noise" in fMRI data. Data from BOLD fMRI are temporally autocorrelated with more noise at some frequencies than at others. The shape of this noise distribution is characterized by a 1/frequency function with increasing noise at lower frequencies *(79)*. Traditional parametric and nonparametric statistical tests assume that the noise is not temporally autocorrelated, that is, each observation is independent. Therefore, any statistical test used in fMRI studies must account for the noise structure of fMRI data. If not, the false-positive rates will inflate *(79, 80)*.

Type II error is rarely considered in functional neuroimaging studies. When a brain map from an fMRI experiment is presented, several areas of activation are typically attributed to some experimental manipulation. The focus of most fMRI studies is on brain activation whereas it is often implicitly assumed that all of the other areas (typically most of the brain) were not activated during the experiment. Power as a statistical concept refers to the probability of correctly rejecting the null hypothesis *(75)*. As the power of an fMRI study to detect changes in brain activity increases, the false-negative rate decreases. Unfortunately, power calculations for particular fMRI experiments are rarely performed, although this methodology is evolving *(81–83)*. Reports that specific brain areas were not active during an experimental manipulation should provide an estimate of the power required for detection of a change in the region. All experiments should be designed to maximize power. Relatively simple strategies can increase power in an fMRI experiment in certain circumstances, such as increasing the amount of imaging data collected or increasing the number of subjects studied. It is also important to note that task designs can affect sensitivity *(84)*. For example, since BOLD fMRI data are temporally autocorrelated, experiments with fundamental frequencies in the lower range (e.g., a boxcar design with 60 s epochs) will have reduced sensitivity, due to the presence of greater noise at these lower frequencies. Finally, in a study that simultaneously measured neural signal via intracortical recording and BOLD signal in a monkey, it was observed that the SNR of the neural signal was on average at least one order of magnitude higher than that of the BOLD signal. The investigators of this study concluded that "the statistical and thresholding methods applied to the hemodynamic responses probably underestimate a great deal of actual neural activity related to a stimulus or task" *(85)*. Thus, the magnitude of type II error in BOLD fMRI may currently be underestimated and warrants further consideration in the interpretation of almost any cognitive neuroscience experiment.

5.2. Altered Hemodynamic Response

When comparing changes in fMRI BOLD signal levels within the brain of an individual subject across different cognitive tasks and making conclusions regarding changes in neural activity and the pattern of activity, numerous assumptions are made regarding the steps comprising neurovascular coupling (stimulus → neural activity → hemodynamic response → BOLD signal) and the regional variability of the metabolic and vascular parameters influencing the BOLD signal. It should be obvious that fMRI studies of cognition of individuals with local vascular compromise or diffuse vascular disease (e.g., patients with strokes or normal elderly) are potentially problematic. For example, many fMRI studies have sought to identify age-related changes in the neural substrates of cognitive processes. Those studies that directly compare changes in fMRI BOLD signal intensity across age groups rely upon the assumption of age-equivalent coupling of neural activity to BOLD signal. However, there is empirical evidence that suggests that this general assumption may not hold true. Extensive research on the aging neurovascular system has revealed that it undergoes significant changes in multiple domains in a continuum throughout the human lifespan, probably as early as the fourth decade (for review *see* **ref.** *86)*. These changes affect the vascular ultrastructure *(87)*, the resting cerebral blood flow *(88, 89)*, the vascular responsiveness of the vessels *(90)*, and the cerebral metabolic rate of oxygen consumption *(91, 92)*. Aging is also frequently associated with co-morbidities such as diabetes, hypertension, and hyperlipidemia, all of which may affect the fMRI BOLD signal by affecting cerebral blood flow and neurovascular coupling *(93)*. Any one of these age-related differences in the vascular system could conceivably produce age-related differences in BOLD fMRI signal responsiveness, greatly affecting the interpretation of results from such studies.

Our laboratory compared the hemodynamic response function (HRF) characteristics in the sensorimotor cortex of young and older subjects in response to a simple motor reaction-time task *(71)*. The provisional assumption was made that there was identical neural activity between the two populations based on physiological findings of equivalent movement-related electrical potentials in subjects under similar conditions *(94)*. Thus, we presumed that any changes that we observed in BOLD fMRI signal between young and older individuals in motor cortex would be due to vascular, and not neural activity changes in normal aging. Several important similarities and differences were observed between age groups. Although, there was no significant difference in the shape of the hemodynamic response curve or peak amplitude of the signal, we found a significantly decreased SNR in the fMRI BOLD signal in older individuals as compared to young individuals. This was attributed to a greater level of noise in the older individuals. We also observed a decrease in the spatial extent of the BOLD signal in older individuals compared to younger individuals in sensorimotor cortex (i.e., the median number of suprathreshold voxels). Similar results have been replicated by two other laboratories *(95, 96)*. These findings suggest that there is some property of the coupling between neural activity and fMRI BOLD signal that changes with age.

The notion that vascular differences among individuals may affect BOLD signal is especially a concern when considering studies of patient populations with known vascular changes such as stroke. A recent fMRI study addressed the issue of the influence of vascular factors on the BOLD signal in a symptomatic stroke population *(97)*. They analyzed the time course of the BOLD HRF in the sensorimotor cortex of patients with an isolated subcortical lacunar stroke compared to a group of age-matched controls. They found a decrease in the rate of rise and the maximal fMRI BOLD HRF to a finger- or hand-tapping task in both the sensorimotor cortex of the hemisphere affected by the stroke and the unaffected hemisphere. These investigators proposed that given the widespread changes of these fMRI BOLD signal differences, the change was unlikely a direct consequence of the subcortical lacunar stroke, but rather a manifestation of pre-existing diffuse vascular pathology.

In summary, comparing BOLD signal in two different groups of individuals that may differ in their vascular system should be done with caution. For example, in one scenario, a comparison of activation of young and elderly individuals during a cognitive task may show less activation by elderly (as compared to young subjects) in some brain regions, but greater activation in other regions *(98)*. In this scenario, it is unlikely that regional variations in the hemodynamic coupling of neural activity to fMRI signal would account for such age-related differences in patterns of activation. In another scenario, a comparison of young and elderly subjects may show less activation by elderly (as compared to young subjects) in some brain regions, but no evidence of greater activation in any other region. In this case, it is possible that the observed age-related differences are not due to differences in intensity of neural activity, but rather to other nonneuronal contributions to the imaging signal, i.e., neurovascular coupling.

Consequently, BOLD contrast methods yield signal changes that result from a complex mix of vascular effects and provide only relative, rather than absolute, measures. One approach to accounting for the influence of purely vascular effects is to directly measure regional and individual variability in vascular reactivity via a breathholding task, which increases carbon dioxide concentration in the blood and leads to vascular dilatation *(99)*. The task-related BOLD signal in each subject can then be corrected for particular region- and subject-specific vascular effects. An alternative functional neuroimaging approach, based on more direct measurements of cerebral blood flow to active brain areas, is known as arterial spin labeling (ASL).

In the various ASL techniques, the MRI scanner selectively magnetizes flowing blood with a particular range of locations and/or velocities, then waits for the appearance of the magnetic "tag" in downstream vessels. It thus becomes possible to obtain absolute measures of cerebral perfusion *(99)*, thereby opening up the possibility of more quantitatively distinguishing between the differential influence of a disease on blood flow, and its effect on brain activity *(100)*. Additionally, relative to BOLD contrast these absolute measurements appear to be more stable over long experiments *(101)*, to show less between-subject and between-session variability *(102)*, and to produce decreased susceptibility artifact in areas such as medial temporal lobe *(103)*. The current major limitation is temporal resolution: one must both wait for the generation of sufficient magnetic label, and also acquire two scans, a reference scan and a postlabeling scan, to produce a single data point. However, efforts are underway to improve this resolution. Research into so-called "Turbo" ASL, for example, is attempting to reduce the time required to apply the magnetic tag, and to optimize image acquisition with respect to the arrival time of the tag in downstream areas *(104)*. A final potential disadvantage somewhat related to the temporal issues is the lower SNR of ASL relative to BOLD, but this decline may be compensated by the observation that ASL methods appear to be less variable across subjects *(100)*.

6. Types of Hypotheses Tested Using fMRI

Functional neuroimaging experiments test hypotheses regarding the anatomical specificity for cognitive processes (functional specialization) or direct or indirect interactions among brain regions (functional integration). The experimental design and statistical analyses chosen will determine the types of questions that can be addressed. Ultimately, the most powerful approach for the testing of theories on brain-behavior relationships is the analysis of converging data from multiple methods.

6.1. Functional Specialization

The major focus of fMRI studies of cognition is testing theories on functional specialization. The concept of functional specialization is based on the premise that functional modules exist within the brain, that is, areas of the cerebral cortex are specialized for a specific cognitive process. For example, facial recognition is a critical primary function likely served by a functional module. Prosopagnosia is the selective inability to recognize faces. Patients with prosopagnosia, however, can recognize familiar faces, such as those of relatives, by other means, such as the voice, dress, or shape. Other types of visual recognition, such as identifying common objects, are normal. Prosopagnosia arises from lesions of the inferomedial temporo-occipital lobe, which are usually due to a stroke within the posterior cerebral artery circulation. No lesion studies have precisely localized the area crucial for facial perception. However, they provide strong evidence that a brain area is specialized for processing faces. Functional imaging studies have provided anatomical specificity for such a module. For example, Kanwisher et al. *(105)* used fMRI to test a group of healthy individuals and found that the fusiform gyrus was significantly more active when the subjects viewed faces than when they viewed assorted common objects. The specificity of a "fusiform face area" was further demonstrated by the finding that this area also responded significantly more strongly to passive viewing of faces than to scrambled two-tone faces, front-view photographs of houses, and photographs of human hands. These elegant experiments allowed the investigators to reject alternative functions of the face area, such as visual attention, subordinate-level classification, or general processing of any animate or human forms, demonstrating that this region selectively perceives faces.

Of course, the existence of brain areas specialized for certain functions does not exclude the strong possibility that those areas are part of larger networks. Recent neuroimaging work has focused on pattern classification methods – that is, on techniques to explore whether a distributed spatial pattern of brain activity corresponds to object (or more abstract) representations. This area of research, still in its infancy, draws on results from physics, computer science, and statistics, among other disciplines, to search for more broadly distributed structure in neuroimaging data. As such, the techniques themselves differ. For example, to distinguish between voxel activity patterns across experimental conditions, various reports have used correlations between the set of activations in visual responses to faces and other objects *(106)*; neural network classifiers to identify particular patterns correlated with particular memories *(107)*; and variants of a matrix algebra transformation known as singular value decomposition to look for distributed spatial correlates of memory storage and search *(108)*. A large number of other techniques – too large to be reviewed here – are also being tested. As such research continues, this type of pattern classification will need to be validated via comparison with behavioral responses, in order to ensure that these patterns are not epiphenomenal *(109)*.

6.2. Functional Integration

Functional neuroimaging experiments can also test hypotheses about interactions between brain regions by focusing on covariances of activation levels between regions *(110, 111)*. These covariances reflect "functional connectivity," a concept that was originally developed in reference to temporal interactions among individual neurons *(112)*.

In addition to providing information about the specialization of various brain regions, functional neuroimaging can also address the interactions between brain regions that underlie cognitive processing. Understanding the various techniques that permit these types of analysis comprises a very active area of current research *(113)*. However, most, if not all, of the techniques used to test for regional interactions are ultimately based on the covariance of activation levels in different brain regions across time – in other words, on the way in which activity levels in different areas of the brain rise or fall in relation to each other. Such statistical techniques are commonly known as "multivariate," both because they rely on interactions between two or more brain areas, and to distinguish them from the "univariate" methods applied in most tests of functional specialization.

The universe of multivariate techniques is further subdivided into two types, determined by whether the method in question is designed to assess connectivity in a model-free ("functional connectivity") or model-based ("effective connectivity") fashion. The former refers simply to methods that measure the temporal covariance in activity between brain areas without a priori notions about which brain areas are relevant or how they should interact. Examples of model-free techniques would include correlation and its frequency-based analogue, coherence, which can be applied irrespective of hypotheses about the neural events that produced them. On the contrary, model-based, or effective connectivity, approaches begin with hypotheses about the interactions between different brain regions, and attempt to support/refute them by evaluating the presence/absence of specific activity covariance patterns. Examples of these techniques would include structural equation modeling and dynamic causal modeling, both of which start by postulating the existence of influences (potentially complex, potentially time-varying) between specific brain regions. Both types of statistical techniques have value, of course; their use is determined by the problem at hand. Model-free approaches are more general, and more easily deployed in exploratory analyses. However, they are not as powerful as model-based methods that address specific hypotheses about how regions interact – but which fail if the model is misspecified. Model-free methods, for example, may be more useful when attempting to determine which networks of brain areas might be involved in a task, whereas model-based methods may be most appropriate when the nodes of the network are known, and specific notions about how they interact need to be tested.

In our own laboratory, we have developed and used functional connectivity techniques to understand how brain interactions change under different task conditions, and over time *(114, 115)*. For example, we have shown that functional connectivity changes as subjects learn a complex finger tapping task *(116)*. In the early phases of learning, the data show that subjects not only activate wide areas of primary sensorimotor cortex, premotor cortex, and the supplementary motor area, but also that the coherence between these areas is increased relative to later stages. Such changes were not observed when subjects performed an already learned motor skill; and more importantly, they were not found in the univariate responses, whose means were unchanged despite the changes in the subjects' facility at the task. Similarly, in a working memory task for faces *(117)*, we have found an interesting dissociation between their univariate and multivariate analyses in the networks that support so-called "delay period" activity (see below). In our protocol, subjects encoded a cue face, maintained the image across a delay of several seconds, and then decided whether a subsequently presented probe face matched the initial one. Interestingly, we found that despite a general decrease in the univariate activity from the cue to the delay period, there was a robust increase in the correlation between activity in the right fusiform face area (a brain region known to be sensitive to face stimuli) and a diffuse set of brain regions including the frontal and parietal cortices as well as the basal ganglia.

In such known networks, effective connectivity techniques can be employed to more specifically evaluate the influence of the nodes of the network on each other. McIntosh et al., for example, were able to exploit their own functional neuroimaging research on working memory networks to formulate a hypothesis about the interactions of the PFC, cingulate cortex, and other brain regions during task performance *(111)*. Using structural equation modeling, the authors found shifting prefrontal and limbic interactions in a working memory task for faces as the retention delay increased (**Fig. 4**). The different interactions between brain regions at short and long delays were interpreted as a functional change. For example, strong corticolimbic interactions were found at short delays, but at longer delays, when the image of the face was more difficult to maintain, strong fronto-cingulate-occipital interactions were found. The investigators postulated that the former finding was due to maintaining an iconic facial representation, and the latter due to an expanded encoding strategy, resulting in more resilient memory. As in our own previous studies, information that was not seen in the univariate analysis was captured by an approach sensitive to regional interactions. In addition to structural equation modeling, other approaches have been applied to fMRI datasets to capture information regarding the relative timing of activation across brain regions such as Granger causality, information analysis, and coherence (*see* **Fig. 5** and **refs.** *114, 115, 118)*.

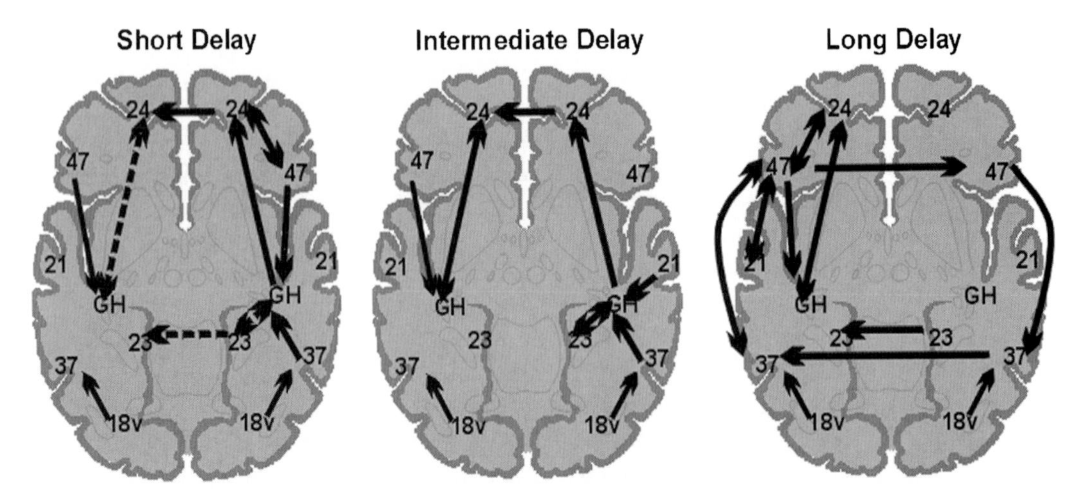

Fig. 4. Network analysis of fMRI data using structural equation modeling during performance of a working memory task cross three different delay periods *(111)*. Areas of correlated increases in activation (*solid lines*) and areas of correlated decreases in activation (*dotted lines*) are shown. Note the different pattern of interactions among brain regions at short and long delays.

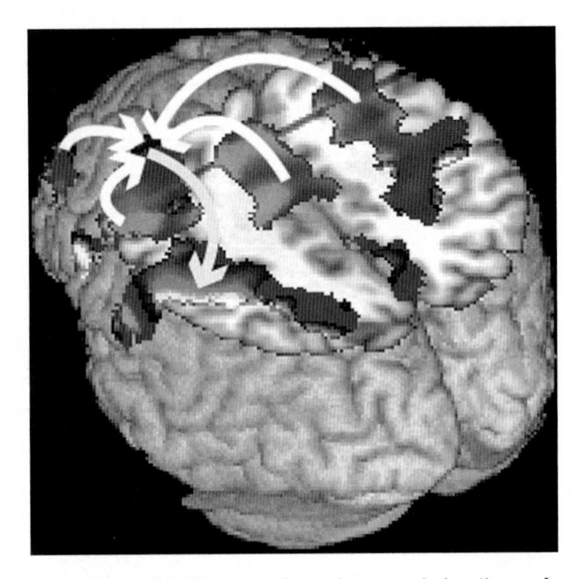

Fig. 5. Network analysis of fMRI data using coherence during the performance of a motor learning task. Activity of some brain regions precedes activity in the region of interest whereas activity in other areas follows in time after activation of the region of interest.

6.3. Cognitive Theory

An exciting new direction for studies using functional neuroimaging are those that test theories of the underlying mechanisms of cognition. For example, an fMRI study *(119)* attempted to answer the question, "To what extent does perception depend on attention?" One hypothesis is that unattended stimuli in the environment receive very little processing *(120)*, but another hypothesis is that the processing load in a relevant task determines the extent to which irrelevant stimuli are processed *(121)*. These alternative hypotheses were tested by asking normal individuals to perform linguistic tasks of low or high load while ignoring irrelevant visual motion in the periphery of a display. Visual motion was used as the distracting stimulus, because it activates a distinct region of the brain (cortical area MT or V5, another functional module in the visual system). Activation of area MT would indicate that irrelevant visual motion was processed. Although task and irrelevant stimuli were unrelated, fMRI of motion-related activity in MT showed a reduction in motion processing during the high-processing load condition in the linguistic task. These findings supported the hypothesis that perception of irrelevant environmental information depends on the information processing load that is currently relevant and being attended to. Thus, by the finding that perception depends on attention, this fMRI experiment provides insight regarding underlying cognitive mechanism.

7. Integration of Multiple Methods

The most powerful approach toward understanding brain-behavior relationships comes from analyzing converging data from multiple methods. There are several ways in which different methods can provide complementary data. For example, one method can provide superior spatial resolution (e.g., fMRI) whereas the other can provide superior temporal resolution (e.g., ERP). Also, the data from one method may allow for different conclusions to be drawn from it such as whether a particular brain region is necessary to implement a cognitive process (i.e., lesion methods) or whether it is only involved during its implementation (i.e., physiological methods). The following sections describe examples of such approaches.

7.1. Combined fMRI/ Lesion Studies

The combined use of functional neuroimaging and lesions studies can be illustrated with studies of the neural basis of semantic memory, the cognitive system that represents our knowledge of the world. Early studies of patients with focal lesions supported the notion that the temporal lobes mediate the retrieval of semantic knowledge *(122)*. For example, patients with temporal lobe lesions may show a disproportionate impairment in the knowledge of living things (e.g., animals) compared with nonliving things. Other patients have a disproportionate deficit in the knowledge of nonliving things *(123)*. These observations led to the notion that the semantic memory system is subdivided into different sensorimotor modalities, that is, living things, compared with nonliving things, are represented by their visual and other sensory attributes (e.g., a banana is yellow), while nonliving things are represented by their function (e.g., a hammer is a tool but comes in many different visual forms). The small number of patients with these deficits, and often large lesions, limits precise anatomical-behavioral relationships. However, functional neuroimaging studies in normal subjects can provide spatial resolution that the lesion method lacks *(124)*.

These original observations regarding the neural basis of semantic memory conflicted with functional neuroimaging studies consistently showing activation of the left IFG during the retrieval of semantic knowledge. For example, an early cognitive activation PET study revealed IFG activation during a verb generation task compared with a simple word repetition task *(61)*. A subsequent fMRI study *(125)* offered a fundamentally different interpretation of the apparent conflict between lesion and functional neuroimaging studies of semantic knowledge: left IFG activity is associated with the need to select some relevant feature of semantic knowledge from competing alternatives, not retrieval of semantic knowledge per se. This interpretation was supported by an fMRI experiment in normal individuals in which selection, but not retrieval, demands were varied across three semantic tasks. In a verb generation task, in a high selection condition, subjects generated verbs to nouns with many appropriate associated responses without any clearly dominant response (e.g., "wheel"), but in a low selection condition nouns with few associated responses or with a clear dominant response (e.g., "scissors") were used. In this way, all tasks required semantic retrieval, and differed only in the amount of selection required. The fMRI signal within the left IFG increased as the selection demands increased (**Fig. 6**). When the degree of semantic processing varied independently of selection demands, there was no difference in left IFG activity, suggesting that selection, not retrieval, of semantic knowledge drives activity in the left IFG.

To determine if left IFG activity was correlated with but not necessary for selecting information from semantic memory, the same task used during the fMRI study was used to examine the ability of patients with focal frontal lesions to generate verbs *(126)*. Supporting the earlier claim regarding left IFG function derived from an fMRI study *(125)*, the overlap of the lesions in patients with deficits on this task corresponded to the site of maximum fMRI activation in healthy young subjects during the verb generation task (**Fig. 6**). In this example, the approach of using converging evidence from lesion and fMRI studies differs in a subtle but important way from the study described earlier that isolated the face processing module. Patients with left IFG lesions do not present with an identifiable neurobehavioral syndrome reflecting the nature of the processing in this region. Guided by the fMRI results from healthy young subjects, the investigators

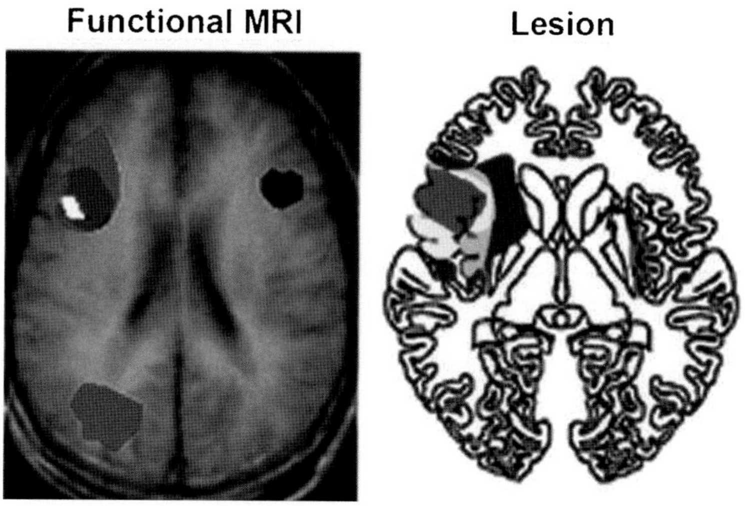

Functional MRI Lesion

Fig. 6. Regions of overlap of fMRI activity in healthy human subjects (*left side of figure*) during the performance of three semantic memory tasks, with the convergence of activity within the left inferior frontal gyrus (*white region*) ***(125)***. Regions of overlap of lesion location in patients with selection-related deficits on a verb generation task (*right side of figure*) with maximal overlap within the left inferior frontal gyrus ***(126)***.

studied patients with left IFG lesions to test a hypothesis regarding the necessity of this region in a specific cognitive process. Coupled with the well-established finding that lesions of the left temporal lobe impair semantic knowledge, these studies further our understanding of the neural network mediating semantic memory.

7.2. Combined fMRI/Transcranial Magnetic Stimulation Studies

Transcranial magnetic stimulation (TMS) is a noninvasive method that can induce a reversible "virtual" lesion of the cerebral cortex in a normal human subject *(127)*. Using both fMRI and TMS provides another means of combining brain activation data with data derived from the lesion method. There are several advantages for using TMS as a lesion method. First, brain injury likely results in brain reorganization after the injury and studies of patients with lesions assume that the nonlesioned brain areas have not been affected, whereas TMS is performed on the normal brain. Another advantage for using TMS is that it has excellent spatial resolution and can target specific locations in the brain whereas lesions in patients with brain injury are markedly variable in location and size across individuals. Such an approach can be illustrated in a recent investigation of the role of the medial frontal cortex in task switching *(128)*. In this study, subjects first performed an fMRI study that identified the regions that were active when they stayed on the current task vs. when they switched to a new task. It was found that medial frontal cortex is activated when switching between tasks. In order to determine if the medial frontal cortex was necessary for the processes involved in task switching, the same paradigm was utilized during inactivation of the medial frontal cortex with TMS. Guided by the locations of activation observed in the fMRI study, and using an MRI guided frameless stereotaxic procedure, it was found that applying a TMS pulse over the medial frontal cortex disrupted performance only during trials during which the subject was required to switch between tasks. TMS over adjacent brain regions did not show this effect. Also, the excellent temporal resolution of TMS allowed the investigators to stimulate during precise periods of the task, determining that the observed effect was during the time when the subjects were presented a cue indicating they must switch tasks prior to the actual performance of the new task. Thus, combining the results from both fMRI and TMS, it was concluded that medial PFC was essential for allowing individuals to intentionally switch to a new task.

Recently, some groups have begun to perform TMS studies not only as an adjunct to, but also concurrently with, fMRI. The advantage of this approach is clear: applying TMS at various times *during* (rather than after) fMRI scans permits it to be causally linked with functional changes in the brain, even independently of behavior. In a recent study employing this technique, Ruff, Driver and colleagues *(129, 130)* examined the influence on early visual cortex of a parietal region (the anterior intraparietal sulcus, or aIPS) implicated in the generation of both covert spatial attention and eye movements. They chose a range of TMS stimulus intensities, all of which were thought to be in an effectively stimulatory rather than inhibitory range, and applied them to the aIPS while subjects fixated the center of a viewing screen. On some trials, a randomly moving visual stimulus was present; subjects had no other task than to maintain fixation. Using this approach, the authors were able to demonstrate a parametric, so-called "top-down" effect from aIPS following TMS – an increase in the BOLD response in early visual cortex with increasing TMS intensity – that could be found only when visual stimuli were absent, and that did not vary with retinotopic eccentricity. In distinction, their previous work (extended here) had shown that TMS of the frontal eye field (FEF) led to a decrease in BOLD response in the central visual field but to an increase in BOLD response in the peripheral visual field, irrespective of the presence or absence of a visual stimulus. The authors were consequently able to conclude that the aIPS and the FEF have distinct top-down effects on visual cortex, a finding that would not have been possible without concurrent TMS.

7.3. Combined fMRI/Event-Related Potential Studies

The strength of combining these two methods is coupling the superb spatial resolution of fMRI with the superb temporal resolution of ERP recording. An example of such a study was reported by Dehaene et al. who asked the question "Does the human capacity for mathematical intuition depend on linguistic competence or on visuospatial representations?" *(131)*. In this study, subjects performed two addition tasks – one in which they were instructed to select the correct sum from two numerically close numbers (exact condition) and one in which they were instructed to estimate the result and select the closest number (approximate condition). During fMRI scanning greater bilateral parietal lobe activation was observed in the approximation condition as compared to the exact condition. Since this activation was outside the perisylvian language zone, it was taken as support that visuospatial processes were engaged during the cognitive operations involved in approximate calculation. Greater left lateralized frontal lobe activation was observed to be greater in the exact condition as compared to the approximate condition, which was taken as evidence for language dependent coding of exact addition facts. In order to consider an alternative explanation of the fMRI findings, the investigators also performed an ERP study. The alternative explanation was that in both the exact and approximate tasks, subjects would compute the exact result using the same representation for numbers but later processing, when they had to make a decision as to the correct choice, was what led to the differences in brain activation. Since fMRI does not offer adequate temporal resolution to resolve these two behavioral events on such a brief time scale, ERP was the appropriate method to test this hypothesis. In the ERP study it was demonstrated that the evoked neural response during exact and approximate trials already differed significantly during the first 400 ms of a trial before subjects had to make a decision.

7.4. Combined fMRI/ Pharmacological Studies

Combining pharmacological challenges during the performance of cognitive tasks during fMRI scanning may yield significantly different information than either method alone. In isolation, fMRI cognitive task paradigms provide little information with respect to the underlying pharmacologic systems involved in cognition. On the contrary, drug administration without a brain measure cannot determine underlying neural mechanisms of the effects of neuromodulatory systems on cognition. Combining the two approaches allows the potential of probing the pharmacologic bases of behavior. One may measure the interactive effects of drug (compared to placebo, or a range of doses) with cognitive task-related modulation of brain activity. It is fair to infer that drug × task interactions reflect modulation of the underlying anatomical and chemical brain systems, and do not simply reflect nonspecific vascular effects. For example, dopaminergic agonists have been shown to have task-specific effects *(132–134)*. For example, different component processes of working memory are differentially affected by a dopaminergic drug, with effects that may differ between individuals depending on their baseline state *(135)*. This study demonstrated that a dopamine agonist improved the flexible updating (switching) of relevant information in working memory. However, the effect only occurred in individuals with low working memory capacity, but not in individuals with higher working memory capacity. This behavioral effect was accompanied by dissociable effects of the dopaminergic agonist on frontostriatal activity. The dopamine agonist modulated the striatum during switching but not during distraction from relevant information in working memory, while the lateral frontal cortex was modulated by the drug during distraction but not during switching.

8. Application of a Cognitive Neuroscience Approach Toward Clinical Studies

8.1. Use of Biomarkers Derived from Cognitive Neuroscience Studies

Cognitive neuroscience studies using fMRI may provide an important foundation for clinical studies. A biomarker is an indicator that reflects a process, event, or condition in a biological system. Biomarkers may be useful for providing a measure of exposure, effect, or susceptibility. Reliable biomarkers of a neural system could reliably quantify how such a neural system is affected by almost any input. The input may be the effects of a drug, the effects of cognitive therapy, or the effects of a disease process. For a measurement to be useful as a biomarker in clinical studies, it needs to have well-defined significance based on preclinical studies. That is, a change in an fMRI measurement would ideally reflect a change in a well-understood process, thus providing a clear a priori hypothesis and interpretation of the findings. Once the processes are established, fMRI biomarkers may then be useful for addressing a number of clinical questions. For any neurophysiologic measurement to be a *surrogate* marker, a stable, reliable relationship between the fMRI measurement and a defined clinical outcome needs to be defined. Only in that scenario would an fMRI measurement provide a suitable surrogate for other clinical outcomes. Cognitive neuroscience studies provide the foundation for fMRI biomarkers, but the studies necessary for defining fMRI surrogate markers have not been done.

Questions regarding the mechanisms of brain function disrupted by pathologic states, processes affected by treatment interventions, or the nature of postinjury reorganization of function are examples of clinical questions that can be tested with fMRI.

For example, attentional modulation of information processing-related activity in visual cortex is a well-established phenomenon, with effects measurable using fMRI. It has been shown that activity in category-selective regions of inferior temporal cortex is modulated based on the target of attention, relatively up-modulated if the target is relevant to the region and down-modulated if not relevant *(136, 137)*. This finding provides a biomarker of attentional control over visual processing, and has been used to test how such control is altered by aging *(138)*. Using this biomarker, it was shown that a mechanism of cognitive aging might be a selective deficit in goal-directed down-modulation of nonrelevant activity.

8.2. Functional MRI for Measuring the Effect of Clinical Interventions

Functional MRI may be useful not only in defining "static" brain–behavioral relationships, but also may be applied to defining changes that occur due to learning, experience, or injury. In order to assess changes, longitudinal or repeated measurements are required. Because fMRI involves no exposure-limiting factors such as radiation, it is suitable for repeated measurements. However, multisession studies are also significantly more complicated to design, analyze, and interpret due to a number of issues discussed below.

There are at least two distinct approaches relevant to assessing changes within an individual. First, fMRI may be used for determining the after-effects of a learning intervention. Functional MRI measures pre- and post-intervention may be used to address this question. For example, after two pieces of information have been strongly associated over repetitive exposures, one may find reduced activation in response to presentation of that information, but increased functional connectivity between regions of the brain that process the two types of information *(110)*. Second, fMRI may be used for determining the processes that occur during an intervention, such as cognitive training. To do this one would need to acquire fMRI data *during* the process of training. An alternative approach is to use a cross-sectional approach to examine differences across individuals rather than within individuals *(139)*. For example, brain activation differences between long-term meditation practitioners and novices may be used to infer the neural effects of meditation training. However, other confounding effects of differences between cohorts are difficult to exclude.

The use of fMRI to define changes over time requires consideration of certain additional methodological issues. Test–retest reliability needs to be considered. Estimates of reliability depend on what is being measured. For example, in statistical parametric mapping, the question may be whether particular brain regions are stably labeled as "active" or not in serial sessions. A handful of studies have addressed this question. For example, one group showed that with a classification learning task, scans 1 year apart resulted in highly concordant results with defined regions of interest *(140)*. Another group showed that maps obtained from a working memory task were similar across time *(141)*, but with a motor task, there appeared to be significant variation over time in volume and spatial location of activation *(142)*. The reliability of interregional functional connectivity fMRI measurements has not yet been systematically tested.

In longitudinal studies, sources of variability may be both physiologic and nonphysiologic (e.g., MRI hardware). In some cases, the magnitudes of activation in specific brain regions of interest are themselves an outcome of interest. In these instances the stability of BOLD signal measurements becomes an even more salient issue. It may be worthwhile to utilize within-session indices that effectively normalize parameters of interest. For example, rather than comparing estimates of the magnitudes of activation, it may be worthwhile utilizing an index of activity for one condition compared to a second controlled condition with each session. An additional statistical approach that could account for potential variability in SNR is to combine data sets across sessions and then "whiten" the noise, effectively normalizing noise contribution across sessions. Other analytic approaches may be taken that are less sensitive to nonphysiologic instabilities. For example, one could test for changes in the spatial *pattern* of activation, which is not necessarily affected by signal magnitude changes. That is, one could test whether the patterns of activity are identical to within a scaling factor *(108)*. Finally, other factors that concurrently change over time can produce confounds to the interpretation of longitudinal studies. For example, performance may change, resulting in changes in reaction time or accuracy. All of these may alter measured responses making determination of the neural bases of the process of interest, such as a treatment intervention, more difficult. These and a number of other issues are discussed in greater depth by Poldrack in consideration of learning-related changes *(139)*.

9. Conclusions

Functional MRI is an extremely valuable tool for studying brain-behavior relationships, as it is widely available, noninvasive, and has superb temporal and spatial resolution. New approaches in fMRI experimental design and data analysis are appearing in the literature at an almost exponential rate, leading to numerous options for testing hypotheses on brain-behavior relationships.

Combined with information from complementary methods, such as the study of patients with focal lesions, healthy individuals with TMS, pharmacological interventions, or ERP, data from fMRI studies provide new insights regarding the organization of the cerebral cortex as well as the neural mechanisms underlying cognition. Moreover, cognitive neuroscience approaches that have been developed for fMRI provide an excellent foundation for its use as a clinical tool.

References

1. Broca P. Remarques sur le siege de la faculte du langage articule suivies d'une observation d'amphemie (perte de al parole). Bull Mem Soc Anat Paris 1861;36:330–57.
2. Buckner RL, Raichle ME, Petersen SE. Dissociation of human prefrontal cortical areas across different speech production tasks and gender groups. J Neurophysiol 1995;74(5):2163–73.
3. Sarter M, Bernston G, Cacioppo J. Brain imaging and cognitive neuroscience: toward strong inference in attributing function to structure. Am Psychol 1996;51:13–21.
4. Gaffan D, Gaffan EA. Amnesia in man following transection of the fornix: a review. Brain 1991;114:2611–8.
5. Feeney DM, Baron JC. Diaschisis. Stroke 1986;17(5):817–30.
6. Fuster JM, Alexander GE. Neuron activity related to short-term memory. Science 1971;173:652–4.
7. Funahashi S, Bruce CJ, Goldman-Rakic PS. Mnemonic coding of visual space in the monkey's dorsolateral prefrontal cortex. J Neurophysiol 1989;61:331–49.
8. Funahashi S, Bruce CJ, Goldman-Rakic PS. Dorsolateral prefrontal lesions and oculomotor delayed-response performance: Evidence for mnemonic "scotomas". J Neurosci 1993;13:1479–97.
9. Watanabe T, Niki H. Hippocampal unit activity and delayed response in the monkey. Brain Res 1985;325(1–2):241–54.
10. Cahusac PM, Miyashita Y, Rolls ET. Responses of hippocampal formation neurons in the monkey related to delayed spatial response and object-place memory tasks. Behav Brain Res 1989;33(3):229–40.
11. Alvarez P, Zola-Morgan S, Squire LR. The animal model of human amnesia: long-term memory impaired and short-term memory intact. Proc Natl Acad Sci U S A 1994;91(12):5637–41.
12. Corkin S. Lasting consequences of bilateral medial temporal lobectomy: clinical course and experimental findings in H.M. SeminNeurol 1984;4:249–59.
13. Ranganath C, D'Esposito M. Medial temporal lobe activity associated with active maintenance of novel information. Neuron 2001;31(5):865–73.
14. Druzgal TJ, D'Esposito M. Activity in fusiform face area modulated as a function of working memory load. Brain Res Cogn Brain Res 2001;10(3):355–64.
15. Henson R. Forward inference using functional neuroimaging: dissociations versus associations. Trends Cogn Sci 2006;10(2):64–9.
16. Cohen MS, Kosslyn SM, Breiter HC, et al. Changes in cortical activity during mental rotation: a mapping study using functional MRI. Brain 1996;119:89–100.
17. D'Esposito M, Ballard D, Aguirre GK, Zarahn E. Human prefrontal cortex is not specific for working memory: a functional MRI study. Neuroimage 1998;8(3):274–82.
18. Poldrack RA. Can cognitive processes be inferred from neuroimaging data? Trends Cogn Sci 2006;10(2):59–63.
19. Hutton C, Bork A, Josephs O, Deichmann R, Ashburner J, Turner R. Image distortion correction in fMRI: a quantitative evaluation. Neuroimage 2002;16(1):217–40.
20. Jezzard P, Clare S. Sources of distortion in functional MRI data. Hum Brain Mapp 1999;8(2–3):80–5.
21. Zeng H, Constable RT. Image distortion correction in EPI: comparison of field mapping with point spread function mapping. Magn Reson Med 2002;48(1):137–46.
22. Grasby PM. Imaging the neurochemical brain in health and disease. Clin Med 2002;2(1):67–73.
23. Bunge SA, Dudukovic NM, Thomason ME, Vaidya CJ, Gabrieli JD. Immature frontal lobe contributions to cognitive control in children: evidence from fMRI. Neuron 2002;33(2):301–11.

24. Casey BJ, Cohen JD, Jezzard P, et al. Activation of prefrontal cortex in children during a nonspatial working memory task with functional MRI. Neuroimage 1995;2(3):221–9.
25. Savoy RL, Ravicz ME, Gollub R. The psychophysiological laboratory in the magnet: stimulus delivery, response recording, and safety. In: Moonen CTW, Bandettini PA, eds. Functional MRI. Berlin: Springer; 1999:347–65.
26. Edminster WB, Talavage TM, Ledden PJ, Weisskoff RM. Improved auditory cortex imaging using clustered volume acquisitions. Hum Brain Mapp 1999;7:88–97.
27. Belin P, Zatorre RJ, Hoge R, Evans AC, Pike B. Event-related fMRI of the auditory cortex. Neuroimage 1999;10(4):417–29.
28. Sobel N, Prabhakaran V, Hartley CA, et al. Blind smell: brain activation induced by an undetected air-borne chemical. Brain 1999;122 (Pt 2):209–17.
29. Sobel N, Prabhakaran V, Desmond JE, Glover GH, Sullivan EV, Gabrieli JD. A method for functional magnetic resonance imaging of olfaction. J Neurosci Methods 1997;78(1–2):115–23.
30. Gitelman DR, Parrish TB, LaBar KS, Mesulam MM. Real-time monitoring of eye movements using infrared video-oculography during functional magnetic resonance imaging of the frontal eye fields. Neuroimage 2000;11(1):58–65.
31. Kimmig H, Greenlee MW, Gondan M, Schira M, Kassubek J, Mergner T. Relationship between saccadic eye movements and cortical activity as measured by fMRI: quantitative and qualitative aspects. Exp Brain Res 2001;141(2):184–94.
32. Palmer ED, Rosen HJ, Ojemann JG, Buckner RL, Kelley WM, Petersen SE. An event-related fMRI study of overt and covert word stem completion. Neuroimage 2001;14(1 Pt 1):182–93.
33. Fu CH, Morgan K, Suckling J, et al. A functional magnetic resonance imaging study of overt letter verbal fluency using a clustered acquisition sequence: greater anterior cingulate activation with increased task demand. Neuroimage 2002;17(2):871–9.
34. Barch DM, Sabb FW, Carter CS, Braver TS, Noll DC, Cohen JD. Overt verbal responding during fMRI scanning: empirical investigations of problems and potential solutions. Neuroimage 1999;10(6):642–57.
35. Goldman RI, Stern JM, Engel J, Jr., Cohen MS. Acquiring simultaneous EEG and functional MRI. Clin Neurophysiol 2000;111(11): 1974–80.
36. Lazeyras F, Zimine I, Blanke O, Perrig SH, Seeck M. Functional MRI with simultaneous EEG recording: feasibility and application to motor and visual activation. J Magn Reson Imaging 2001;13(6):943–8.
37. Mantini D, Perrucci MG, Cugini S, Ferretti A, Romani GL, Del Gratta C. Complete artifact removal for EEG recorded during continuous fMRI using independent component analysis. Neuroimage 2007;34(2):598–607.
38. Otzenberger H, Gounot D, Foucher JR. Optimisation of a post-processing method to remove the pulse artifact from EEG data recorded during fMRI: an application to P300 recordings during e-fMRI. Neurosci Res 2007;57(2):230–9.
39. Gnadt JW, Andersen RA. Memory related motor planning activity in posterior parietal cortex of macaque. Exp Brain Res 1988;70:216–20.
40. Aguirre GK, Zarahn E, D'Esposito M. The variability of human, BOLD hemodynamic responses. Neuroimage 1998;8(4):360–9.
41. Bandettini PA, Wong EC, Hinks RS, Tikofsky RS, Hyde JS. Time course of EPI of human brain function during task activation. Magn Reson Med 1992;25:390–7.
42. Boynton GM, Engel SA, Glover GH, Heeger DJ. Linear systems analysis of functional magnetic resonance imaging in human V1. J Neurosci 1996;16:4207–21.
43. Kim SG, Richter W, Ugurbil K. Limitations of temporal resolution in fMRI. Magn Reson Med 1997;37:631–6.
44. Savoy RL, Bandettini PA, O'Craven KM, et al. Pushing the temporal resolution of fMRI: studies of very brief stimuli, onset of variability and asynchrony, and stimulu-correlated changes in noise. Proc Soc Magn Reson Med 1995;3:450.
45. Zarahn E, Aguirre GK, D'Esposito M. A trial-based experimental design for functional MRI. NeuroImage 1997;6:122–38.
46. Burock MA, Buckner RL, Woldorff MG, Rosen BR, Dale AM. Randomized event-related experimental designs allow for extremely rapid presentation rates using functional MRI. Neuroreport 1998;9(16):3735–9.
47. Clark VP, Maisog JM, Haxby JV. fMRI studies of visual perception and recognition using a random stimulus design. Soc Neurosci Abstr 1997;23:301.
48. Dale AM, Buckner RL. Selective averaging of rapidly presented individual trials using fMRI. Hum Brain Mapp 1997;5:1–12.
49. Miezin FM, Maccotta L, Ollinger JM, Petersen SE, Buckner RL. Characterizing the

hemodynamic response: effects of presentation rate, sampling procedure, and the possibility of ordering brain activity based on relative timing. Neuroimage 2000;11(6 Pt 1):735–59.

50. D'Esposito M, Zarahn E, Aguirre GK. Event-related functional MRI: implications for cognitive psychology. Psychol Bull 1999;125:155–64.
51. Logothetis NK, Guggenberger H, Peled S, Pauls J. Functional imaging of the monkey brain. Nat Neurosci 1999;2(6):555–62.
52. Cheng K, Waggoner RA, Tanaka K. Human ocular dominance columns as revealed by high-field functional magnetic resonance imaging. Neuron 2001;32(2):359–74.
53. Logothetis N, Merkle H, Augath M, Trinath T, Ugurbil K. Ultra high-resolution fMRI in monkeys with implanted RF coils. Neuron 2002;35(2):227–42.
54. Malonek D, Grinvald A. Interactions between electrical activity and cortical microcirculation revealed by imaging spectroscopy: implications for functional brain mapping. Science 1996;272:551–4.
55. Kim SG, Duong TQ. Mapping cortical columnar structures using fMRI. Physiol Behav 2002;77(4–5):641–4.
56. Menon RS, Ogawa S, Strupp JP, Ugurbil K. Ocular dominance in human V1 demonstrated by functional magnetic resonance imaging. J Neurophysiol 1997;77(5):2780–7.
57. Grill-Spector K, Malach R. fMR-adaptation: a tool for studying the functional properties of human cortical neurons. Acta Psychol (Amst) 2001;107(1–3):293–321.
58. Grill-Spector K, Kushnir T, Edelman S, Avidan G, Itzchak Y, Malach R. Differential processing of objects under various viewing conditions in the human lateral occipital complex. Neuron 1999;24(1):187–203.
59. Posner MI, Petersen SE, Fox PT, Raichle ME. Localization of cognitive operations in the human brain. Science 1988;240:1627–31.
60. Sternberg S. The discovery of processing stages: extensions of Donders' method. Acta Psychologica 1969;30:276–315.
61. Petersen SE, Fox PT, Posner MI, Mintun M, Raichle ME. Positron emission tomographic studies of the cortical anatomy of single word processing. Nature 1988;331:585–9.
62. Fuster J. The Prefrontal Cortex: Anatomy, Physiology, and Neuropsychology of the Frontal Lobes. 3rd ed. Raven Press: New York; 1997.
63. Jonides J, Smith EE, Koeppe RA, Awh E, Minoshima S, Mintun MA. Spatial working memory in humans as revealed by PET. Nature 1993;363:623–5.
64. Zarahn E, Aguirre GK, D'Esposito M. Temporal isolation of the neural correlates of spatial mnemonic processing with fMRI. Cogn Brain Res 1999;7(3):255–68.
65. Attwell D, Iadecola C. The neural basis of functional brain imaging signals. Trends Neurosci 2002;25(12):621–5.
66. Heeger DJ, Ress D. What does fMRI tell us about neuronal activity? Nat Rev Neurosci 2002;3(2):142–51.
67. Friston KJ, Josephs O, Rees G, Turner R. Nonlinear event-related responses in fMRI. Magn Reson Med 1998;39(1):41–52.
68. Glover GH. Deconvolution of impulse response in event-related BOLD fMRI. Neuroimage 1999;9(4):416–29.
69. Miller KL, Luh WM, Liu TT, et al. Nonlinear temporal dynamics of the cerebral blood flow response. Hum Brain Mapp 2001;13(1): 1–12.
70. Vazquez AL, Noll DC. Nonlinear aspects of the BOLD response in functional MRI. NeuroImage 1998;7(2):108–18.
71. D'Esposito M, Zarahn E, Aguirre GK, Rypma B. The effect of normal aging on the coupling of neural activity to the bold hemodynamic response. Neuroimage 1999;10(1):6–14.
72. Rosen BR, Buckner RL, Dale AM. Event-related functional MRI: past, present, and future. Proc Natl Acad Sci U S A 1998;95(3):773–80.
73. Donaldson DI, Petersen SE, Ollinger JM, Buckner RL. Dissociating state and item components of recognition memory using fMRI. Neuroimage 2001;13(1):129–42.
74. Mitchell KJ, Johnson MK, Raye CL, D'Esposito M. fMRI evidence of age-related hippocampal dysfunction in feature binding in working memory. Brain Res Cogn Brain Res 2000;10(1–2):197–206.
75. Keppel G, Zedeck S. Data Analysis for Research Design. New York: W.H. Freeman & Company; 1989.
76. Worsley KJ, Friston KJ. Analysis of fMRI time-series revisited–again. Neuroimage 1995;2: 173–82.
77. Everitt BS, Bullmore ET. Mixture model mapping of the brain activation in functional magnetic resonance images. Hum Brain Mapp 1999;7(1):1–14.
78. Genovese CR, Lazar NA, Nichols T. Thresholding of statistical maps in functional neuroimaging using the false discovery rate. Neuroimage 2002;15(4):870–8.

79. Zarahn E, Aguirre GK, D'Esposito M. Empirical analyses of BOLD fMRI statistics. I. Spatially unsmoothed data collected under null-hypothesis conditions. NeuroImage 1997;5:179–97.
80. Aguirre GK, Zarahn E, D'Esposito M. Empirical analyses of BOLD fMRI statistics. II. Spatially smoothed data collected under null-hypothesis and experimental conditions. NeuroImage 1997;5:199–212.
81. D'Esposito M, Ballard D, Zarahn E, Aguirre GK. The role of prefrontal cortex in sensory memory and motor preparation: an event-related fMRI study. Neuroimage 2000;11(5 Pt 1):400–8.
82. Zarahn E, Slifstein M. A reference effect approach for power analysis in fMRI. Neuroimage 2001;14(3):768–79.
83. Van Horn JD, Ellmore TM, Esposito G, Berman KF. Mapping voxel-based statistical power on parametric images. Neuroimage 1998;7(2):97–107.
84. Aguirre GK, D'Esposito M. Experimental design for brain fMRI. In: Moonen CTW, Bandettini PA, eds. Functional MRI. Berlin: Springer Verlag; 1999:369–80.
85. Logothetis NK, Pauls J, Augath M, Trinath T, Oeltermann A. Neurophysiological investigation of the basis of the fMRI signal. Nature 2001;412(6843):150–7.
86. Farkas E, Luiten PG. Cerebral microvascular pathology in aging and Alzheimer's disease. Prog Neurobiol 2001;64(6):575–611.
87. Fang HCH. Observations on aging characteristics of cerebral blood vessels, macroscopic and microscopic features. In: Gerson S, Terry RD, eds. Neurobiology of Aging. New York: Raven Press; 1976.
88. Bentourkia M, Bol A, Ivanoiu A, et al. Comparison of regional cerebral blood flow and glucose metabolism in the normal brain: effect of aging. J Neurol Sci 2000;181(1–2):19–28.
89. Schultz SK, O'Leary DS, Boles Ponto LL, Watkins GL, Hichwa RD, Andreasen NC. Age-related changes in regional cerebral blood flow among young to mid-life adults. Neuroreport 1999;10(12):2493–6.
90. Yamamoto M, Meyer JS, Sakai F, Yamaguchi F. Aging and cerebral vasodilator responses to hypercarbia: responses in normal aging and in persons with risk factors for stroke. Arch Neurol 1980;37(8):489–96.
91. Yamaguchi T, Kanno I, Uemura K, et al. Reduction in regional cerebral rate of oxygen during human aging. Stroke 1986;17:1220–8.
92. Takada H, Nagata K, Hirata Y, et al. Age-related decline of cerebral oxygen metabolism in normal population detected with positron emission tomography. Neurol Res 1992;14(2 Suppl):128–31.
93. Claus JJ, Breteler MM, Hasan D, et al. Regional cerebral blood flow and cerebrovascular risk factors in the elderly population. Neurobiol Aging 1998;19(1):57–64.
94. Cunnington R, Iansek R, Bradshaw JL, Phillips JG. Movement-related potentials in Parkinson's disease. Presence and predictability of temporal and spatial cues. Brain 1995;118 (Pt 4):935–50.
95. Buckner RL, Snyder AZ, Sanders AL, Raichle ME, Morris JC. Functional brain imaging of young, nondemented, and demented older adults. J Cogn Neurosci 2000;12 (Suppl 2):24–34.
96. Huettel SA, Singerman JD, McCarthy G. The effects of aging upon the hemodynamic response measured by functional MRI. Neuroimage 2001;13(1):161–75.
97. Pineiro R, Pendlebury S, Johansen-Berg H, Matthews PM. Altered hemodynamic responses in patients after subcortical stroke measured by functional MRI. Stroke 2002;33(1):103–9.
98. Rypma B, Prabhakaran V, Desmond JE, Gabrieli JD. Age differences in prefrontal cortical activity in working memory. Psychol Aging 2001;16(3):371–84.
99. Wolf RL, Detre JA. Clinical neuroimaging using arterial spin-labeled perfusion magnetic resonance imaging. Neurotherapeutics 2007;4(3):346–59.
100. Brown GG, Clark C, Liu TT. Measurement of cerebral perfusion with arterial spin labeling. Part 2. Applications. J Int Neuropsychol Soc 2007;13(3):526–38.
101. Aguirre GK, Detre JA, Zarahn E, Alsop DC. Experimental design and the relative sensitivity of BOLD and perfusion fMRI. Neuroimage 2002;15(3):488–500.
102. Liu TT, Brown GG. Measurement of cerebral perfusion with arterial spin labeling. Part 1. Methods. J Int Neuropsychol Soc 2007;13(3):517–25.
103. Fernandez-Seara MA, Wang J, Wang Z, et al. Imaging mesial temporal lobe activation during scene encoding: comparison of fMRI using BOLD and arterial spin labeling. Hum Brain Mapp 2007;28(12):1391–400.
104. Lee GR, Hernandez-Garcia L, Noll DC. Functional imaging with Turbo-CASL: transit time and multislice imaging considerations. Magn Reson Med 2007;57(4):661–9.

105. Kanwisher N, McDermott J, Chun MM. The fusiform face area: a module in huma extrastriate cortex specialized for face perception. J Neurosci 1997;17:4302–11.
106. Haxby JV, Gobbini MI, Furey ML, Ishai A, Schouten JL, Pietrini P. Distributed and overlapping representations of faces and objects in ventral temporal cortex. Science 2001;293(5539):2425–30.
107. Polyn SM, Natu VS, Cohen JD, Norman KA. Category-specific cortical activity precedes retrieval during memory search. Science 2005;310(5756):1963–6.
108. Zarahn E, Rakitin BC, Abela D, Flynn J, Stern Y. Distinct spatial patterns of brain activity associated with memory storage and search. Neuroimage 2006;33(2):794–804.
109. Williams MA, Dang S, Kanwisher NG. Only some spatial patterns of fMRI response are read out in task performance. Nat Neurosci 2007;10(6):685–6.
110. Buchel C, Coull JT, Friston KJ. The predictive value of changes in effective connectivity for human learning. Science 1999;283(5407): 1538–41.
111. McIntosh AR, Grady CL, Haxby JV, Ungerleider LG, Horwitz B. Changes in limbic and prefrontal functional interactions in a working memory task for faces. Cereb Cortex 1996;6(4):571–84.
112. Gerstein GL, Perkel DH, Subramanian KN. Identification of functionally related neural assemblies. Brain Res 1978;140(1):43–62.
113. Penny WD, Stephan KE, Mechelli A, Friston KJ. Modelling functional integration: a comparison of structural equation and dynamic causal models. Neuroimage 2004;23 (Suppl 1):S264–74.
114. Sun FT, Miller LM, D'Esposito M. Measuring interregional functional connectivity using coherence and partial coherence analyses of fMRI data. Neuroimage 2004;21(2): 647–58.
115. Sun FT, Miller LM, D'Esposito M. Measuring temporal dynamics of functional networks using phase spectrum of fMRI data. Neuroimage 2005;28(1):227–37.
116. Sun FT, Miller LM, Rao AA, D'Esposito M. Functional connectivity of cortical networks involved in bimanual motor sequence learning. Cereb Cortex 2007;17(5):1227–34.
117. Gazzaley A, Rissman J, Desposito M. Functional connectivity during working memory maintenance. Cogn Affect Behav Neurosci 2004;4(4):580–99.
118. Fuhrmann Alpert G, Sun FT, Handwerker D, D'Esposito M, Knight RT. Spatio-temporal information analysis of event-related BOLD responses. Neuroimage 2007;34(4): 1545–61.
119. Rees G, Frith CD, Lavie N. Modulating irrelevant motion perception by varying attentional load in an unrelated task. Science 1997;278(5343):1616–9.
120. Treisman AM. Strategies and models of selective attention. Psychol Rev 1969;76(3): 282–99.
121. Lavie N, Tsal Y. Perceptual load as a major determinant of the locus of selection in visual attention. Percept Psychophys 1994;56(2): 183–97.
122. McCarthy RA, Warrington EK. Disorders of semantic memory. Philos Trans R Soc Lond B Biol Sci 1994;346(1315):89–96.
123. Warrington EST. Category specific semantic impairments. Brain 1984;107:829–54.
124. Thompson-Schill SL. Neuroimaging studies of semantic memory: inferring "how" from "where". Neuropsychologia 2003;41(3): 280–92.
125. Thompson-Schill SL, D'Esposito M, Aguirre GK, Farah MJ. Role of left inferior prefrontal cortex in retrieval of semantic knowledge: a reevaluation. Proc Natl Acad Sci U S A 1997;94(26):14792–7.
126. Thompson-Schill SL, Swick D, Farah MJ, D'Esposito M, Kan IP, Knight RT. Verb generation in patients with focal frontal lesions: a neuropsychological test of neuroimaging findings. Proc Natl Acad Sci U S A 1998;95(26):15855–60.
127. Pascual-Leone A, Tarazona F, Keenan J, Tormos JM, Hamilton R, Catala MD. Transcranial magnetic stimulation and neuroplasticity. Neuropsychologia 1999;37(2):207–17.
128. Rushworth MF, Hadland KA, Paus T, Sipila PK. Role of the human medial frontal cortex in task switching: a combined fMRI and TMS study. J Neurophysiol 2002;87(5):2577–92.
129. Ruff CC, Bestmann S, Blankenburg F, et al. Distinct causal influences of parietal versus frontal areas on human visual cortex: evidence from concurrent TMS fMRI. Cereb Cortex 2008;18(4):817–27.
130. Ruff CC, Blankenburg F, Bjoertomt O, et al. Concurrent TMS-fMRI and psychophysics reveal frontal influences on human retinotopic visual cortex. Curr Biol 2006;16(15): 1479–88.
131. Dehaene S, Spelke E, Pinel P, Stanescu R, Tsivkin S. Sources of mathematical thinking: behavioral and brain-imaging evidence. Science 1999;284(5416):970–4.

132. Gibbs SE, D'Esposito M. Individual capacity differences predict working memory performance and prefrontal activity following dopamine receptor stimulation. Cogn Affect Behav Neurosci 2005;5(2):212–21.
133. Gibbs SE, D'Esposito M. A functional MRI study of the effects of bromocriptine, a dopamine receptor agonist, on component processes of working memory. Psychopharmacology (Berl) 2005;180(4):644–53.
134. Gibbs SE, D'Esposito M. A functional magnetic resonance imaging study of the effects of pergolide, a dopamine receptor agonist, on component processes of working memory. Neuroscience 2006.
135. Cools R, Sheridan M, Jacobs E, D'Esposito M. Impulsive personality predicts dopamine-dependent changes in frontostriatal activity during component processes of working memory. J Neurosci 2007;27(20):5506–14.
136. Kastner S, Pinsk MA. Visual attention as a multilevel selection process. Cogn Affect Behav Neurosci 2004;4(4):483–500.
137. Gazzaley A, Cooney JW, McEvoy K, Knight RT, D'Esposito M. Top-down enhancement and suppression of the magnitude and speed of neural activity. J Cogn Neurosci 2005;17(3):507–17.
138. Gazzaley A, Cooney JW, Rissman J, D'Esposito M. Top-down suppression deficit underlies working memory impairment in normal aging. Nat Neurosci 2005; 8(10):1298–300.
139. Poldrack RA. Imaging brain plasticity: conceptual and methodological issues - a theoretical review. Neuroimage 2000;12(1): 1–13.
140. Aron AR, Gluck MA, Poldrack RA. Long-term test-retest reliability of functional MRI in a classification learning task. Neuroimage 2006;29(3):1000–6.
141. Wei X, Yoo SS, Dickey CC, Zou KH, Guttmann CR, Panych LP. Functional MRI of auditory verbal working memory: long-term reproducibility analysis. Neuroimage 2004;21(3):1000–8.
142. Yoo SS, Wei X, Dickey CC, Guttmann CR, Panych LP. Long-term reproducibility analysis of fMRI using hand motor task. Int J Neurosci 2005;115(1):55–77.

Chapter 11

fMRI of Language Systems

Jeffrey R. Binder

Summary

Language refers to the uniquely human capacity for communication through productive combination of symbolic representations. Functional neuroimaging studies have in recent decades greatly expanded our knowledge of the brain systems supporting language, producing a dramatic reawakening of interest in this topic and a call to revise and extend the nineteenth century neuroanatomical model formulated by Broca, Wernicke, and others. This chapter presents some theoretical issues regarding functional imaging of language systems, a model of the functional neuroanatomy of language based on recent empirical results in several selected processing domains, and a survey of language mapping paradigms in common clinical use. A central theme is that interpretation of fMRI language studies depends on an informed analysis of the cognitive processes engaged during scanning. This analytic approach can help avoid common pitfalls in task design that limit the sensitivity and specificity of language mapping studies and should encourage the development of a standardized methodological and conceptual framework for such studies.

Key words: fMRI, Language, Semantics, Phonology, Orthography

The central role of language in human culture and social interaction is self-evident. In addition to providing a formal system for overt communication, the symbolic structures of language enable such uniquely human cognitive capacities as the ability to manipulate concepts, plan the future, and invent new technology. Scientific investigation of the neural basis of language began in earnest with the work of Broca, Wernicke, and other nineteenth-century neurologists *(1, 2)*, leading to the classical *Wernicke–Lichtheim* model of language and aphasia that remains with us today *(3, 4)*. Over the past two decades, however, functional imaging methods, particularly fMRI, have greatly expanded our knowledge of the brain systems supporting language, producing a dramatic reawakening of interest in this topic and a call to revise and extend the classical model *(5, 6)*. This chapter provides

M. Filippi (ed.), *fMRI Techniques and Protocols*, Neuromethods, vol. 41
DOI 10.1007/978-1-60327-919-2_11, © Humana Press, a part of Springer Science+Business Media, LLC 2009

a brief survey of some of this work, together with a discussion of theoretical issues central to the design and interpretation of language mapping studies.

1. Language and Language Processes

What is language? One definition often cited is that language processes are those that enable communication. In biological terms, however, this definition is overly inclusive, in that many bodily functions (e.g., cardiac, pulmonary, general arousal, and sustained attention systems) provide critical support for communication but are not linguistic in nature. Communication typically requires neural systems that process auditory or visual sensory information, hold this information in a short-term store, direct attention to specific features or aspects of the information, perform comparisons and other general operations on the information, select a response based on such operations, and carry out the response. The extent to which any of these systems is specialized for language is a matter of debate. Careful consideration of these domain-general systems is especially relevant for interpreting and designing language mapping studies, which often employ relatively complex tasks that engage motor, sensory, attentional, memory, and "central executive" functions in addition to language. Should these other components be considered part of the language system because they are so critical for adequate task performance, or should they be delineated from language processes per se? In this chapter, it is assumed that the goal of language mapping is to identify neural systems involved specifically in language processes, that is, to distinguish these brain networks from early sensory, motor, and general executive systems.

A more precise definition of language is that it is a system of communication based on the symbolic representation and manipulation of information. Languages are also, by definition, *generative*, in that the symbols of a language can be productively combined to make a virtually limitless number of new expressions. In formulating a general definition of this kind, however, it is critical to keep in mind that language is not a unitary process, but rather a collection of processes operating at distinct levels and on distinct types of information. Clinicians working with aphasic patients historically have focused on the dichotomy between "expressive" and "receptive" language functions, but a more useful taxonomy of component language processes is available from the field of linguistics. For spoken languages, these processes include (1) phoneme perception, the processes serving recognition of speech sounds, (2) phonology, the processes by which speech sounds are represented and manipulated in abstract form,

(3) speech articulation, the processes by which speech movements are planned and executed, (4) orthography, the processes by which written characters are represented and manipulated in abstract form, (5) semantics, the processing of word meanings, names, and other declarative knowledge about the world, and (6) syntax, the processes by which words are combined to make sentences and sentences are analyzed to reveal underlying relationships between words. A basic assumption of language mapping is that activation tasks can be designed to make varying demands on these processing subsystems. For example, a task requiring careful listening to word-like nonwords (often called *pseudowords*, e.g., "nurdle") would make great demands on phoneme perception (and on prephonetic auditory processing and attention) but very little demand on semantic or syntactic processing, given that the stimuli have no (or very little) meaning. In contrast, a task requiring semantic categorization of printed words (e.g., "Is it an animal or not?") would make great demands on orthographic and semantic processing but relatively little on phonetic, phonological, or syntactic processing.

On the other hand, the processing subcomponents of language often act together. The extent to which each component can be examined in isolation remains a major methodological issue, as it is not yet clear to what extent the systems responsible for these processes become active "automatically" when presented with linguistic stimuli *(7)*. One familiar example of this interaction is the Stroop effect, in which orthographic and phonological processing of printed words occurs even when subjects are instructed to attend to the color of the print, and even when this processing interferes with task performance *(8)*. Other examples include semantic priming effects during word recognition, picture–word interference effects, lexical effects on phonetic perception, orthographic effects on letter perception, and semantic–syntactic interactions during sentence comprehension *(9–17)*. If linguistic stimuli such as words and pictures evoke obligatory, automatic language processing, these effects need to be considered in the design and interpretation of language activation experiments. Use of such stimuli in a "baseline" condition could result in undesirable subtraction (or partial subtraction) of language-related activation. Because investigators frequently try to match stimuli in control and language tasks very closely, such inadvertent subtraction is relatively a commonplace in functional imaging studies of language processing.

A final theoretical issue is the extent to which language processes occur during "resting" states or states with minimal task requirements (e.g., visual fixation or "passive" stimulation). Language involves interactive systems for manipulating *internally stored knowledge* about words and word meanings. In examining these systems we typically use familiar stimuli or cues to engage processing, yet it seems likely that activity in these systems could occur independently of external stimulation and task demands. The idea that the conscious mind can be internally active independent of external events has a long history in psychology and neuroscience *(18–23)*. When asked, subjects in experimental studies frequently report experiencing seemingly unprovoked thoughts (including words and recognizable images) that are unrelated to the task at hand *(21, 24, 25)*. The precise extent to which such "thinking" engages linguistic knowledge remains unclear *(26, 27)*, but many researchers have demonstrated close parallels between behavior and language content, suggesting that at least some internal thought processes make use of verbally encoded semantic knowledge and other linguistic representations *(26, 28, 29)*. Some authors have argued that "rest" and similar conditions are actually active states in which subjects frequently are engaged in processing linguistic and other information *(30–38)*. Use of such states as control conditions for language imaging studies may thus obscure similar processes that occur during the language task of interest. This is a particularly difficult problem for language studies because the internal processes in question cannot be directly measured or precisely controlled.

2. Functional Neuroanatomy of Component Language Systems

Considering that neuroimaging studies on language processing, now number in the *thousands*, the following review is inevitably incomplete and somewhat cursory. Nor is it possible to cover every topic that might be of interest. This review focuses on single word and sublexical processes, with less attention to studies of sentence comprehension and syntax. Several interesting topics, including bilingualism and sign language processing, are not touched on here, though the interested reader is referred to a number of excellent reviews *(39–42)*.

2.1. Phoneme Perception

Traditional clinical models of aphasia often treat comprehension of speech as a single function *(43)*, but it is important to distinguish at least two processes engaged during speech comprehension. Spoken words not only possess meanings, but they are also composed of very complex and rapidly changing auditory signals. Thus, useful models of auditory word recognition include not only a semantic stage in which words are mapped onto their meanings, but also a stage prior to semantic access in which consonant and vowel sounds – known collectively as *phonemes* – are identified. The distinction between these stages becomes clearer if one considers the differences between listening to a tone (e.g., a note played on a piano), a nonword (such as "dap"), and a word (such as "tap").

The tone has no phonemic value; it cannot be identified as any vowel or consonant. In contrast, the nonword "dap" contains three phonemes - /d/, /æ/, /p/ – although it has no meaning. Finally, the word "tap" conveys both phonemic and semantic information. The importance of the phoneme perception stage is illustrated by the fact that "dap" and "tap" differ at a physical level only in the presence of a brief (typically 20–30 ms) noise at the beginning of "tap" but not "dap," produced by release of the tongue from the roof of the mouth slightly prior to the onset of vocal cord vibration. These and many other subtle acoustic cues must be rapidly and continuously processed in real time for accurate speech comprehension to occur. The absence of this important linguistic process in traditional clinical models of language can be attributed to at least two factors. First, very little was known about the physical acoustic properties of phonemes prior to the mid-twentieth century; the scientific study of speech perception has developed only in the last 50–60 years. Second, cases of isolated phoneme perception difficulty, known as *pure word deafness*, are rare, resulting in a relative lack of familiarity with this field of study on the part of many clinicians.

Over the past 15 years, scientists using functional neuroimaging methods have identified a region in the superior temporal lobes that responds more strongly to speech than to nonspeech sounds such as tones and noise *(44–50)*. These speech-related activations are found consistently in the middle portion of the superior temporal sulcus (STS), that is, the sulcus separating the superior temporal gyrus (STG) from the middle temporal gyrus (MTG). This activation is often found in both the left and right STS, though usually with leftward lateralization. These activations are identical whether the stimuli are words or word-like pseudo-words, thus it is unlikely that they represent processing of word meaning *(49)*. Though some of the activation in this region could be explained by the fact that speech sounds are more acoustically complex than the tones and noises used as nonspeech controls, more recent experiments using acoustically matched speech and nonspeech sounds (e.g., rotated speech, sinewave speech) have shown convincingly that at least some of the activation in this region is due specifically to activation of phoneme codes (or sequences of phoneme codes, also called *auditory word forms*) *(51–54)*.

These observations are fully consistent with localization data from patients with pure word deafness, who typically have lesions restricted to the STG and STS *(55–61)*. Most of these cases have bilateral lesions, though rarely a large left temporal lobe lesion can produce the syndrome *(62, 63)*. These patients show impairment in recognizing speech phonemes and may have other deficits of higher-order auditory perception, especially when the lesions are bilateral, but they have no deficits in written comprehension, naming, or propositional speech production that would indicate any loss of word concepts. Taken together, these functional imaging and lesion data make it clear that the left STG plays a relatively specific role in language processing, that is, that it contains general auditory systems and specialized networks for recognizing speech phonemes. This conceptualization stands in stark contrast to the traditional clinical model of aphasia, which identifies the left STG as "Wernicke's area," the principal site for "language comprehension."

2.2. Grapheme Perception and Orthographic Processing

Traditional neuroanatomical models of written word recognition derive mainly from the late nineteenth century descriptions by Déjerine of alexia with and without agraphia *(64–66)*. According to these models, visual perception of letters occurs in the primary visual cortex of both hemispheres, which then transmit this information to the left angular gyrus, where memories of written words are activated. Lesions of the left angular gyrus destroy these visual word codes, producing both inability to read and inability to write. Alexia without agraphia (also known as pure alexia, peripheral alexia, or letter-by-letter reading) results when an occipital lobe lesion destroys both the left visual cortex and the decussating white matter pathway from right visual cortex to left angular gyrus, effectively disconnecting the angular gyrus from visual input without destruction of the visual word codes themselves *(67)*.

In recent decades it has become clear, however, that pure alexia can result from ventral occipital–temporal lesions that damage neither the primary visual cortex nor the angular gyrus. Though initially ascribed to involvement of white matter pathways projecting to the angular gyrus *(68–70)*, it is now clear that most of these cases are due to focal damage to the left ventral occipital–temporal junction, particularly the midportion of the left fusiform gyrus *(71–75)*. Thus, normal reading requires the participation of a left-lateralized visual association area in or near the mid-fusiform gyrus, which receives input from earlier visual processing stages in both hemispheres.

Recent functional neuroimaging research has strongly confirmed this model. Numerous studies have demonstrated a focal region in the lateral left fusiform gyrus that responds more strongly to words and word-like nonwords than to consonant letter strings or nonsense characters *(76–80)*. This focal region of cortex has consequently been named the "visual word form area." Activation in this area increases as a direct function of how frequently the letter combinations comprising the stimulus occur in the reader's language, and this activation is also correlated with efficiency of letter perception during tachistoscopic presentation *(81)*. It thus appears that during the many hours spent learning to read fluently, neurons in this region become "tuned" to detect familiar letter combinations, resulting in a high degree of perceptual expertise that allows multiletter fragments and even whole words to be processed in parallel. Destruction of these "expert" neurons prevents the patient from recognizing multiletter fragments efficiently, forcing the adoption of a much slower, letter-by-letter decoding process *(82–84)*.

2.3. Phonological Access and Phonological Working Memory

The term *paraphasia* refers to speech production that is fluent but contains errors, such as substitution of incorrect phonemes or words, or rearrangement of the order of phonemes within a word. Paraphasia is characteristic of many forms of aphasia, particularly Wernicke and conduction syndromes, and typically affects both spoken and written output. Paraphasia indicates an inability to retrieve or properly use a mental representation of word sounds – what nineteenth century theorists called "sound images" and what in modern parlance are referred to as *phonological representations*. Patients who cannot access (i.e., activate, compute) correct phonological representations show paraphasic errors on all speech output tasks, such as speaking, naming objects, reading aloud, and repeating, as well as on a variety of other tasks that require phonological access. For example, patients with phonological impairments may be unable to determine whether two printed words rhyme. Writing normally involves a mapping from phonological (sound-based) to orthographic (grapheme-based) representations, which is why patients with impaired phonology also typically show paraphasic errors in their writing.

The brain regions most strongly implicated in phonological access are in the left posterior perisylvian area, especially the posterior STG, posterior STS, and supramarginal gyrus (SMG). For example, patients with conduction aphasia – a relatively isolated disorder of phonological access featuring phonemic paraphasia in naming, reading, and repetition tasks – have lesions confined to this region *(85–90)*, as do patients with phonological deficits in written production *(91, 92)*. A number of functional imaging results also support this localization. For example, contrasts between visual stimuli that can be named and those that cannot (e.g., pictures vs. nonsense shapes, pronounceable vs. unpronounceable letter strings, letters vs. unfamiliar characters) reliably produce activation in the left posterior STS, STG, and inferior SMG *(93–102)* as do silent "word generation" tasks *(103–105)*.

Temporary activation of phonological representations is also central to the concept of *verbal working memory*. The standard model of verbal working memory includes a "phonological loop" responsible for maintaining phonological sequences (whether words or nonwords) in short-term memory *(106)*. The phonological loop is further subdivided into a "phonological store" that represents the phonological information itself and an "articulatory rehearsal" mechanism that reactivates the information before it fades from the store. Several neuroimaging studies have linked the phonological store with the left inferior parietal region, particularly the SMG *(107, 108)*, and with the posterior STG and STS *(100, 109–111)*.

These regions implicated in phonological access and temporary storage of phonological representations overlap partly with those implicated in speech perception, though there is some evidence that the phonological system extends more dorsally than the speech perception system, involving dorsal STG (planum temporale) and SMG to a greater extent *(112)*. These systems cannot be entirely overlapping, since most patients with conduction aphasia (phonemic paraphasia) do not have speech perception deficits. As noted earlier, the speech perception system is also bilaterally represented, which may explain why it is more resistant to left STG damage than is the phonological access system, which is more strongly left-lateralized.

2.4. Semantic Memory and Semantic Processing

The brain has an enormous capacity to acquire knowledge from experience. The characteristic shapes, colors, textures, movements, sounds, smells, and actions associated with objects in the environment, for example, must all be learned from experience. Much of this knowledge is represented symbolically in language and underlies our understanding of word meanings. These relationships between words and the stores of knowledge they signify are known collectively as the *semantics* of a language *(113)*. The term *semantic processing* refers to the cognitive act of accessing stored knowledge about the world through words. The stored knowledge itself is often called *semantic memory*.

Semantic properties of words are readily distinguished from their structural properties. For example, words can have both spoken (phonological) and written (orthographic) forms, but these surface forms are typically related to word meanings only through the arbitrary conventions of a particular language. There is nothing, for example, about the letter sequences D-O-G or C-H-I-E-N that inherently links these sequences to a particular concept. Conversely, it is trivial to construct surface forms (e.g., CHOG) that possess all of the phonological and orthographic properties of words in a particular language, but which have no meaning in that language. A simple, operational distinction can thus be made between the processes involved in analyzing the surface form (phonology, orthography) of words, and semantic processes, which concern access to knowledge that is *not directly represented in the surface form*.

Semantic processing is a defining feature of human behavior, central not only to language, but also to our capacity to access acquired knowledge in reasoning, planning, and problem solving. Impairments of semantic processing figure in a variety of brain disorders, such as Alzheimer's disease, semantic dementia, fluent aphasia, schizophrenia, and autism. The neural basis of semantic processing has been studied extensively by analyzing patterns of brain damage in such patients *(114–122)*. This topic has also been addressed in a large number of functional neuroimaging studies conducted on healthy volunteers, most of which have focused on the neuroanatomical representation of semantic memory [see *(121, 123–125)* for reviews]. Of greatest interest are those studies that have focused specifically on semantic processing by incorporating control tasks that make comparable demands on surface form (phonological or orthographic) processing and on general executive processes such as attention, working memory, and response production.

Results from these neuroimaging studies are remarkably consistent, revealing a very large, distributed network of brain regions underlying semantic memory storage and retrieval. Five major brain regions have been implicated: (1) the ventral temporal lobes, including middle and inferior temporal, anterior fusiform, and anterior parahippocampal gyri, (2) the angular gyrus, (3) the anterior aspect (pars orbitalis) of the inferior frontal gyrus (IFG), (4) dorsal prefrontal cortex, including the superior frontal and portions of the middle frontal gyrus, and (5) the posterior cingulate gyrus *(32, 44, 126–138)*. These are all regions characterized as supramodal, distant from primary sensory and motor areas, and therefore likely to be involved in processing conceptual (i.e., nonperceptual) information. Activation in these regions tends to be strongly left-lateralized, though most studies show at least some activation in homologous regions of the right hemisphere. Recent research suggests that the right hemisphere semantic system is involved primarily in processing concrete, imageable concepts and concrete objects, and plays little or no role in recognizing abstract words *(136, 138, 139)*.

These functional neuroimaging results are fully consistent with pathological data from patients with semantic disorders. For example, lesion localization studies in patients with transcortical sensory aphasia, a syndrome characterized by multimodal semantic impairment with intact phonological processing, implicate widely distributed regions of the left ventral temporal lobe and angular gyrus *(114, 140–142)*. Semantic dementia, a degenerative disorder characterized by gradual loss of semantic knowledge, is associated with progressive neuronal loss in the anterior and ventral temporal lobes bilaterally *(119, 143, 144)*. Other pathological conditions that affect the ventral temporal lobes, such as herpes encephalitis and Alzheimer disease, often produce focal semantic memory loss, particularly loss of knowledge about living things *(118, 145–148)*, while inferior parietal and posterior temporal lobe damage may produce selective loss of knowledge about human-made objects, particularly tools *(118, 123)*. Whereas these temporal and parietal lesions damage the semantic memory store itself, dorsal left prefrontal lesions seem to impair the ability to retrieve information from the semantic store. These latter lesions produce transcortical motor aphasia, a syndrome characterized by inability to initiate spontaneous speech *(141, 149)*.

2.5. Sentence Comprehension and Syntax Processing

The work reviewed so far focused on processing of single word structure and meaning, which can be thought of as the basic building blocks of language. Natural language, however, consists almost entirely of sentences. At least two phenomena distinguish processing at the sentence level from processing of single words. First, in sentence processing, the meanings of individual words are combined to create more complex and context-specific meanings. For example, consider the following:

1. The tigers lost their jungle habitat.
2. The tigers lost in extra innings.

It is the combination of words that specifies in each case the meaning of "tigers" and "lost." This process of *conceptual combination* is a central phenomenon in language production and comprehension that has been insufficiently emphasized. A second distinguishing feature of sentence processing is the use of syntactic information – word order, grammatical function words, and word inflections – to indicate the thematic roles played by constituent content words. In the two illustrative sentences given earlier, for example, "the" marks the beginning of a noun phrase, which can be followed by either a noun or a modifier phrase. The plural inflection of "tigers" then identifies this second word as a noun, which because of its position is likely to be the subject of the sentence, and so on.

One type of neuroimaging study used to examine these processes compares processing of sentences with word lists that do not form a sentence, the latter sometimes created simply by randomly rearranging the order of words in sentences ("scrambled sentences"). Common areas of activation in these contrasts (sentences vs. scrambled sentences) include the left anterior superior temporal lobe, left IFG, and left angular gyrus *(150–156)*. Debate has ensued over whether these activations represent syntactic or semantic processes, as sentences possess both syntactic structure and more complex meaning than lists of isolated words. This question has focused particularly on the anterior temporal lobe, a region often activated in studies using sentence materials but rarely in studies using isolated words. Humphries et al. *(156)* showed that activation in the anterior STS is modulated by the presence of syntactic structure independently of the meaningfulness of constituent content words, suggesting that this region may play a role in early parsing processes (e.g., lexical analysis and role assignment). These authors also examined the processing of combinatorial semantic structure in word lists and sentences by manipulating the degree to which words in the stimuli were thematically related. Remarkably, this contrast showed widespread regions in the ventral left temporal lobe, angular gyrus, and inferior frontal lobe that were activated when words were thematically related (and thus could be combined to form more complex and specific meanings) compared to when words were unrelated *(156, 157)*. This effect of combinatorial semantic structure was largely unaffected by whether the stimuli were syntactically correct sentences or word lists.

Many other fMRI and positron emission tomography (PET) studies have focused on specific syntactic operations, such as repair of syntactic and morphosyntactic violations *(158–163)* and comprehension of object-extracted relative clauses, passive voice, and other noncanonical or derived syntactic structures *(164–173)*. While these studies have generally implicated regions in the left IFG and left superior temporal lobe, the precise localization of specific syntactic operations remains a source of debate. Another ongoing discussion centers on whether these activations reflect operations specific to syntax processing or instead more domain-general working memory and executive processes *(164, 171, 174–177)*. The reader is referred to several excellent reviews that cover this work in detail *(178–181)*.

3. A Survey of Language Mapping Paradigms in Common Use

The variety of possible stimuli and tasks that could be used to induce language processing is vast, and a coherent, concise discussion is difficult. **Table 1** lists some of the broad categories of stimuli that have been used and some of the brain systems they engage. "Auditory nonspeech" refers to noises or tones that are not perceived as speech. Such stimuli can be variably "complex" in their temporal or spectral features and possess to varying degrees the acoustic properties of speech (*see* **refs.** *49, 182–185)*. "Auditory phonemes" are speech sounds that do not comprise words in the listener's language; these may be simple consonant–vowel monosyllables or longer sequences (e.g., pseudowords). "Visual nonletter" refers to any unfamiliar visual shape. Examples include characters from unfamiliar alphabets, nonsense shapes, and "false font." Such stimuli can be variably complex and possess to varying degrees the visual properties of familiar letters. "Visual letter strings" are random strings of letters that do not form familiar or easily pronounceable letter combinations (e.g., FCJVB). "Visual pseudowords" are letter strings that are not words but possess the orthographic and phonological characteristics of real words (e.g., SNADE).

The degree to which these stimuli engage the processes listed in **Table 1** depends partly on the task that the subject is asked to perform, though the processes in **Table 1** are activated

Table 1
Effects of auditory and visual stimuli on sensory and linguistic processing systems

Stimuli	Early sensory	Auditory word form	Visual word form	Object recognition	Syntax
Auditory nonspeech	Aud	–	–	–	–
Auditory phonemes	Aud	+	–	–	–
Auditory words	Aud	+	–	–	–
Auditory sentences	Aud	+	–	–	+
Visual nonletters	Vis	–	–	–	–
Visual letter strings	Vis	–	+/–	–	–
Visual pseudowords	Vis	–	+	–	–
Visual words	Vis	–	+	–	–
Visual sentences	Vis	–	+	–	+
Visual objects	Vis	–	–	+	–

"automatically" to some degree even when subjects are given no explicit task. This is less true for the processing systems listed in **Table 2**, which seem to be strongly task-dependent. The semantic system appears to be partly active even during "rest" or when stimuli are presented "passively" to the subject *(32, 33, 35, 37, 38, 186)*. Other tasks suppress semantic processing by requiring a focusing of attention on perceptual, orthographic, or phonological properties of stimuli *(32, 36, 37, 186)*. Examples include "sensory discrimination" tasks (e.g., intensity, size, color, frequency, and other discriminations based on physical features), "phonetic decision" tasks in which the subject must detect a target phoneme or phonemes, "phonological decision" tasks requiring a decision based on the phonological properties of a stimulus (e.g., detection of rhymes, judgment of syllable number), and "orthographic decision" tasks requiring a decision based on the letters in the stimulus (e.g., case matching, letter identification). Other tasks, such as reading and repeating, make no overt demands on semantic systems but may elicit automatic semantic processing. The extent to which this occurs probably depends on how meaningful the stimulus is: sentences likely elicit more semantic processing than isolated words, which in turn elicit more than pseudowords. Finally, many tasks make overt demands on retrieval and use of semantic knowledge. These include "semantic decision" tasks requiring a decision based on the meaning of the stimulus (e.g., "Is it living or non-living?"), "word generation" tasks requiring retrieval of a word or series of words related in

Table 2
Effects of task states on some linguistic processing systems

Tasks	Semantics	Phonological access	Speech articulation	Working memory	Other language
Rest or "passive"	+	–	–	–	–
Sensory discrimination	–	–	–	+/–	–
Read or repeat covert	+	+	–	+/–	–
Read or repeat overt	+	+	+	+/–	–
Phonetic decision	–	+	–	+	–
Phonological decision	–	+	–	+	–
Orthographic decision	–	+/–	–	–	–
Semantic decision	+	+/–	–	+	Semantic search
Word generation covert	+	+	–	+	Lexical search
Word generation overt	+	+	+	+	Lexical search
Naming covert	+	+	–	–	Lexical search
Naming overt	+	+	+	–	Lexical search

meaning to a cue word, and "naming" tasks requiring retrieval of a verbal label for an object or object description.

As noted earlier, "phonological access" refers to the processes engaged in retrieving a phonological (sound-based) representation of a word (or pseudoword). In addition to speech output and phonological tasks, any task using printed words, including orthographic and semantic tasks, will be accompanied to some degree by obligatory phonological access *(8, 15, 95)*. In contrast, "speech articulation" processes are engaged fully only when an overt spoken response is produced *(105)*. Verbal "working memory" is required whenever a written or spoken stimulus must be held in memory. Some degree of short-term phonological memory is needed for most language tasks, and in particular in cases where the stimulus is relatively long (i.e., sentences more than single words) or has multiple components, or must be held in memory while a response is generated (e.g., word generation tasks involving multiple responses for each cue). Finally, semantic decision, word generation, and naming tasks make strong demands on frontal mechanisms involved in searching for and retrieving information associated with a stimulus *(104, 187, 188)*.

With these somewhat oversimplified stimulus and task characterizations in mind, it is possible to make some general predictions about the processing systems whose level of activation

Table 3
Some task contrasts used for language mapping and the regions in which robust activations are typically observed

	Ventrolateral Prefrontal	Dorsal Prefrontal	Superior Temporal	Ventrolateral Temporal	Ventral Occipital	Angular Gyrus
Hearing words vs. rest			B			
Hearing words vs. nonspeech sounds			L > R			
Word generation vs. rest	L > R			L > R	B	
Word generation vs. reading	L					
Object naming vs. rest	B			L > R	B	
Semantic decision vs. sensory discrimination	L	L	L > R	L		L
Semantic decision vs. phonological decision		L		L		L
Reading sentences vs. letter strings	L > R		L > R	L > R		

L left hemisphere, *R* right hemisphere, *B* bilateral

will differ when two task conditions are contrasted, and thus the likely pattern of brain activation that will be observed in a simple subtraction analysis. Some commonly encountered examples are listed here and in **Table 3**.

Paradigm 1

Language task: Passively listening to words or sentences

Control task: Rest

As shown in **Table 1**, auditory words activate early auditory cortices and auditory word form areas. Since both rest and passive stimulation are accompanied by spontaneous semantic processes and make no other overt cognitive demands, no other language-related activation should appear in this contrast. The resulting activation pattern involves mainly auditory cortex in the STG bilaterally **(Fig. 1a)** *(49, 150, 186, 189, 190)*. The magnitude and extent of this activation increase with rate of word presentation *(191, 192)*. This STG activation is relatively symmetrical and is not correlated with language dominance as measured by Wada testing *(193)*. Although many authors have equated this STG activation with "Wernicke's area for receptive language," most of this activation represents early auditory processing rather than language-specific processes per se.

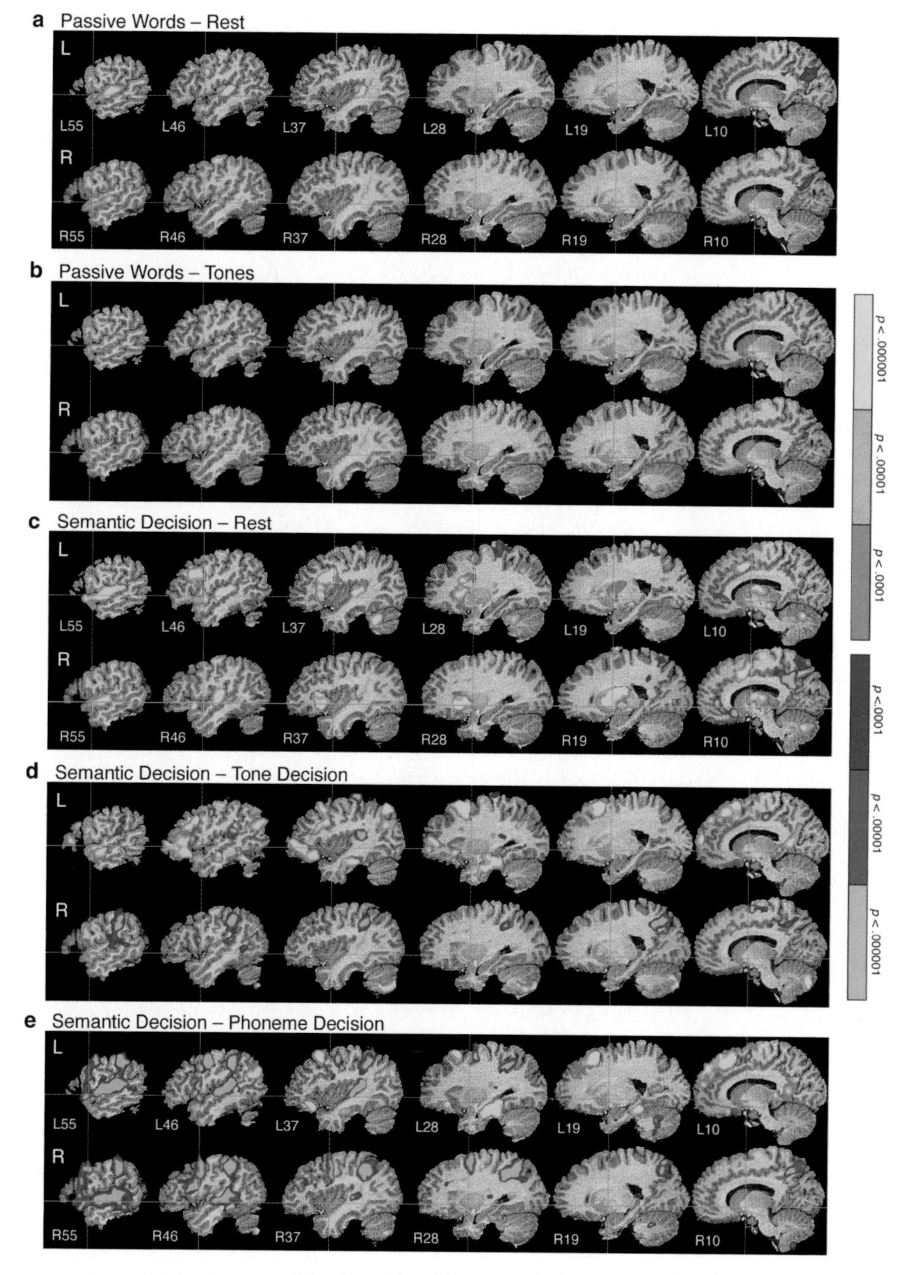

Fig. 1. Group average fMRI activation patterns in 26 neurologically normal, right-handed volunteers during five fMRI language paradigms (*see* **ref. *186*** for details). Auditory word and tone stimuli were equivalent in each of the five paradigms. (**a**) Passive listening to words contrasted with resting. Superior temporal activation occurs bilaterally. (**b**) Passive listening to words contrasted with passive listening to tones. A small region in the left superior temporal sulcus (STS) shows activation specifically related to speech processing. (**c**) Semantic decision on words contrasted with resting. Activation occurs in bilateral auditory (superior temporal gyrus, STG) and attentional/working memory (dorsolateral prefrontal, anterior cingulate, anterior insula, intraparietal sulcus, and subcortical) networks, with left lateralization in the inferior frontal gyrus (IFG). (**d**) Semantic decision on words contrasted with a tone decision task. Activation is strongly left-lateralized in prefrontal, lateral, and ventral temporal, angular, and posterior cingulate cortices. (**e**) Semantic decision on words contrasted with a phoneme decision task on pseudowords. Activation is strongly left-lateralized in dorsal prefrontal, angular, ventral temporal, and posterior cingulate cortices. Data are displayed as serial sagittal sections through the brain at 9-mm intervals. *X*-axis locations for each slice are given in the *top panel*. *Green lines* indicate the stereotaxic *Y* and *Z* origin planes. Hot colors (*red-yellow*) indicate positive activations and cold colors (*blue-cyan*) indicate negative activations for each contrast. All maps are thresholded at a whole-brain corrected $P < 0.05$ using voxel-wise $P < 0.0001$ and cluster extent >200 mm^3.

Paradigm 2

Language task: Passively listening to words or sentences

Control task: Passively listening to auditory nonspeech

Because there are no differences in task requirements, and because semantic processing occurs in all passive conditions, the activation pattern associated with this contrast mainly reflects activation of auditory word forms (**Table 1**). As mentioned earlier, studies employing such contrasts reliably show activation in the STS, with leftward lateralization, and little or no activation elsewhere (**Fig.1b**) *(45, 47–49, 182, 186)*. When sentences are used, this STS activation extends more anteriorly into the dorsal temporal pole region, possibly reflecting early syntactic parsing processes *(150, 155, 156, 182, 194–197)*.

Paradigm 3

Language task: Word generation

Control task: Rest

Because the rest state includes no control for sensory processing, early auditory or visual cortices may be activated bilaterally depending on the sensory modality of the cue stimulus (**Table 1**). The strength of this sensory activation depends, of course, on the rate of stimulus presentation: in some protocols, a single cue (e.g., a letter or a semantic category) is provided only at the beginning of an activation period; in others, a new cue is provided every few seconds. Unlike rest, word generation makes demands on phonological access, working memory, and lexical search systems (**Table 2**). Speech articulation systems will also be activated if an overt spoken response is required. These predictions are confirmed by many studies employing this contrast, which results primarily in activation of the left IFG and left > right premotor cortex, systems thought to be involved in phonological production, working memory, and lexical search *(104, 189, 193, 198–203)*. There may be activation of left posterior temporal regions (posterior MTG and STG) due to engagement of the phonological access system *(103, 105, 204)*.

Paradigm 4

Language task: Word generation

Control task: Reading or repeating

Here, we assume that the same stimulus modality (auditory or visual) is used for both tasks. The stimuli in both cases are single words, thus no difference in activation of sensory or word form systems is expected. Both tasks are accompanied by semantic processing (automatic semantic access in the case of the control task, effortful semantic retrieval in the case of word generation) and by phonological access processes. The word generation task makes greater demands on lexical search and on working memory; consequently greater activation is expected in left inferior frontal areas associated with these processes. These predictions match findings in many studies using this contrast, which show primarily left-lateralized activation in the IFG *(104, 188, 204, 205)*.

Paradigm 5

Language task: Visual object naming

Control task: Rest

Compared to resting, visual object perception activates early visual sensory cortices and higher-level object recognition systems bilaterally **(Table 1)** *(206–208)*. There may be additional, left-lateralized activation in semantic systems of the ventrolateral posterior temporal lobe *(209–213)*. Unlike resting, naming requires lexical search and phonological access, and, when overt, speech articulation **(Table 2)**. These predictions match findings in several studies using this contrast, which show extensive bilateral visual system activation and modest left lateralized inferior frontal activation *(201, 213, 214)*.

Paradigm 6

Language task: Semantic decision

Control task: Sensory discrimination

We again assume that the same stimulus modality is used for both tasks. If the stimuli in the sensory discrimination task are nonlinguistic (e.g., tones or nonsense shapes), then the semantic decision task will produce relatively greater activation in auditory or visual word form systems, depending on the sensory modality. In addition, there will be greater activation of semantic memory and semantic search mechanisms in the semantic decision task. Note that unlike the resting and passive control tasks used in the protocols described so far, effortful sensory discrimination tasks interrupt ongoing semantic processes, providing a control state that is relatively free of conceptual or semantic processing *(32, 35, 37, 38, 186)*. Working memory systems may or may not be activated in this contrast, depending on whether or not the control task also has a working memory component. These predictions match findings in studies using this contrast, which show left lateralized activation of auditory (middle and anterior STS) or visual (mid-fusiform gyrus) word form regions, and extensive activation of left prefrontal, lateral and ventral left temporal, and left posterior parietal systems involved in semantic memory and semantic access **(Fig.1d)** *(5, 44, 127, 186, 215–217)*.

Paradigm 7

Language task: Semantic decision

Control task: Phonological decision

These tasks can also be given in either the visual or auditory modality. Stimuli in the phonological decision task can be either words or pseudowords, and these can be matched to the words used in the semantic task on all structural (physical, orthographic, phonological) variables. Thus, there should be no activation of sensory or word form systems in this contrast. There will be greater activation of semantic memory and semantic search systems in the semantic decision task. These predictions match findings in many studies using this contrast, which show activation of left prefrontal, lateral and ventral left temporal, and left posterior parietal systems believed to be involved in semantic processing (**Fig. 1e**) *(32, 44, 128–130, 132, 134, 186, 218–220)*.

Paradigm 8

Language task: Sentence or word reading

Control task: Passively viewing letter strings

Compared to letter strings, sentences engage visual word form, syntactic, and phonological access systems, and make variable demands on working memory. Both reading and passive viewing probably involve semantic processing. There should be left-lateralized activation of the fusiform gyrus (visual word form system), posterior STG and STS (phonological access), and IFG (orthographic-phonological mapping, working memory, syntax). These predictions are consistent with several studies using this contrast *(95–97, 221, 222)*.

In many clinical settings, the main goal of language mapping is simply to identify as many language-related areas as possible and to assess hemispheric lateralization of language. A review of **Table 3** suggests that the "semantic decision vs. sensory discrimination" paradigm may offer advantages for this purpose in terms of the sheer number of regions activated and leftward lateralization of activation. My colleagues and I put this prediction to a quantitative test by comparing the extent and lateralization of activation produced by five language-related task contrasts, conducted on the same 26 participants during a single scanning session *(186)*. These contrasts included (1) passively listening to words vs. resting, (2) passively listening to words vs. passively listening to tones, (3) performing a semantic decision task with words vs. resting, (4) performing a semantic decision task with words vs. a sensory discrimination task with tones, and (5) performing a semantic decision task with words vs. a phonological task with pseudowords. As shown in **Fig. 2**, the semantic decision-tone decision contrast produced by far the largest activation volume in the left hemisphere, as well as an optimal combination of extensive activation and strong left-lateralization.

The example paradigms discussed here cover but a small sample of all possible language activation protocols. There are also numerous published studies employing designs that do not fit neatly into the schema provided here. Many of these represent

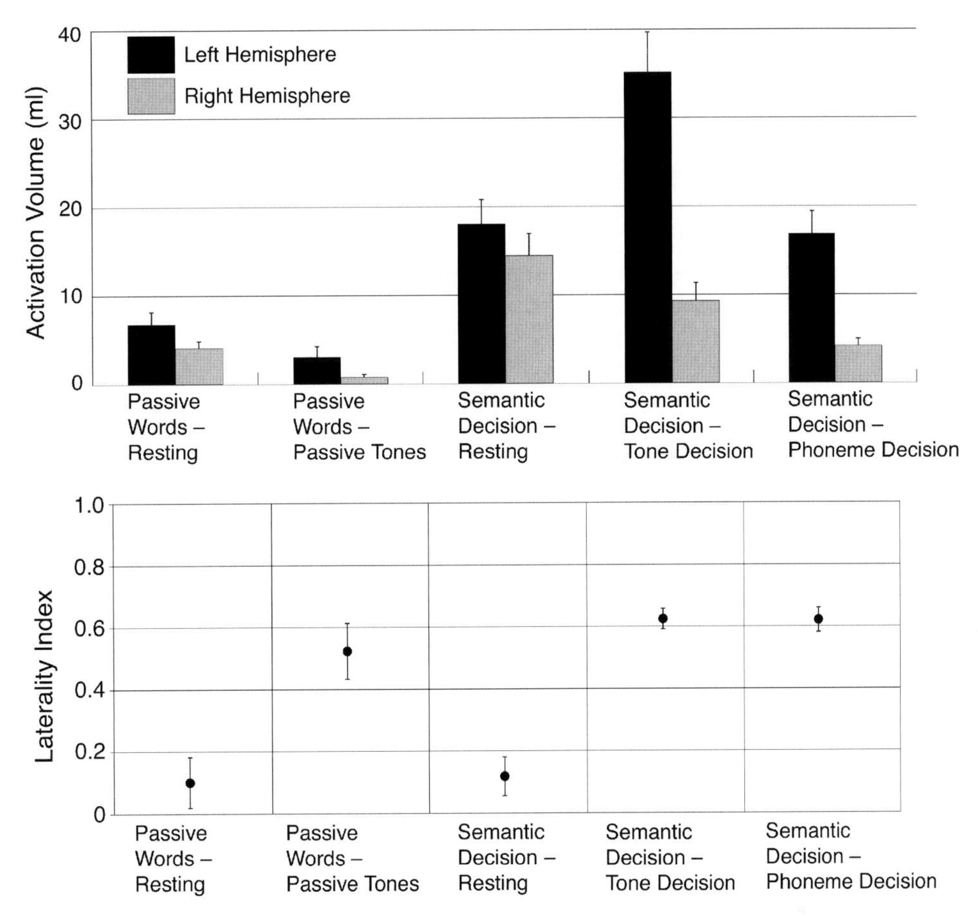

Fig. 2. Group average activation volumes (*top graph*) and laterality indexes (*bottom graph*) for five fMRI language paradigms ***(186)***. Laterality indexes can vary from −1 (all activation in the right hemisphere) to +1 (all activation in the left hemisphere). Error bars represent standard error. The semantic decision–tone decision paradigm produces the greatest left hemisphere activation as well as a strongly left-lateralized pattern.

attempts to further define or fractionate a particular language process, or to define further the functional role of a specific brain region. The reader should appreciate that the review given here is merely a coarse outline of some of the most commonly used types of stimuli and tasks. Above all, it is important to note that activations in a particular part of the language system are seldom "all or none," but vary in a graded way depending on the particular stimuli and tasks used.

4. Conclusions and Future Directions

Functional neuroimaging techniques have enhanced profoundly our understanding of how language processes are implemented in the human brain. This work has led to a number of new discoveries, such as a more precise localization of cortical networks underlying phoneme and grapheme perception, phonological access, and semantic processing. Not all of the claims made here with regard to these component language systems are uncontroversial. In particular, there are ongoing debates and a number of unresolved issues concerning localization of semantic memory and semantic retrieval systems, especially with regard to the role played by the left IFG in semantic processes *(132, 187, 188, 223–226)*. The notion that semantic processes are actively engaged during the resting state, though gaining traction in some quarters, is far from universal acceptance or recognition. For example, many authors continue to regard with suspicion any activation associated with semantic tasks that is not also observed in comparison to a resting baseline *(133, 227, 228)*. As in other areas of cognitive neuroscience, neuroimaging research on language processing has been to some extent clouded by an incomplete understanding of task demands and inadequate recognition of potential confounding factors. For example, many task contrasts are confounded by differences in task difficulty, which are well known to cause differential activation of domain-general networks involved in arousal, attention, working memory, decision, response selection, and error monitoring *(137, 229–235)*. Despite these well-documented effects, many researchers continue to advocate the use of covert tasks and passive conditions that provide no information about task performance, level of attention, or degree of difficulty. These deficiencies are particularly troubling in clinical studies, where the interpretation of brain activation (or lack of activation) can substantially influence clinical decision-making and patient outcome.

As the field of functional neuroimaging continues to mature, it is likely that these potential pitfalls will eventually be universally recognized and that an increasingly standardized methodological and conceptual framework for language mapping studies will emerge. FMRI practitioners, whether working in clinical or research fields, should continue to strive toward these goals.

Acknowledgments

My thanks to Lisa Conant, Rutvik Desai, Thomas Hammeke, Colin Humphries, Einat Liebenthal, Merav Sabri, David Sabsevitz, Sara Swanson and other colleagues in the Language Imaging Laboratory at MCW for many clarifying and thought-provoking discussions on the issues discussed in this chapter.

References

1. Broca P. Remarques sur le siège de la faculté du langage articulé; suivies d'une observation d'aphemie. Bulletin de la Société Anatomique de Paris 1861;6:330–57.
2. Wernicke C. Der aphasische Symptomenkomplex. Breslau: Cohn & Weigert; 1874.
3. Lichtheim L. On aphasia. Brain 1885;7:433–84.
4. Geschwind N. Aphasia. N Eng J Med 1971;284:654–6.
5. Binder JR, Frost JA, Hammeke TA, Cox RW, Rao SM, Prieto T. Human brain language areas identified by functional MRI. J Neurosci 1997;17:353–62.
6. Démonet J-F, Thierry G, Cardebat D. Renewal of the neurophysiology of language: Functional neuroimaging. Physiol Rev 2005;85:49–95.
7. Binder JR, Price CJ. Functional imaging of language. In: Cabeza R, Kingstone A, eds. Handbook of functional neuroimaging of cognition. Cambridge, MA: MIT Press; 2001:187–251.
8. Macleod CM. Half a century of research on the Stroop effect: An integrative review. Psychol Bull 1991;109:163–203.
9. Reicher GM. Perceptual recognition as a function of meaningfulness of stimulus material. J Exp Psychol 1969;81:274–80.
10. Warren RM, Obusek CJ. Speech perception and phonemic restorations. Percept Psychophys 1971;9:358–62.
11. Ganong WF. Phonetic categorization in auditory word perception. J Exp Psychol Hum Percept Perform 1980;6:110–5.
12. Marslen-Wilson WD, Tyler LK. Central processes in speech understanding. Phil Trans Royal Soc London B 1981;295:317–32.
13. Carr TH, McCauley C, Sperber RD, Parmalee CM. Words, pictures, and priming: On semantic activation, conscious identification, and the automaticity of information processing. J Exp Psychol Hum Percept Perform 1982;8:757–77.
14. Marcel AJ. Conscious and unconscious perception: Experiments on visual masking and word recognition. Cognitive Psychol 1983;15: 197–237.
15. Van Orden GC. A ROWS is a ROSE: Spelling, sound, and reading. Mem Cognit 1987;15: 181–98.
16. Burton MW, Baum SR, Blumstein SE. Lexical effects on phonetic categorization of speech: The role of acoustic structure. J Exp Psychol Hum Percept Perform 1989;15:567–75.
17. Glaser WR. Picture naming. Cognition 1992;42:61–105.
18. James W. Principles of psychology. New York: Dover; 1890.
19. Hebb DO. The problem of consciousness and introspection. In: Adrian ED, Bremer F, Jasper HH, eds. Brain mechanisms and consciousness: A symposium. Springfield, IL: Charles C. Thomas; 1954:402–21.
20. Miller GA, Galanter E, Pribram K. Plans and the structure of behavior. New York: Holt; 1960.
21. Pope KS, Singer JL. Regulation of the stream of consciousness: Toward a theory of ongoing thought. In: Schwartz GE, Shapiro D, eds. Consciousness and self-regulation. New York: Plenum; 1976:101–35.
22. Aurell CG. Perception: A model comprising two modes of consciousness. Percept Motor Skills 1979;49:431–44.
23. Picton TW, Stuss DT. Neurobiology of conscious experience. Curr Opin Neurobiol 1994;4:256–65.
24. Antrobus JS, Singer JL, Greenberg S. Studies in the stream of consciousness: Experimental enhancement and suppression of spontaneous cognitive processes. Percept Motor Skills 1966;23:399–417.
25. Teasdale JD, Proctor L, Lloyd CA, Baddeley AD. Working memory and stimulus-independent thought: Effects of memory load and presentation rate. Eur J Cogn Psychol 1993;5:417–33.
26. Révész G, ed. Thinking and speaking: A symposium. Amsterdam: North Holland Publishing; 1954.
27. Weiskrantz L, ed. Thought without language. Oxford: Clarendon; 1988.
28. Vygotsky LS. Thought and language. New York: Wiley; 1962.
29. Karmiloff-Smith A. Beyond modularity: A developmental perspective on cognitive science. Cambridge, MA: MIT Press; 1992.
30. Andreasen NC, O'Leary DS, Cizadlo T, et al. Remembering the past: Two facets of episodic memory explored with positron emission tomography. Am J Psychiatry 1995;152:1576–85.
31. Shulman GL, Fiez JA, Corbetta M, et al. Common blood flow changes across visual tasks. II. Decreases in cerebral cortex. J Cogn Neurosci 1997;9:648–63.
32. Binder JR, Frost JA, Hammeke TA, Bellgowan PSF, Rao SM, Cox RW. Conceptual processing during the conscious resting state:

A functional MRI study. J Cogn Neurosci 1999;11:80–93.

33. Mazoyer B, Zago L, Mellet E, et al. Cortical networks for working memory and executive functions sustain the conscious resting state in man. Brain Res Bull 2001;54:287–98.
34. Raichle ME, McLeod AM, Snyder AZ, Powers WJ, Gusnard DA, Shulman GL. A default mode of brain function. Proc Natl Acad Sci USA 2001;98:676–82.
35. Stark CE, Squire LR. When zero is not zero: The problem of ambiguous baseline conditions in fMRI. Proc Natl Acad Sci USA 2001;98:12760–6.
36. McKiernan KA, Kaufman JN, Kucera-Thompson J, Binder JR. A parametric manipulation of factors affecting task-induced deactivation in functional neuroimaging. J Cogn Neurosci 2003;15:394–408.
37. McKiernan KA, D'Angelo BR, Kaufman JN, Binder JR. Interrupting the "stream of consciousness": An fMRI investigation. Neuroimage 2006;29:1185–91.
38. Mason MF, Norton MI, Van Horn JD, Wegner DM, Grafton ST, Macrae CN. Wandering minds: The default network and stimulus-independent thought. Science 2007;315:393–5.
39. Abutalebi J, Cappa S, Perani D. What can functional neuroimaging tell us about the bilingual brain? In: Kroll J, de Groot AMB, eds. Handbook of bilingualism: Psycholinguistic approaches. New York: Oxford University Press; 2005:497–515.
40. Hernandez A, Li P, MacWhinney B. The emergence of competing modules in bilingualism. Trends Cogn Sci 2005;9:220–5.
41. Corina DP, Knapp H. Sign language processing and the mirror neuron system. Cortex 2006;42:529–39.
42. Campbell R, MacSweeney, Waters D. Sign language and the brain: A review. J Deaf Studies Deaf Educ 2008;13:3–20.
43. Bogen JE, Bogen GM. Wernicke's region – where is it? Ann NY Acad Sci 1976;290:834–43.
44. Démonet J-F, Chollet F, Ramsay S, et al. The anatomy of phonological and semantic processing in normal subjects. Brain 1992;115:1753–68.
45. Zatorre RJ, Evans AC, Meyer E, Gjedde A. Lateralization of phonetic and pitch discrimination in speech processing. Science 1992;256:846–9.
46. Binder JR, Frost JA, Hammeke TA, Rao SM, Cox RW. Function of the left planum temporale in auditory and linguistic processing. Brain 1996;119:1239–47.
47. Mummery CJ, Ashburner J, Scott SK, Wise RJS. Functional neuroimaging of speech perception in six normal and two aphasic subjects. J Acoust Soc Am 1999;106:449–57.
48. Belin P, Zatorre RJ, Lafaille P, Ahad P, Pike B. Voice-selective areas in human auditory cortex. Nature 2000;403:309–12.
49. Binder JR, Frost JA, Hammeke TA, et al. Human temporal lobe activation by speech and nonspeech sounds. Cereb Cortex 2000;10:512–28.
50. Desai R, Liebenthal E, Possing ET, Waldron E, Binder JR. Volumetric vs. surface-based alignment for localization of auditory cortex activation. Neuroimage 2005;26:1019–29.
51. Dehaene-Lambertz G, Pallier C, Serniclaes W, Sprenger-Charolles L, Jobert A, Dehaene S. Neural correlates of switching from auditory to speech perception. Neuroimage 2005;24:21–33.
52. Liebenthal E, Binder JR, Spitzer SM, Possing ET, Medler DA. Neural substrates of phonetic perception. Cereb Cortex 2005;15:1621–31.
53. Benson RR, Richardson M, Whalen DH, Lai S. Phonetic processing areas revealed by sinewave speech and acoustically similar nonspeech. Neuroimage 2006;31:342–53.
54. Mottonen R, Calvert GA, Jaaskelainen IP, et al. Perceiving identical sounds as speech or non-speech modulates activity in the left posterior superior temporal sulcus. Neuroimage 2006;30:563–9.
55. Barrett AM. A case of pure word-deafness with autopsy. J Nerv Ment Dis 1910;37:73–92.
56. Henschen SE. On the hearing sphere. Acta Otolaryngol 1918–1919;1:423–86.
57. Wohlfart G, Lindgren A, Jernelius B. Clinical picture and morbid anatomy in a case of "pure word deafness." J Nerv Ment Dis 1952;116:818–27.
58. Lhermitte F, Chain F, Escourolle R, Ducarne B, Pillon A, Chedru F. Etude des troubles perceptifs auditifs dans les lésions temporales bilatérales. Rev Neurol 1972;24:327–51.
59. Kanshepolsky J, Kelley JJ, Waggener JD. A cortical auditory disorder: Clinical, audiologic and pathologic aspects. Neurology 1973;23:699–705.
60. Buchman AS, Garron DC, Trost-Cardamone JE, Wichter MD, Schwartz D. Word deafness: One hundred years later. J Neurol Neurosurg Psychiatry 1986;49:489–99.
61. Poeppel D. Pure word deafness and the bilateral processing of the speech code. Cogn Sci 2001;25:679–93.
62. Liepmann H, Storch E. Der mikroskopische Gehirnbefund bei dem Fall Gorstelle. Monatsschr Psychiatr Neurol 1902;11:115–20.

63. Stefanatos GA, Gershkoff A, Madigan S. On pure word deafness, temporal processing and the left hemisphere. J Int Neuropsychol Soc 2005;11:456–70.
64. Déjerine J. Sur un cas de cécité verbal avec agraphie, suivi d'autopsie. Comptes Rendus des Séances de la Société de Biologie 1891;3: 197–201.
65. Déjerine J. Contribution à l'étude anatomo-pathologique et clinique des différentes variétés de cécité verbale. Comptes Rendus des Séances de la Société de Biologie 1892;44:61–90.
66. Déjerine J, Vialet N. Contribution a l'étude de la localisation anatomique de la cécité verbale pure. Comptes Rendus des Séances de la Société de Biologie 1893;45:790–3.
67. Geschwind N. Disconnection syndromes in animals and man. Brain 1965;88:237–94, 585–644.
68. Greenblatt SH. Subangular alexia without agraphia or hemianopsia. Brain Lang 1976;3:229–45.
69. Vincent FM, Sadowsky CH, et al. Alexia without agraphia, hemianopia, or color-naming defect: A disconnection syndrome. Neurology 1977;27:689–91.
70. Henderson VW. Anatomy of posterior pathways in reading: A reassessment. Brain Lang 1986;29:119–33.
71. Binder JR, Mohr JP. The topography of transcallosal reading pathways: A case-control analysis. Brain 1992;115:1807–26.
72. Beversdorf DQ, Ratcliffe NR, Rhodes CH, Reeves AG. Pure alexia: Clinical-pathological evidence for a lateralized visual language association cortex. Clin Neuropathol 1997;16:328–31.
73. Sakurai Y, Takeuchi S, Takada T, Horiuchi E, Nakase H, Sakuta M. Alexia caused by a fusiform or posterior inferior temporal lesion. J Neurol Sci 2000;178:42–51.
74. Leff AP, Crewes H, Plant GT, Scott SK, Kennard C, Wise RJS. The functional anatomy of single-word reading in patients with hemianopic and pure alexia. Brain 2001;124:510–21.
75. Cohen L, Martinaud O, Lemer C, et al. Visual word recognition in the left and right hemispheres: Anatomical and functional correlates of peripheral alexias. Cereb Cortex 2003;13:1313–33.
76. Tarkiainen A, Helenius P, Hansen PC, Cornelissen PL, Salmelin R. Dynamics of letter string perception in the human occipitotemporal cortex. Brain 1999;122:2119–31.
77. Cohen L, Dehaene S, Naccache L, et al. The visual word form area. Spatial and temporal characterization of an initial stage of reading in normal subjects and posterior split-brain patients. Brain 2000;123:291–307.
78. Dehaene S, Naccache L, Cohen L, et al. Cerebral mechanisms of word masking and unconscious repetition priming. Nat Neurosci 2001;4:752–8.
79. Cohen L, Lehéricy S, Chochon F, Lemer C, Rivaud S, Dehaene S. Language-specific tuning of visual cortex? Functional properties of the visual word form area. Brain 2002;125:1054–69.
80. Polk TA, Farah MJ. Functional MRI evidence for an abstract, not perceptual, word-form area. J Exp Psychol Gen 2002;131:65–72.
81. Binder JR, Medler DA, Westbury CF, Liebenthal E, Buchanan L. Tuning of the human left fusiform gyrus to sublexical orthographic structure. Neuroimage 2006;33:739–48.
82. Patterson KE, Kay J. Letter-by-letter reading: Psychological descriptions of a neurological syndrome. Q J Exp Psychol A 1982;34: 411–42.
83. Reuter-Lorenz PA, Brunn JL. A prelexical basis for letter-by-letter reading: A case study. Cogn Neuropsychiatry 1990;7:1–20.
84. Behrmann M, Plaut DC, Nelson J. A literature review and new data supporting an interactive activation account of letter-by-letter reading. In: Coltheart M, ed. Pure alexia (letter-by-letter reading). Hove, UK: Pscyhology Press; 1998:7–51.
85. Liepmann H, Pappenheim M. Über einem Fall von sogenannter Leitungsaphasie mit anatomischer Befund. Z Gesamte Neurol Psychiatr 1914;27:1–41.
86. Benson DF, Sheremata WA, Bouchard R, Segarra JM, Price D, Geschwind N. Conduction aphasia. A clinicopathological study. Arch Neurol 1973;28:339–46.
87. Damasio H, Damasio AR. The anatomical basis of conduction aphasia. Brain 1980;103:337–50.
88. Palumbo CL, Alexander MP, Naeser MA. CT scan lesion sites associated with conduction aphasia. In: Kohn SE, ed. Conduction aphasia. Hillsdale, NJ: Lawrence Erlbaum; 1992:51–75.
89. Anderson JM, Gilmore R, Roper S, et al. Conduction aphasia and the arcuate fasciculus: A reexamination of the Wernicke–Geschwind model. Brain Lang 1999;70:1–12.
90. Quigg M, Fountain NB. Conduction aphasia elicited by stimulation of the left posterior superior temporal gyrus. J Neurol Neurosurg Psychiatry 1999;66:393–6.
91. Roeltgen DP, Sevush S, Heilman KM. Phonological agraphia: Writing by the lexical-semantic route. Neurology 1983;33:755–65.

92. Alexander MP, Friedman RB, Loverso F, Fischer RS. Lesion localization of phonological agraphia. Brain Lang 1992;43:83–95.
93. Howard D, Patterson K, Wise R, et al. The cortical localization of the lexicons. Brain 1992;115:1769–82.
94. Price CJ, Wise RJS, Watson JDG, Patterson K, Howard D, Frackowiak RSJ. Brain activity during reading. The effects of exposure duration and task. Brain 1994;117:1255–69.
95. Price CJ, Wise RSJ, Frackowiak RSJ. Demonstrating the implicit processing of visually presented words and pseudowords. Cereb Cortex 1996;6:62–70.
96. Bavelier D, Corina D, Jezzard P, et al. Sentence reading: A functional MRI study at 4 tesla. J Cogn Neurosci 1997;9:664–86.
97. Indefrey P, Kleinschmidt A, Merboldt K-D, et al. Equivalent responses to lexical and nonlexical visual stimuli in occipital cortex: A functional magnetic resonance imaging study. Neuroimage 1997;5:78–81.
98. Tagamets M-A, Novick JM, Chalmers ML, Friedman RB. A parametric approach to orthographic processing in the brain: An fMRI study. J Cogn Neurosci 2000;12:281–97.
99. Hickok G, Erhard P, Kassubek J, et al. A functional magnetic resonance imaging study of the role of left posterior superior temporal gyrus in speech production: Implications for the explanation of conduction aphasia. Neurosci Lett 2000;287:156–60.
100. Hickok G, Buchsbaum B, Humphries C, Muftuler T. Auditory-motor interaction revealed by fMRI: Speech, music, and working memory in area Spt. J Cogn Neurosci 2003;15:673–82.
101. Indefrey P, Levelt WJM. The spatial and temporal signatures of word production components. Cognition 2004;92:101–44.
102. Callan AM, Callan DE, Masaki S. When meaningless symbols become letters: Neural activity change in learning new phonograms. Neuroimage 2005;28:553–62.
103. Fiez JA, Raichle ME, Balota DA, Tallal P, Petersen SE. PET activation of posterior temporal regions during auditory word presentation and verb generation. Cereb Cortex 1996;6:1–10.
104. Warburton E, Wise RJS, Price CJ, et al. Noun and verb retrieval by normal subjects. Studies with PET. Brain 1996;119:159–79.
105. Wise RSJ, Scott SK, Blank SC, Mummery CJ, Murphy K, Warburton EA. Separate neural subsystems within 'Wernicke's area'. Brain 2001;124:83–95.
106. Baddeley AD. Working memory. Oxford: Oxford University Press; 1986.
107. Paulesu E, Frith CD, Frackowiak RSJ. The neural correlates of the verbal component of working memory. Nature 1993;362:342–5.
108. Salmon E, Van der Linden M, Collette F, et al. Regional brain activity during working memory tasks. Brain 1996;119:1617–25.
109. Postle BR, Berger JS, D'Esposito M. Functional and neuroanatomical double dissociation of mnemonic and executive control processes contributing to working memory performance. Proc Natl Acad Sci USA 1999;96:12959–64.
110. Buchsbaum B, Hickok G, Humphries C. Role of left posterior superior temporal gyrus in phonological processing for speech perception and production. Cogn Sci 2001;25: 663–78.
111. Buchsbaum BR, Olsen RK, Koch P, Berman KF. Human dorsal and ventral auditory streams subserve rehearsal-based and echoic processes during verbal working memory. Neuron 2005;48:687–97.
112. Hickok G, Poeppel D. Towards a functional neuroanatomy of speech perception. Trends Cogn Sci 2000;4:131–8.
113. Bréal M. Essai de sémantique (science des significations). Paris: Librairie Hachette; 1897.
114. Alexander MP, Hiltbrunner B, Fischer RS. Distributed anatomy of transcortical sensory aphasia. Arch Neurol 1989;46:885–92.
115. Hart J, Gordon B. Delineation of single-word semantic comprehension deficits in aphasia, with anatomic correlation. Ann Neurol 1990;27:226–31.
116. Chertkow H, Bub D, Deaudon C, Whitehead V. On the status of object concepts in aphasia. Brain Lang 1997;58:203–32.
117. Tranel D, Damasio H, Damasio AR. A neural basis for the retrieval of conceptual knowledge. Neuropsychologia 1997;35:1319–27.
118. Gainotti G. What the locus of brain lesion tells us about the nature of the cognitive defect underlying category-specific disorders: A review. Cortex 2000;36:539–59.
119. Mummery CJ, Patterson K, Price CJ, Ashburner J, Frackowiak RS, Hodges JR. A voxel-based morphometry study of semantic dementia: Relationship between temporal lobe atrophy and semantic memory. Ann Neurol 2000;47:36–45.
120. Hillis AE, Wityk RJ, Tuffiash E, et al. Hypoperfusion of Wernicke's area predicts severity of semantic deficit in acute stroke. Ann Neurol 2001;50:561–6.

121. Damasio H, Tranel D, Grabowski T, Adolphs R, Damasio A. Neural systems behind word and concept retrieval. Cognition 2004;92:179–229.
122. Dronkers NF, Wilkins DP, Van Valin RD, Redfern BB, Jaeger JJ. Lesion analysis of the brain areas involved in language comprehension. Cognition 2004;92:145–77.
123. Martin A, Chao LL. Semantic memory in the brain: Structure and processes. Curr Opin Neurobiol 2001;11:194–201.
124. Bookheimer SY. Functional MRI of language: New approaches to understanding the cortical organization of semantic processing. Annu Rev Neurosci 2002;25:151–88.
125. Thompson-Schill SL. Neuroimaging studies of semantic memory: Inferring "how" from "where." Neuropsychologia 2003;41:280–92.
126. Mummery CJ, Patterson K, Hodges JR, Wise RJS. Generating 'tiger' as an animal name or a word beginning with T: Differences in brain activation. Proc Royal Soc Lond B 1996;263:989–95.
127. Vandenberghe R, Price C, Wise R, Josephs O, Frackowiak RSJ. Functional anatomy of a common semantic system for words and pictures. Nature 1996;383:254–6.
128. Price CJ, Moore CJ, Humphreys GW, Wise RJS. Segregating semantic from phonological processes during reading. J Cogn Neurosci 1997;9:727–33.
129. Cappa SF, Perani D, Schnur T, Tettamanti M, Fazio F. The effects of semantic category and knowledge type on lexical-semantic access: A PET study. Neuroimage 1998;8:350–9.
130. Roskies AL, Fiez JA, Balota DA, Raichle ME, Petersen SE. Task-dependent modulation of regions in the left inferior frontal cortex during semantic processing. J Cogn Neurosci 2001;13:829–43.
131. Binder JR, McKiernan KA, Parsons M, et al. Neural correlates of lexical access during visual word recognition. J Cogn Neurosci 2003;15:372–93.
132. Devlin JT, Matthews PM, Rushworth MFS. Semantic processing in the left inferior prefrontal cortex: A combined functional magnetic resonance imaging and transcranial magnetic stimulation study. J Cogn Neurosci 2003;15:71–84.
133. Rissman J, Eliassen JC, Blumstein SE. An event-related fMRI investigation of implicit semantic priming. J Cogn Neurosci 2003;15:1160–75.
134. Scott SK, Leff AP, Wise RJS. Going beyond the information given: A neural system supporting semantic interpretation. Neuroimage 2003;19:870–6.
135. Ischebeck A, Indefrey P, Usui N, Nose I, Hellwig F, Taira M. Reading in a regular orthography: An fMRI study investigating the role of visual familiarity. J Cogn Neurosci 2004;16:727–41.
136. Binder JR, Westbury CF, Possing ET, McKiernan KA, Medler DA. Distinct brain systems for processing concrete and abstract concepts. J Cogn Neurosci 2005;17:905–17.
137. Binder JR, Medler DA, Desai R, Conant LL, Liebenthal E. Some neurophysiological constraints on models of word naming. Neuroimage 2005;27:677–93.
138. Sabsevitz DS, Medler DA, Seidenberg M, Binder JR. Modulation of the semantic system by word imageability. Neuroimage 2005;27:188–200.
139. Vandenbulcke M, Peeters R, Fannes K, Vandenberghe R. Knowledge of visual attributes in the right hemisphere. Nat Neurosci 2006;9:964–70.
140. Damasio H. Neuroimaging contributions to the understanding of aphasia. In: Boller F, Grafman J, eds. Handbook of neuropsychology. Amsterdam: Elsevier; 1989:3–46.
141. Rapcsak SZ, Rubens AB. Localization of lesions in transcortical aphasia. In: Kertesz A, ed. Localization and neuroimaging in neuropsychology. San Diego: Academic Press; 1994:297–329.
142. Berthier ML. Transcortical aphasias. Hove: Psychology Press; 1999.
143. Chan D, Fox NC, Scahill RI, et al. Patterns of temporal lobe atrophy in semantic dementia and Alzheimer's disease. Ann Neurol 2001;49:433–42.
144. Davies RR, Hodges JR, Krill JJ, Patterson K, Halliday GM, Xuereb JH. The pathological basis of semantic dementia. Brain 2005;128:1984–95.
145. Warrington EK, Shallice T. Category specific semantic impairments. Brain 1984;107: 829–54.
146. Fung TD, Chertkow H, Whatmough C, et al. The spectrum of category effects in object and action knowledge in dementia of the Alzheimer's type. Neuropsychology 2001;15:371–9.
147. Chan AS, Salmon DP, De La Pena J. Abnormal semantic network for "animals" but not "tools" in patients with Alzheimer's disease. Cortex 2001;37:197–217.
148. Gonnerman LM, Andersen ES, Devlin JT, Kempler D, Seidenberg MS. Double dissociation

of semantic categories in Alzheimer's disease. Brain Lang 1997;57:254–79.

149. Alexander MP, Benson DF, Stuss DT. Frontal lobes and language. Brain Lang 1989;37:656–91.
150. Mazoyer BM, Tzourio N, Frak V, et al. The cortical representation of speech. J Cogn Neurosci 1993;5:467–79.
151. Bottini G, Corcoran R, Sterzi R, et al. The role of the right hemisphere in the interpretation of figurative aspects of language. A positron emission tomography activation study. Brain 1994;117:1241–53.
152. Stowe LA, Paans AMJ, Wijers AA, Zwarts F, Mulder G, Vaalburg W. Sentence comprehension and word repetition: A positron emission tomography investigation. Psychophysiology 1999;36:786–801.
153. Friederici AD, Meyer M, von Cramon DY. Auditory language comprehension: An event-related fMRI study on the processing of syntactic and lexical information. Brain Lang 2000;74:289–300.
154. Vandenberghe R, Nobre AC, Price CJ. The response of left temporal cortex to sentences. J Cogn Neurosci 2002;14:550–60.
155. Humphries C, Swinney D, Love T, Hickok G. Response of anterior temporal cortex to syntactic and prosodic manipulations during sentence processing. Hum Brain Mapp 2005;26:128–38.
156. Humphries C, Binder JR, Medler DA, Liebenthal E. Syntactic and semantic modulation of neural activity during auditory sentence comprehension. J Cogn Neurosci 2006;18:665–79.
157. Humphries C, Binder JR, Medler DA, Liebenthal E. Time course of semantic processes during sentence comprehension: An fMRI study. Neuroimage 2007;36:924–32.
158. Kang AM, Constable RT, Gore JC, Avrutin S. An event-related fMRI study of implicit phrase-level syntactic and semantic processing. Neuroimage 1999;10:98–110.
159. Embick D, Marantz A, Miyashita Y, O'Neil W, Sakai KL. A syntactic specialization for Broca's area. Proc Natl Acad Sci USA 2000;97:6150–4.
160. Meyer M, Friederici AD, von Cramon DY. Neurocognition of auditory sentence comprehension: Event related fMRi reveals sensitivity to syntactic violations and task demands. Cogn Brain Res 2000;9:19–33.
161. Ni W, Constable RT, Mencl WE, et al. An event-related neuroimaging study distinguishing form and content in sentence processing. J Cogn Neurosci 2000;12:120–33.
162. Newman AJ, Pancheva R, Ozawa K, Neville HJ, Ullman MT. An event-related fMRI study of syntactic and semantic violations. J Psycholinguist Res 2001;30:339–64.
163. Kuperberg GR, Holcomb PJ, Sitnikova T, Greve D, Dale AM, Caplan D. Distinct patterns of neural modulation during the processing of conceptual and syntactic anomalies. J Cogn Neurosci 2003;15:272–93.
164. Just MA, Carpenter PA, Keller TA, Eddy WF, Thulborn KR. Brain activation modulated by sentence comprehension. Science 1996;274:114–6.
165. Caplan D, Alpert N, Waters GS. Effects of syntactic structure and prepositional number on patterns of regional cerebral blood flow. J Cogn Neurosci 1998;10:541–52.
166. Fiebach CJ, Schlesewsky M, Friederici AD. Syntactic working memory and the establishment of filler-gap dependencies: Insights from ERPs and fMRI. J Psycholinguist Res 2001;30:321–38.
167. Ben-Shachar M, Hendler T, Kahn I, Ben-Bashat D, Grodzinsky Y. The neural reality of syntactic transformations: Evidence form fMRI. Psychol Sci 2003;14:433–40.
168. Friederici AD, Rüschemeyer S-A, Hahne A, Fiebach CJ. The role of left inferior frontal gyrus and superior temporal cortex in sentence comprehension: Localizing syntactic and semantic processes. Cereb Cortex 2003;13:170–7.
169. Ben-Shachar M, Palti D, Grodzinsky Y. The neural correlates of syntactic movement: Converging evidence from two fMRI experiments. Neuroimage 2004;21:1320–36.
170. Wartenburger I, Heekeren HR, Burchert F, Heinemann S, De Bleser R, Villringer A. Neural correlates of syntactic transformations. Hum Brain Mapp 2004;22:72–81.
171. Fiebach CJ, Schlesewsky M, Lohmann G. Revisiting the role of Broca's area in sentence processing: Syntactic integration versus syntactic working memory. Hum Brain Mapp 2005;24:79–91.
172. Chen E, West WC, Waters G, Caplan D. Determinants of BOLD signal correlates of processing object-extracted relative clauses. Cortex 2006;42:591–604.
173. Caplan D, Stanczak L, Waters G. Syntactic and thematic constraint effects on blood oxygenation level dependent signal correlates of comprehension of relative clauses. J Cogn Neurosci 2008;20:643–56.
174. Stowe LA, Broere CA, Paans AM, et al. Localizing components of a complex task: Sentence processing and working memory. Neuroreport 1998;9:2995–9.

175. Caplan D, Waters GS. Verbal working memory and sentence comprehension. Behav Brain Sci 1999;22:77–94.
176. Keller TA, Carpenter PA, Just MA. The neural bases of sentence comprehension: A fMRI examination of syntactic and lexical processing. Cereb Cortex 2001;11:223–37.
177. Cooke A, Zurif EB, DeVita C, et al. Neural basis for sentence comprehension: Grammatical and short-term memory components. Hum Brain Mapp 2002;15:80–94.
178. Caplan D. Functional neuroimaging studies of syntactic processing. J Psycholinguist Res 2001;30:297–320.
179. Friederici AD, Kotz SA. The brain basis of syntactic processes: Functional imaging and lesion studies. Neuroimage 2003;20:S8–S17.
180. Martin RC. Language processing: Functional organization and neuroanatomical basis. Annu Rev Psychol 2003;54:55–89.
181. Grodzinsky Y, Friederici AD. Neuroimaging of syntax and syntactic processing. Curr Opin Neurobiol 2006;16:240–6.
182. Scott SK, Blank C, Rosen S, Wise RJS. Identification of a pathway for intelligible speech in the left temporal lobe. Brain 2000;123:2400–6.
183. Davis MH, Johnsrude IS. Hierarchical processing in spoken language comprehension. J Neurosci 2003;23:3423–31.
184. Specht K, Reul J. Functional segregation of the temporal lobes into highly differentiated subsystems for auditory perception: An auditory rapid event-related fMRI task. Neuroimage 2003;20:1944–54.
185. Uppenkamp S, Johnsrude IS, Norris D, Marslen-Wilson W, Patterson RD. Locating the initial stages of speech-sound processing in human temporal cortex. Neuroimage 2006;31:1284–96.
186. Binder JR, Swanson SJ, Hammeke TA, Sabsevitz DS. A comparison of five FMRI protocols for mapping speech comprehension systems. Epilepsia 2008;49:1980–1997.
187. Thompson-Schill SL, D'Esposito M, Aguirre GK, Farah MJ. Role of left inferior prefrontal cortex in retrieval of semantic knowledge: A reevaluation. Proc Natl Acad Sci USA 1997;94:14792–7.
188. Thompson-Schill SL, D'Esposito M, Kan IP. Effects of repetition and competition on activity in left prefrontal cortex during word generation. Neuron 1999;23:513–22.
189. Wise R, Chollet F, Hadar U, Friston K, Hoffner E, Frackowiak R. Distribution of cortical neural networks involved in word comprehension and word retrieval. Brain 1991;114:1803–17.
190. Price CJ, Wise RJS, Warburton EA, et al. Hearing and saying. The functional neuroanatomy of auditory word processing. Brain 1996;119:919–31.
191. Price C, Wise R, Ramsay S, et al. Regional response differences within the human auditory cortex when listening to words. Neurosci Lett 1992;146:179–82.
192. Binder JR, Rao SM, Hammeke TA, Frost JA, Bandettini PA, Hyde JS. Effects of stimulus rate on signal response during functional magnetic resonance imaging of auditory cortex. Cogn Brain Res 1994;2:31–8.
193. Lehéricy S, Cohen L, Bazin B, et al. Functional MR evaluation of temporal and frontal language dominance compared with the Wada test. Neurology 2000;54:1625–33.
194. Humphries C, Willard K, Buchsbaum B, Hickok G. Role of anterior temporal cortex in auditory sentence comprehension: An fMRI study. Neuroreport 2001;12:1749–52.
195. Crinion JT, Lambon-Ralph MA, Warburton EA, Howard D, Wise RJS. Temporal lobe regions engaged during normal speech comprehension. Brain 2003;126:1193–201.
196. Spitsyna G, Warren JE, Scott SK, Turkheimer FE, Wise RJS. Converging language streams in the human temporal lobe. J Neurosci 2006;26:7328–36.
197. Awad M, Warren JE, Scott SK, Turkheimer FE, Wise RJS. A common system for the comprehension and production of narrative speech. J Neurosci 2007;27:11455–64.
198. Eulitz C, Elbert T, Bartenstein P, Weiller C, Müller SP, Pantev C. Comparison of magnetic and metabolic brain activity during a verb generation task. NeuroReport 1994;6:97–100.
199. Ojemann JG, Buckner RL, Akbudak E, et al. Functional MRI studies of word-stem completion: Reliability across laboratories and comparison to blood flow imaging with PET. Hum Brain Mapp 1998;6:203–15.
200. Yetkin FZ, Swanson S, Fischer M, et al. Functional MR of frontal lobe activation: Comparison with Wada language results. Am J Neuroradiol 1998;19:1095–8.
201. Benson RR, FitzGerald DB, LeSeuer LL, et al. Language dominance determined by whole brain functional MRI in patients with brain lesions. Neurology 1999;52:798–809.
202. Palmer ED, Rosen HJ, Ojemann JG, Buckner RL, Kelley WM, Petersen SE. An event-related fMRI study of overt and covert word stem completion. Neuroimage 2001;14:182–93.
203. Liégois F, Connelly A, Salmond CH, Gadian DG, Vargha-Khadem F, Baldeweg T. A direct test for lateralization of language activation

using fMRI: Comparison with invasive assessments in children with epilepsy. Neuroimage 2002;17:1861–7.

204. Raichle ME, Fiez JA, Videen TO, et al. Practice-related changes in human brain functional anatomy during nonmotor learning. Cereb Cortex 1994;4:8–26.

205. Petersen SE, Fox PT, Posner MI, Mintun M, Raichle ME. Positron emission tomographic studies of the cortical anatomy of single-word processing. Nature 1988;331:585–9.

206. Malach R, Reppas JB, Benson RR, et al. Object-related activity revealed by functional magnetic resonance imaging in human occipital cortex. Proc Natl Acad Sci USA 1995;92:8135–9.

207. Kanwisher N, Woods R, Iacoboni M, Mazziotta J. A locus in human extrastriate cortex for visual shape analysis. J Cogn Neurosci 1996;91:133–42.

208. Grill-Spector K, Kushnir T, Edelman S, Avidian-Carmel G, Itzchak Y, Malach R. Differential processing of objects under various viewing conditions in the human lateral occipital complex. Neuron 1999;24: 187–203.

209. Bookheimer SY, Zeffiro TA, Blaxton T, Gaillard T, Theodore W. Regional cerebral blood flow during object naming and word reading. Hum Brain Mapp 1995;3:93–106.

210. Martin A, Wiggs CL, Ungerleider LG, Haxby JV. Neural correlates of category-specific knowledge. Nature 1996;379:649–52.

211. Price CJ, Moore CJ, Humphreys GW, Frackowiak RSJ, Friston KJ. The neural regions sustaining object recognition and naming. Proc Royal Soc Lond B 1996;263:1501–7.

212. Zelkowicz BJ, Herbster AN, Nebes RD, Mintun MA, Becker JT. An examination of regional cerebral blood flow during object naming tasks. J Int Neuropsychol Soc 1998;4:160–6.

213. Murtha S, Chertkow H, Beauregard M, Evans A. The neural substrate of picture naming. J Cogn Neurosci 1999;11:399–423.

214. Kiasawa M, Inoue C, Kawasaki T, et al. Functional neuroanatomy of object naming: A PET study. Graefes Archives of Clinical and Experimental Ophthalmology 1996;234: 110–5.

215. Carpentier A, Pugh KR, Westerveld M, et al. Functional MRI of language processing: Dependence on input modality and temporal lobe epilepsy. Epilepsia 2001;42:1241–54.

216. Devlin JT, Russell RP, Davis MH, et al. Is there an anatomical basis for category-specificity? Semantic memory studies with PET and fMRI. Neuropsychologia 2002;40:54–75.

217. Xu B, Grafman J, Gaillard WD, et al. Neuroimaging reveals automatic speech coding during perception of written word meaning. Neuroimage 2002;17:859–70.

218. Mummery CJ, Patterson K, Hodges JR, Price CJ. Functional neuroanatomy of the semantic system: Divisible by what? J Cogn Neurosci 1998;10:766–77.

219. Chee MWL, O'Craven KM, Bergida R, Rosen BR, Savoy RL. Auditory and visual word processing studied with fMRI. Hum Brain Mapp 1999;7:15–28.

220. Miceli G. The neural correlates of grammatical gender: An fMRI investigation. J Cogn Neurosci 2002;14:618–28.

221. Herbster AN, Mintun MA, Nebes RD, Becker JT. Regional cerebral blood flow during word and nonword reading. Hum Brain Mapp 1997;5:84–92.

222. Chee MW, Caplan D, Soon CS, et al. Processing of visually presented sentences in Mandarin and English studied with fMRI. Neuron 1999;23:127–37.

223. Démonet JF, Wise R, Frackowiak RSJ. Language functions explored in normal subjects by positron emission tomography: A critical review. Hum Brain Mapp 1993;1:39–47.

224. Fiez JA. Phonology, semantics and the role of the left inferior prefrontal cortex. Hum Brain Mapp 1997;5:79–83.

225. Poldrack RA, Wagner AD, Prull MW, Desmond JE, Glover GH, Gabrieli JDE. Functional specialization for semantic and phonological processing in the left inferior prefrontal cortex. Neuroimage 1999;10:15–35.

226. Gold BT, Buckner RL. Common prefrontal regions coactivate with dissociable posterior regions during controlled semantic and phonological tasks. Neuron 2002;35:803–12.

227. Henson RNA, Price CJ, Rugg MD, Turner R, Friston KJ. Detecting latency differences in event-related BOLD responses: Application to words versus nonwords and initial versus repeated face presentations. Neuroimage 2002;15:83–97.

228. Mechelli A, Gorno-Tempini ML, Price CJ. Neuroimaging studies of word and pseudoword reading: Consistencies, inconsistencies, and limitations. J Cogn Neurosci 2003;15:260–71.

229. Braver TS, Cohen JD, Nystrom LE, Jonides J, Smith EE, Noll DC. A parametric study of prefrontal cortex involvement in human working memory. Neuroimage 1997;5:49–62.

230. Honey GD, Bullmore ET, Sharma T. Prolonged reaction time to a verbal working memory task predicts increased power of posterior parietal cortical activation. Neuroimage 2000;12:495–503.

231. Adler CM, Sax KW, Holland SK, Schmithorst V, Rosenberg L, Strakowski SM. Changes in neuronal activation with increasing attention demand in healthy volunteers: An fMRI study. Synapse 2001;42:266–72.

232. Braver TS, Barch DM, Gray JR, Molfese DL, Snyder A. Anterior cingulate cortex and response conflict: Effects of frequency, inhibition and errors. Cereb Cortex 2001;11:825–36.

233. Ullsperger M, von Cramon DY. Subprocesses of performance monitoring: A dissociation of error processing and response competition revealed by event-related fMRI and ERPs. Neuroimage 2001;14:1387–401.

234. Binder JR, Liebenthal E, Possing ET, Medler DA, Ward BD. Neural correlates of sensory and decision processes in auditory object identification. Nat Neurosci 2004;7:295–301.

235. Desai R, Conant LL, Waldron E, Binder JR. FMRI of past tense processing: The effects of phonological complexity and task difficulty. J Cogn Neurosci 2006;18:278–97.

Chapter 12

Imaging Brain Attention Systems: Control and Selection in Vision

George R. Mangun, Sean P. Fannon, Joy J. Geng, and Clifford D. Saron

Summary

Selective attention is an essential cognitive ability that permits us to effectively process and act upon relevant information while ignoring distracting events. The human capacity to focus attention is at the core of mental functioning. Elucidating the neural bases of human selective attention remains a key challenge for neuroscience and represents an essential aim in translational efforts to ameliorate attentional deficits in a wide variety of neurological and psychiatric disorders. In this chapter, we discuss how functional imaging methods have helped us to understand fundamental aspects of attention: How attention is controlled, and how this control results in the selection of relevant stimuli. Work from our group and from others will be discussed. We will focus on fMRI methods, but where appropriate will include related discussion of electromagnetic recording methods used in conjunction with fMRI.

Key words: Attention, Selection, Control, fMRI, Human, Vision

1. Introduction

Attention is a key cognitive ability that supports our momentary awareness, and affects how we analyze sensory inputs, retain information in memory, process it for meaning, and, finally, act upon it. In this review chapter we will consider the role of attention in sensory processing and perception. First, we must define attention as it will be investigated in the studies to follow, and we start with the definition provided by psychologist William James *(1)*:

> "Everyone knows what attention is. It is the taking possession by the mind, in clear and vivid form, of one out of what seem several simultaneously possible objects or trains of thought. … It implies withdrawal from some things in order to deal effectively with others,…"

M. Filippi (ed.), *fMRI Techniques and Protocols*, Neuromethods, vol. 41
DOI: 10.1007/978-1-60327-919-2_12, © Humana Press, a part of Springer Science+Business Media, LLC 2009

Using introspection, James notes key characteristics of attention that frame the theoretical landscape of attentional mechanisms. He notes the voluntary aspects of attention and its selection of relevant from irrelevant information, as well as its capacity limitations.

Our review will focus on studies of visual attention, and therefore, we also consider the work of James' contemporary, Hermann von Helmholtz *(2)*, who in his studies of visual psychophysics made observations and speculations on the mechanisms of visual attention. He wrote, in studies investigating the limits of visual perception, the following:

> "...by a voluntary kind of intention, even without eye movements, and without changes of accommodation, one can concentrate attention on the sensation from a particular part of our peripheral nervous system and at the same time exclude attention from all other parts."

Helmholtz presages today's knowledge about attention mechanisms, which includes detailed understanding of how attention influences stimulus processing at early and late stages of sensory analysis.

1.1. Varieties of Attention

As suggested earlier, attention is neither a single capability, nor is it supported by a single mechanism or brain system. Theoretically, we can consider two main forms of attention - voluntary attention and reflexive attention *(3)*. Voluntary attention is goal-directed and suggests a top-down influence that is under intentional control. In contrast, reflexive attention is a stimulus-driven process involving bottom-up effects, for example, as when a salient sensory signal grabs our attention. These two general categories of attention differ in their properties and perhaps their neural mechanisms. In this chapter, we will concentrate on studies of voluntary attention to illustrate how functional imaging has been used to elucidate brain attention mechanisms.

Another way of thinking about attention mechanisms is to consider the domain of information processing on which attention operates. For example, attention can operate within or between sensory modalities. That is, we can attend to visual inputs at the expense of auditory ones, or vice versa, or may attend to one aspect of visual input (e.g., stimulus location) at the expense of other stimulus attributes (e.g., color or motion). Thus, one may ask whether the mechanisms of attentional control are similar or different for attention to different sensory modalities compared with attention to different stimulus attributes *(4)*. One may also consider whether the mechanisms supporting stimulus selection are modality and attribute independent or are instead specialized for the particular items to be attended. In order to constrain the discussion in this chapter, for the most part, we will focus on visual attention, emphasizing attentional control and selection for attention based on stimulus location (spatial attention) and elementary stimulus features (e.g., color and motion).

1.2. Early and Late Selection Models

One of the key questions in attention research has been where in information processing attention has its influence. If as Helmholtz suggested, attention could select some information coming from the peripheral nervous system, then it is essential to ask where within the ascending sensory pathways attention can alter stimulus processing to achieve selective processing. In the 1950s, Broadbent *(5)* described the idea of an attentional gate that could be opened for attended information and closed for ignored information. Like Helmholtz, Broadbent suggested that information selection might occur early in sensory processing. This idea has been termed *early selection*, and it is the idea that a stimulus need not be completely perceptually analyzed before it can be selected for further processing or rejected by a gating mechanism.

In contrast, so-called *late selection* models hold that all (attended and unattended) sensory inputs are processed equivalently by the perceptual system to a very high level of coding *(6)* before they are selected by attention. This high level is generally considered to be the level of categorical, or semantic, encoding where the elementary featural codes (e.g., orientation, contrast, color, form) are replaced by conceptual codes (e.g., that is a chair). Later selection models posit that selection takes place on these higher level codes and thereafter representation in awareness may take place. This view, then, holds that selective attention does not influence our perceptions of stimuli by changing the low-level sensory-perceptual processing of the stimulus.

A long debate and many studies have addressed the early versus late selection controversy, and physiological approaches including functional imaging have provided key information about the stages of sensory processing that are influenced by selective attention. We will thus begin our consideration of attention with its role in stimulus selection processes and how functional imaging has been used to investigate these mechanisms. But first, in the next section, some design issues in the study of selective attention deserve consideration as they are of paramount importance for physiological studies of attention, including functional imaging.

1.3. Methodological Issues in Experimental Studies of Selective Attention

The focus of the vast majority of studies of effects of attention on perception involves selective attention, attention to one thing at the expense of another. This is to be contrasted with nonselective attention, which includes generalized behavioral arousal (e.g., the classic orienting response) that does not necessarily involve attending one input while ignoring another. These latter nonselective attention mechanisms are certainly interesting, but selective mechanisms have generated the greatest interest and we review only these studies. Therefore, it is critical to understand the design parameters that permit selective versus nonselective attention to be isolated and studied, and to make note of some of the confounding influences that might contaminate studies of selective attention when nonselective factors (e.g., arousal) are not properly controlled. We turn then to an example from the early history of the physiology of attention.

In the 1950s, Raul Hernandez-Peon et al. *(7)* studied the neuroanatomy and neurophysiology of the ascending sensory pathways and how top-down attention might modulate sensation. Their work in the auditory system was motivated by the well-known neuroanatomical substrate for top-down modulation, the olivocochlear bundle (OCB), which involves centrifugal neural projections from higher levels of the central nervous system downward to earlier processing stages especially the peripheral nervous system out to the level of the cochlea. The OCB provided a very strong neuroanatomical mechanism by which top-down effects of mental processes like attention might gate early auditory processing, in a fashion suggested by Helmholtz and later Broadbent, among others.

Hernandez-Peon's group recorded the activity in neurons in the subcortical auditory pathway in cats while they were either passively listening to the sounds from a speaker or were not attending the sounds. By showing the cats two live mice safely contained in a closed bottle, the cats were induced to ignore the sounds while they attended the mice. The researchers found that the amplitude of activity recorded from electrodes implanted in the cochlear nucleus was larger when the animals attended the sounds versus ignored the sounds while attending the bottled mice. This was interpreted as evidence that selective attention influences stimulus processing as early as the subcortical sensory pathways via the influence of the top-down neural control inputs.

This work, published in the journal *Science*, would however, later be shown to include a fatal flaw. The flaw was that the sounds presented to the cats to evoke auditory activity were delivered by speakers near the cats. When the cats attended the mice, they also oriented their ears and heads toward the mice, and away from the speakers, thereby altering the amplitudes of the auditory sounds at the ears due to simple physical differences in the relation of the ears to the sounds. Changes in the amplitudes of the sounds at the ear lead, in and of themselves, to changes in the amplitudes of auditory responses in the ascending pathways, and this tells us nothing about attention, although it mimics the kind of effect one would expect from an attentional mechanism. This problem can be eliminated by controlling the amplitudes and qualities of the sensory stimuli when attended and ignored so that an ensuing difference in neural responses would be attributable to internal attentional modulations of sensory processing, not differences in the physical stimuli themselves across conditions. Similarly, the two conditions of Hernandez-Peon and colleagues likely also differed in nonspecific behavioral arousal that may also have influenced the neuronal recordings since when cats see mice they are undoubtedly more aroused than when they listen to Beethoven, although the opposite would hopefully hold for at least some human listeners. In either case, differences in behavioral arousal between comparison conditions could confound the effects of selective attention, and therefore, like physical stimulus differences, must be rigorously controlled. Failure to do so properly could lead to changes in, for example, hemodynamic signals that would present a serious confound in functional imaging studies of attention.

2. Imaging Attentional Selection Mechanisms

Studies in animals and humans using electrophysiological recordings have provided evidence that voluntary visual attention can influence the processing of sensory inputs at early stages of neural analysis. Van Voorhis and Hillyard *(8)* used scalp recordings of event-related brain potentials (ERPs) in humans to demonstrate that by 70-ms post stimulus onset, the electrical brain response to a stimulus presented to an attended spatial location was enhanced compared with an identical stimulus presented when that location was ignored. The timing of the ERP method (in the order of milliseconds) provides strong evidence that early stages of information processing were influenced by visual selective attention because the first inputs to the human visual cortex arrive only at about 40 ms after stimulus onset (based on intracranial recordings in patients). However, the scalp recordings are limited in the spatial localization they can provide in humans because the ERPs are recorded outside the skull and are, therefore, distant from the neuroelectric sources generating the signals. The volume conduction of the electrical signal through the brain, skull, and scalp is, on the one hand, an advantage that permits scalp recordings in the first place, but it also means that the electrical currents spread on the scalp (and are also filtered and spatially blurred by the intervening tissues, especially the skull) and are hard to track backward to their three-dimensional intracranial site of generation *(9)*. Subsequent studies in animals showed that area V4 in extrastriate cortex in the macaque monkey showed increased neuronal firing rates to attended-location stimuli as compared with unattended stimuli *(10)*, and more recent studies have extended this to other extrastriate areas *(11, 12)*, as well as striate cortex *(13)*.

In humans, however, studies using functional imaging methods have provided the most detailed information about where in the ascending pathways visual selective attention modulates stimulus processing, because methods such as positron emission tomography (PET) and functional magnetic resonance imaging (fMRI) provide precise information about the neuroanatomical loci of attention effects and attention processes in the brain.

2.1. Functional Imaging of Visual-Spatial Attention

Functional imaging methods provide neuroanatomical information about the mechanisms of human attention. That is, together with information from neurophysiological recordings, imaging approaches help to locate the neuroanatomical stages of information processing influenced by attention. For visual-spatial attention (*see* **ref.** ***14*** for studies of nonspatial attention), we began using O_{15} PET to provide functional anatomical information to complement our earlier ERP studies. Indeed, in these early studies that began in 1991, we first combined ERP and PET methods to provide a spatial-temporal approach to studying human attention ***(15, 16)***. In subsequent studies in the mid-1990s, we incorporated fMRI methods ***(17)***.

The goal of these functional imaging studies was to identify where within the visual system visual-spatial selective attention first influenced sensory analysis. We presented subjects with bilateral stimulus arrays of nonsense symbols (two in each hemifield) flashed at a rapid varying rate, averaging about two stimulus arrays per second (**Fig. 1**). In order to control for physical stimulus differences between conditions, the subjects were required to maintain fixation on a central fixation point and their compliance was ensured using high-resolution infrared photometric monitoring and the horizontal electro-oculograms (EOGs). In order to control for differences in nonspecific behavioral arousal, we included two main attention conditions that were equated in task

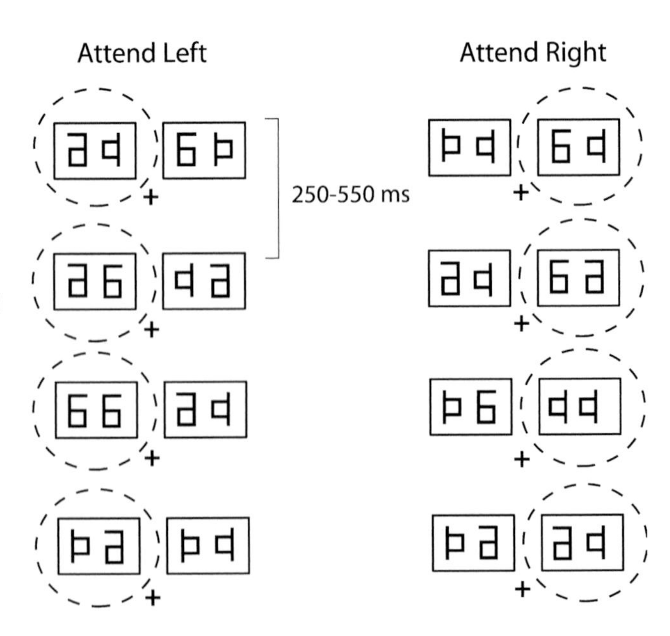

Fig. 1. Stimuli and task used in a functional imaging and event-related potential (ERP) study of spatial attention (***17***). Two conditions of attention are shown: Attend left condition (*left column*) and attend right condition (*right column*). Subjects viewed rapid sequences of arrays (about 2.5/s) of nonsense symbols (flashed for 100 ms) while maintaining fixation of their eyes on a central fixation spot (*plus sign*). There were always two symbols in the left and two in the right visual hemifield in locations each marked by an outline rectangle. The task was to detect and press a button to pairs of symbols at the attended location and to ignore all stimuli in the opposite hemifield. In the figure (but not in the actual experiment) the focus of covert spatial attention is indicated by a *dashed circle*.

difficulty and arousal. The subjects had to covertly pay attention to the left half of the array in one condition and ignore the right (attend left condition), or attend the right half of the array while ignoring the left (attend right condition). Hemodynamic responses (activations) in the brain could be compared for attend-left versus attend-right conditions. So importantly, the stimulus arrays were identical, striking the same regions of the retina in the two attention conditions, and the two attention conditions (attend left and attend right) were also equated for nonspecific arousal because the discrimination task was identical for both conditions. As a result, any changes we observed in visual cortex could not be attributed to either differences in visual stimulation or to nonspecific factors such as arousal, but instead could be attributed to modulations of visual processing with the direction of spatial attention.

We observed that spatial attention led to activations in extrastriate cortex in the cerebral hemisphere that was contralateral to the attended side of the stimulus arrays **(Fig. 2a)**. These activations were highly statistically significant in the posterior fusiform gyrus on the ventral cortical surface and in lateral-ventral occipital regions. In these early studies, no effects of visual-spatial attention could be observed in primary visual cortex (area V1) but as

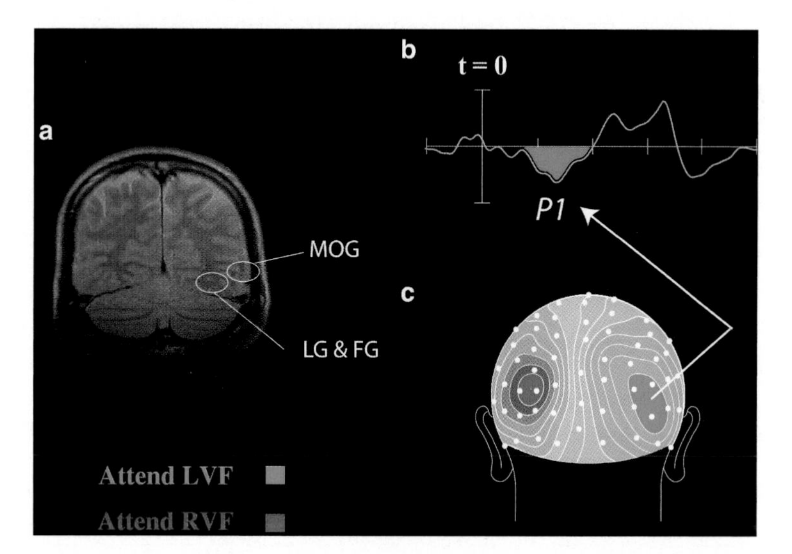

Fig. 2. fMRI and event-related potential (ERP) data from a study of visual-spatial selective attention. (**a**) Horizontal structural scan of single subject showing activations in contralateral visual cortex with spatial attention in the extrastriate cortex (left hemisphere is on the left). The activations were focused in the lingual gyrus (LG) and posterior fusiform gyrus (FG) and the middle occipital gyrus (MOG). (**b**) ERP attention effects shown as difference waves from a single lateral occipital electrode site in the right hemisphere (attend left minus attend right). The vertical scale is 2 μV per side (positive plotted downward). The onset of the array is indicated at time zero ($t = 0$), and the tick marks are 100 ms. (**c**) Topographic voltage attention difference map (110-ms latency) on the scalp surface viewed from the rear (left side of head on left side of figure) [adapted from ***(17)***].

we shall see later, such effects have since been observed using fMRI. The finding of localized brain activations corresponding to the action of spatial attention alone is in accord with evidence from prior ERP studies *(8, 18, 19)*. Also in this study, in a separate session, testing the same subjects, we recorded ERPs for the same stimuli and task. The scalp-recorded ERPs showed the expected P1 attention effects **(Fig. 2b, c)**, and by using neuroelectric dipole modeling we investigated whether intracranial neural generators at the loci of functional activations could produce activity in the model scalp that was similar to that which we actually recorded in the ERPs (not shown in figure). We showed that the ERP and fMRI attention effects in the posterior fusiform gyrus (i.e., extrastriate visual cortex) were strongly related. This combined use of ERPs and functional imaging provide evidence for short-latency (around 100 ms after stimulus onset) changes in responses to visual stimuli as a function of spatial attention that were generated early in extrastriate visual cortex. In several studies we have followed up these effects *(20, 21)* and have observed that these modulations with spatial attention affect multiple stages of visual cortical processing, from V1 toward inferotemporal cortex in the ventral visual stream. The next section reviews related work that also combined structural, functional, and cognitive imaging to detail the structure of spatial attention effects in visual cortex.

2.2. Mapping Spatial Attention in Vision

A beautiful illustration of how spatial attention influences sensory processing in human visual cortex comes from the work of Tootell et al. *(22)*. They used fMRI to identify the borders of the first few visual areas in humans (retinotopic mapping) and then conducted a spatial attention study similar to what was described earlier. In the Tootell study, subjects performed a simple spatial attention task that required subjects to covertly and selectively attend stimuli located in one visual field quadrant while ignoring those in the other quadrants; different quadrants were attended in different conditions. Attentional activations were then mapped onto the flattened representations of the visual cortex, permitting the attention effects to be related directly to the multiple visual areas of human visual cortex, showing that spatial attention led to robust modulations of activity in striate cortex and multiple extrastriate visual areas **(Fig. 3)**.

By combining different methods for recording electrical activity, imaging brain structure, defining functional anatomy (i.e., retinotopic maps), and combining this with functional imaging in carefully controlled studies of selective attention, we can learn a great deal about the effects exerted by attention on sensory processes in humans. Specifically, for spatial attention we now understand that it exerts powerful influences over the processing of visual inputs: Attended stimuli produce greater neural responses than

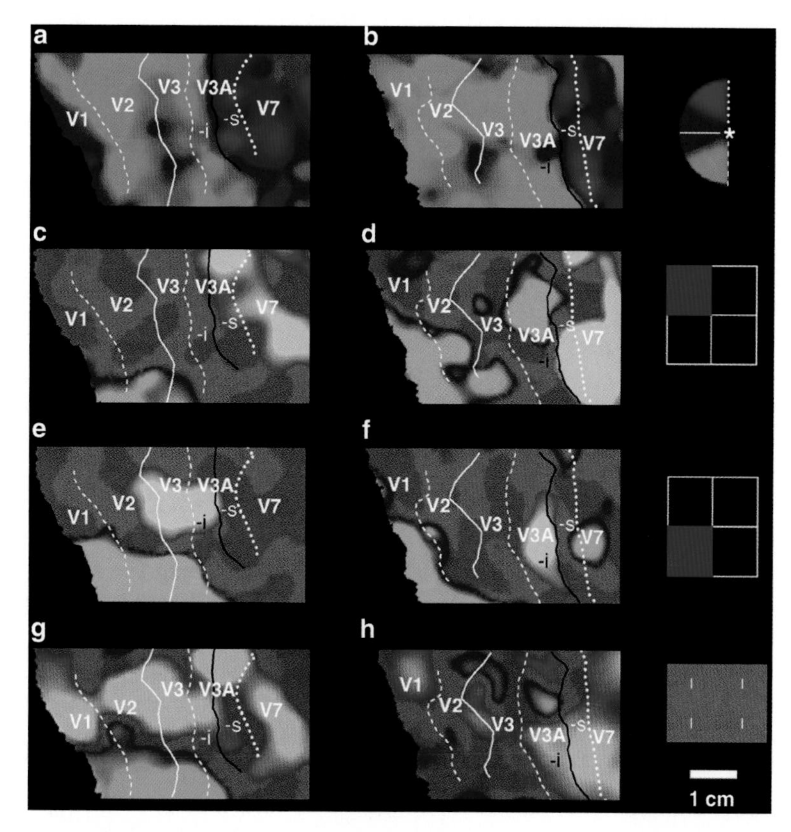

Fig. 3. Spatial attention effects in multiple visual cortical areas as demonstrated by fMRI. Activations with spatial attention to left-field stimuli are shown in the flattened right visual cortices of two subjects (one in the left column and the other on the right). The *white lines* (*dotted and solid*) indicate the borders of the visual areas as defined by representations of the horizontal and vertical meridians; each area is labeled from V1 (striate cortex) through V7, a retinotopic area adjacent to V3A. The *solid black line* is the representation of the horizontal meridian in V3A. Panels (**a**) and (**b**) show the retinotopic mappings of the left visual field for each subject, with colored activations corresponding to the polar angles shown at right (which represent the left visual field). Panels (**c**) and (**d**) show the attention-related modulations (attended vs. unattended) of sensory responses to a target in the upper left quadrant (the quadrant of the stimuli is shown at right). Panels (**e**) and (**f**) show the same for stimuli in the lower left quadrant. In (**c**) through (**f**), the *yellow* to *red colors* indicate areas where activity was greater when the stimulus was attended to than when it was ignored; the *bluish colors* represent the opposite, where the activity was greater when the stimulus was ignored than when attended. The attention effects in (**c**) through (**f**) can be compared to the pure sensory responses to the target bars when passively viewed [(**g**) and (**h**)]. Note the retinotopic pattern of the attention effects in (**c**) through (**f**): The attention effects to targets in the lower left quadrant produced activity in several lower field representations, which included the appropriate half of V3A (inferior V3A labeled with an "i") in both subjects, and V3 and V2 in one subject. In contrast, attention to the upper left quadrant produced activity in the upper field representation of V3A (S) and in the adjacent upper field representation of area V7 [from *(**22**)*].

do unattended stimuli, even when arousal and physical stimulus differences are controlled, and this happens in multiple visual cortical areas beginning at short latencies after stimulus onset (as short as 70 ms in humans). In part, such effects may help to bias competition between attended and ignored sensory inputs at the level of neuronal receptive fields as well as higher order representations of the stimuli *(23)*.

3. Neuroimaging of Attentional Control Networks

3.1. Isolating Attentional Control Mechanisms

A major goal in studies of selective attention is to understand how attention is controlled in the brain. Given that activity in sensory-specific cortex is modulated by attention, and therefore that the neuronal response properties of sensory neurons have been temporarily altered, what mechanisms lead to these changes in information processing? Presumably, some brain systems form executive control systems that, as a function of momentary behavioral goals, are able to alter sensory-neural activity. Some models have referred to attentional control networks as the "sources" of attentional signals, and the sensory (and motor) systems that are affected as the "sites" where attention is implemented *(24)*.

This source versus site dichotomy is useful as it helps to distinguish between the role attention plays in modulating sensory processing in the sensory systems and the neural mechanisms that produce this effect (**Fig. 4**). Presumably, neuronal projections from executive attentional control systems contact and influence neurons in sensory-specific cortical areas in order to alter their excitability. As a result, the response in sensory areas to a stimulus

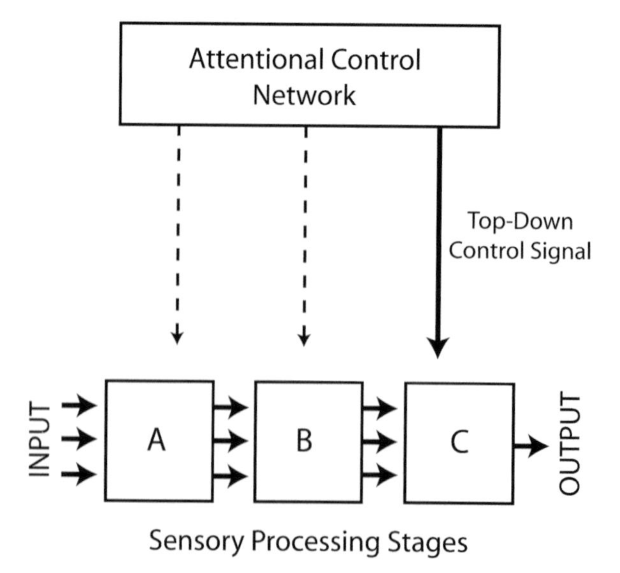

Fig. 4. Diagram of the influence of top-down attention control networks (*top*) on sensory processing (*bottom*). Sensory inputs are transduced at left and processed in multiple stages of analysis (A, B, and C). Top-down influences are shown as vertical arrows from the attentional control network. Here, the *dashed line arrows* indicated no influence on sensory processing stages A or B, and the *heavy solid line arrow* indicates an influence of attention on processing in sensory processing stage C, with the result being selection [represented by three arrows coming into C, but only one arrow leaving as output (*bottom right*)]. If this were to refer to spatial attention, then the *arrows* in the sensory processing stream could correspond to different parallel visual field location inputs in a retinotopic fashion, and the *single output arrow* at C reflecting that selective attention to one location was relatively facilitated (selected) with respect to the other locations.

may be either enhanced if the stimulus is given high priority (i.e., is relevant to the behavioral goal) or attenuated if it is irrelevant to the current goal. A network including the pulvinar nucleus of the thalamus, the posterior parietal cortex, and the dorsolateral and superior prefrontal cortex may mediate cortical excitability in the visual cortex as a function of selective attention. More generally, though, attentional control systems could be involved in modulating thoughts and actions, as well as sensory processes.

Studies of patients with brain damage, animal recordings and lesion studies, and functional imaging converge to show that a large network of cortical and subcortical areas is activated during attentional orienting and selection *(25)*. How can we measure the activity of the different brain regions during the execution of attentional selection tasks in order to determine which networks involve attention control (sources of attention) and which are the sites of selection? In part, the answer is to implement tasks that, at least theoretically, dissociate attentional control processes from selection in time, as do attentional cuing paradigms. In such paradigms, attention is cued at time 1, which is followed by a delay period of several hundred milliseconds or seconds, and then by the target stimuli at time 2 *(26)*. Such designs are different from that described earlier (**Fig. 1**), which used blocked attention conditions and rapid streams of stimuli to study attentional mechanisms, because the cue (e.g., an arrow) triggers the action of the attentional control network, whereas selective stimulus processing would not take place until the target is presented later. Hence, by recording brain responses to the cue and target separately, one can identify the networks for control and selection. Blocked-design fMRI attention studies make this difficult as the activations produced by attention control and selection processes cannot be easily distinguished because they are co-occurring in the images. Although functional brain imaging, which taps hemodynamic changes, is a rather poor method for tracking the time course of brain activity, when used in an event-related design, and appropriate analytic strategies are employed to separate overlapping hemodynamic responses, it is possible to conduct an experiment like that just described; we and others have done so in several studies of selective attention.

In our initial studies *(20)*, we used event-related fMRI to investigate attentional control mechanisms during visual-spatial attention. An arrow, presented at the center of the display, indicated the side to which attention should be directed for that trial. Eight seconds later, a bilateral target display (flickering black and white checkerboards) appeared for 500 ms. The participants' task was to press a button if some of the checks on the cued sides only were gray rather than white. The 8-s gap between the cue and the target stimuli allowed us to extract the hemodynamic responses linked to the attention-directing cues separately from those linked to the subsequent targets. We were thus able to identify a top-down attentional control network triggered by the presentation of the cue.

The attentional control network consisted of regions in the superior frontal cortex, inferior parietal cortex, superior temporal cortex, and portions of the posterior cingulate cortex and insula **(Fig. 5)**. These areas were not involved in the sensory analysis of the cue that was reflected by activity in the visual cortex. Instead, this attentional network of frontal, parietal, and temporal brain areas can be considered the sources of attention control signals in the brain. The result of attentional control on the activity of visual cortex before and during target processing helps us to understand the relationship between control signals and selective sensory effects of spatial attention.

Figure 6 shows coronal sections through the visual cortex at two time points: first in the cue-to-target period, prior to the appearance of the lateral target arrays, and then second, during target processing. Two contrasts are shown for each time point: the case for activity when left cue was greater than right cue (attend left > right), and the inverse (attend right > left). Following the attention-directing cue one can see significant

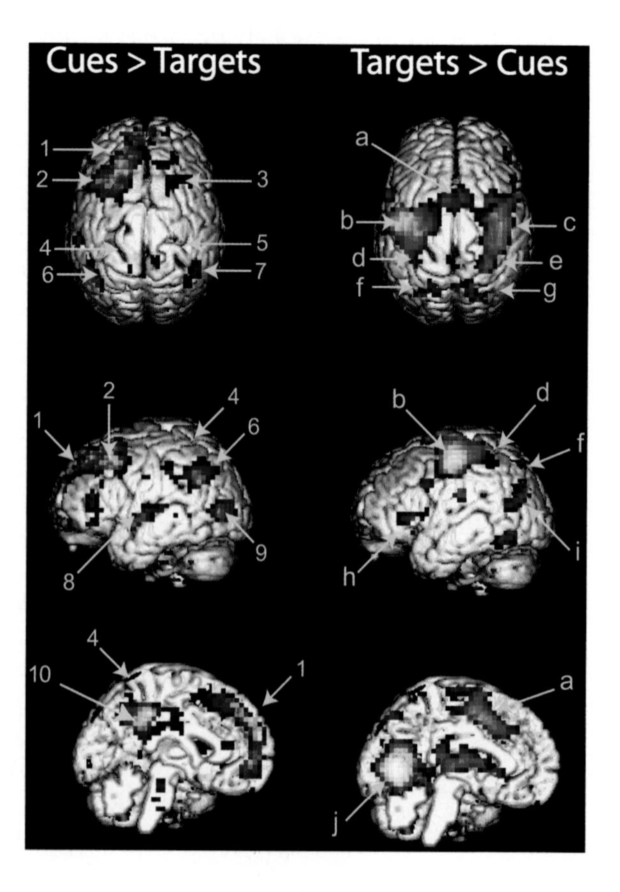

Fig. 5. Activations during attention control and target selection. Group ($N = 6$ subjects) average activations time-locked to the cue (cues > targets contrast) isolating attentional control regions are shown on the left. Those time locked to the targets (targets > cues contrast), indicating target processing and motor responses, are on the right. The activated areas are shown in different colors to reflect the statistical contrasts that revealed the activations: *bluish* for brain regions that were more activated to cues than targets, and *reddish-yellow* to indicate regions that were more active to targets than cues. The *top row* shows a view of the dorsal surface of the brain, the *middle row* shows a lateral view of the left hemisphere, and the *bottom row* shows its medial surface; the activity was the same for the right hemisphere, which is not shown. Attentional control involved the frontal-parietal attention network involving superior and middle frontal gyri (labeled 1–3), and the regions in and around the IPS ***(4–7)***, the superior temporal cortex ***(8)***, and the posterior cingulate cortex ***(10)***. In contrast, during target selection, the main areas of activation were now the supplementary motor area (labeled a), the motor and somatosensory cortex (b-e), posterior superior parietal lobule (f, g), the ventral-lateral prefrontal cortex (h) and the cuneus (i) and the visual cortex (j) [adapted from ***(20)***].

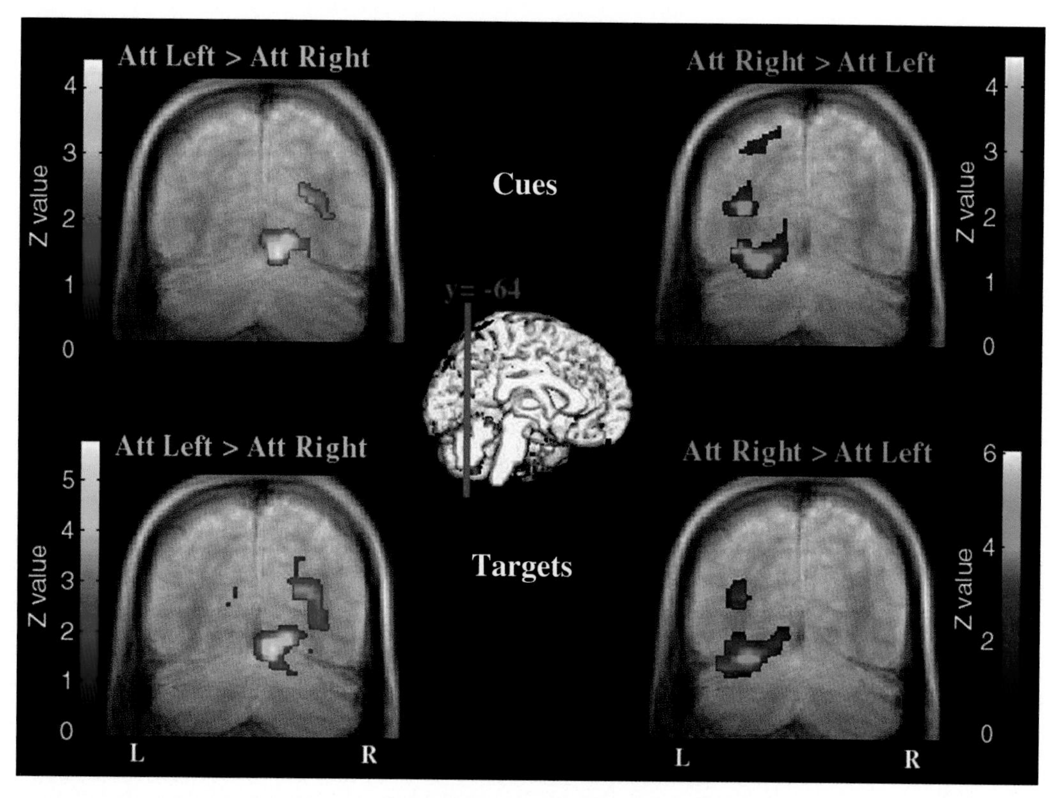

Fig. 6. Priming of visual cortex by spatial attention. The *top row* shows increased baseline activity in six subjects to the cues in visual cortex. The *bottom row* shows the same effects to targets [adapted from *(**20**)*].

contralateral activation in multiple regions of visual cortex. These changes are spatially selective, being in the right visual cortex for leftward attention, and in the left cortex for rightward attention. Importantly, they occur prior to the onset of the targets, which occurred more than 8 s later, and were not related to the simple visual features of the cue, which stimulated different regions of visual cortex. These contralateral activations to the cues represent a kind of attentional priming of sensory cortex, which is thought to form the basis for later selective processing of target inputs. Indeed, as also shown in the bottom half of **Fig. 6**, the selective processing of subsequent target stimuli produced similar contralateral activations in visual cortex. Another way to describe these patterns of activity to cues and targets is to say that regions of visual cortex that code the spatial locations of the expected target stimuli showed increases in background activity levels when attention was directed to those locations by the cues, even before the targets appeared.

Similar effects have been observed in neurophysiological studies in monkeys *(27)*. Computationally, such effects may well be a mechanism for selective sensory processing by changing the baseline gain of sensory neurons such that when later stimulated, they produced enhanced responses *(28)*. This illustrates the mechanism put forward at the beginning of this section that top–down attentional control may lead to selective changes in sensory processing by changing the background firing rates of neurons, thereby improving their sensitivity and/or selectivity.

3.2. Specializations in Attentional Control

One key question about top–down attentional control systems is the extent to which such mechanisms are generalized for the control of attention regardless of the modality or specific stimulus attributes to which selective attention is directed. For example, does preparatory attention for spatial selective attention involve the same or different control networks as would selective attention for color or motion? We have addressed this in studies using similar methods to those described in the foregoing.

In one study *(21)* we undertook a direct test of whether the neural mechanisms for the top-down control of both spatial and nonspatial attention were the same. In this study, we compared spatial selective attention to nonspatial color-selective attention. That is, subjects were either cued to select the target stimulus based on location or color. Our design was as follows. We randomly intermingled trials in which either the location or the color of an upcoming target was cued. The cues were letters (e.g., “L” = attend left and “B” = attend blue) located at fixation in one task or the periphery above fixation in another version of the task. The stimuli to be discriminated were rectangles (outlines) that when following spatial cues were located in the left or right hemifields or when following color cues were overlapping outline rectangles located at fixation in one task version or above fixation in the second task version. The participants were instructed to covertly direct attention to select an upcoming rectangle stimulus based on the cued feature (location or color) with the task of discriminating its orientation. That is, in the spatial cue condition, if the cue signified left, the subjects were to maintain fixation on the central fixation point, but focus covert attention to the left and to discriminate the orientation of the rectangle presented there (vertical or horizontal), and to press the appropriate response key. When the cue indicated blue, however, this meant that the subjects should focus covert attention in preparation for discriminating the blue rectangle location in the same place as the cue (i.e., either at fixation or in the periphery just above fixation).

As in Hopfinger et al. *(20)* described earlier, event-related fMRI measures were obtained to the cues and to the targets, and separately for spatial location and color attention trials. Because we jittered the stimulus-onset-asynchrony between cues and targets from 1,000 to 8,000 ms, it was possible to deconvolve the overlapping hemodynamic responses *(29–31)*. This permitted us to utilize a paradigm that was more similar to cued attention designs in the cognitive psychology literature where the time between cues and targets was not too long for subjects to maintain a strong attentional set (i.e., to sustain the covert allocation of attention to the cued color or location). In our prior work *(20)* and that of others, the need to separate the sluggish hemodynamic responses to adjacent stimuli typically resulted in long stimulus-onset-asynchronies in order to avoid overlap of the responses to cues and targets. By using specially designed stimulus sequences and analytic strategies it is thus possible to design cued attention studies that optimize the design parameters for investigating attentional mechanisms. Indeed, in cued attention designs, the speed of stimulus presentation can be even faster than the 1,000–8,000-ms lag between cues and target described here *(32)*.

In this study, comparing preparatory attention for locations and colors we found that large regions of the frontoparietal network were commonly activated by the spatial and nonspatial cues alike (**Fig. 7**). Such related patterns of activity reflect those aspects of the task that the two attentional control conditions shared, such as low-level sensory processing of the cues, decoding of linguistic information in the cue letter and matching that to the task instruction, establishing the appropriate attentional

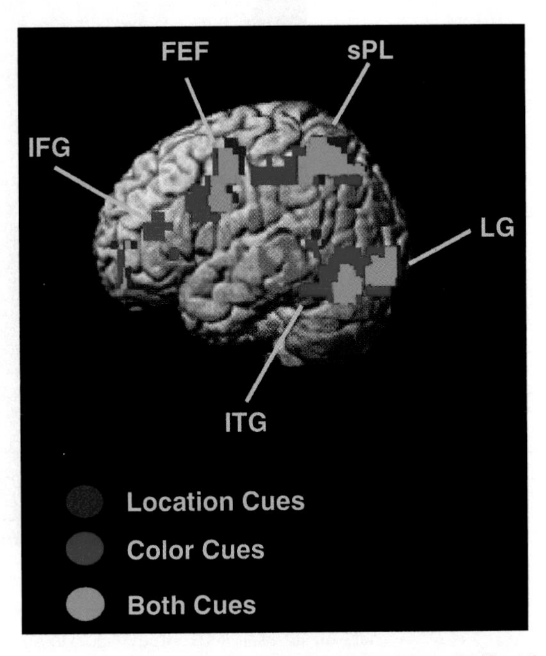

Fig. 7. Cue-related activity. Group-averaged data for brain regions significantly activated to attention-directing cues, overlaid onto a brain rendered in 3D. Areas activated in response to location cues are shown in *blue*, color cues in *red*, and those areas activated by both cues are shown in *green*. Maps are displayed using a height threshold of $p < 0.005$ (uncorrected) and an extent threshold of ten contiguous voxels [modified from ***(21)***].

set and holding this information in working memory during the cue-to-target period, and finally, preparing to respond.

To test whether any of the areas activated in response to the cues were selective for spatial or nonspatial orienting, we directly statistically compared activity in response to location and color cues. The results of this direct comparison are shown in **Fig. 8a** for the activations where locations cues produce more activity than color cues. We found nonoverlapping regions of superior frontal and parietal cortex that were activated during orienting to location versus color. Orienting attention based on stimulus location activated regions of the dorsal frontal cortex [posterior middle frontal gyrus and frontal eye fields (FEF)], posterior parietal cortex [intraparietal sulcus (IPS) and precuneus], and supplementary motor cortex. The inverse contrast (color > location cues) produced no significant activations in the superior frontal or parietal regions, and only showed activity in the ventral occipital

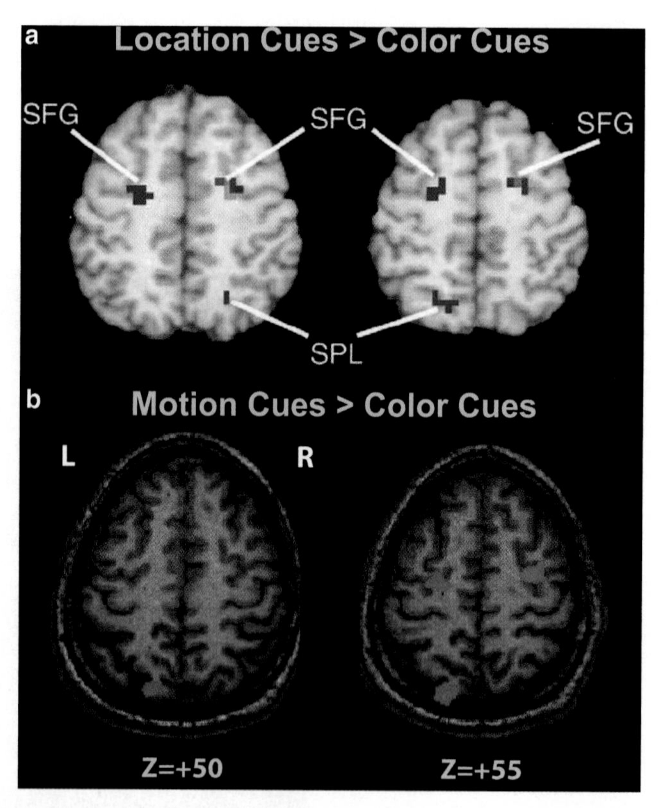

Fig. 8. Attentional control activations for location and motion cues versus color cues. (*Top*) Results of the direct comparison between location and color cue conditions overlaid on an axial slice. Greater activity to location cues than color cues is seen in the superior frontal gyrus (SFG) and the superior parietal lobule (SPL) bilaterally [modified from Giesbrecht et al. *(**21**)*]. (*Bottom*) Results of the direct comparison between motion and color cue conditions overlaid on an axial slice. Greater activity to motion cues than color cues is seen in the superior frontal gyrus (SFG) bilaterally, but the superior parietal lobule (SPL) only on the left [modified from *(**65**)*].

cortex (OCC) (posterior fusiform, posterior middle temporal cortex, and left insula) (not shown in the figure).

The pattern of selectivity in the frontoparietal attentional control network for location cuing suggests that neural specializations exist for the control of orienting attention to locations or for coding some aspect of space tapped in spatial attention tasks *(33)*. But a question that must be addressed is whether these superior frontal and parietal regions are sensitive only for spatial orienting, and we have tested this by investigating specializations in attentional control for other presumably nonspatial features.

We investigated the idea that specializations in superior frontal and parietal cortex for preparatory spatial selective attention might also be involved in other aspects of visual attentional processing. In one study we compared preparatory attention for stimulus motion, a dorsal stream process, to the same nonspatial ventral stream feature, color. In this study each trial began with an auditory word cue that instructed subjects to attend to a target of a particular stimulus feature (i.e., involving either color or motion attention). If cued to a color, then they were to prepare for and detect brief color flashes within a display of randomly moving dots presented during a subsequent test period. If cued to motion, then they were to prepare for and detect the brief coherent motion stimulus in the display of randomly moving dots [the full details of this paradigm can be found *(34)*].

In the study of preparatory attention for motion and color we observed activity in the frontal-parietal attention network for both motion and color cues that was similar to prior work *(20, 21, 32, 35–39)*. As in our study comparing location versus color processing **(Fig. 8a)** we conducted direct statistical contrasts for motion versus color cues **(Fig. 8b)**. Again, as with the location versus color analyses, for motion versus color we found subregions of the attentional control network that were more active for motion than for color, and vice versa. Bilateral posterior superior frontal cortex and the left superior parietal cortex were selectively active to motion cues, but no superior cortical areas were selectively activated for the inverse contrast of color versus motion (color > motion). The activations for the motion versus color attentional control contrast were very similar to three regions of the attention control network we previously found to be selectively activated for preparatory spatial attention. There were also some differences, however, including failure to see activity in the right superior parietal cortex for the motion versus color contrast, suggesting that this right parietal region may indeed be more involved in directing attention to locations.

It is important to point out that in **Fig. 8** we are comparing results across different studies in different populations of volunteers, and indeed using slightly different imaging methods and analyses (although conceptually the same). In order to know whether preparatory attention for location and motion activated identical subregions in the frontal-parietal attention network, one would have to do that experiment, comparing the effects within subjects. Nonetheless, the similar findings across these studies are highly suggestive and require some interpretation. How can one conceptualize these fMRI results of similar specializations in location and motion attention?

One view is that the dorsal visual stream projecting from visual cortex to parietal cortex represents, in part, visual information required for generating actions *(40–42)*, rather than merely coding location per se as originally conceived by Ungerleider and Mishkin *(43)*. In line with this view is the evidence that the dorsal stream attentional control areas overlap with regions implicated in the control of voluntary eye movements *(33, 44–46)*. This opens the possibility that the close correspondence of superior frontal and parietal activity present for location and motion selective attention more than for nonspatial (color) attention is related to the role of both of these forms of attention in preparing actions, specifically those involved in oculomotor output. In the case of attention to locations the activity in oculomotor areas may represent preparation for *unexecuted* eye movements toward the attended location, and in the case of attention to directions of motion the activity may represent analogous preparation for ocular pursuit. Such a view is supported by the study of Astafiev et al. *(47)* who investigated brain activations in cued covert attention, overt saccade tasks, and pointing tasks. They observed overlapping activity in the superior frontal and parietal cortex for all three tasks. As noted, future studies will be required to address the speculations we have provided here and must include studies designed to compare and contrast different forms of preparatory attention.

3.3. Connectivity Analysis in Attentional Control Networks

The network(s) involved in attentional control, predominantly the frontal-parietal attentional control network, are clearly involved in a myriad of cognitive processes, as reflected by the evidence that subregions of the network appear to be specialized for certain aspects of attention in vision, as described in the foregoing *(32, 48, 49)*. In part, this reflects the differing roles of either subportions of the networks, and/or different interactions between nodes in the network in the service of different computational tasks. One way to assess the latter is to conduct analyses of the functional connectivity of the various regions of the attentional control network and their connectivity with sensory cortex with changing stimulus and task demands. One relatively new method is that of dynamic causal modeling (DCM) *(50–53)*.

We have applied DCM to investigate the connectivity between frontal cortex, parietal cortex, and visual cortex as a function of top-down attention and bottom-up stimulus salience in a preliminary study and analysis *(54)*. In our task, participants were cued to attend to one of two locations in the lower left and right visual fields for the duration short miniblocks of six trials; these locations were marked by mask stimuli. On each trial of the miniblock the "masks" were replaced briefly by a target (defined by the direction of the spatial cue) and a distracter in the unattended location in the opposite visual hemifield. We manipulated visual salience of the target and distracter by changing the luminance contrast of the stimuli *(55)* such that salient stimuli were higher in luminance contrast from the background than "masks" and nonsalient stimuli were identical in contrast to masks **(Fig. 9)**. The functional imaging data were acquired using echoplanar imaging on a 3-T Siemens Trio scanner and analyzed using SPM5 [Wellcome Department of Imaging Neuroscience, London, UK; *(56)*]. Based on the existing literature on voluntary spatial attention we identified regions of interest (ROIs) from our functional data reflecting the nodes of the attentional control and selection networks, including the FEF, IPS, and extrastriate OCC. Each functionally defined ROI was defined based on a combination of coordinates from group random-effects analysis and individual anatomy in order to provide uniformity between subjects and generality to the population *(57)*, but locating the ROI center based on individual anatomy resulted in greater precision.

As in prior studies, we found that the frontal-parietal attention network was engaged during the attention task. Both FEF

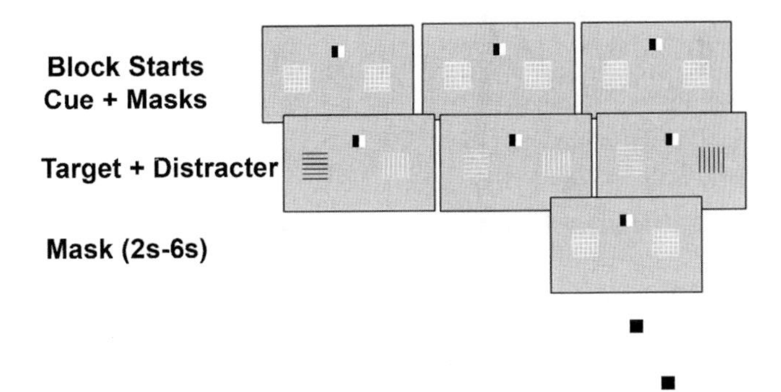

Fig. 9. Stimuli in dynamic causal modeling study of attention. Illustration of possible alternative trial conditions during an attend-left block. Each block began with the onset of bilateral mask stimuli and a fixation square that cued the attentional set for the block (i.e., black left half signals attention to the left). Each trial consisted of "masks" changing into vertical or horizontal lines that remained visible for 300 ms before being replaced by the *masks* again. On some trials, either the left or right line stimulus was visually salient, as defined by a change in color (*black*) and contrast. A variable delay followed each trial. In cued left and right attention blocks, participants responded during the delay with a manual button press to indicate the perceived orientation of lines on the attended side.

and IPS were engaged during attentional control in our task, and extrastriate cortex contralateral to the direction of covert attention was engaged. In addition, IPS showed sensitivity to task-irrelevant visual salience, that is, to the stimuli of increased contrast but only when the visually salient item was a distracter in an unattended location. Hence, FEF and IPS behaved differently with respect to the interaction of attentional control and bottom–up sensory factors. We constructed a series of models that each represents a different hypothesis of how connectivity is modulated during spatial attention and with visual salience. Model comparison was accomplished using a Bayesian selection procedure *(52)*. All models had full intrinsic connectivity and direct stimulation into OCC and were estimated separately for left and right hemisphere ROIs.

We first compared two models that tested whether salience primarily modulated the feedforward or feedback connection between IPS and OCC (**Fig. 10a**, model 1 and model 2). Both models include parameters corresponding to contralateral and ipsilateral spatial attention on the feedback connection from FEF to occipital regions due to the strong a priori evidence for the importance of FEF in modulating visual perception and neuronal activity in visual cortex *(58–61)*. The model (model 2) with feedback modulation by salience from IPS to OCC proved the best

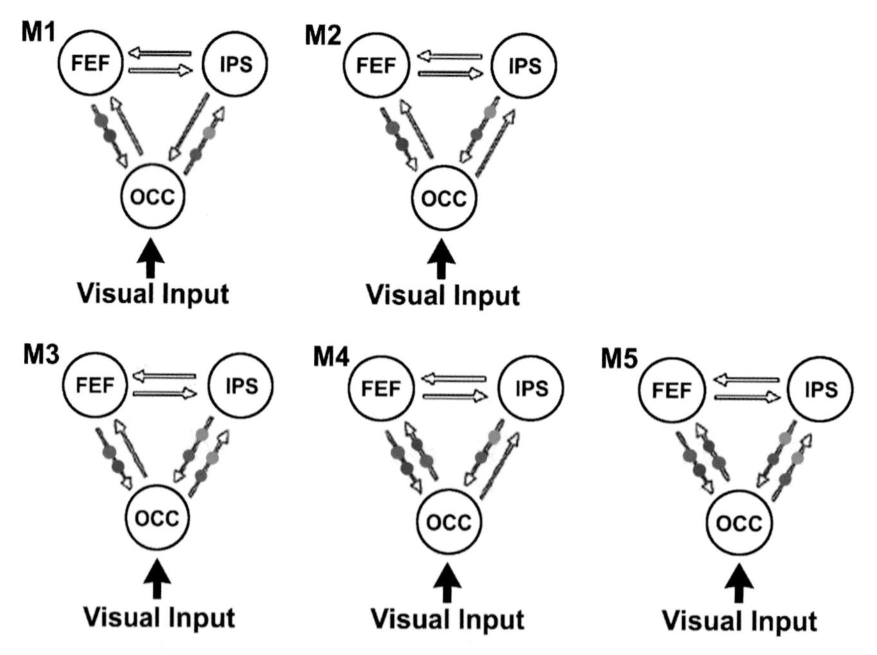

Fig. 10. Models of frontal, parietal, and occipital interactions with attention and stimulus salience. Illustrations of all dynamic casual modeling (DCM) models that were compared for each subject. All models shared full intrinsic connectivity and perturbation by direct visual stimulation to occipital cortex (OCC). The models differed only in terms of coupling parameters that reflect modulation of connectivity by contralateral and ipsilateral attention (*blue, red dots*) and salience (*green, purple dots*).

fit in 8/11 subjects in the left hemisphere (LH) and 9/11 subjects in the right hemisphere (RH). Next, we compared model 2 against two models with greater complexity (**Fig.10a**, models 3 and 4) that had additional modulation parameters for attention from OCC to FEF, or for salience from OCC to IPS. These models tested whether an increase in model accuracy due to the inclusion of feedforward modulation was sufficiently large to overcome any penalization due to a decrease in model parsimony. Model 4 was optimal in 11/11 subjects in the left hemisphere, and in 10/11 subjects in the right hemisphere, providing strong evidence in favor of model 4. Finally, we compared model 4 to a new model, model 5, containing bidirectional modulation of both attention and salience between FEF and OCC, and IPS and OCC, respectively. Model 5 was not found to explain the data well and therefore we concluded that model 4 was the best fit: Model 4 was characterized by bidirectional attentional modulation between FEF and OCC, and unidirectional salience modulation from IPS to OCC.

The specific pattern of change (increased vs. decreased) in connectivity due to our experimental conditions was investigated by taking the maximum a posteriori (MAP) parameter estimates from optimal model 4 for each subject and entering these into group-level analyses separately for each hemisphere (as replications). This permitted evaluation of intersubject consistency in how connection strengths are modulated as a function of spatial attention and stimulus salience and were investigated in two ways: *(1)* by comparing contralateral and ipsilateral parameter estimates on a single connection directly with a paired t test to test for differences between the two conditions and *(2)* by entering the contralateral or ipsilateral parameters separately into one-sample t tests (Bonferroni corrected) to test whether the direction of change in connection strength within each condition is positive or negative.

MAP parameter estimates of contralateral and ipsilateral attention for the connectivity from FEF onto OCC were significantly different such that changes due to contralateral attention were more positive than ipsilateral attention ($p < 0.001$ for both hemispheres). In contrast, contralateral and ipsilateral attention on the connection from OCC onto FEF were also significantly different in a paired t test ($p < 0.001$) but now ipsilateral attention resulted in more positive changes in connectivity. Although this difference is somewhat counterintuitive, it may be that this pattern of connectivity reflects a mechanism of “monitoring” the unattended location *(62)*.

Stimulus salience produced significant modulations of the connection from IPS to OCC. We found that this resulted from a significant decrease in coupling ($p < 0.05–0.01$). This demonstrates that the connectivity between IPS and OCC was reduced when a task-irrelevant, but visually salient stimulus was present. Together with findings from the conventional analyses, this suggests that IPS is sensitive to visual salience as well as the attentional demands of the task by reducing the strength of connectivity with visual cortex when irrelevant salience is present *(63)*.

In sum, we examined the role of the frontal-parietal attentional control network during spatial attention and found the expected pattern of activation in FEF, IPS, and OCC. Interestingly, FEF and IPS showed differential sensitivity to manipulations of spatial attention and stimulus salience; FEF activation was determined by the location of spatial attention, but IPS additionally showed sensitivity to the presence of visually salient distracters. This finding relates well to monkey studies of LIP in which neurons were found to encode not only physical salience, but also behavioral relevance *(62–64)*. Extrastriate visual areas showed greatest activation for attention directed to the contralateral visual hemifield, consistent with much prior work *(15, 17, 22, 32)*. Of main interest though were the results of our functional connectivity analysis using DCM methods.

We tested different models regarding the possible patterns of connectivity between nodes of the attention network during our task and found that in the optimal model, spatial attention modulated the connections between FEF and extrastriate OCC, whereas stimulus salience modulated the connection from IPS to extrastriate OCC. The connection from FEF to extrastriate OCC was strengthened for contralateral compared with ipsilateral attention, as might be expected from studies showing that microstimulation or TMS stimulation of FEF can increase the sensitivity of visual cortex in a spatially specific manner *(58–61)*. Overall activation was also greatest in both FEF and extrastriate OCC during contralateral attention, suggesting that these regions together contribute to maintenance and selective processing of attended contralateral visual information.

Surprisingly, we also found that ipsilateral attention produced a greater increase in the connection strength from extrastriate cortex to FEF, although both attentional conditions increased connectivity overall. Perhaps when a spatial location is being nominally ignored, information from that location is still being monitored by a change in the gain of information flow, possibly for the purpose of reorienting attention if something unexpected but behaviorally relevant occurs. Related to this is the finding that while attention modulates the connections between FEF and OCC, visual salience was primarily involved in modulating the connectivity from IPS to OCC. Connectivity from IPS to OCC was decreased when a salient distracter stimulus was present. Again, this may be related to reorienting attention by reducing the current focus of attention when a salient, potentially relevant event occurs in the visual scene *(63)*.

4. Conclusions

In this chapter we have reviewed the use of functional imaging in the study of attention. We addressed this topic from three perspectives: First, to lay out the theoretical and experimental design issues that are critical for studying attention, especially selective attention. Second, we illustrated that the attention system can be conceptualized as consisting of different components, and, therefore, functional imaging methods that permit these different components to be isolated and studied are necessary. Further, in passing we make note of the use of combined methods to provide temporal information (from electroencephalography) that is not available using functional imaging based on hemodynamic or metabolic methods. Third, we described evidence that specialization in attention systems can be investigated by experimental design approaches that capitalized on comparisons between different forms of attention. Lastly, we presented an application of the analysis of functional connectivity to probe the nodes within attention networks, in this case, using DCM as described in detail in a companion chapter of this volume by Karl Friston. We took our examples primarily from work in our laboratory but also presented the work of others where useful. As well, here we focused on visual attention, principally spatial attention, but the methodological approaches are germane to all studies of attention, across modalities, and including translational efforts in disorders of attention.

Methodological advances have helped to reveal the neural mechanisms of attention. It is clear that the brain attention mechanisms help to refine the processing of sensory information beginning as early as sensory-specific cortex, a concept that was hotly debated until the 1990s. For voluntary attention, a frontal and parietal attentional control system provides biasing signals to sensory cortex that result in selective sensory processing, and functional imaging has proven crucial for understanding these relationships in humans. These control systems and sensory (and decision and motor) systems interact to create specific attention networks in the service of goal-directed behavior in humans and animals.

References

1. James W. Principles of psychology. New York: H Holt, 1890.
2. Von Helmholtz H. Handbuch der Physiologischen Optik. Leipzig: L Vos, Hamburg, Germany, 1867.
3. Posner MI, Cohen Y. Components of visual orienting. In: Bouma H, Bouwhis D, eds. Attention and performance X. Hillsdale, NJ: Erlbaum, 1984:531–556.
4. Harter MR, Aine CJ. Brain mechanisms of visual selective attention. In: Parasuraman R, Davies DR, eds. Varieties of attention. Orlando: Academic Press, 1984:293–321.
5. Broadbent D. Perception and communication. New York: Pergamon, 1958.
6. Deutsch JA, Deutsch D. Attention: Some theoretical considerations. Psychol Rev 1963;70: 80–90.

7. Hernandez-Peon R, Scherrer H, Jouvet M. Modification of electrical activity in cochlear nucleus during attention in unanesthetized cats. Science 1956;123:331–332.
8. Van Voorhis S, Hillyard SA. Visual evoked potentials and selective attention to points in space. Percept Psychophys 1977;22:54–62.
9. Nunez PL, Srinivasan R. Electric Fields of the Brain: The Neurophysics of EEG, 2nd Edition. New York: Oxford University Press, 2006.
10. Moran J, Desimone R. Selective attention gates visual processing in the extrastriate cortex. Science 1985:229:782–784.
11. Chelazzi L, Miller EK, Duncan J, Desimone R. A neural basis for visual search in inferior temporal cortex. Nature 1993;363:345–347.
12. Chelazzi L, Miller EK, Duncan J, Desimon R. Responses of neurons in macaque area V4 during memory-guided visual search. Cereb Cortex 2001;11:761–772.
13. McAdams C, Reid RC. Attention modulates the responses of simple cells in monkey primary visual cortex. J Neurosci 2005;25:11023–11033.
14. Corbetta M, Miezin FM, Dobmeyer S, Shulman GL, Petersen SE. Selective and divided attention during visual discriminations of shape, color, and speed: Functional anatomy by positron emission tomography. J Neurosci 1991;11:2383–2402.
15. Heinze HJ, Mangun GR, Burchert W, et al. Combined spatial and temporal imaging of brain activity during visual selective attention in humans. Nature 1994;372:543–546.
16. Mangun GR, Hopfinger J, Kussmaul C, Fletcher E, Heinze HJ. Covariations in PET and ERP measures of spatial selective attention in human extrastriate visual cortex. Hum Brain Mapp 1997;5:273–279.
17. Mangun GR, Buonocore M, Girelli M, Jha A. ERP and fMRI measures of visual spatial selective attention. Hum Brain Mapp 1998;6:383–389.
18. Hillyard SA, Münte TF. Selective attention to color and location: An analysis with event-related brain potentials. Percept Psychophys 1984;36:185–198.
19. Mangun GR, Hillyard SA. Modulations of sensory-evoked brain potentials indicate changes in perceptual processing during visual-spatial priming. J Exp Psychol Hum Percept Perform 1991;17:1057–1074.
20. Hopfinger JB, Buonocore MH, Mangun GR. The neural mechanisms of top-down attentional control. Nat Neurosci 2000;3:284–291.
21. Giesbrecht B, Woldorff MG, Song AW, Mangun GR. Neural mechanisms of top-down control during spatial and feature attention. Neuroimage 2003;19:496–512.
22. Tootell RB, Hadjikhani N, Hall EK, et al. The retinotopy of visual spatial attention. Neuron 1998;21:1409–1422.
23. Desimone R, Duncan J. Neural mechanisms of selective visual attention. Ann Rev Neurosci 1995;18:193–222.
24. Posner MI, Petersen SE. The attention system of the human brain. Annu Rev Neurosci 1990;13:25–42.
25. Gitelman DR, Nobre AC, Parrish TB, et al. A large-scale distributed network for covert spatial attention: Further anatomical delineation based on stringent behavioural and cognitive controls. Brain 1999;122:1093–1106.
26. Posner MI, Snyder CRR, Davidson BJ. Attention and the detection of signals. J Exp Psychol Gen 1980;109:160–174.
27. Luck SJ, Chelazzi L, Hillyard SA, Desimone R. Neural mechanisms of spatial selective attention in areas V1, V2 and V4 of macaque visual cortex. J Neurophysiol 1997;77:24–42.
28. Chawla D, Rees G, Friston KJ. The physiological basis of attentional modulation in extrastriate visual areas. Nat Neurosci 1999;7:671–676.
29. Burock MA, Buckner RL, Woldorff MG, Rosen BR, Dale AM. Randomized event-related experimental designs allow for extremely rapid presentation rates using functional MRI. Neuroreport 1998;9:3735–3739.
30. Ollinger JM, Corbetta M, Shulman GL. Separating processes within a trial in event-related functional MRI. Neuroimage 2001;13:218–229.
31. Ollinger JM, Shulman GL, Corbetta M. Separating processes within a trial in event-related functional MRI. Neuroimage 2001;13:210–217.
32. Woldorff MG, Hazlett CJ, Fichtenholtz HM, Weissman DH, Anders AM, Song AW. Functional parcellation of attentional control regions of the brain. J Cogn Neurosci 2004;16:149–165.
33. Corbetta M. Frontoparietal cortical networks for directing attention and the eye to visual locations: Identical, independent, or overlapping neural systems? Proc Natl Acad Sci USA 1998;95:831–838.
34. Fannon SP, Saron CD, Mangun GR. Baseline shifts do not predict attentional modulation of target processing during feature-based visual attention. Front Hum Neurosci 2008;1:7. doi:10.3389/neuro.09.007.2007.
35. Corbetta M, Kincade JM, Ollinger JM, McAvoy MP, Shulman GL. Voluntary orienting is dissociated from target detection in human posterior parietal cortex. Nat Neurosci 2000;3:292–297.

36. Kastner S, Ungerleider L. Mechanisms of visual attention in the human cortex. Annu Rev Neurosci 2000;23:315–341.
37. Kincade JM, Abrams RA, Astafiev SV, Shulman GL, Corbetta M. An event-related functional magnetic resonance imaging study of voluntary and stimulus-driven orienting of attention. J Neurosci 2005;25:4593–4604.
38. McMains SA, Fehd HM, Emmanouil TA, Kastner S. Mechanisms of feature and space-based attention: Response modulation and baseline increases. J Neurophysiol 2007;98:2110–2121.
39. Wilson KD, Woldorff MG, Mangun GR. Control networks and hemispheric asymmetries in parietal cortex during attentional orienting in different spatial reference frames. Neuroimage 2005;25:668–683.
40. Cohen YE, Andersen RA. A common reference frame for movement plans in the posterior parietal cortex. Nat Rev Neurosci 2002;3:553–562.
41. Goodale M, Milner A. Separate visual pathways for perception and action. Trends Neurosci 1992;15:20–25.
42. Goodale M, Westwood D. An evolving view of duplex vision: Separate but interacting cortical pathways for perception and action. Curr Opin Neurobiol 2004;14:203–211.
43. Ungerleider LG, Mishkin M. Two cortical visual systems. In: Ingle DJ, Goodale MA, Mansfield RJW, eds. Analysis of Visual Behavior. Cambridge: MIT Press, 1982:549–586.
44. Corbetta M, Tansy AP, Stanley CM, Astafiev SV, Snyder AZ, Shulman GL. A functional MRI study of preparatory signals for spatial location and objects. Neuropsychologia 2005;43:2041–2056.
45. Moore T, Fallah M. Microstimulation of the frontal eye field and its effects on covert spatial attention. J Neurophysiol 2004;91:152–162.
46. Nobre AC, Sebestyen GN, Miniussi C. The dynamics of shifting visuospatial attention revealed by event-related potentials. Neuropsychologia 2000;38:964–974.
47. Astafiev SV, Shulman GL, Stanley CM, Snyder AZ, Van Essen DC, Corbetta M. Functional organization of human intraparietal and frontal cortex for attending, looking, and pointing. J Neurosci 2003;23:4689–4699.
48. Slagter HA, Giesbrecht B, Kok A, et al. fMRI evidence for both generalized and specialized components of attentional control. Brain Res 2007;1177:90–102.
49. Corbetta M, Shulman G. Control of goal-directed and stimulus-driven attention in the brain. Nat Rev Neurosci 2002;3:201–215.
50. Friston KJ. Dynamic causal modeling of brain responses. In: Filippi M, ed. FMRI Techniques and Protocols. New York: Humana Press, 2008.
51. Friston KJ, Harrison L, Penny W. Dynamic causal modelling. Neuroimage 2003;19:1273–1302.
52. Penny WD, Stephan KE, Mechelli A, Friston KJ. Comparing dynamic causal models. Neuroimage 2004;22:1157–1172.
53. Stephan KE, Harrison LM, Penny WD, Friston KJ. Biophysical models of fMRI responses. Curr Opin Neurobiol 2004;14:629–635.
54. Geng JJ, Mangun GR. Spatial attentional selection in the presence of irrelevant stimulus salience. San Diego, CA: Society for Neuroscience Abstracts, 2007; Vol. 32.
55. Reynolds JH, Desimone R. Interacting roles of attention and visual salience in V4. Neuron 2003;37:853–863.
56. Friston KJ, Holmes AP, Worsley KJ, Poline J-P, Frith CD, Frackowiak RSJ. Statistical parametric maps in functional imaging: A general linear approach. Hum Brain Mapp 1995;2:189–210.
57. Friston KJ, Holmes AP, Worsley KJ. How many subjects constitute a study? Neuroimage 1999;10:1–5.
58. Moore T, Armstrong KM. Selective gating of visual signals by microstimulation of frontal cortex. Nature 2003;421:370–373.
59. Moore T, Armstrong KM, Fallah M. Visuomotor origins of covert spatial attention. Neuron 2003;40:671–683.
60. Ruff CC, Blankenburg F, Bjoertomt O, et al. Concurrent TMS-fMRI and psychophysics reveal frontal influences on human retinotopic visual cortex. Curr Biol 2006;16:1479–1488.
61. Silvanto J, Lavie N, Walsh V. Stimulation of the human frontal eye fields modulates sensitivity of extrastriate visual cortex. J Neurophysiol 2006;96:941–945.
62. Bisley JW, Goldberg ME. Neuronal activity in the lateral intraparietal area and spatial attention. Science 2003;299:81–86.
63. Bisley JW, Goldberg ME. Neural correlates of attention and distractibility in the lateral intraparietal area. J Neurophysiol 2006;95:1696–1717.
64. Gottlieb J. From thought to action: The parietal cortex as a bridge between perception, action, and cognition. Neuron 2007;53:9–16.
65. Mangun GR, Fannon SP. Networks for attentional control and selection in spatial vision. In: Mast F, Jäncke L, eds. Spatial Processing in Navigation, Imagery and Perception. Springer: Amsterdam, 2007

Chapter 13

fMRI of Memory

Federica Agosta, Indre V. Viskontas, and Maria Luisa Gorno-Tempini

Summary

Numerous fMRI studies have investigated the network of brain regions critical for memory. Whereas neuropsychological techniques can delineate the brain regions that are necessary for intact memory function, neuroimaging techniques can be used to investigate which regions are recruited during healthy memory formation, storage, and retrieval. For example, fMRI studies have shown that lateral prefrontal cortex (PFC) supports some components of working memory function. However, working memory is not localized to a single brain region but is likely a property of the functional interaction between the PFC and posterior brain regions. The medial temporal lobe (MTL) and its connections with neocortical, prefrontal, and limbic structures are implicated in episodic memory. Semantic memory is mediated by a network of neocortical structures, including lateral and anterior temporal lobes, and inferior frontal cortex, possibly to a greater extent in the left hemisphere. Memory for semantic information benefits from the MTL for only a limited time, and can be acquired, albeit slowly and with difficulty, without it. To date, most of the emphasis has been on exploring the unique aspects of these different types of memory. Some evidence, however, of functional overlap in general retrieval processes does exist.

Key words: Working memory, Encoding, Retrieval, Episodic memory, Semantic memory, Prefrontal cortex, Medial tmporal lobe

1. Introduction

Memory shapes our behavior by allowing us to store, retain, and retrieve past experiences, and thus enables us to imagine the consequences of our actions. These processes influence and are modified by the type of information that is to be remembered, the duration of time over which it must be retained, and the way in which the brain will use the information in the future. The neural circuits underlying these processes are dynamic, reflecting the flexibility of memory itself. fMRI enables detailed

M. Filippi (ed.), *fMRI Techniques and Protocols*, Neuromethods, vol. 41
DOI 10.1007/978-1-60327-919-2_13, © Humana Press, a part of Springer Science+Business Media, LLC 2009

study of the neural networks that support memory function. To delineate the neural circuitry underlying memory, it is helpful to breakdown memory into simpler components. In this chapter, we focus on memory that is consciously accessible, and use the duration of retention to dictate its parcellation: beginning with working memory, which holds information in mind for seconds to minutes, and moving on to long-term memory, which can be further divided into episodic and semantic memory.

2. Working Memory

Working memory (WM) refers to the temporary storage and manipulation of information that was most recently experienced or retrieved from long-term memory but is no longer available in the external environment *(1–3)*. Most models of WM separate two components of WM: temporary stores of information, in the form of "buffers" or "slave systems," that are usually modality-specific, and a central "executive" or set of processes that manipulate the information *(2, 4)*. Items in WM are stored only as long as the information is either being rehearsed (subvocally) or manipulated in some other fashion (i.e., rotated or integrated with existing information in semantic memory). The capacity of WM is limited by attention to about seven (plus or minus two) meaningful "bits" (or chunks) of information – these bits can be manipulated and either discarded or associated and transferred into long-term memory *(5)*. WM is central to everyday functioning and contributes significantly to other areas of cognition. Baddeley and Hitch *(1)* proposed a model of memory that has influenced virtually all subsequent research in the area. Their model is composed of a three-component system: (1) the "phonological loop," comprising a limited capacity phonological store in which verbal information is stored temporarily and maintained by subvocal rehearsal (e.g., repeated subvocal articulation when trying to keep a phone number in mind), (2) the "visuospatial sketch pad," a storage buffer for nonverbal material, such as the visual representations of objects, and (3) the "central executive," which is responsible for strategic manipulation and execution of the aforementioned "slave" systems. The original model has recently been updated *(2)* to include an "episodic buffer" that provides an interface between the subsystems of WM and long-term memory.

The first evidence for a role of prefrontal cortex (PFC) in WM came from lesion and electrophysiological studies in non-human primates *(6, 7)*. The first neurons discovered – showing persistent activity during the delay period of WM – were found in the monkey PFC using single-unit neuron recording techniques

(8–10). This sustained activity is thought to provide a bridge between the stimulus cue, for instance, the location of a flash of light, and its contingent response, such as a saccade to the remembered location. Persistent activity during blank memory intervals is a very powerful observation and established a strong link implicating the PFC as a critical node supporting WM *(11)*. Since then, physiological studies in nonhuman primates have revealed active delay neurons in a large number of brain regions, including the dorsolateral (DL) and ventrolateral (VL) PFC, the intraparietal sulcus (IPS), posterior perceptual areas, and subcortical structures, such as the caudate and the thalamus (for a review, *see* **ref.** *12)*. In humans, however, the advent of modern functional brain imaging techniques, such as positron emission tomography (PET) and fMRI, enabled a more detailed study of the functional neuroanatomy of WM processes. Many of the neuroimaging experiments designed to elucidate the neural underpinnings of WM separate executive control processes from storage. For example, a number of studies have been designed such that WM maintenance is the task of interest (e.g., delayed response, delayed recognition, delayed alternation, and delayed match-to-sample tasks). In a typical delayed recognition trial, the subject is first required to remember a stimulus presented during a *cue* period and then to maintain this information for a brief *delay* interval when the stimulus is absent. Then, the subject responds to a *probe* stimulus to determine whether the information was successfully retained. Thus, the maintenance and recognition decision processes are temporally segregated and can be investigated in relative isolation. In this way, using fMRI, investigators can record neural activity during these distinct stages of WM. Brain regions exhibiting persistent activity above resting baseline during the delay period are often interpreted as being involved in WM maintenance processes *(13–21)*. Functional neuroimaging studies have revealed that the same collection of regions that were shown to be involved in WM in nonhuman primates displays significantly increased activity during delay tasks in humans *(22–30)*. For instance, in one fMRI study, subjects were scanned while they performed an oculomotor delayed matching-to-sample task that required maintenance of the spatial position of single dot of light over a delay period after which a memory-guided saccade was generated *(28)*. Both frontal eye fields and IPS showed activity that spanned the entire delay period. Moreover, the magnitude of the activity correlated positively with the accuracy of the memory-guided saccade that followed later *(28)*. Despite the relative simplicity of the delay period, however, multiple cognitive processes remain engaged concurrently, including information maintenance, suppression of distraction, motor response preparation, mental timing, expectancy, monitoring of internal and external states, and preservation of alertness. As a result,

even the maintenance period of WM is likely to be mediated by a distributed network of distinct brain regions, rather than to be localized to a single brain region *(21)*.

2.1. Organization of the WM Network

The extensive reciprocal connections from the PFC to virtually all cortical and subcortical structures place the PFC in an unique neuroanatomical position to monitor and manipulate diverse cognitive processes *(21)*. Recently, Marklund et al. *(29)* employed a mixed block and event-related design in an fMRI study of episodic, semantic, and working memories contrasted with sustained attention. This approach identified transient activity, particularly in the left DL PFC, that appears to reflect the operation of WM during retrieval from long-term memory *(29)*. Furthermore, it has been found that the PFC shows activity during retention interval of delay task regardless of the type of information (e.g., spatial, faces, objects, words) *(23, 25, 30)*. In fact, there is a critical mass of functional neuroimaging studies investigating the neural activity in the lateral PFC during the delay period in humans (for review, *see* **ref.** *11)*.

A controversial issue in the literature is the extent to which WM can be segregated anatomically according to the type of to-be-retained material (e.g., verbal, space, object, visual, auditory, etc.) or the component processes (e.g., maintenance vs. manipulation of information) (for a review, *see* **refs.** *11, 31–34)*.

2.1.1. Organization of WM by Material Type

First, let us evaluate the evidence supporting anatomical segregation on the basis of stimulus category. Three types of material have been most commonly studied: verbal, spatial, and object information. A prevalent theory of material-type segregation in the frontal cortex suggests that there are dorsal and ventral memory streams for spatial and object information, respectively, similarly to the "where" and "what" pathways of the visual system *(35)*. The dorsal stream projects from the extrastriate cortex to the inferior parietal lobule (IPL) and the IPS and is involved in processing spatial information *(35)*. The ventral stream extends from the extrastriate cortex to the inferior surface of the frontal pole and processes object information *(35)*. Within the frontal cortex, WM for spatial information involves the superior DL PFC or the superior frontal sulcus, whereas object WM relies upon several mid- and inferior frontal regions (VL PFC) (for review, *see* **ref.** *11)*. Furthermore, there is a tendency for verbal and object WM to recruit more left-hemisphere areas, and for spatial tasks to recruit more right-hemisphere areas (for review, *see* **ref.** *11)*.

The processing of nonverbal spatial information is right lateralized and associated with the activation of a fronto-parietal network (e.g., *see* **ref.** *36–41)*. Using event-related fMRI, Courtney et al. *(37)* demonstrated a neuroanatomical dissociation between delay period activity during WM maintenance for either the identity (object memory) or location (spatial memory) of a set of three face stimuli. Greater activity during the delay period on face identity trials was observed in the left inferior frontal gyrus (IFG), whereas greater activity during the delay period of the location task was observed in dorsal frontal cortex (bilateral superior frontal sulcus) *(37)* **(Fig. 1)**.

WM for objects, mostly visually presented faces, houses, and line drawings that are not easily verbalizable, seems to be

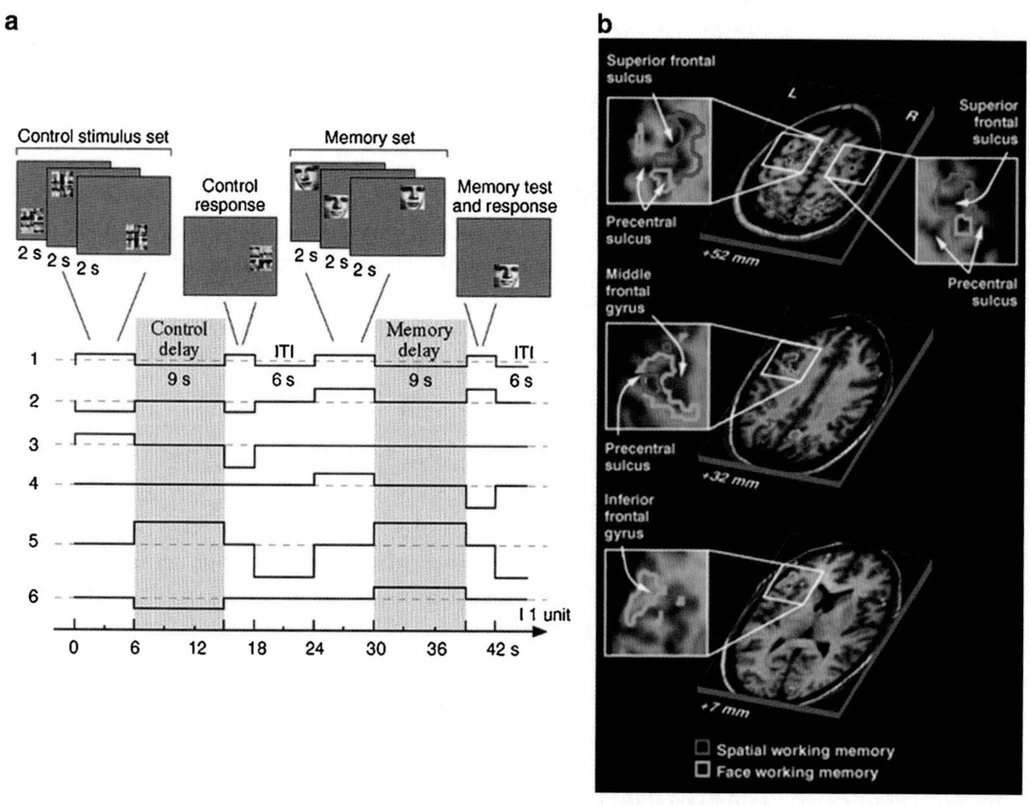

Fig. 1. An event-related fMRI study of delay period activity during working memory (WM) maintenance for either the identity (object memory) or location (spatial memory) of a set of three face stimuli. (**a**) Schematic depiction of the fMRI tasks: subjects saw a series of three faces, each presented for 2 s in a different location on the screen, followed by a 9-s memory delay. Then a single test face appeared in some location on the screen for 3 s, followed by a 6-s intertrial interval. Before each series, subjects were instructed to remember the locations or the identities of the three faces in the memory set. For the spatial task, the subject indicated with a left or right button whether the test location was the same as one of the three locations presented in the memory set, regardless of the face that marked that location. For the face memory task, the subject indicated whether the test face was the same as one of the three faces observed in the memory set, regardless of the location where the face appeared. For the sensorimotor control task, scrambled faces appeared (control stimulus set), and when the fourth scrambled picture appeared after the delay, subjects pressed both buttons (control response). Contrasts between task components are shown below the task diagram: (1) visual stimulation vs. no visual stimulation, (2) memory stimuli vs. control stimuli, (3) control stimulus set vs. control response, (4) memory stimulus set vs. test stimulus and response, (5) delays during anticipation of response vs. intertrial intervals, and (6) memory delays vs. control delays. (**b**) Areas with significant sustained recruitment in a single subject during the WM delay for faces (*blue outline*) and for spatial locations (*red outline*) overlaid onto the subject's Talairach normalized anatomical MR image. Level above the bicommissural plane is indicated for each axial section [from Courtney et al. ***(37)***].

right-lateralized and activates the temporal-occipital regions [Brodmann area (BA) 37] (e.g., *see* **refs**. *22, 27, 30, 36, 39, 42–44)*.

In contrast to object, processing of verbal WM activates regions in the left hemisphere. Broca's area (BAs 44/45), premotor areas (supplementary motor area and premotor cortex) *(45–48)*, and the cerebellum *(25, 47, 49, 50)* are critical for the articulatory subvocal rehearsal. Phonological maintenance has been associated with activity in parietal areas, particularly the inferior parietal lobe (i.e., BAs 39 and 40) *(25, 45, 46, 49–53)*, but also in the superior parietal lobe *(47, 49, 52, 53)*. For instance, using an event-related design, Chein and Fiez *(45)* designed a delayed serial recall task requiring subjects to encode, maintain, and overtly recall sets of verbal items for which phonological similarity, articulatory length, and lexical status were manipulated and reported that Broca's area and BA 40 showed patterns of sustained activity during the delay period of a verbal WM task.

More recently, Raye et al. *(48)* provided evidence that left VL PFC appears associated with subvocal rehearsal of single words, whereas left DL PFC activation is associated with participants "refreshing" (simply thinking about) the visual appearance of a recently presented word *(48)*.

Despite this growing evidence for segregation by material type, there are also several human functional imaging studies that have failed to find evidence for segregation in PFC WM activity (e.g., *see* **refs.** *17, 54, 55)*. For instance, using an event-related design, Postle and D'Esposito *(17)* evaluated the organization of WM for the identity and location of visually presented stimuli (target stimuli for all object trials were 16 abstract polygon stimuli, determined in normative testing to be difficult to associate with real-world objects). Although the task produced considerable delay-period activity in VL PFC, DL PFC, and superior frontal cortex, in no subject, PFC activity was greater for one stimulus domain than for the other *(17)*. Moreover, in a large meta-analysis based on 60 neuroimaging (both PET and fMRI) studies of WM *(55)*, analyses of material type showed the expected dorsal-ventral dissociation between spatial and nonspatial storage in the posterior cortex, but not in the frontal lobe. Some support was found for left frontal dominance in verbal WM, but only for tasks with low executive demand. Executive demand increased right lateralization in the frontal cortex for spatial WM *(55)*.

2.1.2. Organization of WM by Process Type

The other axis along which investigators have suggested that human lateral PFC involvement in WM is segregated is according to the type of operation performed upon the contents of WM, rather than the type of information being maintained. In particular, several fMRI studies have focused on the distinction between two fundamental WM processes, namely the passive *maintenance* of information in short-term memory and the active *manipulation* of this information, within the PFC (for review, *see* **ref.** ***11***). This model received initial support from a PET study by Owen et al. ***(56)*** in which dorsal PFC activation was found during three spatial WM tasks thought to require greater monitoring of remembered information (i.e., a mnemonic variant of modified Tower of London planning task requiring short-term retention and reproduction of problem solutions) than two other WM tasks (i.e., the modified Tower of London planning task, and a control condition that involved identical visual stimuli and motor responses) that activated only ventral PFC.

This model was also tested using fMRI (for review, *see* **ref.** ***11***). The VL PFC has been suggested to be primarily involved only in WM mechanisms that support simple retrieval of information for sensory-guided sequential behavior (maintenance) ***(57–60)***, whereas DL PFC (particularly BA 46) has been found to serve mechanisms of active monitoring and manipulation of information (or generalized executive processing) in WM ***(61–63)***. For instance, in an event-related fMRI study ***(61)***, subjects were presented with two types of trials in random order in which they were required to either maintain a sequence of letters across a delay period or manipulate (alphabetize) this sequence during the delay in order to respond correctly to a probe. The authors found that dorsal PFC activity was greater in trials during which actively maintained information was manipulated, providing further support for a process-specific PFC organization. However, this functional PFC division has not been consistently replicated and seems to vary with materials and difficulty ***(55)***. Moreover, this distinction is not compatible with evidence of continuous DL PFC activity in tasks without any manipulation ***(33, 64)***. A large meta-analysis ***(55)*** showed that tasks requiring executive processing generally produce more dorsal frontal activations than do storage-only tasks, but not all executive processes show this pattern. For instance, superior frontal cortex (BAs 6, 8, and 9) responded most when WM must be continuously updated and when memory for temporal order has to be maintained ***(55)***. Right ventral frontal cortex (BAs 10 and 47) responded more frequently with demand for manipulation (including dual-task requirements or mental operations) ***(55)***. Posterior parietal cortex (BA 7) was found to be involved in all types of executive functions ***(55)***.

2.2. Medial Temporal Lobe Involvement in WM

fMRI studies of WM have also found that the fronto-parietal network is not the only region that is active during the temporary retention of task-relevant information. PFC and parietal cortex do not seem to be sufficient to perform WM for novel stimuli when parahippocampal regions are lesioned ***(65)***, although they are sufficient to maintain normal WM for familiar stimuli ***(66)***. Surprisingly, early fMRI studies of WM did not report activity within parahippocampal regions such as perirhinal (PrC) or entorhinal cortex *(67)*. An fMRI study by Stern et al. *(68)* demonstrated differential activation for novel vs. familiar stimuli during performance of a 2-back WM task. This study showed that WM for a highly familiar set of complex visual images primarily activated prefrontal and parietal cortices, whereas the same task using novel (trial-unique) visual images strongly activated parahippocampal structures in addition to prefrontal and parietal cortices. Activation of parahippocampal structures associated with WM for novel stimuli has also been shown in an event-related fMRI study using novel face stimuli *(26, 69)*.

2.3. Connectivity Analysis Within the WM Network

By varying experimental design [e.g., parametric memory load variation *(24, 43)*], attempts have been made to associate identified brain regions with different processes occurring during the delay. The first fMRI study aimed at characterizing regional interactions in WM has used a block-design fMRI paradigm with a graded *n*-back verbal WM task *(70)*. Changes of effective connectivity within a fronto-parietal WM network related to different levels of WM load were studied. The results revealed enhanced inferior fronto-parietal connectivity and also increased interhemispheric communication between DL PFC regions as correlates of increasing WM load *(70)*. A more recent event-related fMRI study characterized the neural network mediating the on-line maintenance of faces *(71)* (**Fig. 2**). The fusiform face area (FFA)

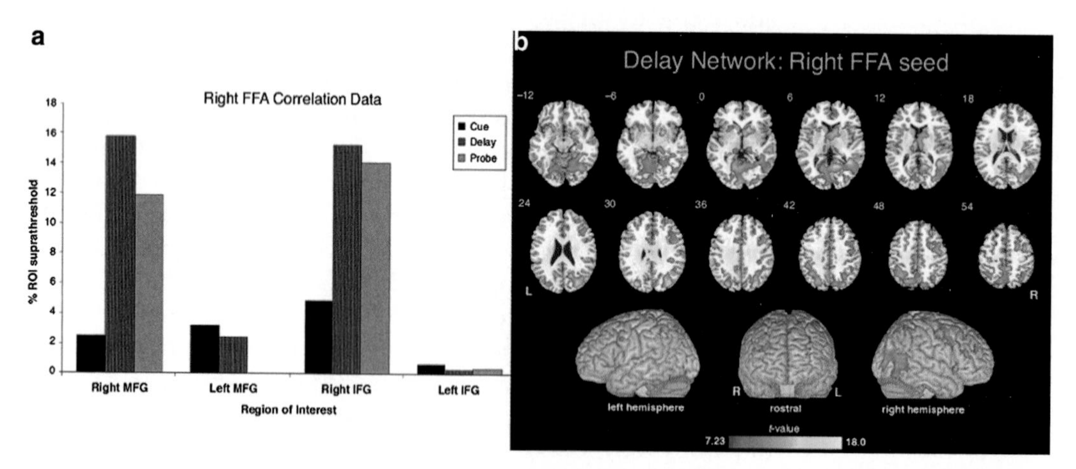

Fig. 2. Functional connectivity analysis between brain regions associated with the maintenance of a representation of a visual stimulus (face recognition task) over a short delay interval. The fusiform face area (FFA) was used as the exploratory seed. (**a**) Quantification of right FFA correlation effects in the prefrontal cortex (PFC). Bars indicate the number of significant voxels present during each stage (cue, delay, and probe) in four prefrontal regions of interest. *MFG* middle frontal gyrus, *IFG* inferior frontal gyrus, *ROI* region of interest. (**b**) Delay period right FFA seed correlation map including bilateral regions in the dorsolateral and ventrolateral PFC, premotor cortex, IPS, caudate nucleus, thalamus, hippocampus, and occipitotemporal regions. Activations are thresholded at $p < 0.05$ (corrected) and are shown overlaid on both axial slices and a three-dimensionally rendered Montreal Neurological Institute (MNI) template brain. The color scale indicates the magnitude of t values [from Gazzaley et al. *(71)*].

was defined as a seed and was then used to generate whole-brain correlation maps. A random effects analysis revealed a network of brain regions exhibiting significant correlations with the FFA seed during the WM delay period. This maintenance network included the DL and VL PFC, the premotor cortex, the IPS, the caudate nucleus, the thalamus, the hippocampus, and occipito-temporal regions *(71)*. These findings support the notion that the coordinated functional interaction between nodes of a widely distributed network underlies the active maintenance of a perceptual representation and provides convergent evidence that fronto-parietal and interhemispheric frontal connectivities are central in WM *(70, 71)*.

3. Long-Term Memory for Consciously Accessible Material: Episodic and Semantic

Within the declarative or consciously accessible long-term memory system, "episodic" and "semantic" memory can be distinguished. Episodic memory (EM) allows the recollection of unique personal experiences: rich, vivid reexperiencing of past events. Semantic memory (SM), in contrast, refers to generic information that is acquired across many different instances and accessed independently of the details of the context in which the information was first encountered *(72)*. This fractionation of declarative memory is supported by evidence that episodic and semantic memory have distinctive anatomical substrates *(73–76)*. Therefore, we will consider each memory type individually. Of note, autobiographical memory can be either semantic, as in one's knowledge of the names of all the schools that one attended, or episodic, as in one's memory for a particular birthday: what binds autobiographical memories to each other is self-awareness.

4. Episodic Memory

EM enables us to access and reexperience the sights, sounds, smells, and other details of a specific event *(72)*. Most EMs are available for several minutes or hours but over time access to their details degrades *(77)*. Others remain accessible with their details for a lifetime *(78)*. This temporal difference in storage highlights the complexity of EM: episodic remembering is composed of several component processes, including the retrieval of information from across sensory domains and the reconstruction of an event from a set of individual details.

4.1. Distinguishing Encoding and Retrieval Processes

One of the major advances in the study of EM has been the application of neuroimaging techniques to distinguish the component processes of encoding and retrieval. In neuropsychological studies, it is often difficult or even impossible to separate failures of EM due to encoding or retrieval processes. Functional imaging affords researchers the opportunity to separate the neural underpinnings of these processes and is particularly helpful in understanding the distinct roles of medial temporal lobe (MTL) subregions, which are often damaged equivocally in patients with lesions *(79)*. Furthermore, neuroimaging evidence has shown that prefrontal and other cortical areas are engaged during episodic remembering, a fact that had not been clear from patient data alone *(80)*.

Neuroimaging studies have indicated that individual elements of EMs may be permanently stored within the same neocortical regions that are involved in initial processing and analyzing of the information *(81, 82)*. Several studies have demonstrated the reactivation of the visual cortex during retrieval of visual details *(81, 83)*, activation in the auditory cortex during auditory memory retrieval *(84)*, and activation in the motor cortex during the retrieval of memory for actions *(85)*. Insofar as episodic remembering involves reexperiencing the details of an event, it is not surprising that brain regions involved in the initial perception of these details are reactivated during their retrieval.

According to several influential memory models, each different cortical region makes a unique contribution to the storage of a given memory and all regions participate together in the creation of a complete memory representation *(86–88)*. The MTL, then, is saddled with the task of binding together these different regional contributions into a coherent memory trace *(86)*. In the MTL, the hippocampal formation receives processed sensory information from association areas in the frontal, parietal, and occipital lobes via the parahippocampal cortex *(89)*. Given its anatomical placement and architecture, the hippocampus has the unique ability to bind "what happened," "when it happened," and "where it happened" together *(86)*. The architecture of the hippocampus includes a circular pathway of neurons from the entorhinal cortex to the dentate gyrus, CA3 and CA1 neurons of the hippocampus to the subiculum, and back to the entorhinal cortex *(90)*. The connections within the hippocampal formation and between the MTL and neocortical regions are formed more rapidly than are the connections between disparate cortical regions *(88)*. Therefore, when a particular cue in the environment or the mental state of the person activates cells in the cortical regions, the MTL network that is associated with that cue is reactivated and the entire neocortical representation is strengthened. As multiple reactivations occur, the connections between the relevant neocortical regions are slowly strengthened until the memory trace no longer depends on the activity of the MTL, but may be entirely represented in the neocortex *(86)*. Recent evidence, however, using both fMRI and neuropsychologal techniques, suggests that the hippocampus remains involved in the retrieval of EMs regardless of the age of the memory *(91)*: several authors, then, have proposed that the hippocampus acts as a pointer or index, recreating the activation in the neocortex that represents the individual elements of the memory *(92)*.

4.1.1. Encoding

Simply perceiving and attending to information in the world is not sufficient to create a lasting long-term memory trace. Some other process or processes are engaged in order to bind elements of an episode into a coherent memory trace. There is some disagreement concerning whether all memories are initially episodic and later become semantic, or whether some memories are semantic in nature from the outset *(93, 94)*. One might argue that every learning episode is encoded as such, and as items from an episode get integrated into the network of semantic information, those items become disassociated from the episode itself and associated instead with the information already in semantic memory *(93)*. In fact, several memory models suggest that episodic and semantic information is learned via different mechanisms: elements of an episode are rapidly bound together by autoassociative processes in the dentate gyrus and CA3 field *(95)*, while semantic information is gradually acquired over many repetitions by reorganization of Hebbian synapses in the neocortex *(87, 96, 97)*.

Pioneering neuroimaging studies in the 1990s implicated the PFC, particularly the left lateral PFC, in semantic or associative encoding, by showing that this region, in addition to the MTL, shows greater neural activity during semantic encoding than during more superficial or perceptual encoding *(98, 99)*. Furthermore, novel stimuli have been shown to elicit greater neural activity in the MTL than familiar stimuli *(100, 101)*, providing more evidence for the involvement of the MTL in encoding processes.

A direct link to episodic encoding processes, however, required the advent of event-related fMRI designs to investigate these cognitive processes *(102)*. Encoding trials were binned according to whether information presented on a given trial is subsequently remembered or forgotten. Subsequent memory studies have shown that activation in the left and right PFC, the parahippocampal gyrus *(103, 104)*, and hippocampus *(105)* during encoding predicts successful memory retrieval (*see* **Fig. 3**). Since these original studies, many more studies have replicated the finding that MTL activity correlates with episodic encoding *(106–109)*. Furthermore, fMRI studies have shown that hippocampal activity correlates with the subsequent retrieval of contextual details presented during encoding, while activity

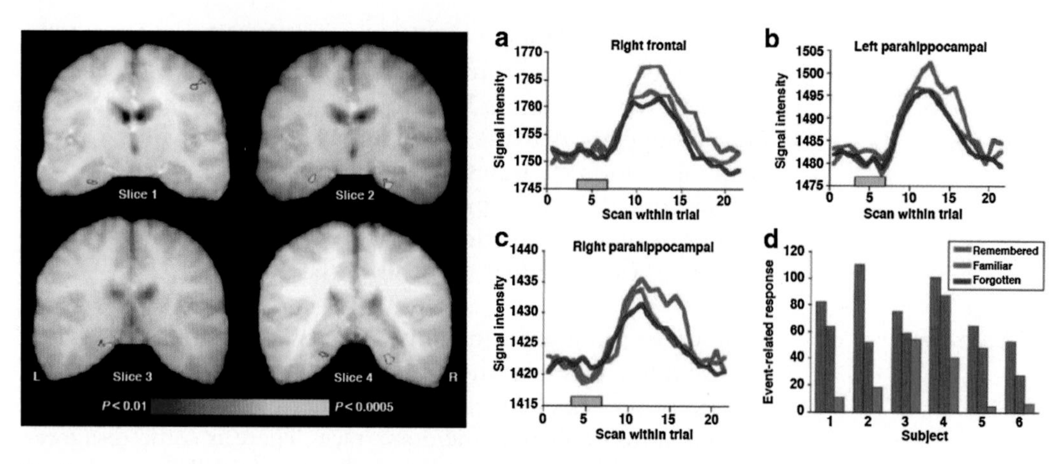

Fig. 3. Composite statistical activation maps displaying voxels with significant positive correlations between event-related activations to pictures and subsequent memory for those pictures. Areas activated are right dorsolateral prefrontal cortex (*upper right* in slice 1) and bilateral parahippocampal cortex (*lower left* in slices 1 and 3; *left* and *right* in slices 2 and 4). Examples of average signal magnitude during study from six subjects in (**a**) right frontal (slice 1), (**b**) left parahippocampal (slice 4), and (**c**) right parahippocampal (slice 2) regions for remembered (*red*), familiar (*green*), and forgotten (*blue*) pictures. *Gray block* depicts onset and offset of picture presentation. (**d**) Mean voxel response in parahippocampal areas showing signficant correlation with subsequent memory in each subject for remembered, familiar, and forgotten pictures [from Brewer et al. ***(103)***].

in the PrC correlates with successful retrieval of an individual item, but not with retrieval of episodic details *(107–111)*. Most recently, Staresina and Davachi *(112)* have shown that since the PrC receives inputs mainly from cortical areas devoted to the processing of visual information, recruitment of this region also correlates with the retrieval of visual features of items, such as the color in which they were presented, but not with other contextual details. These data suggest that while the hippocampus binds elements from all domains, the more specialized regions of the MTL and surrounding cortices may play domain-specific roles in EM encoding. Furthermore, the authors report that the activation in the hippocampus increased stepwise with increasing numbers of successfully encoded associations.

A number of studies, however, have also failed to find reliable MTL activation related to successful episodic encoding *(113–116)*. While neuropsychological data demonstrate that the MTL is necessary for new episodic encoding, processes following encoding such as retrieval, consolidation, interference, and so on may have a larger impact on subsequent remembering than encoding-related MTL activity.

4.1.2. Retrieval

Work on EM retrieval points to a set of highly flexible retrieval operations that are differentially engaged depending on the particular demands of the situation. In general, the network that supports these retrieval operations includes PFC, MTL structures, and posterior sensory cortices. Identifying the contributions of these brain regions to episodic retrieval processes has demanded new methodologies capable of distinguishing subtle dissociations across similar retrieval tasks. This section explores two emerging themes: the role of PFC and posterior sensory cortices and the role of MTL subregions in retrieval success.

PFC and Posterior Sensory Cortices

Early neuroimaging work contrasting encoding and retrieval processes revealed a consistent asymmetry in the activations of prefrontal regions *(117, 118)*: specifically, greater activation of the left PFC during encoding, and relatively greater activation of the right PFC during episodic retrieval *(117, 118)*. This pattern, termed "hemispheric encoding-retrieval asymmetry" (HERA), provided early insights into the neural basis of EM retrieval and a framework for the design of future studies *(118)*.

Further exploration of this relationship, however, demonstrated that the HERA model was insufficient. The current view is that HERA reflects the maintenance of a retrieval mode, or a background cognitive state in which one is mentally attuned to the retrieval process, is sensitive to incoming cues, and is capable of becoming consciously aware of successful retrieval. Lepage et al. *(119)* compared data across four PET studies and concluded that six PFC regions were consistently recruited by episodic recognition. These included bilateral posterior ventrolateral areas, bilateral frontopolar regions, right dorsal PFC, and midline cingulate area near the supplementary motor area. How and when these regions are recruited in EM retrieval depends on task demands. As proposed in the source-monitoring framework *(120)*, retrieval attempts differ in the strategies used to access different cortical representations *depending on the nature and source of those representations.* For example, visuospatial, semantic, or emotional cues are often called upon to aid in episodic retrieval, and each of these cues stimulates a slightly different set of cortical regions.

Using event-related fMRI, Dobbins and Wagner *(121)* showed that the recollection of conceptual or perceptual details of an episode results in greater activation in the left frontopolar and posterior PFC than the detection of novelty. The authors interpret this finding as an evidence that a domain-general control network is engaged during contextual remembering. In contrast, left anterior VL PFC coactivated with a left middle temporal region associated with semantic representation, during conceptual recollection, while right VL PFC and bilateral occipito-temporal cortices were coactivated during recollection of perceptual details. Therefore, whereas left frontopolar and posterior PFC may be involved in domain-general retrieval processes, the middle temporal, right VL PFC, and occipito-temporal regions may be more domain-specific.

Interestingly, emerging data suggest that PFC is not involved in distinguishing correct from incorrect retrieval. Dobbins et al.

(122) found that activation in the left MTL was greater for correct than for incorrect source attributions, or episodic retrieval, but that numerous PFC regions that showed increased activation during source memory retrieval did not distinguish between correct and incorrect trials. Moreover, activation in some of these regions was actually numerically greater for failed retrieval attempts. The authors interpret these findings as evidence that the PFC is involved in elaborative or monitoring operations, which may be enhanced when retrieval fails. Furthermore, Velanova et al. *(123)*, using a mixed block/event-related design to explore sustained and transient control processing during EM retrieval, found that left PFC activation was associated trial-by-trial with retrieval when high control was required, whereas right PFC and several right posterior regions were associated with sustained control processes, which the authors viewed as a reflection of attentional set or retrieval mode. In line with this idea, Kahn et al. *(124)* also failed to find differences in prefrontal responses as a function of source retrieval outcome, and PFC regions were more active during trials for old items than for new ones. In addition, Wheeler and Buckner *(125)* found that activation in PFC correlates with retrieval processes and not with retrieval success: they found that left posterior and mid-VL PFC activity for items that were highly rehearsed and therefore easily identified as old was no different than for new items. By contrast, left PFC activity for items previously encountered only once was greater than for both the old and new items, suggesting that left PFC activity does not correspond to retrieval success. The authors concluded that left VL PFC activity reflects a processing control operation that is selectively engaged during demanding retrieval.

MTL Activation in Episodic Remembering

Investigating MTL activation using fMRI can be difficult not only because the region is susceptible to artifacts attributed to the ear canal, but also because the region seems to be active during a wide variety of tasks, including undirected "rest" *(126)*. Therefore, many studies of episodic remembering do not report greater activation in the MTL, even though this region is known to be involved. In order to observe MTL activity, a baseline task such as an odd/even digit judgment may be used to deactivate the MTL during the control trials *(126)*.

Unlike data from the PFC, activation in the MTL has been found to correlate with episodic retrieval success *(127, 128)* (**Fig. 4**). Several neuroimaging studies investigating the role of MTL subregions in EM have relied on the remember/know procedure to distinguish episodic retrieval, or the processes of "remembering or recollecting" (R), from retrieval based on familiarity, called the "knowing or recalling" (K) *(129)*. This technique was based on the finding that patients with MTL damage, especially those showing selective loss of hippocampal function, show impairments in

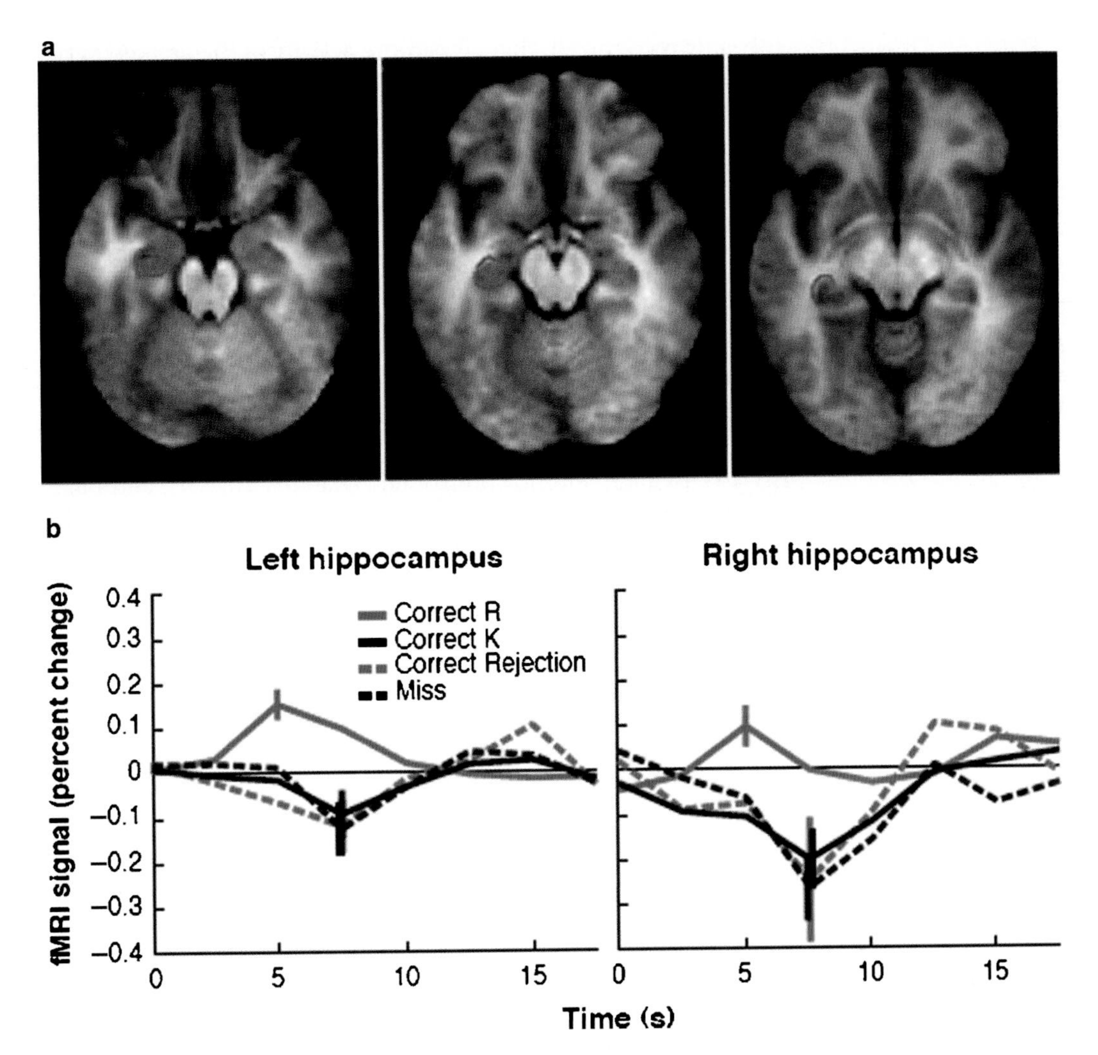

Fig. 4. Results from anatomically defined hippocampal regions of interest. (**a**) Sections from the anatomical template with the left hippocampal region of interest outlined in *red*. (**b**) Averaged event-related responses in the hippocampus from 11 subjects. Error bars represent one standard error (between subjects) of estimated response amplitudes. *Correct R* correct remember (R) response (when subjects could recollect the moment the item was studied), *Correct K* correct know (K) response (when the word is familiar but unaccompanied by the recollection of the specific moment the word was presented), *Correct rejection* correct response for nonrecognized items, *Miss* miss response (in which subjects did not recognize old items) (from Eldridge et al. ***(127)***].

EM and a comparatively intact SM *(76, 130)*. In addition, patients with degeneration of extrahippocampal temporal cortex show the opposite pattern: impaired SM combined with a relatively intact EM *(74)*. Using the remember/know procedure, Eldridge et al. *(127, 128)* found that hippocampal activity is primarily associated with "remembering" rather than "knowing": the hippocampus was more active during retrieval of "R" items than during retrieval of "K" items. In a follow-up study, Eldridge et al. *(131)* found that during encoding of items that were subsequently correctly recognized, there was more activity in the dentate gyrus and CA2/3 fields of the hippocampus than during encoding of items that were subsequently forgotten. During retrieval, activity in the subiculum correlated uniquely with "R" responses. These data are particularly compelling because the dentate gyrus and CA2/3 fields are located early in the hippocampal circuit, while the subiculum is the major output region of the hippocampus.

4.2. Temporal Processing

One somewhat less-studied component of EM is the process of assigning a temporal order to a series of events. While this part of the field is sparse in terms of neuroimaging data, there are a handful of studies that can shed some light on this process. Bilateral middle prefrontal areas near BA 9, left inferior prefrontal (near BA 44/45), left anterior prefrontal (near BA 10/46), and bilateral medial temporal areas show more activation during "high" temporal order retrieval trials (that is, when choosing between two words that were close together on a list: ie., which came first, word # 6 or word #3?) than during "low" trials (that is, when choosing between words that were spaced far apart on a list: i.e., which came first, word #1 or word # 9?) *(132)*. Activation in the middle frontal gyrus (MFG) near BA 9 is especially interesting because there is convergent evidence for the involvement of this region in the human neuropsychological *(133)* and monkey literature *(134)*. There may also be a hemispheric specialization in temporal processing. Suzuki et al. *(135)* asked participants to study pictures during two separate sessions: one in the morning and another in the afternoon. In the scanner, participants were asked to judge whether an item was studied in the morning or in the afternoon, or which of two items in the same list was studied earlier. They found that right prefrontal activity was associated with temporal order judgments of items between lists (morning vs. afternoon) while left prefrontal activity was associated with the retrieval of temporal order information within a list.

4.3. Summary of Episodic Memory

Prior to the advent of fMRI, declarative memory, particularly EM, was thought to be dependent almost exclusively on the MTL. Both animal and human lesion studies provided the bulk of this evidence. Neuroimaging techniques have demonstrated, however, that PFC regions are recruited time and time again to support EM retrieval. Interestingly, PFC does not seem to be sensitive to the outcome of retrieval, but rather seems to support retrieval intention and strategy. The MTL, in contrast, seems to be driven largely by "bottom-up" processes: being more data-driven and less voluntary *(136)*. Whereas details of an event are thought to be eventually stored in the same regions that were initially involved in their perception, the hippocampus creates and stores an index of the memory that can regenerate the original pattern of activation during EM retrieval. Therefore, in healthy adults, EM involves a large network of prefrontal, neocortical, and MTL regions acting in concert to support this complex reconstructive process.

4.4. Future Directions

The difference between EM and SM is the difference between the active reconstruction of an event to extract information specific to that occurrence and the abstraction of statistical regularities and general properties about the world over multiple experiences. The constructive nature of memory, therefore, is most easily observed in EM: the reexperiencing of past events requires the reconstruction of a narrative sequence via the reactivation of stored sensory information. A growing interest in the constructive aspects of EM has recently led to the postulation of the *constructive episodic simulation* hypothesis *(137–139)*, which suggests that the EM system is built, in part, to enable the simulation or imagination of future events. Support for this view comes from several recent neuroimaging studies demonstrating that the network of brain regions involved in EM retrieval overlaps substantially with that supporting the imagination of future events *(140–142)*. Furthermore, as Hassabis and Maguire *(143)* have eloquently described, EM retrieval can thought of as relying heavily on scene construction as a key component process. This approach may help explain why many other cognitive tasks such as imagining a fictitious event seem to employ many of the same regions involved in episodic remembering.

5. Semantic Memory

In contrast to EM, SM corresponds to the general knowledge of objects, words, facts, and people: declarative memories that are laid down independently of the original encoding context *(72)*. This information is often encountered over multiple repetitions, in a variety of contexts, and thus can be retrieved without regenerating details of the original learning event. While initially limited to a memory system for “words and other verbal symbols, their meanings and referents, about relations among them, and about rules, formulas and algorithms” *(72)*, SM currently refers to a broader knowledge set that includes facts, concepts, and beliefs *(144)*.

Reports of patients with focal deficits in SM that were category-specific *(145–148)* led to the hypothesis that SM is organized by taxonomic categories *(149)*. In addition, a highly influential theory proposed by Allport *(150)* suggests that the same sensorimotor areas that are involved in the initial processing of experience are also used to represent abstractions related to the experience. Much like findings from neuroimaging studies of EM, this theory predicts that modality-specific information is represented in the cortex that processes that modality. Finally, studies of patients with progressive neural degeneration leading to deficits in SM suggest that information in SM is also organized hierarchically, such that the more specific a concept is, the more vulnerable it may be to brain damage *(151)*. Therefore, we will consider evidence from neuroimaging studies that address these three components of SM: category specificity, reactivation of sensorimotor areas, and hierarchical organization.

5.1. Organization of SM Network

Before the advent of functional brain imaging, our knowledge of the neural bases of SM was dependent on studies of patients with brain injury. Investigations of semantic impairment arising from brain disease suggested that the anterior temporal lobes (ATLs) are critical for semantic abilities in humans, across all stimulus modalities and for all types of conceptual knowledge *(147, 148, 152–158)*. As mentioned earlier, patients with semantic dementia and progressive degeneration of the anterior temporal cortex are impaired on all tasks requiring knowledge about the meanings of words, objects, and people, although possibly to different degrees for each category depending on the lateralization of atrophy *(155, 157, 159, 160)* Other brain diseases that can affect the ATLs, such as Alzheimer's disease *(161)* and herpes simplex viral encephalitis *(148)*, also often disrupt SM. Finally, it is worth noting that patients with damage to the left PFC often have difficulty in retrieving words in response to specific cues, even in the absence of aphasia *(162)*.

Although ATLs' activation has been associated with a few semantic tasks (i.e., sentence comprehension, and famous face naming or identification, e.g., **refs.** *163, 164)*, the vast majority of the functional imaging experiments on SM have reported posterior temporal, typically stronger in the left than in the right hemisphere, and/or frontal activations in the left VL PFC, with no mention to the ATLs (for review, *see* **refs.** *144, 165–168)*. Intersubject variability and the occurrence of fMRI susceptibility artifacts in the ATLs may be possible reasons for the lack of fMRI activations detected in ATLs *(169)*.

Functional imaging results have also indicated that semantic knowledge is encoded in a widely distributed cortical network, with different regions specialized to represent particular types of information *(170)*, particular categories of objects *(171)*, or both *(172)*, leading to the concept that no single region supports semantic abilities for all modalities and categories. However, in addition to these modality-specific regions and connections, the various different surface representations (such as shape) connect to, and communicate through, a shared, amodal "hub" in the ATLs *(168)*. At the hub stage, associations between different pairs of attributes (such as shape and name, shape and action, or shape and color) are all processed by a common set of neurons and synapses, regardless of the task. Damasio et al. *(173)* were the first to argue for unified conceptual representations that abstract away from modality-specific attributes. They proposed the existence of "convergence zones" that associate different aspects of knowledge clearly articulating the importance of such zones for semantic processing. The convergence-zone hypothesis proposes that there is no multimodal cortical area that would build an integrated and independent semantic representation from its low-level sensory representations. Instead, the representation takes place only in the low-level cortices, with the different parts bound together by a hierarchy of convergence zones. A semantic representation can be recreated by activating its corresponding binding pattern in the convergence zone.

Finally, unlike EM, remote SM is not dependent on the involvement of the MTL *(91)*. The MTL is needed only temporarily, until the knowledge itself is represented permanently by the neocortical structures specialized in processing the acquired information.

5.2. Category-Specific Organization of SM

Functional neuroimaging studies in which the cortical organization for semantic knowledge has been addressed have revealed dissociations in the processing of different object categories (for review, *see* **refs.** ***149, 165, 166, 168, 170, 174)***. The most frequently documented distinction is between "living" and "nonliving" items, body parts and numerals, and the most studied categories have been human faces, houses, animals, and tools. Various theoretical models have been proposed to explain the cognitive mechanisms underlying category specificity. (1) The sensory and functional/motor theory states that categories are defined by the type of information needed to recognize items as belonging to that category. "Living" items require object-related information appreciable through perceptual channels (shape, color, sound, etc.), whereas tools and body parts are more recognizable from information concerning action, activity, or the motor scheme to use them *(147, 148, 153)*, (2) The "domain-specific theory" suggests that evolutionary pressure has led to specific adaptations for recognizing and responding to animals and plants, but not to objects *(149)*. (3) The "correlated-structure principle theory" proposes that conceptual organization reflects the statistical co-occurrence of the properties of objects rather than an explicit division into "living" and "nonliving" categories *(175)*.

An impairment for processing items in the "living things" category has mainly been described in patients with damage to the anterior portions of the temporal lobe bilaterally *(155, 156, 176, 177)*. Nevertheless, the majority of functional neuroimaging studies (both PET and fMRI) failed to show consistent

ATLs' activations for "living item" stimuli (see earlier). PET studies in normal subjects have shown that "living items" tend to activate predominantly posterior visual association cortices *(178–181)*. Only a meta-analysis of seven individual PET studies *(182)* found activations for "living" objects in the temporal poles bilaterally. This large multistudy dataset provided sufficient sensitivity to detect ATLs' activations despite their inconsistency across subjects and lack of significance in each study taken in isolation *(182)*. In contrast, patients with deficits in the "nonliving" category show damage in the left dorsolateral perisylvian regions *(156, 176, 177)*. Consistently, PET studies found activations specific to "nonliving" stimuli in the left posterior middle and superior temporal gyri and in the left inferior frontal cortex *(178, 179, 182–184)*. In an event-related fMRI study in which words belonging to the categories "living" and "nonliving" were presented visually, common areas of activation during processing of both categories included the inferior occipital gyri bilaterally, the left IFG, and the left IPL *(185)*. During processing of "living" minus "nonliving" items, signal changes were present in the right inferior frontal, middle temporal, and fusiform gyri.

Numerous fMRI studies have shown that different object categories elicit activity in different regions of the ventral temporal cortex (for review, *see* **refs.** *149, 165, 166, 168, 170, 174)*. Although it is more likely that these differences are due to perceptual, object recognition process rather than to semantic memory function, we will briefly discuss them here. Perceiving animals showed heightened, bilateral activity in the more lateral region of the fusiform gyrus, whereas tools show heightened, bilateral activity in the medial region of the fusiform gyrus and in the posterior middle temporal gyrus (MTG) *(186)*. A similar pattern of activations was found for viewing faces (in the lateral fusiform) relative to viewing houses (the medial fusiform) *(186)*. The so-called FFA *(187)* responds more strongly to faces than to other object categories, but is not exclusive for faces *(188, 189)* (**Fig. 5**). House-related activity was reported in more medial regions, including the fusiform and lingual gyri *(190)*, and parahippocampal cortex [the parahippocampal place area *(184)*], especially for landmarks *(191)*. More interestingly, a meta-analysis by Joseph *(188)* revealed that the recognition task used (i.e., viewing, matching, or naming) also predicted brain activation patterns. Specifically, matching tasks recruit more inferior occipital regions than either naming or viewing tasks do, whereas naming tasks recruit more anterior ventral temporal sites than either viewing or matching tasks do, thus indicating that the cognitive demands of a particular recognition task are as predictive of cortical activation patterns as is category membership.

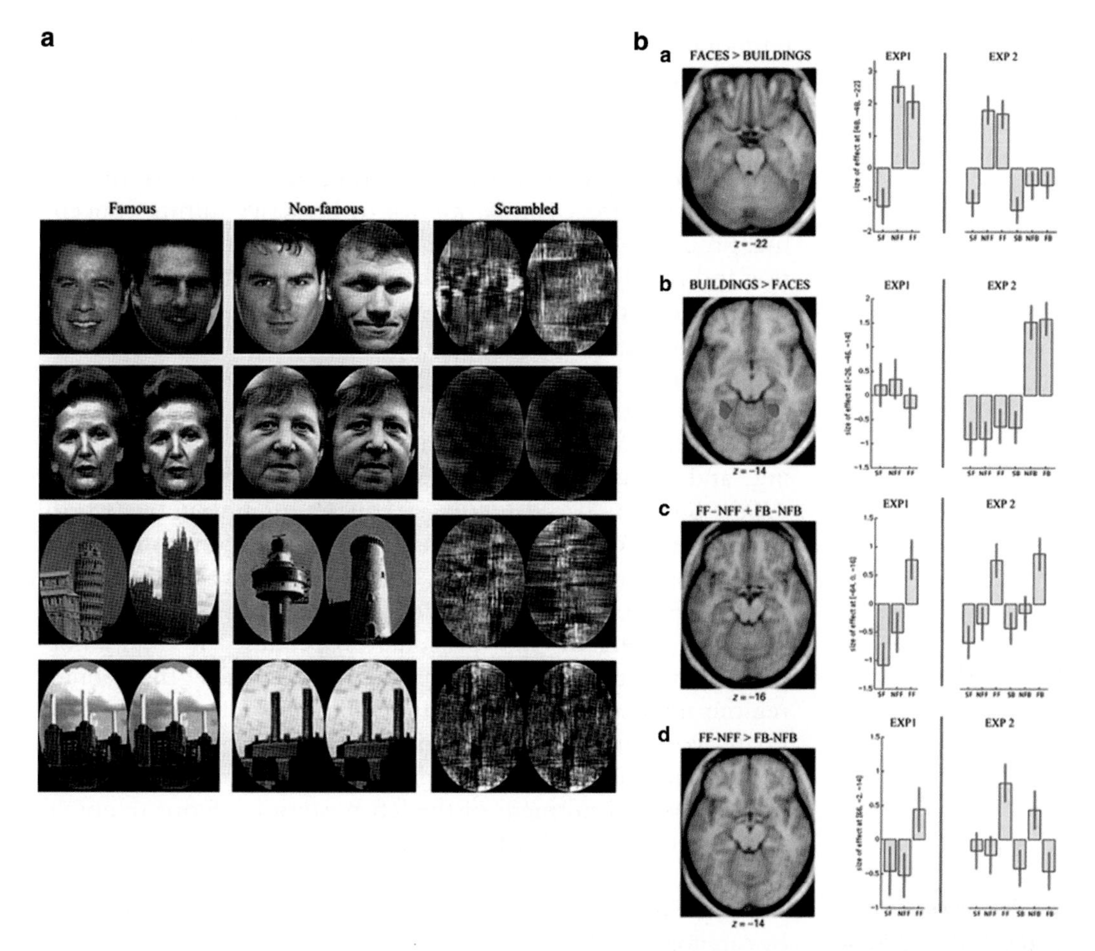

Fig. 5. A PET study investigating brain responses to famous and nonfamous faces and buildings. The results showed category-specific effects in the right fusiform and bilateral parahippocampal/lingual gyri for faces and buildings, respectively, but no effect of fame. In contrast, the left anterior middle temporal gyrus showed an effect of fame for both faces and buildings, but no effect of category. (A) Examples of the stimuli used in experiments 1 and 2 for the face conditions and in experiment 2 for the building conditions. (B) From *top* to *bottom*, this figure illustrates areas of activation and parameter estimates for regions that were more activated for (**a**) faces than for buildings, (**b**) buildings than for faces, (**c**) famous than for nonfamous faces and buildings, and (**d**) famous than for nonfamous faces only. In the left column, all activations are superimposed on axial slices of the mean of the nine subjects' normalized structural MRIs and thresholded at $p < 0.001$ (uncorrected). In the right column, the plots indicate the value of the normalized regional cerebral blood flow at the indicated voxel (*y*-axis) for each of the experimental conditions in experiments 1 and 2 (*x*-axis). *EXP* experiment, *FF* famous faces, *NFF* nonfamous faces, *FB* famous buildings, *NFB* nonfamous buildings (from Gorno-Tempini et al. ***(189)***].

5.3. Modality-Specific Organization of SM

Several of the early PET studies of semantic processes focused on the question of whether words (both auditory and visual) and pictures are interpreted by a common semantic system *(163, 181, 192, 193)*, or whether distinct systems are needed to support these two domains. These studies reached similar conclusions providing evidence for a distributed semantic system that is shared by visual/auditory and verbal modalities *(163, 181, 192, 193)* and that it is distributed throughout inferior temporal and frontal cortices with a few areas uniquely activated by pictures only (left posterior inferior temporal sulcus) or words (left anterior MTG and left inferior frontal sulcus) *(193)*. More recently, fMRI studies have provided additional support for this view by demonstrating that regions of the left posterior temporal cortex, known to be active during conceptual processing of pictures and words (fusiform gyrus and inferior and middle temporal gyri), are also active during auditory sentence comprehension *(194–196)*: activity was modulated by speech intelligibility *(194, 195)* and semantic ambiguity *(196)*. A recent fMRI study used the phenomenon of semantic ambiguity to identify regions within the fronto-temporal language network that are involved in the processes of activating, selecting, and integrating contextually appropriate word meanings *(196)*. Subjects heard sentences containing ambiguous words (e.g., "The shell was fired towards the tank") and well-matched low-ambiguity sentences (e.g., "Her secrets were written in her diary"). Although these sentences had similar acoustic, phonological, syntactic, and prosodic properties, the high-ambiguity sentences required additional processing by those brain regions involved in activating and selecting contextually appropriate word meanings. The ambiguity in these sentences went largely unnoticed, and yet high-ambiguity sentences resulted in increased recruitment of the left posterior inferior temporal cortex and the IFG bilaterally *(196)*.

5.4. Representations of Object Properties

While the neural systems involved in SM may be modulated both by categories and modalities there is also evidence that the sensorimotor regions that are involved in the initial processing of particular information are recruited during SM retrieval (for review, *see* **refs.** *165, 166, 168, 170, 174, 197)*. In particular, semantic decisions involving object properties suggest a broad relationship between perceptual knowledge retrieval and sensory brain mechanisms, though recent work also suggests that this observation may be itself category-specific *(158)*.

Activation of the left or bilateral ventral temporal cortex (fusiform gyrus) when retrieving *color* information, relative to other properties, has been replicated by several fMRI studies *(167, 198, 199)*. Beauchamp et al. *(198)* showed that neural activity is limited to the occipital lobes when color perception was tested by a passive viewing. When the task was made more demanding by requiring subjects to use color information to perform a color-sequencing task, several areas in the ventral cortex were identified: the most posterior, located in the posterior fusiform gyrus, corresponded to an area activated by passive viewing of colored stimuli, and more anterior and medial color-selective areas located in the collateral sulcus and fusiform gyrus *(198)*. These more anterior areas were also most active when visual color information was behaviorally relevant, suggesting that attention influences activity in color-selective areas *(198)*.

Parietal cortex appears to be involved in retrieval of *size(167)*. For instance, in Oliver and Thompson-Schill *(167)*, seven subjects made binary decisions about the shape, color, and size of named objects during fMRI. Bilateral parietal activity was significantly greater during retrieval of shape and size than during retrieval of color *(167)*.

An area in the posterior superior temporal cortex, adjacent to the auditory-association cortex, is activated when participants are asked to judge the *sound* that an object makes *(200)*. *Action knowledge* involves the left lateral temporal cortex, particularly the medial and superior temporal (MT/MST) regions, anterior to an area associated with motion perception *(201–203)*. For instance, in two fMRI experiments Kourtzi and Kanwisher *(202)* found stronger activation of the MT/MST regions during viewing of static photographs with implied motion compared with viewing of photographs without it. Taken together, these data provide strong evidence that information about a particular object property is stored in the same neural system engaged when the property is perceived.

6. Conclusions

Working, episodic, and semantic memory systems engage multiple brain regions and rely upon a number of cognitive processes. WM is likely a property of the functional interaction between the PFC and posterior brain regions. The MTL and its connections with neocortical, prefrontal, and limbic structures are implicated in EM. SM is mediated through a network of neocortical structures, including the lateral and ATLs, and the inferior frontal cortex, possibly to a greater extent in the left hemisphere. Taken together, these memory systems do have overlapping neuroanatomical underpinnings, and even share some of the same component processes, such as item and information retrieval. One emergent similarity between the systems is the finding that those sensory and association areas that are engaged during perceptual and sensorimotor processing of item information are tapped once again when the information is required for memory processing. Future work by researchers using neuroimaging to study memory will require even further refinement in the definitions of memory systems and their components, as exemplified by the current work in EM, where the retrieval component is being further reduced into components such as scene construction.

References

1. Baddeley A, Hitch G. Working memory. In: Bower GH, ed. The Psychology of Learning and Motivation. San Diego: Academic Press; 1974:47–90.
2. Baddeley A. The episodic buffer: A new component of working memory? Trends Cogn Sci 2000;4:417–423.
3. Curtis CE, D'Esposito M. Functional neuroimaging of working memory. In: Cabeza R, Kingstone A, eds. Handbook of Functional Neuroimaging of Cognition. Cambridge: MIT Press; 2006:269–306.
4. Miyake A, Shah P. Models of Working Memory. New York: Cambridge University Press; 1999.
5. Miller GA. The magical number seven, plus or minus two: Some limits on our capacity for processing information. Psychol Rev 1956;63:81–97.
6. Goldman-Rakic PS. Working memory dysfunction in schizophrenia. J Neuropsychiatry Clin Neurosci 1994;6:348–357.
7. Muller NG, Machado L, Knight RT. Contributions of subregions of the prefrontal cortex to working memory: Evidence from brain lesions in humans. J Cogn Neurosci 2002;14:673–686.
8. Fuster JM, Alexander GE. Neuron activity related to short-term memory. Science 1971;173:652–654.
9. Kubota K, Niki H. Prefrontal cortical unit activity and delayed alternation performance in monkeys. J Neurophysiol 1971;34:337–347.
10. Funahashi S, Bruce CJ, Goldman-Rakic PS. Mnemonic coding of visual space in the monkey's dorsolateral prefrontal cortex. J Neurophysiol 1989;61:331–349.
11. Curtis CE, D'Esposito M. Working memory. In: Cabeza R, Kingstone A, eds. Handbook of Functional Neuroimaging of Cognition. Cambridge, MA: MIT Press; 2006.
12. Owen AM, Herrod NJ, Menon DK, et al. Redefining the functional organization of working memory processes within human lateral prefrontal cortex. Eur J Neurosci 1999;11:567–574.
13. Jonides J, Smith EE, Koeppe RA, Awh E, Minoshima S, Mintun MA. Spatial working memory in humans as revealed by PET. Nature 1993;363:623–625.
14. Petrides M, Alivisatos B, Meyer E, Evans AC. Functional activation of the human frontal cortex during the performance of verbal working memory tasks. Proc Natl Acad Sci USA 1993;90:878–882.
15. McCarthy G, Blamire AM, Puce A, et al. Functional magnetic resonance imaging of human prefrontal cortex activation during a spatial working memory task. Proc Natl Acad Sci USA 1994;91:8690–8694.
16. Awh E, Jonides J, Smith EE, et al. Rehearsal in spatial working memory: Evidence from neuroimaging. Psychol Sci 1999;10:433–437.
17. Postle BR, D'Esposito M. "What"-Then-"Where" in visual working memory: An event-related fMRI study. J Cogn Neurosci 1999;11:585–597.
18. Rowe JB, Passingham RE. Working memory for location and time: Activity in prefrontal area 46 relates to selection rather than maintenance in memory. Neuroimage 2001;14:77–86.
19. Corbetta M, Kincade JM, Shulman GL. Neural systems for visual orienting and their relationships to spatial working memory. J Cogn Neurosci 2002;14:508–523.
20. Davachi L, Maril A, Wagner AD. When keeping in mind supports later bringing to mind: Neural markers of phonological rehearsal predict subsequent remembering. J Cogn Neurosci 2001;13:1059–1070.
21. D'Esposito M. Working memory. In: Goldenberg G, Miller B, eds. Handbook of Clinical Neurology Neuropsychology and behavioral Neurology. 3rd series edn. Amsterdam: Elsevier; 2008:237–247.
22. Courtney SM, Ungerleider LG, Keil K, Haxby JV. Transient and sustained activity in a distributed neural system for human working memory. Nature 1997;386:608–611.
23. Courtney SM, Petit L, Haxby JV, Ungerleider LG. The role of prefrontal cortex in working memory: Examining the contents of consciousness. Philos Trans R Soc Lond B Biol Sci 1998;353:1819–1828.
24. Jha AP, McCarthy G. The influence of memory load upon delay-interval activity in a working-memory task: An event-related functional MRI study. J Cogn Neurosci 2000;12 Suppl 2:90–105.
25. Gruber O. Effects of domain-specific interference on brain activation associated with verbal working memory task performance. Cereb Cortex 2001;11:1047–1055.
26. Ranganath C, D'Esposito M. Medial temporal lobe activity associated with active

maintenance of novel information. Neuron 2001;31:865–873.

27. Postle BR, Druzgal TJ, D'Esposito M. Seeking the neural substrates of visual working memory storage. Cortex 2003;39:927–946.
28. Curtis CE, Rao VY, D'Esposito M. Maintenance of spatial and motor codes during oculomotor delayed response tasks. J Neurosci 2004;24:3944–3952.
29. Marklund P, Fransson P, Cabeza R, Petersson KM, Ingvar M, Nyberg L. Sustained and transient neural modulations in prefrontal cortex related to declarative long-term memory, working memory, and attention. Cortex 2007;43:22–37.
30. Druzgal TJ, D'Esposito M. Dissecting contributions of prefrontal cortex and fusiform face area to face working memory. J Cogn Neurosci 2003;15:771–784.
31. D'Esposito M, Aguirre GK, Zarahn E, Ballard D, Shin RK, Lease J. Functional MRI studies of spatial and nonspatial working memory. Brain Res Cogn Brain Res 1998;7:1–13.
32. D'Esposito M, Postle BR, Rypma B. Prefrontal cortical contributions to working memory: Evidence from event-related fMRI studies. Exp Brain Res 2000;133:3–11.
33. Curtis CE, D'Esposito M. Persistent activity in the prefrontal cortex during working memory. Trends Cogn Sci 2003;7:415–423.
34. Passingham D, Sakai K. The prefrontal cortex and working memory: Physiology and brain imaging. Curr Opin Neurobiol 2004;14:163–168.
35. Ungerleider LG, Haxby JV. 'What' and 'where' in the human brain. Curr Opin Neurobiol 1994;4:157–165.
36. McCarthy G, Puce A, Constable RT, Krystal JH, Gore JC, Goldman-Rakic P. Activation of human prefrontal cortex during spatial and nonspatial working memory tasks measured by functional MRI. Cereb Cortex 1996;6:600–611.
37. Courtney SM, Petit L, Maisog JM, Ungerleider LG, Haxby JV. An area specialized for spatial working memory in human frontal cortex. Science 1998;279:1347–1351.
38. Munk MH, Linden DE, Muckli L, et al. Distributed cortical systems in visual short-term memory revealed by event-related functional magnetic resonance imaging. Cereb Cortex 2002;12:866–876.
39. Sala JB, Rama P, Courtney SM. Functional topography of a distributed neural system for spatial and nonspatial information maintenance in working memory. Neuropsychologia 2003;41:341–356.
40. Walter H, Wunderlich AP, Blankenhorn M, et al. No hypofrontality, but absence of prefrontal lateralization comparing verbal and spatial working memory in schizophrenia. Schizophr Res 2003;61:175–184.
41. Leung HC, Seelig D, Gore JC. The effect of memory load on cortical activity in the spatial working memory circuit. Cogn Affect Behav Neurosci 2004;4:553–563.
42. Courtney SM, Ungerleider LG, Keil K, Haxby JV. Object and spatial visual working memory activate separate neural systems in human cortex. Cereb Cortex 1996;6:39–49.
43. Druzgal TJ, D'Esposito M. Activity in fusiform face area modulated as a function of working memory load. Brain Res Cogn Brain Res 2001;10:355–364.
44. Rama P, Sala JB, Gillen JS, Pekar JJ, Courtney SM. Dissociation of the neural systems for working memory maintenance of verbal and nonspatial visual information. Cogn Affect Behav Neurosci 2001;1:161–171.
45. Chein JM, Fiez JA. Dissociation of verbal working memory system components using a delayed serial recall task. Cereb Cortex 2001;11:1003–1014.
46. Gruber O, von Cramon DY. The functional neuroanatomy of human working memory revisited. Evidence from 3-T fMRI studies using classical domain-specific interference tasks. Neuroimage 2003;19:797–809.
47. Ravizza SM, Delgado MR, Chein JM, Becker JT, Fiez JA. Functional dissociations within the inferior parietal cortex in verbal working memory. Neuroimage 2004;22:562–573.
48. Raye CL, Johnson MK, Mitchell KJ, Greene EJ, Johnson MR. Refreshing: A minimal executive function. Cortex 2007;43: 135–145.
49. Chen SH, Desmond JE. Cerebrocerebellar networks during articulatory rehearsal and verbal working memory tasks. Neuroimage 2005;24:332–338.
50. Kirschen MP, Chen SH, Schraedley-Desmond P, Desmond JE. Load- and practice-dependent increases in cerebro-cerebellar activation in verbal working memory: An fMRI study. Neuroimage 2005;24: 462–472.
51. Smith EE, Jonides J. Storage and executive processes in the frontal lobes. Science 1999;283:1657–1661.
52. Henson RN, Burgess N, Frith CD. Recoding, storage, rehearsal and grouping in verbal

short-term memory: An fMRI study. Neuropsychologia 2000;38:426–440.

53. Crottaz-Herbette S, Anagnoson RT, Menon V. Modality effects in verbal working memory: Differential prefrontal and parietal responses to auditory and visual stimuli. Neuroimage 2004;21:340–351.
54. Postle BR. Working memory as an emergent property of the mind and brain. Neuroscience 2006;139:23–38.
55. Wager TD, Smith EE. Neuroimaging studies of working memory: A meta-analysis. Cogn Affect Behav Neurosci 2003;3:255–274.
56. Owen AM, Doyon J, Petrides M, Evans AC. Planning and spatial working memory: A positron emission tomography study in humans. Eur J Neurosci 1996;8:353–364.
57. Thompson-Schill SL, D'Esposito M, Aguirre GK, Farah MJ. Role of left inferior prefrontal cortex in retrieval of semantic knowledge: A reevaluation. Proc Natl Acad Sci USA 1997;94:14792–14797.
58. Thompson-Schill SL, D'Esposito M, Kan IP. Effects of repetition and competition on activity in left prefrontal cortex during word generation. Neuron 1999;23:513–522.
59. D'Esposito M, Postle BR, Jonides J, Smith EE. The neural substrate and temporal dynamics of interference effects in working memory as revealed by event-related functional MRI. Proc Natl Acad Sci USA 1999;96:7514–7519.
60. Jonides J, Smith EE, Marshuetz C, Koeppe RA, Reuter-Lorenz PA. Inhibition in verbal working memory revealed by brain activation. Proc Natl Acad Sci USA 1998;95:8410–8413.
61. D'Esposito M, Postle BR, Ballard D, Lease J. Maintenance versus manipulation of information held in working memory: An event-related fMRI study. Brain Cogn 1999;41:66–86.
62. Postle BR, Berger JS, D'Esposito M. Functional neuroanatomical double dissociation of mnemonic and executive control processes contributing to working memory performance. Proc Natl Acad Sci USA 1999;96:12959–12964.
63. Bunge SA, Klingberg T, Jacobsen RB, Gabrieli JD. A resource model of the neural basis of executive working memory. Proc Natl Acad Sci USA 2000;97:3573–3578.
64. Cohen JD, Perlstein WM, Braver TS, et al. Temporal dynamics of brain activation during a working memory task. Nature 1997;386:604–608.
65. Squire LR, Stark CE, Clark RE. The medial temporal lobe. Annu Rev Neurosci 2004;27:279–306.
66. Corkin S. Lasting consequences of bilateral medial temporal lobectomy: Clinical course and experimental findings in H.M. Semin Neurol 1984;4:249–259.
67. Braver TS, Cohen JD, Nystrom LE, Jonides J, Smith EE, Noll DC. A parametric study of prefrontal cortex involvement in human working memory. Neuroimage 1997;5:49–62.
68. Stern CE, Sherman SJ, Kirchhoff BA, Hasselmo ME. Medial temporal and prefrontal contributions to working memory tasks with novel and familiar stimuli. Hippocampus 2001;11:337–346.
69. Ranganath C, Rainer G. Neural mechanisms for detecting and remembering novel events. Nat Rev Neurosci 2003;4:193–202.
70. Honey GD, Fu CH, Kim J, et al. Effects of verbal working memory load on corticocortical connectivity modeled by path analysis of functional magnetic resonance imaging data. Neuroimage 2002;17:573–582.
71. Gazzaley A, Rissman J, Desposito M. Functional connectivity during working memory maintenance. Cogn Affect Behav Neurosci 2004;4:580–599.
72. Tulving E. Episodic and semantic memory. In: Tulving E, Donaldson W, eds. Organisation of memory. New York: Academic Press; 1972:381–403.
73. Vargha-Khadem F, Gadian DG, Watkins KE, Connelly A, Van Paesschen W, Mishkin M. Differential effects of early hippocampal pathology on episodic and semantic memory. Science 1997;277:376–380.
74. Hodges JR, Graham KS. Episodic memory: Insights from semantic dementia. Philos Trans R Soc Lond B Biol Sci 2001;356:1423–1434.
75. Levine B, Black SE, Cabeza R, et al. Episodic memory and the self in a case of isolated retrograde amnesia. Brain 1998;121:1951–1973.
76. Viskontas IV, McAndrews MP, Moscovitch M. Remote episodic memory deficits in patients with unilateral temporal lobe epilepsy and excisions. J Neurosci 2000;20:5853–5857.
77. Dudukovic NM, Knowlton BJ. Remember-Know judgments and retrieval of contextual details. Acta Psychol 2006;122:160–173.
78. Levine B, Svoboda E, Hay JF, Winocur G, Moscovitch M. Aging and autobiographical

memory: Dissociating episodic from semantic retrieval. 2002;17:677–689.
79. Bayley PJ, Hopkins RO, Squire LR. The fate of old memories after medial temporal lobe damage. J Neurosci 2006;26:13311–13317.
80. Nolde SF, Johnson MK, D'Esposito M. Left prefrontal activation during episodic remembering: An event-related fMRI study. Neuroreport 1998;9:3509–3514.
81. Wheeler ME, Petersen SE, Buckner RL. Memory's echo: Vivid remembering reactivates sensory-specific cortex. Proc Natl Acad Sci USA 2000;97:11125–11129.
82. Heil M, Rosler F, Hennighausen E. Topographically distinct cortical activation in episodic long-term memory: The retrieval of spatial versus verbal information. Mem Cogn 1996;24:777–795.
83. O'Craven KM, Kanwisher N. Mental imagery of faces and places activates corresponding stiimulus-specific brain regions. J Cogn Neurosci 2000;12:1013–1023.
84. Halpern AR, Zatorre RJ. When that tune runs through your head: A PET investigation of auditory imagery for familiar melodies. Cereb Cortex 1999;9:697–704.
85. Nyberg L, Petersson KM, Nilsson LG, Sandblom J, Aberg C, Ingvar M. Reactivation of motor brain areas during explicit memory for actions. Neuroimage 2001;14:521–528.
86. Marr D. Simple memory: A theory for archicortex. Philos Trans R Soc Lond B Biol Sci 1971;262:23–81.
87. Hasselmo ME, McClelland JL. Neural models of memory. Curr Opin Neurobiol 1999;9:184–188.
88. Aggleton JP, Brown MW. Episodic memory, amnesia, and the hippocampal-anterior thalamic axis. Behav Brain Sci 1999;22: 425–444.
89. Insausti R, Amaral DG, Cowan WM. The entorhinal cortex of the monkey. II. Cortical afferents. J Comp Neurol 1987;264: 356–395.
90. Amaral DG, Insausti R, Cowan WM. The entorhinal cortex of the monkey. I. Cytoarchitectonic organization. J Comp Neurol 1987;264:326–355.
91. Moscovitch M, Rosenbaum RS, Gilboa A, et al. Functional neuroanatomy of remote episodic, semantic and spatial memory: A unified account based on multiple trace theory. J Anat 2005;207:35–66.
92. Teyler TJ, Rudy JW. The hippocampal indexing theory and episodic memory: Updating the index. Hippocampus 2007;17: 1158–1169.
93. Squire LR, Knowlton B, Musen G. The structure and organization of memory. Annu Rev Psychol 1993;44:453–495.
94. Baddeley A, Vargha-Khadem F, Mishkin M. Preserved recognition in a case of developmental amnesia: Implications for the acquisition of semantic memory? J Cogn Neurosci 2001;13:357–369.
95. Rolls ET. A theory of hippocampal function in memory. Hippocampus 1996;6:601–620.
96. Gluck MA, Myers CE. Hippocampal mediation of stimulus representation: A computational theory. Hippocampus 1993;3:491–516.
97. Norman KA, O'Reilly RC. Modeling hippocampal and neocortical contributions to recognition memory: A complementary-learning-systems approach. Psychol Rev 2003;110:611–646.
98. Fletcher PC, Shallice T, Dolan RJ. The functional roles of prefrontal cortex in episodic memory. I. Encoding. Brain 1998;121:1239–1248.
99. Shallice T, Fletcher P, Frith CD, Grasby P, Frackowiak RS, Dolan RJ. Brain regions associated with acquisition and retrieval of verbal episodic memory. Nature 1994;368:633–635.
100. Dolan RJ, Fletcher PC. Dissociating prefrontal and hippocampal function in episodic memory encoding. Nature 1997;388: 582–585.
101. Gabrieli JD, Brewer JB, Desmond JE, Glover GH. Separate neural bases of two fundamental memory processes in the human medial temporal lobe. Science 1997;276:264–246.
102. Wagner AD, Schacter DL, Rotte M, et al. Building memories: Remembering and forgetting of verbal experiences as predicted by brain activity. Science 1998;281: 1188–1191.
103. Brewer JB, Zhao Z, Desmond JE, Glover GH, Gabrieli JD. Making memories: Brain activity that predicts how well visual experience will be remembered. Science 1998;281:1185–1187.
104. Wagner AD, Poldrack RA, Eldridge LL, Desmond JE, Glover GH, Gabrieli JD. Material-specific lateralization of prefrontal activation during episodic encoding and retrieval. Neuroreport 1998;9:3711–3717.
105. Fernandez G, Weyerts H, Schrader-Bolsche M, et al. Successful verbal encoding into episodic memory engages the posterior hippocampus: A parametrically analyzed functional magnetic resonance imaging study. J Neurosci 1998;18:1841–1847.

106. Davachi L, Wagner AD. Hippocampal contributions to episodic encoding: Insights from relational and item-based learning. J Neurophysiol 2002;88:982–990.
107. Davachi L, Mitchell JP, Wagner AD. Multiple routes to memory: Distinct medial temporal lobe processes build item and source memories. Proc Natl Acad Sci USA 2003;100:2157–2162.
108. Kirwan CB, Stark CE. Medial temporal lobe activation during encoding and retrieval of novel face-name pairs. Hippocampus 2004;14:919–930.
109. Ranganath C, Yonelinas AP, Cohen MX, Dy CJ, Tom SM, D'Esposito M. Dissociable correlates of recollection and familiarity within the medial temporal lobes. Neuropsychologia 2004;42:2–13.
110. Kensinger EA, Schacter DL. Amygdala activity is associated with the successful encoding of item, but not source, information for positive and negative stimuli. J Neurosci 2006;26:2564–2570.
111. Uncapher MR, Otten LJ, Rugg MD. Episodic encoding is more than the sum of its parts: An fMRI investigation of multifeatural contextual encoding. Neuron 2006;52: 547–556.
112. Staresina BP, Davachi L. Selective and Shared Contributions of the Hippocampus and Perirhinal Cortex to Episodic Item and Associative Encoding. J Cogn Neurosci 2008;20:1478–1489.
113. Baker JT, Sanders AL, Maccotta L, Buckner RL. Neural correlates of verbal memory encoding during semantic and structural processing tasks. Neuroreport 2001;12:1251–1256.
114. Buckner RL, Wheeler ME, Sheridan MA. Encoding processes during retrieval tasks. J Cogn Neurosci 2001;13:406–415.
115. Henson RN, Hornberger M, Rugg MD. Further dissociating the processes involved in recognition memory: An FMRI study. J Cogn Neurosci 2005;17:1058–1073.
116. Otten LJ, Rugg MD. Task-dependency of the neural correlates of episodic encoding as measured by fMRI. Cereb Cortex 2001;11:1150–1160.
117. Nyberg L, McIntosh AR, Cabeza R, Habib R, Houle S, Tulving E. General and specific brain regions involved in encoding and retrieval of events: what, where, and when. Proc Natl Acad Sci USA 1996;93: 11280–11285.
118. Tulving E, Kapur S, Craik FI, Moscovitch M, Houle S. Hemispheric encoding/retrieval asymmetry in episodic memory: Positron emission tomography findings. Proc Natl Acad Sci USA 1994;91:2016–2020.
119. Lepage M, Ghaffar O, Nyberg L, Tulving E. Prefrontal cortex and episodic memory retrieval mode. Proc Natl Acad Sci USA 2000;97:506–511.
120. Johnson MK, Hashtroudi S, Lindsay DS. Source monitoring. Psychol Bull 1993; 114:3–28.
121. Dobbins IG, Wagner AD. Domain-general and domain-sensitive prefrontal mechanisms for recollecting events and detecting novelty. Cereb Cortex 2005;15:1768–1778.
122. Dobbins IG, Rice HJ, Wagner AD, Schacter DL. Memory orientation and success: Separable neurocognitive components underlying episodic recognition. Neuropsychologia 2003;41:318–333.
123. Velanova K, Jacoby LL, Wheeler ME, McAvoy MP, Petersen SE, Buckner RL. Functional-anatomic correlates of sustained and transient processing components engaged during controlled retrieval. J Neurosci 2003;23:8460–8470.
124. Kahn I, Davachi L, Wagner AD. Functional-neuroanatomic correlates of recollection: Implications for models of recognition memory. J Neurosci 2004;24:4172–4180.
125. Wheeler ME, Buckner RL. Functional dissociation among components of remembering: Control, perceived oldness, and content. J Neurosci 2003;23:3869–3880.
126. Stark CE, Squire LR. When zero is not zero: The problem of ambiguous baseline conditions in fMRI. Proc Natl Acad Sci USA 2001;98:12760–12766.
127. Eldridge LL, Knowlton BJ, Furmanski CS, Bookheimer SY, Engel SA. Remembering episodes: A selective role for the hippocampus during retrieval. Nat Neurosci 2000;3:1149–1152.
128. Wheeler ME, Buckner RL. Functional-anatomic correlates of remembering and knowing. Neuroimage 2004;21:1337–1349.
129. Tulving E. Memory and consciousness. Can Psychol 1985;1:1–12.
130. Tulving E, Schacter DL, McLachlan DR, Moscovitch M. Priming of semantic autobiographical knowledge: A case study of retrograde amnesia. Brain Cogn 1988; 8:3–20.
131. Eldridge LL, Engel SA, Zeineh MM, Bookheimer SY, Knowlton BJ. A dissociation of encoding and retrieval processes in the human hippocampus. J Neurosci 2005;25:3280–3286.

132. Konishi S, Uchida I, Okuaki T, Machida T, Shirouzu I, Miyashita Y. Neural correlates of recency judgment. J Neurosci 2002;22:9549–9555.
133. Milner B, Corsi P, Leonard G. Frontal-lobe contribution to recency judgements. Neuropsychologia 1991;29:601–618.
134. Petrides M. Functional specialization within the dorsolateral frontal cortex for serial order memory. Proc Biol Sci 1991;246:299–306.
135. Suzuki M, Fujii T, Tsukiura T, et al. Neural basis of temporal context memory: A functional MRI study. Neuroimage 2002;17:1790–1796.
136. Moscovitch M. Recovered consciousness: A hypothesis concerning modularity and episodic memory. J Clin Exp Neuropsychol 1995;17:276–290.
137. Schacter DL, Addis DR. The cognitive neuroscience of constructive memory: Remembering the past and imagining the future. Philos Trans R Soc Lond B Biol Sci 2007;362:773–786.
138. Dudai Y, Carruthers M. The Janus face of Mnemosyne. Nature 2005;434:567.
139. Suddendorf T, Corballis MC. Mental time travel and the evolution of the human mind. Genet Soc Gen Psychol Monogr 1997;123:133–167.
140. Okuda J, Fujii T, Ohtake H, et al. Thinking of the future and past: The roles of the frontal pole and the medial temporal lobes. Neuroimage 2003;19:1369–1380.
141. Addis DR, Wong AT, Schacter DL. Remembering the past and imagining the future: Common and distinct neural substrates during event construction and elaboration. Neuropsychologia 2007;45: 1363–1377.
142. Szpunar KK, Watson JM, McDermott KB. Neural substrates of envisioning the future. Proc Natl Acad Sci USA 2007;104: 642–647.
143. Hassabis D, Maguire EA. Deconstructing episodic memory with construction. Trends Cogn Sci 2007;11:299–306.
144. Martin A. Functional neuroimaging of semantic memory. In: Cabeza R, Kingstone A, eds. Handbook of Functional Neuroimaging of Cognition. Cambridge: The MIT Press; 2001:153–186.
145. Damasio AR, McKee J, Damasio H. Determinants of performance in color anomia. Brain Lang 1979;7:74–85.
146. McKenna P, Warrington EK. Testing for nominal dysphasia. J Neurol Neurosurg Psychiatry 1980;43:781–788.
147. Warrington EK, McCarthy R. Category specific access dysphasia. Brain 1983;106: 859–878.
148. Warrington EK, Shallice T. Category specific semantic impairments. Brain 1984;107:829–854.
149. Caramazza A, Shelton JR. Domain-specific knowledge systems in the brain the animate-inanimate distinction. J Cogn Neurosci 1998;10:1–34.
150. Allport DA. Distributed memory, modular systems and dysphagia. In: Newman SK, Epstein R, eds. Current Perspectives in Dysphagia. Edinburgh: Churchill Livingstone; 1985:32–60.
151. Hodges JR, Graham N, Patterson K. Charting the progression in semantic dementia: Implications for the organisation of semantic memory. Memory 1995;3:463–495.
152. Warrington EK. The selective impairment of semantic memory. Q J Exp Psychol 1975;27:635–657.
153. Warrington EK, McCarthy RA. Categories of knowledge. Further fractionations and an attempted integration. Brain 1987;110:1273–1296.
154. Hart J Jr, Gordon B. Delineation of single-word semantic comprehension deficits in aphasia, with anatomical correlation. Ann Neurol 1990;27:226–231.
155. Hodges JR, Patterson K, Oxbury S, Funnell F. Semantic dementia. Progressive fluent aphasia with temporal lobe atrophy. Brain 1992;115:1783–1806.
156. Gainotti G. What the locus of brain lesion tells us about the nature of the cognitive defect underlying category-specific disorders: A review. Cortex 2000;36:539–559.
157. Gorno-Tempini ML, Dronkers NF, Rankin KP, et al. Cognition and anatomy in three variants of primary progressive aphasia. Ann Neurol 2004;55:335–346.
158. Brambati SM, Myers D, Wilson A, et al. The anatomy of category-specific object naming in neurodegenerative diseases. J Cogn Neurosci 2006;18:1644–1653.
159. Snowden JS, Thompson JC, Neary D. Knowledge of famous faces and names in semantic dementia. Brain 2004;127: 860–872.
160. Gorno-Tempini ML, Rankin KP, Woolley JD, Rosen HJ, Phengrasamy L, Miller BL. Cognitive and behavioral profile in a case of right anterior temporal lobe neurodegeneration. Cortex 2004;40:631–644.
161. Hodges JR, Patterson K. Is semantic memory consistently impaired early in the course

of Alzheimer's disease? Neuroanatomical and diagnostic implications. Neuropsychologia 1995;33:441–459.

162. Baldo JV, Shimamura AP. Letter and category fluency in patients with frontal lobe lesions. Neuropsychology 1998;12:259–267.
163. Gorno-Tempini ML, Price CJ, Josephs O, et al. The neural systems sustaining face and proper-name processing. Brain 1998;121:2103–2118.
164. Damasio H, Grabowski TJ, Tranel D, Hichwa RD, Damasio AR. A neural basis for lexical retrieval. Nature 1996;380:499–505.
165. Bookheimer S. Functional MRI of language: New approaches to understanding the cortical organization of semantic processing. Annu Rev Neurosci 2002;25:151–188.
166. Martin A. The representation of object concepts in the brain. Annu Rev Psychol 2007;58:25–45.
167. Oliver RT, Thompson-Schill SL. Dorsal stream activation during retrieval of object size and shape. Cogn Affect Behav Neurosci 2003;3:309–322.
168. Patterson K, Nestor PJ, Rogers TT. Where do you know what you know? The representation of semantic knowledge in the human brain. Nat Rev Neurosci 2007;8:976–987.
169. Devlin JT, Russell RP, Davis MH, et al. Susceptibility-induced loss of signal: Comparing PET and fMRI on a semantic task. Neuroimage 2000;11:589–600.
170. Martin A, Chao LL. Semantic memory and the brain: Structure and processes. Curr Opin Neurobiol 2001;11:194–201.
171. Caramazza A, Mahon BZ. The organization of conceptual knowledge: The evidence from category-specific semantic deficits. Trends Cogn Sci 2003;7:354–361.
172. Thompson-Schill SL, Aguirre GK, D'Esposito M, Farah MJ. A neural basis for category and modality specificity of semantic knowledge. Neuropsychologia 1999;37:671–676.
173. Damasio AR. Time-locked multiregional retroactivation: A systems-level proposal for the neural substrates of recall and recognition. Cognition 1989;33:25–62.
174. Thompson-Schill SL. Neuroimaging studies of semantic memory: Inferring "how" from "where." Neuropsychologia 2003;41:280–292.
175. Tyler LK, Moss HE. Towards a distributed account of conceptual knowledge. Trends Cogn Sci 2001;5:244–252.
176. Hillis AE, Caramazza A. Category-specific naming and comprehension impairment: A double dissociation. Brain 1991;114:2081–2094.
177. Tranel D, Damasio H, Damasio AR. A neural basis for the retrieval of conceptual knowledge. Neuropsychologia 1997;35:1319–1327.
178. Martin A, Wiggs CL, Ungerleider LG, Haxby JV. Neural correlates of category-specific knowledge. Nature 1996;379:649–652.
179. Mummery CJ, Patterson K, Hodges JR, Wise RJ. Generating 'tiger' as an animal name or a word beginning with T: Differences in brain activation. Proc Biol Sci 1996;263:989–995.
180. Perani D, Cappa SF, Bettinardi V, et al. Different neural systems for the recognition of animals and man-made tools. Neuroreport 1995;6:1637–1641.
181. Perani D, Schnur T, Tettamanti M, Gorno-Tempini M, Cappa SF, Fazio F. Word and picture matching: A PET study of semantic category effects. Neuropsychologia 1999;37:293–306.
182. Devlin JT, Moore CJ, Mummery CJ, et al. Anatomic constraints on cognitive theories of category specificity. Neuroimage 2002;15:675–685.
183. Gorno-Tempini ML, Cipolotti L, Price CJ. Category differences in brain activation studies: Where do they come from? Proc Biol Sci 2000;267:1253–1258.
184. Cappa SF, Perani D, Schnur T, Tettamanti M, Fazio F. The effects of semantic category and knowledge type on lexical-semantic access: A PET study. Neuroimage 1998;8:350–359.
185. Leube DT, Erb M, Grodd W, Bartels M, Kircher TT. Activation of right fronto-temporal cortex characterizes the 'living' category in semantic processing. Brain Res Cogn Brain Res 2001;12:425–430.
186. Chao LL, Haxby JV, Martin A. Attribute-based neural substrates in temporal cortex for perceiving and knowing about objects. Nat Neurosci 1999;2:913–919.
187. Kanwisher N, McDermott J, Chun MM. The fusiform face area: A module in human extrastriate cortex specialized for face perception. J Neurosci 1997;17:4302–4311.
188. Joseph JE. Functional neuroimaging studies of category specificity in object recognition: A critical review and meta-analysis. Cogn Affect Behav Neurosci 2001;1:119–136.

189. Gorno-Tempini ML, Price CJ. Identification of famous faces and buildings: A functional neuroimaging study of semantically unique items. Brain 2001;124:2087–2097.

190. Aguirre GK, Zarahn E, D'Esposito M. An area within human ventral cortex sensitive to "building" stimuli: Evidence and implications. Neuron 1998;21:373–383.

191. Epstein R, Harris A, Stanley D, Kanwisher N. The parahippocampal place area: Recognition, navigation, or encoding? Neuron 1999;23:115–125.

192. Petersen SE, Fox PT, Posner MI, Mintun M, Raichle ME. Positron emission tomographic studies of the cortical anatomy of single-word processing. Nature 1988;331:585–589.

193. Vandenberghe R, Price C, Wise R, Josephs O, Frackowiak RSJ. Functional anatomy of a common semantic system for words and pictures. Nature 1996;383:254–256.

194. Davis MH, Johnsrude IS. Hierarchical processing in spoken language comprehension. J Neurosci 2003;23:3423–3431.

195. Giraud AL, Kell C, Thierfelder C, et al. Contributions of sensory input, auditory search and verbal comprehension to cortical activity during speech processing. Cereb Cortex 2004;14:247–255.

196. Rodd JM, Davis MH, Johnsrude IS. The neural mechanisms of speech comprehension: fMRI studies of semantic ambiguity. Cereb Cortex 2005;15:1261–1269.

197. Kartsounis LD, Shallice T. Modality specific semantic knowledge loss for unique items. Cortex 1996;32:109–119.

198. Beauchamp MS, Haxby JV, Jennings JE, DeYoe EA. An fMRI version of the Farnsworth-Munsell 100-Hue test reveals multiple color-selective areas in human ventral occipitotemporal cortex. Cereb Cortex 1999;9:257–263.

199. Goldberg RF, Perfetti CA, Schneider W. Perceptual knowledge retrieval activates sensory brain regions. J Neurosci 2006;26:4917–4921.

200. Kellenbach ML, Brett M, Patterson K. Large, colorful, or noisy? Attribute- and modality-specific activations during retrieval of perceptual attribute knowledge. Cogn Affect Behav Neurosci 2001;1:207–221.

201. Puce A, Allison T, Bentin S, Gore JC, McCarthy G. Temporal cortex activation in humans viewing eye and mouth movements. J Neurosci 1998;18:2188–2199.

202. Kourtzi Z, Kanwisher N. Activation in human MT/MST by static images with implied motion. J Cogn Neurosci 2000;12:48–55.

203. Kable JW, Lease-Spellmeyer J, Chatterjee A. Neural substrates of action event knowledge. J Cogn Neurosci 2002;14:795–805.

Chapter 14

fMRI of Emotion

Simon Robinson, Ewald Moser, and Martin Peper

Summary

Recent brain imaging work has expanded our understanding of the mechanisms of perceptual, cognitive, and motor functions in human subjects, but research into the cerebral control of emotional and motivational function is at a much earlier stage. Important concepts and theories of emotion are briefly introduced, as are research designs and multimodal approaches to answering the central questions in the field. We provide a detailed inspection of the methodological and technical challenges in assessing the cerebral correlates of emotional activation, perception, learning, memory, and emotional regulation behavior in healthy humans. fMRI is particularly challenging in structures such as the amygdala as it is affected by susceptibility-related signal loss, image distortion, physiological and motion artifacts and colocalized Resting State Networks (RSNs). We review how these problems can be mitigated by using optimized echo-planar imaging (EPI) parameters, alternative MR sequences, and correction schemes. High-quality data can be acquired rapidly in these problematic regions with gradient compensated multiecho EPI or high resolution EPI with parallel imaging and optimum gradient directions, combined with distortion correction. Although neuroimaging studies of emotion encounter many difficulties regarding the limitations of measurement precision, research design, and strategies of validating neuropsychological emotion constructs, considerable improvement in data quality and sensitivity to subtle effects can be achieved. The methods outlined offer the prospect for fMRI studies of emotion to provide more sensitive, reliable, and representative models of measurement that systematically relate the dynamics of emotional regulation behavior with topographically distinct patterns of activity in the brain. This will provide additional information as an aid to assessment, categorization, and treatment of patients with emotional and personality disorders.

Key words: Emotion, fMRI, Research design, Reliability, Validity, Amygdala, Signal loss, Distortion, Resting state networks

1. Introduction

While recent brain imaging work has expanded our understanding of the mechanisms of perceptual, cognitive, and motor functions in human subjects, research into the cerebral control of emotional

M. Filippi (ed.), *fMRI Techniques and Protocols*, Neuromethods, vol. 41
DOI 10.1007/978-1-60327-919-2_14, © Humana Press, a part of Springer Science+Business Media, LLC 2009

and motivational functions has been less intense. For several years, however, a growing body of fMRI and positron emission tomography (PET) work has been assessing the cerebral correlates of emotional activation, perception, learning and memory, and emotional regulation behavior in healthy humans *(1–6)*.

Current brain imaging work is based on the concepts and hypotheses of the multidisciplinary field of "affective neuroscience" *(7–9)*. The endeavors of the subdisciplines of affective neuroscience have not only complemented but also promoted each other, stimulating a rapid growth of knowledge in the functional neuroanatomy of emotions. It is increasingly recognized that these areas also share similar methodological problems.

The expanding area of emotion neuroimaging has provided new methods for validating neurocognitive models of emotion processing that are crucial for many areas of research and clinical application. Progress is being made in disentangling the cerebral correlates of interindividual differences, personality, as well as of abnormal conditions such as, for example, anxiety, depression, psychoses, and personality disorders *(10–12)*. It has been recognized that psychological assessment, categorization procedures, and psychotherapy treatment may profit from models that integrate functional connectivity information. The relevance and usefulness of valid neurocognitive models of emotion processing have recently been recognized by many fields of applied research such as, for example, psychotherapy research *(13)*, criminology *(14)*, as well as areas such as "neuroeconomics" *(15)* and "neuromarketing" *(16)*.

Several human lesion studies have pointed to the deficits of neurological patients in recognizing emotions in faces, particularly often for the decoding of fearful faces especially after bilateral amygdala damage *(17–20)*. Other studies have reported impairments not only for fear but also for other negative emotions such as anger, disgust, and sadness *(21, 22)*. Recent functional imaging studies have confirmed the importance of the amygdala in emotion processing. Due to the multiple connections between the amygdala and various cortical and subcortical areas, and the fact that the amygdala receives processed input from all the sensory systems, its participation is essential during the initial phase of stimulus evaluation *(23)*. The appraisal function of the amygdala, combining external cues with an internal reaction, reflects the starting point for a differential emotional response and is hence the basis for emotional learning. Involvement of the amygdala during classical conditioning especially during the initial stages of learning *(24, 25)* as well as during processing signals of strong emotions has been documented repeatedly with fMRI. However, a problem in verifying amygdala activation with neuroimaging tools may be the rapid habituation of its responses *(10, 26)*.

Although the need for brain imaging data is not unequivocally acknowledged by all researchers in their specialties, the increasing body of neuroimaging data has value in challenging and constraining existing theories. Followers of cognitive emotion theory must face the fact that their results need to be compatible with or at least not contradict with established neuroscience (neuroimaging) findings *(27)*. However, to appropriately evaluate and integrate this knowledge, it is necessary to deal with the basic methodological problems of the field.

Therefore, this chapter is organized around the two major issues of the neuroimaging of emotional function. First, it addresses underlying conceptual issues and difficulties associated with operationalizing and measuring emotion (for more detailed reviews, *see* **refs.** *28–31)*. The problems and limitations of brain imaging work that are associated with measurement precision, response scaling, reproducibility, as well as validity and generalizability are discussed corresponding to general principles of behavioral research *(32)*.

Second, the complexities of neuroimaging methods are examined to supplement recent quantitative meta-analyses (for a summary of findings of the emotional neuroimaging literature, *see* **refs. 2**, **4**, and **5**). We raise here some grounds for reflection about current measurement in neuroimaging of emotions, and to encourage the adoption of recent methodological advances of fMRI technology. In summary, it is suggested that additional interdisciplinary efforts are needed to advance measurement quality and validity, and to accomplish an integration of brain imaging technology and neuropsychological assessment theory.

2. Psychological Methods

2.1. Emotion Theories and Constructs

2.1.1. Definitions

Emotions have been defined as episodes of temporarily coupled, coordinated changes in component functions as a response of the organism to external or internal events of major significance. These component functions entail subjective feelings, physiological activation processes, cognitive processes, motivational changes, motor expression, and action tendencies *(33, 34)*. Emotions represent functions of fast and flexible systems that provide basic response tendencies for adaptive action *(35)*.

Emotions can be differentiated from mood changes (extended change in subjective feeling with low intensity), interpersonal stances (affective positions during interpersonal exchange), attitudes (enduring, affectively colored beliefs, preferences, and predispositions toward objects or persons), and personality traits (stable dispositions and behavior tendencies) *(29, 34)*.

The frequently used concept of "emotional activation" characterizes a relatively broad class of physiological or mental phenomena (e.g., strain, stress, physiological activation, arousal, etc.). It can be specified with respect to a variety of dimensions such as valence (quality of emotional experience), intensity or arousal (global organismic change), directedness (motivational and orientating functions), and selectivity (specific patterns of change) *(36)*. In contrast, the terms emotional reactivity or arousability, and psychophysical reactivity refer to the ***dispositional*** variability of the above activation processes under defined test conditions *(37, 38)*.

Environmental objects possess a latent meaning structure of emotional information, which is represented by a hierarchy of constructs with relatively fixed intra- and interclass relations *(29, 39, 40)*. Accordingly, physical stimulus properties or surface cues serve as a basis for "universal" emotion categories such as happiness, surprise, fear, anger, sadness, and disgust that originate at a primary level *(41)*. On a secondary level, dimensions such as valence and arousal arise from the preceding levels *(42)*. **Table 1** suggests a potential structure of emotional concepts or domains that integrates both discrete (primary) and secondary emotions *(29)*.

Table 1
Hierarchical organization of emotion concepts (modified from [29])

Emotion concepts or domains	Example constructs	Basis for higher-order grouping
Dimensional concepts	Valence (positive/negative emotions), approach/withdrawal, activity (active/passive), control, etc.	Conceptual or meaning space for subjective experience and verbal labels
Basic, fundamental, discrete, modal emotions or emotion families	Anger, fear, sadness, joy, etc.	Similarity of appraisal, motivational consequences, and response patterns; convenient label for appropriate description and communication
Specific appraisal/response configurations for recurring events/situations	Righteous anger, jealousy, mirth, fright, etc.	Temporal coordination of different response systems for a limited period of time as produced by a specific appraisal pattern
Continuous adaptational changes	Orienting reflex, defense reflex, startle, sympathetic arousal, etc.	Automatic activations and coordination of basic biobehavioral units

Emotional activation has also been characterized as a process with a sequence of stages *(29, 35)*: following an initial evaluation of novelty, familiarity, and self-relevance, a stimulus object or context is fully encoded. This involves detection of physical stimulus features, recognition of object identity, and identification of higher-order emotional dimensions such as pleasantness or need significance. During the subsequent stages, cognitive appraisal processes are initiated to evaluate the significance of the event. These evaluation checks include an appraisal of whether the stimulus is relevant for personal needs or achieving certain goals. Finally, the potential to overcome or cope with the event and the compatibility of behavior with the self or social norms is evaluated *(35)*.

2.1.2. Operationalization

The measurement of emotions crucially depends on an appropriate operationalization of the construct of interest and definition of response parameters. Such considerations have typically been elaborated in the context of psychological assessment theory *(30, 38, 43)*. The latter explains how psychological and physiological measures can be empirically assessed, decomposed, and used as indicators of the psychological constructs of interest. It organizes the assumptions concerning measurement, segmentation, and aggregation of activation measures, and evaluates the distribution characteristics and reliability of the data. It also determines the range of the construct of interest by localizing it according to variables, subjects or settings/situations, or combinations of these sources of variation. Since most current operationalizations are confined to one of these aspects, the range of conclusions to be drawn from the findings is also limited.

A particular problem associated with measuring emotional reactions is a certain lack of covariation of response measures. A frequent finding is that the expected synchronization of verbal, motor, and physiological response systems during an emotional episode is the exception rather than the rule. Although emotional episodes supposedly give rise to a synchronization of central, autonomic, motor, and behavioral variables *(44)*, most emotional response measures only show imperfect coupling *(45)*. This response incoherence may be attributed to a temporary decoupling or dissociation of function *(46)*. This has led authors to suggest a triple response measurement strategy that suggests a multimodal assessment of emotion including responses in the verbal, gross motor, and physiological (autonomic, cortical, neuromuscular) response systems *(47)*.

Research on human emotion has illustrated how the broad concept of emotion is subdivided into several component functions that dynamically interact during an emotional episode. Diverse operationalizations have been suggested to assess these subconstructs, many of which are highly correlated and form clusters or families of similar functions. Emotional activation processes are embedded in a multicomponential system of situational and personal determinants. Factors that shape the level and pattern of the emotional activation process are the following *(29, 48)*: the functional context of the task (e.g., cognitive processing, motor responses, autonomic functions, etc.); the direction and extension of effects (e.g., global versus selective activation); the intensity and the degree of emotional strain (e.g., low, middle, or traumatic intensity; degree of threat; intensity of physical/mental load; stimulus intensities below or above threshold); the time characteristics (e.g., duration, structure, and variability of a stimulus; effects of stimulus repetition or pre-exposure); the informational content (e.g., the degree of information and dimensions inherent in the experimental stimuli such as emotional valence or arousal, preparedness, novelty, safety, predictability, contingency information, etc.); the implications for action (conduciveness, implications for instrumental reactions; artificial vs. realistic nature of the procedure); the coping potential (e.g., active coping vs. passive enduring, degree of controllability, helplessness, social support, specific coping strategies); and, compatibility with self or social norms (e.g., personal relevance).

These different aspects have led to a large number of operationalizations. These include procedures to elicit orienting or startle reactions, basic emotions or "stress," as well as stimulus-response paradigms and conditioning procedures. For example, one such standard procedure is to elicit orienting reactions (OR) by emotionally meaningful stimuli. The OR is a nonassociative process being modulated by excitatory (sensitization) and inhibitory (habituation) mechanisms. Pavlovian (classical) or instrumental conditioning of excitatory or inhibitory reactions has traditionally been investigated in autonomic reactions (cardiovascular, vasomotor, and electrodermal conditioning), motor responses (eye blink), and endocrine or immune system reactions *(1)*.

Emotional experience is strongly influenced by cognitive activities which modulate attention and alertness (avoidance and escape), vigilance processes (information search and problem solving), person-situation interactions (denial, distancing, cognitive restructuring, positive reappraisal, etc.), and actions, which change the person-environment relationship *(49)*. Coping research has identified typical cognitive strategies to regulate arousal during an emotional episode such as rejection (venting, disengagement) and accommodation strategies (relaxation, cognitive work) *(50)*. Cognitive activities subsume engagement (reconceptualization, re-evaluation strategies such as rationalization or reappraisal) and distraction techniques.

These behavioral and cognitive regulation processes have been studied for many decades *(51)*. This research has shown that the outcome of coping processes crucially depends upon the valence, ambiguity, controllability, and changeability of a stressor. Input-related regulation [denial, distraction, defense, or cognitive restructuring; *(52)*] or antecedent-focused regulation [selection, modification or cognitive restructuring of situational antecedents; *(53)*] have been differentiated from response-focused processes [suppression of expressive behavior and physiological arousal; *(53)*].

While the behavioral procedures mentioned above are mostly unstandardized, a vast number of standardized psychometric instruments are available to assess the higher-order emotional processes (for a review *see* **ref. 31**). Questionnaires are the most frequently used method, being followed by behavior ratings by experts or significant others. However, these data assess subjective representations, that is, personal constructs and may be obscured by biased responding

2.2. Research Design and Validity

2.2.1. Research Design

The requirements for experimental research *(32)* are not always fulfilled by many early research designs of emotional neuroimaging work. This is typical for the pilot stage of scientific progress. In many cases, only preliminary or correlational interpretations are possible due to incomplete or missing control conditions [e.g., with respect to the "awareness" of emotional stimuli; *(54)*]. In contrast, more recent work increasingly makes use of full factorial designs or applies parametric variations of the independent variable *(55)*. Moreover, new techniques of covariance analysis are available to explore the causal predictive value of structural data on emotional brain activation. The relationship of structural and functional connectivity data has been explored by means of Structural Equation Modeling *(56–58)* and Dynamic Causal Modeling *(59)*. Moreover, functional brain imaging has been successfully combined with the lesion approach to elucidate the modulating influences of interconnected brain regions *(60)*. Thus, by means of appropriate research plans and advanced techniques of analysis, an "effective connectivity" can be identified that elucidates the causal relations of one neural system to another *(61)*. For example, the functional connectivity of the prefrontal cortex (PFC) that is supposed to modulate amygdala activity *(62)* might thus be better evaluated in terms of causality.

To avoid operationalization errors, the quality of the emotion induction procedure needs to be scrutinized, that is, it must be evaluated whether the intended emotion has actually been elicited. For example, since a variety of emotional and nonemotional stimulus situations may trigger amygdala activations *(5)*, it is necessary to evaluate whether the intended emotion (such as fear) has actually been elicited. Since subjective report is not always an appropriate manipulation check, additional psychophysiological criteria are needed to validate the intended emotion. Sympathetic activity as indexed by electrodermal activity (EDA) has been assessed during imaging procedures for this purpose. Nevertheless, this does not validate fear since skin conductance responses represent the endpoint of many different processes *(63)*.

2.2.2. Construct Validation

Brain imaging work implements specific neuropsychological construct validation strategies by associating behavioral measurement of emotion with functional brain activation data for different localizations *(31)*. Here, functional (physiological) data are related to but still remain categorically distinct from the psychological data that emerge from a particular behavioral paradigm. During the process of construct validation, indicators of connectional or neurophysiological constructs are related to the indicators of psychological constructs. Thus, different operationalizations of a certain psychological construct (procedures or task) are expected to be correlated with activations of a certain area or cluster of areas. A different construct is expected to correlate with another but not the previous area and vice versa. This corresponds to the double dissociation approach, which inspects task by localization interactions. This process of neuropsychological concept formation typically starts at a relatively broad level and proceeds downward in the above hierarchy finally specifying within-systems localization constructs *(64)*.

However, depending on limitations of the measurement device described below, the reliability of psychological or activation data declines at lower levels of structural constructs complicating this validation process. The diverse validation attempts typically draw upon convergent or divergent associations of constructs that are located at quite different levels of generality. However, successful construct validation very much depends upon whether brain activation and psychological measures are analyzed on the same level of generality. In cases of asymmetry, low relationships may result that provoke misinterpretations and confuse the validation process. Thus, successful construct validation in the affective neurosciences requires emotional constructs and brain activation data to be measured on the same (symmetrical) level of generality or aggregation *(31)*.

Emotional neuroimaging is typically guided by neuropsychological construct validation strategies. Here, the constructs are operationally defined by the complementary methods of emotion psychology and of neurophysiology. Both construct types are embedded in hierarchically organized networks with lower- and higher-order levels of generality. Both types of data are associated with each other during validation. However, it is necessary to define neural and emotional constructs on the same level of generality. For example, when a relatively broad behavioral category or set of functions ("emotion regulation") is being associated with isolated cerebral substructures, the relationship is likely to be asymmetrical and disappointing low correlations might result confusing the validation process.

2.2.3. Internal Validity

FMRI is known to be a highly reactive measure because the scanner setting (gradient noise and the supine position) causes the subject to respond to the experimental situation as a stressor. Unless habituation sessions are included in the procedure, tonic stress and arousal effects may be induced that modulate responding as discussed above. For example, a decreasing rate of response of the amygdala to a conditioned stimulus during the late phase of acquisition ***(10, 24, 26, 65)*** may also be attributable to testing effects (sensitization to the setting, acquaintance with the procedure and type of unconditioned stimulation) rather than fast amygdala habituation per se (other factors might also explain reduced amygdala perfusion measures such as potential ceiling effects, baseline dependencies, and regression to the mean). In general, familiarity with emotionally activating procedures in the scanner induces states of expectation, sensitizing or desensitizing effects that may confound follow-up measurement. In addition to these testing effects, history, that is, occurrences other than the treatment and individual experiences between a first and a second measurement are likely to endanger the assessment of emotion (e.g., when assessing psychotherapy effects).

Changes in the observational technique, the measurement device or sequence and other instrumentation effects may also obscure emotion-related treatment variance during an fMRI session or across sessions. From the discussion of MR methods it is clear that longitudinal changes of measurement precision are also to be expected from inconsistent acquisition geometry and shim, as well as system instabilities and hardware changes.

It is well known from psychophysiological research that the interpretation of repeated measurement factors is complicated by initial value dependencies ***(66)***. When the hemodynamic response is fitted relative to the prestimulus baseline, a physiological or statistical dependency of tonic perfusion levels and the phasic reaction may prevail ***(67)***. While the first experimental blocks may show extreme effects, subsequent measurements are likely to be closer to the mean. Moreover, it has been pointed out above that the reliability of blood oxygen level-dependent (BOLD) measurements may be compromised by distortions or signal loss. When emotional paradigms with inconsistent effects are used or when subjects with an extreme variability of emotional responsivity are investigated, experimental effects are likely to show "regression to the mean."

Subjects change as a function of time and these maturation effects may occur during the time range of the experiment (psychophysiological changes of organismic state or psychological stance, in particular during aversive paradigms). State-dependent influences or maturation effects may hamper within-subject replication or evaluations of long-term psychotherapy effects.

Subject groups with an elevated emotionality are more likely to show greater drop out rates in stressful experiments, that is, subjects of one group drop out as a consequence of their specific reactivity to the emotionally strain of the challenge paradigm. If exit from an emotionally activating study is not random, this effect of "experimental mortality" may confound comparison between groups.

Selection effects, that is, group differences from the outset of the study, are likely in functional imaging studies with very small numbers of participants. Selective recruitment of volunteers or drop out of participants may lead to decreased reactivity and lower emotionality in the remaining study group. Poor recruitment techniques (e.g., drafting subjects from the social circle of the lab partially acquainted with the procedures) or lack of random assignment to groups may further limit the validity of emotional fMRI studies.

Interactions of selection with maturation may occur when groups that differ with respect to maturation processes are compared (e.g., administering a social stress test for cortisol stimulation at different times of the day). Gender, personality traits, or psychopathology are all associated with specific individual differences of emotional regulation behavior. When these behaviors change over time as a function of personal development, follow-up measurements may be confounded by this type of effect. Thus, poor randomization or lack of control of personality-specific variance may jeopardize brain activation studies of emotional behavior. Finally, an interaction of selection with instrumentation occurs when experimental subjects and controls show pre-experimental differences with respect to the shape of their responses such as floor or ceiling effects.

In general, emotional responses show an intraindividual instability due to measurement artifacts (*see* **Subheading 3**), state-dependent influences, or characteristics of the subject (age, gender, experience, temperament) all impose additional effects on functional neuroimaging results *(5)*. A considerable degree of within- and between-subject variation in the time course of emotional responding depends on habitual, subject-specific mechanisms. First, the phasic activation pattern reflects the short-term modulation in response to the emotional stimulus. Due to the temporal within-trial variability of BOLD responses in different brain regions, averaging across subjects may obscure the detection of activation in a specific region and reduce effect sizes specifically for higher-level reactions. Second, activation also varies across the time course of the experiment. Most subjects show a constant increase in autonomic arousal depending on the degree of emotional stimulation. This is not only accompanied by a systemic response (tonic increase of sympathetic activation including blood pressure, cardiac contractility, and variability), but also by variations of tonic perfusion. These changes may show divergent trends for cortical and limbic regions imposing an unknown error on the measurement of the phasic BOLD reaction. These tonic and phasic variations appear to reflect the subject-specific mechanisms of emotional regulation behavior.

2.2.4. External Validity and Generalizability

Generalization to Other Procedures and Paradigms

The majority of current paradigms have focused on lower-level perceptual or learning processes pertaining to basic or secondary emotional categories. Since the results depend on the selected task parameters (degree of induced arousal, hedonic strength, and motivational value; degree of involvement of memory processes; reinforcement schedule; conditioning to cues or contexts; etc.), a comparison with and generalization to other operationalizations remains difficult. Systematic neuroimaging approaches to higher-level appraisal processes are still sparse. These involve evaluations of the motivational conditions and coping potential, that is, the ability to overcome obstructions or to adapt to unavoidable consequences *(29)*. An expanded range of constructs would involve an assessment of social communication processes, beliefs, preferences, predispositions, high-level evaluation checks, as well as modulating sociocultural influences. Higher-order appraisal processes involve the evaluation of whether stimulus events are compatible with social standards and values or with the self-concept. Another function to be explored concerns the degree to which a stimulus event may increase, decrease, or even block goal attainment or need satisfaction, and activate a reorientation of the individual's goal/need hierarchy and behavioral planning (goal/need priority setting) *(29)*.

Whereas frontostriatal mechanisms of motor control have been increasingly investigated, recent work has made efforts toward developing an understanding of how emotion and motivation are linked to the frontal mechanisms controlling the preparation and execution of behavior *(68, 69)*. Behavior preparation and execution represent closely integrated components within an emotional episode. Mobilization of energy is required to prepare for a certain class of behavior. Action planning and motor preparation requires sequencing of actions and generation of movements. However, an emotion preceding behavior is only one of a number of factors, including situational pressures, strategic concerns, or instrumentality, involved in eliciting the concrete action. Additional research is needed to trace the information flow from motivational to motor systems.

Another component is the verbal or nonverbal communication of emotions such as facial expression or vocal prosody *(70)*. The ability to verbally conceptualize emotions and to communicate emotional experiences plays an important role in the regulation of an ongoing emotional episode. For example, explicit emotion-labeling tasks have been shown to decrease the activation level of the amygdala *(71, 72)*.

Finally, sociocultural factors may shape attitudes (relatively enduring, affectively colored beliefs, preferences, and predispositions toward objects or persons) as well as interpersonal stances (affective stance taken toward another person in a specific interaction). The ability of the individual to form representations of beliefs, intentions, and affective states of others has a considerable importance for affective and interpersonal interaction. However, the effects of beliefs, preferences, and predispositions on lower levels of emotional responding have attracted little attention. Top-down processes may induce considerable variations of task and stimulus parameters by modulating lower-level automatic processes and by controlling the late behavior preparation stages during the emotional process. Thus, generalization to other paradigms and constructs has limitations because higher-level behavioral and cognitive strategies that are part of the individual emotion regulation system [*(50)*; see later] modulate the emotion process.

Generalization to Other Subjects and Populations

The study groups of many fMRI studies have been relatively small and poorly described with respect to personality dimensions. Since several studies provide evidence for trait-dependent differences in responding *(73–76)*, it remains unclear to what extent the results may have been influenced by interindividual differences of the participating subjects. The representativeness of results is particularly poor if members of the social circle of the lab serve as participants instead of independently recruited participants. Thus, when the effects of an emotional paradigm interact with characteristics of the study groups (such as a low level of emotionality in subjects willing to participate in an activating scanning condition), this selection × treatment effect may endanger generalizations to other populations.

Generalization to Other Times and Settings

The prediction of future emotional or psychopathological disorders on the basis of emotional behavior assessed in the scanner remains difficult *(77)*. Eliciting emotions in the imaging scanner is a highly artificial situation. It remains unclear to what extent these results can be generalized to other settings and, in particular, to real life settings. Small and Nusbaum *(78)* have criticized the unnatural MRI scanner setting and suggested an "ecological functional brain imaging approach" that includes monitoring of natural behaviors using a multimodal assessment and environmental context of presentation or behavior. Nevertheless, in contrast to the scanner, emotion in real settings is not restricted to simple reactions but includes the full range of regulatory actions. By correlating fMRI and field data, such as, for example, generated by emotion monitoring during everyday life *(79)*, the "ecological validity," that is, the predictive value of cerebral perfusion patterns for real-life emotions could be better evaluated.

3. fMRI Methods

3.1. Methodological Challenges

3.1.1. Introduction

A host of fMRI studies have identified the amygdalae as central structures in emotion processing (*see* **Subheading 1** and Zald et al. *(5)*, for example, for a review). The amygdalae lie in the anterior medial temporal lobe (MTL), bounded ventrolaterally by the lateral ventricles and medially by the sphenoid sinuses (**Fig. 1**). The differing magnetic susceptibilities of these tissues cause large deviations in the static magnetic field, B_0. There is also a strong gradient in B_0 in the MTL, and differing precession frequencies lead to dephasing of the bulk magnetization and loss of signal in images. This problem is not restricted to the amygdala, however. Inferior frontal and orbitofrontal regions, likewise involved in emotion processing *(80)*, are also zones of high static magnetic field gradient. In addition to signal loss, static magnetic field gradients also lead to echo times (TE) becoming shifted, so that BOLD sensitivity may be reduced, or signal may not be acquired at all [termed "Type 2" loss *(81)*]. These problems are examined in **Subheading 3.1.2**.

Local variations in the static magnetic field strength confound spatial encoding of the MR signal, leading to image distortion. Particular considerations for the MTL in this regard are discussed in **Subheading 3.1.3**. Even at high field, deviations from B_0 immediately in the amygdala are relatively moderate (**Fig. 1** left; 10 Hz measured at the arrow position, for data acquired at 4.0 T) but the field gradient is high (2 Hz/mm at the same position),

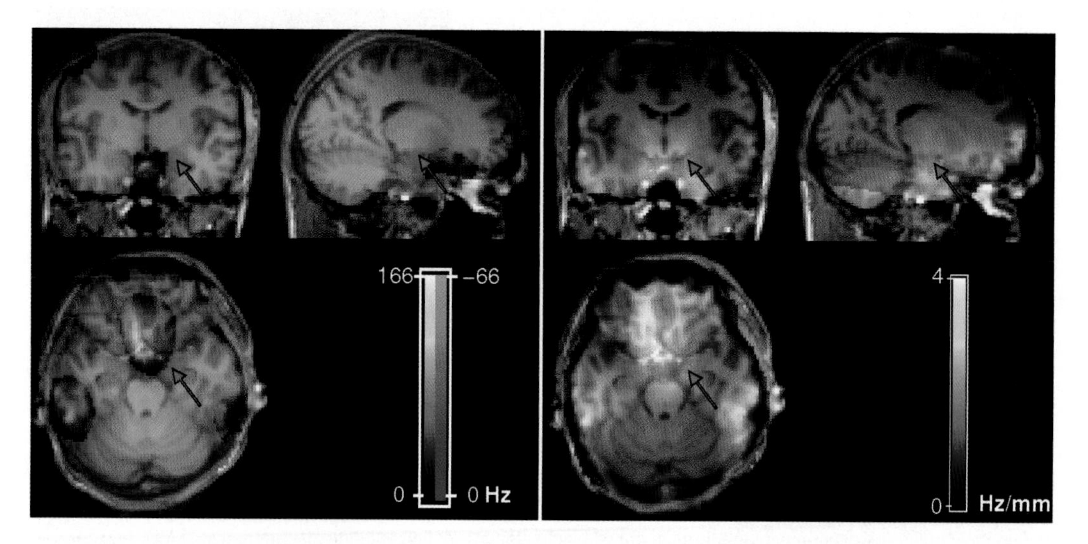

Fig. 1. The amygdalae, central brain structures in emotion processing, lie in a region of moderate deviation from the static magnetic field (*left*) and very high static magnetic field gradients (*right*). The planes intersect in the amygdala at MNI co-ordinate (18, −2, −18), marked by *arrows*. Single subject measurement at 4.0 T.

leading to very large distortions in neighboring structures, which can cause signal to encroach into the amygdalae.

The ventral brain is also prone to physiological artifacts of cardiac and respiratory origin, as described in **Subheading 3.1.4**, which may be mitigated to some extent by simultaneous measurement of cardiac and respiratory processes and the application of postprocessing corrections. In addition to the measurement challenges of ventral brain imaging, the presence of large magnetic field gradients makes the ventral brain susceptible to stimulus-correlated motion (SCM) artifacts, as discussed in **Subheading 3.1.5**. These can lead to the appearance of neuronal activation **(Fig. 2)** arising from subtle head movements which are time locked to stimuli.

A further potential confound is the presence of RSNs which co-localize with regions under study. These show slow fluctuations in the absence of stimuli and constitute sources of unmodeled noise and interrial variation. The existence of a RSN in the amygdalae **(Fig. 3)** offers a possible explanation of why small signal changes are generally recorded in these structures, despite the high neurovascular reactivity of deep gray matter nuclei. This and other RSNs which may involve the amygdala are described in **Subheading 3.1.6**.

In **Subheading 3.1** we expand on the problems outlined here, and go on in **Subheading 3.2** to detail approaches to optimizing conventional single-shot 2D gradient-recalled echo-planar imaging (EPI) to mitigate their effects, alternative sequences which are less sensitive to static magnetic field gradients and, in

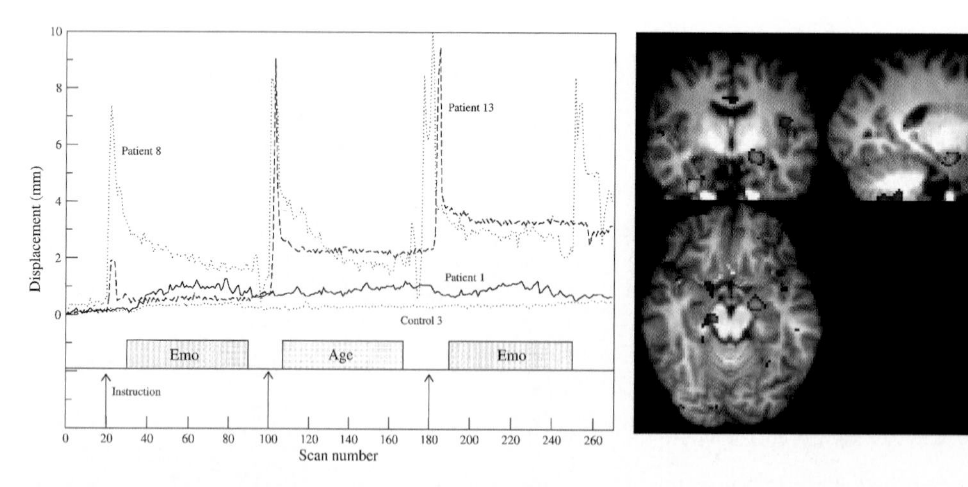

Fig. 2. Large static magnetic field gradients make the amygdala region prone to the artifactual appearance of neuronal activation when stimulus-correlated motion (SCM) is present. *Left*: Observed patterns of SCM of schizophrenic patients and controls in a 3.0-T experiment with three stimulus blocks (facial emotion and age discrimination "EMO" and "AGE"). *Right*: a baseline (no stimulus) study in which a subject executed submillimeter SCM similar to that of Patient 1. The contrast corresponds to the "EMO" periods (uncorrected $p < 0.0001$; t threshold = 5, Montreal Neurological Institute coordinates 22, −6, −16).

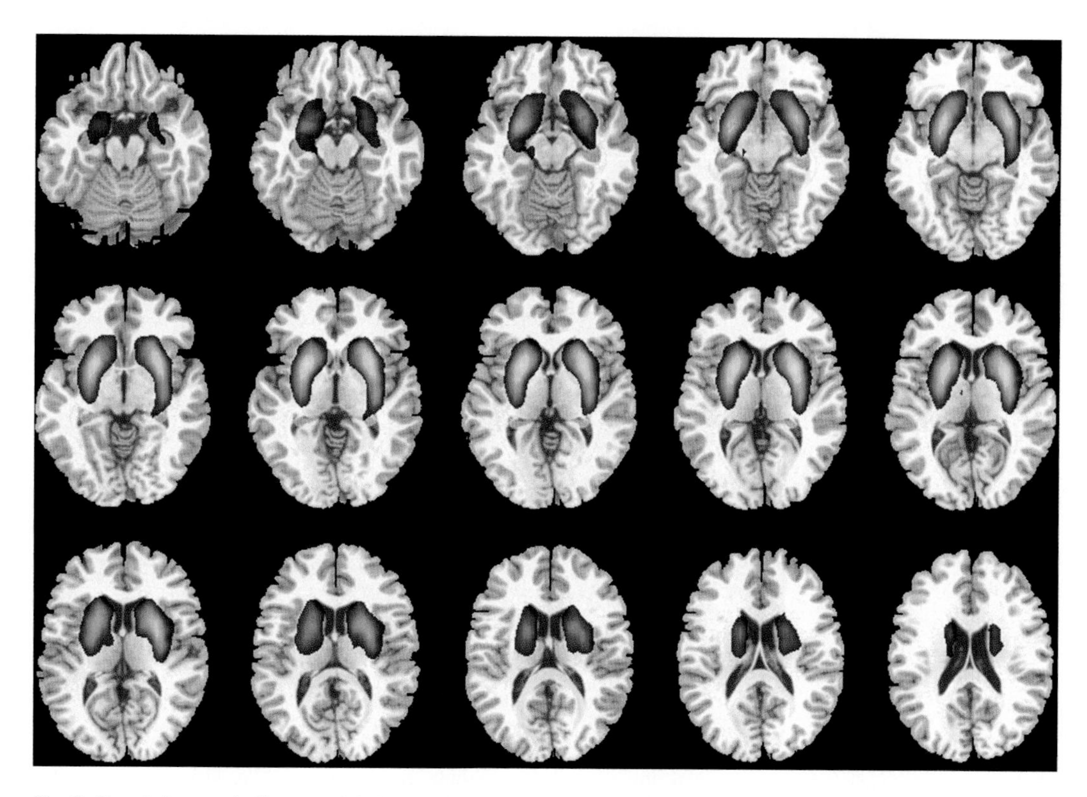

Fig. 3. Signal changes in the amygdala in emotion experiments have to be measured against a background of resting state fluctuations. A resting state network recently been reported, covering the amygdala and basal ganglia (3.0 T, group independent component analysis of 26 young healthy adults). Adapted from ***(106)*** with permission from the ISMRM.

Subheading 3.3, methods to correct for image distortion, physiological noise, and SCM artifacts.

3.1.2. Signal Loss and BOLD Sensitivity Loss

It is worthwhile to briefly review the problem of signal loss from an empirical perspective. A temporal resolution of 1–3 s is usually desirable in fMRI. The whole brain may be covered in this time by acquiring images with voxels of typically 3-mm size (or 27 μl). Relatively long TEs are employed, partly also as a technical necessity – to allow time for gradient switching and echo sampling – but also to confer T_2^* weighting. As well as providing sensitivity to BOLD effects, however, this allows time for dephasing from macroscopic inhomogeneities to develop. The severe signal loss seen in EPI in the anterior MTL with typical parameters is illustrated **Fig. 4** in the lower left two images.

In gradient-echo imaging, the MR signal decays with a time constant T_2^*, comprising the transverse relaxation time, T_2 (reflecting irreversible decay arising from time-varying microscopic spin-spin processes), and T_2', the reversible contribution to the transverse decay rate and the major source of BOLD contrast.

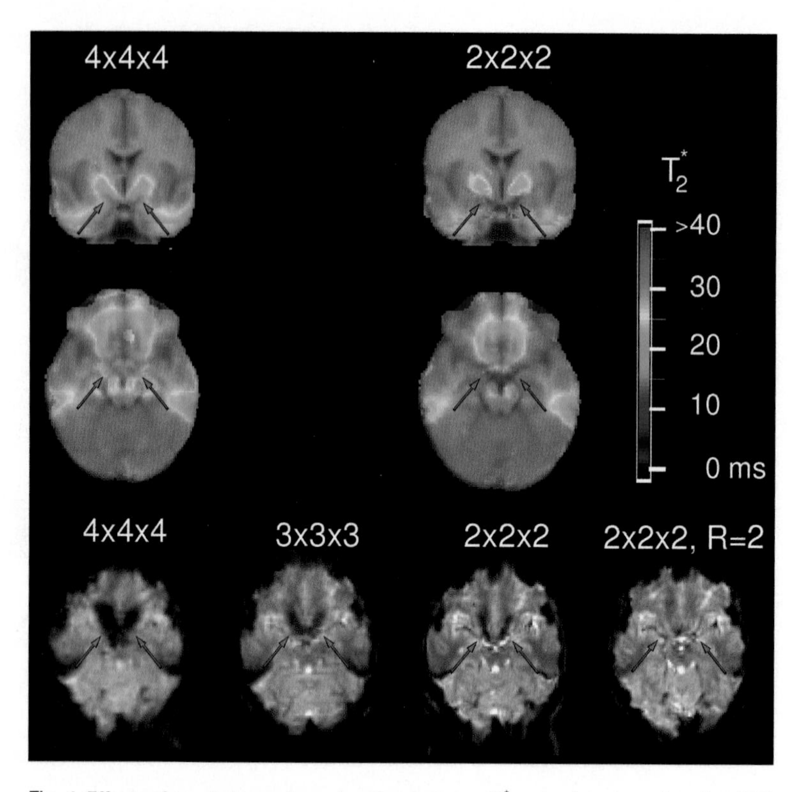

Fig. 4. Effects of voxel size and acceleration factor on T_2^* and echo-planar imaging (EPI) image quality at high field (4.0 T). *Top*: T_2^* in coronal and axial slices through the amygdala at two voxel sizes. *Bottom*: corresponding EPI in slices through the amygdala with acquisition voxel sizes of 4 × 4 × 4 mm, 3 × 3 × 3 mm, 2 × 2 × 2 mm, and 2 × 2 × 2 with GRAPPA acceleration of factor 2, all with echo time (TE) = 32 ms.

T_2' itself can be separated into "mesoscopic" contributions (which operate on a scale smaller than the voxel, e.g., dephasing in the capillary bed), and "macroscopic" contributions (meaning larger than the voxel) which stem from bulk field inhomogeneities and which are dependent on the tissues present, on the quality of shim, and on the scanning parameters such as voxel size and slice orientation. Separating these effects, the MR signal S in a gradient-echo experiment decays such that at the TE it can be expressed *(82)* as:

$$S(\mathrm{TE}) = S(0) \times \exp\left(\frac{-\mathrm{TE}}{\mathrm{T}_2}\right) \times F(\mathrm{TE}), \qquad (1)$$

where an approximation to $F(\mathrm{TE})$ for linear field variations over voxels, ΔBi, in the x, y, and z directions is

$$F(\mathrm{TE}) = \mathrm{sinc}\left(\frac{\gamma \Delta B_x \mathrm{TE}}{2}\right) \times \mathrm{sinc}\left(\frac{\gamma \Delta B_y \mathrm{TE}}{2}\right) \times \mathrm{sinc}\left(\frac{\gamma \Delta B_z \mathrm{TE}}{2}\right). \qquad (2)$$

This illustrates that the signal decay rate may be reduced by decreasing the voxel size – to reduce the gradients across voxels, ΔBi – or by reducing the TE.

The aim of any attempt to optimize an EPI sequence is not just to maximize signal, described above, but also BOLD sensitivity (BS), which is equal to the product of image intensity and TE; for magnetically homogeneous regions is a maximum when the EPI effective TE is equal to the T_2^* of the target region *(83)*. In homogeneous regions, however, the presence of field gradients shifts the location of signal in *k*-space, mainly in the phase-encode direction (because of the low bandwidth), changing the local TE *(81)*. Through-plane field gradients lead to signal loss and reduce BS. If the component of the in-plane susceptibility gradient in the phase-encode direction is antiparallel to the phase-encode gradient "blip" direction, then the TE is also reduced, reducing BS further. Conversely, if it is parallel to the phase-encode "blips" then TE increases. While this increases BS, to some extent compensating for signal loss, if the shift of TE is too large the echo will fall outside the acquisition window, leading to complete signal dropout. This is commonly observed in the anterior MTL for a negative-going phase-encode scheme.

This description motivates the optimization approaches to EPI in susceptibility-affected regions which will be outlined later in this section; compensating through-plane gradients, selecting image orientation and gradient direction to minimize echo shifts, and reducing voxel sizes to reduce field gradients. These techniques will be shown to increase both signal and BS.

3.1.3. Image Distortion

Accurate spatial encoding in MRI is founded upon a homogeneous static magnetic field in the object. The location of signal is deduced from the local field strength under the application of small orthogonal, linear magnetic fields in directions usually referred to as *slice select*, *readout*, and *phase-encode*. The method is confounded if there are regional variations in the static magnetic field, which lead to signal mislocalization (distortion). Typical field offsets are illustrated in **Fig. 1** (left) and lead to EPI distortions of the image shown in **Fig. 4**.

The extent of distortion, expressed as the number of pixels by which signal is mislocalized, is equal to the local magnetic field deviation divided by the bandwidth per pixel (the reciprocal of the time between measuring adjacent points in *k*-space), expressed in the same units. The bandwidth per pixel in the readout direction (rBWread/pix) is equal to the total imaging bandwidth (the signal sampling rate) divided by the image matrix size in the readout direction. In EPI, the pixel bandwidth in the phase-encode direction is smaller than this again by a factor of the image matrix size in the phase-encode direction. The fact that total bandwidth is often increased in proportion with the readout matrix dimension in order to keep rBWread/pix constant means that distortion (in distance rather than number of pixels) is approximately constant as a function of matrix size (and thereby resolution, at constant matrix size). To illustrate the size of expected distortions, in a 64 × 64 matrix acquisition, a typical rBWread might be 1,500 Hz/pixel, giving (as 1,500/64) a rBWphase of 23 Hz/pixel. A value of ΔB_0 of 50 Hz (common at high fields, *see* **Fig. 1**) would lead to a shift of 0.03 voxels in the readout direction, but 2 voxels in the phase-encode direction, or 7 mm for a typical field of view for brain imaging. In a higher resolution acquisition with a 128 × 128 matrix and the same rBWread, rBWphase would be 12 Hz/pixels and the distortion 4 voxels, but also 7 mm because of the proportionately smaller voxel size.

The relationship between EPI distortion and field strength is not simple, depending both on hardware and usage. Susceptibility-induced field changes increase linearly with static magnetic field strength while gradient amplitude (the factor which limits sampling rate) is approximately constant in the standard to high field regimes. While theoretically this leads to an approximate proportionality between distortion and field strength, in practice higher acquisition bandwidths are often used at high field to the achieve shorter effective TEs, to match reduced T_2^* times.

Image distortion frustrates attempts to coregister data from many subjects to a common probabilistic atlas *(84)*, which can reduce significance in fMRI even in relatively homogeneous areas *(85)*. Established methods for correcting image distortion are compared for their performance in the amygdala in **Subheading 3.3.1**.

3.1.4. Physiological Artifacts

A number of physiological processes give rise to fluctuations in the MR signal which are unrelated to neuronal activation, and should therefore be corrected for or modeled in a statistical analysis. The amygdala area is particularly prone to cardiac artifacts due to the proximity of the arteries in the Circle of Willis, and to respiratory artifacts because of the susceptibility gradients.

Respiration leads to head motion, changes in the magnetic field distribution in the head due to changes of gas volume or oxygen concentration in the chest *(86)*, and variation in the local oxyhemoglobin concentration, probably due to flow changes in draining veins *(87)*. Subtle changes in respiration rate and depth are thought to be the origin of spontaneous changes in arterial carbon dioxide level at about 0.03 Hz which have been shown to lead to significant low-frequency variations in BOLD signal *(88)*. The lag of 6 s in this process corresponds to the time taken for blood to transit from the lungs to the brain, and for cerebral blood flow volume to respond to CO_2, a cerebral vasodilator. Magnetic field changes in the head particularly affect ventral brain imaging due high field gradients. Respiration-related artifacts typically affect the image periphery, making them problematic for the amygdala, which is usually at the anterior boundary of the signal-providing region.

Cardiac pulsatility causes expansion of the arteries, bulk motion of the brain, and cerebrospinal fluid flow and leads to the influx of fully relaxed spins into an imaging slice. As a consequence, the signal may increase in many of the arteries that lie close to the amygdala, such as the middle cerebral artery and other elements of the Circle of Willis *(89)*. Cardiac artifacts are particularly complex with regard to emotion studies as the amygdala innervates the autonomic nervous system via the hypothalamus and brainstem, increasing heart rate, as has been shown in fMRI *(90)*, and human depth electrode studies *(91)*. Recently, fluctuations in cardiac rate have been shown to explain almost as much variation in the BOLD signal as the oscillations related to each cardiac cycle, as revealed by shifted cardiac rate regressors *(92)*.

Cardiac and respiratory cycles are connected by a number of processes *(93)*, leading to many regions showing BOLD fluctuations of cardiac origin *(92)* being also observed in studies of respiratory effects *(94)*.

Cardiac and respiratory artifacts may be corrected for by a number of approaches, some of which require additional measurements at the time of imaging. The effectiveness of these techniques in the ventral brain is outlined in **Subheading 3.3.2**.

3.1.5. Motion Artifacts

Motion artifacts affect all regions of the brain, but are particularly problematic in emotion studies because the nature of the task material is prone to induce SCM as a startle, attention, or repulse response. Patients with disorders with emotional components (such as schizophrenia and posttraumatic stress disorder) are less likely to remain still throughout the experiment and the interaction between motion and distortion in regions of high susceptibility gradient produces nonlinear pixel shifts that are not well corrected with rigid-body methods. Partial brain coverage protocols, such as those that may be used to allow *z*-shimming or high spatial and temporal resolution fMRI in the amygdala, are also more prone to partial voluming in the outermost slices and spin history effects, in which motion between the acquisition of adjacent slices leads to some spins being excited twice within one repetition time (TR) while others are not excited at all.

Head motion can be minimized using bite bars, vacuum cushions, thermoplastic masks, or plaster head casts. As well as effective immobilization, casts allow for repositioning in longitudinal studies *(95)*. Such devices are not appropriate for emotion studies, however, due to the added degree of discomfort and distraction they provide.

SCM was originally investigated by Hajnal et al. *(96)* in hybrid simulations with quite large (3 mm) introduced pixel shifts, which led to peripheral correlations. A study by Field et al. *(97)* found that small-amplitude motion can lead to false positive results, particularly in regions of high field gradient. Likewise, larger motions can reduce significance and lead to false negative results. Two distinct patterns of SCM are often observed in fMRI experiments. As in the example of identified motion with sample schizophrenic patients and controls (**Fig. 2**, left), patients may execute large motions at the first presentation of a stimulus, and many patients and controls show very small displacements which endure for entire blocks. Reproducing the submillimeter head motions observed in that experiment in a separate session (without stimuli), these have been shown to lead to highly significant correlations in the amygdala which are difficult to distinguish from genuine activation (**Fig. 2**, right), a problem not mitigated by standard motion correction methods *(98)*.

3.1.6. Colocalized Resting State Networks

An additional methodological confound comes in the form of RSNs, which constitute additional sources of signal fluctuations unrelated to experimental task. In the absence of tasks or stimuli, the brain undergoes slow (0.01–0.1 Hz) fluctuations in functionally related networks of brain regions *(99, 100)*. These endure during task execution, and have been shown to account not only for much of the intertrial variation in the BOLD response in evoked brain response *(101)*, but also to the intertrial variability in behavior *(102)*. Approximately ten such RSNs have been discovered over the past decade *(99, 100, 103–105)* in networks relating to sensory or cognitive function. A network with similar low-frequency characteristics has recently been identified in the amygdala and basal ganglia *(106)*.

The network illustrated in **Fig. 3** shows the results from a group of independent component analysis (ICA), performed with MELODIC *(107)*, of resting state data acquired from 26 subjects. It is continuous, fully incorporating symmetrically the striate nuclei (pallidum, puitamen, and caudate nuclei), extending inferiorly to the amygdaloid complexes. The network is weaker than those previously reported (measured by the amount of variance it explains in the data), but is reproducible across subgroups of subjects, runs, and resting state conditions (fixation and eyes closed) and offers a tantalizing explanation as to why, despite the fact that neurovascular reactivity is high in deep gray nuclei, BOLD signal changes are weaker and less consistent in the amygdalae and basal ganglia than in the cortex.

This may not be the only RSN in which the amygdala is involved. Correlations were observed between the amygdalae, and between the amygdalae and hippocampi and anterior temporal lobes in one of the earliest resting state analyses, using functional connectivity *(100)*. The amygdala was also listed as an element in the "default mode" network *(108)*, when originally reported as regions showing deactivations across a number of tasks in PET *(109)*. The fact that the amygdala has not been observed as part of this network in this context may relate to the technical challenges of measurement discussed in this chapter.

3.2. MR Methods, Sequences, and Protocols

3.2.1. Field Strength

While the signal to noise ratio (SNR), the magnitude of BOLD signal changes, and the specificity of the BOLD response to microvascular contributions all increase with field strength, so do physiological noise, field inhomogeneities, and physiological artifacts which specifically affect the anterior MTL. The advantages of high field for emotion studies are therefore restricted to particular regimes and methods in which these problems are minimized. Human emotion fMRI studies have been carried out at field strengths from 1.0 to 7.0 T. In line with the development of sequences and approaches to EPI in susceptibility-affected area which are discussed in **Subheadings 3.2.2–3.2.8** (high resolution single and multishot EPI, multiecho and spiral acquisitions, gradient compensation, and parallel imaging), emotion fMRI in the high field regime (3.0–4.0 T) has become commonplace, although applied studies have generally used standard sequences and parameters despite the problems which have received attention in the MR literature *(110)* and a number of promising remedies (see the following sections). Ultra-high field strength studies of emotion are still sparse, however, and it is likely that they will be restricted to highly specific questions during the next 5–10 years of hardware and sequence development.

Theoretical gains in SNR at high field are limited by physiological noise, which increases both with field strength and voxel size, and causes time-series SNR (tSNR) to reach as asymptotic limit with voxel volume *(111)*. This limit was found to increase only modestly with field strength, being 65 at 1.5 T, 75 at 3 T, and 90 at 7 T, so that for large (5 × 5 × 3 mm) voxels, tSNR was only 11% higher at 3 T than at 1.5 T, and only 25% higher at 7 T than 1.5 T. The tendency toward asymptotic behavior began at relatively small volume volumes, with 80% of the asymptotic maximum being reached at 28.6 mm^3, 15.0 mm^3, and 11.7 mm^3 at 1.5 T, 3 T, and 7 T, respectively. For small voxels, however, where thermal noise dominates, tSNR gains were almost linear with field strength. In the same study, the authors found that with 1.5 × 1.5 × 3 mm^3 voxels, tSNR increased by 110% at 3 T compared to 1.5 T, and by 245% at 7 T compared to 1.5 T *(111)*. This study clearly shows that tSNR gains are to be made at high field in the small voxel volume regime.

These tSNR results also explain the often modest gains achieved in fMRI studies at higher field, particularly in regions affected by signal dropout. Krasnow et al. *(112)* compared activation in response to perceptual, cognitive, and affective tasks at 1.5 and 3 T with a relatively large voxel protocol (3 × 3 × 4 mm) and observed only moderate increases in activated volume at 3 T for the perceptual and cognitive tasks (23% and 36%, respectively), but no significant improvement in the activated amygdala volume due to increased susceptibility-related signal loss. A high-resolution, high-field approach has been exemplified in the only human study of amygdala function at 7 T to date of which we are aware, which was carried out at submillimeter resolution *(113)*.

These studies define the regime in which field strength gains are to be made, but it is fair to ask why one should move to high resolution measurements if the neuroscience question does not require, for instance, subnuclei of the amygdala to be resolved, but – as is more commonly the case – the study of interactions between the amygdalae and the cortex, for which whole brain coverage is essential. The use of high resolution here is not principally to distinguish activation in small structures, but to reduce both physiological noise and susceptibility artifacts. A number of works have shown the value of averaging thin slices, downsampling and smoothing data acquired at high resolution *(114–116)* and using multichannel coils *(115)* to regain losses in SNR inherent to small voxels generally and yielding net gains in susceptibility affected areas *(115, 117)*.

3.2.2. z-Shimming, Gradient Compensation, Tailored RF Pulses

The effect of signal dephasing arising from through-plane gradients may be reduced by creating a composite image from a number of acquisitions in which different slice-select gradients are applied *(118)*, a process known as *z*-shimming. In each image the applied gradient pulse is appropriate to counteract susceptibility gradients in particular regions. The method is effective in regaining signal in the anterior MTL, but clearly reduces temporal resolution by a factor equal to the number of images acquired, usually a minimum of 3. Alternatively, a single, moderate preparation pulse may be used. This reduces through-plane dephasing in affected areas at limited cost to BS and signal in homogeneous areas, and allows slices to be orientated so that TE shifts are small, reducing signal loss due to in-plane gradients *(119)*. *z*-Shimming and other compensation schemes have been applied in a number of other sequences described in this section.

Spins may also be refocused using tailored radio frequency pulses which create uniform in-plane phase but quadratic phase variation through the slice, allowing dephasing to be "precompensated" *(120)*. Analogous to *z*-shimming, in the original implementation a number of acquisitions with different precompensations were required, suited to different regions. More recently 3D versions have been developed, and while these are promising the pulse lengths are long, and the distribution of susceptibilities must be known *(121)*, or calculated iteratively online *(122)*. These are, however, important steps toward single-shot compensation of susceptibility dropout.

3.2.3. Slice Orientation and Gradient Directions

Divergent findings and recommendations for the optimum slice orientation for amygdala fMRI are due to the absence, until relatively recently, of an adequate description of signal loss and BS in the presence of field gradients *(81, 119, 123)*.

In many early studies, quite nonisotropic voxels were used to achieve short TR while minimizing demands on scanner hardware, with slice thickness being substantially larger than the in-plane voxel size. Gradients across voxels were highest then, and signal loss most severe, if the direction of strongest field gradient was along the slice (through-plane) direction *(124)*. With many studies finding that the direction of the field vector across the amygdala was principally superior-inferior *(125)*, this prescription precluded an axial orientation. As bilateral structures, the amygdala could be imaged in the same slice in the coronal but not the sagittal planes, leading to the coronal orientation being preferred by many *(110)*.

The optimum imaging plane is also dependent on whether gradient compensation is used *(81)*. If so, through-plane gradients may be compensated for with a moderate gradient in the slice direction, although this will lead to a small decrease in BS in unaffected areas. The slice can then be orientated so that in-plane gradients are below the critical threshold for Type 2 signal loss. The value of this has been demonstrated in the orbitofrontal cortex *(119)* but the approach yields lower rewards in the amygdala region *(126)* as gradients are higher (making it more difficult to find a suitable value for compensation), and are more variable between subjects.

The simulations of Chen et al. *(125)* for the amygdala suggested that the maximum BS was to be achieved by orienting the slice direction perpendicular to the maximum gradient vector and the readout direction parallel to it, indicating an (oblique) coronal orientation with superior–inferior readout. The angle between the gradient vector and the superior–inferior direction was shown to vary widely between subjects (from $-7°$ to $+26°$ at 1.5 T, from $-5°$ to $+34°$ at 3 T), meaning that field gradients need to be mapped for each subject before measurement. This scheme also invokes distortions which are asymmetric about the midline (left–right). If erroneous conclusions about lateralization are to be avoided, residual distortions in the amygdala should be symmetric, requiring the phase–encode direction to be superior–inferior for coronal slices or anterior–posterior for axial slices.

As well as the direction of imaging gradients, the sign of phase-encode blips is important for signal loss and BS *(123)*. Encoding in EPI can be either with a large positive phase-encode "prewinder" followed by a succession of small negative "blips," or a negative prewinder followed by positive blips. In homogeneous fields these schemes are equivalent, but we have seen that in the presence of susceptibility gradients echo positions are shifted away from the center of *k*-space, along the phase-encode axis. Positive and negative blip schemes have quite different properties, therefore, depending on whether the component of susceptibility gradient in the phase-encode direction is itself positive or negative *(123)*. The phase-encode direction (PE), slice angle, and *z*-shimming prepulse gradient moments (PP) that lead to maximum BS for EPI with otherwise standard EPI parameters (TE = 50/30 ms at 1.5 T/3 T, 3 × 3 × 2 mm^3 voxels) have been measured throughout the brain by Weiskopf et al. at 1.5 and 3 T *(126)*. They define positive slice angles as being those in which, beginning from the axial plane, the anterior edge is tilted toward the feet, and a positive PE as being that in which the prewinder gradient points from the posterior to the anterior of the brain. In the amygdala they find that the highest BS is achieved with positive PE, a –45° slice tilt and a PP = +0.6 mT/m ms at 3 T, and positive PE, –45° slice tilt and PP = 0.0 mT/m ms at 1.5 T. These values led to a 14% increase in BS at 3 T over a standard acquisition (with positive PE, a –0° slice tilt and a PP = –0.4 mT/m ms) but only 5% at 1.5 T. This indicates that BS can be increased by selecting optimum geometry parameters and compensations gradients, although improvement is more modest than that which has been demonstrated with the more technically challenging or time-consuming strategies described in this chapter. The gradient and geometry values suggested in Weiskopf et al. *(126)* should be adopted for EPI with standard parameters at these field strengths. At other field strengths their analysis could be followed, or interpolated values adopted from the trends evident in that study.

3.2.4. Voxel Size

Among many solutions to the problem of signal loss in the anterior MTL, reduced voxel size was established very early as an effective means of mitigating susceptibility-related signal loss *(127, 128)*. **Equation 2** describes how the rate of signal decay is reduced with voxel size by lowering field gradients across voxels. The effectiveness of this can be seen in the 4-T images of **Fig. 4** over a range of resolutions, with T_2^* in the amygdala (measured with a multiple gradient-echo sequence with the same geometry as the EPI) increasing from 22 to 38 ms when the voxel size is reduced from 64 to 8 mm^3, with corresponding EPI signal increase apparent in the anterior MTL.

Reducing voxel size comes at the expense of temporal resolution (or brain coverage) and SNR. The relationship between image SNR and voxel volume, ΔV, is

$$\mathrm{SNR} = \Delta V \sqrt{\frac{N_x N_y N_z}{\mathrm{rBW}}}, \quad (3)$$

where *Ni* is the number of samples in direction *i* and rBW the receiver bandwidth *(129)*. The commonly held view that voxel volume is simply proportional to SNR is premised on changing the volume via the field of view *(130)*, or that, in addition to increasing *Nx* and *Ny* by a factor *f*, (considering only in-plane resolution) receiver bandwidth is also increased by the same factor. If receiver bandwidth and field of view are held constant, however, then we see from **Eq. 3** (because $N_x N_y N_z = k / \Delta V|_{\text{FOV}}$, where k is the total imaged volume) that SNR is proportional to the square root of the voxel volume, and SNR may be restored by downsampling high resolution images. In this time-consuming scheme, partial *k*-space acquisition may be used to achieve the desired TE, SNR can be increased with multichannel coils, as has been validated for the MTL *(115)* and parallel imaging used to reduce an otherwise long TR.

While this analysis provides the basis for the dependence of image signal on imaging parameters, it neglects the effects of physiological noise. The most important measure of signal in this context is tSNR, which translates into the feasibility of detecting a specified signal change in fMRI *(131)* and has been shown to be useful in assessing the viability of amygdala fMRI in individual subjects *(132)*. In a study of optimum parameters for GE-EPI for 3-T amygdala EPI with a volume coil, a protocol with approximately 2-mm isotropic voxels was found to yield 60% higher tSNR than a protocol with standard parameters (with approximately 4-mm isotropic voxels) *(117)*, despite having been measured at twice the receiver bandwidth. Additional gains with smaller voxels (thinner slices) were not large, because T_2^* had already increased to a value close to that in homogeneous regions. This is in concordance with models calculations which suggest that 2 mm represents the smallest voxel size that should be used for amygdala imaging providing the activated size is itself at least 2 mm *(125)*.

There are many differences between the conditions and metrics of the methodological work cited and typical fMRI studies. It is encouraging, therefore, that these findings have been confirmed in the significance and extent of amygdala activation in fMRI experiments *(133, 134)*.

In summary, small voxels should be used in high field strength studies in order to operate in a regime dominated by technical, rather than physiological noise. In inhomogeneous regions this results in reduced field gradients, reducing signal loss and echo shifts, making BS more uniform in the volume. Time-series SNR may be increased by using multichannel coils and downsampling small voxels.

3.2.5. Echo Time

Taking the simplest approach of matching effective echo time (TE_{eff}) to the T_2^* of the structures of interest in GE-EPI might be seen as being problematic in large voxel size acquisitions, with T_2^*s varying quite widely (e.g., between the amygdala and the fusiform face area). One solution is to use a multiecho sequence, in which the each time of each image is appropriate for regions with particular field gradients, as will be described in more detail in

Subheading 3.2.8. A novel solution to matching TE_{eff} to T_2^* in the amygdala without sacrificing BS in more dorsal slices is to use an axial acquisition with slice-specific TE, demonstrated at 1.5 T with TE_{eff} = 60 ms in dorsal slices, TE_{eff} = 40 ms in ventral slices, and a transition zone with intermediate effective TE *(135)*.

It should be remembered, though, that the maximum of BS is quite flat as a function of TE, and TE is itself not well defined in EPI. In the previous sections, we also saw that in-plane susceptibility gradients change local TE *(81)*. This exposes the limitation of the approach of simply reducing the TE_{eff} of the sequence. In the common, negative blip scheme, signal in the anterior MTL will in fact be shifted to a longer TE. Using a short TE_{eff} makes the sequence more prone to complete (type 2) signal loss.

This explains the experimental findings of Gorno-Tempini et al. *(136)* and Morawetz et al. *(134)*. In 2-T dual-echo EPI with large voxels, Gorno-Tempini et al. found that although signal loss was reduced at the short TE (26 ms) BOLD activation was significantly greater in the hippocampus at the longer TE (40 ms). Morawetz et al. *(134)* studied four EPI protocols in their efficacy at mapping amygdala activation, using variants with two different TE (27 ms and 36 ms) and slices thicknesses (2 mm and 4 mm), all with high in-plane resolution (2 mm). Activation results were poor in the 4-mm protocols, even at the shorter TE.

A more effective approach than reducing TE_{eff} is to reduce susceptibility gradients, and thereby signal dephasing and echo shifts, using the techniques described earlier; gradient compensation, selection of appropriate gradient direction and slice orientation, and the use of smaller voxels. This increases T_2^* in susceptibility-affected regions and, by reducing echo shifts, makes BS more homogeneous throughout the imaging volume. Conditions then approach those with a homogeneous static field, where BS is maximized by using $TE_{eff} = T_2^*$.

The increase in T_2^* in the amygdala with reduced voxel size is illustrated at 4 T in **Fig. 4**; from 22 ms in a 4 × 4 × 4-mm acquisition to 38 ms in 2 × 2 × 2-mm data, consistent with previous results at 3 T *(117)*. Likewise, increase in BS was illustrated in the Morawetz et al. study *(134)*, in which robust amygdala activation was only detectable in the high-resolution acquisition.

3.2.6. Parallel Imaging

The previous sections have shown that many of the techniques which mitigate susceptibility-related signal loss in the amygdala, hypothalamus, and MTL are also time consuming, limiting either temporal resolution or brain coverage. This is undesirable where brain coverage cannot be reduced to the amygdala. Parallel imaging allows acceleration by undersampling *k*-space and using the sensitivity profiles of a number of receiver channel to reconstruct data without image fold-over *(137, 138)*. By this means it is possible to reduce TE_{eff}, which reduces susceptibility loss, and to reduce TR by the acceleration factor. Image distortions and echo shifts are likewise reduced by the acceleration factor so that even at the same effective TE as in a conventional acquisition, signal loss in the amygdala region is lower (**Fig. 4**, bottom right). The noise properties of images reconstructed from parallel acquisition lead to BS reductions of the order of 15–20% in other regions, however *(139)*.

The effectiveness of parallel imaging and suitable acceleration factors for the MTL have been studied by Schmidt et al. *(140)*. Statistical power in the study of MTL activation was higher in the parallel-acquisition data with an acceleration factor of 2 than in the acquisition without acceleration, but neither image quality nor statistical power improved with higher acceleration factors, as noise and reconstruction artifacts reduced tSNR prohibitively. Particular gains in BS can be made in the MTL using parallel imaging with a modest acceleration factor combined with high resolution imaging *(115)*. Combining parallel imaging, high resolution and high field has even allowed differential response of the hypothalamus to be recorded in response to funny as opposed to neutral stimuli at 3 T *(141, 142)*, which could potentially be used to diagnose narcolepsy and cataplexy.

3.2.7. Flip Angle

The following is a consideration which is common to fMRI studies in all brain regions. The flip angle that should be used in a sequence is that which maximizes the signal with a particular experimental TR. In a spoiled gradient-echo sequence this is the Ernst angle, θ_E, given by

$$\theta_E = \arccos\left(e^{\frac{-TR}{T_1}}\right).$$

T_1 values can be taken from the literature, if available, or mapped in a single study of a representative group of subjects, mostly simply using an inversion recovery sequence and a range of inversion times. At high (3.0–4.0 T) and very high field (7.0 T or higher), dielectric effects lead to B_1 inhomogeneity, and flip angles achieved deviate from nominal values. Particularly at 7.0 T it is worthwhile to map the RF field [e.g., using the 180° signal null point using a simple spoiled gradient-echo sequence *(143)* to calibrate nominal flip angles].

3.2.8. Alternatives to 2D, Single-Shot, Gradient-Echo EPI

If multiple echo images are acquired following a single excitation, the range of TE_{eff} in these provides near-optimum BS for a number of regions *(144, 145)*. Images acquired at different TEs may be analyzed separately, or combined to maximize BOLD contrast-to-noise ratio *(145)*. Acquiring multiple images in a single shot also allows additional features to be built into the sequence, such as 3D gradient compensation, in which different combinations of compensation gradients are applied to each echo *(146)*, leading to excellent signal recovery in the amygdala in the combined image *(147)*. Alternatively, the phase-encoding gradient polarity may be reversed to yield images with distortions in opposite directions, allowing for their correction *(148)*.

Similar multiecho and compensation techniques have been applied to spiral acquisitions. A spiral-in trajectory has been shown to reduce signal loss compared to a conventional spiral–out scheme with the same TE, and SNR and BS could be increased with a spiral in-out scheme by combining images optimally from the two acquisitions *(149)*. A number of variants of this have been developed to further reduce susceptibility artifacts, including applying a *z*-shim gradient to the second echo *(150)* or subject-dependent slice-specific *z*-shims to both echoes *(151)*.

A number of segmented methods are being developed to overcome the temporal constraints of multiecho and high-resolution acquisitions. In conventional segmented EPI, subsets of interleaved *k*-space lines are acquired after successive excitations. The higher phase-encode bandwidth leads to reduced distortions and smaller echo shifts, but the method is inherently slow and prone to motion and physiological fluctuations, as each image is built up over a number of TRs. In the MESBAC sequence, navigator echoes are acquired in both the readout and phase-encode directions between each segment. Multiple echoes are acquired with different amounts of compensation for each echo *(152)*, and combined to give impressive signal in inferior frontal areas.

3.2.9. Summary

In the subsections of **Subheading 3.2** we have looked at the influence of field strength, gradient compensation, slice orientation, voxel size, TE, and acquisition acceleration factor on susceptibility-related signal and BS reduction in the anterior MTL, as well as discussing some variants of multiecho and spiral schemes which have been tailored for this region. While the interdependent nature of EPI parameters and changing considerations at different field strength necessarily make some considerations complex, we would like to pick out two lines of approach presented here as being particularly effective, and clarify recommendations.

The first approach is high-field, high-resolution single-shot EPI with gradient compensation and acceleration. BOLD signal changes are greater at high field (3.0–4.0 T), and the tSNR advantages of high field strength are capitalized upon by measuring with small (circa 8-μl voxels), where thermal noise rather than physiological noise dominates. Measuring with small voxels reduces signal dephasing, making T_2^* more homogeneous. Shifts in local TE are also less, reducing Type 2 signal loss and increasing BOLD sensitivity. Moderate slice select gradient compensation and an oblique axial acquisition with a tilt between 20 and

45° (anterior slice edge toward the head) reduces in-plane gradients and echo shifts further. With susceptibility gradients reduced – evidenced by T_2^* values close to those in magnetically homogeneous regions – BS can be maximized by setting the $TE_{eff} = T_2^*$. The TE_{eff} can be reached using parallel imaging acceleration (e.g., factor 2), which further reduces both TE shifts and image distortion. Images acquired with these parameters have high signal in the anterior MTL, low distortion, and quite homogenous BS. Time-series SNR can be increased before statistical analysis by downsampling or smoothing images. This approach is attractive in that it may be achieved on most modern high field systems.

Not only the value of gradient compensation was discussed in **Subheading 3.2.2**, but also the high cost in temporal resolution, if images with a number of compensation gradients are acquired. The second approach we wish to highlight involves the application of a range of compensation gradients to each of a number of echoes acquired after a single excitation, so reducing the time penalty. Both the multiecho echo-planar *(146)* and multiecho spiral acquisitions *(151)* described in **Subheading 3.2.8** have been shown to be effective in reducing susceptibility-related signal loss in the anterior MTL.

3.3. Correction Methods

3.3.1. Distortion Correction with the Field Map and Point-Spread Function Methods

The field map (FM) method was first described by Weisskoff and Davis *(153)* and developed by Jezzard and Balaban *(154)*. In **Subheading 3.1.2** we saw that distortion in EPI is only significant in the phase-encode direction and that the number of pixels by which signal is mislocated is equal to the local field offset divided by the bandwidth per pixel in the phase-encode direction. In the fieldmap method, static magnetic field deviations, ΔB, are calculated from the phase difference, $\Delta\phi$, between two scans with TE separated by $\Delta \mathrm{TE}$ (or a dual-echo scan), using the relation $\Delta B = 2\pi\gamma\,\Delta\mathrm{TE}\,\Delta\varphi$. This map is distorted (*forward-warped*) to provide a map of the voxel shifts required to reverse the distortion at each EPI location. Gaps in the corrected image are filled by interpolation.

While undemanding from the sequence perspective, considerable postprocessing is required to produce FMs that do not contain errors. Phase imaging is only capable of encoding phase values in a 2π range, with values outside this range being aliased, causing "wraps" in the image. These can be removed in the spatial domain using a number of freely available algorithms [e.g., PRELUDE *(155)* or ΠUN *(156)*], or by examining voxel-wise phase evolution in time if three or more echoes are acquired *(157)*. If imaging is being carried out with a multichannel radiofrequency receive coil, phase images created via the sum-of-squares reconstruction *(158)* will show nonphysical discontinuities from arbitrary phase offsets between the coil channels (incongruent wraps) unless these offsets are removed *(159, 160)*. Alternatively, images from channels may be processed separately and individual FMs, weighted by coil sensitivities, combined. In 2D spatial unwrapping, additional global, erroneous 2π phase changes are occasionally inferred between TE when the algorithm begins to unwrap from different sides of a phase wrap at the two TE. In multichannel imaging, these slice phase shifts may be identified by examining the consistency between coil channels *(161)*, as may unreliable voxels at the image edge and in regions of high-field gradient. The FM may finally need to be smoothed to remove high frequency features and dilated to ensure that it extends to the periphery of the brain.

In the point spread function (PSF) approach *(162)* applied to distortion correction *(163)*, the imaging sequence is similar to EPI, but with the initial phase prewinder gradient replaced by a phase gradient table, the values are applied in a loop. The PSF of each voxel is the Fourier transform of the acquired data, and the displacement of the voxel is the shift of the center of the PSF (e.g., if the center of this is at zero additional phase, this corresponds to no local field offset). For one major scanner manufacturer, this method has been robustly implemented with the flexibility to be used for parallel imaging with high acceleration factors *(164)*.

The FM and PSF methods have been compared at 1.5 T *(163)*. The PSF was found to be generally superior, although some conclusions were based on deficiencies in FMs in regions of high field gradient which may be improved upon.

The effectiveness of the two methods in correcting larger distortions at 4.0 T is shown in **Fig. 5**, focusing on a section through the amygdala (top row), and comparing this with the situation in a more dorsal slice (bottom row). Raw and corrected EPIs are compared to a gradient-echo reference which has the same (subvoxel) distortion in the readout direction, but no distortion in the phase-encode direction. The distortion at the anterior boundary of the amygdala (A) is circa 3 mm – moderate compared to the displacement of the ventricles (9 mm at B) and the frontal gray-white matter border indicated at C (12 mm). If the multiplicity of phase information available from multichannel coils is used in the FM method *(161)*, both FM and PSF methods perform very well in all areas, with only minor errors at the periphery of the FM-corrected images due to residual field map inaccuracies at those locations (at D, not present in the PSF-corrected images).

The choice of correction method is often a pragmatic one based on which is more robustly and conveniently implemented.

3.3.2. Correction of Physiological Artifacts

Physiological fluctuation in a sequence of gradient-echo images can be corrected using a navigator echo technique *(165)*. A single echo is acquired before the encoding scheme is begun and used

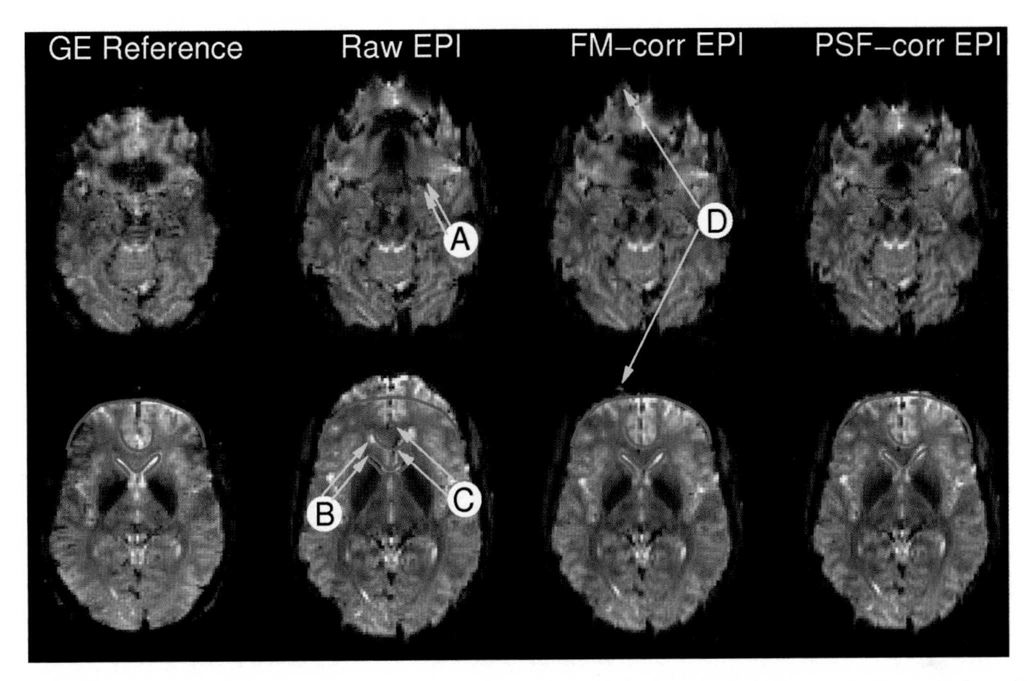

Fig. 5. Distortion correction of echo-planar imaging (EPI) at high field (4.0 T). A comparison of field-map (column 3) and point-spread function (column 4) correction of distortion in EPI (column 2) at the level of the amygdala (*top row*) compared to a more dorsal section (*bottom row*). Salient features have been copied from a gradient-echo geometric reference scan (column 1).

to amend the phase changes in the image data which arise from susceptibility effects. This "global" correction approach, using the central *k*-space point only, can be extended to 1D *(166)* and 2D *(167)*. These methods are effective, but have as drawbacks an increase in TR.

To avoid them being aliased in EPI time series, respiratory fluctuations (circa 0.2–0.3 Hz) and cardiac fluctuations (circa 1 Hz) would need to be sampled at least at 2 Hz. That is, the TR of the sequence would need to be 500 ms or less. Typical TRs in whole-brain fMRI are 1–4 s, and the previous sections have indicated that many of the strategies that should be implemented to improve data quality in fMRI for emotion studies lead to longer repetition times. Respiratory and cardiac fluctuations will normally be aliased, then, and not generally into a particular frequency band *(168)*. Simple band-pass filtering is therefore not generally possible; although a range of alternative correction methods have been developed.

A class of correction methods requires additional physiological measurement to be made concurrent with the fMRI time-series, using a respiration belt to monitor breathing and an electrocardiogram or pulse oximeter to monitor heart rate. Applied in image space, the RETROICOR correction method involves plotting pixels according to their acquisition time within the respiratory cycle (classified also by respiration depth) and subtracting a fit to fluctuations over the cycle *(169)*. Despite the many reasons why physiological artifacts are expected to particularly affect amygdala fMRI, their correction with RETROICOR was found to bring only modest improvements in group fMRI results in an emotion processing task; up to 13% in *t* statistic values depending on the degree of smoothing *(170)*. Those improvements were mostly due to correction of cardiac effects. Recent findings that cardiac rate changes lead to signal changes of similar size to the effect due to cardiac action itself *(88)*, which are not modeled in the RETROICOR approach, suggest that further gains are possible.

Modeling physiological fluctuations *(171)* by including measured signal as "Nuisance Variable Regressors (NVRs)" is a convenient alternative to fitting and removing them. A detailed examination of these and other sources of noise showed respiratory-induced noise particularly at the edge of the brain, larger veins and ventricles, and cardiac-induced noise focused on the middle cerebral artery and Circle of Willis, close to the amygdala *(168)*, which could be well modeled.

A number of image-based methods for physiological artifact correction have been developed, which do not require physiological monitoring data. Physiological fluctuations can be modeled with NVRs based on ventricular and white matter ROI values *(172)*. Alternatively, the data can be decomposed using ICA [e.g., MELODIC *(173)* or GIFT *(174)*] and components relating to physiological processes identified with automated or semiautomated methods. These can be based on experimental thresholds *(175)*, statistical testing *(176)*, automatic thresholding *(177)*, or supervised classifiers *(178)*. Once identified, these components can be removed from the data. While in their infancy, these methods are very promising, particularly for the ventral brain. Tohka et al., for instance, demonstrated marked *Z*-score increases in frontal ventral regions and other areas close to susceptibility artifacts.

3.3.3. Correction of Stimulus-Correlated Motion Artifacts

In patient group studies, Bullmore et al. *(179)* have shown the need to compare the extent to which SCM explains variance between the groups, and suggest that this be identified using an analysis of covariance (ANCOVA). Without this approach, differences between the groups arising from higher SCM in the schizophrenic group in their study would have been attributed to differential activation in response to the task.

In the example of **Fig. 2** (left), realignment of the time series in the motion-only replication did not substantially reduce the amygdala SCM artifact (right), but including identified motion parameters in the model as NVRs was effective *(168, 180)*.

Alternatively, a boxcar NVR corresponding to presentation and response periods can be included in the model *(181)*. This and a number of other studies *(182)* have shown that the temporal shift in response introduced by the hemodynamic response function (HRF) makes it possible to separate motion from activation for short presentation periods, making event-related designs less sensitive to motion than block designs.

4. Summary and Discussion

Emotional neuroimaging is a rapidly expanding area that provides an interface between neurobiological work and psychophysiological emotion research. One important view that has emerged from the area of behavioral neuroscience is that emotional processes play a central role in the adaptive modulation of perceptual encoding, learning and memory, attention, decision-making, and control of action *(9)*. Many of neuroimaging studies have demonstrated that amygdala activation, for example, modulates attention and memory storage in other brain regions such as the hippocampus, striatum, and neocortex. Such interactions may occur as facilitations or modulations of neurocognitive function at several levels of processing. Conversely, recent work has shown that the organism is prevented from excessive emotional activation not only by low-level habituation or negative feedback mechanisms but also as a result of protective inhibition processes. Diverse behavioral and cognitive strategies have been identified that modulate and downregulate the ongoing emotion process *(6)*. The modulating effects on emotional arousal during an emotional episode such as rejection (venting and disengagement) or accommodation (relaxation, distraction, reconceptualization, rationalization, or reappraisal) deserve further inspection with respect to the involved neural mechanisms.

Although important advances have been made in the area of human emotion perception, learning, and autonomic conditioning, research has typically been limited to a small number of primary and mostly negative emotions such as fear, anger, or disgust. Limiting the range of investigated categories (neglecting shame, guilt, interest, etc.), dimensions (neglecting positive emotions such as care, support, etc.) and behavioral procedures does not do justice to the complexity of the multistage emotional appraisal process described above *(29)*. It is equally important but more difficult to identify the correlates of complex emotions such as those resulting from beliefs, preferences, predispositions, or interpersonal exchange. Not only the social dimensions such as untrustworthiness or dishonesty *(183, 184)*, but also positive aspects such as social fairness *(185)*, trust, and supportiveness play a role. Moreover, an understanding of modulating sociocultural influences is essential for a comprehensive conceptualization of human emotion *(29)*.

Current neuroimaging research on emotion can be described as an ongoing construct validation process *(186)*, which draws upon convergent and divergent associations of local activation variables and psychological constructs. The experimental measures (operationalizations of psychological constructs) are expected to be correlated with regional brain activations. It is evaluated whether topographically distinct patterns of activation in a certain region consistently predict engagement of different processes (for an example in the area of cognitive processing, *see* **ref.** *187)*. Indicators of a different construct are expected to correlate with activations of different areas. This corresponds to the well-known double dissociation strategy that inspects task by localization interactions in neuropsychology *(64)*.

This validation process typically starts at a relatively broad construct level and proceeds downward in the hierarchy of constructs to finally specify within-systems constructs. Previous studies have demonstrated a relatively high cross-laboratory repeatability of emotional brain activation patterns at a higher systems level. At lower levels, however, the reliability of psychological or activation data may decline depending on limitations of the instruments.

Nonetheless, high-field fMRI scanners permit an improved discrimination of activations, for example, within the different subnuclei of the amygdala *(5)*. It is evident that increased discrimination on the neural side must be accompanied by a refined technology to assess more fine-grained emotional constructs on the behavioral side.

Neuropsychological construct validation requires additional physiological data to obtain some kind of convergent information about the indicator variable. At the neurophysiological level, the perfusion mechanisms has been elucidated by combining the greater spatial resolution of fMRI with the real-time resolution of intracortical local field ERP (LFP) recordings. The neurophysiological coupling mechanisms of neural activity and the BOLD response can thus be assessed *(188)*. An application of both fMRI methods and electrophysiological approaches (e.g., surface and deep electrode recordings from limbic brain structures) is useful *(189)*. The combination of brain perfusion changes and electrophysiological correlates of oscillatory coupling will foster the understanding of the neural interaction processes within frontal and temporal networks *(190)*.

On the level of the autonomic nervous system, multivariate coregistrations of psychophysiological response patterns including emotion modulated startle, heart rate variability, or cortisol secretion alleviate the validation of experimentally induced emotions or presence of specific emotional disorders.

Emotional neuroimaging has continuously profited from improvement in scanning techniques and the adaptation and standardization of signal processing strategies. However, this area has not only benefited from the diverse contributions of its subdisciplines but also inherited their methodological problems. An inspection of brain imaging studies of emotion showed that measurement quality may be influenced by many factors: by a rapid and differential habituation of responses to emotional stimuli in some regions; by artifacts of certain signal scaling techniques that are applied by default; by situational or state-dependent influences; and by insufficient validation of the emotion to be elicited (manipulation check). Interindividual differences of emotional regulation behavior appear to modulate event-related reactions during the time course of the experiment.

Some of the many approaches to reducing signal loss in EPI in the anterior MTL have been outlined here, as well as some of the methods for identifying and correcting artifacts arising from SCM, distortion, and physiological artifacts. Despite the gravity of the problem and the effectiveness of some of these strategies, the overwhelming majority of fMRI studies of the emotions use the same measurement protocols and analysis methods as have been applied to study cognitive function over the last decade.

Combining many of the simpler strategies described here – high field strength, small voxel volumes, partial *k*-space acquisition with the correction of physiological and SCM artifacts – allows reliable results to be achieved in the anterior MTL *(191)*. **Figure 6** demonstrates such an example; the detection of subtle differences in amygdala activation between explicit and implicit emotion processing *(192)*.

Moreover, new research designs and analysis methods such as Structural Equation Modeling or Dynamic Causal Modeling are now available to inspect the effective or causal connectivity that, for example, permits the PFC to modulate amygdala activity *(62)*. The influences of individual brain regions on each another can also been studied by combining functional brain imaging with the lesion approach or transcranial magnetic stimulation *(193)*.

We have raised a number of caveats that highlight some of the limitations of emotion assessment in a scanner environment. As has been argued above, a lack of representativeness must be noted, that is, emotion includes a much broader conceptual network than currently covered by neuroimaging research. Thus, generalizations to other areas of functioning remain difficult. Representative designs are needed that pay greater attention to high-level strategies that depend on sociocultural factors and initiate, modulate, or regulate emotions. Moreover, the representativeness of results is limited due to small, selected, and poorly

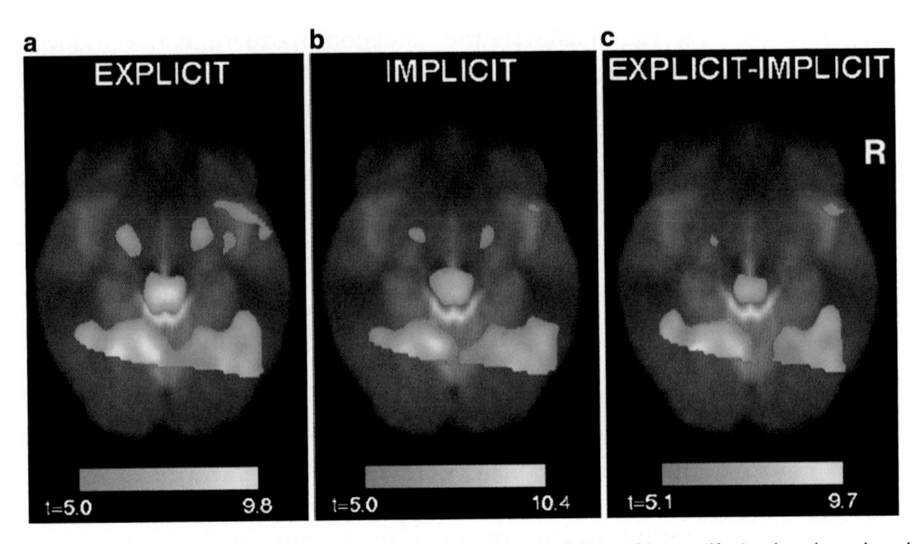

Fig. 6. High resolution imaging detailed in this chapter allows the acquisition of low-artifact echo-planar imaging (EPI) and allows subtle processing effects to be distinguished. Group results from 29 subjects for the conditions (**a**) emotion recognition (**b**) implicit emotion processing (age discrimination) and (**c**) the difference between the two conditions (3.0 T). Results, showing activation in the amygdala and fusiform gyrus (as well as cerebellum and brainstem) are overlaid on mean EPI and thresholded at $p = 0.05$, family-wise error corrected. Reprinted from ***(192)***, with permission.

described study groups. Finally, since emotion elicitation in the scanner has been highly artificial, the power to predict emotions outside the neuroimaging context remains questionable. An ecological functional brain imaging approach that includes natural behaviors and environmental contexts of presentation may help to obtain a more representative view of real-life emotions.

Subject-specific mechanisms regulating the strength and temporal pattern of response to emotional stimuli and the balance of excitatory and inhibitory processes are of particular interest. The variability of the BOLD response between trials and across the time course of the experiment needs to be explained. Future research may therefore examine individual and group differences with a view to resolving inconsistencies in the literature *(5, 12, 77)*. Investigations into personality disorders or psychiatric diseases will provide further insight into the dispositional factors modifying the response to situational stressors. Paradigms specifically adapted to the investigated disorder may help to identify prefrontal dysfunction and associated failure to tonically inhibit amygdala output or to recognize safety signals eventually inducing sympathetic overactivity *(194)*. It may be that – as is the case in motor tasks – a large proportion of the intertrial variation not only in the behavioral response *(102)*, but also in the BOLD signal *(101)* is explained by fluctuations in underlying RSNs.

Eliciting emotions in the environment of an imaging scanner remains a highly artificial process. This raises the question as to the predictive value of current neuroimaging data for explaining the emotional modulations in real-life contexts. This is particularly important for applied areas such as psychotherapy and coping research. Thus, in addition to identifying the neurobiological basis of emotional regulation behavior, the generalizability or predictive validity of imaging data for real-life emotions should be systematically evaluated.

5. Conclusions

Neuroimaging has replicated and extended earlier findings of neuropsychological studies in brain damaged subjects. It has significantly contributed to unraveling the organization of neural systems subserving the different components of emotional stimulus-response mediation along the neuraxis in healthy human subjects. Improved operational definitions and paradigms have contributed to differentiating subcomponents of emotional functions such as, for example, perceptual decoding, anticipation, associative learning, awareness, and response mediation. However, despite obvious advances, a comprehensive model integrating the diverse emotional behaviors on the basis of involved cerebral mechanisms is still unavailable. Moreover, the interpretation of findings is complicated by technical and methodological difficulties.

Research advances not only depend upon the technical refinement of imaging methodology but also on the improvement of behavioral procedures and measurement models. Neuropsychological construct validation procedures imply that an increase of localization precision of the imaging technology would also require an enhanced precision on the side of behavioral operationalizations. However, this seems not to be case as many studies still use unsophisticated stimulus materials or global instructions involving multiple or undefined subfunctions. As much as relatively global operationalizations are applied, however, the obtained neuropsychological correlations (for example, regarding activations of the PFC) will remain incomprehensible.

We have suggested here the framework of a lense-type assessment model, wherein activations in well-characterized neural structures may be used as predictors of particular emotional processes. According to this, a hierarchy of latent constructs constitutes the behavioral level, an idea, which is largely accepted in psychology. On the level of brain activity, patterns or families of topographically distinct activity can be identified in a similar way and used as a predictor of behavioral function. Following the assumptions of a methodological parallelism, neuropsychological construct validation procedures make uses of this framework of activity–behavior associations on different levels of the hierarchy. It can be extrapolated from multivariate personality theory, that the prediction of behavior will only be successful if activation measures and psychological data are analyzed on a similar level of generality or aggregation.

In view of the complexities of emotional regulation behavior in human subjects, it is equally important to advance assessment theory, psychological conceptualization, and behavioral methodology *(29)*. Future work should therefore more closely inspect issues related to model construction, symmetry of neural and behavioral variables, and their aggregation levels. Multidisciplinary approaches that combine improvement in brain activation measurement with enhanced psychological data theory may thus foster construct validity, reliability, and predictive power of emotional neuroimaging.

Knowledge pertaining to the localization of brain activations and its functional connectivity is also an important input to inform and constrain cognitive theories of emotion psychology. Thus, insights from the brain will thus help to explain the incoherences of psychophysiological, behavioral, and subjective indicators of emotion that are so frequently observed in psychophysiological studies. Activation data may also help to establish models that possess a better "breakdown compatibility," that is, power to predict behavioral change as a consequence of brain damage.

The introduction of structural/connectional and functional data has considerably bolstered scientific construct validation processes in the affective neurosciences and emotion psychology. Topographically distinct activity patterns are increasingly identified that possess a certain incremental validity, that is, an increasing power to predict the individual dynamics of emotional regulation behavior. Establishing a representative and valid model of emotional functioning is a necessary precondition for many areas of application such as the categorization of patients with emotional disorders and the assessment of psychotherapy.

Greater attention to methodological issues may help to bring more rigors to experimentation in the field of emotional neuroimaging, promote interdisciplinary research, and alleviate cross-laboratory replication. A wealth of approaches have been presented to countering BS loss in the amygdala, many of which are available as standard on commercial scanners or simply require the adoption of suitable imaging parameters *(117, 125, 134)*. Also, in the absence of a measurement theory that describes validated procedures or instruments for assessing emotional constructs, single findings cannot be trusted. Although absence of validation is acceptable for early stages of the research cycle, current emotional neuroimaging work has only just begun to approach the confirmatory stage. To establish confidence in the suggested models, additional efforts are required to empirically validate assessment strategies and instrumentation.

Acknowledgments

The author's own work reported in this chapter was supported by the Austria FWF grant P16669-B02, grant 11437 from the Austrian National Bank, the government of the Provincia Autonoma di Trento, Italy, the private foundation Fondazione Cassa di Risparmio di Trento e Rovereto, the University of Trento, Italy, and by grant Pe 499/3–2 from the Deutsche Forschungsgemeinschaft to MP. J. Jovicich is thanked for helpful comments.

References

1. Büchel C, Dolan RJ. Classical fear conditioning in functional neuroimaging. Curr Opin Neurobiol 2000;10:219–23.
2. Phan KL, Wager T, Taylor SF, Liberzon I. Functional neuroanatomy of emotion: A meta-analysis of emotion activation studies in PET and fMRI. NeuroImage 2002;16: 331–48.
3. Dolan RJ, Vuilleumier P. Amygdala automaticity in emotional processing. Ann N Y Acad Sci 2003;985:348–55.
4. Wager TD, Phan KL, Liberzon I, Taylor SF. Valence, gender, and lateralization of functional brain anatomy in emotion: A meta-analysis of findings from neuroimaging. NeuroImage 2003;19:513–31.
5. Zald DH. The human amygdala and the emotional evaluation of sensory stimuli. Brain Res Brain Res Rev 2003;41:88–123.
6. Ochsner KN, Gross JJ. The cognitive control of emotion. Trends Cogn Sci 2005;9 :242–9.
7. Panksepp J. Affective neuroscience: The foundations of human and animal emotions. Oxford: Oxford University Press, 1998.
8. Davidson RJ, Jackson DC, Kalin NH. Emotion, plasticity, context, and regulation: Perspectives from affective neuroscience. Psychol Bull 2000;126:890–909.
9. Dolan RJ. Emotion, cognition, and behavior. Science, 2002;298:1191–4.
10. Breiter HC, Rauch SL. Functional MRI and the study of OCD: From symptom provocation to cognitive-behavioral probes of cortico-striatal systems and the amygdala. NeuroImage 1996;4:127–38.
11. Johnson PA, Hurley RA, Benkelfat C, Herpertz SC, Taber KH. Understanding emotion regulation in borderline personality disorder: Contributions of neuroimaging. J Neuropsychiatry Clin Neurosci 2003;15:397–402.
12. Hamann S, Canli T. Individual differences in emotion processing. Curr Opin Neurobiol 2004;14:233–8.
13. Paquette V, Levesque J, Mensour B, et al. Change the mind and you change the brain: Effects of cognitive-behavioral therapy on the neural correlates of spider phobia. Neuroimage 2003;18:401–9.
14. Popma A, Raine A. Will future forensic assessment be neurobiologic? Child Adolesc Psychiatr Clin N Am 2006;15:429–44.
15. Bräutigam S. Neuroeconomics – from neural systems to economic behaviour. Brain Res Bull 2005;67:355–60.
16. Walter H, Abler B, Ciaramidaro A, Erk S. Motivating forces of human actions: Neuroimaging reward and social interaction. Brain Res Bull 2005;67:368–81.
17. Adolphs R, Tranel D, Damasio H, Damasio A. Impaired recognition of emotion in facial

expressions following bilateral damage to the human amygdala. Nature 1994;372: 669–72.

18. Broks P, Young AW, Maratos EJ, et al. Face processing impairments after encephalitis: Amygdala damage and recognition of fear. Neuropsychologia 1998;36:59–70.
19. Sprengelmeyer R, Young AW, Schröder U, et al. Knowing no fear. Proc R Soc Lond B Biol Sci 1999;266:2451–6.
20. Adolphs R, Gosselin F, Buchanan T, Tranel D, Schyns P, Damasio A. A mechanism for impaired fear recognition after amygdala damage. Nature 2005;433:68–72.
21. Adolphs R, Tranel D, Hamann S, et al. Recognition of facial emotion in nine individuals with bilateral amygdala damage. Neuropsychologia 1999;37:1111–7.
22. Schmolck H, Squire LR. Impaired perception of facial emotions following bilateral damage to the anterior temporal lobe. Neuropsychology 2001;15:30–8.
23. LeDoux JE. Emotion: Clues from the brain. Annu Rev Psychol 1995;46:209–35.
24. LaBar KS, Gatenby JC, Gore JC, LeDoux JE, Phelps EA. Human amygdala activation during conditioned fear acquisition and extinction: A mixed-trial fMRI study. Neuron 1998;20:937–45.
25. Morris JS, Büchel C, Dolan RJ. Parallel neural responses in amygdala subregions and sensory cortex during implicit fear. Neurobiol 2002;12:169–77.
26. Büchel C, Morris J, Dolan RJ, Friston KJ. Brain systems mediating aversive conditioning: An event-related fMRI study. Neuron 1998;20:947–57.
27. Krech D. Dynamic systems as open neurological systems. Psychol Rev 1950;57:345–61.
28. Cacioppo JT, Tassinary LG, Berntson GG (Eds.). Handbook of psychophysiology. Cambridge: Cambridge University Press, 2000.
29. Scherer KR, Peper M. Psychological theories of emotion and neuropsychological research. In: G. Gainotti (Ed.), Handbook of neuropsychology, Vol. 5: Emotional behavior and its disorders (2nd edition). Amsterdam: Elsevier, 2001, pp. 17–48.
30. Coan JA, Allen JJB (Eds.). Handbook of emotion elicitation and assessment. New York, NY: Oxford University Press, 2007.
31. Peper M, Vauth R. Socio-emotional processing competences: Assessment and clinical application. In: Vandekerckhove M, von Scheve C, Ismer S, Jung S, Kronast S (Eds.), Regulating emotions (ch. 9). Hoboken, NJ: Wiley, 2008.
32. Kerlinger FN, Lee HB. Foundations of behavioral research (4th edition). Fort Worth, TX: Harcourt College Publishers, 2000.
33. Scherer KR. Neuroscience projections to current debates in emotion psychology. Cogn Emot 1993;7:1–41.
34. Scherer KR. Psychological models of emotion. In J. Borod (Ed.). The neuropsychology of emotion. Oxford: Oxford University Press, 2000, pp. 137–62.
35. Scherer KR. Appraisal theories. In: Dalgleish T, Power M (Eds.). Handbook of cognition and emotion. Chichester: Wiley, 1999, pp. 637–63.
36. Peper M, Fahrenberg J. Psychophysiologie. In: Sturm W, Herrmann M, Münte TF (Eds.). Lehrbuch der Klinischen Neuropsychologie. Heidelberg: Spektrum Akademischer Verlag, 2008.
37. Stemmler G, Fahrenberg J. Psychophysiological assessment: Conceptual, psychometric, and statistical issues. In: G. Turpin (Ed.) Handbook of clinical psychophysiology. Chichester: Wiley, 1989, pp. 71–104.
38. Stemmler G. Differential psychophysiology: Persons in situations. Heidelberg: Springer, 1992.
39. Johnsen BH, Thayer JF, Hugdahl K. Affective judgment of the Ekman faces: A dimensional approach. J Psychophysiol 1995;9:193–202.
40. Peper M, Irle E. The decoding of emotional concepts in patients with focal cerebral lesions. Brain Cogn 1997;34:360–87.
41. Ekman P. Strong evidence for universals in facial expressions. Psychol Bull 1994;115: 268–87.
42. Russell JA. Is there universal recognition of emotion from facial expressions? Psychol Bull 1994;115:102–41.
43. Stemmler G. Emotionen. In: F Rösler (Ed.) Enzyklopädie der Psychologie: Themenbereich C Theorie und Forschung, Serie 1 Biologische Psychologie, Band 5 Ergebnisse und Anwendungen der Psychophysiologie. Göttingen: Hogrefe 1998, pp. 95–163.
44. Scherer KR. On the nature and function of emotion: A component process approach. In: Scherer KR, Ekman P (Eds.) Approaches to emotion. Hillsdale, NJ: Erlbaum, 1984, pp. 203–317.
45. Lacey JI. Somatic response patterning and stress: Some revisions of activation theory. In: Appley MH, Trumbull R (Eds.) Psychological stress: Issues in research. New York, NY: Appleton-Century Crofts, 1967, pp. 14–42.

46. Peper M. Awareness of emotions: A neuropsychological perspective. Adv Consciousness Stud 2000;16:245–70.
47. Lang P, Rice DG, Sternbach RA. The psychophysiology of emotion. In: Greenfield NS, Sternbach RA (Eds.) Handbook of psychophysiology. New York, NY: Holt, 1972, pp. 623–43.
48. Fahrenberg J. Psychophysiologische Methodik. In: Groffman K-J, Michel L (Eds.) Enzyklopädie der Psychologie: Themenbereich B Methodologie und Methoden, Serie II Psychologische Diagnostik, Band 4 Verhaltensdiagnostik. Göttingen: Hogrefe, 1983, pp. 1–192.
49. Lazarus RS. Emotion and adaptation. New York, NY: Oxford University Press, 1991.
50. Parkinson B, Totterdell P. Classifying affect-regulation strategies. Cogn Emot 1999;13: 277–303.
51. Gross JJ, Levenson RW. Emotional suppression: Physiology, self-report, and expressive behavior. J Pers Soc Psychol 1991;64: 970–86.
52. Lazarus RS, Folkman S. Stress, appraisal, and coping. New York, NY: Springer, 1984.
53. Gross JJ. Rev Gen Psychol 1999;2:271–99.
54. Whalen PJ, Rauch SL, Etcoff NL, McInerney SC, Lee MB, Jenike MA. Masked presentations of emotional facial expressions modulate amygdala activity without explicit knowledge. J Neurosci 1998;18:411–8.
55. Pessoa L, Japee S, Sturman D, Ungerleider LG. Target visibility and visual awareness modulate amygdala responses to fearful faces. Cereb Cortex 2006;16:366–75.
56. McIntosh AR, Gonzalez-Lima F. Network interactions among limbic cortices, basal forebrain, and cerebellum differentiate a tone conditioned as a Pavlovian excitor or inhibitor: Fluorodeoxyglucose mapping and covariance structural modeling. J Neurophysiol 1994;72:1717–33.
57. Büchel C, Friston KJ. Modulation of connectivity in visual pathways by attention: Cortical interactions evaluated with structural equation modelling. Magn Reson Imaging 1997;15:763–70.
58. Büchel C, Friston KJ. Assessing interactions among neuronal systems using functional neuroimaging. Neural Netw 2000;13: 871–82.
59. Friston KJ, Harrison L, Penny W. Dynamic causal modelling. NeuroImage 2003;19: 1273–302.
60. Vuilleumier P, Richardson MP, Armony JL, Driver J, Dolan RJ. Distant influences of amygdala lesion on visual cortical activation during emotional face processing. Nat Neurosci 2004;7:1271–8.
61. Friston KJ. Functional and effective connectivity in neuroimaging: A synthesis. Hum Brain Mapp 1994;2:56–78.
62. Ochsner KN, Ray RD, Cooper JC, et al. For better or for worse: Neural systems supporting the cognitive down- and up-regulation of negative emotion. Neuroimage 2004;23: 483–99.
63. Fahrenberg J, Peper M. Psychophysiologie. In: Sturm W, Herrmann M, Wallesch CW (Eds.) Lehrbuch der Neuropsychologie, chapter 1.10. Amsterdam: Swets and Zeitlinger, 2000, pp. 154–68.
64. Passingham RE, Stephan KE, Kötter R. The anatomical basis of functional localization in the cortex. Nat Rev Neurosci 2002;3: 606–16.
65. Büchel C, Dolan RJ, Armony JL, Friston KJ. Amygdala-hippocampal involvement in human aversive trace conditioning revealed through event-related functional magnetic resonance imaging. J Neurosci 1999;19:10869–76.
66. Foerster F. On the problems of initial-value-dependencies and measurement of change. J Psychophysiol 1995;9:324–41.
67. Peper M, Herpers M, Spreer J, Hennig J, Zentner J. Functional neuroimaging studies of emotional learning and autonomic reactions. J Physiol Paris 2006;99:342–54.
68. Cardinal RN, Parkinson JA, Hall J, Everitt BJ. Emotion and motivation: The role of the amygdala, ventral striatum, and prefrontal cortex. Neurosci Biobehav Rev 2002;26:321–52.
69. O'Doherty J, Dayan P, Schultz J, Deichmann R, Friston K, Dolan RJ. Dissociable roles of ventral and dorsal striatum in instrumental conditioning. Science 2004;304:452–4.
70. Grandjean D, Sander D, Pourtois G, et al. The voices of wrath: Brain responses to angry prosody in meaningless speech. Nat Neurosci 2005;8:145–6.
71. Critchley H, Daly E, Phillips M, et al. Explicit and implicit neural mechanisms for processing of social information from facial expressions: A functional Magnetic Resonance Imaging study. Hum Brain Mapp 2000;9:93–105.
72. Hariri S, Bookheiner SY, Mazziotta JC. Modulating emotional responses: Effects of a neocortical network on the limbic system. Neuroreport 2000;11:43–8.
73. Furmark T, Fischer H, Wik G, Larsson M, Fredrikson M. The amygdala and individual differences in human fear conditioning. Neuroreport 1997;8:3957–60.

74. Canli T, Zhao Z, Desmond JE, Kang E, Gross J, Gabrieli JDE. An fMRI study of personality influences on brain reactivity to emotional stimuli. Behav Neurosci 2001;115:33–42.
75. Canli T, Sivers H, Whitfield SL, Gotlib IH, Gabrieli JDE. Amygdala response to happy faces as a function of extraversion. Science 2002;296:2191.
76. Schienle A, Schafer A, Stark R, Walter B, Vaitl D. Relationship between disgust sensitivity, trait anxiety and brain activity during disgust induction. Neuropsychobiology 2005;51: 86–92.
77. Canli T, Amin Z. Neuroimaging of emotion and personality: Scientific evidence and ethical considerations. Brain Cogn 2002;50:414–31.
78. Small SL, Nusbaum HC. On the neurobiological investigation of language understanding in context. Brain Lang 2004;89:300–11.
79. Fahrenberg J, Myrtek M (Eds.) Progress in ambulatory assessment. Seattle, WA: Hogrefe and Huber, 2001.
80. Adolphs R. Neural systems for recognizing emotion. Curr Opin Neurobiol 2002;12: 169–77.
81. Deichmann R, Josephs O, Hutton C, Corfield DR, Turner R. Compensation of susceptibility-induced BOLD sensitivity losses in echo-planar fMRI imaging. NeuroImage 2002;15:120–35.
82. Yablonskiy DA. Quantitation of intrinsic magnetic susceptibility-related effects in a tissue matrix. Phantom study. Magn Reson Med 1998;39:417–28.
83. Lipschutz B, Friston KJ, Ashburner J, Turner R, Price CJ. Assessing study-specific regional variations in fMRI signal. Neuroimage 2001;13: 392–8.
84. Toga AW, Thompson PM. Maps of the brain. Anat Rec 2001;265:37–53.
85. Hutton C, Bork A, Josephs O, Deichmann R, Ashburner J, Turner R. Image distortion correction in fMRI: A quantitative evaluation. NeuroImage 2002;16:217–40.
86. Raj D, Paley DP, Anderson AW, Kennan RP, Gore JC. A model for susceptibility artefacts from respiration in functional echo-planar magnetic resonance imaging. Phys Med Biol 2000;45:3809–20.
87. Windischberger C, Langenberger H, Sycha T, et al. On the origin of respiratory artifacts in BOLD-EPI of the human brain. Magn Reson Imaging 2002;20:575–82.
88. Wise RG, Ide K, Poulin MJ, Tracey I. Resting fluctuations in arterial carbon dioxide induce significant low frequency variations in BOLD signal. Neuroimage 2004;21:1652–64.
89. Dagli MS, Ingeholm JE, Haxby JV. Localization of cardiac-induced signal change in fMRI. Neuroimage 1999;9:407–15.
90. Critchley HD, Rotshtein P, Nagai Y, O'Doherty J, Mathias CJ, Dolan RJ. Activity in the human brain predicting differential heart rate responses to emotional facial expressions. Neuroimage 2005;24:751–62.
91. Frysinger RC, Harper RM. Cardiac and respiratory correlations with unit discharge in human amygdala and hippocampus. Electroencephalogr Clin Neurophysiol 1989;72 :463–70.
92. Shmueli K, van Gelderen P, de Zwart JA, et al. Low-frequency fluctuations in the cardiac rate as a source of variance in the resting-state fMRI BOLD signal. Neuroimage 2007;38:306–20.
93. Cohen MA, Taylor JA. Short-term cardiovascular oscillations in man: Measuring and modelling the physiologies. J Physiol 2002;542: 669–83.
94. Birn RM, Diamond JB, Smith MA, Bandettini PA. Separating respiratory-variation-related fluctuations from neuronal-activity-related fluctuations in fMRI. Neuroimage 2006;31:1536–48.
95. Edward V, Windischberger C, Cunnington R, et al. Quantification of fMRI artifact reduction by a novel plaster cast head holder. Hum Brain Mapp 2000;11:207–13.
96. Hajnal J, Myers R, Oatridge A, Schwieso J, Young I, Bydder G. Artifacts due to stimulus correlated motion in functional imaging of the brain. Magn Reson Med 1994;31: 283–91.
97. Field A, Yen Y, Burdette J, Elster A. False cerebral activation on BOLD functional MR images: Study of low-amplitude motion weakly correlated to stimulus. AJNR Am J Neuroradiol 2000;21:1388–96.
98. Robinson S, Moser E. Positive results in amygdala fMRI: Emotion or head motion? NeuroImage 2004;22:S47, WE 294.
99. Biswal B, Yetkin FZ, Haughton VM, Hyde JS. Functional connectivity in the motor cortex of resting human brain using echo-planar MRI. Magn Reson Med 1995;34:537–41.
100. Lowe MJ, Mock BJ, Sorenson JA. Functional connectivity in single and multislice echoplanar imaging using resting-state fluctuations. Neuroimage 1998;7:119–32.
101. Fox MD, Snyder AZ, Zacks JM, Raichle ME. Coherent spontaneous activity accounts for trial-to-trial variability in human evoked brain responses. Nat Neurosci 2006;9: 23–5.

102. Fox MD, Snyder AZ, Vincent JL, Raichle ME. Intrinsic fluctuations within cortical systems account for intertrial variability in human behavior. Neuron 2007;56:171–84.
103. Beckmann CF, De Luca M, Devlin JT, Smith SM. Investigations into resting-state connectivity using independent component analysis. Philos Trans R Soc Lond B Biol Sci 2005;360:1001–13.
104. Damoiseaux JS, Rombouts SA, Barkhof F, et al. Consistent resting-state networks across healthy subjects. Proc Natl Acad Sci USA 2006;103:13848–53.
105. De Luca M, Beckmann CF, De Stefano N, Matthews PM, Smith SM. fMRI resting state networks define distinct modes of long-distance interactions in the human brain. Neuroimage 2006;29:1359–67.
106. Robinson S, Soldati N, Basso G, et al. A resting state network in the basal ganglia. Proc Intl Soc Magn Res Med 2008;16:746.
107. Beckmann CF, Smith SM. Tensorial extensions of independent component analysis for multisubject FMRI analysis. Neuroimage 2005;25:294–311.
108. Raichle M, MacLeod A, Snyder A, Powers W, Gusnard D, Shulman G. A default mode of brain function. Proc Natl Acad Sci US A 2001;98:676–82.
109. Shulman G, Fiez J, Corbetta M, et al. Common blood flow changes across visual tasks: II. Decreases in cerebral cortex. J Cogn Neurosci 1997;9:648–63.
110. Merboldt KD, Fransson P, Bruhn H, Frahm J. Functional MRI of the human amygdala? Neuroimage 2001;14:253–7.
111. Triantafyllou C, Hoge RD, Krueger G, et al. Comparison of physiological noise at 1.5 T, 3 T and 7 T and optimization of fMRI acquisition parameters. Neuroimage 2005;26: 243–50.
112. Krasnow B, Tamm L, Greicius MD, et al. Comparison of fMRI activation at 3 and 1.5 T during perceptual, cognitive, and affective processing. Neuroimage 2003;18:813–26.
113. Dickerson BC, Wright CI, Miller S, et al. Ultrahigh-field differentiation of medial temporal lobe function: Sub-millimeter fMRI of amygdala and hippocampal activation at 7 Tesla. NeuroImage 2006;31:S154.
114. Merboldt KD, Finsterbusch J, Frahm J. Reducing inhomogeneity artifacts in functional MRI of human brain activation-thin sections vs gradient compensation. J Magn Reson 2000;145:184–91.
115. Bellgowan PS, Bandettini PA, van Gelderen P, Martin A, Bodurka J. Improved BOLD detection in the medial temporal region using parallel imaging and voxel volume reduction. Neuroimage 2006;29:1244–51.
116. Triantafyllou C, Hoge RD, Wald LL. Effect of spatial smoothing on physiological noise in high-resolution fMRI. Neuroimage 2006;32: 551–7.
117. Robinson S, Windischberger C, Rauscher A, Moser E. Optimized 3 T EPI of the amygdalae. NeuroImage 2004;22:203–10.
118. Frahm J, Merboldt KD, Hänicke W. Direct FLASH MR imaging of magnetic field inhomogeneities by gradient compensation. Magn Reson Med 1988;6:474–80.
119. Deichmann R, Gottfried JA, Hutton C, Turner R. Optimized EPI for fMRI studies of the orbitofrontal cortex. Neuroimage 2003;19:430–41.
120. Cho Z, Ro Y. Reduction of susceptibility artifact in gradient-echo imaging. Magn Reson Med 1992;23:193–200.
121. Stenger VA, Boada FE, Noll DC. Three-dimensional tailored RF pulses for the reduction of susceptibility artifacts in T(*)(2)-weighted functional MRI. Magn Reson Med 2000;44:525–31.
122. Yip CY, Fessler JA, Noll DC. Advanced three-dimensional tailored RF pulse for signal recovery in T2*-weighted functional magnetic resonance imaging. Magn Reson Med 2006;56:1050–9.
123. De Panfilis C, Schwarzbauer C. Positive or negative blips? The effect of phase encoding scheme on susceptibility-induced signal losses in EPI. Neuroimage 2005;25:112–21.
124. Ojemann JG, Akbudak E, Snyder AZ, McKinstry RC, Raichle ME, Conturo TE. Anatomic localization and quantitative analysis of gradient refocused echo-planar fMRI susceptibility artifacts. Neuroimage 1997;6: 156–67.
125. Chen N, Dickey CC, Guttman CRG, Panych LP. Selection of voxel size and slice orientation for fMRI in the presence of susceptibility field gradients: Application to imaging of the amygdala. NeuroImage 2003;19:817–25.
126. Weiskopf N, Hutton C, Josephs O, Deichmann R. Optimal EPI parameters for reduction of susceptibility-induced BOLD sensitivity losses: A whole-brain analysis at 3 T and 1.5 T. Neuroimage 2006;33: 493–504.
127. Young IR, Cox IJ, Bryant DJ, Bydder GM. The benefits of increasing spatial resolution as a means of reducing artifacts due to field inhomogeneities. Magn Reson Imaging 1988;6:585–90.

128. Hyde S, Biswal B, Jesmanowicz A. High-resolution fMRI using multislice partial k-space GR-EPI with cubic voxels. Magn Reson Med 2001;46:114–25.
129. Haacke E, Brown R, Thompson M, Venkatesan R. Magnetic resonance imaging: Physical principles and sequence design. New York, NY: Wiley-Liss, 1999.
130. Scouten A, Papademetris X, Constable RT. Spatial resolution, signal-to-noise ratio, and smoothing in multi-subject functional MRI studies. Neuroimage 2006;30:787–93.
131. Parrish T, Gitelman D, LaBar K, Mesulam M. Impact of signal-to-noise on functional MRI. Magn Reson Med 2000;44: 925–32.
132. LaBar K, Gitelman D, Mesulam M, Parrish T. Impact of signal-to-noise on functional MRI of the human amygdala. Neuroreport 2001;12:3461–64.
133. Robinson S, Hoheisel B, Windischberger C, Habel U, Lanzenberger R, Moser E. FMRI of the emotions, towards an improved understanding of amygdala function. Curr Med Imaging Rev 2005;1:115–29.
134. Morawetz C, Holz P, Lange C, et al. Improved functional mapping of the human amygdala using a standard functional magnetic resonance imaging sequence with simple modifications. Magn Reson Imaging 2008;26:45–53.
135. Stocker T, Kellermann T, Schneider F, et al. Dependence of amygdala activation on echo time: Results from olfactory fMRI experiments. Neuroimage 2006;30:151–9.
136. Gorno-Tempini M, Hutton C, Josephs O, Deichmann R, Price C, Turner R. Echo time dependence of BOLD contrast and susceptibility artifacts. NeuroImage 2002;15: 136–42.
137. Sodickson DK, Manning WJ. Simultaneous acquisition of spatial harmonics (SMASH): Fast imaging with radiofrequency coil arrays. Magn Reson Med 1997;38:591–603.
138. Pruessmann K, Weiger M, Scheidegger M, Boesiger P. SENSE: Sensitivity encoding for fast MRI. Magn Reson Med 1999;42: 952–62.
139. Lutcke H, Merboldt KD, Frahm J. The cost of parallel imaging in functional MRI of the human brain. Magn Reson Imaging 2006;24:1–5.
140. Schmidt CF, Degonda N, Luechinger R, Henke K, Boesiger P. Sensitivity-encoded (SENSE) echo planar fMRI at 3 T in the medial temporal lobe. NeuroImage 2005;25:625–41.
141. Fürsatz M, Windischberger C, Karlsson KÆ, Moser E. Successful fMRI of the hypothalamus at 3T. Proc Intl Soc Magn Reson Med 2008;16:2501.
142. Fürsatz M, Windischberger C, Karlsson KÆ, Mayr W, Moser E. Valence-dependent modulation of hypothalamic activity. Neuroimage in press.
143. Dowell NG, Tofts PS. Fast, accurate, and precise mapping of the RF field in vivo using the 180 degrees signal null. Magn Reson Med 20907;58:622–30.
144. Speck O, Hennig J. Functional imaging by I0- and T2*-parameter mapping using multi-image EPI. Magn Reson Med 1998;40: 243–8.
145. Posse S, Wiese S, Gembris D, et al. Enhancement of BOLD-contrast sensitivity by single-shot multi-echo functional MR imaging. Magn Reson Med 1999;42:87–97.
146. Posse S, Shen Z, Kiselev V, Kemna LJ. Single-shot T(2)* mapping with 3D compensation of local susceptibility gradients in multiple regions. Neuroimage 2003;18:390–400.
147. Posse S, Holten D, Gao K, Rick J, Speck O. Evaluation of interleaved XYZ-shimming with multi-echo EPI in prefrontal cortex and amygdala at 4 Tesla. NeuroImage 2006; 31:S154.
148. Weiskopf N, Klose U, Birbaumer N, Mathiak K. Single-shot compensation of image distortions and BOLD contrast optimization using multi-echo EPI for real-time fMRI. Neuroimage 2005;24:1068–79.
149. Glover GH, Law CS. Spiral-in/out BOLD fMRI for increased SNR and reduced susceptibility artifacts. Magn Reson Med 2001;46:515–22.
150. Guo H, Song AW. Single-shot spiral image acquisition with embedded z-shimming for susceptibility signal recovery. J Magn Reson Imaging 2003;18:389–95.
151. Truong TK, Song AW. Single-shot dual-z-shimmed sensitivity-encoded spiral-in/out imaging for functional MRI with reduced susceptibility artifacts. Magn Reson Med 2008;59:221–7.
152. Li Z, Wu G, Zhao X, Luo F, Li SJ. Multi-echo segmented EPI with z-shimmed background gradient compensation (MESBAC) pulse sequence for fMRI. Magn Reson Med 2002;48:312–21.
153. Weisskoff RM, Davis TL. Correcting gross distortion on echo planar images. Paper Presented at the SMRM, Berlin, 1992.
154. Jezzard P, Balaban RS. Correction for geometric distortion in echo planar images

from B0 field variations. Magn Reson Med 1995;34:65–73.

155. Jenkinson M. Fast, automated, N-dimensional phase-unwrapping algorithm. Magn Reson Med 2003;49:193–7.
156. Witoszynskyj S, Rauscher A, Reichenbach JR, Barth M. ΠUN (Πhase UNwrapping) validation of a 2D region-growing phase unwrapping program. Proc Intl Soc Magn Reson Med 2007;15:3436.
157. Windischberger C, Robinson S, Rauscher A, Barth M, Moser E. Robust field map generation using a triple-echo acquisition. J Magn Reson Imaging 2004;20:730.
158. Roemer PB, Edelstein WA, Hayes CE, Souza SP, Mueller OM. The NMR phased array. Magn Reson Med 1990;16:192–225.
159. Bernstein MA, Grgic M, Brosnan TJ, Pelc NJ. Reconstructions of phase contrast, phased array multicoil data. Magn Reson Med 1994;32:330–4.
160. Hammond KE, Lupo JM, Xu D, et al. Development of a robust method for generating 7.0 T multichannel phase images of the brain with application to normal volunteers and patients with neurological diseases. Neuroimage 2008;39:1682–92.
161. Robinson S, Jovicich J. EPI distortion corrections at 4 T: Multi-channel field mapping and a comparison with the point-spread function method. Proc Intl Soc Magn Reson Med 2008;16:3031.
162. Robson MD, Gore JC, Constable RT. Measurement of the point spread function in MRI using constant time imaging. Magn Reson Med 1997;38:733–40.
163. Zeng H, Constable RT. Image distortion correction in EPI: Comparison of field mapping with point spread function mapping. Magn Reson Med 2002;48:137–46.
164. Zaitsev M, Hennig J, Speck O. Point spread function mapping with parallel imaging techniques and high acceleration factors: Fast, robust, and flexible method for echo-planar imaging distortion correction. Magn Reson Med 2004;52:1156–66.
165. Hu X, Kim SG. Reduction of signal fluctuation in functional MRI using navigator echoes. Magn Reson Med 1994;31:495–503.
166. Bruder H, Fischer H, Reinfelder HE, Schmitt F. Image reconstruction for echo planar imaging with nonequidistant k-space sampling. Magn Reson Med 1992;23:311–23.
167. Barry RL, Klassen LM, Williams JM, Menon RS. Hybrid two-dimensional navigator correction: A new technique to suppress respiratory-induced physiological noise in multi-shot echo-planar functional MRI. Neuroimage 2008;39:1142–50.
168. Lund TE, Madsen KH, Sidaros K, Luo WL, Nichols TE. Non-white noise in fMRI: Does modelling have an impact? Neuroimage 2006;29:54–66.
169. Glover GH, Li TQ, Ress D. Image-based method for retrospective correction of physiological motion effects in fMRI: RETROICOR. Magn Reson Med 2000;44:162–7.
170. Windischberger C, Friedreich S, Hoheisel B, Moser E. The importance of correcting for physiological artifacts for functional MRI in deep brain structures. NeuroImage 2004;22:S28.
171. Josephs O, Howseman A, Friston K, Turner R. Physiological noise modelling for multi-slice EPI fMRI using SPM. Proc Intl Soc Magn Reson Med 1997;5:1682.
172. Weissenbacher A, Windischberger C, Lanzenberger R, Moser E. Efficient correction for artificial signal fluctuations in resting-state fMRI-data. Proc Intl Soc Magn Reson Med 2008;16:2467.
173. Beckmann CF, Smith SM. Probabilistic independent component analysis for functional magnetic resonance imaging. IEEE Trans Med Imaging 2004;23:137–52.
174. Calhoun V, Adali T, Stevens M, Kiehl K, Pekar J. Semi-blind ICA of fMRI: A method for utilizing hypothesis-derived time courses in a spatial ICA analysis. Neuroimage 2005;25: 527–38.
175. Thomas CG, Harshman RA, Menon RS. Noise reduction in BOLD-based fMRI using component analysis. Neuroimage 2002;17: 1521–37.
176. Kochiyama T, Morita T, Okada T, Yonekura Y, Matsumura M, Sadato N. Removing the effects of task-related motion using independent-component analysis. Neuroimage 2005;25:802–14.
177. Perlbarg V, Bellec P, Anton JL, Pelegrini-Issac M, Doyon J, Benali H. CORSICA: Correction of structured noise in fMRI by automatic identification of ICA components. Magn Reson Imaging 2007;25:35–46.
178. Tohka J, Foerde K, Aron AR, Tom SM, Toga AW, Poldrack RA. Automatic independent component labeling for artifact removal in fMRI. Neuroimage 2008;39: 1227–45.
179. Bullmore ET, Brammer MJ, Rabe-Hesketh S, et al. Methods for diagnosis and treatment of stimulus-correlated motion in generic brain activation studies using fMRI. Hum Brain Mapp 1999;7:38–48.

180. Morgan VL, Dawant BM, Li Y, Pickens DR. Comparison of fMRI statistical software packages and strategies for analysis of images containing random and stimulus-correlated motion. Comput Med Imaging Graph 2007;31:436–46.
181. Preibisch C, Raab P, Neumann K, et al. Event-related fMRI for the suppression of speech-associated artifacts in stuttering. Neuroimage 2003;19:1076–84.
182. Birn RM, Bandettini PA, Cox RW, Shaker R. Event-related fMRI of tasks involving brief motion. Hum Brain Mapp 1999;7: 106–14.
183. Phelps EA, O'Connor KJ, Cunningham WA, et al. Performance on indirect measures of race evaluation predicts amygdala activation. J Cogn Neurosci 2000;12:729–38.
184. Winston JS, Strange BA, O'Doherty J, Dolan RJ. Automatic and intentional brain responses during evaluation of trustworthiness of faces. Nat Neurosci 2002;5:277–83.
185. Singer T, Kiebel SJ, Winston JS, Dolan RJ, Frith CD. Brain responses to the acquired moral status of faces. Neuron 2004;41: 653–62.
186. Campbell DT, Stanley JC. Experimental and quasi-experimental designs for research. Boston: Houghton Mifflin, 1966.
187. Poldrack RA, Wagner AD. What can neuroimaging tell us about the mind? Insights from prefrontal cortex. Curr Dir Psychol Sci 2004;13:177–81.
188. Logothetis NK, Pauls J, Augath M, Trinath T, Oeltermann A. Neurophysiological investigation of the basis of the fMRI signal. Nature 2001;412:150–7.
189. Janz C, Heinrich SP, Kornmayer J, Bach M, Hennig J. Coupling of neural activity and BOLD fMRI response: New insights by combination of fMRI and VEP experiments in transition from single events to continuous stimulation. Magn Reson Med 2001; 46:482–6.
190. Baas D, Aleman A, Kahn RS. Lateralization of amygdala activation: A systematic review of functional neuroimaging studies. Brain Res Brain Res Rev 2004;4:96–103.
191. Robinson S, Pripfl J, Bauer H, Moser M. Empirical evidence for the minimum voxel size required for reliable 3 T fMRI of the amygdala. NeuroImage 2005;26:S795.
192. Habel U, Windischberger C, Derntl B, et al. Amygdala activation and facial expressions: Explicit emotion discrimination versus implicit emotion processing. Neuropsychologia 2007;45:2369–77.
193. Sack AT, Linden DE. Combining transcranial magnetic stimulation and functional imaging in cognitive brain research: Possibilities and limitations. Brain Res Brain Res Rev 2003;43:41–56.
194. Thayer JF, Brosschot JF. Psychosomatics and psychopathology: Looking up and down from the brain. Sychoneuroendocrinology 2005;30:1050–8.

Chapter 15

fMRI of Pain

Emma G. Duerden and Gary H. Duncan

Summary

The field of pain research has progressed immensely due to the advancement of brain imaging techniques. The initial goal of this research was to expand our understanding of the cerebral mechanisms underlying the perception of pain; more recently the research objectives have shifted toward chronic pain – understanding its origins, developing methods for its diagnosis, and exploring potential avenues for its treatment. While several different neuroimaging approaches have certain advantages for the study of pain, fMRI has ultimately become the most widely utilized imaging technique over the past decade because of its noninvasive nature, high-temporal and spatial resolution, and general availability; thus, the following chapter will focus on fMRI and the special aspects of this technique that are particular to pain research. Subheading 1 begins with a brief review on the spinal pathways and neuroanatomical regions involved in pain processing, and highlights the novel information that has been gained about these structures and their function through the use of fMRI and other neuroimaging techniques. Subheading 2 reviews a few of the aspects associated with the blood-oxygen-level-dependent signal commonly used in fMRI, as they apply to the particular challenges of pain research. Likewise, Subheading 3 summarizes some of the special considerations of experimental design and statistical analysis that are encountered in pain research and their applications to fMRI studies. Subheading 4 reviews special applications of fMRI for the study of higher cognitive processes implicated in pain processing, including pain empathy and cognitive reappraisal of one's own pain perception. The chapter concludes with Subheading 5, exploring some of the future prospects of fMRI techniques and new applications related to pain research.

Key words: Pain, Human, Functional neuroimaging, Brain, Perception

1. Introduction

The history of pain imaging is relatively short, although it has advanced immensely within the last decade due to improvements in imaging techniques, statistical analysis, and specialized equipment for the delivery of painful stimuli. Initially, brain imaging studies sought simply to examine the brain areas that are involved

M. Filippi (ed.), *fMRI Techniques and Protocols*, Neuromethods, vol. 41
DOI 10.1007/978-1-60327-919-2_15, © Humana Press, a part of Springer Science+Business Media, LLC 2009

in pain processing, to make comparisons with the long established neurophysiological studies reported in this field. Many of these initial imaging studies were prompted by electrophysiological data from patients undergoing brain surgery in the early part of the twentieth century *(1)*, which had questioned the role of the cortex in nociceptive processing. It was initially believed that the thalamus was primarily responsible for nociceptive processing as suggested by deficits in pain perception observed in patients with thalamic lesions *(2)*.

In the early 1990s, activation in the human brain evoked by experimental pain stimuli was studied using positron emission tomography (PET) *(3, 4)* and single photon emission tomography (SPECT) *(5)*. Then in 1995 the first fMRI studies examining the cortical representation of pain *(6)* were conducted largely to confirm the findings of previous PET studies and to examine whether the cortical nociceptive signal could be detected using fMRI. In more recent years, the field of pain imaging has expanded immensely, allowing researchers to answer complex questions concerning pain processing, such as how cortical regions are connected and modified during the perception of pain and, most importantly, how the cortex responds during the modulation of pain. These experimental studies were conducted in healthy humans in order to answer broad questions regarding pain processing, with the eventual goal of applying this knowledge to a better understanding and alleviation of pain and suffering associated with chronic pain syndromes. The use of fMRI and other imaging techniques has revealed a number of cortical and subcortical changes that may occur as a result of exposure to chronic pain *(7)*. Indeed, with the advent of high-speed image acquisition and computational processing, not only has the technology of fMRI revealed areas of cortical plasticity associated with chronic pain, but it is also now possible to use fMRI in real-time to furnish feedback to subjects (and patients) to teach them how to modulate their cortical activation in response to chronic pain *(8)*.

This chapter reviews and discusses the various advances in our knowledge of cerebral pain processing that have been achieved using fMRI, the response properties of cortical nociceptive neurons in relation to both imaging techniques and stimuli used to evoke pain, the applications of this research to treat clinical pain in patients, and the future of pain research using fMRI.

2. Background

2.1. Neuroanatomy of Pain Processing

Before describing how fMRI measures the cortical and spinal nociceptive signal, it is important to understand how this signal is transferred to the cortex. In the periphery, a painful stimulus applied to the body is transmitted to the central nervous system (CNS) through nociceptors *(9)*. Myelinated A-delta fibers transmit sharp pricking pain *(10)*, while unmyelinated C-fibers transmit slow burning pain, often referred to as second pain *(11)*. The cell bodies of A-delta and C-fibers are located in the dorsal root ganglia, receiving afferent input from the periphery and then sending the information into the spinal cord to terminate in the dorsal horn *(12, 13)*. Axons from the second-order dorsal horn neurons rise through several ascending pathways that transmit nociceptive information to the thalamus, reticular formation, and cortex *(14–19)*. Pain and temperature information applied to the face is relayed through cranial nerves to the spinal nucleus V terminating in the thalamus via the trigeminothalamic tract which is then relayed to the cortex. A number of spinal and cortical neurons respond to noxious stimuli including nociceptive specific (NS) and wide dynamic range (WDR) projection neurons, the latter of which respond to both noxious and innocuous stimuli. Additionally, the dorsal horns and cortical somatosensory regions contain neurons responsive solely to innocuous stimuli called low threshold mechanical (LTM) neurons and thermoreceptive neurons responsive to temperatures in the warm and cold range. This range of responses is an important consideration when interpreting results from fMRI studies of pain in terms of exactly what the activation pattern is reflecting.

Typically, pain-evoked brain activation is achieved by applying contact thermodes to the skin. This technique involves an increase in temperature at the rate of 1–10°C s^{-1}. Depending on the baseline temperature it can take several seconds to reach perceived pain threshold. In addition to activating NS neurons with noxious heat, contact thermodes may activate both LTM and WDR neurons through innocuous mechanical and thermal stimulation of the skin as the stimulation temperature rises towards pain threshold. Therefore, to examine pain-specific cortical activations, it is necessary to compare pain-related activations to those associated with the presentation of innocuous warm stimuli.

In addition to conductive heating of the skin using contact thermodes, nociceptive afferents can be activated using thermal radiation administered through infrared laser stimulators *(20, 21)*. Lasers can deliver heat stimuli without the need for a contact probe, thus selectively stimulating C-fibers and A-delta fibers without contaminant activation of A-beta fibers that transmit touch information. Additionally, laser stimuli can activate nociceptive nerve endings at rapid rates for short durations (1 ms) *(22, 23)* and are therefore well suited for rapid event-related fMRI studies. However, an important consideration associated with the use of laser stimuli is the difficulty of measuring and controlling skin temperature, which is the primary factor triggering the cascade of neural responses that culminate in the processing of heat-related nociceptive information in the brain and likewise the assessment of pain by the subjects *(24)*. Laser and contact heat stimuli have been shown to produce similar patterns of blood-oxygen-level-dependent (BOLD) activation in the secondary somatosensory cortex (SII), anterior cingulate cortex (ACC), insula, the primary motor cortex, prefrontal cortex (PFC), parahippocampal gyrus, thalamus, basal ganglia, periaquaductal gray (PAG), and cerebellum. However, stronger activation in response to contact heat stimuli was noted in SII, posterior insula, posterior ACC, and regions in parietal and frontal cortices *(25)*. Thus, these two modes of delivering noxious heat stimulation cannot be considered identical in terms of the evoked pain-related BOLD activations, and the advantages and disadvantages of each should be weighed in relation to the research questions and appropriate stimulation paradigms.

2.1.1. Supraspinal Processing of Nociceptive Stimuli

A recent review of 68 pain neuroimaging studies using healthy subjects revealed a homogeneity of reported activations across cortical regions, thus implicating a cerebral network for pain processing *(26)*. Regions most frequently activated by painful stimuli include primary somatosensory cortex (SI), SII, ACC, the insula, the PFC, and the thalamus.

Regions responsible for pain processing are categorized along two functional lines – the first being the sensory-discriminative (lateral pain system) component involved in the perception of temporal, intensity, and localization aspects of pain processing, and the second, the affective-motivational (medial) component associated with the emotional aspects of pain *(27)*. Dissociations between the two systems are made through subjective reports on pain scales. After exposure to noxious stimuli, subjects are asked to quantify separately how intense and how unpleasant is the perceived pain. Subjects' scores are recorded typically using numerical or visual analog scales (VAS) *(28)*. Regions implicated in the lateral pain system include SI, SII, posterior insula, and lateral thalamus, while the medial pain system consists of the medial thalamic nuclei, the ACC, and the PFC. Much of what is known regarding the two components in pain processing was initially explored through single-unit recordings in nonhuman primates and lesion studies in humans. However, the more recent ability to study these functional components noninvasively in humans using fMRI and other brain mapping techniques has allowed pain researchers to advance rapidly in their understanding of the role of these cortical regions in pain processing and how they interact.

2.1.2. Primary Somatosensory Cortex

SI is located in the postcentral gyrus, is composed of four areas (areas 3a, 3b, 1, and 2) *(29)*, and is involved in the processing of both tactile and noxious stimuli *(30)*. SI is the first relay from the principle sensory nuclei in the thalamus *(31–35)* and receives input from nociceptive neurons. While this information, gathered from nonhuman primates, established a role for SI in nociceptive processing, it was long debated whether SI was necessary to perceive pain. Early studies of patients with brain lesions suggested that deficits in nociceptive processing were rather common following lesions to the thalamus, but were very rare when damage was restricted to the area believed to incorporate SI *(2)*. Likewise, later studies, using electrical stimulation of the human cortex during awake brain surgery, reported that direct stimulation of SI rarely evoked any perception of pain in patients *(1)*.

The advent of imaging technology allowed a more global exploration of the role of SI and other cortical regions involved in pain processing, and these studies could be conducted in healthy volunteers, rather than in patients with brain injuries that might alter normal function. The first of these studies involved PET and demonstrated that noxious stimuli applied to the hands were associated with robust activation in SI *(3)*. Several other early studies failed to detect SI activation *(4, 5)*, and subsequent reports, using either PET or fMRI, have resulted in contradictory findings (for SI activation, see for example: *(36–39)*; for absence of SI activation, see **refs**. *40, 41)*.

The inconsistency of SI activation reported across imaging studies could be due to several factors. Wide variations in the location of the central sulcus across subjects may lead to a wash out in signal across averaged group data. In addition, a reduction in SI activity below statistically significant levels could be caused in some paradigms by inhibitory effects induced by noxious stimuli on tactile inputs. This effect has been reported at the cortical level using optical imaging *(42)* and SPECT *(5)* as well as in thalamus *(43)*. In a review discussing the issue of pain-related activation of SI, Bushnell et al. concluded that the BOLD signal in SI largely depends on task design that is likely to influence the attentional state of the subject *(44)*. Results from subsequent studies have likewise indicated that pain-related BOLD activation of SI is increased when subjects attend to pain and decreased when they are distracted *(45, 46)*.

On the contrary, attention may also show a deleterious effect on SI activation as noted by Oshiro et al. *(47)* in their fMRI study examining the neural correlates involved in processing spatial localization of pain. The authors failed to find activation in SI in response to painful stimulation of the calf. However, the authors noted that this lack of activation may have been a result of the response properties of the cortical nociceptive neurons. Nociceptive input to SI is somatotopically organized *(48–51)*, and the small receptive fields of SI *(52)* suggest that this region is well suited to make fine spatial discriminations of noxious stimuli applied to the body. Oshiro et al. *(47)* required subjects to focus on stimulation applied to their calves, and this increased attention on the leg area may have caused a reduction in the receptive field sizes of nociceptive neurons, which would enhance spatial acuity needed to perform the task – but cause deterioration in resulting brain activation. In another study using a discrimination task, Albanese et al. *(53)* explored short-term memory for the spatial location and intensity of painful thermal stimuli applied to the palms. In contrast to the study by Oshiro et al. *(47)*, Albanese et al. *(53)* reported robust pain-related activation in SI/posterior parietal cortex, which was sustained during the memory period of the trial, suggesting that this region has a role in the encoding and retention of noxious stimuli. Differences between the two studies may be due to the larger somatotopic organization of the hand representation of SI. Additionally, subjects in the Albanese study were required to detect the end of each stimulus, a strategy that may have heightened attention toward the stimuli and contributed to a temporal summation of the BOLD signal in SI.

2.1.3. Secondary Somatosensory Cortex (SII)

SII is also considered to be an important region for processing the sensory-discriminative component of pain. SII is located in the parietal operculum in the dorsal bank of the lateral sulcus. Like SI, this region receives projections from the ventroposterior lateral nucleus (VPL) *(54)* of the thalamus, but its major nociceptive input comes directly from the ventroposterior inferior (VPI) nucleus *(55)*. Although few nociceptive neurons have been recorded in SII in nonhuman primates *(56–58)*, this area is nevertheless commonly activated by noxious stimuli in human imaging studies *(26)*. Likewise, studies of patients with lesions that include SII have demonstrated deficits in the perception of pain intensity *(59, 60)*; however, lesions comprised additional cortical regions that may work in concert with SII to process this piece of information. In addition to these clinical findings, converging evidence from a number of studies *(61–63)*, now supports the notion that SII possesses a functional capacity to discriminate between different intensities of noxious stimuli presented to the contralateral side of the body. Furthermore, evidence from PET provides a role for this region in intensity processing in that subjects' ratings of pain intensity in response to thermal heat pain have been shown to be highly correlated with activation of SII *(64)*. Additionally, an fMRI study by Maihofner et al. *(65)* found increased activation in SII in response to painful mechanical stimuli compared to thermal heat pain. In turn, ratings of subjective intensity were correlated with the intensity of mechanical pain. However, dissociative processing was noted in this region as ratings of unpleasantness were not found to correlate with SII activation. Contrary to these findings, evidence from fMRI suggests this region may be involved in some emotional aspects of pain processing. For example, Gracely et al. *(66)* found that fibromyalgia patients who scored higher on a pain catastrophizing questionnaire showed increased activation in both the ACC and

SII in response to noxious stimuli. Catastrophizing (and in turn anxiety about painful stimuli) is inherently linked with pain perception, where the individual's emotional state augments neural processing of these stimuli. In line with these findings are data that show increased activity in SII during the anticipation of painful stimuli, indicative of an enhanced emotional response *(67)*.

2.1.4. Insular Cortex

The insula receives inputs from both SI and SII, and also from thalamic nuclei (VPI, the centromedian-parafasicular, the medial dorsal [MD], and the ventral medial posterior [Vmpo] nuclei); in turn, these nuclei receive nociceptive input via the spinothalamic tract *(31, 54, 68, 69)*. Early clinical reports *(70)*, as well as more recent quantitative studies *(71)*, have indicated that patients with lesions encompassing the insula do not exhibit normal withdrawal or emotional responses to noxious stimuli, indicating an altered or deficient perception of pain affect. Accordingly, fMRI activity in this region in response to noxious stimuli is correlated with subjective ratings of pain unpleasantness *(72)*.

The insula has also been found to process sensory-discriminative features of nociceptive information, making it a likely area of convergence of the two pain systems. Evidence for the role of the insula in sensory-discriminative processing comes from direct electrical stimulation to the region during awake brain surgery, demonstrating evoked painful sensations in the body *(73)*. Furthermore, several other lines of evidence indicate that this region may be involved in the localization of painful stimuli, as it contains a somatotopic map of the body. The dorsal posterior insula receives pain and temperature information from a somatotopically organized region of the thalamus – the VMpo *(74)*, which in turn receives projections from thermoreceptive and nociceptive neurons residing in lamina I of the spinal cord *(75)*.

Neuroimaging studies of pain perception frequently report insular activation, making it difficult to dissociate it from activation seen in adjacent regions of SII *(76)*. Resolving the precise somatotopic organization of the insula using fMRI has only recently become feasible with the availability of high-field strength magnets. As of late, two fMRI studies at 3 T have revealed a nociceptive somatotopic organization in the dorsal posterior insula in response to both cutaneous and muscle pain *(77, 78)*. Henderson et al. *(77)* also reported a distinct somatotopic organization in the right anterior insula ipsilateral to the muscle pain stimuli, and found that activation of this area was greater in comparison to cutaneous stimuli. The authors attributed the increase as a reflection of the enhanced unpleasantness associated with muscular pain.

2.1.5. Anterior Cingulate Cortex (ACC)

The ACC plays a prominent role in pain processing. This region receives thalamocortical input from nociceptive neurons in the thalamus *(79, 80)* and contains nociceptive-specific neurons responsive to noxious stimuli *(81)*. Additionally, the ACC is implicated in mediating antinociceptive responses as it contains high numbers of opiate receptors *(82, 83)*.

Historically, the ACC was considered key to affective processing, as it was classified along with the retrosplenial cortex, the hippocampus, the amygdala, and several basal forebrain structures as part of the limbic lobe, which was considered central in mediating emotion *(84, 85)*. Likewise, the ACC was targeted for surgical lesions to alleviate the suffering of chronic pain *(86–88)*; patients reported that they still experienced the pain they felt prior to surgery, but its emotional unpleasantness was dampened *(89, 90)*.

The ACC is subdivided cytoarchitectonically into several Brodmann areas (BA), namely 24 and 32 *(91)*, with two further subdivisions BA33 located in the perigenual region, and BA25 located in the subcallosal region. The ACC is functionally divided, rather independent of the cytoarchitectonic borders, into a caudal cognitive division involved in attention (BA24 and BA32) and a rostral affective division, which is more involved in emotional processes (BA24, 25, 33) *(92)*. Dissociation between the cognitive division and pain-related processing region was elegantly demonstrated using fMRI by Davis et al. *(93)*, who compared BOLD activation evoked by noxious stimuli to that seen during a demanding cognitive task. Activation associated with the noxious stimuli was found to be inferior and caudal to that produced by the cognitive task.

The first direct evidence for the role of ACC in processing affective components of pain came from a PET study, in which subjects under hypnosis were instructed to modulate the perceived unpleasantness of a painful stimulus while maintaining perceived pain intensity *(94)*. Results showed that activation of the ACC was highly correlated with the subjects' ratings of pain unpleasantness, while activation of the SI was unaltered by emotional processes. Nevertheless, these imaging and lesion data should not be interpreted too rigidly, since the ACC has been shown to have some sensory-discriminative characteristics, such as a crude nociceptive somatotopic organization *(95)*. Furthermore, reductions in both pain intensity and unpleasantness have been described following a neurosurgical capsulotomy – interruption of fiber tracts to the ACC *(96)*.

2.1.6. Prefrontal Cortex (PFC)

Regions of the PFC have been implicated in both pain processing and pain modulation. PFC activation seen in brain imaging studies of pain is believed to reflect attention toward the stimuli *(64, 97)*, but it has also been shown to be directly involved in modulating responses to painful stimuli. Using fMRI, Wager et al. have recently demonstrated increased PFC activity during the anticipation of pain, which was interpreted as a preemptive anticipatory response triggering a descending modulation of the pending nociceptive signals via activation of midbrain structures *(98)*.

2.1.7. Amygdala

The amygdala, buried beneath the uncus and located at the tail of the caudate nucleus, is a key limbic structure involved in the processing of emotional stimuli. The amygdala is suited for such processing as it is the sole subcortical structure to receive projections from every sensory area. Interestingly, projections to the amygdala from visual and auditory areas are greater in primates than in other species (for review *see* **ref.** *99)*.

Functional neuroimaging studies utilizing various types of aversive stimuli including pain, habitually report amygdala activation *(100)*. Studies using fMRI have demonstrated that amygdala activation is associated with extremely unpleasant noxious stimuli, suggesting an involvement of this region in processing the affective component of pain *(101)*. Other evidence from fMRI has implicated the amygdala in processing uncertainty associated with painful stimuli *(102)*.

2.1.8. Brainstem

In addition to cortical regions, a host of midbrain structures are also involved in processing pain affect including the PAG, the superior colliculus, the red nucleus, nucleus cuneiformis, the Edinger-Westphal nucleus, nucleus of Darkschewitsch, pretectal nuclei, the interstitial nucleus, and intercolliculus nucleus. Several of these structures are involved in pain modulation – the best characterized being the PAG. The PAG surrounds the cerebral aqueduct in the midbrain. Inhibitory enkephalin containing neurons in the PAG disinhibit local interneurons and in turn excite neurons in the rostral ventral medulla (RVM) and/or the locus coeruleous (LC). The aminergic projection from the RVM and LC then projects to the spinal cord and dampens pain transmission in dorsal horn neurons through several different mechanisms *(103, 104)*.

Presently, the sensitivity and in-plane resolution of 1.5 and 3.0 T MRI scanners are limited in their ability to resolve fine spatial localization of many brainstem structures. Brainstem functional imaging is also limited by image distortion and is susceptible to local magnetic field inhomogeneities and pulsation artifacts *(105–107)*.

2.1.9. Motor Cortices

A number of other cortical and subcortical regions are commonly activated during fMRI studies of pain including many regions involved in motor processing. Motor regions include the primary motor cortex, the premotor cortex, the supplementary motor area, the cerebellum, and basal ganglia. Frequently, these regions are concomitantly activated along with those involved with affective and sensory aspects of pain processing.

The study of pain–motor interactions is just developing in neuroimaging *(108)*, and our current understanding of this complex interaction is still incomplete. The perception of a painful stimulus involves an orienting response and subsequent retraction of the body part being targeted. Activation of motor areas during functional neuroimaging studies is believed to reflect motor preparatory responses. However, several of these areas, such as the nuclei associated with the basal ganglia, are directly responsive to noxious stimuli *(109)*. Using fMRI, a reliable somatotopic organization has been shown in the putamen *(110)* in response to noxious stimuli, which indicates that this region may be involved in sensory-discriminative processing of pain.

3. Use of fMRI to Study Nociceptive Processing

Compared to other brain mapping techniques currently used to study pain experimentally in humans, such as PET, electroencephalography (EEG), magnetoencephalography (MEG), or optical imaging, fMRI is the tool of choice, given its high spatial resolution, noninvasiveness, and reasonable temporal resolution, which allows the study of rapid dynamic processes involved in pain processing. A number of methodological issues are reviewed below concerning the use of the BOLD signal in research involving cortical, and more recently, spinal mechanisms of pain perception.

3.1. Nociceptive BOLD Signal

fMRI measures local blood flow changes in response to brain activity. Increased neuronal activity causes an increase in oxygen consumption resulting in an increase in local blood flow and volume *(111)*. This occurs after a delay of ~2 s with the hemodynamic response function (HRF) peaking after ~6–9 s after stimulus onset *(112)*. However, for cortical nociceptive processing related to cutaneous heat stimuli, the HRF peaks slightly later and lasts longer in comparison to innocuous stimuli. Chen et al. *(113)* performed a direct comparison of the temporal properties of the HRF in response to noxious thermal heat pain and innocuous brushing stimuli in SI and SII. While both stimuli were of the same duration, the time course for innocuous stimuli peaked ~10 s after the onset of the stimulus and dissipated quickly after its removal. However, noxious thermal heat stimuli produced a time course peaking at ~15 s after the onset of the stimulus and the response was sustained for several seconds **(Fig. 1)**. Similar results have been reported in response to painful electrical stimuli *(114)*; identical trains of noxious and innocuous stimuli produced differential time courses, with the HRF for painful stimulation lasting twice as long as that produced by nonpainful stimuli.

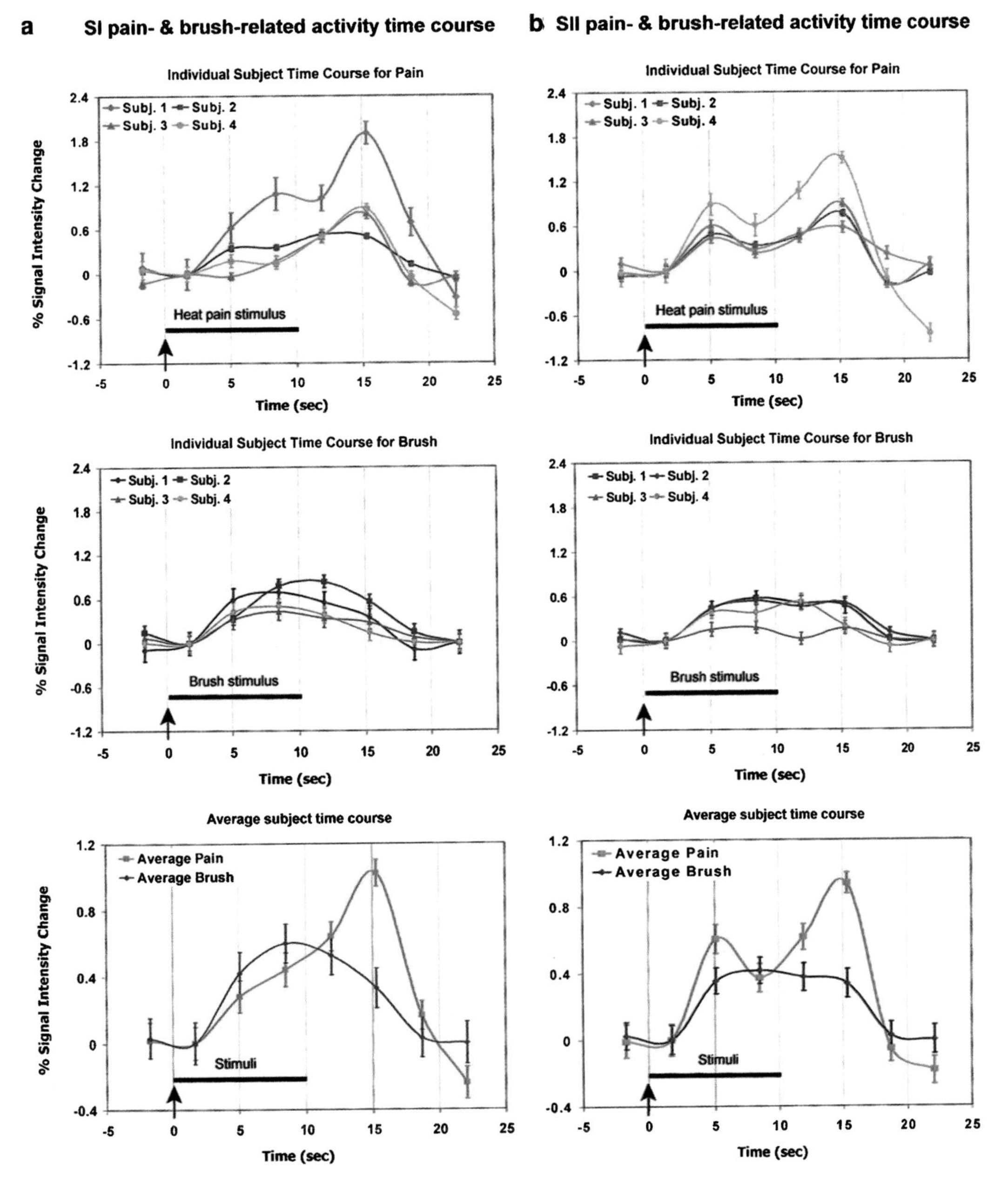

Fig. 1. Time course of the BOLD nociceptive signal. (*Top graphs*) Individual subject data showing percent signal change (±SE) in primary somatosensory cortex (SI) (*left*) and secondary somatosensory cortex (SII) (*right*) in response to thermal heat pain applied to the left inner calf. (*Middle graphs*) Time course information obtained in identical cortical regions and from body part locations in response to mechanical stimuli. (*Bottom graphs*) Averaged time course across subjects for heat pain (*red line*) and mechanical stimuli (*blue line*). The response to the thermal stimuli shows a slow gradual increase that peaks on average 15 s after the onset of the stimuli. In comparison, mechanical stimuli (*blue line*) demonstrate a faster rise time with a peak response occurring on average 5–8 s following stimulus onset and was sustained for ~10 s. Reprinted with permission from ***(113)***.

Time course information on the BOLD response to noxious stimuli is crucial for interpreting data analyzed using the standard canonical HRFs available in the majority of fMRI analysis software, which approximate this time period at ~6 s. Ideally, to establish a more representative model of painful stimuli, a canonical HRF should be created based on data from independent studies employing similar noxious stimulation. The BOLD signal can then be regressed against this canonical HRF to reveal activation more specific to the nociceptive signal.

A related issue in analyzing data recorded during experimental pain studies is the critical importance of considering the rise time of thermal stimuli when establishing time periods in the event design matrix. As the temperature of the thermode gradually increases, warm and pain fibers will become increasingly activated. In order to maximize sensitivity for detection of the pain-related BOLD signal, it is important to enter into the design matrix solely the period of time during which the thermode has exceeded the subjects' pain threshold - not the initial rise-time of the stimulus period, which would be associated with the innocuous warm sensations perceived before the actual onset of pain.

3.2. BOLD fMRI of Spinal Nociceptive Signals

A newly developing field in pain fMRI is spinal cord imaging, which is crucial for a better understanding of CNS pain processing. The spinal cord and brainstem receive input from the periphery before relaying this information on to the cortex. These subcortical regions are involved in the modulation of nociceptive input and the potentially abnormal processing of that input that may lead to chronic pain syndromes. Therefore, knowledge concerning the peripheral mechanisms of nociceptive processing is crucial to understanding a number of pathological pain conditions resulting from nerve injury or inflammation. These factors contribute to the generation and maintenance of two key components of chronic pain, namely hyperalgesia and allodynia. Hyperalgesia is the phenomenon where an exaggerated response occurs after exposure to a noxious stimulus. Allodynia is an exaggerated response toward nonpainful mechanical stimuli. Both occur when nociceptive fibers become sensitized, after exposure to a noxious stimulus, causing the release of "painful" substances in the periphery. Peripheral sensitization can occur due to inflammation of peripheral tissues as a result of a burn or cut. Because of this barrage of input, peripheral nociceptors can become hyperexcitable. This peripheral sensitization can also occur due to ectopic firing of peripheral nerves resulting from an amputation or injury. Central sensitization can occur in the dorsal horns of the spinal cord, when peripheral nerves that were once insensitive to nociceptive input switch their firing patterns and begin to transmit nociceptive information, causing the area of affected skin to become painful to the slightest touch. Much research in this area is focused at the periphery, although these processes have been shown to have supraspinal effects resulting in aberrant cortical activity and the reorganization of body maps in somatosensory cortices.

To fulfill this need to study spinal mechanisms of nociception, experimental models directed toward spinal fMRI have begun in humans *(115–117)*. fMRI of the spinal cord is challenging because of several factors. Most importantly, the small size of the spinal cord makes it difficult to achieve high spatial resolution without loss of signal-to-noise ratio (SNR). High SNR is important for good image quality and will increase in relation to voxel size, the number of image acquisitions, phase encodings, or the number of scans. The spinal cord is at its largest in the cervical segment, measuring ~16 mm × 10 mm. To achieve high spatial resolution of such a small structure, slice thickness, field of view (FOV), and matrix size can be reduced; however, these strategies reduce the SNR. For example, reducing FOV from 340 to 250 mm causes a reduction in signal of ~50%. On the contrary, thinner slices improve image quality since they are less susceptible to partial volume effects, which are inherent in imaging the spinal cord (due to the small size of the spinal cord, different tissue types and pulsating cerebrospinal fluid [CSF] make it difficult to dissociate one from the other). Spinal fMRI is very sensitive to a number of artifacts, such as magnetic field inhomogeneities; differences in the magnetic susceptibility and field gradients of each component of the spinal cord (bone, discs, cartilage, tissue) result in a loss of signal. Other factors causing increased noise in spinal fMRI signal include physiological motion such as CSF pulsation, respiration, and cardiac rhythms. One potential analysis strategy involves recording these physiological parameters during image acquisition, identifying them during postprocessing, and subsequently removing their contribution to the BOLD signal using independent components analysis *(116)*.

To date, only a few reports have assessed the feasibility of studying nociception using fMRI of the spinal cord. One study by Brooks et al. *(116)* examined the spinal nociceptive signal at 1.5 T in response to noxious heat pain stimuli. Using a tailored, high-resolution scanning protocol and postprocessing techniques for controlling physiological noise, they demonstrated reliable pain-related activation in the ipsilateral dorsal horn.

Applications of spinal fMRI to the study of chronic pain could have vast clinical applications. Use of a noninvasive functional imaging modality could shed light on the spinal mechanisms involved in the generation of neuropathic pain, such as dysesthetic pain in patients with spinal cord injury or syringomyelia. In addition to understanding the effects of chronic pain on neuroplasticity of the spinal cord, spinal fMRI could provide insight into the potential mechanisms of medications and their efficacy at treating chronic pain.

4. Methods for fMRI Pain Experiments

4.1. Pain Assessment

A key issue in functional imaging of the cortical nociceptive signal is to ensure that the stimuli delivered to the subjects are perceived as noxious. Pain thresholds are commonly determined during a separate session prior to the scan. This procedure also serves to familiarize participants with the stimuli and reduce anxiety, thereby minimizing anxiety-related fluctuations in cardiovascular activity *(118)*. Stimuli utilized for the scanning session are frequently tailored to each individual's pain threshold; otherwise, all subjects can be administered the same level of noxious stimulation, which has been determined to evoke the perception of pain in all subjects. A corollary to the appropriate choice of noxious stimuli is the confirmation that predetermined levels of stimulation are actually perceived as painful, within the scanning environment. A number of contextual factors can alter the perception of stimuli that were originally considered painful during a prescanning test, including the temperature of the scanning suite, the position of the body in the scanner, and distractions of noise, possible feelings of claustrophobia, and other conditions specific to the scanning paradigm.

It is also important to note that the perception of pain can change during the course of a scanning session, due either to habituation *(119)*, sensitization, or the potential changes in attention during a long scanning experiment. To address this issue, pain assessment ratings can be obtained *during* the fMRI scanning session through subjective reports from participants using a variety of methods. Subjects can rate their perception after each stimulus, continuously during the stimuli, or at the end of the scanning run by giving an average rating of all the stimuli. Ratings can be obtained using electronic VAS scales, verbal, or simple manual reports.

In fMRI experiments, ratings can be obtained during or immediately after the presentation of each stimulus. Conversely, due to methodological issues, with PET studies pain ratings can be taken only at the end of a scanning session several minutes after stimulus presentation. Increased time between stimulus presentation and assessment can cause inaccuracies in subject responses *(120)*. This is a special consideration in studies examining mechanisms of analgesic relief since retrospective ratings can be inflated with increased time after stimulus presentation *(121, 122)*.

In addition to ensuring that the noxious stimuli are actually painful, pain assessment ratings (and other behavioral measures) can be used as regressors in the fMRI design matrix to aid in identifying cortical regions involved in various aspects of pain processing. Behavioral data can be incorporated into the fMRI design matrix as a weighting factor applied to the canonical HRF. Alternatively, continuous pain ratings (recorded during the stimulus presentations) can be modeled in the design matrix (e.g., *see* **ref.** *123)*. The resulting contrasts produce activation sites that are more closely based on the degree to which a region's activity correlates with the perceived intensity of the stimuli rather than with the physical intensity of the stimulus – in other words a "percept-related" activation as opposed to a "stimulus-related" activation *(124)*. This is an important consideration as it has been demonstrated that subjects' continuous ratings of brief (~35 s) thermal heat pain stimuli correlate well with the nociceptive BOLD signal *(123)*.

This experimental approach may have important implications for studying the dissociation that sometimes occurs between the intensity of peripheral stimulation and the perception of pain. For example, presentation of noxious mechanical stimuli over longer durations (~2 min) has been shown to disrupt the relationship between the firing frequency of nociceptive afferents and the perceived intensity of pain evoked by the stimuli *(125, 126)*. This paradoxical relationship may be explained by the process of temporal summation – a disproportionate increase in the firing rate of dorsal horn neurons over time, whereby their response threshold to sensory input is substantially lowered. Additionally, repeated exposure to short-duration heat pain stimuli can cause habituation to both the perceived intensity and unpleasantness of the stimuli *(127)*. Therefore, subjective pain ratings can play a key role in the interpretation of nociceptive processing in the cortex, as opposed to utilizing simply the duration or intensity of the noxious stimuli that may not aptly reflect the resulting activations. Only a few studies, however, have explored the possible cerebral mechanisms underlying habituation or sensitization to painful stimuli *(119, 128, 129)*, and these gave conflicting results concerning any specific association between cerebral activity and ratings of pain intensity. On the whole, however, these ambiguities in the correspondence between stimulus delivery, evoked nociceptive signal, and subjective reports of pain intensity, underscore the importance of accessing the level of perceived pain during scanning sessions, rather than assuming a fixed relationship between stimulation and percept.

A number of advantages and potential disadvantages are associated with obtaining continuous pain ratings of stimuli during fMRI experiments. Clearly, participants' perceptual evaluations will rely less on memory and will tend to be more accurate, compared with evaluations made after the scanning run. In turn, the resulting brain activation will be less reflective of mnemonic or error detection processes. Additionally, continuous ratings can be used to deduce the time lag between the application of the stimulus and the onset of pain perceived by the subject, and to provide further details about the time course of pain perception and the underlying neural activity.

While continuous ratings provide real-time information about a subject's perception of the stimuli, a clear disadvantage to their use is that the motor activity and motor-related activation can produce a confound that complicates interpretation of sensory-related activity. However, this can be accounted for by including the movements as covariates in the fMRI design matrix. This technique was utilized in a recent fMRI study that explored the impact of continuous rating on brain activity by presenting subjects with painful mechanical stimuli to one hand and requiring that they rate the intensity of every second stimulus using the opposite hand *(130)*; as a control, identical scans were performed utilizing innocuous mechanical stimuli. Interestingly, the BOLD signal in somatosensory regions was found to be heavily dependent on the rating of the stimuli, with *t*-values more than doubled for rated stimuli compared to unrated stimuli during both the noxious and innocuous scans. The authors note that the enhanced activity was likely dependent on the greater attention paid to the rated stimuli. On the contrary, virtually no differences were seen between the levels of activity evoked by the two intensities of mechanical stimulation - noxious and innocuous, a finding that is contrary to those of previous neuroimaging studies, which had shown intensity-dependent activation in response to noxious and innocuous stimuli *(37, 102, 131, 146)*. Furthermore, the majority of brain regions activated during the pain task were correlated with the movements associated with the continuous ratings, which the authors attributed to motor planning and attentional effects.

4.2. Statistical Techniques

4.2.1. Conjunction Analysis

Conjunction analysis permits the identification of brain regions commonly activated during separate trial epochs *(132)*. For example, to determine whether brain regions responsible for the perception of pain were activated during an anticipation phase, a conjunction analysis was performed on these two time periods *(133)*. Resulting activations from the conjunction analysis, observed in the PAG, ACC, thalamus, and premotor cortex, suggest the importance of these areas in the anticipation of noxious stimuli and their potential role in the subsequent modulation of pain-related activations within these same areas.

Conjunction analysis has also been applied to understand how pain modulates the cognitive processing of concurrent sensory stimuli. Outside of a controlled experimental setting, acute, or chronic pain is frequently experienced in the presence of competing stimuli. Bingel et al. *(134)* addressed this question by presenting painful stimuli during a visual working memory task and an object visibility task. Activation common to these different tasks was reported in the bilateral lateral occipital cortex, a region previously shown to be modulated by the amount of information processed in working memory *(135)*. Results of this conjunction analysis provide insight into the mechanisms responsible for modulating visual input in the presence of an aversive stimulus.

4.2.2. Connectivity Analysis

In addition to identifying brain activation associated with the different stages involved in the processing of pain-related information, it is also important to understand how these different brain regions interact. Advances in multivariate analysis techniques now allow for a noninvasive examination of relationships between coactivated brain regions to understand how these networks covary during the processing of painful stimuli *(136)*.

Analysis of functional connectivity examines patterns of coactivating brain regions but makes no assumptions about their interrelated anatomical connectivity. The basis of this analysis assumes that brain regions with similar or covarying time courses are likely to interact and are therefore functionally connected during a particular task. Valet et al. *(137)* used a functional connectivity analysis to examine the relationship among regions involved in modulating pain perception during a distracting cognitive task. Results showed that both pain ratings and pain-related activations in medial pain processing regions were lower during the distracter task. However, the distraction period was associated with increased activations of prefrontal, PAG, and thalamic areas. The time course of the significantly activated voxels in the perigenual ACC and orbitofrontal cortex during pain, with and without distraction, was extracted and included as regressors in the design matrix. Applying this time course in the general linear model (GLM) identified other brain regions showing similar patterns of brain activations *(136, 138)*. Activations of the PAG and thalamus were found to significantly covary with that in the cingolo-frontal cortex during pain accompanied by distraction. These results suggest that distraction may reduce pain through activation of prefrontal regions, which trigger descending inhibitory controls via the PAG and thalamus.

One study also applied an analysis of functional connectivity using a partial least square computation *(139)* to assess how pain perception modulates a network of cortical regions involved in a cognitively demanding task *(140)*. Acute phasic pain was found to significantly enhance brain activity in several cortical regions, which are involved in processes related to the cognitive task. Such findings provide insight into the possible mechanisms underlying the detrimental effects of chronic pain on cognitive tasks that demand a high level of attention *(141, 142)*. The protective function of nociceptive processes may require focused attention on pain perception, thus engaging the network of interacting cortical regions involved in general attention. The increased activity of this network during chronic pain states may supersede attentional demands of cognition resulting in apparent deficits in the performance of cognitive tasks.

Another approach that compliments the study of functional connectivity is that of "effective connectivity," which describes the causal relationships that one region exerts on another *(143)*. Additionally, psychological or psychophysical data can be modeled into an analysis of effective connectivity (thus, referred to as a psycho-physiological interaction analysis) in order to measure the influence of one cortical region on another based on the experimental context or the behavioral state of the subject *(138, 144)*. Such an analysis was applied by Bingel et al. *(134)* – as an extension of their findings in the lateral occipital cortex – to examine which brain regions were involved in modulating the activity in the lateral occipital cortex during an object-visibility task. A seed region was placed in the ACC, as this region showed increased activation on high pain intensity compared with low pain intensity trial periods. The time course of the activation in the ACC was extracted and used as the physiological variable, while the degree of visibility of the objects was used as the psychological variable. These variables were then implemented as regressors in the fMRI design matrix. Results showed that modulatory activity in the lateral occipital cortex by pain was driven by the ACC – a finding that is consistent with known anatomical connectivity.

Connectivity analyses of pain processing can be complemented through the use of in vivo mapping of white matter fiber pathways using diffusion tensor (DT) MRI. A recent DT-MRI study examined the role of cortical connectivity in the modulation of pain by the PAG and nucleus cuneiformis *(145)*, areas that previous neuroimaging studies have implicated in pain processing *(105)*. Focused attention on a noxious stimulus has been shown to increase brain activity in PFC, ACC, and thalamus *(146)*. However, during distraction, activations in insular cortex, ACC, and thalamus were found to decrease while increased activity was reported in PAG *(147, 148)*. Furthermore, results from fMRI indicate significant interactions between the PFC and brainstem structures during pain modulation *(98, 137)*. Results from DT-MRI showed separate pathways for the PAG and nucleus cuneiformis connecting with the PFC, amygdala, thalamus, hypothalamus, and the RVM. Interestingly, no correlation was found between the PAG and the ACC, in spite of previous results from fMRI studies indicating a strong correlation between the activities of the two regions during pain modulation *(137)*. These findings highlight the importance of combining emerging noninvasive imaging techniques to deepen our understanding of pain processing and its modulation.

5. fMRI and the Study of Higher Cognitive Pain Processing

5.1. Pain Modulation

fMRI is a useful tool for examining cerebral mechanisms of pain modulation, whereby subjects experience either analgesia or hyperalgesia – a decrease or increase in perceived pain, respectively. Pain modulation can occur through both endogenous mechanisms and as a result of exogenously administered agents. One final common pathway for analgesic mechanisms is believed to be through the release of endogeneous opioids *(149)* acting on sites in the brainstem and midbrain that block the nociceptive signal through their descending pathways; the final effects of this descending modulation are exerted either on the spinal cord and/or at the site of peripheral nerves that transmit the nociceptive stimuli. Additionally, recent research has implicated endocannibinoids in pain modulation, which may act on similar descending pathways *(150)*. fMRI has been applied to study the initial factors triggering these modulatory processes either through endogenous mechanisms utilizing cognitive strategies, such as attention *(147, 151)*, hypnosis *(152–154)*, and placebo analgesia *(155)*, or through exogenous agents, such as pharmacological interventions *(156, 157)*.

While several experimental protocols have been applied using radio-ligands and PET to study neurochemical mechanisms involved in pain modulation – such as in studies of placebo analgesia *(158, 159)*, many of the characteristics of fMRI contribute toward its potential to address questions in pain modulation, as has been suggested in several sections above. First, fMRI offers greater spatial and temporal resolution than PET *(160)*. Thus, fMRI is more suited to accurately localize small brain regions involved in pain modulation, such as the RVM or PAG *(148, 161)*, and is better able to assess the time course of activations of those regions. fMRI is also well suited to study procedures that evoke changes in pain perception since it accommodates the use of parametric data, whereby experimental parameters such as pain ratings (intensity, expectation, unpleasantness) can be correlated with brain activations and thus used to characterize cortical structures according to their response profile to various experimental parameters. fMRI also has the advantage of allowing a larger number of scans within a single session *(112)* and a larger number of experimental conditions during a single experimental paradigm, as opposed to PET studies, which limit the number of measurements that can be taken in order to minimize exposure to radiation. As a corollary of increased temporal resolution, a major advantage of using fMRI to study pain modulation is the possibility of utilizing event-related designs whereby the time course of brain activations over different phases of the modulation period can be studied – the anticipation of the noxious stimulus, the onset of pain perception, changes in pain perception over time, and poststimulus ratings. Anticipation of the painful stimulus is a crucial phase of the pain modulation process, since at this time point neural mechanisms act on descending modulatory systems to diminish or enhance the response to the stimulus *(162)*. Rapid event-related fMRI designs also permit a short stimulus-delivery phase (on the order of seconds for thermal stimuli or milliseconds for laser and electrical stimuli), which can have several advantages for a number of different experimental designs. Namely, short-duration stimuli avoid or minimize sensitization of the skin that may occur with the much longer stimulus presentations that are required in PET studies. Additionally, short-duration stimuli minimize the potential attenuation of the BOLD response to noxious stimuli *(163)* or the reduction in pain sensitivity and activation of antinociceptive responses *(149, 164)* that may be evoked by long-duration tonic stimulation.

5.2. Pain Empathy

Inherent to processing the emotional component of pain is the ability to understand the emotional reactions of other people who are experiencing pain – i.e., pain empathy *(165)*. This rapidly growing field of empathy research is directed toward studying the mental representation of pain – both that which is perceived to be experienced by others, as well as that which is perceived as one's own. Several different types of experimental stimuli implicating other people in pain have been used in these fMRI paradigms, including photographic images *(166–170)*, or short animations *(171)* of body parts in potentially tissue-damaging situations, viewing the faces of actors evoking facial expressions of pain *(172, 173)*, or subjects actually receiving painful stimuli *(174)*, or those of chronic pain patients *(175)*, or being cued that a loved one in the room was receiving painful stimuli *(176)*. A common finding from these studies is that the processing of pain in others recruits brain regions involved in affective processing – namely the ACC and insula.

In a recent meta-analysis, Jackson et al. *(177)* compiled brain activation coordinates from ten studies examining neural correlates of viewing pain in others and compared them with data from ten studies in which pain was evoked in healthy volunteers. Distinct activation was noted in BA24 in response to pain of the self; however, viewing the pain of others primarily produced activity more anteriorly in the perigenual (BA24/33) and subcallosal (BA32/25) regions. A similar pattern was observed in the left and right insula whereby pain of the self was associated with activation in the mid to posterior, dorsal insula, while processing pain in others was more anterior.

While the majority of imaging studies have not reported modulation of sensory-discriminative regions associated with pain empathy, emerging evidence, from the use of transcranial magnetic stimulation (TMS), suggests a somatotopic specificity in the perceived pain of others. In two separate TMS studies, Avenanti et al. *(178, 179)* reported reduced motor-evoked potentials – a reflection of corticospinal activity – in muscles that were homologous to those of other subjects being targeted by painful stimuli, as seen by the viewer. Furthermore, these reductions were correlated with the viewer's subjective ratings of the pain intensity implied by the noxious stimulation, but not with ratings of pain unpleasantness. Contrary to these findings are those of an fMRI study that used a similar experimental protocol; Morrison et al. *(180)* administered painful pinprick stimuli to the fingertips of subjects and then later showed images of others receiving the same stimuli. The authors did not report any changes in somatosensory or motor cortices. The lack of concordance between TMS and fMRI studies may reflect subtle differences in the types of tasks, or changes occurring in the sensorimotor system that are below statistical significance thresholds in the fMRI analyses *(181)*.

fMRI has provided considerable insight into the neural mechanisms of processing pain in others, and suggests a number of interesting clinical implications. Since pain is a sensory and emotional phenomenon that is primarily experienced by the patient – as opposed to an easily measured sign of illness, such as fever or weight loss, for example – health-care professionals who are confronted with patients in pain must be able to infer their discomfort accurately and treat them accordingly. Further understanding of the neural mechanisms underlying how we interpret pain in others is an initial step toward how these neural circuits can change – depending on the clinical context or after years of repeated exposure to those in pain.

6. Future of Pain Imaging

6.1. Increased Sensitivity

The last decade has seen a considerable improvement in the sensitivity of fMRI in both the spatial and temporal localization of regions of activation, and this trend is expected to continue. The shift to higher field strengths of 4.0 and 7.0 T scanners has been shown to significantly enhance the SNR, compared to that observed with the 1.5–3.0 T scanners *(182)*, which have been used in most pain studies. Several imaging centers have begun human fMRI studies at 7 T; however, none to date have applied it to studying pain processing. Pain imaging is poised to benefit from these advances more than other disciplines, because – unlike visual or motor tasks, for example, which produce changes in cerebral blood flow (CBF) on the order of ~40% *(183–186)* – BOLD responses to nociceptive stimuli produce signal changes only in the range of ~5% *(36, 38, 40)*. Improved spatial localization of fMRI pain protocols would provide better information regarding the specificity of somatosensory regions involved in noxious processing and their somatotopic organization; likewise, improved spatial localization and SNR will aid greatly to investigations of small brainstem structures that have been implicated in modulating pain processing at both spinal and supra-spinal levels.

Another burgeoning field in pain imaging is that of arterial spin labeling (ASL) perfusion MRI, which was first described more than a decade ago *(187)*. ASL directly measures CBF by magnetically labeling water molecules in inflowing arteries. Recent application of ASL to study experimental pain in healthy subjects *(188)* has shown that this technique offers several advantages. ASL gives a precise localization of neuronal structures and has demonstrated great interindividual reliability of activation. Additionally, compared with BOLD fMRI, ASL is well suited for pain imaging studies, since it is less susceptible to signal loss and image distortions *(189)* due to magnetic field inhomogeneities at the air–tissue interface around frontal, medial, and inferior temporal lobes *(190–192)*. Although several methods are available to reduce these susceptibility artifacts in the BOLD signal, ASL is nevertheless an attractive alternative for pain studies that target the limbic system where signal loss from susceptibility artifacts is troublesome for such regions as the orbitofrontal cortex and amygdala. ASL also has the additional advantage of permitting longer acquisition times and is thus well suited for studying neuronal processing that may take longer to develop, such as pain modulation through hypnotic induction; fMRI, on the contrary, is limited in terms of its length of acquisition due to drifts in the baseline. However, ASL is limited in that it cannot detect changes occurring faster than 30 s and is therefore not suited for event-related designs. Additionally, the technique is limited by its temporal resolution and slice coverage in which whole brain imaging is not possible using current methods. These issues should be resolved with advances made in fast echo planar imaging sequences.

6.2. Meta-Analysis of Functional Neuroimaging Data

Recent advances in computational techniques have led to more sophisticated tools that can be used for performing meta-analyses on existing brain imaging data. Interpreting the results from individual fMRI studies is limited by factors such as head motion artifacts, small sample size, intersubject variability, low SNR, and reporting false positives. Methods to perform meta-analyses of functional neuroimaging data include region- or labeled-based models *(193)*, the spatial density method *(194)*, and the generation of activation likelihood estimate (ALE) maps *(195)*. The ALE method is proving to be especially useful in that it is automated; it allows for a more precise measurement of both the localization and concordance of peak activation sites across studies; it also permits the generation of significance thresholds based on permutation analysis of randomly generated coordinates *(196)*. Using the ALE method to assess possible differences in imaging techniques for this review chapter, we performed a meta-analysis on 30 fMRI and 30 PET studies of noxious stimuli applied to both arms (**Fig. 2**). Results showed that the probability maps generated for the fMRI and PET studies indicated considerable overlap in a number of cortical regions including SI, SII, ACC, insula, PFC, thalamus, midbrain, and cerebellum. However, fMRI studies appear to have a wider distribution of probabilistic values in cortical regions, namely SI, suggesting an enhanced sensitivity of the technique, compared with that of PET. Findings may be due to the fact that fMRI allows for longer scans and a greater repetition of scans within the same session compared to PET, which both contribute to an increased statistical power. However, at a subcortical level, probabilistic values for fMRI data were localized within the territory of the sensorimotor thalamus, while PET probability values (although similar in magnitude to those of fMRI) were more widely distributed across the thalamus. These findings indicate that the likelihood of detecting pain-related activation in the thalamus using either of the two brain imaging modalities is similar, but that the increased spatial resolution of fMRI may allow a better localization of small nuclei within the thalamus. In summary, the meta-analysis of imaging data collected across many studies provides insights and information that may not be obtained from individual studies, no matter how carefully they were designed and executed.

6.3. Combining fMRI with Morphometry

Recent advances in the analysis of anatomical images derived from MR scanners have led to new research strategies that are expanding the concept of "functional" MRI. Whereas the BOLD signal of an fMRI study is a reliable marker of the current short-term function of an activated region, measures of anatomical variability may hold clues to particular aspects of the long-term function of that region. Our growing understanding of processes like learning and memory, and their influences on neuronal plasticity, leads to a measurable corollary of long-term function – specifically, changes in the anatomical features of specific areas of the brain. Just as the power of fMRI lies in the correlation of the BOLD signal with stimulus, motor tasks, or cognitive events, likewise the potential utility of quantifying macro-anatomical changes in the brain – a morphometric analysis – lies in the correlation of these changes in anatomical structure with the subjects' history of stimulus exposure, practice with motor tasks, and characteristics

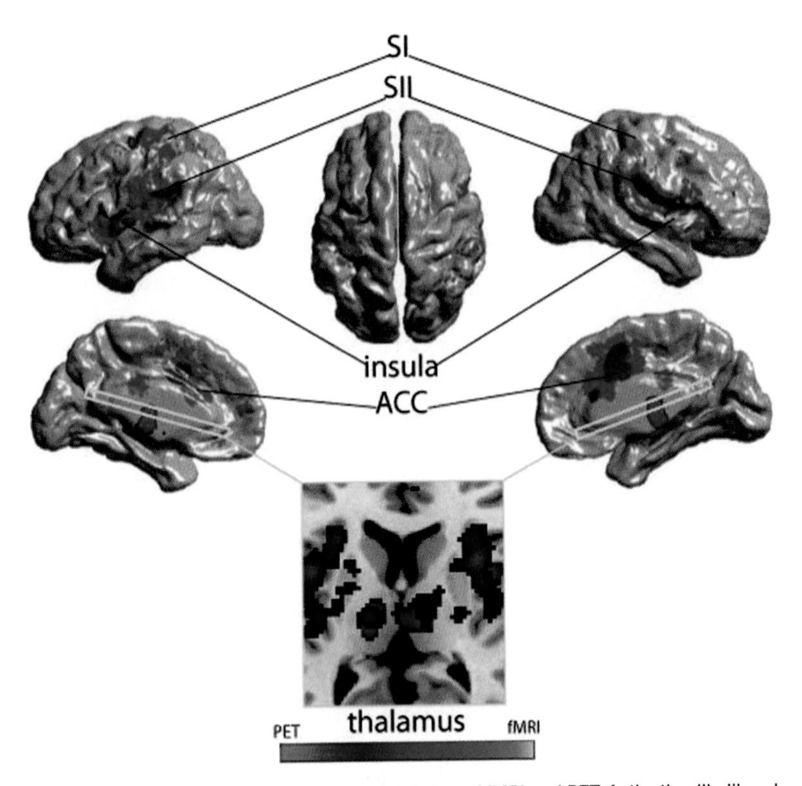

Fig. 2. Comparison of the sensitivity and resolution of fMRI and PET. Activation likelihood estimation (ALE) maps produced from a meta-analysis from 30 fMRI studies compared to an equal number of PET studies whereby noxious stimuli were applied to the arms. Journal articles were selected through an initial Medline search using the key words "fMRI," "PET," "pain," and "experimental." Experiments were then screened to ensure that the noxious stimuli were applied to left or right arms in healthy volunteers. Of the resulting studies, comparable types of stimuli and locations across studies were selected for the final selection. Stimuli included thermal (radiant, contact heat, and cold), electric shock, pressure, impact, injection of capsaicin, and ethanol. Relevant information related to the studies was recorded including imaging modality, size of the blurring kernel, year, subject number, pain stimulus, stimulus location, and activation coordinates. All coordinates points were recorded and converted to a standardized stereotactic space ***(220)***. This yielded 590 foci for fMRI and 554 foci for PET studies. Using the application GingerALE (www.brainmap.org), the data were subjected to a quantitative voxel-level meta-analysis that produced ALE maps for the fMRI and PET activation coordinates ***(195)***. Coordinates were smoothed by 8 mm and then thresholded based on a permutation test ($N = 1000$) and a false discovery rate (FDR) correction of $q = 0.05$. The resulting maps are displayed on an average cortical surface from healthy volunteers registered in Montreal Neurological Institute (MNI) space using SurfStat (http://www.math.mcgill.ca/keith/surfstat/). fMRI studies (shown in *blue*) yielded highest probabilistic values in bilateral secondary somatosensory cortex (SII) (*right*: $p = 0.85$; *left*: $p = 0.07$), anterior cingulate cortex (ACC) (*right*: $p = 0.06$; *left*: $p = 0.08$), insula (*right*: $p = 0.07$; *left*: $p = 0.06$), primary somatosensory cortex (SI) (*right*: $p = 0.01$; *left*: $p = 0.04$), thalamus (*right*: $p = 0.04$; *left*: $p = 0.04$), prefrontal cortices (PFC) (*right*: $p = 0.04$; *left*: $p = 0.03$), left primary motor cortex (MI) ($p = 0.04$), right midbrain ($p = 0.02$) and cerebellum ($p = 0.02$). PET studies (shown in *red*) showed high probabilistic values in similar regions with the largest values in bilateral ACC (*right*: $p = 0.06$; *left*: $p = 0.05$), insula (*right*: $p = 0.06$; *left*: $p = 0.04$), SII (*right*: $p = 0.05$; *left*: $p = 0.05$), thalamus (*right*: $p = 0.04$; *left*: $p = 0.04$), PFC (*right*: $p = 0.03$; *left*: $p = 0.06$), SI (*right*: $p = 0.02$; *left*: $p = 0.02$), and the right cerebellum ($p = 0.06$), right MI ($p = 0.03$), and left periaqueductal gray ($p = 0.02$).

of their personality that may be associated with certain cognitive traits.

Examples of morphometric analyses include voxel-based morphometry (VBM) *(197)* and cortical thickness analysis *(198)*. VBM examines changes in gray or white matter density across the entire brain while cortical thickness analysis measures the surface gray matter. It is generally assumed that at least in disease states decreases in gray matter density and cortical thinning are related to neuronal loss *(199–201)*.

For the study of nociceptive processing and pain perception, MRI-based morphometric analyses can be used to examine neuroanatomical changes that are correlated with particular chronic pain states or to examine differences in the anatomy of specific brain regions that may underlie the variability in pain perception that is observed within a population in healthy volunteers.

A number of recent studies have reported changes in cortical and subcortical brain regions in individuals with chronic pain *(202–205)*. In a recent study examining structural changes in patients with irritable bowel syndrome (IBS), VBM analysis revealed decreases in gray matter density in the anterior medial thalamus *(203)*. This analysis was complemented by cortical thickness analysis and demonstrated cortical thinning in right ACC and bilateral insula. The same group showed in an fMRI study reduced activation in the ACC and anterior insula in response to rectal pain in IBS patients *(206)*. These comparative findings provide a neuroanatomical basis for reduced cortical activity strengthening the importance of relating anatomical structure to physiological function.

To date, a few studies have combined morphometric and functional neuroimaging analysis. However the future of pain fMRI lies in the development of complimentary brain imaging analysis techniques to improve our understanding of pain processing.

6.4. fMRI as a Therapy for Chronic Pain

Recent improvements in the speed of analysis of fMRI data have led to the possibility that "real-time" fMRI (rt-fMRI) can be developed as a potential "therapy" for chronic pain patients. In principle, if patients can be given feedback regarding the level of activity in specific areas of the brain that are associated with the perception of pain or its unpleasantness, then learning to (self)-regulate this activity can allow them to control their own chronic pain – in much the same way as neurosurgeons attempt to control a patient's pain by stimulating a specific area of the brain or by placing a lesion in a targeted area. Self-regulation training with EEG has provided the basis for much of the neurofeedback research; however, due to several methodological limitations EEG offers relatively poor spatial specificity within the brain *(207, 208)*. By contrast, fMRI offers superior spatial resolution, especially for deeper brain regions, and is more suitable for targeting activity in a small, localized brain region *(209, 210)*.

Neurofeedback, using real-time analysis of fMRI data, was initially developed by Cox et al. *(211)*, and several groups have used this technology to study learned control over brain activity during a number of tasks *(212–216)*. Recently, the use of rt-fMRI has been applied to several clinical conditions whose etiology or symptoms might be linked to abnormal activity in known areas of the brain. In one study testing the feasibility of rt-fMRI as a neuroimaging therapy for chronic pain patients *(8)*, normal subjects receiving experimental noxious stimuli were trained to control activity in a targeted region within the ACC – an area previously shown to be strongly associated with the perception of pain unpleasantness *(94)*. Results demonstrated that these subjects were able to use the feedback provided by rt-fMRI to either increase or decrease, on command, ACC activity, and that the level of this activity correlated with their estimates of pain evoked by the experimental stimuli. Likewise, a small cohort of chronic pain patients, following a similar rt-fMRI training paradigm, reported a significant reduction in their level of chronic pain in comparison to that of a control patient group, which received feedback training based only on autonomic measures. Furthermore, the patients in the rt-fMRI group demonstrated a direct correlation between their ability to control ACC activation and their degree of pain reduction. In the future, rt-fMRI could also be applied to modify cortical hyperactivity that has been described for a number of other pain syndromes *(217–219)*.

7. Conclusion

The experience of pain is complex: both sensory and cognitive components depend on a network of neural processing spread throughout many cortical and subcortical regions of the CNS. The advent of noninvasive imaging techniques has allowed us to gain a deep understanding of this multifaceted phenomenon in humans – the experimental preparation that is most relevant to our ultimate goal of understanding, managing, and alleviating pain in patients. Pain is a characteristic common to many diseases and injuries, a consequence of many medical and dental procedures, and chronic pain is essentially a syndrome in its own right – an insufferable sensation that many times has no obvious stimulus. Pain has an enormous impact on society: it costs billions of US dollars annually due to losses in productivity, and strains the health-care systems across the world. fMRI in human subjects is helping us to understand the cerebral mechanisms of pain processing and the modulation of pain by both endogenous and exogenous factors. The results of these studies are making substantial contributions to the development of efficacious interventions for treating and alleviating pain.

References

1. Penfield W, Bouldrey E. Somatic motor and sensory representation in the cerebral cortex of man as studied by electrical stimulation. Brain 1937;60:389–443.
2. Head H, Holmes G. Sensory disturbances from cerebral lesions. Brain 1911;34:102.
3. Talbot JD, Marrett S, Evans AC, Meyer E, Bushnell MC, Duncan GH. Multiple representations of pain in human cerebral cortex. Science 1991;251:1355–1358.
4. Jones AK, Brown WD, Friston KJ, Qi LY, Frackowiak RS. Cortical and subcortical localization of response to pain in man using positron emission tomography. Proc Biol Sci 1991;244:39–44.
5. Apkarian AV, Stea RA, Manglos SH, Szeverenyi NM, King RB, Thomas FD. Persistent pain inhibits contralateral somatosensory cortical activity in humans. Neurosci Lett 1992;140:141–147.
6. Davis KD, Wood ML, Crawley AP, Mikulis DJ. fMRI of human somatosensory and cingulate cortex during painful electrical nerve stimulation. Neuroreport 1995;7: 321–325.
7. Flor H. The functional organization of the brain in chronic pain. Prog Brain Res 2000;129:313–322.
8. deCharms RC, Maeda F, Glover GH, Ludlow D, Pauly JM, Soneji D et al. Control over brain activation and pain learned by using real-time functional MRI. Proc Natl Acad Sci U S A 2005;102(51):18626–18631.
9. Willis WD, Jr. The pain system. The neural basis of nociceptive transmission in the mammalian nervous system. Pain Headache 1985;8:1–346.
10. Adriaensen H, Gybels J, Handwerker HO, Van Hees J. Response properties of thin myelinated (A-delta) fibers in human skin nerves. J Neurophysiol 1983;49:111–122.
11. Ochoa J, Torebjork E. Sensations evoked by intraneural microstimulation of C nociceptor fibres in human skin nerves. J Physiol 1989;415:583–599.
12. Cervero F, Iggo A. The substantia gelatinosa of the spinal cord: a critical review. Brain 1980;103:717–772.
13. Wilson P, Kitchener PD. Plasticity of cutaneous primary afferent projections to the spinal dorsal horn. Prog Neurobiol 1996;48: 105–129.
14. Boivie J. An anatomical reinvestigation of the termination of the spinothalamic tract in the monkey. J Comp Neurol 1979;186:343–369.
15. Craig AD, Bushnell MC, Zhang ET, Blomqvist A. A thalamic nucleus specific for pain and temperature sensation. Nature 1994;372:770–773.
16. Craig AD. Distribution of trigeminothalamic and spinothalamic lamina I terminations in the macaque monkey. J Comp Neurol 2004;477:119–148.
17. Ma W, Peschanski M, Ralston HJ, III. Fine structure of the spinothalamic projections to the central lateral nucleus of the rat thalamus. Brain Res 1987;414:187–191.
18. Applebaum AE, Leonard RB, Kenshalo DR, Jr., Martin RF, Willis WD. Nuclei in which functionally identified spinothalamic tract neurons terminate. J Comp Neurol 1979;188:575–585.
19. Graziano A, Jones EG. Widespread thalamic terminations of fibers arising in the superficial medullary dorsal horn of monkeys and their relation to calbindin immunoreactivity. J Neurosci 2004;24:248–256.
20. Bromm B, Treede RD. Nerve fibre discharges, cerebral potentials and sensations induced by CO2 laser stimulation. Hum Neurobiol 1984;3:33–40.
21. Carmon A, Dotan Y, Sarne Y. Correlation of subjective pain experience with cerebral evoked responses to noxious thermal stimulations. Exp Brain Res 1978;33:445–453.
22. Iannetti GD, Leandri M, Truini A, Zambreanu L, Cruccu G, Tracey I. A[delta] nociceptor response to laser stimuli: selective effect of stimulus duration on skin temperature, brain potentials and pain perception. Clin Neurophysiol 2004;115:2629–2637.
23. Spiegel J, Hansen C, Treede R-D. Clinical evaluation criteria for the assessment of impaired pain sensitivity by thulium-laser evoked potentials. Clin Neurophysiol 2000; 111:725–735.

24. Leandri M, Saturno M, Spadavecchia L, Iannetti GD, Cruccu G, Truini A. Measurement of skin temperature after infrared laser stimulation. Neurophysiologie Clinique/Clin Neurophysiol 2006;36:207–218.
25. Helmchen C, Mohr C, Roehl M, Bingel U, Lorenz J, Buchel C. Common neural systems for contact heat and laser pain stimulation reveal higher-level pain processing. Hum Brain Mapp 2007.
26. Apkarian AV, Bushnell MC, Treede RD, Zubieta JK. Human brain mechanisms of pain perception and regulation in health and disease. Eur J Pain 2005;9:463–484.
27. Melzack R, Casey KL. Sensory, motivational and central control determinants of pain: a new conceptual model. In: Kenshalo DR, editor. The skin senses. Springfield IL: Thomas, 1968: 423–443.
28. Price DD, Bush FM, Long S, Harkins SW. A comparison of pain measurement characteristics of mechanical visual analogue and simple numerical rating scales. Pain 1994;56:217–226.
29. Kaas JH, Nelson RJ, Sur M, Lin CS, Merzenich MM. Multiple representations of the body within the primary somatosensory cortex of primates. Science 1979;204:521–523.
30. Kenshalo DR, Jr., Isensee O. Responses of primate SI cortical neurons to noxious stimuli. J Neurophysiol 1983;50:1479–1496.
31. Jones EG, Burton H. Areal differences in the laminar distribution of thalamic afferents in cortical fields of the insular, parietal and temporal regions of primates. J Comp Neurol 1976;168(2):197–247.
32. Kenshalo DR, Jr., Giesler GJ, Jr., Leonard RB, Willis WD. Responses of neurons in primate ventral posterior lateral nucleus to noxious stimuli. J Neurophysiol 1980;43(6):1594–1614.
33. Jones EG, Friedman DP. Projection pattern of functional components of thalamic ventrobasal complex on monkey somatosensory cortex. J Neurophysiol 1982;48:521–544.
34. Jones EG, Leavitt RY. Retrograde axonal transport and the demonstration of nonspecific projections to the cerebral cortex and striatum from thalamic intralaminar nuclei in the rat, cat and monkey. J Comp Neurol 1974;154:349–377.
35. Rausell E, Jones EG. Chemically distinct compartments of the thalamic VPM nucleus in monkeys relay principal and spinal trigeminal pathways to different layers of the somatosensory cortex. J Neurosci 1991;11:226–237.
36. Casey KL, Minoshima S, Berger KL, Koeppe RA, Morrow TJ, Frey KA. Positron emission tomographic analysis of cerebral structures activated specifically by repetitive noxious heat stimuli. J Neurophysiol 1994;71:802–807.
37. Casey KL, Minoshima S, Morrow TJ, Koeppe RA. Comparison of human cerebral activation pattern during cutaneous warmth, heat pain, and deep cold pain. J Neurophysiol 1996;76:571–581.
38. Coghill RC, Talbot JD, Evans AC, Meyer E, Gjedde A, Bushnell MC et al. Distributed processing of pain and vibration by the human brain. J Neurosci 1994;14:4095–4108.
39. Gelnar PA, Krauss BR, Szeverenyi NM, Apkarian AV. Fingertip representation in the human somatosensory cortex: an fMRI study. Neuroimage 1998;7:261–283.
40. Derbyshire SW, Jones AK. Cerebral responses to a continual tonic pain stimulus measured using positron emission tomography. Pain 1998;76:127–135.
41. Disbrow E, Buonocore M, Antognini J, Carstens E, Rowley HA. Somatosensory cortex: a comparison of the response to noxious thermal, mechanical, and electrical stimuli using functional magnetic resonance imaging. Hum Brain Mapp 1998;6:150–159.
42. Tommerdahl M, Delemos KA, Vierck CJ, Jr., Favorov OV, Whitsel BL. Anterior parietal cortical response to tactile and skin-heating stimuli applied to the same skin site. J Neurophysiol 1996;75:2662–2670.
43. Yen CT, Shaw FZ. Reticular thalamic responses to nociceptive inputs in anesthetized rats. Brain Res 2003;968:179–191.
44. Bushnell MC, Duncan GH, Hofbauer RK, Ha B, Chen JI, Carrier B. Pain perception: is there a role for primary somatosensory cortex? Proc Natl Acad Sci U S A 1999;96: 7705–7709.
45. Seminowicz DA, Mikulis DJ, Davis KD. Cognitive modulation of pain-related brain responses depends on behavioral strategy. Pain 2004;112:48–58.
46. Dunckley P, Aziz Q, Wise RG, Brooks J, Tracey I, Chang L. Attentional modulation of visceral and somatic pain. Neurogastroenterol Motil 2007;19:569–577.
47. Oshiro Y, Quevedo AS, McHaffie JG, Kraft RA, Coghill RC. Brain mechanisms supporting spatial discrimination of pain. J Neurosci 2007;27:3388–3394.
48. Tarkka IM, Treede RD. Equivalent electrical source analysis of pain-related somatosensory evoked potentials elicited by a CO2 laser. J Clin Neurophysiol 1993;10:513–519.
49. Andersson JL, Lilja A, Hartvig P, Langstrom B, Gordh T, Handwerker H et al. Somatotopic

organization along the central sulcus, for pain localization in humans, as revealed by positron emission tomography. Exp Brain Res 1997;117:192–199.
50. DaSilva AF, Becerra L, Makris N, Strassman AM, Gonzalez RG, Geatrakis N et al. Somatotopic activation in the human trigeminal pain pathway. J Neurosci 2002;22:8183–8192.
51. Ogino Y, Nemoto H, Goto F. Somatotopy in human primary somatosensory cortex in pain system. Anesthesiol 2005;103:821–827.
52. Kaas JH. What, if anything, is SI? Organization of first somatosensory area of cortex. Physiol Rev 1983;63:206–231.
53. Albanese MC, Duerden EG, Rainville P, Duncan GH. Memory traces of pain in human cortex. J Neurosci 2007;27:4612–4620.
54. Friedman DP, Murray EA. Thalamic connectivity of the second somatosensory area and neighboring somatosensory fields of the lateral sulcus of the macaque. J Comp Neurol 1986;252:348–373.
55. Apkarian AV, Shi T. Squirrel monkey lateral thalamus. I. Somatic nociresponsive neurons and their relation to spinothalamic terminals. J Neurosci 1994;14:6779–6795.
56. Dong WK, Salonen LD, Kawakami Y, Shiwaku T, Kaukoranta EM, Martin RF. Nociceptive responses of trigeminal neurons in SII-7b cortex of awake monkeys. Brain Res 1989;484:314–324.
57. Dong WK, Chudler EH, Sugiyama K, Roberts VJ, Hayashi T. Somatosensory, multisensory, and task-related neurons in cortical area 7b (PF) of unanesthetized monkeys. J Neurophysiol 1994;72:542–564.
58. Robinson CJ, Burton H. Somatic submodality distribution within the second somatosensory (SII), 7b, retroinsular, postauditory, and granular insular cortical areas of M. fascicularis. J Comp Neurol 1980;192:93–108.
59. Greenspan JD, Lee RR, Lenz FA. Pain sensitivity alterations as a function of lesion location in the parasylvian cortex. Pain 1999;81:273–282.
60. Ploner M, Freund HJ, Schnitzler A. Pain affect without pain sensation in a patient with a postcentral lesion. Pain 1999;81:211–214.
61. Kitamura Y, Kakigi R, Hoshiyama M, Koyama S, Shimojo M, Watanabe S. Pain related somatosensory evoked magnetic fields. Electroencephalogr Clin Neurophysiol 1995;95:463–474.
62. Valeriani M, Le Pera D, Niddam D, Arendt-Nielsen L, Chen AC. Dipolar source modeling of somatosensory evoked potentials to painful and nonpainful median nerve stimulation. Muscle Nerve 2000;23:1194–1203.
63. Opsommer E, Weiss T, Plaghki L, Miltner WH. Dipole analysis of ultralate (C-fibres) evoked potentials after laser stimulation of tiny cutaneous surface areas in humans. Neurosci Lett 2001;298:41–44.
64. Coghill RC, Sang CN, Maisog JM, Iadarola MJ. Pain intensity processing within the human brain: a bilateral, distributed mechanism. J Neurophysiol 1999;82:1934–1943.
65. Maihofner C, Herzner B, Otto Handwerker H. Secondary somatosensory cortex is important for the sensory-discriminative dimension of pain: a functional MRI study. Eur J Neurosci 2006;23:1377–1383.
66. Gracely RH, Geisser ME, Giesecke T, Grant MAB, Petzke F, Williams DA et al. Pain catastrophizing and neural responses to pain among persons with fibromyalgia. Brain 2004;127:835–843.
67. Sawamoto N, Honda M, Okada T, Hanakawa T, Kanda M, Fukuyama H et al. Expectation of pain enhances responses to nonpainful somatosensory stimulation in the anterior cingulate cortex and parietal operculum/posterior insula: an event-related functional magnetic resonance imaging study. J Neurosci 2000;20:7438–7445.
68. Burton H, Jones EG. The posterior thalamic region and its cortical projection in New World and Old World monkeys. J Comp Neurol 1976;168:249–301.
69. Friedman DP, Jones EG, Burton H. Representation pattern in the second somatic sensory area of the monkey cerebral cortex. J Comp Neurol 1980;192:21–41.
70. Schilder P, Stengel E. Asymbolia for pain. Arch Neurol Psychiatry 1932;25:598–600.
71. Berthier M, Starkstein S, Leiguarda R. Asymbolia for pain: a sensory-limbic disconnection syndrome. Ann Neurol 1988;24:41–49.
72. Maihofner C, Handwerker HO. Differential coding of hyperalgesia in the human brain: a functional MRI study. Neuroimage 2005;28:996–1006.
73. Mazzola L, Isnard J, Mauguiere F. Somatosensory and Pain Responses to Stimulation of the Second Somatosensory Area (SII) in Humans. A Comparison with SI and Insular Responses. Cerebral Cortex 2006;16:960–968.
74. Craig AD. New and old thoughts on the mechanisms of spinal cord injury pain. In: Yezierski RP, Burchiel KJ, editors. Spinal Cord Injury Pain: Assessment, Mechanisms, Management. Seattle: IASP Press, 2002:237–264.

75. Blomqvist A, Zhang ET, Craig AD. Cytoarchitectonic and immunohistochemical characterization of a specific pain and temperature relay, the posterior portion of the ventral medial nucleus, in the human thalamus. Brain 2000;123:601–619.
76. Peyron R, Frot M, Schneider F, Garcia-Larrea L, Mertens P, Barral FG et al. Role of operculoinsular cortices in human pain processing: converging evidence from PET, fMRI, dipole modeling, and intracerebral recordings of evoked potentials. Neuroimage 2002;17:1336–1346.
77. Henderson LA, Gandevia SC, Macefield VG. Somatotopic organization of the processing of muscle and cutaneous pain in the left and right insula cortex: a single-trial fMRI study. Pain 2007;128:20–30.
78. Brooks JC, Zambreanu L, Godinez A, Craig AD, Tracey I. Somatotopic organisation of the human insula to painful heat studied with high resolution functional imaging. Neuroimage 2005;27:201–209.
79. Yasui Y, Itoh K, Kamiya H, Ino T, Mizuno N. Cingulate gyrus of the cat receives projection fibers from the thalamic region ventral to the ventral border of the ventrobasal complex. J Comp Neurol 1988;274:91–100.
80. Wang CC, Shyu BC. Differential projections from the mediodorsal and centrolateral thalamic nuclei to the frontal cortex in rats. Brain Res 2004;995:226–235.
81. Hutchison WD, Davis KD, Lozano AM, Tasker RR, Dostrovsky JO. Pain-related neurons in the human cingulate cortex. Nat Neurosci 1999;2:403–405.
82. Jones AKP, Qi LY, Fujirawa T, Luthra SK, Ashburner J, Bloomfield P et al. In vivo distribution of opioid receptors in man in relation to the cortical projections of the medial and lateral pain systems measured with positron emission tomography. Neurosci Lett 1991;126:25–28.
83. Baumgartner U, Buchholz HG, Bellosevich A, Magerl W, Siessmeier T, Höhnemann S et al. High opiate receptor binding potential in the human lateral pain system: A (FEDPN) PET study. Clin Neurophysiol 2007;118:e12.
84. Pessoa L. On the relationship between emotion and cognition. Nat Rev Neurosci 2008;9(2):148–158.
85. Broca P. Anatomie comparée des circonvolutions cérébrales: le grande lobe limbique. Rev Anthropol 1878;1:385–498.
86. Ballantine HT, Jr., Cassidy WL, Flanagan NB, Marino R, Jr. Stereotaxic anterior cingulotomy for neuropsychiatric illness and intractable pain. J Neurosurg 1967;26:488–495.
87. Hassenbusch SJ, Pillay PK, Barnett GH. Radiofrequency cingulotomy for intractable cancer pain using stereotaxis guided by magnetic resonance imaging. Neurosurgery 1990;27:220–223.
88. Pillay PK, Hassenbusch SJ. Bilateral MRI-guided stereotactic cingulotomy for intractable pain. Stereotact Funct Neurosurg 1992;59:33–38.
89. Foltze EL, White LE, Jr. Pain "relief" by frontal cingulumotomy. J Neurosurg 1962;19: 89–100.
90. Gybels JM, Sweet WH. Neurosurgical treatment of persistent pain. Physiological and pathological mechanisms of human pain. Pain Headache 1989;11:1–402.
91. Vogt BA, Nimchinsky EA, Vogt LJ, Hof PR. Human cingulate cortex: surface features, flat maps, and cytoarchitecture. J Comp Neurol 1995;359:490–506.
92. Devinsky O, Morrell MJ, Vogt BA. REVIEW ARTICLE: Contributions of anterior cingulate cortex to behaviour. Brain 1995;118: 279–306.
93. Davis KD, Taylor SJ, Crawley AP, Wood ML, Mikulis DJ. Functional MRI of pain- and attention-related activations in the human cingulate cortex. J Neurophysiol 1997;77: 3370–3380.
94. Rainville P, Duncan GH, Price DD, Carrier B, Bushnell MC. Pain affect encoded in human anterior cingulate but not somatosensory cortex. Science 1997;277:968–971.
95. Arienzo D, Babiloni C, Ferretti A, Caulo M, Del Gratta C, Tartaro A et al. Somatotopy of anterior cingulate cortex (ACC) and supplementary motor area (SMA) for electric stimulation of the median and tibial nerves: an fMRI study. Neuroimage 2006;33:700–705.
96. Talbot JD, Villemure JG, Bushnell MC, Duncan GH. Evaluation of pain perception after anterior capsulotomy: a case report. Somatosens Mot Res 1995;12:115–126.
97. Casey KL. Forebrain mechanisms of nociception and pain: Analysis through imaging. PNAS 1999;96:7668–7674.
98. Wager TD, Rilling JK, Smith EE, Sokolik A, Casey KL, Davidson RJ et al. Placebo-induced changes in FMRI in the anticipation and experience of pain. Science 2004;303: 1162–1167.
99. McDonald AJ. Cortical pathways to the mammalian amygdala. Prog Neurobiol 1998;55: 257–332.

100. Zald DH. The human amygdala and the emotional evaluation of sensory stimuli. Brain Res Rev 2003;41:88–123.
101. Schneider F, Habel U, Holthusen H, Kessler C, Posse S, Muller-Gartner HW et al. Subjective ratings of pain correlate with subcortical-limbic blood flow: an fMRI study. Neuropsychobiology 2001;43: 175–185.
102. Bornhovd K, Quante M, Glauche V, Bromm B, Weiller C, Buchel C. Painful stimuli evoke different stimulus-response functions in the amygdala, prefrontal, insula and somatosensory cortex: a single-trial fMRI study. Brain 2002;125:1326–1336.
103. Mason P. Deconstructing endogenous pain modulations. J Neurophysiol 2005;94: 1659–1663.
104. Fields HL. Pain modulation: expectation, opioid analgesia and virtual pain. Prog Brain Res 2000;122:245–253.
105. Dunckley P, Wise RG, Fairhurst M, Hobden P, Aziz Q, Chang L et al. A comparison of visceral and somatic pain processing in the human brainstem using functional magnetic resonance imaging. J Neurosci 2005;25:7333–7341.
106. Tracey I, Iannetti GD. Brainstem functional imaging in humans. Suppl Clin Neurophysiol 2006;58:52–67.
107. Guimaraes AR, Melcher JR, Talavage TM, Baker JR, Ledden P, Rosen BR et al. Imaging subcortical auditory activity in humans. Hum Brain Mapp 1998;6:33–41.
108. Farina S, Tinazzi M, Le Pera D, Valeriani M. Pain-related modulation of the human motor cortex. Neurol Res 2003;25:130–142.
109. Chudler EH, Dong WK. The role of the basal ganglia in nociception and pain. Pain 1995;60:3–38.
110. Bingel U, Glascher J, Weiller C, Buchel C. Somatotopic representation of nociceptive information in the putamen: an event-related fMRI study. Cereb Cortex 2004;14: 1340–1345.
111. Logothetis NK, Pfeuffer J. On the nature of the BOLD fMRI contrast mechanism. Magn Reson Imaging 2004;22:1517–1531.
112. Kwong KK, Belliveau JW, Chesler DA, Goldberg IE, Weisskoff RM, Poncelet BP et al. Dynamic magnetic resonance imaging of human brain activity during primary sensory stimulation. Proc Natl Acad Sci U S A 1992;89:5675–5679.
113. Chen JI, Ha B, Bushnell MC, Pike B, Duncan GH. Differentiating noxious- and innocuous-related activation of human somatosensory cortices using temporal analysis of fMRI. J Neurophysiol 2002;88: 464–474.
114. Iramina K, Iramina K, Kamei H, Uchida S, Kato T, Ugurbil K et al. Effects of stimulus intensity on fMRI and MEG in somatosensory cortex using electrical stimulation. IEEE Trans Magn 1999;35:4106–4108.
115. Brooks J, Tracey I. From nociception to pain perception: imaging the spinal and supraspinal pathways. J Anat 2005;207:19–33.
116. Brooks JC, Beckmann CF, Miller KL, Wise RG, Porro CA, Tracey I et al. Physiological noise modelling for spinal functional magnetic resonance imaging studies. Neuroimage 2008;39:680–692.
117. Mackey S, Lucca A, Soneji D, Kaplan K, Glover G. FMRI evidence of noxious thermal stimuli encoding in the human spinal cord. J Pain 2006;7(4, Suppl 1):S25.
118. Rollnik JD, Schmitz N, Kugler J. Anxiety moderates cardiovascular responses to painful stimuli during sphygmomanometry. Int J Psychophysiol 1999;33:253–257.
119. Becerra LR, Breiter HC, Stojanovic M, Fishman S, Edwards A, Comite AR et al. Human brain activation under controlled thermal stimulation and habituation to noxious heat: an fMRI study. Magn Reson Med 1999;41:1044–1057.
120. Rainville P, Doucet JC, Fortin MC, Duncan GH. Rapid deterioration of pain sensory-discriminative information in short-term memory. Pain 2004;110:605–615.
121. Charron J, Rainville P, Marchand S. Direct comparison of placebo effects on clinical and experimental pain. Clin J Pain 2006;22:204–211.
122. Price DD, Milling LS, Kirsch I, Duff A, Montgomery GH, Nicholls SS. An analysis of factors that contribute to the magnitude of placebo analgesia in an experimental paradigm. Pain 1999;83:147–156.
123. Apkarian AV, Darbar A, Krauss BR, Gelnar PA, Szeverenyi NM. Differentiating cortical areas related to pain perception from stimulus identification: temporal analysis of fMRI activity. J Neurophysiol 1999;81: 2956–2963.
124. Porro CA, Lui F, Facchin P, Maieron M, Baraldi P. Percept-related activity in the human somatosensory system: functional magnetic resonance imaging studies. Magn Reson Imaging 2004;22:1539–1548.

125. Andrew D, Greenspan JD. Peripheral coding of tonic mechanical cutaneous pain: comparison of nociceptor activity in rat and human psychophysics. J Neurophysiol 1999;82:2641–2648.
126. Adriaensen H, Gybels J, Handwerker HO, Van Hees J. Nociceptor discharges and sensations due to prolonged noxious mechanical stimulation – a paradox. Hum Neurobiol 1984;3:53–58.
127. Gallez A, Albanese MC, Rainville P, Duncan GH. Attenuation of sensory and affective responses to heat pain: evidence for contralateral mechanisms 1. J Neurophysiol 2005;94:3509–3515.
128. Bingel U, Schoell E, Herken W, Buchel C, May A. Habituation to painful stimulation involves the antinociceptive system. Pain 2007;131:21–30.
129. Valeriani M, de Tommaso M, Restuccia D, Le Pera D, Guido M, Iannetti GD et al. Reduced habituation to experimental pain in migraine patients: a CO(2) laser evoked potential study. Pain 2003;105:57–64.
130. Schoedel AL, Zimmermann K, Handwerker HO, Forster C. The influence of simultaneous ratings on cortical BOLD effects during painful and non-painful stimulation. Pain 2008;135:131–141.
131. Buchel C, Bornhovd K, Quante M, Glauche V, Bromm B, Weiller C. Dissociable neural responses related to pain intensity, stimulus intensity, and stimulus awareness within the anterior cingulate cortex: a parametric single-trial laser functional magnetic resonance imaging study. J Neurosci 2002;22:970–976.
132. Friston KJ, Penny WD, Glaser DE. Conjunction revisited. Neuroimage 2005;25: 661–667.
133. Fairhurst M, Wiech K, Dunckley P, Tracey I. Anticipatory brainstem activity predicts neural processing of pain in humans. Pain 2007;128:101–110.
134. Bingel U, Rose M, Glascher J, Buchel C. fMRI reveals how pain modulates visual object processing in the ventral visual stream. Neuron 2007;55:157–167.
135. Rose M, Schmid C, Winzen A, Sommer T, Buchel C. The functional and temporal characteristics of top-down modulation in visual selection. Cerebral Cortex 2005;15: 1290–1298.
136. Friston KJ. Functional and effective connectivity in neuroimaging: A synthesis. Hum Brain Mapp 1994;2:56–78.
137. Valet M, Sprenger T, Boecker H, Willoch F, Rummeny E, Conrad B et al. Distraction modulates connectivity of the cingulo-frontal cortex and the midbrain during pain – an fMRI analysis. Pain 2004;109:399–408.
138. Friston KJ, Buechel C, Fink GR, Morris J, Rolls E, Dolan RJ. Psychophysiological and modulatory interactions in neuroimaging. Neuroimage 1997;6:218–229.
139. McIntosh AR, Bookstein FL, Haxby JV, Grady CL. Spatial pattern analysis of functional brain images using partial least squares. Neuroimage 1996;3:143–157.
140. Seminowicz DA, Davis KD. Pain enhances functional connectivity of a brain network evoked by performance of a cognitive task. J Neurophysiol 2007;97:3651–3659.
141. Glass JM. Cognitive dysfunction in fibromyalgia and chronic fatigue syndrome: new trends and future directions. Curr Rheumatol Rep 2006;8:425–429.
142. Sjogren P, Christrup LL, Petersen MA, Hojsted J. Neuropsychological assessment of chronic non-malignant pain patients treated in a multidisciplinary pain centre. Eur J Pain 2005;9:453–462.
143. Ramnani N, Behrens TE, Penny W, Matthews PM. New approaches for exploring anatomical and functional connectivity in the human brain. Biol Psychiatry 2004;56:613–619.
144. Friston KJ, Frith CD, Frackowiak RSJ. Time-dependent changes in effective connectivity measured with PET. Hum Brain Mapp 1993;1:69–80.
145. Hadjipavlou G, Dunckley P, Behrens TE, Tracey I. Determining anatomical connectivities between cortical and brainstem pain processing regions in humans: a diffusion tensor imaging study in healthy controls. Pain 2006;123:169–178.
146. Peyron R, Garcia-Larrea L, Gregoire MC, Costes N, Convers P, Lavenne F et al. Haemodynamic brain responses to acute pain in humans: sensory and attentional networks. Brain 1999;122:1765–1780.
147. Bantick SJ, Wise RG, Ploghaus A, Clare S, Smith SM, Tracey I. Imaging how attention modulates pain in humans using functional MRI. Brain 2002;125:310–319.
148. Tracey I, Ploghaus A, Gati JS, Clare S, Smith S, Menon RS et al. Imaging attentional modulation of pain in the periaqueductal gray in humans. J Neurosci 2002;22:2748–2752.
149. Levine JD, Gordon NC, Jones RT, Fields HL. The narcotic antagonist naloxone enhances clinical pain. Nature 1978;272:826–827.
150. Hohmann AG, Suplita RL. Endocannabinoid mechanisms of pain modulation. AAPS J 2006;8:E693–E708.

151. Buffington AL, Hanlon CA, McKeown MJ. Acute and persistent pain modulation of attention-related anterior cingulate fMRI activations. Pain 2005;113:172–184.
152. Roder CH, Michal M, Overbeck G, van dV V, Linden DE. Pain response in depersonalization: a functional imaging study using hypnosis in healthy subjects. Psychother Psychosom 2007;76:115–121.
153. Raij TT, Numminen J, Narvanen S, Hiltunen J, Hari R. Brain correlates of subjective reality of physically and psychologically induced pain. Proc Natl Acad Sci U S A 2005;102:2147–2151.
154. Schulz-Stubner S, Krings T, Meister IG, Rex S, Thron A, Rossaint R. Clinical hypnosis modulates functional magnetic resonance imaging signal intensities and pain perception in a thermal stimulation paradigm. Reg Anesth Pain Med 2004;29: 549–556.
155. Bingel U, Lorenz J, Schoell E, Weiller C, Buchel C. Mechanisms of placebo analgesia: rACC recruitment of a subcortical antinociceptive network. Pain 2006;120:8–15.
156. Maihofner C, Ringler R, Herrndobler F, Koppert W. Brain imaging of analgesic and antihyperalgesic effects of cyclooxygenase inhibition in an experimental human pain model: a functional MRI study. Eur J Neurosci 2007;26:1344–1356.
157. Wise RG, Lujan BJ, Schweinhardt P, Peskett GD, Rogers R, Tracey I. The anxiolytic effects of midazolam during anticipation to pain revealed using fMRI. Magn Reson Imaging 2007;25:801–810.
158. Nemoto H, Nemoto Y, Toda H, Mikuni M, Fukuyama H. Placebo analgesia: a PET study. Exp Brain Res 2007;179:655–664.
159. Petrovic P, Kalso E, Petersson KM, Ingvar M. Placebo and opioid analgesia - imaging a shared neuronal network. Science 2002;295:1737–1740.
160. Menon RS, Goodyear BG. Spatial and Temporal Resolution in fMRI. Functional Magnetic Resonance Imaging: An Introduction to Methods. Oxford: Oxford University Press, 2001:149–158.
161. Fields HL, Heinricher MM. Anatomy and physiology of a nociceptive modulatory system. Philos Trans R Soc Lond B Biol Sci 1985;308:361–374.
162. Porro CA. Functional imaging and pain: behavior, perception, and modulation. Neuroscientist 2003;9:354–369.
163. Davis KD, Kwan CL, Crawley AP, Mikulis DJ. Event-related fMRI of pain: entering a new era in imaging pain. Neuroreport 1998;9:3019–3023.
164. Zubieta JK, Smith YR, Bueller JA, Xu Y, Kilbourn MR, Jewett DM et al. Regional mu opioid receptor regulation of sensory and affective dimensions of pain. Science 2001;293:311–315.
165. Thompson E. Empathy and consciousness. J Consc Stud 2001;8:1–32.
166. Jackson PL, Brunet E, Meltzoff AN, Decety J. Empathy examined through the neural mechanisms involved in imagining how I feel versus how you feel pain. Neuropsychologia 2006;44:752–761.
167. Jackson PL, Meltzoff AN, Decety J. How do we perceive the pain of others? A window into the neural processes involved in empathy. Neuroimage 2005;24:771–779.
168. Lamm C, Nusbaum HC, Meltzoff AN, Decety J. What are you feeling? Using functional magnetic resonance imaging to assess the modulation of sensory and affective responses during empathy for pain. PLoS ONE 2007;2:e1292.
169. Moriguchi Y, Decety J, Ohnishi T, Maeda M, Mori T, Nemoto K et al. Empathy and judging other's pain: an fMRI study of alexithymia. Cereb Cortex 2007;17:2223–2234.
170. Morrison I, Lloyd D, di Pellegrino G, Roberts N. Vicarious responses to pain in anterior cingulate cortex: is empathy a multisensory issue? Cogn Affect Behav Neurosci 2004;4:270–278.
171. Morrison I, Peelen MV, Downing PE. The sight of others' pain modulates motor processing in human cingulate cortex. Cereb Cortex 2007;17:2214–2222.
172. Simon D, Craig KD, Miltner WH, Rainville P. Brain responses to dynamic facial expressions of pain. Pain 2006;126:309–318.
173. Chen JI, Simon D, Duncan GH, Rainville P. Brain responses to facial expression of pain and negative emotions. Society for Neuroscience, Washington DC. 2005.
174. Botvinick M, Jha AP, Bylsma LM, Fabian SA, Solomon PE, Prkachin KM. Viewing facial expressions of pain engages cortical areas involved in the direct experience of pain. Neuroimage 2005;25:312–319.
175. Saarela MV, Hlushchuk Y, Williams AC, Schurmann M, Kalso E, Hari R. The compassionate brain: humans detect intensity of pain from another's face. Cereb Cortex 2007;17:230–237.
176. Singer T, Seymour B, O'Doherty J, Kaube H, Dolan RJ, Frith CD. Empathy for pain involves the affective but not sensory

components of pain. Science 2004;303: 1157–1162.

177. Jackson PL, Rainville P, Decety J. To what extent do we share the pain of others? Insight from the neural bases of pain empathy. Pain 2006;125:5–9.
178. Avenanti A, Bueti D, Galati G, Aglioti SM. Transcranial magnetic stimulation highlights the sensorimotor side of empathy for pain. Nat Neurosci 2005;8:955–960.
179. Avenanti A, Paluello IM, Bufalari I, Aglioti SM. Stimulus-driven modulation of motor-evoked potentials during observation of others' pain. Neuroimage 2006;32:316–324.
180. Morrison I, Lloyd D, di Pellegrino G, Roberts N. Vicarious responses to pain in anterior cingulate cortex: is empathy a multisensory issue? Cogn Affect Behav Neurosci 2004;4:270–278.
181. Singer T, Frith C. The painful side of empathy. Nat Neurosci 2005;8:845–846.
182. Norris DG. High field human imaging. J Magn Reson Imaging 2003;18:519–529.
183. Kim SG. Quantification of relative cerebral blood flow change by flow-sensitive alternating inversion recovery (FAIR) technique: application to functional mapping. Magn Reson Med 1995;34:293–301.
184. Fox PT, Raichle ME, Mintun MA, Dence C. Nonoxidative glucose consumption during focal physiologic neural activity. Science 1988;241:462.
185. Ye FQ, Smith AM, Yang Y, Duyn J, Mattay VS, Ruttimann UE et al. Quantitation of regional cerebral blood flow increases during motor activation: a steady-state arterial spin tagging study. Neuroimage 1997;6: 104–112.
186. Ramsey NF, Kirkby BS, van Gelderen P, Berman KF, Duyn JH, Frank JA et al. Functional mapping of human sensorimotor cortex with 3D BOLD fMRI correlates highly with H2(15)O PET rCBF. J Cereb Blood Flow Metab 1996;16:755–764.
187. Detre JA, Leigh JS, Williams DS, Koretsky AP. Perfusion imaging. Magn Reson Med 1992;23:37–45.
188. Owen DG, Bureau Y, Thomas AW, Prato FS, St Lawrence KS. Quantification of pain-induced changes in cerebral blood flow by perfusion MRI. Pain 2008;136(1–2): 85–96.
189. Wang J, Li L, Roc AC, Alsop DC, Tang K, Butler NS et al. Reduced susceptibility effects in perfusion fMRI with single-shot spin-echo EPI acquisitions at 1.5 Tesla. Magn Reson Imaging 2004;22:1–7.
190. Devlin JT, Russell RP, Davis MH, Price CJ, Wilson J, Moss HE et al. Susceptibility-induced loss of signal: comparing PET and fMRI on a semantic task. Neuroimage 2000;11:589–600.
191. Merboldt KD, Fransson P, Bruhn H, Frahm J. Functional MRI of the human amygdala? Neuroimage 2001;14:253–257.
192. Ojemann JG, Akbudak E, Snyder AZ, McKinstry RC, Raichle ME, Conturo TE. Anatomic localization and quantitative analysis of gradient refocused echo-planar fMRI susceptibility artifacts. Neuroimage 1997;6:156–167.
193. Paus T, Koski L, Caramanos Z, Westbury C. Regional differences in the effects of task difficulty and motor output on blood flow response in the human anterior cingulate cortex: a review of 107 PET activation studies. Neuroreport 1998;9:R37–R47.
194. Wager TD, Jonides J, Reading S. Neuroimaging studies of shifting attention: a meta-analysis. Neuroimage 2004;22:1679–1693.
195. Turkeltaub PE, Eden GF, Jones KM, Zeffiro TA. Meta-analysis of the functional neuroanatomy of single-word reading: method and validation. Neuroimage 2002;16:765–780.
196. Laird AR, Fox PM, Price CJ, Glahn DC, Uecker AM, Lancaster JL et al. ALE meta-analysis: controlling the false discovery rate and performing statistical contrasts. Hum Brain Mapp 2005;25:155–164.
197. Ashburner J, Friston KJ. Voxel-based morphometry – the methods. Neuroimage 2000;11:805–821.
198. Lerch JP, Evans AC. Cortical thickness analysis examined through power analysis and a population simulation. Neuroimage 2005;24:163–173.
199. Baron JC, Chételat G, Desgranges B, Perchey G, Landeau B, de la Sayette V et al. In vivo mapping of gray matter loss with voxel-based morphometry in mild Alzheimer's disease. Neuroimage 2001;14:298–309.
200. Thieben MJ, Duggins AJ, Good CD, Gomes L, Mahant N, Richards F et al. The distribution of structural neuropathology in pre-clinical Huntington's disease. Brain 2002;125:1815–1828.
201. Singh V, Chertkow H, Lerch JP, Evans AC, Dorr AE, Kabani NJ. Spatial patterns of cortical thinning in mild cognitive impairment and Alzheimer's disease. Brain 2006;129:2885–2893.
202. Apkarian AV, Sosa Y, Sonty S, Levy RM, Harden RN, Parrish TB et al. Chronic back pain is associated with decreased prefrontal

and thalamic gray matter density. J Neurosci 2004;24:10410–10415.

203. Davis KD, Pope G, Chen J, Kwan CL, Crawley AP, Diamant NE. Cortical thinning in IBS: implications for homeostatic, attention, and pain processing. Neurology 2008;70:153–154.

204. Schmidt-Wilcke T, Leinisch E, Gänssbauer S, Draganski B, Bogdahn U, Altmeppen J et al. Affective components and intensity of pain correlate with structural differences in gray matter in chronic back pain patients. Pain 2006;125:89–97.

205. Draganski B, Moser T, Lummel N, Gänssbauer S, Bogdahn U, Haas F et al. Decrease of thalamic gray matter following limb amputation. Neuroimage 2006;31:951–957.

206. Kwan CL, Diamant NE, Pope G, Mikula K, Mikulis DJ, Davis KD. Abnormal forebrain activity in functional bowel disorder patients with chronic pain. Neurology 2005;65:1268–1277.

207. Vernon DJ. Can neurofeedback training enhance performance? An evaluation of the evidence with implications for future research. Appl Psychophysiol Biofeedback 2005;30:347–364.

208. Tao JX, Ray A, Hawes-Ebersole S, Ebersole JS. Intracranial EEG substrates of scalp EEG interictal spikes. Epilepsia 2005;46: 669–676.

209. Stern JM. Simultaneous electroencephalography and functional magnetic resonance imaging applied to epilepsy. Epilepsy Behav 2006;8:683–692.

210. Lantz G, Spinelli L, Menendez RG, Seeck M, Michel CM. Localization of distributed sources and comparison with functional MRI. Epileptic Disord 2001;Special Issue:45–58.

211. Cox RW, Jesmanowicz A, Hyde JS. Real-time functional magnetic resonance imaging. Magn Reson Med 1995;33:230–236.

212. Yoo SS, Jolesz FA. Functional MRI for neurofeedback: feasibility study on a hand motor task. Neuroreport 2002;13:1377–1381.

213. Yoo SS, O'Leary HM, Fairneny T, Chen NK, Panych LP, Park H et al. Increasing cortical activity in auditory areas through neurofeedback functional magnetic resonance imaging. Neuroreport 2006;17:1273–1278.

214. Weiskopf N, Veit R, Erb M, Mathiak K, Grodd W, Goebel R et al. Physiological self-regulation of regional brain activity using real-time functional magnetic resonance imaging (fMRI): methodology and exemplary data. Neuroimage 2003;19:577–586.

215. Weiskopf N, Scharnowski F, Veit R, Goebel R, Birbaumer N, Mathiak K. Self regulation of local brain activity using real-time functional magnetic resonance imaging (fMRI). J Physiol Paris 2004;98:357–373.

216. Posse S, Fitzgerald D, Gao K, Habel U, Rosenberg D, Moore GJ et al. Real-time fMRI of temporolimbic regions detects amygdala activation during single-trial self-induced sadness. Neuroimage 2003;18: 760–768.

217. Flor H, Braun C, Elbert T, Birbaumer N. Extensive reorganization of primary somatosensory cortex in chronic back pain patients. Neurosci Lett 1997;224:5–8.

218. Diers M, Koeppe C, Diesch E, Stolle AM, Holzl R, Schiltenwolf M et al. Central processing of acute muscle pain in chronic low back pain patients: an EEG mapping study. J Clin Neurophysiol 2007;24:76–83.

219. Apkarian AV, Thomas PS, Krauss BR, Szeverenyi NM. Prefrontal cortical hyperactivity in patients with sympathetically mediated chronic pain. Neurosci Lett 2001;311:193–197.

220. Collins DL, Neelin P, Peters TM, Evans AC. Automatic 3D intersubject registration of MR volumetric data in standardized Talairach space. J Comput Assist Tomogr 1994;18:192–205.

Chapter 16

fMRI of the Sensorimotor System

Massimo Filippi and Maria A. Rocca

Summary

The extensive application of fMRI to the assessment of the human sensorimotor system has disclosed a complexity that is largely beyond our original understanding. From the available data, it is accepted that this system consists of a large, and somewhat yet unknown, number of cortical and subcortical areas, with a precise location and a specialized function. In particular, a large number of regions in the frontal and parietal lobes contribute to different aspects of motor act performance. It is also evident that the properties and potentialities of this network still need to be fully elucidated by further research. Defining how the human sensorimotor system works is of outmost importance for understanding its dysfunction in case of diseases and also to develop potential therapeutic strategies capable to enhance its functional plasticity and reserve.

Key words: Sensorimotor system, Human, Mirror-neuron system, fMRI, Motor training

1. Introduction

During the past 15 years, fMRI has became a valuable tool to study normal brain function, due to the development of revolutionary methods for data acquisition and postprocessing, as well as for paradigm design. Due to its noninvasiveness and relatively high spatial and temporal resolution, fMRI has rapidly substituted other techniques, such as positron emission tomography (PET), in the assessment of brain function. In addition, the combination of fMRI with neurophysiological techniques, such as transcranial magnetic stimulation (TMS), is providing important pieces of information for the understanding of brain function in healthy individuals, which, on turn, is critical for the interpretation of functional changes in diseased people.

M. Filippi (ed.), *fMRI Techniques and Protocols*, Neuromethods, vol. 41
DOI 10.1007/978-1-60327-919-2_16, © Humana Press, a part of Springer Science+Business Media, LLC 2009

This chapter summarizes the major contributions of fMRI for the in vivo assessment of the sensorimotor network in healthy human subjects, with a specific focus on studies of performance of a simple motor task with the dominant upper limb.

2. Sensorimotor Paradigms

Activity of the sensorimotor system has been investigated by using several experimental paradigms. The majority of the studies analyzed the performance of active tasks consisting of movement of the hand, using tasks that require flexion-extension of the hand and/or fingers, tapping the hand or fingers, closing-opening the hand, and squeezing. A few studies investigated the movement of the foot, leg, arm, shoulder, and tongue, with the main goal of defining the somatotopic location and hemispheric lateralization of these body parts *(1, 2)*. Other studies analyzed the fMRI correlates of interlimb coordination *(3, 4)*.

The brain activations associated to the performance of passive tasks have also been evaluated *(2, 5, 6)*. This strategy has mainly been prompted by the need of obtaining meaningful comparisons between controls and patients with neurological affections that might impair the "ability" to perform active tasks correctly *(5, 6)*. The use of passive tasks is also supported by the fact that there are reciprocal projections between the motor and the related sensory cortices; hence, patterns of brain activations that reflect local field potentials from presynaptic activity primarily *(7)*, even with entirely passive movements, may identify those brain regions involved in active voluntary movements. This hypothesis has indeed been confirmed by fMRI studies of healthy controls which have demonstrated that activations associated to active and passive hand movements are similar in localization and size *(5, 6)*. Finally, it is now established that even the imagination of movements activates the motor network *(8, 9)*.

One of the major caveats in the set up of fMRI experiments of the sensorimotor system is an adequate monitoring of subjects' performance during task execution, which might require to be corrected during the statistical analysis. Several variables have been shown to influence the observed patterns of movement-associated cortical activations in healthy subjects during motor task execution, including:

1. Movement rate, which has been positively correlated with the recruitment of the contralateral primary sensorimotor cortex (SMC) *(10)*, supplementary motor area (SMA) *(11)*, and ipsilateral cerebellar cortex *(12)*.

2. Force, as suggested by the load-dependent effect observed in the primary SMC *(10, 13)*.
3. Movement complexity, which has been shown to modulate activity of the primary SMC *(14, 15)*, SMA, and premotor cortex *(16)*, as well as several regions of the parietal lobes *(10)*.

Several strategies can be adopted to minimize these possible confounding factors, including accurate monitoring of task performance during fMRI acquisition either visually or using more sophisticated techniques, such as force transducers.

Other variables that need to be considered when dealing with motor task investigations include:

1. Hemispheric dominance. Approximately 90% of the population has a left-hemispheric dominance for processing motor acts *(17)*. In line with this, fMRI studies have demonstrated that motor-related activations are usually lateralized to the left hemisphere in right-handers and bilateralized or lateralized to the right hemisphere in left-handers *(18–21)*.
2. Gender. Women have been shown to have larger activations of cortical motor areas during motor tasks, while men exhibited significantly stronger activation in the striatal regions *(22)*.
3. Age. There appears to be greater motor task-related brain activity in a wider network of brain regions in older compared to younger subjects *(23)*. A recent study of healthy individuals has demonstrated an age-related increased functional connectivity of motor cortices between the two hemispheres *(24)*. These results are consistent with a more general reduction of functional lateralization of the motor cortex recruitment with aging, which has been interpreted as a compensatory response to increased functional demands **(Fig. 1)** *(25, 26)*.

3. Components of the Human Sensorimotor Network

The control of motor acts is a complex process that involves several motor, sensory, and association areas, including the primary SMC, the secondary sensorimotor cortex (SII), the SMA, the cingulum, the basal ganglia, the cerebellum, and several regions located in the frontal and parietal lobes. Thus, the sensorimotor network is a relatively complex system, with a hierarchic organization. Anatomically, this view is supported by the presence of large, somatotopically organized, primary cortices with converging projections to smaller association areas. The extensive application of functional techniques to the assessment of this system's function in healthy subjects is contributing to increase our knowledge of its behavior and connections. In addition, this has allowed to

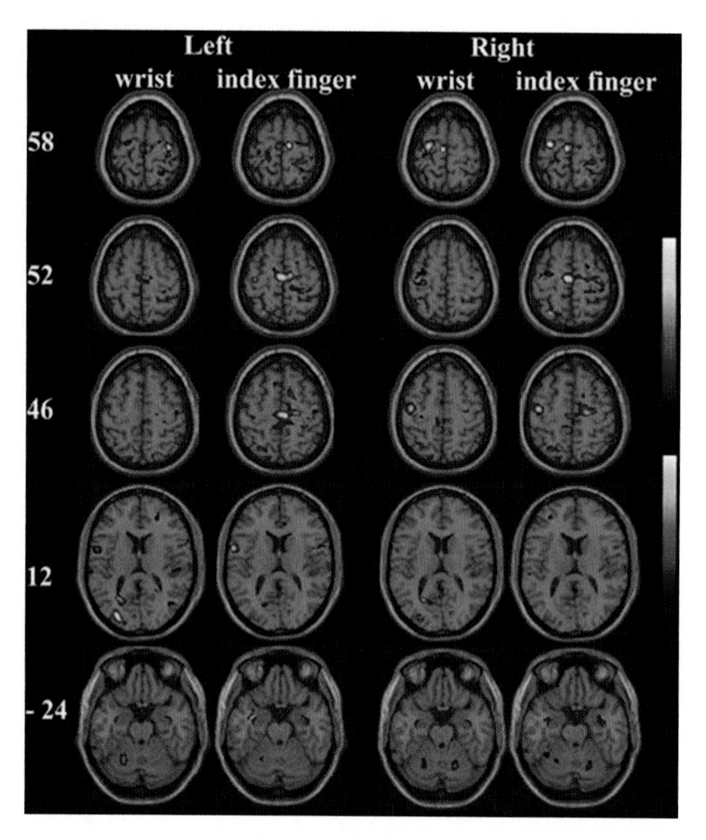

Fig. 1. Comparison of mean activation in old vs. young healthy subjects during the performance of wrist extension/flexion and index finger abduction/adduction with the left and right upper limb, respectively. Areas more significantly activated in old subjects are coded in red spectrum, while areas more significantly activated in young subjects are coded in blue spectrum. Activations have been overlaid on a standard T1 brain image in neurological view. For each motor task, the contralateral primary sensorimotor cortex and the premotor cortex had significantly greater activation in the young group and caudal supplementary motor area had significantly greater activation in the old group. Ipsilateral sensorimotor cortex was more significantly activated in the old group for index finger motor tasks of both hands (From **ref.** ***25***, with permission.).

define the role that the different components of the network have during the performance of a motor act.

3.1. The Primary Sensorimotor Cortex

Anatomically, the primary SMC is the cortex lying within the anterior and posterior banks of the central sulcus *(1)*. In line with clinical and electrical stimulation studies, fMRI studies confirmed the somatotopic organization of the primary SMC of the left hemisphere, with distinct subregions controlling movements of the foot, arm, and face *(27)*. As already mentioned, there is a large body of evidence supporting the prominent role of the primary SMC of the dominant hemisphere in the performance of simple motor acts. In healthy subjects, the role of the ipsilateral primary SMC in the control of movements is still controversial, since several studies have reported conflicting results with respect to the occurrence of ipsilateral primary SMC activation *(10, 18)*. In particular, while there is a general agreement on consistent activation of the primary SMC of the ipsilateral hemisphere with increasing motor task complexity *(10, 16)*, only a few studies reported its activation during simple task performance *(18, 27)*. A mechanism that has been advocated to shed lights on primary SMC behavior is transcallosal inhibition. In healthy individuals, a transcallosal inhibitory pathway between the primary motor cortices of the two hemispheres has been previously shown to exist by neurophysiological studies *(28, 29)* and has been postulated to be responsible for the controls of homologous hand muscles during unilateral movements *(28, 29)*. These data are supported by fMRI studies that have shown a decreased activation of the ipsilateral primary SMC during sequential finger movements *(30–32)*. The reason for the ipsilateral inhibition during unilateral hand movements remains speculative. However, this decreased excitability could improve the capacity to perform fine movements of the fingers, for which a high level of dexterity is needed. Usually, such movements are carried out unilaterally. A suppression of excitability of the ipsilateral SMC would then minimize the risk of contralateral interference, and improve the cortical focus on unilateral activation *(33)*.

3.2. The Supplementary Motor Area

Another important component of the motor network is the SMA, which is the cortex lying above the cingulated sulcus and anteriorly within the paracentral lobule *(1)*. The SMA contributes to the preparation, coordination, temporal course, and execution of movements *(34–36)*. Studies of healthy individuals suggest that the movement-related activity of the primary SMC might be mediated by the extensive input it receives from the SMA and that the SMA recruitment might increase by increasing task difficulty and complexity *(10, 16)*. The extent of SMA activation has been inversely related to the amount of training an individual has gained with that specific task *(34, 36)*. Inter- and intrahemispheric connections between the primary SMC, the premotor cortex, and the SMA are likely to be mediated primarily by the SMA *(37)*. In addition, strong bilateral connections exist between bilateral SMA and the basal ganglia. In agreement with this notion, lesions of the SMA typically result in alterations of bimanually coordinated movements *(38, 39)*. Efferents from the SMA project directly to the brainstem and the cervical cord; as a consequence, an increased SMA activation might represent recruitment of motor pathways that can function in parallel with the contralateral corticospinal tract *(40)*.

Functionally, the SMA can be divided into a pre-SMA (located more rostrally) and a SMA-proper (located more caudally), and event-related fMRI studies have shown that the pre-SMA is activated preferentially during movement preparation *(35, 41)*. In addition, recent work has also shown that pre-SMA recruitment precedes primary SMC activation by several seconds *(42)*.

3.3. The Frontal Cortex

The frontal cortex contains many areas contributing to the motor network *(43, 44)*. In addition to the primary SMC, these areas include the ventral premotor areas (including the inferior frontal gyrus [IFG]), the dorsal premotor cortex (sometimes divided into a caudal and a rostral part) (PMd), and a set of motor areas on the medial wall of the hemispheres, such as the SMA and the cingulate motor area (CMA). The premotor areas in the frontal lobe influence motor output through connections with the primary SMC and direct projections to the spinal cord *(45)*. All the previous premotor areas contain corticospinal neurons that give a substantial contribution to corticospinal projections, which have a high degree of topographic organization *(46)*.

The role of the left inferior frontal lobe (ventral premotor cortex/Broca's area) in motor sequence control is well documented by several studies *(47–50)*. Activation of Broca's area has been reported in various functional imaging studies based on finger movements *(48, 49)*, movement imagination, and motor learning *(47)*. This area is supposed to receive rich sensory information originating from the parietal lobe (including the SII) and to use it for action *(51)*. In addition, modulation of this area's activity by task complexity has been clearly documented *(50)*. Studies in humans have shown that this region is important for encoding hand/object interaction *(51)*.

The PMd has an important role in motor preparation, selection, and initiation of voluntary actions *(52–54)*. Imaging and TMS experiments suggest that the PMd cortex of the left hemisphere is dominant in right-handed people *(55)*. This area is reciprocally connected with the ipsilateral and contralateral primary SMC, as well as with the parietal cortex and the contralateral PMd *(55, 56)*. Using a labeling retrograde strategy, Marconi et al. *(57)* showed transcallosal homotopic and heterotopic connections between different portions of the PMd of the two hemispheres and between the two PMd and the SMA. A recent experiment *(58)* in healthy individuals demonstrated a correlation between preservation of motor performance after disruption of the left PMd activity by means of TMS and increased activation of the right PMd cortex, the SMA, and the cingulate cortex (**Fig. 2**). This pattern was not seen after TMS inhibition of the left SMC, while TMS of the reorganized right PMd disrupted motor performance. These findings suggest that adaptive changes of PMd function might contribute to maintaining motor behavior despite the presence of structural damage *(58)*.

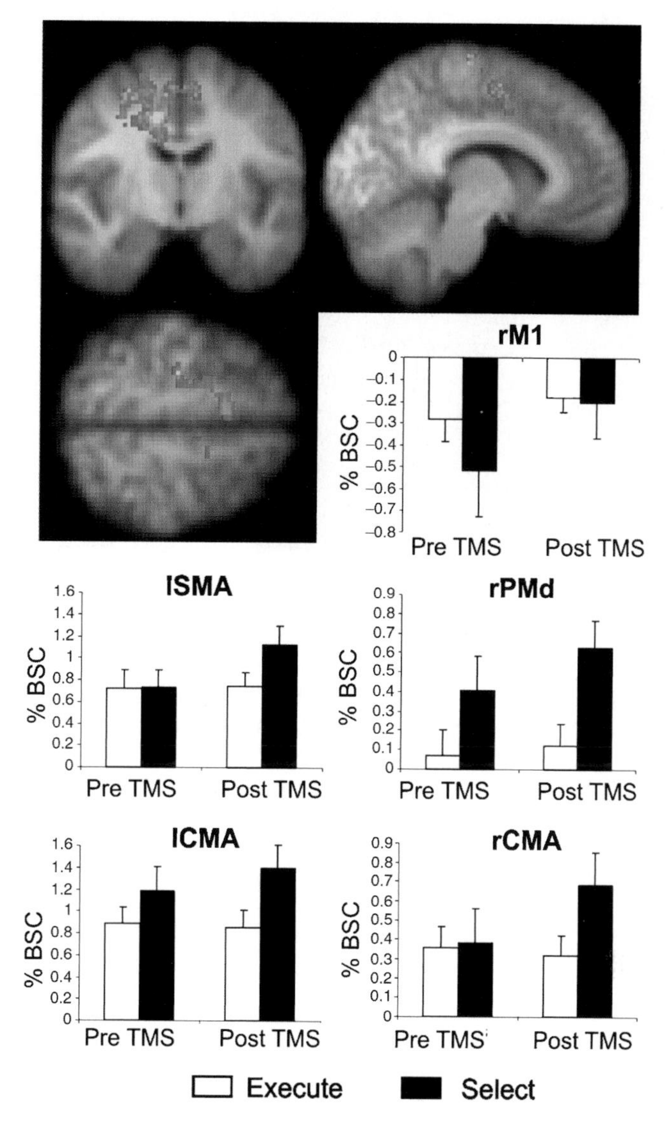

Fig. 2. Compensatory activation increases in the action selection network after transcranial magnetic stimulation (TMS) of the left dorsal premotor cortex (lPMd). A whole-brain random-effects analysis showed that lPMd TMS-induced increased activation that was most prominent in right PMd and the right cingulate motor area (CMA). Additional changes were seen in the right primary motor cortex, left supplementary motor area, and left CMA. For each of these regions, the graphs show mean percent blood oxygen level dependent (BOLD) signal change (%BSC) values for each task (select/execute) and TMS condition (pre-/post-TMS). Note that the TMS-induced activation increases are specific to the process of action selection. White bars = execute; black bars = select (From **ref. *58***, with permission.).

The anatomical variability of the cingulated sulcus in humans hampers functional analysis of this region. The caudal CMA is considered to be primarily involved in movement execution **(Fig. 3)** *(1, 43, 59)*, while the rostral portion of the CMA has

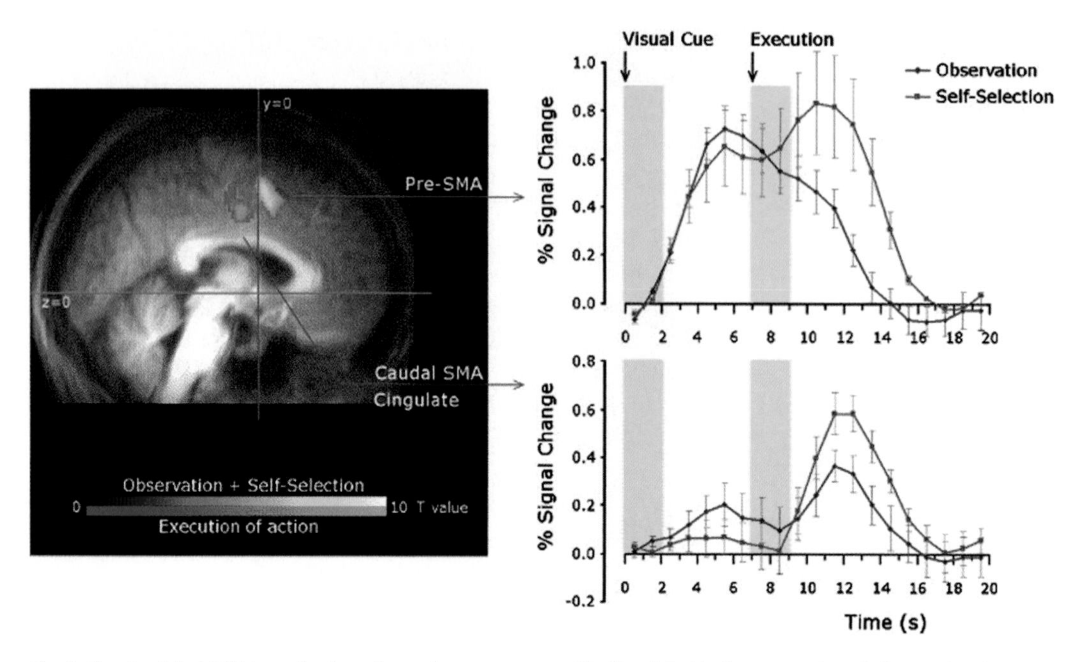

Fig. 3. Event-related fMRI investigation of neural processes specifically related to the processing of observed actions, the selection of our own intended actions, the preparation for movement, and motor response execution. In the mesial frontal cortex, the pre-supplementary motor area (SMA) showed significant activation prior to movement for both the observation and self-selection of action, while the caudal SMA and cingulate motor area showed significantly greater activation during movement execution than during the visual presentation and pre-movement delay period. Overlayed are different regions of activation for the conjunction analysis (observation and self-selection of action) and for the contrast of movement execution vs. the visual presentation period, with mean event-related time courses (plus standard error across subjects) for each region (From **ref. *59***, with permission.).

been shown to have a role in action selection *(60)*, initiation, motivation, and goal-directed behaviors *(61)*. Activation of this region has also been found to be related to the presentation of new motor tasks and perhaps its recruitment reflects relative task difficulty *(16, 62)*. This cortical area is involved in attentional tasks and subserves several executive functions *(63)*. In addition, the CMA has been suggested to play an important role in conflict monitoring *(64, 65)*. The role of the CMA in the execution of spatially complex coordination tasks has been underlined by a study of Wenderoth et al. *(66)*, where an increased CMA activation was detected during the performance of a bimanual task.

3.4. The Parietal Cortex

The parietal cortex is formed by a multiplicity of independent areas, each of which deals with specific aspects of sensory information *(67)*. Physiological and imaging techniques have been extensively applied to define the location and functional specialization of parietal cortex regions in humans. Although this effort resulted in the identification of several areas related to the sensorimotor network, understanding their precise function and relationship still require further experiments.

Among the regions of the parietal cortex, the SII is considered to function as a high-order processing area for somatosensory perception, and its activation seems also to be related to attention, manual dexterity, and coordination *(68, 69)*. SII activity has been associated with processing of the temporal features of somatic sensations, sensorimotor integration *(70)*, tactile recognition, and tactile learning and memory *(71)*. In addition, neurons from SII project directly to the spinal cord *(72)*, indicating that this region might provide alternative pathways for motor control in case of primary SMC injury. SII is known to have extensive connections with the prefrontal cortex, the parietal lobe, and the insula. Similarly to the primary SMC, SII has a somatotopic representation of different body parts, with the upper limb areas located more anteriorly and more inferiorly than the lower limb areas *(73)*.

Numerous areas along the intraparietal sulcus (IPS) have also been associated with processing of sensorimotor tasks. The anterior part of the IPS contains neurons that discharge in response to 3D object presentation and during grasping movements *(74)*, and it is connected to the IFG for control of action in object manipulation *(48)*. This area has a central role for visuomotor integration (crossmodal information process) *(67)*. In addition, increased activity of the IPS has also been described in normal subjects during complex finger movement sequences *(14)*. Activity of the caudal end of the IPS has been associated to object matching and grasping, as well as discrimination of object size *(67)*.

The precuneus has been related to the execution of spatially complex coordination tasks *(66)*, which require shifting attention between different locations in space, while the superior parietal gyrus (SPG) is thought to be involved in the elaboration of somatosensory modalities and has a well-demonstrated hand/finger representation *(48)*.

3.5. The Basal Ganglia, Insula, and Thalamus

Basal ganglia have extensive connections to the motor and somatosensory cortices and are involved in motor programming, execution, and control *(75)*. In particular, basal ganglia activity has been associated with motor program selection and suppression at early stages of motor planning, as well as with control of movement simulation *(76)*. In addition, they are implicated in the formation of motor skills and are also part of subsystems whose activity has been associated with timing of motor acts *(77–79)*.

The thalamus *(75, 80)* and the insula *(81)* also have extensive connections with the motor and somatosensory cortices and are involved in motor execution *(75, 81)*. Interestingly, the thalamus is an important relay station of the complex re-entrant circuitry that links the motor and the prefrontal cortices to the basal ganglia and which is part of the feedback loops of the limbic system able to modulate the cortical motor output *(82)*.

The insular cortex has been shown to play a role in crossmodal transfer of information *(83)*. In addition, the insular cortex is also involved in the synchronization of movement kinematic *(84)*, and has connections with numerous cortical and subcortical motor regions.

3.6. The Cerebellum

The cerebellum integrates sensory information and motor programs to coordinate fine movements. Functionally, the cerebellum is organized in modules arranged in the medio-lateral direction, being the medial part responsible for control of posture and the lateral regions for coordination and movements. Anatomically, the cerebellum is divided along the rostro-caudal axis in the anterior lobe, which contains a somatotopic representation of movement of the ipsilateral side *(85)*, and the posterior lobe, which is thought to be related to motor imagery *(51)* and motor learning *(86, 87)*. The posterior lobe of the cerebellum has projections from and to regions of the parietal cortex, involved in the processing of sensory information *(88, 89)*, which is then used to correct movements.

Several imaging studies have reported a cerebellar recruitment associated to timing of rhythmic movements *(90)*. Some studies also reported increased cerebellar activation corresponding to increase in movement frequency *(34, 91)*. The cerebellum has also been involved in the "automatization" (improvement of motor performance) of learned skills, the establishment of movement strategies, and the consolidation of such a motor knowledge *(92, 93)*. Further evidence supporting the role of the cerebellum in motor learning is based on data from patients with focal cerebellar lesions, who have shown impairment in learning new motor skills *(92, 94, 95)*, and imaging studies of motor learning in healthy individuals, who showed prominent cerebellar recruitment *(96, 97)*.

4. Cortical Reorganization During Motor Training and Motor Skill Learning

Psychophysiological studies have demonstrated that the acquisition of motor skills follows two distinct stages. The first is a fast learning stage during which considerable improvement in performance can be observed within a single training session; the second is a later, slow learning stage, during which further gains can be observed across several sessions of practice *(15)*. Karni et al. *(15)* used a simple finger-opposition task, during which healthy individuals were trained over the course of several weeks and were scanned at weekly interval using fMRI. Repetition of the task after 3 weeks of practice showed that there was a significant larger activation of the contralateral primary SMC as compared with the activation obtained with a control, untrained finger-opposition sequence. These results support the notion that motor practice induces recruitment of additional M1 units into a local network specifically representing the motor trained sequence. These findings are in agreement with the demonstration that, in healthy individuals, the recruitment of the primary SMC can be modified by previous activities, such as playing musical instruments *(91, 98)* or racquets *(99)*. In the previous experiment, changes in primary SMC recruitment were also observed in the early scan session, reflecting a sort of initial habituation-like effect, in which the second sequence performed in a set evoked a smaller response than the first sequence *(15)*.

While there is agreement that learning of relatively complex motor tasks is associated with an increased activation of the contralateral primary SMC, the evaluation of activation patterns associated with repetition of simple movements gave conflicting results, since some studies reported reductions, and others increases of task-related activations *(15, 100–102)*. These discrepancies among studies might be related to variability in number and length of sessions, length of training, as well as monitoring of motor performance. Recent evidence suggests that the repetition of a simple sequence within a brief time window typically results in a reduced recruitment of the primary SMC, due to habituation *(15, 100, 103)*. In addition, a change in the degree of activation of the parietal lobe from healthy volunteers has also been described after motor training *(103)*.

Dynamic activations changes during acquisition of motor skills have also been seen in different regions of the basal ganglia *(104)*. In detail, the dorsal parts of the putamen and the more rostral striatal areas have been shown to be active only during the early learning stage. On the contrary, activations of the posteroventral regions of the putamen and globus pallidus increase with practice **(Fig. 4)** *(104)*.

5. The Mirror-Neuron System

The mirror-neuron system (MNS) is an observation-execution matching system. Several neurophysiological *(105, 106)* and neuroimaging *(51, 107, 108)* studies have demonstrated that MNS neurons discharge not only when an individual performs a specific goal-directed action, but also when an individual observes actions made by other individuals, implying an involvement of this system in imitation and motor learning *(109)*. The main role of the MNS is postulated to be the understanding of actions *(110)*. This system is also thought to be involved in motor

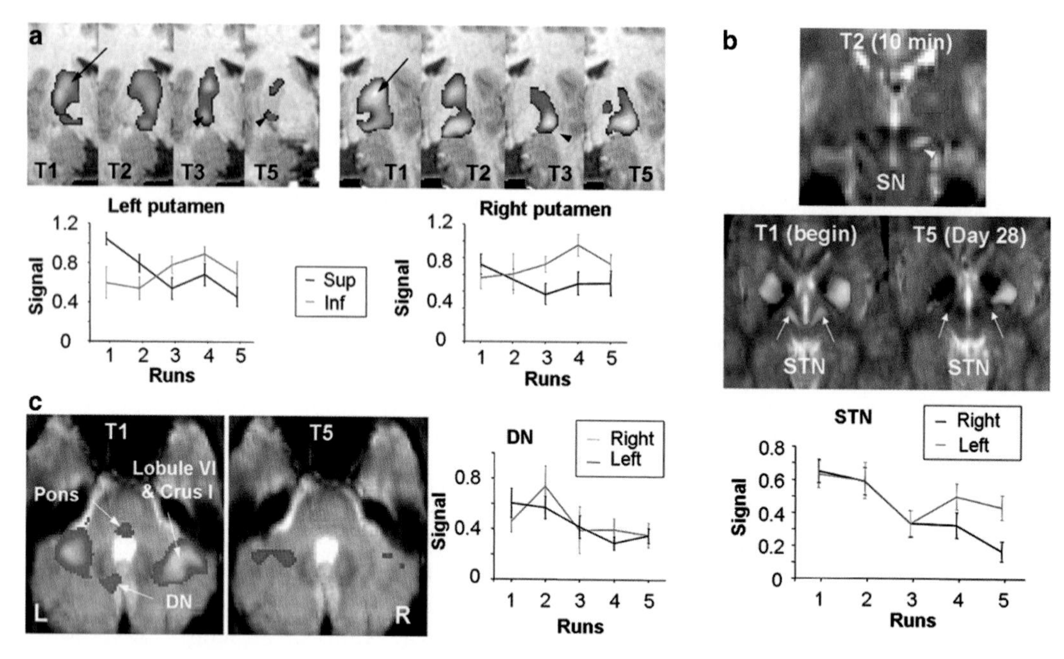

Fig. 4. Activation patterns in the basal ganglia and cerebellum during acquisition of motor skills. (**a** *Upper*) Activation maps obtained in the putamen superimposed on a coronal T1-weighted image. There was a progressive activation decrease in the dorsal part of the putamen (*arrows*) and an increase in a more ventrolateral area (*arrowheads*) bilaterally, which persisted after 4 weeks of training. (**a** *Lower*) Percentage signal increase ± SEM averaged across all subjects for each run of the trained sequence confirmed the activation decrease in the dorsal putamen and increase in the ventral putamen. (**b** *Top*) Activation maps obtained in the substantia nigra (SN) and subthalamic nucleus (STN) superimposed on EPI images. During session 1, STN activation was observed during the first run of T-sequence (T1). After 4 weeks of training, these areas were no more activated during the T-sequence. There was no significant signal change in the SN across runs. (**b** *Bottom*) Signal-to-time curves ± SEM in the STN averaged across all subjects and epochs confirm the activation decrease. (**c** *Left*) Activation maps obtained in the cerebellum during the T-sequence (T1 on day 1 and T5 on day 28). Activation in the lateral cerebellar hemispheres, the left dentate nucleus (DN), and the pons decreased with training. (**c** *Right*) Percentage signal increase ± SEM averaged across all subjects for each run of the trained sequence in the left and right DN. In the right DN, activation increased transiently during T2 (10 min of practice) and returned to pretraining values (From **ref. *104***, with permission.).

imagery *(111)* and empathy *(112)*. In humans, neurons of this system have been described in the IFG, the adjacent premotor cortex, and the rostral part of the inferior parietal lobule *(113)*. The MNS is connected with the superior temporal sulcus (STS) that provides a higher-order visual description of the observed action *(114)*. Mirror neurons are likely to be multimodal, as they respond to both the visual observation of an action as well as the sound associated with specific actions *(115)*. In humans, mirror neurons are part of a system serving the imitation of actions and speech generation. Therefore, the MNS might constitute a bridge between action and language processing and might represent the neuronal substrate from which human language evolved *(116)*. Activity of this system is elicited by both the execution and observation of object-related transitive and intransitive actions

(108, 117, 118). These observations suggest that the MNS is a network that has been preserved during evolution and has developed "new" functions. Therefore, the MNS seems to be an extremely plastic system, with the capacity to adapt to new cognitive, social, or behavioral requirements to which an individual is exposed. This hypothesis assumes that these "evolutionary" changes have occurred over an extremely long time window (phylogenetic plasticity). It is plausible that, as shown for other brain networks, including the motor one, disease-related changes of such a plastic system might occur in case of CNS injury (adaptive plasticity). This hypothesis has indeed been supported by the results of a recent study in patients with multiple sclerosis, which demonstrated that these patients tend to activate regions that are part of the MNS during the performance of a simple motor task *(119)*. Defining the role of the MNS after brain injury may be central to a better understanding of the clinical manifestations of various neurological conditions and, as a consequence, to develop new rehabilitative strategies.

The majority of the MNS studies has been focused on the attempt to better define the exact role of this system and its precise location in healthy individuals. In this perspective, it has been demonstrated that: (1) the MNS has a bilateral representation *(120)*; (2) mirror neurons in the premotor cortex have a somatotopic organization, as shown for the classical motor cortex homunculus *(118)*; and, finally, (3) there are gender differences in this system's function *(121)*. Recently, it has been suggested that there might be a relation between activity of the MNS and handedness **(Fig. 5)***(122)*.

Functional studies have suggested a role of MNS dysfunction, in combination with limbic system impairment, in patients with autism *(123, 124)*, suggesting that this neuronal system may play a role in autistic social impairment. These studies described a reduced activity in the IFG and premotor cortex during action/face imitation and observation in adults *(123)* and children *(124)* with autism spectrum disorders.

6. Conclusions

Functional neuroimaging has dramatically changed our understanding of the human sensorimotor system by showing that it is constituted by a large number of cortical and subcortical areas, with a precise location and a specialized function. It is also evident that this system functions in cooperation with other brain networks in order to integrate all the information coming from the environment and to finalize the performance of motor acts.

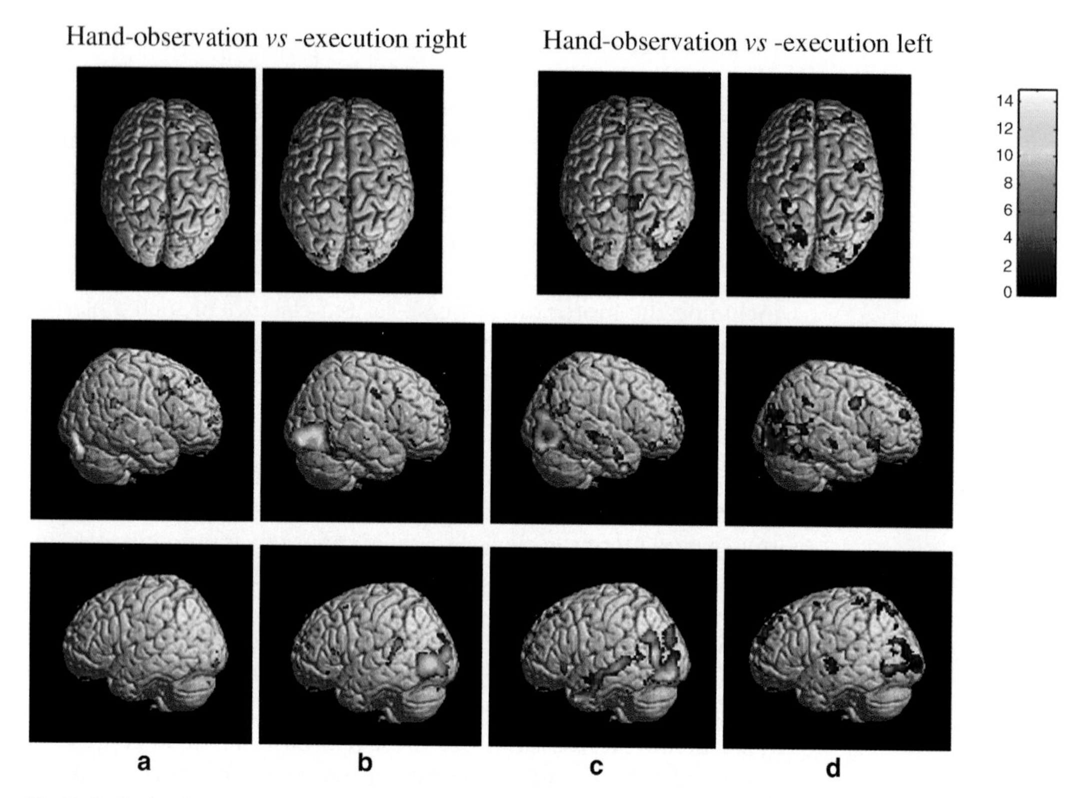

Fig. 5. Cortical activations on a rendered brain from right-handed (RH) (**a**, **c**) and left-handed (LH) healthy subjects (**b**, **d**) during the performance of an observation task involving the right (**a**, **b**) and left (**b**, **d**) upper limbs (within-group analysis, one-sample *t* tests, $p < 0.05$ corrected for multiple comparisons). During hand-observation/right, the activation of areas of the mirror-neuron system (MNS) is mainly located in the left cerebral hemisphere in RH subjects and in the right cerebral hemisphere in LH subjects. During hand-observation/left, the activation of areas of the MNS is mainly located in the right cerebral hemisphere in both groups of subjects. Blue circles identify the inferior frontal gyrus activation. Images are in neurological convention (From **ref.** ***122***, with permission.).

Defining this system's behavior is of the outmost importance for the understanding of its dysfunction in case of disease and to develop potentially successful therapeutic strategies capable to enhance its plasticity.

References

1. Fink GR, Frackowiak RS, Pietrzyk U, Passingham RE. Multiple nonprimary motor areas in the human cortex. J Neurophysiol 1997;77:2164–2174.
2. Ciccarelli O, Toosy AT, Marsden JF, Wheeler-Kingshott CM, Sahyoun C, Matthews PM, Miller DH, Thompson AJ. Identifying brain regions for integrative sensorimotor processing with ankle movements. Exp Brain Res 2005;166:31–42.
3. Debaere F, Swinnen SP, Beatse E, Sunaert S, Van Hecke P, Duysens J. Brain areas involved in interlimb coordination: a distributed network. NeuroImage 2001;14:947–958.
4. Rocca MA, Gatti R, Agosta F, Tortorella P, Riboldi E, Broglia P, Filippi M. Influence of body segment position during in-phase and antiphase hand and foot movements: a kinematic and functional MRI study. Hum Brain Mapp 2007;28:218–227.

5. Reddy H, Floyer A, Donaghy M, Matthews PM. Altered cortical activation with finger movement after peripheral denervation: comparison of active and passive tasks. Exp Brain Res 2001;138:484–491.
6. Reddy H, Narayanan S, Woolrich M, Mitsumori T, Lapierre Y, Arnold DL, Matthews PM. Functional brain reorganization for hand movement in patients with multiple sclerosis: defining distinct effects of injury and disability. Brain 2002;125:2646–2657.
7. Logothetis NK, Pauls J, Augath M, Trinath T, Oeltermann A. Neurophysiological investigation of the basis of the fMRI signal. Nature 2001;412:150–157.
8. Decety J, Perani D, Jeannerod M, Bettinardi V, Tadary B, Woods R, Mazziotta JC, Fazio F. Mapping motor representations with positron emission tomography. Nature 1994; 371:600–602.
9. Porro CA, Francescato MP, Cettolo V, Diamond ME, Baraldi P, Zuiani C, Bazzocchi M, di Prampero PE. Primary motor and sensory cortex activation during motor performance and motor imagery: a functional magnetic resonance imaging study. J Neurosci 1996;16: 7688–7698.
10. Wexler BE, Fulbright RK, Lacadie CM, Skudlarski P, Kelz MB, Constable RT, Gore JC. An fMRI study of the human cortical motor system response to increasing functional demands. Magn Reson Imaging 1997;15:385–396.
11. Deiber MP, Honda M, Ibañez V, Sadato N, Hallett M. Mesial motor areas in self-initiated versus externally triggered movements examined with fMRI: effect of movement type and rate. J Neurophysiol 1999;81:3065–3077.
12. VanMeter JW, Maisog JM, Zeffiro TA, Hallett M, Herscovitch P, Rapoport SI. Parametric analysis of functional neuroimages: application to a variable-rate motor task. NeuroImage 1995;2:273–283.
13. Dettmers C, Fink GR, Lemon RN, Stephan KM, Passingham RE, Silbersweig D, Holmes A, Ridding MC, Brooks DJ, Frackowiak RS. Relation between cerebral activity and force in the motor areas of the human brain. J Neurophysiol 1995;74:802–815.
14. Schlaug G, Knorr U, Seitz R. Inter-subject variability of cerebral activations in acquiring a motor skill: a study with positron emission tomography. Exp Brain Res 1994;98:523–534.
15. Karni A, Meyer G, Jezzard P, Adams MM; Turner R, Ungerleider LG. Functional MRI evidence for adult motor cortex plasticity during motor skill learning. Nature 1995;377: 155–158.
16. Rao SM, Binder JR, Bandettini PA, Hammeke TA, Yetzkin FAZ, Jesmanowicz A, Lisk LM, Morris GL, Mueller WM, Estkowski LD, et al. Functional magnetic resonance imaging of complex human movements. Neurology 1993;43:2311–2318.
17. Annett M. Handedness in families. Ann Hum Genet 1973;37:93–105.
18. Kim SG, Ashe J, Hendrich K, Ellermann JM, Merkle H, Ugurbil K, et al. Functional magnetic resonance imaging of motor cortex: hemispheric asymmetry and handedness. Science 1993;261:615–617.
19. Singh LN, Higano S, Takahashi S, Kurihara N, Furuta S, Tamura H, et al. Comparison of ipsilateral activation between right and left handers: a functional MR imaging study. Neuroreport 1998;9:1861–1866.
20. Solodkin A, Hlustik P, Noll DC, Small SL. Lateralization of motor circuits and handedness during finger movements. Eur J Neurol 2001;8:425–434.
21. Verstynen T, Diedrichsen J, Albert N, Aparicio P, Ivry RB. Ipsilateral motor cortex activity during unimanual hand movements relates to task complexity. J Neurophysiol 2005;93:1209–1222.
22. Lissek S, Hausmann M, Knossalla F, Peters S, Nicolas V, Güntürkün O, Tegenthoff M. Sex differences in cortical and subcortical recruitment during simple and complex motor control: an fMRI study. NeuroImage 2007; 37:912–926.
23. Ward NS. Compensatory mechanisms in the aging motor system. Ageing Res Rev 2006;5:239–254. Review.
24. Taniwaki T, Okayama A, Yoshiura T, Togao O, Nakamura Y, Yamasaki T, Ogata K, Shigeto H, Ohyagi Y, Kira J, Tobimatsu S. Age-related alterations of the functional interactions within the basal ganglia and cerebellar motor loops in vivo. NeuroImage 2007;36:1263–1276.
25. Hutchinson S, Kobayashi M, Horkan CM, Pascual-Leone A, Alexander MP, Schlaug G. Age-related differences in movement representation. NeuroImage 2002;17:1720–1728.
26. Mattay VS, Fera F, Tessitore A, Hariri AR, Das S, Callicott JH, Weinberger DR. Neurophysiological correlates of age-related changes in human motor function. Neurology 2002;58:630–635.
27. Alkadhi H, Crelier GR, Boendermaker SH, Golay X, Hepp-Reymond MC, Kollias SS. Reproducibility of primary motor cortex somatotopy under controlled conditions. AJNR Am J Neuroradiol 2002;23:1524–1532.

28. Netz J, Ziemann U, Hömberg V. Hemispheric asymmetry of transcallosal inhibition in man. Exp Brain Res 1995;104:527–533.
29. Liepert J, Dettmers C, Terborg C, Weiller C. Inhibition of ipsilateral motor cortex during phasic generation of low force. Clin Neurophysiol 2001;112:114–121.
30. Allison JD, Meador KJ, Loring DW, Figueroa RE, Wright JC. Functional MRI cerebral activation and deactivation during finger movement. Neurology 2000;54:135–142.
31. Nirkko AC, Ozdoba C, Redmond SM, Burki M, Schroth G, Hess CW, et al. Different ipsilateral representations for distal and proximal movements in the sensorimotor cortex: activation and deactivation patterns. NeuroImage 2001;13:825–835.
32. Stefanovic B, Warnking JM, Pike GB. Hemodynamic and metabolic responses to neuronal inhibition. NeuroImage 2004;22:771–778.
33. Geffen GM, Jones DL, Geffen LB. Interhemispheric control of manual motor activity. Behav Brain Res 1994;64:131–140. Review.
34. Sadato N, Yonekura Y, Waki A, Yamada H, Ishii Y. Role of the supplementary motor area and the right premotor cortex in the coordination of bimanual finger movements. J Neurosci 1997;17:9667–9674.
35. Lee KM, Chang KH, Roh JK. Subregions within the supplementary motor area activated at different stages of movement preparation and execution. NeuroImage 1999;9: 117–123.
36. Ohara S, Ikeda A, Kunieda T, Yazawa S, Baba K, Nagamine T, Taki W, Hashimoto N, Mihara T, Shibasaki H. Movement-related change of electrocorticographic activity in human supplementary motor area proper. Brain 2000;123:1203–1215.
37. Rouiller EM, Babalian A, Kazennikov O, Moret V, Yu XH, Wiesendanger M. Transcallosal connections of the distal forelimb representations of the primary and supplementary motor cortical areas in macaque monkeys. Exp Brain Res 1994;102:227–243.
38. Brinkman C. Lesions in supplementary motor area interfere with a monkey's performance of a bimanual coordination task. Neurosci Lett 1981;27:267–270.
39. Brinkman C. Supplementary motor area of the monkey's cerebral cortex: short- and long-term deficits after unilateral ablation and the effects of subsequent callosal section. J Neurosci 1984;4:918–929.
40. Martino AM, Strick PL. Corticospinal projections originate from the arcuate premotor area. Brain Res 1987;404:307–312.
41. Humberstone M, Sawle GV, Clare S, Hykin J, Coxon R, Bowtell R, Macdonald IA, Morris PG. Functional magnetic resonance imaging of single motor events reveals human presupplementary motor area. Ann Neurol 1997; 42:632–637.
42. Weilke F, Spiegel S, Boecker H, von Einsiedel HG, Conrad B, Schwaiger M, Erhard P. Time-resolved fMRI of activation patterns in M1 and SMA during complex voluntary movement. J Neurophysiol 2001;85:1858–1863.
43. Picard N, Strick PL. Motor areas of the medial wall: a review of their location and functional activation. Cereb Cortex 1996;6:342–353.
44. Rizzolatti G, Luppino G. The cortical motor system. Neuron 2001;31:889–901. Review.
45. Dum RP, Strick PL. The origin of corticospinal projections from the premotor areas in the frontal lobe. J Neurosci 1991;11:667–689.
46. Dum RP, Strick PL. Motor areas in the frontal lobe of the primate. Physiol Behav 2002;77:677–682. Review.
47. Stephan KM, Fink GR, Passingham RE, et al. Functional anatomy of the mental representation of upper extremity movements in healthy subjects. J Neurophysiol 1995;73:373–386.
48. Binkofski F, Buccino G, Posse S, et al. A fronto-parietal circuit for object manipulation in man: evidence from an fMRI-study. Eur J Neurosci 1999;11:3276–3286.
49. Harrington DL, Rao SM, Haaland KY, et al. Specialized neural systems underlying representations of sequential movements. J Cogn Neurosci 2000;12:56–77.
50. Haslinger B, Erhard P, Weilke F, et al. The role of lateral premotor-cerebellar-parietal circuits in motor sequence control: a parametric fMRI study. Brain Res Cogn Brain Res 2002;13:159–168.
51. Grafton ST, Arbib MA, Fadiga L, Rizzolatti G. Localization of grasp representation in humans by positron emission tomography. 2. Observation compared with imagination. Exp Brain Res 1996;112:103–111.
52. Scott SH, Sergio LE, Kalaska JF. Reaching movements with similar hand paths but different arm orientations. II. Activity of individual cells in dorsal premotor cortex and parietal area 5. J Neurophysiol 1997;78:2413–2426.
53. Grafton ST, Fagg AH, Arbib MA. Dorsal premotor cortex and conditional movement selection: a PET functional mapping study. J Neurophysiol 1998;79:1092–1097.
54. Bestmann S, Swayne O, Blankenburg F, Ruff CC, Haggard P, Weiskopf N, Josephs O, Driver J, Rothwell JC, Ward NS. Dorsal premotor cortex exerts state-dependent causal

influences on activity in contralateral primary motor and dorsal premotor cortex. Cereb Cortex 2007;18:1281–1291.
55. Schluter ND, Krams M, Rushworth MF, Passingham RE. Cerebral dominance for action in the human brain: the selection of actions. Neuropsychologia 2001;39:105–113.
56. Schluter ND, Rushworth MF, Passingham RE, Mills KR. Temporary interference in human lateral premotor cortex suggests dominance for the selection of movements. A study using transcranial magnetic stimulation. Brain 1998;121:785–799.
57. Marconi B, Genovesio A, Giannetti S, Molinari M, Caminiti R. Callosal connections of dorso-lateral premotor cortex. Eur J Neurosci 2003;18:775–788.
58. O'Shea J, Johansen-Berg H, Trief D, Göbel S, Rushworth MF. Functionally specific reorganization in human premotor cortex. Neuron 2007;54:479–490.
59. Cunnington R, Windischberger C, Robinson S, Moser E. The selection of intended actions and the observation of others' actions: a time-resolved fMRI study. NeuroImage 2006;29:1294–1302.
60. Deiber MP, Passingham RE, Colebatch JG, Friston KJ, Nixon PD, Frackowiak RS. Cortical areas and the selection of movement: a study with positron emission tomography. Exp Brain Res 1991;84:393–402.
61. Devinsky O, Morrell MJ, Vogt BA. Contributions of anterior cingulate cortex to behaviour. Brain 1995;118:279–306.
62. Paus T, Petrides M, Evans AC, Meyer E. Role of the human anterior cingulate cortex in the control of oculomotor, manual, and speech responses: a positron emission tomography study. J Neurophysiol 1993;70:453–469.
63. Vogt BA, Finch DM, Olson CR. Functional heterogeneity in cingulate cortex: the anterior executive and posterior evaluative regions. Cereb Cortex 1992;2:435–443. Review.
64. Botvinick M, Nystrom LE, Fissell K, Carter CS, Cohen JD. Conflict monitoring versus selection-for-action in anterior cingulate cortex. Nature 1999;402:179–181.
65. Carter CS, Braver TS, Barch DM, Botvinick MM, Noll D, Cohen JD. Anterior cingulate cortex, error detection, and the online monitoring of performance. Science 1998;280:747–749.
66. Wenderoth N, Debaere F, Sunaert S, Swinnen SP. The role of anterior cingulate cortex and precuneus in the coordination of motor behaviour. Eur J Neurosci 2005;22:235–246.
67. Rizzolatti G, Fogassi L, Gallese V. Parietal cortex: from sight to action. Curr Opin Neurobiol 1997;7:562–567. Review.
68. Karhu J, Tesche CD. Simultaneous early processing of sensory input in human primary (SI) and secondary (SII) somatosensory cortices. J Neurophysiol 1999;81:2017–2025.
69. Hamalainen H, Hiltunen J, Titievskaja I. fMRI activations of SI and SII cortices during tactile stimulation depend on attention. Neuroreport 2000;11:1673–1676.
70. Huttunen J, Wikström H, Korvenoja A, Seppäläinen AM, Aronen H, Ilmoniemi RJ. Significance of the second somatosensory cortex in sensorimotor integration: enhancement of sensory responses during finger movements. Neuroreport 1996;7:1009–1012.
71. Mima T, Nagamine T, Nakamura K, Shibasaki H. Attention modulates both primary and second somatosensory cortical activities in humans: a magnetoencephalographic study. J Neurophysiol 1998;80:2215–2221.
72. Dobkin BH. Functional MRI: a potential physiologic indicator for stroke rehabilitation interventions. Stroke 2003;34:23–28.
73. Del Gratta C, Della Penna S, Ferretti A, Franciotti R, Pizzella V, Tartaro A, Torquati K, Bonomo L, Romani GL, Rossini PM. Topographic organization of the human primary and secondary somatosensory cortices: comparison of fMRI and MEG findings. NeuroImage 2002;17:1373–1383.
74. Culham JC, Kanwisher NG. Neuroimaging of cognitive functions in human parietal cortex. Curr Opin Neurobiol 2001;11:157–163. Review.
75. Parent A, Hazrati LN. Functional anatomy of the basal ganglia. I. The cortico-basal ganglia-thalamo-cortical loop. Brain Res Brain Res Rev 1995;20:91–127. Review.
76. Kessler K, Biermann-Ruben K, Jonas M, Siebner HR, Baumer T, Munchau A, et al. Investigating the human mirror neuron system by means of cortical synchronization during the imitation of biological movements. NeuroImage 2006;33:227–238.
77. Harrington DL, Haaland KY, Knight RT. Cortical networks underlying mechanisms of time perception. J Neurosci 1998;18: 1085–1095.
78. Ivry RB, Keele SW, Diener HC. Dissociation of the lateral and medial cerebellum in movement timing and movement execution. Exp Brain Res 1988;73:167–180.
79. Jantzen KJ, Steinberg FL, Kelso JA. Brain networks underlying human timing behavior are influenced by prior context. Proc Natl Acad Sci U S A 2004;101:6815–6820.
80. Brooks DJ. The role of the basal ganglia in motor control: contributions from PET. J Neurol Sci 1995;128:1–13. Review.

81. Mesulam MM. From sensation to cognition. Brain 1998;121:1013–1052.
82. Chaudhuri A, Behan PO. Fatigue and basal ganglia. J Neurol Sci 2000;179:34–42. Review.
83. Hadjikhani N, Roland PE. Cross-modal transfer of information between the tactile and the visual representations in the human brain: a positron emission tomographic study. J Neurosci 1998;18:1072–1084.
84. Mosier K, Bereznaya I. Parallel cortical networks for volitional control of swallowing in humans. Exp Brain Res 2001;140:280–289.
85. Nitschke MF, Kleinschmidt A, Wessel K, Frahm J. Somatotopic motor representation in the human anterior cerebellum. A high-resolution functional MRI study. Brain 1996;119:1023–1029.
86. Sakai K, Takino R, Hikosaka O, Miyauchi S, Sasaki Y, Putz B, Fujimaki N. Separate cerebellar areas for motor control. Neuroreport 1998;9:2359–2363.
87. Kim JJ, Thompson RF. Cerebellar circuits and synaptic mechanisms involved in classical eyeblink conditioning. Trends Neurosci 1997;20:177–181. Review.
88. Ehrsson HH, Kuhtz-Buschbeck JP, Forssberg H. Brain regions controlling nonsynergistic versus synergistic movement of the digits: a functional magnetic resonance imaging study. J Neurosci 2002;22:5074–5080.
89. Allen GI, Tsukahara N. Cerebrocerebellar communication systems. Physiol Rev 1974;54: 957–1006. Review.
90. Ramnani N, Passingham RE. Changes in the human brain during rhythm learning. J Cogn Neurosci 2001;13:952–966.
91. Jancke L, Shah NJ, Peters M. Cortical activations in primary and secondary motor areas for complex bimanual movements in professional pianists. Brain Res Cogn Brain Res 2000;10:177–183.
92. Doyon J, Laforce RJ, Bouchard G, et al. Role of the striatum, cerebellum and frontal lobes in the automatization of a repeated visuomotor sequence of movements. Neuropsychologia 1998;36:625–641.
93. Jueptner M, Weiller C. A review of differences between basal ganglia and cerebellar control of movements as revealed by functional imaging studies. Brain 1998;121:1437–1449. Review.
94. Sanes JN, Dimitrov B, Hallett M. Motor learning in patients with cerebellar dysfunction. Brain 1990;113:103–120.
95. Bracha V, Zhao L, Irwin KB, Bloedel JR. The human cerebellum and associative learning: dissociation between the acquisition, retention and extinction of conditioned eyeblinks. Brain Res 2000;860:87–94.
96. Jenkins IH, Frackowiak RS. Functional studies of the human cerebellum with positron emission tomography. Rev Neurol 1993; 149:647–653.
97. Jenkins IH, Brooks DJ, Nixon PD, et al. Motor sequence learning: a study with positron emission tomography. J Neurosci 1994; 14:3775–3790.
98. Krings T, Topper R, Foltys H, et al. Cortical activation patterns during complex motor tasks in piano players and control subjects. A functional magnetic resonance imaging study. Neurosci Lett 2000;278:189–193.
99. Pearce AJ, Thickbroom GW, Byrnes ML, Mastaglia FL. Functional reorganisation of the corticomotor projection to the hand in skilled racquet players. Exp Brain Res 2000;130: 238–243.
100. Dirnberger G, Duregger C, Lindinger G, Lang W. Habituation in a simple repetitive motor task: a study with movement-related cortical potentials. Clin Neurophysiol 2004;115:378–384.
101. Loubinoux I, Carel C, Alary F, Boulanouar K, Viallard G, Manelfe C, Rascol O, Celsis P, Chollet F. Within-session and between-session reproducibility of cerebral sensorimotor activation: a test-retest effect evidenced with functional magnetic resonance imaging. J Cereb Blood Flow Metab 2001;21:592–607.
102. Tracy JI, Faro SS, Mohammed F, Pinus A, Christensen H, Burkland D. A comparison of 'Early' and 'Late' stage brain activation during brief practice of a simple motor task. Brain Res Cogn Brain Res 2001;10:303–316.
103. Morgen K, Kadom N, Sawaki L, Tessitore A, Ohayon J, Frank J, McFarland H, Martin R, Cohen LG. Kinematic specificity of cortical reorganization associated with motor training. NeuroImage 2004;21:1182–1187.
104. Lehéricy S, Benali H, Van de Moortele PF, Pélégrini-Issac M, Waechter T, Ugurbil K, Doyon J. Distinct basal ganglia territories are engaged in early and advanced motor sequence learning. Proc Natl Acad Sci U S A 2005;102:12566–12571.
105. Fadiga L, Fogassi L, Pavesi G, Rizzolatti G. Motor facilitation during action observation: a magnetic stimulation study. J Neurophysiol 1995;73:2608–2611.
106. Hari R, Forss N, Avikainen S, Kirveskari E, Salenius S, Rizzolatti G. Activation of human primary motor cortex during action observation: a neuromagnetic study. Proc Natl Acad Sci U S A 1998;95:15061–15065.
107. Grezes J, Armony JL, Rowe J, Passingham RE. Activations related to "mirror" and "canonical" neurones in the human brain: an fMRI study. NeuroImage 2003;18:928–937.

108. Rizzolatti G, Fadiga L, Matelli M, Bettinardi V, Paulesu E, Perani D, Fazio F. Localization of grasp representations in humans by PET: 1. Observation versus execution. Exp Brain Res 1996;111:246–252.
109. Buccino G, Vogt S, Ritzl A, Fink GR, Zilles K, Freund HJ, Rizzolatti G. Neural circuits underlying imitation learning of hand actions: an event-related fMRI study. Neuron 2004;42:323–334.
110. Rizzolatti G, Craighero L. The mirror-neuron system. Annu Rev Neurosci 2004;27:169–192. Review.
111. Johnson SH, Rotte M, Grafton ST, Hinrichs H, Gazzaniga MS, Heinze HJ. Selective activation of a parietofrontal circuit during implicitly imagined prehension. NeuroImage 2002;17:1693–1704.
112. Leslie KR, Johnson-Frey SH, Grafton ST. Functional imaging of face and hand imitation: towards a motor theory of empathy. NeuroImage 2004;21:601–607.
113. Rizzolatti G, Fogassi L, Gallese V. Neurophysiological mechanisms underlying action understanding and imitation. Nat Rev Neurosci 2001;2:661–670.
114. Iacoboni M. Neural mechanisms of imitation. Curr Opin Neurobiol 2005;15:632–637. Review.
115. Kohler E, Keysers C, Umilta MA, Fogassi L, Gallese V, Rizzolatti G. Hearing sounds, understanding actions: action representation in mirror neurons. Science 2002;297:846–848.
116. Rizzolatti G, Arbib MA. Language within our grasp. Trends Neurosci 1998;21:188–194. Review.
117. Iacoboni M, Woods RP, Brass M, Bekkering H, Mazziotta JC, Rizzolatti G. Cortical mechanisms of human imitation. Science 1999;286:2526–2528.
118. Buccino G, Binkofski F, Fink GR, Fadiga L, Fogassi L, Gallese V, Seitz RJ, Zilles K, Rizzolatti G, Freund HJ. Action observation activates premotor and parietal areas in a somatotopic manner: an fMRI study. Eur J Neurosci 2001;13:400–404.
119. Rocca MA, Tortorella P, Ceccarelli A, Falini A, Tango D, Scotti G, Comi G, Filippi M. The "mirror-neuron system" in MS: a 3 tesla fMRI study. Neurology 2008;70:255–262.
120. Aziz-Zadeh L, Koski L, Zaidel E, Mazziotta J, Iacoboni M. Lateralization of the human mirror neuron system. J Neurosci 2006;26:2964–2970.
121. Cheng YW, Tzeng OJ, Decety J, Imada T, Hsieh JC. Gender differences in the human mirror system: a magnetoencephalography study. Neuroreport 2006;17:1115–1119.
122. Rocca MA, Falini A, Comi G, Scotti G, Filippi M. The mirror-neuron system and handedness: A "right" world? Hum Brain Mapp 2008;29:1243–1254.
123. Theoret H, Halligan E, Kobayashi M, Fregni F, Tager-Flusberg H, Pascual-Leone A. Impaired motor facilitation during action observation in individuals with autism spectrum disorder. Curr Biol 2005;15:R84–R85.
124. Dapretto M, Davies MS, Pfeifer JH, Scott AA, Sigman M, Bookheimer SY, Iacoboni M. Understanding emotions in others: mirror neuron dysfunction in children with autism spectrum disorders. Nat Neurosci 2006;9:28–30.

Chapter 17

Functional Imaging of the Human Visual System

Guy A. Orban and Zoe Kourtzi

Summary

The human visual system consists of a large, yet unknown number of cortical areas. We summarize the efforts made to identify these areas, using the macaque visual cortex as a guide. So far, retinotopic mapping has identified several regions and study of functional properties such as motion and shape has revealed further expanses of visual cortex. Macaques and humans share early areas (V1, V2, and V3) and a motion-sensitive middle temporal (MT/V5) region, but the intervening cortex has considerably developed in humans with the appearance of new areas. The kinetic occipital region is located in this part of cortex between V3A and the human MT/V5 complex. Several regions sensitive to motion and even higher order motion have been described in parietal cortex. On the other hand, both dorsal and ventral regions are sensitive to shape, which is most pronounced in the lateral occipital complex (LOC). The anterior part of this complex represents visual objects rather than image properties.

Key words: Vision, Retinotopy, Cortical area, Visual field, Motion, 2D and 3D shape, Depth

1. Introduction

The human visual system is located in the occipital lobe and extends forward into the parietal and temporal lobes. It is estimated to encompass 30% of human cortex *(1)*. Functional imaging gives us direct access to the function of this important part of human cortex. One way to study this system is to consider a number of perceptual or visual cognitive functions and to localize their neural correlates. An alternative is to consider the visual system as an anatomically organized collection of cortical areas and subcortical centers that process retinal information

M. Filippi (ed.), *fMRI Techniques and Protocols*, Neuromethods, vol. 41
DOI 10.1007/978-1-60327-919-2_17, © Humana Press, a part of Springer Science+Business Media, LLC 2009

and transform it into messages appropriate for processing in the nonvisual cerebral regions to which the visual system projects. The critical aim in visual neuroscience is to define the different cortical areas that make up the human visual system. In other species, such as the nonhuman primates, cortical areas are defined by the combination of four criteria: (1) cyto and myeloarcitectonics, (2) anatomical connections with other (known) areas, (3) topographic organization, i.e., retinotopic organization, and (4) functional properties. It is important to note that while not all criteria may apply to each area, it is critical to obtain as much converging information as possible. In the nonhuman primate, 30 or more visual cortical areas have been identified with these criteria, although it is fair to state that even in these species there is discussion about the exact definition of areas, especially those at the higher levels in the system *(1)*. The definition of the visual cortical areas is only a first step in understanding the visual system; next is the investigation of the type of processing performed by these areas and the flow of information through the areas as a function of the task context and demands.

Recent advances in brain imaging have provided powerful tools for the definition and mapping of cortical areas. Functional magnetic resonance imaging (fMRI) provides insights into the functional characteristics of cortical areas by means of specific contrasts of brain activity that isolate a functional property. For example, in the monkey in which a number of visual areas have been identified using anatomical and neurophysiological measurements, fMRI has shown that a small number of functional characteristics, defined by a few subtractions, allows the definition of six motion-sensitive regions in the monkey superior temporal sulcus (STS) *(2)*. fMRI can also provide evidence for retinotopic organization. It actually is more powerful than single cell studies in this respect, as it is less biased in its sampling and the measure required is simply responsiveness. It has been suggested that the topology of an area, that is, its localization with respect to neighboring areas, might be a valuable addition for the identification of areas *(3)*. Imaging has not yet provided clear means to obtain histological structure, although at high field (7.0 T) the stria of Gennari becomes visible. The situation is slightly better for anatomical connections, as diffusion tensor imaging (DTI) *(4, 5)* is increasingly seen as a potential measure of connectivity between areas, although the methodological issues remain formidable. In the present chapter we will provide an overview of how these two strategies, functional specialization and retinotopic organization, have been used for defining cortical areas.

Despite all its strengths functional imaging has severe limitations due to its limited temporal and spatial resolution. With the present 3T systems a few millimeters can be resolved. While this is ample to define cortical regions it is a long way from the resolution of the single neuron. In fact, fMRI signals are only indirect reflections of average activity of thousands of neurons. Hence, fMRI is very sensitive at detecting average activity levels, but it has great difficulty in measuring neuronal selectivity. It has been proposed that repetition suppression can be used to measure neuronal tuning, but the case for it might be overstated *(6)*. Recent developments using multivoxel pattern analysis (MVPA) *(7)* provide sensitive tools for studying neural representations beyond the resolution of conventional fMRI approaches. Yet the estimation provided by this analysis depends heavily on the clustering of neurons with similar properties, like those in cortical columns, and the discrimination provided falls quite short of what single neurons can achieve. For example, single V1 neurons can signal orientation differences of 5°–10° with an 84% chance of success *(8)*. MVPA of human V1 has so far yielded values of 35° *(9)*. Therefore, much can be gained by combining functional imaging in humans with knowledge derived from invasive studies, such as single cell recordings in nonhuman primates. The combination has become possible with the advent of fMRI in the awake monkey *(10)*. Indeed this allows parallel imaging experiments leading to the definition of cortical regions and their characteristics in the two species, paving the way for establishing homologies. Once a homology is established, one can test whether the neuronal properties in that area apply to the human homolog. Indeed, comparison of the single cell recordings and fMRI in the monkey using similar stimuli allows one to derive an fMRI signature of a neuronal property. One can then verify that the human homolog also exhibits this fMRI signature. Hence, the definition of cortical areas in both species is a critical step for knowledge transfer from animal models to the human visual system.

2. Methodological Issues

2.1. Stimulus Definition

Definition of the visual stimulus is important as it determines to a large degree the brain activation pattern and thus the experimental findings reported. It is important to note that precise stimulus description is crucial for repeating an experiment and replicating the results. For example, very different stimuli are used for defining motion-responsive areas. A motion localizer used to localize human middle temporal (hMT/V5) region often consists of random dots, but may also consist of gratings, either rectangular or circular. Random dots may have different densities, sizes, luminance, etc., or the whole pattern may be of a different size. Random dots may translate in one or several directions, but may also rotate or move radially. All these paradigms, using very different stimuli, are referred to as motion localizers, but because of their differences they may result in activation of different cortical regions, reducing the value of the localization.

2.2. Tasks

One of the main challenges in brain imaging is investigating the link between neural activity and human behavior. Recent studies using parametric stimulus manipulation employ detection or discrimination tasks *(11–13)* rather than passive viewing of the stimuli. These paradigms allow correlation between behavioral data (psychometric functions) and fMRI activations. This approach is important for discerning the functional role of different cortical areas and evaluating their contribution to behavior. Further, attentionally demanding tasks (e.g., detection of changes in the fixation target, 1-back matching task) are used during scanning to ensure that observers pay attention across all stimulus conditions and that activation differences across conditions are not simply due to differences in the general arousal of the participants or the task difficulty across conditions. For example, when mapping the lateral occipital complex (LOC), participants view intact and scrambled images of objects. It is possible that higher activations for intact images of objects are due to the fact that these images attract the participants' attention more than scrambled images. To control for this potential confound observers are instructed to perform a task on different properties of the fixation target or the stimulus (e.g., dimming of the fixation point or part of the shape) *(14)* that entail similar attention across all stimulus conditions. Another task that has been adopted for controlling attentional confounds is the 1-back matching task (detect a repeat of an intact or scrambled image) *(15, 16)*. This task is more demanding for scrambled than intact images, thus excluding the possibility that higher activations for intact images of objects are due to attentional differences.

2.3. Control of Eye Movements

Control of fixation is mandatory in motion response studies, retinotopic mapping experiments, and in spatial attention studies. Although in the past it was acceptable to show that the subjects fixated well based on off-line measurements, standards have evolved. In addition, precise eye movement records, provided by infrared corneal reflection methods, allow one to remove the effect of residual eye movements that occur despite fixation. In general, in all visual experiments, control of fixation will ensure that the part of visual field stimulated is known and will remove eye movements as a source of unwanted and uncontrolled activations.

2.4. fMRI Designs and Paradigms

The conventional fMRI approach for identifying cortical areas involved in different processes and cognitive tasks entails a subtraction of activations between different stimulus types that are presented in blocked or event-related designs.

One of the limitations of these fMRI paradigms is that they average across neural populations that may respond homogeneously across stimulus properties or may be differentially tuned to different stimulus attributes. Thus, in most cases, it is impossible to infer the properties of the underlying imaged neural populations. fMRI adaptation (or repetition suppression) paradigms *(17–22)* have recently been employed to study the properties of neuronal populations beyond the limited spatial resolution of fMRI. These paradigms capitalize on the reduction of neural responses for stimuli that have been presented for prolonged time or repeatedly *(23, 24)*. A change in a specific stimulus dimension that elicits increased responses (i.e., rebound of activity) identifies neural populations that are tuned to the modified stimulus attributes (**Fig. 1**). fMRI adaptation paradigms have been used in both monkey and human fMRI studies as a sensitive tool that allows us to investigate: (a) the sensitivity of the neural populations to stimulus properties, and (b) the invariance of their responses within the imaged voxels. Adaptation across a change between two stimuli suggests a common neural representation invariant to that change, while recovery from adaptation suggests neural representations sensitive to specific stimulus properties. For example, recent imaging studies tested whether fMRI measurements can reveal neural populations in early visual areas sensitive to elementary visual features, e.g., orientation, color, and direction of motion *(25–29)*.

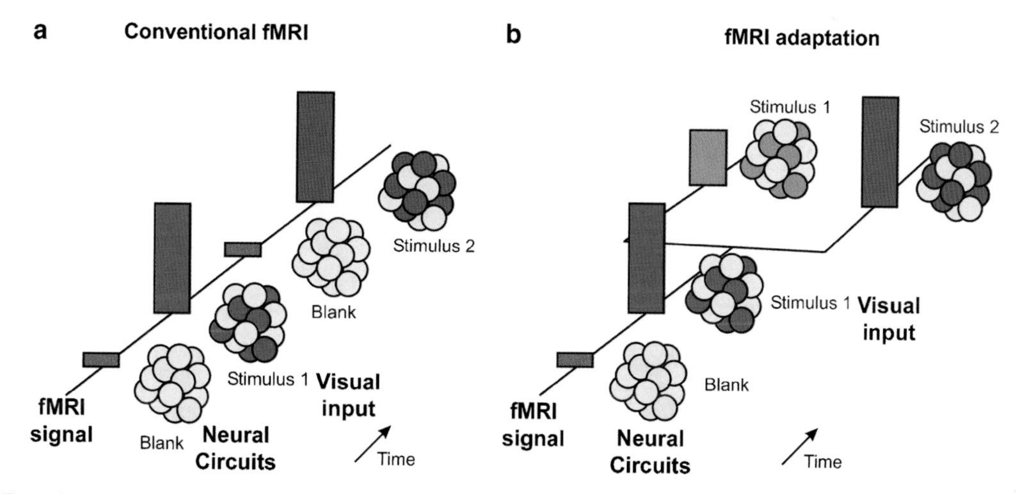

Fig. 1. fMRI paradigms. (a) Conventional fMRI: fMRI responses to two stimulus conditions 1 and 2 are compared. If neural subpopulations that encode the two stimuli are intermixed in the measured voxel, it is possible that the strength of the blood oxygen level-dependent (BOLD) signal will be the same under these two conditions failing to reveal differences in the representations of these neural populations. (b) fMRI adaptation paradigm: stimulus 1 is shown for a prolonged time or repeatedly resulting in adaptation of the BOLD signal. If different neural subpopulations encode stimulus 1 and 2, then after presentation of stimulus 2 the signal shows a rebound, that is, release from adaptation. If the same neural subpopulations encode stimulus 1 and 2, then the responses for stimulus 2 remain adapted after adaptation to stimulus 1.

Consider the case of motion direction: after prolonged exposure to the adapting motion direction, observers were tested with the same stimulus in the same or in an orthogonal motion direction. Decreased fMRI responses were observed in MT when the test stimuli were at the same motion direction as the adapting stimulus. However, recovery from this adaptation effect was observed for stimuli presented at an orthogonal direction. These studies suggest that the neural populations in human MT are sensitive to direction of motion *(26, 29)*. Using the same procedure in the monkey, Nelissen et al. *(30)* indeed observed adaptation in MT/V5 but also in other motion-sensitive regions, such as the medial superior temporal (MST) region. Similarly, recent studies have shown stronger adaptation in human hMT/V5+ for coherently than transparently moving plaid stimuli. These findings provide evidence that fMRI adaptation responses are linked to the activity of pattern-motion rather than component-motion cells in MT/MST *(27)*. Thus, these studies suggest that the fMRI signal can reveal neural selectivity consistent with the selectivity established by neurophysiological methods. However, recent studies comparing fMRI adaptation and neurophysiology in monkeys call for cautious interpretation of the relationship between fMRI adaptation effects and neural selectivity or invariance *(6)*. In particular, fMRI adaptation in a given cortical area may be the result of adaptation at earlier or later stages of processing that is propagated along the visual areas.

Interestingly, novel MVPA methods *(31–33)* provide an alternative approach for investigating neural selectivity based on fMRI signals. Unlike conventional univariate analysis, MVPA takes advantage of the information across multiple voxels in a cortical area and allows us to characterize neural representations of features that are encoded at a higher spatial resolution in the brain than the typical resolution of fMRI. These classification analyses have been used successfully for the decoding of elementary visual features [e.g., orientation *(9, 34)*, motion direction *(35)*, and object categories *(36–39)*].

2.5. Whole Brain Versus Region of Interest Analyses

The statistical evaluation of activation differences between stimulus and tasks is typically conducted by comparing responses for each voxel using the general linear model. Analysis of activation patterns across the whole brain (whole brain analysis) reveals clusters of activations in different anatomical regions that show significant differences in their functional processing. This approach has allowed researchers to identify and localize cortical regions with different functions and evaluate their involvement in various cognitive tasks. In contrast, region of interest (ROI) analysis focuses on specific cortical areas identified anatomically or functionally following standard mapping procedures (e.g., retinotopic mapping). The advantage of this approach is that it allows us to zoom in specific cortical regions and investigate their neural computations using parametric stimulus manipulations. Such manipulations result in fine stimulus variations and differences in behavioral performance. Identifying fMRI activations that reflect these fine differences in neural processing may require the high signal-to-noise ratio that is possible when scanning and analyzing smaller regions of cortex. However, ROI analyses are limited in two respects: (a) the ROI may be outside the volume scanned or analyzed, (b) the voxels of interest (i.e., voxels that show differential activations across conditions) may cover a smaller cortical volume than the ROI; as a result, the differential activations may be averaged out within the ROI. Taken together, whole brain and ROI analyses can be used as complementary tools for studying the functional roles of cortical regions. Whole brain analyses search the entire brain for regions involved in the analysis of a given stimulus or a cognitive task, while ROI methods are more appropriate for finer investigation of the neural processing in these cortical regions *(40, 41)*.

3. Retinotopic Organization

3.1. Early Visual Areas (V1, V2, V3)

Initially, positron emission tomography (PET) studies have concentrated on the retinotopy of V1 *(42)*, which is a large area of known localization in the calcarine sulcus. With the advent of fMRI, mapping was extended to areas neighboring V1 *(43)* [but see also *(44)*]. An additional step was the introduction of angular and eccentricity periodic sweeping stimuli that generate eccentricity and polar angle maps based on phase encoding of stimulus position *(45)*. This allowed the mapping of all three early areas (V1, V2, V3, **Fig. 2**) *(46–48)*, in which polar angle and eccentricity vary along orthogonal directions on the cortical surface. These three areas all have a large, complete representation of the contralateral hemifield, with the upper quadrant projecting ventrally and lower quadrant dorsally. The representation of the vertical meridian (VM) constitutes the boundary between V1 and V2 as well as the anterior boundary of V3. The representations of the horizontal meridian (HM) split the V1 representation and constitute the boundary between V2 and V3. The central representations of the three areas are fused. The coordinates of these landmarks are listed in **Table 1**. This retinotopic organization is very similar in humans and macaques **(Fig. 3)**. This is not surprising as the presence of three early visual areas is a feature of primates *(49, 50)*. In all three areas the central representation is magnified compared to that of the periphery *(51)*. Duncan and Boynton *(52)* observed a correlation between magnification

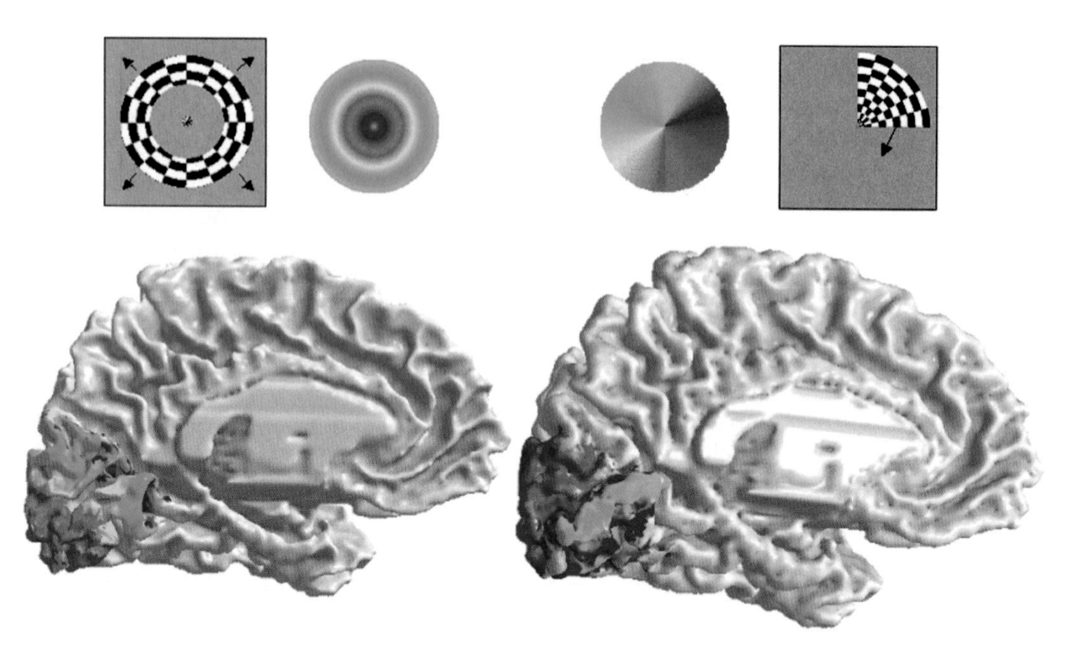

Fig. 2. The sweeping stimulus retinotopy paradigm. Two stimuli are used to measure the retinotopic maps in the cortex. Expanding ring stimuli map eccentricity, and rotating wedge stimuli map polar angle. The phase of the best-fitting sinusoid for each voxel indicates the position in the visual field that produces maximal activation for that voxel. Thus, these pseudo-color phase maps are used to visualize the retinotopic maps. Data area is shown for the left hemisphere (medial view) of one subject. Because of the heavy folding of human cortex, these retinotopic maps are best seen on flattened hemispheres [from Dougherty et al. ***(53)***].

Table 1
Talairach coordinates (X, Y, Z) of the visual areas

Visual area		Left			Right		
		X	Y	Z	X	Y	Z
V1, V2, V3 *(53)*							
	Confluence	-29	-78	-11	25	-80	-9
	V1 12° HM	-13	-63	3	9	-67	5
	V2/3v 12° HM	-14	-59	-4	10	-59	-6
	V2/3d 12° HM	-12	-75	13	7	-74	15
hV4, VO1/2 *(77)*							
	hV4	-26	-79	-16	23	-77	-14
	VO1	-27	-67	-14	21	-66	-10
	VO2	-27	-58	-13	23	-55	-10
LO1/2 *(71)*							
	LO1	-31	-90	1	32	-89	3
	LO2	-38	-83	0	38	-82	1
V3A/B, V7 *(83)*							
	V3A	-20	-90	23	20	-90	23
	V3B	-33	-85	16	33	-85	16
	V7	-26	-79	31	26	-79	31

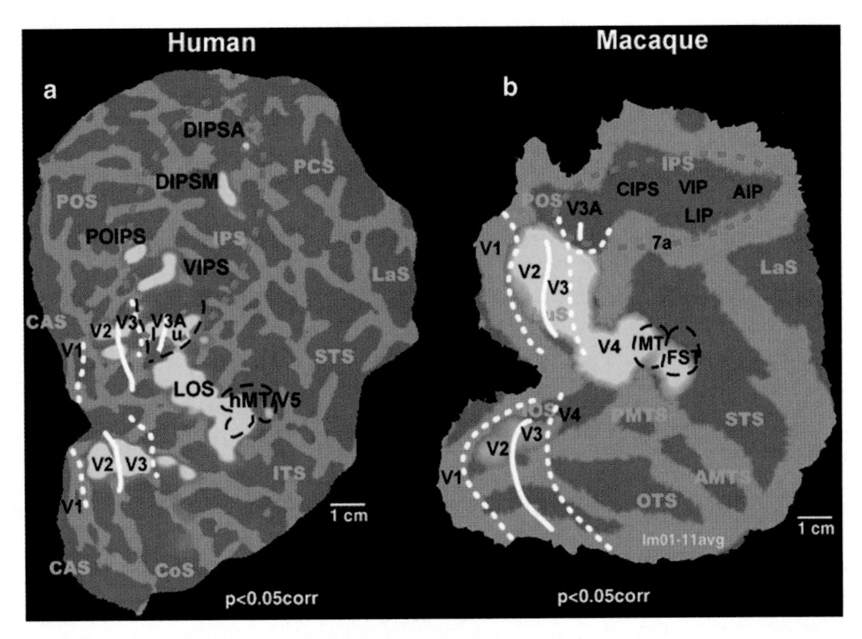

Fig. 3. Visual cortical regions sensitive to 3D shape from motion in human and macaque. Statistical parametric maps (SPMs) for the subtraction viewing of 3D rotating lines minus viewing of 2D translating lines ($p < 0.05$, corrected) of a single human (a) and monkey (M4) (b) subject projected on the posterior part of the flattened right hemisphere. *White stippled and solid lines*: vertical and horizontal meridian projections (from separate retinotopic mapping experiments); *black stippled lines*: motion-responsive regions from separate motion localizing tests; *purple stippled lines*: region of interspecies difference encompassing V3 and intraparietal sulcus. *PCS* post central sulcus, *IPS* intraparietal sulcus, *LaS* lateral sulcus, *POS* parieto-occipital sulcus, *CAS* calcarine sulcus, *STS* superior temporal sulcus, *ITS* inferior temporal sulcus, *CoS* collateral sulcus, *IOS* inferior occipital sulcus, *OTS* occipito-temporal sulcus, *PMTS* posterior middle temporal sulcus, *AMTS* anterior middle temporal sulcus [modified from ***(91)***].

factor in V1 of human subjects and Vernier acuity but not grating acuity. The surface of V1 has been estimated from histological specimens to range between 2,000 and 4,500 mm^2, while the central 12° occupy 2,200 mm^2 according to one imaging study *(53)*. Comparison between histological and fMRI estimates is difficult because of the difficulty of estimating the shrinkage in the histological specimens and the portion of V1 occupied by the central representation *(54)*. In both types of studies large variation between individuals (a factor of 2) were observed. A similar range of variation has been observed in the macaque, in which the average surface of V1 is roughly half the size of its human counterpart *(55)*. The surface of human V2 is estimated to be 80% of that of V1 – that of V3 60%. Hence, cortical magnification is somewhat lower in V2 and V3 than in V1 *(51, 53)*, but magnification factors decrease with eccentricity at similar rates in V1, V2, and V3 *(53)*.

3.2. Two Middle Level Areas: hMT/V5+ and V3A

V5 or MT area in humans was initially localized in the ascending branch of the inferior temporal sulcus (ITS) *(56, 57)*. This identification was supported by the fMRI study of Tootell et al. *(58)*, showing that this region of human cortex has properties, such as luminance and color contrast sensitivity, similar to those of macaque MT/V5. Subsequently this region has been referred to as human MT/V5+ *(47)* to indicate that probably it corresponds not just to MT/V5 of the macaque but also to several of its satellites. It has proven difficult to demonstrate a retinotopic organization in this region. Huk et al. *(59)* have suggested that the MT/V5 complex in humans contains a posterior retinotopic part, considered the homolog of MT/V5., and an anterior part driven by ipsilateral stimuli *(60)*, considered the homolog of MST. It is at present unclear whether the human complex also contains an homolog of the fundus of the superior temporal (FST) area, although the ventral part of the complex responsive to 3D shape from motion might correspond to FST (**Fig. 3**). Also the retinotopic organization of what is believed to be MT in humans *(59, 61)* seems opposite to that of macaque MT in which the lower visual field projects in the dorsal part of MT *(62, 63)*. The reasons why it is so difficult to map the retinotopy of human MT/V5 are unclear. Indeed the sweeping technique has proven to map MT and its satellites very well in the macaque *(64)*. There is at present little consensus on the criteria to define human MST *(59, 65)*.

In humans, V3A has a similar retinotopic organization as in macaque: it is defined by hemifield representation in which the representations of the two quadrants, separated by the HM, are neighbors and occupy the banks of the transverse sulcus *(66)*. The posterior quadrant is the lower quadrant, separated from that of V3d by a lower VM. In contrast to macaque V3A, hV3A is motion sensitive *(10, 66, 67)*. In the initial mapping study *(66)* the central representation of V3A was considered to be fused with that of V1–V2–V3. Subsequent studies *(68–70)* have shown that the central representation is separated from and located more dorsal than that of the V1–3 confluence, as it generally is in monkeys [5/8 hemispheres in *(63)*]. It has also been noted in humans that this foveal projection, which V3A shares with V3B (see later), can vary considerably in clarity, being well defined in about half (13/30) hemispheres *(71)*. It is noteworthy that in all primates the visual cortex includes an MT area, but that the presence of an area V3A in new world monkeys in unclear *(50, 72)*.

3.3. The Fate of V4 in Human Visual Cortex

In their 1995 study, Sereno et al. *(46)* reported an upper quadrant representation anterior to V3v, that they labeled V4v as it occupied the same position as ventral V4 in macaque. Many studies have replicated that finding of a lower quadrant in front of V3v, but it has proven difficult to identify a corresponding dorsal V4 quadrant in front of dorsal V3 *(73)*. One possible explanation was that standard mapping technique locating meridians did not apply.

Indeed, in the macaque the horizontal meridian, which represents the anterior border of ventral V4, forms the boundary of dorsal V4 only over a short distance, as it curves to join the HM splitting MT/V5 into two halves *(63, 74)*. Since the region between V3/V3A and hMT/V5+, which we refer to as LOS *(14)*, is located in a position similar to that of dorsal V4 and has functional properties relatively similar to those of macaque dorsal V4, for example, is sensitive to 3D shape from motion (**Fig. 3**), to 2D shape *(14)*, and kinetic boundaries *(75, 76)*; we *(3)* and others *(73)* have suggested that this region is the homolog of macaque dorsal V4.

Recent mapping studies concentrating on the central 6° of the visual field suggest that the two halves of macaque V4 have become separated in humans and are each integrated into a separate representation of the contralateral hemifield. Brewer et al. *(77)* have shown that a lower quadrant was located in front of the upper quadrant initially labeled V4v, with the eccentricity running at right angle to the polar variations. They proposed that this hemifield, located in front of V3v (**Fig. 4**) should be considered human V4. They went on to describe two additional

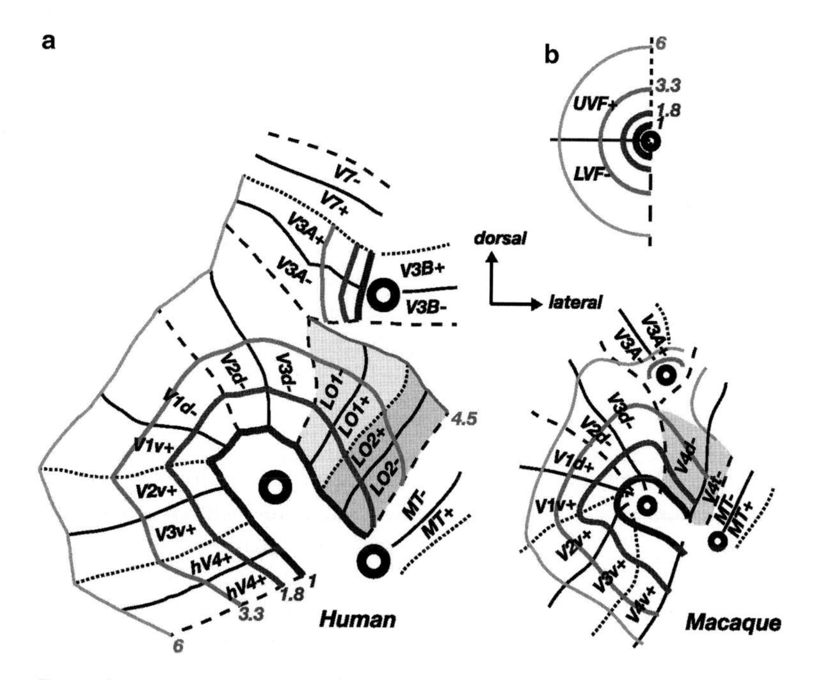

Fig. 4. Schematic summary of visual retinotopic regions according to Larsson and Heeger ***(71)***. (a) Human visual cortex. Topography and location of LO1 and LO2 relative to other retinotopic areas, shown on flattened right hemisphere, averaged across 30 hemispheres; (b) The organization of the corresponding region of macaque visual cortex is shown for comparison (not to scale). Adapted from ***(63, 74, 142)***.

maps located in front of hV4: ventral occipital (VO)1 and VO2, each supposedly containing a hemifield representation. Interestingly, the two face areas, the fusiform and occipital face areas are located just lateral to hV4 and VO2, respectively. In the same vein, Larsson and Heeger *(71)* have described a complete hemifield representation in front of V3d, which they refer to as lateral occipital (LO)1. The posterior half of this region is a lower quadrant that was initially described by Smith et al. *(78)* as V3B. Thus, the posterior parts of hV4 and LO1 apparently seem more responsive, as they were discovered first, and may be the remnants of the macaque V4 quadrants. Just as is the case ventrally, a second hemifield representation has been described in front of LO1: LO2, of which the anterior border is close to hMT/V5+ **(Fig. 4)**. Thus, while in macaque hemifield representations seemed the exception and split representations with a separate dorsal and ventral quadrants the rule, the opposite seems to hold true in humans, in which beyond the early V1–3 areas all areas seem to have hemifield arrangements **(Fig. 4)**. What is the benefit of this arrangement? As noticed earlier the dorsal region between V3/V3A and hMT/V5, in macaque as well as in human, has some particular functional characteristics, such as 3D shape from motion sensitivity. The advantage of the human arrangement is that this sensitivity applies to the whole visual field, while in macaque it applies only to the lower field. This might be an evolutionary advantage explaining the changes in this region, which has expanded considerably in humans.

After completion of the present review, two paper appeared: one entitled 'Topographic organization in and near human visual area V4' by Hansen, Kay and Gallant (J Neurosci 2007, 27:11896–11911), which presents evidence that human visual cortex includes a visual area similar to macaque V4, a second entitled 'The processing of three-dimensional shape from disparity in the human brain' by Geogieva, Peeters, Kolster and Orban (J Neurosci 2009, 29: 727–741), which supports the LO1, LO2, hV4 scheme.

3.4. Dorsal Occipital and Intraparietal Areas

Human V3A shares its central representation with an area referred to as V3B, located in front of V3A and dorsally from the LO1/LO2 pair *(69, 71)*. V3B occupies a position initially referred to as V7 *(79)*. V7 is now instead described as an area dorsal of the pair V3A/V3B, which also contains a complete hemifield **(Fig. 4)**. Area V7 seems to correspond to the motion-sensitive region, ventral intraparietal sulcus (VIPS) *(67, 70, 80)*, located in the most ventral part of the occipital part of human intraparietal sulcus (IPS) *(81)*.

In the human parieto-occipital sulcus *(82)* has described V6, which borders the dorsal parts of V2 and V3 and seems to be similar in both species. It represents the contralateral hemifield, but with an emphasis on the periphery of the visual field rather than the center.

Finally, several attempts have been made to parcel visual regions in human IPS. Using standard retinotopic mapping, Swisher et al. *(83)* described four retinotopic maps, labeled IPS1–4, separated by VM representations. Responses to standard retinotopic stimuli are weak in this region, and within anterior parts of IPS moving stimuli are more appropriate to map retinotopic organization than are black and white flickering checkerboards (**Fig. 5**). Others have used attentional stimuli *(84)* or delayed saccade stimuli *(85–87)* to map retinotopic organisation, but the relationship between the various maps described in human IPS is still unclear.

4. Motion-Sensitive Regions

4.1. Low-Level Motion Regions

The two most prominent motion-sensitive regions in human visual cortex are human MT/V5+ and V3A (see earlier). They display the highest z scores in a contrast between moving and static random dots. Their activation remains significant at low stimulus contrasts typical of the magnocellular stream *(66)*. In the occipital cortex motion responses have also been noted in lingual gyrus, probably corresponding to ventral V2, V3, and in parts of LOS *(14, 67, 88, 89)*.

In the early studies *(57, 66)* it was noted that some parietal regions were also responsive to motion in a contrast between moving and static random dots. Sunaert et al. *(67)* described four motion-sensitive regions in the IPS. The VIPS region is located at the bottom of the IPS near hV3A. This region, we believe corresponds to V7. The parieto-occipital IPS (POIPS) region is located dorsally with respect to VIPS, at the junction of the parieto-occipital sulcus and IPS, in the vicinity of hV6. Not surprisingly, it represents mainly the peripheral visual field *(70)* (**Fig. 5**). The dorsal IPS medial and anterior (DIPSM and DIPSA) regions are located in the horizontal part of IPS, and both represent mainly the central visual field *(70)* (**Fig. 5**). They are considered the homolog of anterior part of lateral intraparietal (LIP) region (DIPSM) and posterior part of anterior intraparietal (AIP) region (DIPSA), and indeed DIPSA is located just behind the region referred to as human homolog of AIP based on activation by grasping actions *(90)*. All these regions are also activated by 3D shape from motion *(81)*, which just as motion itself has a much more extensive representation in human IPS than in macaque IPS *(91, 92)*. We have speculated that this might in part be due to the more extensive tool use in humans than in monkeys, and using a tool indeed activates DIPSM and DIPSA *(93)*. These different parietal regions may be engaged in different visuo-motor control circuits, for example, in the control of heading *(94)*. Furthermore,

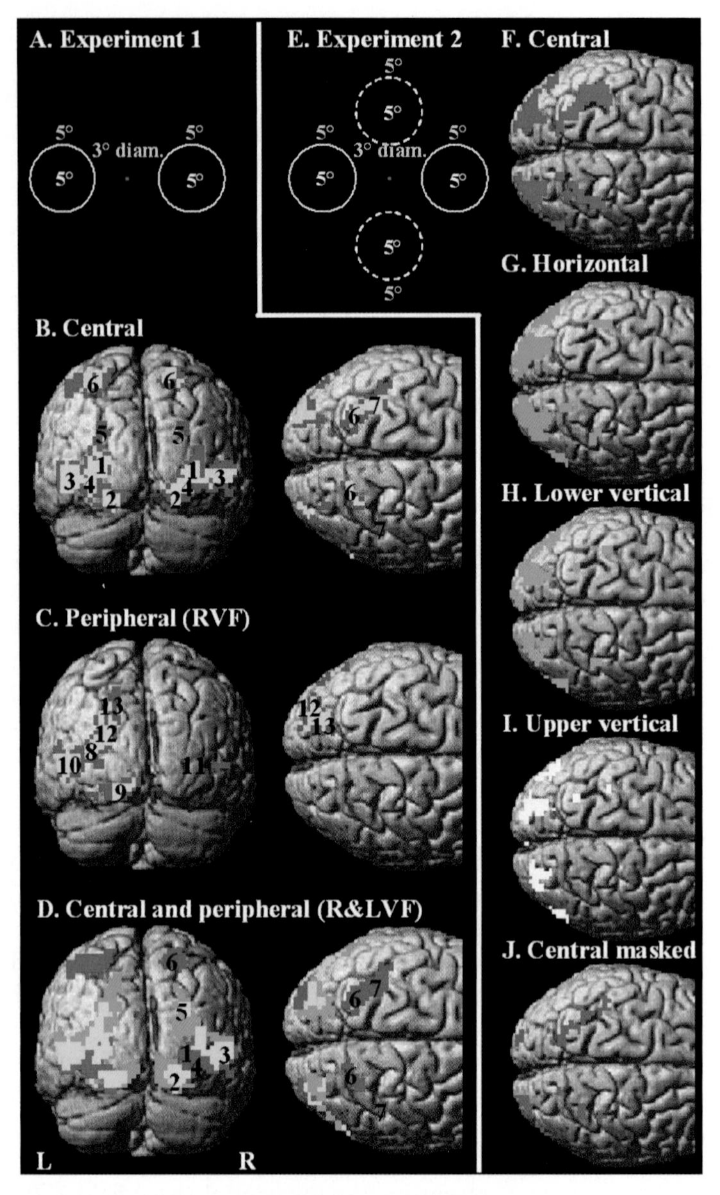

Fig. 5. Human motion-sensitive regions: distinction between central and peripheral visual field. Central and peripheral motion processing in humans. (a) Stimulus configuration in experiment 1: the randomly textured pattern (RTP) was positioned either centrally or 5° into left and right visual field (*red dot* indicates fixation point). (b, c) Statistical parametric maps (SPMs) showing voxels significant (*yellow*: $p < 0.0001$ uncorrected for multiple comparisons, corresponding to a false discovery rate of less than 5% false positives; *red*: $p < 0.001$ uncorrected) in the group random-effects analysis (experiment 1, $n = 16$) for the subtraction moving minus stationary conditions for the centrally (b) and peripherally (right visual field) (c) positioned stimulus, rendered on the posterior and superior views of the standard human brain. Further statistical testing revealed that the interaction between type of stimulus (motion, stationary) and location (center, periphery) was significant (random effects analysis) in DIPSA ($Z = 3.12$, $p < 0.001$ uncorrected and $Z = 3.58$, $p < 0.001$ uncorrected for right and left, respectively), DIPSM ($Z = 3.58$, $p < 0.001$ uncorrected and $Z = 4.35$, $P < 0.0001$ uncorrected for right and left, respectively) and weakly in POIPS ($Z = 2.69$, $P < 0.01$ uncorrected and $Z = 2.24$, $p < 0.01$ uncorrected for right and left, respectively). (d) Overlap of voxels ($p < 0.001$ uncorrected; yellow) in the group random-effects analysis for the subtraction moving minus stationary conditions for the centrally (*red*) and peripherally (right and left visual field; *green*) positioned stimulus (experiment 1), rendered on the posterior and superior views of the standard human brain.

it has been shown that flicker is rejected gradually from hMT/V5+ to the more anterior IPS regions *(67, 80, 95)*.

Further regions sensitive to motion, but not to 3D shape from motion, are premotor regions corresponding to the frontal eye field (FEF) *(67, 81)*, as well as a region in the posterior insula, which we refer to as posterior insular cortex (PIC) region *(67, 92)*, and which might be the homolog of a visual region located next to the posterior insular vestibular cortex (PIVC) in macaques *(10, 96)*.

4.2. The Kinetic Occipital (KO) Region

Using kinetic gratings, that is, stimuli in which random dots move in opposite directions in alternate stripes, and comparing them to luminance gratings or uniform motion, our group *(75, 97, 98)* discovered a region located between V3/V3A and hMT/V5+ that appeared selective for kinetic boundaries and that we referred to as the kinetic occipital (KO) region. Recent work by Zeki et al. *(100)* has proposed that KO responds to boundaries defined by other cues (e.g., colors). These findings do not dispute the responsiveness of KO to kinetic gratings as several groups have observed these responses *(71, 99)*. Although they have been presented differently, these findings are in fact consistent with our PET *(98)* and fMRI studies *(75)* showing responses in KO for both kinetic and luminance gratings, suggesting that KO responds to contours of different nature, not just kinetic contours. However, it is important to emphasize that in contrast with responses in hMT/V5+ and other motion-sensitive regions, KO is selective for kinetic contours as opposed to uniform motion. Thus, we meant selectivity in the motion domain, not in the domain of cues defining contours, when we stated *(75)* that KO is selective for kinetic boundaries. In the Van Oostende et al. study *(75)* we observed overlap of the KO region with response to the LO localizer. Indeed, Larsson and Heeger *(71)* in their study identifying LO1/2 showed that the maximal response to kinetic gratings compared to transparent motion, the contrast most sharply defining KO *(75)*, was strongest in LO1 and V3A/B. The coordinates of LO1 (**Table 1**) are very similar

Fig. 5. (continued) (e) Stimulus configuration in experiment 2: RTP was positioned centrally or at 5° eccentricity on upper or lower vertical or horizontal meridian. (f-i) SPMs showing voxels significant ($p < 0.05$ corrected) in experiment 2 ($n = 3$) for the subtraction moving minus stationary conditions for the stimuli positioned in the central visual field (f, *red*), peripherally left and right on the horizontal meridian (g, *green*), and on the lower (h, *blue*) and upper vertical meridian (i, *white*), rendered on the superior view of the standard human brain (posterior part). (j) SPM showing voxels that are active only in the central condition (obtained by exclusive masking of the subtraction in (f) with those in (g-i). The opposite procedure, subtractions (g–i) masked by that in (f), yielded no active voxels. *R* right, *L* left, *VF* visual field. *White and yellow numbers* in (a) and (e) indicate eccentricity and diameter (diameter), respectively. Numbers in (b-d) correspond to the activation sites listed: 1 and 8: hV3A; 2 and 9: lingual gyrus; 3, 10, and 11: hMT/V5+; 4: LOS; 5 and 12: VIPS; 13: POIPS; 6: DIPSM; and 7: DIPSA [from Orban et al.*(70)*].

to those of KO [±31, −91, 0, and −32, −92, 0, *(75)*]. Since the coordinates of KO were those of the local maxima, there is some argument for identifying LO1 as the core region of KO.

4.3. High-Level Motion Area

All these motion-sensitive regions are low-level motion regions in the sense that they are driven by motion of light over the retina. Claeys et al. *(80)* provided evidence for an attention-based motion-sensitive region in the inferior parietal lobule (IPL). This region has activated equiluminant color gratings in which one of the colors is more salient than the other, a paradigm tapping third order motion *(101, 102)*. In addition this region has a bilateral representation of the visual field, while all other motion-sensitive areas have mainly a contralateral representation.

5. Shape-Sensitive Regions

There is accumulating evidence that neuronal processes supporting object recognition are coarsely localized in the ventral visual stream *(103)* that contains a hierarchy of cortical processing stages (V1 → V2 → V4 → IT). The highest stages of this stream [i.e., anterior inferior temporal cortex, AIT or anterior TE in the monkey, and LOC in the human *(14, 15, 104–106)*] are thought to be involved in shape processing and support object recognition (**Fig. 6**). But how are these neuronal representations that support object recognition constructed in the brain? In the monkey, the visual system has been suggested to recruit a hierarchical network of areas across the ventral visual pathway *(103, 107)* with selectivity for features of increasing complexity from early to later stages of processing *(108)*. Recent neuroimaging studies suggest a similar organization in the human brain. That is, local image features (e.g., position, orientation) are shown to be processed at the first stage of cortical processing (V1) *(32, 34)* while complex shapes and even abstract object categories (faces, bodies, places) are represented toward the end of the pathway in the LOC *(109–112)*. Recently, combined monkey and human fMRI studies showed that the perception of global shapes involves both early (retinotopic) and higher (occipitotemporal) visual areas that may integrate local elements to global shapes at different spatial scales *(113, 114)*. However, unlike neurons in early visual areas that integrate local information about global shapes within the neighborhood of their receptive fields, neural populations in the LOC represent the perceived global form of objects. In particular, recent imaging studies *(115)* have shown fMRI adaptation in LOC when the perceived shape of visual stimuli was identical but the image contours differed (because occluding bars occurred in

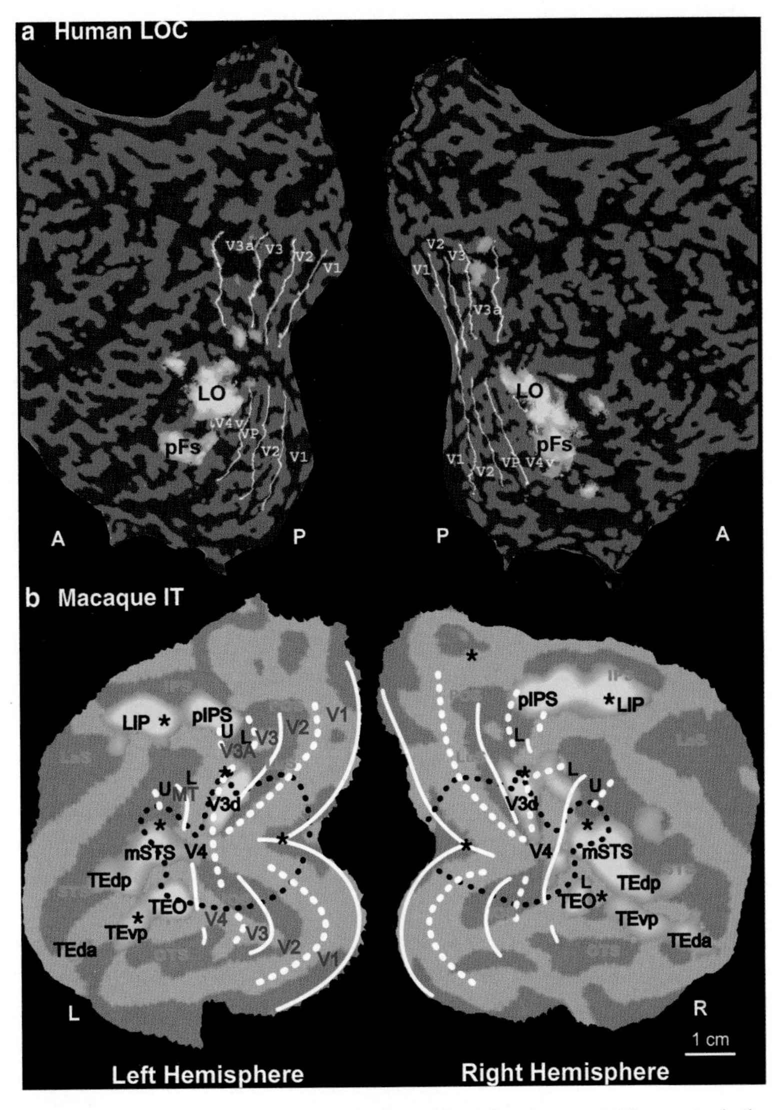

Fig. 6. The lateral occipital complex (LOC) in the human brain and the infero-temporal (IT) complex in the macaque brain: (a) Functional activation maps for one subject showing the early retinotopic regions and the lateral occipital complex (LOC). The functional activations are superimposed on flattened cortical surfaces of the right and left hemispheres. The sulci are coded in darker gray than the gyri and the anterior-posterior orientation is noted by A and P. Major sulci are labeled: *STS* superior temporal sulcus, *ITS* inferior temporal sulcus, *OTS* occipitotemporal sulcus, *CoS* collateral sulcus. The borders (shown by lines) of the early visual regions were defined with standard retinotopic techniques. The LOC was defined as the set of all contiguous voxels in the ventral occipitotemporal cortex that were activated more strongly ($p < 10^{-4}$) by intact than by scrambled images of objects. The posterior (LO) and anterior regions (pFs) of the LOC were identified based on anatomical criteria. These two parts are referred to as posterior ITG or LO and mid FG or LOa by Denys et al. ***(14)***. (b) Shape-sensitive regions in monkeys. Statistical parametric maps (T-score maps) indicating that voxels (colored *yellow to red*) are significantly more active when viewing intact than scrambled images in the group of four subjects superimposed on the flattened (FreeSurfer) left and right hemispheres (*A*) of subject M3. Threshold is $p < 0.05$ corrected (fixed effect) in all panels. The labels of the shape-sensitive regions are indicated in *black*. The *white and black lines* indicate retinotopic organization: *solid white lines*, horizontal median; *dashed white lines*, vertical median; *dashed black lines and stars*, central visual field (1.5°); U and L indicate upper and lower visual fields, respectively ***(63)***. *Blue labels* indicate retinotopic regions and MT/V5 identified from motion localizer. Green labels indicate sulci. *LaS* lateral sulcus, *OTS* occipitotemporal sulcus, *IOS* inferior occipital sulcus, *LuS* lunate sulcus, *POS* parieto-occipital sulcus [from Denys et al. ***(14)***]. Notice the parietal activation in the monkey brain maps. This activation does not reflect a species difference, as it was also observed in humans in that study ***(14)***. It most likely reflects the different task performed by the subjects. In monkeys, in contrast to humans, it is possible to record neurons located in the activated regions. This approach has revealed 2D shape-selective neurons in the anterior parietal activation site ***(143, 144)***.

front of the shape in one stimulus and behind the shape in the other). In contrast, recovery from adaptation was observed when the contours were identical but the perceived shapes were different (because of a figure-ground reversal).

Importantly, neural representations for global shapes and object categories are characterized by three main principles. First, representations for a large range of object categories (animate and inanimate) have been shown to be distributed across neural populations in the LOC, but specialized regions for the processing of faces, bodies, and places have also been identified *(36, 116)*. Second, object representations show a retinotopic organization *(71, 117)* and a center-periphery organization; that is, objects that entail central-vision processing (e.g., faces, words) show center-biased representations, while objects that entail integration across the visual field (e.g., places) show periphery-biased representations *(118, 119)*. Third, unlike earlier visual areas, patterns of neuronal activity in the LOC contain information about object identity independent of image changes (e.g., object position, size, orientation) *(109, 110, 120, 121)*. In particular, representations in the anterior subregion of the LOC in the fusiform gyrus (pFs) were shown to be largely invariant to size and position, but not invariant to the direction of illumination and rotation around the vertical axis. In contrast, representations in the posterior subregion of the LOC in the lateral occipital (LO) cortex did not show size or position invariance *(19, 122)*. These findings suggest a progression of visual object representations from posterior temporal regions that contain information about local image properties (e.g., position, size) to anterior regions that show a higher degree of tolerance to changes to these image properties.

6. Depth Processing and 3D Shape Perception

Neurophysiological studies have revealed selectivity for binocular disparity at multiple levels of the visual hierarchy in the monkey brain from early visual areas, to object- and motion-selective areas and the parietal cortex [for reviews: *(123–126)*]. Imaging studies have identified multiple human brain areas in the visual, temporal, and parietal cortex that show stronger activations for stimuli defined by binocular or monocular depth cues than for 2D versions of these stimuli. In particular, areas V3A *(127–130)* and V3B/LO1/KO *(75, 99, 100, 131)* have been implicated in the analysis of disparity-defined surfaces and boundaries. Furthermore, studies have employed parametric manipulations to investigate the neural correlates of surface depth (i.e., near vs. far) judgments *(129, 132)* and 3D shape perception *(13)*. Finally, several recent studies suggest that areas involved in disparity processing, primarily in the temporal and parietal cortex, are also engaged in the processing of monocular cues to depth (e.g., texture, motion, shading) *(14, 81, 91, 133–140)* and the combination of binocular and monocular cues for depth perception *(141)*.

7. Conclusions

It is likely that the human visual system includes about 40–50 cortical areas. Only a quarter of these have been identified so far, using retinotopic mapping or functional properties. Further progress can be expected from mapping retinotopic organization with functionally more specific stimuli than black and white checkerboards and from studying higher order visual attributes, such as 3D shape or actions.

References

1. Van Essen DC. Organization of visual areas in macaque and human cerebral cortex. In: Chalupa LM, Werner JS, editors. The Visual Neurosciences, Vol. 1. Cambridge, MA: MIT Press; 2004, pp. 507–521.
2. Nelissen K, Luppino G, Vanduffel W, Rizzolatti G, Orban G. Representation of observed actions in macaque occipitotemporal and parietal cortex. Soc Neurosci Abstr USA 2006 (abstract) 306–312.
3. Orban GA, Van Essen D, Vanduffel W. Comparative mapping of higher visual areas in monkeys and humans. Trends Cogn Sci 2004; 8(7):315–324.
4. Dougherty RF, Ben-Shachar M, Deutsch G, Potanina P, Bammer R, Wandell BA. Occipital-callosal pathways in children: Validation and atlas development. Ann N Y Acad Sci 2005; 1064:98–112.
5. Schmahmann JD, Pandya DN, Wang R et al. Association fibre pathways of the brain: Parallel observations from diffusion spectrum imaging and autoradiography. Brain 2007;130 (Part 3): 630–653.
6. Sawamura H, Orban GA, Vogels R. Selectivity of neuronal adaptation does not match response selectivity: A single-cell study of the FMRI adaptation paradigm. Neuron 2006; 49(2):307–318.
7. Haynes JD, Rees G. Decoding mental states from brain activity in humans. Nat Rev Neurosci 2006;7(7):523–534.
8. Vogels R, Orban GA. How well do response changes of striate neurons signal differences in orientation: A study in the discriminating monkey. J Neurosci 1990;10(11):3543–3558.
9. Haynes JD, Rees G. Predicting the orientation of invisible stimuli from activity in human primary visual cortex. Nat Neurosci 2005;8(5):686–691.
10. Vanduffel W, Fize D, Mandeville JB et al. Visual motion processing investigated using contrast agent-enhanced fMRI in awake behaving monkeys. Neuron 2001;32(4):565–577.
11. Boynton GM, Demb JB, Glover GH, Heeger DJ. Neuronal basis of contrast discrimination. Vision Res 1999;39(2):257–269.
12. Zenger-Landolt B, Heeger DJ. Response suppression in v1 agrees with psychophysics of surround masking. J Neurosci 2003;23(17):6884–6893.
13. Chandrasekaran C, Canon V, Dahmen JC, Kourtzi Z, Welchman AE. Neural correlates of disparity-defined shape discrimination in the human brain. J Neurophysiol 2007;97(2):1553–1565.
14. Denys K, Vanduffel W, Fize D et al. The processing of visual shape in the cerebral

cortex of human and nonhuman primates: A functional magnetic resonance imaging study. J Neurosci 2004;24(10):2551–2565.

15. Kourtzi Z, Kanwisher N. Cortical regions involved in perceiving object shape. J Neurosci 2000;20:3310–3318.
16. Kourtzi Z, Kanwisher N. Cortical regions involved in perceiving object shape. J Neurosci 2000;20(9):3310–3318.
17. Buckner RL, Goodman J, Burock M et al. Functional-anatomic correlates of object priming in humans revealed by rapid presentation event-related fMRI. Neuron 1998;20: 285–296.
18. Grill-Spector K, Kushnir T, Edelman S, Avidan G, Itzchak Y, Malach R. Differential processing of objects under various viewing conditions in the human lateral occipital complex. Neuron 1999;24:187–203.
19. Grill-Spector K, Malach R. fMR-adaptation: A tool for studying the functional properties of human cortical neurons. Acta Psychol 2001;107(1–3):293–321.
20. Grill-Spector K, Henson R, Martin A. Repetition and the brain: Neural models of stimulus-specific effects. Trends Cogn Sci 2006;10(1):14–23.
21. Koutstaal W, Wagner AD, Rotte M, Maril A, Buckner RL, Schacter DL. Perceptual specificity in visual object priming: Functional magnetic resonance imaging evidence for a laterality difference in fusiform cortex. Neuropsychologia 2001;39(2):184–199.
22. Vuilleumier P, Henson RN, Driver J, Dolan RJ. Multiple levels of visual object constancy revealed by event-related fMRI of repetition priming. Nat Neurosci 2002;5(5):491–499.
23. Lisberger SG, Movshon JA. Visual motion analysis for pursuit eye movements in area MT of macaque monkeys. J Neurosci 1999;19(6):2224–2246.
24. Muller JR, Metha AB, Krauskopf J, Lennie P. Rapid adaptation in visual cortex to the structure of images. Science 1999;285(5432):1405–1408.
25. Tootell RBH, Reppas JB, Dale AM et al. Visual motion aftereffect in human cortical area MT revealed by functional magnetic resonance imaging. Nature 1995;375:139–141.
26. Huk AC, Ress D, Heeger DJ. Neuronal basis of the motion aftereffect reconsidered. Neuron 2001;32:161–172.
27. Huk AC, Heeger DJ. Pattern-motion responses in human visual cortex. Nat Neurosci 2002;5(1):72–75.
28. Engel SA, Furmanski CS. Selective adaptation to color contrast in human primary visual cortex. J Neurosci 2001;21(11):3949–3954.
29. Tolias AS, Smirnakis SM, Augath MA, Trinath T, Logothetis NK. Motion processing in the macaque: Revisited with functional magnetic resonance imaging. J Neurosci 2001;21:8594–8601.
30. Nelissen K, Vanduffel W, Orban GA. Charting the lower superior temporal region, a new motion-sensitive region in monkey superior temporal sulcus. J Neurosci 2006;26(22):5929–5947.
31. Cox DD, Savoy RL. Functional magnetic resonance imaging (fMRI) "brain reading": Detecting and classifying distributed patterns of fMRI activity in human visual cortex. Neuroimage 2003;19(2, Part 1):261–270.
32. Haynes JD, Rees G. Decoding mental states from brain activity in humans. Nat Rev Neurosci 2006;7(7):523–534.
33. Norman KA, Polyn SM, Detre GJ, Haxby JV. Beyond mind-reading: Multi-voxel pattern analysis of fMRI data. Trends Cogn Sci 2006;10(9):424–430.
34. Kamitani Y, Tong F. Decoding the visual and subjective contents of the human brain. Nat Neurosci 2005;8(5):679–685.
35. Kamitani Y, Tong F. Decoding seen and attended motion directions from activity in the human visual cortex. Curr Biol 2006;16(11):1096–1102.
36. Haxby JV, Gobbini MI, Furey ML, Ishai A, Schouten JL, Pietrini P. Distributed and overlapping representations of faces and objects in ventral temporal cortex. Science 2001;293(5539):2425–2430.
37. Hanson SJ, Matsuka T, Haxby JV. Combinatorial codes in ventral temporal lobe for object recognition: Haxby (2001) revisited: Is there a "face" area? 2004;23(1):156–166.
38. O'Toole AJ, Jiang F, Abdi H, Haxby JV. Partially distributed representations of objects and faces in ventral temporal cortex. J Cogn Neurosci 2005;17(4):580–590.
39. Williams MA, Dang S, Kanwisher NG. Only some spatial patterns of fMRI response are read out in task performance. Nat Neurosci 2007;10(6):685–686.
40. Friston KJ, Rotshtein P, Geng JJ, Sterzer P, Henson RN. A critique of functional localisers. Neuroimage 2006;30(4):1077–1087.
41. Saxe R, Brett M, Kanwisher N. Divide and conquer: A defense of functional localizers. Neuroimage 2006;30(4):1088–1096.
42. Fox PT, Mintun MA, Raichle ME, Miezin FM, Allman JM, Van Essen DC. Mapping human visual cortex with positron emission tomography. Nature 1986;323:806–809.
43. Schneider W, Noll DC, Cohen JD. Functional topographic mapping of the cortical ribbon in

human vision with conventional MRI scanners. Nature 1993;365:150–153.

44. Shipp S, Watson JDG, Frackowiak RSJ, Zeki S. Retinotopic maps in human prestriate visual cortex: The demarcation of areas V2 and V3. Neuroimage 1995;2:125–132.
45. Engel SA, Rumelhart DE, Wandell BA et al. fMRI of human visual cortex. Nature 1994;369:525.
46. Sereno MI, Dale AM, Reppas JB et al. Borders of multiple visual areas in humans revealed by functional MRI. Science 1995;268:889–893.
47. DeYoe EA, Carman GJ, Bandettini P et al. Mapping striate and extrastriate visual areas in human cerebral cortex. Proc Natl Acad Sci USA 1996;93:2382–2386.
48. Engel SA, Glover GH, Wandell BA. Retinotopic organization in human visual cortex and the spatial precision of functional MRI. Cerebr Cortex 1997;7:181–192.
49. Lyon DC, Kaas JH. Evidence for a modified V3 with dorsal and ventral halves in macaque monkeys. Neuron 2002;33:453–461.
50. Rosa MG, Tweedale R. Brain maps, great and small: Lessons from comparative studies of primate visual cortical organization. Philos Trans R Soc Lond B Biol Sci 2005;360(1456): 665–691.
51. Sereno MI, Dale AM, Reppas JB et al. Borders of multiple visual areas in humans revealed by functional magnetic resonance imaging. Science 1995;268(5212):889–893.
52. Duncan RO, Boynton GM. Cortical magnification within human primary visual cortex correlates with acuity thresholds. Neuron 2003;38(4):659–671.
53. Dougherty RF, Koch VM, Brewer AA, Fischer B, Modersitzki J, Wandell BA. Visual field representations and locations of visual areas V1/2/3 in human visual cortex. J Vis 2003;3(10):586–598.
54. Adams DL, Sincich LC, Horton JC. Complete pattern of ocular dominance columns in human primary visual cortex. J Neurosci 2007;27(39):10391–10403.
55. Van Essen DC, Newsome WT, Maunsell JH. The visual field representation in striate cortex of the macaque monkey: Asymmetries, anisotropies, and individual variability. Vision Res 1984;24:429–448.
56. Zeki S, Watson JDG, Lueck CJ, Friston KJ, Kennard C, Frackowiak RSJ. A direct demonstration of functional specialization in human visual cortex. J Neurosci 1991;11:641–649.
57. Watson JDG, Myers R, Frackowiak RSJ et al. Area V5 of the human brain: Evidence from a combined study using positron emission tomography and magnetic resonance imaging. Cereb Cortex 1993;3:79–94.
58. Tootell RBH, Reppas JB, Kwong KK et al. Functional analysis of human MT/V5 and related visual cortical areas using magnetic resonance imaging. J Neurosci 1995;15:3215–3230.
59. Huk AC, Dougherty RF, Heeger DJ. Retinotopy and functional subdivision of human areas MT and MST. J Neurosci 2002;22(16):7195–7205.
60. Dukelow SP, DeSouza JF, Culham JC, van-den-Berg AV, Menon RS, Vilis T. Distinguishing subregions of the human MT+ complex using visual fields and pursuit eye movements. J Neurophysiol 2001;86:1991–2000.
61. Smith AT, Wall MB, Williams AL, Singh KD. Sensitivity to optic flow in human cortical areas MT and MST. Eur J Neurosci 2006; 23(2):561–569.
62. Van Essen DC, Maunsell JHR, Bixby JL. The middle temporal visual area in the macaque: Myeloarchitecture, connections, functional properties and topographic organization. J Comp Neurol 1981;199:293–326.
63. Fize D, Vanduffel W, Nelissen K et al. The retinotopic organization of primate dorsal V4 and surrounding areas: A functional magnetic resonance imaging study in awake monkeys. J Neurosci 2003;23(19):7395–7406.
64. Kolster H, Ekstrom LB, Mandeville JB, Wald LL, Dale AM, Vanduffel W. Can we distinguish MT from neighboring areas V4t and MST using awake monkey fMRI procedures at 7T? Soc Neurosci Abstr USA 2006 (abstract) 114–118.
65. Morrone MC, Tosetti M, Montanaro D, Fiorentini A, Cioni G, Burr DC. A cortical area that responds specifically to optic flow, revealed by fMRI. Nat Neurosci 2000;3:1322–1328.
66. Tootell RB, Mendola JD, Hadjikhani NK et al. Functional analysis of V3A and related areas in human visual cortex. J Neurosci 1997;17:7060–7078.
67. Sunaert S, Van Hecke P, Marchal G, Orban GA. Motion-responsive regions of the human brain. Exp Brain Res 1999;127(4):355–370.
68. Press WA, Brewer AA, Dougherty RF, Wade AR, Wandell BA. Visual areas and spatial summation in human visual cortex. Vision Res 2001;41(10–11):1321–1332.
69. Wandell BA, Brewer AA, Dougherty RF. Visual field map clusters in human cortex. Philos Trans R Soc Lond B Biol Sci 2005;360(1456):693–707.
70. Orban GA, Claeys K, Nelissen K et al. Mapping the parietal cortex of human and

non-human primates. Neuropsychologia 2006;44(13):2647–2667.

71. Larsson J, Heeger DJ. Two retinotopic visual areas in human lateral occipital cortex. J Neurosci 2006;26(51):13128–13142.

72. Lyon DC, Kaas JH. Evidence from V1 connections for both dorsal and ventral subdivisions of V3 in three species of New World monkeys. J Comp Neurol 2002;449(3):281–297.

73. Tootell RB, Hadjikhani N. Where is 'dorsal V4' in human visual cortex? Retinotopic, topographic and functional evidence. Cereb Cortex 2001;11:298–311.

74. Gattass R, Sousa AP, Gross CG. Visuotopic organization and extent of V3 and V4 of the macaque. J Neurosci 1988;8:1831–1845.

75. Van Oostende S, Sunaert S, Van Hecke P, Marchal G, Orban GA. The kinetic occipital (KO) region in man: An fMRI study. Cereb Cortex 1997;7(7):690–701.

76. Nelissen K, Vanduffel W, Sunaert S, Janssen P, Tootell RB, Orban GA. Processing of kinetic boundaries investigated using fMRI and double-label deoxyglucose technique in awake monkeys. Soc Neurosci Abstr USA 2000;26:1584.

77. Brewer AA, Liu J, Wade AR, Wandell BA. Visual field maps and stimulus selectivity in human ventral occipital cortex. Nat Neurosci 2005;8(8):1102–1109.

78. Smith AT, Greenlee MW, Singh KD, Kraemer FM, Hennig J. The processing of first- and second-order motion in human visual cortex assessed by functional magnetic resonance imaging (fMRI). J Neurosci 1998; 18(10):3816–3830.

79. Tootell RB, Tsao D, Vanduffel W. Neuroimaging weighs in: Humans meet macaques in "primate" visual cortex. J Neurosci 2003; 23(10):3981–3989.

80. Claeys KG, Lindsey DT, De SE, Orban GA. A higher order motion region in human inferior parietal lobule: Evidence from fMRI. Neuron 2003;40(3):631–642.

81. Orban GA, Sunaert S, Todd JT, Van HP, Marchal G. Human cortical regions involved in extracting depth from motion. Neuron 1999;24(4):929–940.

82. Pitzalis S, Galletti C, Huang RS et al. Wide-field retinotopy defines human cortical visual area v6. J Neurosci 2006;26(30):7962–7973.

83. Swisher JD, Halko MA, Merabet LB, McMains SA, Somers DC. Visual topography of human intraparietal sulcus. J Neurosci 2007;27(20):5326–5337.

84. Silver MA, Ress D, Heeger DJ. Topographic maps of visual spatial attention in human parietal cortex. J Neurophysiol 2005;94(2): 1358–1371.

85. Sereno MI, Pitzalis S, Martinez A. Mapping of contralateral space in retinotopic coordinates by a parietal cortical area in humans. Science 2001;294(5545):1350–1354.

86. Schluppeck D, Glimcher P, Heeger DJ. Topographic organization for delayed saccades in human posterior parietal cortex. J Neurophysiol 2005;94(2):1372–1384.

87. Schluppeck D, Curtis CE, Glimcher PW, Heeger DJ. Sustained activity in topographic areas of human posterior parietal cortex during memory-guided saccades. J Neurosci 2006;26(19):5098–5108.

88. Sunaert S, Van Hecke P, Marchal G, Orban GA. Attention to speed of motion, speed discrimination, and task difficulty: An fMRI study. Neuroimage 2000;11(6, Part 1):612–623.

89. Rees G, Friston K, Koch C. A direct quantitative relationship between the functional properties of human and macaque V5. Nat Neurosci 2000;3:716–723.

90. Binkofski F, Dohle C, Posse S et al. Human anterior intraparietal area subserves prehension: A combined lesion and functional MRI activation study. Neurology 1998;50(5):1253–1259.

91. Vanduffel W, Fize D, Peuskens H et al. Extracting 3D from motion: Differences in human and monkey intraparietal cortex. Science 2002;298(5592):413–415.

92. Orban GA, Fize D, Peuskens H et al. Similarities and differences in motion processing between the human and macaque brain: Evidence from fMRI. Neuropsychologia 2003;41(13):1757–1768.

93. Stout D, Chaminade T. The evolutionary neuroscience of tool making. Neuropsychologia 2007;45(5):1091–1100.

94. Peuskens H, Sunaert S, Dupont P, Van HP, Orban GA. Human brain regions involved in heading estimation. J Neurosci 2001;21(7):2451–2461.

95. Braddick OJ, O'Brien JM, Wattam-Bell J, Atkinson J, Turner R. Form and motion coherence activate independent, but not dorsal/ventral segregated, networks in the human brain. Curr Biol 2000;10(12):731–734.

96. Grüsser O-J, Guldin WO, Mirring S, Salah-Eldin A. Comparative physiological and anatomical studies of the primate vestibular cortex. In: Albowitz B, Albus K, Kuhnt U, Nothdurft H-C, Wahle P, editors. Structural and functional organization of the neocortex.

Proceedings of a Symposium in the Memory of Otto D. Creutzfeldt, May 1993, Exp. Brain Res. Series 24. 1994, pp. 358–371.
97. Orban GA, Dupont P, De Bruyn B, Vogels R, Vandenberghe R, Mortelmans L. A motion area in human visual cortex. Proc Natl Acad Sci USA 1995;92(4):993–997.
98. Dupont P, De Bruyn B, Vandenberghe R et al. The kinetic occipital region in human visual cortex. Cereb Cortex 1997;7(3):283–292.
99. Tyler CW, Likova LT, Kontsevich LL, Wade AR. The specificity of cortical region KO to depth structure. Neuroimage 2006;30(1):228–238.
100. Zeki S, Perry RJ, Bartels A. The processing of kinetic contours in the brain. Cereb Cortex 2003;13 (2):189–202.
101. Lu Z-L, Sperling G. Attention-generated apparent motion. Nature 1995;377:237–239.
102. Lu ZL, Lesmes LA, Sperling G. The mechanism of isoluminant chromatic motion perception. Proc Natl Acad Sci USA 1999;96:8289–8294.
103. Ungerleider LG, Mishkin M. Two cortical visual systems. In: Ingle DJ, Mansfield RJW, Goodale MS, editors. The analysis of visual behavior. Cambridge, MA: MIT Press;1982, pp. 549–586.
104. Malach R, Reppas JB, Benson RR et al. Object-related activity revealed by functional magnetic resonance imaging in human occipital cortex. Proc Natl Acad Sci USA 1995;92(18):8135–8139.
105. Kanwisher N, Chun MM, McDermott J, Ledden PJ. Functional imagining of human visual recognition. Brain Res Cogn Brain Res 1996;5(1–2):55–67.
106. Sawamura H, Georgieva S, Vogels R, Vanduffel W, Orban GA. Using functional magnetic resonance imaging to assess adaptation and size invariance of shape processing by humans and monkeys. J Neurosci 2005;25(17):4294–4306.
107. Felleman DJ, Van Essen DC. Distributed hierarchical processing in the primate cerebral cortex. Cereb Cortex 1991;1:1–47.
108. Tanaka K, Saito H, Fukada Y, Moriya M. Coding visual images of objects in the inferotemporal cortex of the macaque monkey. J Neurophysiol 1991;66:170–189.
109. Grill-Spector K, Malach R. The human visual cortex. Annu Rev Neurosci 2004;27: 649–677.
110. Quiroga RQ, Reddy L, Kreiman G, Koch C, Fried I. Invariant visual representation by single neurons in the human brain. Nature 2005;435(7045):1102–1107.
111. Reddy L, Kanwisher N. Coding of visual objects in the ventral stream. Curr Opin Neurobiol 2006;16(4):408–414.
112. Privman E, Nir Y, Kramer U et al. Enhanced category tuning revealed by intracranial electroencephalograms in high-order human visual areas. J Neurosci 2007; 27(23):6234–6242.
113. Altmann CF, Bulthoff HH, Kourtzi Z. Perceptual organization of local elements into global shapes in the human visual cortex. Curr Biol 2003;13(4):342–349.
114. Kourtzi Z, Tolias AS, Altmann CF, Augath M, Logothetis NK. Integration of local features into global shapes: Monkey and human FMRI studies. Neuron 2003;37(2):333–346.
115. Kourtzi Z, Kanwisher N. Representation of perceived object shape by the human lateral occipital complex. Science 2001;293: 1506–1509.
116. Spiridon M, Kanwisher N. How distributed is visual category information in human occipito-temporal cortex? An fMRI study. Neuron 2002;35(6):1157–1165.
117. Wade AR, Brewer AA, Rieger JW, Wandell BA. Functional measurements of human ventral occipital cortex: Retinotopy and colour. Philos Trans R Soc Lond B Biol Sci 2002; 357:963–973.
118. Levy I, Hasson U, Avidan G, Hendler T, Malach R. Center–periphery organization of human object areas. Nat Neurosci 2001; 4(5):533–539.
119. Hasson U, Levy I, Behrmann M, Hendler T, Malach R. Eccentricity bias as an organizing principle for human high-order object areas. Neuron 2002;34(3):479–490.
120. Rolls ET. Functions of the primate temporal lobe cortical visual areas in invariant visual object and face recognition. Neuron 2000;27(2):205–218.
121. Hung CP, Kreiman G, Poggio T, DiCarlo JJ. Fast readout of object identity from macaque inferior temporal cortex. Science 2005;310(5749).
122. Grill-Spector K, Kushnir T, Edelman S, Avidan G, Itzchak Y, Malach R. Differential processing of objects under various viewing conditions in the human lateral occipital complex. Neuron 1999;24:187–203.
123. Cumming BG, DeAngelis GC. The physiology of stereopsis. Annu Rev Neurosci 2001;24:203–238.

124. Parker AJ. Binocular depth perception and the cerebral cortex. Nat Rev Neurosci 2007;8(5):379–391.
125. Neri P, Bridge H, Heeger DJ. Stereoscopic processing of absolute and relative disparity in human visual cortex. J Neurophysiol 2004;92(3):1880–1891.
126. Orban GA, Janssen P, Vogels R. Extracting 3D structure from disparity. Trends Neurosci 2006;29(8):466–473.
127. Gulyas B, Roland PE. Processing and analysis of form, clour and binocular disparity in the human brain: Functional anatomy by positron emision tomography. Eur J Neurosci 1994;6:1811–1828.
128. Mendola JD, Dale AM, Fischl B, Liu AK, Tootell RB. The representation of illusory and real contours in human cortical visual areas revealed by functional magnetic resonance imaging. J Neurosci 1999;19:8560–8572.
129. Backus BT, Fleet DJ, Parker AJ, Heeger DJ. Human cortical activity correlates with stereoscopic depth perception. J Neurophysiol 2001;86(4):2054–2068.
130. Tsao DY, Vanduffel W, Sasaki Y et al. Stereopsis activates V3A and caudal intraparietal areas in macaques and humans. Neuron 2003;39(3):555–568.
131. Brouwer GJ, van ER, Schwarzbach J. Activation in visual cortex correlates with the awareness of stereoscopic depth. J Neurosci 2005;25(45):10403–10413.
132. Gilaie-Dotan S, Ullman S, Kushnir T, Malach R. Shape-selective stereo processing in human object-related visual areas. Hum Brain Mapp 2002;15(2):67–79.
133. Shikata E, Hamzei F, Glauche V et al. Surface orientation discrimination activates caudal and anterior intraparietal sulcus in humans: An event-related fMRI study. J Neurophysiol 2001;85(3):1309–1314.
134. Taira M, Nose I, Inoue K, Tsutsui K. Cortical areas related to attention to 3D surface structures based on shading: An fMRI study. Neuroimage 2001;14(5):959–966.
135. James TW, Humphrey GK, Gati JS, Servos P, Menon RS, Goodale MA. Haptic study of three-dimensional objects activates extrastriate visual areas. Neuropsychologia 2002;40(10):1706–1714.
136. Sereno ME, Trinath T, Augath M, Logothetis NK. Three-dimensional shape representation in monkey cortex. Neuron 2002;33:635–652.
137. Kourtzi Z, Erb M, Grodd W, Bulthoff HH. Representation of the perceived 3-D object shape in the human lateral occipital complex. Cereb Cortex 2003;13(9):911–920.
138. Murray SO, Olshausen BA, Woods DL. Processing shape, motion and three-dimensional shape-from-motion in the human cortex. Cereb Cortex 2003;13(5):508–516.
139. Orban GA. Three-dimensional shape: Cortical mechanisms of shape extraction. In: Masland RHATD, editor. Handbook of The Senses, Vol. 5: Vision. Amsterdam: Elsevier;2007.
140. Durand JB, Nelissen K, Joly O et al. Anterior regions of monkey parietal cortex process visual 3D shape. Neuron 2007;55(3):493–505.
141. Welchman AE, Deubelius A, Conrad V, Bulthoff HH, Kourtzi Z. 3D shape perception from combined depth cues in human visual cortex. Nat Neurosci 2005;8(6):820–827.
142. Brewer AA, Press WA, Logothetis NK, Wandell BA. Visual areas in macaque cortex measured using functional magnetic resonance imaging. J Neurosci 2002;22(23):10416–10426.
143. Sereno AB, Maunsell JH. Shape selectivity in primate lateral intraparietal cortex. Nature 1998;395:500–503.
144. Janssen P, Srivastava S, Ombelet S, Orban GA. Coding of shape and position in macaque area LIP. J Neurosci 2008;28:6679–6690.

Chapter 18

fMRI of the Central Auditory System

Deborah Ann Hall and Aspasia Eleni Paltoglou

Summary

Over the years, blood oxygen level-dependent (BOLD) fMRI has made important contributions to the understanding of central auditory processing in humans. Although there are significant technical challenges to overcome in the case of auditory fMRI, the unique methodological advantage of fMRI as an indicator of population neural activity lies in its spatial precision. It can be used to examine the neural basis of auditory representation at a number of spatial scales, from the micro-anatomical scale of population assemblies to the macro-anatomical scale of cortico-cortical circuits. The spatial resolution of fMRI is maximised in the case of mapping individual brain activity, and here it has been possible to demonstrate known organisational features of the auditory system that have hitherto been possible only using invasive electrophysiological recording methods. Frequency coding in the primary auditory cortex is one such example that we shall discuss in this chapter. Of course, non-invasive procedures for neuroscience are the ultimate aim and as the field moves towards this goal by recording in awake, behaving animals so human neuroimaging techniques will be increasingly relied upon to provide an interpretive link between animal neurophysiology at the multi-unit level and the operation of larger neuronal assemblies, as well as the mechanisms of auditory perception itself. For example, the neural effects of intentional behaviour on stimulus-driven coding have been explored both in animals, using electrophysiological techniques, and in humans, using fMRI. While the feature-specific effects of selective attention are well established in the visual cortex, the effect of auditory attention in the auditory cortex has generally been examined at a very coarse spatial scale. Ongoing research in our laboratory has started to address this question and here we present preliminary evidence for frequency-specific effects of attentional enhancement in the human auditory cortex. We end with a brief discussion of several future directions for auditory fMRI research.

Key words: Technical challenges, Frequency coding, Selective attention, Perceptual representation, Task specificity

1. Challenges of Auditory fMRI

The construction of a brain image using MR imaging depends upon the magnetic properties of hydrogen ions that, when placed in a static magnetic field, can absorb pulses of radiowave energy

M. Filippi (ed.), *fMRI Techniques and Protocols*, Neuromethods, vol. 41
DOI 10.1007/978-1-60327-919-2_18, © Humana Press, a part of Springer Science+Business Media, LLC 2009

of a specific frequency. The time taken for the ion alignments to return to equilibrium after the radiofrequency (RF) pulse differs according to the surrounding tissue, thus providing the image contrast, for example between grey matter, white matter, cerebrospinal fluid, and bone. The use of MR techniques for detecting functional brain activation relies on two factors: first that local neural activity is a metabolically demanding process that is closely associated with a local increase in the supply of oxygenated blood to those active parts of the brain, and second that the different paramagnetic properties of oxygenated and deoxygenated blood produce measurable effects on the MR signal. The functional signal detected during fMRI is known as the blood oxygen level-dependent (BOLD) response. Essentially, the functional image represents the spatial distribution of blood oxygenation levels in the brain, and the small fluctuations in these levels over time are correlated with the stimulus input or cognitive task.

MR scanners operate using three different types of electromagnetic fields: a very high static field generated by a superconducting magnet, time-varying gradient magnetic fields, and pulsed RF fields. The latter two fields are much weaker than the first, but all pose a number of unique and considerable technical challenges for conducting auditory fMRI research within this hostile environment. In the first place, the static and time-varying magnetic fields preclude the use of many types of electronic sound presentation equipment, as well as preventing the safe scanning of patients who are wearing listening devices such as hearing aids or implants. Additionally, the high levels of scanner noise generated by the flexing of the gradient coils in the static magnetic field can potentially cause hearing difficulties. The scanner noise masks the perception of the acoustic stimuli presented to the subject in the scanner making it difficult to calibrate audible hearing levels and adding to the difficulty of the listening task. And finally, the scanner noise not only activates parts of the auditory brain, but also interacts with the patterns of activity evoked by experimental stimuli. Auditory fMRI poses a number of other challenges, not related to the hostile environment of the MR scanner, but related instead to the nature of the neural coding in the auditory cortex. The response of auditory cortical neurons to a particular class of sound is determined not only by the acoustic features of that sound, but also by its presentation context. For example, neurons respond strongly to the onset of sound events and thereafter tend to show rapid adaptation to that sound in terms of a reduction in their firing rate. Thus, the result of any particular auditory fMRI experiment will depend not only on the physical attributes of a stimulus, but also on the way in which the stimuli are presented. In this first section, we shall take each one of these issues in turn, introducing the problems in more detail as well as proposing some solutions.

1.1. Use of Electronic Equipment for Sound Presentation in the MR Scanner

1.1.1. Problems

The ideal requirement is a sound presentation system that produces a range of sound levels [up to 100-dB sound pressure level (SPL)], with low distortion, a flat frequency response, and a smooth and predictable phase response. The first commercially available solution utilised loudspeakers, placed away from the high static magnetic field, from which the sound was delivered through plastic tubes inserted into the ear canal **(Fig. 1a)** through a protective ear defender **(Fig. 1b)**. One general disadvantage of the tube phone system is that the tubing distorts both the phase and amplitude of the acoustic signal, for example, by imposing a severe ripple on the spectra and reducing sound level, especially at higher frequencies. Another limitation is the leak of the scanner noise through the pipe walls to the pipe inner and hence the ear. Despite alternative systems now being readily available, tube phone systems are still commercially manufactured (e.g. Avotec Inc. Stuart, Florida, USA, www.avotec.org/). The Avotec system has been specifically designed for fMRI use and boasts an equaliser to provide a reasonably flat audio output (±5 dB) across its nominal bandwidth (150 Hz to 4.5 kHz) and a procedure for acoustic calibration that feeds a known electrical input signal to the audio system input and makes a direct acoustic output measurement at the headset.

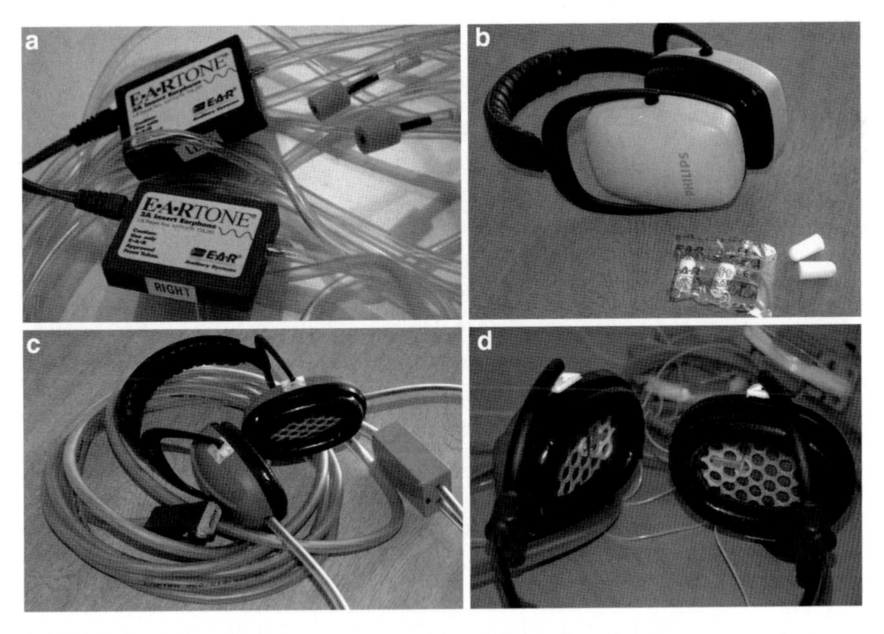

Fig. 1. MR compatible headsets for sound delivery and noise reduction: (**a**) tube phones system with foam ear inserts, (**b**) circum-aural ear defenders, plus foam ear plugs for passive noise reduction, (**c**) MRC IHR sound presentation headset combining commercially available electrostatic transducers in an industry standard ear defender, and (**d**) modified MRC IHR headset for sound presentation and for active noise cancellation (ANC), including an optical error microphone positioned underneath the ear defender.

Alternative electronic systems often used for psychoacoustical research deliver high-quality signals, but these systems are generally unsuitable for use in the MR environment because most headphones use an electromagnet to push and pull on a diaphragm to vibrate the air and generate sound. Of course, this electromagnet is rendered inoperable by the magnetic fields in the MR scanner. Headphone components constructed from ferromagnetic material also disrupt the magnetic fields locally and induce signal loss or spatial distortion in areas close to the ears. In addition, the electronic components can be damaged by the static magnetic field, while electromagnetic interference generated by the equipment is detected by the MR receiver head coil. Electronic sound delivery systems for use in auditory fMRI research have been designed specifically to overcome these difficulties.

1.1.2. Solutions

Despite the restriction on the materials that can be used in a scanner, a number of different MR-compatible active headphone driving units have been produced. An ingenious system has been developed and marketed by one auditory neuroimaging research group (MR confon GmbH, Magdeburg, Germany, www.mr-confon.de). This system incorporates a unique, electrodynamic driver that uses the scanner's static magnetic field in place of the permanent magnets that are found in conventional headphones and loudspeakers. It produces a wide frequency range (less than 200 Hz to 35 kHz) with a flat frequency response (±6 dB). Another company manufactures and supplies high-quality products for MRI, with a special focus on the fast-growing field of functional imaging (NordicNeuroLab AS, Bergen, Norway, www.nordicneurolab.com/). Their audio system uses electrostatic transducers to ensure high performance. Electrostatic headphones generate sound using a conductive diaphragm placed next to a fixed conducting panel. A high voltage polarises the fixed panel and the audio signal passing through the diaphragm rapidly switches between a positive and a negative signal, attracting or repelling it to the fixed panel and thus vibrating the air. Their technical specification claims a flat frequency response from 8 Hz to 35 kHz. The signal is transferred from the audio source to the headphones in the RF screened scanner room using either filters through a filter panel or fibre-optic cable through the waveguide.

Here at the MRC Institute of Hearing Research, we became engaged in auditory fMRI research well before such commercial systems were widely available and so, for our own purposes, we developed an MR-compatible headset **(Fig. 1c)** based on commercially available electrostatic headphones, modified to remove or replace their ferromagnetic components, and combined with standard industrial ear defenders to provide good acoustic isolation *(1)*. Our custom-built system delivers a flat frequency response (±10 dB) across the frequency range 50 Hz to 10 kHz and has an output level capability up to 120-dB SPL. Again, the digital audio source, electronics, and power supply that drive the system are housed outside the RF screened scanner room to avoid electromagnetic interference with MR scanning, and all electrical signals passing into the screened scanner room are RF filtered.

1.2. Risk to Patients Who Are Wearing Listening Devices in the MR Scanner

1.2.1. Problems

No ferromagnetic components can be placed in the scanner bore as they would experience a strong attraction by the static magnetic field and potentially cause damage not only to the scanner and the listening device, but also to the patient. Induced currents in the electronics, caused directly by the time-varying gradient magnetic fields or the RF pulses, are an additional hazard to the electronic devices themselves, while some materials can also absorb the RF energy causing local tissue heating and even burns if in contact with soft tissue. For these reasons, there are restrictions on scanning people who have electronic listening devices. These include hearing aids, cochlear implants, and brainstem implants. Hearing aids amplify sound for people who have moderate to profound hearing loss. The aid is battery-operated and worn in or around the ear. Hearing aids are available in different shapes, sizes, and types, but they all work in a similar way. They all have a built-in microphone that picks up sound from the environment. These sounds are processed electronically and made louder, either by analogue circuits or digitally, and the resulting signals are passed to a receiver in the hearing aid where they are converted back into audible sounds. In contrast, cochlear and brainstem implants are both small, complex electronic devices that can help to provide a sense of sound to people who are profoundly deaf or severely hard-of-hearing. Cochlear implants bypass damaged portions of the inner ear (the cochlea) and directly stimulate the auditory nerve, while auditory brainstem implants bypass the vestibulocochlear nerve in cases when it is damaged by tumours or surgery and directly stimulate the lower part of the auditory brain (the cochlear nucleus). In general, both types of implant consist of an external portion that sits behind the ear and a second portion that is surgically placed under the skin. They contain a microphone, a sound processor (which converts sounds picked up by the microphone into an electrical code), a transmitter and receiver/stimulator (which receive signals from the processor and convert them into electric impulses), and finally an electrode array (which is a set of electrodes that collect the impulses from the stimulator and stimulate groups of auditory neurons). Coded information from the sound processor is delivered across the skin via electromagnetic induction to the implanted receiver/stimulator, which is surgically placed on a bone behind the ear.

1.2.2. Solutions

Official approval for the manufacture of implant devices requires rigorous testing for susceptibility to electromagnetic fields, radiated electromagnetic fields and electrical safety testing (including susceptibility to electrical discharge). However, such tests are conducted under normal conditions, not in the magnetic fields of an MR scanner. Some implant designs have been proven to be MR compatible *(2–5)*, but they are not routinely supplied in clinical practice. Standard listening devices do not meet MR compatibility criteria and, for the patient, risks include movement of the device and localised heating of brain tissue, whereas, for the device, the electronic components may be damaged. Magnetic Resonance Safety Testing Services (MRSTS) is a highly experienced testing company that conducts comprehensive evaluations of implants, devices, objects, and materials in the MR environment (MRSTS, Los Angeles, CA, www.magneticresonancesafetytesting.com/). Testing includes approved assessment of magnetic field interactions, heating, induced electrical currents, and artefacts. A database of the devices and results of implant testing is accessible to the interested reader (www.mrisafety.com/). However, auditory devices have generally been tested only at low magnetic fields (up to 1.5 T) because most clinical MR systems operate at this field strength. Since research systems typically operate at 3.0 T (for improved BOLD signal-to-noise ratio, BOLD SNR) it may be necessary for individual research teams to ensure the safety of their patients. For example, here at the MRC Institute of Hearing Research, we have recently assessed the risks of movement and localised tissue heating for two middle ear piston devices *(6)*. For the safety reasons discussed in this subsection, listeners who normally wear hearing aids could be scanned without their aid but, to compensate, have been presented with sounds amplified to an audible level. Given that implanted devices cannot be removed without surgical intervention, clinical imaging research of implantees has generally used other brain imaging methods, namely positron emission tomography *(7)*.

1.3. Intense MR Scanner Noise and Its Effects on Hearing

1.3.1. Problems

The scanning sequence used to measure the BOLD fMRI signal requires rapid on and off switching of electrical currents through the three gradient coils of wire in order to create time-varying magnetic fields that are required for selecting and encoding the three-dimensional image volume (in the x, y, and z planes). This rapid switching in the static magnetic field induces bending and buckling of the gradient coils during MRI. As a result, the gradient coils act like a moving coil loudspeaker to produce a compression wave in the air, which is heard as acoustic noise during the image acquisition. Scanner noise increases non-linearly with static magnetic field strength, such that ramping from 0.5 to 2 T could account for a rise in sound level of as much as 11-dB SPL *(8)*. A brain scan is composed of a set of two-dimensional 'slices' through the brain. Gradient switching is required for each slice acquisition and so an intense scanner 'ping' occurs each time a brain slice is collected. Each ping lasts about 50 ms and so during fMRI, each scan is audible as a rapid sequence of such 'pings' (see inset in **Fig. 2** for an example of the amplitude envelope of the scanner noise).

The dominant components of the noise spectrum are composed of a peak of sound energy at the gradient switching frequency plus its higher harmonics. Most of the energy lies below 3 kHz. Secondary acoustic noise can be produced if the vibration of the coils and the core on which they are wound conducts through the core supports to the rest of the scanner structure. These secondary noise characteristics depend more on the mechanical resonances of the coil assemblies than on the type of imaging sequence and they tend to be the dominant contributor to the bandwidth and the spectral envelope of the noise. In this example of the frequency spectrum captured from a BOLD fMRI scanning sequence that was run on a Philips 3 Tesla Intera **(Fig. 2)**, the spectrum has a peak component at 600 Hz with several other prominent pseudo-harmonics at 300, 1,080, and 1,720 Hz. The sound level measured in the bore of the scanner is typically 99-dB SPL [98 dB(A) using an A-weighting], measured using the maximum 'fast' root-mean-square (RMS) time constant (125 ms). Clearly, exposure to such an intense sound levels without protection is likely to cause a temporary threshold shift in hearing and

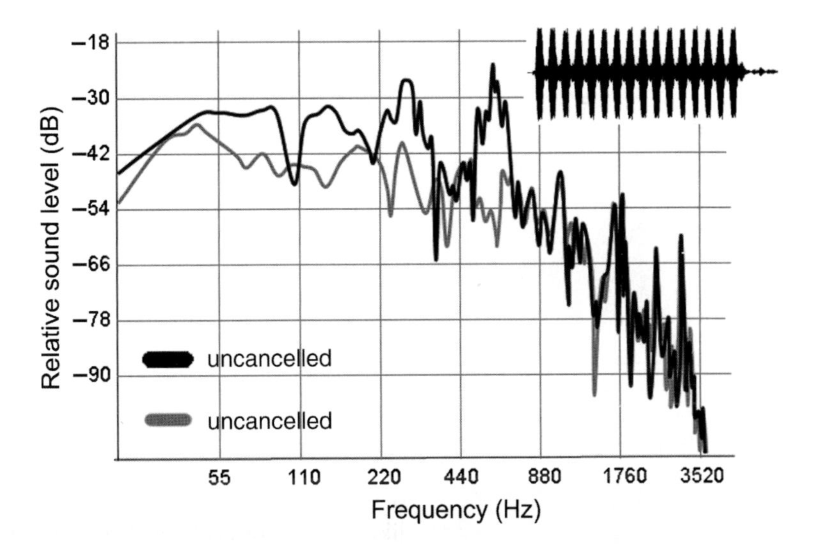

Fig. 2. Typical frequency spectrum of the scanner noise generated during blood oxygen level-dependent (BOLD) fMRI. This example was measured in the bore of a Philips Intera 3.0 Tesla scanner. The *black line* (uncancelled) indicates the acoustic energy of the noise recorded under normal scanning conditions. The *grey line* (cancelled) indicates the residual acoustic energy at the ear when the active noise cancellation (ANC) system is operative. The inset (*upper right*) shows an example of the amplitude envelope of the scanner noise for a brain scan consisting of 16 slices corresponding to a sequence of 16 intense 'pings'.

tinnitus, and it could be permanently damaging over a prolonged dosage *(9)*.

1.3.2. Solutions

The simplest way to treat the intense noise is to use ear protection in the form of ear defenders and/or ear plugs (shown in **Fig. 1b**). Foam ear plugs can compromise the acoustic quality of the experimental sounds delivered to the subject and so ear defenders are preferable. Typically transducers are fitted into sound attenuating earmuffs to reduce the ambient noise level at the subject's ears. Attenuation of the external sound by up to 40 dB can be achieved in this manner, although the level of reduction drops off at the high-frequency end of the spectrum. Commercial sound delivery systems all incorporate passive noise attenuation of this sort. An additional method of noise reduction is to line the bore of the scanner with a sound-energy absorbing material [*(10)*; see also www.ihr.mrc.ac.uk/research/technical/soundsystem/]. The results of a set of measurements directly comparing the sound intensity of the scanner noise with and without the foam lining are shown in **Fig. 3**. However, this strategy does not provide a feasible solution because neither the design of the scanner bore nor the automated patient table are suited to the permanent installation of a foam lining and some types of acoustic foam can present risks of noxious fumes if they catch fire.

Some manufacturers have attempted to minimise scanner sound levels by modifying the design of the scanner hardware. For example, MR scanners manufactured by Toshiba (Toshiba America Medical systems, Inc., www.medical.toshiba.com/) incorporate Pianissimo technology - employing a solid foundation for gradient

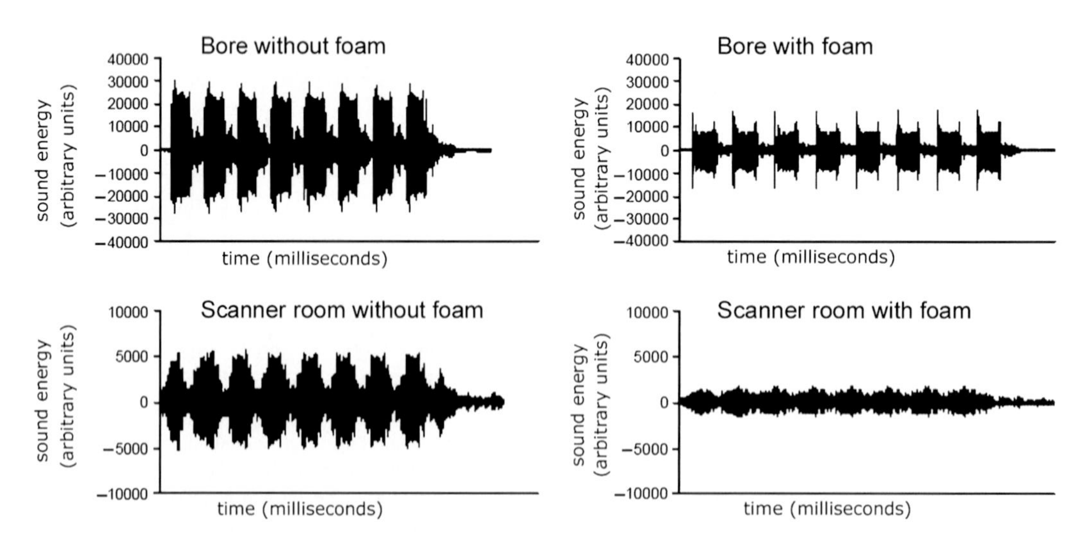

Fig. 3. Acoustic waveforms of the scanner noise measured with and without a lining of acoustic damping foam in the bore of the scanner. Our data demonstrate that the foam reduces the sound pressure level (SPL) at the position of the subject's head and in scanner room by a significant margin (about 8 dB). The segment of scanner noise that is illustrated here has a duration of approximately 1 s.

support, integrating sound dampening material in the gradient coils and enclosing them in a vacuum to reduce acoustic noise, even at full gradient power. This technology claims to reduce scanner noise by up to 90% *(11)*. Subjects are reported to hear sounds at the volume of gentle drumming instead of the jack-hammer noise level of other MR systems.

Another solution is to run modified pulse sequences that reduce acoustic noise by slowing down the gradient switching. This approach is based on the premise that the spectrum of the acoustic noise is determined by the product of the frequency spectrum of the gradient waveforms and the frequency response function of the gradient system *(12)*. The frequency response function is generally substantially reduced at low frequencies (i.e. below 200 Hz) and so the sound level can be reduced by using gradient pulse sequences whose spectra are band limited to this low-frequency range using pulse shapes with smooth onset and offset ramps *(13)*. A low-noise fast low-angle shot (FLASH) sequence can be modified to have a long gradient ramp time (6,000 μs) and it generates a peak sound level of 48-dB SPL measured at the position of the ear. This type of sequence has been used for mapping central auditory function *(14)*. However, the low noise is achieved at the expense of slower gradient switching, extending the acquisition time. Low-noise sequences are not suitable for rapid BOLD imaging in which the fundamental frequency of the gradient waveform is greater than 200 Hz.

1. 4. The Effect of Scanner Noise on Stimulus Audibility

1.4.1. Problems

Not only is the intense scanner noise a risk for hearing, but it also masks the perception of the acoustic stimuli presented to the subject. The exact specification of the acoustic signal-to-scanner-noise ratio (acoustic SNR) in fMRI studies using auditory stimuli is a potentially complicated matter. Nevertheless, we have sought to establish the relative difference between the stimulus level and the scanner noise level at the ear, by measuring these signals using a reference microphone placed inside the cup of the ear defender while participants perform a signal detection in noise task. Detection thresholds for a narrow band noise centred at the peak frequency of the scanner noise (600 Hz) are elevated when the target coincides with the scanner noise. We have demonstrated an average 11-dB shift in the 71% detection threshold for the 600-Hz target when we modulate the perceived level of the scanner noise using active noise cancellation (ANC) methods (see later).

This evidence suggests that even with hearing protection, whenever the scanner noise coincides with the presented sound stimulus it produces changes in task performance and probably also increases the attentional demands of the listening task. The frequency range of the scanner acoustic noise is crucial for speech intelligibility, and speech experiments can be particularly compromised by a noisy environment [*(15)*; for review, see *(16)*].

A recent study has quantified the effect of acoustic SNR using four listening tasks: pitch discrimination of complex tones, same/different judgments of minimal-pair nonsense syllables, lexical decision, and judgement of sentence plausibility *(17)*. Across these tasks, performance was assessed in silence (acoustic SNR = infinity) and in a background of MR scanner noise at the three acoustic SNR levels (−6, −12, and −18 dB). Performance of normally hearing listeners significantly decreased as a function of the noise **(Fig. 4)**.

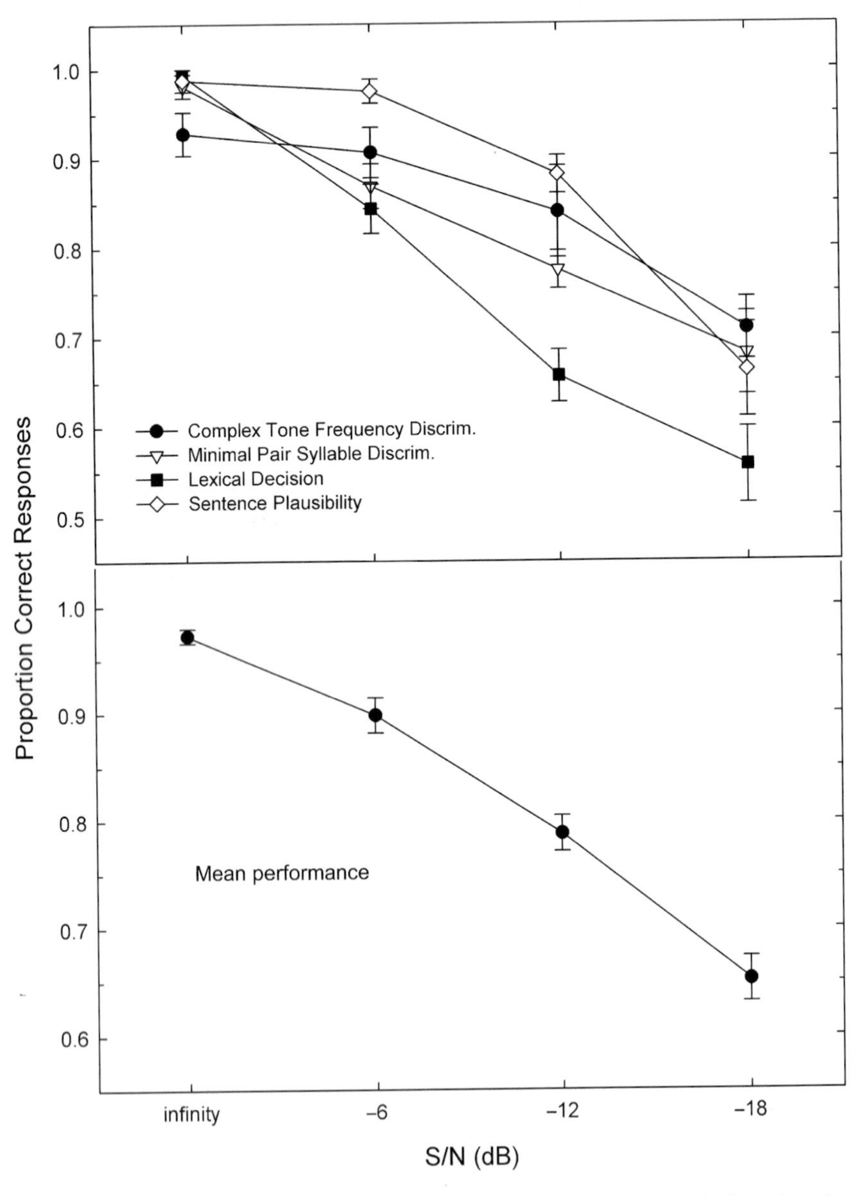

Fig. 4. Mean performance in a simulated scanning environment across four acoustic signal-to-noise ratios *(17)*. The *top panel* plots the proportion of correct responses on the individual tasks, while the *bottom panel* shows the overall mean performance (*SNR* signal-to-noise ratio, *dB* decibels).

Even at –6 dB acoustic SNR, participants made many more errors than in quiet listening conditions ($p < 0.01$). Thus, across a range of auditory tasks that vary in linguistic complexity, listeners are highly susceptible to the disruptive impact of the intense noise associated with fMRI scanning.

1.4.2. Solutions

The aggregate noise dosage can be reduced by acquiring either a single or at least very few brain slices, but at the expense of only a partial view of brain activity *(18)*. For whole brain fMRI, other strategies are required.

One novel method that has been developed and evaluated at our Institute combines optical microphone technology with an active noise controller for significant attenuation of ambient noise received at the ears *(19)*. The canceller is based upon a variation of the single channel feed-forward filtered-x adaptive controller and uses a digital signal processor to achieve the noise reduction in real time. The canceller minimises the noise pressure level at a specific control point in space that is defined by the position of the error microphone, positioned underneath the circum-aural ear defender of the headset (*see* **Fig. 1d**). In 2001, we published a psychophysical assessment of the system using a prototype system built in the laboratory that utilised a loudspeaker as the noise generator *(19)*. This system produced 10–20 dB of subjective noise reduction between 250 Hz and 1 kHz and smaller amounts at higher frequencies. More recently, we have obtained psychophysical threshold data in a Philips 3 Tesla scanner confirming that the same level of cancellation is achieved in the real scanner environment (**Fig. 5**; *(20)*). Again, the subjective impression of the scanner noise is the volume of gentle drumming when the

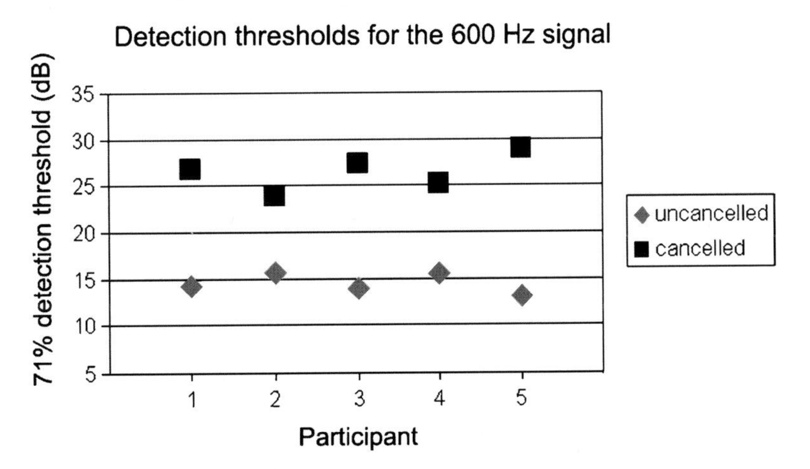

Fig. 5. Performance on a signal detection in noise task measured in a real scanning environment [Philips 3.0 Tesla MR scanner during blood oxygen level-dependent (BOLD) fMRI]. The data show that when the noise canceller was operative, the sound level of the signal could be 8–16 dB softer (depending upon the listener) in order to achieve the same detection performance.

sound system is operating in its cancelled mode. Thus, it is possible to achieve a high level of noise attenuation by combining both passive and active methods.

A much more common strategy for reducing the masking influence of the concomitant scanner noise combines a passive method of ear protection with an experimental protocol that carefully controls the timing between stimulus presentation and image acquisition so that sound stimuli can be delivered during brief periods of quiet in between successive brain scans *(21)*. Specific details of several pulse sequence protocols that reduce the masking effects of scanner noise are discussed in more detail in the next subsection.

1.5. The Effect of Scanner Noise on Sound-Related Activation in the Brain

1.5.1. Problems

To increase the BOLD SNR, it is necessary to acquire a large number of scans in each condition in an fMRI experiment. Typically, an experimenter would collect many hundreds of brain scans in a single session, with the time in between each scan chosen to be as short as the scanner hardware and software will permit. Remember that, for fMRI, an intense 'ping' is generated for each slice of the scan and so of course this means that the participant can easily be subjected to several thousand repeated 'pings' of noise during the experiment. Not only does this scanner noise acoustically mask the presented sound stimuli, but the elevated baseline of sound-evoked activation due to the ambient scanner noise also makes the experimentally induced auditory activation more difficult to detect statistically. Much of the work examining the influence of acoustic scanner noise has been directed toward its capacity to interfere with the study of audition or speech perception by producing activation of various brain regions, especially the auditory cortex *(22–25)*. Several studies highlight the reduced activation signal (i.e. the difference between stimulation and baseline conditions) in the auditory cortex when the amount of prior scanner noise is increased, demonstrating that the scanner noise effectively masks the detection of auditory activation *(22, 26, 27)*. In another example, taken from one of the early fMRI experiments conducted at the MRC Institute of Hearing Research, we used a specially tailored scanning protocol to measure the amplitude and the time course of the BOLD response to a high-quality recording of a single burst of scanner noise presented to participants over headphones *(24)*. Our results revealed a reliable transient increase in the BOLD signal across a large part of the auditory cortex. As in many other brain regions, the evoked response to this single brief stimulus event was smoothed and delayed in time. It rose to a peak by 4–5 s after stimulus onset and decayed by 5–8 s after stimulus offset *(24)*. Its amplitude reached about 1.5% of the overall signal change, which is considerable considering that stimulus-related activation usually accounts for a BOLD signal change of approximately 2–5%. **Figure 6** illustrates the canonical BOLD response to a noise onset.

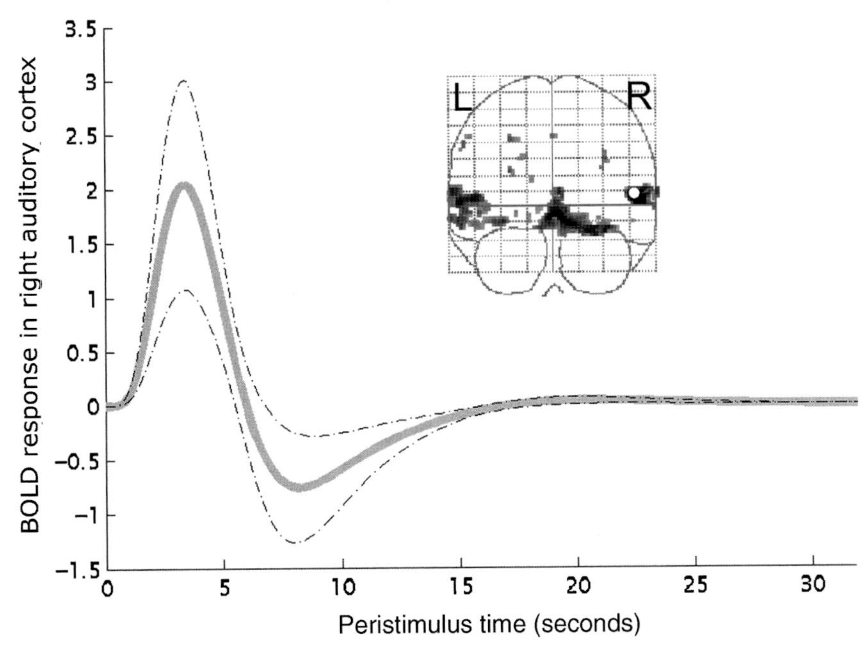

Fig. 6. Transient blood oxygen level-dependent (BOLD) response to a noise onset. The graph shows the fitted response and the 90% confidence interval. This example illustrates all the characteristic features of the transient BOLD response –a peak at 4-s post-stimulus onset followed by an undershoot and then return to baseline at 16 s.

In many fMRI experimental paradigms, regions of stimulus-evoked activation are detected by comparing the BOLD scans acquired during one sound condition with the BOLD scans acquired during another condition, which could be either a condition in which a different type of sound was presented or no sound (known as a baseline 'silent' condition) was presented. Activation is defined as those parts of the brain that demonstrate a statistically significant difference between the two conditions. For example, let us consider the simplest case in which one condition contains a sound and the other does not. Since the scanner noise is present throughout, the sound condition effectively contains both stimulus and scanner noise, while the baseline condition also contains the scanner noise (i.e. it is not silent). Given the spectrotemporal characteristics of the scanner noise, it generates widespread sound-related activity across the auditory cortex. Thus, the subtraction analysis for detecting activation is sensitive only to whatever is the small additional contribution of the sound stimulus to auditory neural activity.

1.5.2. Solutions

A number of different scanning protocols have been used to minimise the effect of the scanner acoustic noise on the measured patterns of auditory cortical activation. In this section, we will describe two of these, but before we do, we need to consider some important details about the time course of the BOLD response to the scanner noise and introduce some new terms.

During an fMRI experiment, the BOLD response to the scanner noise spans two different temporal scales. First, the 'ping' generated by the acquisition of one slice early in the scan may induce a BOLD response in a slice, which is acquired later in the same scan if that later scan is positioned over the auditory cortex. We shall call this inter-slice interference. Inter-slice interference is maximally reduced when all slices in the scan are acquired in rapid succession and the total duration of the scan is not more than 2 s *(26)*. A common term for the scanning protocol that uses a minimum inter-slice interval is a clustered-acquisition sequence. Edmister et al. *(28)* found that the clustered-acquisition sequence provides an advantageous auditory BOLD SNR compared with a conventional scanning protocol. The second form of interference is called inter-scan interference. This occurs when the scanner noise evokes an auditory BOLD response that extends across time to subsequent scans, predominantly when the interval between scans is as short as the MR system will permit. Reducing the inter-scan interference can easily be achieved by extending the period between scans (the inter-scan interval). By separately manipulating the timing between slices and between scans, we can reduce the inter-slice and inter-scan interference independently of one another. When the clustered-acquisition sequence is combined with a long (e.g. 10 s) inter-scan interval, the activation associated with the experimental sound can be separated from the activation associated with the scanner sound **(Fig. 7a)**. Furthermore, because the scanner sound is temporally offset, it does not produce acoustical masking and does not distract the listener. This scanning protocol is commonly known as sparse

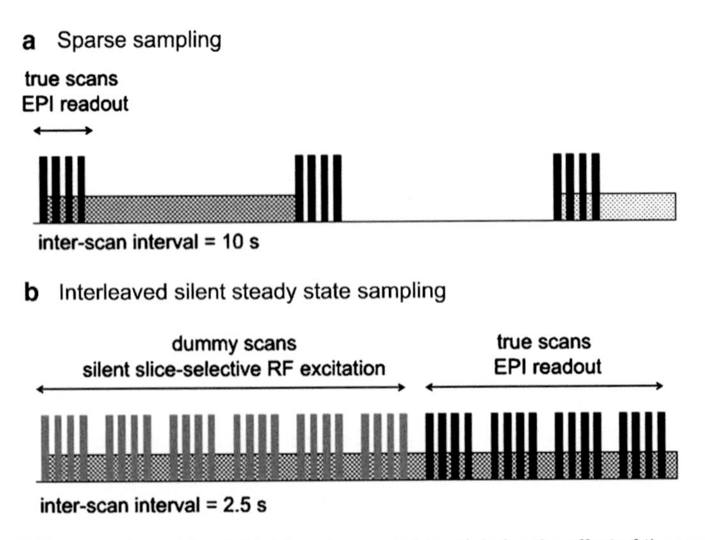

Fig. 7. Two scanning protocols that have been used to minimise the effect of the scanner acoustic noise on the measured patterns of auditory cortical activation. See text for further explanation (*s* seconds, *EPI* echo-planar imaging, *RF* radiofrequency).

sampling *(21)*. Sparse sampling is often the scanning protocol of choice for identifying auditory cortical evoked responses in the absence of scanner noise [*see* e.g. *(29–33)*]. However, it requires a scanning session that is longer than that of conventional 'continuous' protocols in order to acquire the same amount of imaging data, and participants can be intolerant of long sessions. It also relies upon certain assumptions about the time to peak of the BOLD response after stimulus onset and a sustained plateau of evoked activity for the duration of the stimulus.

A second type of scanning protocol acquires a rapid set of scans following each silent period in order to avoid some of the aforementiioned difficulties - 'interleaved silent steady state' sampling *(34)*. The increased number of scans permits a greater proportion of scanning time to be used for data acquisition and at least partial mapping of the time course of the BOLD response (**Fig. 7b**). However, some pulse programming is required to avoid T1-related signal decay during the data acquisition, hence ensuring that signal contrast is constant across successive scans. The software modification maintains the longitudinal magnetisation in a steady state throughout the scanning session by applying a train of slice-selective excitation pulses (quiet dummy scans) during each silent period.

1.6. The Effect of Stimulus Context: Neural Adaptation to Sounds

1.6.1. Problems

The acoustic environment is typically composed of one or more sound sources that change over time. Over the years, both psychophysical and electrophysiological studies have amply demonstrated that stimulus context strongly influences the perception and neural coding of individual sounds, especially in the context of stream segregation and grouping *(35–37)*. A simple example of the influence of stimulus context is forward masking, which occurs when the presence of one sound increases the detection threshold for the subsequent sound. The perceptual effects of forward masking are strongest when the spectral content of the first sound is similar to the second sound, when there is no delay between the two sounds, and when the masker duration is long *(38)*. Forward inhibition typically lasts from 70 to 200 ms. This type of suppression has not only been demonstrated in anaesthetised preparations, but also in awake primates. In the latter case, suppression was seen to extend up to 1 s in time *(39)*. As well as tone-tone interactions, neural firing rate is sensitive to stimulus duration. Neurons respond strongly to the onset of a sound and their response decays thereafter. Many illustrative examples can be found in the literature, especially in cases where longer duration sounds are presented [e.g. 750–1,500 ms in the case of Bartlett and Wang *(38)*, see their **Fig. 4**].

By transporting these well-established paradigms into a neuroimaging experiment, researchers are beginning to address the context dependency of neural coding in humans. One way in which the effect of sound context on the auditory BOLD fMRI signal has been examined is in terms of different repetition rates *(19, 40)*. This is conceptually analogous to the presentation rate manipulations of the forward masking studies described earlier, but goes beyond the simple case of two-tone interactions. In the fMRI studies, stimuli were long trains of noise bursts presented at different rates. The slowest rate was 2 Hz and the fastest rate was 35 Hz, with intermediate rates being 10 and 20 Hz. Noise bursts at each repetition rate were presented in prolonged blocks of 30 s, each followed by a 30-s 'silent' period. During sound presentation, scans were acquired at a short inter-scan interval (approximately 2 s) so that the experimenters could reconstruct the 30-s time course of the BOLD response to each of the different repetition rates, hence determining the multi-second time pattern of neural activity. The scans were positioned so that a number of different auditory sites in the ascending auditory system could be measured: (1) the inferior colliculus in the midbrain, (2) the medial geniculate nucleus in the thalamus, and (3) Heschl's gyrus and the superior temporal gyrus in the cortex. The plots of the BOLD time course demonstrated a systematic change in its shape from midbrain up to cortex. In the inferior colliculus, the amplitude of the BOLD response increased as a function of repetition rate while its shape was sustained throughout the 30-s stimulus period. In the medial geniculate body, increasing rate also produced an increase in BOLD amplitude with a moderate peak in the BOLD shape just after stimulus onset. Repetition rate exerted its largest effect in the auditory cortex. The most striking change was in the shape of the BOLD response. The low repetition rate (2 Hz) elicited a sustained response, whereas the high rate (35 Hz) elicited a phasic response with prominent peaks just after stimulus onset and offset. The follow-up study *(40)* confirmed that it was the temporal envelope characteristics of the acoustic stimulus, not its sound level or bandwidth, that strongly influenced the shape of the BOLD response. The authors offer a perceptual interpretation of the neural response to different repetition rates. The shift in the shape of the cortical BOLD response from sustained to phasic corresponds to a shift from a stimulus in which component noise bursts are perceptually distinct to one in which successive noise bursts fuse to become individually indistinguishable. The onset and offset responses of the phasic response coincide with the onset and offset of a distinct, meaningful event. The logical conclusion to this argument is that the succession of individual perceptual events in the low repetition rate conditions defines the sustained BOLD response observed at the 2-Hz rate. It is clear from these results that while the amplitude of the BOLD response to sound can inform us about the tuning properties of the underlying neural population (e.g. sensitivity to repetition rate), other properties of the BOLD response, such as its shape, provide different information about neural coding (e.g. segmentation of the auditory environment into perceptual events).

It is crucial that these contextual influences on the BOLD signal are accounted for in the design and/or interpretation of auditory fMRI experiments. To illustrate this case in point, I use a set of our own experimental data *(41)*. In this experiment, one of the sound conditions was a diotic noise (identical signal at the two ears) presented continuously for 32 s at a constant sound level (~86-dB SPL) and at a fixed location in the azimuthal plane. Scans were acquired every 4 s throughout the stimulus period. When the scans acquired during this sound condition were combined together and contrasted against the scans acquired during the 'silent' baseline condition, no overall significant activation was obtained ($p > 0.001$). We interpret this lack of activation as evidence that the auditory response had rapidly habituated to a static signal. This conclusion is confirmed by plotting out the time course of the response at one location within the auditory cortex. The initial transient rise in the BOLD response at the onset of the sound begins to decay at about 4 s and this reduction continues across the stimulus epoch. The end of the epoch is characterised by a further rise in the BOLD response, elicited by the other types of sound stimuli that were presented in the experiment **(Fig. 8a)**.

1.6.2. Solutions

It is common for auditory fMRI experiments to use a blocked design in which a sound condition is presented over a prolonged time period that extends over many seconds, even tens of seconds. Indeed as we described in **Subheading 1.5**, the blocked design is at the core of the sparse sampling protocol, and so the risk of neural adaptation is a legitimate one. The BOLD signal detection problem caused by neural adaptation is often circumvented by presenting the stimulus of interest as a train of stimulus bursts at a repetition rate that elicits the sustained cortical response (e.g. 2 Hz). Many of the auditory fMRI experiments that have been conducted over the years in our research group have taken this form *(30, 31, 42–44)*. Alternatively, if the stimulus contains dynamic spectrotemporal changes, then it is not always necessary to pulse the stimulus on and off. To illustrate this case in point, I return to a set of our own experimental data *(41)*. In this experiment, one of the sound conditions was a broadband noise convolved with a generic head-related transfer function to give the perceptual impression of a sound source that was continuously rotating around the azimuthal plane of the listener. Although the sound was presented continuously for 32 s, the filter functions of the pinnae imposed a changing frequency spectrum and the head shadow effect imposed low-rate amplitude modulations in the sound envelope presented to each ear. When the scans acquired during this sound condition were combined together

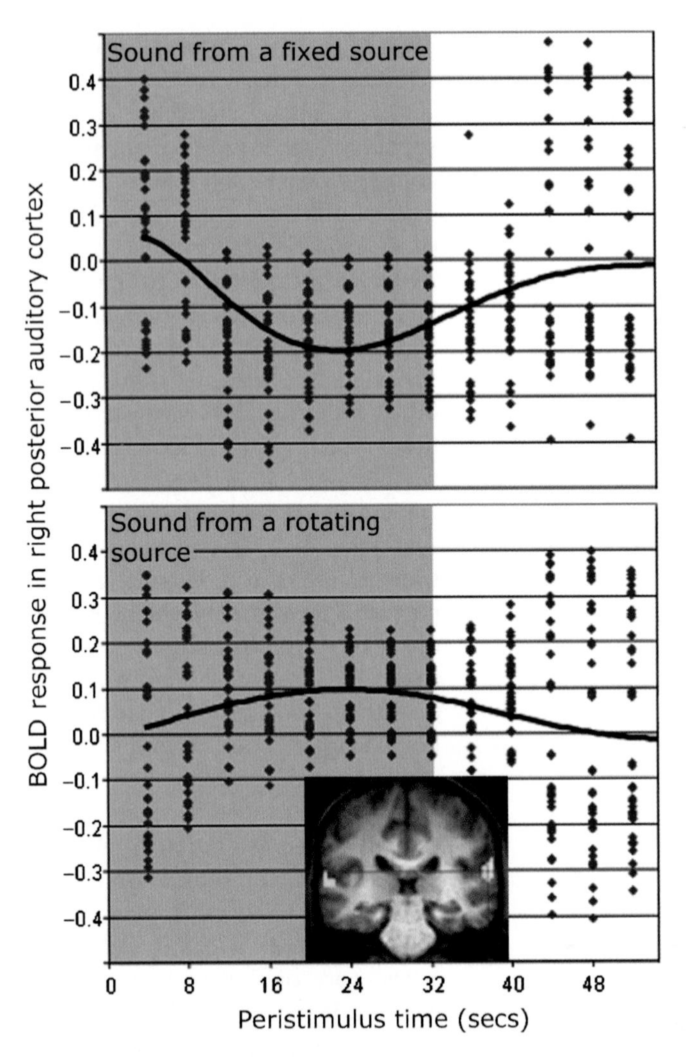

Fig. 8. Adjusted blood oxygen level-dependent (BOLD) response (measured in arbitrary units) across the 32-s stimulus epoch shaded in grey (**a**) for a sound from a fixed source and (**b**) for a sound from a rotating source. Adjusted values are combined for all six participants and the trend line is indicated using a polynomial sixth order) function. The response for both stimulus types is plotted using the same voxel location in the planum temporale region of the right auditory cortex (coordinates x 63, y –30, z 15 mm). The position of this voxel is shown in the inserted panel. The activation illustrated in this insert represents the subtraction of the fixed sound location from the rotating sound conditions ($p < 0.001$).

and contrasted against the scans acquired during the 'fixed sound source' condition, widespread activation was obtained ($p < 0.001$) across the posterior auditory cortex (planum temporale): an area traditionally linked with spatial acoustic analysis. The time course of activation demonstrates a sustained BOLD response across the entire duration of the epoch **(Fig. 8b)**. The sustained response contrasts with the transient response observed for the fixed sound source condition **(Fig. 8a)**.

2. Examples of Auditory Feature Processing

2.1. The Representation of Frequency in the Auditory Cortex

Within the inner ear, an incoming sound is separated into its individual frequency components by the way in which the energy at different frequencies travels along the cochlear partition *(45)*. High-frequency tones maximally stimulate those nerve fibres near the base of the cochlea while low-frequency tones are best coded towards the apex. This cochleotopic representation persists throughout the auditory pathway where it is referred to as a tonotopic map. Within the mammalian auditory cortex, electrophysiological recordings have revealed many tonotopic maps *(46, 47)*. Within each map, neurons tuned to the same sound frequency are co-localised in a strip across the cortical surface, with an orderly progression of frequency tuning across adjacent strips. Frequency tuning is sharper in the primary auditory fields than it is in the surrounding non-primary fields, and so the most complete representations of the audible frequency range are found in the primary fields. Primates have at least two tonotopic maps in primary auditory cortex, adjacent to one another and with mirror-reversed frequency axes. It is possible to demonstrate tonotopy by fMRI as well as by electrophysiology, even though frequency selectivity deteriorates at the moderate to high sound intensities required for fMRI sound presentation. As a recent example, mirror-symmetric frequency gradients have been confirmed across primary auditory fields using high-resolution fMRI at 4.7 T in anesthetised macaques and at 7.0 T in awake behaving macaques *(48)*. This section describes results from several fMRI experiments that have sought to demonstrate tonotopy in the human auditory cortex.

fMRI is an ideal tool for exploring the spatial distribution of the frequency-dependent responses across the human auditory cortex because it provides good spatial resolution and the analysis requires few a priori modelling assumptions [see *(49)* for a review]. In addition, it is possible to detect statistically significant activation using individual fMRI analysis. This is important when determining fine-grained spatial organisation because averaging data across different listeners would inevitably blur the subtle distinctions. A number of recent studies have sought to determine the organisation of human tonotopy *(29, 33, 50–52)*. To avoid the problem of neural adaptation discussed in **Subheading 1.6**, experimenters chose stimuli that would elicit robust auditory cortical activation. For example, Talavage et al. *(51, 52)* presented amplitude-modulated signals, while Schönwiesner et al. *(50)* presented sine tones that were frequency modulated across a narrow bandwidth. Langers et al. *(33)* used a signal detection task in which the tone targets at each frequency were briefly presented (0.5 s). In agreement with the primate literature, evidence for the presence of tonotopic organisation is at its most apparent within the primary auditory cortex while frequency preferences in the surrounding non-primary areas are more erratic *(33)*. Thus, we shall consider in more detail the precise arrangement of tonotopy in the primary region.

In their first study, Talavage et al. *(51)* contrasted pairs of low (<66 Hz) and high (>2,490 Hz) frequency stimuli of moderate intensity and sufficient spectral separation to produce spatially resolvable differences in activation (low > high and high > low) across the auditory cortical surface. These activation foci were considered to define the endpoints of a frequency gradient. In total, Talavage et al. identified eight frequency-sensitive sites across Heschl's gyrus (HG, the primary auditory cortex) and the surrounding superior temporal plane (STP, the non-primary auditory cortex). Each site was reliably identified across listeners and the sites were defined by a numerical label *(1–8)*.

Foci 1–4 occurred around the medial two-thirds of HG and are good candidates for representing frequency coding within the primary auditory cortex **(Fig. 9)**. Finding several endpoints does not provide direct confirmation of tonotopy because tonotopy

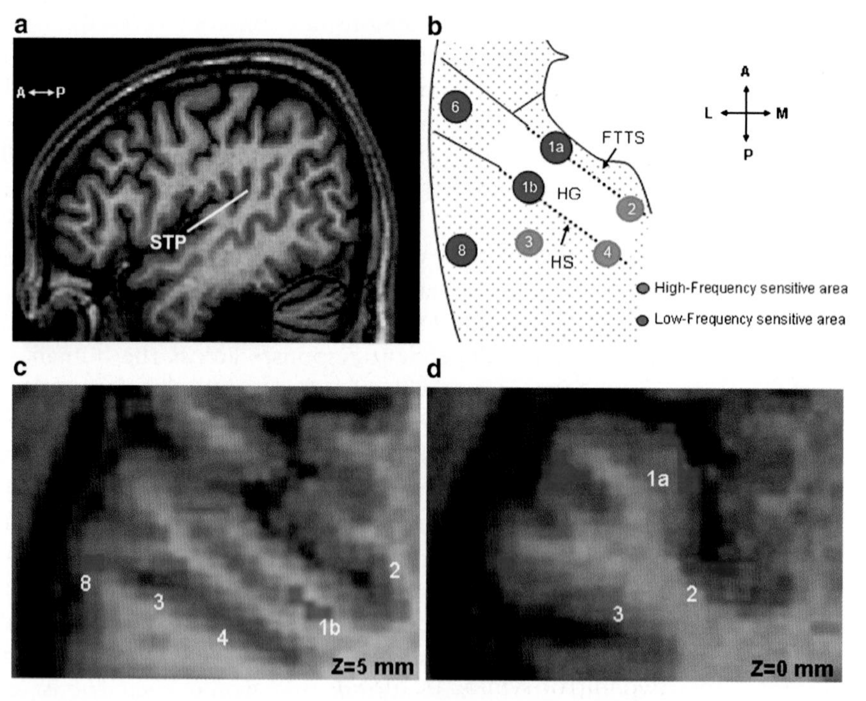

Fig. 9. (**a**) Sagittal view of the brain with the *oblique white line* denoting the approximate location and orientation of the schematic view shown in panel (**b**) along the supratemporal plane. (**b**) Schematic representation of the most consistently found high (*red*) and low (*blue*) frequency-sensitive areas across the human auditory cortex reported by Talavage et al. *(50, 51)*. The primary area is shown in *white* and the non-primary areas are shown by *dotted shading*. Panels (**c**) and (**d**) illustrate the high- (*red*) and low- (*blue*) frequency sensitive areas across the left auditory cortex of one participant (unpublished data). Two planes in the superior-inferior dimension are shown ($z = 5$ mm and $z = 0$ mm above the CA-CP line). *A* anterior, *P* posterior, *M* medial, *L* lateral, *HG* Heschl's gyrus, *HS* Heschl's sulcus, *FTTS* first transverse temporal sulcus, *STP* supratemporal plane.

necessitates a linear gradient of frequency sensitivity. Nevertheless, Talavage et al. argued that the foci 1–3 were at least consistent with predictions from primate electrophysiology. The arrangement of the three foci encompassed the primary auditory cortex, suggested a common low-frequency border, and had a mirror-image reversed pattern. This interpretation was criticised by Schönwiesner et al. *(50)* who stated that it was wrong to associate these foci with specific tonotopic fields because pairs of low- and high-frequency foci could not clearly be attributed to specific frequency axes nor to anatomically defined fields. Indeed, in their own study, Schönwiesner et al. *(50)* did not observe the predicted gradual decrease in frequency-response amplitude at locations away from the best-frequency focus, but instead found a rather complex distribution of response profiles. Their explanation for this finding was that the regions of frequency sensitivity reflected not tonotopy, but distinct cortical areas that each preferred different acoustic features associated with a limited bandwidth signal.

Increasing the BOLD SNR might be necessary for characterising some of the more subtle changes in the response away from best frequency and more recent evidence using more sophisticated scanning techniques does support the tonotopy viewpoint. Frequency sensitivity in the primary auditory cortex was studied using a 7-T ultra-high field MR scanner to improve the BOLD SNR and to provide reasonably fine-grained (1 mm^3) spatial resolution *(29)*. Formisano et al. *(29)* sought to map the progression of activation as a smooth function of tone frequency across HG. Frequency sensitivity was mapped by computing the locations of the best response to single frequency tones presented at a range of frequencies (0.3, 0.5, 0.8, 1, 2, and 3 kHz). Flattened cortical maps of best frequency revealed two mirror-symmetric gradients (high-to-low and low-to-high) travelling along HG from an anterolateral point to the posteromedial extremity. In general, the amplitude of the BOLD response decreased as the stimulating tone frequency moved away from the best frequency tuning characteristics of the voxel. A receiver coil placed close to the scalp over the position of the auditory cortex is another way to achieve a good BOLD SNR and this was the method used by Talavage et al. *(52)*. Talavage et al. measured best-frequency responses to an acoustic signal that was slowly modulated in frequency across the range 0.1–8 kHz. Again, the results confirmed the presence of two mirror-symmetric maps that crossed HG (extending from the anterior first transverse temporal sulcus to the posterior Heschl's sulcus) and shared a low-frequency border.

Although more evidence will be required before a clear consensus is established, the studies presented in this section have made influential contributions to the understanding of frequency representation in the human auditory cortex and its correspondence to primate models of auditory coding.

2.2. The Influence of Selective Attention on Frequency Representations in Human Auditory Cortex

We live in a complex sound environment in which many different overlapping auditory sources contribute to the incoming acoustical signal. Our brains have a limited processing capacity and so one of the most important functions of neural coding is to separate out these competing sources of information. One way to achieve this is by filtering out the uninformative signals (the 'ground') and attending to the signal of interest (the 'figure'). Competition between incoming signals can be resolved by a bottom-up, stimulus-driven process (such as a highly salient stimulus that evokes an involuntary orienting response), or it can be resolved by a top-down, goal-directed process (such as selective attention). Selective attention provides a modulatory influence that enables a listener to focus on the figure and to filter out or attenuate the ground *(53)*.

Visual scientists have shown that attention can be directed to the features of the figure [feature-based attention, for a review see *(54)*] or to the entire figure [object-based attention, for a review see *(55)*]. Given that so little is known about the mechanisms by which auditory objects are coded *(56)*, we shall focus on those studies of auditory feature-based attention. A sound can be defined according to many different feature dimensions including frequency spectrum, temporal envelope, periodicity, spatial location, sound level, and duration. The experimenter can instruct listeners to attend to any feature dimension in order to investigate the effect of selective attention on the neural coding of that feature. Different listening conditions have been used for comparison with the 'attend' condition. The least controlled of these is a passive listening condition in which participants are not given any explicit task instructions *(30, 57, 58)*. Even if there are cases where a task is required, but the cognitive demand of that task is low, participants are able to divide their attention across both relevant and irrelevant stimulus dimensions [see *(59)* for a review on attentional load]. Again, this leads to an uncontrolled experimental situation. For greater control, some studies have employed a visual distractor task to compete for attentional resources and pull selective attention away from the auditory modality *(60, 61)*. However, there is some evidence that the mere presence of a visual stimulus exerts a significant influence on auditory cortical responses *(62, 63)* and hence modulation related to selective attention might interact with that related to the presence of visual stimuli in a rather complex manner. This can make comparison between the results from bimodal studies *(60, 61)* and unimodal auditory studies *(32, 64)* somewhat problematic.

One paradigm that has been commonly used to examine feature-based attention manipulates two different feature dimensions independently within the same experimental session and listeners are required to make a discrimination judgement to one feature or the other. Studies have compared attention to spatial features such as location, motion, and ear of presentation with attention to non-spatial features such as pitch and phonemes *(60, 64)*. Results typically demonstrate a response enhancement in non-primary auditory regions. For example, Degerman et al. *(60)* found auditory enhancement in left posterior non-primary regions, but only for attending to location relative to pitch and not the other way round. Ahveninen et al. *(64)* used a novel paradigm in which they measured the effect of attention on neural adaptation. Their fMRI results showed smaller adaptation effects in the right posterior non-primary auditory cortex when attending to location (relative to phonemes), but again not the other way round. Both studies reported enhancement for attending to location in additional non-auditory regions, notably the prefrontal and right parietal areas. This asymmetry in the effects observed across spatial and non-spatial attended domains is worthy of further exploration since spatial analysis is well known to engage the right posterior auditory and right parietal cortex *(65)*.

Another experimental design that has been used to examine feature-based attention presents concurrent visual and auditory stimuli and participants are required to make a discrimination judgement to stimuli in one modality or the other. One example of this design used novel melodies and geometric shapes, and participants were required to respond to either long note targets or vertical line targets *(57, 58)*. When 'attending to the shapes' was subtracted from 'attending to the melodies' the results revealed relative enhancement bilaterally in the lower boundary of the superior temporal gyrus. This finding supports the view that there is sensory enhancement when attending to the auditory modality. In addition, it was shown that when 'attending to the shapes', the auditory response was suppressed relative to a bimodal passive condition. This is tentative evidence for neural suppression when ignoring the auditory modality. A novel feature of the experiment by Degerman et al. *(66)* was that in one selective attention condition, participants had to respond to a target defined by a particular combination of cross-modal features (e.g. high pitch and red circle). The conventional general linear analysis did not show any significant difference in the magnitude of the auditory response in the cross-modal condition compared with a condition in which participants simply attended to the high- and low-pitch targets in the audiovisual stimulus. However, a region of interest analysis (defining a region in the posterolateral superior temporal gyrus) did suggest some enhancement for the audiovisual attention condition compared with the auditory attention condition. Thus, it is possible that non-primary auditory regions are involved in attention–dependent binding of synchronous auditory and visual events into coherent audio–visual objects.

In audition, it has long been established behaviourally that when participants expect a tone at a specific frequency, their ability to detect a tone in a noise masker is significantly better when the tone is at the expected frequency than when it is at an unexpected frequency [the probe-signal paradigm *(67)*]. The benefit of selective attention for signal detection thresholds can be plotted as a function of frequency. The ability to detect tones at frequencies close to the expected frequency is also enhanced, and this benefit drops off smoothly with the distance away from the expected frequency *(67, 68)*. The width of this attention-based listening band is comparable to the width of the critical band related to the frequency-tuning curve, which can be measured psychophysically using notched noise maskers *(68)*. This equivalence suggests that selective attention might be operating at the level of the sensory representation of tone frequency.

Evidence from electrophysiological recordings demonstrates frequency-specific attentional modulation at the level of the primary auditory cortex, consistent with a neural correlate of the psychophysical phenomena found in the probe-signal paradigm. In a series of experiments, awake behaving ferrets were trained to perform a number of spectral tasks including tone detection and frequency discrimination *(69)*. In the tone detection task, ferrets were trained to identify the presence of a tone against a background of broadband rippled noise. The spectro-temporal receptive fields measured during the noise for frequency-tuned neurons showed strong facilitation around the target frequency that persisted for 30–40 ms. In the two-tone discrimination task, ferrets performed an oddball task in which they responded to an infrequent target frequency. Again, the spectro-temporal receptive fields showed an enhanced and persistent response for the target frequency, plus a *decreased* response for the reference frequency. These opposite effects serve to magnify the contrast between the two centre frequencies, and thus facilitate the selection of the target. The results of these two tasks confirm that the acoustic filter properties of auditory cortical neurons can dynamically adapt to the attentional focus of the task.

Recently, we have addressed the question of attentional enhancement for selective attention to frequency using a high-resolution scanning protocol ($1.5\ mm^2 \times 2.5\ mm$) (unpublished data). To control for the demands on selective attention, we presented two concurrent streams (low- and high-frequency tones). Participants were requested to attend to one frequency stream or the other and these attend conditions were presented in an interleaved manner throughout the experiment. Behavioural testing confirmed that performance significantly deteriorated when these sounds were presented in a divided attention task. To be able to identify high- and low-frequency sensitive areas around the primary auditory cortex we designed two types of stimuli using different rhythms for each of the two streams. For example, one stimulus contained a 'fast' high-frequency rhythm and a 'slow'

low-frequency rhythm so that the stimulus contained a majority of high-frequency tones. The other stimulus was the converse. Areas of high-frequency sensitivity were identified by subtracting the low-frequency majority stimulus from the high-frequency majority stimulus, and vice-versa (**Fig. 9c, d**). For each of the three participants, we selected those frequency-specific areas that best corresponded to areas 1–4 [defined by Talavage et al. *(51, 52)*; see **Subheading 2.1**]. Within these areas, we extracted the BOLD signal time course for every voxel and performed a log transform to standardise the data. We collapsed the data across low- and high-frequency sensitive areas *(1–4)* according to their 'best frequency' (BF). The best frequency of an area corresponds to the frequency that evokes the largest BOLD response. A univariate ANOVA showed response enhancement when participants were attending to the BF of that area, compared with attending to the other frequency ($p < 0.01$). In addition, response enhancement was also found when attending to the BF of that area, compared with passive listening ($p < 0.05$) (**Fig. 10**). Note that for these results area 4 was excluded from the analysis, because it showed different pattern of attentional modulation. The response profile of area 4 might differ from that of areas 1–3 in other ways because it is not consistently present in all listeners *(51)*. Our finding of frequency-specific attentional enhancement in primary auditory regions contrasts with that of Petkov et al. *(61)*,

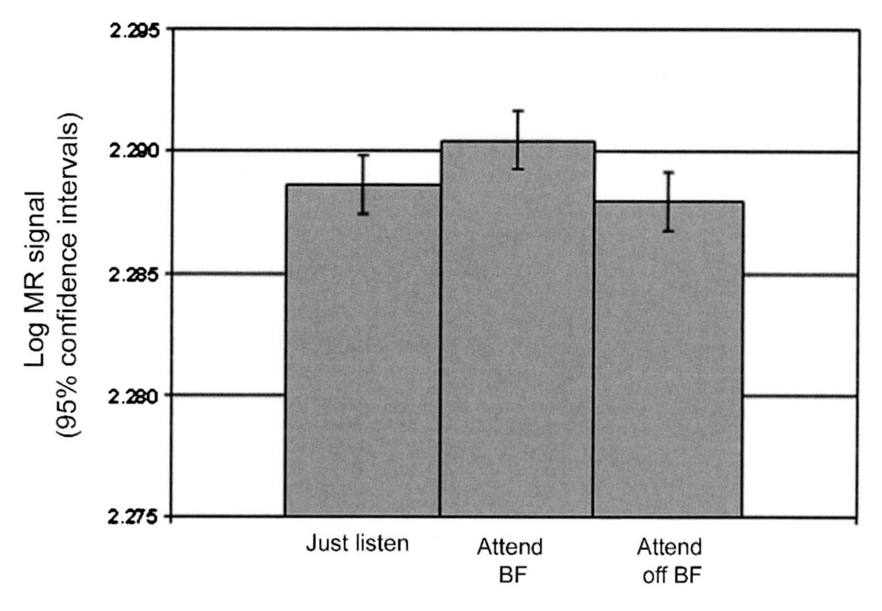

Fig. 10. Response to the three listening conditions: just listen, attend to the best frequency (BF) tones, and attend OFF BF. The data shown are for those stimuli in which BF tones formed the majority (80%) of the tones in the sound sequence, combining responses across areas 1–3. The error bars denote the 95% confidence intervals.

who reported attention-related modulation to be independent of stimulus frequency and to engage mainly the non-primary auditory areas. However, our result is more in keeping with the predictions made by the neurophysiological data reported by Fritz et al. *(69)*.

3. Future Directions

It is increasingly likely that auditory cortical regions compute aspects of the sound signal that are more complicated in their nature than the simple physical acoustic attributes of the sound. Thus, the encoded features of the sound reflect an increasingly abstract representation of the sound stimulus. We have already presented some evidence for this in terms of the way in which the auditory cortical response is exquisitely sensitive to the temporal context of the sound, particularly the way in which the time course of the BOLD response represents the temporal envelope characteristics of the sound, including sound onsets and offsets [*(18, 40)*, see **Subheading 1.4**]. However, there are many other ways in which neural coding reflects higher level processing. In this final section, we shall introduce two important aspects of the listening context that determine the auditory BOLD response: the perceptual experience of the listener and the operational aspects of the task. A number of fMRI studies have demonstrated ways in which activity within the human auditory cortex is modulated by auditory sensations, including loudness, pitch, and spatial width. Other studies have revealed that task relatedness is also a significant determining factor for the pattern of activation. These findings highlight how future auditory fMRI studies could usefully investigate these contributory factors in order to provide a more complete picture of the neural basis of the listening process.

3.1. Cortical Activation Reflects Perceptually Relevant Coding

One approach used in auditory fMRI to investigate perceptually relevant coding imposes systematic changes to the listener's perception of a sound signal by parametrically manipulating certain acoustic parameters and subsequently correlating the perceptual change with the variation in the pattern of activation. For example, by increasing sound intensity (measured in SPL), one also increases its perceived loudness (measured in phons). Loudness is a perceptual phenomenon that is a function of the auditory excitation pattern induced by the sound, integrated across frequency. Sound intensity and loudness are measures of different phenomena. For example, if the bandwidth of a broadband signal is increased while its intensity is held constant, then loudness nevertheless increases because the signal spans a greater number of frequency channels. In an early fMRI study, Hall et al. *(31)* presented single-frequency tones and harmonic-complex tones that were matched either in intensity or in loudness. The results showed that the complex tones produced greater activation than did the single-frequency tones, irrespective of the matching scheme. This result indicates that bandwidth had a greater effect on the pattern of auditory activation than sound level. Nevertheless, when the data were collapsed across stimulus class, the amount of activation was significantly correlated with the loudness scale, not with the intensity scale.

In people with elevated hearing thresholds, the perception of sound level is distorted. They typically experience the same dynamic range of loudness as normally hearing listeners despite having a compressed range of sensitivity to sound level. The BOLD response to sound level is reflected in a disproportionate increase in loudness with intensity. A recent study has characterised the BOLD response to frequency-modulated tones presented at a broad range of intensities (0–70 dB above the normal hearing threshold) *(33)*. Both normally hearing and hearing impaired groups showed the same steepness in the linear increase in auditory activation as a function of loudness, but not of intensity **(Fig. 11)**. The results from this study clearly demonstrate that the BOLD response can be interpreted as a correlate of the subjective strength of the stimulus percept.

Pitch can be defined as the sensation whose variation is associated with musical melodies. Together with loudness, timbre, and spatial location, pitch is one of the primary auditory sensations. The salience of a pitch is determined by several physical properties

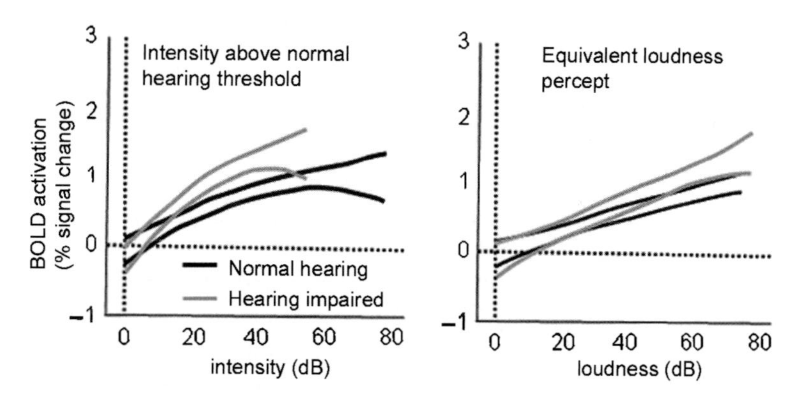

Fig. 11. Growth in the blood oxygen level-dependent (BOLD) response (measured in % signal change) for high (4–8 kHz) frequency-modulated tone presented across a range of sound levels for ten normally hearing participants (*black lines*) and ten participants with a high-frequency hearing loss (*grey lines*). The *left hand panel* shows the growth as a function of sound intensity, while the *left hand panel* plots the same data as a function of the equivalent loudness percept. The *upper and lower lines* denote the 95% confidence interval of the quadratic polynomial fit to the data. This graph summarises data presented by Langers et al.

of the pitch signal, one being the numbered harmonic components comprising a harmonic-complex tone. The cochlea separates out the frequency components of sounds to a limited extent, so that the first eight harmonics of a harmonic-complex tone excite distinct places in the cochlea and are said to be 'resolved', whereas the higher harmonics are not separated and are said to be 'unresolved'. Pitch discrimination thresholds for unresolved harmonics are substantially higher than those for resolved harmonics, consistent with the former type of stimulus evoking a less salient pitch *(70)*. A pairwise comparison between the activation patterns for resolved (strong pitch) and unresolved (weak pitch) harmonic-complex tones has identified differential activation in a small, spatially localised region of non-primary auditory cortex, overlapping the anterolateral end of Heschl's gyrus *(71)*. The authors claim that this finding reflects the cortical representation for pitch salience. Another way to determine the salience of a pitch is by the degree of fine temporal regularity in the stimulus (i.e. the monaural repeating pattern within frequency channels). This is true even for signals in which there are no distinct frequency peaks in the cochlear excitation pattern from which to calculate the pitch. A range of pitch saliencies can be created by parametrically varying the degree of temporal regularity in an iterated-ripple noise stimulus (using 0, 1, 2, 4, 8, and 16 add-and-delay iterations during stimulus generation) *(72)*. Again the anterolateral end of Heschl's gyrus appeared highly responsive to the change in pitch salience, in a linear manner.

Spatial location is another important auditory sensation that is determined by the fine temporal structure in the signal, this time it being the binaural temporal characteristics across the two ears. The interaural correlation (IAC) of a sound represents the similarity between the signals at the left and right ears. Changes in the IAC of a wideband signal result in changes in sound's perceived 'width' when presented through headphones. A noise with an IAC of 1.0 is typically perceived as sound with a compact source located at the centre of the head. As the IAC is reduced the source broadens. For an IAC of 0.0, it eventually splits into two separate sources, one at each ear *(73)*. Again the parametric approach has been employed to measure activation across a range of IAC values (1.00, 0.93, 0.80, 0.60, 0.33, and 0.00) *(74)*. The authors found a significant positive relationship between BOLD activity and IAC, which was confined to the anterolateral end of Heschl's gyrus, the region that is also responsive to pitch salience. The slope of the function was not precisely linear but the BOLD response was more sensitive to changes in IAC at values near to unity than at values near zero. This response pattern is qualitatively compatible with previous behavioural measures of sensitivity to IAC *(75)*.

There is some evidence to support the claim that the neural representations of auditory sensations (including loudness, pitch, and spatial width) evolve as one ascends the auditory pathway. Budd et al. *(74)* examined sensitivity to values of IAC associated with spatial width within the inferior colliculus, the medial geniculate nucleus, as well as across different auditory cortical regions, but the effects were significant only within the non-primary auditory cortex. Griffiths et al. *(72, 76)* also examined sensitivity to the increases in temporal regularity associated with pitch salience within the cochlear nucleus, inferior colliculus, medial geniculate nucleus, as well as across different auditory cortical regions. Some degree of sensitivity to pitch salience was found at all sites, but the preference appeared greater in the higher centres than in the cochlear nucleus *(76)*. Thus, the evidence supports the notion of an increasing responsiveness to percept attributes of sound throughout the ascending auditory system, culminating in the non-primary auditory cortex. These findings are consistent with the hierarchical processing of sound attributes.

Encoding the perceptual properties of a sound is integral to identifying the object properties of that sound source. The non-primary auditory cortex probably plays a key role in this process because it has widespread cortical projections to frontal and parietal brain regions and is therefore ideally suited to access distinct higher level cortical mechanisms for sound identification and localisation. Recent trends in auditory neuroscience are increasingly concerned with auditory coding beyond the conventional limits of the auditory cortex (the superior temporal gyrus in humans), particularly with respect to the hierarchical organisation of sensory coding via dorsal and ventral auditory processing routes. At the top of this hierarchy stands the brain's representation of an auditory 'object'. The concept of an auditory object still remains controversial *(56)*. Although it is clear that the brain needs to code information about the invariant properties of a sound source, research in this field is considerably underdeveloped. Future directions are likely to begin to address critical issues such as the definition of an auditory object, whether the concept is informative for auditory perception, and optimal paradigms for studying object coding.

3.2. Cortical Activation Also Reflects Behaviourally Relevant Coding

Listeners interact with complex auditory environments that, at any one time point, contain multiple auditory objects located at dynamically varying spatial locations. One of the primary challenges for the auditory system is to analyse this external environment in order to inform goal-directed behaviour. In **Subheading 2.2** we introduced some of the neurophysiological evidence for the importance of the attentional focus of the task in determining the pattern of auditory cortical activity *(69)*. Here, we consider the contribution of human auditory fMRI research to this question. In particular, we present the interesting findings of one group who have started to address how the auditory cortex responds to the context and the procedural and cognitive demands of the listening task [*see (77)* for a review].

In that review, Scheich and colleagues report a series of research studies in which they suggest that the function of different auditory cortical areas is not determined so much by stimulus features (such as timbre, pitch, motion, etc.), but rather by the task that is performed. For example, one study reported the results of two fMRI experiments in which the same frequency-modulated stimuli were presented under different task conditions *(78)*. Top-down influences strongly affected the strength of the auditory response. When a pitch-direction categorisation task was compared with passive listening, a greater response was found in right posterior non-primary auditory areas (planum temporale). Moreover, hemispheric differences were also found when comparing the response to two different tasks. The right non-primary areas responded more strongly when the task required a judgement about pitch direction (rising or falling), whereas the left non-primary areas responded more strongly when the task required a judgement about the sound duration. It is not the case that the right posterior non-primary areas were *only* engaged by sound categorisation because this region was more responsive to the critical sound feature (frequency modulation) during passive listening than were surrounding auditory areas (see also **ref.** *32)*. These results broadly indicate an interaction between the stimulus and the task, which influences the pattern of auditory cortical activity. The precise characteristics of this interaction are worthy of future studies.

References

1. Palmer AR, Bullock DC, Chambers JD. A high-output, high-quality sound system for use in auditory fMRI. Neuroimage 1998;7:S357.
2. Chou CK, McDougall JA, Chan KW. Absence of radiofrequency heating from auditory implants during magnetic-resonance imaging. Bioelectromagnetics 1995;16(5):307–316.
3. Heller JW, Brackmann DE, Tucci DL, Nyenhuis JA, Chou CK. Evaluation of MRI compatibility of the modified nucleus multichannel auditory brainstem and cochlear implants. Am J Otol 1996;17(5):724–729.
4. Shellock FG, Morisoli S, Kanal E. MR procedures and biomedical implants, materials, and devices – 1993 update. Radiology 1993; 189(2):587–599.
5. Weber BP, Neuburger J, Battmer RD, Lenarz T. Magnetless cochlear implant: Relevance of adult experience for children. Am J Otol 1997;18(6):S50–S51.
6. Wild DC, Head K, Hall DA. Safe magnetic resonance scanning of patients with metallic middle ear implants. Clin Otolaryngol 2006; 31(6):508–510.
7. Giraud AL, Truy E, Frackowiak R. Imaging plasticity in cochlear implant patients. Audiol Neurootol 2001;6(6):381–393.
8. Moelker A, Piotr A, Wielopolski, Pattynama PM. Relationship between magnetic field strength and magnetic-resonance-related acoustic noise levels. MAGMA 2003;16:52–55.
9. Foster JR, Hall DA, Summerfield AQ, Palmer AR, Bowtell RW. Sound-level measurements and calculations of safe noise dosage during fMRI at 3T. J Magn Reson Imaging 2000;12:157–163.
10. Ravicz ME, Melcher JR. Isolating the auditory system from acoustic noise during functional magnetic resonance imaging: Examination of noise conduction through the ear canal, head, and body. J Acoust Soc Am 2001;109(1): 216–231.
11. Price DL, De Wilde JP, Papadaki AM, Curran JS, Kitney RI. Investigation of acoustic noise on 15 MRI scanners from 0.2 T to 3 T. J Magn Reson Imaging 2001;13(2):288–293.
12. Hedeen RA, Edelstein WA. Characterization and prediction of gradient acoustic noise in MR imagers. Magn Reson Med 1997;37(1):7–10.

13. Hennel F, Girard F, Loenneker T. "Silent" MRI with soft gradient pulses. Magn Reson Med 1999;42:6–10.
14. Brechmann A, Baumgart F, Scheich H. Sound-level-dependent representation of frequency modulations in human auditory cortex: A low-noise fMRI study. J Neurophysiol 2002; 87:423–433.
15. Sumby WH, Pollack I. Visual contribution to speech intelligibility in noise. J Acoust Soc Am 1954;26(2):212–215.
16. Assmann P, Summerfield Q. Perception of speech under adverse conditions. In: S. Greenberg, W. A. Ainsworth, A. N. Popper, R. R. Fay, eds. Speech processing in the auditory system. New York: Springer; 2004:231–308.
17. Healy EW, Moser DC, Morrow-Odom KL, Hall DA, Fridriksson J. Speech perception in MRI scanner noise by persons with aphasia. J Speech Lang Hear Res 2007;50:323–334.
18. Harms MP, Melcher JR. Sound repetition rate in the human auditory pathway: Representations in the waveshape and amplitude of fMRI activation. J Neurophysiol 2002;88:1433–1450.
19. Chambers JD, Akeroyd MA, Summerfield AQ, Palmer AR. Active control of the volume acquisition noise in functional magnetic resonance imaging: Method and psychoacoustical evaluation. J Acoust Soc Am 2001;110(6): 3041–3054.
20. Hall DA, Chambers J, Foster J, Akeroyd MA, Coxon R, Palmer AR. Acoustic, psychophysical, and neuroimaging measurements of the effectiveness of active cancellation during auditory functional magnetic resonance imaging. The Journal of the Acoustical Society of America 2009;125(1):347–359.
21. Hall DA, Haggard MP, Akeroyd MA, Palmer AR, Summerfield AQ, Elliott MR, Gurney EM, Bowtell RW. 'Sparse' temporal sampling in auditory fMRI. Hum Brain Mapp 1999;7:213–223.
22. Bandettini PA, Jesmanowicz A, Van Kylen J, Birn RM, Hyde JS. Functional MRI of brain activation induced by scanner acoustic noise. Magn Reson Med 1998;39:410–416.
23. Bilecen D, Scheffler K, Schmid N, Tschopp K, Seelig J. Tonotopic organization of the human auditory cortex as detected by BOLD-FMRI. Hear Res 1998;126:19–27.
24. Hall DA, Summerfield AQ, Gonçalves MS, Foster JR, Palmer AR, Bowtell RW. Time-course of the auditory BOLD response to scanner noise. Magn Reson Med 2000;43:601–606.
25. Shah NJ, Jäncke L, Grosse-Ruyken M-L, Müller-Gärtner HW. Influence of acoustic masking noise in fMRI of the auditory cortex during phonetic discrimination. J Magn Reson Imaging 1999;9(1):19–25.
26. Talavage TM, Edmister WB, Ledden PJ, Weisskoff RM. Quantitative assessment of auditory cortex responses induced by imager acoustic noise. Hum Brain Mapp 1999;7(2):79–88.
27. Elliott MR, Bowtell RW, Morris PG. The effect of scanner sound in visual, motor, and auditory functional MRI. Magn Reson Med 1999;41(6):1230–1235.
28. Edmister WB, Talavage TM, Ledden PJ, Weisskoff RM. Improved auditory cortex imaging using clustered volume acquisitions. Hum Brain Mapp 1999;7:89–97.
29. Formisano E, Kim DS, Di Salle F, van de Moortele PF, Ugurbil K, Goebel R. Mirror-symmetric tonotopic maps in human primary auditory cortex. Neuron 2003;40(4):859–869.
30. Hall DA, Haggard MP, Akeroyd MA, Summerfield AQ, Palmer AR, Elliott MR, Bowtell RW. Modulation and task effects in auditory processing measured using fMRI. Hum Brain Mapp 2000;10(3):107–119.
31. Hall DA, Haggard MP, Summerfield AQ, Akeroyd MA, Palmer AR, Bowtell RW. Functional magnetic resonance imaging measurements of sound-level encoding in the absence of background scanner noise. J Acoust Soc Am 2001;109(4):1559–1570.
32. Hart HC, Palmer AR, Hall DA. Different areas of human non-primary auditory cortex are activated by sounds with spatial and non-spatial properties. Hum Brain Mapp 2004;21: 178–190.
33. Langers DRM, Backes WH, Van Dijk P. Representation of lateralization and tonotopy in primary versus secondary human auditory cortex. Neuroimage 2007;34:264–273.
34. Schwarzbauer C, Davis MH, Rodd JM, Johnsrude I. Interleaved silent steady state (ISSS) imaging: A new sparse imaging method applied to auditory fMRI. Neuroimage 2006;29(3): 774–782.
35. Bregman AS. Auditory scene analysis: The perceptual organisation of sound. MIT, Cambridge, MA; 1990.
36. Fishman YI, Arezzo JC, Steinschneider M. Auditory stream segregation in monkey auditory cortex: Effects of frequency separation, presentation rate, and tone duration. J Acoust Soc Am 2004;116(3):1656–1670.
37. Fishman YI, Reser DH, Arezzo JC, Steinschneider M. Neural correlates of auditory stream segregation in primary auditory cortex of the awake monkey. Hear Res 2001;151:167–187.
38. Brosch M, Schreiner CE. Time course of forward masking tuning curves in cat primary

auditory cortex. J Neurophysiol 1997;77: 923–943.
39. Bartlett EL, Wang X. Long-lasting modulation by stimulus context in primate auditory cortex. J Neurophysiol 2005;94:83–104.
40. Harms MP, Guinan JJ, Sigalovsky IS, Melcher JR. Short-term sound temporal envelope characteristics determine multisecond time patterns of activity in human auditory cortex as shown by fMRI. J Neurophysiol 2005;93:210–222.
41. Palmer AR, Hall DA, Sumner C, Barrett DJK, Jones S, Nakamoto K, Moore DR. Some investigations into non-passive listening. Hear Res 2007;229:148–157.
42. Hall DA, Edmondson-Jones M, Fridriksson J. Periodicity and frequency coding in human auditory cortex. Eur J Neurosci 2006;24: 3601–3610.
43. Hall DA, Johnsrude IS, Haggard MP, Palmer AR, Akeroyd MA, Summerfield AQ. Spectral and temporal processing in human auditory cortex. Cereb Cortex 2002;12:140–149.
44. Hart HC, Hall DA, Palmer AR. The sound-level-dependent growth in the extent of fMRI activation in Heschl's gyrus is different for low- and high-frequency tones. Hear Res 2003;179(1–2):104–112.
45. Von Békésy G. The variations of phase along the basilar membrane with sinusoidal vibrations. J Acoust Soc Am 1947;19:452–460.
46. Kosaki H, Hashikawa T, He J, Jones EG. Tonotopic organization of auditory cortical fields delineated by parvalbumin immunoreactivity in macaque monkeys. J Comp Neurol 1997;386:304–316.
47. Merzenich MM, Brugge JF. Representation of the cochlear partition on the superior temporal plane of the macaque monkey. Brain Res 1973;50:275–296.
48. Petkov CL, Kayser C, Augath M, Logothetis NK. Functional imaging reveals numerous fields in the monkey auditory cortex. PLoS Biol 2006;4(7):213–226.
49. Hall DA, Hart HC, Johnsrude IS. Relationships between human auditory cortical structure and function. Audiol Neurootol 2003;8(1):1–18.
50. Schönwiesner M, Von Cramon DY, Rubsamen R, Is it tonotopy after all?Neuroimage 2002;17:1144–1161.
51. Talavage TM, Ledden PJ, Benson RR, Rosen BR, Melcher JR. Frequency-dependent responses exhibited by multiple regions in human auditory cortex. Hear Res 2000; 150:225–244.
52. Talavage TM, Sereno MI, Melcher JR, Ledden PJ, Rosen BR, Dale AM. Tonotopic organization in human auditory cortex revealed by progressions of frequency sensitivity. J Neurophysiol 2004;91:1282–1296.
53. Kastner S, Ungerleider LG. Mechanisms of visual attention in the human cortex. Annu Rev Neurosci 2000;23(1):315–341.
54. Maunsell JHR, Treue S. Feature-based attention in visual cortex. Trends Neurosci 2006; 29(6):317–322.
55. Scholl BJ. Objects and attention: The state of the art. Cognition 2001;80(1–2):1–46.
56. Griffiths TD, Warren JD, . What is an auditory object?Nat Rev Neurosci 2004;5(11):887–892.
57. Johnson JA, Zatorre RJ. Attention to simultaneous unrelated auditory and visual events: Behavioral and neural correlates. Cereb Cortex 2005;15(10):1609–1620.
58. Johnson JA, Zatorre RJ. Neural substrates for dividing and focusing attention between simultaneous auditory and visual events. Neuroimage 2006;31(4):1673–1681.
59. Lavie N. Distracted and confused? Selective attention under load. Trends Cogn Sci 2005; 9(2):75–82.
60. Degerman A, Rinne T, Salmi J, Salonen O, Alho K. Selective attention to sound location or pitch studied with fMRI. Brain Res 2006;1077(1):123–134.
61. Petkov CI, Kang X, Alho K, Bertrand O, Yund EW, Woods DL. Attentional modulation of human auditory cortex. Nat Neurosci 2004;7(6):658–663.
62. Kayser C, Petkov CI, Augath M, Logothetis NK. Functional imaging reveals visual modulation of specific fields in auditory cortex. J Neurosci 2007;27(8):1824–1835.
63. Lehmann C, Herdener M, Esposito F, Hubl D, di Salle F, Scheffler K, Bach DR, Federspiel A, Kretz R, Dierks T, Seifritz E. Differential patterns of multisensory interactions in core and belt areas of human auditory cortex. Neuroimage 2006;31(1):294–300.
64. Ahveninen J, Jaaskelainen IP, Raij T, Bonmassar G, Devore S, Hamalainen M, Levanen S, Lin F-H, Sams M, Shinn-Cunningham BG, Witzel T, Belliveau JW. Task-modulated "what" and "where" pathways in human auditory cortex. Proc Natl Acad Sci USA 2006;103(39):14608–14613.
65. Lewald J, Meister IG, Weidemann J, Topper R. Involvement of the superior temporal cortex and the occipital cortex in spatial hearing: Evidence from repetitive transcranial magnetic stimulation. J Cogn Neurosci 2004;16(5):828–838.
66. Degerman A, Rinne T, Pekkola J, Autti T, Jaaskelainen IP, Sams M, Alho K. Human brain

activity associated with audiovisual perception and attention. Neuroimage 2007;34(4):1683–1691.

67. Greenberg GZ, Larkin WD. Frequency-response characteristic of auditory observers detecting signals of a single frequency in noise: The probe-signal method. J Acoust Soc Am 1968;44(6):1513–1523.
68. Schlauch RS, Hafter ER. Listening bandwidths and frequency uncertainty in pure-tone signal detection. J Acoust Soc Am 1991;90(3):1332–1339.
69. Fritz JB, Elhilali M, David SV, Shamma SA, . Does attention play a role in dynamic receptive field adaptation to changing acoustic salience in A1?Hear Res 2007;229:186–203.
70. Shackleton TM, Carlyon RP. The role of resolved and unresolved harmonics in pitch perception and frequency modulation discrimination. J Acoust Soc Am 1994;95:3529–3540.
71. Penagos H, Melcher JR, Oxenham AJ. A neural representation of pitch salience in nonprimary human auditory cortex revealed with functional magnetic resonance imaging. J Neurosci 2004;24(30):6810–6815.
72. Griffiths TD, Büchel C, Frackowiak RSJ, Patterson RD. Analysis of temporal structure in sound by the human brain. Nat Neurosci 1998;1:422–427.
73. Blauert J, Lindemann W. Spatial mapping of intracranial auditory events for various degrees of interaural coherence. J Acoust Soc Am 1986;79(3):806–813.
74. Budd TW, Hall DA, Goncalves MS, Akeroyd MA, Foster JR, Palmer AR, Head K, Summerfield AQ. Binaural specialisation in human auditory cortex: An fMRI investigation of interaural correlation sensitivity. Neuroimage 2003;20(3):1783–1794.
75. Culling JF, Colburn HS, Spurchise M. Interaural correlation sensitivity. J Acoust Soc Am 2001;110(2):1020–1029.
76. Griffiths TD, Uppenkamp S, Johnsrude I, Josephs O, Patterson RD. Encoding of the temporal regularity of sound in the human brainstem. Nat Neurosci 2001;4:633–637.
77. Scheich H, Brechmann A, Brosch M, Budinger E, Ohl FW. The cognitive auditory cortex: Task-specificity of stimulus representations. Hear Res 2007;229:213–224.
78. Brechmann A, Scheich H. Hemispheric shifts of sound representation in auditory cortex with conceptual listening. Cereb Cortex 2005;15(5):578–587.

Part III

fMRI Clinical Application

Chapter 19

Application of fMRI to Multiple Sclerosis and Other White Matter Disorders

Massimo Filippi and Maria A. Rocca

Summary

The variable effectiveness of reparative and recovery mechanisms following tissue damage is among the factors that might contribute to explain, at least partially, the paucity of the correlation between clinical and magnetic resonance imaging (MRI) findings in patients with white matter disorders. Among the mechanisms of recovery, brain plasticity is likely to be one of the most important with several possible different substrates (including increased axonal expression of sodium channels, synaptic changes, increased recruitment of parallel existing pathways or "latent" connections, and reorganization of distant sites). The application of fMRI has shown that plastic cortical changes do occur after white matter injury of different aetiology, that such changes are related to the extent of white matter damage, and that they can contribute in limiting the clinical consequences of brain damage. Conversely, the failure or exhaustion of the adaptive properties of the cerebral cortex might be among the factors responsible for the accumulation of "fixed" neurological deficits in patients with white matter disorders.

Key words: Multiple sclerosis, Functional magnetic resonance imaging, White matter, Adaptation, Maladaptation, Myelitis, Vasculitides

1. Introduction

Over the past decade, modern structural magnetic resonance imaging (MRI) techniques have been extensively used to study patients with white matter disorders with the ultimate goal of increasing the understanding of the mechanisms responsible for the accumulation of irreversible disability *(1–3)*. Although the application of these techniques has provided important insight into the pathobiology of many of these disorders, the magnitude of the correlation between MRI and clinical findings remains suboptimal *(1–3)*. This might be explained, at least partially, by the

M. Filippi (ed.), *fMRI Techniques and Protocols*, Neuromethods, vol. 41
DOI 10.1007/978-1-60327-919-2_19, © Humana Press, a part of Springer Science+Business Media, LLC 2009

variable effectiveness of reparative and recovery mechanisms following tissue damage. Cortical reorganization has been recently suggested as a potential contributor to the recovery or to the maintenance of function in the presence of irreversible white matter damage *(4, 5)*. Brain plasticity is a well-known feature of the human brain, which is likely to have several different substrates (including increased axonal expression of sodium channels, synaptic changes, increased recruitment of parallel existing pathways or "latent" connections, and reorganization of distant sites) *(6)*. The application of functional MRI (fMRI) has shown that plastic cortical changes do occur after central nervous system (CNS) white matter injury of different aetiology, that such changes are related to the extent of WM damage, and that they can contribute in limiting the clinical consequences of widespread disease-related tissue damage *(4, 5)*. Conversely, the failure or the exhaustion of the adaptive properties of the cerebral cortex might be among the factors responsible for the accumulation of "fixed" neurological deficits in patients affected by white matter disorders (WMD) *(4, 5)*.

This chapter summarizes the major contributions of fMRI for the in vivo monitoring of several white matter diseases. Since fMRI has been mostly applied to improve our understanding of the pathophysiology of multiple sclerosis (MS), a special focus is devoted to this condition and allied WMD.

2. fMRI in MS

2.1. General Considerations

The main problem in the interpretation of fMRI studies in diseased people is that the observed changes might be biased by differences in task performance between patients and controls. Clearly, this is a major issue in MS, which typically causes impairment of various functional systems. Therefore, despite providing several important pieces of information, the value of the earliest fMRI studies of patients with MS *(7–13)* has to be weighted against this background. For this reason, the most recent fMRI studies in MS have been based on larger and more selected patients' groups than the seminal studies. These studies have investigated the brain patterns of cortical activations during the performance of a number of motor, visual, and cognitive tasks in patients with all the major clinical phenotypes of the disease. One of the most solid conclusion that can be drawn from fMRI studies of MS is that cortical reorganization does occur in patients affected by this condition. The correlation between various measures of structural MS damage and the extent of cortical activations also suggests an adaptive role of such cortical changes in contributing to clinical recovery and maintaining a normal level of functioning in patients with MS, despite the presence of irreversible axonal/neuronal loss.

2.2. Visual System

The method usually applied to investigate the visual system consists of the application of a 8 Hz photic stimulation to one or both eyes *(8, 12, 13–18)*. A study of the visual system *(12)* in patients who had recovered from a single episode of acute unilateral optic neuritis demonstrated that these patients, relative to healthy volunteers, had an extensive activation of the visual network, including the claustrum, lateral temporal and posterior parietal cortices, and thalamus, in addition to the primary visual cortex, when the clinically affected eye was studied. Conversely, when the unaffected eye was stimulated, only activations of the visual cortex and the right insula/claustrum were observed. Furthermore, a strong correlation was found in these patients between the volume of the extra-occipital activation and the latency of the visual evoked potential (VEP) P100, suggesting that the functional reorganization of the cortex might represent an adaptive response to a persistently abnormal visual input. The results of this preliminary study have been confirmed and extended by subsequent studies *(14, 15, 17)*. Toosy et al. *(14)* replicated the previous study *(12)* using a longer photic stimulation epoch to better elucidate the nature of the abnormal extra-occipital response observed, which had a peak response during the OFF phase of the stimulation paradigm. The results of this study confirmed the original findings of a phase-dependent increase of the blood oxygen level dependent (BOLD) signal in the extra-occipital regions during the baseline condition. Russ et al. *(15)* used fMRI and VEP to monitor the functional recovery after an acute unilateral optic neuritis and found a strong relationship between fMRI and VEP latencies, suggesting that, in MS, fMRI might contribute to the assessment of the temporal evolution of the visual deficits during recovery, either natural or modified by treatment. Levin et al. *(17)* showed reduced activation of the primary visual cortex and increased activation of the lateral occipital complex (LOC) in eight subjects who recovered clinically from an episode of optic neuritis, but who still had prolonged VEP latencies in comparison with healthy controls. More recently, in a 1-year follow-up study, Toosy et al. *(14)*, using a novel technique that modeled the fMRI response and optic nerve structure together with clinical function, demonstrated a potential adaptive role of cortical reorganization within the extra-striate visual areas, which are regions involved in higher-order visual processing, early after optic neuritis. In addition, in this study, an increased optic nerve gadolinium-enhanced lesion length at baseline was associated with a reduced functional activation within the visual cortex and poorer vision. At 3 months, more severe optic nerve damage was associated with an increased fMRI response in the bilateral temporal cortices, whereas at 1 year, the right temporal cortex correlation reversed. These results illustrate how the same regions may play different roles at different times during recovery, reflecting the complexity of brain plasticity and the MS process. This notion has been supported by a recent region-of-interest longitudinal study *(18)* that demonstrated dynamic changes in the fMRI response following visual stimulation not only in V1, V2, and the LOC, but also in the lateral geniculate nucleus (LGN) in patients with isolated acute optic neuritis. In this study, abnormal LGN response was found not only following stimulation of the affected eye, but also after stimulation of the unaffected one, indicating that the visual pathways undergo early functional changes following tissue injury (**Fig. 1**).

In patients with established MS and a relapsing-remitting (RR) course with a unilateral optic neuritis, a reduced recruitment of the visual cortex after stimulation of the affected and the unaffected eyes was found when compared with healthy subjects. On average, patients with optimal clinical recovery showed increased visual cortex activation than those with poor or no recovery, although the extent of the activation remained reduced compared with controls *(8)*. A more recent study *(13)* of nine

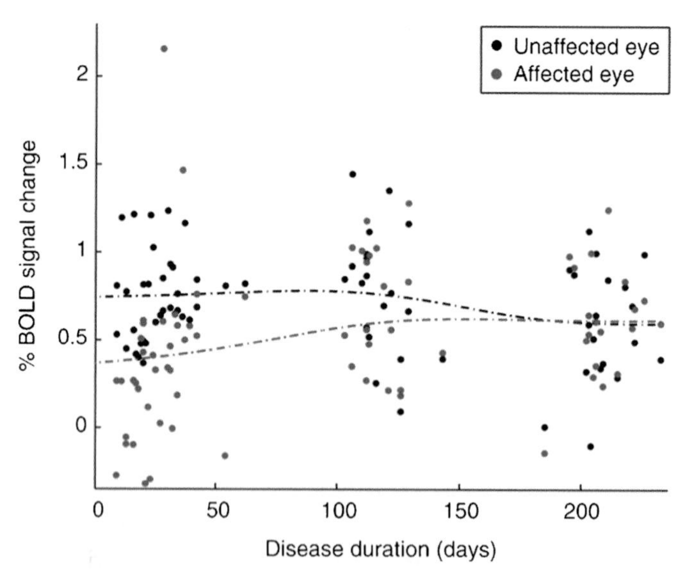

Fig. 1. Mean lateral geniculate nucleus (LGN) activation during stimulation of the affected and unaffected eye over time. The *dash-dotted lines* indicate weighted moving averages for the affected eye (*red*) and unaffected eye (*black*). During recovery there is a significant increase in LGN blood oxygenated level dependent (BOLD) signal from stimulation of the affected eye. The LGN BOLD signal from stimulation of the unaffected eye decreases over time in a stepwise fashion. Post hoc analysis showed this decrease to be significant. The BOLD signal changes for the affected and unaffected eye level off at a similar positive value as time increases (from **ref.** ***18***).

patients with previous optic neuritis confirmed the results of a previous study *(8)* and showed that these patients not only have a reduced activation of the primary visual cortex, but also a reduced fMRI percentage signal change in this region, again suggesting an abnormality of the synaptic input.

2.3. Motor System

The investigation of the motor system in patients with MS has mainly focused on the analysis of the performance of simple motor tasks with the dominant right upper limbs *(9–11, 19–38)*. Such tasks were either self-paced or paced by a metronome. A few studies assessed the performance of simple motor tasks with the dominant right lower limbs *(22, 25, 31)*, while even fewer studies have investigated the performance of more complex tasks, including phasic movements of dominant hand and foot *(25, 31)*, object manipulation *(39)*, and visuo-motor integration tasks *(40)*.

An altered brain pattern of movement-associated cortical activations, characterized by an increased recruitment of the contralateral primary sensorimotor cortex (SMC) during the performance of simple tasks *(21, 25)* and by the recruitment of additional "classical" and "higher-order" sensorimotor areas during the performance of more complex tasks *(25)*, has been demonstrated in patients with clinically isolated syndrome (CIS) suggestive of MS. The clinical and conventional MRI follow-up of these patients has shown that, at disease onset, CIS patients with a subsequent evolution to clinically definite MS tend to recruit a more widespread sensorimotor network than those without short-term disease evolution *(26)*. These findings suggest that in CIS patients the extent of early cortical reorganization might be a factor associated with a different clinical evolution. This would support the notion that, whereas increased recruitment of a widespread sensorimotor network contributes to limiting the impact of structural damage during the course of MS, its early activation might be counterproductive, as it might result in an early exhaustion of the adaptive properties of the brain. This notion is also supported by studies of stroke patients, where a persistent over-recruitment of a widespread cortical network has been related to an unfavorable clinical outcome *(41)*.

An increased recruitment of several sensorimotor areas, mainly located in the cerebral hemisphere ipsilateral to the limb that performed the task, has also been demonstrated in patients with early RRMS and a previous episode of hemiparesis *(27)*. In patients with similar characteristics, but who presented with an episode of optic neuritis, this increased recruitment involved sensorimotor areas that were mainly located in the contralateral cerebral hemisphere *(28)*.

In patients with established MS and a RR course, functional cortical changes, mainly characterized by an increased recruitment of "classical" motor areas, including the primary SMC, the supplementary motor area (SMA), and the secondary sensorimotor cortex (SII), have been shown during the performance of simple motor *(9–11, 20)* and visuo-motor integration tasks *(40)*. Movement-associated cortical changes, characterized by the activation of highly specialized cortical areas, have also been described in patients with secondary progressive (SP) MS *(22)* during the performance of a simple motor task and in patients with primary progressive (PP) MS during the performance of active *(19, 31)* and passive *(42)* motor experiments.

The concept that movement-associated cortical reorganization varies across patients at different stages of the disease has been shown by a recent fMRI study of patients with different disease phenotypes *(34)*. The study compared data from 16 patients with a CIS suggestive of MS, 14 with RRMS and no disability, 15 with RRMS and mild clinical disability, and 12 with SPMS, acquired during the performance of a simple motor task with their unimpaired dominant hand. CIS patients had an increased activation of the contralateral primary SMC when compared with those with RRMS and no disability, whereas patients with RRMS and no disability had an increased activation of the SMA when compared with those with CIS (**Fig. 2**). Patients with RRMS and no disability had an increased activation of the primary SMC, bilaterally, and ipsilateral SMA when compared with patients with RRMS and mild clinical disability. Conversely, patients with RRMS and mild clinical disability had an increased activation of the contralateral SII, inferior frontal gyrus (IFG), and ipsilateral precuneus. Patients with RRMS and mild clinical disability had an increased activation of the contralateral thalamus and ipsilateral SII when compared with those with SPMS. The opposite contrast showed that patients with SPMS had an increased activation of the IFG, bilaterally, middle frontal gyrus (MFG), bilaterally, contralateral precuneus, and ipsilateral cingulate motor area (CMA) and inferior parietal lobule. This study suggests that early in the disease course more areas typically devoted to motor tasks are recruited, then a bilateral activation of these regions is seen, and late in the disease course, areas that healthy people recruit to perform novel or complex tasks are activated *(34)*, perhaps in an attempt to limit the functional consequences of accumulating tissue damage.

As described in another chapter of this book, fMRI has been recently applied to the investigation of cervical cord neuronal activity during a proprioceptive and a tactile stimulation of the right upper limb from patients with relapsing MS *(43)*. During the application of both stimulations, healthy controls and MS patients had significant activations of the cervical cord between C5 and C8. On average, MS patients had 20% higher cord fMRI signal changes during either stimulations, suggesting an abnormal cord function in these patients.

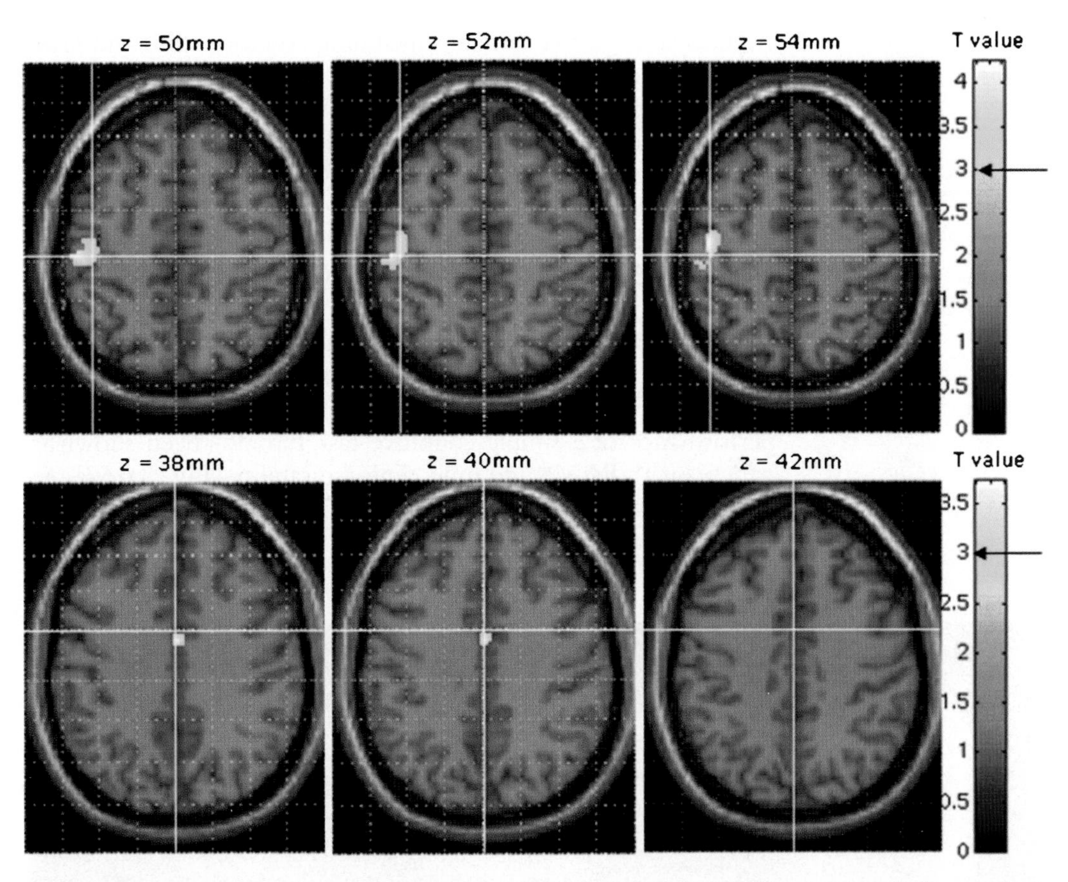

Fig. 2. Comparisons of patients at presentation with clinically isolated syndrome (CIS) suggestive of MS and patients with relapsing-remitting (RR) MS and no disability during a simple, right-hand, motor task. Patients with CIS showed an increased activation of the contralateral primary sensorimotor cortex when compared with patients with RRMS and no disability (*top row*). Patients with RRMS and no disability had a more significant activation of the supplementary motor area, bilaterally, when compared with patients with a CIS (*bottom row*). Images are color-coded for activation and arrows show *t* cut-off values. Activations were superimposed on a high-resolution T1-weighted scan obtained from one healthy individual and normalized into a standard statistical parametric mapping space (neurological convention) (from **ref.** ***34***).

2.4. Cognition

Recent fMRI studies have suggested that functional cortical changes might have an adaptive role also in limiting MS-related cognitive impairment *(44–60)*. Therefore, brain plasticity might, in part, explain the weak relationship found in MS between neuropsychological deficits and conventional MRI measures of disease burden *(61)*.

Several cognitive domains have been investigated in MS patients with fMRI. Working memory has been the most extensively studied by means of the Paced Auditory Serial Addition Test (PASAT) or the Paced Visual Serial Addition Task (PVSAT) *(44, 45–47, 52, 56)* (which also involve sustained attention, information processing speed, and simple calculation), the *n*-back task *(51, 53–55, 57)*, or a task adapted from the Sternberg paradigm *(49)*.

Additional cognitive domains including attention *(50)* and planning *(60)* have also been interrogated.

In patients at presentation with CIS suggestive of MS, an altered pattern of cortical activations has been described during the performance of the PASAT *(47, 48)*, confirming the presence of cortical reorganization at the earliest clinical stage of the disease. Staffen et al. *(44)* found that, during the performance of the PVSAT, MS patients with intact task performance had an increased activation of several regions located in the frontal and parietal lobes, bilaterally, compared with healthy volunteers, suggesting the presence of functional compensatory mechanisms. An increased recruitment of several cortical areas during the performance of a simple cognitive task has also been shown in patients with RRMS and mild clinical disability **(Fig. 3)** *(52)*. An increased activation of regions exclusively located in the right cerebral hemisphere (in particular in the frontal and temporal

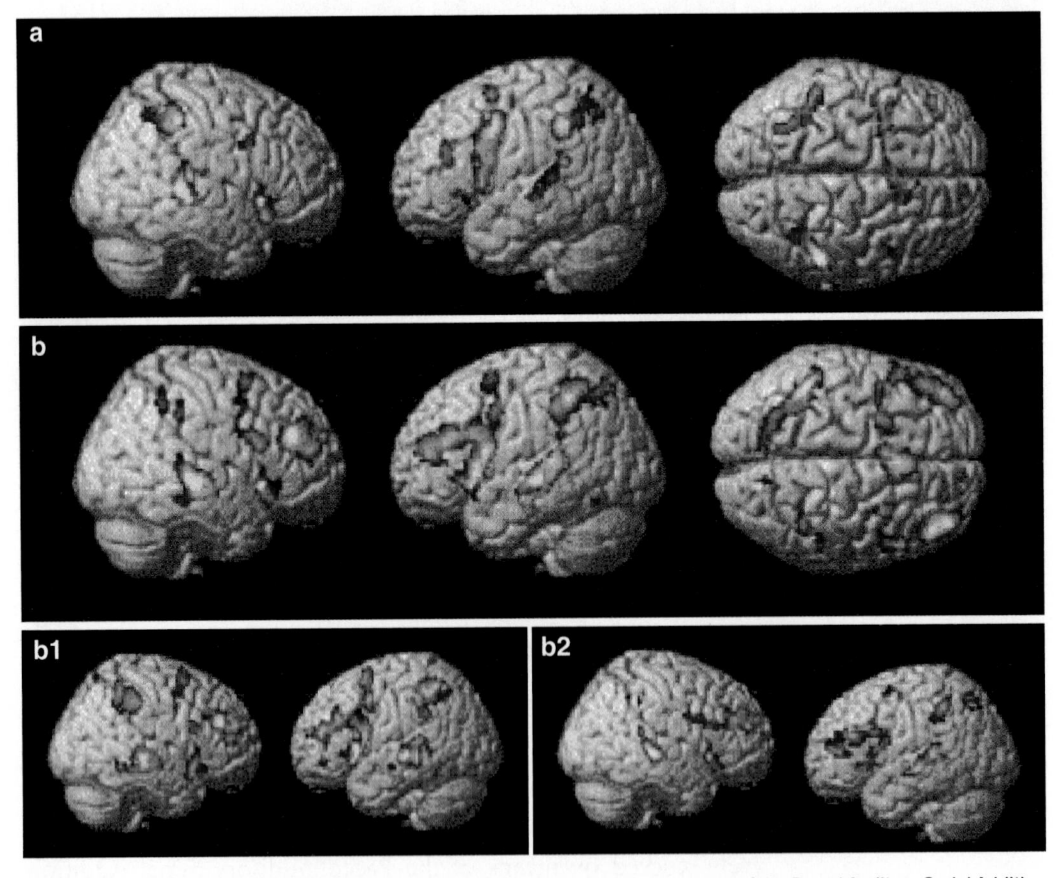

Fig. 3. Brain patterns of cortical activations on a rendered brain during the execution of the Paced Auditory Serial Addition Task (PASAT) in (**a**) 22 healthy controls and in (**b**) 22 patients with MS. (**b**1, **b**2) Rendered images for patients with MS subgrouped according to their performance at the PASAT during fMRI showing significant activated foci in (**b**1) for 12 patients whose performance was similar to that of healthy controls and in (**b**2) the activations found in the ten patients who exhibited lower scores (from **ref.** ***52***).

lobes) has also been found in MS patients when testing rehearsal within working memory *(49)*. The degree of right hemisphere recruitment was strongly related to patient neuropsychological performance *(49)*. In patients with RRMS and no cognitive deficits, using fMRI during an *n*-back test, a reduced activation of the "core" areas of the working memory circuitry (including prefrontal and parietal regions) and an increased activation of other regions within and beyond the typical working memory circuitry (including areas in the frontal, parietal, temporal, and occipital lobes) have been found *(55)*. This shift of activation was most prominent with increased working memory demands. These findings suggest that, as shown for motor and visual tasks, dynamic changes of brain activation patterns can occur in RRMS patients during cognitive tasks. Other studies *(53, 54, 58)*, which also investigated working memory performance in MS patients, demonstrated: (1) an increased recruitment of regions related to sensorimotor functions and anterior attentional/executive components of the working memory system in patients compared with healthy controls, and (2) a reduced recruitment of several regions in the right cerebellar hemisphere in patients compared with healthy individuals *(58)*, thus suggesting that the cerebellum might play a role in the working memory impairment of MS.

In a recent fMRI study *(57)*, working memory was investigated with an *n*-back task and functional connectivity analysis in a group of 21 RRMS patients and 16 age- and sex-matched healthy controls. With similar task performances, activations were found in similar regions for both groups. However, patients had relatively reduced activations of the superior frontal and anterior cingulate gyri. Patients also showed a variable, but generally substantially smaller increase of activation than healthy controls with greater task complexity, depending on the specific brain regions assessed. These findings suggest that, despite similar brain regions were recruited in both groups, patients have a reduced functional reserve for cognition relevant to memory. The functional connectivity analysis revealed increased correlations between right dorsolateral prefrontal and superior frontal/anterior cingulate activations in controls, and increased correlations between activations in the right and left prefrontal cortices in patients (**Fig. 4**), suggesting that altered interhemispheric interactions between dorsal and lateral prefrontal regions may yet be an additional adaptive mechanism distinct from recruitment of novel processing regions *(57)*.

Using a 3 T scanner, more significant activations of several areas of the cognitive network involved in the performance of the Stroop test have also been demonstrated in a group of 15 cognitively preserved patients with benign MS (BMS) when compared with 19 healthy controls (**Fig. 5**) *(59)*. BMS patients also showed an increased connectivity of several cortical areas of the sensorimotor

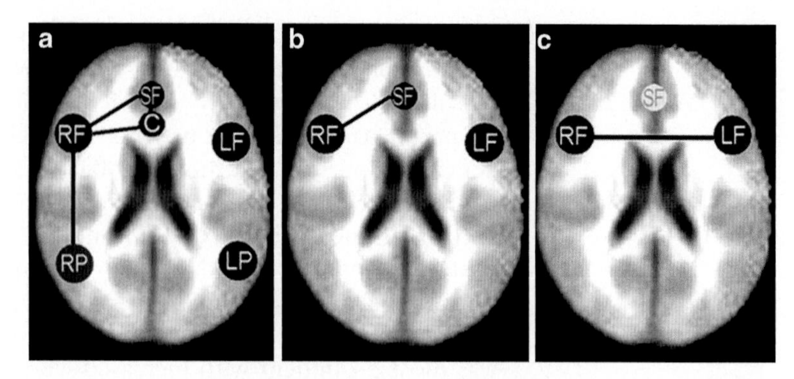

Fig. 4. Analysis of functional connectivity during the performance of the *n*-back task with different levels of difficulty in healthy individuals and patients with MS. The most significant correlations between activation in regions involved in processing increasing task demand are indicated in (**a**). In (**b**) are those connections more significant for controls ($p < 0.05$). The image (**c**) shows connections that were more significant in patients than controls ($p < 0.05$). *C* cingulate, *SF* superior medial frontal, *RF* right dorsolateral prefrontal, *LF* left dorsolateral prefrontal, *RP* right parietal, *LP* left parietal (from **ref. *57***).

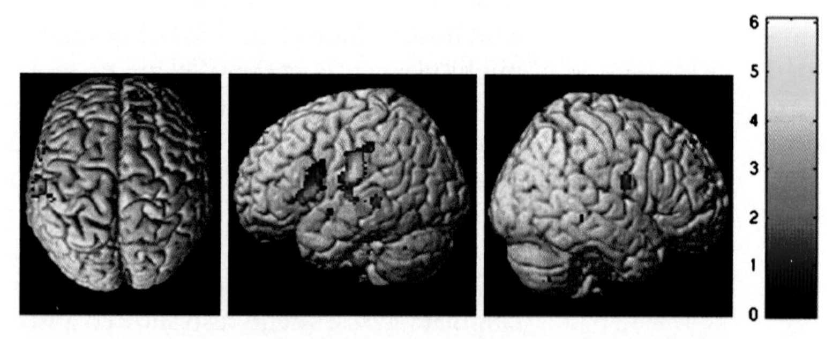

Fig. 5. Areas showing increased activations in patients with BMS in comparison with healthy controls during the analysis of the Stroop facilitation condition (random effect interaction analysis, ANOVA, $p < 0.05$ corrected for multiple comparisons). BMS patients had increased activations of several areas located in the frontal and parietal lobes, bilaterally, including the anterior cingulate cortex, the superior frontal sulcus, the inferior frontal gyrus, the precuneus, the secondary sensorimotor cortex, the bilateral visual cortex, and the cerebellum, bilaterally. Note that the color-encoded activations have been superimposed on a rendered brain and normalized into standard SPM space (neurological convention) (*see* **ref. *59***).

network, including the left IFG, the anterior cingulated cortex and the left SII, with the right IFG and the right cerebellum, as well as a decreased connectivity between some areas (including the left SII, the prefrontal cortex, and the right cerebellum), and the anterior cingulate cortex. These results suggest an altered interhemispheric balance in favor of the right hemisphere in BMS patients in comparison with healthy controls, when performing cognitive tasks.

2.5. Correlations Between the Extent of Functional Cortical Reorganization and the Extent of Brain Damage in MS

In addition to an abnormal pattern of functional activations, the majority of the previous studies described a variable relationship between the extent of fMRI activation and several measures of tissue damage *(6–8, 10, 11, 19–23, 28, 31, 40, 45, 48, 50, 52)*.

An increased recruitment of several brain areas with increasing T2 lesion load has been shown in patients with RR *(6–8, 45, 52)* and PP *(31)* MS. The severity of intrinsic T2-visible lesion damage, measured using T1-weighted images *(28)*, magnetization transfer (MT), and diffusion tensor (DT) MRI *(20)*, has been found to modulate the activity of some cortical areas in these patients. The severity of normal appearing brain tissue (NABT) injury, measured using proton MR spectroscopy *(10, 11, 21)*, MT MRI *(19, 20, 45, 48)*, and DT MRI *(19, 21, 23)*, is another important factor associated to an increased recruitment of motor- and cognitive-related brain regions, as shown by studies of patients at presentation with CIS suggestive of MS *(21, 45, 48)*, patients with RRMS and variable degrees of clinical disability *(10, 11, 20, 23)*, patients with PPMS *(19)*, and with SPMS *(10, 11)*. Finally, subtle GM damage, which goes undetected when using conventional MRI, may also influence functional cortical recruitment, as demonstrated, for the motor system, in patients with RRMS *(40)*, SPMS *(22)*, and patients with clinically definite MS and nonspecific (less than three focal white matter lesions) conventional MRI findings *(23)*. In cognitively intact MS patients, the increased activation of a left prefrontal region during the counting Stroop task has been correlated with the normalized brain parenchymal volume *(50)*.

2.6. Functional Cortical Reorganization and Regional Damage in MS

Structural damage of white matter pathways that connect functional relevant areas for a given task has been shown to modify the observed brain patterns of cortical activations in patients with MS. Damage to the corticospinal tract *(28, 29)* (**Fig. 6**) as well as damage to the corpus callosum (CC) *(62, 63)* has been related to a more bilateral movement-associated brain pattern of cortical activations. The role of the CC in interhemispheric connectivity and in eliciting functional cortical changes has been underpinned by a study by Lowe et al. *(30)*, who showed, by measuring low-frequency BOLD fluctuations, a reduced functional connectivity between the right and the left hemisphere primary motor cortices in MS patients.

The recent development of diffusion-based tractography methods that allow to define with precision the pathways connecting different CNS structures and their application to patients with MS resulted in an improvement of the correlation between structural and functional abnormalities. In particular, recent works combined measures of abnormal functional connectivity with DT MR measures of damage within selected white matter fiber bundles in patients with RRMS *(38)* and BMS *(59)*.

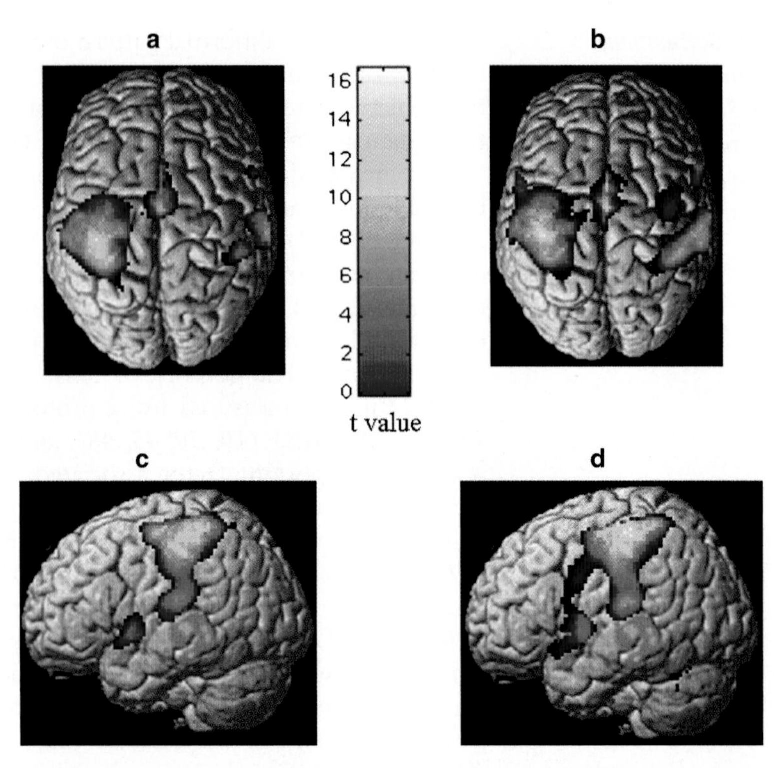

Fig. 6. Brain patterns of cortical activations on a rendered brain from MS patients without (**a**, **c**) and with (**b**, **d**) lesions in the left corticospinal tracts, during the performance of a simple motor task with their clinically unimpaired and fully normal functioning, dominant right hands. In patients with corticospinal tract lesions, a more bilateral pattern of activations is visible. Note that the activations are color-coded according to their *t* values. Images are in neurological convention (*see* **ref.** ***29***).

In patients with RRMS and no clinical disability *(38)*, measures of abnormal connectivity inside the motor network were correlated with structural MRI metrics of tissue damage of the corticospinal and the dentatorubrothalamic tracts, while no correlation was found with measures of damage within "not-motor" white matter fiber bundles. These findings suggest an adaptive role of functional connectivity changes in limiting the clinical consequences of structural damage to selected white matter pathways in RRMS patients *(38)*. In patients with BMS *(59)*, measures of abnormal connectivities inside the cognitive network were moderately correlated with structural MRI metrics of tissue damage within intra- and interhemispheric cognitive-related white matter fiber bundles, while no correlations were found with the remaining fiber bundles studied, suggesting that functional cortical changes in patients with BMS might represent an adaptive response driven by damage to specific white matter structures *(59)*.

In patients with PPMS *(19)*, a relationship has been demonstrated between the severity of spinal cord pathology, measured using MT MRI, and the extent of movement-associated cortical activations.

2.7. Adaptive Role of Functional Cortical Reorganization in MS

Although the actual role of cortical reorganization on the clinical manifestations of MS remains to be established, there are several pieces of evidence which suggest that cortical adaptive changes are likely to contribute in limiting the clinical consequences of MS-related structural damage. In nondisabled patients with RRMS *(20)*, an increased activation of several motor regions, mainly located in the contralateral cerebral hemisphere, has been seen during the performance of a simple motor task. The correlations found in this study *(20)* between the extent of fMRI activations and several MT and DT MRI metrics of structural brain damage suggested that an increased recruitment of movement-associated cortical network contributes to limiting the functional impact of MS-related damage.

The notion that an increased recruitment of areas that are usually activated by healthy individuals when performing different/more complex motor tasks might be one of the cortical reorganization mechanisms playing a role in MS recovery/maintenance of function has been highlighted by the results of two recent experiments *(37, 39)*. The first showed that MS patients, during the performance of a simple motor task, activate some regions that are part of a fronto-parietal circuit, whose recruitment occurs typically in healthy subjects during object manipulation (**Fig. 7**) *(39)*. The second, which assessed the fMRI patterns of activation during the performance of a simple motor task and of a task aimed at investigating the mirror-neuron system, demonstrated activations of regions that are part of the mirror-neuron system in patients with MS during the performance of the simple motor task *(37)*.

The compensatory role of cortical reorganization has also been demonstrated by studies investigating the cognitive domains, which showed increased recruitment of several cortico-subcortical areas in cognitively preserved MS patients *(44–50)*. In patients complaining of fatigue, when compared with matched nonfatigued MS patients *(64)*, a reduced activation of a complex movement-associated cortical/subcortical network, including the cerebellum, the thalamus, and regions in the frontal lobes, has been shown. The correlation found in these patients between the reduction of thalamic activity and the clinical severity of fatigue indicates that a "pseudoreduction" of brain functional recruitment might be associated with the appearance of MS symptomatology. Additional work has shown that the pattern of movement-associated cortical activations in MS is determined by both the extent of brain injury and disability and that these changes are distinct *(24, 31)*.

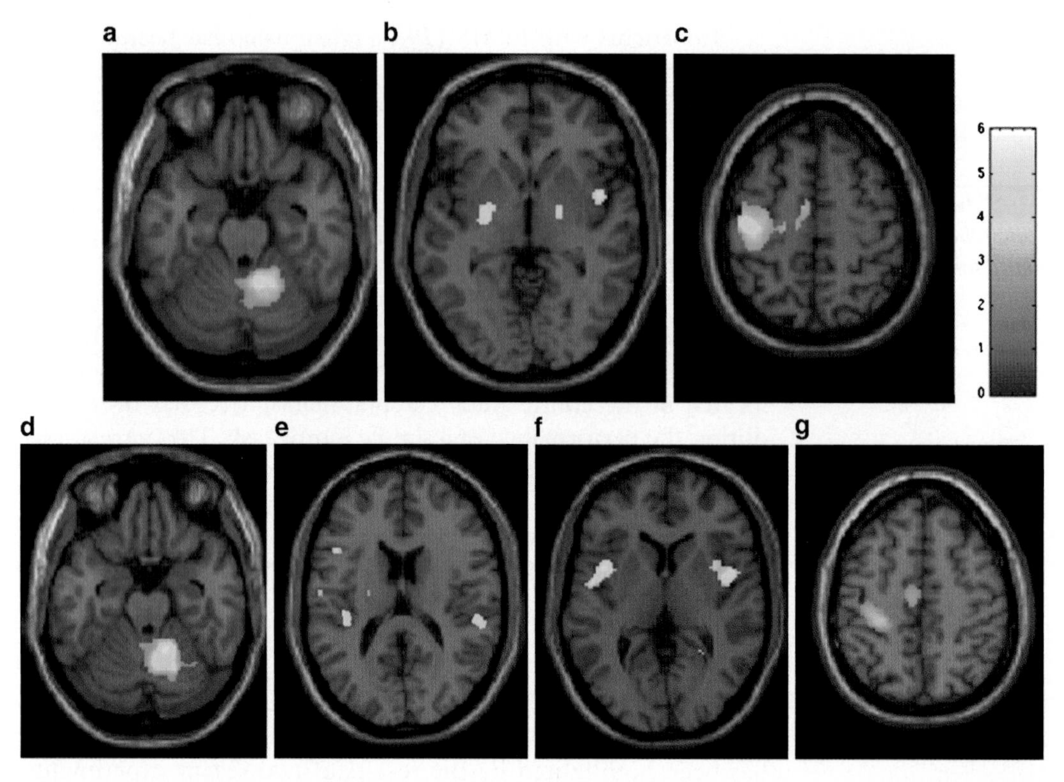

Fig. 7. Comparison of simple vs. complex task with the dominant right hands in healthy subjects (*top row*; **a–c**) and patients with MS (*bottom row*; **d–g**) (paired *t* test for each group, corrected *p* value <0.05). The ipsilateral anterior lobe of the cerebellum (**a**, **d**), bilateral insula/basal-ganglia (**b**, **e**, **f**) and contralateral primary sensorimotor cortex and supplementary motor area (**c**, **g**) were identified in both groups. Compared with healthy subjects, MS patients also had a significant activation of the contralateral inferior frontal gyrus and bilateral secondary sensorimotor cortex (**e**). Note that the activations are color-coded according to their *t* values. Images are in neurological convention (*see* **ref.** ***39***).

2.8. Maladaptive Role of Functional Cortical Reorganization

The results of several studies suggest that an increased cortical recruitment might not always be beneficial for patients with MS. As already mentioned, disease progression and accrual of disability has been observed in patients with SPMS, despite the widespread activations of regions in the frontal and parietal lobes during the performance of simple motor tasks *(22, 34)*. Three fMRI studies of the motor system *(19, 31, 42)* of patients with PPMS suggested a lack of "classical" adaptive mechanisms as a potential additional factor contributing to the accumulation of disability. In these patients during the performance of different motor tasks with the nonimpaired dominant limbs, a recruitment of a widespread movement-associated cortical network usually considered to function in motor, sensory, and multimodal integration processing (i.e., the frontal and temporal lobes, and the insula) was detected *(19, 31, 42)*. The absence of a concomitant recruitment of the "classical" motor areas, including the primary SMC, the SMA, the infraparietal sulcus, and the SII, was interpreted as a failure of part of the adaptive capacity of the cerebral cortex in this severely disabling phenotype of the disease *(19, 31)*. The notion that multimodal integration areas might have a critical role in PPMS patients has been strengthened by another study which showed increased activations of these regions in PPMS patients, in comparison with healthy controls, during passive movements *(42)*. Similar findings have led to similar conclusions in patients with cognitive decline *(51, 56)*, in whom a "poor" pattern of cortical activations in the expected areas *(51)* and the activation of regions that are not normally devoted to the performance of the investigated task *(56)* were detected during the performance of the administered tasks.

The comparison of the movement-associated brain patterns of cortical activations between RRMS patients complaining of reversible fatigue after weekly interferon (IFN) beta-1a administration and those without fatigue suggested an association between the presence of fatigue and an increased activation of several areas of the motor network, including the thalamus, the cingulum, and several regions located in the frontal lobes, including the SMA and the primary SMC bilaterally *(35)* (**Fig. 8**). These results suggest that the overrecruitment of brain networks in MS might, at least to some degree, have a detrimental effect.

2.9. Use of fMRI to Assess Longitudinal Changes of Cortical Reorganization

Dynamic functional changes have been described in an MS patient following an acute relapse *(10)*. These results have been confirmed and extended by a recent study which assessed the early cortical changes following acute motor relapses secondary to pseudotumoral lesions in 12 MS patients and the evolution over time of cortical reorganization in a subgroup of these patients *(36)*. In this study *(36)*, short-term cortical changes were mainly characterized by the recruitment of pathways in the unaffected hemisphere. A recovery of function of the primary SMC of the affected hemisphere was found in patients with clinical improvement, while in patients without clinical recovery, there was a persistent recruitment of the primary SMC of the unaffected hemisphere, suggesting that the restoration of function of motor areas of the affected hemisphere might be a critical factor for a favorable recovery (**Fig. 9**).

A longitudinal (time interval of 15–26 months) fMRI study of the motor system has been conducted in a group of patients with early RRMS *(65)*. Patients exhibited greater bilateral activations than controls in both fMRI studies. Although no significant differences between the two fMRI scans were observed in controls, a reduction of the functional activity of the ipsilateral SMC and the contralateral cerebellum was seen in patients at follow-up. Moreover, activation changes in ipsilateral motor areas correlated inversely with age, extent and progression of T1 lesion load, and occurrence of a new relapse, suggesting that younger patients with less structural brain damage and a favorable clinical

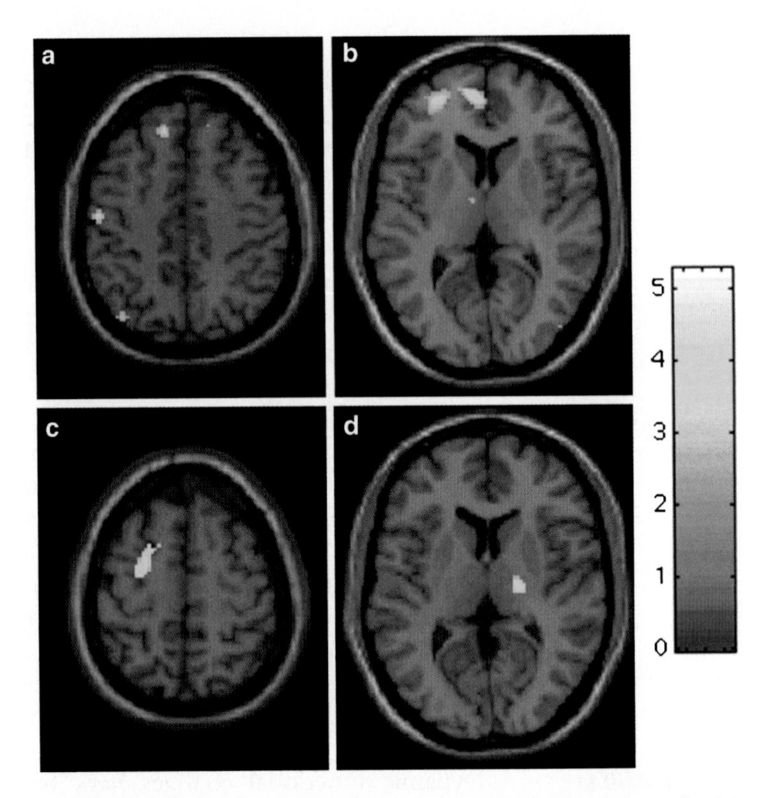

Fig. 8. Relative cortical activations of MS patients with reversible fatigue after interferon (IFN) beta-1a injection during the performance of a simple motor task with their clinically unimpaired and fully normal functioning, dominant right hands. At entry (**a**, **b**) (when they did not complain of fatigue), compared with MS patients without reversible fatigue, these patients showed an increased recruitment of the contralateral primary SMC (**a**), the thalamus (**b**), the superior frontal sulcus (**a**, **b**), and the cingulate motor area (**b**). At day 1 (**c**, **d**) (after IFN beta-1a administration, when fatigue was present), compared with MS patients without fatigue, these patients showed increased recruitment of the ipsilateral thalamus (**d**), and contralateral middle frontal gyrus (**c**). Note that the color-coded activations have been superimposed on a high-resolution T1-weighted scan obtained from a single, healthy subject and normalized into standard statistical parametric mapping space (neurological convention) (from **ref. *35***).

course demonstrate brain plasticity that follows a more lateralized pattern of brain activations *(65)*.

2.10. fMRI to Monitor Treatment

Only a few fMRI studies have been performed to monitor the effect of treatments in MS *(50, 66, 67)*. A preliminary study, conducted in five patients, tested the effects of acute administration of rivastigmine, a central cholinesterase inhibitor, on the patterns of brain activations during the performance of a cognitive task (Stroop task) *(50)*. After treatment administration, a relative normalization of the abnormal Stroop-associated brain activations was observed in patients, while no change in the pattern of brain activations was found in any of the four healthy controls studied. In MS patients, increased activation in the ipsilateral

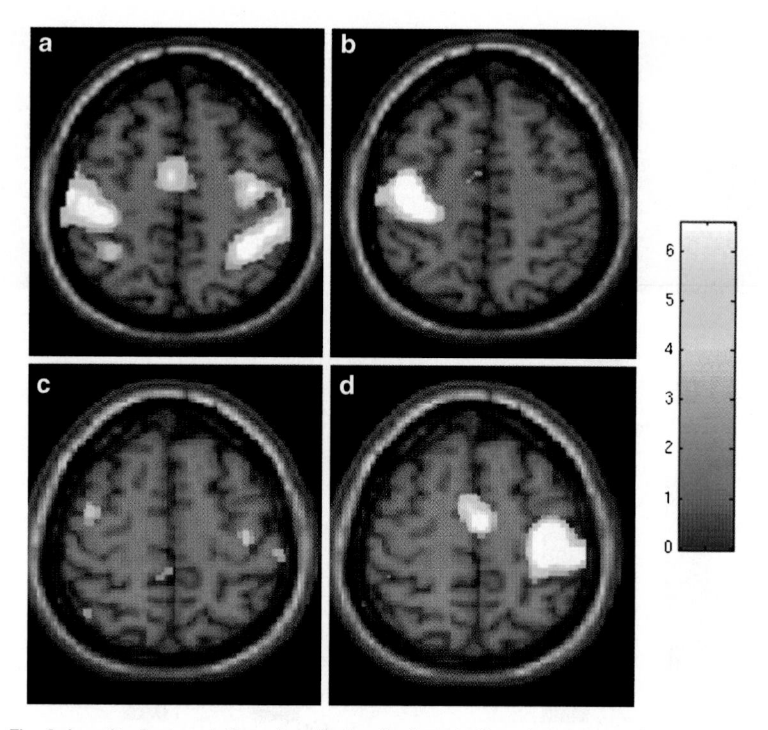

Fig. 9. Longitudinal evolution of cortical activations in the primary sensorimotor cortex (SMC), bilaterally, during task performance with impaired hand compared with unimpaired hand in one patient with good clinical recovery during follow-up (**a**, **b**) and in one patient with poor/absent clinical recovery (**c**, **d**). Scans obtained during left hand motor task have been flipped to keep the left hemisphere contralateral to movement. At baseline, both patients showed an increased activation of the primary SMC of the unaffected (ipsilateral) hemisphere (**a**, **c**). During follow-up, the patient with good clinical recovery showed an increased recruitment of the primary SMC of the affected hemisphere (**b**), while the patient with poor/absent clinical recovery continued to show an increased recruitment of the primary SMC of the unaffected hemisphere (**d**). Note that the activations are color-coded according to their *t* values (*see* **ref.** ***36***).

primary SMC and SMA has been observed after a single dose of 3,4-diaminopyridine (a potassium channel blocker), suggesting that this treatment may modulate brain motor activity in patients with MS, probably by enhancing excitatory synaptic transmission *(66)*. A recent study analyzed how the motor network responds to motor training in MS patients with mild motor impairment of the right upper extremity *(67)*. Before training, MS patients had a more prominent activation of the contralateral dorsal premotor cortex during thumb movements when compared with controls. After training, unlike the control group, MS patients did not exhibit task-specific reductions in activation of the contralateral primary SMC and adjacent parietal association cortices (**Fig. 10**). The absence of training-dependent reductions in activations supports the notion that MS patients have a decreased capacity to optimize recruitment of the motor network with practice.

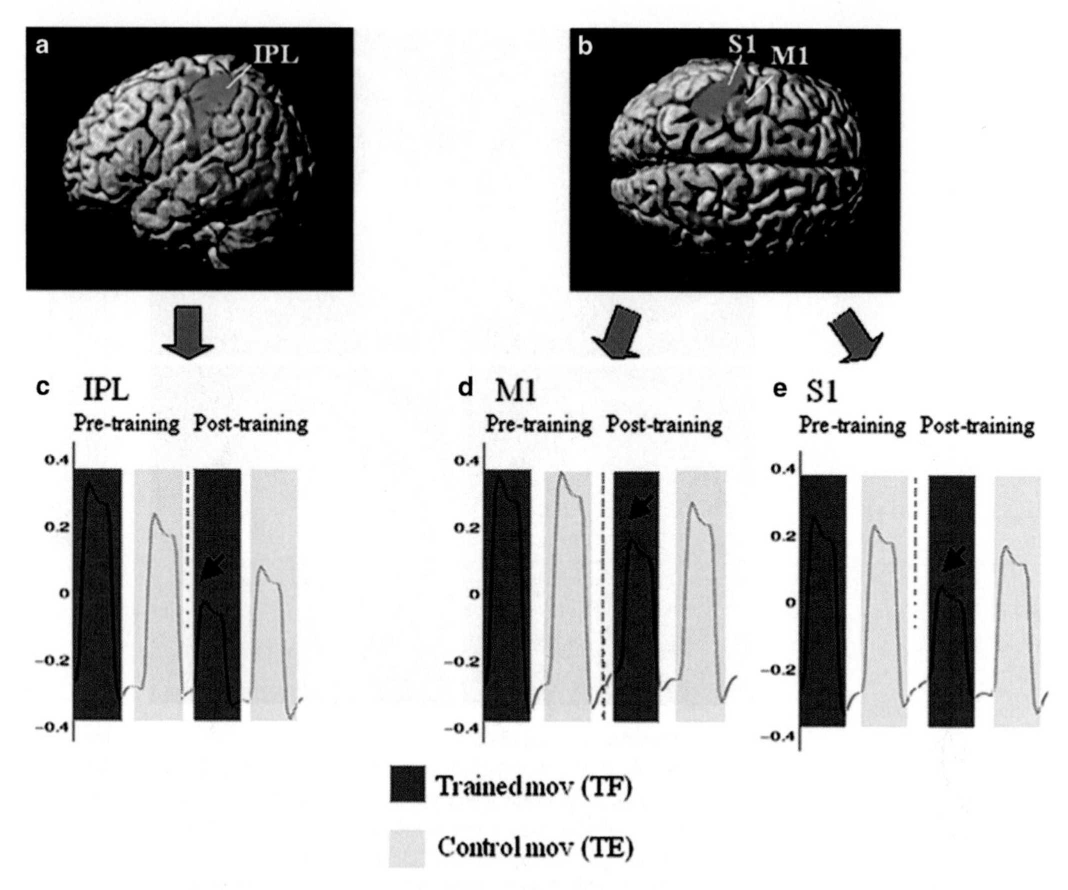

Fig. 10. Three-dimensional rendering (**a**, **b**) of task-specific reductions of activations in contralateral inferior parietal lobule (IPL) (BA 40; **a**, **c**), primary motor cortex (M1) (**b**, **d**), and primary somatosensory cortex (S1) (**b**, **e**) in healthy volunteers after training. **c**, **d**, and **e** illustrate average signal intensities in voxels representing cluster maxima in IPL (BA 40), M1 and S1 for 30 s blocks of thumb flexion (trained movement, *grey bars*), extension (control movement, *blue bars*), and rest. Note that training led to a decrease in signal intensity predominantly for the trained movement (*black arrows*). Images are in neurological convention (from **ref.** ***75***).

3. fMRI in MS-Allied Conditions

Among the known MS-allied conditions, only patients with neuromyelitis optica (NMO) have been assessed using fMRI. Rocca et al. *(68)* investigated the performance of a simple motor task with the dominant and nondominant upper limbs in ten patients with NMO and found that, compared with matched controls, NMO patients had an increased recruitment of several regions of the sensorimotor network (primary SMC, postcentral gyrus, MFG, rolandic operculum, SII, precuneus, and cerebellum) and of several other regions mainly in the temporal and occipital lobes, such as MT/V5, the fusiform gyrus, the cuneus, and the parahippocampal gyrus during the performance of both tasks. For both tasks, strong correlations were found between relative activations of cortical sensorimotor areas and the severity of cervical cord damage, suggesting that the observed functional cortical changes might have an adaptive role in limiting the clinical outcome of NMO structural pathology.

4. fMRI in Other WMD and Conditions Associated with "Significant" White Matter Damage

4.1. Isolated Spinal Cord Injury

Studies of patients with spinal cord injury of different etiology (i.e., traumatic and/or demyelinating) with no or only partial clinical recovery have shown movement-associated cortical changes, consisting of an abnormal location of the activated areas and in a more widespread recruitment of motor areas, mainly located in the hemisphere contralateral to the limb used to perform the task *(69–71)*.

In patients with isolated myelitis of probable demyelinating origin and normal function in the investigated limbs, an abnormal pattern of movement-associated cortical activation has been described *(72–74)* and has been related to the degree of daily hand use *(72)*, the severity of cervical cord damage *(73, 74)*, and the level of spinal cord involvement *(74)*.

4.2. CNS Vasculitides

In patients with neuropsychiatric systemic lupus erythematosus (NPSLE) without overt motor impairment, movement-associated functional cortical changes, characterized by more significant activations of the contralateral primary SMC, putamen, dentate nucleus, several regions located in the frontal and parietal lobes, MT/V5, and the middle occipital gyrus, bilaterally, have been observed when compared with matched healthy controls *(75)* (**Fig. 11**). The correlations found in these patients between relative activations of sensorimotor areas and the extent and severity of brain damage suggest that also in these patients functional cortical changes might contribute to the maintenance of their normal functional capacities *(75)* (**Fig. 11**).

4.3. Migraine

Brain T2-weighted white matter abnormalities in patients with migraine are a relatively common finding, which is thought to be the consequence of ischemic damage secondary to the repeated regional blood flow reductions known to occur during headache attacks *(76)*. In a group of 15 migraine patients with white matter abnormalities on conventional MRI scans of the brain, Rocca et al. *(77)* described an increased recruitment of the contralateral primary SMC and a rostral displacement of the SMA during the performance of a simple motor task with the dominant, right upper limb, when compared with healthy controls. The shapes

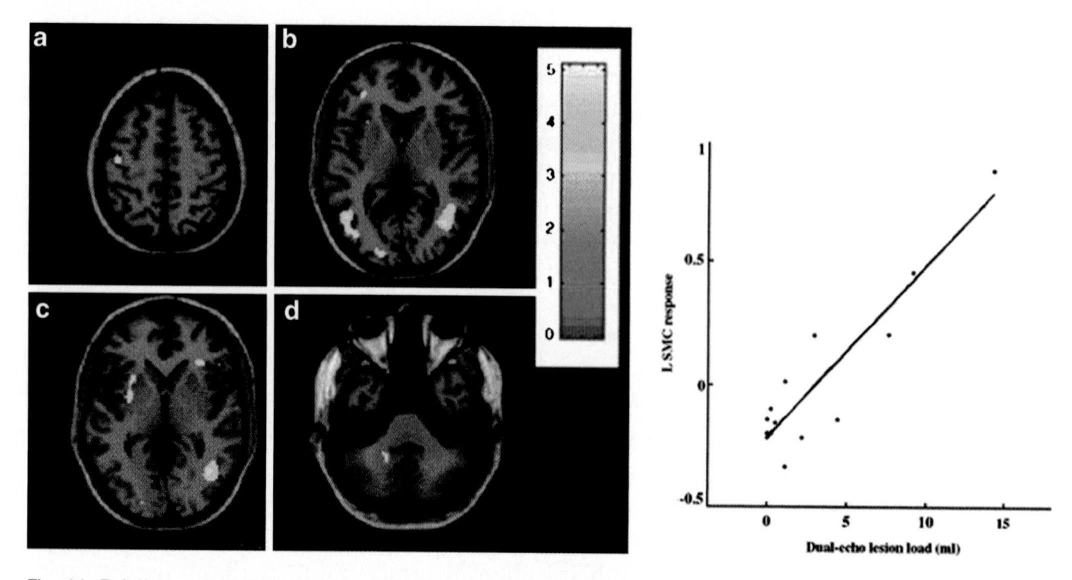

Fig. 11. Relative cortical activations in patients with neuropsychiatric systemic lupus erythematosus during a simple motor task with the right hand in comparison with healthy volunteers (color-coded *t* values). (**a**) Contralateral primary sensorimotor cortex. (**b**) Contralateral putamen, contralateral middle frontal gyrus (MFG), bilateral MT/V5 complex, contralateral middle occipital gyrus (MOG). (**c**) Contralateral putamen, ipsilateral inferior frontal gyrus, bilateral MT/V5 complex, ipsilateral MOG. (**d**) Contralateral dentate nucleus. The relative activation of the contralateral primary sensorimotor cortex was significantly correlated with brain dual-echo lesion load (**e**) ($p < 0.001$, $r = 0.79$) (from **ref.** ***75***).

of the curves reflecting the time course for fMRI signal intensity changes were similar between migraine patients and controls for all of the cortical areas studied. A significant correlation was found between the extent of displacement of the SMA and DT MRI measures of normal-appearing white matter damage. This study suggests that functional cortical changes occur in patients with migraine and brain MRI abnormalities and that they might be secondary to the extent of subcortical structural damage.

5. Conclusions

Taken all together, fMRI studies of patients with various white matter disorders demonstrate the potential of this technique to provide important insights into the mechanisms of cortical reorganization following white matter injury. As a consequence, fMRI holds promise to improve our understanding of the factors associated with the accumulation of irreversible disability in MS and other white matter conditions. Although the role of cortical reorganization in limiting the functional impact of white matter structural damage is still not proved definitively, the available data support the concept that cortical adaptive responses may have an important role in compensating for irreversible tissue damage, such as axonal loss. Thereby, it can be concluded that the presently available fMRI data suggest that the rate of accumulation of disability in MS and other white matter disorders might be a function not only of tissue loss, but also of the progressive failure of the adaptive capacity of the brain with increasing tissue damage.

References

1. Filippi M, Rocca MA. Magnetization transfer magnetic resonance imaging in the assessment of neurological diseases. J Neuroimaging 2004;14:303–313. Review.
2. Filippi M, Rocca MA, Comi G. The use of quantitative magnetic-resonance-based techniques to monitor the evolution of multiple sclerosis. Lancet Neurol 2003;2:337–346.
3. Hesselink JR. Differential diagnostic approach to MR imaging of white matter diseases. Top Magn Reson Imaging 2006;17:243–263. Review.
4. Rocca MA, Filippi M. Functional MRI to study brain plasticity in clinical neurology. Neurol Sci 2006;27(Suppl 1):S24–S26. Review.
5. Rocca MA, Filippi M. Functional MRI in multiplesclerosis. J Neuroimaging 2007;17(Suppl 1): 36S–41S. Review.
6. Waxman SG. Demyelinating diseases: new pathological insights, new therapeutic targets. N Engl J Med 1998;338:323–326.
7. Clanet M, Berry I, Boulanouar K. Functional imaging in multiple sclerosis. Int MS J 1997;4:26–32.
8. Rombouts SA, Lazeron RH, Scheltens P, Uitdehaag BM, Sprenger M, Valk J, Barkhof F. Visual activation patterns in patients with optic neuritis: an fMRI pilot study. Neurology 1998;50:1896–1899.
9. Lee MA, Reddy H, Johansen-Berg H, Pendlebury S, Jenkinson M, Smith S, Palace J, Matthews PM. The motor cortex shows adaptive functional changes to brain injury from multiple sclerosis. Ann Neurol 2000;47:606–613.
10. Reddy H, Narayanan S, Matthews PM, Hoge RD, Pike GB, Duquette P, Antel J, Arnold DL. Relating axonal injury to functional recovery in MS. Neurology 2000;54:236–239.
11. Reddy H, Narayanan S, Arnoutelis R, Jenkinson M, Antel J, Matthews PM, Arnold DL. Evidence for adaptive functional changes in the cerebral cortex with axonal injury from multiple sclerosis. Brain 2000;123: 2314–2320.
12. Werring DJ, Bullmore ET, Toosy AT, Miller DH, Barker GJ, MacManus DG, Brammer MJ, Giampietro VP, Brusa A, Brex PA, Moseley IF, Plant GT, McDonald WI, Thompson AJ. Recovery from optic neuritis is associated with a change in the distribution of cerebral response to visual stimulation: a functional magnetic resonance imaging study. J Neurol Neurosurg Psychiatry 2000;68:441–449.
13. Langkilde AR, Frederiksen JL, Rostrup E, Larsson HB. Functional MRI of the visual cortex and visual testing in patients with previous optic neuritis. Eur J Neurol 2002;9:277–286.
14. Toosy AT, Werring DJ, Bullmore ET, Plant GT, Barker GJ, Miller DH, Thompson AJ. Functional magnetic resonance imaging of the cortical response to photic stimulation in humans following optic neuritis recovery. Neurosci Lett 2002;330:255–259.
15. Russ MO, Cleff U, Lanfermann H, Schalnus R, Enzensberger W, Kleinschmidt A. Functional magnetic resonance imaging in acute unilateral optic neuritis. J Neuroimaging 2002;12:339–350.
16. Toosy AT, Hickman SJ, Miszkiel KA, Jones SJ, Plant GT, Altmann DR, Barker GJ, Miller DH, Thompson AJ. Adaptive cortical plasticity in higher visual areas after acute optic neuritis. Ann Neurol 2005;57:622–633.
17. Levin N, Orlov T, Dotan S, Zohary E. Normal and abnormal fMRI activation patterns in the visual cortex after recovery from optic neuritis. NeuroImage 2006;33: 1161–1168.
18. Korsholm K, Madsen KH, Frederiksen JL, Skimminge A, Lund TE. Recovery from optic neuritis: an ROI-based analysis of LGN and visual cortical areas. Brain 2007;130:1244–1253.
19. Filippi M, Rocca MA, Falini A, Caputo D, Ghezzi A, Colombo B, Scotti G, Comi G. Correlations between structural CNS damage and functional MRI changes in primary progressive MS. NeuroImage 2002;15:537–546.
20. Rocca MA, Falini A, Colombo B, Scotti G, Comi G, Filippi M. Adaptive functional

changes in the cerebral cortex of patients with non-disabling MS correlate with the extent of brain structural damage. Ann Neurol 2002;51:330–339.
21. Rocca MA, Mezzapesa DM, Falini A, Ghezzi A, Martinelli V, Scotti G, Comi G, Filippi M. Evidence for axonal pathology and adaptive cortical reorganisation in patients at presentation with clinically isolated syndromes suggestive of MS. NeuroImage 2003;18:847–855.
22. Rocca MA, Gavazzi C, Mezzapesa DM, Falini A, Colombo B, Mascalchi M, Scotti G, Comi G, Filippi M. A functional magnetic resonance imaging study of patients with secondary progressive multiple sclerosis. NeuroImage 2003;19:1770–1777.
23. Rocca MA, Pagani E, Ghezzi A, Falini A, Zaffaroni M, Colombo B, Scotti G, Comi G, Filippi M. Functional cortical changes in patients with MS and non-specific conventional MRI scans of the brain. NeuroImage 2003;19:826–836.
24. Reddy H, Narayanan S, Woolrich M, Mitsumori T, Lapierre Y, Arnold DL, Matthews PM. Functional brain reorganization for hand movement in patients with multiple sclerosis: defining distinct effects of injury and disability. Brain 2002;125:2646–2657.
25. Filippi M, Rocca MA, Mezzapesa DM, Ghezzi A, Falini A, Martinelli V, Scotti G, Comi G. Simple and complex movement-associated functional MRI changes in patients at presentation with clinically isolated syndromes suggestive of MS. Hum Brain Mapp 2004;21:108–117.
26. Rocca MA, Mezzapesa DM, Ghezzi A, Falini A, Martinelli V, Scotti G, Comi G, Filippi M. A widespread pattern of cortical activations in patients at presentation with clinically isolated symptoms is associated with evolution to definite multiple sclerosis. AJNR Am J Neuroradiol 2005;26:1136–1139.
27. Pantano P, Iannetti GD, Caramia F, Mainero C, Di Legge S, Bozzao L, Pozzilli C, Lenzi GL. Cortical motor reorganization after a single clinical attack of multiple sclerosis. Brain 2002;125:1607–1615.
28. Pantano P, Mainero C, Iannetti GD, Caramia F, Di Legge S, Piattella MC, Pozzilli C, Bozzao L, Lenzi GL. Contribution of corticospinal tract damage to cortical motor reorganization after a single clinical attack of multiple sclerosis. NeuroImage 2002;17:1837–1843.
29. Rocca MA, Gallo A, Colombo B, Falini A, Scotti G, Comi G, Filippi M. Pyramidal tract lesions and movement-associated cortical recruitment in patients with MS. NeuroImage 2004:23:141–147.
30. Lowe MJ, Phillips MD, Lurito JT, Mattson D, Dzemidzic M, Mathews VP. Multiple sclerosis: low-frequency temporal blood oxygen level-dependent fluctuations indicate reduced functional connectivity initial results. Radiology 2002;224:184–192.
31. Rocca MA, Matthews PM, Caputo D, Ghezzi A, Falini A, Scotti G, Comi G, Filippi M. Evidence for widespread movement-associated functional MRI changes in patients with PPMS. Neurology 2002;58:866–872.
32. Rocca MA, Mezzapesa DM, Ghezzi A, Falini A, Agosta F, Martinelli V, Scotti G, Comi G, Filippi M. Cord damage elicits brain functional reorganization after a single episode of myelitis. Neurology 2003;61:1078–1085.
33. Rocca MA, Agosta F, Mezzapesa DM, Falini A, Martinelli V, Salvi F, Bergamaschi R, Scotti G, Comi G, Filippi M. A functional MRI study of movement-associated cortical changes in patients with Devic's neuromyelitis optica. NeuroImage 2004;21:1061–1068.
34. Rocca MA, Colombo B, Falini A, Ghezzi A, Martinelli V, Scotti G, Comi G, Filippi M. Cortical adaptation in patients with MS: a cross-sectional functional MRI study of disease phenotypes. Lancet Neurol 2005;4: 618–626.
35. Rocca MA, Agosta F, Colombo B, Mezzapesa DM, Falini A, Comi G, Filippi M. fMRI changes in relapsing-remitting multiple sclerosis patients complaining of fatigue after IFNβ-1a injection. Hum Brain Mapp 2007;28:373–382.
36. Mezzapesa DM, Rocca MA, Rodegher M, Comi G, Filippi M. Functional cortical changes of the sensorimotor network are associated with clinical recovery in multiple sclerosis. Hum Brain Mapp 2008;29:562–573.
37. Rocca MA, Tortorella P, Ceccarelli A, Falini A, Tango D, Scotti G, Comi G, Filippi M. The "mirror-neuron system" in MS: a 3 Tesla fMRI study. Neurology 2008;70:255–262.
38. Rocca MA, Pagani E, Absinta M, Valsasina P, Falini A, Scotti G, Comi G, Filippi M. Altered functional and structural connectivities in patients with MS: a 3T fMRI study. Neurology 2007;69:2136–2145.
39. Filippi M, Rocca MA, Mezzapesa DM, Falini A, Colombo B, Scotti G, Comi G. A functional MRI study of cortical activations associated with object manipulation in patients with MS. NeuroImage 2004;21:1147–1154.
40. Cerasa A, Fera F, Gioia MC, Liguori M, Passamonti L, Nicoletti G, Vercillo L, Paolillo A, Clodomiro A, Valentino P, Quattrone A. Adaptive cortical changes and the functional correlates of visuo-motor integration in

relapsing-remitting multiple sclerosis. Brain Res Bull 2006;69:597–605.

41. Calautti C, Baron JC. Functional neuroimaging studies of motor recovery after stroke in adults: a review. Stroke 2003;34:1553–1566. Review.
42. Ciccarelli O, Toosy AT, Marsden JF, Wheeler-Kingshott CM, Miller DH, Matthews PM, Thompson AJ. Functional response to active and passive ankle movements with clinical correlations in patients with primary progressive multiple sclerosis. J Neurol 2006;253: 882–891.
43. Agosta F, Valsasina P, Sala S, Caputo D, Stroman PW, Filippi M. Functional MRI of the spinal cord in patients with relapsing-remitting multiple sclerosis. Proc Int Soc Magn Reson Med 2007;15:108.
44. Staffen W, Mair A, Zauner H, Unterrainer J, Niederhofer H, Kutzelnigg A, Ritter S, Golaszewski S, Iglseder B, Ladurner G. Cognitive function and fMRI in patients with multiple sclerosis: evidence for compensatory cortical activation during an attention task. Brain 2002;125:1275–1282.
45. Au Duong MV, Audoin B, Boulanouar K, Ibarrola D, Malikova I, Confort-Gouny S, Celsis P, Pelletier J, Cozzone PJ, Ranjeva JP. Altered functional connectivity related to white matter changes inside the working memory network at the very early stage of MS. J Cereb Blood Flow Metab 2005;25:1245–1253.
46. Au Duong MV, Boulanouar K, Audoin B, Treseras S, Ibarrola D, Malikova I, Confort-Gouny S, Celsis P, Pelletier J, Cozzone PJ, Ranjeva JP. Modulation of effective connectivity inside the working memory network in patients at the earliest stage of multiple sclerosis. NeuroImage 2005;24:533–538.
47. Audoin B, Ibarrola D, Ranjeva JP, Confort-Gouny S, Malikova I, Ali-Cherif A, Pelletier J, Cozzone P. Compensatory cortical activation observed by fMRI during a cognitive task at the earliest stage of MS. Hum Brain Mapp 2003;20:51–58.
48. Audoin B, Au Duong MV, Ranjeva JP, Ibarrola D, Malikova I, Confort-Gouny S, Soulier E, Viout P, Ali-Cherif A, Pelletier J, Cozzone PJ. Magnetic resonance study of the influence of tissue damage and cortical reorganization on PASAT performance at the earliest stage of multiple sclerosis. Hum Brain Mapp 2005;24:216–228.
49. Hillary FG, Chiaravalloti ND, Ricker JH, Steffener J, Bly BM, Lange G, Liu WC, Kalnin AJ, DeLuca J. An investigation of working memory rehearsal in multiple sclerosis using fMRI. J Clin Exp Neuropsychol 2003;25:965–978.
50. Parry AM, Scott RB, Palace J, Smith S, Matthews PM. Potentially adaptive functional changes in cognitive processing for patients with multiple sclerosis and their acute modulation by rivastigmine. Brain 2003;126: 2750–2760.
51. Penner IK, Rausch M, Kappos L, Opwis K, Radu EW. Analysis of impairment related functional architecture in MS patients during performance of different attention tasks. J Neurol 2003;250:461–472.
52. Mainero C, Caramia F, Pozzilli C, Pisani A, Pestalozza I, Borriello G, Bozzao L, Pantano P. fMRI evidence of brain reorganization during attention and memory tasks in multiple sclerosis. NeuroImage 2004;21:858–867.
53. Sweet LH, Rao SM, Primeau M, Mayer AR, Cohen RA. Functional magnetic resonance imaging of working memory among multiple sclerosis patients. J Neuroimaging 2004;14:150–157.
54. Sweet LH, Rao SM, Primeau M, Durgerian S, Cohen RA. Functional magnetic resonance imaging response to increased verbal working memory demands among patients with multiple sclerosis. Hum Brain Mapp 2006;27: 28–36.
55. Wishart HA, Saykin AJ, McDonald BC, Mamourian AC, Flashman LA, Schuschu KR, Ryan KA, Fadul CE, Kasper LH. Brain activation patterns associated with working memory in relapsing-remitting MS. Neurology 2004;62:234–238.
56. Chiaravalloti N, Hillary F, Ricker J, Christodoulou C, Kalnin A, Liu WC, Steffener J, DeLuca J. Cerebral activation patterns during working memory performance in multiple sclerosis using FMRI. J Clin Exp Neuropsychol 2005;27:33–54.
57. Cader S, Cifelli A, Abu-Omar Y, Palace J, Matthews PM. Reduced brain functional reserve and altered functional connectivity in patients with multiple sclerosis. Brain 2006;129: 527–537.
58. Li Y, Chiaravalloti ND, Hillary FG, Deluca J, Liu WC, Kalnin AJ, Ricker JH. Differential cerebellar activation on functional magnetic resonance imaging during working memory performance in persons with multiple sclerosis. Arch Phys Med Rehabil 2004;85:635–639.
59. Rocca MA, Valsasina P, Ceccarelli A, Absinta M, Ghezzi A, Riccitelli G, Pagani E, Falini A, Comi G, Scotti G, Filippi M. Structural and functional MRI correlates of Stroop control in benign MS. Hum Brain Mapp 2009;30: 276–290.
60. Lazeron RH, Rombouts SA, Scheltens P, Polman CH, Barkhof F. An fMRI study of

planning-related brain activity in patients with moderately advanced multiple sclerosis. Mult Scler 2004;10:549–555.

61. Comi G, Rovaris M, Leocani L, Martinelli V, Filippi M. Clinical and MRI assessment of brain damage in MS. Neurol Sci 2001;22: S123–S127.
62. Lenzi D, Conte A, Mainero C, Frasca V, Fubelli F, Totaro P, Caramia F, Inghilleri M, Pozzilli C, Pantano P. Effect of corpus callosum damage on ipsilateral motor activation in patients with multiple sclerosis: a functional and anatomical study. Hum Brain Mapp 2007;28:636–644.
63. Manson SC, Palace J, Frank JA, Matthews PM. Loss of interhemispheric inhibition in patients with multiple sclerosis is related to corpus callosum atrophy. Exp Brain Res 2006;174: 728–733.
64. Filippi M, Rocca MA, Colombo B, Falini A, Codella M, Scotti G, Comi G. Functional magnetic resonance imaging correlates of fatigue in multiple sclerosis. NeuroImage 2002;15:559–567.
65. Pantano P, Mainero C, Lenzi D, Caramia F, Iannetti GD, Piattella MC, Pestalozza I, Di Legge S, Bozzao L, Pozzilli C. A longitudinal fMRI study on motor activity in patients with multiple sclerosis. Brain 2005;128: 2146–2153.
66. Mainero C, Inghilleri M, Pantano P, Conte A, Lenzi D, Frasca V, Bozzao L, Pozzilli C. Enhanced brain motor activity in patients with MS after a single dose of 3,4-diaminopyridine. Neurology 2004;62:2044–2050.
67. Morgen K, Kadom N, Sawaki L, Tessitore A, Ohayon J, McFarland H, Frank J, Martin R, Cohen LG. Training-dependent plasticity in patients with multiple sclerosis. Brain 2004;127:2506–2517.
68. Rocca MA, Agosta F, Mezzapesa DM, Falini A, Martinelli V, Salvi F, Bergamaschi R, Scotti G, Comi G, Filippi M. A functional MRI study of movement-associated cortical changes in patients with Devic's neuromyelitis optica. NeuroImage 2004;21:1061–1068.
69. Mikulis DJ, Jurkiewicz MT, McIlroy WE, Staines WR, Rickards L, Kalsi-Ryan S, Crawley AP, Fehlings MG, Verrier MC. Adaptation in the motor cortex following cervical spinal cord injury. Neurology 2002;58:794–801.
70. Curt A, Alkadhi H, Crelier GR, Boendermaker SH, Hepp-Reymond MC, Kollias SS. Changes of non-affected upper limb cortical representation in paraplegic patients as assessed by fMRI. Brain 2002;125:2567–2578.
71. Sabbah P, de Schonen, Leveque C, Gay S, Pfefer F, Nioche C, Sarrazin JL, Barouti H, Tadie M, Cordoliani YS. Sensorimotor cortical activity in patients with complete spinal cord injury: a functional magnetic resonance imaging study. J Neurotrauma 2002;19: 53–60.
72. Cramer SC, Fray E, Tievsky A, Parker RA, Riskind PN, Stein MC, Wedeen V, Rosen BR. Changes in motor cortex activation after recovery from spinal cord inflammation. Mult Scler 2001;7:364–370.
73. Rocca MA, Mezzapesa DM, Ghezzi A, Falini A, Agosta F, Martinelli V, Scotti G, Comi G, Filippi M. Cord damage elicits brain functional reorganization after a single episode of myelitis. Neurology 2003;61:1078–1085.
74. Rocca MA, Agosta F, Martinelli V, Falini A, Comi G, Filippi M. The level of spinal cord involvement influences the pattern of movement-associated cortical recruitment in patients with isolated myelitis. NeuroImage 2006;30:879–884.
75. Rocca MA, Agosta F, Mezzapesa DM, Ciboddo G, Falini A, Comi G, Filippi M. An fMRI study of the motor system in patients with neuropsychiatric systemic lupus erythematosus. NeuroImage 2006;30:478–484.
76. Kurth T, Slomke MA, Kase CS, Cook NR, Lee IM, Gaziano JM, Diener HC, Buring JE. Migraine, headache, and the risk of stroke in women: a prospective study. Neurology 2005;64:1020–1026.
77. Rocca MA, Colombo B, Pagani E, Falini A, Codella M, Scotti G, Comi G, Filippi M. Evidence for cortical functional changes in patients with migraine and white matter abnormalities on conventional and diffusion tensor magnetic resonance imaging. Stroke 2003;34:665–670.

Chapter 20

fMRI in Cerebrovascular Disorders

Nick S. Ward

Summary

Stroke is a major cause of long-term disability worldwide. One of the key factors underpinning recovery of function is reorganization of surviving neural networks. Noninvasive techniques such as fMRI allow this reorganization to be studied in humans. However, the design of experiments involving patients with impairment requires careful consideration and is often constrained. Difficulty with some tasks can lead to a number of performance confounds, and so tasks and task parameters that avoid or minimize this should be selected. Furthermore, when studying patients with cerebrovascular disease, it is important to consider the possibility that the blood oxygen level dependent signal may be altered and affect interpretation of results. Despite these potential problems, careful experimental design can provide real insights into system-level reorganization after stroke and how it is related to functional recovery. Currently, results suggest that functionally relevant reorganization does occur in cerebral networks in human stroke patients. For example, it is apparent that initial attempts to move a paretic limb following stroke are associated with widespread activity within the distributed motor system in both cerebral hemispheres. This reliance on nonprimary motor output pathways is unlikely to support full recovery, but improved efficiency of the surviving networks is associated with behavioral gains. This reorganization can only occur in structurally and functionally intact brain regions. Understanding the dynamic process of system-level reorganization will allow greater understanding of the mechanisms of recovery and potentially improve our ability to deliver effective restorative therapy.

Key words: fMRI, Stroke, Blood oxygen level dependent, Motor cortex, Premotor cortex, Plasticity, Rehabilitation

1. Introduction

Studying patients who have suffered from stroke with functional brain imaging is difficult for a variety of reasons. The motivation behind such studies is a desire to understand and subsequently improve the process of functional recovery. Stroke and other forms of neurological damage account for nearly half of all

M. Filippi (ed.), *fMRI Techniques and Protocols*, Neuromethods, vol. 41
DOI 10.1007/978-1-60327-919-2_20, © Humana Press, a part of Springer Science+Business Media, LLC 2009

severely disabled adults *(1–3)*. Longitudinal studies of recovery suggest that only 50% of stroke survivors with significant initial upper limb paresis recover useful function of the limb *(4)*. Furthermore, those with poor recovery of arm function have dramatically impaired quality of life and sense of well being *(5, 6)*. It is clear that effective treatment of motor impairment after stroke is critically important to many people.

The mainstay of treatment is neurorehabilitation. The overall approach is effective and the benefit of strategies aimed at helping patients *adapt* to impairment well proven *(7)*. Treatments aimed at *reducing* impairment, however, are poorly developed. There is an overriding assumption that one way to tackle impairment in those patients with focal brain damage is to attempt to promote functionally relevant reorganization within surviving neural networks *(8)*. Over the last decade, advances in our understanding of how the normal brain is organized at the molecular, cellular, and systems level have improved enormously. Advances in our understanding of the mechanisms of impairment after brain injury, including stroke, are way behind. Translating findings from proof-of-principle studies into real treatments is proving problematic and requires urgent attention *(9)*. Thus, the clinical neurosciences have the potential to make a unique contribution toward developing rehabilitation strategies designed to reduce impairment.

The tools available for studying the working human brain are different to those used in animal models. In human subjects, experiments are performed at the level of neural systems rather than single cells or molecules. This chapter will concentrate on the way that fMRI can be used to contribute. The first half will consider the specific difficulties involved in performing fMRI experiments in stroke patients, in particular how fMRI signals might be affected in cerebrovascular disease and how studying patients with impairment should influence experimental design. In the second half, examples of studies using fMRI in stroke patients will be discussed to illustrate advances in our understanding of post-stroke functional brain reorganization. Many studies have been performed in the somatosensory *(10–12)* and language systems *(13–15)*, but studies of the motor system are particularly numerous and will be used to illustrate how fMRI may be used.

2. Blood Oxygen Level Dependent Signal in Cerebrovascular Disease

fMRI relies on the blood oxygen level dependent (BOLD) signal. In brief, the BOLD signal relies on the close coupling between blood flow and metabolism. During an increase in neuronal activation there is an increase in local cerebral blood flow, but only a small proportion of the greater amount of oxygen delivered locally to the tissue is used. This results in a net increase in the tissue concentration of oxyhemoglobin and a net reduction in paramagnetic deoxyhemoglobin in the local capillary bed. The magnetic properties of hemoglobin depend on its level of oxygenation so that this change results in an increase in local tissue-derived signal intensity on T2*-weighted MR images *(16)*.

The mechanism of neurovascular signalling to the blood vessels controlling cerebral blood flow is still unclear, although it may involve metabolic *(17, 18)* or neurochemical *(19, 20)* mechanisms. In addition, the generation of BOLD signal is still reliant on venous blood volume, blood flow, blood oxygenation, and oxygen consumption. Thus, it is possible that any disease state that changes these parameters will potentially modify the BOLD signal. It is therefore legitimate to be concerned whether the BOLD signal is reliable in patients who have suffered stroke and in subjects with evidence of both large and small vessel atherosclerosis. The potential problem arises because in general, the BOLD signal is assumed to have the same shape in all subjects and in all brain regions. In one case, the canonical hemodynamic response function (HRF) has been derived by principle component analysis of empirical data with a peak magnitude occurring 6 s after the neuronal activity *(21)*.

There is evidence to suggest that the shape of the hemodynamic response might be altered after stroke. Newton et al. *(22)* demonstrated a greater time to peak BOLD response in ipsilateral compared with contralateral M1 in controls. In three chronic stroke patients, the time to peak BOLD response was increased in ipsilesional (contralateral) M1 compared with controls. Interestingly, in these patients the time to peak BOLD response in contralesional M1 was equivalent or less than that for ipsilesional M1, representing a finding opposite to that seen in healthy controls. Pineiro et al. *(23)* have also described a slower time to peak BOLD response in sensorimotor cortex bilaterally in 12 chronic stroke patients with lacunar infarcts. Thus, modeling the BOLD response with a canonical HRF might be less efficient in stroke patients. It is worth considering what the effect of this would be in the context of a standard functional imaging analysis using the general linear model approach. If the canonical HRF was a poor fit for the actual response, then the residual error of the analytical model would be greater (than if the fit was good), thus *lowering t-* and *Z*-scores and depressing sensitivity to detection of differences. In fact, in general, most studies of stroke patients have found increased activation in a number of brain regions over and above healthy controls, so it might be the case that these overactivations have been *underestimated*. However, modeling differences in measured HRF from the canonical is likely to be beneficial. The use of temporal and dispersion derivatives of the canonical HRF to specifically capture differences in the timing or duration of the peak response, for example, is likely to increase sensitivity.

In addition to changes in the shape of the HRF, there is evidence that in patients with impaired cerebrovascular reserve or advanced narrowing of the cerebral arteries, the BOLD fMRI signal may be reduced, or even become negative *(24–26)*. Röther et al. *(27)* describe a single patient, which illustrates the point. The patient was found to have bilateral occluded internal carotid arteries and an occluded vertebral artery. The cerebrovascular reactivity, as determined by reduced change in T2* signal during hypercapnia, was severely impaired in the left hemisphere. The finding of importance was that during a motor task with the right hand, the BOLD response in the left motor cortex was negative for the duration of the task. This suggests that the initial dip in BOLD signal due to a relative decrease in oxyhemoglobin was not followed by the normal vascular response (which would have resulted in an increase in oxygenated and decrease in deoxygenated hemoglobin). This subject had previously suffered from a transient ischemic attack involving the right arm. It is likely that these symptoms were related to hemodynamic insufficiency, and it is interesting to speculate that the presence of a prolonged negative dip in BOLD signal represents a marker for those at risk from such symptoms. Others have made the point that this impaired cerebrovascular reactivity might be due to either large or small vessel disease *(26)*. Further investigation will reveal whether this idea has genuine potential as a clinical tool.

Several studies have now suggested that impairment of normal vasodilatation in response to hypercapnia is associated with diminished magnitude of BOLD signal *(25, 26, 28, 29)*. Thus, in patients with severely impaired cerebrovascular reactivity neuronal activation may not translate into a BOLD response in the conventional sense, and standard models using the canonical HRF may not be sufficient.

However, the scale of the problem is not yet clear. For example, the cerebrovascular reactivity in the right hemisphere of the patient studied by Röther et al. *(27)* was moderately impaired and the BOLD response during a motor task with the left hand was entirely normal. Patients with hemodynamic symptoms are rare, and are likely to be excluded from fMRI studies. In addition, patients with severe stenosis of ipsilesional internal carotid arteries are usually also excluded, although it is not clear that this is necessary. It may also be the case that small vessel disease may also make a significant contribution to impaired cerebrovascular reactivity.

Thus, although there is evidence that impaired cerebrovascular reactivity can diminish the BOLD response, there is no evidence that the BOLD signal is erroneously detected in these patients, i.e., this is largely a problem of false negative results. In general, the literature concerning differences between stroke patients and healthy controls is dominated by the finding of overactivity in patients compared with controls, and once again, it is possible that the issue of cerebrovascular disease has led to an underestimation of changes in cortical organization after stroke. It is clear that when examining for differences between a group of patients and a group of healthy controls, the nonneural factors that can influence the BOLD response will contribute significantly to whether a difference is found or not. However, several studies have begun to use a correlation approach. That is to say, to attempt to explain variability in the task-related BOLD signal with some other parameters, such as a measure of recovery *(30)* or corticospinal tract integrity *(31, 32)*. It is unlikely that changes in cerebrovascular responsiveness will correlate with recovery or an anatomical measure of corticospinal tract integrity, and so it is unlikely to account for any significant (and biologically plausible) results. As we have already discussed, however, the ability to detect a real finding is likely to be diminished by alterations in nonneural contributors to the BOLD signal. A multimodal approach using different imaging techniques (BOLD, perfusion, hypercapnic challenge) and concurrent neurophysiological methods (electroencephalography [EEG], magnetoencephalography [MEG], transcranial magnetic stimulation [TMS]) may be useful when addressing the influence of multiple physiological variables. These issues will require further empirical study.

The discussion regarding differences in hemodynamic coupling is also of relevance when considering the effects of age, given that stroke is commoner with advancing age, and that often age-matched controls are used in studies of stroke patients. D'Esposito et al. *(33)* examined the effects of age on the BOLD signal generated during a button press task in response to a visual cue, using a sparse event-related design. Using a standard fixed effects analysis, task-related activation was detected in primary motor cortex (M1) above the chosen threshold in only 75% of the older subjects but 100% of the younger subjects. Furthermore, for those in whom activation was detected, four times the number of suprathreshold voxels was present in the younger compared with the older subjects. Thus, on the face of it, M1 appears to be less active during a button press task in older subjects. However, the most important finding was in relation to the signal-to-noise ratio (SNR), which was reduced in elderly subjects. Results from single subject or group fixed effects analyses of functional imaging data are generally presented as t statistics for each voxel (volume element) of the brain. The result is therefore dependent on both the magnitude of the signal change and the residual variance after this has been accounted for. Thus, an increased SNR will lead to a lower *t* statistic, and therefore fewer suprathreshold voxels. In fact, D'Esposito et al. *(33)* found no difference in the magnitude of task-related signal change in M1, supporting the notion that the diminished number of suprathreshold voxels was largely attributable to the decreased SNR in the older subjects.

The problem of reduced SNR can be effectively dealt with by employing the statistical technique of random effects analysis as opposed to fixed effects analysis. Random effects analysis of functional imaging data treats each subject as a random variable. The experimental variance is dominated by between subjects variability (as opposed to within subject variability in the case of fixed effects models). The data for each subject comprise the voxel-wise parameter estimate for the task under consideration, which reflects the magnitude of the signal change in each voxel. Appropriate statistics can be performed on these data, which are less likely to be influenced by differences in SNR *(33)*. Using a random effects analysis, and employing both temporal and dispersion derivatives of the HRF, Ward et al. *(34)* demonstrated no change in the shape of the hemodynamic response during a hand grip task with advancing age, in keeping with the findings described earlier.

3. Issues in Experimental Design

The results from any functional imaging study are only as reliable as the care with which the experiment is constructed and executed, but studies involving patients who have had a stroke raise some specific issues. The selection of patients, choice of experimental paradigm, within scanner monitoring of performance, and the approaches to data analysis all require careful consideration.

3.1. Subject Selection

In general, stroke patients are a heterogeneous group differing in several important ways, not least the site and size of infarct, patency of the vascular system, age, co-morbidities, and concurrent medication. The criteria for patient selection will to an extent depend on the experimental question. It is unlikely that averaging the results from a wide variety of patient types will prove useful because of this variability, but there are two other ways of approaching the experimental design. First, it may be desirable to use a group of patients highly selected on the basis of lesion location, for example. Results from this type of controlled study are powerful, but do not generalize outside the subgroup selected.

Alternatively, it might be more useful to study a group of patients who vary in a specific factor of interest (e.g., outcome), to explore the relationship between this factor and task-related brain activity. Results from studies using this approach can be generalized more easily.

3.2. Performance Confounds

The choice of experimental task is critical and is dependent on the experimental question. For example, a study of the relationship between brain activation and outcome after stroke will by necessity involve patients with different performance abilities. Similarly, a longitudinal study will require that patients are studied at different stages of recovery. For an active motor or language task this can result in the problem of performance confounds, because the ability to perform the task is not the same across patients or sessions. A change in experimental task performance can have significant effects upon the pattern of brain activation. In other words, comparison across patients or time points is made difficult if the patients are performing the task differently. Thus, each patient must perform the *same task* during the fMRI experiment, so that a meaningful comparison can be made across subjects or scanning sessions. Maintaining a consistent task is therefore of great importance, but in stroke recovery studies equality of task may be interpreted in a number of ways. In particular, a task may be consistent across patients with different abilities in terms of *absolute* or *relative* parameters.

This can be illustrated by considering a simple motor experiment. The absolute task parameters can be fixed (by setting the same target force and rate for each subject or session), but performing a task may be experienced as more or less effortful depending on the level of recovery. Consequently, any differences in results between subjects or sessions could be attributed to differences in "effort" exerted. Alternatively, the relative task parameters (i.e., the level of task difficulty) can be fixed across subjects/sessions. In this scenario, patients will perform the task at different absolute forces and rates, and so differences in results across subjects/sessions could be attributed to differences in the absolute task parameters. When using an "active task," these factors must always be considered and results interpreted with these confounds in mind. Increased effort is a potentially useful strategy for overcoming motor, language, or cognitive impairment in a real world setting. As described earlier, some patients may use less effortful as their performance improves. Is the focus of interest the reorganization that might be the substrate for recovery, or is it the strategy that each patient uses to perform a task to a certain level, given the constraints of their impairment? Both may be of interest, but the choice of experimental design has an impact on which process is being studied. The problem of performance confounds is avoided with passive tasks (e.g., passive limb movements and passive listening), but these are complementary approaches to active tasks, not substitutes for them.

3.3. Task Frequency

The rate of task performance is also something that has implications for both data analysis and interpretation of results. Consider once again a simple motor task such as finger tapping or hand squeezing. The rate of performance of a repetitive task will influence how effortful the task is, in the same way as the target force. Most experiments are conducted in a "block design"; that is to say a period of activity (usually for 16–30 s) followed by a period of rest. If subjects are asked to perform at the same rate (even if the target force is scaled according to each subjects own performance abilities), for example, finger tapping at 1 Hz for 20 s, then differences across subjects/sessions could be due to differences in perceived effort, just as with equal absolute target forces. Some investigators have varied the rate at which subjects are asked to perform a task to try to control for effort exerted. However, comparing blocks with different numbers of "events" within them is problematic because the BOLD signal summates depending on how many events there are. The BOLD response needs approximately 10 s and longer to return to baseline, but "events" are usually more frequent than this (e.g., 1 Hz finger tapping). It is usually assumed that there is a summation of the overlapping BOLD responses, which is largely (but not entirely) linear *(35, 36)*. The basis function (boxcar design) in the general linear model will have the same "height," and so more frequent events will result in a larger parameter estimate, for the same amount of event-related activity. In fact, it is the quantity of events, not their frequency, which will modulate BOLD signal in motor cortices *(37)*. Thus, if a subject performs the task at 1 Hz and then at 0.5 Hz in both cases for 20 s, and each period is modeled with the same boxcar basis function, then the parameter estimate (or magnitude of activity) will appear to be roughly twice as much during the more frequent task. This reflects the quantity of "events," but not a change in the way the brain is organized. Thus, if patients perform a task more slowly than healthy controls to control for effort, then the brain activity associated with that movement will be underestimated in comparison to those subjects who perform the task at a faster rate. One way around this problem is to use an event-related design in which the intertrial interval is long enough for the task to be performed repetitively without increasing the sense of effort. This design may be less efficient in terms of fMRI design, but avoids the confounds described earlier *(31, 32)*.

3.4. Task Complexity

Investigators are often tempted to use more complex tasks when studying patients with impairment in the hope that this will maximize differences between patients and control subjects. This is sometimes done in the hope of exploring a more *ecologically valid* task, i.e., one which is relevant to function in the real world. However, it is never possible to study the neural correlates of a task that a subject cannot themselves perform. By introducing more complexity into the task, patients with significant impairment are more likely to adopt new operational strategies toward these experimental tasks in an attempt to adapt to their impairment. These differences in strategy could therefore account for differences between subject groups. Although of clinical interest, differences in strategy across a group represent a potential experimental confound if they are unexpected and not measured. One approach is therefore to use a simple task that minimizes difference in strategic approach to the task so that valid comparisons can then be made across subjects/sessions.

3.5. Task Monitoring

Once a paradigm has been selected it is important that task performance is monitored during the experiment. Intersubject variability may be greater after stroke and new sources of variability can arise, such as mirror or associated movements. To take account of this, some investigators record behavior during a prescan rehearsal, whilst others incorporate the increasingly available instrumentation that is compatible with the MRI setting. Prescan rehearsal provides some idea of whether a task can be performed correctly, or whether mirror movements are present, for example. However, in-scanner recordings allow this information to be incorporated into image analysis as a covariate, and thus improving statistical power by accounting for correlated variance in the measured scan signal.

The experimental approach is therefore dictated by the experimental question. Not all investigators will have the same question, but the issues discussed earlier need to be considered in all cases. For most questions, this approach is entirely appropriate and standardization of experimental paradigms, patient selection, and method of analysis across experiments is not required. In the case of experimental questions that require a multicenter approach that is technically feasible, standardization of such factors would be required.

4. Reorganization in the Motor System After Stroke

Early studies of motor system organization after stroke compared brain activation during movement in well-recovered patients and normal controls. Early group studies of stroke patients with subcortical lesions described greater activation within a number of motor-related cortical regions compared with controls during a finger tapping task *(38–42)*. It was suggested that nonprimary (or secondary) cortical motor regions were thus responsible for recovery of motor function in these patients. Strick *(43)* had proposed this as a potential mechanism of restoration of function, some years before based on an understanding of the organization of the cortical motor system in primates. Normal distal motor function is facilitated largely through the corticospinal pathway, from the cortical motor system to the spinal cord motor neurons. The majority of corticospinal fibers originate in the M1, but there are contributions from other cortical regions *(44)*. In primates, the M1, arcuate (or lateral) premotor cortex (PM), and supplementary motor area (SMA) are each part of parallel, independent motor networks with (i) separate projections to spinal cord motoneuron and (ii) interactions at the level of the cortex *(43)*. There is some similarity between the corticospinal projections from the hand regions of M1, PM, and SMA. Thus, it seemed feasible that a number of motor networks acting in parallel could generate an output to the spinal cord necessary for movement, and that damage in one of these networks could be at least partially compensated for by activity in another *(45, 46)*. Subsequently, many studies have demonstrated that the performance of a simple motor task with the affected limb is associated with greater bilateral brain activation in a number of cortical motor-related areas compared with healthy volunteers, including dorsal PM (PMd) and ventral PM (PMv), SMA, and cingulate motor areas (CMA) *(23, 38–42, 47–52)*.

A critical question is whether these differences are related to recovery. As discussed previously in this chapter, this question requires that the group of patients examined have a wide variety of outcomes, or else longitudinal studies should be performed. In the first such cross-sectional study, a group of chronic stroke patients with infarcts sparing M1 were scanned during a hand grip with visual feedback task using fMRI *(30)*. The target forces used were always a proportion of each subject's own maximum grip force, so that any differences were unlikely to be due to differences in effort. The more affected patients had greater task-related activity in secondary motor regions in both affected and unaffected hemispheres, whereas patients with the best motor scores had activation patterns that were indistinguishable from healthy age-matched volunteers. A similar result was observed in a group of patients studied at approximately 10 days post-stroke *(53)*. It was hypothesized that, secondary motor areas are recruited in response to damage to corticospinal output. A subsequent study demonstrated a strong positive correlation between secondary motor area recruitment in both hemispheres and corticospinal system damage **(Fig. 1)** *(31)*. A more "normal" corticospinal system was associated with greater task-related activity in contralesional M1 (hand area), suggesting a progressive shift

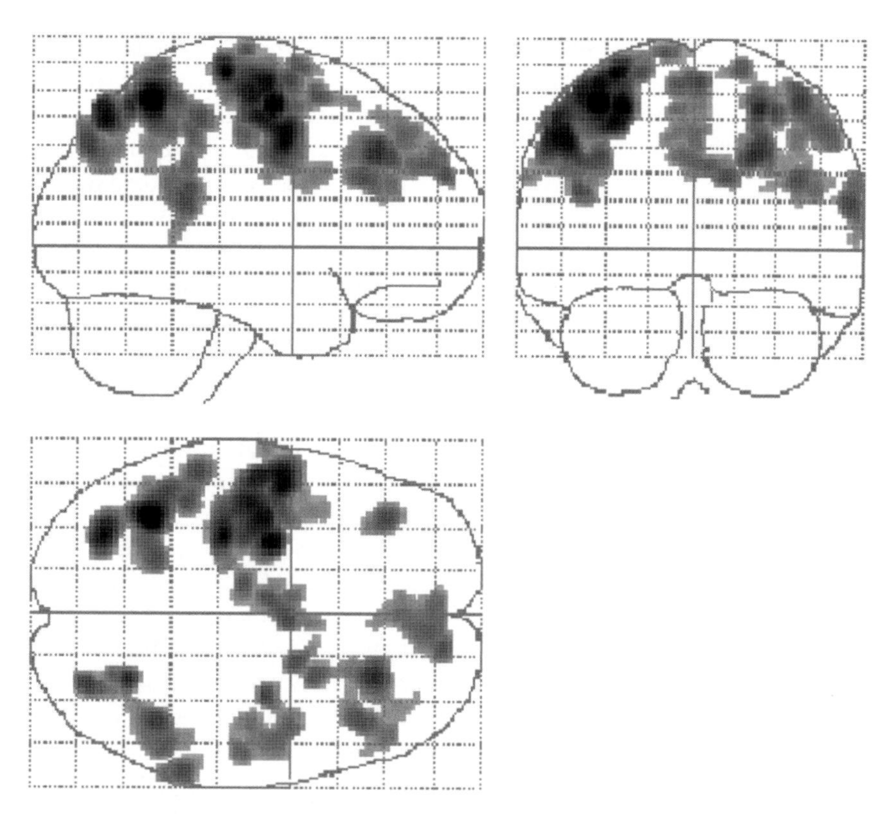

Fig. 1. Brain regions in which there is a negative correlation between corticospinal system integrity (as assessed with transcranial magnetic stimulation) and task-related signal change during hand grip with the affected hand. Increasing task-related activity is seen in a number of secondary motor areas including premotor regions and supplementary motor area as damage to the corticospinal system increases. The affected hand was on the left side. Results are displayed on a "glass brain" shown from the right side (*top left image*), from behind (*top right image*), and from above (*bottom left image*). Voxels are significant at $P < 0.001$ (uncorrected), and clusters are significant at $P < 0.05$ (corrected) (Reproduced from **ref. *31***, Oxford University Press.).

away from primary to secondary motor areas with increasing disruption to corticospinal system. This has also been described as an increase in either bilateral or contralesional activity, but the exact pattern is likely to depend on the anatomy of the damage. Furthermore, several secondary motor areas have bilateral projections to motor output systems *(54, 55)*. Thus, after stroke, the brain will use what is available (i.e., what is intact and connected so that motor output can be influenced) in an attempt to generate motor output to spinal cord motoneurons.

These results do not immediately support the idea that secondary motor areas are the substrate for motor recovery. Labeling corticospinal neurons with retrograde tracers has revealed multiple nonprimary corticospinal output zones in both the lateral and the medial areas of the frontal lobe (SMA, CMA, PMd, and PMv) *(56–58)*. These output zones contain large numbers of corticospinal neurons that project to the intermediate zone and ventral horn of the spinal cord suggesting their potential for direct control of spinal motoneurons in a way paralleling corticospinal output from M1 *(45)*. In primates, however, projections from secondary motor areas to spinal cord motor neurons are less numerous and less efficient at exciting spinal cord motoneurons than those from M1 *(59, 60)*. Moreover, unlike M1, facilitation of distal muscles from SMA, PMd, and PMv is not significantly stronger than facilitation of proximal muscles. The pathways through which (bilateral) secondary cortical motor regions generate this motor output are not clear. The fact that descending pathways from secondary motor areas are thought not to be able to efficiently generate distal limb movements suggests that cortico-cortical interactions, presumably with surviving M1 output, play an important role.

What is the evidence for the idea that the secondary motor areas of both hemispheres are contributing to recovered function? There are two ways to investigate the functional relevance of secondary motor region recruitment. One is to measure how task-related activity co-varies with modulation of task parameters. In healthy humans, for example, increasing force production is associated with linear increases in BOLD signal in contralateral M1 and medial motor regions, implying that they have a functional role in force production *(61–63)*. A recent study examined specifically for regional changes in the control of force modulation after stroke *(32)*. In patients with greater corticospinal system damage, force-related signal changes were seen mainly in contralesional dorsolateral PM, bilateral ventrolateral premotor cortices, and contralesional cerebellum, but not ipsilesional M1 (**Fig. 2**).

Thus, not only do premotor cortices become increasingly active as corticospinal system integrity diminishes *(31)*, but they can take on a new "M1-like" role during modulation of force output, which implies a new and functionally relevant role in motor control.

Second, experiments in which premotor activity is transiently disrupted with TMS can lead to worsening of recovered motor behaviors in patients with no effect on the performance of control subjects *(51, 64, 65)*, again implying new and functionally relevant roles. Furthermore, TMS to contralesional PMd is more disruptive in patients with greater impairment *(51)*, whereas TMS to ipsilesional PMd is more disruptive in less impaired patients *(64)* in keeping with a general shift toward functionally relevant activity in the contralesional hemisphere of patients with greater damage to motor output pathways.

These results are important because they tell us that the response to focal injury does not involve simple substitution of one cortical region for another. It is clear that nodes within remaining motor networks can take on new functional roles.

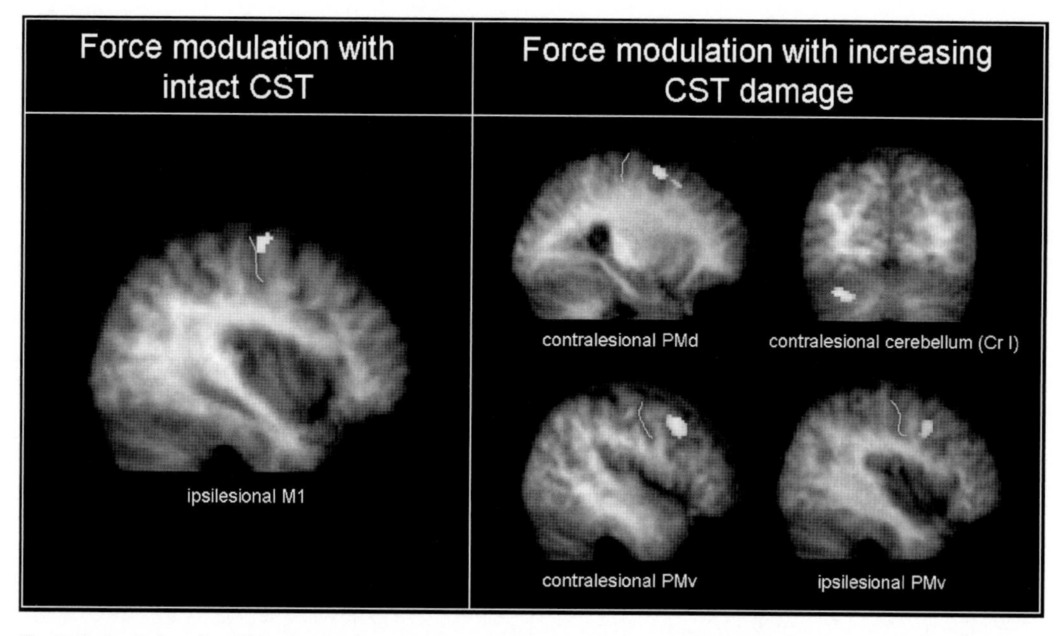

Fig. 2. Brain regions in which the blood oxygen level dependent (BOLD) signal varies linearly with force exerted during hand grip change as a function of corticospinal tract (CST) integrity (as assessed with transcranial magnetic stimulation). The affected hand was on the left side. In the *left panel*, increasing force leads to greater modulation of BOLD signal in ipsilesional M1 in patients with less damage to CST. In the *right panel*, increasing force leads to greater modulation of BOLD signal in contralesional dorsal premotor cortex, contralesional cerebellum, contralesional, and ipsilesional ventral premotor cortex. This demonstrates that brain regions involved in force modulation shift away from primary motor cortex to premotor regions with increasing CST damage. Results are overlaid onto the average T1-weighted structural scan obtained from all stroke patients in the study (Adapted from **ref.** ***32***, Blackwell Publishing.).

In summary, in the chronic stroke brain, there is a reconfiguration of the cerebral motor system. Task-related brain activation varies across chronic stroke patients in a way that appears to be predictable. It is important to stress that this reorganization is often not successful in returning motor function to normal. It is less effective than that in the intact brain but will nevertheless attempt to generate some form of motor signal to spinal cord motoneurons in the most efficient way. The exact configuration of this new motor system will be determined most obviously by the extent of the anatomic damage. This includes the extent to which the damage affects cortical motor regions, white matter pathways, and even which hemisphere is affected *(66)*. The more of the normal functional architecture that survives, the greater will be the potential for full recovery. In patients with damage to primary sensorimotor cortex, for example, tests of fractionated finger movement correlated more strongly with the proportion of surviving "normal" sensorimotor cortex (as defined by functional activation maps in normal controls) than with total infarct volume *(67)*.

This anatomic explanation accounts for why some patients do better than others, but it does not account for the recovery of function that occurs over weeks and months in individual patients. How does the reorganized state evolve? Longitudinal fMRI studies have shed further light on the process *(47, 50, 68–70)*, although only a handful have studied patients on more than two occasions. One study scanned subcortical stroke patients on average eight times over 6 months after stroke *(70)* and demonstrated an early overactivation in primary and many nonprimary motor regions. Thereafter, functional recovery was associated with a focusing of task-related brain activation patterns toward a "normal" lateralized pattern. In general, longitudinal studies have demonstrated a focusing of activity toward the lesioned hemisphere motor regions that is associated with improvement in motor function *(47, 68)*, with some patients showing persistent recruitment *(50)*.

5. Conclusions

In summary, the brain activation pattern of an individual patient represents the state of reorganization within that system at the time of study. This pattern is highly influenced by a number of methodological factors as previously discussed. However, in appropriately controlled experiments, these activation patterns tell us something about how that brain is functionally organized. Functional improvement with treatment is likely to be associated with changes within this network. The potential for functionally relevant change to occur will depend on a number of other factors, not least the biologic age of the subject and the premorbid state of their brain, but also current drug treatments. Furthermore, levels of neurotransmitters and growth factors that are able to influence the ability of the brain to respond to afferent input (i.e., how plastic it is) might be determined by their genetic status *(71)*. All of these factors will influence the potential for activity driven change within the intact motor networks, the putative mechanism of therapy driven improvements in motor performance.

References

1. Hoffman C, Rice D, Sung HY. Persons with chronic conditions. Their prevalence and costs. JAMA 1996;276(18):1473–1479.
2. Office of Population Censuses and Surveys. OPCS surveys of disability in Great Britain. I. The prevalence of disability among adults. London: HMSO. 1988.
3. Wade DT, Hewer RL. Epidemiology of some neurological diseases with special reference to work load on the NHS. Int Rehabil Med 1987; 8(3):129–137.
4. Wade DT. Measuring arm impairment and disability after stroke. Int Disabil Stud 1989; 11(2):89–92.

5. Nichols-Larsen DS, Clark PC, Zeringue A, Greenspan A, Blanton S. Factors influencing stroke survivors' quality of life during subacute recovery. Stroke 2005;36(7):1480–1484.
6. Wyller TB, Sveen U, Sodring KM, Pettersen AM, Bautz-Holter E. Subjective well-being one year after stroke. Clin Rehabil 1997;11(2):139–145.
7. Stroke Unit Trialists' Collaboration. Organised inpatient (stroke unit) care for stroke (Cochrane Review). In: The Cochrane Library, Issue 2.Oxford: Update Software. 2000.
8. Ward NS, Cohen LG. Mechanisms underlying recovery of motor function after stroke. Arch Neurol 2004;61(12):1844–1848.
9. The Academy of Medical Sciences. Restoring neurological function: putting the neurosciences to work in neurorehabilitation. London: Academy of Medical Sciences. 2004.
10. Loubinoux I, Carel C, Pariente J, Dechaumont S, Albucher JF, Marque P et al. Correlation between cerebral reorganization and motor recovery after subcortical infarcts. Neuroimage 2003;20(4):2166–2180.
11. Tombari D, Loubinoux I, Pariente J, Gerdelat A, Albucher JF, Tardy J et al. A longitudinal fMRI study: in recovering and then in clinically stable sub-cortical stroke patients. Neuroimage 2004;23(3):827–839.
12. Ward NS, Brown MM, Thompson AJ, Frackowiak RS. Longitudinal changes in cerebral response to proprioceptive input in individual patients after stroke: an FMRI study. Neurorehabil Neural Repair 2006;20(3):398–405.
13. Lee A, Kannan V, Hillis AE. The contribution of neuroimaging to the study of language and aphasia. Neuropsychol Rev 2006;16(4): 171–183.
14. Price CJ, Crinion J. The latest on functional imaging studies of aphasic stroke. Curr Opin Neurol 2005;18(4):429–434.
15. Wise RJ. Language systems in normal and aphasic human subjects: functional imaging studies and inferences from animal studies. Br Med Bull 2003;65:95–119.
16. Buxton RB. An introduction to functional magnetic resonance imaging: principles and techniques. Cambridge: Cambridge University Press, 2002.
17. Magistretti PJ, Pellerin L. Cellular mechanisms of brain energy metabolism and their relevance to functional brain imaging. Philos Trans R Soc Lond B Biol Sci 1999;354(1387):1155–1163.
18. Magistretti PJ, Pellerin L, Rothman DL, Shulman RG. Energy on demand. Science 1999;283(5401):496–497.
19. Iadecola C. Neurovascular regulation in the normal brain and in Alzheimer's disease. Nat Rev Neurosci 2004;5(5):347–360.
20. Attwell D, Iadecola C. The neural basis of functional brain imaging signals. Trends Neurosci 2002;25(12):621–625.
21. Friston KJ, Josephs O, Rees G, Turner R. Nonlinear event-related responses in fMRI. Magn Reson Med 1998;39(1):41–52.
22. Newton J, Sunderland A, Butterworth SE, Peters AM, Peck KK, Gowland PA. A pilot study of event-related functional magnetic resonance imaging of monitored wrist movements in patients with partial recovery. Stroke 2002;33(12):2881–2887.
23. Pineiro R, Pendlebury S, Johansen-Berg H, Matthews PM. Functional MRI detects posterior shifts in primary sensorimotor cortex activation after stroke: evidence of local adaptive reorganization? Stroke 2001;32(5): 1134–1139.
24. Carusone LM, Srinivasan J, Gitelman DR, Mesulam MM, Parrish TB. Hemodynamic response changes in cerebrovascular disease: implications for functional MR imaging. AJNR Am J Neuroradiol 2002;23(7): 1222–1228.
25. Hamzei F, Knab R, Weiller C, Rother J. The influence of extra- and intracranial artery disease on the BOLD signal in fMRI. Neuroimage 2003;20(2):1393–1399.
26. Rossini PM, Altamura C, Ferretti A, Vernieri F, Zappasodi F, Caulo M et al. Does cerebrovascular disease affect the coupling between neuronal activity and local haemodynamics? Brain 2004;127(Pt 1):99–110.
27. Rother J, Knab R, Hamzei F, Fiehler J, Reichenbach JR, Buchel C et al. Negative dip in BOLD fMRI is caused by blood flow-oxygen consumption uncoupling in humans. Neuroimage 2002;15(1):98–102.
28. Krainik A, Hund-Georgiadis M, Zysset S, von Cramon DY. Regional impairment of cerebrovascular reactivity and BOLD signal in adults after stroke. Stroke 2005;36(6): 1146–1152.
29. Murata Y, Sakatani K, Hoshino T, Fujiwara N, Kano T, Nakamura S et al. Effects of cerebral ischemia on evoked cerebral blood oxygenation responses and BOLD contrast functional MRI in stroke patients. Stroke 2006;37(10):2514–2520.
30. Ward NS, Brown MM, Thompson AJ, Frackowiak RS. Neural correlates of outcome after stroke: a cross-sectional fMRI study. Brain 2003;126(Pt 6):1430–1448.

31. Ward NS, Newton JM, Swayne OB, Lee L, Thompson AJ, Greenwood RJ et al. Motor system activation after subcortical stroke depends on corticospinal system integrity. Brain 2006;129(Pt 3):809–819.
32. Ward NS, Newton JM, Swayne OB, Lee L, Frackowiak RS, Thompson AJ et al. The relationship between brain activity and peak grip force is modulated by corticospinal system integrity after subcortical stroke. Eur J Neurosci 2007;25(6):1865–1873.
33. D'Esposito M, Zarahn E, Aguirre GK, Rypma B. The effect of normal aging on the coupling of neural activity to the bold hemodynamic response. Neuroimage 1999;10(1):6–14.
34. Ward NS, Swayne OB, Newton JM. Age-dependent changes in the neural correlates of force modulation: an fMRI study. Neurobiol Aging 2008;29(9):1434–1446.
35. Pollmann S, Dove A, Yves von Cramon D, Wiggins CJ. Event-related fMRI: comparison of conditions with varying BOLD overlap. Hum Brain Mapp 2000;9(1):26–37.
36. Wager TD, Vazquez A, Hernandez L, Noll DC. Accounting for nonlinear BOLD effects in fMRI: parameter estimates and a model for prediction in rapid event-related studies. NeuroImage 2005;25(1):206–218.
37. Kim JA, Eliassen JC, Sanes JN. Movement quantity and frequency coding in human motor areas. J Neurophysiol 2005;94(4):2504–2511.
38. Cao Y, D'Olhaberriague L, Vikingstad EM, Levine SR, Welch KM. Pilot study of functional MRI to assess cerebral activation of motor function after poststroke hemiparesis. Stroke 1998;29(1):112–122.
39. Chollet F, DiPiero V, Wise RJ, Brooks DJ, Dolan RJ, Frackowiak RS. The functional anatomy of motor recovery after stroke in humans: a study with positron emission tomography. Ann Neurol 1991;29(1):63–71.
40. Cramer SC, Nelles G, Benson RR, Kaplan JD, Parker RA, Kwong KK et al. A functional MRI study of subjects recovered from hemiparetic stroke. Stroke 1997;28(12):2518–2527.
41. Weiller C, Chollet F, Friston KJ, Wise RJ, Frackowiak RS. Functional reorganization of the brain in recovery from striatocapsular infarction in man. Ann Neurol 1992;31(5):463–472.
42. Weiller C, Ramsay SC, Wise RJ, Friston KJ, Frackowiak RS. Individual patterns of functional reorganization in the human cerebral cortex after capsular infarction. Ann Neurol 1993;33(2):181–189.
43. Strick PL. Anatomical organization of multiple motor areas in the frontal lobe: implications for recovery of function. Adv Neurol 1988;47:293–312.
44. Porter R, Lemon RN. Corticospinal function and voluntary movement. Oxford, UK: Oxford University Press. 1993.
45. Dum RP, Strick PL. Spinal cord terminations of the medial wall motor areas in macaque monkeys. J Neurosci 1996;16(20):6513–6525.
46. Rouiller EM, Moret V, Tanne J, Boussaoud D. Evidence for direct connections between the hand region of the supplementary motor area and cervical motoneurons in the macaque monkey. Eur J Neurosci 1996;8(5):1055–1059.
47. Calautti C, Leroy F, Guincestre JY, Baron JC. Dynamics of motor network overactivation after striatocapsular stroke: a longitudinal PET study using a fixed-performance paradigm. Stroke 2001;32(11):2534–2542.
48. Calautti C, Leroy F, Guincestre JY, Baron JC. Displacement of primary sensorimotor cortex activation after subcortical stroke: a longitudinal PET study with clinical correlation. Neuroimage 2003;19(4):1650–1654.
49. Cramer SC, Shah R, Juranek J, Crafton KR, Le V. Activity in the peri-infarct rim in relation to recovery from stroke. Stroke 2006;37(1):111–115.
50. Feydy A, Carlier R, Roby-Brami A, Bussel B, Cazalis F, Pierot L et al. Longitudinal study of motor recovery after stroke: recruitment and focusing of brain activation. Stroke 2002;33(6):1610–1617.
51. Johansen-Berg H, Rushworth MF, Bogdanovic MD, Kischka U, Wimalaratna S, Matthews PM. The role of ipsilateral premotor cortex in hand movement after stroke. Proc Natl Acad Sci U S A 2002;99(22):14518–14523.
52. Seitz RJ, Hoflich P, Binkofski F, Tellmann L, Herzog H, Freund HJ. Role of the premotor cortex in recovery from middle cerebral artery infarction. Arch Neurol 1998;55(8):1081–1088.
53. Ward NS, Brown MM, Thompson AJ, Frackowiak RS. The influence of time after stroke on brain activations during a motor task. Ann Neurol 2004;55(6):829–834.
54. Dancause N, Barbay S, Frost SB, Plautz EJ, Stowe AM, Friel KM et al. Ipsilateral connections of the ventral premotor cortex in a new world primate. J Comp Neurol 2006;495(4):374–390.
55. Dancause N, Barbay S, Frost SB, Mahnken JD, Nudo RJ. Interhemispheric connections of the ventral premotor cortex in a new

world primate. J Comp Neurol 2007;505(6): 701–715.
56. Dum RP, Strick PL. The origin of corticospinal projections from the premotor areas in the frontal lobe. J Neurosci 1991;11(3): 667–689.
57. He SQ, Dum RP, Strick PL. Topographic organization of corticospinal projections from the frontal lobe: motor areas on the lateral surface of the hemisphere. J Neurosci 1993;13(3):952–980.
58. He SQ, Dum RP, Strick PL. Topographic organization of corticospinal projections from the frontal lobe: motor areas on the medial surface of the hemisphere. J Neurosci 1995;15(5 Pt 1):3284–3306.
59. Boudrias MH, Belhaj-Saif A, Park MC, Cheney PD. Contrasting properties of motor output from the supplementary motor area and primary motor cortex in rhesus macaques. Cereb Cortex 2006;16(5):632–638.
60. Maier MA, Armand J, Kirkwood PA, Yang HW, Davis JN, Lemon RN. Differences in the corticospinal projection from primary motor cortex and supplementary motor area to macaque upper limb motoneurons: an anatomical and electrophysiological study. Cereb Cortex 2002;12(3):281–296.
61. Dettmers C, Fink GR, Lemon RN, Stephan KM, Passingham RE, Silbersweig D et al. Relation between cerebral activity and force in the motor areas of the human brain. J Neurophysiol 1995;74(2):802–815.
62. Thickbroom GW, Phillips BA, Morris I, Byrnes ML, Sacco P, Mastaglia FL. Differences in functional magnetic resonance imaging of sensorimotor cortex during static and dynamic finger flexion. Exp Brain Res 1999;126(3):431–438.
63. Ward NS, Frackowiak RS. Age-related changes in the neural correlates of motor performance. Brain 2003;126(Pt 4):873–888.
64. Fridman EA, Hanakawa T, Chung M, Hummel F, Leiguarda RC, Cohen LG. Reorganization of the human ipsilesional premotor cortex after stroke. Brain 2004;127(Pt 4):747–758.
65. Lotze M, Markert J, Sauseng P, Hoppe J, Plewnia C, Gerloff C. The role of multiple contralesional motor areas for complex hand movements after internal capsular lesion. J Neurosci 2006;26(22):6096–6102.
66. Zemke AC, Heagerty PJ, Lee C, Cramer SC. Motor cortex organization after stroke is related to side of stroke and level of recovery. Stroke 2003;34(5):e23–e28.
67. Crafton KR, Mark AN, Cramer SC. Improved understanding of cortical injury by incorporating measures of functional anatomy. Brain 2003;126(Pt 7):1650–1659.
68. Marshall RS, Perera GM, Lazar RM, Krakauer JW, Constantine RC, DeLaPaz RL. Evolution of cortical activation during recovery from corticospinal tract infarction. Stroke 2000;31(3):656–661.
69. Small SL, Hlustik P, Noll DC, Genovese C, Solodkin A. Cerebellar hemispheric activation ipsilateral to the paretic hand correlates with functional recovery after stroke. Brain 2002;125(Pt 7):1544–1557.
70. Ward NS, Brown MM, Thompson AJ, Frackowiak RS. Neural correlates of motor recovery after stroke: a longitudinal fMRI study. Brain 2003;126(Pt 11):2476–2496.
71. Kleim JA, Chan S, Pringle E, Schallert K, Procaccio V, Jimenez R et al. BDNF val66met polymorphism is associated w ith modified experience-dependent plasticity in human motor cortex. Nat Neurosci 2006;9(6):735–737.

Chapter 21

fMRI in Psychiatric Disorders

Erin L. Habecker, Melissa A. Daniels, and Perry F. Renshaw

Summary

Functional neuroimaging has become an important tool for clinical research, with the potentiality to provide information on psychiatric disease pathology and treatment response. We review functional magnetic resonance imaging (fMRI) research findings for six psychiatric disorders: schizophrenia, major depressive disorder, bipolar disorder, obsessive-compulsive disorder, posttraumatic stress disorder, and Alzheimer's disease. Brain functional abnormalities and possible underlying mechanisms for disease symptoms are discussed, with a focus on future clinical implications for fMRI in psychiatric disease.

Key words: fMRI, Blood oxygen level dependent, Psychiatric disorders, Schizophrenia, Major depressive disorder, Bipolar disorder, Obsessive-compulsive disorder, Posttraumatic stress disorder, Alzheimer's disease

1. Introduction

1.1. Overview of fMRI

fMRI is a unique, noninvasive method of measuring neural activation through changes in oxidation and regional blood flow. An important clinical research tool that has been used more and more frequently in recent years, fMRI is able to indirectly detect cortical activity in the working brain, allowing for the assessment of psychiatric disease physiology and treatment effects. fMRI does not involve exposure to radioactive tracers, thus allowing patients and subjects to undergo multiple scans over a short period of time, if necessary. Most fMRI studies involve the measurement of signal arising from hydrogen nuclei *(1, 2)*. Common types of fMRI used in psychiatric neuroimaging include blood oxygen level dependent (BOLD) and arterial spin labeling (ASL).

M. Filippi (ed.), *fMRI Techniques and Protocols*, Neuromethods, vol. 41
DOI 10.1007/978-1-60327-919-2_21, © Humana Press, a part of Springer Science+Business Media, LLC 2009

Instead of incorporating a radioactive tracer as in positron emission tomography (PET) or single photon emission computed tomography (SPECT), fMRI makes use of the unique properties of hemoglobin (BOLD and BOLD contrast methods) or the water molecules of flowing blood (ASL) to produce images of neural activation. Most fMRI studies today are BOLD studies that make use of T_2^* mechanisms. ASL fMRI depends on T_1 effects, which will be further described below. Hemoglobin is present in the body in two forms: the oxygenated form, oxyhemoglobin; and the deoxygenated form, deoxyhemoglobin. T_2^* weighed images of each form of hemoglobin are distinctive because the two have different magnetic properties. Neuronal activity results in greater cerebral blood flow (CBF) to the specific brain areas involved in the processing of a particular task, leading to an increase in T_2^* signal and a more intense MR signal on the images created. Tasks and stimuli are used during fMRI to elicit a predicted brain response - they are intended to alter neural activity in brain regions thought to be impacted by the disorder in question.

1.2. Bold

The advantages of BOLD fMRI – the most common psychiatric fMRI modality – as compared with other functional imaging techniques, such as PET, include the greater sensitivity of the fMRI signal to event-related changes in neuronal blood flow and the increased spatial resolution of fMRI images *(3)*. Temporal resolution, which was historically poor in previous imaging methods, has been improved drastically through the use of high-speed MR scanners with the ability to perform echo planar imaging, acquiring single image planes in 50–100 ms *(4)*. However, one distinct disadvantage of this mode of functional imaging that must be taken into consideration during experimental construction is the inability of BOLD signal to differentiate between changes in CBF that are correlated with neuronal activity and changes that are independent of it. Such changes include activity-related signal changes in draining veins away from the brain activity *(5)*, incidental neural activations that are unrelated to the task at hand *(6)*, and changes in CBF caused by changes in respiration. Even small respiration changes can alter blood arterial carbon dioxide tension (PCO_2), which has a large effect on CBF *(7, 8)*. Subjects with an anxiety disorder, or state anxiety induced by the MRI environment, are particularly susceptible to variations in respiration, and this must be taken into account during experiment planning using BOLD. The effects of respiration changes on PCO_2 may be managed by acquiring continuous measurements of PCO_2, or end-tidal CO_2, during the experimental protocol *(3)*. When this function is available, investigators have the option of either acquiring data only during steady-state CO_2 levels *(9)*, or attempting to adjust for the effect of PCO_2 on global CBF *(10, 11)* and integrate this modification into fMRI indices.

1.3. ASL

ASL differs from BOLD in that it depends on T_1 mechanisms and the magnetic labeling of water molecules to generate images. Water molecules in flowing blood are tagged through the saturation or inversion of the longitudinal component of the MR signal *(12)*; these molecules then diffuse from capillaries into brain tissue where they alter the magnetization of the local tissue *(1)*. As blood flow into the imaging slice increases, there is a more significant difference between the magnetized condition and the control condition, during which the magnetization of arterial blood is fully relaxed *(1)*. Control and tagged images are then taken, and the difference between them is proportional to the CBF. ASL can be used to measure global CBF changes dynamically; its use of water molecules as an endogenous blood flow tracer means that the images generated by ASL are not susceptible to neurovascular changes that are not related to neuronal activation. ASL has another advantage over BOLD in that effects of frequency drifts tend to be minimized in ASL, making this method more suitable for longer duration scans *(12)*. However, BOLD acquisitions tend to have greater temporal resolution, greater maximum number of slices, and appear to be more sensitive to parametric manipulations of task demands *(1, 13, 14)*. BOLD maps also usually have larger activation areas than ASL maps *(15, 16)*, which could either be due to the decreased sensitivity or improved signal localization inherent in ASL *(13)*. There are three classes of ASL methods: pulsed ASL, continuous ASL, and velocity selective ASL *(1)*. A discussion of the relative methodologies and merits of the three techniques is beyond the scope of this review.

2. fMRI in Psychiatry

2.1. Clinical Disorders

A wide range of neuropsychiatric disorders have now been investigated using fMRI techniques and protocols. This review will explore the paradigms employed, imaging results, and future research opportunities in six mental disorders: schizophrenia, major depression, bipolar disorder (BD), obsessive-compulsive disorder (OCD), posttraumatic stress disorder (PTSD), and Alzheimer's disease (AD). These particular disorders were selected due to the fact that they represent a subset of psychotic, mood, anxiety, and late life disorders; have a significant prevalence in the general population (0.4–14% depending on age and gender of the sample); are popular candidates for fMRI research; and have each been the subject of research on diagnosis, disease progression, and treatment using imaging. **Table 1** summarizes the nature, range, and prevalence of the selected disorders in the general population.

Table 1
Summary of psychiatric disorders discussed in this chapter

Disorder	Type	Characteristic symptoms	Subtypes	Prevalence	Prognosis	Typical Treatment
Schizophrenia	Psychotic disorder	Delusions, hallucinations, disorganized speech, grossly disorganized or catatonic behavior, negative symptoms (affective flattening, alogia, aviolotion)	Paranoid, disorganized, catatonic, undifferentiated, residual	0.5–1.5% among adults	Complete remission uncommon. Some individuals display exacerbations and remissions of symptoms while others remain chronically ill, either on a stable course or a progressively worsening one	Antipsychotic medications, including clozapine, risperidone, olanzapine, quetiapine, ziprasidone, and aripiprazole
Major depressive disorder	Mood disorder	One or more major depressive episodes accompanied by changes in appetite, weight, or sleep; decreased energy; feelings of worthlessness or guilt; difficulty thinking, concentrating, or making decisions; or recurrent thoughts of death or suicide	Single episode, recurrent	5–9% among adult women, 2–3% among adult men	Major depressive episodes may end completely (2/3 of cases) or partially or not at all (1/3 of cases). Up to 15% of affected individuals die by suicide	SSRIs, tricyclics, and MAOIs
Bipolar disorder	Mood disorder	One or more manic or mixed episodes (Bipolar I) or one or more hypomanic episodes (Bipolar II) accompanied by one or more major depressive episodes. Cyclothymia characterized by periods of hypomanic and depressive symptoms	Bipolar I, bipolar II, cyclothymia, bipolar disorder not otherwise specified	0.4–5% a mong adults	Recurrent disorder: interval between episodes tends to decrease with age. 20–30% of affected individuals do not return to full functionality between episodes. Up to 15% of affected individuals die by suicide	Mood stabilizers, including lithium, valproate, carbamazepine, lamotrigine, gabapentin, and topiramate

Obsessive-compulsive disorder	Anxiety disorder	Recurrent obsessions (persistant ideas, thoughts, impulses, or images) or compulsions (repetitive behaviors) that are time consuming or cause marked distress or significant impairment. Affected adults realize that obsessions or compulsions are excessive or unreasonable	Obsessive-compulsive disorder, obsessive compulsive disorder with poor insight	1.5–2.5% among adults	Majority of individuals have chronic disease course with waxing and waning symptoms. 15% show progressive deterioration of occupational and social functioning. 5% have episodic course with minimal symptoms between episodes	SSRIs, tricyclics (clomipramine only), MAOIs, and benzodiazepines
Posttraumatic stress disorder	Anxiety disorder	Development of characteristic symp-toms following an extreme traumatic stressor. Symptoms include persistent reexperiencing of the trauma, avoidance of stimuli associated with the trauma, numbing of general responsiveness, and increased arousal	Acute, chronic, delayed onset	1–14% among adults; up to 58% for at risk populations (combat victims, victims of criminal violence, etc.)	Symptomatic onset usually occurs within 3 months after trauma. Complete recovery occurs within 3 months in 50% of cases; symptoms persist longer than 12 months in many cases	SSRIs, including sertraline and paroxetine
Alzheimer's disease	Disorder of late life	Multiple cognitive defects, including memory impairment and at least one of the following cognitive disturbances: aphasia, apraxia, agnosia, or disturbance in executive functioning	Early onset, late onset, with delirium, with delusions, with depressed mood, uncomplicated	2–4% among individuals over the age of 65	Terminal disease, with symptoms worsening over time. Typical course is early deficits in recent memory, followed by development of aphasia, apraxia, and agnosia after several years. Average length of illness from onset of symptoms to death is 8–10 years.	Cholinesterase inhibitors, including galantamine, rivastigmine, donepezil, and tacrine. One NMDA antagonist, memantine, used in moderate to severe Alzheimer's disease

MAOIs monoamine oxidase inhibitors; *NMDA* *N*-methyl D-aspartate; *SSRIs* selective serotonin reuptake inhibitors

Modern imaging techniques have been crucial to the delineation of the brain structures and functions that are negatively impacted in psychiatric disorders such as the ones reviewed here. Traditionally, such disorders have been characterized primarily via clinical psychiatric evaluation of abnormal symptoms, and treatments consist of a trial-and-error strategy combined with patient self-selection of treatment options or option combinations *(17)*. The use of fMRI to evaluate the underlying cognitive disturbances present across a heterogeneous psychiatric disorder, or even a range of psychiatric disorders, is important in that it allows for the investigation of core dysfunctions that might highlight more effective treatment options. Studying differences in neural response between psychiatric patients and normal subjects with respect to affected brain regions, and incorporating a variety of emotional and cognitive challenges to examine localized activation, investigators have the opportunity to evaluate subtle differences in the ways that the brains of patients process different types of information and perform tasks. In addition, functional imaging can be used to evaluate the efficacy of a psychiatric medication, especially in conjunction with the usual clinical symptom assessments. New findings are also highlighting the ways in which fMRI could be utilized to aid in the diagnosis of psychiatric disease *(17–19)*.

3. Psychotic Disorders

3.1. Schizophrenia

Schizophrenia is a lifelong illness associated with a high rate of morbidity and disability due to the severity and neurologically disruptive nature of its symptoms. It is often thought of as the most serious psychiatric disorder, and the afflicted population (about 1% of the general population) is impaired in one or more major areas of functioning: interpersonal relations, work or education, or self-care (American Psychiatric Association, Diagnostic and Statistical Manual of Mental Disorders, 4th edition, 1994 – DSM-IV 1994). The disorder, which typically manifests sometime in an individual's mid-twenties, is diagnosed through a number of characteristic symptoms falling into positive (an excess or distortion of normal function) or negative (attenuation or loss of normal function) categories. Positive symptoms include delusions, hallucinations, disorganized speech, and disorganized or catatonic behavior, while the negative symptoms encompass affective flattening, alogia, and aviolation (DSM-IV, 1994). Because schizophrenia is a heterogeneous disorder with a wide range of associated impairments, the range of tasks employed during fMRI studies has been similarly broad. **Table 2** summarizes the most common paradigms, which include verbal fluency *(20)*, affective pictures

Table 2
Summary of fMRI research in schizophrenia

Authors	Subjects	fMRI paradigm	Results
		Schizophrenia	
Manoach et al. *(25)*	9 schizophrenic subjects and 9 healthy controls	Working memory: Sternberg Item Recognition Paradigm adapted to include monetary reward for correct responses	Schizophrenic patients exhibited deficiencies in working memory with equal activity in the DLPFC as compared to healthy controls. Schizophrenic patients activated the basal ganglia and thalamus during the task while the healthy controls did not
Rubia et al. *(27)*	6 male medicated patients with schizophrenia and 7 matched healthy controls	Inhibitory control: "stop" and "go/no-go" tasks	Schizophrenic patients exhibited reduced left prefrontal activation as compared to healthy controls
Hempel et al. *(21)*	10 partially remitted schizophrenic patients and 10 healthy controls	Facial affect discrimination and labeling	Schizophrenic patients showed a significantly decreased activation in the anterior cingulate during the discrimination portion of the trial, and decreased activity in the amygdala-hippocampal complex bilaterally during labeling
Hofer et al. *(24)*	10 male outpatients with schizophrenia and 10 male healthy controls	Episodic encoding/recognition of words	Patients with schizophrenia demonstrated bilateral impairments in activation in the DLPFC and lateral temporal cortices
Kubicki et al. *(35)*	9 male schizophrenic subjects and 9 control subjects	Semantic encoding	Schizophrenic patients had left inferior prefrontal cortex underactivation and left superior temporal gyrus overactivation as compared to healthy controls
Habel et al. *(34)*	13 male patients with schizophrenia, 13 of their nonaffected brothers ($N = 13$), and 26 unrelated matched healthy controls	Positive and negative mood evocation	Schizophrenic patients and brothers of schizophrenic patients showed reduced amygdala activity as compared to healthy controls
Ragland et al. *(26)*	14 patients with schizo-phrenia and 15 healthy comparison subjects	Word encoding/recognition	Schizophrenic patients demonstrated prefrontal cortex underactivation and parahippocampal overactivation

(continued)

Table 2 (continued)

Schizophrenia

Authors	Subjects	fMRI paradigm	Results
Takahashi et al. *(22)*	15 schizophrenics and 15 healthy volunteers	Affective pictures	Patients with schizophrenia had decreased activity in the right amygdala and medial prefrontal cortex
Williams et al. *(23)*	27 schizophrenic patients and 22 matched healthy controls	Facial expressions of fear	Schizophrenic subjects had enhanced arousal responses coupled with reduction in amygdala and medial prefrontal activity
Honey et al. *(33)*	12 healthy volunteers	Episodic memory task. 100 ng/ml plasma ketamine or placebo	Left frontal activation was increased by ketamine when semantic processing was required at encoding. Successful encoding was supplemented by additional nonverbal processing in subjects on ketamine
Morey et al. *(32)*	52 subjects: 16 healthy controls, 10 ultra-high risk for schizophrenia, 15 with early schizophrenia, 11 with chronic schizophrenia	Visual oddball task	Frontostriatal in the ACC, inferior frontal gyrus, and the medial frontal gyrus activation associated with target stimuli in the early and chronic groups was significantly lower than the control group, while the ultra-high-risk group showed a trend toward the early group.
Yurgelun-Todd et al. *(20)*	12 schizophrenic patients	Word fluency task. 8 weeks of D-cycloserine treatment or placebo	Patients receiving D-cycloserine showed a significant increase in temporal lobe activation as compared to patients receiving placebo. This increased activation was associated with a reduction in negative symptoms
Juckel et al. *(141)*	10 male schizophrenic patients and 10 age-matched male healthy controls	Incentive monetary delay task	Schizophrenic patients showed reduced ventral striatal activation during the presentation of reward-indicating cues as compared to healthy controls. Decreased activation of the left ventral striatum was inversely correlated with the severity of symptoms
Vink et al. *(28)*	21 schizophrenic patients, 15 unaffected siblings, and 36 matched healthy controls	Inhibitory control: stop cues	Control subjects activated the striatum when responding to motor cues while schizophrenic patients and unaffected siblings did not

ACC anterior cingulate cortex; *DLPFC* dorsolateral prefrontal cortex

(21–23), working memory (WM) *(24–26)*, and inhibitory control *(27, 28)*. These studies have reported attenuation and deactivation of fMRI signal, as compared to healthy control groups, in prefrontal and temporal lobe structures including the amygdala, hippocampus, and parahippocampal gyrus. In addition, increased activation of the basal ganglia and striatum has been observed during WM and inhibitory control tasks *(25, 28)*.

fMRI assessment of cognitive verbal and memory tasks in schizophrenic patients allows for an analysis of cognitive deficits and enables investigation of the abnormal language functionality seen in some individuals with the disorder. WM performance, as commonly measured by encoding and recognition tasks, continuous performance paradigms, and the Sternberg test, to name a few, appears to be impacted by schizophrenia. Difficulties in encoding and free recall are common, indicating the possible involvement of the prefrontal cortex and hippocampus, both of which have been shown to participate in the neural mechanisms of working, episodic, and semantic memory *(29–31)*. Ragland et al. *(26)*, using a word encoding and recognition task, observed dorsolateral prefrontal cortex (DLPFC) dysfunction manifested by bilateral defects during encoding, left hemisphere hypoactivity during recognition, and right side signal attenuation during successful retrieval, as compared to a healthy control group. This finding has been replicated using a visual oddball continuous task *(32)*. In addition, marked increase in parahippocampal activation in schizophrenic patients as compared to healthy controls has been observed during task performance, suggesting a core deficit in the reciprocal connections between the hippocampus and the neocortex *(26)*. Other WM fMRI studies of schizophrenia have reported attenuated activations in the anterior cingulate cortex (ACC) and cerebellum *(33)*, and significantly increased activity in the basal ganglia and thalamus *(25)*. It has been hypothesized that frontostriatal circuitry defects could account for these deviations, and the anatomical as well as functional normality of this circuitry in schizophrenic patients has been targeted for further study *(25)*.

Investigations of emotional processing, long a part of psychiatric research into mood disorders, have begun in recent years to be used in the study of schizophrenia. fMRI protocols have made use primarily of affective facial expressions, as well as emotional pictures and words, to evoke cortical responses in the temporal and frontal lobes. Attenuation of amygdala response, as well as that of the amygdala-hippocampal complex, has been noted by several investigators in response to emotional faces, discrimination of facial affect, and the evocation of negative mood *(21–23, 34)*. Some of these same studies have also shown attenuation in the anterior cingulate and medial prefrontal cortex (mPFC) as compared to healthy controls *(21, 22)*. This reduced activation is often accompanied by increased activation in another area, such as the middle frontal gyrus (MFG) in one study *(21)*. This invites the hypothesis that the observed increased activations are secondary mechanisms evoked to compensate for the dysfunction in related areas.

Other fMRI studies of schizophrenia have focused on the neural underpinnings of the episodic memory, language, and learning deficits common in the disorder. Hofer et al. *(24)* found decreased activations in the bilateral DLPFC and lateral temporal cortices in schizophrenic patients, despite the fact that recognition performance in the schizophrenic patients was intact. A language processing task highlighted underactivity in the temporal lobe which improved after treatment with d-cycloserine, in conjunction with negative symptom improvement *(20)*. Reduced activation of the left inferior prefrontal cortex was observed, accompanied by exaggerated activation in the left superior temporal gyrus, during a semantic encoding learning task *(35)*. Findings such as these are particularly relevant in a clinical sense, as encoding, learning, and language skills are so important for normal social interaction and functioning. These studies increase our level of understanding of the neuropsychological dysfunctions that underlie some of the most disruptive symptoms of schizophrenia, and point to a need for further study to determine optimal treatment options.

4. Mood Disorders

4.1. Major Depression

Major depression is diagnosed in patients who experience one or more major depressive episodes without any associated mania. A major depressive episode occurs when a patient presents with persistent feelings of deep despair and loss of pleasure or interest in nearly all activities for at least 2 weeks, accompanied by at least four of the following symptoms: sleep disturbances, disruption of appetite, lethargy, feelings of hopelessness or worthlessness, difficulty concentrating, or suicidal thoughts (DSM IV, 1994). Individuals with major depressive disorder may present a range of heterogeneous symptoms within this framework, implying that the disease impacts more than one brain region or neurotransmitter system. Studying mood and cognition-induced brain activations in affected individuals represents a powerful way to unlock the functional discrepancies between the depressed and normal nervous system. fMRI studies have revealed a wide-range network of limbic and paralimbic neural regions and circuitry whose interactions appear to be disrupted in major depressive disorder *(36–38)*.

The limbic-cortical model of depression advanced by Mayberg et al. hypothesizes that major depression as a dysfunction among discrete, but functionally integrated pathways in the dorsal, ventral, and rostral compartments of the brain *(17)*. Respective dysfunctions in components of each of these compartments, which include the dorsolateral and dorsomedial prefrontal cortex, dorsal anterior cingulate, and posterior cingulate (dorsal compartment); subgenual anterior cingulate, ventral prefrontal cortex, insula, hippocampus, and amygdala (ventral compartment); and rostral ACC (rostral compartment), can all be associated with the collection symptoms seen in major depression **(Fig. 1)** *(38)*. In addition, it has been theorized that the failure of the healthy elements in the system to maintain homeostasis of emotionality during times of stress when a part of the system is compromised is a contributor to major depressive episodes *(17)*. Depressed subjects demonstrate abnormalities in regional cerebral blood flow (rCBF) and regional cerebral glucose metabolism (regional cerebral metabolic rate for glucose, rCMRglc) in the dorsal and ventral compartments. Decreases rCBF and rCMRglc that have been observed in the dorsolateral prefrontal cortex *(37, 39)*, dorsomedial and dorsal anterolateral prefrontal cortex, as well as the dorsal anterior cingulate *(39, 40)* in depressed subjects during PET and SPECT studies have highlighted these areas for exploration with fMRI BOLD paradigms – as do the observed increases in rCBF and rCMRglc that have been found in components of the ventral compartment, including the ventrolateral, ventromedial, orbitofrontal cortex (OFC), the sugenal prefrontal cortex, the amygdala, and the insular cortex *(41, 42)*.

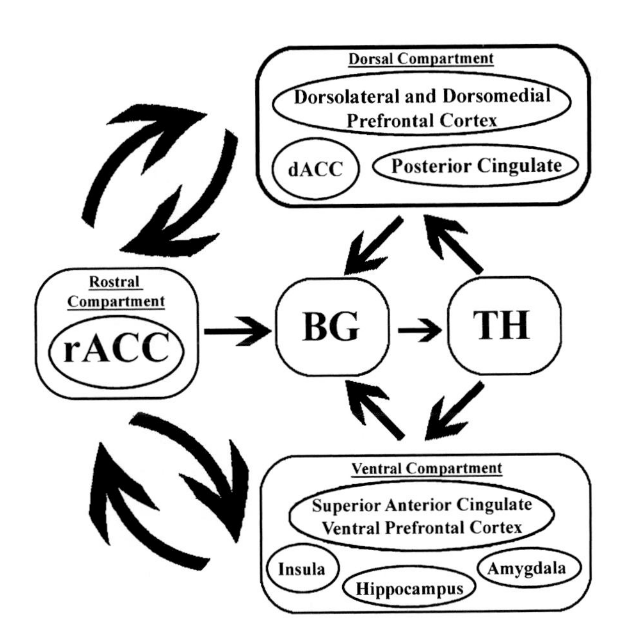

Fig. 1. A limbic-cortical model of depression adapted and modified from Mayberg et al. *(38)*. It involves three compartments: a dorsal, a ventral, and a rostral compartment. *dACC* dorsal anterior cingulate; *rACC* rostral anterior cingulated.

The amygdala has been extensively investigated for its involvement in the processing of emotional stimuli and abnormal response in individuals with major depressive disorder. Emotional paradigms used in these fMRI studies include the exhibition of emotional film clips, facial photographs, or the presentation of audio cues during the scan to induce feelings of sadness, happiness, or fear *(4)*. During negative emotional tasks (negative words and sad faces), it has been shown that the amygdala in the depressed brain displays abnormally sustained activations as compared to the amygdala in the normal brain *(19, 43, 44)*. Other studies utilizing emotional facial expression stimuli with depressed subjects have also shown increased activation in the hippocampus, left parahippocampal gyrus, and other regions of the left brain **(Table 3)** *(19, 45)*, as well as increased dynamic range bilaterally in the cerebellum and anterior cingulate gyrus extending to the rostral prefrontal cortex in response to sad faces, as compared to healthy controls *(19)*.

Cognitive disturbances associated with depression, such as concentration difficulties, explicit memory impairment, and impairment in executive functioning, have been examined with fMRI paradigms that incorporate various WM and executive control tasks. In comparison to healthy control subjects, depressed individuals demonstrate slower reaction times and decreased accuracy with regards to executive control challenges in conjunction with decreased dorsolateral prefrontal cortex activity *(44)*. Interestingly, depressed subjects were also shown, in another study, to have increased activation in the dorsolateral prefrontal cortex in response to cognitive load in a WM task *(46)*.

Ideally, identifying brain regions and circuitry impacted by depressive disorder will lead to a greater understanding of the effects of drug treatment and yield important information regarding the feasibility and success of various treatment options. In order to delineate the neurocircuitry involved in processing emotional cues and gather information about the pharmacodynamics of various antidepressants, studies have been undertaken that examine the effects of antidepressant treatment in a variety of scenarios. PharmacoMRI (pMRI) studies of a single dose of an antidepressant with healthy control subjects allow any focal changes in brain activity induced by the drug to be observed during BOLD scanning. Pre- and posttreatment studies of antidepressants incorporate structural MRI and fMRI to combine activation paradigms with antidepressant treatment and identify brain functional correlates of antidepressant treatment and symptomatic response *(19, 47)*. Two studies investigating, respectively, the effects of citalopram and mirtazapine on the healthy nervous system in a pMRI format found that each antidepressant enhanced activations right OFC during a Go/No-go task *(48, 49)*. In addition, one study incorporating the emotional faces paradigm found

Table 3
Summary of fMRI research in mood disorders

Major depressive disorder			
Authors	**Subjects**	**fMRI paradigm**	**Results**
Siegle et al. *(43)*	7 depressed and 10 never-depressed subjects	Alternating 15 s emotional processing tasks and nonemotional processing trials	Depressed subjects displayed sustained amygdalar responses to negative words that lasted throughout the following nonemotional processing trial (25 s later). Never-depressed subjects displayed amygdalar responses to all emotional stimuli, decaying within 10 s
Fu et al. *(19)*	19 medication free, acutely depressed subjects and 19 matched healthy volunteers	Facial expressions of sadness: low, medium and high intensity. Subjects asked to identify sex of face; accuracy and reaction time measured	Depressed subjects had increased capacity in regions of the left brain: hippocampus, amygdala, parahippocampal gyrus, insula, caudate nucleus, thalamus, dorsal cingulate gyrus, inferior parietal cortex. Depressed subjects had increased dynamic range in bilateral cerebellum and anterior cingulate gyrus extending bilaterally to rostral prefrontal cortex
Surguladze et al. *(45)*	14 healthy controls and 16 individuals with major depressive disorder	Facial expressions: happy and sad	Healthy individuals only displayed linear increases in response in bilateral fusiform gyri and right putamen to expressions of increasing happiness. Depressed subjects demonstrated linear increases in response in left putamen, left parahippocampal gyrus/amygdala, and right fusiform gyrus to expressions of increasing sadness
Del-Ben et al. *(48)*	12 healthy male subjects, single blind crossover design	Go/No-go, Loss/No-loss, covert (averse) face emotion recognition. 7.5 mg IV citalopram or placebo	Prefrontal and subcortical regions activated by tasks consistent with previous studies (BA47). Citalopram enhanced activations in the right BA47 during Go/No-go but attenuated BA47 response to aversive faces. Right amygdala response to aversive faces attenuated by citalopram. BA11 response to Loss/No-loss attenuated by citalopram
Wagner et al. *(142)*	16 patients with unipolar depression; 16 matched healthy controls	Adapted version of Stroop task	No differences in reaction time and accuracy between groups. Hyperactivity in the rostral anterior cingulate gyrus and left dorsolateral prefrontal cortex seen in depressed patients during interference condition

(continued)

Table 3 (continued)

Major depressive disorder

Authors	Subjects	fMRI paradigm	Results
Siegle et al. *(44)*	27 unmedicated, unipolar depressive subjects and 25 never-depressed healthy controls	Executive control task: digit sorting; emotional information processing: personal relevance rating of words	Depressed subjects displayed sustained amygdala reactivity on emotional tasks and decreased dorsolateral prefrontal cortex activity on the digit-sorting task
Vollm et al. *(49)*	45 healthy male subjects, double blind, placebo controlled	Go/No-go, Reward/No-reward, Loss/No-loss. Mirtazapine or placebo	Task activations consistent with previous findings. During behavioral inhibition, enhanced activations in the right orbitofrontal cortex with mirtazapine during Go/No-go and Reward/No-reward. Increased activations in bilateral parietal cortex with mirtazapine during Reward/No-reward
Walter et al. *(46)*	12 partially remitted, medicated inpatients with major depressive disorder; 17 healthy control	Working memory: delayed match to sample	Depressed patients slower and less accurate in task. Depressed patients exhibited more activation in the left dorsolateral prefrontal cortex during highest cognitive load; more activation in the ventromedial prefrontal cortex during control condition
Bipolar disorder			
Yurgelun-Todd et al. *(62)*	14 BD subjects and 10 healthy controls	Faces: happy and fearful affect recognition paradigm	BD subjects had reduced DLPFC activation and increased amygdala response to fearful facial affect
Blumberg et al. *(53)*	36 BD subjects: 11 elevated mood state, 10 depressed mood state, 15 euthymic mood state; 20 matched healthy controls	Color-word Stroop task	Elevated mood group: small signal increase in right frontal cortex. Depressed mood group: large signal increase in the left frontal cortex. BD subjects showed somewhat increased activation in rostral region of left VPFC, independent of mood state
Adler et al. *(57)*	12 euthymic BD patients (8 medicated) and 10 healthy controls	Working Memory: Two-back task, zero-back control/attention task	BD patients performed more poorly than healthy controls. BD patients had increased activation in fronto-polar prefrontal cortex, temporal cortex, basal ganglia, thalamus, and posterior parietal cortex. No attenuated activations observed

Chang et al. *(58)*	12 young (12–18 years old) male BD subjects off medication for 24 h; 10 age-matched healthy controls	Working Memory: two-back task	BD subjects had greater activations in bilateral ACC, left putamen, left thalamus, left DLPFC, and right inferior frontal gyrus than control group
Gruber et al. *(59)*	11 BD patients on stable pharmacotherapy regimen; 10 healthy controls	Stroop test	BD patients had reduced activations in the right AAA subdivision of the ACC and an increase in DLPFC activity during task as compared to healthy controls
Lawrence et al. *(61)*	12 euthymic BD patients, 9 major depressive disorder patients, and 11 healthy controls	Faces: fear, happiness, and sadness	BD patients had increased activations in the ventral striatal, thalamic, hippocampal, and ventral prefrontal cortical areas in response to intense fear, mild happiness, and mild sadness
Malhi et al. *(65)*	10 hypomanic-state female subjects with BD, 10 matched healthy controls	Negative, positive, or neutral captioned pictures	BD patients had increased activations in the caudate and thalamus in response to negative-captioned pictures
Malhi et al. *(66)*	10 depressed-state female subjects with BD, 10 matched healthy controls	Negative, positive or neutral captioned pictures	BD patients had increased reactions in the amygdala, thalamus, hypothalamus, and medial globus pallidus in response to positive-captioned pictures
Monks et al. *(55)*	12 male BD subjects on lithium carbonate monotherapy and 12 healthy controls	Working Memory: Two-back task and Sternberg test	BD group had attenuated activity in bilateral frontal, temporal, and parietal regions; increased activity in left precentral, right medial frontal and left supramarginal gyri as compared to healthy controls during two-back task
Blumberg et al. *(63)*	17 BD patients, 5 unmedicated, and 17 healthy controls	Faces: fear, happiness, sadness, and neutral	Increased amygdala activation in all BD patients, greatest activations observed in unmedicated patients. Rostral anterior cingulate activations attenuated in unmedicated patients as compared to healthy controls and medicated patients
Frangou et al. *(56)*	7 euthymic BD subjects (five women, two men) on monotherapy and seven healthy controls	Working Memory: N-back task and Iowa Gambling task	BD patients exhibited decreased activity in VPFC bilaterally and in the left DPFC

(continued)

Table 3 (continued)

Major depressive disorder			
Authors	Subjects	fMRI paradigm	Results
Strakowski et al. *(51)*	16 euthymic BD patients and 16 healthy controls	Stroop test	Impaired task performance in BD subjects. BD patients had greater activation as compared to healthy controls in medial occipital cortex; less activation in temporal cortical regions, middle frontal gyrus, putamen, and midline cerebellum
Pavuluri et al. *(64)*	Ten euthymic, unmedicated BD subjects and ten matched healthy controls	Faces: angry, happy, and neutral	BD patients had reduced right rostral ventrolateral prefrontal cortex activation and increased activity in right pregenual anterior cingulate, amygdala, and paralimbic cortex in response to angry and happy faces as compared to healthy control group
Lagopoulos et al. *(54)*	Ten euthymic BD patients and ten healthy controls	Working memory	BD patients exhibited attenuated or absent activations of frontal brain regions, including the DLPFC, superior frontal gyri, anterior cingulate gyri, and intraparietal sulcus. BD patients activated the inferior frontal gyrus during task conditions and had slower reaction times

AAA anterior amygdaloid area; *ACC* anterior cingulate cortex; *BD* bipolar disorder; *DLPFC* dorsolateral prefrontal cortex; *VPFC* ventral prefrontal cortex that the administration of a serotonergic drug attenuated right amygdala response to aversive faces *(48)*.

Pre- and posttreatment studies of depressed subjects have posed the question of whether clinical response to antidepressant drugs can be predicted by indicators present at the baseline scan. One study found that more positive activation in the ACC at the baseline scan (in response to facial affect processing) was associated with faster rates of symptom improvement as measured by Hamilton Depression Rating Scale *(50)*. This finding is in need of replication, but clear differences between pre- and post-treatment have been shown in the fMRI results: significantly attenuated activations after treatment were seen in limbic-subcortical systems, including the amygdala, that had shown enhanced activations in depressed subjects at baseline in response to an emotional faces task *(19)*. In addition, as the treatment decreased this capacity for over-activation in the limbic-subcortical region, a corresponding increase was seen in prefrontal cortex activation in response to the highest levels of affective load *(19)*. This could be explained by the fact that treatment exposure induced changes in mood state have a proposed association with reciprocal changes in limbic-subcortical systems and frontoparietal circuitry: as limbic-subcortical regions activations to sadness are selectively lowered by drug treatment, greater dynamic range is available for high levels of affective load and increased activation in the prefrontal cortex is seen *(17)*. These results have great implications for the future of fMRI studies in major depressive disorder: in theory, it should be possible in the years to come to use quantitative measurements of brain function to determine optimal treatment and predict treatment response patterns for a person presenting with a major depressive episode, thereby increasing positive outcomes and chances of eventual recovery *(17)*.

4.2. Bipolar Disorder

BD, a prevalent neuropsychiatric illness manifesting as depressed and manic episodes in affected individuals, is among the leading worldwide causes of disability (DSM IV, 1994). Bipolar depressed patients exhibit symptomatology that overlaps with that of unipolar depressed patients: feelings of despair, lack of motivation and goal-setting behavior, social isolation lethargy, and sleep disturbances. Bipolar patients in the manic state, meanwhile, experience elevated mood, heightened energy levels, altered thought processes, and sometimes irritability, while bipolar euthymic individuals show neither depressive or manic symptoms (DSM IV, 1994).

Functional imaging studies coupled with analysis of bipolar clinical manifestations have led to the hypothesis that, much like depression, the mechanism of the disorder involves abnormalities in the limbic, frontal, and subcortical cortical areas, perhaps caused by a dysfunction in the neural networks present in these regions *(51, 52)*. Anterior limbic networks, incorporating the amygdala, ACC, DLPFC, and midline cerebellum, have been implicated in this disorder, as these areas contribute to behavioral functions seen to be abnormal in individuals with BD *(51)*. Functional imaging studies have been performed on bipolar patients in the euthymic state, the manic state, and the depressed state to compare cortical activations across groups, and found comparatively increased activations in prefrontal and dorsal ACC in all states that were not consistent bilaterally *(53)*. There was evidence of a small increase in signal on the right side of the ventral prefrontal cortex for patients in the manic state, while patients in the depressed state showed much greater signal increases as compared to healthy controls and manic patients in the left ventral prefrontal cortex *(53)*. This indicates that differential signal changes across brain hemispheres may be connected with the type of mood episode experienced by the bipolar individual. Traditionally, two types of tasks have been used to generate neural activations thought to be altered in BD: fronto-executive function cognitive studies and emotional processing studies *(51–53)*, as bipolar patients are known to have emotional regulatory impairments and impaired cognitive control.

Cognitive tasks employed in fMRI study of BD include tests of WM, interference tasks, encoding tasks, and other performance related paradigms *(52)*. Lagopoulos et al. *(54)*, noting that deficits in WM seem to be of particular significance in bipolar individuals, examined cortical activations in euthymic bipolar patients and healthy controls during a parametric WM task with three load conditions. As compared to the healthy subjects, subjects with bipolar exhibited attenuation of activation across several frontal brain regions. The DLPFC, which was activated across all WM conditions in healthy controls, failed to activate under the same conditions in bipolar patients, although subjects with bipolar did not have significantly poorer task performance. As the group noted that bipolar subjects recruited the inferior frontal gyrus during all WM components, while healthy subjects did not, it was theorized that this could represent a compensatory mechanism for normal DLPFC performance *(54)*. This group also noted a failure of BD patients to activate the parahippocampal gyrus during the delay condition. It should be noted that these cognitive defects were observed in euthymic bipolar patients, suggesting that BD-associated cognitive defects are not restricted to state conditions of mania or depression. Similarly, Monks et al. *(55)* found that euthymic bipolar patients on lithium therapy showed reduced activations bilaterally in frontal, parietal, and temporal regions during two WM tasks, coupled with increased activations as compared to the control group in the left precentral, right medial frontal, and left supramarginal gyri. This and similar data *(56)* suggest that fronto-executive region function is compromised during WM tasks in bipolar patients, which could lead to the recruitment of other neural areas in task performance. Two other studies, also using WM, both found exaggerated task-induced activations in the left DLPFC, ACC, and thalamus of bipolar patients *(57, 58)*. Unification of this data will require further fMRI study with increased, heterogeneous sample sizes, but all the studies consistently point to an underlying dysfunction in the region of the prefrontal-subcortical circuitry *(58)*.

Stroop tasks have been utilized in several fMRI studies of BD to examine variations in local activations induced by cognitive interference. As mentioned above, Blumberg et al. *(53)* found bilateral differences in the cortical activations of bipolar patients that were state-dependent and significantly different from those of healthy controls. In addition, Gruber et al. showed that stable bipolar patients had reduced signal intensity in the right anterior amygdaloid area (AAA) subdivision of the ACC, accompanied by an increase in DLPFC activation during the interference condition in what was hypothesized to be a compensatory manner *(59)*. In a study of medicated bipolar subjects, unmedicated bipolar subjects, and healthy controls, both groups of patients were seen to exhibit relatively increased activations as compared to the control group in the medial occipital cortex, as well as reduced activations in the temporal cortical regions, MFG, putamen, and midline cerebellum *(60)*. These studies illustrate the various differences in neural structures involved in interference processing in the bipolar vs. healthy brain, as well as differences in magnitude of MR signal intensity in identical areas.

Emotional processing studies of BD involve the evocation of transient mood reactions by presenting subjects in the scanner with cues, such as charged facial expressions or auditory stimuli. Four fMRI paradigms involving the presentation of some combination of happy, fearful, sad, and neutral faces yielded varied results. Lawrence et al. *(61)* noted increased subcortical and ventrolateral PFC activation to all categories of emotional expression, as compared to the healthy group and a group with major depressive disorder. Yurgelun-Todd et al. *(62)* found increased amygdala activation in BD patients in response to fearful facial affect, accompanied by a reduction in DLPFC signal. Individuals with BD also demonstrated an impaired ability to identify fearful faces as compared to their ability to identify faces carrying positive emotion. An increased amygdala response was also found by Blumberg et al. *(63)*, this time in response to happy faces, as well as decreased rostral anterior cingulate activation in unmedicated BD patients–medicated BD patients, meanwhile, exhibited an attenuation of emotional response differences across the two groups, demonstrating that mood-stabilizing medications have the ability to ameliorate BD-induced functional abnormalities. Increased activity in the right amygdala, right pregenual anterior cingulate, and paralimbic cortex was noted by Pavuluri et al. *(64)*

in pediatric BD in response to faces displaying both a positive and a negative emotional state. Face stimuli presentation to bipolar patients, then, appears to elicit a range of abnormal frontotemporal responses, with consistent overactivation of the amygdala seen across several studies.

State-dependent differences in brain activation in response to emotional cues have been investigated using charged pictures as well as facial affect paradigms. Malhi et al. *(65)* showed positive, negative, and reference captioned pictures to hypomanic and depressed female patients, finding that the hypomanic patients restricted response to the negative-captioned pictures to subcortical regions while healthy controls displayed a more widespread pattern of cortical activation. In depressed-state patients, positive-captioned pictures significantly increased similar subcortical region reactions, including activations in the thalamus and amygdala *(66)*. The depressed patients also showed relatively increased right-side brain activity as compared to the healthy control group *(66)*. These results suggest that subcortical limbic systems are involved to a much greater extent in emotional processing in the bipolar hypomanic and the bipolar depressed individual, as well as further illustrating the need to study euthymic, manic, and depressed bipolar subjects as separate groups.

BD involves dysfunction in several key limbic and cortical networks, evidence for which is summarized above. One other additional feature reflected by BD fMRI research is the consistent findings of abnormal PFC activations across state and trait boundaries. Similar findings in major depressive disorder and schizophrenia could indicate that the dysfunction caused by BD may share certain underlying characteristics with other psychiatric disorders *(52)*.

5. Anxiety Disorders

5.1. Obsessive-Compulsive Disorder

OCD is a complex and clinically heterogeneous disorder characterized by obsessions (intrusive, unwanted, and repetitive thoughts) which prompt compulsions (repetitive behaviors) intended to neutralize the distress generated by the obsessions (DSM IV, 1994). The etiology of OCD consists of four common symptom dimensions which can exist independently of one another or overlap within the same patients: contamination/washing, obsessions/checking, hoarding, and symmetry/ordering (DSM IV, 1994). The intensity of symptoms is generally varied throughout a patient's lifetime, but complete and spontaneous remission is rare *(67)*. Some studies have pointed to an association between different dimensions of the disorder and different treatment responses: in particular, the hoarding impulse has been associated with poorer behavioral and pharmacologic treatment response *(68)*. These findings illustrate the heterogeneity of the disease; however, fMRI and other neuroimaging studies have tended to group together patients with a range of symptoms, allowing for the delineation of underlying neural mechanisms present across all categories.

Resting-state studies of functional neuroanatomy using PET and SPECT have uncovered elements of the neurobiology of OCD that should be taken into consideration when undertaking and analyzing results of fMRI research. Specifically, examinations of CBF in OCD patients during a resting state have revealed that these patients exhibit increased metabolism in the OFC and head of the caudate nucleus compared to healthy controls *(67)*. In addition, one study involving OCD patients with comorbid major depression showed reduced CBF in the hippocampus, caudate, and thalamus as compared to both the group of "pure" OCD patients and the control group *(69)*. These resting state variations should be taken into account when conducting BOLD research on this disorder, as the sensitivity of fMRI to changes in brain metabolism that are independent of task-evoked activation can be a significant confounding factor *(3)*. Accordingly, fMRI techniques have been used to investigate a number of states in patients with OCD (**Table 4**): studies comparing local activations in the brains of OCD patients and healthy controls during cognitive tasks; pre- and posttreatment studies; and symptom-provocation studies during which transient OCD-related anxiety symptoms are incited through pictures or contact with "contaminated" objects *(70, 71)*. This research has generally supported the involvement of frontal-striatal-thalamic-cortical circuitry in OCD symptomatology, with OCD patients demonstrating functional deviations from healthy controls in the affected brain regions *(70, 72, 73)*.

Cognitive challenge studies examine abnormal activations in the brains of OCD patients as compared to healthy controls during a variety of learning and inhibition control tasks. The proposal of the frontal cortex and striatum as possible sites of dysfunction in the disease suggests the use of tasks that have previously been found to require processing by the frontal and subcortical systems during performance by healthy subjects *(74)*. Using a Tower of London task, van den Heuvel et al. *(75)* found decreased frontal-striatal responsiveness in OCD patients as compared to the control group, noting that this was accompanied by increased involvement of the ACC, the ventrolateral prefrontal cortex, and the parahippocampal cortex in a possibly "compensatory" mechanism. Roth et al., using a response inhibition "Go/no-go" task, demonstrated that the OCD group had reduced activations in the right thalamus during response inhibition *(73)*,

Table 4
Summary of fMRI research in anxiety disorders

Obsessive-compulsive disorder

Authors	Subjects	fMRI paradigm	Results
van den Heuvel et al. *(75)*	22 nonmedicated OCD patients and 22 healthy controls	Tower of London	Decreased frontal-striatal responsiveness observed in OCD patients, mainly in dorsolateral prefrontal cortex and caudate nucleus. OCD patients showed increased involvement of anterior cingulate, ventrolateral prefrontal, and parahippocampal cortices as compared to healthy controls
Remijnse et al. *(72)*	20 nonmedicated OCD patients and 27 healthy controls	Reversal learning task	Patients with OCD had a reduced number of correct responses relative to control subjects but had similar responses to receipt of punishment and demonstrated normal affective switching. Patients had reduced activations in the right medial and lateral OFC and in right caudate nucleus. Patients recruited the left posterior OFC, bilateral insular cortex, bilateral dorsolateral and bilateral anterior prefrontal cortex to a lesser extent than control subjects
Roth et al. *(73)*	12 adults with OCD and 14 healthy control subjects	Response inhibition: Go/No-go	During response inhibition, healthy controls demonstrated right-hemisphere activation while the patient group showed more diffuse and bilateral pattern of activation. The OCD group had less activation than comparison group during response inhibition in several right-hemisphere regions. Severe OCD symptoms were positively correlated with thalamic and posterior cortical activations and inversely correlated with right orbitofrontal and anterior cingulate gyri activations
Posttraumatic stress disorder			
Rauch et al. *(89)*	Eight (male, right handed) Vietnam combat veterans with PTSD and eight Vietnam combat veterans free from PTSD	Masked-fearful vs. masked-happy faces	Subjects with PTSD had an increased amygdalar response to the masked-fearful faces as compared to control group and to the PTSD group's responses to the masked-happy faces

(continued)

Table 4 (continued)

Obsessive-compulsive disorder

Authors	Subjects	fMRI paradigm	Results
Lanius et al. *(93)*	Nine traumatized subjects with PTSD, nine traumatized subjects without PTSD	Script-driven symptom provocation	PTSD subjects had less activation of the thalamus, anterior cingulate gyrus, and medial frontal gyrus as compared to healthy controls
Shin et al. *(104)*	Eight Vietnam veterans with PTSD, eight Vietnam veterans without PTSD	Emotional counting stroop task	PTSD group exhibited diminished response in rostral anterior cingulate cortex
Hendler et al., 2003 *(96)*	21 male veterans, 10 with PTSD, and 11 without	Parametric factorial design with combat slides and noncombat slides	Increased activation in amygdala in PTSD group in response to all images, increased activation in visual cortex of PTSD group in response to combat content
Lanius et al. *(103)*	Ten traumatized subjects with PTSD, ten traumatized subjects without PTSD	Script-driven symptom provocation	PTSD subjects had less activation of the thalamus and the anterior cingulate gyrus as compared to healthy controls
Driessen et al. *(97)*	12 traumatized female patients with BPD, 6 with PTSD and 6 without PTSD	Autobiographical recall of traumatic vs. negative but nontraumatic events	Subgroup without PTSD: predominant bilateral activation of orbitofrontal cortex and Broca's area. Subjects with PTSD: predominant activation of right anterior temporal lobes, mesiotemporal areas, amygdala, posterior cingulate gyrus, occipital areas, and cerebellum
Protopopescu et al. *(98)*	11 patients with assault-related PTSD, 21 healthy controls	Trauma and nontrauma-related emotional words	PTSD patients had increased initial amygdala response to trauma-related emotional words. PTSD patients did not become habituated to negative stimuli
Shin et al. *(105)*	13 traumatized subjects with PTSD, 13 traumatized subjects without PTSD	Emotional faces	The PTSD group exhibited increased amygdala response and diminished medial prefrontal cortex response to fearful facial expressions. The PTSD group showed decreased ability to habituate right amygdala response to fearful faces

BPD borderline personality disorder; *OCD* obsessive-compulsive disorder; *OFC* orbitofrontal cortex; *PTSD* posttraumatic stress disorder a finding consistent with a significant body of literature reporting structural *(76, 77)* and functional *(69, 78)* thalamus abnormalities in OCD patients. This study also reported a reduced activation of the right OFC and dorsal cingulate gyrus in patients with the most severe symptoms during response inhibition, while another group also observed reduced activations in the right OFC and in the right caudate nucleus during a reversal learning task *(72)*. These findings support the involvement of the OFC, thalamus, and cortical circuitry in the abnormal patterns of response inhibition that characterize OCD, as well as indicating that some of the behavioral impairments associated with the disorder may be attributed to dysfunction in this region *(72)*.

Treatment studies in OCD incorporating fMRI to track local activation changes across a course of medication have traditionally combined symptom-provocation tasks with longitudinal study designs *(70)*. The caudate nucleus has shown decreased glucose metabolism following treatment with serotonin reuptake inhibitors (SRIs) such as clomipramine and fluoxetine, suggesting that the right anteriolateral OFC plays a role in the mediation of OCD symptoms and response of OCD patients to pharmacotherapy *(79)*. During symptom-provocation experiments, increased activations have been observed in the OFC, cingulate cortex, striatum, thalamus, lateral prefrontal cortex, amygdala, caudate, and insula among unmedicated OCD patients *(72, 79)*. These results add support to the theory that dysfunctions of the OFC and frontal-subcortical circuitry are responsible for a great extent of OCD symptomatology. One theory with the potential to unify a great deal of functional imaging data to date involves the orbitofrontal-subcortical circuitry, which has classically been described as having a "direct" and an "indirect" pathway (**Fig. 2**) *(80)*. It has been hypothesized that OCD symptoms could be caused by a captured signal in the direct pathway creating a positive feedback loop and increasing activity in the OFC, ventromedial caudate, and medial dorsal thalamus – leading to an excessive fixation on issues of hygiene, order, danger, violence, and sex coupled with an inability to distract oneself from these thoughts or change behavior patterns *(79)*. Interventions that alter and functional imaging experiments that study the direct–indirect pathway balance within the orbitofrontal-subcortical circuits would be particularly beneficial to the future of OCD research, as they could directly test these theories and help to advance the understanding of OFC functionality in the brains of OCD patients *(79)*.

5.2. Posttraumatic Stress Disorder

PTSD is an anxiety disorder caused by the onset of an extreme stressor such as combat, childhood physical/sexual abuse, motor vehicle accidents, rape, and natural disasters. Symptoms vary across subtype of disorder and can include one or several of the following: hypervigilance, sleep disturbance, intrusive memories,

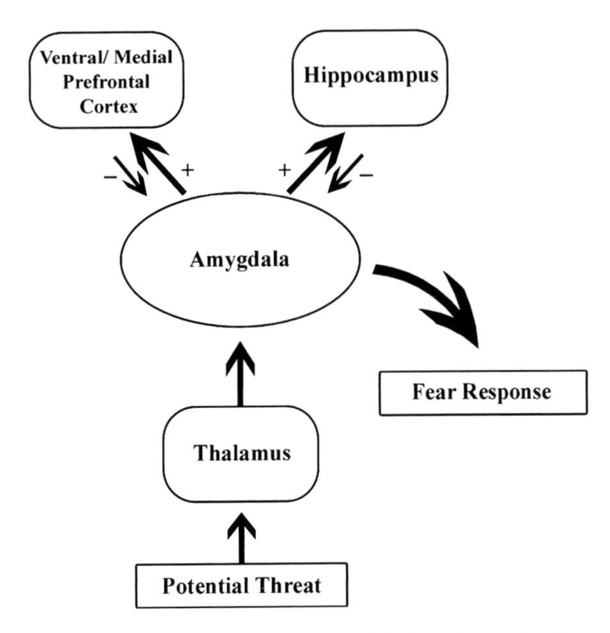

Fig. 2. The amygdalocentric neurocircuitry model of posttraumatic stress disorder (adapted from **ref.** ***80***). Excitatory connections are labeled "+"; inhibitory connections are labeled "–".

flashbacks, avoidance of traumatic stimuli, numbing of emotions, and social dysfunction (DSM IV, 1994).

There are four basic neural mechanisms that appear to function abnormally in the brains of patients with PTSD: the fear response, fear extinction, behavioral sensitization, and memory *(81)*. Fear response in PTSD patients appears greatly exaggerated: in the normal brain, the pairing of potentially dangerous stimuli with fear is an important survival mechanism, decreasing response time and initiating fight or flight mechanisms when such a stimuli is presented. In the brain of a PTSD patient, it seems that there is an overgeneralization of danger cues such that nonthreatening stimuli is seen as dangerous and can then be linked to past traumatic memories, which come hand-in-hand with intrusive memories and flashbacks. As a result of such nonthreatening yet triggering stimuli, patients with PTSD often exhibit avoidance of such stimuli or numbing of emotional reactions *(82)*. In functional neuroimaging studies of PTSD, symptom-provocation paradigms measure brain activity when subjects are exposed to visual or auditory stimulation reminiscent of their past experienced trauma to determine the mechanism of the abnormal response *(2)*. The main structure involved in the response to fearful stimuli is the central nucleus of the amygdala, with additional involvement by the sensory cortex, thalamus, and the mPFC *(82)*. PTSD patients also often exhibit a failure of fear extinction: normally, the brain is able to process stimuli from a dangerous situation from which there were no adverse outcomes such that the representation of these stimuli elicits a smaller fear response than the initial situation. However, in patients with PTSD, repeated encounters with fearful stimuli can continue to result in a consistent, heightened fear response, regardless of the actual danger of the situation *(2)*. The main neuroanatomical structures involved in the extinction of fear response overlap with those involved in fear conditioning, and their functionality may be studied concurrently *(83)*.

PTSD patients often have increased sensitivity to stress, which leads to an increase in responses such as arousal and vigilance in response to stressful stimuli *(81)* known as behavioral sensitization. The neuroanatomy of the stress response is not centralized but involves a wide range of structures and mechanisms – many of which overlap with those involved in fear conditioning and fear response *(84)*. Many of those afflicted with PTSD also exhibit memory deficiencies thought to be connected to hippocampal function and reduction in hippocampal volume *(85)*. It is unclear whether a stressful event leading to PTSD causes volume reductions in the hippocampus through exposure to elevated glucocorticoids accompanied by a reduction of brain-derived neurotrophic factor *(86, 87)*, or if persons with reduced hippocampal volume from birth are simply more prone to developing PTSD *(88)*. However, hippocampal volume and functionality is an important area of PTSD study.

FMRI research into PTSD has focused on the examination of the brain structures outlined above using symptom-provocation paradigms to examine fear response and stress sensitivity, as well as cognitive task paradigms to investigate memory deficiencies and other cognitive problems potentially associated with the disorder. It has been theorized that neurocircuitry links the amygdala to the mPFC, the hippocampus, and the thalamus through excitatory and inhibitory connections that are dysfunctional in PTSD *(80)* (**Fig. 3**). Consistent with this, fMRI studies have found alterations in BOLD signal in OFC, ACC, anterior temporal cortex, and amygdala *(89, 90)*, as well as in the hippocampus, parahippocampus, and thalamus *(2, 91, 92)*. Symptom-provocation studies find exaggerated amygdala responses and decreased activation within medial frontal areas *(93–95)*. Rauch et al. *(89)* used an fMRI paradigm incorporating a happy vs. fearful faces task in healthy combat veterans and combat veterans suffering from PTSD. The study found increased activation of the amygdala in the PTSD group in response to fearful faces, which could be positively correlated with PTSD symptom severity. Similarly, Hendler et al. *(96)* presented combat and noncombat related slides to PTSD and non-PTSD Israeli soldiers, and found that activity in the amygdala was significantly increased, although their results demonstrated this increased activity regardless of whether the

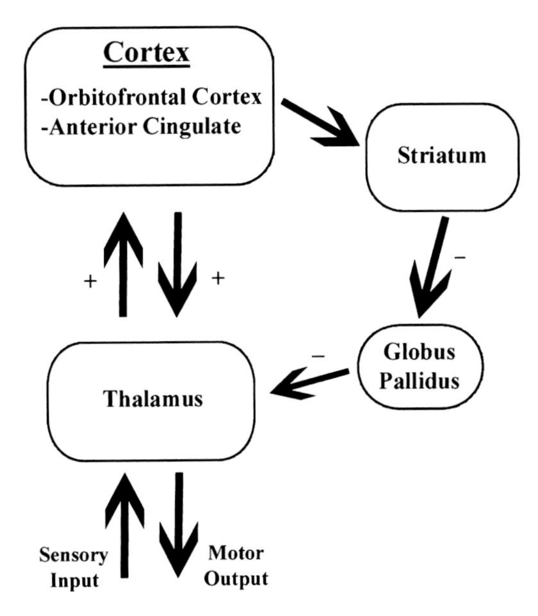

Fig. 3. The cortico-striatal model of obsessive-compulsive disorder (adapted from **ref.** ***80***). The striatum projects via direct and indirect pathways through the globus pallidus to the thalamus, which, in turn, projects to the neocortex. Excitatory connections are labeled "+"; inhibitory connections are labeled "−".

slide presented was traumatic (combat-related) or nontraumatic (noncombat-related). One study presenting autobiographical cues to borderline personality disorder patients with and without PTSD found increased activation in the amygdala in the group with PTSD only *(97)*, and another showed increased amygdala activation in response to traumatic stimuli correlating with PTSD symptom severity *(98)*.

The inappropriate fear responses associated with PTSD may be linked to dysfunction of the mPFC, as well as the amygdala. The mPFC is associated with fear response and fear extinction *(99–101)*, and it has been reported that mPFC lesions will disrupt emotional regulation in animals *(102)*. A number of studies have reported diminished activations in the mPFC as compared to groups of healthy control subjects *(93, 103–106)*. These abnormal activations have been reported during such symptom-provocation tasks as traumatic narratives, combat-related stimuli, emotional word tasks, and emotional Stroop tasks. For example, using an autobiographical script, two studies by Lanius et al. *(93, 103)* found reduced activity as compared to healthy controls in both the mPFC and thalamus of PTSD subjects, suggesting an alteration of normal neurocircuitry in those regions. In an fMRI paradigm incorporating a happy vs. fearful faces task, Shin et al. *(105)* noted that the observed decrease in mPFC activity in the PTSD group as compared to controls was accompanied by increased amygdala activations. The theory has been advanced that PTSD symptoms are caused by an overactive amygdala in PTSD patients accompanied by a failure of the mPFC, including the ACC, and the hippocampus to inhibit this abnormal activation *(107–110)*. The fact that the functionality and structure of the hippocampus is also often altered in this disorder contributes evidence to this hypothesis. In one study combining fMRI and PET with verbal declarative memory tasks, the hippocampus of PTSD patients failed to activate entirely during the same tasks that caused activations in two control groups *(87)*. Other PET studies have noted hypoactivations of the hippocampus during memory tasks *(111, 112)*, but one other fMRI study found increased hippocampal activation during a Stroop task *(47)*, and another PET study noted increased resting state blood flow in the hippocampus and parahippocampal gyrus that was positively correlated with symptom severity *(113)*. Overall, reduced volume and increased resting blood flow in the hippocampus have been seen in many PTSD patients during both PET and fMRI studies. More study is needed to bring a greater degree of consistency to the imaging data with regards to this structure, incorporating subjects that are consistent for age, gender, and subtype of PTSD (primarily manifesting as flashback symptoms or primarily involving dissociation) *(103)*.

The majority of functional PTSD fMRI studies to date provide evidence for hyperactivation of the amygdala coupled with a relative decrease in mPFC activity. It is hypothesized that the hippocampus, parahippocampus, and thalamus also represent participatory areas and that dysfunction in regional neurocircuitry leads to the symptoms of the disorder *(80)*. More research with larger subject groups and standardized study guidelines is needed to unify the fMRI data and uncover more consistent trends in PTSD imaging research.

6. Disorders of Late Life

6.1. Alzheimer's Disease

AD is the leading cause of dementia in the elderly. It is a progressive neurodegenerative disorder characterized by memory loss and cognitive decline, with a prevalence of approximately 5.7% in people aged 65 years and more *(114)*. Age is the dominant risk factor in AD. The current total direct costs can approach upwards of $60,000 per patient annually, with formal care accounting for approximately $30,000 *(115)*. There are also many indirect costs, including lost productivity of patients and caregivers. Statistics from 1998 show AD to be the ninth highest cause of death in people in the USA aged 65 years and older, totaling 22,725 deaths annually *(116)*.

The progressive nature of neurodegeneration suggests an age-dependent process that ultimately leads to synaptic failure and neuronal damage *(117)* in cortical areas of the brain essential for memory and higher mental functions *(118)*. Despite the considerable amount that is known regarding the mechanisms of the disease, there is no cure and, at the present time, definitive diagnosis can only be made post-mortem.

Criteria for making diagnosis of AD are based on those published by the National Institute of Neurologic Disorders and Stroke - AD and Related Disorders Association (NINDS –ADRDA) Work Group *(119)*, and the Diagnostic and Statistical Manual, 4th edition (DSM-IV, 1994). The American Academy of Neurology recommends NINDS-ADRDA criteria for "probable AD" or DSMIIIR criteria for "dementia of Alzheimer's type (DAT)" based on studies regarding the accuracy of clinical diagnoses, correlated with neuropathologic confirmation, which show an average sensitivity of 81% and average specificity of 70% when using a DSM-IIIR definition of DAT or a NINDS-ADRDA definition of "probable AD" *(114, 120)*.

Because there is a lack of sensitive and specific biomarkers for early AD, early diagnosis is very difficult. Imaging is being used more and more frequently for biomarker quantitation and there is an active search for an imaging biomarker *(121–125)*. A number of pathologic features have been identified in the brains of patients with AD, including metabolic disturbances, structural disturbances, blood flow disturbances, and extracellular neuritic plaques. fMRI has been utilized in the study of AD to study the functional pathology of this disorder (*see* **Table 5**).

Ross et al. *(126)* have reported that BOLD signal responses to photic stimulation decreased with aging. Similar, age-dependent results have been reported for odor-stimulated fMR responses *(127)* and for activations that accompany a simple reaction time task *(128)*. These observations suggest that BOLD fMRI studies of older adults with brain disorders such as AD may be compromised by the limited contrast: noise ratios for the detection and quantitation of task-specific activations, especially if the expected change is a decrease in signal intensity.

Despite these limitations, many fMRI studies have been consistent in their findings. According to a review by Prvulovic *(129)*, healthy aging has been associated with hypofunction of sensory or motor cortex and occipital cortex *(127, 128, 130, 131)*, while pathologic conditions such as AD have been associated with decreased activation in temporal and parietal lobes *(132, 133)*.

A number of fMRI studies have looked at tasks such as learning and memory (which are known to be disturbed) in AD *(134–136)*. Kato et al. *(134)* used fMRI to examine the learning ability of

Table 5
Summary of fMRI research in Alzheimer's disease

Alzheimer's disease

Authors	Subjects	fMRI paradigm	Results
Kato et al. *(134)*	Eight young adult controls, eight late middle-aged controls and seven patients with early AD	Memory and recall of geometric figures	AD subjects showed increased activation during stimulus presentation in visual association area and failure of activation in either temporal lobe or prefrontal region as compared to the two control groups
Smith et al. *(133)*	Healthy adults females at risk for AD	Letter fluency task	The at-risk group showed increased activation in the left parietal region
Grady et al. *(139)*	12 adults with probable AD and 12 healthy adults	Semantic and episodic memory tasks	AD subjects use additional neural resources in prefrontal cortex that mediate executive functions
Sperling et al. *(138)*	Ten young adult controls, ten elderly controls, and seven patients with AD	Face-name associative encoding task; block design with three conditions: novel face-name pairs, repeated face-name pairs, and visual fixation	AD subjects showed significantly less hippocampal activation and greater activation in medial parietal and posterior cingulated regions as compared to elderly controls; elderly controls showed less activation in superior and inferior prefrontal cor tices but greater activation in parietal regions than young controls during encoding
Dickerson et al. *(137)*	Ten healthy adults, nine adults with MCI, ten adults with AD	Face-name associative encoding task	Greater hippocampal activation in MCI group as compared to controls; hippocampal and entrohinal hypoactivation and atrophy in AD as compared to controls
Pariente et al. *(140)*	17 healthy adults and 12 adults with early state AD	Face-name associative encoding and recognition tasks	Hippocampal hypoactivation during encoding and recognition in AD subjects as compared to controls; bilateral hyperactivation of parts of parietal and frontal lobes in AD subjects as compared to controls

AD Alzheimer's disease; *MCI* mild cognitive impairment

patients with early AD. While learning complex, geometric figures, reduced signals were found in the medial and temporal regions. These regions could potentially be markers for AD *(123, 134–137)*. Additionally, there have been many studies that have found reduced activation in hippocampal and parahippocampal regions as compared to healthy control subjects *(137, 138)*. There has also been evidence of increased activation in some neocortical regions in AD patients, indicating compensation for hippocampal deactivation *(138–140)*.

7. Summary and Future Directions

Table 6 summarizes the overall neural activation differences between affected psychiatric patients and healthy subjects during a sampling of common fMRI paradigms designed to test the normality of the patients' emotional, memory, inhibitory, learning, language, and executive functionality. As expected, decreased activity in cortical regions is common in the diseased brain, but neural structure overactivity is just as prevalent in certain disorders. Many researchers have theorized about the possibility of secondary effects, whereby the functional deficits that underlie the symptomatology of certain disorders are compensated for by involvement of accessory structures or overactivation in another area *(17, 21)*. Clearly, the effects of these disorders cannot be summarized by a single regional deficit, and current fMRI research is moving toward investigation of neural circuitry and abnormal systemic interactions. It is hoped that the documentation of these disorders' underlying dysfunctionality will yield common threads that connect the ranges of heterogeneous symptoms and indicate new clinical strategies. However, the use of fMRI as a diagnostic imaging method in psychiatric practice has thus far been limited. Although some studies are beginning to hone in on early indicators of disease, as well as the potential treatment response of the newly diagnosed, clinical method development is hampered by cost, complexity of typical research paradigms and protocols, and the known and unknown effects of drug treatments of CBF *(4)*.

An important future direction for fMRI research into these disorders is the development of simple paradigms that can be relied upon to produce distinct and consistent patterns of activation in healthy and affected brains. In addition, the integration of current trends in clinical and molecular genetics, cellular biology, and social health factors such as population vulnerability factors and life events into the study of these disorders has the potential to delineate disease causality, risk factors, and course to an increasingly accurate extent.

Table 6
Summary of fMRI paradigms and protocols in psychiatric disorders

Function	Stimulation paradigms	Observed activation differences in affected group	Disorder
Emotional processing	Evocation of happy and sad mood (autobiographical scripts, positive, and negative words, etc.), affective pictures, facial expressions	↓ Anterior cingulate	Schizophrenia
		↓ Amygdala– hippocampal complex (Hempel et al. *(21)*)	
		↓ Amygdala (Habel et al. (*34*); Williams et al. *(23)*)	
		↓ Right amygdala	
		↓ mPFC (Takahashi et al. *(22)*)	
		↑ Left amygdala	Major depressive disorder
		↑ Ventral striatum	
		↑ Left hippocampus, insula, caudate nucleus, thalamus, dorsal cingulate gyrus, inferior parietal cortex (Fu et al. *(19)*)	
		↑ Amygdala (Siegle et al.) *(43, 44)*)	
		↓ DLPFC (Yurgulen-Todd et al. *(62)*)	Bipolar disorder
		↑ Amygdala (Yurgulen-Todd et al. *(62)*; Malhi et al. *(66)*; Blumberg et al. *(63)*; Pavuluri et al. *(64)*)	
		↑ Thalamus (Lawrence et al. *(61)*; Malhi et al. *(65)*; Malhi et al. *(66)*)	
		↑ Ventral striatum, ventral PFC	
		↑ Hippocampus (Lawrence et al. *(61)*)	

		↑ Caudate (Malhi et al. *(65)*)	
		↑ Hypothalamus, medial globus pallidus (Malhi et al. *(66)*)	
		↓ Rostral anterior cingulate (Blumberg et al. *(63)*)	
		↓ Right rostral ventrolateral PFC (Pavuluri et al. *(64)*)	
		↑ Right pregenual anterior cingulate, paralimbic cortex (Pavuluri et al. *(64)*)	
		↑ Amygdala (Rauch et al. *(89)*; Hendler et al. *(96)*; Driessen et al. *(97)*; Protopopescu et al. *(98)*; Shin et al. *(105)*)	Posttraumatic stress disorder
		↓ Thalamus, anterior cingulate gyrus (Lanius et al. *(103)*)	
		↑ Right anterior temporal lobes, mesiotemporal areas, posterior cingulate gyrus, occipital areas, cerebellum (Driessen et al. *(97)*)	
		↓ Medial prefrontal cortex (Shin et al. *(105)*)	
Working memory/attention	Word encoding/recognition, delayed match to sample, Sternberg test, N-back, continuous performance	↑ Basal ganglia	Schizophrenia
		↑ Thalamus (Manoach et al. *(25)*)	
		↑ Parahippocampus (Ragland et al. *(26)*)	
		↓ Prefrontal cortex (Ragland et al. *(26)*); Morey et al. *(32)*)	
		↓ Anterior cingulate cortex	
		↓ Cerebellum (Honey et al. *(33)*)	
		↑ Left DLPFC	Major depressive disorder
		↑ Ventromedial PFC (Walter et al. *(46)*)	
		↓ Left DLPFC (Rubia et al. *(27)*); Chang et al. *(58)*; Frangou et al. *(56)*)	Bipolar disorder
		↑ Fronto-polar PFC, temporal cortex, temporal cortex, posterior parietal cortex	

(continued)

Table 6 (continued)

Function	Stimulation paradigms	Observed activation differences in affected group	Disorder
		↑ Thalamus, basal ganglia (Adler et al. *(57)*)	
		↑ Bilateral ACC, left thalamus, left putamen, right inferior frontal gyrus (Chang et al. *(58)*)	
		↓ Frontal, temporal, parietal regions (Monks et al. *(55)*)	
		↑ Left precentral, right medial frontal, left supramarginal gyri (Monks et al. *(55)*)	
		↓ Bilateral VPFC (Frangou et al. *(56)*)	
		↓ DLPFC (Lagopoulos et al. *(54)*)	
		↑ Inferior frontal gyrus (Lagopoulos et al. *(54)*)	
		↓ Temporal lobes, prefrontal region (Kato et al. *(134)*)	Alzheimer's disease
		↑ Visual association area (Kato et al. *(134)*)	
		↑ Medial parietal cortex, posterior cingulate cortex (Sperling et al. *(138)*)	
		↓ Hippocampus (Dickerson et al. *(137)*; Sperling et al. *(138)*; Pariente et al. *(140)*)	
		↓ Entrohinal cortex (Dickerson et al. *(137)*)	
		↓ Parietal and frontal lobes (Pariente et al. *(140)*)	
Inhibitory control	Go/No-go, stop cues, Stroop task	↑ Rostral anterior cingulate gyrus	Major depressive disorder
		↑ Left DLPFC (Wagner et al. *(142)*)	
		↑ Rostral left VPFC	Bipolar disorder

		↑ Left frontal cortex (depressed mood group)	
		↑ Right frontal cortex (elevated mood group) (Blumberg et al. *(53)*)	
		↓ ACC (Gruber et al. *(59)*)	
		↑ DLPFC (Gruber et al. *(59)*)	
		↓ Temporal cortical regions, medial frontal gyrus, putamen, midline cerebellum (Strakowski et al., *(60)*)	
		↑ Medial occipital cortex (Strakowski et al. *(60)*)	
		↓ Right inferior and medial frontal gyri	Obsessive-compulsive disorder
		↓ Right orbitofrontal cortex, dorsal anterior cingulate gyrus (Roth et al. *(73)*)	
		↓ Rostral anterior cingulate cortex (Shin et al. *(104)*)	Posttraumatic stress disorder
Executive function	Digit sorting, Tower of London	↓ DLPFC (Siegle et al. *(44)*)	Major depressive disorder
		↓ DLPFC	Obsessive-compulsive disorder
		↓ Caudate nucleus (van den Heuvel, *(75)*)	
		↑ Anterior cingulate, ventrolateral PFC, parahippocampal cortex (van den Heuvel, *(75)*)	
Episodic memory	Episodic encoding/ recognition	↓ Bilateral DLPFC	Schizophrenia
		↓ Lateral temporal cortex (Hofer et al. *(24)*)	
		↑ Bilateral DLPFC	Alzheimer's disease
		↑ Bilateral posterior cortices	

(continued)

Table 6 (continued)

Function	Stimulation paradigms	Observed activation differences in affected group	Disorder
Language processing	Word fluency, letter fluency	↓ Temporal lobe (Yurgelun-Todd et al. *(20)*)	Schizophrenia
		↑ Left parietal region (Smith et al. *(133)*)	Group at risk for Alzheimer's disease
Reward and punishment	Loss/No loss, reward/ no reward, incentive monetary tasks	↓ Ventral striatum (Juckel et al. *(141)*)	Schizophrenia
Learning	Picture encoding, semantic encoding	↓ Left inferior PFC ↑ Left superior temporal gyrus (Kubicki et al. *(35)*)	Schizophrenia
Affective switching	Reversal learning task	↓ Right medial and right lateral OFC, right caudate nucleus ↓ Left posterior OFC, insular cortex, DLPFC, anterior PFC (Remijnse et al. *(72)*)	Obsessive-compulsive disorder

References

1. Brown GG, Perthen JE, Liu TT, Buxton RB. A primer on functional magnetic resonance imaging. Neuropsychol Rev 2007;**17**: 107–125.
2. Francati V, Vermetten E, Bremner JD. Functional neuroimaging studies in posttraumatic stress disorder: review of current methods and findings. Depress Anxiety 2007;**24**: 202–218.
3. Giardino ND, Friedman SD, Dager SR. Anxiety, respiration, and cerebral blood flow: implications for functional brain imaging. Compr Psychiatry 2007;**48**:103–112.
4. Yurgelun-Todd DA, Femia LA. Applications of fMRI to psychiatry. In: Functional MRI: basic principles and clinical applications, Faro SH M.F.,Editor. 2006, Springer Science + Business Media, Inc.: New York. pp 183–220.
5. Lai S, Hopkins AL, Haacke EM, et al. Identification of vascular structures as a major source of signal contrast in high resolution 2D and 3D functional activation imaging of the motor cortex at 1.5T: preliminary results. Magn Reson Med 1993;**30**: 387–392.
6. Saad ZS, Ropella KM, DeYoe EA, Bandettini PA. The spatial extent of the BOLD response. Neuroimage 2003;**19**:132–144.
7. Poulin MJ, Liang PJ, Robbins PA. Dynamics of the cerebral blood flow response to step changes in end-tidal PCO_2 and PO_2 in humans. J Appl Physiol 1996;**81**:1084–1095.
8. Ide K, Poulin MJ. The relationship between middle cerebral artery blood velocity and end-tidal PCO_2 in the hypocapnichypercapnic range in humans. J Appl Physiol 2003;**95**: 129–137.
9. Rostrup E, Knudsen GM, Law I, Holm S, Larsson HB, Paulson OB. The relationship between cerebral blood flow and volume in humans. Neuroimage 2005;**24**:1–11.
10. Grubb RL, Jr., Raichle ME, Eichling JO, Ter-Pogossian MM. The effects of changes in $PaCO_2$ on cerebral blood volume, blood flow, and vascular mean transit time. Stroke 1974;**5**:630–639.
11. Reiman EM, Raichle ME, Robins E, et al. The application of positron emission tomography to the study of panic disorder. Am J Psychiatry 1986;**143**:469–477.
12. Aguirre GK, Detre JA, Wang J. Perfusion fMRI for functional neuroimaging. Int Rev Neurobiol 2005;**66**:213–236.
13. Liu TT, Brown GG. Measurement of cerebral perfusion with arterial spin labeling. Part 1. Methods. J Int Neuropsychol Soc 2007;**13**: 517–525.
14. Rao SM, Salmeron BJ, Durgerian S, et al. Effects of methylphenidate on functional MRI blood-oxygen-level-dependent contrast. Am J Psychiatry 2000;**157**:1697–1699.
15. Mildner T, Zysset S, Trampel R, Driesel W, Möller HE. Towards quantification of blood-flow changes during cognitive task activation using perfusion-based fMRI. Neuroimage 2005;**27**:919–926.
16. Tjandra T, Brooks JC, Figueiredo P, Wise R, Matthews PM, Tracey I.Quantitative assessment of the reproducibility of functional activation measured with BOLD and MR perfusion imaging: implications for clinical trial design. Neuroimage 2005;**27**:393–401.
17. Mayberg HS. Modulating dysfunctional limbic-cortical circuits in depression: towards development of brain-based algorithms for diagnosis and optimised treatment. Br Med Bull 2003;**65**:193–207.
18. Deckersbach T, Dougherty DD, Rauch SL. Functional imaging of mood and anxiety disorders. J Neuroimaging 2006;**16**:1–10.
19. Fu CH, Williams SC, Cleare AJ, et al. Attenuation of the neural response to sad faces in major depression by antidepressant treatment: a prospective, event-related functional magnetic resonance imaging study. Arch Gen Psychiatry 2004;**61**:877–889.
20. Yurgelun-Todd DA, Coyle JT, Gruber SA, et al. Functional magnetic resonance imaging studies of schizophrenic patients during word production: effects of D-cycloserine. Psychiatry Res 2005;**138**:23–31.
21. Hempel A, Hempel E, Schönknecht P, Stippich C, Schröder J. Impairment in basal limbic function in schizophrenia during affect recognition. Psychiatry Res 2003;**122**:115–124.
22. Takahashi H, Koeda M, Oda K, Matsuda T, et al. An fMRI study of differential neural response to affective pictures in schizophrenia. Neuroimage 2004;**22**:1247–1254.
23. Williams LM, Das P, Harris AW, et al. Dysregulation of arousal and amygdala-prefrontal systems in paranoid schizophrenia. Am J Psychiatry 2004;**161**:480–489.
24. Hofer A, Weiss EM, Golaszewski SM, et al. An FMRI study of episodic encoding and recognition of words in patients with schizophrenia in remission. Am J Psychiatry 2003;**160**: 911–918.
25. Manoach DS, Gollub RL, Benson ES, et al. Schizophrenic subjects show aberrant fMRI

activation of dorsolateral prefrontal cortex and basal ganglia during working memory performance. Biol Psychiatry 2000;**48**: 99–109.

26. Ragland JD, Gur RC, Valdez J, et al. Event-related fMRI of frontotemporal activity during word encoding and recognition in schizophrenia. Am J Psychiatry 2004;**161**:1004–1015.
27. Rubia K, Russel T, Bullmore ET, et al. An fMRI study of reduced left prefrontal activation in schizophrenia during normal inhibitory function. Schizophr Res 2001;**52**:47–55.
28. Vink M., et al. Striatal dysfunction in schizophrenia and unaffected relatives. Biol Psychiatry 2006;**60**(1):32–39.
29. Braver TS, et al. Direct comparison of prefrontal cortex regions engaged by working and long-term memory tasks. Neuroimage 2001;**14**(1 Pt 1):48–59.
30. Cohen NJ, et al. Hippocampal system and declarative (relational) memory: summarizing the data from functional neuroimaging studies. Hippocampus 1999;**9**(1):83–98.
31. Nyberg L, et al. Common prefrontal activations during working memory, episodic memory, and semantic memory. Neuropsychologia 2003;**41**(3):371–377.
32. Morey RA, et al. Imaging frontostriatal function in ultra-high-risk, early, and chronic schizophrenia during executive processing. Arch Gen Psychiatry 2005;**62**(3):254–262.
33. Honey GD, et al. Ketamine disrupts frontal and hippocampal contribution to encoding and retrieval of episodic memory: an fMRI study. Cereb Cortex 2005;**15**(6):749–759.
34. Habel U, et al. Genetic load on amygdala hypofunction during sadness in nonaffected brothers of schizophrenia patients. Am J Psychiatry 2004;**161**(10):1806–1813.
35. Kubicki M, et al. An fMRI study of semantic processing in men with schizophrenia. Neuroimage 2003;**20**(4):1923–1933.
36. Drevets WC. Neuroimaging studies of mood disorders. Biol Psychiatry 2000;**48**(8): 813–829.
37. Mayberg HS. Limbic-cortical dysregulation: a proposed model of depression. J Neuropsychiatry Clin Neurosci 1997; **9**(3):471–481.
38. Mayberg HS, et al. Reciprocal limbic-cortical function and negative mood: converging PET findings in depression and normal sadness. Am J Psychiatry 1999;**156**(5):675–682.
39. Baxter LR, Jr., et al. Reduction of prefrontal cortex glucose metabolism common to three types of depression. Arch Gen Psychiatry 1989;**46**(3):243–250.
40. Bench CJ, et al. The anatomy of melancholia – focal abnormalities of cerebral blood flow in major depression. Psychol Med 1992;**22**(3): 607–615.
41. Drevets WC, et al. Subgenual prefrontal cortex abnormalities in mood disorders. Nature 1997;**386**(6627):824–827.
42. Liotti M, et al. Differential limbic–cortical correlates of sadness and anxiety in healthy subjects: implications for affective disorders. Biol Psychiatry 2000;**48**(1):30–42.
43. Siegle GJ, et al. Can't shake that feeling: event-related fMRI assessment of sustained amygdala activity in response to emotional information in depressed individuals. Biol Psychiatry 2002;**51**(9):693–707.
44. Siegle GJ, et al. Increased amygdala and decreased dorsolateral prefrontal BOLD responses in unipolar depression: related and independent features. Biol Psychiatry 2007;**61**(2):198–209.
45. Surguladze SA, et al. Recognition accuracy and response bias to happy and sad facial expressions in patients with major depression. Neuropsychology 2004;**18**(2): 212–218.
46. Walter H, et al. Increased left prefrontal activation in patients with unipolar depression: an event-related, parametric, performance-controlled fMRI study. J Affect Disord 2007;**101**(1–3):175–185.
47. Sheline YI, et al. Increased amygdala response to masked emotional faces in depressed subjects resolves with antidepressant treatment: an fMRI study. Biol Psychiatry 2001;**50**(9): 651–658.
48. Del-Ben CM, et al. The effect of citalopram pretreatment on neuronal responses to neuropsychological tasks in normal volunteers: an FMRI study. Neuropsychopharmacology 2005;**30**(9):1724–1734.
49. Vollm B, et al. Serotonergic modulation of neuronal responses to behavioural inhibition and reinforcing stimuli: an fMRI study in healthy volunteers. Eur J Neurosci 2006;**23**(2):552–560.
50. Chen CH, et al. Brain imaging correlates of depressive symptom severity and predictors of symptom improvement after antidepressant treatment. Biol Psychiatry 2007;**62**(5): 407–414.
51. Strakowski SM, Delbello MP, Adler CM. The functional neuroanatomy of bipolar disorder: a review of neuroimaging findings. Mol Psychiatry 2005;**10**(1):105–116.
52. Yurgelun-Todd DA Ross AJ. Functional magnetic resonance imaging studies in bipolar disorder. CNS Spectr 2006;**11**(4):287–297.

53. Blumberg HP, et al. A functional magnetic resonance imaging study of bipolar disorder: state- and trait-related dysfunction in ventral prefrontal cortices. Arch Gen Psychiatry 2003;**60**(6):601–609.
54. Lagopoulos J, Ivanovski B Malhi GS. An event-related functional MRI study of working memory in euthymic bipolar disorder. J Psychiatry Neurosci 2007;**32**(3):174–184.
55. Monks PJ, et al. A functional MRI study of working memory task in euthymic bipolar disorder: evidence for task-specific dysfunction. Bipolar Disord 2004;**6**(6):550–564.
56. Frangou S. The Maudsley bipolar disorder project. Epilepsia 2005;**46** (Suppl 4):19–25.
57. Adler CM, et al. Changes in neuronal activation in patients with bipolar disorder during performance of a working memory task. Bipolar Disord 2004;**6**(6):540–549.
58. Chang K, et al. Anomalous prefrontal-subcortical activation in familial pediatric bipolar disorder: a functional magnetic resonance imaging investigation. Arch Gen Psychiatry 2004;**61**(8):781–792.
59. Gruber SA, Rogowska J, Yurgelun-Todd DA. Decreased activation of the anterior cingulate in bipolar patients: an fMRI study. J Affect Disord 2004;**82**(2):191–201.
60. Strakowski SM, et al. Abnormal FMRI brain activation in euthymic bipolar disorder patients during a counting Stroop interference task. Am J Psychiatry 2005;**162**(9):1697–1705.
61. Lawrence NS, et al. Subcortical and ventral prefrontal cortical neural responses to facial expressions distinguish patients with bipolar disorder and major depression. Biol Psychiatry 2004;**55**(6):578–587.
62. Yurgelun-Todd DA, et al. fMRI during affect discrimination in bipolar affective disorder. Bipolar Disord 2000;**2**(3 Pt 2):237–248.
63. Blumberg HP, et al. Preliminary evidence for medication effects on functional abnormalities in the amygdala and anterior cingulate in bipolar disorder. Psychopharmacology (Berl) 2005;**183**(3):308–313.
64. Pavuluri MN, et al. Affective neural circuitry during facial emotion processing in pediatric bipolar disorder. Biol Psychiatry 2007;**62**(2): 158–167.
65. Malhi GS, et al. Cognitive generation of affect in hypomania: an fMRI study. Bipolar Disord 2004;**6**(4):271–285.
66. Malhi GS, et al. Cognitive generation of affect in bipolar depression: an fMRI study. Eur J Neurosci 2004;**19**(3):741–754.
67. Whiteside SP, Port JD, Abramowitz JS. A meta-analysis of functional neuroimaging in obsessive-compulsive disorder. Psychiatry Res 2004;**132**(1):69–79.
68. Mataix-Cols D, Rosario-Campos MC, Leckman JF. A multidimensional model of obsessive-compulsive disorder. Am J Psychiatry 2005;**162**(2):228–238.
69. Saxena S, et al. Cerebral metabolism in major depression and obsessive-compulsive disorder occurring separately and concurrently. Biol Psychiatry 2001;**50**(3):159–170.
70. Mitterschiffthaler MT, et al. Applications of functional magnetic resonance imaging in psychiatry. J Magn Reson Imaging 2006;**23**(6): 851–861.
71. Saxena S, et al. Neuroimaging and frontal-subcortical circuitry in obsessive-compulsive disorder. Br J Psychiatry Suppl 1998(35): 26–37.
72. Remijnse PL, et al. Reduced orbitofrontal-striatal activity on a reversal learning task in obsessive-compulsive disorder. Arch Gen Psychiatry 2006;**63**(11):1225–1236.
73. Roth RM, et al. Event-related functional magnetic resonance imaging of response inhibition in obsessive-compulsive disorder. Biol Psychiatry 2007;**62**(8):901–909.
74. Purcell R, et al. Cognitive deficits in obsessive-compulsive disorder on tests of frontal-striatal function. Biol Psychiatry 1998;**43**(5): 348–357.
75. van den Heuvel OA, et al. Frontal-striatal dysfunction during planning in obsessive-compulsive disorder. Arch Gen Psychiatry 2005;**62**(3):301–309.
76. Gilbert AR, et al. Decrease in thalamic volumes of pediatric patients with obsessive-compulsive disorder who are taking paroxetine. Arch Gen Psychiatry 2000;**57**(5): 449–456.
77. Kim JJ, et al. Grey matter abnormalities in obsessive-compulsive disorder: statistical parametric mapping of segmented magnetic resonance images. Br J Psychiatry 2001;**179**: 330–334.
78. Lacerda AL, et al. Elevated thalamic and prefrontal regional cerebral blood flow in obsessive-compulsive disorder: a SPECT study. Psychiatry Res 2003;**123**(2):125–134.
79. Saxena S Rauch SL. Functional neuroimaging and the neuroanatomy of obsessive-compulsive disorder. Psychiatr Clin North Am 2000;**23**(3):563–586.
80. Rauch SL, et al. The functional neuroanatomy of anxiety: a study of three disorders using positron emission tomography and symptom provocation. Biol Psychiatry 1997;**42**(6): 446–452.

81. Charney DS, et al. Psychobiologic mechanisms of posttraumatic stress disorder. Arch Gen Psychiatry 1993;**50**(4):295–305.
82. Charney DS. Psychobiological mechanisms of resilience and vulnerability: implications for successful adaptation to extreme stress. Am J Psychiatry 2004;**161**(2):195–216.
83. Quirk GJ, Gehlert DR. Inhibition of the amygdala: key to pathological states? Ann N Y Acad Sci 2003;**985**:263–272.
84. Stein MB, et al. Increased amygdala and insula activation during emotion processing in anxiety-prone subjects. Am J Psychiatry 2007;**164**(2):318–327.
85. Geuze E, Vermetten E, Bremner JD. MR-based in vivo hippocampal volumetrics. 2. Findings in neuropsychiatric disorders. Mol Psychiatry 2005;**10**(2):160–184.
86. Bremner JD. Neuroimaging of childhood trauma. Semin Clin Neuropsychiatry 2002;7(2):104–112.
87. Bremner JD, et al. MRI and PET study of deficits in hippocampal structure and function in women with childhood sexual abuse and posttraumatic stress disorder. Am J Psychiatry 2003;**160**(5):924–932.
88. Gilbertson MW, et al. Smaller hippocampal volume predicts pathologic vulnerability to psychological trauma. Nat Neurosci 2002;**5**(11):1242–1247.
89. Rauch SL, et al. Exaggerated amygdala response to masked facial stimuli in posttraumatic stress disorder: a functional MRI study. Biol Psychiatry 2000;**47**(9):769–776.
90. Whalen PJ, et al. Masked presentations of emotional facial expressions modulate amygdala activity without explicit knowledge. J Neurosci 1998;**18**(1):411–418.
91. Lanius RA, et al. Functional connectivity of dissociative responses in posttraumatic stress disorder: a functional magnetic resonance imaging investigation. Biol Psychiatry 2005;**57**(8):873–884.
92. Vermetten E, et al. Long-term treatment with paroxetine increases verbal declarative memory and hippocampal volume in posttraumatic stress disorder. Biol Psychiatry 2003;**54**(7):693–702.
93. Lanius RA, et al. Neural correlates of traumatic memories in posttraumatic stress disorder: a functional MRI investigation. Am J Psychiatry 2001;**158**(11):1920–1922.
94. Pitman RK, Shin LM, Rauch SL. Investigating the pathogenesis of posttraumatic stress disorder with neuroimaging. J Clin Psychiatry 2001;**62** (Suppl 17):47–54.
95. Villarreal G, King CY. Brain imaging in posttraumatic stress disorder. Semin Clin Neuropsychiatry 2001;**6**(2):131–145.
96. Hendler T, et al. Sensing the invisible: differential sensitivity of visual cortex and amygdala to traumatic context. Neuroimage 2003;**19**(3):587–600.
97. Driessen M, et al. Posttraumatic stress disorder and fMRI activation patterns of traumatic memory in patients with borderline personality disorder. Biol Psychiatry 2004;**55**(6):603–611.
98. Protopopescu X, et al. Differential time courses and specificity of amygdala activity in posttraumatic stress disorder subjects and normal control subjects. Biol Psychiatry 2005;**57**(5):464–473.
99. Morgan MA, LeDoux JE. Differential contribution of dorsal and ventral medial prefrontal cortex to the acquisition and extinction of conditioned fear in rats. Behav Neurosci 1995;**109**(4):681–688.
100. Quirk GJ, et al. The role of ventromedial prefrontal cortex in the recovery of extinguished fear. J Neurosci 2000;**20**(16):6225–6231.
101. Santini E, et al. Consolidation of fear extinction requires protein synthesis in the medial prefrontal cortex. J Neurosci 2004;**24**(25):5704–5710.
102. Morgan MA, Romanski LM, LeDoux JE. Extinction of emotional learning: contribution of medial prefrontal cortex. Neurosci Lett 1993;**163**(1):109–113.
103. Lanius RA, et al. Recall of emotional states in posttraumatic stress disorder: an fMRI investigation. Biol Psychiatry 2003;**53**(3):204–210.
104. Shin LM, et al. An fMRI study of anterior cingulate function in posttraumatic stress disorder. Biol Psychiatry 2001;**50**(12):932–942.
105. Shin LM, et al. A functional magnetic resonance imaging study of amygdala and medial prefrontal cortex responses to overtly presented fearful faces in posttraumatic stress disorder. Arch Gen Psychiatry 2005;**62**(3):273–281.
106. Williams LM, et al. Trauma modulates amygdala and medial prefrontal responses to consciously attended fear. Neuroimage 2006;**29**(2):347–357.
107. Golier JA, et al. Memory performance in Holocaust survivors with posttraumatic stress disorder. Am J Psychiatry 2002;**159**(10):1682–1688.
108. Rauch SL, Shin LM, Phelps EA. Neurocircuitry models of posttraumatic stress disorder and extinction: human neuroimaging research – past, present, and future. Biol Psychiatry 2006;**60**(4):376–382.

109. Shin LM, et al. Hippocampal function in posttraumatic stress disorder. Hippocampus 2004;**14**(3):292–300.
110. Vermetten E, Bremner JD. Circuits and systems in stress. II. Applications to neurobiology and treatment in posttraumatic stress disorder. Depress Anxiety 2002;**16**(1):14–38.
111. Bremner JD, et al. Neural correlates of memories of childhood sexual abuse in women with and without posttraumatic stress disorder. Am J Psychiatry 1999;**156**(11): 1787–1795.
112. Shin LM, et al. Regional cerebral blood flow during script-driven imagery in childhood sexual abuse-related PTSD: a PET investigation. Am J Psychiatry 1999;**156**(4): 575–584.
113. Semple WE, et al. Higher brain blood flow at amygdala and lower frontal cortex blood flow in PTSD patients with comorbid cocaine and alcohol abuse compared with normals. Psychiatry 2000;**63**(1):65–74.
114. Sair HI, Doraiswamy PM, Petrella JR. In vivo amyloid imaging in Alzheimer's disease. Neuroradiology 2004;**46**(2):93–104.
115. Rice DP, et al. Prevalence, costs, and treatment of Alzheimer's disease and related dementia: a managed care perspective. Am J Manag Care 2001;7(8):809–818.
116. Murphy SL Deaths: final data for 1998. Natl Vital Stat Rep 2000;**48**(11):1–105.
117. Masters CL, Beyreuther K. Alzheimer's disease. BMJ 1998;**316**(7129):446–448.
118. Masters CL, et al. Molecular mechanisms for Alzheimer's disease: implications for neuroimaging and therapeutics. J Neurochem 2006;**97**(6):1700–1725.
119. McKhann G, et al. Clinical diagnosis of Alzheimer's disease: report of the NINCDS-ADRDA Work Group under the auspices of Department of Health and Human Services Task Force on Alzheimer's Disease. Neurology 1984;**34**(7):939–944.
120. Knopman D, et al. Cardiovascular risk factors and cognitive decline in middle-aged adults. Neurology 2001;**56**(1):42–48.
121. Burns A, Russell E, Page S. New drugs for Alzheimer's disease. Br J Psychiatry 1999;**174**:476–479.
122. Coimbra A, Williams DS, Hostetler ED. The role of MRI and PET/SPECT in Alzheimer's disease. Curr Top Med Chem 2006;**6**(6): 629–647.
123. Lee BC, et al. Imaging of Alzheimer's disease. J Neuroimaging 2003;**13**(3):199–214.
124. Mosconi L, et al. Magnetic resonance and PET studies in the early diagnosis of Alzheimer's disease. Expert Rev Neurother 2004;**4**(5):831–849.
125. Schmidt B, Braun HA, Narlawar R. Drug development and PET-diagnostics for Alzheimer's disease. Curr Med Chem 2005;**12**(14): 1677–1695.
126. Ross MH, et al. Age-related reduction in functional MRI response to photic stimulation. Neurology 1997;**48**(1):173–176.
127. Yousem DM, et al. The effect of age on odor-stimulated functional MR imaging. AJNR Am J Neuroradiol 1999;**20**(4):600–608.
128. D'Esposito M, Weksler ME. Brain aging and memory: new findings help differentiate forgetfulness and dementia. Geriatrics 2000;**55**(6):55–58, 61–62.
129. Prvulovic D, et al. Functional activation imaging in aging and dementia. Psychiatry Res 2005;**140**(2):97–113.
130. Cerf-Ducastel B, Murphy C. FMRI brain activation in response to odors is reduced in primary olfactory areas of elderly subjects. Brain Res 2003;**986**(1–2):39–53.
131. Grady CL, et al. Subgroups in dementia of the Alzheimer type identified using positron emission tomography. J Neuropsychiatry Clin Neurosci 1990;**2**(4):373–384.
132. Bookheimer SY, et al. Patterns of brain activation in people at risk for Alzheimer's disease. N Engl J Med 2000;343(7):450–456.
133. Smith CD, et al. Women at risk for AD show increased parietal activation during a fluency task. Neurology 2002;**58**(8): 1197–1202.
134. Kato T, Knopman D, Liu H. Dissociation of regional activation in mild AD during visual encoding: a functional MRI study. Neurology 2001;**57**(5):812–816.
135. Petrella JR, et al. Prefrontal activation patterns in subjects at risk for Alzheimer disease. Am J Geriatr Psychiatry 2002;**10**(1): 112–113.
136. Small SA, et al. Differential regional dysfunction of the hippocampal formation among elderly with memory decline and Alzheimer's disease. Ann Neurol 1999;**45**(4): 466–472.
137. Dickerson BC, et al. Increased hippocampal activation in mild cognitive impairment compared to normal aging and AD. Neurology 2005;**65**(3):404–411.
138. Sperling RA, et al. fMRI studies of associative encoding in young and elderly controls and mild Alzheimer's disease. J Neurol Neurosurg Psychiatry 2003;**74**(1):44–50.
139. Grady CL, et al. Evidence from functional neuroimaging of a compensatory prefrontal

network in Alzheimer's disease. J Neurosci 2003;**23**(3):986–993.

140. Pariente J, et al. Alzheimer's patients engage an alternative network during a memory task. Ann Neurol 2005;**58**(6):870–879.
141. Juckel G, et al. Dysfunction of ventral striatal reward prediction in schizophrenic patients treated with typical, not atypical, neuroleptics. Psychopharmacology (Berlin) 2006;**187** (2): 222–228.
142. Wagner G, et al. Cortical inefficiency in patients with unipolar depression: an event-related fMRI study with the Stroop task. Biol Psychiatry 2006;**59**(10):958–965.

Chapter 22

fMRI in Neurodegenerative Diseases: From Scientific Insights to Clinical Applications

Bradford C. Dickerson

Summary

fMRI is a technology with great promise as a tool to probe abnormalities of brain activity in neurodegenerative diseases. The detection of functional brain abnormalities may be useful, in the appropriate clinical context, for early diagnosis, differential diagnosis, or prognostication. Prediction of response to treatment or therapeutic monitoring may also be possible with fMRI. In addition, fMRI has the potential to provide a variety of scientific insights that may have clinical relevance, including compensatory hyperactivation of brain circuits or genetic modulation of functional brain activity.

Key words: Alzheimer's disease, Amyotrophic lateral sclerosis, Functional MRI, Huntington disease, Magnetic resonance imaging, Neurodegenerative diseases, Parkinson's disease

1. Introduction

The neurodegenerative diseases are a major medical and social burden in many societies, particularly with the growth of older population segments. Neurodegenerative diseases include many dementias, movement disorders, cerebellar, and motor neuron diseases. In many cases, these diseases involve the pathologic accumulation of abnormal protein forms. As the biology of these diseases is elucidated, hope is beginning to emerge for specific treatments targeted at modification of fundamental pathophysiologic processes *(1, 2)*. For this hope to be realized, methods for early detection of specific disease processes need to be identified. Furthermore, reliable methods for monitoring the progression of the diseases will likely be critical in demonstrating the effects of putative disease-modifying therapies. Magnetic resonance

M. Filippi (ed.), *fMRI Techniques and Protocols*, Neuromethods, vol. 41
DOI 10.1007/978-1-60327-919-2_22, © Humana Press, a part of Springer Science+Business Media, LLC 2009

imaging (MRI) and other neuroimaging tools offer great potential for these purposes *(3)*. Given the growing body of evidence that alterations in synaptic function are present very early in the course of neurodegenerative disease processes, possibly long before the development of clinical symptoms and even significant neuropathology *(4, 5)*, fMRI may be particularly useful for detecting alterations in brain function that may be present very early in the trajectory of neurodegenerative diseases. The uses of fMRI in neurodegenerative diseases will be reviewed, with a focus on Alzheimer's disease (AD) to illustrate many points of relevance to other neurodegenerative diseases.

Before specifically discussing fMRI, though, it is worth considering the current concepts of clinicopathologic constructs of neurodegenerative diseases, since the interpretation of imaging data in patients depends critically on a detailed understanding of the clinical characteristics of the patient population(s) being studied.

2. Constructs of Neurodegenerative Disease: Clinical, Prodromal, and Presymptomatic Phases

Many neurodegenerative diseases are thought to arise from pathophysiologic processes that take place over years, possibly decades, prior to the development of symptoms. For example, the clinical diagnosis of AD is made after a patient has developed impairment in multiple cognitive domains that is substantial enough to interfere with routine social and/or occupational function (dementia). It is only after this point that FDA-approved medications are currently indicated – that is, clinically probable AD. By this time, substantial neuronal loss and neuropathologic change have damaged many brain regions. Although data from animal models suggest that it may be possible to impede this process as it is developing *(6, 7)*, and potentially reverse some aspects of it *(8)*, it is not clear whether the pathology typically present when patients are clinically diagnosed with AD can be reversed. Thus, it would be ideal to initiate treatment with neuroprotective medications at a time when – or even before – AD is mildly symptomatic *(3)*. This scenario is true for many neurodegenerative diseases and is even more compelling in diseases in which known genetic abnormalities can be identified that predict future disease, as in Huntington's disease.

To approach the goal of early intervention in neurodegenerative diseases, we must improve our capability to identify individuals in the earliest symptomatic phases of the diseases prior to significant functional impairment. For example, individuals are categorized as having mild cognitive impairment (MCI) when symptoms suggestive of AD are present but mild enough that traditional diagnostic criteria (which require functional impairment consistent with dementia) are not fulfilled. This gradual transitional state may last for a number of years. Diagnostic criteria for MCI have been developed *(9)* and operationalized *(10)* in a manner that suggests that cohorts of such individuals can be reliably identified for clinical trials. Recently, new diagnostic criteria for AD have been proposed with the goal of diagnosing probable AD prior to dementia *(11)*. If the pathophysiologic process of AD can be slowed at this stage of the disease, then it may be possible to preserve cognitive function and delay the ultimate development of dementia for a period of time, which is clearly clinically meaningful. Therefore, MCI and other patients in the prodromal phases of neurodegenerative diseases present excellent target populations for clinical trials of disease-modifying therapies.

Finally, the presymptomatic phase of neurodegenerative diseases is the phase when pathologic alterations are developing but symptoms are not yet apparent. In the case of AD, this phase may best be studied through the identification of cohorts with particular risk factors, such as genetic determinants [e.g., amyloid precursor protein (APP) or presenilin mutations, Down syndrome] or susceptibility factors [e.g., apolipoprotein E (APOE) ε4]. Ideally, it would ultimately be possible to initiate disease-modifying therapies at this point based on the presence of risk factors, much as is done in the case of primary preventive measures for cerebrovascular disease. Yet given that some of these therapies may not be benign, it would be best to have a panel of biomarkers that could be used to help guide the timing of such therapies, such that individuals at elevated risk for AD could be followed over time. When changes in biomarkers indicate the earliest phase of active pathophysiology, treatment could be initiated.

3. Strengths and Weaknesses of fMRI As a Tool to Probe Brain Activity in Neurodegenerative Diseases

Since functional neuroimaging tools assess inherently dynamic processes that may change over short time intervals in relation to a host of factors, these measures have unique characteristics that may offer both strengths and weaknesses as potential biomarkers of neurologic disease. Functional neuroimaging measures may be affected by transient brain and body states at the time of imaging, such as arousal, attention, sleep deprivation, sensory processing of irrelevant stimuli, or the effects of substances with pharmacologic central nervous system activity. Imaging measures of brain function may also be more sensitive than structural measures to constitutional or chronic differences between individuals, such as genetics, intelligence or educational level, learning, mood, or medication use. While these may be effects of interest in certain experimental settings, they need to be controlled when the focus is on disease-related changes and differences between subject groups or within individuals over time.

Among functional neuroimaging techniques, fMRI has many potential advantages in studying patients with neurodegenerative disorders, as it is a noninvasive imaging technique that does not require the injection of a contrast agent. It can be repeated many times over the course of a longitudinal study and thus lends itself well as a measure in clinical drug trials. It has relatively high spatial and temporal resolution, and the use of event-related designs enables the hemodynamic correlates of specific behavioral events, such as successful memory formation *(12)*, to be measured.

A caveat essential to the interpretation of task-related functional neuroimaging data is that healthy individuals of any age demonstrate differences in brain activation depending on how well they are able to perform the particular task. For example, when cognitively intact individuals learn new information during fMRI scanning, the strength of this signal is related to subsequent ability to remember the information *(13–17)*. AD patients typically perform less well on the memory tasks, which complicates the interpretation of these data *(18)*. Conversely, the recruitment of additional brain regions during task performance by patients with neurodegenerative or other neurologic disease may indicate the presence of processes attempting to compensate for damaged networks *(19, 20)*. While the task performance factor is important to consider when designing or interpreting functional neuroimaging studies of neurodegenerative diseases, it also indicates that these imaging biomarkers may be particularly sensitive to changes in cognitive or sensorimotor function, which not only provides face validity for these measures but also supports their potential use in short-term, early proof-of-concept drug trials.

There are additional challenges to performing fMRI studies in patients with neurodegenerative diseases. The technique is particularly sensitive to even small amounts of head motion. Finally, it is critical to complete further reliability experiments if fMRI is to be used in longitudinal or pharmacologic studies. Although there are now a few studies of fMRI test-retest reliability in young subjects *(21–23)*, reproducibility studies are only beginning to be performed in patients with neurodegenerative diseases.

4. Clinical Applications of fMRI in Neurodegenerative Diseases

fMRI has been applied in a number of ways in studies of patients with neurodegenerative diseases. fMRI has been used to identify abnormal patterns of brain activity during the performance of a variety of tasks in patients with neurodegenerative diseases; these abnormal patterns may reveal new insights into the disruption of brain circuits by such diseases. They may also be useful in differential diagnosis. Compensatory hyperactivation has been identified in many of these studies, which is a fascinating area of ongoing research. fMRI has been used to assess the modulatory effects of genetic factors on brain activation in patients with or at risk for neurodegenerative disorders, and it has been used to monitor the effects of therapeutic agents and is beginning to be used to try to help predict the course of the diseases.

4.1. Patterns of Abnormal Regional Brain Activation During Task Performance

fMRI has been used to investigate abnormalities in patterns of regional brain activation during a variety of cognitive tasks in patients diagnosed with mild AD compared with control subjects. It is important to keep in mind that the particular abnormalities found in an fMRI study of an AD or other patient group are heavily dependent on the type of behavioral task used in the study – if the task does not engage a particular circuit, functional abnormalities will not likely be observed. Also, the nature of functional abnormalities may depend on whether the activated brain regions are directly affected by the disease, are indirectly affected via connectivity, or are not pathologically affected. Tools are now available to directly investigate the overlap of disease-related alterations in brain structure and task-related functional activity (**Fig. 1**). Yet it should also be kept in mind that even brain regions not usually thought to be affected by a particular neurodegenerative disease (e.g., sensorimotor areas in AD) have been shown to exhibit abnormal function *(24, 25)*.

In addition to memory, which is discussed next, aspects of language and attention have been studied. Altered patterns of frontal and temporal activation have been observed in AD patients performing language tasks *(26–28)*. Similarly, although temporoparietal activation was diminished in AD during performance of semantic memory task, increased activation in temporal and frontal regions was also observed, suggesting possible compensatory processes *(29)*. During performance of a visual attention task, AD patients showed alterations in parietal activation; increased prefrontal activation was also observed compared with controls, again suggesting possible compensatory mechanisms *(30)*.

With respect to memory, a number of fMRI studies in patients with clinically diagnosed AD, using a variety of visually presented stimuli, have identified decreased activation in hippocampal and

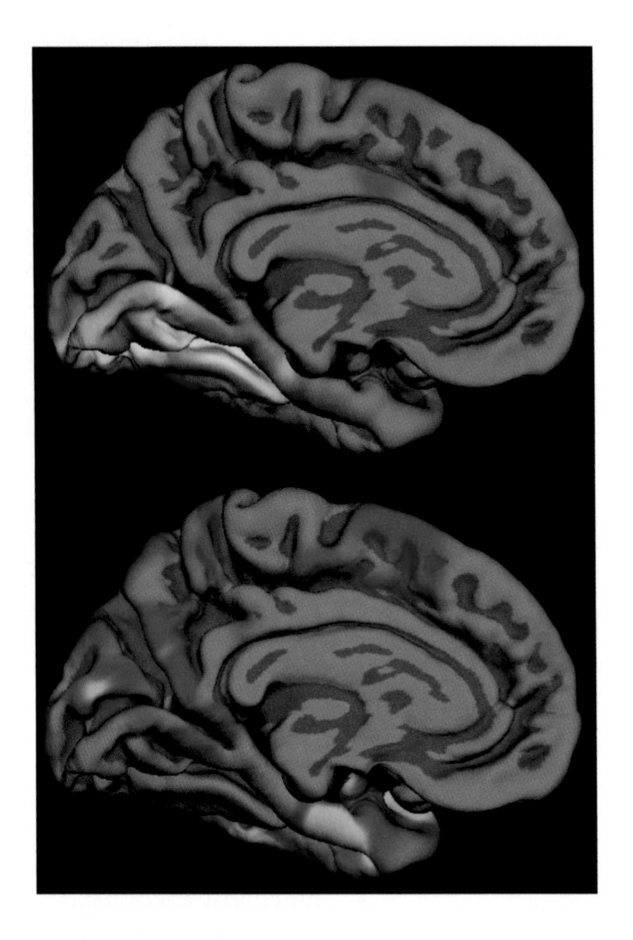

Fig. 1. The localization, magnitude, and extent of abnormalities observed in fMRI studies of patients with neurologic diseases depend on both localization and severity of pathology and on functional networks engaged by the particular fMRI task, as well as participant performance on the task. In this illustration, regions of cortical thinning in Alzheimer's disease from structural MRI (bottom; *brighter blue colors* indicate greater degree of thinning) are compared with cortical areas activated, as measured with fMRI, in normals during an event-related study of successful learning of new information that was able to later be freely recalled (*top, yellow colors* indicate greater blood oxygen level-dependent (BOLD) signal in the contrast of recalled items vs. fixation).

parahippocampal regions than in control subjects during episodic encoding tasks *(31–35)*. Neocortical abnormalities in AD have also been demonstrated using fMRI, including decreased activation in temporal and prefrontal regions. In addition to AD-related differences in task-related blood oxygen level-dependent (BOLD) signal amplitude or spatial extent, the temporal dynamics of activation appear to be altered in patients with AD *(36)*. Increased activation in prefrontal and other regions has also been found in AD patients performing memory tasks *(35)*.

While memory-task related fMRI data regarding medial temporal lobe (MTL) activation in individuals with MCI are less consistent than data from patients diagnosed with AD, with reports of both decreased and increased activation *(20, 31, 34, 37–42)*, they indicate that differences are present in comparison to older controls. Some of the variability in fMRI data in MTL activation appears to relate to degree of impairment along the spectrum of MCI, which suggests that fMRI may be sensitive to relatively subtle clinical differences in disease severity *(43)*.

fMRI has been used to study abnormal patterns of brain activation in a variety of tasks in patients with Parkinson's disease (PD), particularly with respect to sensorimotor cortical hypoactivation during motor task performance *(44)*. One study of a motor task demonstrated hypoactivation of supplementary motor areas (SMAs) accompanied by hyperactivation in both the primary and premotor cortices *(45)*. In addition, cognitive tasks have been employed, such as a working memory paradigm that demonstrated hypoactivation in fronto-striatal regions in PD patients with cognitive impairment compared with those patients who were not cognitively impaired *(46)*. Very little work using fMRI has been performed in patients with diffuse Lewy body disease (DLB), with one recent study demonstrating a complex set of differences between DLB and AD patients in visual cortical activation during face, color, and motion perceptual tasks, many of which were explainable by differences in task performance *(47)*.

In Huntington's disease (HD), attentional tasks have been used to probe fronto-parietal and fronto-striatal systems using fMRI. Findings have included reduced activation in multiple cortical regions in HD patients performing a serial reaction time task than in controls *(48)*. On a visual attention/response inhibition task, HD patients showed reduced prefrontal-anterior cingulate interhemispheric functional connectivity, and reduced connectivity predicted slower reaction times and increased numbers of errors on the task *(49)*.

In amyotrophic lateral sclerosis (ALS), fMRI has been used to demonstrate abnormalities in motor cortical activation. Furthermore, it has been used to investigate whether there are different patterns of cortical activation during a simple motor task in ALS patients with a primarily upper motor neuron (UMN) pattern of clinical deficits vs. those with a lower motor neuron (LMN) pattern *(50)*. The UMN patient subgroup showed relatively greater activation of the anterior cingulate and caudate than the LMN subgroup, suggesting that fMRI may provide insights into the differentiation of disease subtypes.

4.2. Compensatory Hyperactivation: A Universal Adaptation Response to Brain Injury?

Aside from neocortical hyperactivation in AD, recent data suggest that there is a phase of increased MTL activation in MCI (**Fig. 2**). This increase, which also may be present in cognitively intact carriers of the APOE-ε4 allele [for review, *see (51)*], may represent an attempted compensatory response to AD neuropathology, given that some MCI individuals with smaller hippocampal volume perform similarly on memory tasks to MCI individuals with larger hippocampal volume but have relatively greater MTL activation *(20, 39)*. Additional studies employing event-related fMRI paradigms *(12, 17, 42)* will be very helpful in determining whether increased MTL activation in MCI patients is specifically associated with successful memory, as opposed to a general effect that is present regardless of success (possibly indicating increased effort). It is possible that MTL hyperactivation reflects cholinergic or other neurotransmitter upregulation in MCI patients *(52)*. Alternatively, increased regional brain activation may be a marker of the pathophysiologic process of AD itself, such as aberrant

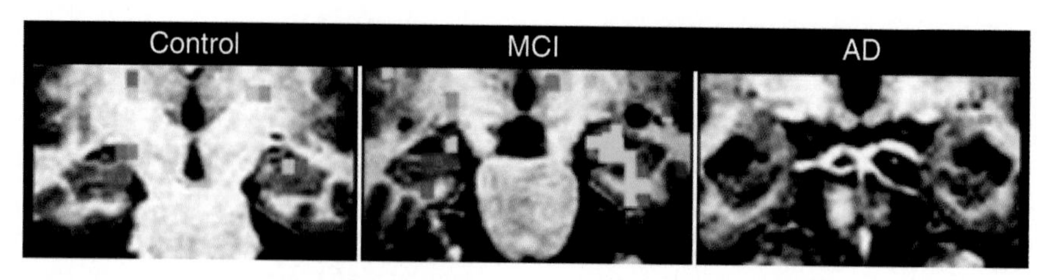

Fig. 2. A phase of compensatory hyperactivation appears to occur in the medial temporal lobe (MTL) in mild cognitive impairment (MCI), prior to the clinical onset of Alzheimer's disease (AD) dementia. Representative single subjects from each group, showing normal memory-related MTL activation measured with fMRI in normal older controls, hyperactivation and very mild atrophy in MCI, and hypoactivation and more prominent atrophy in mild AD.

sprouting of cholinergic fibers *(53)* or inefficiency in synaptic transmission *(54)*. It is important, however, to acknowledge that multiple nonneural factors may confound the interpretation of changes in the hemodynamic response measured by BOLD fMRI, such as age- and disease-related changes in neurovascular coupling *(24, 25)*, AD-specific alterations in vascular physiology *(55)*, and resting hypoperfusion and metabolism in MCI and AD *(56)*, which may result in an amplified BOLD fMRI signal during activation *(57, 58)*. Further research to determine the specificity of hyperactivation with respect to particular brain regions and behavioral conditions will be valuable to better characterize this phenomenon.

A number of authors have hypothesized that MTL and other cortical hyperactivation during the performance of memory and other cognitive tasks may play, at least in part, a compensatory role for neuropathologic abnormalities in MCI/mild AD *(59–61)*. "Compensation" is typically defined as greater regional brain activity (hyperactivation) in an MCI/AD group in the setting of task performance accuracy that is similar to that of a matched control group. Regional hyperactivation may involve greater magnitude of activity in brain regions typically active during performance of the task (when performed by controls), or the recruitment of additional brain regions not normally engaged by controls. However, it is also clear that greater task difficulty may provoke similar alterations in regional brain activity in healthy individuals *(62–65)*. It is challenging to know to what degree MCI/AD groups find memory tasks to be "more difficult" than they would in the absence of disease. This has led some investigators to attempt to match task difficulty between MCI/AD patients and controls *(61)*. It is also possible that different cognitive strategies during memory task performance (e.g., semantic elaborative encoding strategies vs. visualization strategies) may contribute to differences in the recruitment of particular brain regions *(66)* and that this may vary between patient and control groups. Further work in this area, including longitudinal studies in MCI/AD patients, ideally including detailed behavioral measures of reaction time as well as accuracy and possibly self-report of task difficulty, will be important to better clarify the situations in which activity increases can be reasonably interpreted as compensatory for brain disease.

In PD, a potential compensatory response was demonstrated in an fMRI study showing that maintenance of movement is accompanied by hyperactivation in lateral premotor areas *(45)*. In the same subjects, dopaminergic therapy normalized these activation patterns in the setting of constant motor performance [see later for additional discussion of pharmacologic fMRI (phMRI)]. Monchi et al. *(67)* showed that, during a set-shifting task, PD patients have reduced ventrolateral prefrontal activity relative to controls, but greater dorsolateral prefrontal activity, suggesting not only that frontostriatal circuits are dysfunctional in PD but that there also may be attempted compensatory activity. As shown in **Fig. 3**, an elegant study of motor imagery in PD patients with strongly lateralized symptoms demonstrated that when patients judged the laterality of hand images in different orientations, occipito-parietal cortex hyperactivation was most prominent for imagery employing the affected hand compared with the unaffected hand *(68)*.

In HD, hyperactivation was observed in multiple regions during a visual attention/interference task *(69)*. Notably, greater premotor activation correlated with a greater degree of clinical impairment, supporting the conjecture that response is at least attempting to compensate for the disease. The authors suggest that the HD patients may have required increased effort to

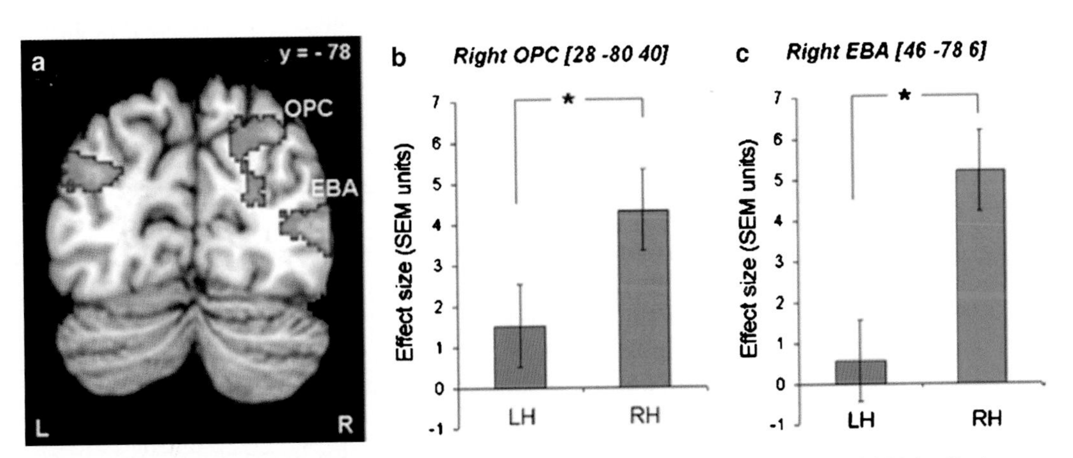

Fig. 3. Cerebral activity related to performance of a motor imagery task in patients with largely right-lateralized symptoms of Parkinson's disease. (**a**) Map of cerebral activity that increased as a function of stimulus rotation. Note relatively larger extent of activation in right occipitoparietal (OPC) and extrastriate body (EBA) regions. Bar graphs show relatively larger magnitude of activation in right hand (RH) than left hand (LH) in both regions. Figure reprinted from ***(68)*** with permission from Elsevier.

inhibit inappropriate motor responses. As in most studies identifying possible compensatory hyperactivation, there are a number of other interpretations of how hyperactivation may relate to severity of illness.

In ALS, sensorimotor cortical hyperactivation is present in comparison to both healthy normal controls and to controls with peripheral motor weakness, indicating that the hyperactivation is not purely a reflection of weakness *(70)*. Schoenfeld et al. *(71)* used a button-press sequencing task to investigate whether task difficulty level was primarily responsible for greater activation within motor circuits. Although during the simple task ALS patients showed motor hyperactivation compared with controls, when the task was manipulated such that controls had to respond more rapidly and thus made more errors (equivalent to those of ALS patients in the simpler task), motor activation was similar between the two groups.

Despite the caveats mentioned earlier with regard to many of the studies of hyperactivation in neurodegenerative diseases, accumulating evidence suggests that task-related regional brain hyperactivation may be a universal neural response to insult, as it occurs in sleep deprivation *(72)*, aging *(73)*, and a variety of neuropsychiatric disorders and conditions, including AD/MCI, PD, HD, ALS, cerebrovascular disease *(74, 75)*, multiple sclerosis *(76, 77)*, traumatic brain injury *(78)*, human immunodeficiency virus (HIV) *(79)*, alcoholism *(80)*, and schizophrenia *(81)*. In many of these studies, task-related regional brain hyperactivation was associated with the relative preservation of performance on the task, suggesting that hyperactivation may be serving, at least in part, a compensatory role for neurologic insult. The evidence discussed earlier also indicates that increased MTL activation can be seen in MCI in the setting of minimal MTL atrophy *(38)*, which provides in vivo support for laboratory and animal data suggesting that physiologic alterations may precede significant structural abnormalities very early in the course of a neurodegenerative disease such as AD *(4, 82)* and may represent inefficient neural circuit function *(54)*. Thus, fMRI may provide a means to detect changes in human brain circuit function that underlie the earliest symptoms of neurodegenerative diseases and may be useful in identifying groups of subjects at high risk for future decline prior to a clinical diagnosis of these diseases (**Fig. 4**).

It is possible, however, that hyperactivation reflects inefficient function of neural circuits in the face of injury, and that such a response may be deleterious in the long run. Thus, it will be critical to elucidate the relationships between behavioral performance, neural circuit function, and clinical course of disease, with the ultimate goal of determining how best to use these fMRI measures as biomarkers of putative therapeutic response in clinical trials.

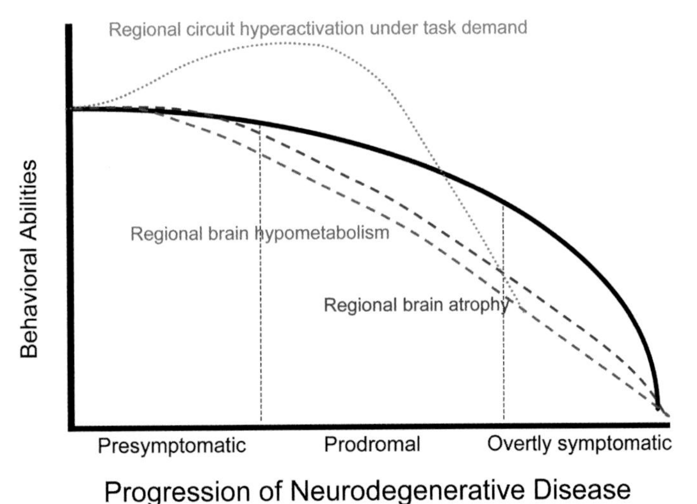

Fig. 4. Illustration of degenerative-compensatory model, proposed as a universal response to brain insult. In the case of neurodegenerative diseases, the model proposes that there is a phase of task-related hyperactivation of regional brain circuits subserving task performance, followed by the gradual development of regional hypometabolism and atrophy as the disease progresses from presymptomatic to prodromal to overtly symptomatic phases.

4.3. The Modulatory Effects of Genetic Risk Factors for Neurologic Disease on Brain Activation

In the last few years, there has been an explosion in literature on imaging and genetics, primarily in psychiatric disorders *(83)* and the basic science of genetic modulators of brain function *(84)*. This is an area that is ripe for study in neurologic disease, with a number of studies having been done in populations at elevated genetic risk for AD.

The *APOE* ε4 allele is a major genetic susceptibility factor associated with increased risk for AD *(85)*. Several fMRI studies have investigated regional brain activation during task performance in cognitively intact subjects stratified by their *APOE* allele status. Smith et al. reported decreased activation in inferior temporal regions on a visual naming and a letter fluency fMRI paradigm (there was no hippocampal or other medial temporal activation reported with these tasks) in *APOE* ε4 carriers *(86)*. In a subsequent report, this group reported increased parietal activation in women with an *APOE* ε4 allele *(87)*. Bookheimer et al. reported increased activation in left hippocampal, parietal, and prefrontal regions among *APOE* ε4 carriers, compared with noncarriers, using a word-pair associative memory paradigm *(88)*. In addition, an increased number of activated regions in the left hemisphere at baseline was associated with a decline in memory at the 2-year follow-up among the *APOE* ε4 carriers. The authors hypothesized that this increase in activation in the *APOE* ε4 carriers might represent the additional cognitive effort or neuronal recruitment required to adequately perform the task. Similarly increased activation in multiple brain regions was recently reported in cognitively intact older *APOE* ε4 carriers compared with ε3 carriers, although the effect was lateralized to the right MTL region (left hippocampal activation was greater in ε3 carriers) *(89)*. Among a group of 29 controls, MCI subjects, and AD patients, we recently reported that 13 *APOE* ε4 carriers demonstrated greater entorhinal activation than noncarriers, in the absence of genotype-related differences in the volumes of these regions *(38)*. Other studies suggest that decreased medial temporal activation may also be seen in APOE ε4 carriers *(90)*.

Five members of a family with familial AD were recently studied using fMRI memory tasks; in this family, an autosomal-dominant presenilin mutation leads to AD, typically by age 48 *(91)*. In a young (age, 20) mutation carrier, increased hippocampal and frontotemporal activation was observed during an episodic-memory task, while decreased activation in these regions was found in a middle-aged (age, 45) mutation carrier, compared with activation in the three family members without the mutation (one middle-aged and two young adults) and in a group of young, unrelated controls (mean age, 22). The authors infer that these findings are consistent with two phases of predementia functional brain change in AD: first, a hyperactivation phase, followed by a hypoactivation phase.

HD provides an excellent opportunity to study functional brain alterations in individuals who are genetically destined to develop the disease but who are not yet manifesting symptoms. During an attentional interference task, presymptomatic HD patients demonstrated reduced anterior cingulate activation compared with controls, suggesting a functional basis for subtle performance abnormalities *(92)*.

4.4. fMRI in the Differential Diagnosis of Neuropsychiatric Syndromes

The early detection and differential diagnosis of disorders causing cognitive impairment is a promising aim for further work using fMRI. Since clinical evaluation and neuropsychological testing are currently the most sensitive approaches to diagnosis, and fMRI is sensitive to both cognitive performance and clinical status, it seems reasonable to hope that the potential capability of fMRI to detect alterations in the pattern and degree of regional brain activation during task performance may provide additional useful data to complement clinical and psychometric evaluations. For example, fMRI may be of value in demonstrating regional brain activation differences in distinct clinical subtypes of disease, as described earlier in ALS *(50)*. However, relatively little fMRI data have been published on differential diagnosis or disease subtyping to date.

In elderly individuals with cognitive symptoms, it can be difficult to distinguish a neurodegenerative process from depression – fMRI may be helpful in this setting. In a study of older individuals who had sought clinical evaluation for memory-related symptoms, Gron et al. *(93)* investigated the utility of fMRI to differentiate patterns of regional brain activation in those diagnosed with depression vs. AD (as well as a control group). Hippocampal activation during the memory task was decreased in AD patients than in controls and depressed patients. In contrast, orbitofrontal and cingulate activation were greater in depressed patients than in AD subjects and controls.

Furthermore, different forms of neurodegenerative dementias may be challenging to diagnose specifically early in their course. fMRI may provide helpful data to assist in differential diagnosis of the dementias. Rombouts et al. *(94)* compared regional brain activation during a working memory task in patients with frontotemporal dementia (FTD) to that of AD patients. Although both groups activated similar fronto-parietal-thalamic regions, frontoparietal activation was diminished in FTD patients than in AD patients.

Further insights into the utility of fMRI in assisting with differential diagnosis may potentially be gained through prospective studies of patients presenting for clinical evaluation with subtle symptoms consistent with a degenerative dementia who do not yet have a clear clinical diagnosis. If such individuals are scanned using tasks they can still perform and then followed clinically, it may be possible to determine whether fMRI has predictive power in differential diagnosis.

4.5. fMRI As a Predictive Biomarker

Very little work has been done on the use of fMRI as a predictive biomarker for prognosis. We recently pursued such a study of a group of 25 senior citizens spanning the spectrum of MCI, none of whom were demented at the time of baseline assessment, but who exhibited varying degrees of mild symptoms of cognitive impairment clinically [as measured using the CDR sum-of-boxes (CDR-SB)] *(95)*. At baseline, subjects performed a visual scene-encoding task during fMRI scanning and were clinically followed longitudinally after scanning. Over about 4 years of follow-up after scanning, subjects demonstrated a wide range of cognitive decline, with some showing no change and others progressing to dementia (change in CDR-SB ranged from 0 to 4.5). The degree of cognitive decline was predicted by hippocampal activation at the time of baseline scanning, with greater hippocampal activation predicting greater decline ($p < 0.05$) **(Fig. 5)**. This finding was present even after controlling for baseline degree of impairment (CDR-SB), age, education, and hippocampal volume. These data suggest that fMRI may provide a physiologic imaging biomarker useful for identifying the subgroup of MCI individuals at highest risk of cognitive decline for potential inclusion in disease-modifying clinical trials.

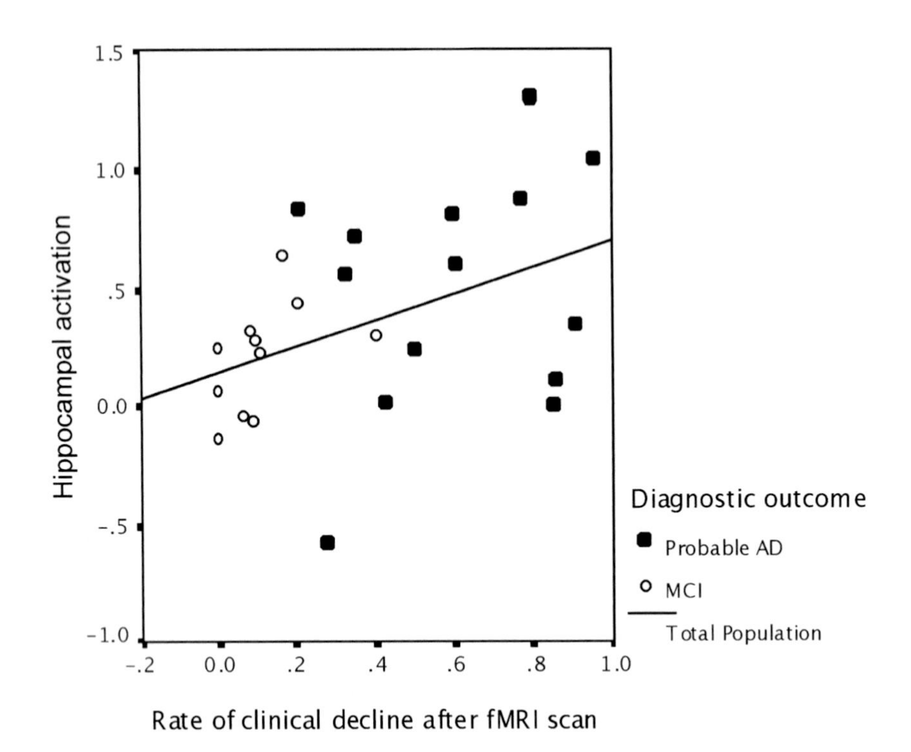

Fig. 5. fMRI as a predictive quantitative imaging biomarker. In a group of mild cognitive impairment patients, hippocampal activation at baseline predicts the degree of cognitive decline over 4 years after scanning. *Y*-axis indicates percent blood oxygen level-dependent (BOLD) signal change within hippocampal formation; *X*-axis indicates rate of clinical decline as measured by change in CDR sum-of-boxes (CDR-SB) per year, with higher numbers indicating greater decline.

4.6. Links Between Task-Related Brain Function Abnormalities and Altered Resting Brain Activity

Recent fMRI studies are beginning to reveal a link between disease-related hemodynamic alterations and the well-described resting perfusion/metabolic abnormalities in AD. Hypoperfusion/metabolism is typically seen with nuclear medical imaging techniques [such as fluorodeoxyglucose positron emission tomography (FDG-PET) or single photon emission computed tomography (SPECT)] in temporo-parietal/posterior cingulate cortical regions in AD patients during the "resting" state. The medial parietal/posterior cingulate cortex, along with medial frontal and lateral parietal regions, appears to compose a "default mode" network that is more active when individuals are not engaged in particular tasks and is thought to play a role in vigilance, readiness, or monitoring - these regions "deactivate" (BOLD signal amplitude falls below baseline) during cognitive task performance *(96)*. Several recent studies in AD patients have demonstrated alterations in the deactivation and functional connectivity of these regions, suggesting that this *default mode* network is disrupted by the disease *(97–100)*. Substantial overlap is present between these *default mode* areas and the localization of PET amyloid tracer binding *(101)*.

4.7. Uses of fMRI in Understanding and Monitoring Neurotherapeutics

fMRI may be particularly valuable in evaluating acute and subacute effects of medications on neural activity. Alterations in memory-related activation related to the administration of pharmacologic agents known to impair memory can be detected with pharmacological MRI (phMRI) *(23, 102, 103)*. The effects of cognitive enhancing drugs on brain activation during cognitive task performance have shown that fMRI can detect changes after administration of cholinesterase inhibitors in patients with AD and MCI *(104,105)*. Rombouts et al. *(104)* found that, after receiving a single dose of galanthamine, AD patients demonstrated increased fusiform activity during a face encoding task and increased prefrontal activity during a working memory task. Saykin et al. *(105)* reported that, compared with the control group of healthy older individuals, MCI patients who received 6 weeks of donepezil showed increased prefrontal activation after this course of medication, which related to improvements in performance of a working memory task. In a recent study of galanthamine in patients with MCI, Goekoop et al. *(106)* found that, after patients received galanthamine for approximately 1 week, performance on a working memory task was improved in conjunction with increased activation in precuneus and middle frontal regions. In addition, increases in hippocampal, prefrontal, cingulate, and occipital regions were seen during an episodic encoding task, although performance did not improve on this memory task. Although these pilot studies did not include placebo-control groups to reduce potential confounding factors, such as learning effects, they indicate that fMRI is sensitive to both acute and subacute medication effects, some of which relate to behavioral change.

Parkinson's disease (PD) is an excellent clinical scenario in which to apply phMRI, given the typical rapid responsiveness of symptoms to dopaminergic therapy. In drug-naïve patients with mild PD, SMA and contralateral motor cortex hypoactivation during a simple finger movement task normalized with l-dopa therapy *(107)*. Motor performance was constant across conditions, suggesting that any change could be ascribed to pharmacological modulation within basal ganglia–thalamocortical loops.

In addition to motor behavior, the modulatory effects of dopaminergic therapy on cognition and emotion have also been studied in PD. In one well-controlled study, the modulatory effects of dopamine replacement were studied using both sensorimotor and working memory tasks *(108)*. The cortical motor regions activated during the motor task showed greater activation during the dopamine-replete state, but the cortical regions subserving working memory displayed greater activation during the hypodopaminergic state. Interestingly, the greater cortical activation during the working memory task in the hypodopaminergic state correlated with errors in task performance, while the increased activation in the cortical motor regions during the dopamine-replete state was correlated with improvement in motor function. These results are consistent with evidence that the hypodopaminergic state is associated with decreased efficiency of prefrontal cortical information processing and that dopaminergic therapy improves the physiological efficiency of this region, and also indicate that hyperactivation does not necessarily reflect better behavioral performance.

In a study of emotional face processing in PD, amygala activation was reduced compared with controls during a hypodopaminergic state, and partly normalized with dopaminergic therapy *(109)*.

In other clinical populations, phMRI has been applied to investigations of cocaine dependence *(110)*, depression *(111)*, and schizophrenia *(112)*. A recently emerging, exciting application of fMRI is in the prediction of clinical response to pharmacologic treatment – e.g., patients with depression who are treated with antidepressant medications demonstrate changes in activation of affective circuitry that are associated with clinical response to pharmacologic treatment *(113)* and may enable prediction of response *(114)*. The addition of genetic factors to these investigations holds fascinating and powerful promise to illuminate fundamental pharmacogenetic mechanisms of central nervous

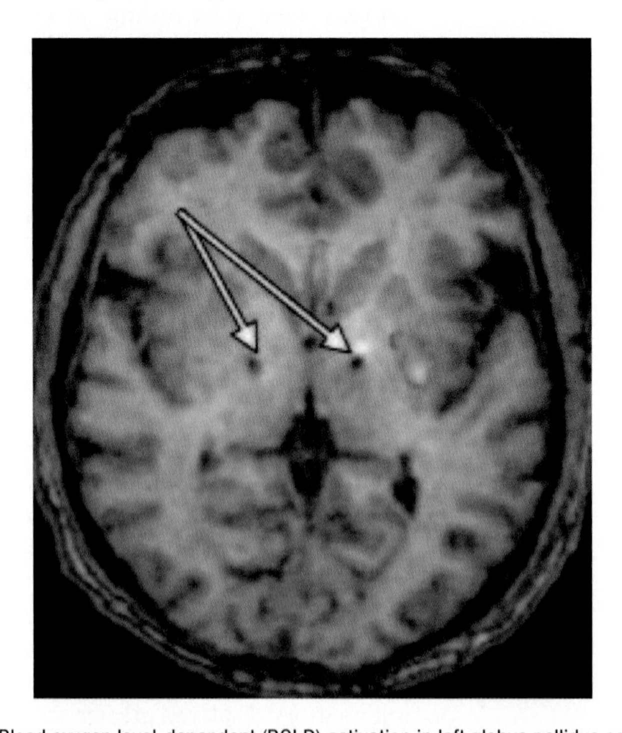

Fig. 6. Blood oxygen level-dependent (BOLD) activation in left globus pallidus as measured by fMRI in subthalamic nucleus in patient with Parkinson's disease during stimulation of deep brain stimulator electrode. Bilateral electrodes are indicated with arrows; stimulation took place only in left electrode. Figure reprinted from ***(117)*** with permission from the Radiologic Society of North America and Professor Phillips.

system drug effects – e.g., in schizophrenic patients, there is a catechol-*o*-methyl-transferase polymorphism-dependent effect of atypical antipsychotic therapy on prefrontal activation during working memory task performance *(115)*.

Moving beyond its uses in pharmacologic studies, fMRI has recently begun to be used to study the effects of deep brain stimulation (DBS). DBS has become a well-accepted therapeutic modality for PD and is finding a growing number of applications in other neurologic and psychiatric disorders. Yet there remain many questions about the fundamental basis through which DBS modulates neural circuits. These questions have begun to be pursued using functional neuroimaging methods including fMRI, with initial studies focusing on safety and the measurement of BOLD signal changes during on versus off stimulator activity **(Fig. 6)** *(116, 117)*.

5. Beyond Exclusion: The Use of Imaging Measures As Disease Biomarkers

At present, the potential efficacy of disease-modifying therapies for AD and other neurodegenerative diseases is evaluated primarily using clinical measures of cognition, movement, and other behaviors. In animal models, traditional behavioral assessments are often used, such as the rate at which rodents learn to navigate a maze. In clinical trials, outcome measures are typically performance-based instruments, such as the Alzheimer's Disease Assessment Scale (ADAS-Cog) *(118)*, or structured surveys of clinician/ caregiver impression of change *(119)*. Although the efficacy of disease-modifying treatments for AD and other neurodegenerative diseases must ultimately be demonstrated using clinically meaningful outcome measures such as the slowing of decline in progression of symptoms or functional impairment, such trials will likely require hundreds of patients studied for a minimum of 1–2 years. Thus, surrogate markers of efficacy with less variability than clinical assessments are desperately needed to reduce the number of subjects. These markers may also prove particularly valuable in the early phases of drug development to detect a preliminary "signal of efficacy" over a shorter time period.

Since the pathophysiologic process underlying cognitive decline in AD and other neurodegenerative diseases involves the progressive degeneration of particular brain regions, repeatable in vivo neuroimaging measures of brain anatomy, chemistry, physiology, and pathology hold promise as an important class of potential biomarkers *(3)*. A growing body of data indicates that the natural history of gradually progressive cognitive decline in AD can be reliably related to changes in such imaging measures. Furthermore, regionally specific changes in brain anatomy, chemistry, and physiology can be detected by imaging prior to the point at which the disease is symptomatic enough to make a typical clinical diagnosis. Finally, evidence is accumulating that alterations in synaptic function are present very early in the disease process, possibly long before the development of clinical symptoms and significant cell loss, which may relate closely to symptomatic progression in manifest disease *(4, 5, 120)*. Thus, potential disease-modifying therapies may act by impeding the accumulation of neuropathology, slowing the loss of neurons, altering neurochemistry, or preserving synaptic function; neuroimaging modalities exist to measure each of these putative therapeutic goals, and fMRI could potentially play a valuable role in the development of new scientific insights into functional brain abnormalities early in the course of these diseases and in the development of therapeutic agents.

As advances in research provide data to support the use of specific imaging measures as biomarkers, it becomes apparent that these measures could be useful in diagnosis and eventually may find applications in routine clinical practice. As neurologic diagnosis of neurodegenerative diseases moves beyond the simple use of imaging for the exclusion of mass lesions or other "potentially reversible" causes of dementia or other symptoms, these tools will become increasingly more important in routine practice *(121)*. In fact, the American Academy of Neurology recently proposed that a neuroimaging study should be performed in the workup of all cases of dementia *(122)*. An increasing number of diagnostic criteria sets, including frontotemporal, Lewy body, and cerebrovascular dementias, are including neuroimaging evidence as a core or supportive component *(11, 123–126)*. Thus, research motivated toward improving our understanding of the natural history of neurodegenerative diseases and toward the development of new therapies is also assisting in the translation of diagnostic tools from bench to beside.

Although measures of brain structure (MRI) and resting brain function (e.g., SPECT or PET) are already clinically useful and will likely become the first accepted quantitative neuroimaging biomarkers for neurodegenerative diseases, fMRI recently received its first approvals for current procedural technology (CPT) codes by the American Medical Association, one application of which is toward the assessment of complex cognitive and sensorimotor function in patients with neuropsychiatric disorders *(127)*. This is likely the first step toward the translation of fMRI technology from the bench to the bedside. It will be extremely valuable, as this begins to occur, to collect fMRI data in the context of structural MRI, resting functional (e.g., FDG-PET), and other multimodal imaging data to begin to understand the relationships between these data types.

6. Conclusions

fMRI is a particularly attractive method for use by clinical investigators to study task-related brain activation in patients with neurodegenerative illness. Despite the relative infancy of the field, there have already been a number of promising fMRI studies in various neurodegenerative disorders that highlight the potential uses of fMRI in both basic and clinical spheres of investigation. fMRI may provide novel insights into the neural correlates of cognitive and sensorimotor abilities and how they are altered by neurologic disease and by medications. The technique may help elucidate fundamental aspects of brain-behavior relationships, such as the genetic influences on task-related brain physiology. fMRI measures hold promise for multiple clinical applications, including the early detection and differential diagnosis, predicting future change in clinical status, and as a marker of alterations in brain physiology related to neurotherapeutic agents. The greatest potential of fMRI likely lies in the study of very early and preclinical stages of progressive neurologic diseases, at the point of subtle neuronal dysfunction prior to overt anatomic pathology. There is a need for further validation and reliability studies and continued technical advances to fully realize the potential of fMRI.

References

1. Trojanowski JQ. Protein mis-folding emerges as a "drugable" target for discovery of novel therapies for neuropsychiatric diseases of aging. Am J Geriatr Psychiatry 2004;12: 134–135.
2. Bertram L, Tanzi RE. The genetic epidemiology of neurodegenerative disease. J Clin Invest 2005;115:1449–1457.
3. DeKosky ST, Marek K. Looking backward to move forward: early detection of neurodegenerative disorders. Science 2003;302: 830–834.
4. Selkoe DJ. Alzheimer's disease is a synaptic failure. Science 2002;298:789–791.
5. Coleman P, Federoff H, Kurlan R. A focus on the synapse for neuroprotection in Alzheimer disease and other dementias. Neurology 2004;63:1155–1162.
6. Schenk D, Barbour R, Dunn W, et al. Immunization with amyloid-beta attenuates Alzheimer-disease-like pathology in the PDAPP mouse. Nature 1999;400:173–177.
7. Weiner HL, Lemere CA, Maron R, et al. Nasal administration of amyloid-beta peptide decreases cerebral amyloid burden in a mouse model of Alzheimer's disease. Ann Neurol 2000;48:567–579.
8. Lombardo JA, Stern EA, McLellan ME, et al. Amyloid-beta antibody treatment leads to rapid normalization of plaque-induced neuritic alterations. J Neurosci 2003;23: 10879–10883.
9. Petersen RC, Smith GE, Waring SC, Ivnik RJ, Tangalos EG, Kokmen E. Mild cognitive impairment: clinical characterization and outcome. Arch Neurol 1999;56: 303–308.
10. Grundman M, Petersen RC, Ferris SH, et al. Mild cognitive impairment can be distinguished from Alzheimer disease and normal aging for clinical trials. Arch Neurol 2004;61:59–66.
11. Dubois B, Feldman HH, Jacova C, et al. Research criteria for the diagnosis of Alzheimer's disease: revising the NINCDS-ADRDA criteria. Lancet Neurol 2007;6:734–746.
12. Dickerson BC, Miller SL, Greve DN, et al. Prefrontal-hippocampal-fusiform activity during encoding predicts intraindividual differences in free recall ability: an event-related functional-anatomic MRI study. Hippocampus 2007;17:1060–1070.
13. Brewer JB, Zhao Z, Desmond JE, Glover GH, Gabrieli JD. Making memories: brain

activity that predicts how well visual experience will be remembered. Science 1998;281: 1185–1187.
14. Wagner AD, Schacter DL, Rotte M, et al. Building memories: remembering and forgetting of verbal experiences as predicted by brain activity. Science 1998;281:1188–1191.
15. Kirchhoff BA, Wagner AD, Maril A, Stern CE. Prefrontal-temporal circuitry for episodic encoding and subsequent memory. J Neurosci 2000;20:6173–6180.
16. Daselaar SM, Veltman DJ, Rombouts SA, Raaijmakers JG, Jonker C. Neuroanatomical correlates of episodic encoding and retrieval in young and elderly subjects. Brain 2003;126:43–56.
17. Sperling R, Chua E, Cocchiarella A, et al. Putting names to faces: successful encoding of associative memories activates the anterior hippocampal formation. Neuroimage 2003;20:1400–1410.
18. Price CJ, Friston KJ. Scanning patients with tasks they can perform. Hum Brain Mapp 1999;8:102–108.
19. Grady CL, McIntosh AR, Beig S, Keightley ML, Burian H, Black SE. Evidence from functional neuroimaging of a compensatory prefrontal network in Alzheimer's disease. J Neurosci 2003;23:986–993.
20. Dickerson BC, Salat DH, Bates JF, et al. Medial temporal lobe function and structure in mild cognitive impairment. Ann Neurol 2004;56:27–35.
21. Machielsen WC, Rombouts SA, Barkhof F, Scheltens P, Witter MP. FMRI of visual encoding: reproducibility of activation. Hum Brain Mapp 2000;9:156–164.
22. Manoach DS, Halpern EF, Kramer TS, et al. Test–retest reliability of a functional MRI working memory paradigm in normal and schizophrenic subjects. Am J Psychiatry 2001;158:955–958.
23. Sperling R, Greve D, Dale A, et al. Functional MRI detection of pharmacologically induced memory impairment. Proc Natl Acad Sci USA 2002;99:455–460.
24. Buckner RL, Snyder AZ, Sanders AL, Raichle ME, Morris JC. Functional brain imaging of young, nondemented, and demented older adults. J Cogn Neurosci 2000;12 Suppl 2: 24–34.
25. D'Esposito M, Deouell LY, Gazzaley A. Alterations in the BOLD fMRI signal with ageing and disease: a challenge for neuroimaging. Nat Rev Neurosci 2003;4:863–872.
26. Saykin AJ, Flashman LA, Frutiger SA, et al. Neuroanatomic substrates of semantic memory impairment in Alzheimer's disease: patterns of functional MRI activation. J Int Neuropsychol Soc 1999;5:377–392.
27. Johnson SC, Saykin AJ, Baxter LC, et al. The relationship between fMRI activation and cerebral atrophy: comparison of normal aging and Alzheimer disease. Neuroimage 2000;11:179–187.
28. Grossman M, Koenig P, DeVita C, et al. Neural basis for verb processing in Alzheimer's disease: an fMRI study. Neuropsychology 2003;17:658–674.
29. Grossman M, Koenig P, Glosser G, et al. Neural basis for semantic memory difficulty in Alzheimer's disease: an fMRI study. Brain 2003;126:292–311.
30. Thulborn KR, Martin C, Voyvodic JT. Functional MR imaging using a visually guided saccade paradigm for comparing activation patterns in patients with probable Alzheimer's disease and in cognitively able elderly volunteers. AJNR Am J Neuroradiol 2000;21: 524–531.
31. Small SA, Perera GM, DeLaPaz R, Mayeux R, Stern Y. Differential regional dysfunction of the hippocampal formation among elderly with memory decline and Alzheimer's disease. Ann Neurol 1999;45:466–472.
32. Rombouts SA, Barkhof F, Veltman DJ, et al. Functional MR imaging in Alzheimer's disease during memory encoding. AJNR Am J Neuroradiol 2000;21:1869–1875.
33. Kato T, Knopman D, Liu H. Dissociation of regional activation in mild AD during visual encoding: a functional MRI study. Neurology 2001;57:812–816.
34. Machulda MM, Ward HA, Borowski B, et al. Comparison of memory fMRI response among normal, MCI, and Alzheimer's patients. Neurology 2003;61:500–506.
35. Sperling RA, Bates JF, Chua EF, et al. fMRI studies of associative encoding in young and elderly controls and mild Alzheimer's disease. J Neurol Neurosurg Psychiatry 2003;74: 44–50.
36. Rombouts SARB, Goekoop R, Stam CJ, Barkhof F, Scheltens P. Delayed rather than decreased BOLD response as a marker for early Alzheimer's disease. Neuroimage 2005; 26:1078–1085.
37. Johnson SC, Baxter LC, Susskind-Wilder L, Connor DJ, Sabbagh MN, Caselli RJ. Hippocampal adaptation to face repetition in healthy lderly and mild cognitive impairment. Neuropsychologia 2004;42:980–989.
38. Dickerson BC, Salat DH, Greve DN, et al. Increased hippocampal activation in mild

cognitive impairment compared to normal aging and AD. Neurology 2005;65: 404–411.

39. Hamalainen A, Pihlajamaki M, Tanila H, et al. Increased fMRI responses during encoding in mild cognitive impairment. Neurobiol Aging 2007;28:1889–1903.
40. Johnson SC, Schmitz TW, Moritz CH, et al. Activation of brain regions vulnerable to Alzheimer's disease: the effect of mild cognitive impairment. Neurobiol Aging 2006;27:1604–1612.
41. Petrella JR, Krishnan S, Slavin MJ, Tran TT, Murty L, Doraiswamy PM. Mild cognitive impairment: evaluation with 4-T functional MR imaging. Radiology 2006;240: 177–186.
42. Kircher T, Weis S, Freymann K, et al. Hippocampal activation in MCI patients is necessary for successful memory encoding. J Neurol Neurosurg Psychiatry 2007;78:812–818.
43. Dickerson BC, Sperling RA. Functional abnormalities of the medial temporal lobe memory system in mild cognitive impairment and Alzheimer's disease: insights from functional MRI studies. Neuropsychologia 2008;46:1624–1635.
44. Elsinger CL, Rao SM, Zimbelman JL, Reynolds NC, Blindauer KA, Hoffmann RG. Neural basis for impaired time reproduction in Parkinson's disease: an fMRI study. J Int Neuropsychol Soc 2003;9:1088–1098.
45. Haslinger B, Erhard P, Kampfe N, et al. Event-related functional magnetic resonance imaging in Parkinson's disease before and after levodopa. Brain 2001;124: 558–570.
46. Lewis SJ, Dove A, Robbins TW, Barker RA, Owen AM. Cognitive impairments in early Parkinson's disease are accompanied by reductions in activity in frontostriatal neural circuitry. J Neurosci 2003;23:6351–6356.
47. Sauer J, ffytche DH, Ballard C, Brown RG, Howard R. Differences between Alzheimer's disease and dementia with Lewy bodies: an fMRI study of task-related brain activity. Brain 2006;129:1780–1788.
48. Kim JS, Reading SA, Brashers-Krug T, Calhoun VD, Ross CA, Pearlson GD. Functional MRI study of a serial reaction time task in Huntington's disease. Psychiatry Res 2004;131:23–30.
49. Thiruvady DR, Georgiou-Karistianis N, Egan GF, et al. Functional connectivity of the prefrontal cortex in Huntington's disease. J Neurol Neurosurg Psychiatry 2007;78: 127–133.
50. Tessitore A, Esposito F, Monsurro MR, et al. Subcortical motor plasticity in patients with sporadic ALS: an fMRI study. Brain Res Bull 2006;69:489–494.
51. Wierenga CE, Bondi MW. Use of functional magnetic resonance imaging in the early identification of Alzheimer's disease. Neuropsychol Rev 2007;17:127–143.
52. DeKosky ST, Ikonomovic MD, Styren SD, et al. Upregulation of choline acetyltransferase activity in hippocampus and frontal cortex of elderly subjects with mild cognitive impairment. Ann Neurol 2002;51: 145–155.
53. Hashimoto M, Masliah E. Cycles of aberrant synaptic sprouting and neurodegeneration in Alzheimer's and dementia with Lewy bodies. Neurochem Res 2003;28:1743–1756.
54. Stern EA, Bacskai BJ, Hickey GA, Attenello FJ, Lombardo JA, Hyman BT. Cortical synaptic integration in vivo is disrupted by amyloid-beta plaques. J Neurosci 2004;24:4535–4540.
55. Mueggler T, Sturchler-Pierrat C, Baumann D, Rausch M, Staufenbiel M, Rudin M. Compromised hemodynamic response in amyloid precursor protein transgenic mice. J Neurosci 2002;22:7218–7224.
56. El Fakhri G, Kijewski MF, Johnson KA, et al. MRI-guided SPECT perfusion measures and volumetric MRI in prodromal Alzheimer disease. Arch Neurol 2003;60: 1066–1072.
57. Davis TL, Kwong KK, Weisskoff RM, Rosen BR. Calibrated functional MRI: mapping the dynamics of oxidative metabolism. Proc Natl Acad Sci USA 1998;95: 1834–1839.
58. Cohen ER, Ugurbil K, Kim SG. Effect of basal conditions on the magnitude and dynamics of the blood oxygenation level-dependent fMRI response. J Cereb Blood Flow Metab 2002;22:1042–1053.
59. Becker JT, Mintun MA, Aleva K, Wiseman MB, Nichols T, DeKosky ST. Compensatory reallocation of brain resources supporting verbal episodic memory in Alzheimer's disease. Neurology 1996;46:692–700.
60. Backman L, Andersson JL, Nyberg L, Winblad B, Nordberg A, Almkvist O. Brain regions associated with episodic retrieval in normal aging and Alzheimer's disease. Neurology 1999;52:1861–1870.
61. Stern Y, Moeller JR, Anderson KE, et al. Different brain networks mediate task performance in normal aging and AD: defining compensation. Neurology 2000;55: 1291–1297.

62. Gur RC, Gur RE, Skolnick BE, et al. Effects of task difficulty on regional cerebral blood flow: relationships with anxiety and performance. Psychophysiology 1988;25:392–399.
63. Grasby PM, Frith CD, Friston KJ, et al. A graded task approach to the functional mapping of brain areas implicated in auditory-verbal memory. Brain 1994;117:1271–1282.
64. Grady CL. Age-related changes in cortical blood flow activation during perception and memory. Ann N Y Acad Sci 1996;777:14–21.
65. Rypma B, D'Esposito M. The roles of prefrontal brain regions in components of working memory: effects of memory load and individual differences. Proc Natl Acad Sci USA 1999;96:6558–6563.
66. Kirchhoff BA, Buckner RL. Functional-anatomic correlates of individual differences in memory. Neuron 2006;51:263–274.
67. Monchi O, Petrides M, Doyon J, Postuma RB, Worsley K, Dagher A. Neural bases of set-shifting deficits in Parkinson's disease. J Neurosci 2004;24:702–710.
68. Helmich RC, de Lange FP, Bloem BR, Toni I. Cerebral compensation during motor imagery in Parkinson's disease. Neuropsychologia 2007;45:2201–2215.
69. Georgiou-Karistianis N, Sritharan A, Farrow M, et al. Increased cortical recruitment in Huntington's disease using a Simon task. Neuropsychologia 2007;45:1791–1800.
70. Stanton BR, Williams VC, Leigh PN, et al. Altered cortical activation during a motor task in ALS: evidence for involvement of central pathways. J Neurol 2007;254;1260–1267.
71. Schoenfeld MA, Tempelmann C, Gaul C, et al. Functional motor compensation in amyotrophic lateral sclerosis. J Neurol 2005;252:944–952.
72. Drummond SP, Brown GG, Gillin JC, Stricker JL, Wong EC, Buxton RB. Altered brain response to verbal learning following sleep deprivation. Nature 2000;403: 655–657.
73. Cabeza R, Anderson ND, Locantore JK, McIntosh AR. Aging gracefully: compensatory brain activity in high-performing older adults. Neuroimage 2002;17:1394–1402.
74. Carey JR, Kimberley TJ, Lewis SM, et al. Analysis of fMRI and finger tracking training in subjects with chronic stroke. Brain 2002;125:773–788.
75. Johansen-Berg H, Dawes H, Guy C, Smith SM, Wade DT, Matthews PM. Correlation between motor improvements and altered fMRI activity after rehabilitative therapy. Brain 2002;125:2731–2742.
76. Reddy H, Narayanan S, Arnoutelis R, et al. Evidence for adaptive functional changes in the cerebral cortex with axonal injury from multiple sclerosis. Brain 2000;123: 2314–2320.
77. Morgen K, Kadom N, Sawaki L, et al. Training-dependent plasticity in patients with multiple sclerosis. Brain 2004;127: 2506–2517.
78. McAllister TW, Saykin AJ, Flashman LA, et al. Brain activation during working memory 1 month after mild traumatic brain injury: a functional MRI study. Neurology 1999;53:1300–1308.
79. Ernst T, Chang L, Jovicich J, Ames N, Arnold S. Abnormal brain activation on functional MRI in cognitively asymptomatic HIV patients. Neurology 2002;59: 1343–1349.
80. Desmond JE, Chen SH, DeRosa E, Pryor MR, Pfefferbaum A, Sullivan EV. Increased frontocerebellar activation in alcoholics during verbal working memory: an fMRI study. Neuroimage 2003;19:1510–1520.
81. Callicott JH, Mattay VS, Verchinski BA, Marenco S, Egan MF, Weinberger DR. Complexity of prefrontal cortical dysfunction in schizophrenia: more than up or down. Am J Psychiatry 2003;160: 2209–2215.
82. Walsh DM, Selkoe DJ. Deciphering the molecular basis of memory failure in Alzheimer's disease. Neuron 2004;44:181–193.
83. Winterer G, Hariri AR, Goldman D, Weinberger DR. Neuroimaging and human genetics. Int Rev Neurobiol 2005;67:325–383.
84. Hariri AR, Weinberger DR. Functional neuroimaging of genetic variation in serotonergic neurotransmission. Genes Brain Behav 2003;2:341–349.
85. Saunders AM. Apolipoprotein E and Alzheimer disease: an update on genetic and functional analyses. J Neuropathol Exp Neurol 2000;59:751–758.
86. Smith CD, Andersen AH, Kryscio RJ, et al. Altered brain activation in cognitively intact individuals at high risk for Alzheimer's disease. Neurology 1999;53:1391–1396.
87. Smith CD, Andersen AH, Kryscio RJ, et al. Women at risk for AD show increased parietal activation during a fluency task. Neurology 2002;58:1197–1202.
88. Bookheimer SY, Strojwas MH, Cohen MS, et al. Patterns of brain activation in people at risk for Alzheimer's disease. N Engl J Med 2000;343:450–456.
89. Bondi MW, Houston WS, Eyler LT, Brown GG. fMRI evidence of compensatory mech-

anisms in older adults at genetic risk for Alzheimer disease. Neurology 2005;64: 501–508.

90. Johnson SC, Schmitz TW, Trivedi MA, et al. The influence of Alzheimer disease family history and apolipoprotein E epsilon4 on mesial temporal lobe activation. J Neurosci 2006;26: 6069–6076.
91. Mondadori CR, Buchmann A, Mustovic H, et al. Enhanced brain activity may precede the diagnosis of Alzheimer's disease by 30 years. Brain 2006;129:2908–2922.
92. Reading SA, Dziorny AC, Peroutka LA, et al. Functional brain changes in presymptomatic Huntington's disease. Ann Neurol 2004;55:879–883.
93. Gron G, Bittner D, Schmitz B, Wunderlich AP, Riepe MW. Subjective memory complaints: objective neural markers in patients with Alzheimer's disease and major depressive disorder. Ann Neurol 2002;51: 491–498.
94. Rombouts SA, van Swieten JC, Pijnenburg YA, Goekoop R, Barkhof F, Scheltens P. Loss of frontal fMRI activation in early frontotemporal dementia compared to early AD. Neurology 2003;60:1904–1908.
95. Miller S, Bates J, Blacker D, Sperling RA, Dickerson BC. Hippocampal activation in MCI predicts subsequent cognitive decline. In: International Conference on Alzheimer's Disease, Madrid, Spain, 2006.
96. Raichle ME, MacLeod AM, Snyder AZ, Powers WJ, Gusnard DA, Shulman GL. A default mode of brain function. Proc Natl Acad Sci USA 2001;98:676–682.
97. Lustig C, Snyder AZ, Bhakta M, et al. Functional deactivations: change with age and dementia of the Alzheimer type. Proc Natl Acad Sci USA 2003;100:14504–14509.
98. Greicius MD, Srivastava G, Reiss AL, Menon V. Default-mode network activity distinguishes Alzheimer's disease from healthy aging: evidence from functional MRI. Proc Natl Acad Sci USA 2004;101: 4637–4642.
99. Rombouts SA, Barkhof F, Goekoop R, Stam CJ, Scheltens P. Altered resting state networks in mild cognitive impairment and mild Alzheimer's disease: an fMRI study. Hum Brain Mapp 2005;26:231–239.
100. Celone KA, Calhoun VD, Dickerson BC, et al. Alterations in memory networks in mild cognitive impairment and Alzheimer's disease: an independent component analysis. J Neurosci 2006;26:10222–10231.
101. Buckner RL, Snyder AZ, Shannon BJ, et al. Molecular, structural, and functional characterization of Alzheimer's disease: evidence for a relationship between default activity, amyloid, and memory. J Neurosci 2005;25: 7709–7717.
102. Leslie RA, James MF. Pharmacological magnetic resonance imaging: a new application for functional MRI. Trends Pharmacol Sci 2000;21:314–318.
103. Thiel CM, Henson RN, Dolan RJ. Scopolamine but not lorazepam modulates face repetition priming: a psychopharmacological fMRI study. Neuropsychopharmacology 2002;27:282–292.
104. Rombouts SA, Barkhof F, Van Meel CS, Scheltens P. Alterations in brain activation during cholinergic enhancement with rivastigmine in Alzheimer's disease. J Neurol Neurosurg Psychiatry 2002;73:665–671.
105. Saykin AJ, Wishart HA, Rabin LA, et al. Cholinergic enhancement of frontal lobe activity in mild cognitive impairment. Brain 2004;127:1574–1583.
106. Goekoop R, Rombouts SA, Jonker C, et al. Challenging the cholinergic system in mild cognitive impairment: a pharmacological fMRI study. Neuroimage 2004;23:1450–1459.
107. Buhmann C, Glauche V, Sturenburg HJ, Oechsner M, Weiller C, Buchel C. Pharmacologically modulated fMRI – cortical responsiveness to levodopa in drug-naive hemiparkinsonian patients. Brain 2003;126:451–461.
108. Mattay VS, Tessitore A, Callicott JH, et al. Dopaminergic modulation of cortical function in patients with Parkinson's disease. Ann Neurol 2002;51:156–164.
109. Tessitore A, Hariri AR, Fera F, et al. Dopamine modulates the response of the human amygdala: a study in Parkinson's disease. J Neurosci 2002;22:9099–9103.
110. Breiter HC, Gollub RL, Weisskoff RM, et al. Acute effects of cocaine on human brain activity and emotion. Neuron 1997;19: 591–611.
111. Kalin NH, Davidson RJ, Irwin W, et al. Functional magnetic resonance imaging studies of emotional processing in normal and depressed patients: effects of venlafaxine. J Clin Psychiatry 1997;58 Suppl 16:32–39.
112. Honey GD, Bullmore ET, Soni W, Varatheesan M, Williams SC, Sharma T. Differences in frontal cortical activation by a working memory task after substitution of risperidone for typical antipsychotic drugs in patients

with schizophrenia. Proc Natl Acad Sci USA 1999;96:13432–13437.

113. Fu CH, Williams SC, Cleare AJ, et al. Attenuation of the neural response to sad faces in major depression by antidepressant treatment: a prospective, event-related functional magnetic resonance imaging study. Arch Gen Psychiatry 2004;61:877–889.
114. Davidson RJ, Irwin W, Anderle MJ, Kalin NH. The neural substrates of affective processing in depressed patients treated with venlafaxine. Am J Psychiatry 2003;160:64–75.
115. Bertolino A, Caforio G, Blasi G, et al. Interaction of COMT (Val(108/158)Met) genotype and olanzapine treatment on prefrontal cortical function in patients with schizophrenia. Am J Psychiatry 2004;161:1798–1805.
116. Arantes PR, Cardoso EF, Barreiros MA, et al. Performing functional magnetic resonance imaging in patients with Parkinson's disease treated with deep brain stimulation. Mov Disord 2006;21:1154–1162.
117. Phillips MD, Baker KB, Lowe MJ, et al. Parkinson disease: pattern of functional MR imaging activation during deep brain stimulation of subthalamic nucleus – initial experience. Radiology 2006;239:209–216.
118. Rosen WG, Mohs RC, Davis KL. A new rating scale for Alzheimer's disease. Am J Psychiatry 1984;141:1356–1364.
119. Schneider LS, Olin JT, Doody RS, et al. Validity and reliability of the Alzheimer's Disease Cooperative Study – Clinical Global Impression of Change. The Alzheimer's Disease Cooperative Study. Alzheimer Dis Assoc Disord 1997;11 Suppl 2:S22–S32.
120. Scheff SW, Price DA, Schmitt FA, Mufson EJ. Hippocampal synaptic loss in early Alzheimer's disease and mild cognitive impairment. Neurobiol Aging 2006;27:1372–1384.
121. Scheltens P, Fox N, Barkhof F, De Carli C. Structural magnetic resonance imaging in the practical assessment of dementia: beyond exclusion. Lancet Neurol 2002;1:13–21.
122. Knopman DS, DeKosky ST, Cummings JL, et al. Practice parameter: diagnosis of dementia (an evidence-based review). Report of the Quality Standards Subcommittee of the American Academy of Neurology. Neurology 2001;56:1143–1153.
123. Neary D, Snowden JS, Gustafson L, et al. Frontotemporal lobar degeneration: a consensus on clinical diagnostic criteria. Neurology 1998;51:1546–1554.
124. O'Brien JT, Erkinjuntti T, Reisberg B, et al. Vascular cognitive impairment. Lancet Neurol 2003;2:89–98.
125. McKeith IG, Dickson DW, Lowe J, et al. Diagnosis and management of dementia with Lewy bodies. Third report of the DLB consortium. Neurology 2005;65:1863–1872.
126. Neary D, Snowden J, Mann D. Frontotemporal dementia. Lancet Neurol 2005;4:771–780.
127. Bobholz JA, Rao SM, Saykin AJ, Pliskin N. Clinical use of functional magnetic resonance imaging: reflections on the new CPT codes. Neuropsychol Rev 2007;17:189–191.

Chapter 23

fMRI in Epilepsy

Rachel Thornton, Robert Powell, and Louis Lemieux

Summary

This chapter provides an overview of the application of functional MRI applied to the field of Epilepsy and is divided into two sections, covering cognitive mapping and imaging of paroxysmal activity, respectively. In addition to a review of the most scientifically and clinically relevant findings, technical and methodological background information is provided to help the reader better understand the data acquisition process. We show how both approaches may play a role in the pre-surgical evaluation of patients with drug-resistant focal epilepsy and provide opportunities for new insights into the neuropathological processes that underlie both focal and generalised epilepsy.

Key words: Epilepsy, Focal epilepsy, Generalised epilepsy, Interictal, Ictal, Imaging, Functional magnetic resonance imaging, fMRI, Electroencephalography, EEG, Multi-modal imaging, EEG-correlated fMRI, Cognitive mapping, Functional mapping, Brain activity mapping, Language lateralization, Memory mapping, Pre-surgical evaluation

1. Cognitive fMRI in Epilepsy

The commonest surgical procedure for patients with drug resistant temporal lobe epilepsy (TLE) is anterior temporal lobe resection (ATLR). Complications of this operation include a decline in language and memory abilities, and an important part of the presurgical assessment lies in the careful selection of patients to minimise these adverse cognitive sequelae. This has traditionally been the role of neuropsychology and the intracarotid amytal test (IAT). Since the advent of functional MRI (fMRI), however, there has been much interest in its possible role in the presurgical assessment of those with epilepsy, principally in the identification of eloquent cortex to be spared during surgery.

M. Filippi (ed.), *fMRI Techniques and Protocols*, Neuromethods, vol. 41
DOI 10.1007/978-1-60327-919-2_23, © Humana Press, a part of Springer Science+Business Media, LLC 2009

Neuropsychology has played a prominent role throughout the modern era of epilepsy surgery, mainly because of the importance of the temporal lobes in memory function. The principal role of baseline neuropsychological assessments is in predicting the impact of surgery on memory, providing data on lateralisation and localisation of cerebral disturbance, and providing evidence for cerebral reorganisation.

The IAT plays a role in the presurgical assessment of TLE in some centres. Its uses are in assessing the capacity of the contralateral temporal lobe to maintain useful memory functions, thus guarding against a severe post-operative amnesic syndrome, and as a means of lateralising language function. The procedure involves the injection of sodium amytal into one carotid artery, inactivating the corresponding hemisphere for around 10 min, and thus crudely mimicking the effects of surgery on the medial temporal lobe (MTL) structures. During this time, the patient's language and memory abilities are tested. Although still commonly used, the IAT has considerable disadvantages, not least the fact that it is an expensive, invasive procedure with potentially serious complications. Doubts also exist about its reliability and validity in predicting post-operative amnesia. In contrast to the traditional neuropsychological assessment, which relies on standardised tests of cognitive abilities and yields results that are easily validated, IAT procedures vary significantly between institutions with respect to the testing protocol used, choice of behavioural stimuli, dosage, and administration of amytal, all of which can lead to variations in the results *(1)*. The IAT is also poor at predicting verbal memory decline as deactivation of the language dominant hemisphere will cause increased errors on verbal memory testing *(2)*.

fMRI has the potential for replacing the IAT and for providing additional information to that provided by baseline neuropsychological assessment in the lateralisation and localisation of language and memory function. Practically, fMRI is cheaper than the IAT, non-invasive, and repeatable. There are, however, important potential caveats when considering the role of fMRI. First, areas activated by a particular fMRI paradigm are not necessarily crucial for the performance of that task. Second, it does not necessarily follow that all areas involved in a task will be activated by a particular fMRI paradigm. Third, the extent of activation seen in a task, in terms of both the area activated and the magnitude of the peak, may bear no relation to the competence with which that task is performed. Caution will also be needed in the interpretation of results bearing in mind that fMRI techniques, while useful for the localisation of cognitive function, may not reliably indicate the capacity of unilateral temporal lobe structures.

In the following section, we review how fMRI is used to lateralise language function. We then discuss the current state of research efforts to localise brain regions involved in language and memory, and study the effects of epilepsy upon these. Furthermore, the efforts to assess the reliability of fMRI in the prediction of post-operative language and memory deficits following ATLR are discussed.

1.1. Language fMRI

1.1.1. Paradigm Design and Analysis

The aims of pre-operative language fMRI are primarily to lateralise and localise language functions and to use this information to predict and avert postoperative complications. The most widely used tasks in language fMRI experiments are verbal fluency tasks including word generation and verb generation, and semantic decision tasks. These are generally strongly lateralising and reliably identify 'expressive language functions' in the dominant inferior frontal gyrus (IFG) (Brodmann Areas [BA], 44, 45). Although these tasks are usually covert (i.e. performed silently without performance monitoring), they have been reliably replicated in numerous studies in both normal and patient populations. Their within-subject reproducibility has been demonstrated, although frontal activations have been shown to be more reliable than temporoparietal ones *(3)*. In addition they can be applied to patients with a wide range of cognitive abilities, with language lateralisation results appearing to be relatively unaffected by patients' performance levels *(4)*. Although tasks of verbal fluency do show language-related activations, they are not pure language tasks, containing substantial components of executive processing and of working and verbal memory. These activations are typically seen in the middle frontal gyrus (MFG) (BA 46, 49) (**Fig. 1**).

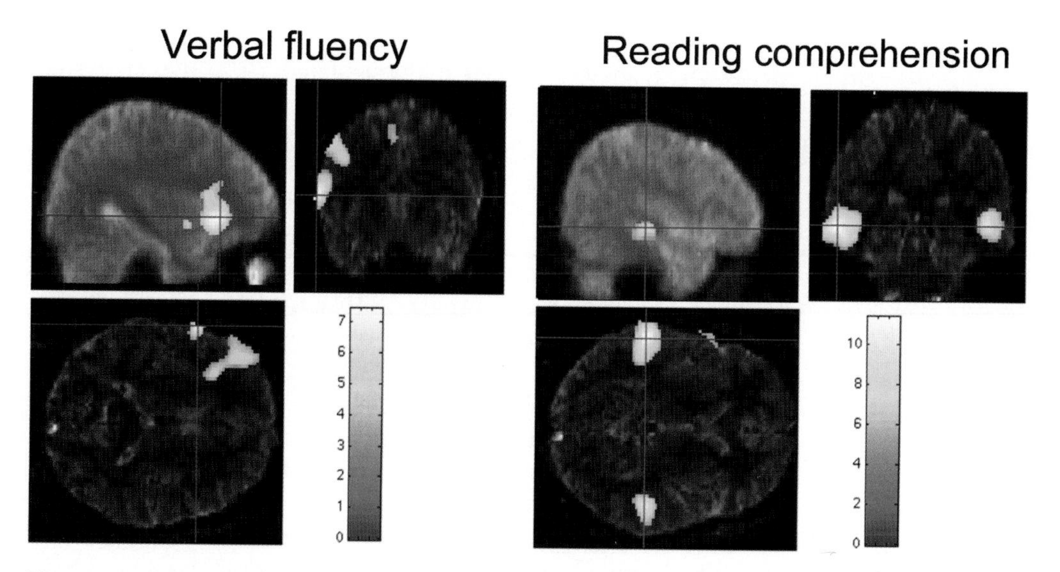

Fig. 1. Verbal fluency and reading comprehension: Typical fMRI findings in a patient performing tasks of verbal fluency (*left*) and reading comprehension (*right*) showing activation in the dominant frontal lobe and bilateral superior temporal lobes, respectively. Areas of activation are overlaid on a distortion matched high resolution echo-planar image.

Fluency tasks are also less reliable in identifying 'receptive' language areas located in the dominant temporal lobe. These processing areas are best assessed by tasks that probe language comprehension such as reading sentences or stories, which tend to activate superior temporal cortex extending to supramarginal gyrus (BA 20, 21, 39) *(5, 6)*, but are less strongly lateralizing than verbal fluency tasks (**Fig. 1**). Using a panel of fMRI tasks (verbal fluency, reading comprehension, and auditory comprehension) was demonstrated to be helpful in reducing inter-rater variability and helped in the evaluation of language laterality in patients with focal epilepsy *(7)*.

These functional imaging experiments have used block design paradigms to detect regions of the brain showing greater activation during task blocks when compared with rest blocks. The advantage of block designs over event-related designs is that they are efficient in detecting differences between two conditions; however, they offer less flexibility in the experimental design required for studying complex cognitive functions.

The degree of lateralisation is often quantified using an asymmetry index, (AI) = $(L - R)/(L + R)$, where L and R represent the strength of activation for the left (L) and right (R) sides, respectively, based on the number of activated voxels for the whole hemisphere or using regions of interest (ROIs) targeted to known as language areas *(8)*. A positive value represents left language lateralisation and a negative, right-sided dominance, although AI values between –0.2 and +0.2 are often classified as bilateral. This can be determined by counting the number of voxels exceeding a specified threshold of significance. This type of AI calculation has some problems, in particular, the fact that the AI can differ according to the significance threshold chosen for the activation map from which it is calculated *(9)*. One suggested solution to this is to calculate AIs from all voxels that correlated positively with a task, but with each weighted by their own statistical significance *(10)*. Using this method, AIs were less variable than those calculated from suprathreshold voxels only. Alternative approaches to estimate the degree of asymmetry include measuring the mean signal intensity change induced by the task within a brain volume of interest *(9)*, and performing a statistical comparison of the magnitude of task-induced activation in homotopic regions of the two hemispheres *(11)*. These methods measure the magnitude of the mean signal change and have the advantage of not being threshold dependent. Other studies have suggested that visual rating appears to work as well as calculating AIs *(12)*.

The simplest forms of study designs (including those described earlier) employ cognitive subtraction designs. These involve selecting a task that activates the cognitive process of interest and a baseline task that controls for all but the process of interest. One problem of this type of design is that it depends on an assumption known as pure insertion, which supposes that a new cognitive component can be inserted without affecting those processes that are also engaged by the baseline task *(13)*. Another problem with cognitive subtraction is in finding baseline tasks that activate all but the process of interest. These problems can be overcome by using more complex experimental designs, such as factorial designs and cognitive conjunctions.

Factorial designs use two or more variables (e.g. sentence vs. word presentation and auditory vs. visual presentation) and allow the effect one variable has on the other to be measured explicitly. The analysis of this type of design involves calculating the main effects of each variable and the interaction between them *(13)*. Cognitive conjunctions are an extension of cognitive subtraction paradigms. Cognitive subtraction looks for activation differences between a single pair of tasks, while cognitive conjunction looks at two or more task pairs, which share a common processing difference *(14)*. The advantages of this approach are that it allows greater freedom in selecting the baseline task as it is not necessary to control for all but the component of interest, and that it does not depend on the assumption of pure insertion.

1.1.2. Language Lateralisation in Epilepsy

Focal epilepsy may be associated with disrupted lateralisation and localisation of language regions; therefore, one would expect a higher probability of abnormal language lateralisation. Nevertheless, significant differences have been reported between centres in the relative proportions of right and left hemisphere dominant patients using the IAT, some of which may be due to the different criteria used for assessing dominance. The percentage of left hemisphere dominant right-handed patients has ranged from 63 to 96% *(15)* while for left handers a similar variation has been reported between 38 and 70% *(16, 17)*. Results of fMRI studies have also shown greater atypical language dominance in patients. In a comparison between 100 right-handed healthy subjects and 50 right-handed epilepsy patients, 94% of the normal subjects were considered as left hemisphere dominant and 6% had bilateral representation. The epilepsy group showed greater variability of language dominance, with 78% showing left hemisphere dominance, 16% symmetric activation, and 6% showing right hemisphere dominance. Atypical language dominance was associated with an earlier age of brain injury and with weaker right hand dominance *(18)*.

The localisation of the epileptogenic lesion and epileptic activity *(19)* has also been shown to influence language organisation. In a retrospective study of patients with hippocampal sclerosis (HS) who had undergone presurgical evaluation, atypical speech dominance occurred in 24% of those with left-sided HS, whereas all those with right-sided HS had left-sided speech dominance. In addition, atypical speech representation was associated with higher spiking frequency and in those with sensory auras suggesting ictal involvement of the lateral temporal structures. No association was demonstrated between either age at epilepsy onset or age at initial precipitating injury and atypical speech representation *(20)*.

Comparing the degree of reorganisation of frontal and temporal lobe language functions has shown a significantly more left lateralised pattern of language activation in controls and right TLE patients compared with that of left TLE patients *(21)*. In patients with atypical language representation, the degree of reorganisation towards the right hemisphere was greater in the temporal lobes than in the frontal lobes. In a study of 50 patients with focal epilepsy, greater atypical language dominance was seen in those with left hemisphere seizure focus *(22)*. Left TLE patients who did not have atypical language also had lower asymmetry indices in both frontal and temporal ROIs, mainly because of greater activation in homologous right hemisphere regions.

The degree of language lateralisation has also been related to the nature of the epileptogenic lesion with early acquired lesions, such as HS, considered to be associated with greater incidence of atypical language lateralisation compared with developmental lesions originating in utero, such as malformations of cortical development (MCDs). A higher degree of atypical language dominance, in both frontal and temporal language areas, has been demonstrated in patients with left HS compared with patients with left frontal and lateral temporal lesions *(23)*, suggesting that the hippocampus itself may play an important role in the establishment of language dominance. A further study, however, demonstrated no difference in the frequency of atypical language lateralisation between left TLE patients with HS and those with developmental tumours *(24)*.

It is interesting to speculate on how TLE affects language lateralisation, and it is possible that strong connectivity between inferior frontal and temporal areas make frontal lobe functions particularly sensitive to temporal pathology. The increased incidence of atypical language dominance in epilepsy illustrates the importance of establishing language dominance prior to performing surgical resection and as a consequence much of the work on fMRI in epilepsy has been directed towards trying to replace the IAT as a means of doing this.

1.1.3. Comparison of fMRI, IAT, and Electrocortical Stimulation Findings

Studies comparing fMRI and the IAT are summarised in **Table 1**. Just as IAT protocols differ between centres, a number of fMRI paradigms to determine language dominance have been employed but overall agreement of approximately 90% is seen between the two techniques. The remaining cases generally exhibit partial disparity where one method shows bilateral

Table 1
Concordancage between fMRI language lateralisation and the IAT

Authors	Sample size[a]	fMRI language lateralisation tasks	Concordance
Desmond et al. (1995) *(41)*	7	Semantic decision task	100%
Binder et al (1996) *(42)*	22	Semantic decision task	$r = 0.96$
Hertz-Pannier et al. (1997) *(43)*	6[b]	Verbal fluency paradigm	100%
Yetkin et al. (1998) *(44)*	13	Word generation task	$r = 0.93$
Benson et al. (1999) *(45)*	12	Verb generation task	100%
Lehericy et al. (2000) *(6)*	10	Semantic fluency Sentence repetition Story listening	Semantic fluency > story listening > sentence repetition. Greater concordance between IAT results and activation asymmetry in frontal than temporal lobes
Carpentier et al. (2001) *(46)*	10	Identification of syntactic/ semantic errors in target sentences	80%
Gaillard et al. (2002) *(8)*	21	Reading paradigm	85%
Woermann et al. (2003) *(12)*	100	Word generation	91%
Sabbah et al. (2003) *(47)*	20[c]	Word generation Semantic decision	95%
Benke et al. (2006) *(25)*	68	Semantic decision	89% – right TLE 72% – left TLE

IAT intracarotid amytal test, *TLE* temporal lobe epilepsy
[a]Some of these studies report fMRI data on larger samples. However, only the patients with fMRI and IAT data are included here
[b]Age range 8–18
[c]Patients with suspected atypical language lateralisation were selected

language representation and the other lateralised language dominance and outright disagreement between fMRI and IAT is rare. One study suggested that fMRI may be less reliable in left-sided neocortical epilepsy (25% disparity) in comparison with left-sided medial TLE (3% disparity) *(12)*. Another showed that concordance between fMRI-based laterality and IAT was much lower in left TLE patients than in patients with right TLE *(25)*. One interesting case of false lateralisation of language function in a post-ictal patient with left HS also illustrates the need for caution in the interpretation of results in individual patients. No activation was seen in the left temporal lobe during multiple language tasks after a cluster of left temporal lobe seizures but in a repeat fMRI experiment 2 weeks later activation was seen predominantly over the left temporal region *(26)*. However, bearing in mind the previously mentioned limitations of the IAT, and the doubts about its gold standard status, it is even debatable whether fMRI and IAT are directly comparable as they probe different aspects of language.

Comparisons have also been performed between fMRI activation maps and regions showing disruption of function during intraoperative electrocortical stimulation (ECS). In order for fMRI to be used instead of ECS, it must demonstrate a high predictive power for the presence as well as the absence of critical language function in regions of the brain. As with IAT, these studies show strong, but incomplete agreement with fMRI, with high sensitivity but lower specificity *(27–29)*. Although false-positive activation (fMRI activation but no ECS disruption) is relatively common, this is not surprising given that fMRI activates whole networks of regions, not all of which are essential for the task in question. False-negative findings (regions showing disruption by ECS but no fMRI activation) are more critical when planning a surgical resection, and these were identified in 2 patients out of 21 reported in two series. Activation and disruption was typically within 5 mm in frontal regions and 10 mm in temporal areas.

A combination of four different language tasks has shown more reliable and robust lateralisation in normal subjects by targeting brain regions common to different tasks, thereby focusing on areas critical to language function. Regions of activation detected in this way corresponded well with ECS findings in the temoroparietal region *(30)*. Sensitivity was 100% in all but one patient. This high negative predictive value suggested that areas where no significant fMRI activity was present could be safely resected without using ECS. fMRI activity, however, was not always absent at noncritical language areas limiting its positive predictive value for the presence of critical language. Although this suggests that fMRI is not yet ready to replace ECS, it could be used to speed up intracranial mapping procedures and to guide the extent of the craniotomy.

1.1.4. Language Localisation and Prediction of Post-Operative Language Deficits

Selective language deficits have been reported following language-dominant ATLR, with naming the most commonly affected function *(31, 32)*. It has also been suggested that the risk for post-operative decline in naming abilities increases with age of seizure onset and the extent of lateral temporal neocortex resected *(33)*. Pre-operative cortical stimulation via subdural grid electrodes has been used to localise language function, suggesting that early onset of dominant temporal lobe seizure foci leads to a more widespread or atypical distribution of language areas, particularly naming and reading areas *(34)*. A subsequent study also reported that markers of early left hemisphere damage (such as early seizure onset, poor verbal IQ, left handedness, and right hemisphere memory dominance) increase the chances of essential language areas being located in more anterior temporal regions. Again these areas were identified using naming and reading tasks *(35)*.

These findings suggest that naming and reading abilities are the language skills most at risk following dominant temporal lobe surgery. Although the IAT may provide a useful index of language laterality, it does not provide detailed information on the localisation of these specific language skills. As these may also vary in location between individuals, the role IAT can play in the prediction of post-operative deficits in individual patients is therefore limited. Designing fMRI paradigms that specifically probe naming and reading skills would provide a useful clinical tool for mapping relevant language skills that could be used in the prediction of post-operative deficits.

One study has used pre-operative functional neuroimaging to predict language deficits following left ATLR. Temporal lobe fMRI asymmetry was found to be predictive of deficits seen on a post-operative naming test with a greater degree of language lateralisation toward the left hemisphere related to poorer naming outcome and language lateralisation towards the right hemisphere associated with less or no decline. The correlation between temporal lobe fMRI AI and naming deficits was stronger than that seen in the frontal lobes and also stronger than that between IAT and naming deficits. *(36)*.

Interestingly, many patients do not suffer any language deficits following ATLR, suggesting that multiple sets of neural systems may exist that are capable of performing the same cognitive function, and that some of these may be engaged following focal brain injuries. In a study of patients who had undergone left ATLR but did not have deficits in sentence comprehension, decreased activation was demonstrated in undamaged areas of the normal left hemisphere system but increased activation was seen in several right frontal and temporal regions not usually engaged by normal subjects *(37)*. This suggests that there is more than one neural system capable of sustaining sentence comprehension. This study was, however, unable to tell whether this functional reorganisation to the right IFG occurred pre or post-operatively. A separate study looked at the role of the right IFG by comparing its functional activation on a verbal fluency task in controls with left TLE patients *(38)*. The patients were shown to activate a more posterior right IFG region compared with controls, although left IFG activation did not differ significantly between the two groups. Further, verbal fluency-related activation in the right IFG was not anatomically homologous to left IFG activation in either patients or controls. This suggests that reorganisation takes place pre-operatively in patients with chronic left TLE, and that the prediction of language outcome following left ATLR may depend not only on the extent of preoperative right hemisphere activation, but also its location.

1.1.5. Combination of fMRI and MR-Tractography

Complex behaviours such as language and memory rely upon networks of neurons, which integrate the functions of spatially remote brain regions. The combination of fMRI to identify cortical regions involved in language function and MR-tractography to visualise white matter pathways connecting these regions offers an opportunity to study the relationship between structure and function in the language system **(Fig. 2)**. Studies have revealed structural asymmetries in controls, with greater left-sided fronto-temporal connections in the dominant hemisphere *(39)*. Patients with left TLE had reduced left-sided and greater right-sided connections than both controls and right TLE patients, reflecting the altered functional lateralisation seen in left TLE patients, and significant correlations were demonstrated between structure and function in controls and patients, with subjects with more highly

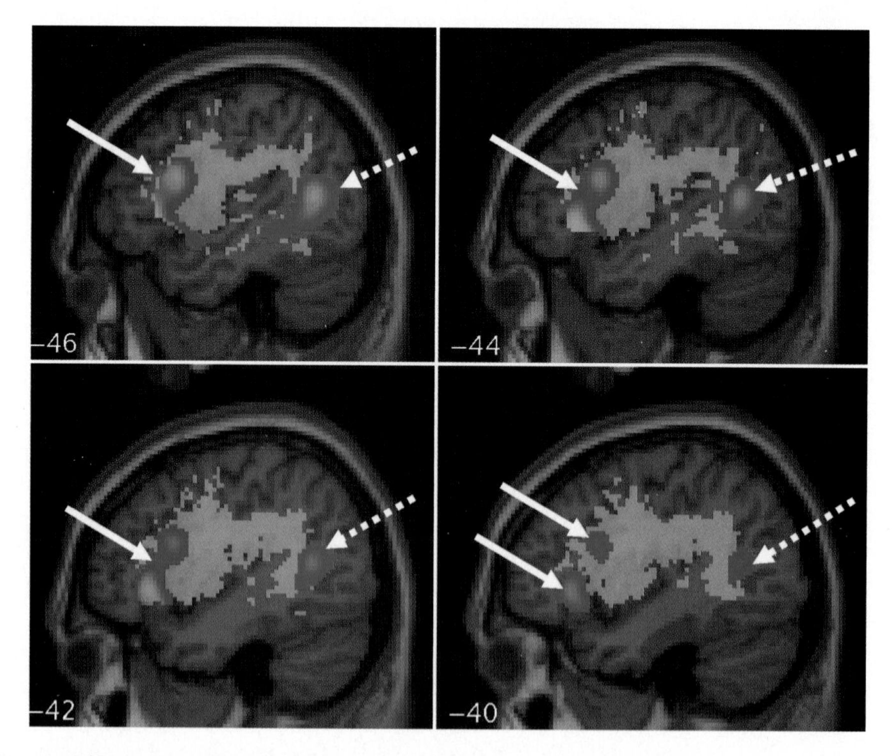

Fig. 2. Combined MR tractography and functional mapping. Frontal lobe connections overlaid on a structural template along with group fMRI effects for word generation (*solid arrows*) and reading comprehension (*dashed arrows*), showing how the tracts connect together the frontal and temporal lobe functionally active regions.

lateralised language function having a more lateralised pattern of connections *(40)*. The combination of fMRI with information on the structural connections of these normally and abnormally functioning areas offers the opportunity to improve understanding of the relationship between brain structure and function and may improve the planning of surgical resections to maximise the chance of seizure remission and to minimise the risks of cognitive impairment.

1.2. Memory fMRI

MTL structures are associated with memory functioning and surgical resection is known to cause reduced memory function in some cases. The study of patients following temporal lobe surgery has provided considerable evidence supporting the critical role that the hippocampi play in memory functioning. Bilateral injury to these areas leads to a characteristic amnesic syndrome *(48)*, while unilateral lesions lead to material-specific deficits, and a decline in verbal memory following surgery to the language-dominant hemisphere has been consistently reported and studied *(49)*, along with deficits in topographical memory following non-dominant ATLR *(50)*. Although rare, some patients have sustained a severe anterograde amnesic syndrome following a unilateral ATLR. Most of these, however, have subsequently been found to have evidence of contralateral hippocampal pathology, either on post-operative electroencephalography (EEG) *(51)*, post-mortem pathological findings *(52)*, or post-operative volumetric MRI *(53)*.

Two different models of hippocampal function have been proposed to explain memory deficits following unilateral ATLR: hippocampal reserve and functional adequacy *(54)*. According to the hippocampal reserve theory, post-operative memory decline depends on the capacity or reserve of the contralateral hippocampus to support memory following surgery, while the functional adequacy model suggests that it is the capacity of the hippocampus that is to be resected that determines whether changes in memory function will be observed. Evidence from baseline neuropsychology *(55)*, the IAT *(56)*, histological studies of hippocampal cell density *(57)*, and MRI volumetry *(58)* has suggested that of the two, it is the functional adequacy of the ipsilateral MTL, rather than the functional reserve of the contralateral MTL that is most closely related to the typical material-specific memory deficits seen following ATLR.

The assessment of ability to sustain memory is critical for planning ATLR as memory decline is not an inevitable consequence of temporal lobe surgery. Accurate prediction of likelihood and severity of post-operative memory decline is necessary to make an informed decision regarding surgical treatment. Much work has been focused on the identification of prognostic indicators for risk of memory loss after ATLR. The severity of

HS on MRI is an important determinant, being inversely correlated with a decline in verbal memory following left ATLR, with less severe HS increasing the risk of memory decline *(58–60)*. Pre-operative memory performance has been related to degree of post-operative memory impairment, with better performance increasing the risk of memory decline *(55, 61, 62)*. These risk factors reflect the functional integrity of the resected temporal lobe and suggest that patients with residual memory function in the pathological hippocampus are at greater risk of memory impairment post-operatively. Recently, fMRI has also been shown to be a potential predictor of post-operative material-specific memory decline following ATLR.

1.2.1. The Difficulty in Seeing Anterior Hippocampal Activation

Impairment in memory encoding following ATLRs suggests that anterior MTL regions are critical for successful memory encoding, and in the patient HM, who was rendered amnesic following bilateral temporal lobe resections, more posterior MTL structures remained intact *(63)*. Intracranial electrophysiological recordings during verbal encoding tasks have also shown greater responses in anterior hippocampal and parahippocampal regions for words remembered than those forgotten *(64)*. However, functional imaging studies have proved contradictory, with many showing encoding-related activations in posterior hippocampal and parahippocampal regions, which would be left intact following ATLRs.

One possible explanation for this apparent conflict is that anterior temporal regions are subject to signal loss during fMRI sequences. It has been demonstrated that signal loss due to susceptibility artefact is most prominent in the inferior frontal and inferolateral temporal regions *(65)*, and as the hippocampus rises from anterior to posterior, one would expect greater susceptibility-induced signal loss in the anterior (inferior) relative to posterior (superior) hippocampus. This may have been one reason for the relative lack of anterior hippocampal activation in early fMRI studies of memory *(66)*. One study has directly examined the effects of susceptibility artefact on hippocampal activation by demonstrating its differential effect on the anterior vs. the posterior hippocampus. The averaged resting voxel intensity in an anterior hippocampal ROI was significantly less than in a posterior hippocampal ROI and intensity decreases were substantial enough to leave many voxels below the threshold at which BOLD effects could be detected *(66)*. On top of this, it has been shown that the sensitivity to BOLD changes is proportional to signal intensity at rest so that voxels with a lower baseline signal (such as those in anterior hippocampal regions) would be more difficult to activate than those with higher baseline signals *(67)*.

An alternative explanation for the lack of anterior hippocampal activation seen in many early memory fMRI experiments is that the paradigms used were not optimal for detecting subsequent memory effects. The use of fMRI in studying memory function is more challenging than for language. This is partly due to the different components involved in memory processing, such as encoding and retrieval, and the fact that the nature of the material being encoded or retrieved influences which brain areas are activated. A further difficulty is how to separate brain activity related specifically to memory from that related to other cognitive processes. In consequence, more complex paradigms are required when studying memory than for examining language function.

Standard fMRI experiments initially used block design paradigms looking for regions of the brain showing greater activation during task blocks compared with rest blocks. A problem when designing memory fMRI experiments was how to separate brain activity specifically due to memory from that due to other cognitive processes being used in the task. Early fMRI studies of memory encoding employed block experimental designs to contrast tasks promoting differing memory performance, using the 'depth of encoding' principle *(68)*. This states that if you manipulate material in a 'deep' way (e.g. make a semantic decision about a word), then it is more likely to be recalled successfully than material manipulated in a 'shallow' way (e.g. make a decision of whether the first letter of a word is alphabetically before the last letter). These studies tended to show consistent activation in left prefrontal cortical regions along with less reliable MTL activation *(69–72)*. Similar assumptions underlie the use of 'novelty' paradigms in probing memory encoding. During these experiments, alternating blocks of novel and repeated stimuli are presented, with the hypothesis being that more memory encoding takes place while viewing a block of novel stimuli than when viewing the same repeated stimulus *(73)*.

The advantage of block designs is that they are generally the most efficient in detecting differences between two conditions. The main problem in their interpretation, however, lies in the inference that the effects shown by these contrasts reflect differences in memory encoding, rather than any other differences between the two conditions (e.g. response to novelty and semantic processing) that are independent from differences in memory encoding. Attempts were made to overcome this problem using parametric block designs but were soon superseded by the advent of event-related studies.

Event-related fMRI is defined as the detection of transient haemodynamic responses to brief stimuli or tasks. This technique, derived from those used by electrophysiologists to study event-related potentials, enables trial-based rather than block-based experiments to be carried out. Trial-based designs have a number of methodological advantages, in particular that trials can be categorised post-hoc according to a subject's performance on a subsequent test to obtain fMRI data at the individual item level. Therefore, when studying memory encoding, activations for individual items presented can be contrasted according to whether they are remembered or forgotten in a subsequent memory test. This type of analysis allows the identification of brain regions showing greater activation during the encoding of items that are subsequently remembered compared with items subsequently forgotten (subsequent memory effects), which are then taken as candidate neural correlates of memory encoding *(74)*. Although event-related designs are less powerful than block designs at detecting differences between two brain states and may be more vulnerable to alterations in the haemodynamic response function (e.g. due to pathology), they have the advantage of permitting specifically the detection of subsequent memory effects due to successful encoding.

One study looking at encoding of words, pictures, and faces in healthy controls employed an experimental design, which allowed data analysis either as a block design, or as an event-related design of successful encoding *(75)*. The results demonstrated a functional dissociation between anterior and posterior hippocampus. The main effects of memory encoding, demonstrated specifically using an event-related analysis, were seen in the anterior hippocampus **(Fig. 3)**, with the main effects of viewing stimuli, demonstrated using a block analysis, being located in more posterior regions.

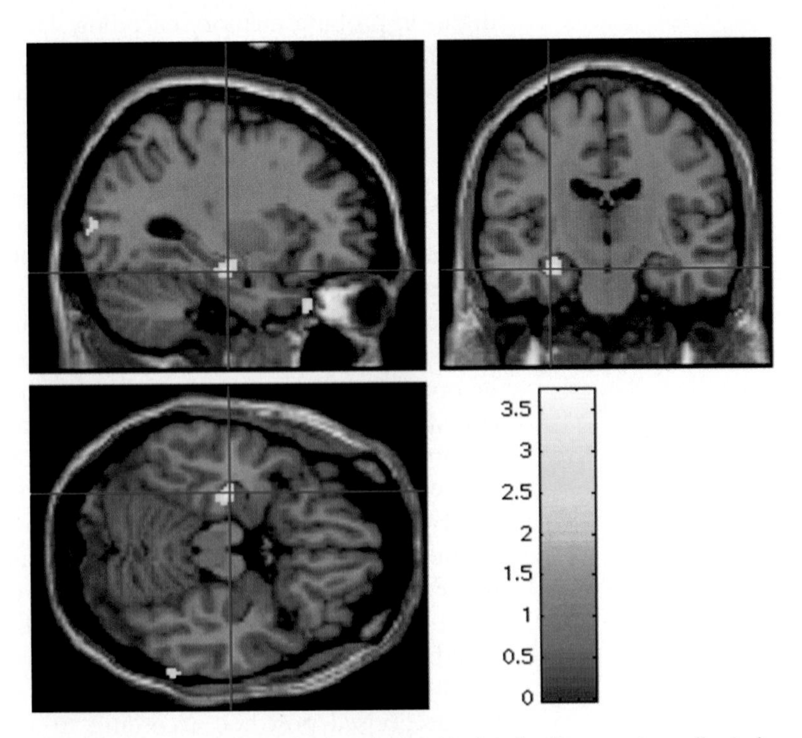

Fig. 3. Left hippocampal activation in a single subject performing a word encoding task.

1.2.2. The Effect of TLE on Memory Processes

Deficits in verbal memory following left and topographical memory following right ATLR suggest a material-specific lateralisation of function in MTL structures. Functional imaging studies have been used to look for lateralisation of cerebral activation patterns during episodic memory processes. Many have shown material-specific lateralisation in prefrontal regions but this has been more difficult to demonstrate in the MTL *(69, 73, 75, 76)*. A number of studies have used fMRI to look at the lateralisation of memory in patients with TLE compared with that seen in normal subjects, and also compared the findings with the results of the IAT. These are summarised in **Table 2**. These employed block design studies, demonstrating predominantly posterior MTL activation, and therefore cannot claim that subsequent memory effects have been specifically examined.

Table 2
fMRI memory studies in TLE

Authors	Sample size	fMRI tasks	Findings
Detre et al. (1998) *(76)*	Controls n = 8 Patients n = 10	Block design Complex visual scenes vs. abstract pictures	Symmetric MTL activation in normal subjects. Lateralization of memory concordant with IAT in 9/10 subjects
Bellgowan et al. (1998) *(79)*	Patients n = 28, 14 left TLE, 14 right TLE	Block design Semantic decision vs auditory perception task	Greater activation in the left MTL in right TLE compared to left TLE group
Dupont et al. (2000) *(79)*	Controls n = 10 Patients n = 7, left HS	Block design Verbal encoding and retrieval vs. fixation on the letter A	Left occipitotemporoparietal network activated in controls. Reduced MTL activation and increased activation in left dorsolateral frontal cortex in patients
Jokeit et al. (2001) *(80)*	Controls n = 17 Patients n = 30	Block design Roland's Hometown Walking vs. baseline	No asymmetry of MTL activation in controls, greater activation in the MTL contralateral to seizure focus in 90% of patients
Golby et al. (2002) *(81)*	Patients n = 9	Block design comparing novel vs. repeated stimuli 4 encoding stimuli used – patterns, faces, scenes and words	Group level – greater activation in the MTL contralateral to seizure focus for all encoding stimuli. Single subjects – lateralization of memory concordant with IAT in 8/9 subjects

Studies performed in patients with TLE showing patient groups studied, experimental design employed, and principal findings *HS* hippocampal sclerosis, *IAT* intracarotid amytal test, *MTL* medial temporal lobe, *TLE* temporal lobe epilepsy

More recently event-related studies have demonstrated a material-specific lateralisation of memory encoding within anterior MTL regions that would be resected during standard ATLR (**Fig. 4**) *(75)*. In addition a reorganisation of function has been demonstrated in patients with unilateral TLE due to HS, with reduced ipsilateral activation, and increased contralateral activation in patients compared with controls *(77, 78)* (**Fig. 5**). Comparing groups of patients with controls demonstrated a functional reorganisation away from the pathological hemisphere; however, it is not clear whether this represents an effective way of maintaining memory function in individual patients. By correlating fMRI activation and performance on standard neuropsychological memory tests, it has been shown that MTL activation ipsilateral to the pathology is correlated with better performance while contralateral, compensatory activation correlates with poorer performance *(78)*. The conclusion that memory function in unilateral TLE is better when sustained by the activation within the damaged hippocampus is consistent with the observation that pre-operative memory performance is a predictor of post-operative memory decline, with better performance predicting worse decline *(61, 62)*, and adds further support to the functional adequacy model of hippocampal function.

1.2.3. The Prediction of Post-Operative Memory Changes

Prediction of post-operative memory decline is necessary to make an informed decision regarding surgical treatment. To date a small number of studies have used fMRI to predict the effect of left or right ATLR on verbal and non-verbal memory. In patients with left HS, greater verbal memory encoding activity in the left hippocampus compared with the right hippocampus predicted the extent of verbal memory decline following left ATLR *(82)*.

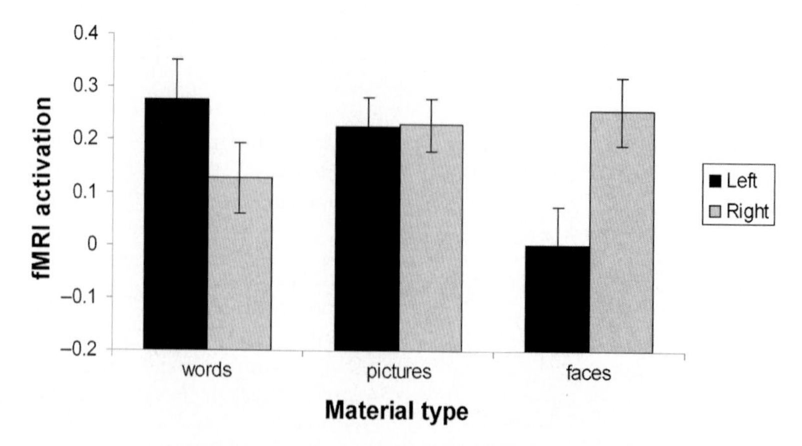

Fig. 4. Material-specific lateralisation of memory encoding in the anterior hippocampus. fMRI activation within left and right hippocampal ROIs in healthy controls demonstrating left lateralised activation for word encoding, right-lateralised activation for face encoding and bilateral activation for picture encoding.

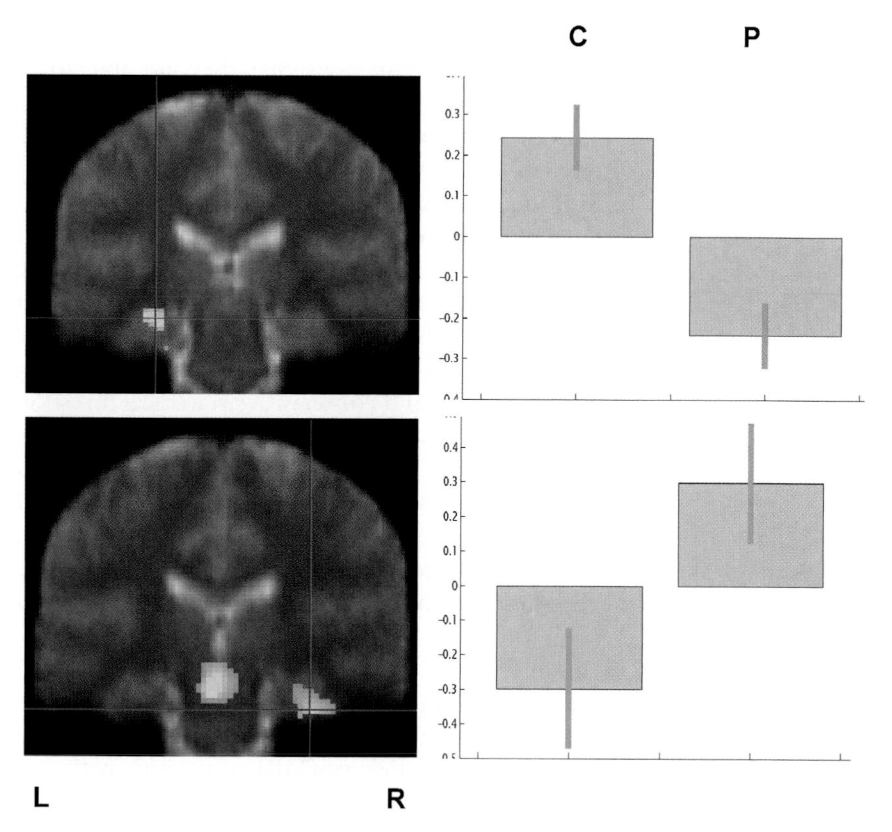

Fig. 5. fMRI memory encoding experiment: Left TLE patients vs. healthy controls. Regions showing significant differences in activation between left temporal lobe epilepsy (TLE) patients and controls are highlighted. Contrast estimates are shown on the right of the images. Controls (C) are on the *left* and patients (P) on the *right*. A reorganisation of function is seen in the left TLE patients with reduced activation in the left hippocampus, and greater activation in the right hippocampus, compared with healthy controls.

In a further analysis of the same patients, it was demonstrated that greater activation within the left hippocampus predicted a greater postoperative decline in verbal memory *(83)*. These findings have since been replicated and extended to patients undergoing right ATLR *(84)*. Other groups have demonstrated correlations between MTL activation asymmetry ratios and post-surgical memory outcome in patients with both left and right TLE, with increased activation ipsilateral to the seizure focus correlating with greater memory decline *(85, 86)*.

As discussed earlier, two different models of hippocampal function have been proposed to explain memory deficits following unilateral ATLR: hippocampal reserve and functional adequacy *(54)*. Studies using asymmetry indices are unable to address this important issue; however, the findings of some of the above studies that greater pre-operative activation within the ipsilateral, to-be-resected hippocampus, correlated with greater post-operative decline in memory support the functional adequacy theory *(83, 84)*.

1.3. Challenges of Clinical Cognitive fMRI

When designing paradigms for patients with neurological deficits, it is important to use tasks that they are able to perform. A differential pattern of activation between patients and normal subjects is only interpretable if patients are performing the task adequately *(87)*.

In addition, one must be aware of differences in the questions being asked by cognitive neuroscientists and clinicians, which can lead to different approaches to data analysis. Generally, neuroscientists look at groups of matched controls performing the same task and determine which brain regions are commonly activated across the group. The emphasis is on avoiding false-positive results (Type I errors) and conservative statistical thresholds need to be used, which may lead to an under representation of brain areas truly involved. Conversely, clinicians are considering individual patients where the priority is to identify all brain regions involved in a task, i.e., avoiding false negatives (Type II errors). As a result, less stringent statistical thresholds are required and indeed thresholds used may need to vary on an individual basis.

1.4. Cognitive fMRI in Epilepsy: Summary

fMRI is a non-invasive and widely available tool, which has had a dramatic impact on cognitive neuroscience. Much of the progress made will benefit clinical neuroimaging, although some problems exist in the application of fMRI to patients with neurological deficits. fMRI allows the non-invasive assessment of language function to be performed and offers a valid alternative to the IAT for establishing language dominance. By tailoring paradigms towards the localisation of the specific language skills most at risk following temporal and frontal resections, it will be possible to map relevant language functions in the epilepsy surgery population. This in turn will allow better assessment of the risks posed by surgery in each individual patient.

Considerable effort is also being made in the development of memory paradigms that can lateralise MTL functions and provide meaningful data at the single subject level. This information, in combination with structural MRI to evaluate hippocampal pathology and baseline neuropsychology, will enable pre-operative prediction of likely material specific memory impairments seen following unilateral ATLR to be made with greater accuracy. In consequence, it will be possible to modify surgical approaches in those patients most at risk and to improve pre-operative patient counselling.

Clinically it is what happens to individual patients that is important and the next step in the validation of these techniques will involve similar studies with larger numbers of patients. These should include more heterogeneous samples, including both left and right TLE undergoing ATLR. As well as showing group level correlations either at the voxel-level or within a predefined ROI, it will be important to establish methods for using this data to predict language and memory changes in individual cases. Investigating how the brain sustains memory post-operatively also requires further investigation. Longitudinal fMRI studies with pre and post-operative imaging, including correlations with neuropsychological measures of language and memory, will be required to look at functional reorganisation following surgery, and it is anticipated that these will offer valuable insights into brain plasticity. As experience grows in the interpretation of patterns of pre-operative language and memory fMRI results associated with good and bad post-operative outcomes, there will be a reduction in need for invasive procedures such as the IAT.

2. fMRI of Paroxysmal Activity

Despite major developments in the field of neuroimaging over the last two decades, the localisation of the brain regions involved in seizure onset is problematic in a significant proportion of patients with focal epilepsy, thereby precluding surgical treatment. Furthermore, our understanding of the neurobiological mechanisms underlying epileptogenic networks in focal and generalised epilepsies is incomplete.

Scalp EEG and magnetoencephalography (MEG) are comparable in their ability to detect and measure synchronised neuronal activity taking place mainly over relatively superficial parts of the brain with exquisite temporal resolution. Although both remain extremely active areas of investigation, the interpretation of EEG and MEG data and in particular their utility in localising brain generators is severely limited as a consequence of the principle of superposition and its corollary, the non-unicity of the inverse solution *(88, 89)*. This is in contrast to tomographic functional imaging modalities such as positorn emission tompgraphy (PET) or fMRI, which do not suffer from the problem of non-unicity and sampling bias is relatively minor. Although much superior to PET, the temporal resolution of fMRI, which is essentially governed by the local haemodynamic response, remains inferior to that of EEG by roughly three orders of magnitude. Nonetheless, fMRI allows haemodynamic changes linked to brief (~ms) neuronal events to be detected and localised with a fair degree of reliability. Although our understanding of the blood oxygen level-dependent (BOLD) fMRI signal is constantly improving, in part due to combined EEG and fMRI experimental data, as a general rule it remains an indirect and relative measure of neuronal activity. Although combined MEG-MRI seems a distant prospect, combined EEG-fMRI experiments were performed only a few years following the advent of fMRI *(90)*.

Although often presented as combining the advantages of its constituent parts, a concept that motivated the technique's pioneers, we will show that inevitably combined EEG-fMRI also suffers from some of their individual limitations. Whatever the technique's pros and cons, it will soon become clear to the reader that EEG-fMRI is unique in allowing the haemodynamic correlates of brief, unpredictable bursts of neuronal activity observed on scalp EEG, such as interictal spikes, to be investigated.

Prior to the possibility of EEG-fMRI experiments, studies of paroxysmal activity using fMRI were limited to ictal events and often relied on the correlation of the image time-series with observed clinical manifestations but sometimes did not ***(91–94)***.

The first studies of paroxysmal brain activity using fMRI were predominantly in patients with focal epilepsy, clinically motivated by the possibility of non-invasively localising seizure focus. This continues to be an important source of motivation for this rapidly moving field, but much current research focuses its attention on the understanding of the networks underlying the generation of seizures in both focal and generalised epilepsies.

Although an exciting development with potential clinical value, the technique currently remains within the realm of advanced, exploratory imaging modalities that require resources either not available in most epilepsy clinics. We therefore begin this review by discussing some of the technique's key technological and methodological aspects. We will then present an overview of the state of EEG-fMRI applied to the investigation of focal and generalised epilepsies.

The analysis and interpretation of fMRI data acquired from patients lying in the resting state with simultaneous EEG recording differs fundamentally from that of paradigm-driven fMRI in at least two ways: a lack of a prior experimental control and uncertainty about the nature of the relationship between EEG event and putative haemodynamic effects. This important topic will be the subject of a discussion.

2.1. EEG-Correlated fMRI in Epilepsy: Technical Issues

The recording of EEG inside the MR scanner still presents safety, image data quality and EEG data quality challenges. Historically, the issue of EEG data quality has been the determining factor in the technique's evolution, from interleaved to simultaneous EEG-fMRI. This reflects in part the fact that a gradual degradation in EEG quality mainly linked to cardiac activity can be readily observed in most subjects as they are moved inside the MR scanner, posing an immediate challenge ahead of any other considerations such as safety or the effect of scanning on EEG quality. In the following, we provide an overview of the state of EEG-fMRI technology, which remains an active area of research in particular in the area of EEG quality, although mostly for the purpose of evoked response recordings. The focus will be on the implications for studies in epilepsy and in particular at field strengths commonly used in neurological studies (≤3.0 T).

2.1.1. Physical Principles of EEG-MR System Interactions

The electro-magnetic processes that take place during MR image acquisition and that are susceptible to interactions with the EEG system are: strong static magnetic field (~1.0–3.0 T), switching magnetic gradient fields (~100 T/m/s), and radio frequency (RF) pulses (~10 μT and 100 MHz). In addition, although MR scanners are designed to optimise the magnetic component of the RF pulses, an electrical component is unavoidable. This may lead to linear antenna effects with possible safety implications *(95)*. EEG recording, on the other hand, requires electrodes and leads to be placed within the imaging field of view and electronic components, depending on the exact equipment and setup, in proximity to the scanner coils and antenna(s).

Four main mechanisms are at the origin of EEG-MR instrumentation interactions:

1. Magnetic induction: any change in magnetic flux (essentially the component of the magnetic field that is perpendicular to a surface) over time through a conducting medium (loop, surface, volume) gives rise to an electromotive force in the material and hence an induced current. This phenomenon is governed by Faraday's law of induction. Changes in magnetic flux, and the associated induced currents, can be caused by movement (change of position, orientation, or shape) of the conducting medium in a magnetic field or change in the magnetic field to which the conducting medium is exposed.
2. Magnetic susceptibility differences: static interactions due to the magnetic properties of the components of the EEG recording system
3. RF radiation emanating from active components of the EEG recording system
4. Magnetic force on ferro- or para-magnetic components; in the following we will assume that all usual design and manipulation precautions have been taken to avoid projectile effects in MR scanners

Magnetic induction can result in additional health hazards and EEG quality degradation in the form of pulse-related and image acquisition (gradient-switching and RF)-related artefacts. Magnetic susceptibility differences and RF radiation linked to the EEG system can give rise to image artefacts.

2.1.2. Safety

Health hazards not normally encountered when MR or EEG are performed separately can arise due to induced currents flowing trough loops or the heating of EEG components in proximity or contact with the subject. For a specific 1.5 T scanner, and based on a worst case scenario, this study recommended that one 10 kΩ current-limiting resistor be inserted serially at each electrode lead and the possibility of large (EEG lead-electrode-head-electrode-lead-amplifier circuit) loops being formed reduced to a minimum by lead twisting. In experiments using a different custom-made EEG system, no significant heating was observed *(96)*. An important general consideration when placing wires in contact with the body is the type of RF transmit coil used and length of wire exposed to the electrical component of the field *(97)*. A number of MR-compatible EEG system or electrode cap vendors have incorporated current-limiting resistors in their product design. To the authors' knowledge, no adverse incident linked specifically to EEG-fMRI data acquisition has been reported to date.

2.1.3. Image Quality

Image quality remains an important issue throughout the field of MRI and the subject of investigation, particularly for echo-planar imaging (EPI), which is particularly prone to distortion and local signal dropout *(98)*. Artefacts caused by electrodes and leads were observed in early EEG-fMRI experiments *(90)*. Therefore, one must consider carefully the choice of materials and components placed within the field-of-view *(96, 99–102)*.

2.1.4. EEG Quality: Pulse Artefact, Reduction, and Correction Methods

The first attempts at recording EEG inside MR scanners revealed the presence of pulse-related artefacts delayed in relation to the QRS complexes on ECG *(90)*. This effect has been shown to be common across subjects and has a slight frontal emphasis *(103)*. The pulse artefacts can have amplitude of the order 50 μV (at 1.5 T) and resemble epileptic spikes. Because of natural heart beat variability, it is considered a more challenging problem than that of image acquisition artefacts. EEG artefacts linked to subject movement are also amplified in the scanner's strong static magnetic field.

The precise mechanism through which the circulatory system exposed to a strong magnetic field gives rise to these artefacts remains uncertain, but it is thought to represent a combination of the motion of the electrodes and leads (induction) and the Hall effect (voltage induced by flow of conducting blood in proximity of electrodes) *(104)*. Electrode motion can result from local arterial pulsation, brain and head motion or whole-body motion (ballistocardiogram, or BCG, in the later case) *(105, 106)*.

Methods to reduce artefacts at the source include: careful laying out and immobilisation of the leads, twisting of the leads, bipolar electrode chain arrangement *(107)*, head vacuum cushion *(108)*. Such measures do not eliminate the problem completely resulting in degraded EEG quality, impeding the identification of epileptiform discharges. The first pulse artefact reduction algorithm published, and to this day still the gold standard against which most methods are compared, is based on subtraction of a running average estimate of the artefact based on automatic

QRS detection, and is commonly referred to as the average artefact subtraction (AAS) method *(103)*. Using this method, the residual artefact is of the order of a few microvolts. The reliance of the algorithm on ECG is a common, though not universal, feature among subsequently developed techniques (some of which use the signal from the standard scanner pulse oxymeter). The method has been and continues to be used successfully in our lab allowing the satisfactory identification of ictal and interictal epileptiform discharges (IED) in real time (at 1.5 T) *(109)* and for the purpose of source analysis *(110)*, and has been implemented in widely used commercial MR-compatible EEG recording systems (*see* **Fig. 6**).

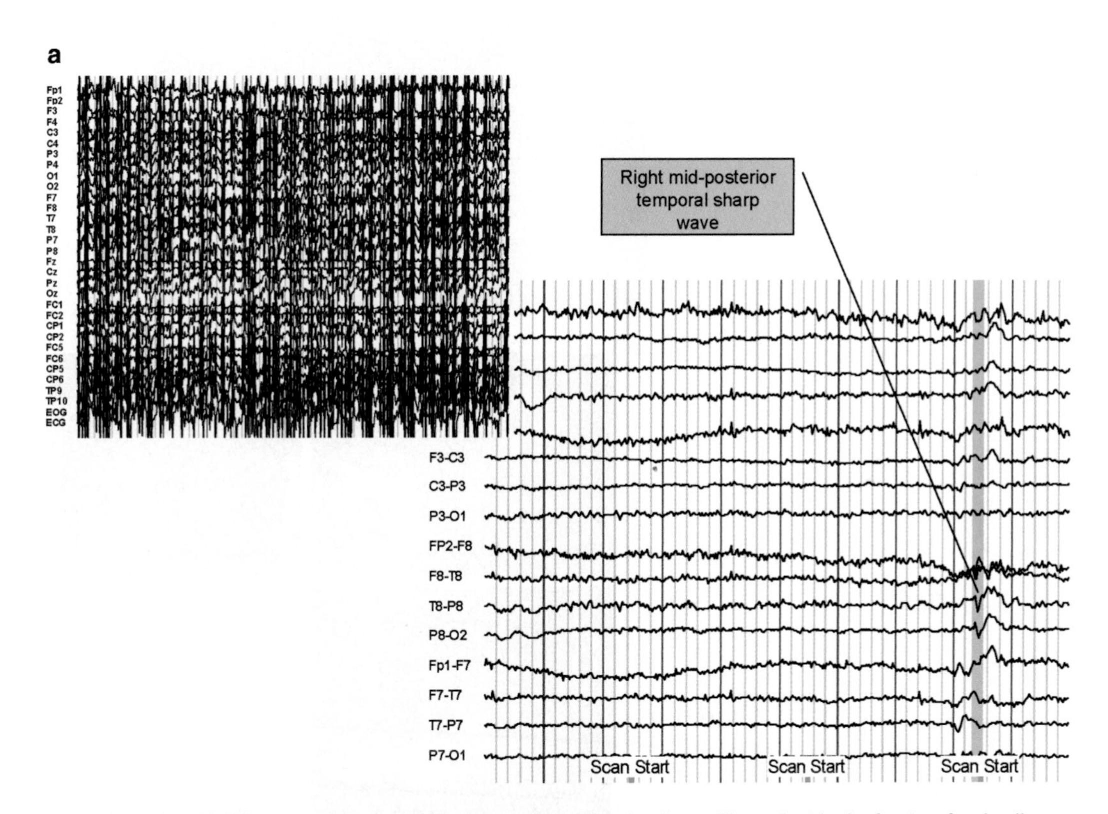

Fig. 6. IED-related BOLD pattern in patient with drug-resistant focal epilepsy. The patient had refractory focal epilepsy, lateralized to the right with a normal structural MRI. Frequent mid and posterior temporal sharp waves were recorded on EEG.(**a**) Representative segment of 32-channel EEG showing a sharp wave, maximum at the right mid-posterior temporal region, recorded during two 20-min fMRI sessions. *Top left*: EEG prior to artefact correction ***(103,126)***. (**b**) Design matrix: BOLD signal changes related to 40 sharp waves were modelled by convolution of the EEG event onsets with a canonical HRF and its time-derivative. Signal changes linked to head motion and heartbeat were modelled as nuisance effects ***(222, 223)***.(**c**) *Top*: SPM showing significant sharp wave-related BOLD response in glass brain display ($p < 0.05$ corrected for multiple comparisons). The *red arrow* marks the global maximum, located in the BA 28 (superior temporal gyrus). *Bottom*: BOLD response overlaid onto the patient's normalised T1-weighted volumetric scan. Intracranial recording confirmed a right posterior temporal lobe onset. No significant sharp wave-related deactivation was revealed. The activation clusters were labelled using the Talairach Daemon, http://ric.uthscsa.edu/project/talairachdaemon.html.

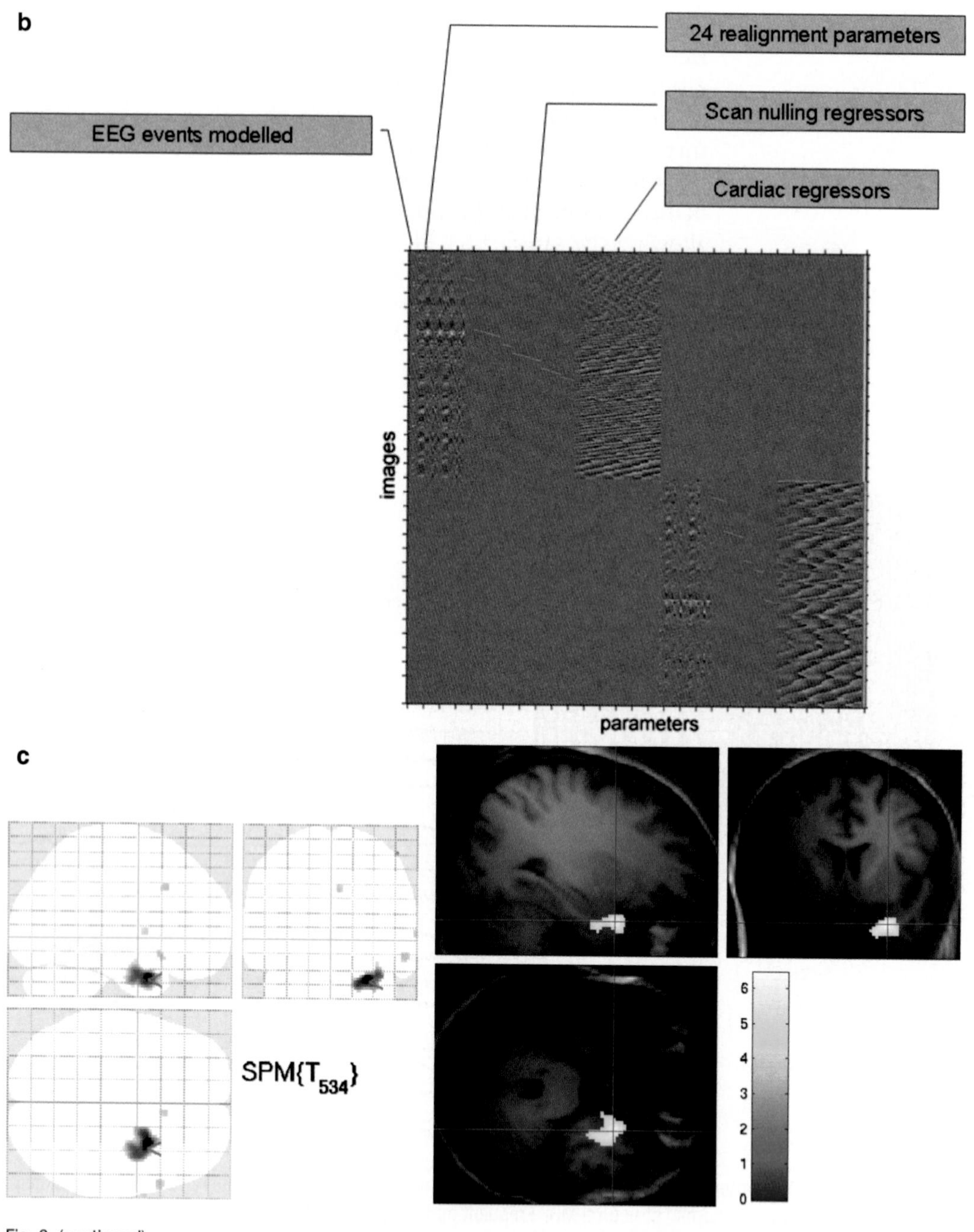

Fig. 6. (continued)

The artefact amplitude is theoretically directly proportional to the scanner static field strength (B_0). This phenomenon, and an increasing interest in recording evoked potentials in the MR scanner, has motivated an important research effort towards improving existing pulse-related artefact reduction methods and the development of new ones. Variants of the AAS method have been proposed, ranging from different ways of estimating the artefact waverform, for example to account for a greater degree of inter-beat variability *(107, 111–113)*, more general motion effects *(101, 114, 115)*, to improving QRS detection *(116)* and removing the need for ECG recording *(117)*.

Spatial EEG filtering methods have been proposed based on temporal principal components analysis (PCA) or independent components analysis (ICA) *(102, 108, 118–124)*. It is important to note that some PCA and ICA-based correction methods may not be applicable in real time, making it difficult to visualise epileptiform activity during the experiments with possible practical and safety implications. In studies in which residual noise was quantified, improvements of the order of 0.1–1 μV compared with various implementations of the AAS method have been demonstrated *(117, 124, 125)*.

2.1.5. EEG Quality: Image Acquisition Artefact, Special EEG Recording Equipment, and Correction Methods

In the absence of any special measures, the EEG recorded inside the MR scanner becomes un-interpretable during image acquisition because of the presence of repetitive artefact waveforms superimposed on the physiological signal *(126, 127)*.

One way of circumventing this problem is to leave gaps in the fMRI acquisition (e.g. between EPI volumes) of sufficient duration to capture the EEG features of interest (assuming sufficient data quality, e.g., following pulse artefact removal); this is *interleaved EEG-fMRI(107, 128, 129)*. This approach relies on artefact not persisting following each acquisition (e.g. due to amplifier saturation). Interleaved EEG-fMRI can be most useful to study predictable events (evoked responses) or slowly varying phenomena, such as brain rhythms. EEG-triggered fMRI and in particular spike-triggered fMRI, which involves limiting fMRI acquisition to single or multi-volume blocks, each triggered following the identification of an EEG event of interest is a form of interleaved EEG-fMRI with obvious relevance to epilepsy *(127, 130–133)*.

Although interleaved EEG-fMRI is capable of providing useful data in many circumstances, it imposes a limit on experimental efficiency due to EEG quality degradation during scanning.

We now review the technical developments that have made it possible to record EEG of sufficient quality throughout fMRI acquisition (so-called *continuous EEG-fMRI*), by the image acquisition artefact to be corrected. For all practical purposes, and assuming that the time gap between volumes is the same as between slices, the artefact's spectral signature ranges from 1/TR (TR: slice acquisition repetition time) to around 1 kHz (corresponding to the readout gradient). In fact it extends into the mega-hertz (RF) range, well beyond the recording capability of any EEG equipment. It can appear artificially benign when captured using standard EEG equipment *(90)*. Only using equipment with sufficient bandwidth, sampling rate and dynamic range, can one capture the artefact with adequate accuracy *(126, 134)*.

Experiments have shown that the gradient switching-related effects generally dominate over RF in terms of amplitude and extent in time, although the balance between the two mechanisms will vary depending on the specific MR sequence used *(116, 126, 134, 135)*. For standard EPI sequences, the pattern of gradients is repeated exactly across slices. Compared with the problem of cardiac-related artefact reduction, this determinism greatly facilitates the task of image acquisition artefact removal; however, the induced waveforms will be subjected to variations in time due to changes in the electrode/lead configuration caused by subject motion.

Before discussing artefact reduction post-processing techniques, let us review some of the hardware modifications and other measures that can facilitate the recording of good quality EEG during fMRI. First, some of the tricks described previously to reduce the pulse-related artefact at the source, and in particular those to limit the area of loops formed by EEG leads and head motion, can also help to lessen the image acquisition artefact problem. Second, low-pass filtering at the front end of the EEG system may be used to reduce the artefact significantly, although not sufficiently to result in adequate EEG quality *(126)*. Third, a scheme has been devised to reduce the amplitude of the artefact at the source by modification of the MR sequence and careful synchronization with EEG sampling *(134)*.

In the studies by Allen et al. *(126)* and Anami et al. *(134)*, custom-built EEG recording systems with high sampling rates (1–20 KHz) and large dynamic range (~20 mV), based on the notion that the artefact must be captured accurately to be understood, were measured and eliminated. Specially designed 'MR-compatible' EEG recording hardware has now become the norm in the field, with a number of commercial products now available on the market (see also *(107, 135, 136)* for a description of other modified or purpose-built apparatus).

Post-processing methods to reduce image acquisition artefacts can be categorised as filtering, template subtraction methods or PCA/ICA. Filtering based on the identification and subsequent suppression of frequencies linked to the image acquisition process can lead to an improvement in EEG quality and be used for simultaneous EEG-fMRI *(113, 136, 137)*. However, it is severely limited by the spectral overlap between the artefact and physiological signals, ringing effects and has been shown to be inferior to template subtraction *(108)*.

The most commonly used image acquisition artefact reduction method is based on average template artefact subtraction (sometimes referred to as AAS) method *(126)* (*see* **Fig. 6**). It relies on the lack of correlation between physiological signals and the artefacts, enabling the latter to be estimated by averaging the EEG over a number of epochs, corresponding to individual scan repetitions, for example. The success of the artefact (template) estimation and its subsequent subtraction from the ongoing EEG depend critically on the sampling rate, the number of averaging epochs, and the precision of their timing. In Allen's original implementation, this is addressed by the use of the scanner's scan trigger pulse to mark each scan acquisition and interpolation. Following subtraction, residual artefacts are reduced using Adaptive Noise Cancellation. The method can be used in real time, allowing continuous EEG-fMRI studies in patients with epilepsy *(138–141)*. Possibly the most important practical development has been the demonstration that synchronised MR acquisition and EEG digitisation lead to significantly improved EEG quality, and in particular over a wider frequency range, when combined with an AAS-like method *(142)*. As noted previously, changes in the artefact waveform due to subject motion will lead to sub-optimal template estimation. To address this, refinements of Allen's method which incorporate PCA of the residual artefact have been proposed *(116, 143)*. The shape of the image acquisition artefact may be captured in a separate experiment for subsequent subtraction *(135)*. In the study by Wan et al. *(144)*, a method designed to bypass the requirement for a slice acquisition signal from the scanner is proposed. As is commonly the case for artefact reduction methods based on ICA, identification of the components containing artefact is mainly done visually *(124)*. A difficulty encountered when comparing the various methods available for artefact reduction is the range of methodologies used. This has been addressed to some degree in a recent comparative study *(145)*.

2.2. Application of fMRI to the Study of Paroxysmal Activity

fMRI can be used to investigate the haemodynamic correlates of paroxysmal activity, and in particular to reveal regional changes in the BOLD signal thereby potentially providing new localising information. *See* **Table 3** for a list of the main published studies.

The conventional approach to the analysis of fMRI data is predicated on the correlation of the fMRI time series with experimentally determined stimuli within the framework of the general linear model (GLM). This methodology allows voxel-by-voxel testing of the degree of fit of predicted and observed BOLD time courses, and subsequent inferences. The application of fMRI for the assessment of spontaneous paroxysmal activity in epilepsy offers a number of additional challenges, namely: the lack of experimental paradigm (subjects scanned in the resting state), the identification of paroxysmal activity in relation to the fMRI time series, and the representation and translation of this activity into a GLM. The latter point is crucial and as we will see, has been an area of continuing investigation in the field.

In addition to the difficulties associated with the observation of subjects within the confined space of an MR scanner, and perhaps

Table 3
EEG-fMRI studies of paroxysmal activity: early milestones and important series

Study	Number of subjects/no. in which IED recorded	Results	Conclusion	Comment
Focal Epilepsy				
Warach et al. 1996 *(127)*	1/1 frequent IED	Bilateral activation where EEG suggested left temporal localization and anterior cingulate activation in relation to generalized epileptiform activity.	'We cannot make conclusions about the source of the discharge from the present data.'	
Seeck et al. 1998 *(159)*	1/1 frequent IED	Multiple areas of signal enhancement on fMRI. Confirmed on 3D-EEG source localization with evidence of a focal onset. Focus later confirmed on subdural recordings.	The combination of EEG-triggered fMRI and 3D EEG source analysis, represents a promising additional tool for presurgical epilepsy evaluation allowing precise non-invasive identification of the epileptic foci.	
Symms et al. 1999 *(154)*	1/1 frequent IED	Reproducible and concordant activation across 4 sessions		
Patel et al. 1999 *(153)*	20/10 frequent IED	9/10 overall reported as showing 'activation corresponding to the EEG focus'		
Krakow et al. 1999 *(130)*	10 frequent IED	Reproducible activations (same lobe and overlapping) obtained in 6/10 patients in close spatial relation to EEG focus		
Krakow et al. 1999 *(185)*	1 frequent IED	Focal activation within a large malformation of cortical development in response to focal epileptiform discharges		
Lazeyras et al. 2000 *(158)*	11 frequent IED	Activation confirmed clinical diagnosis in 7/11. In 5/6 intracranial EEG confirmed result		

Lazeyras et al. 2000 *(149)*	1/1 frequent IED	Area of signal enhancement concordant with hyperintensity seen on ictal FLAIR images in a patient with non-lesional partial epilepsy		
Krakow et al. 2001 *(132)*	24/14 frequent IED	12/24 patients showed activations concordant with EEG focus, 7/12 of which also had concordant structural lesions. 2/24 were discordant and 10/24 showed no significant activation.		
Lemieux et al. 2001 *(139)*	1/1 frequent, stereotyped IED	In a case with stereotyped frequent IED, BOLD signal change concordant with the seizure onset zone was recorded	Localisation of BOLD activation was consistent with previous findings and EEG source modelling.	First description of application of continuous EEG-fMRI
Jager et al. 2002 *(225)*	10/5 frequent IED, focal epilepsy	Focal activation in 5/5 patients, concordant with EEG amplitude mapping. Mean signal increase was 15 ± 9%. Spike amplitude correlated with volume of activation		
Benar et al. 2002 *(226)*	4/4 frequent IED	The average HRF presented a wider positive lobe in three patients and a longer undershoot in two.	There was no clear correlation between the amplitudes of individual BOLD responses and EEG spikes	
Al-Asmi et al. 2003 *(133)*	48/31 frequent IED	BOLD activation in 39% of studies. Concordant with seizure focus in almost all. 4 patients had concordant intracranial recording (by lobe).	Combining EEG and fMRI in focal epilepsy yields regions of activation that are presumably the source of spiking activity and these are high	
Benar et al. 2006 *(181)*	5/5 presurgical candidates having sEEG	When an intracranial electrode is in the vicinity of an EEG or fMRI peak, it usually includes one active contact.		Largest series of intracranial EEG correlated fMRI activation

(continued)

Table 3 (continued)

Study	Number of subjects/no. in which IED recorded	Results	Conclusion	Comment
Aghakhani et al. 2006 *(208)*	64/40 focal epilepsy with frequent unilateral or bilateral IED	A positive thalamic response was seen in 12.5% of studies with unilateral and 55% with bilateral spikes. Cortical acitvation was more concordant with focus than deactivation.	The thalamus is involved in partial epilepsy during interictal discharges. This involvement and also cortical deactivation are more commonly seen with bilateral spikes than focal discharges.	
Salek-Hadaddi et al. 2006 *(141)*	63/34 focal epilepsy frequent IED	Significant hemodynamic correlates were detectable in over 68% of patients and were highly, but not entirely, concordant with site of presumed seizure onset.	These findings provide important new information on the optimal use and interpretation of EEG-fMRI in focal epilepsy	
Ziljmans et al. 2007 *(184)*	29/15 focal epilepsy, declined for surgery	8/15 subjects: IED correlated BOLD response at site of focus. Multifocal in 4, unifocal in 4. Concordant with IC data in 2.	EEG-fMRI provides additional information about the epileptic source in the presurgical work-up of complex cases.	First evaluation of impact on pre-surgical evaluation
Generalised Epilepsy Salek-Haddadi et al. 2003 *(227)*	1/1 prolonged GSW epochs (ictal)	Thalamic activation and widespread cortical deactivation	Supports thalamo-cortical model of GSW	
Baudewig et al. 2001 *(160)*	1/1 frequent GSW	Unilateral insular activation shown in relation to generalized epileptiform discharges.	Strategy resulted in robust BOLD MRI responses to epileptic activity that resemble those commonly observed for functional challenges	

Hamandi et al. 2006 *(138)*	46/30 interictal GSW in IGE and SGE	Thalamic activation and cortical deactivation observed at group level. Deactivation in the default brain areas. Cortical pattern mixed at individual level.	Observed cortical deactivation may represent correlate of clinical absence seizure.	
Aghakhani et al. 2004 *(228)*	15/14 interictal GSW	Bilateral thalamic activation in 80% of BOLD response. Cortical deactivation in 93%.	Cortical deactivation mediated by hyperpolarization of the thalamus.	
Hamandi et al. 2007 *(205)*	4/4 interictal GSW	Qualitatively reproducible BOLD and blood perfusion patterns; Cortical deactivation corresponds to decrease in blood flow.	Consistent with preserved neurovascular coupling in GSW and decreased cortical activity.	
Children				
De Tiege et al. 2007 *(196)*	6/6 focal epilepsy (lesional and non-lesional)	Concordant activation with presumed focus in 4 cases. IC recording corroborative in 1.	EEG-fMRI is a promising tool to noninvasively localize epileptogenic regions in children with pharmacoresistant focal epilepsy.	First series specifically addressing use of EEG-fMRI in children
Jacobs et al. 2007 *(195)*	9/9, mixed focal epilepsy, sedated	All had BOLD activation or deactivation concordant with seizure focus. Deactivation appears more common in adults.	EEG-fMRI at 3T could be useful to localize focus in children with focal epilepsies. Nature and origin of the negative BOLD requires investigation.	
Jacobs et al. 2007 *(229)*	13/13 symptomatic (lesional) epilepsy), sedated	Activation corresponding with the lesion was seen in 20% and deactivation in 52% of the studies.	'Good results could be obtained from the EEG-fMRI recordings, performed in sedated children.'	

more importantly the use of behaviourally-derived time markers when possible, although crucial, does not provide as complete a picture of the event as one would wish given the importance of putative concomitant EEG abnormalities.

The advent of EEG-correlated fMRI has been a major advance in this respect, providing an established, albeit imperfect, marker of paroxysmal activity and more generally brain state. Importantly, it allows the study of interictal activity, particularly IED only manifest on EEG. Although this type of data allows the experimentalist to study haemodynamic changes linked to specific EEG events, it presents a number of challenges linked to the subjective nature of EEG interpretation. The investigator is also soon confronted with a 'chicken and egg' type problem of not knowing precisely what the time course of the changes is, and in particular whether the haemodynamic 'response' function associated with paroxysmal discharges deviates from normality, a necessary element of the modelling, or its spatial substrate, the latter being precisely the motivation for undertaking the experiment given its clinical implication.

2.2.1. Ictal fMRI in Focal Epilepsy

As mentioned previously, a number of case studies of ictal events captured using fMRI alone were published prior to the advent of EEG-fMRI *(91–94)*. Despite the fact that these contain interesting observations, particularly with regard to the signal change around the time of seizure onset, the availability of simultaneously recorded EEG would have added important information. The development of the ability to record good quality EEG inside the MR scanner has offered the possibility of improved models of ictal fMRI signal changes by the inclusion of precisely timed EEG-derived information in the model, and also by providing an indication of head motion due to artefacts on the EEG which may assist interpretation. Clearly the problem of motion is particularly pertinent to the acquisition of seizure data, as it can severely degrade image data quality and consequently the ability to detect and map regional haemodynamic changes accurately in relation to those manifestations *(146–148)*.

Given the above limitations and concerns regarding patient safety, EEG-fMRI data of seizures is usually acquired incidentally in the course of experiments aimed at studying interictal activity. In a patient with simple partial status, epilepticus multiple seizures were captured in a multi-modal MR imaging and spectroscopic study, showing a consistent area of abnormality *(149)*. Salek-Haddadi et al. showed large (up to 6%) focal signal increases in relation to a clinical seizure and concomitant EEG changes, consistent with a normal haemodynamic response, and were able to control for possible motion-related effects *(140)*. In a patient with multiple simple partial seizures, significant BOLD activation concordant with the electroclinically determined seizure focus was revealed. Widespread deactivation was observed on the contralateral side and in other areas of the brain and the haemodynamic change continued beyond the ictal activity observed on scalp EEG *(150)*.

Detection of changes prior to the onset of ictal EEG is of great interest in understanding the generation of seizures and this is a relatively new area of research. PET and single photon emission computed tomography data suggest that there are pre-ictal changes in cerebral metabolism and studies of cerebral blood flow (CBF) support this *(151)*. A series of three patients who had seizures during scanning has been studied with fMRI and changes in the BOLD signal minutes before the onset of seizures was noted, although concurrently recorded EEG was available in only one case and unhelpful *(152)*.

2.2.2. Interictal fMRI: EEG-fMRI

Although the study of ictal events may be useful in pursuing the aim of non-invasively identifying the seizure onset zone and thereby contributing most to pre-surgical evaluation, its acquisition is fraught with difficulty and analysis is subject to the limitations described earlier. Attention has, therefore, focused on interictal activity and the insight gained from it into the function of epileptic networks. Spike-triggered fMRI acquisition was used in the first case reports and series which largely studied subjects pre-selected for the large number of IED observed in prior routine surface EEG or as part of pre-surgical assessment. The acquisition and analysis of spike-triggered fMRI data generally rests on the assumption that interictal spikes are associated with a BOLD signal change pattern similar to the so-called canonical HRF. In these series, regions of positive BOLD signal change associated with IED were observed in approximately 50% of the cases overall, and occasional negative changes *(127, 153, 154)(130, 132, 155–159)*. Using bursts of BOLD EPI scans, Krakow et al. made an initial attempt at estimating the shape of the IED-related HRF *(132)*. We note that in the work by Seeck and colleagues, Clonazepam was used to suppress interictal discharges, thereby creating a control scan state *(158, 159)*. Interleaved EEG-fMRI, whereby fMRI data are acquired in blocks with inter-block gaps of a sufficient duration to allow interpretation of part of the EEG was employed in a patient with IGE for the mapping of BOLD changes related to spike-wave complexes *(160)*.

Following the implementation of image acquisition artefact removal *(126)*, the technique of continuous, simultaneous EEG-fMRI acquisition was demonstrated along with estimation of the shape of the IED-related HRF in a subject with Raussmussen's encephalitis and stereotyped high amplitude sharp waves on the EEG *(161)*. *See* **Fig. 6** for example of EEG-fMRI in patient with focal epilepsy. When applied to relatively large case series of subjects with predominantly focal epilepsy selected on frequent interictal discharges on routine EEG *(141, 157)*, a yield (proportion of significant BOLD activations or deactivations) of 60–70% was observed where IED were recorded. This result has been reproduced in further studies despite variation in the analysis strategies and patient groups. Positive BOLD changes tended to correspond with the site of the presumed seizure onset zone (based on electroclinical localisation), but occasionally appeared at sites distant from this region, the significance of which remains unclear. Negative BOLD changes were more often observed remote from the presumed focus, with a striking pattern of retrosplenial deactivation in a significant proportion of cases *(141)*. It was shown that activations were more likely when there was good electroclinical localization, frequent stereotyped spikes, less head motion, and less background EEG abnormality. Furthermore, the findings suggest that significant activation is more likely for runs of spikes than for isolated discharges, when the event duration is taken into account in the modelling *(141, 162)*. The increase in yield from continuous EEG-fMRI compared with spike-triggered studies, although not formally assessed, seems relatively modest despite the potentially greater number of events observed. This may reflect the more selective approach to data collection (i.e., event used for triggering) in the latter.

In TLE, IED-related activation of the MTL ipsi-lateral to the presumed focus was found to be common, and is reminiscent of the thalamic pattern observed in generalized spike and wave discharges (GSW) *(163)*.

The possible explanations for lack of activation include: combination of insufficient number of events and limited fMRI sensitivity, incorrect model due to sub-optimal EEG event identification and classification, choice of HRF and limited extent to which scalp EEG reflects ongoing activity (throughout the brain). Automated and semi-automated spike detection methods have been proposed to attempt to reduce the level of subjectivity of EEG event identification *(110)*. Nonetheless, although the ability to record EEG during fMRI is necessary to study the haemodynamic correlates of EEG events observed on the scalp, total reliance on scalp EEG can also be seen as a limitation when attempting to interpret the BOLD signal throughout the brain and in particular the part of which may not be linked to activity reflected on the scalp. In the absence of clear epileptiform discharges, the presence of slow activity may provide useful localising information *(164, 165)*. Data-driven or region-based fMRI analysis techniques may offer a way forward in this respect, although interpretation of the observed patterns in the former is significantly impaired by the lack of an a prior model and numerous possible confounds *(166, 167)*.

The HRF in Focal Epilepsy

The normal haemodynamic response associated with neuronal activity arising from brief external stimuli in humans has a characteristic shape with a peak at around 5–6 s following the event, the so-called canonical HRF, with a significant degree of inter-subject variability *(168)*. The shape of the HRF is a key element of fMRI signal modelling with an important impact on sensitivity, and it is therefore important to attempt to characterise it. In epilepsy, particularly in relation to deactivations it has been proposed neurovascular coupling might be abnormal with possible consequences on the shape of the HRF, and this was suggested to be one possible explanation for the significant widespread BOLD deactivations observed in both focal and generalised epilepsy *(169, 170)*.

This has lead to increased interest in estimating the shape of the HRF in epilepsy. An efficient way of estimating the shape of the haemodynamic changes linked with epileptiform discharges is by using a set of functions (sometimes referred to as 'basis set') that can, by linear superposition, fit signal changes with almost any time course. Various such schemes have been used in studies of epilepsy including sets of gamma response functions, Fourier basis sets and a linear combination of canonical HRF, its time derivative and a dispersion derivative *(139, 139, 141, 171, 171–173)*. Although a degree of variability has been observed, the HRF linked to IED was found to be principally canonical in shape. In some cases deviation from the norm at locations distant to the IED may reflect artefacts, while in others, deviant activation in proximity to the presumed focus, early responses may reflect brain activity that systematically precedes the event captured on the scalp or propagation *(171, 174)*. Animal studies provide corroborative evidence for these findings. In a recent study, the development of penicillin-induced epileptic activity was correlated with increase in BOLD signal prior to the onset of IED *(175)*. These studies coupled with those combining EEG-fMRI with multiple source analysis *(176)* suggest that in some cases the maximum BOLD response may be detected just prior to seizure or even IED onset. It may be, however, that this reflects BOLD changes linked to interictal abnormalities, which are not included in the model by virtue of the fact they are not seen on the scalp EEG.

2.2.3. Clinical Relevance of fMRI of Paroxysmal Activity in Focal Epilepsy

Having shown that EEG-fMRI is capable of providing a unique form of localising information, the issue of its clinical value arises. The evaluation of new non-invasive imaging modalities in the pre-surgical assessment of patients with drug-resistant epilepsy is a complex issue in part due to the lack of an established methodological consensus (baseline) across centres. Added value and clinical relevance are a function of the new test's sensitivity and specificity. In the case of EEG-fMRI, neither has been properly assessed to date. To date, the field has focused on proof of principle demonstrations, usually in patients selected based on high rates of EEG abnormalities, with a success rate of roughly 50–60% *(127, 130, 132, 141, 153, 157, 157, 158)*. Therefore, one may anticipate a lower success rate in the most clinically challenging cases for which the need for non-invasive assessment is most pressing.

When available, the new localising information must be evaluated relative to a surgically confirmed irritative and epileptogenic zone *(177)* using the current gold standard of invasive recording and outcome data. In practice, a gradual approach to validation is often taken, whereby the face validity of new localising information is tested against other existing techniques in cases with well characterised syndromes, such as mesial TLE.

In our laboratory, we have assessed the value of the fMRI findings by comparing the localisation of the BOLD cluster containing the most significant activation to the seizure onset zone defined electro-clinically, when possible, and found a very good degree of concordance at the lobar level *(130, 132, 141)*. Additional regions of activation were observed in roughly 50% of the cases with significant activation. The finding of localized BOLD activation in cases with poor electroclinical localisation suggests a means of obtaining target areas for intra-cranial EEG *(141)*. In TLE, the yield has been characterised as relatively high *(178)* and the degree of concordance of BOLD activations with the presumed focus generally good in one study *(141)*, but more varied in another *(178)*.

Comparison of EEG-fMRI with source localisation suggests that IED-related BOLD activation are often in proximity to those detected by conventional EEG source localisation, but some studies have suggested a distance of up to 50 mm. The possible sources of discrepancy between BOLD and electrical (or magnetic) source localisation include: differences in the nature of the observed phenomena and neuro-vascular coupling, vascular architecture and scanner field-related effects on sensitivity, instrumental and physiological noise, source reconstruction limitations, fMRI sensitivity limitations *(179–182)*.

Pre-surgically, intra-cerebral EEG data are widely recognised as the gold standard.

Comparison of scalp EEG-fMRI against the gold standard of invasive EEG recording has been performed in small groups and using widely varying fMRI analysis techniques and comparison criteria *(133, 159, 169)*, noting a degree of concordance between the epileptogenic zone and the area of maximal BOLD activation in some, but not all subjects *(181, 183)*. In the largest study to date and the first to systematically investigate the relationship between EEG-fMRI and intracranial recording, a group of 5 patients in which significant IED-correlated BOLD activation is observed, at least one intracranial contact showing epileptiform discharges was within proximity of the activated region concordant with the putative seizure onset zone *(181)*.

An initial assessment of the potential role of EEG-fMRI in pre-surgical evaluation was recently published *(184)*. The authors considered a series of patients with focal epilepsy in whom surgery was not offered following conventional electroclinical evaluation. The impact of the EEG-fMRI findings was assessed in eight cases with unclear foci or suspected mutifocality (based on the centre's usual battery of tests) in whom significant IED-related BOLD activation. IED-related changes suggested a more restricted seizure onset zone in four, and multifocality in a further patient. The EEG-fMRI finding was concordant with intracranial recordings in two. The authors suggest that this supports a possible role for EEG-fMRI in pre-surgical evaluation where a potential, but possibly widespread focus is identified.

Relationship with Pathology in Focal Epilepsy

Focal epilepsy can be divided by pathological subtype. Given the knowledge that the irritative and epileptogenic zones may extend beyond the area of pathological abnormality and that animal models suggest abnormal sub-populations of neurones within dysplastic areas, there have been studies of EEG-fMRI aimed at evaluating the haemodynamic response in abnormal tissue revealed on MRI generally demonstrating *(164, 185–189)*. In a series of 14 cases with heterotopia notable variability of the BOLD response across abnormal tissue was observed, but the area of BOLD signal increase was often concordant with the area of pathology, with a more mixed pattern of deactivations *(189)*. In MCDs and particularly Taylor type focal cortical dysplasia, BOLD signal increase was observed within the lesion while peri-close and distant from the lesional activity displayed a negative BOLD response in 4 out of 6 cases *(152)*. Other studies in cases with MCD have supported these findings. A recent study of IED-related BOLD signal change have been observed in patients with cavernomas *(190)* close and distant from the lesion. The frequently observed negative BOLD responses, particularly in MCD have been attributed to loss of neuronal inhibition (in the presence of normal neurovascular coupling) in the regions surrounding the abnormality or abnormalities in neurovascular coupling itself. The significance of these deactivations will be discussed in more detail later.

2.2.4. fMRI in the Investigation of Epilepsy Syndromes

EEG-fMRI has been used to attempt to localise sources in specific epilepsy syndromes, in addition to adding evidence to the understanding of the differences in sub-types of focal epilepsy. TLE is of particular interest in this context as surface EEG may not detect the deep sources involved in TLE and surgery for TLE, where it is correctly localised, is associated with excellent outcome *(191)*.

A large series of subjects with temporal and extra-temporal lobe epilepsy showed that temporal lobe spikes are seen at the presumed seizure focus, but also unsurprisingly on the opposite homologous cortex, but did not give further localising information *(192)*.

A further more recent study compared IED-related BOLD responses between subjects with temporal and extra-temporal lobe epilepsy and found that those brain areas involved in the so-called 'default mode network' *(193)* were commonly deactivated during temporal IED whereas other areas were involved in BOLD activation and deactivation in extra-temporal lobe epilepsy *(163)*.

Other sub-types of epilepsy studied include benign rolandic epilepsy of childhood *(182, 194)* and other focal epilepsies in children, which have demonstrated concordant IED-correlated BOLD activation in approximately 60% of cases *(195, 196)*. A study of lesional epilepsy in children not only revealed positive BOLD response concordant with the seizure onset zone in a significant number of subjects, but also a higher number of deactivations than those observed in adult studies *(195)*. However, sedation was used in this study, the effect of which has not been investigated in detail to date. EEG-fMRI is a particularly attractive option for the study of childhood epilepsies as it is non-invasive, but experiments are long and require a high degree of subject cooperation.

2.2.5. EEG-fMRI in Generalised Epilepsy

The generalised epilepsies (that is the syndromes of Absences, Myoclonic epilepsies, and primary generalised tonic clonic seizures coupled with generalised spike and wave discharges on the EEG *(197)*) are not currently amenable to surgical treatment. Therefore, the primary focus of EEG-fMRI investigations of generalised epilepsy has been the exploration of hypotheses developed in vitro and animal models *(138, 180, 198–202)*. In fact, most ictal EEG-fMRI data have been recorded in patients who have brief absence seizures. A striking pattern of thalamic activation and widespread cortical deactivation has been observed linked to four prolonged runs of GSW *(202)*. Evidence from animal models *(203, 204)* and limited intracranial data in humans regarding the generation of generalised spike and wave discharges (GSW) supports the existence of a cortico-thalamic loop. In addition, there is mounting evidence for the existence of a frontal origin of the spike component *(203)*.

EEG-fMRI is able to show significant BOLD changes in roughly 80% of cases in which interictal GSW is captured *(138, 198, 200)*. The increased yield compared with studies of focal discharges may reflect the fact that there are more commonly runs of discharges, which appear to produce more significant responses than single events. There was a considerable degree of inter-subject variability particularly in the cortical patterns of BOLD activation and deactivation, making it difficult to envisage the possibility of using these patterns as a basis to classify patients according to syndrome *(138)*. Additionally, focal cortical activation may reflect an initial focus (*see* **Fig.** 7).

The pattern of cortical deactivation is particularly interesting, corresponding in many cases to the so-called default mode network *(193)*, and it has been proposed that this may be a subclinical representation of the loss of consciousness associated with GSW in absence seizures *(192, 200, 201)*

Simultaneous EEG-Arterial Spin Labelling (ASL) MRI in patients with spike and wave discharges revealed degrees

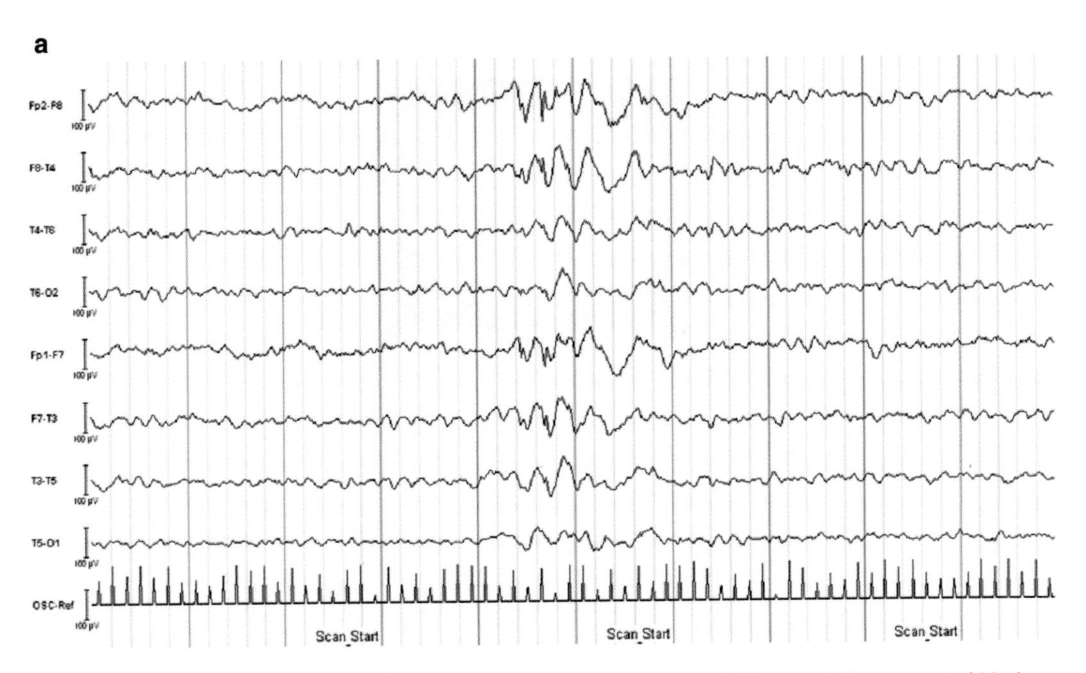

Fig. 7. GSW-related BOLD pattern in patient with secondary generalised epilepsy. (**a**) Representative segment of 10-channel EEG recording during scanning after image and pulse artefact subtraction ***(103, 126)***, displayed on common average montage showing a run of generalized spike and wave activity (2–2.5 Hz) with frontal predominance arising from a slow background. The patient (case no. 33 in ***(138))*** was treated with Lamotrigine (LTG) and Sodium Valproate (VPA). Twenty-five discharges were recorded during the 35-min experiment (mean duration 1.1 s; range 0.3–2.0 s). For the purpose of fMRI modelling, the EEG events of interest were represented by variable-duration blocks and convolved with the canonical HRF, its time derivative (TD) and dispersion derivative (DD) to account for variations in response timing ***(138)***. Motion and cardiac effects were included in the model as confounds ***(222, 223)***.(**b**) *Top*: SPM showing significant GSW-related BOLD response in glass brain display ($p < 0.05$ corrected for multiple comparisons). The *red arrow* marks the global maximum located in the precuneus. Signal changes in the precuneus, frontal cortex (bilateral medial frontal gyrus), bilateral inferior parietal lobuli and right superior frontal gyrus correspond to deactivations. The signal in the thalamus is positive. *Bottom*: BOLD response overlaid onto mean EPI image. The activation clusters were labelled using the Talairach Daemon, http://ric.uthscsa.edu/project/talairachdaemon.html.

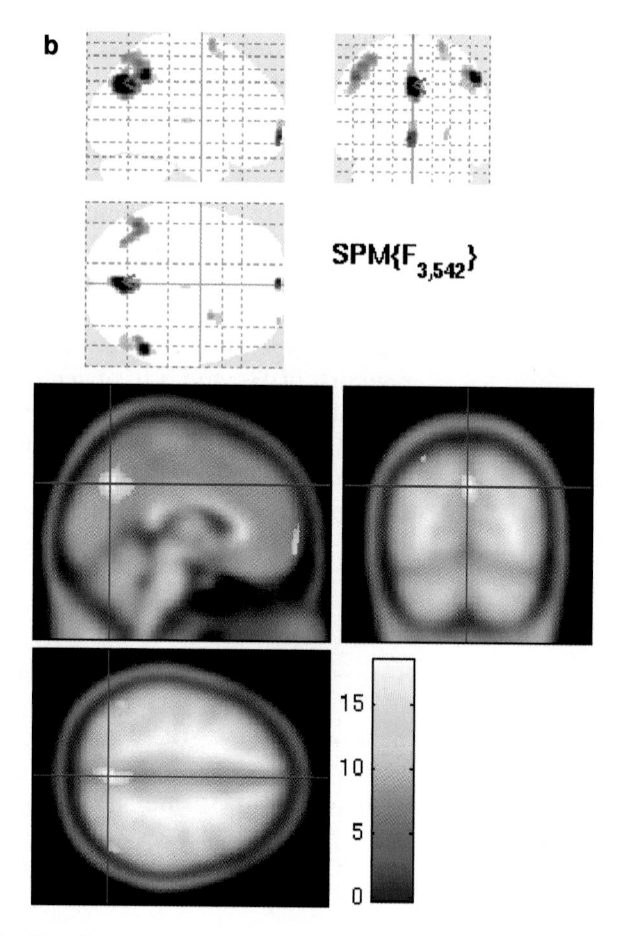

Fig. 7. (continued)

of correlation between BOLD and CBF consistent with normal neuro-vascular coupling *(205)*. This study also revealed a remarkable degree of within subject, inter-session (and inter-scanner) reproducibility, albeit in a small group.

2.2.6. The Significance of BOLD Deactivation in Epilepsy

BOLD signal deactivation has been observed in the monkey visual system and found to have a linear relationship with CBF in the same way as positive BOLD signal change in normal physiological conditions *(206)*. The pattern of cortical deactivation commonly observed in relation to GSW using EEG-fMRI in humans has been discussed earlier. Gotman et al. proposed that the reason for observing the widespread cortical deactivation may be hypersynchronisation of the thalamus, supporting Avoli's proposed model of the thalamus driving the cortex during GSW *(200)*. The results of Hamandi et al. using EEG-ASL are consistent with GSW-related deactivation reflecting decreased cortical activity *(205)*.

Focal IED-related deactivations are less common than activations and seem particularly linked to the presence of activation, possibly reflecting a smaller haemodynamic effect of individual events *(141, 170)*. In cases with MCD IED-related negative BOLD signal change adjacent to the seizure onset zone or area of dysplastic cortex, leading to several possible explanations including 'vascular steal' from more metabolically active regions, abnormal neuronal coupling, or perhaps more likely, the hypothesis is that there may be loss of inhibitory neuronal activity in these regions supporting the link between loss BOLD deactivation and underlying decrease in neuronal activity *(170, 187)*.

Deactivation of the default mode areas is a typical feature of IED-related BOLD changes in TLE in contrast to cases with extra-TLE *(163)*. The observed pattern may reflect IED-related effects in areas involved in cognition, with a possible link to transient cognitive impairment *(207)*.

2.2.7. fMRI and the Neurobiology of Epileptic Networks

Although limited by the sluggish BOLD response, we have seen that EEG-fMRI is able to reveal multiple regions more or less simultaneously activated or deactivated in relation to EEG events *(138, 141, 163)*. The involvement of the thalamus in GSW has been discussed earlier, but it has also been observed that sub-cortical BOLD activation can be seen in focal epilepsy *(192, 208, 209)*. A spike-triggered EEG-fMRI study of subjects with MCD identified BOLD activation concordant with seizure onset in all patients studied, and sub-cortical activations in various structures were observed in 50% of subjects *(187)*, and in Agakhani et al.'s study of focal epilepsies, thalamic activations were observed correlated with interictal IED, particularly where these were bilateral and synchronous *(208)*.

2.2.8. Limitations, Challenges, and Future Work

The study of spontaneous, pathological brain activity using fMRI presents the experimentalist numerous challenges. The advent of simultaneously recorded EEG greatly facilitates this endeavour. EEG-fMRI was originally held to be an excellent tool incorporating the spatial coverage and resolution of imaging techniques and the temporal resolution of EEG. However, a number of challenges remain.

Technically, post-processing methods are now available to offer EEG quality sufficient to allow a degree of abnormality identification reliability comparable to that of routine clinical EEG. Although EEG quality is sufficient to detect most IED with a good degree of certainty (at 1.5 T) *(210)*, pulse-related artefacts can sometimes interfere with EEG interpretation particularly at 3 T. Improved artefact correction and the use of techniques based on EEG source reconstruction may yield more localised and reproducible results *(110, 176)*. In addition, very little work has been done on this aspect of EEG interpretation and in particular the differences between the clinical and experimental approaches, with the former using a summary of the abnormalities observed over the whole EEG rather than categorisation and quantification of all IED required for fMRI.

The technique's reliance on scalp EEG is both a strength, as it allows to answer the question 'what, if any, are the BOLD correlates of EEG pattern X?', and weakness because it is burdened with certain limiting aspects of scalp EEG, such as sensitivity bias subjectivity in interpretation, and the unpredictable nature of the phenomena of interest. This presents the investigator with significant challenges in terms of unpredictable experimental efficiency (and consequently yield) and EEG interpretation for the purpose of GLM building. The former may require a more aggressive approach possibly using drug management as a tool for modulating EEG activity *(159)*. Although EEG event classification remains problematic, possible solutions may give the opportunity of using fMRI to inform EEG interpretation *(110)*.

The 'yield' of EEG-fMRI studies is to some extent a function of the previous points. A particular problem is that subjects are scanned for a limited period of time, and interictal discharges are not always observed during this period. A significant BOLD response is also only present in 50% of cases giving an overall 'yield' of the order of only 25% in many studies.

Attempts at interpreting resting state fMRI patterns in patients with epilepsy without reference to EEG (or clinical manifestations) using data-driven analysis techniques have had mixed success *(167)*, but one can envisage the possibility of being able to reliably identify sets of patterns that are potentially related to epileptic activity, with 100% yield *(211)*.

Despite providing whole-brain coverage of putative BOLD changes, the technique is effectively limited to the study of the events seen on the scalp EEG. We know from intra-cranial EEG that a large amount of pathological activity is not seen on the scalp, despite originating from a fixed location *(212)*. Therefore, the possible mismatch between scalp-derived model and ongoing activity raises the question of the nature of the baseline, with implications for sensitivity. This may be reflected in event onset time offset reflected in 'early responses' *(171, 174)*. Furthermore, this means that EEG-fMRI cannot be used to exclude regions from epileptogenic zone.

An additional degree of modelling uncertainty arises with regards to the IED-related haemodynamic response function linked with the potential effects of pathology and the nature of scalp EEG *(169)*. However, the latest evidence points to a preserved neurovascular coupling and shape of the HRF in line with the physiological response in healthy brains *(171, 205)*.

The sluggishness of the BOLD response in relation to EEG means that activation patterns effectively represent a time averaged picture of a sequence of neuronal events, such as propagation and loops. The causal relationship between the activity in these regions on the one hand and the EEG on the other cannot be untangled based on the correlation-based machinery that is the GLM. Nonetheless, these sets of regions may be thought of as a static or averaged picture of evolving networks, which may be subjected to further investigation. This may be particularly relevant to spontaneous brain activity compared with experimentally controlled experiments because of the greater uncertainty in the origin and sequence of neuronal events. The neurophysiology underlying the BOLD response is not fully understood. Much remains to be elucidated on the spatial-temporal relationship between IED and time-locked BOLD signal changes, for example the significance of responses distant from the presumed focus for which extensive comparison of the fMRI with the gold standards of intra-cranial EEG and post-surgical outcome will be required. The possibility of using MRI to directly detect local changes in magnetic field is an exciting prospect *(213)*.

The interpretation of BOLD activation patterns consisting of multiple clusters in relation to focal IED remains an area of active investigation. Recently, an attempt has been made to increase the technique's ability to temporally resolve multiple regions of BOLD activation by combining EEG-fMRI with multiple source analysis *(182)*.

Finally, EEG-fMRI as implemented and analysed currently may be considered purely as a haemodynamic imaging technique, with EEG simply acting as a time marker for scan classification. However, we envisage a more symmetrical approach whereby the two forms of data are used to infer neuronal activity, which must be the ultimate aim of the entire functional imaging enterprise *(214)*.

2.2.9. fMRI of Paroxysmal Activity: Conclusion

The possibility of recording EEG of good quality during fMRI has created a new instrument, which to date has been mainly used in an exploratory fashion.

EEG-fMRI of ictal and interictal activity in focal epilepsy has demonstrated the capability to provide new localisation information in a large proportion of cases. In focal epilepsy, varied patterns have been identified often suggestive of 'functional lesions' homologous to abnormalities seen on structural imaging, but additionally the involvement of possibly less disease-specific regions. The clinical value of this information remains uncertain and is the subject of ongoing investigations. At the very least, EEG-fMRI may provide complementary data useful for planning of invasive recordings where conventional localisation of the seizure focus is unsuccessful. However, there are signs that it may be able to offer more.

In generalised epilepsy, EEG-fMRI has revealed activation and deactivation patterns that are mainly suggestive of less specific effects linked to generalised spike-wave, rather than reflecting syndrome, the spatial distribution of the EEG generators or more specifically a putative focus responsible for initiating GSW due in part to the averaging effect of BOLD.

2.3. Data Acquisition and Modelling Challenges in Clinical fMRI

Some of the brain regions of most interest in epilepsy are subject to susceptibility artefact, leading to geometric distortions and signal loss during fMRI acquisition. Ideally in the absence of an applied gradient, the magnetic field would be homogenous throughout the bore of an MRI scanner. Unfortunately, the different magnetic properties of bone, tissue, and air introduce inhomogeneities in the field when a head is introduced into the bore. Brain regions closest to borders between sinuses and brain or bone and brain, for example the inferior frontal and MTLs, are most affected, and therefore especially likely to suffer geometric distortions or signal loss *(215)*. This can result in reduced sensitivity and anatomical uncertainties when interpreting the images. Most epilepsy studies have been performed on 1.5 T clinical MRI scanners. Scanning at higher field strength improves signal-to-noise ratio but increases distortions and dropout *(216)*.

Geometric distortions of the EPI data make it difficult to directly overlay fMRI activations on co-registered high-resolution scans. They can be unwarped using techniques that map the local field in the head *(217)*, though it has been shown that approaches of this kind can introduce extra noise into the corrected EPI data *(218)*. Alternative acquisition sequences that do not experience geometric distortions are available *(219)* though these rarely have the temporal resolution or high signal to noise ratio (SNR) per unit time of EPI.

The second artefact in EPI data is more serious, as signal loss leads to sensitivity loss, which is unrecoverable by image processing techniques. Some of these artefacts can be corrected by shimming, a process whereby the static magnetic field is made more homogenous over the region of interest *(215)*. Some will remain, however, leading to distortion and dropout in echo-planar images. Other approaches to removing dropout often involve acquiring extra images, leading to a loss of temporal resolution *(220)*, but more recent work has shown that dropouts and distortions can be reduced without incurring time penalties if regions of reduced spatial extent are imaged *(221)*. The use of high-performance gradients and thin slice acquisitions ameliorate these problems and improve fMRI quality. fMRI is also extremely sensitive to motion, although generally epilepsy patients are familiar with MRI scanners and may move less than control subjects *(8)*.

Motion remains problematic for all applications of fMRI *(148)*, but the problem may be particularly harmful in patient studies in view of two factors: first, patients may be more prone to movement than selected and highly motivated healthy subjects;

second, each patient dataset may have greater value than that from inter-changeable healthy subjects. Therefore, measures have been proposed to extract as much information as possible from what are effectively sub-optimal fMRI studies *(222–224)*.

Acknowledgments

Thanks to Philip Allen for his comments on parts of the manuscript and to Dr. Anna Vaudano and Dr. Serge Vulliemoz for supplying some of the illustrations. Some of the work reported in this chapter was funded through a grant from the Medical Research Council (MRC grant number G0301067) and by the Wellcome Trust. We are grateful to the Big Lottery Fund, Wolfson Trust, and National Society for Epilepsy for supporting the NSE MRI scanner. This work was carried out under the auspices of the UCL/UCLH Biomedical Comprehensive Research Centre.

References

1. Baxendale S. The role of functional MRI in the presurgical investigation of temporal lobe epilepsy patients: a clinical perspective and review. J Clin Exp Neuropsychol 2002; 24(5):664–676.
2. Kirsch HE, Walker JA, Winstanley FS, Hendrickson R, Wong ST, Barbaro NM et al. Limitations of Wada memory asymmetry as a predictor of outcomes after temporal lobectomy. Neurology 2005;65(5):676–680.
3. Fernandez G, Specht K, Weis S, Tendolkar I, Reuber M, Fell J et al. Intrasubject reproducibility of presurgical language lateralization and mapping using fMRI. Neurology 2003; 60(6):969–975.
4. Weber B, Wellmer J, Simone S, Dinkelacker V, Ruhlmann J, Mormann F et al. Presurgical language fMRI in patients with drug-resistant epilpesy: effects of task performance. Epilepsia 2006;47(5):880–886.
5. Schlosser MJ, Aoyagi N, Fulbright RK, Gore JC, McCarthy G. Functional MRI studies of auditory comprehension. Hum Brain Mapp 1998;6(1):1–13.
6. Lehericy S, Cohen L, Bazin B, Samson S, Giacomini E, Rougetet R et al. Functional MR evaluation of temporal and frontal language dominance compared with the Wada test. Neurology 2000;54(8):1625–1633.
7. Gaillard WD, Balsamo L, Xu B, McKinney C, Papero PH, Weinstein S et al. fMRI language task panel improves determination of language dominance. Neurology 2004;63(8):1403–1408.
8. Gaillard WD, Balsamo L, Xu B, Grandin CB, Braniecki SH, Papero PH et al. Language dominance in partial epilepsy patients identified with an fMRI reading task. Neurology 2002;59(2):256–265.
9. Adcock JE, Wise RG, Oxbury JM, Oxbury SM, Matthews PM. Quantitative fMRI assessment of the differences in lateralization of language-related brain activation in patients with temporal lobe epilepsy. Neuroimage 2003; 18(2):423–438.
10. Branco DM, Suarez RO, Whalen S, O'Shea JP, Nelson AP, da Costa JC et al. Functional MRI of memory in the hippocampus: laterality indices may be more meaningful if calculated from whole voxel distributions. Neuroimage 2006;32(2):592–602.
11. Liegeois F, Connelly A, Cross JH, Boyd SG, Gadian DG, Vargha-Khadem F et al. Language reorganization in children with early-onset lesions of the left hemisphere: an fMRI study. Brain 2004;127(Pt 6):1229–1236.
12. Woermann FG, Jokeit H, Luerding R, Freitag H, Schulz R, Guertler S et al. Language lateralization by Wada test and fMRI in 100 patients with epilepsy. Neurology 2003; 61(5):699–701.
13. Friston KJ, Price CJ, Fletcher P, Moore C, Frackowiak RS, Dolan RJ. The trouble with

cognitive subtraction. Neuroimage 1996; 4(2):97–104.
14. Price CJ, Friston KJ. Cognitive conjunction: a new approach to brain activation experiments. Neuroimage 1997;5(4 Pt 1):261–270.
15. Risse GL, Gates JR, Fangman MC. A reconsideration of bilateral language representation based on the intracarotid amobarbital procedure. Brain Cogn 1997;33(1):118–132.
16. Serafetinides EA, Hoare RD, Driver M. Intracarotid sodium amylobarbitone and cerebral dominance for speech and consciousness. Brain 1965;88:107–130.
17. Rasmussen T, Milner B. The role of early left-brain injury in determining lateralization of cerebral speech functions. Ann N Y Acad Sci 1977;299:355–369.
18. Springer JA, Binder JR, Hammeke TA, Swanson SJ, Frost JA, Bellgowan PS et al. Language dominance in neurologically normal and epilepsy subjects: a functional MRI study. Brain 1999;122(Pt 11):2033–2046.
19. Janszky J, Mertens M, Janszky I, Ebner A, Woermann FG. Left-sided interictal epileptic activity induces shift of language lateralization in temporal lobe epilepsy: an fMRI study. Epilepsia 2006;47(5):921–927.
20. Janszky J, Jokeit H, Heinemann D, Schulz R, Woermann FG, Ebner A. Epileptic activity influences the speech organization in medial temporal lobe epilepsy. Brain 2003;126(Pt 9):2043–2051.
21. Thivard L, Hombrouck J, du Montcel ST, Delmaire C, Cohen L, Samson S et al. Productive and perceptive language reorganization in temporal lobe epilepsy. Neuroimage 2005;24(3):841–851.
22. Berl MM, Balsamo LM, Xu B, Moore EN, Weinstein SL, Conry JA et al. Seizure focus affects regional language networks assessed by fMRI. Neurology 2005;65(10):1604–1611.
23. Weber B, Wellmer J, Reuber M, Mormann F, Weis S, Urbach H et al. Left hippocampal pathology is associated with atypical language lateralization in patients with focal epilepsy. Brain 2006;129:346–351.
24. Briellmann RS, Labate A, Harvey AS, Saling MM, Sveller C, Lillywhite L et al. Is language lateralization in temporal lobe epilepsy patients related to the nature of the epileptogenic lesion? Epilepsia 2006;47(5):916–920.
25. Benke T, Koylu B, Visani P, Karner E, Brennais C, Bartha L et al. Language lateralisation in temporal lobe epilepsy: a comparison between fMRI and the Wada Test. Epilepsia 2006;47:1308–1309.
26. Jayakar P, Bernal B, Santiago ML, Altman N. False lateralization of language cortex on functional MRI after a cluster of focal seizures. Neurology 2002;58(3):490–492.
27. FitzGerald DB, Cosgrove GR, Ronner S, Jiang H, Buchbinder BR, Belliveau JW et al. Location of language in the cortex: a comparison between functional MR imaging and electrocortical stimulation. AJNR Am J Neuroradiol 1997;18(8):1529–1539.
28. Schlosser MJ, Luby M, Spencer DD, Awad IA, McCarthy G. Comparative localization of auditory comprehension by using functional magnetic resonance imaging and cortical stimulation. J Neurosurg 1999;91(4):626–635.
29. Pouratian N, Bookheimer SY, Rex DE, Martin NA, Toga AW. Utility of preoperative functional magnetic resonance imaging for identifying language cortices in patients with vascular malformations. J Neurosurg 2002;97(1):21–32.
30. Rutten GJ, Ramsey NF, van Rijen PC, Noordmans HJ, van Veelen CW. Development of a functional magnetic resonance imaging protocol for intraoperative localization of critical temporoparietal language areas. Ann Neurol 2002;51(3):350–360.
31. Davies KG, Bell BD, Bush AJ, Hermann BP, Dohan FC, Jr., Jaap AS. Naming decline after left anterior temporal lobectomy correlates with pathological status of resected hippocampus. Epilepsia 1998;39(4):407–419.
32. Saykin AJ, Stafiniak P, Robinson LJ, Flannery KA, Gur RC, O'Connor MJ et al. Language before and after temporal lobectomy: specificity of acute changes and relation to early risk factors. Epilepsia 1995;36(11):1071–1077.
33. Hermann BP, Perrine K, Chelune GJ, Barr W, Loring DW, Strauss E et al. Visual confrontation naming following left anterior temporal lobectomy: a comparison of surgical approaches. Neuropsychology 1999; 13(1):3–9.
34. Devinsky O, Perrine K, Llinas R, Luciano DJ, Dogali M. Anterior temporal language areas in patients with early onset of temporal lobe epilepsy. Ann Neurol 1993;34(5):727–732.
35. Schwartz TH, Devinsky O, Doyle W, Perrine K. Preoperative predictors of anterior temporal language areas. J Neurosurg 1998; 89(6):962–970.
36. Sabsevitz DS, Swanson SJ, Hammeke TA, Spanaki MV, Possing ET, Morris GL, III et al. Use of preoperative functional neuroimaging to predict language deficits from epilepsy surgery. Neurology 2003;60(11):1788–1792.

37. Noppeney U, Price CJ, Duncan JS, Koepp MJ. Reading skills after left anterior temporal lobe resection: an fMRI study. Brain 2005; 128(Pt 6):1377–1385.
38. Voets NL, Adcock JE, Flitney DE, Behrens TE, Hart Y, Stacey R et al. Distinct right frontal lobe activation in language processing following left hemisphere injury. Brain 2006; 129(Pt 3):754–766.
39. Powell HW, Parker GJ, Alexander DC, Symms MR, Boulby PA, Wheeler-Kingshott CA et al. Hemispheric asymmetries in language-related pathways: a combined functional MRI and tractography study. Neuroimage 2006; 32(1):388–399.
40. Powell HW, Parker GJ, Alexander DC, Symms M, Boulby P, Wheeler-Kingshott CA et al. Abnormalities of language networks in temporal lobe epilepsy. Neuroimage 2007; 36(1):209–221.
41. Desmond JE, Sum JM, Wagner AD, Demb JB, Shear PK, Glover GH et al. Functional MRI measurement of language lateralization in Wada-tested patients. Brain 1995;118(Pt 6):1411–1419.
42. Binder JR, Swanson SJ, Hammeke TA, Morris GL, Mueller WM, Fischer M et al. Determination of language dominance using functional MRI: a comparison with the Wada test. Neurology 1996;46(4):978–984.
43. Hertz-Pannier L, Gaillard WD, Mott SH, Cuenod CA, Bookheimer SY, Weinstein S et al. Noninvasive assessment of language dominance in children and adolescents with functional MRI: a preliminary study. Neurology 1997;48(4):1003–1012.
44. Yetkin FZ, Swanson S, Fischer M, Akansel G, Morris G, Mueller W et al. Functional MR of frontal lobe activation: comparison with Wada language results. AJNR Am J Neuroradiol 1998;19(6):1095–1098.
45. Benson RR, FitzGerald DB, LeSueur LL, Kennedy DN, Kwong KK, Buchbinder BR et al. Language dominance determined by whole brain functional MRI in patients with brain lesions. Neurology 1999;52(4):798–809.
46. Carpentier A, Pugh KR, Westerveld M, Studholme C, Skrinjar O, Thompson JL et al. Functional MRI of language processing: dependence on input modality and temporal lobe epilepsy. Epilepsia 2001;42(10):1241–1254.
47. Sabbah P, Chassoux F, Leveque C, Landre E, Baudoin-Chial S, Devaux B et al. Functional MR imaging in assessment of language dominance in epileptic patients. Neuroimage 2003; 18(2):460–467.
48. Scoville WB, Milner B. Loss of recent memory after bilateral hippocampal lesions. 1957. J Neuropsychiatry Clin Neurosci 2000; 12(1):103–113.
49. Ivnik RJ, Sharbrough FW, Laws ER, Jr. Effects of anterior temporal lobectomy on cognitive function. J Clin Psychol 1987;43(1): 128–137.
50. Spiers HJ, Burgess N, Maguire EA, Baxendale SA, Hartley T, Thompson PJ et al. Unilateral temporal lobectomy patients show lateralized topographical and episodic memory deficits in a virtual town. Brain 2001;124(Pt 12): 2476–2489.
51. Penfield W, Milner B. Memory deficit produced by bilateral lesions in the hippocampal zone. AMA Arch Neurol Psychiatry 1958;79(5):475–497.
52. Warrington EK, Duchen LW. A re-appraisal of a case of persistent global amnesia following right temporal lobectomy: a clinicopathological study. Neuropsychologia 1992; 30(5):437–450.
53. Loring DW, Hermann BP, Meador KJ, Lee GP, Gallagher BB, King DW et al. Amnesia after unilateral temporal lobectomy: a case report. Epilepsia 1994;35(4):757–763.
54. Chelune GJ. Hippocampal adequacy versus functional reserve: predicting memory functions following temporal lobectomy. Arch Clin Neuropsychol 1995;10(5):413–432.
55. Chelune GJ, Naugle RI, Luders H, Awad IA. Prediction of cognitive change as a function of preoperative ability status among temporal lobectomy patients seen at 6-month follow-up. Neurology 1991;41(3):399–404.
56. Kneebone AC, Chelune GJ, Dinner DS, Naugle RI, Awad IA. Intracarotid amobarbital procedure as a predictor of material-specific memory change after anterior temporal lobectomy. Epilepsia 1995;36(9):857–865.
57. Sass KJ, Spencer DD, Kim JH, Westerveld M, Novelly RA, Lencz T. Verbal memory impairment correlates with hippocampal pyramidal cell density. Neurology 1990;40(11): 1694–1697.
58. Trenerry MR, Jack CR, Jr, Ivnik RJ, Sharbrough FW, Cascino GD, Hirschorn KA et al. MRI hippocampal volumes and memory function before and after temporal lobectomy. Neurology 1993;43(9):1800–1805.
59. Hermann BP, Wyler AR, Somes G, Berry AD, III, Dohan FC, Jr. Pathological status of the mesial temporal lobe predicts memory outcome from left anterior temporal lobectomy. Neurosurgery 1992;31(4):652–656.

60. Sass KJ, Westerveld M, Buchanan CP, Spencer SS, Kim JH, Spencer DD. Degree of hippocampal neuron loss determines severity of verbal memory decrease after left anteromesiotemporal lobectomy. Epilepsia 1994; 35(6):1179–1186.
61. Jokeit H, Ebner A, Holthausen H, Markowitsch HJ, Moch A, Pannek H et al. Individual prediction of change in delayed recall of prose passages after left-sided anterior temporal lobectomy. Neurology 1997;49(2):481–487.
62. Helmstaedter C, Elger CE. Cognitive consequences of two-thirds anterior temporal lobectomy on verbal memory in 144 patients: a three-month follow-up study. Epilepsia 1996;37(2):171–180.
63. Corkin S, Amaral DG, Gonzalez RG, Johnson KA, Hyman BT. H. M.'s medial temporal lobe lesion: findings from magnetic resonance imaging. J Neurosci 1997;17(10):3964–3979.
64. Fernandez G, Effern A, Grunwald T, Pezer N, Lehnertz K, Dumpelmann M et al. Real-time tracking of memory formation in the human rhinal cortex and hippocampus. Science 1999; 285(5433):1582–1585.
65. Ojemann JG, Akbudak E, Snyder AZ, McKinstry RC, Raichle ME, Conturo TE. Anatomic localization and quantitative analysis of gradient refocused echo-planar fMRI susceptibility artifacts. Neuroimage 1997;6(3):156–167.
66. Greicius MD, Krasnow B, Boyett-Anderson JM, Eliez S, Schatzberg AF, Reiss AL et al. Regional analysis of hippocampal activation during memory encoding and retrieval: fMRI study. Hippocampus 2003;13(1):164–174.
67. Lipschutz B, Friston KJ, Ashburner J, Turner R, Price CJ. Assessing study-specific regional variations in fMRI signal. Neuroimage 2001; 13(2):392–398.
68. Craik FIM, Lockhart RS. Levels of processing: a framework for memory. J Verbal Learn Verbal Behav 1972;11:671–684.
69. Kelley WM, Miezin FM, McDermott KB, Buckner RL, Raichle ME, Cohen NJ et al. Hemispheric specialization in human dorsal frontal cortex and medial temporal lobe for verbal and nonverbal memory encoding. Neuron 1998;20(5):927–936.
70. Demb JB, Desmond JE, Wagner AD, Vaidya CJ, Glover GH, Gabrieli JD. Semantic encoding and retrieval in the left inferior prefrontal cortex: a functional MRI study of task difficulty and process specificity. J Neurosci 1995; 15(9):5870–5878.
71. Wagner AD, Schacter DL, Rotte M, Koutstaal W, Maril A, Dale AM et al. Building memories: remembering and forgetting of verbal experiences as predicted by brain activity. Science 1998;281(5380):1188–1191.
72. Buckner RL, Kelley WM, Petersen SE. Frontal cortex contributes to human memory formation. Nat Neurosci 1999;2(4):311–314.
73. Golby AJ, Poldrack RA, Brewer JB, Spencer D, Desmond JE, Aron AP et al. Material-specific lateralization in the medial temporal lobe and prefrontal cortex during memory encoding. Brain 2001;124(Pt 9):1841–1854.
74. Wagner AD, Koutstaal W, Schacter DL. When encoding yields remembering: insights from event-related neuroimaging. Philos Trans R Soc Lond B Biol Sci 1999;354(1387): 1307–1324.
75. Powell HW, Koepp MJ, Symms MR, Boulby PA, Salek-Haddadi A, Thompson PJ et al. Material-specific lateralization of memory encoding in the medial temporal lobe: blocked versus event-related design. Neuroimage 2005;27(1):231–239.
76. Detre JA, Maccotta L, King D, Alsop DC, Glosser G, D'Esposito M et al. Functional MRI lateralization of memory in temporal lobe epilepsy. Neurology 1998;50(4): 926–932.
77. Richardson MP, Strange BA, Duncan JS, Dolan RJ. Preserved verbal memory function in left medial temporal pathology involves reorganisation of function to right medial temporal lobe. Neuroimage 2003;20(Suppl 1):S112–S119.
78. Powell HW, Richardson MP, Symms MR, Boulby PA, Thompson PJ, Duncan JS et al. Reorganisation of verbal and non-verbal memory in temporal lobe epilepsy due to unilateral hippocampal sclerosis. Epilepsia 2007; 48(8):1512–1525.
79. Bellgowan PS, Binder JR, Swanson SJ, Hammeke TA, Springer JA, Frost JA et al. Side of seizure focus predicts left medial temporal lobe activation during verbal encoding. Neurology 1998;51(2):479–484.
80. Jokeit H, Okujava M, Woermann FG. Memory fMRI lateralizes temporal lobe epilepsy. Neurology 2001;57(10):1786–1793.
81. Golby AJ, Poldrack RA, Illes J, Chen D, Desmond JE, Gabrieli JD. Memory lateralization in medial temporal lobe epilepsy assessed by functional MRI. Epilepsia 2002;43(8):855–863.
82. Richardson MP, Strange BA, Thompson PJ, Baxendale SA, Duncan JS, Dolan RJ. Pre-operative verbal memory fMRI predicts post-operative memory decline after left temporal lobe resection. Brain 2004;127(Pt 11): 2419–2426.

83. Richardson MP, Strange BA, Duncan JS, Dolan RJ. Memory fMRI in left hippocampal sclerosis: optimizing the approach to predicting postsurgical memory. Neurology 2006; 66(5):699–705.
84. Powell HW, Richardson MP, Symms MR, Boulby PA, Thompson PJ, Duncan JS et al. Preoperative fMRI predicts memory decline following anterior temporal lobe resection. J Neurol Neurosurg Psychiatry 2008; 79(6):686–693.
85. Rabin ML, Narayan VM, Kimberg DY, Casasanto DJ, Glosser G, Tracy JI et al. Functional MRI predicts post-surgical memory following temporal lobectomy. Brain 2004; 127(Pt 10):2286–2298.
86. Janszky J, Jokeit H, Kontopoulou K, Mertens M, Ebner A, Pohlmann-Eden B et al. Functional MRI predicts memory performance after right mesiotemporal epilepsy surgery. Epilepsia 2005;46(2):244–250.
87. Price CJ, Friston KJ. Scanning patients with tasks they can perform. Hum Brain Mapp 1999;8(2–3):102–108.
88. Von Helmholtz HLF. Some laws concerning the distribution of electric currents in volume conductors with applications to experiments on animal electricity (Reprinted from Poggendorff's Annals, vol 89, pp. 211–233, 353–377, 1853). Proc IEEE 2004;92(5):868–870.
89. Geselowitz DB. Introduction to some laws concerning the distribution of electric currents in volume conductors with applications to experiments on animal electricity. Proc IEEE 2004;92(5):864–867.
90. Ives JR, Warach S, Schmitt F, Edelman RR, Schomer DL. Monitoring the patient's EEG during echo planar MRI. Electroencephalogr Clin Neurophysiol 1993;87(6):417–420.
91. Detre JA, Alsop DC, Aguirre GK, Sperling MR. Coupling of cortical and thalamic ictal activity in human partial epilepsy: demonstration by functional magnetic resonance imaging. Epilepsia 1996;37(7):657–661.
92. Krings T, Topper R, Reinges MHT, Foltys H, Spetzger U, Chiappa KH et al. Hemodynamic changes in simple partial epilepsy: a functional MRI study. Neurology 2000;54(2):524–527.
93. Connelly A. Ictal imaging using functional magnetic resonance. Magn Reson Imaging 1995;13(8):1233–1237.
94. Jackson GD, Connelly A, Cross JH, Gordon I, Gadian DG. Functional magnetic resonance imaging of focal seizures. Neurology 1994; 44(5):850–856.
95. Lemieux L, Allen PJ, Franconi F, Symms MR, Fish DR. Recording of EEG during fMRI experiments: patient safety. Magn Reson Med 1997;38(6):943–952.
96. Mirsattari SM, Lee DH, Jones D, Bihari F, Ives JR. MRI compatible EEG electrode system for routine use in the epilepsy monitoring unit and intensive care unit. Clin Neurophysiol 2004;115(9):2175–2180.
97. Konings MK, Bartels LW, Smits HFM, Bakker CJG. Heating around intravascular guidewires by resonating RF waves. J-Magn-Reson-Imaging 2000;12:79–85.
98. Fischer H, Ladebeck R. Echo-planar imaging image artifacts. In: Schmitt F, Stehling MK, Turner R, editors. Echo-Planar Imaging: Theory, Technique, and Application. Berlin: Springer-Verlag, 1998;179–200.
99. Krakow K, Allen PJ, Symms MR, Lemieux L, Josephs O, Fish DR. EEG recording during fMRI experiments: image quality. Hum Brain Mapp 2000;10(1):10–15.
100. Bonmassar G, Anami K, Ives JR, Belliveau JW. Visual evoked potential (VEP) measured by simultaneous 64-channel EEG and 3T fMRI. Neuroreport 1999;10(9): 1893–1897.
101. Bonmassar G, Purdon PL, Jaaskelainen IP, Chiappa KH, Solo V, Brown EN et al. Motion and ballistocardiogram artifact removal for interleaved recording of EEG and EPs during MRI. NeuroImage 2002; 16(4):1127–1141.
102. Scarff CJ, Reynolds A, Goodyear BG, Ponton CW, Dort JC, Eggermont JJ. Simultaneous 3-T fMRI and high-density recording of human auditory evoked potentials. NeuroImage 2004;23(3):1129–1142.
103. Allen PJ, Polizzi G, Krakow K, Fish DR, Lemieux L. Identification of EEG events in the MR scanner: the problem of pulse artifact and a method for its subtraction. Neuroimage 1998;8(3):229–239.
104. Wendt RE, Rokey R, Vick GW, Johnston DL. Electrocardiographic gating and monitoring in NMR imaging. Magn Reson Imaging 1988;6(1):89–95.
105. Poncelet BP, Wedeen VJ, Weiskoff RM, Cohen MS. Brain parenchyma motion: measurement with cine echo-planar MR imaging. Radiology 1992;185:645–651.
106. Tenforde TS, Gaffey CT, Moyer BR, Budinger TF. Cardiovascular alterations in Macaca monkeys exposed to stationary magnetic fields: experimental observations and theoretical analysis. Bioelectromagnetics 1983;4(1):1–9.
107. Goldman RI, Stern JM, Engel Jr, JE, Cohen MS. Acquiring simultaneous EEG and

functional MRI. Clin-Neurophysiol 2000; 111:1974–1980.

108. Benar C, Aghakhani Y, Wang Y, Izenberg A, Al Asmi A, Dubeau F et al. Quality of EEG in simultaneous EEG-fMRI for epilepsy. Clin Neurophysiol 2003;114(3):569–580.

109. Salek-Haddadi A, Lemieux L, Merschhemke M, Diehl B, Allen PJ, Fish DR. EEG quality during simultaneous functional MRI of interictal epileptiform discharges. Magn Reson Imaging 2003;21(10):1159–1166.

110. Liston AD, De Munck JC, Hamandi K, Laufs H, Ossenblok P, Duncan JS et al. Analysis of EEG-fMRI data in focal epilepsy based on automated spike classification and Signal Space Projection. Neuroimage 2006; 31(3):1015–1024.

111. Ellingson ML, Liebenthal E, Spanaki MV, Prieto TE, Binder JR, Ropella KM. Ballistocardiogram artifact reduction in the simultaneous acquisition of auditory ERPS and fMRI. Neuroimage 2004;22(4):1534–1542.

112. Kruggel F, Wiggins CJ, Herrmann CS, von Cramon DY. Recording of the event-related potentials during functional MRI at 3.0 Tesla field strength. Magn Reson Med 2000;44:277–282.

113. Sijbers J, Van Audekerke J, Verhoye M, Van der LA, Van Dyck D. Reduction of ECG and gradient related artifacts in simultaneously recorded human EEG/MRI data. Magn Reson Imaging 2000;18(7):881–886.

114. Kim KH, Yoon HW, Park HW. Improved ballistocardiac artifact removal from the electroencephalogram recorded in fMRI. J Neurosci Methods 2004;135(1–2):193–203.

115. Wan X, Iwata K, Riera J, Ozaki T, Kitamura M, Kawashima R. Artifact reduction for EEG/fMRI recording: nonlinear reduction of ballistocardiogram artifacts. Clin Neurophysiol 2006;117(3):668–680.

116. Niazy RK, Beckmann CF, Iannetti GD, Brady JM, Smith SM. Removal of FMRI environment artifacts from EEG data using optimal basis sets. Neuroimage 2005; 28(3):720–737.

117. In MH, Lee SY, Park TS, Kim TS, Cho MH, Ahn YB. Ballistocardiogram artifact removal from EEG signals using adaptive filtering of EOG signals. Physiol Meas 2006; 27(11):1227–1240.

118. Eichele T, Specht K, Moosmann M, Jongsma ML, Quiroga RQ, Nordby H et al. Assessing the spatiotemporal evolution of neuronal activation with single-trial event-related potentials and functional MRI. Proc Natl Acad Sci USA 2005;102(49): 17798–17803.

119. Otzenberger H, Gounot D, Foucher JR. P300 recordings during event-related fMRI: a feasibility study. Brain Res Cogn Brain Res 2005;23(2–3):306–315.

120. Otzenberger H, Gounot D, Foucher JR. Optimisation of a post-processing method to remove the pulse artifact from EEG data recorded during fMRI: an application to P300 recordings during e-fMRI. Neurosci Res 2007;57(2):230–239.

121. Srivastava G, Crottaz-Herbette S, Lau KM, Glover GH, Menon V. ICA-based procedures for removing ballistocardiogram artifacts from EEG data acquired in the MRI scanner. NeuroImage 2005;24(1):50–60.

122. Nakamura W, Anami K, Mori T, Saitoh O, Cichocki A, Amari S. Removal of ballistocardiogram artifacts from simultaneously recorded EEG and fMRI data using independent component analysis. IEEE Trans Biomed Eng 2006;53(7):1294–1308.

123. Briselli E, Garreffa G, Bianchi L, Bianciardi M, Macaluso E, Abbafati M et al. An independent component analysis-based approach on ballistocardiogram artifact removing. Magn Reson Imaging 2006;24(4): 393–400.

124. Mantini D, Perrucci MG, Cugini S, Ferretti A, Romani GL, Del Gratta C. Complete artifact removal for EEG recorded during continuous fMRI using independent component analysis. Neuroimage 2007; 34(2):598–607.

125. Debener S, Strobel A, Sorger B, Peters J, Kranczioch C, Engel AK et al. Improved quality of auditory event-related potentials recorded simultaneously with 3-T fMRI: removal of the ballistocardiogram artefact. NeuroImage 2007;34(2):587–597.

126. Allen PJ, Josephs O, Turner R. A method for removing imaging artifact from continuous EEG recorded during functional MRI. Neuroimage 2000;12(2):230–239.

127. Warach S, Ives JR, Schlaug G, Patel MR, Darby DG, Thangaraj V et al. EEG-triggered echo-planar functional MRI in epilepsy. Neurology 1996;47:89–93.

128. Bonmassar G, Schwartz DP, Liu AK, Kwong KK, Dale AM, Belliveau JW. Spatiotemporal brain imaging of visual-evoked activity using intervleaved EEG and fMRI recordings. Neuroimage 2001;13:1035–1043.

129. Huang-Hellinger FR, Breiter HC, McCormack G, Cohen MS, Kwong KK, Sutton JP et al. Simultaneous functional magnetic resonance imaging and electrophysiological recording. Hum Brain Mapp 1995;3: 13–23.

130. Krakow K, Woermann FG, Symms MR, Allen PJ, Lemieux L, Barker GJ et al. EEG-triggered functional MRI of interictal epileptiform activity in patients with partial seizures. Brain 1999;122(Pt 9):1679–1688.
131. Krakow K, Allen PJ, Lemieux L, Symms MR, Fish DR. Methodology: EEG-correlated fMRI. Adv Neurol 2000;83:187–201.
132. Krakow K, Lemieux L, Messina D, Scott CA, Symms MR, Duncan JS et al. Spatio-temporal imaging of focal interictal epileptiform activity using EEG-triggered functional MRI. Epileptic Disord 2001;3(2):67–74.
133. Al Asmi A, Benar CG, Gross DW, Khani YA, Andermann F, Pike B et al. fMRI activation in continuous and spike-triggered EEG-fMRI studies of epileptic spikes. Epilepsia 2003;44(10):1328–1339.
134. Anami K, Mori T, Tanaka F, Kawagoe Y, Okamoto J, Yarita M et al. Stepping stone sampling for retrieving artifact-free electroencephalogram during functional magnetic resonance imaging. NeuroImage 2003; 19(2):281–295.
135. Garreffa G, Carni M, Gualniera G, Ricci GB, Bozzao L, De Carli D et al. Real-time MR artifacts filtering during continuous EEG/fMRI acquisition. Magn Reson Imaging 2003;21(10):1175–1189.
136. Hoffmann A, Jager L, Werhahn KJ, Jaschke M, Noachtar S, Reiser M. Electroencephalography during functional echo-planar imaging: detection of epileptic spikes using post-processing methods. Magn Reson Med 2000;44(5):791–798.
137. Sijbers J, Michiels I, Verhoye M, Van Audekerke J, Van der Linden A, Van Dyk D. Restoration of MR-induced artifacts in simultaneously recorded MR/EEG data. Magn Reson Imaging 1999;17(9):1383–1391.
138. Hamandi K, Salek-Haddadi A, Laufs H, Liston A, Friston K, Fish DR et al. EEG-fMRI of idiopathic and secondarily generalized epilepsies. Neuroimage 2006;31(4):1700–1710.
139. Lemieux L, Salek-Haddadi A, Josephs O, Allen P, Toms N, Scott C et al. Event-related fMRI with simultaneous and continuous EEG: description of the method and initial case report. Neuroimage 2001;14(3):780–787.
140. Salek-Haddadi A, Merschhemke M, Lemieux L, Fish DR. Simultaneous EEG-correlated ictal fMRI. Neuroimage 2002; 16(1):32–40.
141. Salek-Haddadi A, Diehl B, Hamandi K, Merschhemke M, Liston A, Friston K et al. Hemodynamic correlates of epileptiform discharges: an EEG-fMRI study of 63 patients with focal epilepsy. Brain Res 2006; 1088(1):148–166.
142. Mandelkow H, Halder P, Boesiger P, Brandeis D. Synchronization facilitates removal of MRI artefacts from concurrent EEG recordings and increases usable bandwidth. NeuroImage 2006;32(3):1120–1126.
143. Negishi M, Abildgaard M, Nixon T, Constable RT. Removal of time-varying gradient artifacts from EEG data acquired during continuous fMRI. Clin Neurophysiol 2004; 115(9):2181–2192.
144. Wan X, Iwata K, Riera J, Kitamura M, Kawashima R. Artifact reduction for simultaneous EEG/fMRI recording: adaptive FIR reduction of imaging artifacts. Clin Neurophysiol 2006;117(3):681–692.
145. Ritter P, Becker R, Graefe C, Villringer A. Evaluating gradient artifact correction of EEG data acquired simultaneously with fMRI. Magn Reson Imaging 2007;25(6):923–932.
146. Friston KJ, Williams S, Howard R, Frackowiack RSJ, Turner R. Movement-related effects in fMRI time-series. Magn Reson Med 1996;35:346–355.
147. Hajnal JV, Myers R, Oatridge A, Schwieso JE, Young IR, Bydder GM. Artifacts due to stimulus correlated motion in functional imaging of the brain. Magn Reson Med 1994;31:283–291.
148. Lund TE, Norgaard MD, Rostrup E, Rowe JB, Paulson OB. Motion or activity: their role in intra- and inter-subject variation in fMRI. Neuroimage 2005;26(3):960–964.
149. Lazeyras F, Blanke O, Zimine I, Delavelle J, Perrig SH, Seeck M. MRI, (1)H-MRS, and functional MRI during and after prolonged nonconvulsive seizure activity. Neurology 2000;55(11):1677–1682.
150. Kobayashi E, Hawco CS, Grova C, Dubeau F, Gotman J. Widespread and intense BOLD changes during brief focal electrographic seizures. Neurology 2006;66(7):1049–1055.
151. Baumgartner C, Serles W, Leutmezer F, Pataraia E, Aull S, Czech T et al. Preictal SPECT in temporal lobe epilepsy: regional cerebral blood flow is increased prior to electroencephalography-seizure onset. J Nucl Med 1998;39(6):978–982.
152. Federico P, Abbott DF, Briellmann RS, Harvey AS, Jackson GD. Functional MRI of the pre-ictal state. Brain 2005;128(Pt 8):1811–1817.
153. Patel MR, Blum A, Pearlman JD, Youssuf N, Ives JR, Saeteng S et al. Echo-planar functional MR imaging of epilepsy with concurrent EEG monitoring. Am J Neuroradiol 1999;20:1916–1919.

154. Symms MR, Allen PJ, Woermann FG, Polizzi G, Krakow K, Barker GJ et al. Reproducible localization of interictal epileptiform discharges using EEG-triggered fMRI. Phys Med Biol 1999;44(7):N161–N168.
155. Archer JS, Briellman RS, Abbott DF, Syngeniotis A, Wellard RM, Jackson GD. Benign epilepsy with centro-temporal spikes: spike triggered fMRI shows somato-sensory cortex activity. Epilepsia 2003;44(2): 200–204.
156. Archer JS, Briellmann RS, Syngeniotis A, Abbott DF, Jackson GD. Spike-triggered fMRI in reading epilepsy: involvement of left frontal cortex working memory area. Neurology 2003;60(3):415–421.
157. Al Asmi A, Benar CG, Gross DW, Khani YA, Andermann F, Pike B et al. fMRI Activation in continuous and spike-triggered EEG-fMRI studies of epileptic spikes. Epilepsia 2003;44(10):1328–1339.
158. Lazeyras F, Blanke O, Perrig S, Zimine I, Golay X, Delavelle J et al. EEG-triggered functional MRI in patients with pharmacoresistant epilepsy. J Magn Reson Imaging 2000;12:177–185.
159. Seeck M, Lazeyras F, Michel CM, Blanke O, Gericke CA, Ives JR et al. Non-invasive epileptic focus localization using EEG-triggered functional MRI and electromagnetic tomography. Electroenceph Clin Neurophysiol 1998;106:508–512.
160. Baudewig J, Bittermann HJ, Paulus W, Frahm J. Simultaneous EEG and functional MRI of epileptic activity: a case report. Clin Neurophysiol 2001;112:1196–1200.
161. Lemieux L, Salek-Haddadi A, Josephs O, Allen P, Toms N, Scott C et al. Event-Related fMRI with Simultaneous and Continuous EEG: description of the method and initial case report. Neuroimage 2001;14(3): 780–787.
162. Bagshaw AP, Hawco C, Benar CG, Kobayashi E, Aghakhani Y, Dubeau F et al. Analysis of the EEG-fMRI response to prolonged bursts of interictal epileptiform activity. Neuroimage 2005;24(4):1099–1112.
163. Laufs H, Hamandi K, Salek-Haddadi A, Kleinschmidt AK, Duncan JS, Lemieux L. Temporal lobe interictal epileptic discharges affect cerebral activity in "default mode" brain regions. Hum Brain Mapp 2007; 28:1023–1032.
164. Diehl B, Salek-Haddadi A, Fish DR, Lemieux L. Mapping of spikes, slow waves, and motor tasks in a patient with malformation of cortical development using simultaneous EEG and fMRI. Magn Reson Imaging 2003;21(10):1167–1173.
165. Laufs H, Hamandi K, Walker MC, Scott C, Smith S, Duncan JS et al. EEG-fMRI mapping of asymmetrical delta activity in a patient with refractory epilepsy is concordant with the epileptogenic region determined by intracranial EEG. Magn Reson Imaging 2006;24(4):367–371.
166. Morgan VL, Price RR, Arain A, Modur P, Abou-Khalil B. Resting functional MRI with temporal clustering analysis for localization of epileptic activity without EEG. Neuroimage 2004;21(1):473–481.
167. Hamandi K, Salek-Haddadi A, Liston A, Laufs H, Fish DR, Lemieux L. fMRI temporal clustering analysis in patients with frequent interictal epileptiform discharges: comparison with EEG-driven analysis. Neuroimage 2005;26(1):309–316.
168. Aguirre GK, Zarahn E, D'Esposito M. The variability of human, BOLD hemodynamic responses. Neuroimage 1998;8:360–369.
169. Salek-Haddadi A, Friston KJ, Lemieux L, Fish DR. Studying spontaneous EEG activity with fMRI. Brain Res Rev 2003; 43(1):110–133.
170. Kobayashi E, Bagshaw AP, Grova C, Dubeau F, Gotman J. Negative BOLD responses to epileptic spikes. Hum Brain Mapp 2006; 27(6):488–497.
171. Lemieux L, Laufs H, Carmichael D, Paul JS, Walker MC, Duncan JS. Noncanonical spike-related BOLD responses in focal epilepsy. Hum Brain Mapp 2008;29(3):329–345.
172. Benar CG, Gross DW, Wang Y, Petre V, Pike B, Dubeau F et al. The BOLD response to interictal epileptiform discharges. Neuroimage 2002;17(3):1182–1192.
173. Lu Y, Bagshaw AP, Grova C, Kobayashi E, Dubeau F, Gotman J. Using voxel-specific hemodynamic response function in EEG-fMRI data analysis. Neuroimage 2006; 32(1):238–247.
174. Hawco CS, Bagshaw AP, Lu Y, Dubeau F, Gotman J. BOLD changes occur prior to epileptic spikes seen on scalp EEG. Neuroimage 2007;35(4):1450–1458.
175. Makiranta M, Ruohonen J, Suominen K, Niinimaki J, Sonkajarvi E, Kiviniemi V et al. BOLD signal increase preceeds EEG spike activity--a dynamic penicillin induced focal epilepsy in deep anesthesia. Neuroimage 2005;27(4):715–724.
176. Siniatchkin M, Moeller F, Jacobs J, Stephani U, Boor R, Wolff S et al. Spatial filters and

automated spike detection based on brain topographies improve sensitivity of EEG-fMRI studies in focal epilepsy. Neuroimage 2007;37(3):834–843. Corrected Proof:-901.

177. Rosenow F, Luders H. Presurgical evaluation of epilepsy. Brain 2001;124:1683–1700.
178. Kobayashi E, Bagshaw AP, Benar CG, Aghakhani Y, Andermann F, Dubeau F et al. Temporal and extratemporal BOLD responses to temporal lobe interictal spikes. Epilepsia 2006;47(2):343–354.
179. Lemieux L, Krakow K, Fish DR. Comparison of spike-triggered functional MRI BOLD activation and EEG dipole model localization. Neuroimage 2001;14(5):1097–1104.
180. Bagshaw AP, Kobayashi E, Dubeau F, Pike GB, Gotman J. Correspondence between EEG-fMRI and EEG dipole localisation of interictal discharges in focal epilepsy. Neuroimage 2006;30(2):417–425.
181. Benar CG, Grova C, Kobayashi E, Bagshaw AP, Aghakhani Y, Dubeau F et al. EEG-fMRI of epileptic spikes: Concordance with EEG source localization and intracranial EEG. Neuroimage 2006;30(4):1161–1170.
182. Boor R, Jacobs J, Hinzmann A, Bauermann T, Scherg M, Boor S et al. Combined spike-related functional MRI and multiple source analysis in the non-invasive spike localization of benign rolandic epilepsy. Clin Neurophysiol 2007;118(4):901–909.
183. Lazeyras F, Blanke O, Perrig S, Zimine I, Golay X, Delavelle J et al. EEG-triggered functional MRI in patients with pharmacoresistant epilepsy. J Magn Reson Imaging 2000;12(1):177–185.
184. Zijlmans M, Huiskamp G, Hersevoort M, Seppenwoolde JH, van Huffelen AC, Leijten FS. EEG-fMRI in the preoperative work-up for epilepsy surgery. Brain 2007;130(Pt 9):2343–2353.
185. Krakow K, Wieshmann UC, Woermann FG, Symms MR, McLean MA, Lemieux L et al. Multimodal MR imaging: functional, diffusion tensor, and chemical shift imaging in a patient with localization-related epilepsy. Epilepsia 1999;40(10):1459–1462.
186. Salek-Haddadi A, Lemieux L, Fish DR. Role of functional magnetic resonance imaging in the evaluation of patients with malformations caused by cortical development. Neurosurg Clin N Am 2002;13(1):63–9, viii.
187. Federico P, Archer JS, Abbott DF, Jackson GD. Cortical/subcortical BOLD changes associated with epileptic discharges: an EEG-fMRI study at 3 T. Neurology 2005; 64(7):1125–1130.
188. Kobayashi E, Bagshaw AP, Jansen A, Andermann F, Andermann E, Gotman J et al. Intrinsic epileptogenicity in polymicrogyric cortex suggested by EEG-fMRI BOLD responses. Neurology 2005;64(7):1263–1266.
189. Kobayashi E, Bagshaw AP, Grova C, Gotman J, Dubeau F. Grey matter heterotopia: what EEG-fMRI can tell us about epileptogenicity of neuronal migration disorders. Brain 2006;129(Pt 2):366–374.
190. Kobayashi E, Bagshaw AP, Gotman J, Dubeau F. Metabolic correlates of epileptic spikes in cerebral cavernous angiomas. Epilepsy Res 2007;73(1):98–103.
191. Wiebe S, Blume WT, Girvin JP, Eliasziw M. A randomized, controlled trial of surgery for temporal-lobe epilepsy. N Engl J Med 2001;345(5):311–318.
192. Kobayashi E, Bagshaw AP, Benar CG, Aghakhani Y, Andermann F, Dubeau F et al. Temporal and extratemporal BOLD responses to temporal lobe interictal spikes. Epilepsia 2006;47(2):343–354.
193. Raichle ME, MacLeod AM, Snyder AZ, Powers WJ, Gusnard DA. A default mode of brain function. Proc Natl Acad Sci USA 2001;98(2):676–682.
194. Lengler U, Kafadar I, Neubauer BA, Krakow K. FMRI correlates of interictal epileptic activity in patients with idiopathic benign focal epilepsy of childhood: a simultaneous EEG-functional MRI study. Clin Neurophysiol 2007;118(4):e70–e51.
195. Jacobs J, Jacobs J, Boor R, Jansen O, Wolff S, Siniatchkin M et al. Localization of epileptic foci in children with focal epilepsies using 3-Tesla simultaneous EEG-fMRI recordings. Clin Neurophysiol 2007; 118(4):e50–e51.
196. De Tiege X, Laufs H, Boyd SG, Harkness W, Allen PJ, Clark CA et al. EEG-fMRI in children with pharmacoresistant focal epilepsy. Epilepsia 2007;48(2):385–389.
197. Engel J. A proposed diagnostic scheme for people with epileptic seizures and with epilepsy: report of the ILAE task force on classification and terminology. Epilepsia 2001; 42(6):796–803.
198. Aghakhani Y, Bagshaw AP, Benar CG, Hawco C, Andermann F, Dubeau F et al. fMRI activation during spike and wave discharges in idiopathic generalized epilepsy. Brain 2004;127(5):1127–1144.
199. Archer JS, Abbott DF, Waites AB, Jackson GD. fMRI "deactivation" of the posterior cingulate during generalized spike and wave. Neuroimage 2003;20(4):1915–1922.

200. Gotman J, Grova C, Bagshaw A, Kobayashi E, Aghakhani Y, Dubeau F. Generalized epileptic discharges show thalamocortical activation and suspension of the default state of the brain. Proc Natl Acad Sci USA 2005; 102(42):15236–15240.
201. Laufs H, Lengler U, Hamandi K, Kleinschmidt A, Krakow K. Linking generalized spike-and-wave discharges and resting state brain activity by using EEG/fMRI in a patient with absence seizures. Epilepsia 2006;47(2):444–448.
202. Salek-Haddadi A, Lemieux L, Merschhemke M, Friston KJ, Duncan JS, Fish DR. Functional magnetic resonance imaging of human absence seizures. Ann Neurol 2003; 53(5):663–667.
203. Meeren HKM, Pijn JP, Van Luijtelaar ELJM, Coenen AML, Lopes da Silva FH. Cortical focus drives widespread corticothalamic networks during spontaneous absence seizures in rats. J Neurosci 2002;22(4):1480–1495.
204. Steriade M, Dossi RC, Nunez A. Network modulation of a slow intrinsic oscillation of cat thalamocortical neurons implicated in sleep delta waves: cortically induced synchronization and brainstem cholinergic suppression. J Neurosci 1991;11(10):3200–3217.
205. Hamandi K, Laufs H, Noth U, Carmichael DW, Duncan JS, Lemieux L. BOLD and perfusion changes during epileptic generalised spike wave activity. Neuroimage 2008;39(2):608–618.
206. Shmuel A, Augath M, Oeltermann A, Logothetis NK. Negative functional MRI response correlates with decreases in neuronal activity in monkey visual area V1. Nat Neurosci 2006;9(4):569–577.
207. Binnie CD. Cognitive impairment during epileptiform discharges: is it ever justifiable to treat the EEG? Lancet Neurol 2003; 2(12):725–730.
208. Aghakhani Y, Kobayashi E, Bagshaw AP, Hawco C, Benar CG, Dubeau F et al. Cortical and thalamic fMRI responses in partial epilepsy with focal and bilateral synchronous spikes. Clin Neurophysiol 2006; 117(1):177–191.
209. Federico P, Archer JS, Abbott DF, Jackson GD. Cortical/subcortical BOLD changes associated with epileptic discharges: an EEG-fMRI study at 3 T. Neurology 2005; 64(7):1125–1130.
210. Salek-Haddadi A, Lemieux L, Merschhemke M, Diehl B, Allen PJ, Fish DR. EEG quality during simultaneous functional MRI of interictal epileptiform discharges. Magn Reson Imaging 2003;21(10):1159–1166.
211. Rodionov R, De Martino F, Laufs H, Carmichael DW, Formisano E, Walker M et al. Independent component analysis of interictal fMRI in focal epilepsy: comparison with general linear model-based EEG-correlated fMRI. Neuroimage 2007; 38(3):488–500.
212. Merlet I, Gotman J. Dipole modeling of scalp electroencephalogram epileptic discharges: correlation with intracerebral fields. Clin Neurophysiol 2001;112(3): 414–430.
213. Liston AD, Salek-Haddadi A, Kiebel SJ, Hamandi K, Turner R, Lemieux L. The MR detection of neuronal depolarization during 3-Hz spike-and-wave complexes in generalized epilepsy. Magn Reson Imaging 2004; 22(10):1441–1444.
214. Daunizeau J, Grova C, Marrelec G, Mattout J, Jbabdi S, Pelegrini-Issac M et al. Symmetrical event-related EEG/fMRI information fusion in a variational Bayesian framework. NeuroImage 2007;36(1):69–87.
215. Jezzard P, Clare S. Sources of distortion in functional MRI data. Hum Brain Mapp 1999;8(2–3):80–85.
216. Bagshaw AP, Torab L, Kobayashi E, Hawco C, Dubeau F, Pike GB et al. EEG-fMRI using z-shimming in patients with temporal lobe epilepsy. J Magn Reson Imaging 2006; 24(5):1025–1032.
217. Jezzard P, Balaban RS. Correction for geometric distortion in echo planar images from B0 field variations. Magn Reson Med 1995; 34(1):65–73.
218. Hutton C, Bork A, Josephs O, Deichmann R, Ashburner J, Turner R. Image distortion correction in fMRI: a quantitative evaluation. Neuroimage 2002;16(1):217–240.
219. Niendorf T. On the application of susceptibility-weighted ultra-fast low-angle RARE experiments in functional MR imaging. Magn Reson Med 1999;41(6):1189–1198.
220. Deichmann R, Josephs O, Hutton C, Corfield DR, Turner R. Compensation of susceptibility-induced BOLD sensitivity losses in echo-planar fMRI imaging. Neuroimage 2002;15(1):120–135.
221. Deichmann R, Gottfried JA, Hutton C, Turner R. Optimized EPI for fMRI studies of the orbitofrontal cortex. Neuroimage 2003;19(2 Pt 1):430–441.
222. Lemieux L, Salek-Haddadi A, Lund TE, Laufs H, Carmichael D. Modelling large motion events in fMRI studies of patients with epilepsy. Magn Reson Imaging 2007; 25(6):894–901.

223. Liston AD, Lund TE, Salek-Haddadi A, Hamandi K, Friston KJ, Lemieux L. Modelling cardiac signal as a confound in EEG-fMRI and its application in focal epilepsy studies. Neuroimage 2006;30(3):827–834.
224. Glover GH, Li TQ, Ress D. Image-based method for retrospective correction of physiological motion effects in fMRI: RETROICOR. Magn Reson Med 2000; 44(1):162–167.
225. Jager L, Werhahn KJ, Hoffmann A, Berthold S, Scholz V, Weber J et al. Focal epileptiform activity in the brain: detection with spike-related functional MR imaging - preliminary results. Radiology 2002;223:860–869.
226. Benar CG, Gross DW, Wang Y, Petre V, Pike B, Dubeau F et al. The BOLD response to interictal epileptiform discharges. Neuroimage 2002;17(3):1182–1192.
227. Salek-Haddadi A, Lemieux L, Merschhemke M, Friston KJ, Duncan JS, Fish DR. Functional magnetic resonance imaging of human absence seizures. Ann Neurol 2003; 53(5):663–667.
228. Aghakhani Y, Bagshaw AP, Benar CG, Hawco C, Andermann F, Dubeau F et al. fMRI activation during spike and wave discharges in idiopathic generalized epilepsy. Brain 2004;127:1127–1144.
229. Jacobs J, Kobayashi E, Boor R, Muhle H, Stephan W, Hawco C et al. Hemodynamic responses to interictal epileptiform discharges in children with symptomatic epilepsy. Epilepsia 2007;48(11):2068–78.

Chapter 24

fMRI in Neurosurgery

Oliver Ganslandt, Christopher Nimsky, Michael Buchfelder, and Peter Grummich

Summary

Functional magnetic resonance imaging has evolved from a basic research application to a useful clinical tool that also has found its place in modern neurosurgery. The localization of functional important brain areas as language and sensorimotor cortex has been the focus of numerous investigations and can now be implemented in neurosurgical planning. Since the neurosurgeon must have detailed knowledge about the individual anatomy and related neurological function to resect a brain tumor with the highest safety, the need for individualized maps of brain function is essential. Advanced fMRI techniques and modern imaging methods contribute significantly to brain mapping as do already established concepts of electrophysiological monitoring and the Wada test. The implementation of functional maps into neuronavigation systems enables the surgeon to superimpose anatomy and function to the surgical site. This chapter describes our experience with the use of fMRI in neurosurgery.

Key words: fMRI, Neurosurgery, Functional neuronavigation, Magnetoencephalography, Language, Somatosensory cortex

1. Introduction

The concept of using information about functionally important brain areas (also known as “eloquent cortex”) to safely guide neurosurgical procedures has been established in the middle of the twentieth century by use of electrical stimulation in awake craniotomies. Based on the seminal work of Penfield *(1)*, modern neurosurgeons used the technique of electrical stimulation to meticulously map the cerebral cortex of their patients, for instance, to delineate the borders of resection in epilepsy and tumor surgery. Today, electrical stimulation is still regarded as the

M. Filippi (ed.), *fMRI Techniques and Protocols*, Neuromethods, vol. 41
DOI 10.1007/978-1-60327-919-2_24, © Humana Press, a part of Springer Science+Business Media, LLC 2009

"gold standard" for neurosurgical functional brain localization *(2, 3)*. However, these invasive direct cortical stimulation methods are not available for preoperative decision making and surgical planning. They are also time consuming and demand special resources. Therefore, in the past decade new efforts have been made to overcome the limitations of using electrical stimulation in awake craniotomy by using new techniques of brain imaging and the implementation of these data into the neurosurgical workspace. In recent years, two noninvasive techniques have been found especially suitable for presurgical localization of the eloquent cortex: magnetoencephalography (MEG) and fMRI. Studies using these techniques successfully localized functional activity *(4, 5)*.

One of the most interesting application was the merge of functional brain imaging with frame-based and frameless stereotaxy, also known as functional neuronavigation *(6, 7)*. There is initial evidence that the use of functional neuronavigation for lesions adjacent to eloquent brain areas may favor clinical outcome *(8, 9)*, but large controlled studies to support this assumption are still needed.

If surgery near eloquent brain areas is planned, a detailed knowledge about the topographic relation of a lesion to the adjacent functional brain area is crucial to avoid postoperative neurological deficits. In neurosurgery, the primary sensorimotor cortex and the cortical areas subserving language comprehension and production are considered to be the main risk structures. These structures usually cannot be depicted from conventional structural imaging techniques. Other reasons that warrant a detailed evaluation are the individual representation of these eloquent areas and the phenomenon of cortical reorganization of these areas from their original positions *(10, 11)*. Furthermore, normal sulcal anatomy is not often discernible because of a space occupying lesion. These situations require methods for localizing functional areas prior to surgery for decision making, planning, and avoiding crippling postoperative results.

fMRI has become indispensable in neurosurgery to easily gain knowledge about the topographic relation of a given lesion to the functional brain area at risk and thus to plan the surgical approach. Furthermore, fMRI-derived information about the extent of cortical involvement in function can be used in conjunction with image-guided surgery during resection of lesions adjacent to eloquent brain areas under general anesthesia for navigation. The almost ubiquitous availability of modern MR scanners favors the use of fMRI over other modalities as MEG or positron emission tomography (PET) that demand resources not commonly available. In addition, its noninvasiveness gives the opportunity to repeat the examinations and conduct follow-up studies on reorganization of cortical function. Advances in MRI technology, such as the introduction of higher field strengths, will undoubtedly improve signal acquisition and processing *(12)*. Over the last year, a substantial number of publications have described the usefulness of clinical fMRI for neurosurgical applications *(13, 14)*. The use of fMRI for the presurgical localization of the sensorimotor cortex is now widely appreciated and has been investigated by several groups *(15–17)*, which also performed comparisons with direct motor stimulation. Language fMRI has been found to be an alternative to the invasive Wada test *(18–20)* for language lateralization. Furthermore, fMRI has been used to predict memory localization *(21)*. Concerning the reliability of fMRI-localization of speech areas in the fronto-temporal cortex, as compared with direct electrical stimulation, there is not enough data that support its application as a substitute for electrical stimulation, since inconsistent agreement has been found between activation sites by fMRI naming and verb generation tasks and cortical stimulation *(5)*. However, more studies are needed to fully understand the mechanisms of language activation and whether the use of fMRI adds complementary information that is essential for neurosurgeons.

2. Methods

In our department, all neurosurgical fMRI measurements are acquired on a 1.5 T MR scanner by echo-planar imaging (Magnetom Sonata, Siemens, Erlangen, Germany).

Measurements for localization of motor and sensory activity are performed with 16 slices of 3 mm thickness, a TR = 1580, and a TE = 60. Stimulation is done in a block paradigm with 120 stimulus presentations in six blocks. Twenty measurements during rest alternated with 20 measurements during stimulation. During the motor activation blocks, the patient is asked to perform a motor task: in particular, we are interested in localizing the cortical representation of the toes, foot, leg, fingers, hand, arm, tongue, lips, and eye lid.

Our selection of the motor tasks for each patient depends on the tumor location. Attention is paid that the patient does not move the opposite limb. With this, the possibility to detect reorganization of functional areas to the contralateral hemisphere is ensured. Each patient is also instructed not to touch anything during movement. For this reason, the motor task is usually not a finger tapping task. Only in cases where we are interested in localizing the supplementary motor area we conduct a finger tapping task.

For the localization of the sensory cortex, we use a tactile stimulation of different parts whose cortical representations we wish to localize.

Measurements for language are done with 25 slices of 3 mm thickness, a TR = 2470, and a TE = 60. Stimulation is done in a block paradigm with 180 measurements in 6 blocks. We perform 30 measurements in an activation condition during which the patient is instructed to perform a language task; we alternate these with 30 measurements in a resting condition. By means of a mirror that is attached to the head coil, the patient is able to observe words, numbers, or pictures projected onto a screen.

We developed several stimulation paradigms for localizing the Broca's and the Wernicke's areas. Usually, each patient is asked to undergo four different paradigms during fMRI measurements. The paradigms are selected according to tumor location and are adapted to the abilities of each patient. The length of the interstimulus interval is also adjusted according to the patient's abilities, varying between 900 and 2000 ms. The duration of the stimulus presentation lies between 600 and 1700 ms (300 ms less than the interstimulus interval). We ask the patients to perform the tasks as quickly as possible immediately after stimulus presentation and to perform the task silently to avoid artifacts from mouth movement.

Paradigms are chosen on the basis of: (a) tumor location, (b) patient cognitive abilities:

(a) In case of tumor location in the inferior parietal area close to the intraparietal sulcus, we use an arithmetic task so that besides Wernicke's area (activated by reading, adding numbers, and formulating the result) the cortex for calculation in the intraparietal sulcus is also activated and so can be spared during surgery.
    In case of tumor location close to Broca's area, we select language tasks that are also expressive and demand grammatical abilities, because these may increase activity in Broca's area. This happens during the verb generation task, but also the verb conjugation task gives suitable Broca's area activations.

(b) For patients with reduced abilities, simple tasks are selected to obtain reliable results. Especially in patients who suffer from word finding disorders, we avoid the picture naming task. For patients with better cognitive performance, we select one or more complex tasks such as verb generation task, because these are reported to show a more clear lateralization, whereas in patients who have difficulties in this complex task the activation is usually worse than with a simple paradigm.

For motion correction, we apply an image-based prospective acquisition correction by applying interpolation in the k-space *(22)*. We produce activation maps by analyzing the correlation between signal intensity and a square wave reference function for each pixel according to the paradigm. Pixels exceeding a significance threshold (typical correlations above a threshold of 0.3 with $p < 0.000045$) are displayed, if at least six contiguous voxels constitute a cluster, to eliminate isolated voxels. We align the functional slices to magnetization prepared rapid acquisition gradient echo (MPRAGE) images (160 slices of 1 mm slice thickness).

3. Results

3.1. Localizations

Since 2002, we have investigated preoperative fMRI with motor or sensory stimulation in 177 cases. Of these patients, 110 underwent tumor resection and 28 had stereotactic brain biopsy. In one additional patient, invasive electrodes were implanted by fMRI guidance for chronic recording of epileptic discharges.

For language testing, we examined 186 cases and used additional information from MEG studies. Of them, 124 underwent tumor resection and 21 had stereotactic brain biopsy. The remaining patients either obtained radiation therapy, endovascular treatment, or were just enrolled in a "wait-and-see" protocol.

It was possible to localize the primary motor and sensory cortex as well as the supplementary motor area (SMA) in all examined cases (**Figs. 1** and **2**). Only in one case, the motor activity of the toe was not detectable by fMRI because of tumor infiltration. However, in this patient it was possible to obtain motor activation from nearby muscle representations of the motor homunculus.

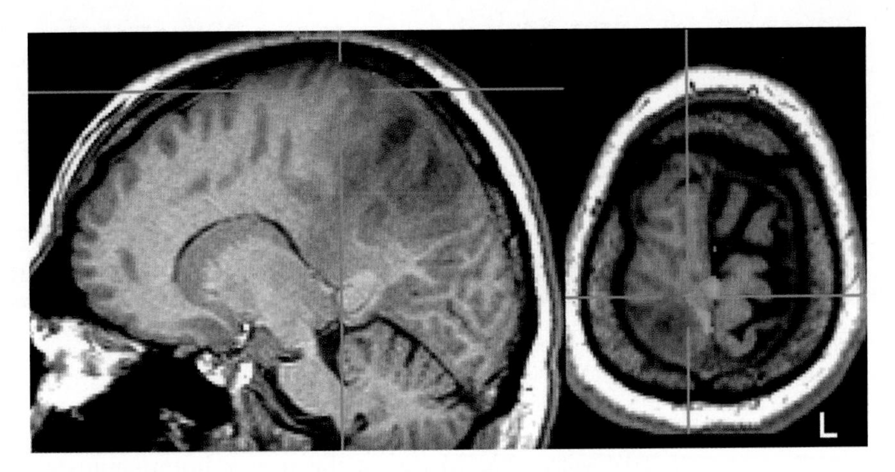

Fig. 1. fMRI activations during movement of left foot in a patient with an oligoastrocytoma (WHO III) in the right parietal lobe. In front of the activation of the motor cortex (posterior wall of precentral gyrus), activation of the supplementary motor area (SMA) is also evident.

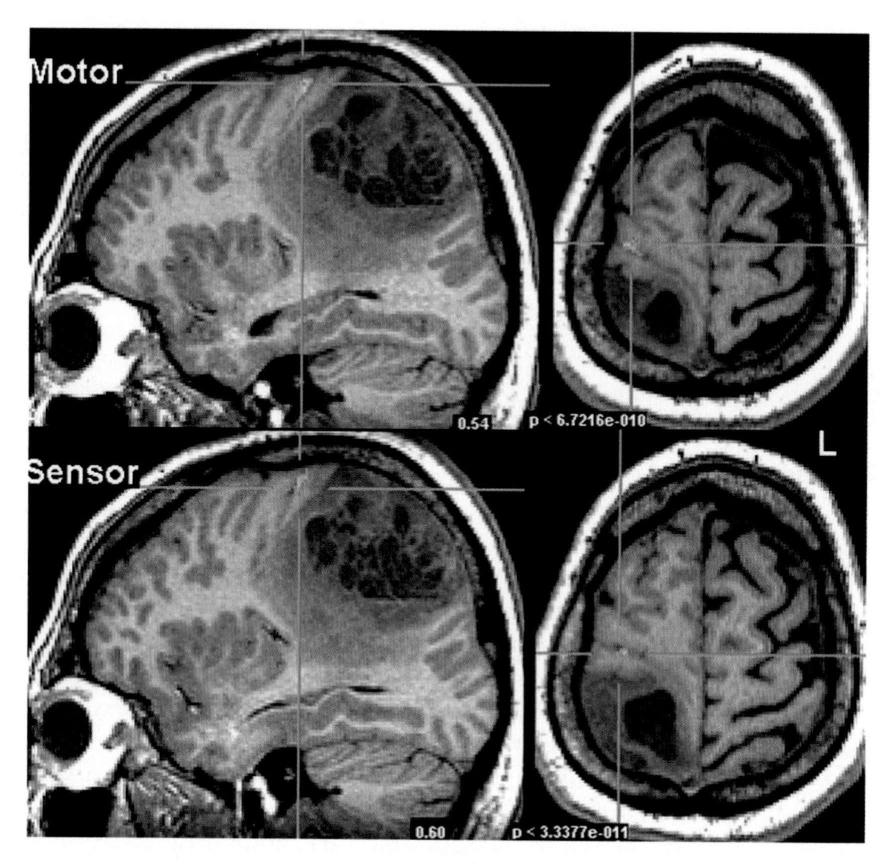

Fig. 2. Comparison of fMRI motor activations during arm movement and sensory stimulation of the arm (oligoastrocytoma WHO III, same patient as **Fig. 1**).

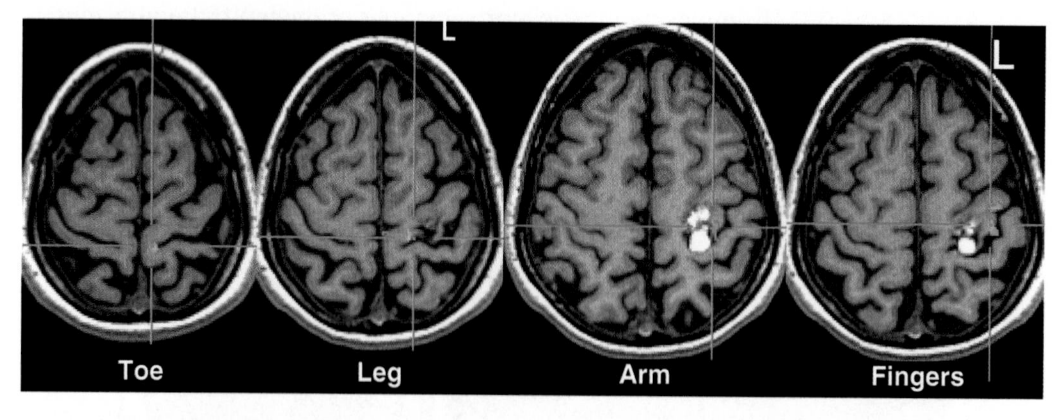

Fig. 3. fMRI activations during movement of toe, leg, arm, and fingers (note that the lesion, a cavernoma in the left motor cortex, is located between the cortical representation of the arm and that of the leg in the precentral gyrus).

We were able to define the motor homunculus along the central sulcus in the posterior wall of the precentral gyrus with fMRI by motor activation of the respective muscle groups **(Fig. 3)**. We localized toe, foot, leg, arm, hand, finger, thumb, tongue, lip, and eye movements. Additionally, sometimes we found activity in the ipsilateral homotopic cortex. Especially when the motor task was more complex there was also activity at the frontal wall of the precentral gyrus. Additional activity can be detected in the SMA in the interhemispheric sulcus.

Activity at the posterior wall of the postcentral gyrus (Brodman area 2) can be found and sometimes in the gyrus posteriorly, which is likely to represent proprioceptive activation due to the positions of the limbs.

Sensory activity was seen in the anterior wall (Brodman area 3) of the postcentral gyrus. Sometimes there was also blood oxygen level dependent (BOLD) activation in the homotopic cortex of the ipsilateral hemisphere that may indicate the presence of mechanisms of cortical plasticity.

With verbal stimulation tasks, we were able to localize language, calculation, and memory activity. In the frontal lobe, we found activity in the following cortical areas: (1) at the bottom of the opercular part of the inferior frontal gyrus anteriorly to the precentral gyrus (Brodmann area 44, classical Broca's area); (2) in the adjacent part of the frontal cranial edge of the insular cortex; and (3) close to the upper end of the inferior frontal gyrus, which sometimes extends into the medial frontal gyrus. Here, there are usually three cortical areas that extend from the pars triangularis to pars opercularis (from anterior to posterior).

In the temporo-parietal region, we found language activity in the superior temporal gyrus at its bottom in the temporal sulcus and at its lateral side. Activity was also found at the top of the temporal gyrus in the planum temporale. Additional activity was found in the supramarginal gyrus in the frontal part of the intraparietal sulcus.

In 53% of our patients with high-grade glioma, language activity was not clearly detectable by fMRI alone, because of changes in vascular function (*see* below).

3.2. Laterality

Although it is generally accepted that the majority of people has left hemispheric language dominance, the true number of atypical (right) dominance is unknown. Studies using the Wada test showed an incidence of left hemispheric dominance in right handers in a range of 63–96% and a right hemispheric dominance for left handers and ambidextrous patients in 48–75% *(23)*. Furthermore, it is thought that there is varying degrees of language dominance in the population.

In certain circumstances, activity can be located in both hemispheres or reorganization to the other hemisphere could have been occurred. FMRI is a useful method to clarify this. If activity is only found in one hemisphere or the activity on one side is much stronger than the activity on the other side, then it is clear that the active area has to be spared during surgery.

It is important to know that certain stimulation tasks and modalities show more lateralized activations than others. In case of complex motor tasks, the ipsilateral hemisphere may also show activation.

For the localization of language activity, we found that visual stimulation shows a more accentuated lateralization than acoustic stimulation *(24)*. Stimulation with words, especially in a complex task, shows a stronger lateralization than a picture naming task, a finding that was also described by Herholz et al. *(25, 26)*. In rare cases, it can occur that not all language areas are located on the same side.

3.3. Surgery

We perform fMRI-guided surgery by coregistering the activation maps onto a 3D MRI data set that can be used with a navigation system. Targets and areas at risk determined by fMRI are segmented and made visible for the surgeon through a navigation microscope (**Figs. 4–6**). Thus functional data are visualized in the operation field throughout the whole surgery. Neuronavigation support is provided by the VectorVision Sky Navigation System (BrainLab AG, Heimstetten, Germany). A fiber optic connection ensures MR-compatible integration into the radiofrequency-shielded room of our intraoperative MR suite. A ceiling-mounted camera is used to monitor the positions of the operating microscope (Pentero, Zeiss, Oberkochen, Germany), which is placed outside the 5 G line, and other instruments.

A 1.0 mm isotropic 3D MPRAGE dataset (TE: 4.38 ms, TR: 2020 ms, slice thickness: 1.0 mm, FOV: 250 × 250 mm, measurement time: 8 min 39 s) is acquired prior to surgery with the head already fixed in the MR-compatible headholder as navigational reference dataset. For registration, five adhesive

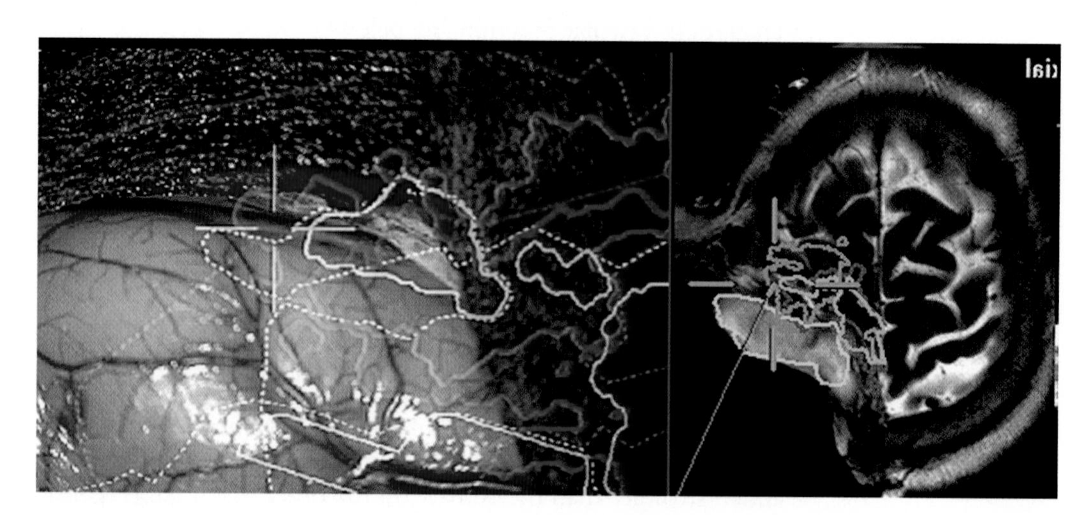

Fig. 4. Microscopic view with neuronavigation markers showing the sensory activation of the arm area in light blue and the pyramidal tract in purple (oligoastrocytoma WHO III, same patient as **Fig. 1**).

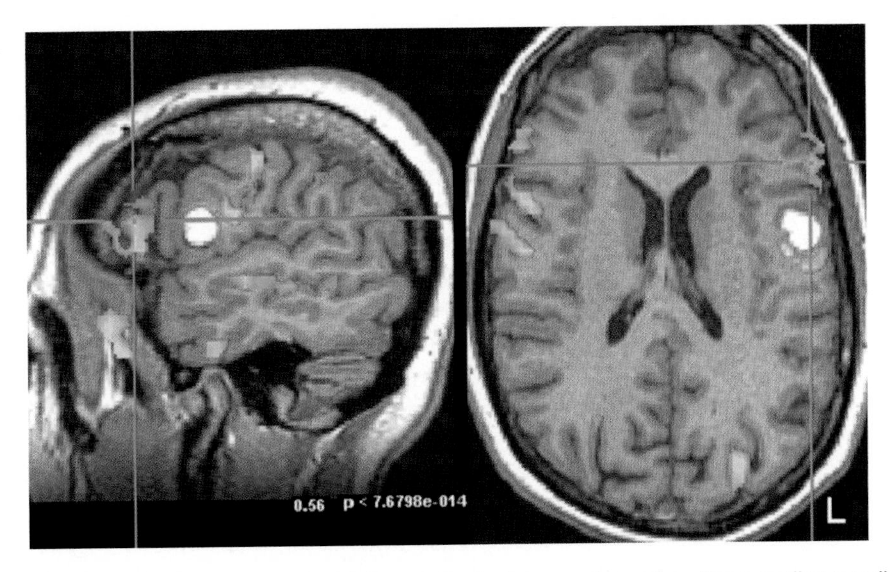

Fig. 5. fMRI activation of Broca's and Wernicke's areas and primary motor cortex after a reading paradigm. The functional mapping was requested to plan surgery of a cavernoma, which was located between Broca's area and the motor cortex.

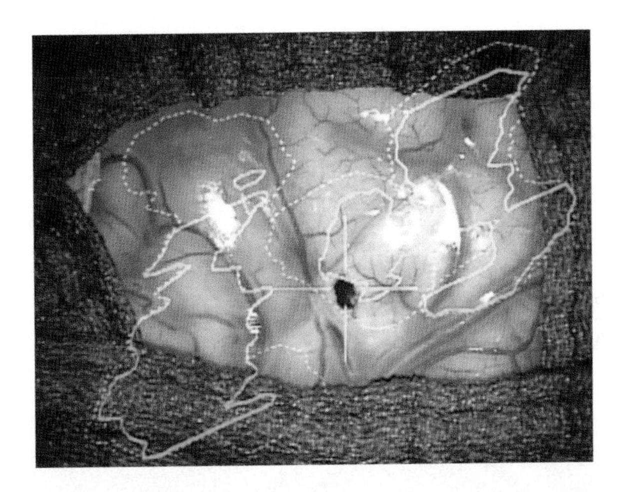

Fig. 6. Same patient as **Fig. 5**. Segmentation lines indicating Broca's area (*left*) and motor cortex of tongue (*right*). The figure shows the beginning of the corticotomy on a trajectory that spared the eloquent cortices (*cross*). Postoperatively the patient was neurologically intact.

skin fiducial markers are placed in a scattered pattern on the head surface prior to imaging and registered with a pointer after their position is defined in the 3D dataset **(Fig. 5)**. Functional data from MEG and fMRI, which were acquired preoperatively, are integrated into the 3D dataset. Furthermore, data from diffusion tensor imaging (DTI) depicting the course of major white matter tracts are integrated as well as in selected cases metabolic maps from proton magnetic resonance spectroscopy (^{1}H-MRS) are co-registered to the navigational dataset. In addition, further standard anatomical datasets, such as T2-weighted images, are co-registered. Repeated landmark checks are performed to ensure overall accuracy. In case intraoperative imaging depicts some remaining tumor, which should be further removed, intraoperative image data are used for updating the navigation system **(Fig. 7)**. After a rigid registration of pre- and intraoperative images (ImageFusion software, BrainLAB, Heimstetten, Germany), all data are transferred to the navigation system and then the initial patient registration file is restored, so that no repeated patient registration procedure is needed.

In our series with a surgical resection close to the motor cortex, only 5 out of 110 patients (4.5%) had postoperative neurologic dysfunction. One of them recovered within 2 days. In the other cases, the condition improved over several months. One patient, who had a hemiplegia prior to surgery, was able to move the affected side after surgery.

No permanent postoperative deterioration of speech was observed in our patients with surgery close to the language areas (N = 124 patients). However, in 20 patients, in whom surgery was conducted very close to the language areas, a transitory deterioration was observed (16% of all patients with surgery neighboring language areas). Patients were not able to name part of the shown objects; this impairment lasted from 1 day to few weeks, but they all resolved completely.

No patient had suffered from global aphasia after surgery. The result of having no permanent speech disorder in our patients indicates that our language mapping is reliable. The presence of patients with transitory disturbances suggests that resection was conducted close to the boundaries of functional areas. This is in accordance to other series that evaluated the outcome of glioma surgery in eloquent areas with direct cortical stimulation.

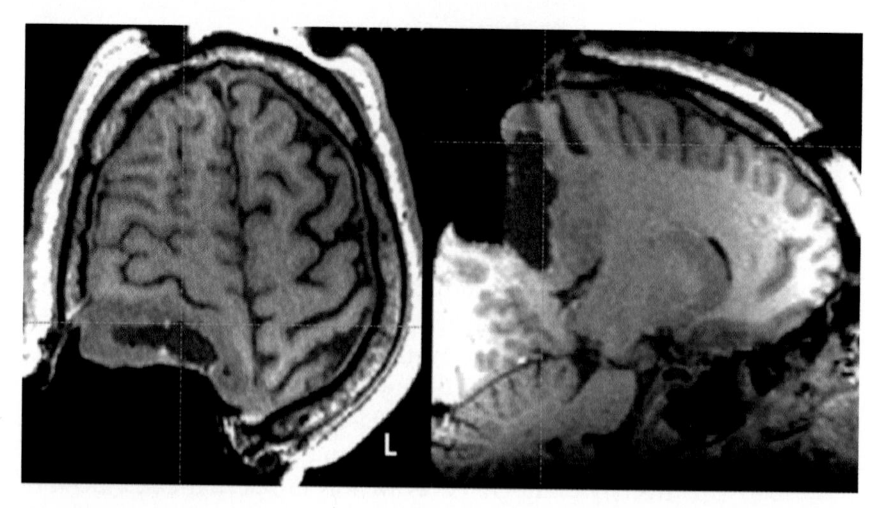

Fig. 7. Intraoperative MRI showing the outcome of the fMRI-guided tumor resection. Note that the tumor was removed sparing the sensory cortex (oligoastrocytoma WHO III *right*, same patient as **Fig. 1**).

In a recent publication by Duffau et al., the rate of severe neurological deficits was 6.5% *(27)*.

The safety margin that should be kept to preserve the functional areas depends from several factors: (1) the kind of functional center and the situation of reorganization; (2) the situation of blood supply; and (3) the status of the connectivity fibers. At present, no recommendations can be given for the exact distance to avoid the risk of neurologic deficits. Neurosurgeons who use fMRI-guided neuronavigation have to keep in mind that the fMRI-activation does not represent the actual extent of the functional brain areas, but rather a "center of gravity" of the functional units that are measured. Also one has to take into account that descending pathways (e.g., the pyramidal tract) have also to be spared. A recent study that investigated the accuracy between the actual location of the pyramidal tract and subcortical electric stimulation with stereotactic navigation found a mean difference in distance of 8.7 ± 3.1 mm (standard deviation) *(28)*. Nevertheless, there are functional areas that can be compensated for, if destroyed. These are the SMA and the area in the fusiform gyrus for word recognition.

For the language areas, Haglund et al. *(29)* described in an electrical stimulation study that above a resection distance of 10 mm from the eloquent areas they observed no permanent language deficits. When surgery was 7–10 mm close to the language area, they found 43% patients who suffered from permanent language deficits (severe or mild aphasia). The 9% of patients had no language deficit at all and the remaining 48% experienced transitory language deficits, which resolved within 4 weeks *(29)*. Two of our patients showed an amelioration of language function after surgery. One patient, who was not able to talk before surgery, was able to talk afterwards. Another patient, who had severe naming problems, showed an improvement after surgery.

3.4. Problems with the BOLD Effect

Sometimes the BOLD activations are not clearly visible in spite of the fact that the function is there, as confirmed by MEG measurements. In our experience, such a discrepancy between MEG and fMRI occurred only in the case of large tumors. Previous reports indicated similar effects of vascular conditions on the BOLD effect *(30–32)*. A reduction of the BOLD effect in the vicinity of a glioma but not in the vicinity of nonglial tumors was described by Schreiber et al. *(33)*. These findings are in agreement with our results. In our series, we found that in 53% of the patients with high-grade gliomas the fMRI maps did not give clear indications of language areas in their vicinity.

Because of the impact of gliomas on the BOLD effect, the dominant hemisphere sometimes is more easily found by MEG measurements. This is seen for a patient with an astrocytoma (WHO Grade II) in **Fig. 8**. Here MEG localizations of

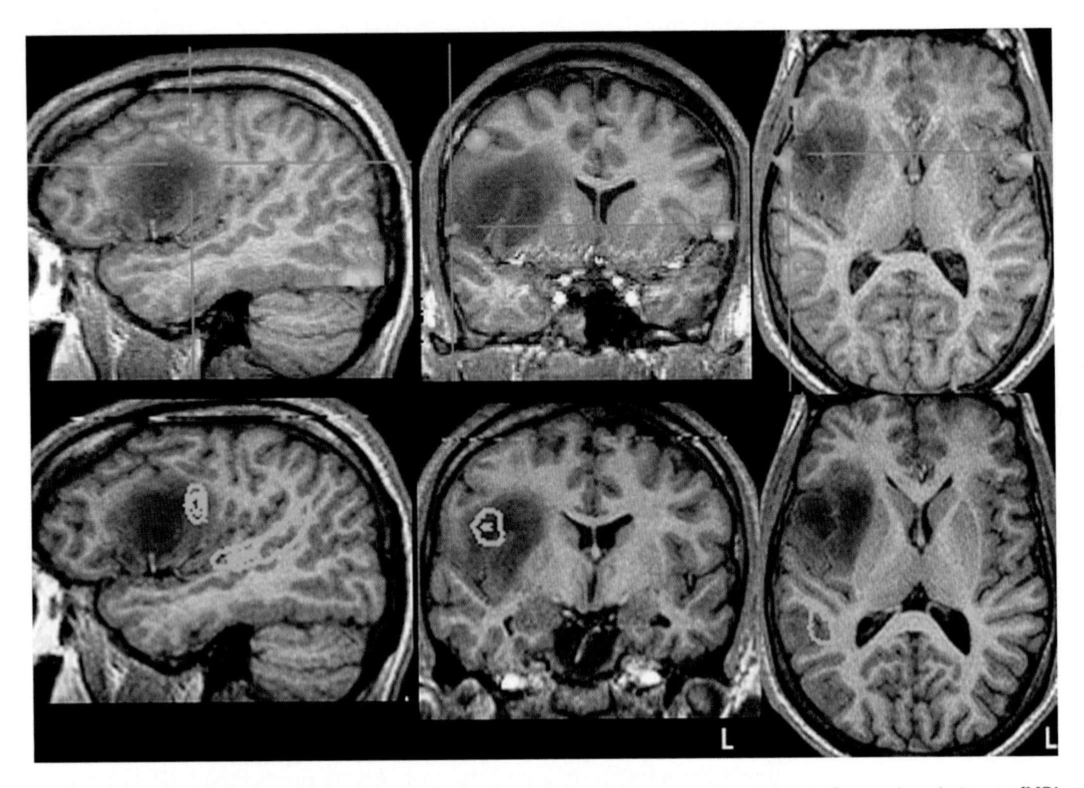

Fig. 8. Wernicke and Broca activity during reading of fragmentary sentences with mistakes. Comparison between fMRI (*orange*) and MEG beamformer localizations at 500 ms (*light blue*). With fMRI a bilateral activation in the operculum frontale and in the superior temporal sulcus can be seen. With MEG, activity is only seen in the right hemisphere. In the first and second image in the lower row, activity of the insula can be seen with MEG only. Left-handed patient with astrocytoma WHO II.

Wernicke's and Broca's area were only on the right side. This was in accordance with the Wada test that showed right hemispheric language dominance in this left-handed patient. In this patient, fMRI localizations of Wernicke's activity were similar on both sides, in MEG they were only found in the right hemisphere. The activity detected by MEG in the right insula was not found by fMRI.

Other reasons that might lead to suboptimal fMRI results are continuing brain activation during rest or a very short activation of brain areas. This might be the reason why memory activity in the hippocampus is rarely found by fMRI.

4. Conclusions

The use of preoperative fMRI brain mapping provides important information for: (1) estimating the risk of a surgical procedure; (2) planning the surgical approach; (3) indicating hemispheric dominance; and (4) revealing whether reorganization of brain function took place and at what degree. The integration of the functional markers into the navigation system is a good tool to continuously track the locations of the functional areas during surgery and enables a resection close to the eloquent areas to be performed. Thus, fMRI-guided functional navigation increases the amount of radical surgery and decreases morbidity. When using fMRI in neurosurgery, it is important to know that, in certain circumstances, the BOLD effect can be suppressed, which may lead to wrong conclusions. Beside integration of fMRI data the additional use of fiber tracking of the descending pathways as well as other paraclinical investigations (PET, proton magnetic resonance spectroscopy, MEG, etc.) should lead to a comprehensive understanding of the options and limitations of glioma surgery adjacent to important functional brain areas.

References

1. Penfield W, Rasmussen T. The Cerebral Cortex of Man. A Clinical Study of Localization of Function. New York: Macmillan; 1950.
2. Berger MS, Rostomily RC. Low grade gliomas: functional mapping resection strategies, extent of resection, and outcome. J Neurooncol 1997;34:85–101.
3. Duffau H, Capelle L, Denvil D, et al. Usefulness of intraoperative electrical subcortical mapping during surgery for low-grade gliomas located within eloquent brain regions: functional results in a consecutive series of 103 patients. J Neurosurg 2003;98:764–78.
4. Kober H, Moller M, Nimsky C, Vieth J, Fahlbusch R, Ganslandt O. New approach to localize speech relevant brain areas and hemispheric dominance using spatially filtered magnetoencephalography. Hum Brain Mapp 2001;14:236–50.
5. Roux FE, Boulanouar K, Lotterie JA, Mejdoubi M, LeSage JP, Berry I. Language functional magnetic resonance imaging in preoperative assessment of language areas: correlation with direct cortical stimulation. Neurosurgery 2003;52:1335–45; discussion 45–7.
6. Nimsky C, Ganslandt O, Kober H, et al. Integration of functional magnetic resonance imaging supported by magnetoencephalography in functional neuronavigation. Neurosurgery 1999; 44:1249–55; discussion 55–6.
7. Rutten GJ, Ramsey N, Noordmans HJ, et al. Toward functional neuronavigation: implementation of functional magnetic resonance imaging data in a surgical guidance system for intraoperative identification of motor and language cortices. Technical note and illustrative case. Neurosurg Focus 2003;15:E6.
8. Gralla J, Ganslandt O, Kober H, Buchfelder M, Fahlbusch R, Nimsky C. Image-guided removal of supratentorial cavernomas in critical brain areas: application of neuronavigation and intraoperative magnetic resonance imaging. Minim Invasive Neurosurg 2003;46: 72–7.
9. Pirotte B, Voordecker P, Neugroschl C, et al. Combination of functional magnetic resonance imaging-guided neuronavigation and intraoperative cortical brain mapping improves targeting of motor cortex stimulation in neuropathic pain. Neurosurgery 2005;56:344–59; discussion 59.
10. Duffau H, Denvil D, Capelle L. Long term reshaping of language, sensory, and motor maps after glioma resection: a new parameter to integrate in the surgical strategy. J Neurol Neurosurg Psychiatry 2002;72:511–6.
11. Grummich P, Nimsky C, Fahlbusch R, Ganslandt O. Observation of unaveraged giant MEG activity from language areas during speech tasks in patients harboring brain lesions very close to essential language areas: expression of brain plasticity in language processing networks? Neurosci Lett 2005; 380:143–8.
12. Tieleman A, Vandemaele P, Seurinck R, Deblaere K, Achten E. Comparison between functional magnetic resonance imaging at 1.5 and 3 Tesla: effect of increased field strength on 4 paradigms used during presurgical work-up. Invest Radiol 2007;42:130–8.
13. Matthews PM, Honey GD, Bullmore ET. Applications of fMRI in translational medicine and clinical practice. Nat Rev Neurosci 2006;7:732–44.

14. Tharin S, Golby A. Functional brain mapping and its applications to neurosurgery. Neurosurgery 2007;60:185–201; discussion 2.
15. Majos A, Tybor K, Stefanczyk L, Goraj B. Cortical mapping by functional magnetic resonance imaging in patients with brain tumors. Eur Radiol 2005;15:1148–58.
16. Matthews PM, Jezzard P. Functional magnetic resonance imaging. J Neurol Neurosurg Psychiatry 2004;75:6–12.
17. Roux FE, Boulanouar K, Ibarrola D, Tremoulet M, Chollet F, Berry I. Functional MRI and intraoperative brain mapping to evaluate brain plasticity in patients with brain tumours and hemiparesis. J Neurol Neurosurg Psychiatry 2000;69:453–63.
18. Desmond JE, Sum JM, Wagner AD, et al. Functional MRI measurement of language lateralization in Wada-tested patients. Brain 1995;118 (Pt 6):1411–9.
19. Lehericy S, Cohen L, Bazin B, et al. Functional MR evaluation of temporal and frontal language dominance compared with the Wada test. Neurology 2000;54:1625–33.
20. Stippich C, Rapps N, Dreyhaupt J, et al. Localizing and lateralizing language in patients with brain tumors: feasibility of routine preoperative functional MR imaging in 81 consecutive patients. Radiology 2007;243:828–36.
21. Branco DM, Suarez RO, Whalen S, et al. Functional MRI of memory in the hippocampus: laterality indices may be more meaningful if calculated from whole voxel distributions. Neuroimage 2006;32:592–602.
22. Thesen S, Heid O, Mueller E, Schad LR. Prospective acquisition correction for head motion with image-based tracking for real-time fMRI. Magn Reson Med 2000;44:457–65.
23. Springer JA, Binder JR, Hammeke TA, et al. Language dominance in neurologically normal and epilepsy subjects: a functional MRI study. Brain 1999;122 (Pt 11):2033–46.
24. Grummich P, Nimsky C, Pauli E, Buchfelder M, Ganslandt O. Combining fMRI and MEG increases the reliability of presurgical language localization: a clinical study on the difference between and congruence of both modalities. Neuroimage 2006;32:1793–803.
25. Herholz K, Reulen HJ, von Stockhausen HM, et al. Preoperative activation and intraoperative stimulation of language-related areas in patients with glioma. Neurosurgery 1997;41:1253–60; discussion 60–2.
26. Lazar RM, Marshall RS, Pile-Spellman J, et al. Interhemispheric transfer of language in patients with left frontal cerebral arteriovenous malformation. Neuropsychologia 2000;38:1325–32.
27. Duffau H, Lopes M, Arthuis F, et al. Contribution of intraoperative electrical stimulations in surgery of low grade gliomas: a comparative study between two series without (1985–96) and with (1996–2003) functional mapping in the same institution. J Neurol Neurosurg Psychiatry 2005;76:845–51.
28. Berman JI, Berger MS, Chung SW, Nagarajan SS, Henry RG. Accuracy of diffusion tensor magnetic resonance imaging tractography assessed using intraoperative subcortical stimulation mapping and magnetic source imaging. J Neurosurg 2007;107:488–94.
29. Haglund MM, Berger MS, Shamseldin M, Lettich E, Ojemann GA. Cortical localization of temporal lobe language sites in patients with gliomas. Neurosurgery 1994;34:567–76; discussion 76.
30. Hamzei F, Knab R, Weiller C, Roether J. Intra- und extrakranielle Gefäßstenosen beeinflussen BOLD Antwort. Aktuelle Neurologie 2002;29:231.
31. Holodny AI, Schulder M, Liu WC, Wolko J, Maldjian JA, Kalnin AJ. The effect of brain tumors on BOLD functional MR imaging activation in the adjacent motor cortex: implications for image-guided neurosurgery. AJNR Am J Neuroradiol 2000;21:1415–22.
32. Holodny AI, Schulder M, Liu WC, Maldjian JA, Kalnin AJ. Decreased BOLD functional MR activation of the motor and sensory cortices adjacent to a glioblastoma multiforme: implications for image-guided neurosurgery. AJNR Am J Neuroradiol 1999;20:609–12.
33. Schreiber A, Hubbe U, Ziyeh S, Hennig J. The influence of gliomas and nonglial space-occupying lesions on blood-oxygen-level-dependent contrast enhancement. AJNR Am J Neuroradiol 2000;21:1055–63.

Chapter 25

Pharmacological Applications of fMRI

Paul M. Matthews

Summary

Modern drug development presents new challenges by the unmet medical needs of chronic neurological and psychiatric disease. Imaging provides a potentially powerful tool for more efficiently translating pre-clinical and clinical studies and enhancing confidence in progression through early phase clinical development. Pharmacological MRI (phMRI) refers specifically to the applications of fMRI methods for direct or indirect measures of drug action. phMRI can be coupled to advanced structural methods to relate pharmacological effects and functional anatomy. Current and potential applications of phMRI to target stratification, patient validation, and pharmacodynamic studies are described. While great new opportunities could arise from extension of these methods as surrogate markers of clinical responses, this review highlights the substantial additional work that would need to be done to make this feasible. The review concludes that there is a strong rational for investment in phMRI for early phase clinical development, but that the short- to medium-term impact on late phase clinical development likely will be modest.

Key words: Pharmacological, fMRI, Pharmacodynamic

1. Introduction

The pharmaceutical industry is searching for better ways of developing drugs *(1)*. The traditional model for development – which involves a sharp transition between early biological evaluation of targets and drugs in pre-clinical model systems and later clinical development assessing efficacy and toxicities with respect to conventional clinical outcome measures – is too slow and too expensive. The problems are particularly acute as (based on recent experience) less than 1/10 new molecules entering clinical development reach the market.

Brain imaging provides a powerful set of tools to bridge this transition and potentially make drug development more efficient.

M. Filippi (ed.), *fMRI Techniques and Protocols*, Neuromethods, vol. 41
DOI 10.1007/978-1-60327-919-2_25, © Humana Press, a part of Springer Science+Business Media, LLC 2009

It already is making a significant contribution to drug development. Almost 30% of new molecular entities approved for neuropsychiatric indications by the Food and Drug Administration (FDA) between 1995 and 2004 were developed with contributions from imaging *(2)*. The majority of these were for studies of receptor occupancy or for proof-of-mechanism (e.g., testing for drug–receptor interactions). Most applications involved positron emission tomography (PET). However, there is a growing appreciation for a potential role for fMRI in new pharmaceuticals development *(3)*.

New ways of thinking about clinical development are putting a premium on integration of early biology and clinical development in the context of "experimental medicine". Experimental medicine can be defined an approach that uses (ethical) human experimentation to address mechanistic questions similar to those asked with pre-clinical studies. It involves *biologically*-driven therapeutics development involving hypothesis-led research often performed across levels of biological complexity (e.g. cells to the whole human). Experimental medicine is based on the premise that, although pre-clinical studies can inform about biological mechanisms and their modulation, humans must be studied to understand drug effects unequivocally relevant to disease. With this thinking, the traditional *unidirectional* "critical path" for drug development (**Fig. 1**) is enabled by tools (e.g. from genetics, proteomics, and imaging) that can be used *bidirectionally* (e.g. from pre-clinical to clinical applications and "back again").

This reviewer believes that imaging can best be applied in drug development through a question-based experimental strategy. Major problems of early drug development provide a framework for these:

1. Target validation: does the chosen therapeutic target potentially play a central role in determining the disease or symptom of interest?
2. Patient stratification: how can the most responsive patient population be identified?

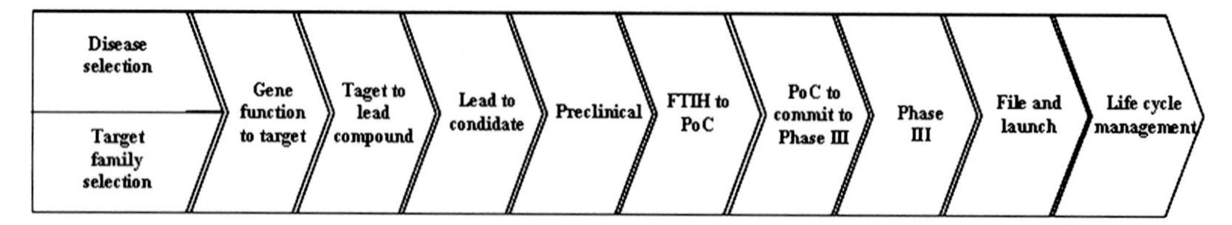

Fig. 1. The "critical path" for drug development. Pharmacological MRI (phMRI) has the potential to enhance the efficiency of early clinical development with better translation of biological concepts from pre-clinical to clinical studies, providing a new pharmacodynamic measure and enhancing potential in proof-of-mechanism studies (*FTIH* first time in human study, *PoC* proof of concept study).

3. Pharmacodynamics: what are the effects of the drug and how long do they last?

Answering these questions, in turn, can be used to enable drugs to be distinguished from each other mechanistically, *differentiating* them in ways that may guide their optimal application *(1)*. Ultimately, as the relationships between imaging measures of drug action and clinical outcomes become established, imaging potentially could provide "surrogate markers" of drug response (i.e. imaging measures that are generally accepted to be predictive of clinical outcomes) to enable later phase trials that are smaller, faster, and cheaper than with conventional clinical endpoints. Although even an optimist would be cautious in suggesting that imaging surrogate markers will be developed either easily or for the majority of applications, the impact of such developments (e.g. as demonstrated already for multiple sclerosis *(4, 5)*) could be huge.

fMRI could provide an especially widely applicable tool for such applications in neurology and psychiatry *(6)*. MRI is a widely disseminated platform. fMRI methodology has matured rapidly over the last decade and a half. Similar experimental types of data can be acquired from pre-clinical models and from human subjects. As a measure of physiological responses, fMRI can define pathological characteristics of disease, validate targets for therapy, and define pharmacodynamic responses. This review will focus on applications of fMRI in the context of applications in pharmacological therapy development (**Table 1**), where it often is referred to as pharmacological MRI (phMRI).

Table 1
Applications of fMRI in drug development

Applications of fMRI for target validation
Endophenotypes for genetic characterisation of disease
In vivo assays for functional polymorphisms
Potential applications of fMRI for patient stratification
Identification of pre-symptomatic subjects, e.g., Alzheimer's disease
Distinguishing specific disease state, e.g., schizophrenia
Enrichment of study populations with treatment responders
Differentiating strong placebo responders
Applications of phMRI for drug response
Relating molecular targets to behaviours, e.g., addiction
Predicting treatment response
Assessing risk of relapse after treatment
Pharmacodynamic markers
Pharmacokinetic markers
Measures of treatment response potentially more sensitive than clinical outcomes, e.g., in analgesia development

2. Principles Common to Both fMRI and phMRI

Both fMRI and phMRI are based on indirect measures of neuronal response by being sensitive to changes in relative blood oxygenation. Increased neuronal activity is associated with a local haemodynamic response involving both increased cerebral blood flow and blood volume. This *neurovascular coupling* appears to be a consequence predominantly of presynaptic neurotransmitter release and thus reflects local signalling *(7)*. The haemodynamic response has a magnitude and a time course that depend on both relative inhibitory and excitatory input *(8)*. Therefore, it can be considered as a measure of local information transfer. Mechanistically, it is regulated by neuronal-glial interactions and mediated by more than one signalling mechanism *(9,10)*. Pharmacological applications need to consider the potential impact of compounds on the coupling mechanisms, as well as on the downstream neuronal effects.

The sensitivity of MRI to blood oxygenation arises because deoxyhaemoglobin is paramagnetic and therefore locally modulates an applied static magnetic field. In the MRI magnet, where a highly *homogeneous* (i.e., spatially invariant) magnetic field is generated, small magnetic field *inhomogeneities* at a very local level are found around blood vessels. Their magnitude increases with the amount of paramagnetic deoxyhaemoglobin. These inhomogeneities reduce the MRI signal acquired with a gradient echo image acquisition sequence (echo planar imaging, EPI). A relationship between neuronal activation and blood oxygenation is observed because blood flow increases with greater neuronal activity, and this increase in blood flow is greater than is needed simply for increased oxygen delivery with greater tissue demands: the local oxygen extraction fraction decreases with synaptic signally. The decrease in blood ratio of deoxy to oxyhaemoglobin is large enough to be associated with an increase in the EPI MRI signal, although this is small (0.5–5% typically at 3 T).

The most commonly used fMRI (or phMRI) imaging method of blood oxygen level-dependent (BOLD) contrast relies on this phenomenon *(11, 12)*. A typical experiment would involve acquisition of a series of brain images during infusion of a drug or over the course of a changing cognitive state (e.g. performing a visually-presented working memory task vs. attending to a simple visual stimulus). Regions of significant signal change with drug infusion or between cognitive states then are defined by statistical analysis of the time series of data. Quantitative measurement of these changes allows measures relevant to drug action on the brain to be defined.

3. Target Validation

The traditional progression of drug development through target validation in pre-clinical models, e.g., by demonstration of a phenotype plausibly related to the human disease, arguably is particularly inappropriate for "complex genetic diseases" that are determined by the interaction of multiple genes and the environment and particularly for those that are uniquely human (which constitute the bulk of neurological and psychiatric disorders). An alternative concept for target validation in humans involves tests for modulation of disease-related systems by allelic variation at candidate target loci *(13)*.

An exciting, emerging opportunity with this approach applies structural MRI and phMRI outcomes as *endophenotypes*-heritable quantitative traits *(14)*. Consider, for example, a complex genetic disease such as schizophrenia, in which both structural and functional differences can be defined relative to the healthy brain. In an early step towards target validation, an endophenotype-based target validation approach should facilitate the distinction of *causative* from simply (possibly incidental or non-specific) *associated* features. Candidate genes *DISC1*, *GRM3*, and *COMT* all have been related to imaging endophenotype for schizophrenia in this way and associated with altered hippocampal structure and function *(15)*, glutamatergic fronto-hippocampal function *(16)*, and prefrontal dopamine responsiveness *(17)*, respectively.

Understanding how genetic risk factors contribute to expression traits that are related to symptoms in a complex genetic disease can be used to validate markers of therapeutic response for phMRI. For example, patients with a history of affective disorders, who are the carriers of the S allele, of the common 5-HTTLPR polymorphism in the serotonin transporter gene (*SLC6A4*) have an exaggerated fMRI response to environmental threat in the amygdala (the endophenotype) relative to L allele homozygotes *(18, 19)*. S allele carriers also have lower amygdala and perigenual cingulate volumes, and correlations between activity in these regions are reduced relative to healthy controls *(20)*, while the functional connectivity between the amygdala and ventromedial prefrontal cortex is increased *(21)*. Elucidation of MRI-based functional–anatomical features could be used in some cases to support further patient stratification *(22)*.

Application of fMRI approaches that define neurobiological bases for general cognitive processes (such as, in the context of psychiatric disease, motivation or reward) facilitate more holistic views of targets that may be relevant to more than one disease. For example, fMRI approaches have contributed to the current appreciation for neural mechanisms common to addictive behaviours across a wide range of substances abuse states. Studies of cue-elicited craving have defined similar activities of the mesolimbic reward circuit in addictions to nicotine *(23)*, alcohol *(24)*, gambling *(25)*, amphetamine *(26)*, cocaine *(27)*, and opiates *(28)*. Combination of fMRI with PET receptor mapping can be used to relate systems-level dysfunction directly with the molecular targets of drug therapies in ways that enhance target validation for new pharmacological treatments faster and more cheaply than conventional clinical designs allow *(29)*. These studies very directly suggest phMRI approaches (e.g. with interventions targeting the dopaminergic system) that could have an impact on drug development across all of these types of addictions. Similar examples can be defined for other disease areas.

4. Patient Stratification

A critical issue in early drug development is to establish an appropriate level of confidence in the potential of a new molecule to become a therapy. One way in this can be done is by better controlling for the substantial variations in therapeutic responses between individuals in early phase studies. As well demonstrated in oncology *(30)*, stratification of patients based on specific disease characteristics can enable more powerful trial designs. Consider, hypothetically, the difference in outcome of trials first for a population in which a new molecule has a 50% treatment effect in 20% of patients (10% net treatment effect) and then in a stratified population enriched so that 70% are responders (net 35% treatment effect). By predicting potential responders, imaging also can suggest ways of best selecting optimal patient groups for applications of new molecules.

Establishing fMRI measures for stratification of patients *(31, 32)* also ultimately can aid in clinical diagnosis and management. Predicting clinical course is a major concern on first presentation with psychosis, for example. Structural MRI may be able to reliably distinguish patients with schizophrenia from those who present with other forms of acute psychosis, allowing more confident selection of patients in an early intervention schizophrenia treatment trial *(33)*. More specifically, structural MRI measures can be related to specific, symptom-related functional systems (e.g. the auditory cortex for auditory hallucinations in schizophrenia) (**Fig. 2**). Establishing plausible relations between

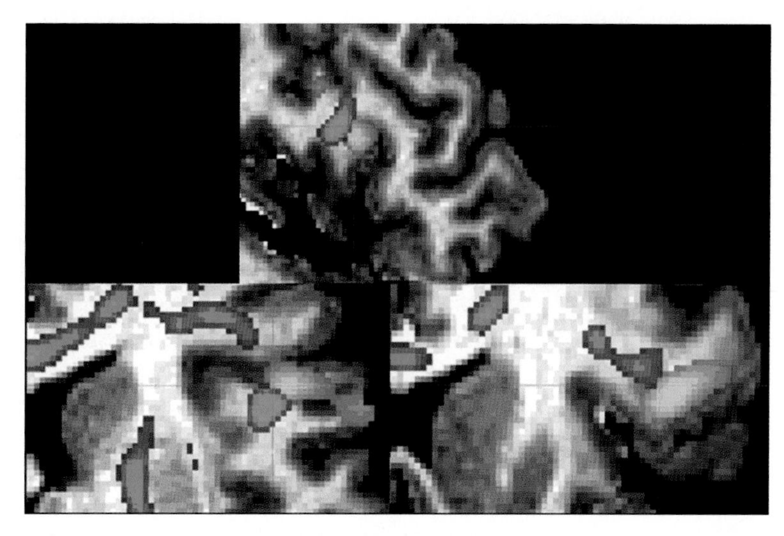

Fig. 2. Conventional structural MRI and new structural imaging methods such as diffusion tensor imaging can provide specific functional anatomical hypotheses for testing using phMRI approaches. Here are shown specific examples from a contrast between adolescent schizophrenia patients and healthy controls ***(33)*** showing decreased grey matter volume (*orange-yellow*) and decreased white matter fractional anisotropy (FA) (*green*) in regions of the left hemisphere associated with perception of language. The three images define decreases of FA in the left arcuate fasculus (*green*) and grey matter loss in Broca's area (*red*). Results from group analysis were overlaid on the brain image of a single control subject for this representation (Images courtesy of Dr. G. Daoud, Oxford).

neuropathological changes measured by MRI and functional anatomy helps to validate fMRI measures proposed to distinguish disease trait specifically *(34–37)*.

Stratification of patients using neurophysiological measures based on fMRI also could be important for identification of individuals at high risk of disease who have not yet manifested symptoms clinically. For example, like the patients themselves, the relatives of patients with schizophrenia can show abnormal prefrontal fMRI activation *(38–40)* or reduced functional connectivity in fronto-thalamo-cerebellar and fronto-parietal networks *(41)*. fMRI changes also differentiated individuals with high genetic risk of Alzheimer's disease who progressed to clinical disease expression over the observation period from those who did not *(42)*. In both instances, the pre or minimally symptomatic subjects can be easier to recruit or to study in clinical trials and also extend the range of behavioural-neurophysiological correlations that can be tested in phMRI-based proof-of-mechanism studies.

Where alternative treatment approaches are available that have potentially significant individual variation in response across a population, selection of the optimal treatment for an individual patient could be assisted by phMRI. At present, such *personalized medicine* applied in any general way remains an aspiration only. However, if it could be made practical, it would have great impact. In depression, for example, responses are highly variable, e.g., only about 70% of patients respond well to a given antidepressant *(43)*. Higher BOLD signal in the amygdala at baseline, therefore, may be predictive of treatment response *(44)*. However, considerable experience is needed to relate phMRI-derived measures directly to clinical outcomes. Aggregate neuronal activity in a single brain region (e.g. the amygdala)-at the level of resolution currently available using phMRI (mm^3) - is unlikely to be unique. Other fMRI signals that change with treatment in depression also are candidate phMRI markers, e.g., signal change in the ventromedial prefrontal and anterior cingulate cortices *(45)* or as a modulation of cortico-limbic functional connectivity *(46)* suggesting the utility of multivariate methods.

5. Pharmacodynamics

Applications of fMRI to the direct assessment of drug action are expanding *(47)*. *Pharmacodynamic* data (e.g. testing whether a drug at the chosen dose has an effect on brain function) can be obtained from analysis of brain imaging changes induced simply by the administration of a drug (nicotine) *(48)*. Additional information can come from correlation of brain activity with any measurable (e.g. behavioural) effects of drug administration (metamphetamine) *(49)* (**Fig. 3**) or with characterisation of the way in which activity of a probe-task is modulated by a drug *(50–53)*. This information can inform clinical dose-ranging studies. As noted earlier, correlations between fMRI measures of brain functional system response and drug receptor or receptor occupancy measurements by PET are possible *(29, 54)*. A number of examples are available already in which phMRI has been able to define effects of treatment in populations too small to allow behavioural effects to be discerned or where usual clinical measures are simply insensitive to drug effects *(55–58)*.

A potential risk of phMRI-derived pharmacodynamic markers is that they may not be specific for (or predictive of) clinically-relevant parameters. One way of minimising this risk is to frame the specific phMRI measures in terms of important disease symptoms based on the relationship between fMRI measures and individual symptoms.

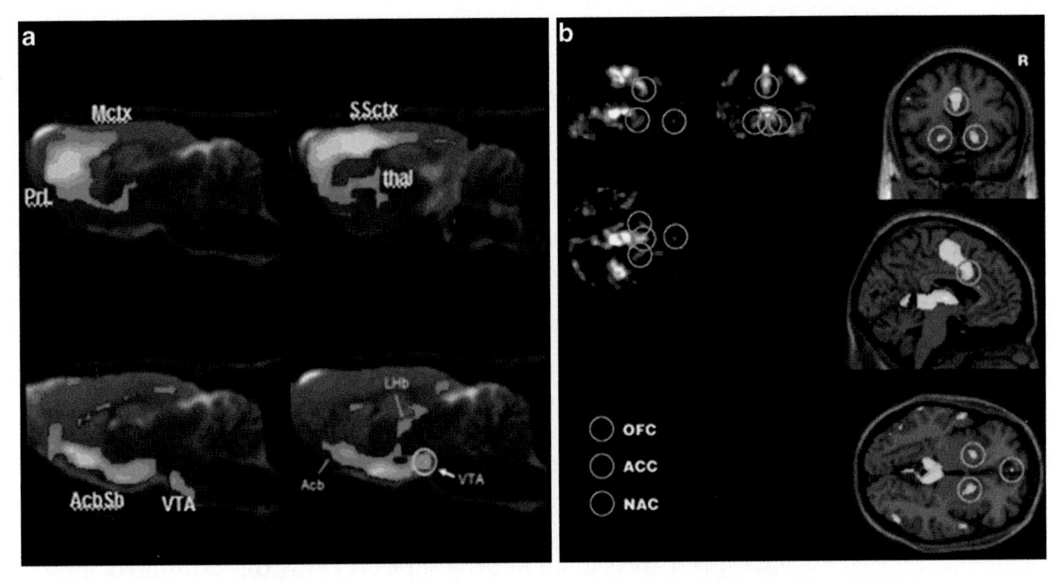

Fig. 3. phMRI can be performed in both animals and humans to assess correspondences in tests of mechanisms. (**a**) phMRI results with metamphetamine challenge of a rodent, identifying major regions in the monamine network as sites of direct or indirect action (*Mctx* motor cortex, *PrL* pre-limbic medial prefrontal cortex, *thal* thalamus, *SSctx* somatosensory cortex, *AcbSh* shell of the nucleus accumbens, *VTA* ventral tegmental area) (Images courtesy of Dr. A. Bifone, GSK, Verona). (**b**) A similar phMRI experiment with acute amphetamine infusion in human subjects performed using "mind racing" as a behavioural index of drug effects identified comparable elements of the core response network (*OFC* orbitofrontal cortex, *ACC* anterior cingulate cortex, *NAC* nucleus accumbens).

Two broad experimental approaches have been adopted for symptom-related fMRI studies, involving measurement of brain activity (1) while the symptom of interest is experienced, or (2) during performance of tasks that engage cognitive processes putatively related to the symptom. Applications to understanding mechanisms contributing to pain and psychosis, respectively, provide good examples of these complementary approaches.

FMRI has allowed dissection of the *subjective experience* of pain into anatomically distinct activities of different functional systems (including arousal and the somatosensory and limbic systems) *(59)*. The clinical importance of such a dissection is that it rationally defines distinct targets for therapeutic modulation. FMRI also has suggested that common physiological mechanisms are shared between directly experienced or imagined pain and between exteroreceptive and affective pain *(60, 61)*. These types of studies, which have identified objective, fMRI-based measures for neurophysiological mechanisms of pain, provide a rational basis to explore applications of phMRI measures to build confidence in early phase therapeutic trials *(59, 62)*. Inter-individual differences in pain responses are reflected in differences in brain responses in the primary somatosensory, anterior cingulate and prefrontal cortices, for example *(63)*. Building on these, phMRI studies have demonstrated *reductions* of responses in these regions with analgesic interventions following noxious stimuli *(62, 64, 65)*. At this point, only preliminary data from one phMRI study of pain in a patient population has been reported to our knowledge *(66)*, but applications of fMRI to exploratory therapeutic and Phase II trials of new analgesics are certain to expand. Additional potential may lie in better definition of the general anticipatory system responses distinguishing placebo *(67–69)* from active treatment responses and in developing understanding of how the brain encodes differences in qualities of pain *(70, 71)*.

A different approach can be taken with symptoms related to perceptions that are qualitatively abnormal, such as auditory hallucinations. Increases in primary auditory *(35–37)* and potentially language-related *(37)* cortical activity are evident from fMRI studies during auditory hallucinations in schizophrenic patients. By contrasting these responses in schizophrenic patients with responses to "inner speech" in healthy subjects (thoughts attributed appropriately to an internal source), specific hypotheses concerning mechanisms of brain dysfunction in schizophrenia responsible for psychotic features can be made *(72)*. Differences in activity of the auditory cortex and association areas between healthy subjects and patients could provide new, short-term candidate phMRI measures of response to anti-psychotic medication.

As highlighted in the introduction to this review, imaging has the potential to bridge directly between pre-clinical and clinical studies. While many behaviours cannot be translated across species, functional-anatomical correlations allow direct drug responses elicited in the brain for translation of underlying neurobiology. For example, phMRI experiments in which unstimulated brain responses to acute compound challenges can be used to define brain regions where activity is modulated by the same compound in animals (**Fig. 3**). Pre-clinically, these observations can be linked to results from more invasive studies, e.g., direct measurements of neurotransmitter release that distinguish direct and indirect effects of the compound *(73)*. Similar observations of drug modulation of brain acitivity can be made in human volunteers, providing a way of confirming mechanism (**Fig. 3**) *(49)*. State-dependent modulation of these regions can further contribute to this *(74)*. By relating plasma concentrations to brain responses, similar approaches could be used to define dose, for example.

Application of more sophisticated analytical methods to phMRI can define not just the way in which individual regions of the brain respond to a drug, but also ways in which interactions between brain regions are changed by a drug. fMRI images show signal fluctuations at rest that occur at low frequencies (0.01–0.05 Hz) with coherent changes between widely-separated brain regions *(75, 76)*. These consistent spatio-temporal coherence patterns define common "default" activity in "networks" that function even in the resting state *(77, 78)*.

These appear to be a direct consequence of slow coherences in faster neuronal activity *(79–81)*. Changes in coherence can be used to distinguish disease and healthy control states *(82)*. More recent information has defined the potential for these and other *(83)* coherence measures to define drug effects.

Information in addition to functional-anatomical modulations can be acquired in this way. The kinetics of the phMRI signal change reflects the convolution of neuronal responses with the much slower haemodynamic response *(84)*. Although this complicates their interpretation, the relatively fast time-course of response (occurring primarily over several seconds) and fundamental dependence of neuronal activation allows them to provide information relevant to ***pharmacokinetics*** (PK) (i.e. data describing the time course of drug action) *(64)* and pharmacodynamics. The approach can be distinguished from more conventional PK methods by providing information on the dynamics of the target organ response.

6. Current Limitations and Some Future Extensions of phMRI

Although there is real promise for phMRI, major challenges to the meaningful quantitative interpretations of phMRI measures remain. First, the relationship of blood flow changes with altered presynaptic activity depends on the physiological (and, potentially, pharmacologic) context *(85–89)*. Even the relative direction relative activation in disease states may be difficult to interpret precisely. For example, reduced activation may reflect brain functional impairment *(90)* or improved efficiency *(91)*. Experimental designs need to accommodate this, e.g., by studying dose-response relations and behavioural correlates in individual studies.

Future methodological developments will make fMRI more informative. Computational advances already allow robust analyses in real time *(92, 93)*. In the context of phMRI, these could enable full quality control during an examination or more precise tailoring of the protocol to the question being asked about an individual patient. Some limitations to interpretation of the BOLD response can be addressed with use of complementary forms of MRI contrast or through the integration of BOLD MRI and other measures in simultaneous data acquisitions. For example, direct measures of brain blood flow can be made using non-invasive "arterial spin labelling" MRI methods, which have greater stability over time for better assessment of slow (on the order of a minutes or more) changes in brain responses *(94)*. With care for safety issues and correction of the artifacts induced by the shifting magnetic field gradients used for MRI, high quality electroencephalography (EEG) now can be obtained simultaneously during an fMRI examination *(95)*, allowing simultaneous pharmaco-EEG studies. In the near future, advances in positron detection methods should support commercial availability of combined human PET/MRI scanners *(54)*.

7. Conclusions

The first step towards translation of fMRI from basic cognitive neuroscience to drug development has been taken. phMRI shows promise for *specific* clinical applications. In these areas, the impact could be significant. The promise is enhanced by the potential to integrate phMRI with structural MRI measures (e.g. for target validation or patient stratification) and ancillary functional measures such as EEG (e.g. for pharmacodynamics and related kinetic studies). However, to this reviewer, it seems unlikely that practical or scientific considerations will allow phMRI to be used routinely in any general way. The challenge is to define those applications for which:

1. Alternative, less expensive or technically demanding approaches other than phMRI are not available.
2. The phMRI experimental design allows shorter, faster or more informative that possible with a clinical endpoint.

The primary immediate applications likely will remain in Phase I or Phase II trials (or in Phase III sub-studies designed to increase understanding of mechanism, stratification or other specific factors) where experimental medicine employing phMRI could inform decision-making in early development, rather than independently contributing evidence for efficacy in the context of a pivotal trial.

The hurdles that would need to be overcome before phMRI could be used as a surrogate marker for later phase development are immense. It may be anticipated that any phMRI measure would need to be evaluated for more than one class of drug and is related clearly to the relevant clinical outcomes if it were to be used as direct support in a regulatory submission for a new molecular entity. Specific issues that need to be considered are how analysis methods can be validated for use in conjunction with phMRI paradigms and, ultimately, how fully regulatory compliant IT environments can be created in clinical centres for such imaging studies.

Despite these challenges posed by the late phase, the short-term promise of phMRI for early phase experimental medicine should compel the therapeutics development community to evaluate potential applications with vigour. Developments in basic and cognitive neurosciences have laid a solid foundation for a broad range of paradigms *(6)*. There is an immediate opportunity to adapt them for pharmacologic applications, set clear criteria to define the added value of phMRI and find those applications for which value can be realised.

Acknowledgments

The author is a full-time employee of GlaxoSmithKline, but acknowledges with gratitude support from the Medical Research Council (UK), the Stroke Association and the MS Society of Great Britain and Northern Ireland for personally-directed research activities referred to in this review. He is grateful to Dr. A. Bifone, GSK, Verona, for helpful discussions and for sharing early data.

References

1. Trusheim MR, Berndt ER, Douglas FL. Stratified medicine: strategic and economic implications of combining drugs and clinical biomarkers. Nat Rev Drug Discov 2007; 6(4):287–93.
2. Uppoor RS, Mummaneni P, Cooper E, et al. The use of imaging in the early development of neuropharmacological drugs: a survey of approved NDAs. Clin Pharmacol Ther 2008;84(1):69–74.
3. Pangalos MN, Schechter LE, Hurko O. Drug development for CNS disorders: strategies for balancing risk and reducing attrition. Nat Rev Drug Discov 2007;6(7):521–32.
4. De Stefano N, Filippi M, Miller D, et al. Guidelines for using proton MR spectroscopy in multicenter clinical MS studies. Neurology 2007;69(20):1942–52.
5. Rocca MA, Filippi M. Functional MRI in multiple sclerosis. J Neuroimaging 2007;17 Suppl 1:36S–41S.
6. Matthews PM, Honey GD, Bullmore ET. Applications of fMRI in translational medicine and clinical practice. Nat Rev Neurosci 2006;7(9):732–44.
7. Logothetis NK. The underpinnings of the BOLD functional magnetic resonance imaging signal. J Neurosci 2003;23(10):3963–71.
8. Caesar K, Thomsen K, Lauritzen M. Dissociation of spikes, synaptic activity, and activity-dependent increments in rat cerebellar blood flow by tonic synaptic inhibition. Proc Natl Acad Sci USA 2003;100(26):16000–5.
9. Buerk DG, Ances BM, Greenberg JH, Detre JA. Temporal dynamics of brain tissue nitric oxide during functional forepaw stimulation in rats. Neuroimage 2003;18(1):1–9.
10. St Lawrence KS, Ye FQ, Lewis BK, Frank JA, McLaughlin AC. Measuring the effects of indomethacin on changes in cerebral oxidative metabolism and cerebral blood flow during sensorimotor activation. Magn Reson Med 2003;50(1):99–106.
11. Kwong KK, Belliveau JW, Chesler DA, et al. Dynamic magnetic resonance imaging of human brain activity during primary sensory stimulation. Proc Natl Acad Sci USA 1992;89(12):5675–9.
12. Ogawa S, Lee TM. Magnetic resonance imaging of blood vessels at high fields: in vivo and

in vitro measurements and image simulation. Magn Reson Med 1990;16(1):9–18.
13. Meyer-Lindenberg A, Weinberger DR. Intermediate phenotypes and genetic mechanisms of psychiatric disorders. Nat Rev Neurosci 2006;7(10):818–27.
14. Gottesman, II, Gould TD. The endophenotype concept in psychiatry: etymology and strategic intentions. Am J Psychiatry 2003;160(4):636–45.
15. Callicott JH, Straub RE, Pezawas L, et al. Variation in DISC1 affects hippocampal structure and function and increases risk for schizophrenia. Proc Natl Acad Sci USA 2005;102(24):8627–32.
16. Egan MF, Straub RE, Goldberg TE, et al. Variation in GRM3 affects cognition, prefrontal glutamate, and risk for schizophrenia. Proc Natl Acad Sci USA 2004; 101(34):12604–9.
17. Egan MF, Goldberg TE, Kolachana BS, et al. Effect of COMT Val108/158 Met genotype on frontal lobe function and risk for schizophrenia. Proc Natl Acad Sci USA 2001;98(12):6917–22.
18. Hariri AR, Drabant EM, Munoz KE, et al. A susceptibility gene for affective disorders and the response of the human amygdala. Arch Gen Psychiatry 2005;62(2):146–52.
19. Hariri AR, Mattay VS, Tessitore A, et al. Serotonin transporter genetic variation and the response of the human amygdala. Science 2002;297(5580):400–3.
20. Pezawas L, Meyer-Lindenberg A, Drabant EM, et al. 5-HTTLPR polymorphism impacts human cingulate-amygdala interactions: a genetic susceptibility mechanism for depression. Nat Neurosci 2005;8(6):828–34.
21. Heinz A, Braus DF, Smolka MN, et al. Amygdala-prefrontal coupling depends on a genetic variation of the serotonin transporter. Nat Neurosci 2005;8(1):20–1.
22. Smith KA, Ploghaus A, Cowen PJ, et al. Cerebellar responses during anticipation of noxious stimuli in subjects recovered from depression. Functional magnetic resonance imaging study. Br J Psychiatry 2002;181:411–5.
23. David SP, Munafo MR, Johansen-Berg H, et al. Ventral striatum/nucleus accumbens activation to smoking-related pictorial cues in smokers and nonsmokers: a functional magnetic resonance imaging study. Biol Psychiatry 2005;58(6):488–94.
24. Myrick H, Anton RF, Li X, et al. Differential brain activity in alcoholics and social drinkers to alcohol cues: relationship to craving. Neuropsychopharmacology 2004;29(2):393–402.
25. Reuter J, Raedler T, Rose M, Hand I, Glascher J, Buchel C. Pathological gambling is linked to reduced activation of the mesolimbic reward system. Nat Neurosci 2005;8(2):147–8.
26. Paulus MP, Tapert SF, Schuckit MA. Neural activation patterns of methamphetamine-dependent subjects during decision making predict relapse. Arch Gen Psychiatry 2005;62(7):761–8.
27. Kaufman JN, Ross TJ, Stein EA, Garavan H. Cingulate hypoactivity in cocaine users during a GO-NOGO task as revealed by event-related functional magnetic resonance imaging. J Neurosci 2003;23(21):7839–43.
28. Forman SD, Dougherty GG, Casey BJ, et al. Opiate addicts lack error-dependent activation of rostral anterior cingulate. Biol Psychiatry 2004;55(5):531–7.
29. Heinz A, Siessmeier T, Wrase J, et al. Correlation between dopamine D(2) receptors in the ventral striatum and central processing of alcohol cues and craving. Am J Psychiatry 2004;161(10):1783–9.
30. Engel RH, Kaklamani VG. HER2-positive breast cancer: current and future treatment strategies. Drugs 2007;67(9):1329–41.
31. Honey GD, Pomarol-Clotet E, Corlett PR, et al. Functional dysconnectivity in schizophrenia associated with attentional modulation of motor function. Brain 2005;128(Pt 11): 2597–611.
32. Honey GD, Sharma T, Suckling J, et al. The functional neuroanatomy of schizophrenic subsyndromes. Psychol Med 2003;33(6): 1007–18.
33. Douaud G, Smith S, Jenkinson M, et al. Anatomically related grey and white matter abnormalities in adolescent-onset schizophrenia. Brain 2007;130(Pt 9):2375–86.
34. Allen P, Amaro E, Fu CH, et al. Neural correlates of the misattribution of speech in schizophrenia. Br J Psychiatry 2007;190:162–9.
35. Dierks T, Linden DE, Jandl M, et al. Activation of Heschl's gyrus during auditory hallucinations. Neuron 1999;22(3):615–21.
36. Shergill SS, Brammer MJ, Amaro E, Williams SC, Murray RM, McGuire PK. Temporal course of auditory hallucinations. Br J Psychiatry 2004;185:516–7.
37. Shergill SS, Brammer MJ, Williams SC, Murray RM, McGuire PK. Mapping auditory hallucinations in schizophrenia using functional magnetic resonance imaging. Arch Gen Psychiatry 2000;57(11):1033–8.
38. Callicott JH, Egan MF, Mattay VS, et al. Abnormal fMRI response of the dorsolateral prefrontal cortex in cognitively intact siblings

of patients with schizophrenia. Am J Psychiatry 2003;160(4):709–19.

39. Whalley HC, Simonotto E, Flett S, et al. fMRI correlates of state and trait effects in subjects at genetically enhanced risk of schizophrenia. Brain 2004;127(Pt 3):478–90.
40. Morey RA, Inan S, Mitchell TV, Perkins DO, Lieberman JA, Belger A. Imaging frontostriatal function in ultra-high-risk, early, and chronic schizophrenia during executive processing. Arch Gen Psychiatry 2005;62(3):254–62.
41. Whalley HC, Simonotto E, Marshall I, et al. Functional disconnectivity in subjects at high genetic risk of schizophrenia. Brain 2005;128(Pt 9):2097–108.
42. Bookheimer SY, Strojwas MH, Cohen MS, et al. Patterns of brain activation in people at risk for Alzheimer's disease. N Engl J Med 2000;343(7):450–6.
43. Baghai TC, Moller HJ, Rupprecht R. Recent progress in pharmacological and non-pharmacological treatment options of major depression. Curr Pharm Des 2006;12(4):503–15.
44. Canli T, Cooney RE, Goldin P, et al. Amygdala reactivity to emotional faces predicts improvement in major depression. Neuroreport 2005;16(12):1267–70.
45. Killgore WD, Yurgelun-Todd DA. Ventromedial prefrontal activity correlates with depressed mood in adolescent children. Neuroreport 2006;17(2):167–71.
46. Anand A, Li Y, Wang Y, et al. Antidepressant effect on connectivity of the mood-regulating circuit: an FMRI study. Neuropsychopharmacology 2005;30(7):1334–44.
47. Honey G, Bullmore E. Human pharmacological MRI. Trends Pharmacol Sci 2004;25(7):366–74.
48. Stein EA, Pankiewicz J, Harsch HH, et al. Nicotine-induced limbic cortical activation in the human brain: a functional MRI study. Am J Psychiatry 1998;155(8):1009–15.
49. Vollm BA, de Araujo IE, Cowen PJ, et al. Methamphetamine activates reward circuitry in drug naive human subjects. Neuropsychopharmacology 2004;29(9):1715–22.
50. Gerdelat-Mas A, Loubinoux I, Tombari D, Rascol O, Chollet F, Simonetta-Moreau M. Chronic administration of selective serotonin reuptake inhibitor (SSRI) paroxetine modulates human motor cortex excitability in healthy subjects. Neuroimage 2005;27(2):314–22.
51. Pariente J, Loubinoux I, Carel C, et al. Fluoxetine modulates motor performance and cerebral activation of patients recovering from stroke. Ann Neurol 2001;50(6):718–29.
52. Goekoop R, Duschek EJ, Knol DL, et al. Raloxifene exposure enhances brain activation during memory performance in healthy elderly males; its possible relevance to behavior. Neuroimage 2005;25(1):63–75.
53. Goekoop R, Rombouts SA, Jonker C, et al. Challenging the cholinergic system in mild cognitive impairment: a pharmacological fMRI study. Neuroimage 2004;23(4):1450–9.
54. Farahani K, Slates R, Shao Y, Silverman R, Cherry S. Contemporaneous positron emission tomography and MR imaging at 1.5 T. J Magn Reson Imaging 1999;9(3):497–500.
55. Wilkinson D, Halligan P. The relevance of behavioural measures for functional-imaging studies of cognition. Nat Rev Neurosci 2004;5(1):67–73.
56. Parry AM, Scott RB, Palace J, Smith S, Matthews PM. Potentially adaptive functional changes in cognitive processing for patients with multiple sclerosis and their acute modulation by rivastigmine. Brain 2003;126:2750–60.
57. Abu-Omar Y, Cader S, Guerrieri Wolf L, Pigott D, Matthews PM, Taggart DP. Short-term changes in cerebral activity in on-pump and off-pump cardiac surgery defined by functional magnetic resonance imaging and their relationship to microembolization. J Thorac Cardiovasc Surg 2006;132(5):1119–25.
58. Matthews PM, Johansen-Berg H, Reddy H. Non-invasive mapping of brain functions and brain recovery: applying lessons from cognitive neuroscience to neurorehabilitation. Restor Neurol Neurosci 2004;22(3–5):245–60.
59. Tracey I. Nociceptive processing in the human brain. Curr Opin Neurobiol 2005;15(4):478–87.
60. Gundel H, O'Connor MF, Littrell L, Fort C, Lane RD. Functional neuroanatomy of grief: an FMRI study. Am J Psychiatry 2003;160(11):1946–53.
61. Saarela MV, Hlushchuk Y, Williams AC, Schurmann M, Kalso E, Hari R. The compassionate brain: humans detect intensity of pain from another's face. Cereb Cortex 2007;17(1):230–7.
62. Borsook D, Ploghaus A, Becerra L. Utilizing brain imaging for analgesic drug development. Curr Opin Investig Drugs 2002;3(9):1342–7.
63. Coghill RC, McHaffie JG, Yen YF. Neural correlates of interindividual differences in the subjective experience of pain. Proc Natl Acad Sci USA 2003;100(14):8538–42.
64. Wise RG, Rogers R, Painter D, et al. Combining fMRI with a pharmacokinetic model

to determine which brain areas activated by painful stimulation are specifically modulated by remifentanil. Neuroimage 2002;16(4): 999–1014.
65. Rogers R, Wise RG, Painter DJ, Longe SE, Tracey I. An investigation to dissociate the analgesic and anesthetic properties of ketamine using functional magnetic resonance imaging. Anesthesiology 2004;100(2):292–301.
66. Koeppe C, Schneider C, Thieme K, et al. The influence of the 5-HT3 receptor antagonist tropisetron on pain in fibromyalgia: a functional magnetic resonance imaging pilot study. Scand J Rheumatol Suppl 2004(119):24–7.
67. Lieberman MD, Jarcho JM, Berman S, et al. The neural correlates of placebo effects: a disruption account. Neuroimage 2004; 22(1):447–55.
68. Petrovic P, Dietrich T, Fransson P, Andersson J, Carlsson K, Ingvar M. Placebo in emotional processing--induced expectations of anxiety relief activate a generalized modulatory network. Neuron 2005;46(6):957–69.
69. Wager TD, Rilling JK, Smith EE, et al. Placebo-induced changes in FMRI in the anticipation and experience of pain. Science 2004;303(5661):1162–7.
70. Henderson LA, Bandler R, Gandevia SC, Macefield VG. Distinct forebrain activity patterns during deep versus superficial pain. Pain 2006;120(3):286–96.
71. Singer T, Seymour B, O'Doherty J, Kaube H, Dolan RJ, Frith CD. Empathy for pain involves the affective but not sensory components of pain. Science 2004;303(5661): 1157–62.
72. Hunter MD, Griffiths TD, Farrow TF, et al. A neural basis for the perception of voices in external auditory space. Brain 2003;126(Pt 1): 161–9.
73. Schwarz AJ, Gozzi A, Reese T, Bifone A. In vivo mapping of functional connectivity in neurotransmitter systems using pharmacological MRI. Neuroimage 2007;34(4): 1627–36.
74. Batterham RL, ffytche DH, Rosenthal JM, et al. PYY modulation of cortical and hypothalamic brain areas predicts feeding behaviour in humans. Nature 2007;450(7166):106–9.
75. Schwindack C, Siminotto E, Meyer M, et al. Real-time functional magnetic resonance imaging (rt-fMRI) in patients with brain tumours: preliminary findings using motor and language paradigms. Br J Neurosurg 2005; 19(1):25–32.
76. Hoge RD, Pike GB. Oxidative metabolism and the detection of neuronal activation via imaging. J Chem Neuroanat 2001;22(1–2): 43–52.
77. Gasser T, Ganslandt O, Sandalcioglu E, Stolke D, Fahlbusch R, Nimsky C. Intraoperative functional MRI: implementation and preliminary experience. Neuroimage 2005;26(3):685–93.
78. Hoge RD, Atkinson J, Gill B, Crelier GR, Marrett S, Pike GB. Linear coupling between cerebral blood flow and oxygen consumption in activated human cortex. Proc Natl Acad Sci USA 1999;96(16):9403–8.
79. Biswal B, Yetkin FZ, Haughton VM, Hyde JS. Functional connectivity in the motor cortex of resting human brain using echo-planar MRI. Magn Reson Med 1995;34(4):537–41.
80. Lowe MJ, Dzemidzic M, Lurito JT, Mathews VP, Phillips MD. Correlations in low-frequency BOLD fluctuations reflect cortico-cortical connections. Neuroimage 2000;12(5):582–7.
81. Raichle ME, MacLeod AM, Snyder AZ, Powers WJ, Gusnard DA, Shulman GL. A default mode of brain function. Proc Natl Acad Sci USA 2001;98(2):676–82.
82. Rombouts SA, Damoiseaux JS, Goekoop R, et al. Model-free group analysis shows altered BOLD FMRI networks in dementia. Hum Brain Mapp 2009;30(1):256–66.
83. Cader S. Cholmergic agonism alters cognitive processing and enhances functional connectivity in patients with multiple sclerosis. J. Psychopharmacol 2008.
84. Buxton RB, Uludag K, Dubowitz DJ, Liu TT. Modeling the hemodynamic response to brain activation. Neuroimage 2004;23 Suppl 1:S220–S233.
85. Smith SM, Beckmann CF, Ramnani N, et al. Variability in fMRI: a re-examination of inter-session differences. Hum Brain Mapp 2005;24(3):248–57.
86. Pineiro R, Pendlebury S, Johansen-Berg H, Matthews PM. Altered hemodynamic responses in patients after subcortical stroke measured by functional MRI. Stroke 2002; 33(1):103–9.
87. Laurienti PJ, Field AS, Burdette JH, Maldjian JA, Yen YF, Moody DM. Relationship between caffeine-induced changes in resting cerebral perfusion and blood oxygenation level-dependent signal. AJNR Am J Neuroradiol 2003;24(8):1607–11.
88. Lawrence NS, Ross TJ, Stein EA. Cognitive mechanisms of nicotine on visual attention. Neuron 2002;36(3):539–48.
89. Austin VC, Blamire AM, Grieve SM, et al. Differences in the BOLD fMRI response to

direct and indirect cortical stimulation in the rat. Magn Reson Med 2003;49(5):838–47.

90. Rombouts SA, van Swieten JC, Pijnenburg YA, Goekoop R, Barkhof F, Scheltens P. Loss of frontal fMRI activation in early frontotemporal dementia compared to early AD. Neurology 2003;60(12):1904–8.

91. Floyer-Lea A, Matthews PM. Changing brain networks for visuomotor control with increased movement automaticity. J Neurophysiol 2004;92(4):2405–12.

92. Leopold DA, Logothetis NK. Spatial patterns of spontaneous local field activity in the monkey visual cortex. Rev Neurosci 2003;14(1–2): 195–205.

93. Montague PR, Berns GS, Cohen JD, et al. Hyperscanning: simultaneous fMRI during linked social interactions. Neuroimage 2002; 16(4):1159–64.

94. Tjandra T, Brooks JC, Figueiredo P, Wise R, Matthews PM, Tracey I. Quantitative assessment of the reproducibility of functional activation measured with BOLD and MR perfusion imaging: implications for clinical trial design. Neuroimage 2005;27(2):393–401.

95. Lemieux L. Electroencephalography-correlated functional MR imaging studies of epileptic activity. Neuroimaging Clin N Am 2004; 14(3):487–506.

Chapter 26

Application of fMRI to Monitor Motor Rehabilitation

Steven C. Cramer

Summary

Motor deficits contribute to disability in a number of neurological conditions. A wide range of emerging restorative therapies has the potential to reduce this by favorably modifying function. In many medical contexts, a study of target organ function improves efficacy of a therapeutic intervention. However, the optimal methods to prescribe a restorative therapy in the setting of central nervous system (CNS) disease are not clear. Brain mapping studies have the potential to provide useful insights in this regard. Examples of restorative therapies are provided, and human trials are summarized whereby brain mapping data have proven useful in promoting motor improvements in subjects with a neurological condition. In some cases, brain mapping findings that correlate with better outcome with spontaneous behavioral recovery correspond to findings that predict better treatment response in the context of a clinical trial. Similarities across CNS conditions, such as stroke and multiple sclerosis, are discussed. Further studies are needed to understand which methods have the greatest value to monitor, predict, triage, and dose restorative therapies in trials that aim to reduce motor, and other neurological, deficits.

Key words: Functional MRI, Brain mapping, Stroke, Motor system, Recovery, Repair, Plasticity, Treatment, Multiple sclerosis, Spinal cord injury

1. Motor Deficits and Restorative Therapies

Motor deficits are a major contributor to disability in the setting of a number of neurological diseases marked by focal central nervous system (CNS) injury, such as stroke, multiple sclerosis (MS), spinal cord injury (SCI), and traumatic brain injury *(1)*. In general, motor deficits show some degree of spontaneous improvement in the weeks following the insult. Spontaneous recovery is generally incomplete, however.

A number of therapies are in development to promote recovery in patients with motor deficits after a CNS insult. Some

M. Filippi (ed.), *fMRI Techniques and Protocols*, Neuromethods, vol. 41
DOI 10.1007/978-1-60327-919-2_26, © Humana Press, a part of Springer Science+Business Media, LLC 2009

target the acute phase of injury, when the brain is galvanized and produces growth-related substances at levels reminiscent of development. Other therapies target patients in the chronic phase. Regardless, the goal of such therapies is not to salvage injured tissue, rather to promote repair and restore function.

Many forms of restorative therapy are under study. Examples include small molecules *(2–5)*, immune approaches such as via neutralization of the axon growth inhibitor Nogo-A with monoclonal antibodies *(6)*, growth factors *(7–15)*, cell-based methods *(16–19)*, electromagnetic stimulation *(20–24)*, neuroprosthetics *(25, 26)*, and methods based on various forms of therapy and practice *(27–37)*.

A key thesis of this chapter is that optimal prescription of such restorative therapies will be achieved by probing the state of the brain. Clinical trials often enroll patients based on demographic or behavioral measures. However, these are only an approximation of the type of brain state information that is important to promoting repair and recovery.

There are examples in other medical specialities. A measure of target organ function is obtained in addition to behavior and demographic data in order to maximize therapeutic gains. For example, hypothyroidism is optimally treated not by serial behavioral examination, rather by serial measures of pituitary–thyroid axis via serum TSH. Treatment of myeloproliferative and related hematological syndromes is ultimately dosed not by behavioral or demographic measures, but at least in part on the basis of serial measure of the cell population of interest. Cardiac arrhythmias and coronary artery disease are often assessed by evaluating cardiac function, e.g., in the setting of electrophysiological studies, exercise, or a sympathomimetic challenge.

These practices suggest the general principle that some form of study of the therapy's target organ might be useful for optimizing therapy. In the setting of focal CNS injury, a technique such as fMRI might therefore be useful for a restorative therapy to assist with study entry criteria, to define optimal therapy dose or duration, or to serve as a biological marker of treatment effect. This issue is considered below with respect to three conditions characterized by an acute focal neurological insult.

2. Stroke

The motor system is among the most frequently affected domains by stroke *(38, 39)*. Duncan et al. *(40, 41)* found that the most dramatic improvements occurred in the first 30 days poststroke, though significant improvement continued to occur up to 90 days after stroke in patients with more severe deficits. Nakayama et al. *(42)* measured arm disability and found that maximum arm function was achieved by 80% of patients within 3 weeks, and by 95% of patients within 9 weeks. Wade et al. *(43)* also found that significant improvement was mainly seen across the first 3 months after stroke. Despite these improvements, residual motor deficits remain in approximately half of patients in the chronic phase of stroke *(38, 39)*.

A number of studies *(44–46)* have examined the brain events underlying the spontaneous recovery of motor behavior that does arise after stroke. In sum, stroke-related injury is associated with reduced activation, function, and neurophysiological responsiveness in injured (or for deep strokes, the overlying/corresponding) primary cortex. The best spontaneous return of behavior is associated with resolution of these reductions, i.e., return of activity in primary cortex, sometimes with particular shifts in the site of activation. Several compensatory responses may also contribute to spontaneous behavioral recovery, including increased activation in secondary areas that are normally connected to the injured zones in a distributed network, as well as a shift in interhemispheric laterality towards the contralesional hemisphere. The larger the injury or greater the deficits, the more these compensatory events are seen. These compensatory responses are tricks of the desperate, but in patients with injury-related deficits they are better present than absent *(47–51)*. These events that underlie spontaneous recovery are important because in some cases they are the same measures important to measure to guide optimization of therapy-derived recovery.

2.1. Use of Functional Neuroimaging to Guide a Restorative Intervention in Patients with Stroke

One study used functional neuroimaging in a clinical trial of a restorative intervention to extract data from an fMRI scan in order to guide details of decision-making during therapy *(24, 52)*. An fMRI scan was used to identify the centroid of ipsilesional primary motor cortex activation when patients with stroke moved the affected hand. This information then guided neurosurgical placement of an investigational epidural cortical stimulation device over ipsilesional motor cortex. Using this approach, patients receiving stimulation plus rehabilitation therapy showed significantly greater arm motor gains than patients receiving rehabilitation therapy alone. A similar approach was used in studies based on transcranial magnetic stimulation (TMS) to identify the optimal physiological representation site for hand motor function. These studies found repetitive TMS to be useful for improving motor function after stroke *(20, 21)*.

2.2. Use of Functional Neuroimaging to Predict Response to a Restorative Intervention in Patients with Stroke

An additional application of functional neuroimaging in the setting of restorative therapy is to predict behavioral response to treatment. Several studies have examined this issue *(53–58)*. For example, Koski et al. *(56)*, using TMS, and Dong et al. *(54)*, using fMRI, have found that changes in brain function early into therapy predict behavioral gains measured at the end of therapy. Note that in both cases, the findings that predicted better treatment response (improved motor evoked response in affected hand with TMS of ipsilesional hemisphere, and increased laterality of fMRI activation, i.e., towards the ipsilesional hemisphere, with movement of the affected hand, respectively) correspond to the findings correlating with better outcome associated to a spontaneous behavioral recovery. This latter pair of studies also hint at the potential use of human brain mapping measures to identify dose of a restorative therapy, in individual patients. For example, could a TMS or an fMRI measure of brain function inform a clinician of the likelihood that the brain is receptive to further change that supports behavioral gains? In this regard, note that a probe of brain plasticity, such as might be used to predict treatment response to a restorative intervention, can be developed even in the setting of severe deficits, such as complete plegia *(59)*.

A recent study *(57)* found that fMRI had independent value for predicting treatment response in a restorative therapy trial in patients with chronic stroke. This study used a multivariate model to examine the specific ability of a baseline fMRI to predict trial-related behavioral gains, and compared this fMRI predictive ability directly to a number of other baseline measures. Patients in this study each underwent baseline clinical and functional MRI assessments, received 6 weeks of rehabilitation therapy with or without investigational motor cortex stimulation, then had repeat assessments. Across all patients, univariate analyses found that several baseline measures had predictive value for trial-related gains. However, multiple linear regression modeling found that only two variables remained significant predictors: degree of motor cortex activation on fMRI (lower motor cortex activation predicted larger gains) and arm motor function (greater arm function predicted larger gains). This study emphasizes that an assessment of brain function can be a unique source of information for clinical decision-making in the setting of restorative therapy after stroke. Interestingly, clinical gains during study participation were paralleled by boosts in motor cortex activity, the latter detected via serial fMRI scanning, suggesting that lower baseline cortical activity in some patients likely represents under-use of an available cortical resource. There are a number of important variables that differ across patients, study designs, fMRI acquisition and analysis methods, and more. As such, further studies are needed to understand the extent to which the above findings generalize across other stroke studies.

2.3. Use of Functional Neuroimaging to Investigate the Biological Mechanisms of Restorative Intervention Effects in Patients with Stroke

Carey et al. *(60)* found that a population of subjects with chronic stroke, when performing a finger tracking task with the stroke-affected hand, had activation within contralesional brain regions, i.e., regions that were primarily ipsilateral to movement. After training at this task, the normal pattern of laterality of brain activation was restored, with activation shifting to ipsilesional brain regions, i.e., contralateral to movement, and thereby more closely resembling findings in healthy control subjects. In this landmark study, functional neuroimaging provided insights into the mechanistic effects of treatment.

A recent meta-analysis *(61)* extended these results by examining studies that have used functional neuroimaging as a biological marker of treatment effects targeting the motor system after stroke. Review of 13 studies of 121 patients permitted drawing a number of conclusions. Motor deficits have been most often studied, in part because of their substantial contribution to overall disability after stroke, and in part because of their relatively high prevalence. Most published studies have focused on patients with good to excellent outcome at baseline since they were more able to perform the motor tasks required to probe brain function. Consequently, less is known about the functional anatomy of therapy-induced recovery processes in the large population of patients with more severe deficits after stroke despite the great need for further study of restorative interventions in this population. Very few studies have used functional imaging to examine treatment effects during the first few months after stroke, when spontaneous behavioral recovery is at its greatest. The effects that many key variables such as lesion site, recovery level, gender, and age have on the performance of functional neuroimaging in this context requires further study. Studies could be improved by incorporating measures of injury and/or physiology.

Baseline differences in the stroke population under study can have a significant impact on the informative value of functional neuroimaging measures in the setting of a clinical trial. One successful constraint-induced motor therapy was associated with *decreased* interhemispheric laterality in a study of weaker patients *(62)*, while a second study found *increased* laterality in a study of stronger patients *(63)* with chronic stroke.

This divergence in findings emphasizes how differences in a single variable, such as baseline motor status, might influence the utility of brain mapping in the setting of a clinical trial, and highlights the need for further studies in this regard.

3. Multiple Sclerosis

Motor deficits are common in MS. For example, across a broad population of subjects with MS, the median time to reach irreversible limited walking ability for more than 500 m without aid or rest is 8 years, to walk with unilateral support no more than 100 m without rest is 20 years, and to walk no more than 10 m without rest while leaning against a wall or support is

30 years *(64)*. Upper extremity motor deficits, such as those related to ataxia and paresis, are also a common source of disability.

Brain plasticity is an important determinant in MS in at least two contexts. First, steady destruction of myelin and of axons over years results in disability. During this period, reorganization of brain function can reduce the impact of such injury on behavioral status. Second, approximately 85% of patients with MS have a relapsing, remitting course *(65)*, in which a relapse peaks over 1–2 months and then improves over a similar time period. The resolution of these MS flares has been attributed to a number of brain events, such as neurological reserve and resolution of inflammatory insult *(65)*, and a number of studies suggest that brain plasticity is also important *(66)*. Note too that there are numerous asymptomatic brain lesions for each symptomatic one in most patients with MS, a fact that might further support the importance of brain plasticity in maintenance of behavioral status in this condition.

Brain plasticity thus is likely important to motor status in MS, by minimizing the debilitating effects of MS injury accrual over time, and by promoting recovery from silent or symptomatic MS flares. A number of studies have provided insights into the brain events important in this regard, with substantial overlap as compared to findings in patients with stroke. This information gains importance in the current discussion because events important to maximizing behavioral status in the natural course of the disease are likely to be many of the same measures whose measurement can guide optimization of therapy-derived recovery.

Studies of brain plasticity in MS have found that, early in the course of the disease, brain activation is larger and more widespread as compared with that in healthy controls. Later in the disease, laterality of activation is reduced (i.e., activation is more bilateral) *(67–69)*, akin to stroke patients who have larger infarcts or greater deficits *(70, 71)*. Bilateral sensorimotor cortical regions are activated to a greater extent in the setting of MS-related white matter injury *(72, 73)*. This increased degree of bilateral organization persists to the greatest extent in subjects with persistent deficits after an acute MS relapse, and returns to a normal, lateralized (i.e., contralateral-predominant) form of organization in subjects with the least degree of persistent disability *(74, 75)*. The pattern of brain activation during performance of a simple motor task in subjects recovered from stroke has been considered similar to the pattern seen in healthy subjects during performance of a complex task *(76)*; a similar analogy has been made in subjects with MS *(77)*.

3.1. Use of Functional Neuroimaging to Investigate the Biological Mechanism of Restorative Intervention Effects in Patients with MS

The extent to which these spontaneous changes in brain function after MS can be used to monitor therapeutic interventions has been assessed in several small studies. One study tested the effects of increased cholinergic tone on the pattern of fMRI activation during performance of a cognitive task, the Stroop test. At baseline, patients with MS and moderate disability had similar behavioral performance as compared with controls, but on fMRI showed increased left medial prefrontal, and decreased right frontal, activation. Treatment with the cholinesterase inhibitor rivastigmine normalized both of these fMRI abnormalities in patients, but had little effect on a small cohort of healthy control subjects *(78)*.

In another study, administration of 3,4-diaminopyridine to patients with MS was associated with increased activation in sensorimotor cortex and SMA ipsilateral to movement. This pattern is the reverse of the laterality pattern seen in normals but might correspond to effects of increased injury *(70, 71)* or task complexity *(79, 80)*. Interestingly, TMS measures were also affected by treatment, showing a drug-induced reduction in intracortical inhibition and increase in intracortical facilitation *(81)*. The relationship that these changes had with behavioral effects of drug administration was not reported.

Subjects with MS were also studied with fMRI before and after 30 min of thumb flexion training. Across the training period, fMRI during thumb movements showed reduced activation in primary sensorimotor and parietal association cortex contralateral to movement in healthy controls, but not in subjects with MS *(82)*.

4. Spinal Cord Injury

Though SCI can be associated with a range of injury patterns, motor deficits are generally a prominent feature. At the time of discharge from initial SCI, the most frequent neurologic category is incomplete tetraplegia (34.1%), followed by complete paraplegia (23.0%), complete tetraplegia (18.3%), and incomplete paraplegia (18.5%). Less than 1% of persons experience complete neurologic recovery by hospital discharge. By 10 years after SCI, 68% of persons with paraplegia, and 76% of those with tetraplegia, are unemployed *(83)*.

Subjects with SCI generally show modest spontaneous sensory and motor improvement in the first 3–6 months following injury *(84, 85)*, although significant improvement beyond the first year post-SCI is uncommon *(86)*. Motor deficits are thus common and persistent after SCI, and these impact a number of health, quality of life, and other issues in subjects with SCI *(87–89)*.

4.1. Use of Functional Neuroimaging to Investigate the Biological Mechanism of Restorative Intervention Effects in Patients with SCI

There has been limited study of the CNS mechanisms underlying spontaneous motor improvement during the months following SCI. Studies to date have more been focused on the nature of brain motor systems function in the chronic state, with some divergence of results to date. Some studies have found a broad decrease in activation *(90–92)*, particularly in primary sensorimotor cortex, whereas others have found supranormal activation *(93)*. The basis for these discrepancies remains unclear but could be due to differences in age or injury pattern of the population studied, years post-SCI at time of study, or the nature of the task used to probe motor system function, some uncovering deficient processing and others emphasizing supranormal efforts to compensate *(91, 94)*. A commonly described feature is a change in somatotopic organization within primary sensorimotor cortex contralateral to sensory or motor events, with representation of supralesional body regions expanding at the expense of infralesional body regions *(95–99)*. Spontaneous changes in laterality, so prominent in studies of stroke or MS, as above, are generally not prominent after SCI, perhaps because in part injury typically affects the CNS bilaterally or perhaps because in part SCI spares brain commisural fibers whose integrity helps maintain normal hemispheric balance. As such, laterality is unlikely to be a useful variable to examine in brain mapping studies of treatment effects in the setting of SCI.

At least two studies have evaluated changes in brain function in relation to therapy after SCI. Winchester et al. *(100)* studied body weight supported treadmill training in four patients with motor incomplete SCI. These authors compared fMRI during attempted unilateral foot and toe movement before vs. after training. This therapy was associated with increased activation within several bilateral areas, including primary sensorimotor cortex and cerebellum, though to a variable extent. The authors observed that, although all participants demonstrated a change in the blood oxygen level dependent signal following training, only those patients who demonstrated a substantial increase in activation of the cerebellum demonstrated an improvement in their ability to walk over ground, suggesting that this measure in this brain region, at least when examined using this task during fMRI, might be useful as a biological marker of successful treatment effect.

Another form of intervention that has been evaluated after SCI is motor imagery. Motor imagery normally activates many of the same brain regions as motor execution, and has been associated with improvements in motor performance *(101, 102)*. The effects of 1 week of motor imagery training to tongue and to foot were evaluated in ten subjects with chronic, complete tetra-/paraplegia plus ten healthy controls *(103)*. The behavioral outcome measure was speed of performance of a complex sequence. Motor imagery training was associated with a significant improvement in this behavior in nonparalyzed muscles (tongue for both groups, right foot for healthy subjects). In both the healthy controls and the subjects with SCI, serial fMRI scanning (before vs. after training) during attempted right foot movement was associated with increased fMRI activation in left putamen, an area associated with motor learning, despite foot movements being present in controls and absent in subjects with SCI.

Training effects on brain plasticity can thus be measured independent of behavior effects, a finding that might be important for designing biological markers in trials targeting severely disabled patient populations. Note that this fMRI change was absent in a second healthy control group serially imaged without training. The main conclusion from this study is that motor imagery training improves brain function whether or not sensorimotor function is present in the trained limb. An additional conclusion is that motor imagery, by virtue of its favorable effects on brain motor system organization, might have value as an adjunct motor restorative therapy. Another key point from this study is that brain plasticity related to plegic limbs can be studied in subjects with chronic SCI.

One hypothesis suggested by this study's findings is that the results of a short-term brain plasticity probe such as this motor imagery training intervention will predict response to a longer-term treatment *(17, 18)*, for example, patients who show the greatest extent of brain plasticity with such a 1 week motor imagery intervention might be those who are most likely to respond to a more intensive intervention such as stem cell injections. Thus, at least in chronic SCI, some measure of the capacity for the brain to adapt in the short term might predict likelihood of response to a more intense intervention.

5. Conclusions

Motor deficits are a major source of disability, across a number of conditions. A number of restorative therapies are under study to improve motor function in this regard. Optimal prescription of such therapies might benefit from an assessment of the function of the target organ, in addition to assessment of behavior or demographics. This is an approach that has often proven fruitful in general medical practice, and given the added complexities related to the CNS, is likely to be particularly important in for application of CNS restorative therapies.

Towards this goal, establishment of standardized protocols, such as for measuring motor cortex plasticity *(104)*, to extract measures of brain function might help maximize the extent to which functional neuroimaging can be effectively applied. Dynamic protocols that incite a CNS response, such as over 30 min *(105)* or days *(54, 56)* of activity, might have particular value as compared to a single cross-sectional behavioral probe. Also, studies that provide a greater understanding of the underlying neurobiologic principles related to spontaneous recovery will also aid application of restorative therapies given that brain changes important to spontaneous recovery likely overlap substantially with changes whose measurement can effectively guide trials to maximize treatment-induced gains.

Further studies are needed to better characterize the measures that have the potential for monitoring, predicting, and dosing in the setting of a restorative trial of patients with motor deficits. Also, a minority of studies has examined language, neglect, and other domains injured in CNS disease, and further studies in these areas are also needed. Some similarities exist across diseases, such as those discussed between stroke and MS above, and further investigation of such points of similarity might prove fruitful in a broader sense to advancing restorative therapeutics. Finally, this review has focused on fMRI as a means of probing the state of the CNS. Other investigative methods might also prove useful, including functional, anatomical, and other forms of probing the CNS. Examples include positron emission tomography, diffusion tensor imaging, proton MR spectroscopy, TMS, electroencephalography, and measures of anatomy, injury, or perfusion. These can measure white matter integrity *(58)*, injury in relation to normal anatomy *(106, 107)*, metabolic state *(108, 109)*, and more might prove equally useful in models that aim to inform therapeutic approaches to restoring motor function in the setting of neurological disease.

References

1. Dobkin B. The Clinical Science of Neurologic Rehabilitation. New York: Oxford University Press; 2003.
2. Chen J, Cui X, Zacharek A, et al. Niaspan increases angiogenesis and improves functional recovery after stroke. Ann Neurol 2007;62(1):49–58.
3. Li L, Jiang Q, Zhang L, et al. Angiogenesis and improved cerebral blood flow in the ischemic boundary area detected by MRI after administration of sildenafil to rats with embolic stroke. Brain Res 2007;1132(1):185–92.
4. Chen P, Goldberg D, Kolb B, Lanser M, Benowitz L. Inosine induces axonal rewiring and improves behavioral outcome after stroke. Proc Natl Acad Sci USA 2002;99(13):9031–6.
5. Freret T, Valable S, Chazalviel L, et al. Delayed administration of deferoxamine reduces brain damage and promotes functional recovery after transient focal cerebral ischemia in the rat. Eur J Neurosci 2006;23(7):1757–65.
6. Papadopoulos CM, Tsai SY, Cheatwood JL, et al. Dendritic plasticity in the adult rat following middle cerebral artery occlusion and Nogo-a neutralization. Cereb Cortex 2006;16(4): 529–36.
7. Kawamata T, Dietrich W, Schallert T, et al. Intracisternal basic fibroblast growth factor (bFGF) enhances functional recovery and upregulates the expression of a molecular marker of neuronal sprouting following focal cerebral infarction. Proc Natl Acad Sci 1997;94:8179–84.
8. Kawamata T, Ren J, Chan T, Charette M, Finklestein S. Intracisternal osteogenic protein-1 enhances functional recovery following focal stroke. NeuroReport 1998;9(7):1441–5.

9. Schabitz WR, Berger C, Kollmar R, et al. Effect of brain-derived neurotrophic factor treatment and forced arm use on functional motor recovery after small cortical ischemia. Stroke 2004;35(4):992–7.
10. Wang L, Zhang Z, Wang Y, Zhang R, Chopp M. Treatment of stroke with erythropoietin enhances neurogenesis and angiogenesis and improves neurological function in rats. Stroke 2004;35(7):1732–7.
11. Tsai PT, Ohab JJ, Kertesz N, et al. A critical role of erythropoietin receptor in neurogenesis and post-stroke recovery. J Neurosci 2006;26(4):1269–74.
12. Schneider UC, Schilling L, Schroeck H, Nebe CT, Vajkoczy P, Woitzik J. Granulocyte-macrophage colony-stimulating factor-induced vessel growth restores cerebral blood supply after bilateral carotid artery occlusion. Stroke 2007;38(4):1320–8.
13. Kolb B, Morshead C, Gonzalez C, et al. Growth factor-stimulated generation of new cortical tissue and functional recovery after stroke damage to the motor cortex of rats. J Cereb Blood Flow Metab 2007;27(5): 983–97.
14. Zhao LR, Berra HH, Duan WM, et al. Beneficial effects of hematopoietic growth factor therapy in chronic ischemic stroke in rats. Stroke 2007;38(10):2804–11.
15. Ehrenreich H, Hasselblatt M, Dembowski C, et al. Erythropoietin therapy for acute stroke is both safe and beneficial. Mol Med 2002;8(8):495–505.
16. Savitz SI, Dinsmore JH, Wechsler LR, Rosenbaum DM, Caplan LR. Cell therapy for stroke. NeuroRx 2004;1(4):406–14.
17. Keirstead HS, Nistor G, Bernal G, et al. Human embryonic stem cell-derived oligodendrocyte progenitor cell transplants remyelinate and restore locomotion after spinal cord injury. J Neurosci 2005;25(19):4694–705.
18. Cummings BJ, Uchida N, Tamaki SJ, et al. Human neural stem cells differentiate and promote locomotor recovery in spinal cord-injured mice. Proc Natl Acad Sci USA 2005;102(39):14069–74.
19. Shen LH, Li Y, Chen J, et al. Therapeutic benefit of bone marrow stromal cells administered 1 month after stroke. J Cereb Blood Flow Metab 2007;27:6–13.
20. Khedr EM, Ahmed MA, Fathy N, Rothwell JC. Therapeutic trial of repetitive transcranial magnetic stimulation after acute ischemic stroke. Neurology 2005;65(3):466–8.
21. Kim YH, You SH, Ko MH, et al. Repetitive transcranial magnetic stimulation-induced corticomotor excitability and associated motor skill acquisition in chronic stroke. Stroke 2006;37(6):1471–6.
22. Malcolm MP, Triggs WJ, Light KE, et al. Repetitive transcranial magnetic stimulation as an adjunct to constraint-induced therapy: an exploratory randomized controlled trial. Am J Phys Med Rehabil/Assoc Acad Physiatrists 2007;86(9):707–15.
23. Hummel F, Celnik P, Giraux P, et al. Effects of non-invasive cortical stimulation on skilled motor function in chronic stroke. Brain 2005;128(Pt 3):490–9.
24. Brown JA, Lutsep HL, Weinand M, Cramer SC. Motor cortex stimulation for the enhancement of recovery from stroke: a prospective, multicenter safety study. Neurosurgery 2006;58(3):464–73.
25. Ring H, Rosenthal N. Controlled study of neuroprosthetic functional electrical stimulation in sub-acute post-stroke rehabilitation. J Rehabil Med 2005;37(1):32–6.
26. Sheffler LR, Chae J. Neuromuscular electrical stimulation in neurorehabilitation. Muscle Nerve 2007;35(5):562–90.
27. Kwakkel G, Kollen BJ, Krebs HI. Effects of Robot-assisted therapy on upper limb recovery after stroke: a systematic review. Neurorehabil Neural Repair 2008;22(2):111–21.
28. Volpe BT, Ferraro M, Lynch D, et al. Robotics and other devices in the treatment of patients recovering from stroke. Curr Neurol Neurosci Rep 2005;5(6):465–70.
29. Reinkensmeyer D, Emken J, Cramer S. Robotics, motor learning, and neurologic recovery. Annu Rev Biomed Eng 2004;6:497–525.
30. Deutsch JE, Lewis JA, Burdea G. Technical and patient performance using a virtual reality-integrated telerehabilitation system: preliminary finding. IEEE Trans Neural Syst Rehabil Eng 2007;15(1):30–5.
31. Duncan P, Studenski S, Richards L, et al. Randomized clinical trial of therapeutic exercise in subacute stroke. Stroke 2003;34(9):2173–80.
32. Woldag H, Hummelsheim H. Evidence-based physiotherapeutic concepts for improving arm and hand function in stroke patients: a review. J Neurol 2002;249(5):518–28.
33. French B, Thomas L, Leathley M, et al. Repetitive task training for improving functional ability after stroke. Cochrane Database of Systematic Reviews (Online) 2007;(4): CD006073.
34. Kwakkel G, Wagenaar R, Twisk J, Lankhorst G, Koetsier J. Intensity of leg and arm training after primary middle-cerebral-artery stroke: a randomised trial. Lancet 1999;354(9174): 191–6.

35. Van Peppen RP, Kwakkel G, Wood-Dauphinee S, Hendriks HJ, Van der Wees PJ, Dekker J. The impact of physical therapy on functional outcomes after stroke: what's the evidence? Clin Rehabil 2004;18(8):833–62.
36. Luft A, McCombe-Waller S, Whitall J, et al. Repetitive bilateral arm training and motor cortex activation in chronic stroke: a randomized controlled trial. JAMA 2004;292(15): 1853–61.
37. Wolf SL, Winstein CJ, Miller JP, et al. Effect of constraint-induced movement therapy on upper extremity function 3 to 9 months after stroke: the EXCITE randomized clinical trial. JAMA 2006;296(17):2095–104.
38. Rathore S, Hinn A, Cooper L, Tyroler H, Rosamond W. Characterization of incident stroke signs and symptoms: findings from the atherosclerosis risk in communities study. Stroke 2002;33(11):2718–21.
39. Gresham G, Duncan P, Stason W, et al. Post-Stroke Rehabilitation. Rockville, MD: U.S. Department of Health and Human Services, Public Health Service, Agency for Health Care Policy and Research; 1995.
40. Duncan P, Goldstein L, Horner R, Landsman P, Samsa G, Matchar D. Similar motor recovery of upper and lower extremities after stroke. Stroke 1994;25(6):1181–8.
41. Duncan P, Goldstein L, Matchar D, Divine G, Feussner J. Measurement of motor recovery after stroke. Stroke 1992;23:1084–9.
42. Nakayama H, Jorgensen H, Raaschou H, Olsen T. Recovery of upper extremity function in stroke patients: the Copenhagen Stroke Study. Arch Phys Med Rehabil 1994;75(4):394–8.
43. Wade D, Langton-Hewer R, Wood V, Skilbeck C, Ismail H. The hemiplegic arm after stroke: measurement and recovery. J Neurol Neurosurg Psychiatry 1983;46(6):521–4.
44. Yozbatiran N, Cramer SC. Imaging motor recovery after stroke. NeuroRx 2006;3(4): 482–8.
45. Ward NS, Cohen LG. Mechanisms underlying recovery of motor function after stroke. Arch Neurol 2004;61(12):1844–8.
46. Baron J, Cohen L, Cramer S, et al. Neuroimaging in stroke recovery: a position paper from the First International Workshop on Neuroimaging and Stroke Recovery. Cerebrovasc Dis (Basel, Switzerland) 2004;18(3):260–7.
47. Lotze M, Markert J, Sauseng P, Hoppe J, Plewnia C, Gerloff C. The role of multiple contralesional motor areas for complex hand movements after internal capsular lesion. J Neurosci 2006;26(22):6096–102.
48. Winhuisen L, Thiel A, Schumacher B, et al. Role of the contra lateral inferior frontal gyrus in recovery of language function in poststroke aphasia: a combined repetitive transcranial magnetic stimulation and positron emission tomography study. Stroke 2005;36(8):1759–63.
49. Johansen-Berg H, Rushworth M, Bogdanovic M, Kischka U, Wimalaratna S, Matthews P. The role of ipsilateral premotor cortex in hand movement after stroke. Proc Natl Acad Sci USA 2002;99(22):14518–23.
50. Werhahn K, Conforto A, Kadom N, Hallett M, Cohen L. Contribution of the ipsilateral motor cortex to recovery after chronic stroke. Ann Neurol 2003;54(4):464–72.
51. Fridman E, Hanakawa T, Chung M, Hummel F, Leiguarda R, Cohen L. Reorganization of the human ipsilesional premotor cortex after stroke. Brain 2004;127(Pt 4):747–58.
52. Cramer S, Benson R, Himes D, et al. Use of functional MRI to guide decisions in a clinical stroke trial. Stroke 2005;36(5):e50–2.
53. Platz T, Kim I, Engel U, Kieselbach A, Mauritz K. Brain activation pattern as assessed with multi-modal EEG analysis predict motor recovery among stroke patients with mild arm paresis who receive the Arm Ability Training. Restor Neurol Neurosci 2002;20 (1–2):21–35.
54. Dong Y, Dobkin BH, Cen SY, Wu AD, Winstein CJ. Motor cortex activation during treatment may predict therapeutic gains in paretic hand function after stroke. Stroke 2006;37(6):1552–5.
55. Fritz SL, Light KE, Patterson TS, Behrman AL, Davis SB. Active finger extension predicts outcomes after constraint-induced movement therapy for individuals with hemiparesis after stroke. Stroke 2005;36(6):1172–7.
56. Koski L, Mernar T, Dobkin B. Immediate and long-term changes in corticomotor output in response to rehabilitation: correlation with functional improvements in chronic stroke. Neurorehabil Neural Repair 2004;18(4): 230–49.
57. Cramer S, Parrish T, Levy R, et al. An assessment of brain function predicts functional gains in a clinical stroke trial. Stroke 2007;38: 520 (abstract).
58. Stinear CM, Barber PA, Smale PR, Coxon JP, Fleming MK, Byblow WD. Functional potential in chronic stroke patients depends on corticospinal tract integrity. Brain 2007;130(Pt 1):170–80.
59. Cramer SC, Orr EL, Cohen MJ, Lacourse MG. Effects of motor imagery training after

chronic, complete spinal cord injury. Exp Brain Res 2007;177(2):233–42.
60. Carey J, Kimberley T, Lewis S, et al. Analysis of fMRI and finger tracking training in subjects with chronic stroke. Brain 2002;125(Pt 4):773–88.
61. Hodics T, Cohen LG, Cramer SC. Functional imaging of intervention effects in stroke motor rehabilitation. Arch Phys Med Rehabil 2006;87(12 Suppl):36–42.
62. Schaechter J, Kraft E, Hilliard T, et al. Motor recovery and cortical reorganization after constraint-induced movement therapy in stroke patients: a preliminary study. Neurorehabil Neural Repair 2002;16(4):326–38.
63. Johansen-Berg H, Dawes H, Guy C, Smith S, Wade D, Matthews P. Correlation between motor improvements and altered fMRI activity after rehabilitative therapy. Brain 2002;125(Pt 12):2731–42.
64. Vukusic S, Confavreux C. Natural history of multiple sclerosis: risk factors and prognostic indicators. Curr Opin Neurol 2007; 20(3):269–74.
65. Vollmer T. The natural history of relapses in multiple sclerosis. J Neurol Sci 2007;256(Suppl 1):S5–S13.
66. Rocca MA, Filippi M. Functional MRI in multiple sclerosis. J Neuroimaging 2007; 17(Suppl 1):36S–41S.
67. Rocca MA, Colombo B, Falini A, et al. Cortical adaptation in patients with MS: a cross-sectional functional MRI study of disease phenotypes. Lancet Neurol 2005;4(10):618–26.
68. Wang J, Hier DB. Motor reorganization in multiple sclerosis. Neurol Res 2007;29(1):3–8.
69. Pantano P, Mainero C, Caramia F. Functional brain reorganization in multiple sclerosis: evidence from fMRI studies. J Neuroimaging 2006;16(2):104–14.
70. Ward N, Brown M, Thompson A, Frackowiak R. Neural correlates of outcome after stroke: a cross-sectional fMRI study. Brain 2003;126(Pt 6):1430–48.
71. Cramer SC, Crafton KR. Somatotopy and movement representation sites following cortical stroke. Exp Brain Res Experimentelle Hirnforschung 2006;168(1/2):25–32.
72. Lenzi D, Conte A, Mainero C, et al. Effect of corpus callosum damage on ipsilateral motor activation in patients with multiple sclerosis: a functional and anatomical study. Hum Brain Mapping 2007;28(7):636–44.
73. Rocca MA, Gallo A, Colombo B, et al. Pyramidal tract lesions and movement-associated cortical recruitment in patients with MS. NeuroImage 2004;23(1):141–7.
74. Reddy H, Narayanan S, Matthews P, et al. Relating axonal injury to functional recovery in MS. Neurology 2000;54(1):236–9.
75. Mezzapesa DM, Rocca MA, Rodegher M, Comi G, Filippi M. Functional cortical changes of the sensorimotor network are associated with clinical recovery in multiple sclerosis. Hum Brain Mapping 2008;29(5):562–73.
76. Cramer S, Nelles G, Benson R, et al. A functional MRI study of subjects recovered from hemiparetic stroke. Stroke 1997;28(12):2518–27.
77. Filippi M, Rocca MA, Mezzapesa DM, et al. A functional MRI study of cortical activations associated with object manipulation in patients with MS. NeuroImage 2004;21(3):1147–54.
78. Parry AM, Scott RB, Palace J, Smith S, Matthews PM. Potentially adaptive functional changes in cognitive processing for patients with multiple sclerosis and their acute modulation by rivastigmine. Brain 2003;126(Pt 12):2750–60.
79. Sadato N, Campbell G, Ibanez V, Deiber M, Hallett M. Complexity affects regional cerebral blood flow change during sequential finger movements. J Neurosci 1996;16(8):2691–700.
80. Verstynen T, Diedrichsen J, Albert N, Aparicio P, Ivry RB. Ipsilateral motor cortex activity during unimanual hand movements relates to task complexity. J Neurophysiol 2005;93(3):1209–22.
81. Mainero C, Inghilleri M, Pantano P, et al. Enhanced brain motor activity in patients with MS after a single dose of 3,4-diaminopyridine. Neurology 2004;62(11):2044–50.
82. Morgen K, Kadom N, Sawaki L, et al. Training-dependent plasticity in patients with multiple sclerosis. Brain 2004;127(Pt 11):2506–17.
83. www.spinalcord.uab.edu. Facts and figures at a glance – June 2006. 2007.
84. Geisler F, Dorsey F, Coleman W. Recovery of motor function after spinal-cord injury – a randomized, placebo-controlled trial with GM-1 ganglioside. New Eng J Med 1991;324(26):1829–38.
85. Ditunno J, Stover S, Freed M, Ahn J. Motor recovery of the upper extremities in traumatic quadriplegia: a multicenter study. Arch Phys Med Rehabil 1992;73(5):431–6.
86. Kirshblum S, Millis S, McKinley W, Tulsky D. Late neurologic recovery after traumatic spinal cord injury. Arch Phys Med Rehabil 2004;85(11):1811–7.
87. Jayaraman A, Gregory CM, Bowden M, et al. Lower extremity skeletal muscle function

in persons with incomplete spinal cord injury. Spinal Cord 2006;44(11):680–7.

88. DeVivo MJ, Richards JS. Community reintegration and quality of life following spinal cord injury. Paraplegia 1992;30(2):108–12.
89. Frankel HL, Coll JR, Charlifue SW, et al. Long-term survival in spinal cord injury: a fifty year investigation. Spinal Cord 1998;36(4): 266–74.
90. Cramer SC, Lastra L, Lacourse MG, Cohen MJ. Brain motor system function after chronic, complete spinal cord injury. Brain 2005;128(Pt 12):2941–50.
91. Jurkiewicz MT, Mikulis DJ, McIlroy WE, Fehlings MG, Verrier MC. Sensorimotor cortical plasticity during recovery following spinal cord injury: a longitudinal fMRI study. Neurorehabil Neural Repair 2007;21(6): 527–38.
92. Sabbah P, de Schonen S, Leveque C, et al. Sensorimotor cortical activity in patients with complete spinal cord injury: a functional magnetic resonance imaging study. J Neurotrauma 2002;19(1):53–60.
93. Alkadhi H, Brugger P, Boendermaker S, et al. What disconnection tells about motor imagery: evidence from paraplegic patients. Cereb Cortex 2005;15(2):131–40.
94. Humphrey D, Mao H, Schaeffer E. Voluntary activation of ineffective cerebral motor areas in short- and long-term paraplegics. Soc Neurosci (abstract) 2000.
95. Topka H, Cohen L, Cole R, Hallett M. Reorganization of corticospinal pathways following spinal cord injury. Neurology 1991;41(8):1276–83.
96. Bruehlmeier M, Dietz V, Leenders K, Roelcke U, Missimer J, Curt A. How does the human brain deal with a spinal cord injury? Eur J Neurosci 1998;10(12):3918–22.
97. Mikulis D, Jurkiewicz M, McIlroy W, et al. Adaptation in the motor cortex following cervical spinal cord injury. Neurology 2002;58(5):794–801.
98. Turner J, Lee J, Martinez O, Medlin A, Schandler S, Cohen M. Somatotopy of the motor cortex after long-term spinal cord injury or amputation. IEEE Trans Neural Syst Rehabil Eng 2001;9(2):154–60.
99. Corbetta M, Burton H, Sinclair R, Conturo T, Akbudak E, McDonald J. Functional reorganization and stability of somatosensory-motor cortical topography in a tetraplegic subject with late recovery. Proc Natl Acad Sci USA 2002;99(26):17066–71.
100. Winchester P, McColl R, Querry R, et al. Changes in supraspinal activation patterns following robotic locomotor therapy in motor-incomplete spinal cord injury. Neurorehabil Neural Repair 2005;19(4):313–24.
101. Lacourse MG, Turner JA, Randolph-Orr E, Schandler SL, Cohen MJ. Cerebral and cerebellar sensorimotor plasticity following motor imagery-based mental practice of a sequential movement. J Rehabil Res Dev 2004;41(4):505–24.
102. Sharma N, Pomeroy VM, Baron JC. Motor imagery: a backdoor to the motor system after stroke? Stroke 2006;37(7):1941–52.
103. Cramer SC, Orr EL, Cohen MJ, Lacourse MG. Effects of motor imagery training after chronic, complete spinal cord injury. Exp Brain Res 2007;177(2):233–42.
104. Kleim J, Kleim E, Cramer SC. Systematic assessment of training-induced changes in corticospinal output to hand using frameless stereotaxic transcranial magnetic stimulation. Nat Protocols 2007;2:1675–84.
105. Kleim JA, Chan S, Pringle E, et al. BDNF val66met polymorphism is associated with modified experience-dependent plasticity in human motor cortex. Nat Neurosci 2006; 9(6):735–7.
106. Newton JM, Ward NS, Parker GJ, et al. Non-invasive mapping of corticofugal fibres from multiple motor areas – relevance to stroke recovery. Brain 2006;129(Pt 7):1844–58.
107. Crafton K, Mark A, Cramer S. Improved understanding of cortical injury by incorporating measures of functional anatomy. Brain 2003;126(Pt 7):1650–9.
108. Heiss W, Emunds H, Herholz K. Cerebral glucose metabolism as a predictor of rehabilitation after ischemic stroke. Stroke 1993;24(12):1784–8.
109. Cappa S, Perani D, Grassi F, et al. A PET follow-up study of recovery after stroke in acute aphasics. Brain Lang 1997;56(1):55–67.

Part IV

Future fMRI Development

Chapter 27

Integration of Measures of Functional and Structural MRI

Heidi Johansen-Berg, Timothy E.J. Behrens, Saad Jbabdi, and Kate E. Watkins

Summary

Recent years have seen a renewed interest in brain anatomy in the neuroimaging community. Developments in techniques for structural MR acquisition and analysis have opened new opportunities for structural mapping of the living human brain. For example, high-resolution MR imaging can be used for "in vivo histology", techniques such as voxel-based morphometry can be used to localise structural variation across populations, diffusion imaging provides information on system-level anatomical connectivity. fMRI studies are increasingly making use of the information provided by such structural mapping techniques in order to discover the anatomical substrate for observed functional effects.

Key words: Structural MRI, Diffusion tensor imaging, Tractography, Voxel-based morphometry, Functional MRI, Individual differences

1. Introduction

While the twenty-first century neuroscientist is most likely to measure brain function, much of the study of the brain has its roots in neuroanatomy. In the seventeenth century, for example, Thomas Willis performed careful study of the gross anatomy of the brain. In the early twentieth century, scientists such as Ramon Y Cajal studied the microscopic architecture of the neuron, while Brodmann, Walker, and others used these cytoarchitectonic features to build up atlases of human and animal brains. They were not able to simultaneously investigate the functions of these regions, but their speculation was that areas that differed in their anatomical features were likely to perform distinct functional roles. Subsequent studies of the function of the brain have proved this notion largely correct.

M. Filippi (ed.), fMRI Techniques and Protocols, Neuromethods, vol. 41
DOI 10.1007/978-1-60327-919-2_27, © Humana Press, a part of Springer Science + Business Media, LLC 2009

Functional and anatomical specialisations are inextricably linked and information about one will greatly inform study of the other.

The rapid development of functional imaging technology, and in particular fMRI, has arguably led to neglect of questions concerning neuroanatomy. The vast majority of fMRI studies focus only on functional responses and have limited interest in the anatomical features of the regions they are mapping functionally. Yet an understanding of brain anatomy is critical to building up a detailed picture of brain function. The anatomical features of a brain region will determine its functional specialisation. For example, the cytoarchitecture of an area, providing an indication of cell density, size, and shape, will reflect the cell types present and therefore the processing operations that can be performed. The anatomical connection patterns of a region, which dictates the information available to it and the influence that it can have on other brain areas, will closely constrain its functional specialisation.

Recent years, however, have seen an increasing interest within the neuroimaging community in questions of neuroanatomy. As is often the case, much of this expansion is technology-led. As more sophisticated techniques are developed for imaging and analysing brain structure, neuroimagers are increasingly likely to add a structural mapping component to complement their functional studies. A multi-modal functional and structural approach is likely to become increasingly common in future.

The preceding chapters provide a comprehensive overview of methods and applications of fMRI. Here, we provide a brief introduction to selected structural MRI methods, and highlight recent applications that have attempted to integrate structural and functional approaches.

2. Introduction to Structural MRI Techniques

Development of new structural imaging techniques and automated analysis approaches has resulted in growing use of structural imaging methods as a complement to functional imaging studies. Interpretation of such structural data is aided by an understanding of the methodology. The following section focuses on two specific and commonly used quantitative approaches to structural brain analysis – voxel-based morphometry (VBM) and diffusion MR modelling and tractography. Other automated approaches to interrogating structural MRI data, such as analysis of sulcal morphology *(1, 2)* or cortical thickness *(3)*, can provide complementary information, but will not be discussed in detail here.

2.1. Voxel-Based Morphometry

The study of variation in brain structure associated with features of the normal population such as age, gender, handedness, or with disease states was greatly advanced by the availability of in vivo imaging. Compared to earlier imaging methods, such as CT, structural MR images offer excellent tissue contrast and high resolution (typically 1 mm^3). Because the original studies of brain images relied mainly on labour-intensive manual measurements, they often limited analyses to focus either on gross morphological measures - widths and lengths of hemispheres – or on one or two regions. Intra- and inter-rater reliabilities were often low because of difficulties in defining tissue and regional boundaries; gross morphological landmarks such as cerebral sulci are often variable in their presence and location. There was a strong need, therefore, for a whole-brain imaging analysis that was both reproducible and automated. One such method is VBM *(4, 5)*. Developed in the mid-1990s and originally implemented in SPM (http//:www.fil.ion.ucl.ac.uk/spm), VBM is a fully automated analysis of tissue "density" that allows assessment of the relative amounts of grey or white matter on a voxel-by-voxel basis throughout the whole brain. Many packages now offer tools to complete each of the stages of a VBM analysis described below, allowing the user to tailor an analysis pipeline specific to their needs and preferences.

Briefly, VBM is a tool that allows the investigator to perform a statistical test for structural effects at each brain voxel. It has four stages: *(1)* classification of each voxel in the image into different tissue types; *(2)* nonlinear spatial transformation of the image to match a template, thereby normalising brain size and shape; *(3)* spatial smoothing to remove residual differences in the exact location of sulci and gyri across brains and to meet statistical criteria for analysis; *(4)* statistical analysis on a voxel-by-voxel basis throughout the image. These steps are illustrated in **Fig. 1**. We elaborate on more specific details of each of these stages below.

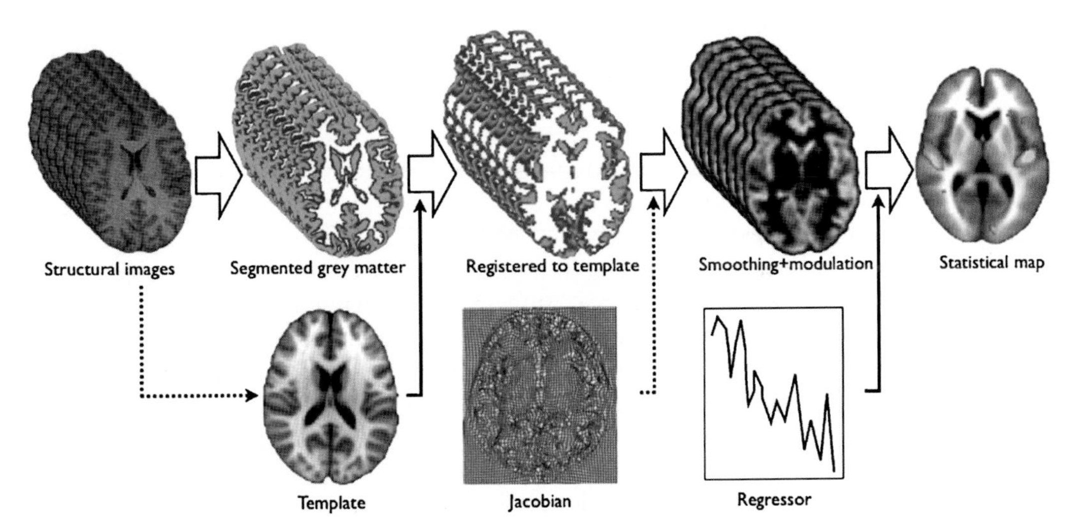

Fig. 1. Summary of the different stages of a voxel-based morphometry analysis: The structural images are segmented, then nonlinearly registered to a (potentially study-specific) template, smoothed (and eventually modulated by the Jacobian of the nonlinear registration), and then statistical regression is performed on a voxel-by-voxel basis.

Typically, the data used for a VBM study is T1-weighted. However, additional contrast images, such as T2-weighted or proton-density images, can improve tissue classification. There are a number of software packages available that allow nonlinear registration between a structural MRI image and a template. Choice of template can influence this registration and most VBM studies therefore prefer to build a study- or population-specific template that reflects specific features of the data being analysed, e.g., image contrast, disease state, race, or age of subjects. Automated tissue classification algorithms are used to segment the image into separate images reflecting tissue class. The data in the resulting segmented image can be binary, reflecting the presence or absence of a tissue type, or probabilistic, reflecting the likelihood that the voxel represents a specific tissue type or combination of types on the basis of either its signal or its location or both. Furthermore, the data in these segmented images nonlinearly registered to a template, may be modulated by the Jacobian determinant of the deformation field to reflect their original volume. That is, the signal in each voxel is increased or decreased depending on whether the area was reduced or increased in volume to match the template. Finally, the data are smoothed using a Gaussian smoothing kernel with a specified full-width at half-maximum (FWHM). For whole-brain analyses, the choice of smoothing kernel is usually between 8- and 14-mm FWHM. The smaller smoothing kernel is often applied to the modulated data sets as these are smoother than unmodulated ones. Smoothing is required to improve the distribution of data values from almost binary into a range of continuous data, which is a requirement for the statistical analyses using Gaussian Random Field theory. It also weights the signal at each voxel to reflect the amount of tissue in surrounding voxels, thereby reflecting regional density or concentration of tissue in an area. Finally, smoothing serves to reduce the effect of spatial differences between subjects in the exact location of gyri and sulci. It should be noted, however, that smoothing the data also increases the detection of differences whose spatial extent matches the width of the smoothing kernel. Previous studies have shown that smaller smoothing kernels can be more sensitive in specific regions of interest such as the hippocampus *(6)*. The smoothed segmented data are then subjected to statistical analysis using standard imaging statistical packages to compare groups of subjects (e.g., *(7)* or hemispheric asymmetries in the same subjects *(8)* or to examine a correlation with another variable such as age *(9)*, or a behavioural measure *(10)*).

Caution needs to be exercised in interpreting the results of a VBM analysis. The manipulation of the data prior to analysis has favoured a description that a difference in a VBM analysis reflects a difference in grey or white matter "density". This term is meant to reflect the relative amounts of tissue in a particular area rather than a precise quantity such as cell-packing density. To reduce confusion, terms such as "concentration" have been used. In analyses using modulation, the term volume is often used. It should be noted, however, that because of smoothing, tissue from both sides of a sulcus can contribute to the apparent "volume" of cortex in a region. One can only conclude with confidence that there is a difference in the region, say between two groups, and that this difference may reflect volume or shape (gyrification) differences or some combination of the two (*see* **Subheading 2.3**).

VBM analysis has primarily been applied to grey or white matter segmentations based on T1-weighted MR data. However, voxelwise values derived from other MR images can also be compared using VBM-like techniques. For example, white matter properties have been explored using VBM-like techniques on values derived from diffusion-weighted MR data, as these images contain information on white matter microstructure (*see* **Subheading 2.2**). However, just as caution is required in interpreting VBM-derived differences in grey matter density, it is also important to establish whether observed differences in white matter values derived from diffusion images are due to genuine changes in white matter microstructure or rather to variations in brain shape or ventricular variations. Alternatives to VBM-style analyses, such as Tract-Based Spatial Statistics *(11)* reduce the likelihood of some of these potential confounds by restricting statistical comparisons to the centres of white matter pathways after non-linear registration of different subject brains to a common tract "skeleton".

2.2. Diffusion Imaging

Diffusion imaging is a catch-all phrase for different MR techniques that rely on sensitizing the MR signal to the motion of water molecules. Crucially, features of this motion are influenced by properties of the local tissue in which the molecules are diffusing, and diffusion distances for water in a typical experiment are on the micrometer scale. Diffusion MRI can therefore give us a microscopic view on tissue architecture that is not available with any other technique *(12)*.

Although the idea that diffusion might influence the MR signal was around in the 1950s *(13)*, and techniques for quantifying this effect from the 1960s *(14)*, the first diffusion-weighted MR image did not appear until the late 1980s *(15)*. It soon became clear that diffusion-weighted imaging had a great deal of promise as a clinical tool: the local apparent diffusion coefficient of water was found to drop by as much as a factor of two immediately after an ischemic stroke *(16)*, hours before the change was visible using any other imaging technique. However, the crucial observation for brain anatomists was, instead, that the diffusion properties of

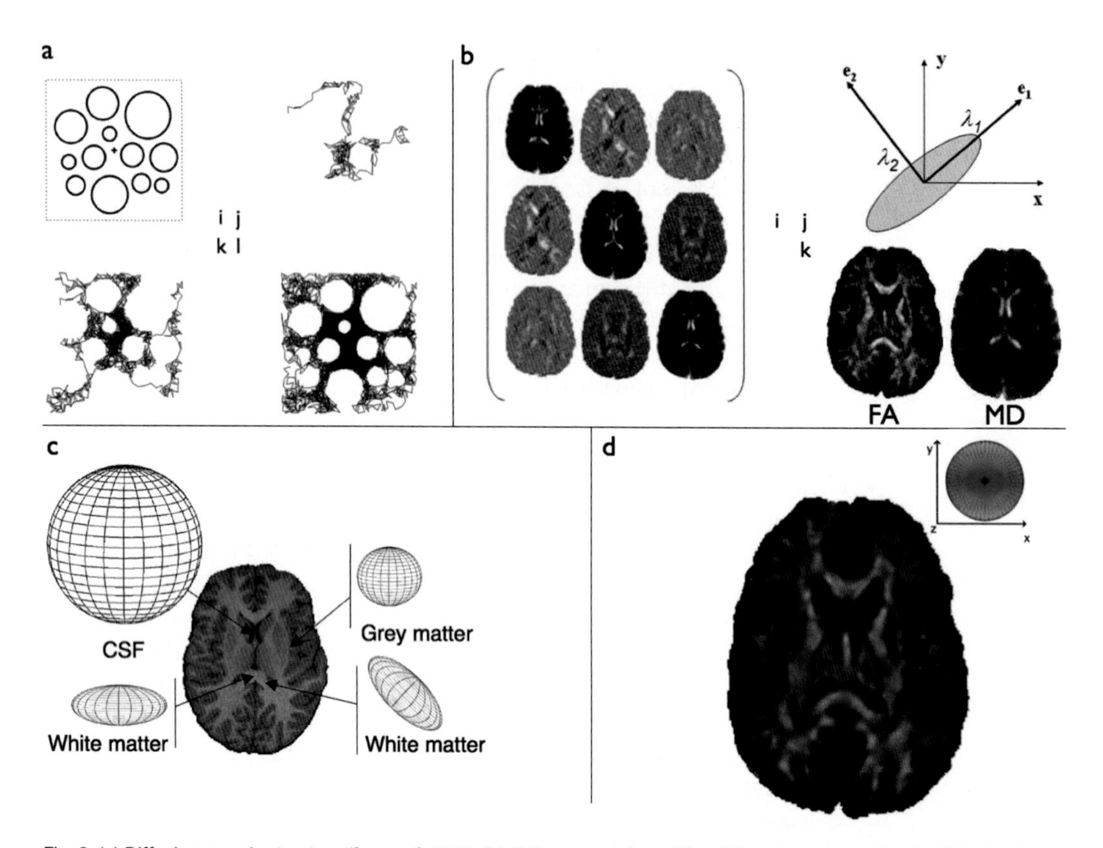

Fig. 2. (**a**) Diffusion reveals structure (from **ref.** *100*). (**b**) [i] Representation of the different components of a diffusion tensor in every brain voxel; [j] ellipsoid representation of a diffusion tensor showing the principal axis and eigenvalues; [k] example of fractional anisotropy (FA) and mean diffusivity (MD) maps. (**c**) Shape, size, and orientation of the diffusion tensor in different tissues. (**d**) RGB Colormap of the tensor principal orientation, showing orientational contrast in the white matter.

water were influenced by the orientational structure of the surrounding tissue (**Fig. 2a**). In brain tissue, diffusion is hindered by the cellular structures, which act as barriers to diffusion, but in white matter diffusion is hindered more when measured perpendicular than parallel to the long axis of the axons *(17)*. This phenomenon is known as diffusion anisotropy – the diffusion coefficient is different when measured in different orientations - and is central to all modern diffusion imaging techniques that focus on white matter anatomy. Not only does diffusion anisotropy allow us to infer the average orientation of axons in an imaging voxel, but it also allows us to quantify the degree to which biological barriers are hindering diffusion. Diffusion anisotropy depends on the myelination of the axons (e.g., **refs.** *18, 19)*, but is present even in the absence of myelin (e.g., **refs.** *20, 21)*. It depends on the diameter of the axons *(22, 23)*, and on the packing density *(23)*. Therefore, diffusion anisotropy is sensitive to many features that might contribute to the ease of passage of an electrical signal along an axon bundle.

To measure diffusion anisotropy and the preferred diffusion orientation in many voxels, it is not sufficient to simply measure parallel and perpendicular diffusion coefficients. Orientations that are parallel to the axon in one voxel will be off-axis in others. The 1990s saw the development and popularisation of diffusion tensor imaging (DTI) *(24)*, a technique that is capable of quantifying diffusion characteristics from apparent diffusion coefficient measurements along a general set of (at least 6) orientations that do not need to be parallel and perpendicular to the crucial axis. Despite recent advances, DTI is still the most popular technique for quantifying diffusion in the in vivo brain. The central premise of DTI is that diffusion in a voxel conforms to a 3D Gaussian distribution. The diffusion tensor can be written mathematically as a symmetric 3×3 matrix, but is often represented graphically as an ellipsoid **(Fig. 2b)**. More anisotropic diffusion results in a change in the relative values of different elements of the tensor, and the diffusion ellipsoid deviates further from a sphere **(Fig. 2c)**. Once a diffusion tensor is fit to the data at each voxel, a number of useful parameters can be calculated from this tensor. In particular, the mean diffusivity (MD) quantifies the average level of diffusion across all orientations; and the fractional anisotropy (FA) quantifies the degree to which the diffusion is directionally dependent or anisotropic **(Fig. 2b)**. The principal diffusion direction (PDD) corresponds to the main axis of the tensor (its principal eigenvector) while the perpendicular diffusion directions correspond to the second and third orthogonal axes. The amount of diffusion along each of these axes can also be calculated.

The estimates of PDD are used to perform what is known as diffusion tractography. In coherent fibre bundles, the PDD corresponds to the underlying fibre direction **(Fig. 2d)**. Tractography algorithms follow these PDD estimates to reconstruct entire white matter trajectories and infer anatomical connectivity in the living human brain *(25–29)*. One limitation of simple tractography algorithms is the assumption that each voxel contains a single, coherent fibre bundle. In fact, we know that many voxels in the brain will contain more than one fibre population, oriented in different directions. Recently, techniques for uncovering these multiple fibre orientations have been proposed *(30)* and used in tractography routines *(31–33)*. The ability to trace paths through these regions of crossing fibres provides improved performance in tracking certain bundles *(31)*.

2.3. Microstructural and Functional Bases of Observed Structure

A key question in MR-based structural analyses such as VBM and diffusion imaging is how to interpret the results of an analysis relative to the underlying neuroanatomy. As both techniques use MR data to study the structure, they are faced with two major limitations: *(1)* they are indirect measures of the quantities of interest, in that they look at the effect of tissue structures on protons' magnetic properties, and *(2)* they have limited spatial resolution, looking at microscopic tissue structure at a millimetric scale. Other technical limitations may also add-up to further complicate the analysis and interpretation of the data, such as noise or imaging artefacts. In interpreting VBM and diffusion imaging, one has to take into account these limiting factors.

2.3.1. Interpretation of VBM Measures

The differences revealed by a VBM analysis of grey matter could reflect differences in cell numbers, their size, and density or the amount of myelin in the cortex. Electronmicrograph studies of mouse cerebral cortex reveal that fibres (axons and dendrites) constitute 60% of the volume, whereas cell bodies and blood vessels constitute only 14% and extracellular space is 5% *(34)*. It is therefore likely that actual cortical volume as well as white matter volume is highly influenced by myelin content. On the other hand, increased myelin content could weaken the apparent "greyness" of grey matter and cortex may be inaccurately classified and its volume *under*estimated (*see* **ref.***35)*. For example, in a DTI study of development we found that increasing FA along the superior corona radiata over adolescence was associated with a correlated decrease in grey matter density in prefrontal areas to which the pathway connects *(36)*. The increasing FA and apparently decreasing grey matter density could both be driven by increasing myelination along this pathway. Combining grey matter and white matter analyses could aid interpretation of VBM changes in other contexts. For example, if Wallerian degeneration occurs in a white matter tract, it is likely that the grey matter region where the tract originates from would show demyelination, and hence potentially an observable T1 effect.

Other methodological issues might also lead to possible confounds in a VBM analysis (**Fig. 3a**). For example, the limitation of acquiring data in box-shaped voxels might lead to different

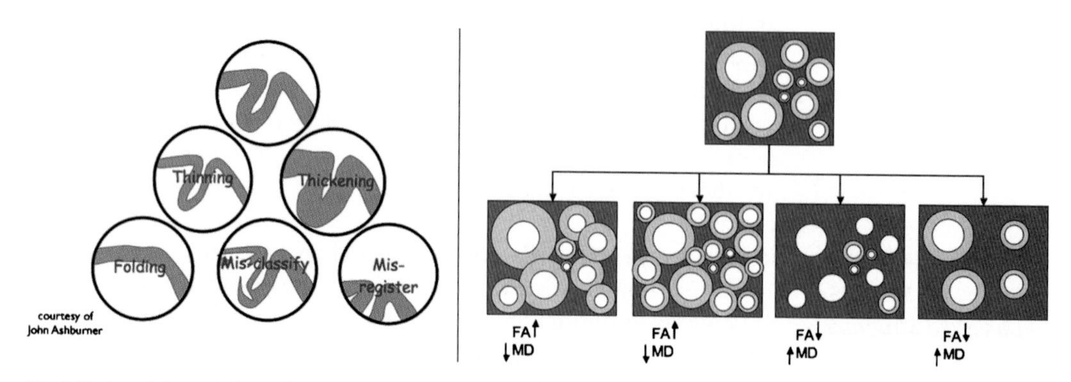

Fig. 3. Various interpretations of voxel-based morphometry (VBM) (*left*) and fractional anisotropy/mean diffusivity (FA/MD) changes in diffusion imaging (*right*).

folding patterns, for different scanned brains, within a single voxel. More folding can then be interpreted as greater cortical thickness. Tissue-type classification and misregistration can also affect the results and lead to erroneous interpretations.

2.3.2. Interpretation of Diffusion Measures

Brain tissues act as barriers that hinder water diffusion. Measuring diffusion hence gives an idea about how tissues are structured *(20)*. For example, more axonal packing makes extracellular diffusion more tortuous, and decreases the mean diffusivity while increasing diffusion anisotropy. It is then tempting to interpret differences in FA and MD between subjects, in terms of variations in axonal bundle organisation. This organisation obviously closely constrains the way brain areas interact. For example, the axonal myelin sheath provides electrical insulation that makes the conduction speed higher. Axonal density, on the other side, might reflect a higher degree of connectivity between two areas, and lead to a lower "activation" threshold. There is, however, not a one to one relationship between diffusion measurements and tissue properties and so different tissue configurations might lead to the same values for diffusion properties. **Figure 3b** illustrates examples of FA/MD changes for different tissue properties. In this example, FA increases with more myelin, or more axons, and decreases with less myelin or fewer axons. To disentangle these two effects, more detailed models *(37)* or combination with other sources of data must be used.

3. Integration of Structure and Function

Information processing in the brain obeys two contradictory, but complementary, principles: functional segregation and integration. The nodes of a brain functional network consist of areas whose functional specialisation is reflected by their neuronal cell composition and intrinsic connectivity. On the other hand, the external connectivity of brain areas constrains the nature and amount of information an area has access to. The combination of these two features in brain networks reflects their complexity and constrains their dynamics *(38)*.

This chapter reviews some recent advances in neuroimaging research that concern the study of regional segregation and integration, at a systems level, both in terms of structure and function. By combining diffusion data with functional data, neuroscientists are now able to identify functional units at an individual level, and in vivo. It is also possible to study the dynamics of these units, and their large-scale interactions, as well as their variations across subjects, and breakdown in disease.

3.1. Relating Functional Borders to Anatomical Borders

A goal of early neuroanatomists, such as Brodmann and Walker, was to define regional borders between anatomically distinct brain regions. This was achieved by microscopic study of the cytoarchitecture of post-mortem brain sections. These microstructural borders are not visible to us in vivo. Until recently, the only anatomical characteristics visible using brain imaging of living brains were gross features such as sulci and gyri. Although these features can provide useful landmarks in some circumstances (e.g., the calcarine fissure for primary visual cortex), there are brain regions in which the correspondence between gross anatomy and underlying microstructural boundaries is poor (e.g., medial and lateral prefrontal areas) *(39)*. However, structural imaging approaches are now beginning to develop methods for defining anatomical boundaries in the living human brain. Parallel fMRI studies have demonstrated that these borders have functional significance.

The ability to define functionally relevant, microstructural borders in vivo opens up new possibilities in the study and treatment of the human brain. First, it allows us to perform parallel studies of anatomy and function in the same individuals. This can be useful for the study of normal brain functional anatomy, and can also be critical for localising altered responses in clinical populations. In brain areas with high interindividual anatomical variation, which do not overlap well across subjects, such approaches enable us to define anatomically accurate regions of interest in individual subjects, without the need to perform cross-subject registration. The ability to accurately identify brain structures in this way could also have significant application in neurosurgery, where pre-operative functional and structural MRI could be used to optimise targeting of specific structures of interest.

3.1.1. High-Resolution Structural MRI

The clear contrast between grey and white matter that is visible on conventional T1-weighted MRI is due to the greater presence of myelin in the white matter. However, there are also variations in myelin content within the grey matter, and this myeloarchitecture can provide useful information about the location of anatomical boundaries. In the visual system, for example, the primary visual cortex, V1, is identified histologically by the presence of a highly myelinated stripe, the "stria of Genari" within the cortical layers, hence the name "striate cortex". Although this feature is not visible on conventional MRI, by using high-resolution, contrast-optimised T1-weighted imaging, it is possible to visualise the stria of Genari in vivo *(40, 41)*. Independently acquired fMRI data from the same subjects confirms that borders identified using functional criteria co-localise extremely well with those identified using this structural information *(41)*. The visual motion area, V5, can also be defined based on its characteristic myeloarchitecture, and again, parallel fMRI studies confirm the functional significance of borders defined in this way *(42)*.

There have been recent suggestions that quantitative relaxometry may be particularly sensitive to detection of microstructural boundaries, e.g., within the thalamus *(43)*. Future work should assess the functional significance of such boundaries.

3.1.2. Diffusion MRI

Diffusion images can also be used to define regional borders within subcortical or cortical grey matter *(44)*. Areas that differ in their functional properties and cytoarchitecture are also likely to have different patterns of anatomical connectivity. As diffusion imaging can provide information on anatomical connections, we can use this information to define borders between regions. One subcortical structure in which such borders are very clear is the thalamus *(45, 46)*. The thalamus is a deep grey matter structure that is made up of multiple, cytoarchitectonically distinct nuclei. Each nucleus relays a different type of information to a distinct area of the cortex. Borders between different nuclei cannot be visualised using conventional MRI, but as each nucleus has a different pattern of anatomical connectivity, we can use diffusion imaging to define these borders. Simply by considering the estimated fibre direction at each location within the thalamus, distinct clusters, with different local fibre directions, can be identified, and are thought to correspond to nuclei *(46)*. Using tractography, it is possible to trace these directions beyond the thalamus to their cortical targets, and create a parcellation of the thalamus based on these differing patterns of cortical connectivity *(45)*. Thalamic subregions identified in this way have been shown to have functional relevance: a meta-analysis of fMRI studies reporting functional activations within the thalamus found that activations during hand movement were located within the thalamic region that connected with sensorimotor cortices, whereas activations during tasks of memory or executive control were located within the thalamic region that connected with the prefrontal cortex *(47)* (**Fig. 4a**).

Diffusion imaging and tractography can also be used to define regional boundaries within cortical grey matter in vivo. Again, the principle is to define boundaries based on detecting changes in anatomical connectivity patterns. This principle was first applied in the medial frontal cortex, where Brodmann's Area 6 can be divided into at least 2 cytoarchitectonically distinct areas – the supplementary motor area (SMA) and pre-supplementary motor area (pre-SMA) *(48)*.These areas have distinct functional roles and different patterns of anatomical connectivity. The SMA is important in motor control *(49)* and connects with motor cortices *(50)*, while the pre-SMA is involved in cognitive functions *(51)*, and has strong connections with prefrontal and cingulate cortices *(52)*. There is no local landmark (e.g., a sulcus) that can be used to define the border between these two regions of the human brain in vivo. However, diffusion imaging can be used to

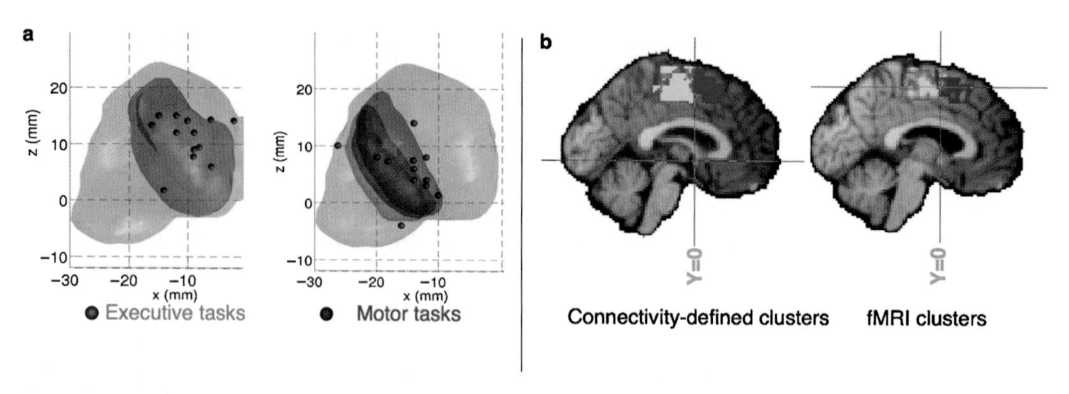

Fig. 4. Relating anatomical borders to functional borders. (**a**) Thalamus: The pale grey shape represents the thalamus in a single hemisphere. Within the thalamus, coloured spheres represent peaks of activation from functional imaging studies and coloured volumes represent areas defined as having high probability of connection with prefrontal cortex (dark grey, *left*) or with premotor (red), primary motor (blue), and primary sensory (green) cortices. Activations during executive tasks co-localise with the region showing high probability of connection to the prefrontal cortex while activations during motor tasks co-localise with the regions showing high probability of connection with sensorimotor cortical areas. Data from *(47)*; (**b**) supplementary motor area. The image on the *left* shows a connectivity based parcellation of supplementary motor area (SMA) (yellow) and pre-SMA (blue). The image on the *right* shows the results of an fMRI study selectively activating SMA with a movement task (red to yellow) or pre-SMA with a serial subtraction task (blue to light blue). Data from *(48)*.

define this border in living human subjects. As with the thalamus, it is possible to define this boundary either based on local diffusion properties, or based on the results of tractography which traces all pathways from the area of interest. By examining the PDD at each voxel within this medial frontal area, it is possible to define the location of a sudden change in these directions *(53)*. Using probabilistic tractography, it is possible to trace all pathways from this region and define a boundary at a location where these pathways suddenly change *(54)*. In both cases, independently acquired fMRI data from the same subjects were used to demonstrate the functional significance of these boundaries. Subjects were asked to perform a finger tapping task (expected to activate the SMA), and a serial subtraction task (expected to activate the pre-SMA). Task-related activations were used to define functional boundaries which co-localised well with the borders defined using anatomical criteria alone **(Fig. 4b)**.

A similar approach has been taken in the lateral part of Brodmann's Area 6, which consists of cytoarchitectonically distinct subregions, including the dorsal premotor cortex and ventral premotor cortex. Again, there is no local gross anatomical landmark that reliably identifies the border between these two regions, but studies in non-human primates suggest that they have distinct patterns of anatomical connectivity, consistent with their different functional specialisations *(55)*. Tractography reliably defines the border between these regions anatomically and demonstrates that the two regions form part of distinct parieto-prefrontal circuits

(56). A functional meta-analysis of fMRI activations within premotor cortex demonstrates the location of a functional border between these regions *(57)*, and Tomassini et al. *(56)* find that this functional border is located close to the border defined using anatomical connectivity.

3.2. Combining Functional and Structural Data for Network Studies

A simple example of integration of functional and structural imaging is the use of fMRI mapping to define "seed points" for diffusion tractography. If both fMRI and DTI data are acquired in the same individuals, an initial fMRI study can be used to locate functional responses in the visual cortex. The regions of increased blood oxygen level dependent (BOLD) signal can then be used to seed diffusion tractography to reconstruct the visual pathways *(58, 59)*. A similar approach has been taken in the motor system *(60)*. There can be considerable disparity between the location of sulcal and gyral landmarks and the position of underlying cytoarchitectonic boundaries. It might therefore be expected that the use of fMRI activation to seed tractography can prove more sensitive than use of anatomical criteria (typically the location of sulci and gyrus) to define seed regions. The usefulness of functional localisation of tractography seeds has been shown particularly convincingly in the context of neurosurgery. Tractography can provide useful information on the location of critical pathways that should be avoided to limit functional deficits post-surgery *(61)*. However, the presence of a tumor or other pathology can distort fibre pathways and therefore render tractography challenging, potentially leading to inability to trace pathways that are still present. By seeding tractography from areas defined using functional imaging, rather than anatomical criteria, it is possible to improve tracing of fibre tracts that have been displaced by a tumor *(62)*, or that pass through low anisotropy regions of perifocal oedema *(63)*.

Using functional imaging to provide functional "nodes" for investigating an anatomical network has also been used to test cognitive hypotheses. Aron and colleagues used a triangulation technique to show that a cognitive network composed of the inferior prefrontal cortex, subthalamic nucleus, and pre-SMA is anatomically interconnected and has a specific role in stopping motor responses *(64)*. The triangulation approach consisted of seeding anatomical networks using probabilistic tractography from each of the three activation sites. Each "node" of the network showed a distinct pattern of anatomical connectivity, but the area of overlap in connections from any two areas among those three always corresponded to the third node of the network. This strengthened the idea of an anatomical-functional network strongly supported by both functional and structural data.

Other approaches to anatomical–functional network studies use explicit models of network structure, in contrast to the more descriptive and exploratory approaches described before. Several methods for modelling functional networks have been proposed in the literature, and some of them have been commented on in previous chapters (e.g., structural equation modelling, dynamic causal modelling). Using in vivo data, such as fMRI or magento- or electro-encephalography (M/EEG), one can study interactions among brain areas during the execution of a given task. Usually, such models require the specification of an anatomical substrate, i.e., the connectivity pattern of the functional network. Ramnani et al. propose using diffusion data to constrain the functional network's connections, as direct functional interaction is contingent on the existence of a physical connection *(65)*.

The benefit of having a functional network model is that, given some functional data, one can test for different network configurations, and assess which one is best supported by the functional data. For example, Kumar et al. used dynamic causal modelling to compare different anatomical models for the connectivity of a functional network responsible for sound processing *(66)*. Two types of models were compared, implying either parallel or serial processing of sounds, with a modulation of the functional connection by the frequency content of the sounds. The favoured model in this study was a serial connection from the primary auditory cortex, to the superior temporal sulcus, through the planum temporale, with a frequency modulation of the first functional connection. This finding was consistent across subjects, and supports the idea that dynamic causal modelling can capture the functional "structure" of a given task.

The limitation of such a method is that the support for one network configuration rather than another is solely driven by the functional data, which is obviously task-dependent. A different paradigm would therefore possibly favour another type of network. There is also a limitation in using fMRI data for studying functional neuronal networks. fMRI is an indirect measure of the neuronal activity, through hemodynamic processes. This means that the temporal resolution of such data does not allow us to distinguish between direct and indirect connections between systems. In other words, the functional structure inferred from the data during a given task might not reflect the underlying true anatomical structure.

Assessing the results of the functional network analysis against diffusion tractography would partly address the question, or one can use Ramnani's suggestion in constraining the structure of functional networks using tractography, but diffusion data itself is not gold standard, and tractography results are difficult to interpret as ground truth. Furthermore, a given anatomical connection is not necessarily used during the functional task under investigation, and as we have already said, functional connections might not be supported by diffusion data if they are indirect. Future research directions are heading towards a symmetric fusion between functional and anatomical models. In such network models, connections would therefore have to explain both functional and diffusion data *(67)*.

3.3. Relating Functional Variation to Structural Variation

There is large interindividual variation in brain structure. This is present at the gross morphological level, seen in differences in sulcal folding patterns, and at the microstructural level, reflected in variation in quantitative measures such as FA or VBM-derived metrics such as grey matter density. Such variations can reflect individual differences, development and ageing, or pathology. Recent studies have demonstrated that these structural differences in brain anatomy are linked to functional variations in brain physiology.

3.3.1. Anatomical and Functional Variation Within the Healthy Population

We have previously discussed the use of diffusion imaging in concert with functional experiments to investigate the relationship between regional brain connectivity and functional localisation. Anatomical connections constrain the type of information available to a brain region, and therefore the type of functional processing it can perform *(68)*. A similar logic might be used to predict the relationship between the quality of information delivered to a particular brain region and the functional response of that region. The relationship between such an ill-defined notion as the quality of information, and the gross measures that are commonly taken from diffusion MRI experiments is necessarily opaque. However, as we have previously discussed in this chapter, many biophysical mechanisms that may ease the passage of an electrical signal between brain regions are also likely to increase the anisotropy of diffusion in the connecting white matter. In cases where this can be shown, not only does it suggest that differences in functional responses are mediated by the underlying brain anatomy, but it also provides a direct functional relevance for the diffusion tractography data that go some way towards providing (albeit indirect) validation.

Despite the logic presented above, it is by no means clear that such a relationship should generally hold. Event-related potentials and transcranial magnetic stimulation data suggest that diffusion properties of the mediating white matter underlie both the signal transfer time between connected cortical regions *(69)* and the influence that one brain region can have on the excitability of another *(70)*. However, there are many factors that might influence the regional BOLD response that have no obvious dependence on the integrity of the incoming white matter pathways. For example, differences in grey matter density *(71)*, cortical thickness, and neurovascular coupling *(72)* have been shown to have measurable effects on the fMRI signal. An important step therefore, is to test the hypothesised structure-function relationship in experimental settings where the BOLD response is robust and well-characterised. Toosy et al. chose a simple visual paradigm to test the relationship *(59)*. They performed an fMRI experiment with visual stimulation, and a separate diffusion MRI experiment in the same subjects. The diffusion experiment allowed the authors to trace the route of the optic radiations (the pathways that carry visual information into the primary visual cortex), and to generate measurements of mean tract FA in each subject. The functional response in the primary visual cortex showed a compelling correlation across subjects with the FA measure reflecting the integrity (or quality) of the optic radiation – the most relevant white matter projection.

Toosy's experiment leads us to believe that we can expect a relationship between functional responses and FA measurements. However, it does not provide insight about the functional specificity of the relationship. By selecting only the FA from the optic radiation and correlating it only with a response to visual stimulation it leaves open the possibility that particular subjects might have a general increase in anisotropy that is related to generally increased functional responses. Olesen et al. addressed this issue in the context of the simultaneous maturation of structure and function *(73)*. Using a standard working memory paradigm, the authors activated a familiar fronto-parietal network in children aged between 8 and 18 years, and acquired diffusion data in the same set of subjects. Within this fronto-parietal network, BOLD activity correlated with individual subject's working memory performance, and when the same measure was used to interrogate the FA data, it correlated with white matter regions that mediate the connections in this network (medial superior longitudinal fasciculus, and the genu of the corpus callosum). Given that each of these imaging measures correlates with working memory performance in crucial parts of the network, it is perhaps unsurprising that they also correlated with one another (even after factoring out the effect of age). However, what is notable is that the effect was specific to the network in question. It cannot represent differential general maturation speeds between subjects, as the BOLD response from the fronto-parietal network correlated with the FA only in regions of white matter mediating the same network.

The functional response of a set of brain regions can therefore be related to features of the subserving white matter that can be measured with diffusion imaging. If it survives the test of time, this idea will prove to be a powerful tool. Different brain networks underlie different brain functions, and many everyday tasks are solved by the combination of such functional networks. By simply examining the relative integrity of the white matter in different brain systems, we may, for example, be able to predict the degree to which subjects will use one brain system over another in performing a particular task. The language system provides us with a richly studied test bed for such an idea. Language tasks tend to engage brain regions in the left hemisphere more than those in the right in most individuals, and the degree of this functional asymmetry will vary among individuals. Powell et al. showed that this functional asymmetry was mirrored by an asymmetry in the FA of the arcuate fasciculus – a fibre bundle connecting Broca's and Wernicke's area in the left hemisphere and their homologues in the right, and the pathway most likely to carry relevant signals *(74)*. Moreover, the degree to which individual subjects relied on the left over the right hemisphere in functional processing of language was predicted by the degree of asymmetry in the integrity of their arcuate fasciculi, as measured by FA. Catani et al. recently showed that asymmetry in these language pathways vary widely across the population, with a tendency for male brains to have stronger lateralisation than do female brains. Further, this study showed a behavioural correlate of the anatomical variation: less asymmetric brains provide an advantage in performing a verbal memory task *(75)*.

Variation in the gross morphology of the language system has also been found to have behavioural and physiological relevance. The primary auditory cortex is located on the transverse gyrus of Heschl (HG), which lies diagonally across the superior surface of the human temporal lobe. Anteriorly it is bounded by the transverse sulcus and posteriorly by Heschl's sulcus. It is common, however, for multiple transverse gyri to occur in either or both hemispheres, a phenomenon noted as early as 1920 *(76, 77)*. There are two common types of duplication (more rarely triplication) of HG: the "posterior duplication" occurs when an additional sulcus extends medially and laterally resulting in two complete transverse gyri and the "common stem" occurs when an intermediate sulcus forms an indentation on the surface of HG but does not extend its full length, causing the duplicated HG to appear heart-shaped when viewed in the sagittal plane *(78)*. In the normal population, duplications in HG are most common in the right rather than left hemisphere. Duplications on the left have been associated with developmental disorders such as autism, dyslexia, and schizophrenia *(79)*. To explore the functional effects of this gyral variation in the normal population, we undertook a small pilot study of 19 individuals. Of these, 6 showed the typical configuration of one HG in the left hemisphere and two in the right, 6 had one HG in each hemisphere, and 5 had duplications in both hemispheres; the remaining 2 had two HG in the left and one on the right. Each subject was tested using a mismatch negativity EEG paradigm that used deviants graded in terms of magnitude of deviation from a standard stimulus separately in the time (duration) and frequency (pitch) domains. The most typical pattern of event-related potentials was observed in the individuals with a single HG in the left and a duplication in the right

(Jamison, Bishop, Matthews and Watkins, under review). These individuals also exhibited the strongest left-lateralisation in grey matter volume in primary auditory cortex. These results suggest that individual variation in auditory cortex morphology has a significant effect on measures of auditory function.

3.3.2. Anatomical and Functional Variation in Clinical Populations

Variation in brain structure across individuals can also relate to functional differences reflecting brain disease or disorder. Below we describe a few specific examples to illustrate such differences.

A number of studies have shown that recovery from brain damage, such as stroke, is associated with functional plasticity, reflected by altered fMRI responses *(80–85)*. One recent study relates this functional plasticity to co-localised structural changes: the somatosensory cortex of patients who had recovered from stroke showed increased BOLD response relative to healthy control subjects; this same cortical area was also found to be thicker in patients than in controls. The increase in cortical thickness observed in the somatosensory cortex was not found in any control cortical regions. This was the first observation of such specific structural changes in the human brain post-stroke and suggests a potential anatomical substrate for the altered functional response *(86)*. These gross changes in the human brain might potentially reflect underlying changes in cellular morphology or cortico-cortical connections that have been described following experimental stroke in animal models *(87, 88)*.

Large-population VBM studies confirm known structural asymmetries in the healthy human brain, such as a rightwards frontal and leftwards occipital asymmetry as well as a leftwards asymmetry in the planum temporale *(8)*. As discussed above, structural asymmetries in the language system relate to functional asymmetry. VBM was used to examine structural hemispheric asymmetries in patients who underwent sodium amytal testing to determine hemispheric dominance for language prior to epilepsy surgery *(89)*. Patients with left- or right-hemisphere dominance for language or bilateral representation did not differ in the degree of asymmetry in the planum temporale. In HG, three subjects with right-hemisphere speech showed a right-larger-than-left volume asymmetry; the other eight subjects with right-hemisphere speech had the same left-larger-than-right volume asymmetry as those with left-hemisphere speech. This suggests that the anatomical asymmetry does not necessarily follow the direction of the functional asymmetry. In other words the structure-function relationship is not obligatory in HG. The fact that deaf subjects also show the expected left-larger-than-right volume differences in the absence of auditory stimulation *(90, 91)* provides further evidence to suggest that morphological asymmetries in the auditory cortex regions may be predetermined at an early developmental stage. Nonetheless, these auditory regions may still interact to some extent with environmental input. For example, Schneider et al. *(92)* showed that professional musicians displayed larger HG than did nonmusicians. On the other hand, for the pars opercularis, it was seen that every one of the subjects with right-hemisphere speech had a right-greater-than-left difference in grey-matter volume. Thus, the anatomical asymmetries in this region seem to be more closely related to the functional asymmetry. This relationship may reflect a use-dependent reorganization. The left pars opercularis is related to speech production and is close to motor and premotor cortices. Many VBM studies have found volume increases in motor regions related to use-dependent factors. For example, Schlaug *(93)* showed that the motor cortex region of musicians revealed grey matter volume changes corresponding to the hand use of these musicians. Also, Penhune et al. *(91)* found that deaf subjects, who use the right hand for signing, showed an increase of grey-matter density in their left motor hand region. Thus we speculate that the morphological asymmetries observed here in the subjects with right-hemisphere speech may also be use-dependent. In other words, patients with right-hemisphere speech may employ more the right motor-related regions in the expression of speech functions. In conclusion, HG morphological asymmetry may be predetermined and more resistant to change, whereas the pars opercularis morphological asymmetry may be related to use-dependent factors. Asymmetries in these two regions appear to bear a stronger relationship to language lateralization than asymmetry in the planum temporale.

Using functional and diffusion imaging, we examined brain structure and function in the motor and language areas in a group of young people who stutter *(94)*. During speech production, irrespective of fluency or auditory feedback, the people who stuttered showed overactivity relative to controls in the anterior insula, cerebellum and midbrain bilaterally, consistent with suggestions that the function of the basal ganglia or dopamine system is abnormal in people who stutter. People who stutter also showed significantly less activity relative to controls in the ventral premotor, Rolandic opercular and sensorimotor, cortex bilaterally and Heschl's gyrus on the left. Analysis of the diffusion data revealed that the integrity of the white matter underlying the underactive areas in ventral premotor cortex was reduced in people who stutter (**Fig. 5**). The white matter tracts in this area via connections with posterior superior temporal and inferior parietal cortex provide a substrate for the integration of articulatory planning and sensory feedback, and via connections with primary motor cortex, a substrate for execution of articulatory movements. The difference in activity of this cortex might be explained, therefore, by a difference in the organisation or integrity of inputs to this region.

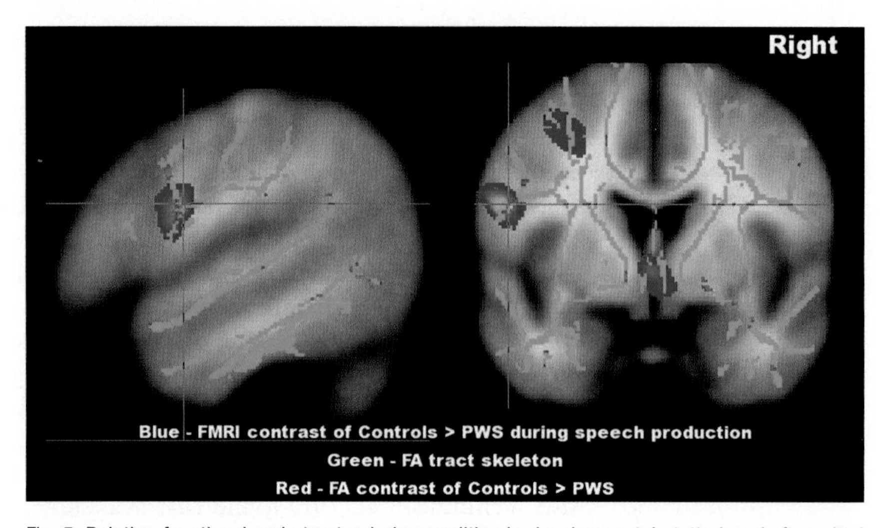

Fig. 5. Relating functional and structural abnormalities in developmental stuttering. *Left*, sagittal slice through the left hemisphere; *right*, coronal slice. Blue areas indicate significant differences in FMRI signal obtained during speech production in people who stutter (PWS) and controls (PWS < Controls). The green skeleton indicates the centre of white matter tracts from a group average of the diffusion data. The red/orange regions indicate significantly reduced FA in PWS relative to controls. The cross-hairs indicated the white matter voxels which showed a significant reduction in fractional anisotropy (FA) in PWS, which underlies the functional underactivation in the cortex in the same subjects (*see* **ref. *94***).

As we have discussed, there are good reasons why we might expect a relationship between the diffusion properties of the white matter and the functional response in cortex, and early studies suggest that such a relationship will hold. However, other measures from fMRI may be more directly related to the white matter integrity. More sophisticated analyses of fMRI data can tell us about the degree of coherence between functional responses (functional connectivity) *(95, 96)* and may even give us some insight into the influence of one brain region on another *(97)*. We may expect such measures to be directly related to the integrity of the specific white matter pathways mediating the functional interactions, as we have previously shown using transcranial magnetic stimulation measures of functional connectivity in the context of an action-selection task in healthy volunteers *(70)*. A recent experiment by He et al. *(98)* follows this logic using fMRI measures of functional connectivity. They analysed a fronto-parietal network during an attentional processing task, on a population of patients with neglect. They found that functional connectivity, as measured by coherence between regional fMRI signals, was disrupted in the ventral part of the network. This functional disruption strongly correlated with impaired attentional processing across different subjects. Interestingly, they also found that the integrity of the white matter tracts connecting the parietal to the frontal cortex correlated both with the severity of the neglect and also with the disrupted functional connectivity. Larger lesions in the white matter tracts were associated with greater disruption of the functional connectivity between the areas connected by that tract. This approach of relating tract damage to functional disruptions and neurological impairments could be generally useful in the study of patients with brain damage, allowing the role of "disconnection" to be explicitly addressed *(99)*.

4. Conclusions

By combining information from structural and functional imaging, we can gain a far richer picture of the anatomical substrate of observed functional responses. Diffusion imaging provides insights into structural networks that can be explicitly related to functional interactions across brain circuits. Structural MRI supplies information on gross brain morphology and can potentially provide measures of tissue microstructure. Relating variation in these anatomical features to observed functional responses will provide powerful insights into the relationship between brain and behaviour and its breakdown in disease.

References

1. Lohmann G: Extracting line representations of sulcal and gyral patterns in MR images of the human brain. IEEE Trans Med Imaging 1998;17:1040–1048.
2. Lohmann G, von Cramon DY, Steinmetz H: Sulcal variability of twins. Cereb Cortex 1999;9:754–763.
3. Fischl B, Dale AM: Measuring the thickness of the human cerebral cortex from magnetic resonance images. Proc Natl Acad Sci USA 2000;97:11050–11055.
4. Ashburner J, Friston KJ: Voxel-based morphometry - the methods. Neuroimage 2000;11:805–821.
5. Wright IC, McGuire PK, Poline JB, Travere JM, Murray RM, Frith CD, Frackowiak RS, Friston KJ: A voxel-based method for the statistical analysis of gray and white matter density applied to schizophrenia. Neuroimage 1995;2:244–252.
6. Gadian DG, Mishkin M, Vargha-Khadem F: Early brain pathology and its relation to cognitive impairment: the role of quantitative magnetic resonance techniques. Adv Neurol 1999;81:307–315.
7. Watkins KE, Vargha-Khadem F, Ashburner J, Passingham RE, Connelly A, Friston KJ, Frackowiak RS, Mishkin M, Gadian DG: MRI analysis of an inherited speech and language disorder: structural brain abnormalities. Brain 2002;125:465–478.
8. Watkins KE, Paus T, Lerch JP, Zijdenbos A, Collins DL, Neelin P, Taylor J, Worsley KJ, Evans AC: Structural asymmetries in the human brain: a voxel-based statistical analysis of 142 MRI scans. Cereb Cortex 2001;11:868–877.
9. Paus T, Zijdenbos A, Worsley K, Collins DL, Blumenthal J, Giedd JN, Rapoport JL, Evans AC: Structural maturation of neural pathways in children and adolescents: in vivo study. Science 1999;283:1908–1911.
10. Mechelli A, Friston KJ, Frackowiak RS, Price CJ: Structural covariance in the human cortex. J Neurosci 2005;25:8303–8310.
11. Smith SM, Jenkinson M, Johansen-Berg H, Rueckert D, Nichols TE, Mackay CE, Watkins KE, Ciccarelli O, Cader MZ, Matthews PM, Behrens TE: Tract-based spatial statistics: voxelwise analysis of multi-subject diffusion data. Neuroimage 2006;31:1487–1505.
12. Le Bihan D: Looking into the functional architecture of the brain with diffusion MRI. Nat Rev Neurosci 2003;4:469–480.

13. Hahn EL: Spin echoes. Phys Rev 1950; 80:580–594.
14. Stejskal EO, Tanner JE: Spin diffusion measurements: spin echoes in the presence of a time-dependent field gradient. J Chem Phys 1965;42:288–292.
15. Le Bihan D, Breton E, Lallemand D, Grenier P, Cabanis E, Laval-Jeantet M: MR imaging of intravoxel incoherent motions: application to diffusion and perfusion in neurologic disorders. Radiology 1986;161:401–407.
16. Moseley ME, Butts K, Yenari MA, Marks M, de Crespigny A: Clinical aspects of DWI. NMR Biomed 1995;8:387–396.
17. Moseley M, Cohen Y, Kucharczyk J, Mintorovitch J, Asgari H, Wendland M, Tsuruda J, Norman D: Diffusion-weighted MR imaging of anisotropic water diffusion in cat central nervous system. Radiology 1990;176:439–445.
18. Nair G, Tanahashi Y, Low HP, Billings-Gagliardi S, Schwartz WJ, Duong TQ: Myelination and long diffusion times alter diffusion-tensor-imaging contrast in myelin-deficient shiverer mice. Neuroimage 2005;28:165–174.
19. Tyszka JM, Readhead C, Bearer EL, Pautler RG, Jacobs RE: Statistical diffusion tensor histology reveals regional dysmyelination effects in the shiverer mouse mutant. Neuroimage 2006;29:1058–1065.
20. Beaulieu C: The basis of anisotropic water diffusion in the nervous system – a technical review. NMR Biomed 2002;15:435–455.
21. Song SK, Sun SW, Ramsbottom MJ, Chang C, Russell J, Cross AH: Dysmyelination revealed through MRI as increased radial (but unchanged axial) diffusion of water. Neuroimage 2002;17:1429–1436.
22. Beaulieu C, Allen PS: Determinants of anisotropic water diffusion in nerves. Magn Reson Med 1994;31:394–400.
23. Takahashi M, Hackney DB, Zhang G, Wehrli SL, Wright AC, O'Brien WT, Uematsu H, Wehrli FW, Selzer ME: Magnetic resonance microimaging of intraaxonal water diffusion in live excised lamprey spinal cord. Proc Natl Acad Sci USA 2002;99:16192–16196.
24. Basser PJ, Mattiello J, LeBihan D: Estimation of the effective self-diffusion tensor from the NMR spin echo. J Magn Reson B 1994;103:247–254.
25. Behrens TE, Woolrich MW, Jenkinson M, Johansen-Berg H, Nunes RG, Clare S, Matthews PM, Brady JM, Smith SM: Characterization and propagation of uncertainty in diffusion-weighted MR imaging. Magn Reson Med 2003;50:1077–1088.
26. Catani M, Howard RJ, Pajevic S, Jones DK: Virtual in vivo interactive dissection of white matter fasciculi in the human brain. Neuroimage 2002;17:77–94.
27. Conturo TE, Lori NF, Cull TS, Akbudak E, Snyder AZ, Shimony JS, McKinstry RC, Burton H, Raichle ME: Tracking neuronal fiber pathways in the living human brain. Proc Natl Acad Sci USA 1999;96:10422–10427.
28. Mori S, Crain BJ, Chacko VP, van Zijl PC: Three-dimensional tracking of axonal projections in the brain by magnetic resonance imaging. Ann Neurol 1999;45:265–269.
29. Parker GJ, Alexander DC: Probabilistic Monte Carlo based mapping of cerebral connections utilising whole-brain crossing fibre information. Inf Process Med Imaging 2003;18:684–695.
30. Alexander DC: Multiple-fiber reconstruction algorithms for diffusion MRI. Ann NY Acad Sci 2005;1064:113–133.
31. Behrens TE, Berg HJ, Jbabdi S, Rushworth MF, Woolrich MW: Probabilistic diffusion tractography with multiple fibre orientations: what can we gain? Neuroimage 2007;34:144–155.
32. Hosey T, Williams G, Ansorge R: Inference of multiple fiber orientations in high angular resolution diffusion imaging. Magn Reson Med 2005;54:1480–1489.
33. Parker GJ, Alexander DC: Probabilistic anatomical connectivity derived from the microscopic persistent angular structure of cerebral tissue. Philos Trans R Soc Lond B Biol Sci 2005;360:893–902.
34. Braitenberg V: Brain size and number of neurons: an exercise in synthetic neuroanatomy. J Comput Neurosci 2001;10:71–77.
35. Paus T: Mapping brain maturation and cognitive development during adolescence. Trends Cogn Sci 2005;9:60–68.
36. Giorgio A, Watkins KE, Douaud G, James AC, James S, De Stefano N, Matthews PM, Smith SM, Johansen-Berg H: Changes in white matter microstructure during adolescence. Neuroimage 2008;39:52–61.
37. Sen PN, Basser PJ: A model for diffusion in white matter in the brain. Biophys J 2005;89:2927–2938.
38. Tononi G, Sporns O, Edelman GM: A measure for brain complexity: relating functional segregation and integration in the nervous system. Proc Natl Acad Sci USA 1994;91:5033–5037.
39. Amunts K, Schleicher A, Burgel U, Mohlberg H, Uylings HB, Zilles K: Broca's region revisited: cytoarchitecture and intersubject variability. J Comp Neurol 1999;412:319–341.

40. Barbier EL, Marrett S, Danek A, Vortmeyer A, van Gelderen P, Duyn J, Bandettini P, Grafman J, Koretsky AP: Imaging cortical anatomy by high-resolution MR at 3.0T: detection of the stripe of Gennari in visual area 17. Magn Reson Med 2002;48:735–738.
41. Bridge H, Clare S, Jenkinson M, Jezzard P, Parker AJ, Matthews PM: Independent anatomical and functional measures of the V1/V2 boundary in human visual cortex. J Vis 2005;5:93–102.
42. Walters NB, Egan GF, Kril JJ, Kean M, Waley P, Jenkinson M, Watson JD: In vivo identification of human cortical areas using high-resolution MRI: an approach to cerebral structure-function correlation. Proc Natl Acad Sci USA 2003;100:2981–2986.
43. Deoni SC, Josseau MJ, Rutt BK, Peters TM: Visualization of thalamic nuclei on high resolution, multi-averaged T1 and T2 maps acquired at 1.5 T. Hum Brain Mapp 2005;25:353–359.
44. Behrens TE, Johansen-Berg H: Relating connectional architecture to grey matter function using diffusion imaging. Philos Trans R Soc Lond B Biol Sci 2005;360:903–911.
45. Behrens TE, Johansen-Berg H, Woolrich MW, Smith SM, Wheeler-Kingshott CA, Boulby PA, Barker GJ, Sillery EL, Sheehan K, Ciccarelli O, Thompson AJ, Brady JM, Matthews PM: Non-invasive mapping of connections between human thalamus and cortex using diffusion imaging. Nat Neurosci 2003;6:750–757.
46. Wiegell MR, Tuch DS, Larsson HB, Wedeen VJ: Automatic segmentation of thalamic nuclei from diffusion tensor magnetic resonance imaging. Neuroimage 2003;19:391–401.
47. Johansen-Berg H, Behrens TE, Sillery E, Ciccarelli O, Thompson AJ, Smith SM, Matthews PM: Functional-anatomical validation and individual variation of diffusion tractography–based segmentation of the human thalamus. Cereb Cortex 2005;15:31–39.
48. Johansen-Berg H, Behrens TE, Robson MD, Drobnjak I, Rushworth MF, Brady JM, Smith SM, Higham DJ, Matthews PM: Changes in connectivity profiles define functionally distinct regions in human medial frontal cortex. Proc Natl Acad Sci USA 2004;101:13335–13340.
49. Picard N, Strick PL: Motor areas of the medial wall: a review of their location and functional activation. Cereb Cortex 1996;6:342–353.
50. Luppino G, Matelli M, Camarda R, Rizzolatti G: Corticocortical connections of area F3 (SMA-proper) and area F6 (pre-SMA) in the macaque monkey. J Comp Neurol 1993;338:114–140.
51. Rushworth MF, Hadland KA, Paus T, Sipila PK: Role of the human medial frontal cortex in task switching: a combined fMRI and TMS study. J Neurophysiol 2002;87:2577–2592.
52. Wang Y, Shima K, Sawamura H, Tanji J: Spatial distribution of cingulate cells projecting to the primary, supplementary, and pre-supplementary motor areas: a retrograde multiple labeling study in the macaque monkey. Neurosci Res 2001;39:39–49.
53. Behrens TE, Jenkinson M, Robson MD, Smith SM, Johansen-Berg H: A consistent relationship between local white matter architecture and functional specialisation in medial frontal cortex. Neuroimage 2006;30:220–227.
54. Johansen-Berg H, Behrens TE, Robson MD, Drobnjak I, Rushworth MF, Brady JM, Smith SM, Higham DJ, Matthews PM: Changes in connectivity profiles define functionally distinct regions in human medial frontal cortex. Proc Natl Acad Sci USA 2004;101:13335–13340.
55. Rizzolatti G, Luppino G: The cortical motor system. Neuron 2001;31:889–901.
56. Tomassini V, Jbabdi S, Klein JC, Behrens TE, Pozzilli C, Matthews PM, Rushworth MF, Johansen-Berg H: Diffusion-weighted imaging tractography-based parcellation of the human lateral premotor cortex identifies dorsal and ventral subregions with anatomical and functional specializations. J Neurosci 2007;27:10259–10269.
57. Mayka MA, Corcos DM, Leurgans SE, Vaillancourt DE: Three-dimensional locations and boundaries of motor and premotor cortices as defined by functional brain imaging: a meta-analysis. Neuroimage 2006; 31:1453–1474.
58. Werring DJ, Clark CA, Parker GJ, Miller DH, Thompson AJ, Barker GJ: A direct demonstration of both structure and function in the visual system: combining diffusion tensor imaging with functional magnetic resonance imaging. Neuroimage 1999;9:352.
59. Toosy AT, Ciccarelli O, Parker GJ, Wheeler-Kingshott CA, Miller DH, Thompson AJ: Characterizing function-structure relationships in the human visual system with functional MRI and diffusion tensor imaging. Neuroimage 2004;21:1452–1463.
60. Guye M, Parker GJ, Symms M, Boulby P, Wheeler-Kingshott CA, Salek-Haddadi A, Barker GJ, Duncan JS: Combined functional MRI and tractography to demonstrate the connectivity of the human primary motor cortex in vivo. NeuroImage 2003;19:1349.
61. Powell HW, Parker GJ, Alexander DC, Symms MR, Boulby PA, Wheeler-Kingshott CA, Barker GJ, Koepp MJ, Duncan JS: MR

tractography predicts visual field defects following temporal lobe resection. Neurology 2005;65:596–599.
62. Schonberg T, Pianka P, Hendler T, Pasternak O, Assaf Y: Characterization of displaced white matter by brain tumors using combined DTI and fMRI. NeuroImage 2006;30: 1100–1111.
63. Bartsch AJ, Homola G, Biller A, Solymosi L, Bendszus M: Diagnostic functional MRI: Illustrated clinical applications and decision-making. J Magn Reson Imaging 2006;23(6):921–932.
64. Aron AR, Behrens TE, Smith S, Frank MJ, Poldrack RA: Triangulating a cognitive control network using diffusion-weighted magnetic resonance imaging (MRI) and functional MRI. J Neurosci 2007;27:3743–3752.
65. Ramnani N, Behrens TE, Penny W, Matthews PM: New approaches for exploring anatomical and functional connectivity in the human brain. Biol Psychiatry 2004;56:613–619.
66. Kumar S, Stephan KE, Warren JD, Friston KJ, Griffiths TD: Hierarchical processing of auditory objects in humans. PLoS Comput Biol 2007;3:e100.
67. Jbabdi S, Woolrich MW, Andersson JL, Behrens TE: A Bayesian framework for global tractography. Neuroimage 2007;37:116–129.
68. Passingham RE, Stephan KE, Kotter R: The anatomical basis of functional localization in the cortex. Nat Rev Neurosci 2002;3:606–616.
69. Westerhausen R, Kreuder F, Woerner W, Huster RJ, Smit CM, Schweiger E, Wittling W: Interhemispheric transfer time and structural properties of the corpus callosum. Neurosci Lett 2006;409:140–145.
70. Boorman ED, O'Shea J, Sebastian C, Rushworth MF, Johansen-Berg H: Individual differences in white-matter microstructure reflect variation in functional connectivity during choice. Curr Biol 2007;17:1426–1431.
71. Oakes TR, Fox AS, Johnstone T, Chung MK, Kalin N, Davidson RJ: Integrating VBM into the General Linear Model with voxelwise anatomical covariates. Neuroimage 2007;34:500–508.
72. Hoge RD, Atkinson J, Gill B, Crelier GR, Marrett S, Pike GB: Investigation of BOLD signal dependence on cerebral blood flow and oxygen consumption: the deoxyhemoglobin dilution model. Magn Reson Med 1999;42:849–863.
73. Olesen PJ, Nagy Z, Westerberg H, Klingberg T: Combined analysis of DTI and fMRI data reveals a joint maturation of white and grey matter in a fronto-parietal network. Brain Res Cogn Brain Res 2003;18:48–57.
74. Powell HW, Parker GJ, Alexander DC, Symms MR, Boulby PA, Wheeler-Kingshott CA, Barker GJ, Noppeney U, Koepp MJ, Duncan JS: Hemispheric asymmetries in language-related pathways: a combined functional MRI and tractography study. Neuroimage 2006;32:388–399.
75. Catani M, Allin MP, Husain M, Pugliese L, Mesulam MM, Murray RM, Jones DK: Symmetries in human brain language pathways correlate with verbal recall. Proc Natl Acad Sci USA 2007;104:17163–17168.
76. Pfeifer R: Myelogenetisch-anatomische Untersuchungen über das kortikale Ende der Hörleitung. Abh Math-Physik Kl sächs Akad Wiss Leipzig 1920;37:1–54.
77. Von Economo C, Horn L: Über Windungsrelief Maße und Rindenarchitektonik der Supratemporalfläche, ihre individuellen und ihre Seitenunterschiede. Z Neurol Psychiatrie 1930;130:678–757.
78. Leonard CM, Puranik C, Kuldau JM, Lombardino LJ: Normal variation in the frequency and location of human auditory cortex landmarks. Heschl's gyrus: where is it? Cereb Cortex 1998;8:397–406.
79. Leonard CM, Eckert MA, Lombardino LJ, Oakland T, Kranzler J, Mohr CM, King WM, Freeman A: Anatomical risk factors for phonological dyslexia. Cereb Cortex 2001;11:148–157.
80. Baron JC, Cohen LG, Cramer SC, Dobkin BH, Johansen-Berg H, Loubinoux I, Marshall RS, Ward NS: Neuroimaging in stroke recovery: a position paper from the First International Workshop on Neuroimaging and Stroke Recovery. Cerebrovasc Dis 2004;18:260–267.
81. Cramer SC, Nelles G, Benson RR, Kaplan JD, Parker RA, Kwong KK, Kennedy DN, Finklestein SP, Rosen BR: A functional MRI study of subjects recovered from hemiparetic stroke. Stroke 1997;28:2518–2527.
82. Weiller C, Chollet F, Friston KJ, Wise RJ, Frackowiak RS: Functional reorganization of the brain in recovery from striatocapsular infarction in man. Ann Neurol 1992; 31:463–472.
83. Johansen-Berg H, Dawes H, Guy C, Smith SM, Wade DT, Matthews PM: Correlation between motor improvements and altered fMRI activity after rehabilitative therapy. Brain 2002;125:2731–2742.

84. Matthews PM, Johansen-Berg H, Reddy H: Non-invasive mapping of brain functions and brain recovery: applying lessons from cognitive neuroscience to neurorehabilitation. Restor Neurol Neurosci 2004;22: 245–260.

85. Ward NS, Brown MM, Thompson AJ, Frackowiak RS: Neural correlates of motor recovery after stroke: a longitudinal fMRI study. Brain 2003;126:2476–2496.

86. Schaechter JD, Moore CI, Connell BD, Rosen BR, Dijkhuizen RM: Structural and functional plasticity in the somatosensory cortex of chronic stroke patients. Brain 2006;129:2722–2733.

87. Biernaskie J, Corbett D: Enriched rehabilitative training promotes improved forelimb motor function and enhanced dendritic growth after focal ischemic injury. J Neurosci 2001;21:5272.

88. Dancause N, Barbay S, Frost SB, Plautz EJ, Chen D, Zoubina EV, Stowe AM, Nudo RJ: Extensive cortical rewiring after brain injury. J Neurosci 2005;25:10167–10179.

89. Dorsaint-Pierre R, Penhune VB, Watkins KE, Neelin P, Lerch JP, Bouffard M, Zatorre RJ: Asymmetries of the planum temporale and Heschl's gyrus: relationship to language lateralization. Brain 2006;129:1164–1176.

90. Emmorey K, Allen JS, Bruss J, Schenker N, Damasio H: A morphometric analysis of auditory brain regions in congenitally deaf adults. Proc Natl Acad Sci USA 2003;100:10049–10054.

91. Penhune VB, Cismaru R, Dorsaint-Pierre R, Petitto LA, Zatorre RJ: The morphometry of auditory cortex in the congenitally deaf measured using MRI. Neuroimage 2003;20:1215–1225.

92. Schneider P, Scherg M, Dosch HG, Specht HJ, Gutschalk A, Rupp A: Morphology of Heschl's gyrus reflects enhanced activation in the auditory cortex of musicians. Nat Neurosci 2002;5:688–694.

93. Schlaug G: The brain of musicians. A model for functional and structural adaptation. Ann NY Acad Sci 2001;930:281–299.

94. Watkins KE, Smith SM, Davis S, Howell P: Structural and functional abnormalities of the motor system in developmental stuttering. Brain 2008;131:50–59.

95. Friston KJ, Buechel C, Fink GR, Morris J, Rolls E, Dolan RJ: Psychophysiological and modulatory interactions in neuroimaging. Neuroimage 1997;6:218–229.

96. Salvador R, Suckling J, Schwarzbauer C, Bullmore E: Undirected graphs of frequency-dependent functional connectivity in whole brain networks. Philos Trans R Soc Lond B Biol Sci 2005;360:937–946.

97. Friston KJ, Harrison L, Penny W: Dynamic causal modelling. Neuroimage 2003;19: 1273–1302.

98. He BJ, Snyder AZ, Vincent JL, Epstein A, Shulman GL, Corbetta M: Breakdown of functional connectivity in frontoparietal networks underlies behavioral deficits in spatial neglect. Neuron 2007;53:905–918.

99. Catani M, ffytche DH: The rises and falls of disconnection syndromes. Brain 2005;128:2224–2239.

100. Nicholson C: Diffusion and related transport mechanisms in the brain. Rep Prog Phys 2001;64:815–884.

Chapter 28

Functional MRI of the Spinal Cord

Patrick W. Stroman and Massimo Filippi

Summary

Evidence to date shows that fMRI of the spinal cord (spinal fMRI) can reliably demonstrate regions involved with sensation of tactile, thermal, and painful stimuli, and with motor tasks. The spin-echo-based spinal fMRI method with "signal enhancement by extravascular protons" contrast has been developed more extensively than the BOLD (blood oxygen level dependent)-based method. Results have demonstrated good localization to areas of activity within the spinal cord cross-section and to the spinal cord segmental level, in both the cervical and lumbar spinal cord, with a range of thermal stimuli as well as tactile, vibration, and motor stimuli. The method has also demonstrated the first results in the injured spinal cord with thermal and motor stimuli, and in people with multiple sclerosis with proprioceptive and tactile stimuli. The image quality obtained with this method has also resulted in the ability to obtain 3D data in thin sagittal slices spanning 20 cm, to spatially normalize the results, and thereby apply group analysis methods of partial least-squares, analysis of effective connectivity, and automatic determination of voxel-by-voxel repeatability across studies or volunteers. The availability of essentially automated analysis, large extent coverage of the spinal cord, and spatial normalization to permit comparisons with reference results and labelling of active regions are essential elements for developing the method into a practical clinical assessment tool.

Key words: Spinal fMRI, Signal enhancement by extravascular protons Blood oxygen level dependent, Multiple sclerosis, cord trauma

1. Introduction

Evidence to date shows that fMRI of the spinal cord (spinal fMRI) can reliably demonstrate regions involved with sensation of tactile, thermal, and painful stimuli, and with motor tasks. There is also reliable evidence of the descending modulation of activity in the spinal cord. While spinal fMRI has not yet been applied or verified in a clinical setting, its value is expected to be in its ability to discriminate changes in response to motor and sensory stimuli,

M. Filippi (ed.), *fMRI Techniques and Protocols*, Neuromethods, vol. 41
DOI 10.1007/978-1-60327-919-2_28, © Humana Press, a part of Springer Science + Business Media, LLC 2009

regardless of whether the patient can feel the stimulus or is even conscious, although results obtained to date show that attention to the stimulus or awareness of it can influence the activity that is detected, and will be discussed in detail below. Preliminary studies have been carried out with patients with cord trauma, and in people with multiple sclerosis (MS) to investigate the clinical utility of the results. Robust methods for analysis, and for displaying the results in an effective manner to facilitate their interpretation, are also necessary. At present, the usefulness and reliability of spinal fMRI as a research tool has been demonstrated, analysis and display methods have been developed, and further improvements are forthcoming. The research completed so far indicates that spinal fMRI will be able to demonstrate where the neuronal activity is altered at any level (cervical, thoracic, lumbar, or sacral), whether or not information is reaching the cord from the periphery, and whether or not there is descending modulation of the response. It may also be able to provide an objective measure of pain, and to demonstrate the extent and mechanism of changes over time after an injury.

In the following paragraphs, we discuss the current evidence for the most effective spinal fMRI methods and the points that are still under debate, the applications that have been carried out to date and the degree of reliability and sensitivity these studies demonstrate, and the proposed future developments and applications.

2. Background of Spinal fMRI

As with any fMRI method, spinal fMRI requires alternated periods of stimulation and of a baseline reference condition, while a time series of images is acquired over several minutes. Neuronal activity is detected only in gray matter regions, and is revealed by the local MR signal intensity having a component of signal change that corresponds with the stimulation paradigm. Unlike brain fMRI, unique challenges are encountered in the heterogeneous tissues of the spine and spinal cord, because differences in magnetic susceptibilities between tissues produce spatial variations in the magnetic field. The spinal cord itself lies within the spinal canal surrounded by cerebrospinal fluid (CSF), and averages 45 cm in length, with cross-sectional dimensions of roughly 15 mm × 8 mm in the largest regions of the cervical and lumbar enlargements (**Fig. 1**). The entire cord is therefore in close proximity to the heart and lungs, and has been observed to move with each heart beat, presumably as a result of the pulsatile CSF flow around it *(1, 2)*. An inherent challenge of fMRI is that it

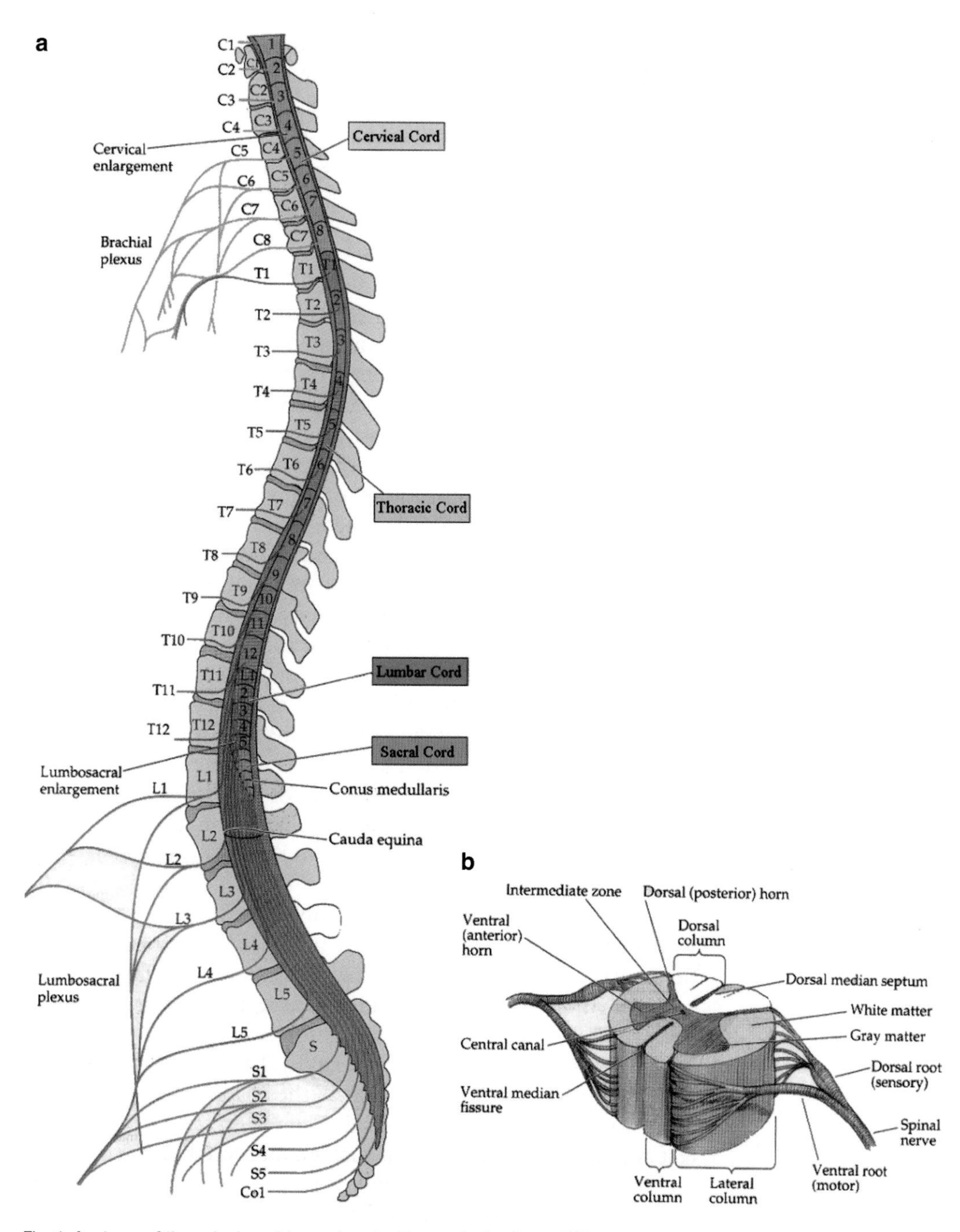

Fig. 1. Anatomy of the spinal cord (reproduced with permission from (***66***)).

is necessary to monitor the signal intensity changes in a specific tissue volume, and this tissue may move between adjacent voxels over the course of the fMRI time series. Moreover, other sources of signal intensity change such as random noise and motion artefacts can obscure the signal changes related to neuronal activity.

Although these challenges may seem daunting, most have been overcome by adapting methods for analysis and motion correction that have been developed for brain fMRI.

3. The Spinal fMRI Method: Evidence of Reliability

The first published example of fMRI in the spinal cord was by Yoshizawa et al. *(3)* This work and the earliest attempts by other groups applied the established brain fMRI methods of the time to the spinal cord *(4, 5)*. The consistent features of the studies by Yoshizawa et al. *(3)*, Stroman et al. *(6)*, Madi et al. *(7)*, and Backes et al. *(8)* were that they were carried out with healthy volunteers and used a hand motor task with imaging of the cervical spinal cord. All used gradient-echo methods with echo times (TE) of 40–50 ms at 1.5 T, and 31 ms at 3 T, as is typical with brain fMRI, and all compared data obtained with transverse and sagittal slices. The areas of activity in the spinal cord that were demonstrated by these studies corresponded to areas of neuronal activity that were expected with the stimuli applied. The conclusions reported from each of these studies included the point that spinal fMRI is a feasible method for assessing neuronal activity in the cord. However, the results obtained also demonstrated variability in the areas of activity, and that it is difficult to obtain high-quality fMRI data in the spinal cord with gradient-echo methods and sensitivity to the BOLD (blood oxygen level dependent) effect.

The debate over the choice of imaging method was started with a study designed to verify that the BOLD effect occurred in the spinal cord *(9)*, by comparing data obtained with gradient-echo and spin-echo methods at the same echo times. This was followed by a study to characterize the signal changes detected with spin-echo methods *(10)*. The BOLD theory shows that with the comparison in the first of these studies the gradient-echo method should produce signal intensity changes between rest and stimulation conditions that are three to four times higher than those produced by the spin-echo method *(11)*. The results however showed that the two methods produced signal intensity changes of approximately equal magnitudes, the image quality obtained with the spin-echo method was superior, and that the spin-echo method may therefore be superior for spinal fMRI.

3.1. BOLD-Based Methods

As mentioned above, the earliest results with BOLD methods *(3, 6–9, 12)* showed that spinal fMRI appears to be feasible. Yoshizawa et al. *(3)* demonstrated areas of activity with a hand motor task which, when combined across subjects, the consistent areas corresponded with spinal cord gray matter and the rostral-caudal distribution corresponded well with the neuroanatomy, and signal changes averaged 4.8%. This was the first work to provide evidence that spinal fMRI is feasible, and set the standard for the studies that followed by Stroman et al. *(6, 9)*, Madi et al. *(7)*, and Backes et al. *(8)*, which had a number of similarities. As in the study by Yoshizawa et al. *(3)*, each of these used gradient-echo imaging (fast gradient-echo or echo-planar encoding), relatively thick (5–10 mm) transverse slices, with the echo time set for BOLD sensitivity. All of these studies investigated activity with motor tasks, two investigated activity with sensory stimuli as well *(6, 9)*, and three of them *(6–8)* compared results obtained with sagittal (4–8 mm) and transverse slices.

The results of these studies showed a number of consistent features **(Fig.2)**. Signal intensity changes with hand motor tasks were consistently in the range of 4.3–4.8% by Yoshizawa et al. *(3)* and Stroman et al. *(6)*, and 0.5–7.5% with graded force tasks by Madi et al. *(7)*, and 8–12% by Backes et al. *(8)*, although there were differences in field strength (1.5 T vs. 3 T) and image resolution which complicate any direct comparison. Each of these studies showed a rostral-caudal dependence of the areas with the task being performed, but only about half observed apparent laterality (left vs. right) of the active regions *(3, 6)*. Backes et al. *(8)* pointed out that there may be problems with the sensitivity to draining veins and the small anatomy, and the differences in spatial localization may simply be attributed to the extent of the draining vein field for the area being stimulated (ventral vs. dorsal). Another significant feature introduced by Backes et al., was the use of cardiac gating, which has an importance that is just

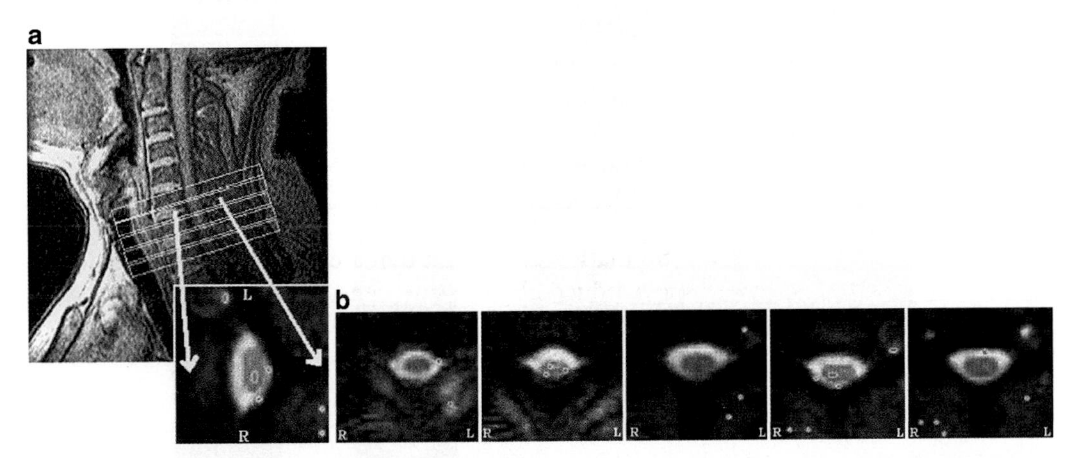

Fig. 2. Example of spinal fMRI results obtained with gradient-recalled echo echo-planar imaging (EPI) in transverse slices of the cervical spinal cord (reproduced from *(9)*). (a) The left side of the body is at the top of this image and the red marks indicate the locations which underwent intensity changes in response to sensory stimulation of the right hand. The location of the slice is indicated in relation to the sagittal view of the cervical spine shown in the larger image. (b) Five transverse slices corresponding to the slices indicated in the sagittal view.

recently being appreciated. In their results the cardiac gating can be expected to have significantly reduced the effects of CSF flow and spinal cord motion, and their results give a good demonstration of the sensitivity of the BOLD methods to the draining veins leading from the gray matter to cord surface.

Another consistent feature of many of these studies was that they used echo-planar imaging (EPI), and made efforts to reduce the data readout time in order to reduce image distortion, including high bandwidth and sampling on the gradient ramps *(7)* or multi-shot acquisitions *(8)*. Recently, Maieron et al. *(13)* used the most advanced methods of parallel imaging (SENSE encoding) with EPI to reduce the distortion effects for spinal fMRI (**Fig.3**). This method showed improvements over previous methods but spatial distortions and variations in sensitivity in the rostral-caudal direction along the spinal cord were clearly visible. Govers et al. *(14)* instead used a 180° refocusing pulse to reduce signal dropout by making the center of k-space T_2-weighted. However, this makes the entire resulting image data T_2-weighted, with less sensitivity to the BOLD effect than would have been obtained with T_2^*-weighted images. The previous "standards" for fMRI resolution no longer hold, even for brain fMRI, as the capabilities of MRI systems have improved, higher resolution is possible even with EPI methods, and the optimal BOLD sensitivity in the

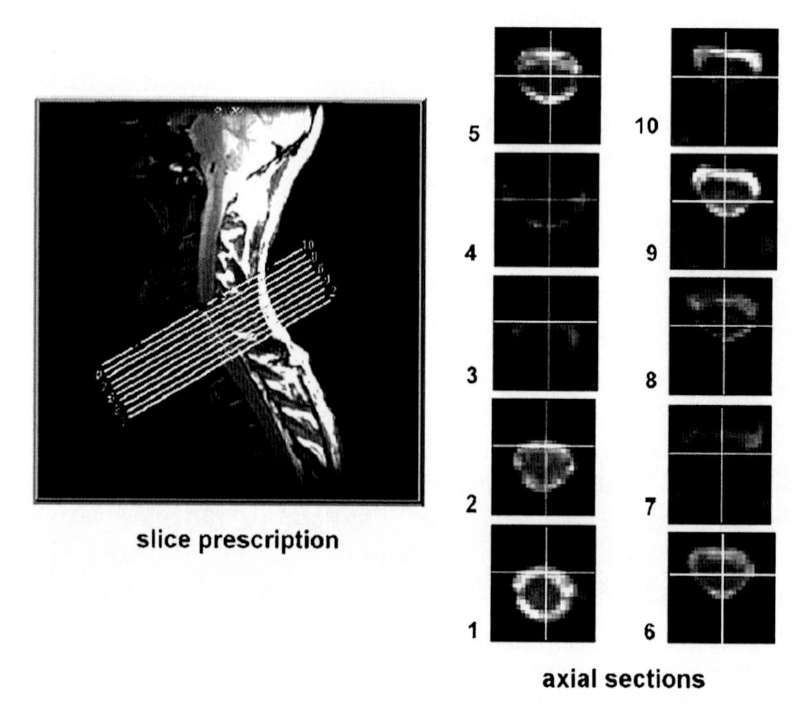

Fig. 3. Example of functional MR image data obtained in the cervical spinal cord with gradient-recalled echo EPI with SENSE encoding to reduce distortion effects (Reproduced with permission from (*13*)).

brain has been shown to be achieved with 1.5-mm cubic voxels *(15)*. What has evolved into standards for brain fMRI does not need to be applied to fMRI of the spinal cord, as has been demonstrated in the majority of the published studies using BOLD fMRI methods in the spinal cord and have in-plane resolutions of less that 2 mm *(3, 6–9, 13, 16)*.

The consistent finding regarding the ability to spatially localize areas of neuronal activity in the spinal cord with these BOLD-based spinal fMRI studies is that the localization is restricted to the spinal cord level, and more detailed localization within the cord cross-section is not possible *(6–8, 13, 14)*. The later studies by Govers et al. *(14)* and Maieron et al. *(13)* demonstrate that the BOLD signal changes are not localized to the gray matter; instead they appear to be dominant in the white matter. There does appear to be a dependence on the side being stimulated, and so the signal changes have some correspondence with neuronal activity. Evidence for the correspondence with neuronal activity was also provided by Madi et al. *(7)* as they showed a graded response to a graded isometric exercise task. Nonetheless, based on the neuroanatomy and vascular anatomy, it appears that the BOLD signal changes show primarily the veins draining the areas of activity, not the actual sites of activity.

3.2. SEEP-Based Methods

The first spinal fMRI study, carried out by Stroman et al. *(6)* in 1999, raised the question as to whether or not the BOLD effect occurred in the spinal cord as in the brain, and had the same properties. This is because with typical BOLD fMRI parameters, and fast gradient-recalled echo acquisition (not EPI), the active regions had signal changes averaging 7.0% at 3 T. This was in line with the 4.8% observed previously by Yoshizawa et al. *(3)*, but was nonetheless large compared with typical brain fMRI using similar methods.

Therefore, in a subsequent study *(9)*, a comparison of T_2-weighted and T_2^*-weighted acquisitions for spinal fMRI was carried out. The intent was to investigate whether the BOLD effect occurred in the spinal cord as in the brain. EPI was used and the slices were aligned with the centers of the vertebral bodies or the intervertebral discs to obtain the best through-slice field homogeneity. Results with both methods, with sensory stimulation (air puffs) and a motor task, showed activity corresponding with neuroanatomy and magnitudes of signal changes at ~5–6%, with a dependence on the duration of the stimulus. The T_2-weighted data had signal changes that were as large, or larger, than the T_2^*-weighted data at approximately the same echo time, and so the observations were not consistent with the BOLD model. It was proposed that the results are consistent with a combination of the BOLD effect and baseline intensity changes, possibly related to changes in extravascular fluid volume at the site of neuronal activity, and indicate that spinal fMRI is feasible with both motor and sensory stimulation at 1.5 T. This conclusion was based on a concurrent study to investigate the nature of T_2-weighted and T_2^*-weighted fMRI data in the brain (visual cortex) at 1.5 and 3 T, over a range of echo times *(17)*. This study demonstrated that in the brain the T_2-weighted fMRI data do not extrapolate to zero at an echo time of zero, which is also inconsistent with the BOLD theory. It was therefore proposed that there is a second contributing factor, in addition to the BOLD effect, which was then termed signal enhancement by extravascular water protons (SEEP). Moreover, contrary to the expected outcome, it was observed that the use of spin-echo imaging methods for fMRI of the spinal cord had sensitivity to neuronal activity that was equal or better than that obtained with gradient-echo methods, and the image quality was superior.

This was followed by a study focused on the use of spin-echo imaging methods for spinal fMRI, with the purpose of characterizing the signal intensity changes observed at different echo times with a motor task and thermal sensory stimulation *(10)*. The results showed that signal intensity changes increased with increasing echo time as predicted by the accepted BOLD model, but again there was an additional offset so that the signal change extrapolated to zero echo time was consistently around 2.5%. To test whether the response was nonlinear, with the signal change approaching zero more rapidly as the echo time approached zero *(18)*, the signal changes in fMRI of the spinal cord were measured using fast spin-echo imaging with a long repetition time (3 s), over a range of echo times from 11 to 66 ms. The results demonstrated that the echo time dependence is nonlinear but does not approach zero as the echo time approaches zero. At an echo time of only 11 ms a signal intensity change of 3.3% was observed. As the data are not T_1-weighted, have negligible T_2-weighting, and are not sensitive to changes in water self-diffusion, but all MRI data are sensitive to changes in proton-density, the only conclusion that can be reached from these studies is that the non-zero intercept arises from a proton-density change. Because the signal intensity change vs. echo time relationship is nonlinear, there must be two different water components contributing, and a two-component model was proposed. This model was separable into two terms, one of which is essentially identical to the accepted BOLD model, and the other which shows the SEEP effect.

3.2.1. The Biophysical Basis of SEEP

At sites of neuronal activity there are well-known physiological processes that occur that are unrelated to the BOLD effect. Astrocytes make contact with both blood vessels and neurons and play a role in providing metabolites to neurons and in maintaining the extracellular concentration of glutamate *(19–21)*. As a result, these cells have been shown to play an important role in effective neuronal signalling *(22, 23)*. When glutamate is released from vesicles in the axon terminal and travels across the synaptic cleft to trigger the depolarization of the adjoining neuron, the remaining glutamate is rapidly absorbed by astrocytes by means of high-affinity sodium-dependent transporters *(22)*. This process helps to maintain the extracellular glutamate concentration at the essentially low level required for effective neuronal function. When glutamate is actively taken up by astrocytes, each molecule is accompanied by three Na^+ and one H^+ transported into the cell, and one K^+ transported out, and so is also accompanied by water entering the cells *(22)*. In addition, astrocytes have been shown to be depolarized during neuronal activity as a result of uptake of potassium in proportion to the number of active neurons in the vicinity and the frequency at which they fire *(20)*. As a result of these effects, the extracellular/intracellular volume ratio has been shown to change significantly *(24, 25)*. Brain cell swelling has been observed in relation to increased neuronal depolarization during seizure *(26)*, anoxia *(27)*, electrical stimulation *(28, 29)*, and spreading depression *(30)*. Activity-dependent cell swelling has been consistently implicated by independent physiological techniques that include volume measurement of the shrinking extracellular space *(28, 31)*, increased light transmittance *(29, 32, 33)*, and increased extracellular resistance *(26)*. In each case, Na^+, Ca^{2+}, and Cl^- influx to discharging neurons, together with K^+ and Cl^- uptake by adjacent astrocytes, draw water intracellularly from the extracellular space. In vivo, this water is drawn in turn from the regional capillaries. The resultant cell swelling increases local tissue water content *(34)* and light transmittance through the tissue *(33)*, providing a means of observing elevated neuronal activity in tissue slices.

The validity of using SEEP contrast to detect neuronal activity with fMRI was therefore tested by combining the fMRI method developed for use in the spinal cord with established methods for imaging of cell swelling in rat cerebral slices based on light transmittance microscopy *(35)*. The tissue slice preparation and methods for imaging with light microscopy were adapted from Anderson and Andrew *(36)*, and methods for fMRI were adapted from Stroman et al. *(37)*. Cerebral tissue slices were superfused with artificial CSF (aCSF) and were stimulated by increasing the concentration of K^+ with no change in osmolality. In separate studies the cortical slices were superfused with hyper- or hypo-osmotic aCSF in order to elicit cell shrinking and swelling, respectively.

fMRI results and supporting measurements of light transmittance changes confirmed that MR signal intensity changes occurred as a result of tissue water changes at sites of neuronal activity in the cerebral slices, therefore verifying the proposed SEEP theory. Within 1 min of stimulation with high K^+ in aCSF, the fMRI time course series demonstrated a reversible increase in signal intensity across the cerebral tissue slices. MR signal intensities averaged across cortical gray matter regions increased during the 2-min exposure and then abruptly decreased upon superfusion with control aCSF to below baseline followed by a slow recovery to control levels. Corresponding measurements of light transmittance changes supported the fMRI results as they revealed a large initial peak in cortical gray matter, previously shown to represent cell swelling along the front of a spreading depression event evoked by similar high K^+ exposure *(36)*. In addition, in control experiments with osmotic challenges the fMRI signal intensity and the light transmittance through the slices were observed to be reversibly elevated in cortical gray matter upon exposure to hypo-osmotic aCSF and reversibly decreased in hyperosmotic aCSF.

The results therefore demonstrate that *(1)* MR signal changes occurred in primarily proton-density weighted images that corresponded with neuronal activity, *(2)* cell swelling occurred corresponding with neuronal activity, and *(3)* cell swelling and shrinking result in corresponding MR signal changes. Therefore, the observed MR signal changes corresponded with neuronal activity, arising as a result of cellular swelling, demonstrating the biophysical basis of SEEP contrast.

3.2.2. Results of SEEP-Based Spinal fMRI Studies

Significant advances in the validation of the spinal fMRI method based on SEEP contrast have been provided by later studies by Ng et al. *(38, 39)*, and Li et al. *(40)*, as they carried out detailed spinal fMRI studies at 0.2 T using proton-density weighted spin-echo methods. At this low field and imaging method the BOLD contribution is negligible. Ng et al. carried out spinal fMRI studies in the cervical spinal cord with a fast spin-echo method (TE = 24 ms) with 10-mm-thick transverse slices and 1.25-mm in-plane spatial resolution. The stimulus paradigm was a bimanual hand gripping task, and corresponding activity was detected in the gray matter in 12 of 14 volunteers, with the peak of activity at C6 and C7. The areas of activity in the spinal cord cross-section corresponded well with the neuroanatomy. In related experiments with the same methods, Li et al. *(40)* then carried out spinal fMRI at 0.2 T with electrical stimulation of acupoints. The results showed localization of activity to expected regions in over 73% of volunteers and the average signal change was 4%. More recently Ng et al. *(39)* have also carried out spinal fMRI of 28 healthy volunteers at 0.2 T using the same methods, with a hand gripping task to elicit activity. Results showed that 11 of 14 people had positive activity in the dorsal gray matter in the cervical cord, and the average signal changes were again 4%, and the overall consistency across volunteers was over 70%.

With evidence in place to identify the biophysical basis of the signal changes it remained to prove that the signal changes that were being detected did indeed correspond with neuronal activity in the spinal cord. This was tested by comparing the activity detected in the cervical spinal cord with the expected activity based on the neuroanatomy, with thermal sensory stimulation of various dermatomes *(41)*. Again a proton-density weighted (TE = 38 ms) single-shot fast spin-echo imaging sequence was used to have low sensitivity to the BOLD effect, and primarily SEEP-weighted fMRI data. During the time series acquisition thermal stimulation was applied alternately to dermatomes overlying the thumb side of the hand (C6), little finger side of the hand (C8), and the forearm (C5). The results showed that the activity within the spinal cord gray matter with each stimulus was predominantly on the side of the body being stimulated, and was distributed across several spinal cord segments, but with the peak of activity having a center that corresponded with the dermatome being stimulated. The results also demonstrated variability and false-positive results occurring primarily in the surrounding CSF and at the surface of the cord, which were reduced by means of clustering based on signal intensity time courses of active voxels. This helps to show that this method for spinal fMRI is able to show neuronal activity.

A related study was then carried out with fMRI of the lumbar spinal cord and thermal stimuli applied to the leg, with the temperature of the stimulus ramped between 32°C and 10°C at various rates, and data were also obtained from a small number of patients with cord trauma for the first time *(37)*. Areas of activity in the lumbar spinal cord were consistently detected and the magnitude of signal intensity changes varied with the stimulus temperature. The areas of activity corresponded well with the neuroanatomy, showing dorsal areas of activity on the side of stimulation, as well as some ventral areas, showing precision of the spatial localization. There was a clear transition in the response between cool sensation between 29°C and 15°C, and more intense sensations becoming noxious below 15°C. The cooler stimuli elicited a significantly larger response (~7% compared to ~2–3%). There was also a difference detected when looking at only the later response to the ramped temperature. This was a passive sensory stimulus and so the only connection between the temperature on the leg and the signal changes in the lumbar spinal cord is the neural input; so the response demonstrates that the signal changes correspond with neuronal activity. Data were also obtained from a small sample (n = 6) of patients with cord trauma. The areas of activity were different in patients with cord trauma, but the areas that were active had the same temperature response as that seen in healthy volunteers.

This study was then extended to include a much larger population of 27 patients with cord trauma *(42)*. Eighteen of the subjects had complete injuries ("American Spinal Injury Association"

[ASIA] impairment scale = A), and nine had incomplete injuries (ASIA = B to D). The results showed well-localized areas of activity, laterality dependence on the side of stimulation, and patterns of activity that varied with the injury and whether the person could feel the thermal stimulus on the leg. The magnitude of signal changes also varied with the stimulation temperature. In all subjects the injury site was rostral to the lumbar spinal cord. Areas of activity were consistently detected caudal to the injury site with complete injuries, showing for the first time the neuronal activity in the spinal cord where self-report techniques provide no information. This is a significant step towards making spinal fMRI a clinical assessment tool.

The applications for spinal fMRI were expanded with a study of activity in the lumbar spinal cord during a lower limb motor task, again using the same spin-echo imaging methods at 1.5 T as the previous studies *(43)*. Activity was mapped during active and passive alternating rhythmic motor task involving ankle movements in a group of healthy volunteers, with the purpose of verifying that leg motor studies were feasible and to establish reference data for subsequent studies of patients with cord trauma. The results demonstrated areas of activity consistent with these motor tasks, and differences between active and passive tasks. Signal changes were larger than those seen with thermal sensory stimuli, at around 12%. The study that followed *(44)* involved 12 people with cervical or thoracic spinal cord injures, and passive and active movements were studied according to each volunteer's abilities. Some people had incomplete injuries and could do the task actively. This study was done with the sagittal slice imaging method to give large coverage of the spinal cord with a resolution of 2.8 mm × 0.9 mm × 0.9 mm (R/L × A/P × S/I). Areas of activity were detected in the lumbar cord corresponding to the motor task. Signal changes were relatively large again at 13.6% and 15.0% during active and passive pedaling. Differences in the areas of activity and the numbers of active voxels were detected depending on the extent of injury (ASIA rating) and the reported sensations and ability to move at the time of the studies. Individual results were also assessed. This is the first time that motor activity has been mapped in the human spinal cord caudal to a site of spinal cord injury, and demonstrates the clinical and research potential of the method.

4. Recent Developments

Recent developments in the analysis methods are discussed in detail in **Subheading 5**, but were used in the following studies and impact on the sensitivity and reliability of the results. These advances include the definition of a normalized coordinate system for the spinal cord and brainstem, characterization and modelling of sources of confounding signal changes, and analysis methods for removing these confounds.

Spinal fMRI studies have now been carried out in children (ages 6–13 years) using these most recent methods with cold thermal stimulation of the hand at 17 and 27°C *(45)*. The results showed a good degree of consistency across subjects, as well as consistency with the neuroanatomy and previous studies in healthy adults. The magnitudes of the signal intensity changes in regions spanning the brainstem and cervical spinal cord were also similar, or marginally higher, than those observed in adults with similar stimuli. Brain fMRI data based on BOLD contrast demonstrated lower signal changes in children than in adults *(46, 47)*, while these data demonstrated no such difference in the spinal cord with SEEP contrast that might confound interpretation of results for clinical assessment or research.

To obtain baseline reference data for studies of neuropathic pain (allodynia and hyperalgesia), spinal fMRI studies have also been carried out with light touch (2- and 15-g von-Frey monofilaments) and light brush (unpublished data). This study required the high sensitivity that has resulted from the ability to normalize the data and to model the confounding effects of physiological motion driven by the cardiac cycle. The results showed consistent well-localized activity in specific brainstem regions (cuneate and gracile nuclei) in the brainstem with 2-g von-Frey stimulation, and little activity in the cervical spinal cord, whereas with the 15-g von-Frey hair there was consistent activity in the right dorsal gray matter and spreading into intermediate or ventral gray matter regions at the seventh cervical spinal cord segment, as well as in the olivary nucleus, reticular formation, periaqueductal gray matter, and raphe nucleus in the rostral pons and midbrain (**Figs. 4** and **5**). The results clearly reflect the difference in stimulation and

Fig. 4. Consistent areas of neuronal activity detected at the seventh cervical spinal cord segment with light touch of the dorsal aspect of the right hand, with 2- and 15-g von-Frey monofilaments. Data were obtained with signal enhancement by extravascular protons (SEEP) contrast with a proton-density weighted fast spin-echo (HASTE) imaging method. Results are combined across 11 volunteers, and are shown as contiguous 1-mm-thick transverse slices in radiological orientation. The colours indicate the number of people with activity at each voxel, or within an immediate neighbour: red 10/11, orange 9, yellow 8, green 7. Areas with negative signal changes upon light touch are also shown, in shades of blue (darker means more repeatable).

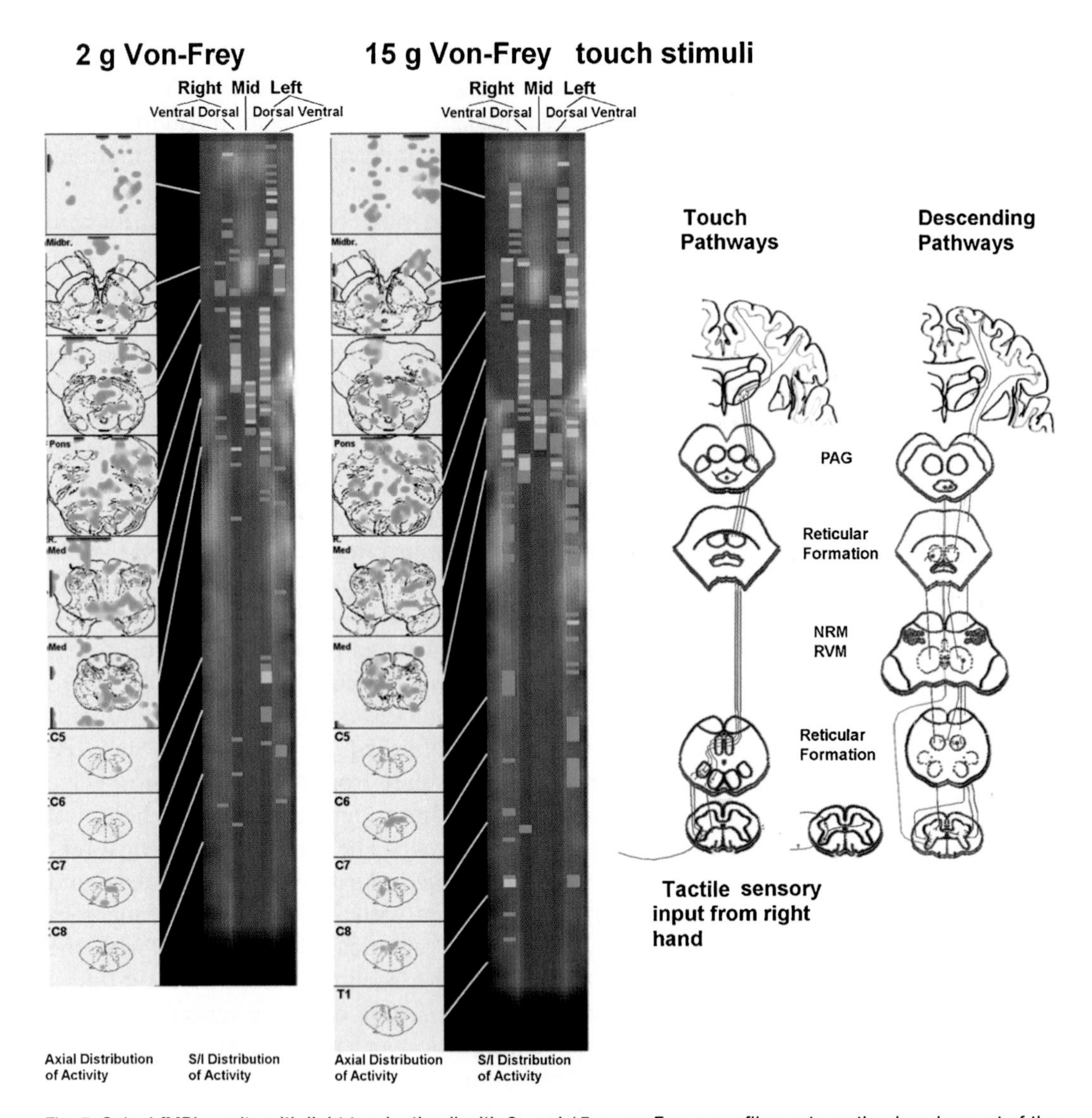

Fig. 5. Spinal fMRI results with light touch stimuli with 2- and 15-g von-Frey monofilaments on the dorsal aspect of the right hand, as shown in **Fig. 4**. In this figure, the activity is shown in all areas spanning the entire brainstem and cervical spinal cord. The distribution of activity in transverse planes is shown for each spinal cord segment, and for selected levels of the brainstem, projected onto schematic drawings of each level. The corresponding rostral-caudal distribution is also shown, with columns representing each of right/left and dorsal/ventral regions.

the different receptors that these activate. An additional interesting feature that was identified with 2-g von-Frey stimulation was consistent areas of negative signal change in the ipsilateral intermediate or ventral gray matter at the level of C7, at precisely the same locations where 15-g von-Frey hair caused positive signal changes. Calculations of effective connectivity between regions *(48)* confirm that the negative signal changes are likely the result of decreased input to the region from the midbrain. It is presumed that the signal changes are most closely related to the pre-synaptic input to a region, as was shown by Logothetis et al. *(49)* for BOLD fMRI data in the brain, and this conclusion is supported by our observations in this study. As a result, we are able to assess sensory responses in the spinal cord and brainstem in both ascending and descending pathways.

Studies of pain pathways in the brainstem and spinal cord have also been carried out with thermal stimuli at 42°C (warm) and 46°C (hot) of the palm of the hand *(50)*. These results demonstrated distinct differences between the warm and hot stimuli that are consistent with the neuroanatomy for sensation vs. pain responses, as areas of activity were detected in the ipsilateral dorsal gray matter in areas of the spinal cord with both stimuli, but were notably increased in the ventral gray matter regions of the spinal cord and in the rostral ventromedial medulla, reticular formation in the pons, raphe nucleus, pariaqueductal gray matter, locus coeruleus, red nucleus, and in the contralateral thalamus, with the hot stimulus. This serves to show the ability of the method to discriminate sensation from pain and to reliably identify specific areas of activity that can be used in future studies of neuropathic pain.

Over the course of the previous study it was also noted that there is variability in the patterns of activity and that it appears to depend on the order of the experiments as well as on whether or not the volunteers are shown pictures during the experiments. It has been shown that subjective ratings of pain sensations are increased when subjects focus their attention on a painful sensation, and fMRI studies of the brain and brainstem have shown altered activity depending on attention. A systematic study was therefore carried out on the effects of directing the subjects' attention to a noxious thermal stimulus or elsewhere, to determine whether or not changes in attentiveness can also influence fMRI results in the cervical spinal cord, and can therefore be a source of variability, although without reflecting errors (unpublished data). Subjects were instructed by means of a visual display to focus on the sensation on their hand and to provide a 1–4 rating of the sensation every 15 s throughout the fMRI time series, in order to have them focus their attention on the sensation. In the same imaging session, the subject was instead instructed to ignore the sensation while he/she watched a movie, and was asked to press any button on a key pad each time a new character with a speaking role appeared in the movie, in order to draw their attention away from the sensation. Two of each type of experiment were carried out in an alternating order. The results showed consistently more activity in the periaqueductal gray matter and raphe nuclei in the medulla when the subject's attention was focussed on the movie, as has been shown previously *(51, 52)*, and also showed increased activity in the sixth cervical spinal cord segment in the ipsilateral dorsal gray matter and ventral gray matter regions. However, there was also systematic variability between repeated identical experiments with the same attention focus. The latter experiments more strongly reflected the attentional modulation that was imposed during the experiments, whereas the earlier experiments appeared to be confounded by other factors. These other factors were reasoned to be the influences of being put into the MRI system, the experience of the enclosed space, the potential anxiety over the impending painful sensations, etc., which decrease over time as the volunteer becomes accustomed to the experimental setup – therefore also factors influencing attention and emotion. The results of this study demonstrate attentional modulation of activity in the cervical spinal cord as a result of descending input from the brainstem, and reveal true physiological variation of neuronal activity. Variation of activity in the spinal cord across repeated studies cannot be assumed to reflect errors. Instead, the variability observed in relation to the volunteers' attention focus confirms the sensitivity of spinal fMRI to true neuronal activity.

More recently, spinal fMRI has been carried out on a 1.5 T clinical scanner using a proton-density-weighted fast spin-echo sequence (TE = 11 ms) in patients with relapsing MS and in age-matched controls to assess the extent and define the role of neuronal activity in the cervical cord during a proprioceptive and a tactile stimulation of the right upper limb *(53)*. The proprioceptive stimulation consisted of a passive flexion and extension of the right wrist, whereas the tactile stimulation consisted of a repeated tapping of the center of the subject's right palm. FMRI data were analyzed using a custom-made software written in MatLab and a general linear model (GLM) approach in which fMRI time-courses at each voxel are modelled as a linear combination of explanatory variables (which constitute the so-called basis set) and a residual error term. In healthy subjects and MS patients, positive activations in the posterior, middle, and anterior cervical cord between C5 and C8, bilaterally, were found during the application of both stimuli. MS patients also showed, on average, a 20% higher cord fMRI signal change during either proprioceptive (3.4% vs. 2.7%, $p = 0.03$) or tactile (3.9% vs. 3.2%, $p = 0.02$) stimulation than did normal individuals, thus suggesting an abnormal cord function in these patients, which in turn is likely to be secondary to the dysfunction of spinal cord interneurons. Moreover, during tactile stimulation of the right palm, unlike controls, MS patients did not show differences between right and left cord average fMRI signal changes. The reduced functional lateralization of the cord activity following tactile stimulation in MS patients was also confirmed by the more frequent evidence of fMRI activity in the cord side contralateral to the stimulus compared with controls. In an attempt to define the role of cord fMRI changes in these patients, the magnitude of the correlation between the extent of prorioceptive-associated cord functional activity and the severity

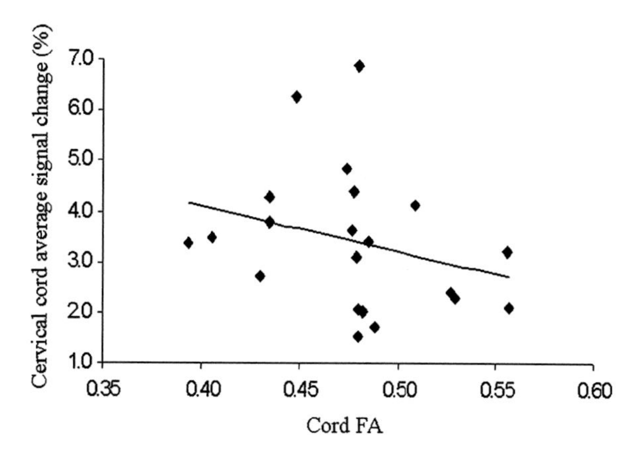

Fig. 6. Scatterplot of the correlation between cervical cord average fMRI signal change during a proprioceptive stimulation and cord fractional anisotropy (FA) in patients with MS ($r = -0.48$, $p = 0.04$).

of brain and cervical cord structural damage has also been investigated. There was a significant correlation between the extent of cord fMRI recruitment and the severity of brain and cord damage (**Fig. 6**). This suggests that, as shown for the brain *(54)*, also in the spinal cord, fMRI changes might be a reflection of functional reorganization which might contribute in limiting the clinical consequences of diffuse disease-related damage.

5. Data Acquisition Details and Analysis Methods

Although the basic principle of all fMRI is to observe neuronal-activity MR signal changes over time, it is essential that each voxel of image data represents the same volume of tissue at all times, and that there are no sources of signal change that are not related to neuronal activity. Analysis of the fMRI data therefore involves *(1)* correcting for any motion or other non-neuronal sources of signal changes (confounds) as much as possible, *(2)* detecting the neuronal-activity-related signal changes, allowing for the possibility of residual contributions from confounding effects, and *(3)* displaying the results in a manner to facilitate their interpretation. How these methods can be applied can often depend on the data that were acquired in terms of temporal resolution, spatial extent, and spatial resolution. The two issues of data acquisition and data analysis are therefore discussed together in the following paragraphs.

5.1. Acquisition Methods: Image Orientation and Resolution, Spatial Normalization

The issues of whether to use a fast spin-echo imaging method, or a fast gradient-echo imaging method (such as with EPI) are discussed extensively above. There are, however, a number of other imaging parameters that are critically important, such as the image resolution, orientation, and methods to reduce motion effects. As with all fMRI it is desirable to reduce partial-volume effects, in order to obtain the highest sensitivity to neuronal-activity-related signal changes. The anatomy of the spinal cord therefore requires relatively high resolution in the spinal cord cross-section, and lower resolution is tolerable parallel to the long axis of the cord. The spinal fMRI studies that have been published to date typically had voxel dimensions of 2 mm or less in the right-left and anterior-posterior directions, and a number have used transverse slices as thick as 10 mm *(3, 7–10, 18, 37–43)*.

Although the use of transverse slices provided relatively good coverage of the cord with low partial-volume effects and high signal (with thick slices), they were limited in the extent of the cord that could be covered in a reasonable amount of time for fMRI. To make the method more practical for routine use, a new method was developed based on thin contiguous sagittal slices *(55)*. In the first application of this method, data were acquired with a resolution of 2.8 mm × 0.94 mm × 0.94 mm (R/L × A/P × S/I) in an effectively 3D volume of the cord spanning 12 cm. The data had lower signal than in previous studies because of the smaller voxel volume, and this was addressed by smoothing the data only parallel to the long axis of the cord. A reference line was drawn manually along the anterior edge of the cord, the volume was interpolated to 0.5-mm cubic voxels, and was then resliced transverse to the reference line. This enabled smoothing only across the slices, and therefore parallel to the long axis of the cord. The results showed consistent activity in the cervical spinal cord in response to a 15°C stimulus on the palm of the hand, with an apparent pattern of activity within spinal cord segments, showing detail that has not been seen before. Signal changes were quite high at 10%. Anatomical details surrounding the cord were also visible, providing position reference information.

The continuous 3D data provided by the sagittal slice method made it possible to define a coordinate system and method for mapping the spinal cord image data to a consistent shape and size across different people (unpublished data). The coordinate system is simply defined with one axis parallel to the long axis of the cord, and the other two axes being right-left and anterior-posterior. The normalization was applied in a study of warm and hot thermal stimuli applied to the hand, spanning the entire cervical spinal cord and brainstem, with a resolution of 2 mm × 1 mm × 1 mm. The caudal edge of the pons (the pontomedullary junction) and the C7/T1 disc were chosen as reference points, and based on data from a small number of subjects, were fixed at being 140 mm apart. Data from subsequent studies were therefore scaled and shifted to align these reference points and the anterior edge of the spinal cord to produce a normalized reference volume. Analysis of the degree of correspondence across eight people showed that 92% of the voxels were aligned to within 2 mm or less of the combined reference volume.

With a normalized reference volume thus defined, it is now also possible to align the normalized fMRI data with the reference volume at each point of the time series, and thus correct for small changes in position over time, including right–left or anterior–posterior flexion. It has also enabled the creation of a region mask, to automatically label regions of the cervical spinal cord and brainstem, or to mask areas outside of these regions of interest. These tools are useful for automated identification of areas of activity and combination of group results. The normalized volumes can now be further refined to assess and improve the consistency of the alignment of gray matter regions.

5.2. Problems Arising from Physiological Motion

Functional MRI of the spinal cord presents unique challenges because of the motion of the cord within the spinal canal, and its proximity to sources of confounding signal changes as well (CSF, heart, lungs). The motion of nearby tissues can create subtle changes in the magnetic field within the spinal cord, or can create motion artefacts which contribute, in error, to the signal within a voxel spanning a volume of the spinal cord. However, means to reduce and/or model these effects have been developed.

It has been shown that some motion of the spinal cord may arise from blood flow in the radicular arteries *(56)* but the bulk of the motion is expected to be caused by CSF flow *(57, 58)*. The magnitude of displacement depends on the rostral-caudal position with the cord, and on its curvature, as can be observed in serial images of the spinal cord with sagittal slices.

Various means of compensating for these sources of motion have been applied in virtually every spinal fMRI study reported. Reduction of image artefacts arising from moving tissues and fluids has been successfully achieved with the use of flow-compensation gradients applied in the rostral-caudal direction, and spatial saturation pulses to eliminate all signal arising from anterior to the spine *(18, 37, 59, 60)*. These serve to eliminate sources of signal from areas outside of the spinal cord that could contribute to voxels within the spinal cord as a result of spatial encoding errors created by motion. The study carried out by Moffitt et al. *(61)* appears to demonstrate that the application of fluid attenuated inversion recovery to eliminate the MR signal from CSF, or the use of spatial saturation pulses to eliminate the signal from CSF rostral and caudal to the imaging region, is not effective at reducing confounding signal changes. The main impact of the CSF flow appears to be the motion it imparts to the spinal cord itself, whereas artifacts from the CSF movement can be reduced with flow-compensation gradients applied in the head-foot direction. Respiratory gated acquisition and breath-hold during acquisition have also been used to reduce or eliminate the effects of lung motion *(6, 9)*. However, in these studies there was no apparent benefit or improvement in data quality obtained from respiratory gating or breath-holding. In one of the earliest spinal fMRI studies reported, Backes et al. *(8)* applied cardiac-gated acquisition and this has not been investigated again until a more recent work reported by Brooks et al. *(62)* In the latter study, a comparison was made to determine the effects of retrospective cardiac gating and the results demonstrated that the majority of noise in spinal fMRI appears to be cardiac in origin. Given that many of the causes of spinal cord motion within the spinal canal (CSF and blood flow) are driven by the cardiac motion, this conclusion seems highly plausible.

Recent studies have succeeded at characterizing the motion of the cervical spinal cord, and have confirmed that the motion is a function of the cardiac cycle *(1, 2)*. This study also demonstrated that there is no quiescent period during the cardiac cycle during which the spinal cord is stationary, and concluded that cardiac gating methods may therefore never be fully effective. Another study has demonstrated that using recordings of the peripheral pulse to model the spinal cord motion in a GLM analysis can significantly reduce errors and improve the sensitivity of spinal fMRI *(63)*. The combination of these two findings is currently being developed, and is expected to yield a very sensitive spinal fMRI method.

Finally, to assess the reproducibility of spinal fMRI results, the same data set from 12 healthy subjects has been analyzed by calculating the cross-correlation coefficient between the stimulus and the time course of every voxel, by using the GLM and the independent component analysis (ICA) *(64)* (**Fig. 7**). The first two methods require the definition of a model for the stimulation and calculate the coupling between model paradigm and every fMRI time series. On the contrary, ICA is a model-free approach, which decomposes the fMRI time series into a set of spatial components having an associated time course in a multivariate manner, i.e., considering all time series at once. The fMRI data set of 12 healthy subjects was acquired during proprioceptive and tactile stimulation of the right upper limb. Model-based approaches (cross-correlation coefficient and GLM) revealed similar patterns of neuronal cord activity both for proprioceptive and tactile tasks. ICA was also able to identify a component related to fMRI stimulation, although with a lower statistical threshold than model-based approaches. Moreover, ICA found a set of components, consistent across subjects, which, because of their spatial and frequency profiles, could be related to artefacts. This study prompts

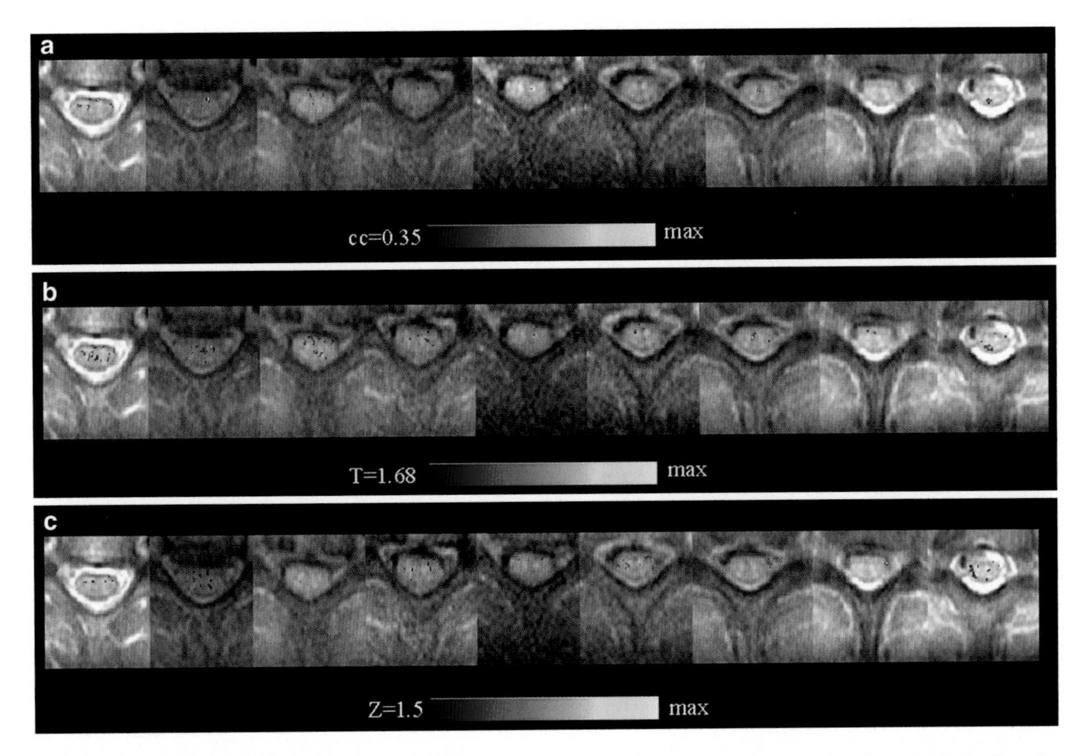

Fig. 7. Illustrative activation maps (color coded for *t* values) of cervical cord on axial proton-density-weighted spin-echo images from C5 to C8 from a healthy control during a tactile stimulation of the right palm hand, obtained with cross-correlation analysis (**a**), general linear model (**b**), and independent component analysis (**c**).

a future combined use of ICA for artefacts estimation, and GLM for activity estimation, which might lead to both an improvement in the sensitivity to fMRI spinal changes and a more accurate estimation of artefact components, thus allowing a better mapping of the cord neuronal function with spinal fMRI.

6. Conclusions and Future Directions

Spin-echo imaging methods provide optimal image quality, and data based on SEEP contrast have so far demonstrated the highest sensitivity and specificity. The main sources of errors arise from physiological motion (primarily cardiac), and can be reduced with flow compensation gradients, spatial saturation pulses to eliminate signal arising from anterior to the spine, and modelling of spinal cord motion to predict the time courses of confounding signal changes for use in GLM analysis. FMRI data obtained in thin contiguous slices can span a large extent of the spinal cord, provide 3D coverage, and can be transformed into a normalized coordinate system for automated analysis and identification of active regions. The advantages of BOLD fMRI methods appear to be speed and temporal resolution, as good precision has been demonstrated for assessing the temporal response to stimuli of various durations *(65)*, at the expense of lower spatial precision and increased sensitivity to physiological motion. There are therefore advantages to each method, and the optimal method to use may depend on the application.

The remaining challenges include the development of *(1)* practical means of analyzing the data in a clinical setting, *(2)* methods for displaying the results in a useful format, *(3)* creation of reference results from healthy volunteers to facilitate interpretation of altered function, and *(4)* an increased use of spinal fMRI in cross-sectional and longitudinal studies of patients with diseases where cord is considered to be the "target" organ. All of this should result in a better understanding of the pathobiology of several neurological conditions, and also set the stage for the use of spinal fMRI as a routine clinical tool.

References

1. Figley CR, Stroman PW. Characterization of spinal cord motion: a source of errors in spinal fMRI? In: Proceedings of the International Society for Magnetic Resonance in Medicine, 14th annual meeting, Seattle, Washington, May 4–9, 2006.
2. Figley CR, Stroman PW. Investigation of human cervical and upper thoracic spinal cord motion: implications for imaging spinal cord structure and function. Magn Reson Med 2007;58(1):185–9.
3. Yoshizawa T, Nose T, Moore GJ, Sillerud LO. Functional magnetic resonance imaging of motor activation in the human cervical spinal cord. NeuroImage 1996;4(3 Pt 1):174–82.
4. Menon RS, Ogawa S, Kim SG, Ellermann JM, Merkle H, Tank DW, Ugurbil K. Functional brain mapping using magnetic resonance imaging. Signal changes accompanying visual stimulation. Invest Radiol 1992;27 (Suppl 2):S47–S53.
5. Ogawa S, Tank DW, Menon R, Ellermann JM, Kim SG, Merkle H, Ugurbil K. Intrinsic signal changes accompanying sensory stimulation: functional brain mapping with magnetic resonance imaging. Proc Natl Acad Sci USA 1992;89(13):5951–5.
6. Stroman PW, Nance PW, Ryner LN. BOLD MRI of the human cervical spinal cord at 3 tesla. Magn Reson Med 1999;42(3):571–6.
7. Madi S, Flanders AE, Vinitski S, Herbison GJ, Nissanov J. Functional MR imaging of the human cervical spinal cord. AJNR Am J Neuroradiol 2001;22(9):1768–74.
8. Backes WH, Mess WH, Wilmink JT. Functional MR imaging of the cervical spinal cord by use of median nerve stimulation and fist clenching. AJNR Am J Neuroradiol 2001;22(10):1854–9.
9. Stroman PW, Ryner LN. Functional MRI of motor and sensory activation in the human spinal cord. Magn Reson Imaging 2001;19(1):27–32.
10. Stroman PW, Krause V, Malisza KL, Frankenstein UN, Tomanek B. Characterization of contrast changes in functional MRI of the human spinal cord at 1.5 T. Magn Reson Imaging 2001;19(6):833–8.
11. Bandettini PA, Wong EC, Jesmanowicz A, Hinks RS, Hyde JS. Spin-echo and gradient-echo EPI of human brain activation using BOLD contrast: a comparative study at 1.5 T. NMR Biomed 1994;7(1/2):12–20.
12. Komisaruk BR, Mosier KM, Liu WC, Criminale C, Zaborszky L, Whipple B, Kalnin A. Functional localization of brainstem and cervical spinal cord nuclei in humans with fMRI. AJNR Am J Neuroradiol 2002;23(4):609–17.
13. Maieron M, Iannetti GD, Bodurka J, Tracey I, Bandettini PA, Porro CA. Functional responses in the human spinal cord during willed motor actions: evidence for side- and rate-dependent activity. J Neurosci 2007;27(15):4182–90.
14. Govers N, Beghin J, Van Goethem JW, Michiels J, van den HL, Vandervliet E, Parizel PM.

Functional MRI of the cervical spinal cord on 1.5 T with fingertapping: to what extent is it feasible? Neuroradiology 2007;49(1):73–81.

15. Hyde JS, Biswal BB, Jesmanowicz A. High-resolution fMRI using multislice partial k-space GR-EPI with cubic voxels. Magn Reson Med 2001;46(1):114–25.
16. Stracke CP, Pettersson LG, Schoth F, Moller-Hartmann W, Krings T. Interneuronal systems of the cervical spinal cord assessed with BOLD imaging at 1.5 T. Neuroradiology 2005;47(2):127–33.
17. Stroman PW, Krause V, Frankenstein UN, Malisza KL, Tomanek B. Spin-echo versus gradient-echo fMRI with short echo times. Magn Reson Imaging 2001;19(6):827–31.
18. Stroman PW, Krause V, Malisza KL, Frankenstein UN, Tomanek B. Extravascular proton-density changes as a non-BOLD component of contrast in fMRI of the human spinal cord. Magn Reson Med 2002;48(1):122–7.
19. Bouzier-Sore AK, Merle M, Magistretti PJ, Pellerin L. Feeding active neurons: (re)emergence of a nursing role for astrocytes. J Physiol Paris 2002;96(3/4):273–82.
20. Nicholls JG, Martin AR, Wallace BG. Properties and Functions of Neuroglial Cells. From Neuron to Brain, 3rd ed. Sunderland, MA: Sinauer Associates; 1992, p 146–83.
21. Pellerin L, Magistretti PJ. Food for thought: challenging the dogmas. J Cereb Blood Flow Metab 2003;23(11):1282–6.
22. Nedergaard M, Takano T, Hansen AJ. Beyond the role of glutamate as a neurotransmitter. Nat Rev Neurosci 2002;3(9):748–55.
23. Piet R, Vargova L, Sykova E, Poulain DA, Oliet SH. Physiological contribution of the astrocytic environment of neurons to inter-synaptic crosstalk. Proc Natl Acad Sci USA 2004;101(7):2151–5.
24. Sykova E, Vargova L, Kubinova S, Jendelova P, Chvatal A. The relationship between changes in intrinsic optical signals and cell swelling in rat spinal cord slices. NeuroImage 2003;18(2):214–30.
25. Sykova E. Diffusion properties of the brain in health and disease. Neurochem Int 2004;45(4):453–66.
26. Traynelis SF, Dingledine R. Role of extracellular-space in hyperosmotic suppression of potassium-induced electrographic seizures. J Neurophysiol 1989;61(5):927–38.
27. Anderson TR, Jarvis CR, Biedermann AJ, Molnar C, Andrew RD. Blocking the anoxic depolarization protects without functional compromise following simulated stroke in cortical brain slices. J Neurophysiol 2005;93(2):963–79.
28. Svoboda J, Sykova E. Extracellular space volume changes in the rat spinal cord produced by nerve stimulation and peripheral injury. Brain Res 1991;560(1/2):216–24.
29. MacVicar BA, Hochman D. Imaging of synaptically evoked intrinsic optical signals in hippocampal slices. J Neurosci 1991;11(5):1458–69.
30. Somjen GG. Osmotic Stress and the Brain. Ions in the Brain: Normal Function, Seizures and Stroke. New York: Oxford University Press; 2004, p 63–74.
31. Krizaj D, Rice ME, Wardle RA, Nicholson C. Water compartmentalization and extracellular tortuosity after osmotic changes in cerebellum of Trachemys scripta. J Physiol 1996;492 (Pt 3):887–96.
32. Andrew RD, MacVicar BA. Imaging cell volume changes and neuronal excitation in the hippocampal slice. Neuroscience 1994;62(2):371–83.
33. Andrew RD, Jarvis CR, Obeidat AS. Potential sources of intrinsic optical signals imaged in live brain slices. Methods 1999;18(2):185–96, 179.
34. Andrew RD, Labron MW, Boehnke SE, Carnduff L, Kirov SA. Physiological evidence that pyramidal neurons lack functional water channels. Cereb Cortex 2007;17(4):787–802.
35. Stroman PW, Andrew RD. Functional magnetic resonance imaging of cortical tissue slices by means of signal enhancement by extravascular water protons (SEEP) contrast. In: Proceedings of the International Society for Magnetic Resonance in Medicine, 15th annual meeting, Berlin, Germany, May 19–25, 2007.
36. Anderson TR, Andrew RD. Spreading depression: imaging and blockade in the rat neocortical brain slice. J Neurophysiol 2002; 88(5): 2713–25.
37. Stroman PW, Tomanek B, Krause V, Frankenstein UN, Malisza KL. Mapping of neuronal function in the healthy and injured human spinal cord with spinal fMRI. NeuroImage 2002;17:1854–60.
38. Ng MC, Wong KK, Li G, Ma QY, Yang ES, Hu Y, Luk KDK. Verification of proton density change in spinal cord fMRI. In: Proceedings of the Fourth IASTED International Conference on Visualization, Imaging, and Image Processing, Marbella, Spain, Sept 6–8, 2004, p 926–30.
39. Ng MC, Wong KK, Li G, Lai S, Yang ES, Hu Y, Luk KD. Proton-density-weighted spinal fMRI with sensorimotor stimulation at 0.2 T. NeuroImage 2006; 29: 995–9.
40. Li G, Ng MC, Wong KK, Luk KD, Yang ES. Spinal effects of acupuncture stimulation

assessed by proton density-weighted functional magnetic resonance imaging at 0.2 T. Magn Reson Imaging 2005;23:995–9.

41. Stroman PW, Krause V, Malisza KL, Frankenstein UN, Tomanek B. Functional magnetic resonance imaging of the human cervical spinal cord with stimulation of different sensory dermatomes. Magn Reson Imaging 2002; 20(1):1–6.
42. Stroman PW, Kornelsen J, Bergman A, Krause V, Ethans K, Malisza KL, Tomanek B. Noninvasive assessment of the injured human spinal cord by means of functional magnetic resonance imaging. Spinal Cord 2004;42(2):59–66.
43. Kornelsen J, Stroman PW. fMRI of the lumbar spinal cord during a lower limb motor task. Magn Reson Med 2004;52(2):411–4.
44. Kornelsen J, Stroman PW. Detection of the neuronal activity occurring caudal to the site of spinal cord injury that is elicited during lower limb movement tasks. Spinal Cord 2007;45:485–90.
45. Stroman PW, Lawrence J, Kornelsen J. Functional MRI of the brainstem and spinal cord of children based on SEEP contrast. In: Proceedings of the International Society for Magnetic Resonance in Medicine, 15th annual meeting, Berlin, Germany, May 19–25, 2007; p 1967.
46. Schapiro MB, Schmithorst VJ, Wilke M, Byars AW, Strawsburg RH, Holland SK. BOLD fMRI signal increases with age in selected brain regions in children. NeuroReport 2004;15(17):2575–8.
47. Thomason ME, Burrows BE, Gabrieli JD, Glover GH. Breath holding reveals differences in fMRI BOLD signal in children and adults. NeuroImage 2005;25(3):824–37.
48. Friston KJ, Buechel C, Fink GR, Morris J, Rolls E, Dolan RJ. Psychophysiological and modulatory interactions in neuroimaging. NeuroImage 1997;6(3):218–29.
49. Logothetis NK, Pauls J, Augath M, Trinath T, Oeltermann A. Neurophysiological investigation of the basis of the fMRI signal. Nature 2001;412(6843):150–7.
50. Stroman PW, Cahill CM. Functional magnetic resonance imaging of the human spinal cord and brainstem during heat stimulation. In: Proceedings of the International Society for Magnetic Resonance in Medicine, 14th annual meeting, Seattle, USA, May 6–12, 2006.
51. Dunckley P, Wise RG, Fairhurst M, Hobden P, Aziz Q, Chang L, Tracey I. A comparison of visceral and somatic pain processing in the human brainstem using functional magnetic resonance imaging. J Neurosci 2005;25(32): 7333–41.
52. Fairhurst M, Wiech K, Dunckley P, Tracey I. Anticipatory brainstem activity predicts neural processing of pain in humans. Pain 2007; 128(1/2):101–10.
53. Agosta F, Valsasina P, Sala S, Caputo D, Stroman PW, Filippi M. Functional MRI of the spinal cord in patients with relapsing-remitting multiple sclerosis. In: Proceedings of the International Society for Magnetic Resonance in Medicine, 15th annual meeting, Berlin, Germany, May 19–25, 2007; p 108.
54. Filippi M, Rocca MA. Cortical reorganisation in patients with MS. J Neurol Neurosurg Psychiatry 2004;75(8):1087–9.
55. Stroman PW, Kornelsen J, Lawrence J. An improved method for spinal functional MRI with large volume coverage of the spinal cord. J Magn Reson Imaging 2005;21(5):520–6.
56. Matsuzaki H, Wakabayashi K, Ishihara K, Ishikawa H, Kawabata H, Onomura T. The origin and significance of spinal cord pulsation. Spinal Cord 1996;34(7):422–6.
57. Schumacher R, Richter D. One-dimensional Fourier transformation of M-mode sonograms for frequency analysis of moving structures with application to spinal cord motion. Pediatr Radiol 2004;34:793–7.
58. Mikulis DJ, Wood ML, Zerdoner OA, Poncelet BP. Oscillatory motion of the normal cervical spinal cord. Radiology 1994;192(1):117–21.
59. Stroman PW, Krause V, Malisza KL, Kornelsen J, Bergman A, Lawrence J, Tomanek B. Spinal fMRI of spinal cord injury in human subjects. In: Proceedings of the International Society of Magnetic Resonance in Medicine, 11th annual meeting, Toronto, Canada, July 10–16, 2005; p 13.
60. Stroman PW, Krause V, Malisza KL, Frankenstein UN, Tomanek B. Functional magnetic resonance imaging of the human cervical spinal cord with stimulation of different sensory dermatomes. Magn Reson Imaging 2002;20:1–6.
61. Moffitt MA, Dale BM, Duerk JL, Grill WM. Functional magnetic resonance imaging of the human lumbar spinal cord. J Magn Reson Imaging 2005;21(5):527–35.
62. Brooks J, Robson M, Schweinhardt P, Wise R, Tracey I. Functional magnetic resonance imaging (fMRI) of the spinal cord: a methodological study. In: American Pain Society, 23rd annual meeting, Vancouver, May 6–9, 2004; p 667.
63. Stroman PW. Discrimination of errors from neuronal activity in functional magnetic resonance imaging in the human spinal cord by means of general linear model analysis. Magn Reson Med 2006;56:452–6.
64. Valsasina P, Agosta F, Caputo D, Stroman P, Filippi M. Spinal fMRI during proprioceptive

and tactile tasks in healthy subjects: activity detected using cross-correlation, general linear model and independent component analysis. Neuroradiology 2008;50(10):895–902.

65. Giulietti G, Giove F, Garreffa G, Venditti E, Colonnese C, Maraviglia B. Spinal Cord fMRI: functional response and linear model assessment. In: Proceedings of the International Society for Magnetic Resonance in Medicine, 15th annual meeting, Berlin, Germany, May 19–25, 2007; p 3201.

66. Blumenfeld H. Neuroanatomy Through Clinical Cases. Sunderland, MA: Sinauer Associates 2002; p 22.

INDEX

D

E

F

G

H

I

Q

R

S

T

U

V

W

Z

Printed in the United States of America